Materials for Biomedical Engineering

Materials for Biomedical Engineering

Fundamentals and Applications

Mohamed N. Rahaman
Missouri University of Science and Technology
Rolla, Missouri, United States

Roger F. Brown
Missouri University of Science and Technology
Rolla, Missouri, United States

The American Ceramic Society

WILEY

Published by John Wiley & Sons, Inc., Hoboken, New Jersey.
Published simultaneously in Canada.

Registered Office
John Wiley & Sons, Inc., 111 River Street, Hoboken, NJ 07030, USA

Editorial Office
111 River Street, Hoboken, NJ 07030, USA

For details of our global editorial offices, customer services, and more information about Wiley products visit us at www.wiley.com.
Wiley also publishes its books in a variety of electronic formats and by print-on-demand. Some content that appears in standard print versions of this book may not be available in other formats.

Library of Congress Cataloging-in-Publication Data

Names: Rahaman, M. N., 1950- author. | Brown, Roger F., author.
Title: Materials for biomedical engineering : fundamentals and applications / Mohamed N. Rahaman, Missouri University of Science and Technology, Rolla, Missouri, United States, Roger F. Brown, Missouri University of Science and Technolog, Rolla, Missouri, United States.
Description: First edition. | Hoboken, NJ : Wiley-American Ceramic Society, 2022. | Includes bibliographical references and index.
Identifiers: LCCN 2021030672 (print) | LCCN 2021030673 (ebook) | ISBN 9781119551089 (cloth) | ISBN 9781119551119 (adobe pdf) | ISBN 9781119551096 (epub)
Subjects: LCSH: Biomedical materials. | Biomedical engineering.
Classification: LCC R857.M3 R34 2022 (print) | LCC R857.M3 (ebook) | DDC 610.28–dc23
LC record available at https://lccn.loc.gov/2021030672
LC ebook record available at https://lccn.loc.gov/2021030673

Cover Design: Wiley
Cover Image: © Mohamed N. Rahaman

MIX
Paper from responsible sources
FSC
www.fsc.org
FSC® C013604

Contents

Preface

Materials for Biomedical Applications places an emphasis on the fundamentals of biomaterials science and how these fundamentals are used, in practice, to modify and adapt various classes of materials for use as biomaterials in a variety of biomedical applications. Over the last several decades, advances in materials science, biological sciences, physical sciences, and engineering, along with an evolution in medical treatment, have led to the creation of biomaterials with a greater degree of sophistication and a considerable increase in range of use and efficacy. Millions of lives are now being improved or saved by the use of biomaterials in a wide variety of medical devices. As this importance of biomaterials to society continues to grow, so too will the significance of materials for biomedical applications, both as an academic field, an area of research to develop new or improved devices, and as an industry.

Our intention has been to write a textbook suitable for a one-semester (or two-quarter) course in biomaterials at the junior or senior undergraduate level, a course that is commonly a fixture in most biomedical engineering curricula at many universities. The book should also provide a useful introductory or reference resource for graduate students and practicing scientists and engineers in the field of biomedical engineering. Biomaterials is an interdisciplinary field and practitioners of biomaterials science often enter the field from a number of other fields. It is hoped that the emphasis of the book on fundamentals should provide a common ground for equipping students and practitioners with a background for understanding how the properties of biomaterials arise and how to design and create new biomaterials. The book assumes a knowledge of introductory physics and chemistry courses normally taught at most universities. Knowledge from introductory materials science and biological sciences courses is useful but it not a requirement, since appropriate principles of materials science, engineering, and biological sciences are considered in the earlier chapters of the book.

The overall arrangement of the topics covered in the book begins with an introductory overview of biomaterials and the biomaterials field. Following this, we discuss the relationship between the composition, structure and properties of materials, which is fundamental to the field of materials science and is critical to understanding how to select and design biomaterials for specific applications. Then specific classes of materials used as biomaterials, metals, ceramics, polymers (including synthetic and natural polymers and hydrogels) and composites, are considered from the point of view of how their composition and structure can be modified and manipulated to control their properties. As the ability of a biomaterial to degrade or not is a key property that influences its performance *in vivo*, we discuss factors that influence the degradation of biomaterials in the physiological environment. This is followed by a discussion of biocompatibility phenomena, that is, the effect of appropriate responses from cells and tissues of the body on the ability of an implanted biomaterial to perform, which is the most critical factor that determines the biomaterial's success *in vivo*. The final part of the book, discussing the applications of biomaterials, is where we see how the principles of materials science, engineering, and biology covered in the earlier chapters are used, in practice, to modify and adapt the various classes of materials for use as biomaterials in a variety of biomedical applications. Because a wide range of applications cannot be adequately covered at a level compatible with the intention of this book, we limit coverage to applications broadly grouped into four areas. These areas, which represent a significant cross section of the biomedical field and biomedical industry, are the repair of hard tissues, repair of soft tissues, tissue engineering and regenerative medicine, and drug delivery.

Each chapter is replete with data and illustrations to augment the discussion, and concludes with a set of problems and a list of references for further reading. These problems may be suitable for classroom tests and discussions, as well as for reviewing key topics covered in each chapter. Overall, the references have been selected to provide sources of additional information related to specific studies discussed in the chapter and to include seminal publications enlarging upon major topics discussed in the chapter.

We would like to thank the many publishers and authors who have given us permission to reuse their figures and tables in this book. We would also like to express our appreciation to the many students we have taught in our biomaterials and tissue engineering courses over the years and to our students who have contributed to our research and publications. The structure of this book and the body of knowledge incorporated within it have benefitted from their contributions. Finally, we would like to express our gratitude to our wives, Vashanti Rahaman and Peggy Brown, for their support during the lengthy period of preparing this book manuscript.

About the Companion Website

This book is accompanied by a companion website:

www.wiley.com/go/rahaman/biomedicalengineering

The website includes the following materials:

- Figures, Tables and Solutions Manual.

Part I

General Introduction

1

Biomaterials – An Introductory Overview

1.1 Introduction

Materials have been used to treat, replace, or augment tissues and organs in the human body since antiquity but their use and degree of sophistication have increased significantly over time, particularly over the last several decades. Advances in materials science, biological sciences, physical sciences, and engineering along with an evolution in medical treatment over the last several decades have led to the creation of biomaterials with more reproducible properties, better performance, and increased functionality. These advances have resulted in a considerable increase in the range of use and the efficacy of biomaterials. Nowadays, millions of lives are being improved or saved by the use of biomaterials in fracture fixation plates, implants for total hip and knee joint replacement, dental implants and restorations, heart valves, vascular grafts and stents, contact and intraocular lenses, skin substitutes, and wound healing materials, for example (Figure 1.1).

The applications of biomaterials are many. Table 1.1 provides a list of selected applications and the types of materials used in these applications. Many of the biomaterials used in these applications were selected from durable, chemically inert materials that were available off the shelf, and they were designed to serve, mainly, a mechanical (or physical) function. The last few decades have seen a shift in emphasis in which the biological sciences are playing a role in the design of biomaterials of significance comparable to that of materials science. Biomaterials are now no longer designed to be chemically inert or to just serve a mechanical function. Instead, advances in biological sciences are being used to design biomaterials to regenerate tissues and organs and to direct the response of specific cells and tissues. In doing so, these biomaterials stimulate the body to heal itself. An example is the creation of functional skin substitutes to treat patients with severe burns (Figure 1.11).

The importance of biomaterials to society has been increasing significantly in the last several decades, both as an academic field, an area of research to develop new or improved devices, and as an industry (Figure 1.2). The number of biomedical engineering (or bioengineering) departments in academic institutions has increased rapidly over the last few decades. There are over 75 biomedical engineering (or bioengineering) departments in the United States alone. Biomaterials are an important area of teaching and research in these departments and they are often emphasized in engineering disciplines such as materials science and engineering, chemical engineering, and mechanical engineering.

Outside of academic institutions, biomaterials are also an important area of research, development, and manufacturing in several industries, such as in the production of medical devices, dental restorations, and devices for drug delivery. The size of companies that manufacture biomaterials and devices for these applications cover a wide range, from small start-up companies to large established companies that are among the so-called Fortune 500 companies in the United States. The commercial market for devices that incorporate biomaterials is immense and it is projected to grow at a healthy rate. Estimates of the commercial market and the number of implantable medical devices that incorporate biomaterials annually are presented in Table 1.1.

1.2 Definition and Meaning of Common Terms

Several terms are commonly used in the biomaterials literature and throughout this book. This section provides a description of the meaning and definition of some of these common terms that are particularly relevant to this chapter.

Biomaterial

Our understanding of what a biomaterial is and, consequently, proposed definitions of the term biomaterial have evolved over time in response to advances in science and

Materials for Biomedical Engineering: Fundamentals and Applications, First Edition. Mohamed N. Rahaman and Roger F. Brown.

Companion website: www.wiley.com/go/rahaman/biomedicalengineering

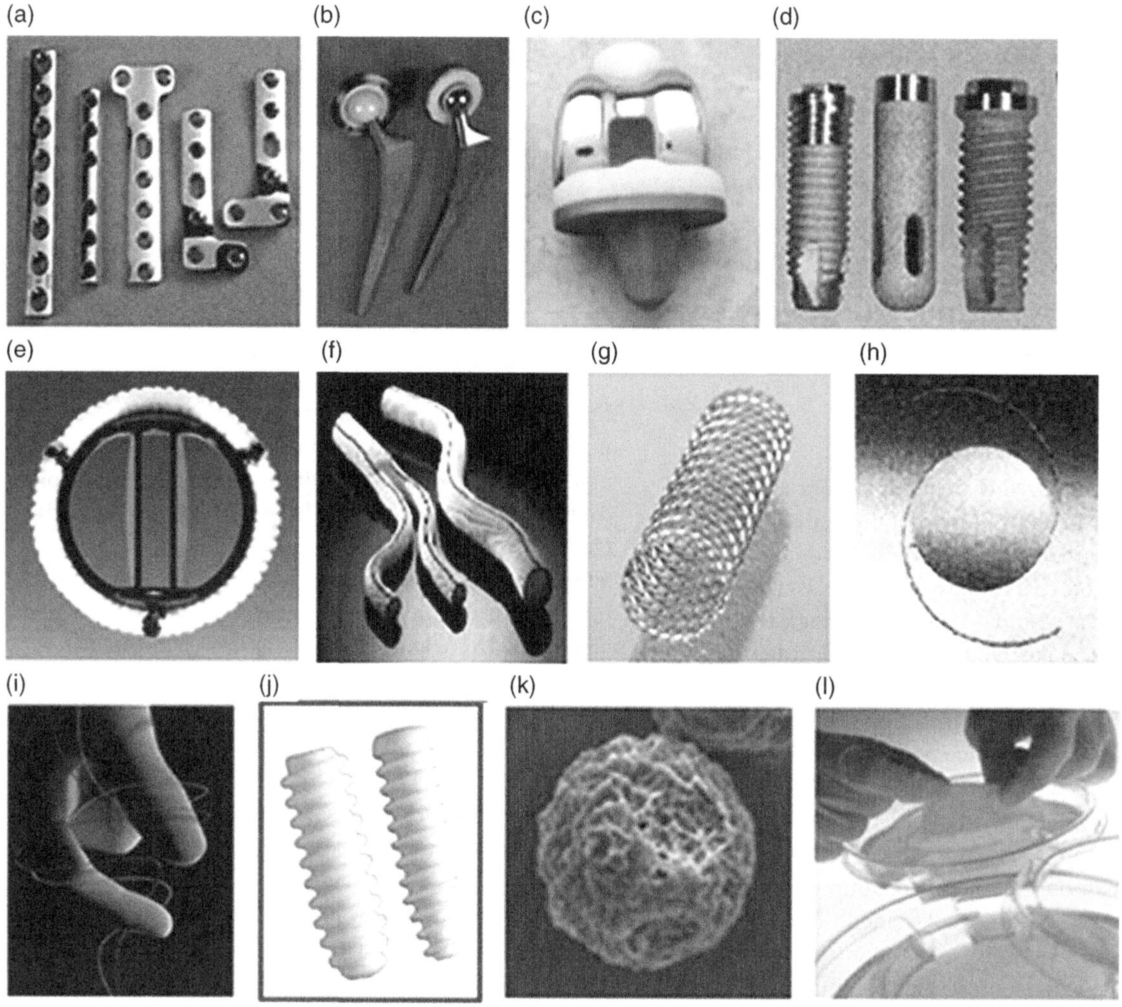

Figure 1.1 Examples of biomaterials in use for medical and dental applications. (a) Fracture fixation plate; (b) implant for total hip replacement; (c) implant for total knee replacement; (d) dental implant; (e) heart valve; (f) vascular graft; (g) intravascular stent; (h) intraocular lens; (i) degradable suture; (j) degradable screw for fracture fixation; (k) degradable polymer microsphere for delivery of therapeutics; (l) functional skin substitute.

technology. An early use of the term and the beginnings of the biomaterials field emerged from symposia held in the United States around the late 1960s and early 1970s. One definition, endorsed by a consensus of experts in the field, is

- A biomaterial is a nonviable material used in a medical device, intended to interact with biological systems.

As biomaterials evolved, another definition was proposed:

- A biomaterial is a material intended to interface with biological systems to evaluate, treat, augment, or replace any tissue, organ or function in the body (Williams 1999).

While these early definitions reflected our understanding and use of biomaterials at that time, they are now too restrictive because of the evolution of biomaterials over the last two to three decades. Biological sciences, as noted earlier, are now playing an increasing role in designing biomaterials. At the same time, advances in engineering have made it possible to create biomaterials with increasing sophistication. Biomaterials are now being studied, developed or created to target specific responses by cells and tissues in the body, regenerate functional tissues and organs, deliver molecules or drugs to target specific cells or tissues, and improve diagnostic imaging of cells and tissues. A recently proposed definition that attempts to capture these more recent advances and broader meaning of the term is:

- A biomaterial is a substance that has been engineered to take a form which, alone or as part of a complex system, is used to direct, by control of interactions with components of living systems, the course of any therapeutic or diagnostic procedure (Williams 2014).

Table 1.1 Key applications of synthetic materials and modified natural materials in medicine[a].

Application	Biomaterial	Number used per year worldwide (or market in US$)
Skeletal system		
Joint replacement (hip; knee; shoulder)	Titanium; stainless steel; polyethylene	2.5 million
Bone fixation plates and screws	Metals; polylactic acid	1.5 million
Spine repair	Titanium; polyether ether ketone; silicon nitride	800 000
Bone cement	Polymethylmethacrylate	($600 million)
Bone defect repair	Calcium phosphates	—
Artificial tendon or ligament	Polyester fibers	—
Dental implants	Titanium	($4 billion)
Cardiovascular system		
Blood vessel prosthesis	Dacron; expanded polyethylene	200 000
Heart valve	Dacron; carbon; metal; treated natural tissue	400 000
Pacemaker	Titanium; polyurethane	600 000
Implantable defibrillator	Titanium; polyurethane	300 000
Stents	Stainless steel; cobalt–chromium alloy; nickel–titanium alloy	1.5 million
Catheters	Teflon; silicone; polyurethane	1 billion ($20 billion)
Organs		
Heart assist devices	Polyurethane; titanium; stainless steel	4000
Hemodialysis	Polysulfone; silicone	1.8 million patients ($70 billion)
Blood oxygenator	Silicone	1 million
Skin substitute	Collagen; cadaver skin; nylon; silicone	($1 billion)
Ophthalmologic		
Contact lens	Acrylate, methacrylate and silicone polymers	150 million
Intraocular lens	Acrylate and methacrylate polymers	7 million
Corneal bandage lens	Hydrogel	—
Glaucoma drain	Silicone; polypropylene	($200 million)
Other		
Cochlear prosthesis	Platinum; platinum–iridium; silicone	250 000 users
Breast implant	Silicone	700 000
Hernia mesh	Silicone; polypropylene; teflon	200 000 ($4 billion)
Sutures	Polylactic acid; polydioxanone; polypropylene; stainless steel	($2 billion)
Blood bags	Polyvinyl chloride	—
Ear tubes (Tympanostomy)	Silicone; teflon	1.5 million
Intrauterine device	Silicone; copper	1 million

[a] Data compiled from many sources – these numbers should be considered rough estimates that are growing with changing markets and new technologies. Where only US numbers are available, world usage is estimated at 2.5 times the US usage.
Source: Modified from Ratner (2013).

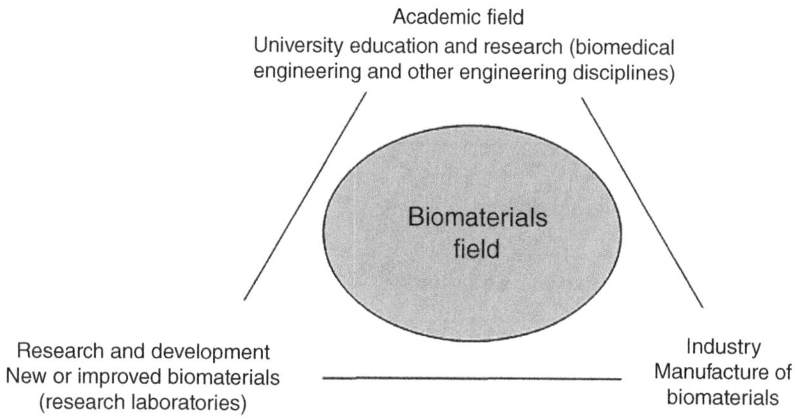

Figure 1.2 Schematic showing the major components of the biomaterials field.

Biocompatibility

A widely accepted definition of biocompatibility is:

- Biocompatibility refers to the ability of a material to perform with an appropriate host response in a specific application (Williams 1999).

Biocompatibility is the most critical factor that determines the success of implanted biomaterials (Chapter 16), but the term is often used loosely in the literature. Biocompatibility is not simply a property of the material alone. As biocompatibility refers to the ability of the biomaterial to perform in a specific application, it is a property of the material–tissue system. It should also be recognized that the biocompatibility phenomena of a given material can change, depending on the situation and the implantation time.

Host Response

The host response to a biomaterial refers to the response of cells and tissues of the body, either local or systemic, to the implanted biomaterial.

Categories of Biomaterials

Biomaterials can be categorized in a number of different ways, depending on their source or the property being emphasized. As they are widely used in the literature, we provide a description of the meaning and definition of the more common terms related to these categories.

Natural and Synthetic Biomaterials

Biomaterials can be classified as natural materials or synthetic materials, depending on their source. Natural biomaterials are materials obtained from human and other vertebrate sources, invertebrate organisms, and from plants. They include bone taken from live humans or cadavers, collagen isolated from animal skin and bones, cellulose produced by certain types of bacteria or obtained from plants, silk from spiders, chitin extracted from the shells of crustaceans, alginate produced by seaweed, and hyaluronic acid extracted from rooster combs or from genetically engineered bacteria. Natural biomaterials are typically similar or identical to substances in living organisms.

In comparison, synthetic biomaterials are produced from industrial materials that have commonly undergone some degree of refinement. As illustrated in Figure 1.3, they are composed of:

- Metals, including their alloys, such as gold; stainless steels; titanium (Ti) and its alloys;
- Ceramics, including glasses and glass-ceramics, such as aluminum oxide (Al_2O_3); the calcium phosphates hydroxyapatite and β-tricalcium phosphate; bioactive glasses;
- Polymers such as polyethylene (PE); polymethyl methacrylate (PMMA); nylon; silicone (polydimethylsiloxane);
- Composites composed of one or more of the abovementioned classes of materials.

Metals, ceramics, and polymers are said to make up the primary classes of synthetic biomaterials. Semiconductors are sometimes described as a fourth class of synthetic materials but the few semiconductors that find use as biomaterials typically have structures similar to ceramics. Consequently, in this book, they are included, where appropriate, within the category of ceramics.

Degradable, Nondegradable and Resorbable Biomaterials

Biomaterials can also be classified in terms of their ability to degrade, typically in an aqueous medium such as a phosphate-buffered saline or the physiological fluid. Degradation refers to the conversion of the material into

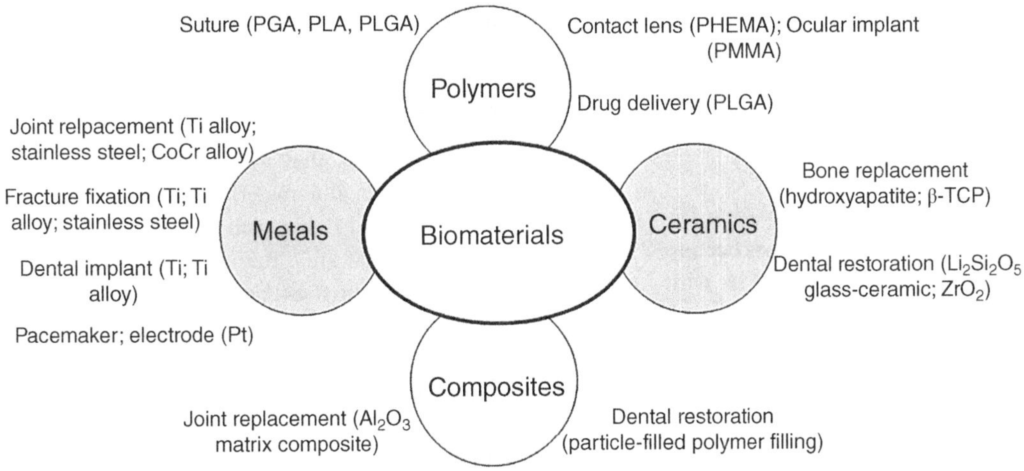

Figure 1.3 Schematic of the classes of materials used as biomaterials, along with examples of specific biomaterials in each class.

smaller constituents, such as ions, molecules or fine particles. Degradation of biomaterials in an aqueous environment can occur by chemical attack by water molecules (hydrolysis), enzyme-mediated hydrolysis, dissolution, oxidative reaction, or some combination of these mechanisms. For example, the synthetic polymer polyglycolic acid has a composition that makes it susceptible to degradation by hydrolysis into smaller molecules composed of glycolic acid.

The term biodegradable is often used in the literature instead of, or interchangeably with, degradable. Sometimes it may be used in a more restrictive sense to describe degradation by enzyme-mediated reactions in the physiological environment, and so would not include unmediated hydrolysis. Another term is resorbable that, like biodegradable, is often used instead of, or interchangeably with, degradable. This term may be used in a more restrictive sense to emphasize what eventually happens to the biomaterial degradation product composed of ions, molecules, or particles. Commonly, the degradation product (or most of it) is removed from the body via renal clearance, hepatic processing, phagocytosis, or by other processes. Alternatively, the degradation product of some biomaterials can be taken up into the body tissues, and it can then be said to be resorbed back into the body tissues. For example, calcium and phosphate ions produced by the degradation of some calcium phosphate biomaterials can be resorbed into bone as these ions are constituents of bone.

In this book, we will simply use the term degradable regardless of the mechanism of degradation or the fate to the degradation products. The terms biodegradable or resorbable will be used only when necessary to emphasize a distinction from the general term degradable.

Whereas all materials show some degree of degradation, no matter how small, over some appropriate duration, the term nondegradable is often used in a qualitative manner to describe a biomaterial that is normally not susceptible to chemical attack in an aqueous environment. Qualitatively, these nondegradable biomaterials can also be said to be chemically inert, chemically stable, or resistant to degradation. In this sense, then, the synthetic polymer PE can be described as nondegradable or chemically stable, even though it can undergo some degree of degradation by oxidative reactions under certain conditions, such as the presence of highly oxidizing molecules in the aqueous environment (Chapter 15).

The term bioinert has often been used to describe these nondegradable biomaterials. However, it is generally recognized that no biomaterial is totally inert in the biological environment. Although they may not degrade themselves when implanted, all biomaterials elicit a response from ions, molecules, cells, or tissues in the body. Consequently, a biomaterial cannot be said to be truly bioinert.

Bioactivity

An original definition of bioactivity is:

> "Bioactivity refers to the ability of a material to elicit a specific reaction at its surface when implanted in the body, leading to the formation of a strong bond at its interface with bone or soft tissue."

The calcium phosphate ceramic hydroxyapatite and certain compositions of glasses, referred to as bioactive glasses, satisfy this definition (Chapter 7). If the requirement of forming a strong bond with bone and soft tissues is omitted, the definition becomes broader. The term is now used more commonly in this broader sense.

Tissue Engineering and Regenerative Medicine

Since antiquity, the conventional approach to repairing diseased or damaged tissues or organs in the body has been to replace them with transplanted tissues or organs, or with durable synthetic implants composed of metals, ceramics or polymers. While this approach has worked well, it suffers from limitations related to factors such as supply of donor tissue or organs, or the inadequate lifetime of the implanted biomaterial. A radical shift in this conventional approach was proposed approximately three decades ago. Instead of focusing on repairing or replacing tissues and organs with transplanted tissues or organs, or with durable materials, this new approach, referred to as tissue engineering, is based on regenerating functional tissues and organs. A definition of the term is:

> "Tissue engineering is an interdisciplinary field that applies the principles of engineering and the life sciences toward the development of biological substitutes that restore, maintain, or improve tissue function" (Langer and Vacanti 1993).

The tissue engineering approach involves harvesting cells, incorporating them into a suitable biomaterial, and stimulating them *ex vivo* (outside the body). Then the resulting construct is implanted into the patient at the appropriate time. Alternatively, the cell-seeded biomaterial can be implanted directly into the body (Chapter 25). More recently, the term regenerative medicine has been used to describe a broader approach to regenerating the patient's own tissues or organs. Regenerative medicine includes tissue engineering as well as two other approaches based on cell therapy and gene therapy. Biomaterials form an important component of the tissue engineering approach, whereas cell therapy and gene therapy involve biomaterials only minimally.

In Vivo, *Ex Vivo*, and *In Vitro*

In vivo refers to procedures performed inside the body whereas *ex vivo* refers to procedures or manipulations performed outside the body. *In vitro* refers to studies, such as cell culture, evaluating the degradation rate of a biomaterial, or evaluating the release profile of molecules from a biomaterial, that are performed outside the body, typically in a science laboratory. Commonly, biomaterials are evaluated *in vitro* prior to implantation or use *in vivo*.

1.3 Biomaterials Design and Selection

Biomaterials are normally designed to have some desirable combination of properties that depend on the intended application. For example, a biomaterial designed for use in healing a defect in a hard tissue such as bone would most likely have properties that are different from one designed for use in healing a defect in a soft tissue such as cartilage. A biomaterial designed to deliver drugs or molecules to a specific site in the body should be capable of being formed into a particulate form, normally of spherical shape and size less than several tens of microns, which is vastly different from a three-dimensional (3D) form for healing a bone defect. The biomaterial may also have to degrade at a desirable rate to release drugs or biomolecules in a controllable manner.

The approach to designing biomaterials has seen a radical shift in the last few decades. Biological sciences are now playing a significantly increasing role, while at the same time, materials science and engineering are being used to design improved biomaterials having properties more comparable to those of natural tissues.

1.3.1 Evolving Trend in Biomaterials Design

In the past, materials science and engineering provided the foundation for designing and selecting biomaterials. Biomaterials were normally selected from durable, chemically inert materials that were available off the shelf, and they were engineered to serve a mainly mechanical (or physical) function. Adequate strength and stiffness, for example, are required for the biomaterial to support physiological stresses without deforming permanently into a nonoptimal shape or without fracturing. Chemical inertness is important to minimize the potential for rejection of the biomaterial.

This approach based on materials science and engineering worked well for some applications and it provided a foundation for more recent developments in the biomaterials field. On the other hand, clinical data show that a high percentage of these biomaterials failed within a certain duration, necessitating a second surgical procedure (sometimes referred to as a revision surgery). Implants used for total hip joint replacement are a good example of this approach (Section 1.5). While the basic design of the hip implant has changed very little in over 50 years, improvements in materials selection over this time have resulted in the creation of implants having better performance and longer lifetime *in vivo*. However, ~25% of these implants still fail within 10–20 years, necessitating revision surgery.

The last few decades have seen a radical shift in the approach to designing and selecting biomaterials. Biological sciences are now playing a significant role, often comparable to that of materials science and engineering. Advances in biological sciences are being utilized to design biomaterials that serve not just a mechanical (or physical) function. Instead, biomaterials are now being designed to serve, in addition, biological functions such as eliciting or directing specific cellular responses. At the same time, advances in the physical sciences, materials science and engineering are being used to design and create improved biomaterials with greater complexity in structural and chemical characteristics to better mimic biological tissues.

As an example of the shift in biomaterials design over the last few decades, we can select the intravascular stent (Figure 1.1g). This device is among the most widely used of all implantable medical devices, with over 1.5 million stents implanted annually in patients worldwide. Stents are used when a blood vessel such as an artery becomes blocked (a process referred to as stenosis) due to the accumulation of plaque, a condition that, in many cases, can lead to a heart attack. Stents are tubular in shape and they are inserted into the artery to compress the plaque and to remain at the site to keep the artery open (Chapter 24).

Until recently, stent designs focused strongly on the need to provide a mechanical function. This called for the use of a biocompatible metal with the attractive mechanical properties characteristic of this class of materials and corrosion resistance to minimize the chances of rejection *in vivo*. Early stent designs employed durable metals that were available off the shelf, such as a stainless steel or a cobalt–chromium (Co–Cr) alloy (Figure 1.4). Although these so-called bare-metal stents work well initially, they themselves can become blocked after a few years (a process referred to as restenosis). A problem with bare-metal stents is that they can irritate the endothelial lining of the artery. This triggers an inflammatory response, which leads to the proliferation of certain cells (vascular smooth muscle cells) and the growth of extracellular fibrotic tissue

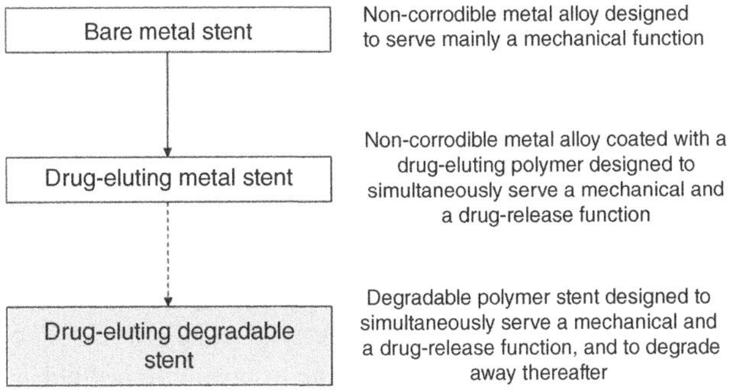

Figure 1.4 Stages in the evolution of the intravascular stent, used as an example to illustrate the considerable shift in biomaterials design over the last few decades.

into the lumen of the stent, resulting in restenosis. The result is that the procedure may have to be repeated at regular intervals.

Modern metal stents are designed to serve the mechanical function of bare-metal stents but to also provide a biological function to reduce restenosis. The biological function of controlling cell proliferation and the growth of fibrotic tissue into the stent is achieved by incorporating a drug-eluting function into the stent. In one approach, a thin coating of a polymer, degradable or nondegradable, in which the drug is dispersed, is applied to the stent surface. Commonly used drugs include paclitaxel and sirolimus, both of which are powerful antiproliferative drugs that inhibit cell proliferation. The use of these drug-eluting stents has produced a significant lowering of restenosis rates, but they have not been entirely successful. A problem is that the stent can provide sites for aggregation of platelets (from the blood), resulting in thrombosis, which normally necessitates prolonged antiplatelet therapy.

Questions have been raised about the need for the continued presence of the stent after stenosis has been resolved. It has been claimed that the benefit of the drug-eluting stent is achieved within a finite time, such as a few months, and the continued presence of the stent contributes to adverse biological effects due to its ability to irritate the endothelium of the artery. Consequently, a more recent approach to the design of modern stents is the use of a degradable stent, composed of a degradable polymer in which the drug is dispersed or a degradable metal coated with a degradable polymer in which the drug is dispersed. The stent serves a mechanical function initially to open up the blocked blood vessel and, as it degrades, the drug is released at a controllable rate to destroy the deposit responsible for stenosis. After resolution of stenosis, the stent degrades completely. While drug-eluting degradable stents provide a useful alternative approach to treating stenosis; a few questions related to their effectiveness remain unresolved at the present time. One is the degree of inflammatory response by cells to the degradation products of the polymer stent and the effect of this response on the endothelium of the blood vessel.

1.3.2 Factors in Biomaterials Design and Selection

The design of a biomaterial for a specific application starts with the selection of a material with certain inherent properties. Then materials science, biological and engineering principles and procedures are utilized to design and create an implant with the requisite properties for the intended application. Important factors that should be considered in biomaterials design and selection are summarized in the following sections.

- *Chemical composition*: It is well known in materials science that the chemical composition (or atomic structure) of a material controls its intrinsic properties, such as mechanical strength and stiffness, whether the material is brittle or not, its ability to degrade in an aqueous environment, whether the material is electrically conducting or not, and so on. Consequently, the chemical composition is perhaps the first factor to be considered in selecting a material for use as a biomaterial.
- *Biocompatibility*: The ability of a material to perform with an appropriate host response in the intended application and its continued function is a critical factor in the selection of a biomaterial. Biocompatibility, as described earlier, is a property of both the biomaterial and the tissue system in which it is implanted, or with which it interfaces. The biomaterial should not, for example, be toxic or release ions or molecules that are toxic locally or systemically.
- *Mechanical properties*: All tissues (and organs) in the body have some characteristic mechanical properties, such as strength and stiffness. Consequently, the mechanical properties of a biomaterial are important in determining its performance *in vivo*. A useful guideline is that a biomaterial used to repair a diseased or damaged tissue should have mechanical properties that are comparable to the host tissue. However, this is often difficult to achieve. A large mismatch in mechanical properties between an implant and the host tissue can lead to adverse biological effects. The use of a strong metal implant to repair a bone defect can, for example, lead to resorption and weakening of the surrounding bone and, eventually, to bone fracture. The mechanical properties of a material can also influence the response of cells and, thus, they can determine the ability to regenerate a specific tissue. For example, a material that is optimal for regenerating bone typically would not be suitable for regenerating a soft tissue such as cartilage.
- *Stability in the biological environment*: A biomaterial should be nondegradable or should degrade at a desirable rate for the intended application. For a degradable biomaterial, a guideline often found in the literature is that the implant should degrade at a rate comparable to the rate at which new tissue is being formed. However, this can often be difficult to achieve in practice.
- *Ease of fabrication*: Typically, a biomaterial will have an external shape (or geometry), complex or simple, that is similar to the tissue or organ to be replaced or regenerated. Biomaterials can be 3D objects, fibers, coatings, films, or particles, depending on the application. The biomaterial might be required to have internal structural features that are also important. These internal structural features relate to the way in which the components or phases, such as the solid phase and porosity, are arranged within the biomaterial. The structure at a microscale and nanoscale, referred to as the "microstructure" and "nanostructure," respectively, are often the major structural features of interest. The microstructure (or nanostructure) can be simple or complex, depending on the application. Materials used to create biomaterials should be capable of being formed economically into the desired external shape and internal microstructure (and/or nanostructure). In common with other technologies, the use of additive manufacturing to create biomaterials, particularly with complex shape and microstructure, has been increasing rapidly.
- *Ability to be sterilized*: A biomaterial should be capable of being sterilized by one of the sterilization methods that employ heat (steam autoclaving or dry heat), gas, radiation, or electron beam treatment prior to their use in studies *in vitro* or implantation *in vivo*. Lack of sterility will invariably lead to infection and destruction of cells, followed by potentially catastrophic failure *in vivo*.
- *Other physicochemical properties*: The applications of biomaterials are many and, consequently, one or more additional properties can be particularly important in certain applications. The capacity to absorb water can be crucial to the use of biomaterials in applications such as drug delivery and tissue engineering. It depends on the composition and the structure of the biomaterial. As water is composed of polar molecules, the capacity to interact with, and absorb, water molecules is favored by the presence of ionic charges, polar groups, or a combination of the two, in the molecular structure of the biomaterial. The biomaterial should also have an interconnected but rather expandable molecular structure to allow the migration of water molecules into the structure.

Whether a material is electrically conducting or insulting is important to the function of some medical devices. Electrically conducting metals are, for example, important for use as electrodes in pacemakers and neural stimulators. In comparison, electrically insulating materials are typically used as coatings to isolate or insulate sensitive electronic devices. The ability to conduct electrical signals is quantified by the electrical conductivity of the material or, less commonly, by the electrical resistivity, which is the inverse of the electrical conductivity. The capacity of a material to respond to a magnetic field is important for its ability to function in some treatments such as hyperthermia treatment of tumors and in diagnostic imaging.

The use of materials in devices such as contact lenses and intraoptical lenses is crucially dependent on their ability to transmit light (i.e. their transparency). Whether a material can conduct heat or not is quantified by its thermal conductivity. The thermal expansion coefficient quantifies the expansion or contraction of a material upon heating or cooling. These thermal properties are normally important for biomaterials that are subjected to sizable temperature changes during manufacture or use. As some biomaterials, particularly natural materials, can deteriorate when heated, their maximum processing or use temperature can also be important.

1.4 Properties of Materials

Biomaterials, as noted earlier, are designed to have some desirable combination of properties, depending on the application. In the selection of a biomaterial for a specific application, it is worth recognizing at the outset that:

- The primary classes of synthetic materials, metals, ceramics and polymers, and natural materials have vastly different intrinsic (or inherent) properties;
- Within each class, the measured properties can cover a wide range.

A useful starting point is a qualitative understanding of the intrinsic properties of these different classes of materials. Then, we will present the intrinsic properties of these classes of synthetic materials and compare them with those of some natural materials. In Chapter 2, the basic principles for understanding the intrinsic properties of these classes of materials in terms of their atomic structure are presented.

1.4.1 Intrinsic Properties of Metals

Metals are composed of single elements (such as Ti) or a combination of elements, forming alloys such as brass, which is an alloy of copper and zinc. Most metals show excellent mechanical properties, such as high strength, high stiffness, high ductility, and good fatigue resistance. Strength refers to the ability of a material to support an applied load (or mechanical stress) without breaking. While stiffness refers to the ability to resist deformation when subjected to an applied load, the elastic modulus is a more effective and more widely used measure of a material's stiffness. Ductility refers to the ability to deform rather than shatter catastrophically, particularly when the applied stress becomes sufficiently high. Fatigue resistance refers to the ability to withstand repeated cyclic loading without fracturing. Most metals generally show moderate hardness and moderate resistance to abrasion or wear, somewhere between ceramics and polymers. The majority of metals have a high density, higher than ceramics and much higher than polymers. The excellent electrical and thermal conductivity of metals is well known.

Except for the noble metals such as gold, silver, and platinum, most pure metals corrode in an aqueous environment, such as the physiological environment. Consequently, most metals cannot be used as implantable biomaterials. On the other hand, a protective oxide surface layer forms rapidly on some metals upon exposure to an oxidizing environment, which passivates them from corrosion. These passivated metals, such as Ti, certain Ti alloys, and stainless steel, have a high resistance to corrosion in the normal physiological environment. Because of their excellent mechanical properties and corrosion resistance, they find considerable use in a variety of orthopedic and dental applications, such as fracture fixation plates, total joint replacement, and dental implants.

Ease of fabrication, as noted earlier, is also an important factor in the selection of a material for use as a biomaterial. Metals can be formed with reasonable ease into 3D objects, coatings and films using conventional fabrication methods that are widely used in the metallurgical industry. Additive manufacturing, also referred to as 3D printing, now provides another method to produce metals with the requisite external shape and microstructure for use as biomaterials.

Overall, metals are normally selected for use as biomaterials when excellent mechanical properties, high electrical conductivity, or a combination of both must be guaranteed. Suitable metals have a high resistance to corrosion in the physiological environment, such as certain noble metals or metals passivated by a protective oxide surface layer.

1.4.2 Intrinsic Properties of Ceramics

Except for carbon, ceramics are compounds of metallic and nonmetallic elements, such as Al_2O_3 or silicon nitride (Si_3N_4), for example. Although ceramics used in technological or engineering applications typically show better strength and elastic modulus than most metals, a distinctive drawback is their inherent brittleness. Brittleness refers to the catastrophic shattering of an object into two or more pieces under sufficiently high mechanical stresses. Despite their brittleness, ceramics can be engineered to function safely and reliably over long durations under high stresses if these stresses are compressive in nature. Ceramics are not recommended for use in high-stress applications when the applied stress has an appreciable tensile or bending component. In general, ceramics show high hardness and wear resistance.

The density of ceramics is lower than that of most metals, often by a factor of two or more, which means that for the same geometry, a ceramic implant will not be as heavy as a metal implant. Ceramics are generally poor conductors of electricity and heat, and those ceramics used in technological applications normally have a high degree of chemical inertness. On the other hand, certain compositions, such as hydroxyapatite and β-tricalcium phosphate, composed of the same ions as the mineral constituent of bone, show some reactivity in the physiological environment.

Ceramics are generally more difficult to fabricate than metals and far more difficult to fabricate than polymers. Forming ceramics into useful objects having the requisite external shape and microstructure often requires high fabrication temperatures, approximately several hundred degrees Celsius. Due to their high hardness, ceramics are expensive and difficult to machine into the desired shape and surface smoothness after their fabrication.

Overall, major limitations of ceramics are their brittleness, and the difficulty and cost of fabricating them into useful objects. On the other hand, ceramics have high compressive strength, stiffness, hardness, and wear resistance. Consequently, ceramics generally find use in applications where the applied stress is much lower than their strength or mainly compressive in nature, and where high wear resistance and chemical inertness are required.

1.4.3 Intrinsic Properties of Polymers

Polymers are generally composed of long-chain molecules formed by repeated bonding of a large number of small molecules. The simplest example is polyethylene (abbreviated PE) whose molecular chains consist a large number (several hundred to several thousand) of ethylene ($H_2C{=}CH_2$) molecules bonded together. Polymers show low strength and low elastic modulus. In contrast to metals and ceramics, they show a time-dependent mechanical response. This means that the measured mechanical properties of polymers, such as strength and elastic modulus, are dependent on the duration or rate of the mechanical testing procedure, or on the temperature at which the test is performed (Chapter 4). A given polymer, such as PMMA, for example, can show a range of mechanical behavior, from ductile to brittle, depending on the rate or the temperature of the mechanical test. Unless they are brittle, polymers typically show good fatigue resistance. Due to their low hardness, the resistance of polymers to abrasive wear is low.

Polymers generally have low density (~1 g/cm^3), low electrical conductivity, and low thermal conductivity. A clear advantage of polymers over metals and ceramics is their ease of fabrication. Polymers can be easily formed into the requisite shape and microstructure using conventional processing or additive manufacturing (3D printing) methods. Another advantage of polymers is their compositional flexibility. Polymers can not only be synthesized with the requisite composition but their composition can also be easily modified to achieve more desirable properties, such as degradation rate.

Overall, the human body, except for bone and teeth, is composed of soft tissues (and organs). When compared to metals and ceramics, polymers can be more easily designed to approximate the structure and properties of these soft tissues. Consequently, polymers find considerable use as biomaterials. Because of the ease in synthesizing compositions with a controllable degradation rate, polymers also find considerable use as biomaterials for drug delivery.

1.4.4 Properties of Composites

Composites are composed of two or more physically distinct materials or phases (Chapter 12). Synthetic composites used as biomaterials are composed of one or more of the primary classes of materials (metals, ceramics, and polymers), and consist of a continuous phase (the matrix) and a dispersed phase (the reinforcing phase). While composites are abundant in nature, synthetic composites find only limited use as biomaterials. A well-known example of a natural composite is bone, produced by embedded cells and composed of an inorganic (ceramic) phase of fine lath-like particles of composition approximating that of hydroxyapatite and an organic phase composed of collagen.

Synthetic composites are often of interest when a single material cannot provide the desired combination of properties. For example, polymers have an advantage of ease of fabrication but, because of their weak mechanical properties, they are not suitable as implants to heal defects in structural bone that have to support a significant physiological stress. To better approximate the mechanical properties of bone, polymers can be reinforced with a strong material, such as a ceramic in the form of particles or fibers. Use of particles composed of hydroxyapatite or bioactive glass can also enhance the functionality of the polymer matrix, such as its bioactivity.

As ceramics are brittle, they are sometimes reinforced with another phase, typically strong ceramic particles or fibers, to improve their resistance to fracture, a property quantified by a measurable parameter called the fracture toughness. For example, zirconium oxide (ZrO_2) is incorporated into alumina (Al_2O_3) to form a composite, referred to as zirconia-toughened alumina (ZTA), an improved form of which is called alumina matrix composite (AMC). Because of its better fracture resistance (higher fracture toughness) and better wear resistance, AMC has now replaced Al_2O_3 as a ceramic femoral head material in hip implants.

1.4.5 Representation of Properties

Charts provide a succinct way to show a direct comparison of material properties. As a mechanical function is often a major consideration in many biomedical applications, charts that compare mechanical properties can be useful at the outset of materials selection and design. Several types of mechanical properties are available and, depending on the application, some properties can be more important than others. Figure 1.5 shows one type of material property chart, which depicts the measured strength versus elastic modulus for the primary classes of synthetic materials and for selected natural materials. The properties of the synthetic materials shown in Figure 1.5 (and in most mechanical property charts) are those for the dense materials, that is, the materials have zero or near zero porosity. As the presence of porosity lowers the mechanical properties of a material, one approach to modifying the properties of strong synthetic materials in order to achieve more optimal properties for use as biomaterials is to incorporate a controlled amount of porosity into them. In comparison, incorporation of a strong phase into polymers to form a composite provides a way to improve their mechanical properties.

1.5 Case Study in Materials Design and Selection: The Hip Implant

Modern implants for total hip replacement (Figure 1.1b) provide a useful example of the design and selection of biomaterials. While these implants do not reflect the aforementioned trend in which biological sciences are playing a more significant role in biomaterials design, they take into account some advances, such as the response of cells and tissues to implanted biomaterials. Although the hip implant has not changed considerably in its overall geometry since Sir John Charnley first implanted it into patients around 1960, the incorporation of advances in materials science, engineering, and biological principles into materials selection has resulted in the creation of longer-lasting implants having better performance. Currently, approximately half a million total hip joint replacement surgeries are performed annually in the United States and elsewhere in the world, which provide an improved quality of life for this significant number of people.

The human hip joint consists essentially of a femoral head with an approximately spherical surface, which articulates within a socket in the acetabular bone (Figure 1.6a). A prosthetic hip implant attempts to reproduce these geometrical features (Figure 1.6b). It consists of three major parts, a femoral stem, a femoral head (also called a ball), and an acetabular cup (socket) composed of a liner and a shell (Figure 1.7). The stem is inserted into the femur to provide stability for the femoral head, which articulates in the liner of the acetabular cup that is secured to the pelvic

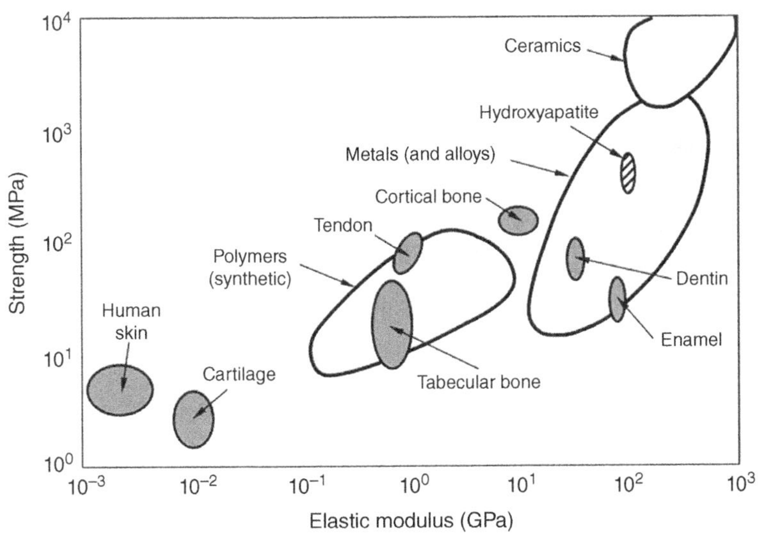

Figure 1.5 Strength versus elastic modulus for the three major classes of synthetic materials used as biomaterials (ceramics, metals, and polymers). The range shown for each class of material is approximate and for the dense (or almost dense) material (zero or close to zero porosity). The presence of porosity in these materials will lead to a reduction in these and other mechanical properties. The properties of some human tissues are shown for comparison.

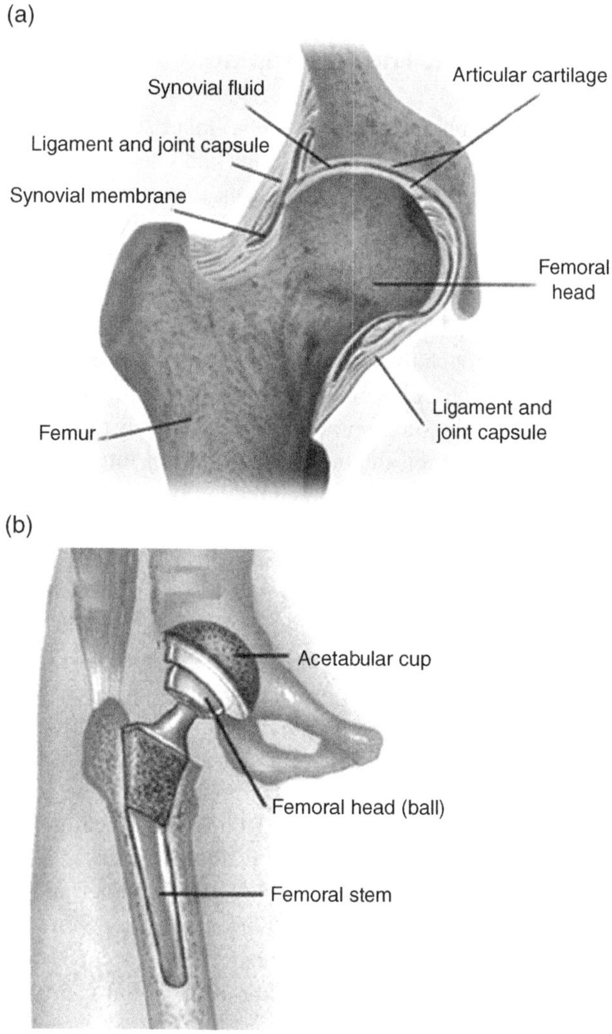

Figure 1.6 Illustration of (a) the human hip joint and (b) an artificial hip joint used to repair a damaged or diseased hip joint. *Source:* From Rahaman et al. (2007) / with permission of John Wiley & Sons.

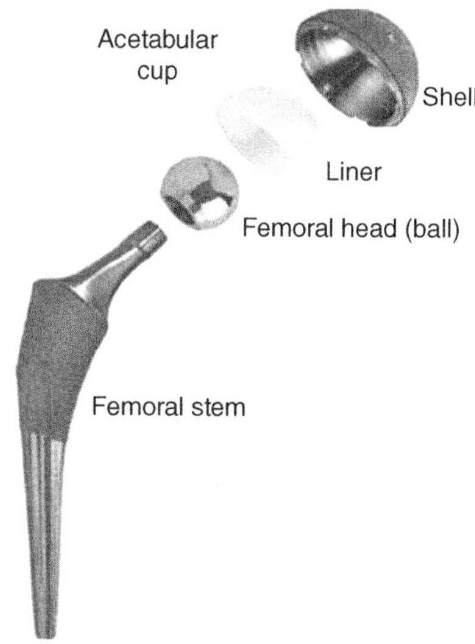

Figure 1.7 Image showing the components of an artificial implant used for total hip joint replacement.

bone. Important factors in materials design and selection for these three components of the hip implant are discussed in the following sections.

Femoral Stem

Major factors in materials design and selection for the femoral stem are mechanical properties, biocompatibility, and ease of fabrication. Mechanical properties are important because the stem supports physiological stresses imposed on the femoral head due to walking, running, and jumping, composed predominantly of repeated cyclic stresses of a compressive and bending nature. The peak force on the hip joint has been measured as three and five times the body weight during walking and running, respectively, resulting in a stress on the femur of an adult of ~10–15 MPa. Additionally, the number of gait cycles in one year is ~10^6 for patients after hip joint replacement. At a minimum, therefore, the selected material should have compressive and bending strengths far in excess of the maximum stress on the femur, high elastic modulus to prevent permanent deformation, and good fatigue resistance to withstand repeated cyclic loading under physiological stresses. These requirements eliminate the use of ceramics, polymers, and polymer matrix composites.

Because hip implants are designed to have a long lifetime, the stem should maintain its mechanical properties and biocompatibility with the surrounding physiological environment over this period. This means that the selected metal should not corrode in the physiological environment. Because the noble metals do not have the requisite mechanical properties, are expensive, or a combination of the two, we are left with the passivated metals. Candidate metals are Ti, certain Ti alloys, stainless steel, and a Co–Cr alloy composed of cobalt (Co), chromium (Cr) and other elements. While high purity or even commercial purity Ti does not have the requisite mechanical properties for this application, a Ti alloy that is widely used in the aerospace and other industries is known to have far better mechanical properties and it is also corrosion resistant. This Ti alloy, containing ~6% aluminum (Al) and ~4% vanadium (V) by weight, is commonly abbreviated Ti6Al4V. While stainless steel and Co–Cr have desirable mechanical properties and corrosion resistance, their densities are approximately twice the value for Ti6Al4V (~4.4 g/cm^3). Based on a combination of mechanical property requirements, corrosion resistance, and weight, femoral stems are now manufactured almost exclusively formed from Ti6Al4V.

Upon implantation, the stem is stabilized (that is, fixed in position) to prevent motion relative to the femur. One method of fixation involves using a polymer (PMMA) cement, often referred to as bone cement. While this cement works well and is commonly used in hip joint surgery, it suffers from a few limitations such as minor toxicity, temperature increase upon setting which can have an adverse effect on the surrounding tissue, and some degree of brittleness, particularly after an extended period.

An alternative is to roughen the surface of the stem to allow bone to grow into the rough regions and stabilize the implant. Another method is to coat the roughened surface with a layer of hydroxyapatite particles. Hydroxyapatite is bioactive and stimulates cells to create new bone, leading to the formation of a strong bond at its interface with host bone. However, due to the difference in thermal expansion coefficient between hydroxyapatite and Ti6Al4V, a strongly adherent and long lasting hydroxyapatite coating on Ti6Al4V has been difficult to achieve. Overall, femoral stems that are not cemented in place often have a rough surface, or a rough surface coated with a hydroxyapatite layer.

Femoral Head

Major requirements for the femoral head (ball) are that it should articulate in the liner of the acetabular cup with as little friction as possible and should have the requisite strength and stiffness to withstand physiological stresses, as noted for the femoral stem. From a mechanics point of view, the ball is retained within the hemispherical acetabular cup and, thus, is subjected to stresses that are predominantly compressive in nature. When compared to the stem, bending stresses are not as serious a consideration but the biocompatibility requirement is still valid. An added requirement is that the ball should be capable of being polished as smoothly as possible to reduce friction and wear when articulating against the liner of the acetabular cup.

Based on the aforementioned mechanical property requirements, a polymer is unsuitable for use as the ball. From a fabrication standpoint, producing the stem and ball together as a single component and from the same material is an attractive option. A serious problem is that Ti6Al4V is difficult to polish to a very smooth finish, is rather sticky and has poor sliding properties when articulating against itself or other materials. Consequently, Ti6Al4V is not suitable as a femoral head material.

Candidate materials for the femoral head are stainless steel and Co–Cr alloys, which were considered in the materials selection for the femoral stem. Ceramics are now also candidate materials due to the absence of significant bending stresses on the femoral head. Stainless steel generally has properties that are inferior to Co–Cr alloys but it is cheaper to manufacture. It is used when cost is an important factor and when the implant is not required to have an average lifetime of longer than ~10 years. The use of Co–Cr results in a longer average lifetime (~10–15 years) of the implant due to its better mechanical properties. Consequently, Co–Cr is the most commonly used femoral head material.

Al_2O_3, a well-known ceramic material, has superior strength and stiffness to stainless steel and Co–Cr alloys, higher hardness and the ability to be polished to a very smooth finish. Although it shows a brittle mechanical response, Al_2O_3 created in the last few decades has shown excellent mechanical reliability due to the use of better quality starting materials and improved fabrication techniques. Al_2O_3 has been replaced by a composite composed of Al_2O_3 reinforced with zirconia (ZrO_2) particles, referred to as ZTA, which has a higher resistance to brittle fracture and a higher wear resistance. More recently, ZTA is being replaced by a composite referred to as AMC, which has similar composition to ZTA but slightly improved mechanical properties due to a modification of the ZTA microstructure. While the potential for brittle failure of AMC (or ZTA) *in vivo* should be recognized, the actual incidence of failure under the physiological stresses of the hip is now very low.

Another ceramic material that has been used as an articulating bearing material in hip implants is yttria (Y_2O_3)-stabilized zirconia (abbreviated YSZ). While YSZ can have comparable or somewhat better mechanical properties than AMC, its manufacturing conditions have to be carefully controlled to achieve these properties. Some compositions of YSZ can also be chemically unstable in an aqueous environment, which can lead to a degradation in mechanical properties and wear resistance *in vivo*. Based on these limitations, YSZ is not currently approved by the US Food and Drug Administration for use as femoral head devices *in vivo*.

When articulating against the liner of the acetabular cup, AMC femoral heads generate a lower amount of wear particles than Co–Cr. Consequently, hip implants produced with AMC femoral heads can have a longer lifetime *in vivo* when compared to similar implants with Co–Cr femoral heads. This improvement in implant lifetime can be a significant factor in the selection of hip implants, particularly for younger patients. However, despite this advantage, femoral heads composed of AMC are used far less frequently than Co–Cr due to their higher manufacturing cost and lingering concerns about their brittle mechanical response.

Acetabular Cup

The femoral head articulates against the liner of the acetabular cup whereas the shell provides mechanical support for the liner and stability of the acetabular cup within the pelvic bone. In addition to a consideration of biocompatibility phenomena, friction, and wear are properties of primary importance for the liner, whereas strength and stiffness are important properties for the shell. Articulation of the femoral head against the liner leads to the production of wear debris, which has been shown to cause adverse responses by tissues surrounding the implant, such as an inflammatory response (Chapter 24). This adverse response, in turn, can lead to loosening of the stem within the femur, necessitating replacement of the implant (revision surgery).

Sir John Charnley selected polytetrafluoroethylene (PTFE) as the liner material in his early implants because it was available off the shelf and was known to have the capacity to be machined and polished to a smooth surface finish with low friction. However, the use of PTFE resulted in a high incidence of implant failure and Sir John Charnley tried another available polymer, PE, which proved to be more successful. Advances in materials science in the last several decades have resulted in the production of PE having better mechanical properties and higher wear resistance. PE with high molecular weight, referred to as ultrahigh molecular weight polyethylene (UHMWPE) or a highly cross-linked UHMWPE is now the most commonly used liner material.

Other candidate liner materials include the metals and ceramics discussed for the femoral head material, which can be polished to a smooth low-friction surface and have a much higher hardness than UHMWPE. Based on their mechanical properties, Co–Cr alloys are worthy of selection and they have been used as the liner in hip implants. However, articulation of Co–Cr femoral heads against Co–Cr liners (sometimes referred to as metal-on-metal bearing couples) have been associated with adverse responses *in vivo*. These bearing couples are prone to scratching and sticking, which reduces the ease of articulation. Of greater importance is the release of metal particles and ions, which has been shown to cause adverse immune responses in patients, such as metal ion sensitivity (Chapter 19). Based on these adverse effects, Co–Cr liners are no longer approved for *in vivo* use by the US Food and Drug Administration.

Ceramic liners composed AMC have been used for articulating against AMC femoral heads over the last few decades. Implants composed of these ceramic-on-ceramic articulating couples produce a much smaller amount of wear particles *in vivo* when compared to the commonly used couple consisting of Co–Cr and UHMWPE. AMC bearing couples show the lowest amount of wear of all the articulating couples currently used in hip implants, and a longer average lifetime when compared to the commonly used Co–Cr on UHMWPE couples. On the other hand, ceramic liners suffer from the same disadvantages described earlier for ceramic femoral heads. They are more difficult to manufacture and more costly than UHMWPE liners, there is a lingering concern about their brittle mechanical response, and they are sometimes prone to developing a squeaking noise when articulating against a ceramic femoral head.

Selection criteria for the shell of the acetabular cup, such as mechanical properties and biocompatibility, are, in general, similar to those already described for the stem. Consequently, Ti6Al4V is the commonly used shell material. The shell can be stabilized in the pelvic bone using Ti6Al4V metal screws or bone cement. Press fitting the shell (with its liner) into the pelvic bone is now also a commonly used procedure. Bone ingrowth into the roughened surface of the shell, as described for the stem, serves to stabilize the acetabular cup in place.

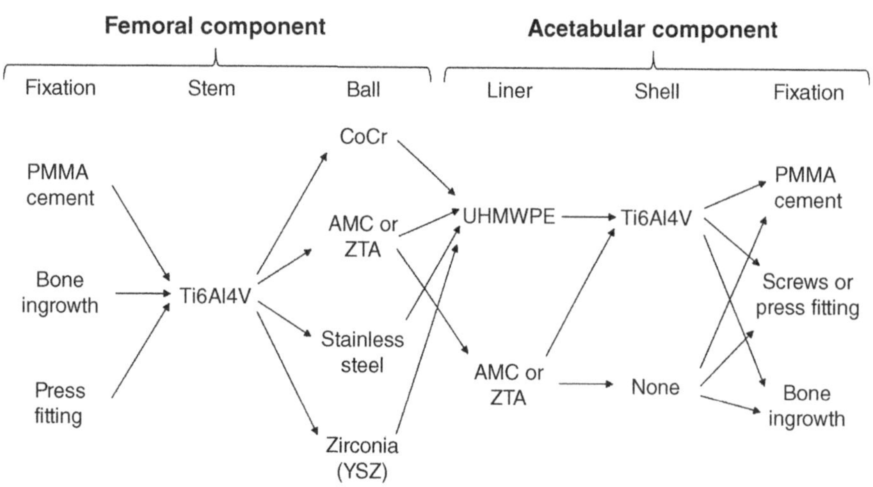

Figure 1.8 Various combinations of materials currently used in implants for total hip replacement.

Modern Hip Implants

Based on the foregoing discussion, we can summarize the materials currently used in implants for total joint surgery (Figure 1.8). Ti6Al4V is used almost exclusively for the femoral stem and the shell of the acetabular cup, while Co–Cr and UHMWPE are the most commonly used combination for the femoral head (ball) and liner of the acetabular cup, respectively. Stainless steel balls are used far less often than Co–Cr. The ball is impacted on to a tapered end of the stem to hold it firmly in place and to allow its removal, if necessary. AMC finds use for both the ball and the liner but far less frequently when compared to the Co–Cr on UHMWPE couple. While AMC bearing couples show the lowest wear rates and the longest average lifetime when compared to all articulating couples used in hip implants, they are expensive to manufacture, their brittleness can be a concern, and a small fraction can suffer from a squeaking noise *in vivo*. AMC balls articulating against UHMWPE liners are also used. They show lower wear than Co–Cr on UHMWPE bearing couples but higher wear than AMC on AMC bearing couples.

1.6 Brief History of the Evolution of Biomaterials

Materials have been used to heal the human body since the early civilizations, although they were not described by the term biomaterials until more recently. Biomaterials have evolved enormously over time, particularly over the last several decades, and they are continuing to evolve with developments in materials science and engineering and a greater emphasis on incorporating biological principles and advances into their design (Section 1.3). While various systems have been used to describe this evolutional history of biomaterials, they all reflect the enormous increase in the sophistication, performance, and application of biomaterials over time. In order to highlight some of the major advances, we provide a brief description of the evolution of biomaterials over time, divided into a few historical periods (Ratner 2013). These periods cover the years prior to World War II, a few decades after World War II, and what can be described as a more contemporary period, covering approximately the last 30–40 years (Figure 1.9).

Prior to World War II

A prosthetic toe formed from wood and leather (both naturally available materials), has been found in an Egyptian mummy dating back to around 1000 BC, while one of the earliest prosthetic arms, composed of iron, is reported to have been used by a Roman general around 200 BC. Archeologists have reported the use of linen sutures by the early Egyptians and catgut sutures in the Middle Ages in Europe. Artificial eyeballs made from gold several centuries ago to replace eyeballs lost as a result of disease or injury evolved into the use of glass by the Venetians in the sixteenth century and, later, in other parts of Europe. Prosthetic teeth, made from gold or carved from dog or calf teeth, were anchored with gold wire or gold plate to form dental restorations which have been found in tombs from the Etruscan civilization dating back to 300–800 BC. Dental implants composed of iron are reported to have been found in a human skeleton dating back to approximately 200 BC in France.

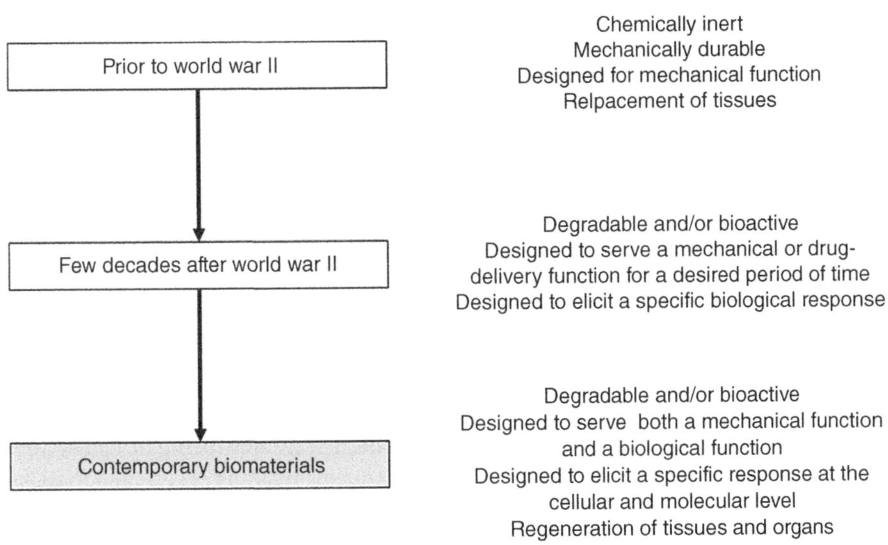

Figure 1.9 Illustrative summary of the evolution of biomaterials classified in terms of historical periods.

A Few Decades After World War II

With the introduction of plastics (synthetic polymers) in the early twentieth century and, as materials designed for use in World War II became more widely available, the first few decades following the World War II saw a rapid growth in the use of synthetic materials as biomaterials. Based on observations showing that fragments of PMMA from shattered canopies of fighter planes in World War II were well tolerated in the eyes of pilots, intraocular lenses were created and implanted in humans by Sir Harold Ridley around 1949. Synthetic arteries (referred to as vascular grafts) made from parachute fabric (silk or nylon), were implanted in a patient by Arthur Voorhees in 1952 to replace a ruptured aneurysm. Sir John Charnley pioneered the hip implant, composed of synthetic metals and polymers, in the late 1950s. An improved version of the implant, composed of stainless steel and high molecular weight PE, showed favorable success rates when implanted in patients the early 1960s. Improved versions of intraocular lenses, vascular grafts, and hip implants, along with several devices created from these chemically inert biomaterials are now implanted in millions of patients.

Contemporary Period

The last 30–40 years have seen a rapid advance in the development and application of biomaterials. Prior to this period, biomaterials were selected from mechanically durable and chemically stable materials that were available off the shelf. These biomaterials were designed to serve a mainly mechanical (or physical) function. The last few decades have seen a significant shift in emphasis in the design of biomaterials (Section 1.3). Whereas materials science played a dominant role in biomaterials design and selection previously, this contemporary period is marked by biological sciences playing a role of significance comparable to that of materials science. Advances in biological sciences are utilized in designing biomaterials that serve not just a mechanical function but also a biological function.

Contemporary biomaterials are degradable or bioactive. Instead of remaining in their original chemically inert form in the body, implants are now being designed to degrade at a controllable rate, be replaced by newly forming tissue, and eventually disappear completely. Additionally, degradable implants can deliver ions and biomolecules to target specific cells and tissues *in vivo*, to control the behavior of cells, or to stimulate cells to create new tissues and organs in the body. In this way, these biomaterials can stimulate the body to heal itself.

Examples of contemporary biomaterials include composites composed of polylactic acid and β-tricalcium phosphate for fracture fixation (Figure 1.1j) and calcium phosphate cements that set *in situ* for healing bone defects. The incorporation of a drug-eluting function in a stent is now becoming common to reduce the tendency for endothelial proliferation and the adherence and clumping of blood cells, thereby improving the long-term efficacy of the stent in treating vascular disease (Section 1.3.1).

Biomaterials are an important component in some tissue engineering (or regenerative medicine) approaches to create functional tissues and organs. At the same time, advances in science and technology are being used to design these biomaterials with greater complexity in structural and chemical characteristics to better mimic the structure of biological tissues. In addition to degradable or bioactive solids, water-filled polymers known as hydrogels, synthetic, or natural, find considerable use in tissue engineering approaches, particularly for creating soft tissues. The stiffness of hydrogels and their structural characteristics can be modified to better mimic those of soft tissues. Functional skin substitutes for healing burns are already available commercially, while there are considerable research and development efforts to create bone, cartilage, liver, and pancreas for implantation in the body.

In addition to medical applications, contemporary biomaterials are playing a significant role in the pharmaceutical industry. In the form of 3D objects or particles of size ~1 μm to a few hundred microns (referred to as microparticles), degradable biomaterials have been used for several decades as devices to deliver drugs in a controlled manner to a particular site in the body. Hydrogels that respond to environmental conditions such as pH or temperature are another type of biomaterials that find use as carriers in drug delivery. Emphasis has been shifting in the last couple of decades toward more precise drug targeting in which the drug molecules are targeted specifically to the cells or tissues where they will exert their desired effect. This is achieved by combining the drug molecules with a suitable biomaterial to which is added a cell-specific antibody or other targeting molecule that has an affinity for the targeted cells or tissues. Increasingly, the biomaterial in these applications is in the form of fine particles, of size smaller than 50–100 nm, referred to as nanoparticles. The combination of chemotherapy drugs and nanoparticles is under development as an alternative technique to radiotherapy or chemotherapy to treat cancer cells. Techniques that rely on the magnetic or optical properties of specially designed nanoparticles are also under development for treating tumors.

Contemporary biomaterials are now being designed and developed for use in diagnostic imaging of cells and tissues. One area of particular interest is the detection and real-time imaging of cancer cells. Here again, biomaterials in the form of nanoparticles, particularly nanoparticles that become highly fluorescent when irradiated with light, are playing a significant role. Nanoparticles that simultaneously combine a drug delivery function and a diagnostic imaging function are under development to provide a technique for real-time diagnosis and treatment of cancer.

1.7 Biomaterials – An Interdisciplinary Field

The significant advances made in the development of contemporary biomaterials, described in the previous section, would not have been possible without the collaboration among researchers from a variety of fields and this collaboration will grow in future. Scientists, engineers, and clinicians working within their respective fields do not possess the range of skills and expertise that are often required to design, create, and evaluate new or improved biomaterials with better performance and functionality for use in medicine and dentistry. The field of biomaterials is a truly interdisciplinary (or multidisciplinary) field that encompasses collaborations from more than one of the following fields: materials science, biological sciences, chemistry, biomedical engineering (or bioengineering), traditional engineering fields such as chemical engineering, mechanical engineering, and electrical engineering, biochemistry, and the medical and dental fields.

1.8 Concluding Remarks

In this chapter, we presented an introductory overview of biomaterials and the biomaterials field. Materials have been used to repair the human body since antiquity, but their use and degree of sophistication have increased significantly over time, particularly over the last several decades:

- In recent years, there has been a radical shift in the approach to designing and selecting biomaterials, in which biological sciences are now playing a significant role, often comparable to materials science and engineering. Instead of the previous approach in which biomaterials were normally selected from durable materials available off the shelf to serve a mainly mechanical function, biomaterials are now being designed to be degradable or bioactive, to deliver biomolecules at targeted sites in the body, and to elicit specific responses from cells and tissues to stimulate the body to heal itself;
- Biomaterials have become a truly interdisciplinary (multidisciplinary) field that serves an important role in society as an academic field, as a research area for developing new or improved devices, and as a manufacturing industry.

Problems

1.1 Distinguish between the terms degradable and bioactive as used to describe biomaterials.

1.2 Discuss the materials design characteristics required for intravascular stents and how the design of the intravascular stent has evolved over the last few decades.

1.3 Discuss the materials design and selection issues for the femoral stem of an artificial hip joint.

1.4 Bearing couples in artificial hip joints are composed of (i) a metal ball articulating against a polyethylene liner (metal-on-polyethylene) or (ii) a ceramic ball articulating against a ceramic liner (ceramic-on-ceramic). Compare the engineering and biological issues related to the use of these two bearing couples.

1.5 Describe the major ways in which the so-called contemporary biomaterials differ from those previously used.

References

Langer, R. and Vacanti, J.P. (1993). Tissue engineering. *Science* 260: 920–926.

Rahaman, M.N., Yao, A., Bal, B.S. et al. (2007). Ceramics for prosthetic hip and knee joint replacement. *Journal of the American Ceramic Society* 90: 1965–1988.

Ratner, B.D. (2013). A history of biomaterials. In: *Biomaterials Science: An Introduction to Materials in Medicine*, 3e (eds. B.D. Ratner, A.S. Hofmann, F.J. Schoen and J.E. Lemons), xli–liii. New York: Elsevier.

Williams, D.F. (1999). *The Williams Dictionary of Biomaterials*. Liverpool, UK: Liverpool University Press.

Williams, D. (2014). *Essential Biomaterials Science*. Cambridge University Press; Chapter 1.

Part II

Materials Science of Biomaterials

2

Atomic Structure and Bonding

2.1 Introduction

A guiding principle in materials science is that the structure of a material controls its properties that, in turn, determine its applications. Consequently, an understanding of the structure of materials is crucial to understanding how their properties arise and how to create materials with optimal properties for specific applications. As biomaterials are typically three-dimensional solids composed of an enormous number of atoms, we are concerned with their structural characteristics at different levels or length scales. In general, two types of structures have the most significant effect on the properties of solids.

The first type of structure is at the atomic level. Atomic structure refers to the type of bonding between the atoms to form molecules. In this level of structure, we also include the way in which the atoms pack together to form a controllable structure. If the atoms pack together in a regularly repeating, ordered manner, the structure is described as crystalline. On the other hand, if the atoms pack together in a random non-ordered manner, the structure is said to be amorphous.

The second type of structure is at a larger scale, referred to as the microstructure. This refers to the nature, quantity, and distribution of the structural components or phases in the solid. A phase is a region of material with uniform physical and chemical properties. Water, for example is composed of a single phase whereas a mixture of ice and water consists of two phases, a solid and a liquid. A pure metal such as titanium is composed of a single phase. On the other hand, at room temperature, the titanium alloy Ti6Al4V is composed of two solid phases. A porous Ti6Al4V implant is composed of three phases, two solid phases and porosity that is considered a gaseous phase.

Table 2.1 shows the main types of structure in materials and their scales. The atomic structure determines the intrinsic properties of the material (Section 1.4), such as the melting point, coefficient of thermal expansion, whether the material is strong or weak, ductile or brittle, electrically conducting or not, and so on. In comparison, many of the properties crucial to the engineering or technological applications of a material are strongly dependent on its microstructure. Most ceramics, for example, are intrinsically strong due to their atomic structure. On the other hand, the actual measured strength relevant to their design for specific engineering applications is strongly dependent on their microstructure, such as the presence of pores within the solid phase, the size of crystals that make up the solid phase, and the presence of microcracks in the solid phase.

The scale of the crystalline or amorphous structure represents the level of the smallest controllable structure in which the atoms are packed to give a solid structure. When present, structural features at the nanoscale (referred to as the nanostructure) often lead to a significant improvement in the engineering properties of a material and can modulate cell and tissue response to the material *in vivo*. In situations where it is not emphasized, the nanostructure is often included within the more general and widely used term microstructure. The macrostructure, such as the external shape or geometry of the material, is often incorporated within the design criteria for the intended application while flaws at the macroscopic level, such as large observable cracks, should be eliminated by proper processing methods.

In this chapter, we will discuss the atomic structure of materials, focusing on the bonding between atoms and molecules. Atomic packing resulting in either crystalline or amorphous structures will be discussed in the next chapter along with the microstructure. Together, these two chapters provide the fundamentals that underlie how the properties of materials are dependent on their structure and how this structure can be modified or manipulated to produce desirable properties for the intended application.

2.2 Interatomic Forces and Bonding Energies

The strength of the bond between two atoms can be described in terms of a force between the atoms which

Materials for Biomedical Engineering: Fundamentals and Applications, First Edition. Mohamed N. Rahaman and Roger F. Brown.

Companion website: www.wiley.com/go/rahaman/biomedicalengineering

binds them together. A simple way of visualizing this force is in terms of a tiny spring that links one atom to the other. However, unlike a mechanical force due to a spring, the force between two atoms is more complex in nature and depends on the particular type of bonding between the atoms. Interatomic forces refer to these forces between atoms. Intermolecular forces refer to forces between molecules whereas intramolecular forces refer to forces within individual molecules. Metals and ceramics, for example, are characterized by strong interatomic forces. In comparison, polymers are characterized by strong interatomic forces between neighboring atoms in their long-chain molecules but weak intermolecular forces between neighboring polymer chains.

Table 2.1 Scales of structure in solids.

Structure	Scale (m)
Atomic structure	10^{-10}
Crystal or amorphous structure	10^{-9}
Nanostructure	10^{-9}–10^{-7}
Microstructure	10^{-7}–10^{-4}
Macrostructure	$>10^{-4}$

Consider two isolated atoms as they approach each other. When their distance of separation x, taken as the distance between the two nuclei, is very large, often stated as infinitely large, the interaction between the atoms is negligible. As the atoms approach and x becomes smaller, an attractive force F_a develops between the atoms. The nature of this attractive force depends on the type of bonding (Figure 2.1).

For two atoms with discrete charges (that is, ions) of opposite sign such as the sodium ion (Na^+) and the chlorine ion (Cl^-) that form an ionic bond in the sodium chloride (NaCl) molecule, for example, this is simply an electrostatic or Coulombic attraction. The attractive force is given by

$$F_a = \frac{q_1 q_2}{4\pi\varepsilon_o x^2} \quad (2.1)$$

where, q_1 and q_2 are the electrical charges of the two ions and ε_o is the permittivity of free space.

On the other hand, for atoms that are bound together by covalent bonds, F_a results from sharing of electrons. The simplest example of covalent bonding is between two hydrogen (H) atoms to form a hydrogen molecule (H_2) (Figure 2.2). Proximity of the two atoms leads to the creation

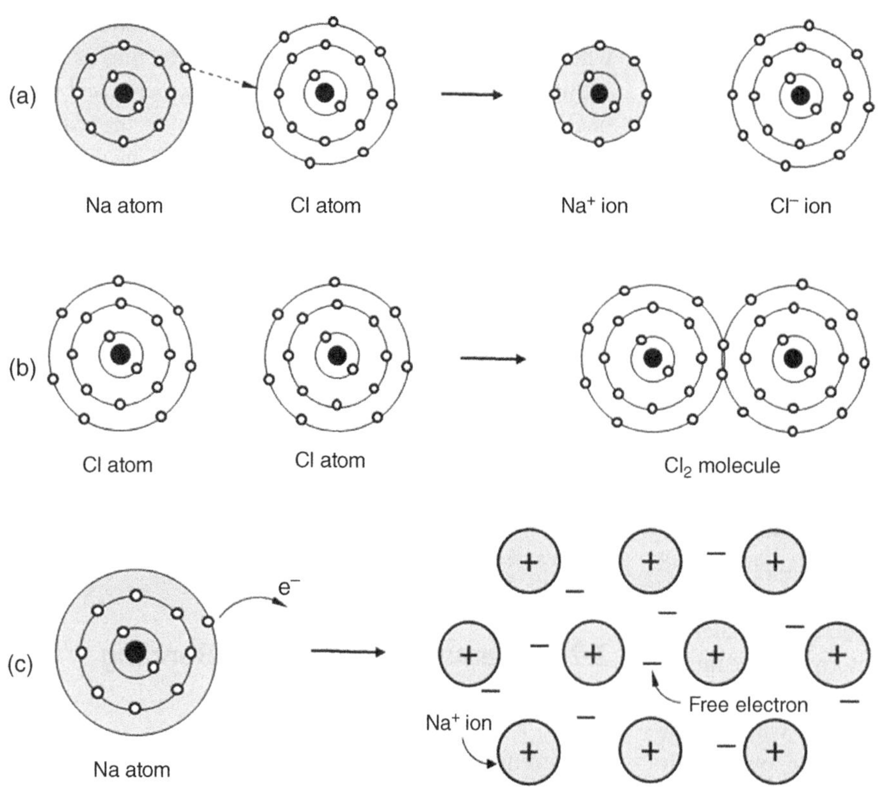

Figure 2.1 Schematic illustration of the formation of (a) ionic bond, (b) covalent bond, and (c) metallic bond.

of a new electron orbital shared by the two atoms into which the two electrons go. This sharing of electrons leads to a reduction in energy and to the formation of a bond. The reduction of energy can be described in terms of an attractive force between the nuclei and shared electrons of the two atoms. In the metallic bond, the attractive force results from the attraction between the positive metal ions and a "sea" of freely moving valence electrons.

For all types of bonding, F_a increases as x decreases (Eq. (2.1)) but when x becomes sufficiently small and the electron cloud surrounding each atom starts to overlap, a repulsive force F_r develops between the atoms. This repulsive force is more than just an electrostatic repulsion between the electron clouds of the two atoms. It has a quantum mechanical component that is rooted in the Pauli exclusion principle and another component that arises from repulsion between the atomic nuclei due to reduced screening of their charge by the overlapping electron cloud. These forces are illustrated in Figure 2.2 for the hydrogen molecule.

Overall, then, the interatomic force consists of two types of forces, an attractive force F_a and a repulsive force F_r, both of which increase in magnitude with decreasing separation distance x between the two atoms. Analyses show that F_r increases more strongly with decreasing x than F_a. The net or resultant interatomic force F is the sum of F_a and F_r. Figure 2.3a is a schematic of F_a, F_r, and F as a function of x. Conventionally, the attractive force and repulsive force are taken to be positive and negative in sign, respectively. Whereas their exact shape depends on the type of bonding (that is, ionic, covalent, or metallic bonding), the curves show the same trend qualitatively regardless of the bond type. As x decreases, F increases, reaches a maximum, and decreases rapidly thereafter.

Instead of the interatomic force, it is often more convenient to describe bonding between two atoms in terms of their potential energy of interaction. This potential energy represents the work done in bringing the two atoms from infinity to a separation distance x. Representation of interatomic bonding in terms of potential energy U is equivalent to that of force F as they are related by the expression

$$U = \int_x^\infty F\,dx \tag{2.2}$$

Substituting for F from Eq. (2.1), we find that for the ionic bond, the attractive potential energy is

$$U_a = \frac{q_1 q_2}{4\pi\varepsilon_o x} \tag{2.3}$$

As shown in Figure 2.3b, U decreases as x decreases (that is, it becomes more negative), reaches a minimum, and

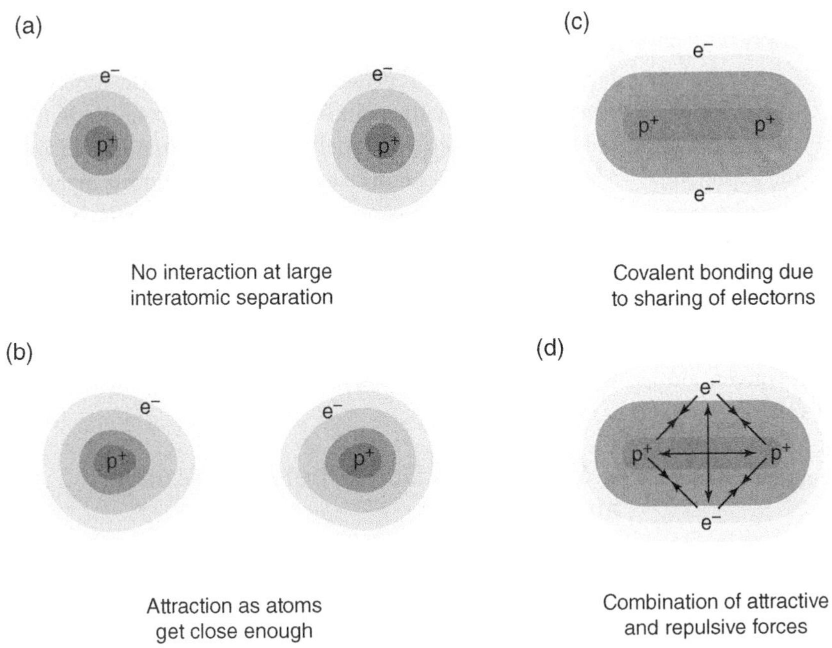

Figure 2.2 Schematic illustration of covalent bonding in the hydrogen molecule (H_2), showing the combination of attractive and repulsive forces.

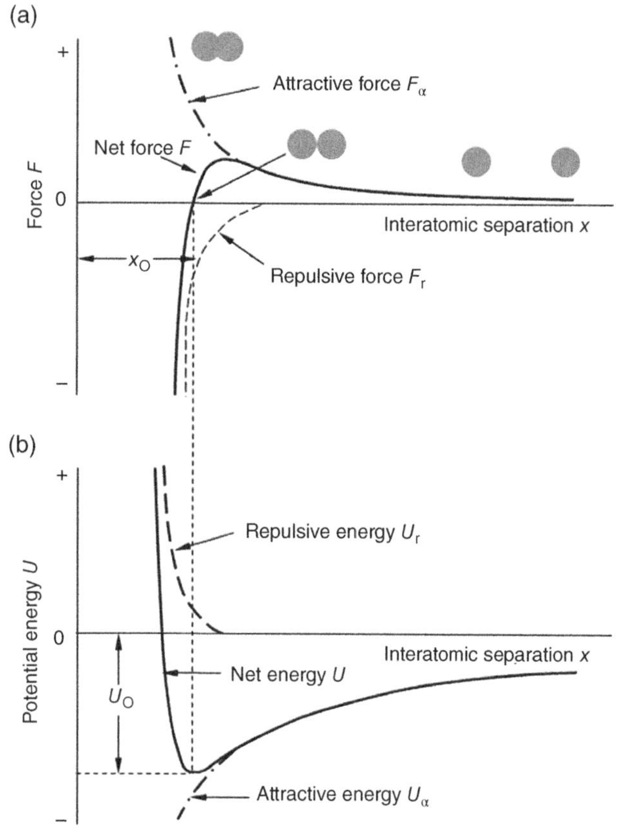

Figure 2.3 Formation of an interatomic bond viewed in terms of (a) interatomic force and (b) potential energy of interaction.

increases rapidly as x continues to decrease. Analyses show that U can be described by an equation of the form

$$U = -\frac{A}{x^m} + \frac{B}{x^n} \tag{2.4}$$

where, A and B are numerical constants and the exponent m is often far smaller than n. In ionic bonding, the first term on the right hand side of Eq. (2.4) is given by Eq. (2.3). For this type of bonding, a correction should be made to take into account the energy required to form the ions from the atoms in the first place, but this is neglected in Figure 2.3b. It is also noticed that the sign convention for potential energy is the opposite of that for the force, that is, the potential energy is taken to be negative and positive in sign, respectively, for attractive and repulsive interactions.

The parameter x_0, the interatomic separation when the net force $F = 0$, is the equilibrium separation between the two atoms and represents the bond length (Figure 2.3). The minimum energy U_0 is the bonding energy of the two atoms and is a measure of the bond strength. It gives the energy required to separate the atoms from their equilibrium spacing to an infinite distance apart. The deeper the well, the more stable the molecule and the larger the amount of energy required to separate the two atoms.

Relationship of Interatomic Force and Bonding Energy to Properties of Materials

As we go from two atoms to solid materials composed of a large number of atoms, the bond length, interatomic force, and bonding energy will be modified by the way the atoms arrange themselves and by their nearest atomic neighbors. Nevertheless, the representations of the interaction between two atoms in Figure 2.3 in terms of interatomic force and bonding energy provide a useful way to understand how many of the intrinsic properties of a material arise. The shape and depth of the potential energy curve, in particular, define various properties.

Molecules with large bonding energy typically exist as solids. Furthermore, the melting point of these solids typically increase as the depth of the potential well increases. This is because a larger amount of energy, in the form of heat, is required to break the interatomic bonds and convert the material from a solid to a liquid (Table 2.2). The elastic modulus is a measure of a material's stiffness, that is, its ability to initially deform when subjected to a mechanical load. It is proportional to the slope of the F versus x curve at $x = x_0$ (Figure 2.4). A steep slope represents a strongly bonded material with high elastic modulus whereas a gradual slope represents a weakly bonded material with low elastic modulus (Table 2.3).

As F and U are related through Eq. (2.2), an alternative view is that a deeper potential energy well (that is, a larger bonding energy), represents a higher elastic modulus. Another property that can be understood from the potential energy curve is the coefficient of thermal expansion. The shape of the potential energy well provides a measure of the relative expansion coefficient of materials. A symmetrical potential energy well, for example, means that there is no change in the average interatomic separation and, thus, no expansion of the material when it is heated. On the other hand, the more asymmetric potential energy well, the larger the thermal expansion coefficient.

2.3 Types of Bonds between Atoms and Molecules

The types of bonds in materials can be classified broadly into two main types: primary bonds and secondary bonds. Primary bonds are interatomic bonds, that is, chemical bonds between atoms. They are strong bonds characterized by a high bond strength and a high bonding energy in the range of ~20–250 kJ/mol (Table 2.2). Primary bonds are

Table 2.2 Bonding energy and melting temperature of various substances.

		Bonding energy		
Type of bond	**Substance**	**kJ/mol**	**eV per atom, ion or molecule**	**Melting temperature (°C)**
Ionic	NaCl	640	3.3	801
	MgO	1000	5.2	2800
Covalent	Si	450	4.7	1410
	C (diamond)	713	7.4	>3550
Metallic	Hg	68	0.7	−39
	Al	324	3.4	660
	Fe	406	4.2	1538
	W	849	8.8	3410
Van der Waals	Ar	7.7	0.08	−189
	Cl_2	31	0.32	−101
Hydrogen	NH_3	35	0.36	−78
	H_2O	51	0.52	0

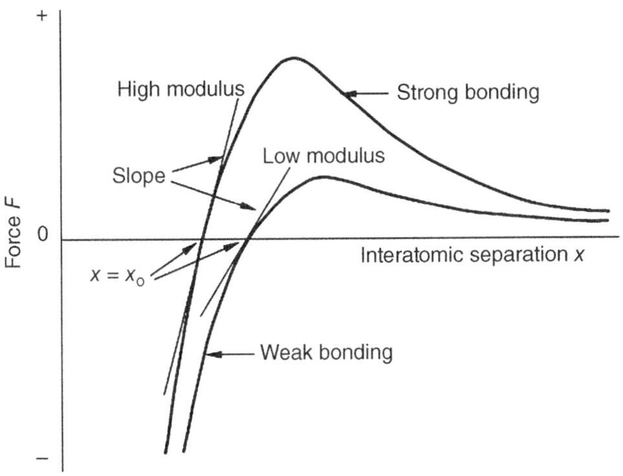

Figure 2.4 Relationship between interatomic force versus displacement curve and the Young's modulus of a solid.

Table 2.3 Calculated stiffness and Young's modulus (the elastic modulus in tension or compression) for various types of bonds.

Type of bond	**Stiffness (N/m)**	**Young's modulus (GPa)**
Covalent	50–180	200–1000
Ionic	8–24	30–90
Metallic	15–75	60–300
Hydrogen	2–3	8–12
Van der Waals	0.5–1	2–4

formed by transfer of one electron or more from one atom to another which leads to the formation of ions with discrete charges, or by sharing of electrons between atoms. There are three types of primary bonds: ionic, covalent, and metallic bonds.

Secondary bonds are intermolecular bonds. They are physical bonds formed by attraction of molecules that are nonpolar or that have a permanent dipole moment and, thus, they do not involve chemical transfer or sharing of electrons. Secondary bonds are weak, with a low bonding energy of ~2–12 kJ/mol (Table 2.2). Common types of secondary bonds are the van der Waals bond and the hydrogen bond.

2.4 Primary Bonds

The type of primary bond formed between two atoms depends on their position in the periodic table of elements. Ionic bonding involves transfer of electrons from one atom to the other, resulting in the formation of a positive ion (cation) and a negative ion (anion) as, for example, the creation of sodium (Na^+) and chlorine (Cl^-) ions to form a sodium chloride molecule (NaCl) (Figure 2.1a). The transferred electrons reside in the outermost shell of the anion and, thus, are almost bound to that ion. Covalent bonding, on the other hand, involves sharing of electrons between two atoms but the shared electrons are bound between the atoms as, for example, in the chlorine molecule (Cl_2) (Figure 2.1b). In the metallic bond, electrons tend to leave the parent atoms, making them cations, and combine to form a sea of highly mobile electrons, as illustrated in Figure 2.1c for sodium (Na). In contrast to the ionic bond, the electrons that leave the metal atoms are not transferred to any one ion. The electrons involved in the formation of primary bonds are typically those in the outermost orbital of the atoms, referred to as valence electrons.

Guidelines have been proposed to predict the type of primary bond that two atoms will form. We provide a description of a few of these guidelines.

The Octet Rule

A useful guideline is the octet rule. The basis of this rule is that the unreactive nature of the inert gases in Group VIII of the periodic table is related to their outermost electron shell composed of eight electrons, except for helium that has two electrons in its outer shell. The electron configuration of argon (Ar), for example, written as $1s^22s^22p^63s^23p^6$, has two electrons in the 3s subshell and six in the 3p subshell, making eight electrons in the outermost shell. According to the octet rule, when two atoms combine to form a molecule, they do so in a manner to form a complete octet in their outermost shell as far as possible. Then the electron configuration of each atom will correspond to that of its closest inert gas in the periodic table. The octet rule is illustrated for the sodium chloride (NaCl) and methane (CH_4) molecules.

NaCl Molecule

The sodium (Na) and chlorine (Cl) atoms have the following electronic configurations:

Na: $1s^22s^22p^63s^1$ which can be written [Ne]$3s^1$, where [Ne] represents the electronic configuration of the inert gas neon (Ne)
Cl: $1s^22s^22p^63s^23p^5$ which can be written [Ne]$3s^23p^5$

According to the octet rule, in combining to form an NaCl molecule, the sodium atom can lose an electron (the $3s^1$ electron) to from a stable octet in the outermost shell while the chlorine atom can gain an electron in its outermost shell (Figure 2.1). This results in the formation of Na^+ and Cl^- ions with an electronic configuration corresponding to Ne and Ar, respectively. Electrostatic attraction between the cations and anions leads to the formation of an ionic bond.

CH_4 Molecule

The carbon (C) and hydrogen (H) atoms have the following electronic configurations:

C: $1s^22s^22p^2$
H: $1s^1$

The formation of a CH_4 molecule can be achieved by the carbon atom sharing each of its four electrons in the outermost shell with an H atom. As described later (Section 2.7), the 2s orbital and three 2p orbitals in the carbon atom combine to form four hybrid orbitals (called sp^3 orbitals), each occupied by one valence electron. By sharing one electron in each orbital with an electron from the hydrogen atom, the carbon atom achieves an outer shell of eight electrons, corresponding to the neon configuration, while the hydrogen atom has two electrons in its shell, corresponding to the helium configuration.

The octet rule is a useful guideline but there are exceptions to it. Furthermore, it does not consider electrons in the d and f shells of atoms. As only s and p electrons are considered, the octet rule is a useful rule for predicting the type bonding in the main group elements in the periodic table, that is, elements not in the transition metal blocks.

Another guiding principle in interatomic bonding is that when atoms lose, gain, or share electrons to form an octet in their outermost shell, they do so in a manner to minimize the number of charges that have to be transferred or shared. An alternative way to form an ionic bond between sodium and chlorine atoms, for example, is by the transfer of seven electrons from the outer shell of the chlorine atom to the sodium atom, forming Cl^{7+} and Na^{7-} ions, respectively. However, this does not occur because the energy required to create the Cl^{7+} and Na^{7-} ions with these many charges in the first place is exorbitant.

Electronegativity of Atoms

The ability of an atom to lose, gain, or share electrons is crucial to the type of bond that it forms with other atoms. This has been quantified in terms of a measureable property called the electronegativity, defined as the ability of an atom in a particular molecule to attract electrons to itself. Linus Pauling originally developed an electronegativity scale for atoms in the 1930s, referred to as the Pauling electronegativity scale, based on measurements of the strengths of covalent bonds between different elements. In this scale, atoms have an electronegativity value between 0 and 4, a range selected arbitrarily by Pauling. A limitation of Pauling's method is that many elements do not form stable covalent compounds with other elements and, thus, their electronegativity cannot be measured. Other methods have since been developed which address this problem but the Pauling electronegativity scale is still widely used. For the main group elements, that is, elements not in the transition metal blocks, the electronegativity values increase as the atomic number decreases in a particular column of the periodic table and increase with atomic number along a particular row.

Empirical relations based on electronegativity values have been proposed to predict or compare the type of bond but they provide only an approximation and involve some degree of arbitrariness. One simple relation, based on the Pauling electronegativity scale, considers the difference in electronegativity, ΔEN, as follows:

- $\Delta EN > 2.0$: ionic bond
- $\Delta EN < 0.4$: covalent bond
- $0.4 < \Delta EN < 2.0$: combination of ionic and covalent character

While these ΔEN values are useful for predicting whether a bond is predominantly ionic or covalent, there is no sharp distinction between an ionic bond and a bond that has a combination of ionic and covalent character, and between a covalent bond and a bond that has a combination of ionic and covalent character. Empirical equations have been developed to predict the percentage of ionic character I of a bond between two atoms with a difference in electronegativity ΔEN. One such equation is

$$I(\%) = \left\{1 - \exp\left[-0.25(\Delta EN)^2\right]\right\} \times 100 \tag{2.5}$$

However, the exact nature of a bond that has a certain percentage of ionic and covalent character is unclear.

Polarity of Covalent Bond

The formation of an ionic bond involves the transfer of one electron or more from one atom to the other, creating ions with discrete positive and negative charges (Figure 2.1a). On the other hand, the covalent bond involves sharing of electrons between two atoms (Figure 2.1b). When the atoms forming the covalent bond are the same, such as two hydrogen atoms or two chlorine atoms to form a hydrogen molecule (H_2) or a chlorine molecule (Cl_2), the bonding electrons are shared equally between the two atoms (Figure 2.5a). However, when the two atoms are different, such as a hydrogen atom and a chlorine atom forming a hydrogen chloride molecule (HCl), the bonding electrons in the covalent bond are not shared equally between the atoms. The chlorine atom has a greater capacity to attract the shared electrons to itself, so due to its higher electronegativity, the shared electrons are attracted more strongly to the chlorine atom than the hydrogen atom (Figure 2.5b). Because of this unequal sharing, the chlorine atom develops a slightly negative character whereas the hydrogen atom develops a slightly positive character. The covalent bond is said to be polar, to consist of a dipole or to have a dipole moment. When appropriate to emphasize the polar nature of the molecule, the more electronegative atom in the bond is shown with a partial negative charge $\delta-$ symbol while the less electronegative atom is shown with a $\delta+$ symbol (Figure 2.5).

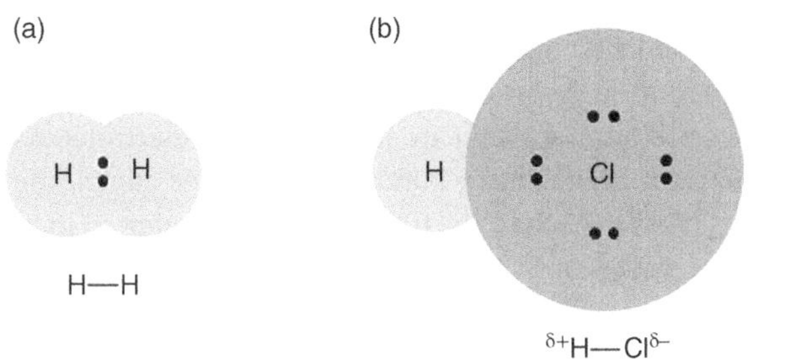

Figure 2.5 Illustration of (a) nonpolar bond in hydrogen molecule (H_2) and (b) polar bond in hydrogen chloride molecule (HCl).

2.4.1 Ionic Bonding

The electrons in each ion occupy orbitals of complicated shape but, at an approximate level, we can assume that the ions have a spherical shape. Thus, in ionic bonding, one ion can be attracted to the other in any direction and, for the same interatomic distance, the force of attraction does not depend on the direction. We say that that ionic bond lacks directionality or is nondirectional. In packing to form a three-dimensional structure, one type of ion will surround itself with as many ions of the other type as far as possible to maximize the attraction between cations and anions and to minimize the repulsion between ions of the same charge (Chapter 3). In this way, the attractive potential energy of the system is maximized. Another feature of ionic bonding is that in the creation of ions, an electron transferred from one atom to the other atom remains localized at this second atom and cannot move freely. These features of the ionic bond provide a basis for understanding the characteristic properties of ionic solids (Section 1.4):

- High strength, high elastic modulus, high melting point, and chemically unreactive due to the strong interatomic bonding
- Low electrical and thermal conductivity due to the localized electrons
- Brittleness: Permanent deformation of an ionic solid due to a large enough mechanical force or stress requires motion of ions from their equilibrium sites to adjacent equilibrium sites. For this to occur, ions in one plane of the solid must move over or near to ions of the same sign in the adjacent plane. This leads to a strong repulsion and to catastrophic fracture of the solid into two or more pieces
- High hardness: As hardness is related to the resistance of the material to be permanently deformed, ionic solids typically show high hardness

Ceramics are held together by ionic or covalent bonds, or by bonds with a mixture of ionic and covalent character. Although they are not useful as biomaterials, sodium chloride (NaCl), and magnesium oxide (MgO) are common examples of ceramics that show ionic bonding. Ceramic biomaterials that show predominantly ionic bonding include aluminum oxide (Al_2O_3) used as bearings in some hip implants, and the calcium phosphates hydroxyapatite, $Ca_{10}(PO_4)_6(OH)_2$, and tricalcium phosphate, $Ca_3(PO_4)_2$, used to repair bone defects (Table 1.1).

Example 2.1 Sodium chloride (NaCl) and magnesium oxide (MgO) are both ionic solids and they have the same crystal structure. However, MgO has a melting temperature (2800 °C) which is far higher than that of NaCl (801 °C). Explain why.

Solution:
Two factors are mainly responsible: (i) the charge of the ions and (ii) the size of the ions, taken as the ionic radius, which determines the equilibrium interatomic spacing.

i) *Charge of the ions*: The sodium atom (atomic number = 11; ground state configuration = $1s^2 2s^2 2p^6 3s^1$) loses one electron to become a sodium ion (Na^+) while the chlorine atom (atomic number = 17; ground state configuration = $1s^2 2s^2 2p^6 3s^2 3p^5$) gains one electron to become a chlorine ion (Cl^-). In comparison, the magnesium atom (atomic number = 12; ground state configuration = $1s^2 2s^2 2p^6 3s^2$) loses two electrons to become a magnesium ion (Mg^{2+}) while the oxygen atom (atomic number = 8; ground state configuration = $1s^2 2s^2 2p^4$) gains two electrons to become an oxygen ion (O^{2-}). The attractive potential energy (Eq. (2.3)) between the ions depends on the product of the ionic charges. For NaCl, this is (1×1) whereas for MgO, this is (2×2). Thus, for a given distance of separation, the attractive potential energy for the MgO bond is four times that for the NaCl bond.
ii) *Size of the ions*: The electronic configuration of the Na^+ and Mg^{2+} ions are the same ($1s^2 2s^2 2p^6$). On the other hand, the Mg^{2+} ion has a higher nuclear charge than the Na^+ ion and, thus, it has a smaller ionic radius due to the greater attraction between the nucleus and the electrons remaining in the ion. The O^{2-} ion has an electronic configuration ($1s^2 2s^2 2p^6$) composed of one less shell than the Cl^- ion ($1s^2 2s^2 2p^6 3s^2 3p^6$). Although the oxygen atom gains two electrons to form an O^{2-} ion whereas the chlorine atom gains one electron to form a Cl^- ion, it might be expected that the O^{2-} ion will have a somewhat smaller radius due to it having one less electron shell. Thus, the equilibrium interatomic separation of the molecule, taken approximately as the sum of the ionic radii, is smaller in MgO than in NaCl.

We see from Eq. (2.3) that the potential energy of attraction is proportional to the product of the charges and inversely proportional to the interatomic distance. Thus, based on the ionic charges and interatomic separation, the attractive potential energy at a separation approximating the equilibrium separation distance for MgO is at least four times higher than for NaCl. Although we have not taken into account the repulsive potential energy, Figure 2.2 indicates that it is far smaller than the attractive potential energy. Overall, it will take a much larger amount of energy to disrupt the MgO bond and, thus, the melting point of MgO is much higher than that of NaCl.

2.4.2 Covalent Bonding

Covalent bonding involves sharing of electrons between two atoms but the shared electrons are localized between the two atoms and do not move freely. Depending on the number of valence electrons, atoms can each share one or more electrons, giving single, double or triple bonds. Common examples include hydrogen (H_2) and chlorine (Cl_2) molecules that share one electron each (single bond), oxygen (O_2) sharing two electrons each, and nitrogen (N_2) molecules sharing three electrons each.

Hybrid Orbitals

Some atoms such as those with electrons in their s, p, and d subshells, for example, can also form hybrid orbitals, each orbital being equivalent and having the same energy. As carbon, nitrogen and oxygen atoms are common constituents of many biomaterials, they are used here to illustrate the formation of hybrid orbitals. The 2s orbital and the three 2p orbitals in the carbon atom, for example, can combine to form four sp^3 orbitals, each containing one electron (Figure 2.6). Formation of these hybrid orbitals can lead to an overall reduction in energy and, thus, to a more stable compound when the atom with the hybridized orbitals forms a bond with other atoms. Nitrogen and oxygen atoms can also form sp^3 hybrid orbitals but one orbital in nitrogen and two orbitals in oxygen are filled with two electrons. As they are not available for bonding with other atoms, the two electrons in each of the filled orbitals are referred to as lone-pair electrons.

Hybrid orbitals are formed at specific angles to one another and, as bonds are formed in the directions of greatest orbital overlap, these orbitals give directionality to the bond that, in turn, determine the arrangement of the atoms in three dimensions and how the atoms pack to form solids.

Hybridization of s and p orbitals, for example, can occur in the following ways (Figure 2.7):

- One s orbital and one p orbital to form two sp orbits that are diametrically opposed to each other. The angle between the orbitals is 180°, giving a linear arrangement of the orbitals
- One s orbital and two p orbitals to form three sp^2 orbitals at 120° to one another. The orbitals form a triangular arrangement in a single plane and they are said to form a trigonal planar arrangement
- One s orbital and three p orbitals to form four sp^3 orbitals at 109.5° to one another. The orbitals point to the corners of a tetrahedron.

When a carbon atom bonds with four hydrogen atoms to form a methane molecule (CH_4), for example, the shared electrons point to the corners of a tetrahedron (Figure 2.8). On the other hand, when oxygen or nitrogen atoms with sp^3 hybridized orbitals bond with other atoms such as hydrogen atoms to form water (H_2O) or ammonia (NH_3) molecules, respectively, the angle between the OH bonds in H_2O or between the NH bonds in NH_3 differs slightly from 109.5° due to the presence of the lone-pair electron orbitals. The bonds and lone-pair electron orbitals form a slightly distorted tetrahedral arrangement, with the angle

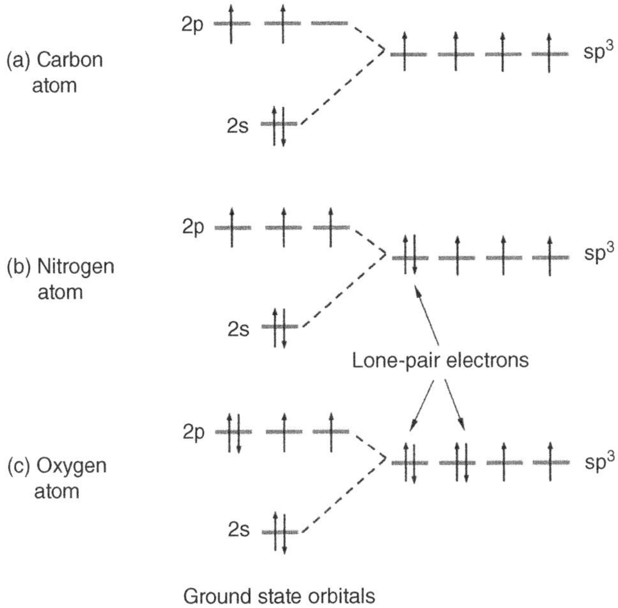

Figure 2.6 Schematic illustration of the formation of four sp^3 orbitals in (a) the carbon atom, (b) the nitrogen atom, and (c) the oxygen atom. Each hybrid orbital in carbon can form a covalent bond by sharing an electron from another atom. In comparison, one orbital in nitrogen and two orbitals in oxygen are filled with two electrons, called lone-pair electrons, which are not available for covalent bonding but available for hydrogen bonding.

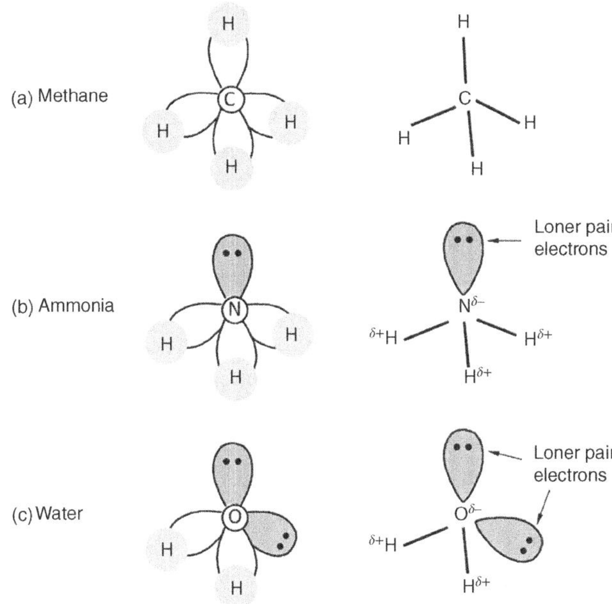

Figure 2.8 Schematic illustration of covalent bonding in methane, ammonia, and water molecules. The polarity of the bonds and lone pair electrons in ammonia and water are also shown.

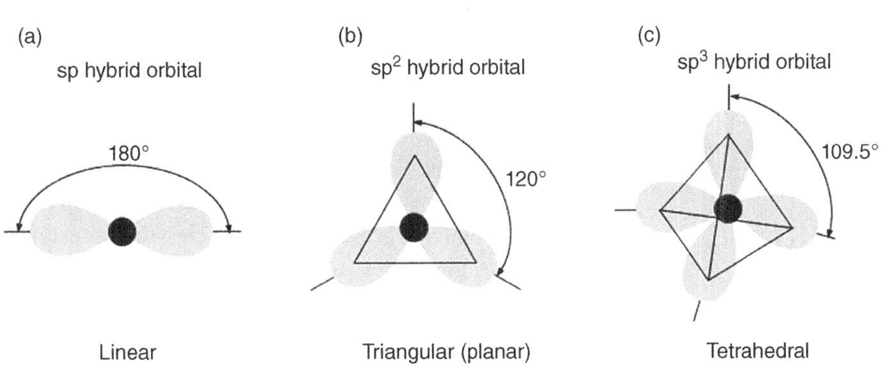

Figure 2.7 Schematic comparison of the directionality of sp, sp^2, and sp^3 hybrid orbitals.

between the OH bonds in H_2O and between the NH bonds in NH_3 having a value of 104.5° and 107°, respectively.

The hybridized electron orbitals in atoms such as carbon, nitrogen, and oxygen can also lead to the formation of double and triple bonds when these atoms bond to the same or other atoms. The oxygen molecule (O_2), for example, has a double bond between the atoms while the nitrogen molecule (N_2) has a triple bond between the atoms. To visualize the formation of single, double, and triple bonds between atoms, we can compare the structures of the hydrocarbon molecules ethane (C_2H_6), ethylene (C_2H_4), and acetylene (C_2H_2) (Figure 2.9). The formation of the ethane molecule, for example, can be viewed in terms of bonding among two carbon atoms, each having four sp^3 hybrid orbitals, and hydrogen atoms. Each bond in the molecule shares two electrons and points directly at each other along the internuclear axis. This type of bond is called a sigma (σ) bond and it is the most common type of covalent bond.

In the ethylene molecule, each carbon atom has three sp^2 hybrid orbitals and one unhybridized p orbital. Each sp^2 orbital forms a σ bond with another carbon atom and with a hydrogen atom. In contrast, the two p orbitals, one from each carbon atom, can share their electrons in a side-by-side manner, perpendicular to the internuclear axis, to complete the valence shell of the carbon atom. This type of bond is called a pi (π) bond. The combination of the σ and π bonds results in a double bond between the carbon atoms. In the acetylene molecule, we have one σ bond between the two carbon atoms, due to sp hybrid orbitals in each carbon atom, plus two π bonds from the two unhybridized p orbitals, making a triple bond between the carbon atoms.

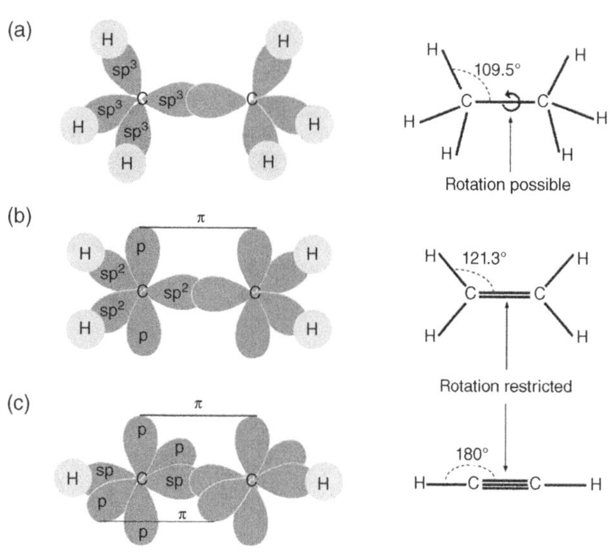

Figure 2.9 Illustration of the formation and geometry of single, double, and triple bonds in (a) ethane, (b) ethylene, and (c) acetylene.

In the case of the hydrocarbons, such as ethane, ethylene, and acetylene, for example, a triple bond between two carbon atoms has a higher bonding energy and a shorter interatomic distance than a double that, in turn, has a higher bonding energy and shorter interatomic distance than a single bond. Another important distinction between the single, double, and triple bonds is the freedom of one carbon atom to rotate about the internuclear axis with the other carbon atom. As the energy barrier for rotation is low, molecules can rotate freely around a single carbon–carbon bond. On the other hand, rotational motion is highly restricted around a double or a triple carbon–carbon bond. The ethylene molecule has a planar shape whereas the acetylene molecule is linear. Similar trends hold for bonds between other pairs of atoms, such as between carbon and nitrogen atoms. Overall, then, the presence of double and triple bonds has important consequences for the structure and, thus, the properties and function of covalent-bonded molecules.

Covalent Bonding in Ceramics

Covalent-bonded ceramics are characterized by intrinsic properties similar in nature to those described for ionic-bonded ceramics (Section 2.6) but as the covalent bond can be stronger than the ionic bond, many covalent-bonded ceramics have better mechanical properties, such as strength, elastic modulus and hardness, and a higher melting point.

Covalent bonding appears in diamond, a crystalline form of carbon, often regarded as a prototypical example of covalent-bonded ceramics. The atoms bonded to a given atom point to the corners of tetrahedron due to sp^3 hybridization of the electron orbitals of the carbon atoms, giving a directional form of bonding (Chapter 3). Covalent bonding is the dominant type of bonding in many other ceramics, including carbides such as silicon carbide (SiC), and nitrides such as silicon nitride (Si_3N_4), and in the network structure of glasses (Chapter 3). It contributes to the bonding in some oxides such as Al_2O_3 and high melting point metals such as tungsten and tantalum.

Covalent-bonded ceramics find some use as biomaterials. Examples include silicon nitride used as implants in spinal repair, a few nondegradable glasses in some medical devices and a few bioactive glasses used in healing bone defects and skin wounds (Chapter 7). Carbon materials such as diamond and graphite are often classified as ceramics but their properties are more relevant to industrial applications (Chapter 3). Pyrolytic carbon is an important material in the manufacture of heart valves whereas diamond-like carbon (DLC) has been investigated as coatings for articulating bearings in hip and knee implants. More recently discovered allotropes of carbon such as fullerenes, carbon nanotubes, and graphene, have been the subject of an enormous amount of investigations for use in technological applications and are now receiving considerable interest for potential biomedical applications.

Covalent Bonding in Polymers

Covalent bonding occurs in synthetic and natural polymers, linking atoms in the backbone of the long-chain molecule, often referred to as a macromolecule. The macromolecules in the synthetic polymer polyethylene, a material used as articulating bearings in hip and knee implants, has the simplest structure, built up of $-H_2C-CH_2-$ repeating units (Figure 2.10). These macromolecules show strong covalent bonds between the carbon atoms in the chain backbone and a tetrahedral arrangement of the covalent bonds at each carbon atom in the chain resulting from sp^3 hybridization of the electron orbitals of the carbon atom. The single bond between the carbon atoms in the chain backbone allows one carbon atom to rotate more or less freely, relative to a neighboring carbon atom and this, coupled with weak intermolecular forces between neighboring chains, often lead to a random arrangement of the chains in three dimensions. On the other hand, if the repeating unit has a simple structure as, for example, in polyethylene, some regions of the macromolecule can arrange themselves in an ordered three-dimensional pattern. When formed under appropriate conditions, such as cooling sufficiently slowly from the molten state, polyethylene is composed of crystalline regions interspersed with amorphous regions, a structure often described as semicrystalline (Chapter 3).

Other synthetic polymers have a structure that is more complex than polyethylene but similar principles are applicable (Chapter 8). Consequently, most solid polymers produced by conventional methods show low strength and low elastic modulus, become soft or molten at low to moderate temperatures, and are insulating electrically and thermally. On the other hand, alignment of the polymer chains in a certain direction and hydrogen bonding, a strong form of intermolecular bonding (Section 2.5.2), can lead to high strength and high elastic modulus in the direction of alignment as, for example, in fibers of the synthetic polymer nylon.

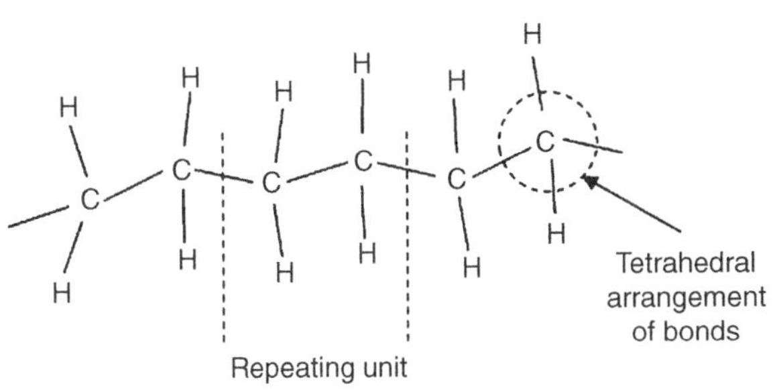

Figure 2.10 Illustration of covalent bonds linking carbon atoms (C) in the chain backbone of the polyethylene molecule and between the carbon atoms and hydrogen (H) atoms.

2.4.3 Metallic Bonding

The metallic bond is the dominant bond in metals and their alloys. Because the metal atoms have low electronegativity, the valence electrons tend to leave the parent atom, making the atom a cation, and form a "sea" of rather free electrons that surrounds the ions (Figure 2.1c). The valence electrons are shared by the cations and are not localized at any one cation. Bonding arises from the strong attraction between the cations and the sea of delocalized electrons. Consequently, the bond has no directionality and the cations arrange themselves to form simple densely packed crystalline structures in solids (Chapter 3). The packing of the cations in the solid can be visualized in a simple manner as the ordered packing of tiny ball bearings.

Except when there is a covalent contribution, as in the high-melting point metals such as tungsten and tantalum, the metallic bond is somewhat weaker than the ionic or covalent bond. Thus, many metals have a lower strength, lower elastic modulus, and lower melting point than most ceramics (Figure 1.5). The sea of mobile electrons is responsible for the high electrical conductivity of metals and this sea of electrons, together with a greater ability of the cations to undergo thermal vibration, is responsible for their high thermal conductivity. Another attractive property of metals is their ductility, the ability to deform when subjected to sufficiently high mechanical stresses instead of fracturing in a brittle manner, characteristic of ceramics. This ductility is due to the ease with which the metallic bond can be broken and reformed when compared to the ionic or covalent bond, coupled with the ease with which the sea of electrons can adjust itself to accommodate the change in position of the cations (Chapter 3). Metals typically show moderate hardness that is lower than that of most ceramics due to their ability to deform in a ductile manner. Corrosion resistant metals find considerable use as biomaterials in a variety of devices such as fracture fixation plates, implants for total joint repair, and dental implants because of their attractive mechanical properties, and electrodes in pacemakers and neural stimulators because of their high electrical conductivity (Table 1.1).

To summarize at this stage, features of the three types of primary bonds are illustrated in Figure 2.11.

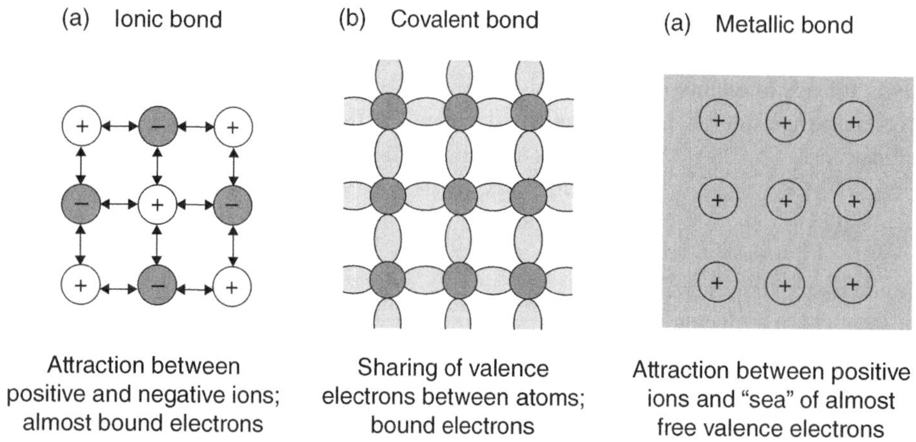

Figure 2.11 Schematic comparison of the attractive interactions in the ionic, covalent, and metallic bond.

2.5 Secondary Bonds

Although considerably weaker than the primary bonds (Table 2.2), secondary bonds still have a significant effect on the properties of some materials. Secondary bonds are typically intermolecular bonds. Intermolecular bonding between neighboring chains, for example, is responsible for polyethylene and many other polymers existing as a solid at room temperature. Without intermolecular bonding between its polar molecules, water would boil at approximately −80 °C instead of existing as a liquid at room temperature. Proteins are the most versatile macromolecules in living organisms and they serve crucial functions in essentially all biological processes. As described in Section 2.6, intermolecular bonding in these macromolecules controls the shape (conformation) that these macromolecules take up and, thus, their functions. Secondary bonds are often divided into two main types: van der Waals bonds and hydrogen bonds.

2.5.1 Van der Waals Bonding

Van der Waals bonding is the weakest type of intermolecular bonding. It exists between all atoms and molecules regardless of what other interactions might be present. There are three types of van der Waals bonds, classified in terms of the type of attractive force between the molecules.

- Dispersion forces, sometimes referred to as London forces, are the type of force that exists between nonpolar atoms such as argon, or between nonpolar molecules such as nitrogen. To see how this type of attractive force arises, we will consider the argon atom (Figure 2.12a). Although the argon atom (Atom 1) has a symmetrical distribution of electrons surrounding its nucleus so that is has no dipole, this situation is only true as an average over time. At any instant, the electron cloud undergoes local fluctuations in charge density, making the argon atom an instantaneous dipole. This instantaneous dipole produces an electric field that polarizes the electron distribution in a neighboring argon atom (Atom 2), so that the neighboring atom itself acquires a dipole. The electrostatic interaction between these two dipoles leads to an attractive force.
- Debye forces, sometimes described as dipole-induced dipole forces: In this type of force, a polar molecule such as hydrogen chloride (HCl), for example, can induce a dipole in a neighboring nonpolar atom or molecule, such as an argon atom, for example, which results in an attractive electrostatic force between them (Figure 2.12b).
- Keesom forces, sometimes described as dipole–dipole forces: A polar molecule such as HCl, for example, can interact electrostatically with another polar molecule, such as HCl, to produce an attractive force (Figure 2.12c).

The attractive force between two dipoles can be calculated from Coulomb's law of electrostatics. A simple case is the force between two parallel dipoles, that is, when their dipole moment is in the same direction, as illustrated in Figure 2.12. A dipole of charge magnitude δq and distance of separation r between its ends has a dipole moment given by

$$p = \delta q \times r \tag{2.6}$$

The attractive force between two parallel dipoles of moment p_1 and p_2, respectively, separated by a distance x that is much larger than r, is

$$F_a = \frac{3p_1p_2}{2\pi\varepsilon_0 x^4} \tag{2.7}$$

where, ε_0 is the permittivity of free space. According to Eq. (2.7), the attractive force is equal to the product of the two dipole moments and inversely proportional to the fourth

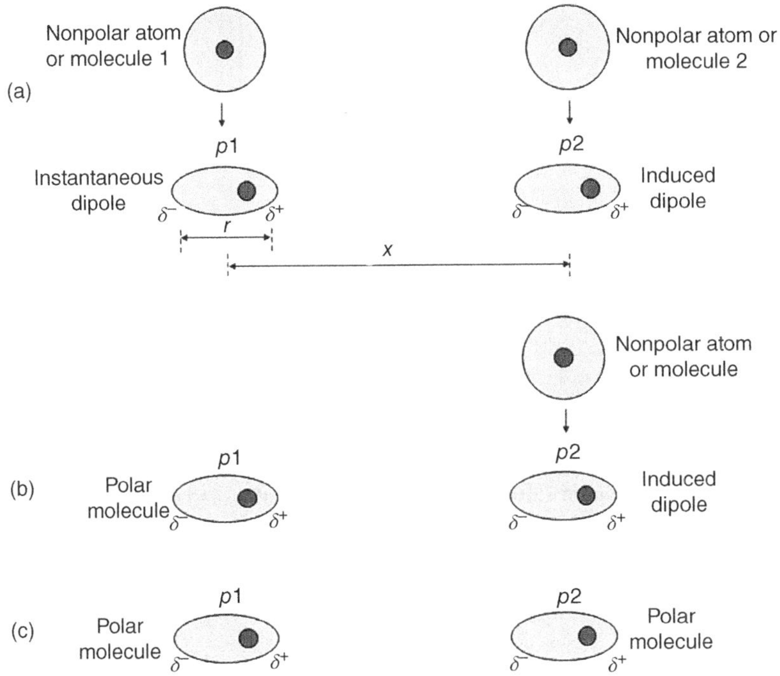

Figure 2.12 Illustration of the interactions in van der Waals bonding: (a) dispersion force between nonpolar atoms or molecules; (b) Debye force between a polar molecule with a permanent dipole moment and a nonpolar atom or molecule; (c) Keesom force between two polar molecules, each having a permanent dipole moment.

power of their separation distance. The larger the dipole moments, the larger the attractive force.

Keesom forces between permanent dipoles typically provide the strongest van der Waals attraction whereas dispersion forces provide the weakest. Figure 2.13 shows an illustration of the structure of trichloromethane ($CHCl_3$), a molecule with a permanent dipole due to the difference in electronegativity of the atoms in the molecule. In comparison, the tetrachloromethane molecule (CCl_4), where the H in $CHCl_3$ is replaced with Cl, is nonpolar because there is no net dipole moment.

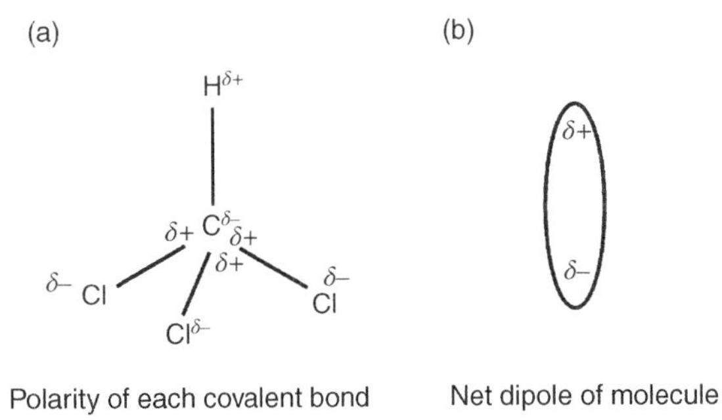

Figure 2.13 Illustration of (a) polarity of each covalent bond in the trichloromethane molecule ($CHCl_3$), and (b) overall net dipole of the molecule.

2.5.2 Hydrogen Bonding

Although much weaker than the primary bonds, hydrogen bonds are a strong type of intermolecular bonding (Table 2.2). Hydrogen bonds result from Keesom forces, the bonding between molecules that have a permanent dipole, but in addition they have the following characteristics:

- A hydrogen atom is covalently bonded to a highly electronegative atom, thus the hydrogen atom has a slight positive character while the highly electronegative atom has a slight negative character.
- The highly electronegative atom to which the hydrogen atom is bonded has at least one active set of lone pair electrons.

A well-known example of hydrogen bonding is between H_2O molecules in water (Figure 2.14). By sharing one electron in its outer shell with each hydrogen atom, the oxygen atom achieves an outer shell containing eight electrons, but the four unshared electrons make up two orbitals, each containing two electrons. The two unshared electrons in each orbital are referred to as lone pair electrons. The hydrogen bond involves intermolecular bonding between a hydrogen atom of one molecule and a lone pair electron in the oxygen atom of a neighboring molecule. When water freezes, the tetrahedral arrangement of sp^3 orbitals of the oxygen atom coupled with hydrogen bonding between the water molecules results in an ordered three-dimensional pattern of the molecules in ice (Figure 2.15). The hydrogen

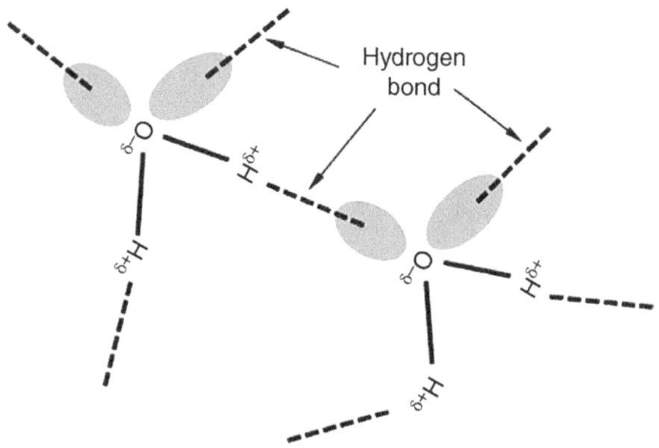

Figure 2.14 Schematic illustration of hydrogen bonds between water molecules.

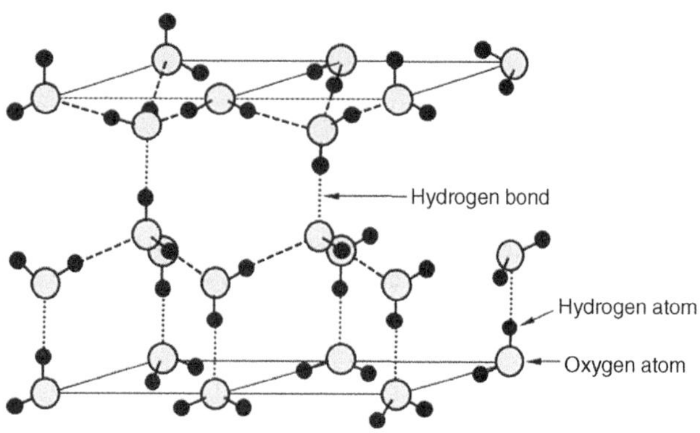

Figure 2.15 Illustration of the arrangement of water (H_2O) molecules in ice showing the hydrogen bonds.

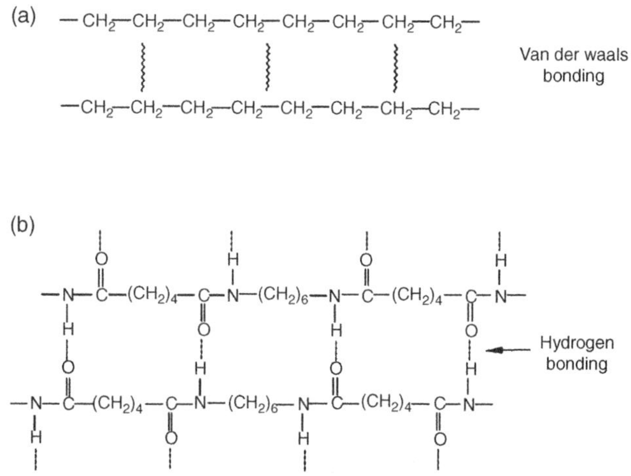

Figure 2.16 Schematic comparison of (a) van der Waals bonding between nonpolar polyethylene macromolecules and (b) hydrogen bonding between polar nylon 6.6 macromolecules.

bonds keep the molecules well apart, which is why ice has a lower density than water.

Secondary bonds also play an important role in polymers, providing links between their constituent macromolecules that make them solids at ordinary temperatures. In polyethylene composed of nonpolar macromolecules, secondary bonding consists of weak van der Waals bonds (Figure 2.16a). On the other hand, polymers composed of macromolecules that have polar groups, such as OH and NH groups, form stronger hydrogen bonds. When the macromolecules of the synthetic polymer nylon 6.6 are aligned in the formation of fibers, hydrogen bonding between carbonyl oxygen and amide hydrogen atoms of neighboring molecules serve to stabilize the structure, thus contributing to the high mechanical properties of the fibers (Figure 2.16b).

2.6 Atomic Bonding and Structure in Proteins

Proteins are the most versatile macromolecules in living organisms and they serve crucial functions in essentially all biological processes (Alberts et al. 2014). They are not only long-chain molecules but they can fold in a complex manner to give a variety of three-dimensional shapes. The functionality of a protein depends critically on its conformation and, thus, an understanding of how atomic bonding controls the overall three-dimensional shape of a protein is important. Because of its complexity, the conformation of proteins is often described at multiple levels, referred to as primary, secondary, tertiary, and quaternary structure, each emphasizing different features of the overall three-dimensional structure.

2.6.1 Primary Structure

The building blocks of simple proteins are approximately 20 naturally occurring α-amino acids (Figure 2.17; Table 2.4). These amino acids have a characteristic composition that includes an amine (NH_2) group at one end of the molecule and a carboxyl (C=O)OH group at the other end. The molecules also have a side-group, denoted generally as R, attached to the carbon atom between the NH_2 and (C=O)OH groups, which can be nonpolar, polar, positively charged or negatively charged. The simplest amino acid, glycine has R = H, the hydrogen atom. Proteins form by a condensation reaction involving the carboxyl group of one amino acid with the amine group of another amino acid, resulting in an amide bond between the two amino acids (Figure 2.18a). Conventionally, in drawing such structures, the NH_2 group of the amino acid or the N terminus of the protein is on the left whereas the (C=O)OH group of the amino acid or C terminus of the protein is on the right.

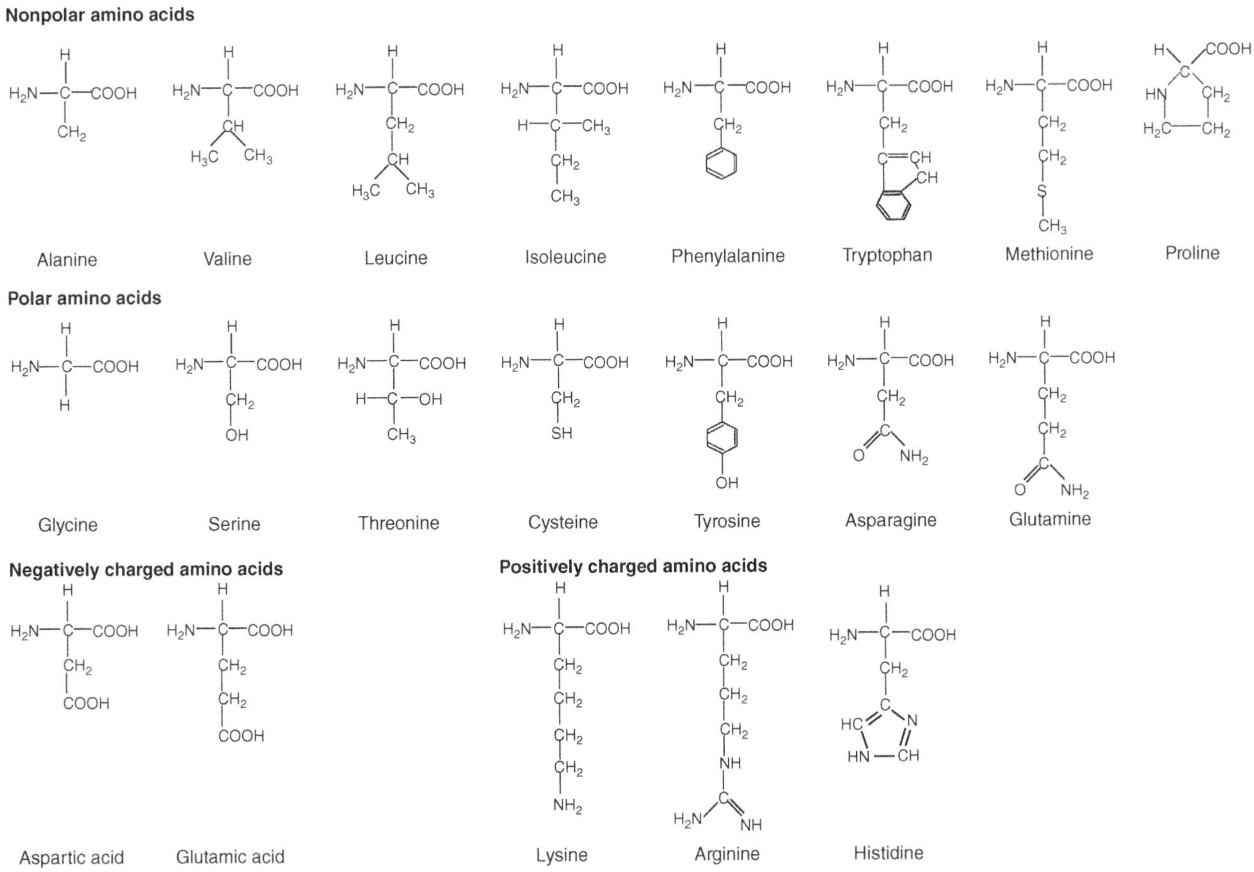

Figure 2.17 Side groups in 20 naturally occurring α-amino acids which can be nonpolar, polar, negatively charged, or positively charged.

A key feature of protein molecules is the presence of amide bonds in the chain backbone. These bonds are present in other macromolecules as well, such as in the synthetic polymer nylon 6.6 and other nylons, for example (Figure 2.16b). However, when it involves two natural amino acids, it is sometimes called a peptide bond. While there is no clear agreement on terminology, a protein molecule is often taken as composed of over 50–60 amino acid groups often called amino acid residues. Smaller molecules composed of less than 10–20 residues are called peptides whereas molecules composed of more than 10–20 residues are often called polypeptides. Consequently, a protein molecule is often called a polypeptide also.

Proteins are large molecules, composed of ~50–2000 amino acid residues, but the peptide illustrated in Figure 2.18b, composed of five residues, shows the key features of the primary bonds in the chain backbone and examples of side groups attached to the chain backbone. The sequence of the amino acid residues in the chain backbone is called the primary structure of the protein. This primary structure dictates the higher-order structure of proteins.

2.6.2 Secondary Structure

The term secondary structure describes the geometry of the long-chain backbone of the individual protein macromolecule. This geometry can consist of randomly coiled chain along with regions composed of a regular repeating pattern. It is strongly influenced by two key factors

- The stereochemistry of the amide bond
- Hydrogen bonding between the oxygen atom of the carbonyl (C=O) group and the hydrogen atom of the amino (N–H) group in the chain backbone.

Stereochemistry of the Amide Bond

A key feature of the amide bond is that the carbonyl (C=O) double bond resonates between the C–O and C–N positions, which gives the C–N bond partial double bond character (Figure 2.19a). The N atom has trigonal planar geometry, in a manner similar to carbon atoms in the ethylene molecule due to sp^2 hybridized electron orbitals (Section 2.4.2). Because rotation about a double bond is restricted, the amide bond has a planar geometry

Table 2.4 The α-amino acids of proteins.

Amino acid	Three-letter symbol	One-letter symbol	pK_a value of side chain
Alanine	Ala	A	
Valine	Val	V	
Leucine	Leu	L	
Isoleucine	Ile	I	
Phenylalanine	Phe	F	
Tryptophan	Trp	W	
Methionine	Met	M	
Proline	Pro	P	
Glycine	Gly	G	
Serine	Ser	S	
Threonine	Thr	T	
Cysteine	Cys	C	
Tyrosine	Tyr	Y	
Asparagine	Asn	N	
Glutamine	Gln	Q	
Aspartic acid	Asp	D	3.9
Glutamic acid	Glu	E	4.3
Lysine	Lys	K	10.5
Arginine	Arg	R	12.5
Histidine	His	H	6.0

(Figure 2.19b). This has important consequences for the secondary and tertiary structure of the protein molecule.

Hydrogen Bonding

Due to electronegativity differences between the atoms (Section 2.4), the carbonyl C=O and amino N–H bonds are polarized. Thus, the O and H atoms in these bonds have a partial negative (δ^-) and positive charge (δ^+), respectively. As the O atom also has lone pair electrons, this leads to hydrogen bonding between the O and H atoms of different amide groups that can occur within a chain (intrachain bonding) and between neighboring chains (interchain bonding). Hydrogen bonding is an important feature of both the secondary and higher order structures of proteins but the way it occurs within the chain controls the secondary structure.

While the overall conformation of a protein is complex and unique, a feature of the chains themselves is that they often contain regions with two regular repeating patterns called the α-helix and the β-sheet. These two structures result from hydrogen bonding between the O atoms of the C=O group and the H atoms of the N–H group of the amide bond in the chain backbone, without involving the side chains.

An α-helix is generated when a single protein chain twists around itself in the form of a helical geometry (Figure 2.20). Hydrogen bonds between every fourth peptide bond stabilizes the structure, giving a regular helix with a complete turn every 3.6 amino acid residues. To reduce steric hindrance to this type of intrachain bonding, the side groups attached to the chain backbone point outward from the chains. Other helices occur but they are less common.

The β-sheet forms as a result of layering of the protein chain, one part on top of the other. This structure is stabilized by hydrogen bonds between neighboring parts of the chain (Figure 2.21). These β-sheets can form in two ways, either from neighboring protein chains that run in the same orientation (called parallel chains) or from an individual protein chain that runs back and forth (antiparallel chains). This arrangement of the chain backbone along with hydrogen bonding results in both types of β-sheet having a rather rigid structure. The side groups are at right angles to the sheets, minimizing steric hindrance between the layers. Regions of β-sheets are common in many proteins because of their efficient packing.

In forming an antiparallel β-sheet, the chain backbone of the protein changes its direction, turning back and forth upon itself, making what is described as a β-turn. This is an important feature of protein folding, which gives a more compact shape. A hydrogen bond between the first and third peptide residues of the β-turn and the amino acid proline are frequently found at bends in the chain backbone (Figure 2.21c). Of the 20 α-amino acids (Figure 2.17), proline is the only one in which the side chain is connected to the chain backbone twice, forming a five-membered nitrogen-containing ring. Because ring structures typically have limited flexibility, proline is unable to occupy many of the main-chain conformations easily adopted by other amino acids. Proline is also nonpolar and, thus, cannot take part in hydrogen bonding. Together, these factors make proline a useful residue for occupying tight turns in a protein structure such as the β-turn.

2.6.3 Tertiary Structure

The tertiary structure of a protein refers to the overall three-dimensional shape (conformation) of its polypeptide chain. Many proteins do not have the regular structure of the α-helix or β-sheet but, instead, rely for their properties on their tertiary structure. A polypeptide chain can fold in a complex manner to give a variety of three-dimensional shapes, but two broad classes of structures, referred to as globular and fibrous, are often distinguished. In globular proteins, the polypeptide chain is more compactly folded like a ball of string with an irregular surface. In comparison, the polypeptide chain in fibrous proteins have a more elongated or

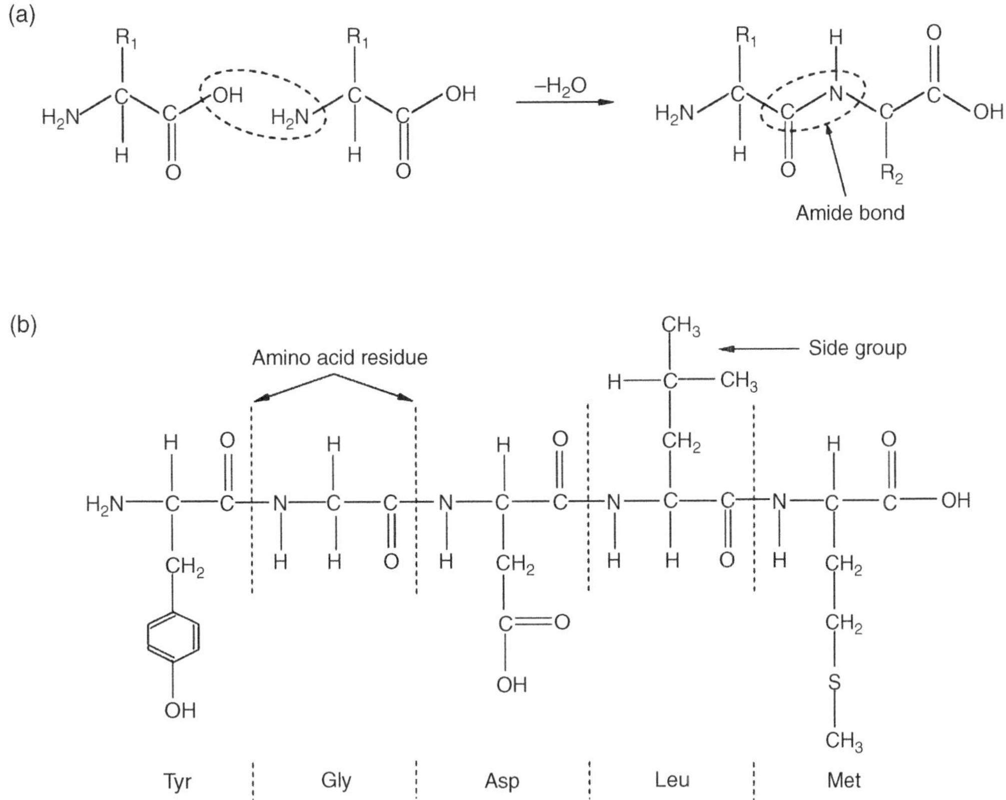

Figure 2.18 Illustration of (a) condensation reaction between two amino acids resulting in the formation of an amide (peptide) bond, and (b) peptide composed of five amino acid residues, showing the atomic bonds in the chain backbone and side groups. The three-letter symbol of each amino acid is given below its residue.

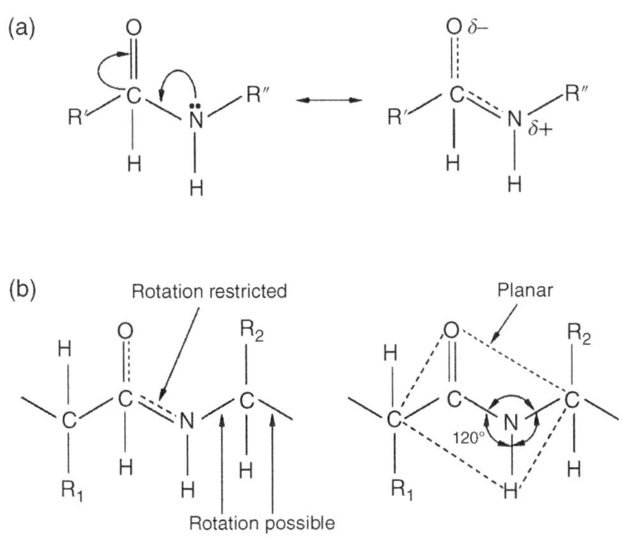

Figure 2.19 Schematic illustration of the stereochemistry of the amide bond. (a) Resonance of the carbonyl (C=O) double bond, resulting in partial double bond character of the amine (N–H) bond. (b) Planar geometry of amide bond.

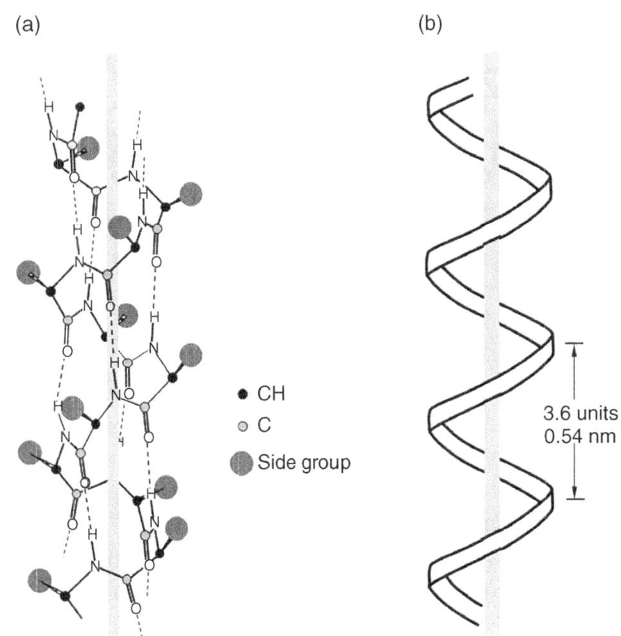

Figure 2.20 Illustration of (a) α-helix structure generated by intrachain hydrogen bonding in polypeptides, and (b) geometrical parameters of α-helix. The side groups point outward from the chain.

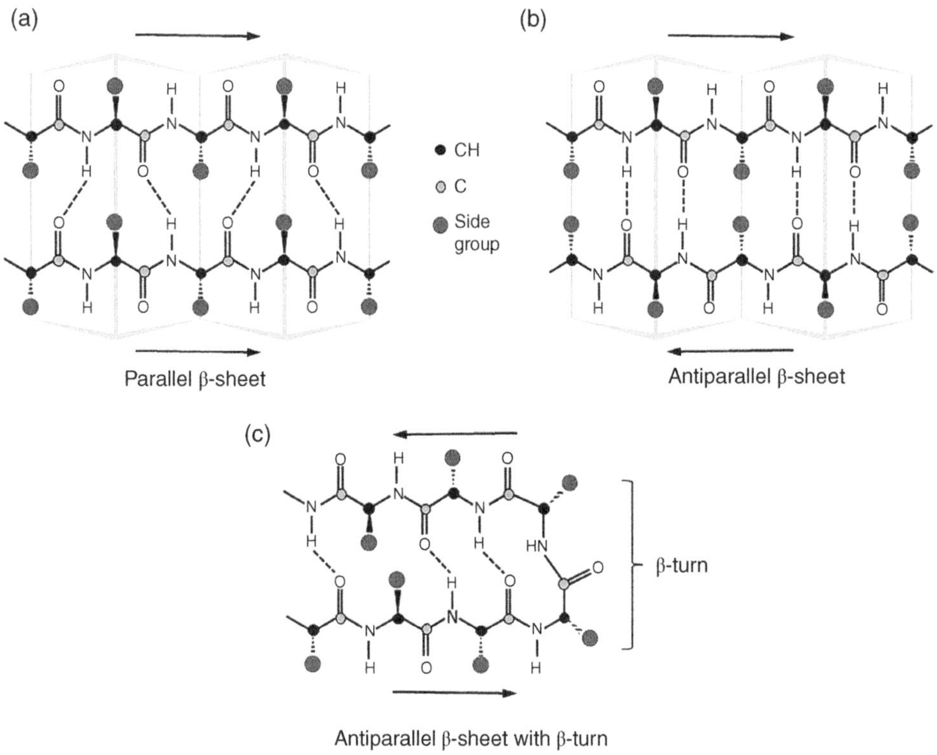

Figure 2.21 Illustration of β-sheet structure generated by layering of polypeptide chains and hydrogen bonding between the chains: (a) parallel β-sheet, (b) antiparallel β-sheet, and (c) β-turn in antiparallel β-sheet.

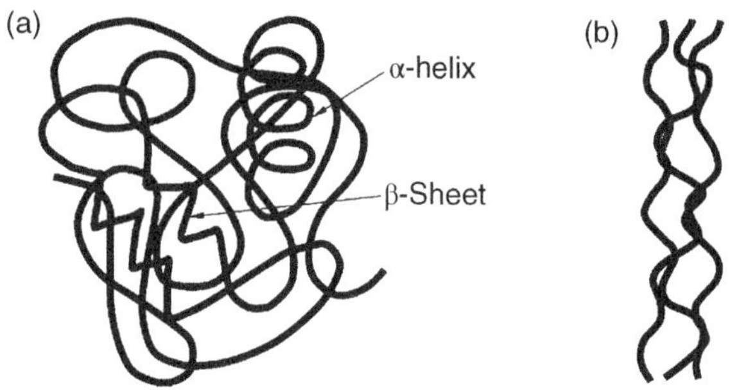

Figure 2.22 Schematic illustration in two dimensions of the overall three-dimensional shape of (a) globular proteins and (b) fibrous proteins.

filamentous three-dimensional structure (Figure 2.22). A fibrous protein often has a length that spans a longer distance than the average diameter of a globular protein.

Globular Proteins

In globular proteins, atoms in the side groups attached to the chain backbone can form covalent bonds and noncovalent bonds, such as ionic, hydrogen and van der Waals bonds, with atoms of other side groups. Unlike hydrogen bonding in the secondary structure that involves the amide bond in the chain backbone without involving the side chains, these bonds can involve side groups of the same chain (intrachain bonding) as well as different chains (interchain bonding). As a result of these interactions, the chain can fold up into various complex three-dimensional shapes that include lengths of randomly coiled chain along with regions having the regular repeating pattern of the β-sheet or α-helix.

Of the approximately twenty naturally occurring α-amino acids, only one, cysteine, offers the possibility of forming a covalent bond while approximately five have side groups than can form ionic bonds (Figure 2.17). Consequently, hydrogen and van der Waals bonds often have the largest effect in determining the tertiary structure of a protein. Although these bonds are much weaker than the covalent and ionic bonds (Table 2.2), the combined effect resulting from a large number of such noncovalent bonds determines and stabilizes the globular three-dimensional shape of the protein. Each type of protein has its own particular conformation but the conformation adopted by any protein is typically the one with the minimum free energy.

Examples of the main types of interactions between the side groups that can determine the conformation of a protein chain are:

- Covalent bonds: Two residues of the amino acid cysteine, for example, can be oxidized at their thiol (S–H) groups to form a disulfide (S–S) bond, sometimes called a disulfide bridge (Figure 2.23a). Reduction reverses this process.

- Ionic bonds: The side groups of some amino acid residues (Figure 2.17) contain weakly acidic or weakly basic groups that are deprotonated or protonated at the physiological pH (~7.4), giving rise to atoms in these molecules that have a negative or positive charge, respectively. Electrostatic attraction between oppositely charged atoms results in an ionic bond (Figure 2.23b).

A common group is the carboxyl group (C=O)OH in which the H atom dissociates to give a (C=O)O$^-$ species with a negatively charged oxygen atom. This group is found in the side groups of the negatively charged amino acids (Figure 2.17). On the other hand, weakly basic groups are composed of nitrogen-containing species such as the amine (NH_2) group. These groups are found in the side groups of the positively charged amino acids. The nitrogen atom in these species become protonated to form a positively charged nitrogen atom.

It should be noted that in these weakly acidic or basic groups, not all of the (C=O)OH or nitrogen-containing groups become deprotonated or protonated. Instead, there is an equilibrium reaction that can be expressed generally by the equations

$$\mathrm{A-COOH} \leftrightarrows \mathrm{A-COO^-} + \mathrm{H^+}$$
$$K_a = [\mathrm{A-COO^-}][\mathrm{H^+}]/[\mathrm{A-COOH}] \tag{2.8}$$

$$\mathrm{B-N^+H_3} \leftrightarrows \mathrm{B-NH_2} + \mathrm{H^+}$$
$$K_a = [\mathrm{B-NH_2}][\mathrm{H^+}]/[\mathrm{B-N^+H_3}] \tag{2.9}$$

where A–COOH and B–NH_2 represent the side groups, the square brackets denote the concentration of the respective species, and K_a is the equilibrium constant for the reaction. A well-known indicator of the acidity or basicity of these weak acids or bases is their pK_a value, defined as

$$pK_a = -\log_{10}(K_a) \tag{2.10}$$

In general, the higher the pK_a value, the weaker the acid. Table 2.4 gives the pK_a values of the molecules in the acidic or basic side groups. As the pK_a value for the side group of aspartic acid is ~3.9, the dissociated side groups (C=O)O$^-$ have a negative charge at the physiological pH. On the other hand, the pK_a value for the side group of lysine is ~10.5 and, thus, the protonated side groups (N^+H_3) in this amino acid have a positive charge at the physiological pH.

- *Hydrogen bonds*: The amide bonds in the protein chain backbone contain polar C=O and N–H bonds while the side group of some amino acid residues contain these bonds or polar hydroxyl (O–H) bonds. These polar species give rise to hydrogen bonding which can again be intrachain or interchain (Figure 2.23c). Although far weaker than the covalent and ionic bonds, hydrogen bonds are numerous in proteins and, thus, have a significant influence on the conformation of proteins.

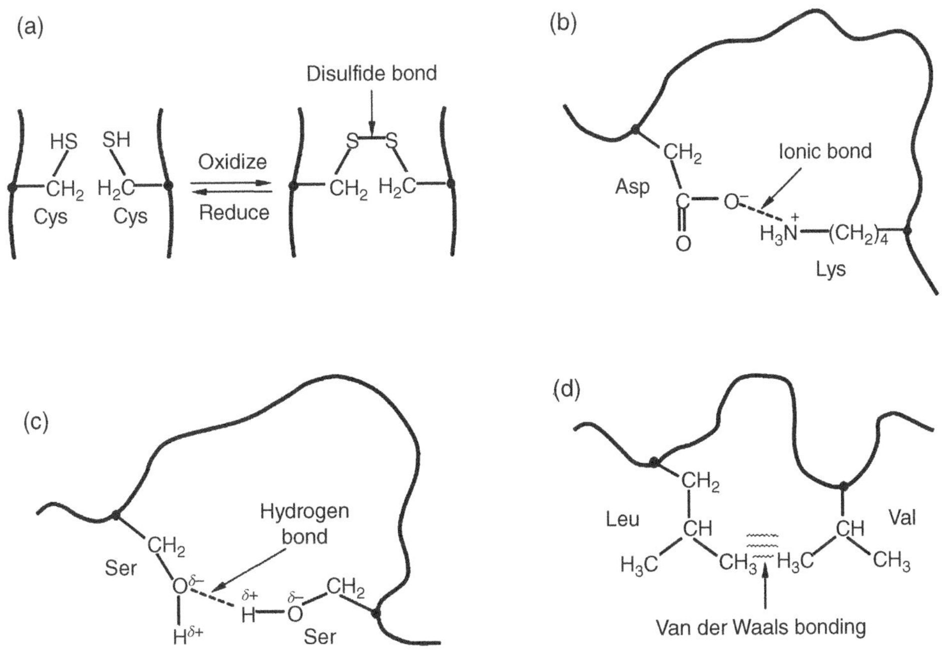

Figure 2.23 Illustration of the main types of interactions between side groups of proteins which can determine the conformation of a protein, using examples of specific amino acid residues. (a) Covalent bond between cysteine (Cys) residues; (b) ionic bond between aspartic acid (Asp) and lysine (Lys) residues; (c) hydrogen bond between serine (Ser) residues; (d) van der Waals bonding between leucine (Leu) and valine (Val) residues.

- *Van der Waals bonds*: These weak bonds involve electrostatic interactions between polar species in the side groups that do not take part in hydrogen bonding due to a lack of lone pair electrons in the relevant atoms or between nonpolar side groups (Figure 2.23d).

In an aqueous physiological environment, van der Waals bonding involving the side groups plays an additional role in determining the conformation of proteins. In general, nonpolar molecules are hydrophobic whereas polar and ionic molecules are hydrophilic. Consequently, as the protein chain folds to form a globular three-dimensional shape, it does so in a manner such that side groups with polar or ionizable molecules gather on the outside of the protein chain backbone whereas side groups with nonpolar molecules are buried on the inside of the chain (Figure 2.24). This manner of chain folding enhances the interaction of the polar or ionizable side groups with the polar water molecules and shields the nonpolar side groups from the aqueous environment. The nonpolar groups cluster together by van der Waals attraction and this, together with the interactions of the polar and ionizable side groups with the water molecules, has a strong influence on the overall three-dimensional shape of the protein. The protein undergoes conformational changes until the sum of its interaction energy is minimized.

In nonaqueous environments, such as the lipid bilayer of the cell membrane, the arrangement of the polar and nonpolar groups is opposite to that in aqueous environments. The nonpolar groups predominantly gather on the outside

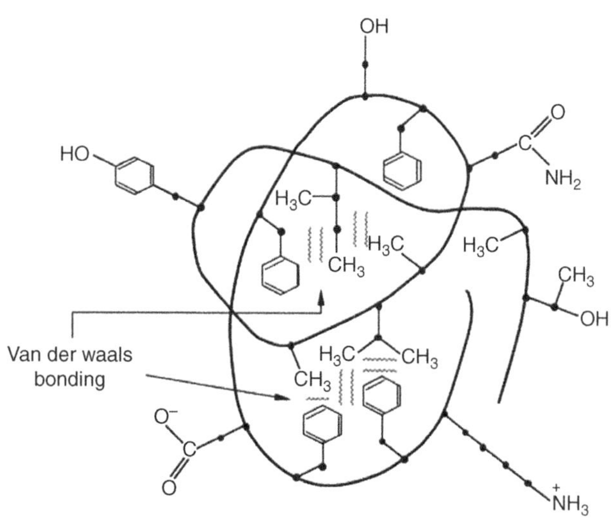

Figure 2.24 Illustration of chain folding of protein to form a globular three-dimensional shape in an aqueous environment. Side groups with polar or ionizable molecules gather on the outside of the protein chain backbone whereas side groups with nonpolar molecules are buried on the inside of the chain. Some side groups are shown as examples.

of the protein chain where they can interact with the nonpolar (hydrophobic) side chains of membrane fatty acids whereas the polar groups arrange themselves anywhere they can contact water. For transmembrane proteins, for example, which project through both sides of the membrane, the polar groups are found at each point where the polypeptide chain emerges from the membrane.

Fibrous Proteins

Fibrous proteins are abundant in mammalian tissues, particularly in the extracellular matrix, a complex structural network of proteins and other substances surrounding and supporting cells. Extracellular matrix proteins are secreted by cells into their surroundings where they often assemble into sheets or fibers. The most abundant of these proteins is a family called collagen. A major component of skin and bone, collagen is the most abundant protein in mammals, comprising approximately 25% of the total dry protein mass.

Collagen is structurally distinct from other proteins in that the molecule consists of three polypeptide chains, called α-chains, that wind around each other to form a long triple-helix structure (Figure 2.25). Glycine, the smallest amino acid, occurs at every third position in the chain backbone. This small size and regular repeating pattern of glycine allows the three chains to wind around each other closely to generate a compact triple-helix structure. Procollagen molecules secreted from cells are cleaved at their ends to form collagen molecules, each ~1.5 nm in diameter and ~300 nm in length. Side-to-side and end-to-end bonding between many of these collagen molecules lead to the formation of much larger fibrils that, in turn, can then assemble to form fibers (Chapter 10). This fibrous morphology is a significant factor in endowing tissues such as tendons and ligaments with their high tensile strength and elastic modulus, in a manner similar to aligned macromolecules that contribute to the high mechanical properties of synthetic fibers such as nylon and polyethylene.

As glycine (Gly) occurs at every third position, the primary structure of each chain has the sequence Gly-X-Y, where X and Y represent the one-letter symbol of the other amino acid residues. X and Y can be any amino acid but, often, they consist of the amino acids proline (Pro) and hydroxyproline (three-letter symbol Hyp). Typically, approximately one-third of the amino acid residues in the chain consists of glycine and approximately one-quarter consists of a combination proline and hydroxyproline, as exemplified by the amino acid content of human tendon (Table 2.5). Hydroxyproline and another modified amino acid, hydroxylysine (three-letter symbol Hyl), which often occur in proteins, are formed by enzymatic reactions involving proline and lysine residues in the chain prior to triple-helix formation within the cell (Figure 2.26).

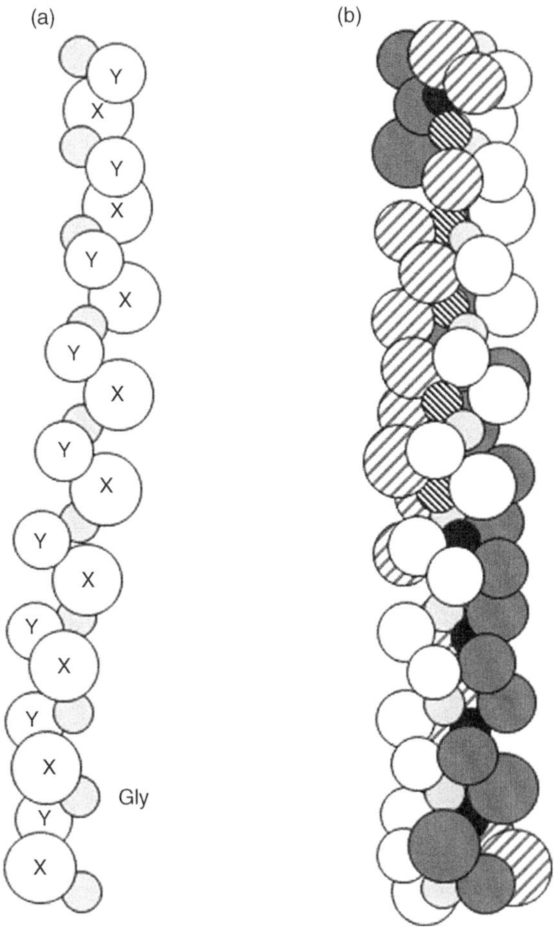

Figure 2.25 (a) Illustration of single α-chain composed of the amino acid sequence Gly-X-Y where Gly is the three letter symbol for glycine that occurs at every third position in the sequence, and X and Y are the one-letter symbol of any amino acid. (b) Illustration of triple-helix structure of collagen formed by winding of three α-chains around each other.

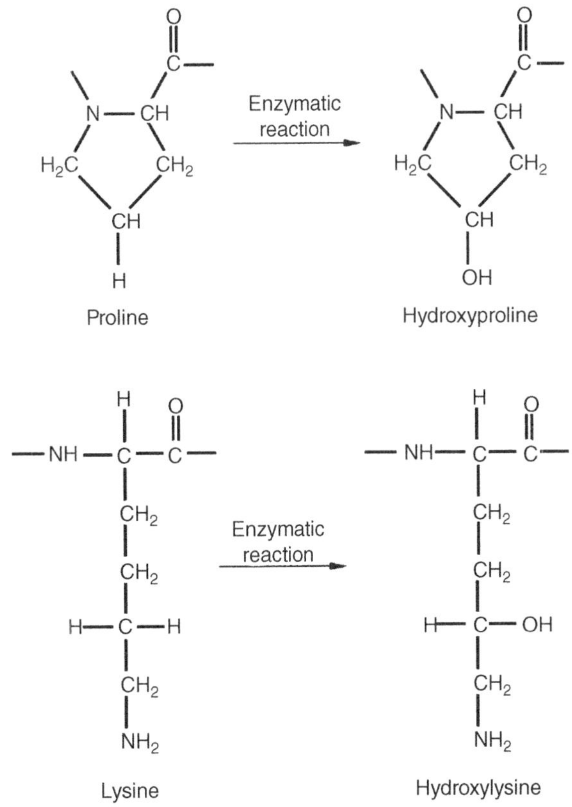

Figure 2.26 Schematic illustration of the structure of the modified amino acid residues hydroxyproline and hydroxylysine formed from proline and lysine residues by enzymatic reaction.

Table 2.5 Measured amino acid content of human tendon.

Amino acid	**Number of residues per 1000 residues**
Glycine	334
Proline	122
Hydroxyproline	96
Acidic or polar (Asn; Glu; Asn)	124
Basic or polar (Lys; Arg; His)	91
Others	233

Interchain hydrogen bonding between the carbonyl oxygen (C=O) and amine hydrogen (N–H) stabilizes the triple-helix structure. Proline also plays a role in stabilizing the structure because it has a limited ability to change its conformation due to the ring structure of its side chain. As the triple-helix already has a structure with the minimum energy due to the closely wound hydrogen-bonded chains and the regular pattern of the amino acid residues, changes to its conformation to provide an entropic contribution to its free energy are not required. Thus, collagen molecules maintain a stable triple-helix structure, unlike globular proteins that can undergo conformational changes.

2.6.4 Quaternary Structure

Some globular proteins, often called oligomeric proteins, are composed of two or more different polypeptide chains. The quaternary structure of these proteins emphasizes the way in which the individual folded protein chains fit together to form an overall three-dimensional structure (Figure 2.27). A well-studied example is hemoglobin, the protein that carries oxygen in red blood cells. This protein has a nearly spherical shape, composed of two identical α-globin subunits and two identical β-globin subunits arranged symmetrically. The same noncovalent bonds that stabilize the tertiary structure stabilize the quaternary structure of proteins.

Figure 2.27 Illustrative example of quaternary structure of a protein composed of four polypeptide chains. In this example, there are two types of protein, labeled Protein 1 and Protein 2, each with two chains.

Table 2.6 Summary of the four levels of protein structure and major atomic interactions that stabilize each level.

Structure	Description	Geometry	Major atomic interactions
Primary	Sequence of amino acid residues in chain backbone	—	Covalent bonding between amino acid residues to form amide (peptide bond)
Secondary	Arrangement of chain backbone of individual polypeptide	Random arrangement Regular helical structure (α-helix) Regular sheet structure (β-sheet)	Intrachain hydrogen bonding
Tertiary	Three-dimensional shape (conformation) of polypeptide chain	Globular Fibrous (elongated)	Interchain and intrachain hydrogen, van der Waals, ionic, and covalent bonding[a]
Quaternary	Overall three-dimensional structure of oligomeric proteins	Approximately spherical or irregular	Interchain hydrogen and van der Waals

[a] Hydrogen and van der Waals bonding often dominate over ionic and covalent bonding.

Table 2.6 summarizes the four levels of protein structure and the main atomic interactions.

2.7 Concluding Remarks

Bonding between atoms and molecules and the way in which the atoms or molecules pack together, whether in an ordered crystalline structure or a disordered amorphous structure, are crucial in understanding the intrinsic properties of materials. In this chapter, we discussed bonding between atoms and molecules, which can be divided into two types: strong primary (interatomic) bonds between atoms and weak secondary (intermolecular) bonds between molecules.

- Ionic and covalent bonds, common in ceramics, and metallic bonding, common in metals and metal alloys, are the three types of primary bonds. Polymers have strong interatomic covalent bonds between atoms in the chain backbone but weak intermolecular bonding between the chains.
- We discussed representations of the interaction between two atoms, in terms of a force or potential energy as a function of interatomic distance, which provide a useful way to understand how many of the intrinsic properties of a material arise
- Van der Waals and hydrogen bonds are the main types of intermolecular bonding
- Proteins, the most versatile macromolecules in living organisms, have a complex three-dimensional shape (conformation) that is determined by intrachain and interchain covalent and noncovalent bonds. We discussed how the bonding in proteins determines their overall three-dimensional shape, which, in turn, has a critical effect on their functionality.

Problems

2.1 Predict the type of bond (ionic, covalent, metallic, or a combination of these) that will be present in each of the following molecules. Explain your answer.

A NaF
B CH_4
C Al_2O_3
D SiC
E $CHCl_3$

2.2 Distinguish between intrinsic properties of materials and their engineering properties, indicating how these properties are dependent on the structure of a material.

2.3 Given the ionic radii of the following ions: sodium = 0.102 nm; chlorine = 0.181 nm; magnesium = 0.072 nm; oxygen = 0.140 nm, determine the potential energy of attraction in (kJ/mol) for (i) the NaCl molecule, and (ii) the MgO molecule at a separation distance equal to the sum of the ionic radii. Comment on the difference between these values and the bonding energy in Table 2.2 for these molecules.

2.4 Sodium chloride (NaCl) and silicon dioxide (SiO_2) are respectively among the prototype ionic and covalent solids. The electronegativities of the atoms in the Pauling electronegativity scale are as follows: sodium = 0.93; chlorine = 3.16; silicon = 1.90; oxygen = 3.44. Determine the percentage ionic and covalent character of the bonds in NaCl and SiO_2 according to Eq. (2.5).

2.5 Carbon and silicon both have four electrons in their outermost shell but carbon dioxide is a gas at room temperature whereas silicon dioxide is a solid with a high melting point. Explain.

2.6 Explain why hydrogen fluoride (HF) has a higher boiling point (19.4 °C) than that (−85 °C) of hydrogen chloride (HCl) even though its molecular weight is lower.

2.7 The trichloromethane ($CHCl_3$) molecule is polar whereas the tetrachloromethane molecule is nonpolar but, despite its polarity, trichloromethane has a boiling point (61.2 °C) that is lower than that of tetrachloromethane (76.7 °C). Explain.

2.8 The pK_a values of the side groups in the amino acids glutamic acid, arginine and histidine are 4.3, 12.5, and 6.0, respectively. With the aid of a sketch of each amino acid residue, explain whether the side group will have a positive or negative charge at the physiological pH, and indicate the position of the charge on the side group.

Reference

Alberts, B., Johnson, A.D., Lewis, J. et al. (2014). *Molecular Biology of the Cell*, 6e. New York: W W Norton & Company Chapter 3.

Further Reading

Callister, W.D. (2007). *Materials Science and Engineering: An Introduction*, 7e. New York: Wiley Chapter 2.

Ebbing, D.D., Gammon, S.D., and Ragsdale, R.O. (2006). *Essentials of General Chemistry*, 2e. Boston, MA: Houghton Mifflin Company Chapter 9.

Shoulders, M.D. and Raines, R.T. (2009). Collagen structure and stability. *Annual Review of Biochemistry* 78: 929–958.

3

Structure of Solids

3.1 Introduction

Biomaterials are typically three-dimensional solids composed of an enormous number of atoms. A titanium (Ti) implant of volume 1 cm^3, for example, contains ~5×10^{22} atoms. Consequently, we are concerned with the structural characteristics at different levels or length scales. Bonding between atoms to form molecules and the way in which the atoms pack together to form a controllable structure determine the intrinsic properties of a solid (Chapter 2). If the atoms pack together in a regularly repeating three-dimensional pattern, the structure is described as crystalline. On the other hand, if the atoms pack together in a random non-ordered pattern, the structure is said to be amorphous. Metals and ceramics in their pure state have a crystalline structure. On the other hand, when cooled at a rapid enough rate from the molten state, some metals and ceramics develop an amorphous structure and the solid is said to be a glass or to have a glassy structure. Whereas glasses are, by definition, amorphous, some materials derived from glasses, called glass-ceramics, consist of a mixture of crystalline and amorphous phases. Many polymers have an amorphous structure but some have a structure composed of a combination of crystalline and amorphous regions, sometimes described as a semicrystalline structure. Whether a material is crystalline or not is important because a crystalline phase has properties that are different and often better than an amorphous phase of the same composition.

A perfect crystal is an idealization. Imperfections, often called defects, occur in the crystal due to packing irregularities of the atoms or the addition of other types of atoms. There are two main types of defects in crystals, classified in terms of whether they occur at one or a few atomic positions, called point defects, or over a more extended one-dimensional line of atomic positions, called line defects or, more commonly, dislocations. These defects control the rate at which atoms can migrate through the crystal in response to a stimulus, such as mechanical stress or temperature. The attractive mechanical property of ductility in metals arises from the presence of dislocations in the crystals that make up the metal.

Unless they are formed by special and, often, expensive methods, crystalline solids are not composed of just one crystal. Instead, they are composed of a large number of small crystals, called grains, of size smaller than 1 μm to several tens of micrometers. These solids are said to be polycrystalline. The grains and, if present, amorphous phase and porosity make up the microstructure that determines the engineering properties of a material relevant to its application (Chapter 2).

In this chapter, we will discuss the following structural features that, in addition to atomic bonding, are important for the design, properties, and applications of biomaterials:

- Ways in which atoms pack to form a crystalline or an amorphous solid
- Defects that are present in crystals
- Microstructure of materials

3.2 Packing of Atoms in Crystals

Packing of identical atoms, such as those of a pure metal, to form crystals can be visualized in two successive steps (Figure 3.1):

- Ordered packing of atoms in two-dimensions to form a plane or layer of atoms
- Packing of these layers on top of one another to form a three-dimensional structure.

Metal atoms can be viewed as tiny ball bearings of the same radius but, for conciseness, to more easily show their packing arrangement in three-dimensions, we can show their positions in space by points located at the center of each sphere. These points represent the crystal lattice, defined as the regular three-dimensional pattern in which the atoms are arranged. As the pattern of lattice points forms an ordered array, it is not necessary to show the

Materials for Biomedical Engineering: Fundamentals and Applications, First Edition. Mohamed N. Rahaman and Roger F. Brown.

Companion website: www.wiley.com/go/rahaman/biomedicalengineering

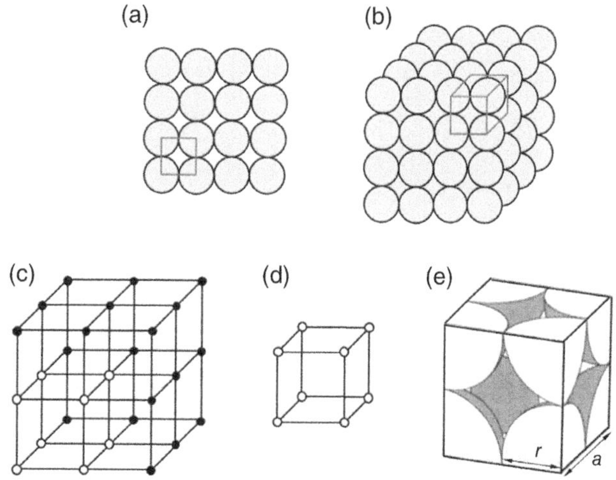

Figure 3.1 Packing of atoms to give a simple cubic structure. (a) Single "square" layer of atoms; (b) packing of layers directly on top of preceding layers; (c) part of a simple cubic lattice; (d) simple cubic unit cell; (e) illustration of atoms within a unit cell.

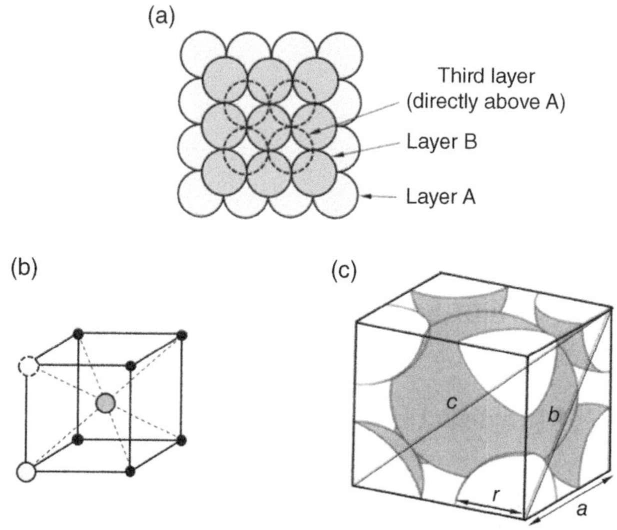

Figure 3.2 (a) Packing of "square" layers of atoms in ABAB pattern to give a body-centered cubic (BCC) structure; (b) illustration of ABAB packing pattern in a unit cell; (c) illustration of atoms within a unit cell.

three-dimensional structure of the entire crystal. Instead, a unit cell, defined as smallest unit in space which, when repeated, completely describes the crystal lattice, is used.

The structure illustrated in Figure 3.1, described as a simple cubic structure, is the simplest type of crystal structure. In this structure, the distance between two lattice points in the x, y, and z directions from a given lattice point (which we can take arbitrarily as the origin), called the unit cell length, is the same and the angle between these lines is 90°. A simple cubic structure can be created by taking the first layer of atoms, composed of this "square" arrangement of the atoms, and placing an identical layer directly on top of it. Then the third layer is placed directly on top of the second layer, and so on.

Another way of forming a cubic structure using this square layer is to take the first layer of atoms A and place the second layer of atoms B into the depressions (hollow regions) between the atoms of the first layer (Figure 3.2). Then the third layer is placed in the depressions of the second layer, which means that it is directly above the first layer. This gives an ABAB pattern of stacking the layers. The unit cell consists of a simple cubic structure with an additional atom located at its center. The structure is referred to as body-centered cubic, abbreviated BCC.

Instead of a square layer of identical atoms, we can use a "triangular" layer in which the centers of any three touching atoms from a triangle (Figure 3.3). This triangular layer takes up the least amount of space in a plane and is called a close-packed plane. To build a regular repeating three-dimensional pattern, we can start with the first layer A and add a second layer by placing the atoms into the depressions of layer A. Then a third layer can be added into the depressions of the second layer, and so on. However, there are two alternative ways of stacking these additional layers. One way is to place the atoms of the second layer B into the depressions, labeled B, of the first layer. Then the third layer C can be added on top of the second layer over the depressions labeled C. By the time we get to the fourth layer, the atoms of this layer are directly above the first layer A. We call this packing sequence an ABCABC pattern and the structure formed in this way is called face-centered cubic (FCC). The unit cell corresponds to a simple cubic structure but with additional atoms located at the center of each face.

Another way to stack the triangular layer is to add the second layer B by placing the atoms into the depressions labeled B of the first layer A (Figure 3.4). Then the atoms of the third layer are placed into the depressions of layer B which means that these atoms are directly above those of layer A. This packing sequence, called an ABAB pattern, results in a hexagonal structure referred to as close-packed hexagonal (CPH).

The atomic packing fraction, sometimes referred to as the packing density, of a crystal structure is defined as the fractional volume of the unit cell that is occupied by the atoms themselves, that is, the volume of the atoms divided by the volume of the unit cell. Table 3.1 shows that the atomic packing fraction of the BCC structure is considerably higher than that of the simple cubic while those of the FCC and CPH structures are equal to one another and higher still.

While the majority of metals and a significant number of ceramics have a BCC, FCC, or CPH structure, these are not

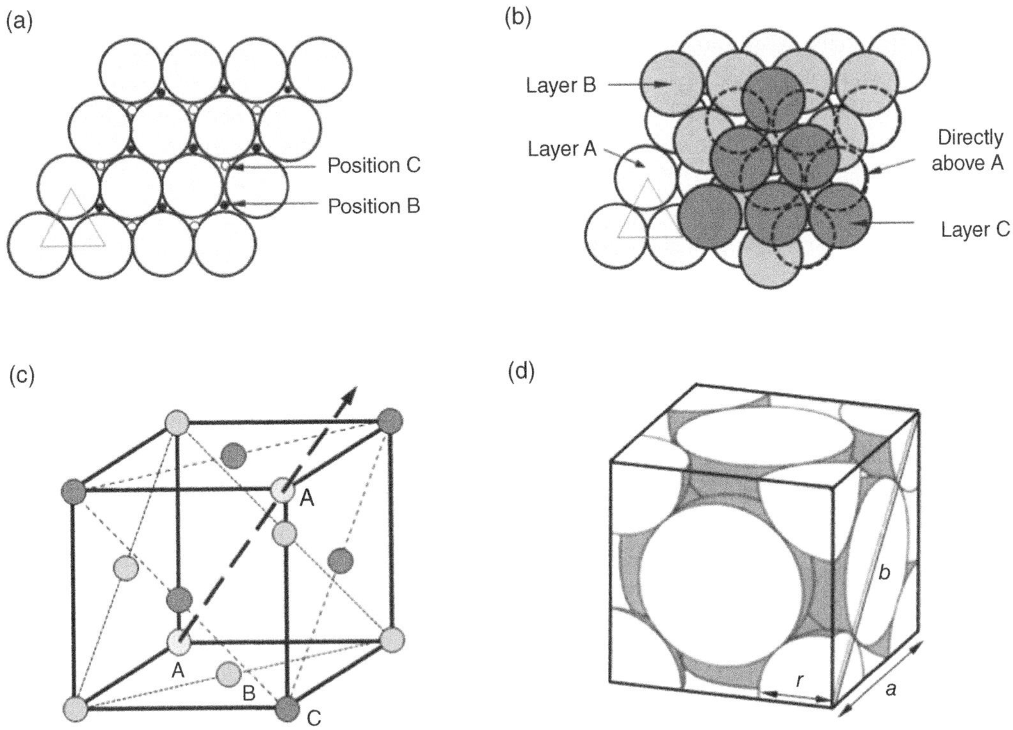

Figure 3.3 (a) Single "triangular" layer of atoms showing interstitial positions (depressions) B and C between the atoms; (b) Packing of triangular layers of atoms in ABCABC pattern to give a face-centered cubic (FCC) structure; (c) atomic positions in a unit cell (resulting from ABCABC packing of the layers in planes perpendicular to the direction of the arrow); (d) illustration of atoms within a unit cell.

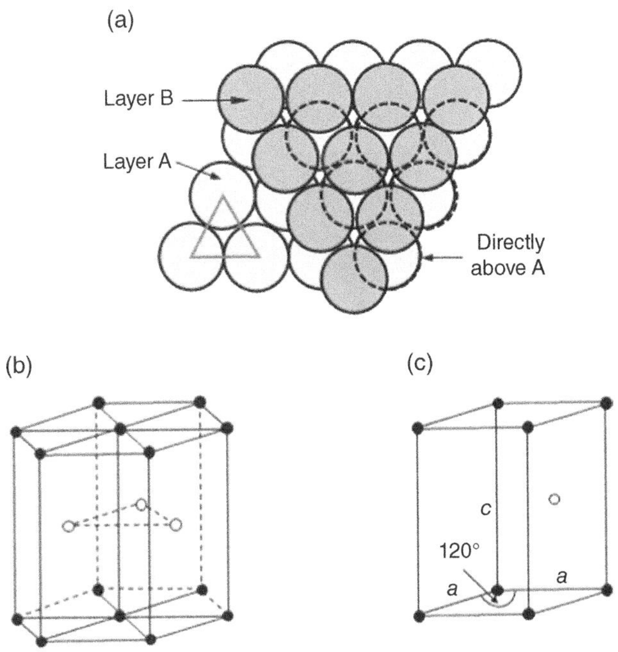

Figure 3.4 (a) Packing of triangular layers of atoms in ABAB pattern to give a close-packed hexagonal (CPH) structure; (b) atomic positions in a hexagonal unit cell; (c) unit cell commonly used for CPH structure.

Table 3.1 Atomic packing fractions of the simple cubic, body-centered cubic (BCC), face-centered cubic (FCC), and close-packed hexagonal (CPH) structures.

Crystal structure	Simple cubic	BCC	FCC	CPH
Atomic packing fraction	0.52	0.68	0.74	0.74

the only types of crystal structure in materials. In general, the crystal structure that a material adopts is the one that gives the minimum energy. This structure does not have to be close-packed or geometrically simple but it must have a regular-repeating ordered three-dimensional pattern.

3.2.1 Unit Cells and Crystal Systems

A standard nomenclature is used to succinctly describe the geometrical arrangement of the atoms in a unit cell and, thus, in a crystal.

Unit Cell Parameters

The unit cell of a crystal is defined in terms of the length of the cell edges from a given lattice point, taken as the origin, and the angle between the cell edges (Figure 3.5). The general convention in orienting a crystal is to use a coordinate system in which the x-axis points toward us as we look directly at the page, the y-axis points to the right, the z-axis points upward, and the origin is the point of intersection of the three axes. Then, the unit cell dimensions are taken as the edge lengths a, b, and c, respectively, in the x, y, and z directions. The angle between the edges b and c, that is, opposite the edge a, is taken as α, the angle between the edges a and c, opposite the edge b, is β, and the angle between the edges a and b, opposite the edge c, is γ. The lattice constants are the edge lengths a, b, and c, and the angles α, β, and γ, which define the crystal structure. The cubic structure, for example, has $a = b = c$ and $\alpha = \beta = \gamma = 90°$.

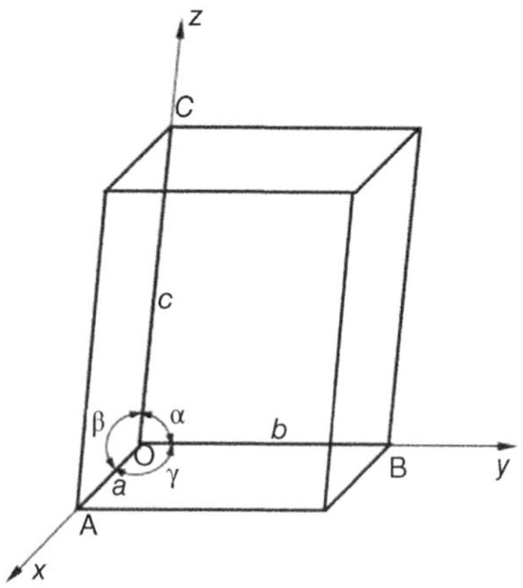

Figure 3.5 Geometry and parameters of a unit cell.

Crystal Systems and Bravais Lattices

There are seven different unit cell shapes, called crystal systems, defined by their unit cell lengths and interaxial angles (Figure 3.6). When packed together, each of these seven unit cell shapes fills up three-dimensional space completely. The unit cell contents are specified in terms of where the atoms are located. Consequently, depending on the location of the atoms in the unit cell, some crystal systems can generate more than one crystal lattice. The cubic system, for example, can generate three different crystal lattices, the simple cubic, BCC and FCC lattice. In total, there are 14 different crystal lattices, called Bravais lattices (Figure 3.7).

Different planes and directions in crystalline solids can have a different arrangement of atoms as, for example, in a pure metal such as Ti. Depending on the composition of the solid, different planes and directions can also have a

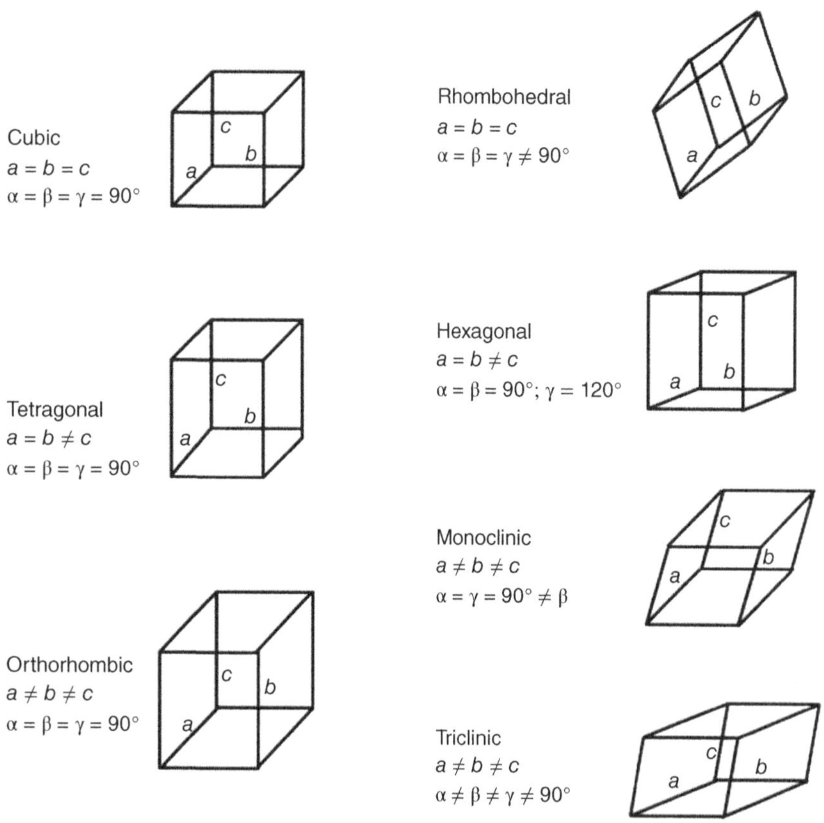

Figure 3.6 The seven crystal systems and their parameters.

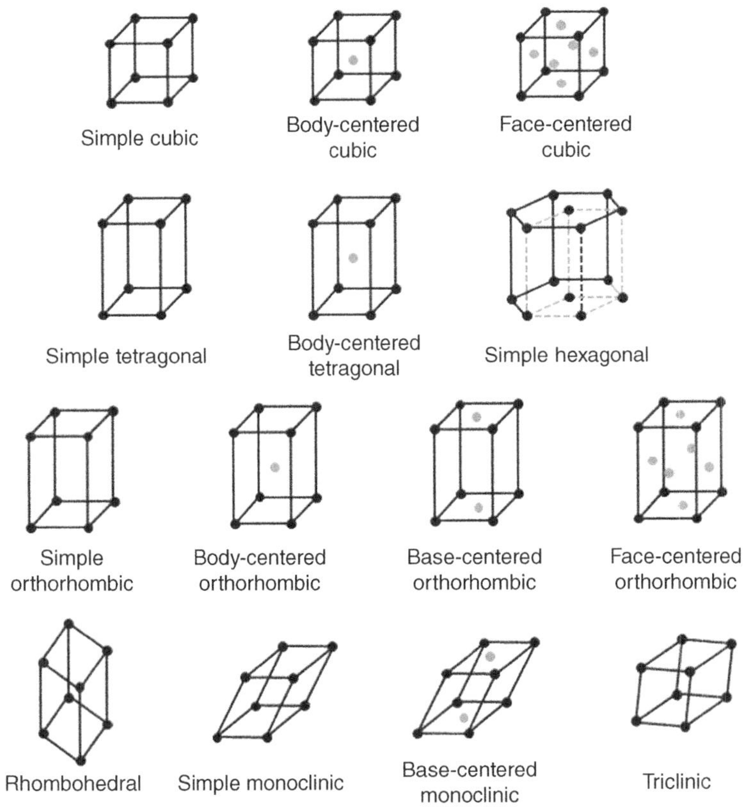

Figure 3.7 The 14 Bravais lattices.

different combination of atoms. Some planes in sodium chloride (NaCl), for example, have different ratios of sodium to chlorine atoms. Consequently, some solids can show different properties in different planes or directions. Often, it is necessary to describe specific atomic positions, lattice planes or lattice directions in a succinct way. How this is done is described in Section 3.6.

3.3 Structure of Solids Used as Biomaterials

As pure metals are composed of a single type of atom, their crystal structures are rather simple. In comparison, the crystal structures of ceramics are more complex and they depend on the type of bonding, ionic, or covalent.

3.3.1 Crystal Structure of Metals

Metallic bonding is nondirectional and the metal cations within the sea of almost free electrons can be viewed as tiny ball bearings (Chapter 2). The metal ions pack together as tightly as possible, subject only to geometric constraints. Consequently, the majority of metals have an FCC, CPH, or BCC structure, as illustrated in Table 3.2 for some common metals.

Example 3.1 Iron (Fe) has a BCC structure and an atomic radius of 0.1241 nm. Determine (i) the number of atoms in a unit cell of Fe, (ii) the unit cell length, and (iii) the theoretical density of Fe (that is, the density of the pore-free material). (Atomic weight of Fe = 55.85; Avogadro number = 6.023×10^{23}/mol.)

Solution:

a) In the BCC structure, each unit cell has one atom at its corners and one atom at its center (Figure 3.2). As an atom at each corner is shared by eight unit cells, only one-eighth of this atom effectively belongs to a given unit cell. Thus, the total number of atoms within a BCC unit cell is 2 (that is, $8 \times 1/8$ at the corners + 1 in the center).
b) The unit cell length can be found from the geometry of the unit cell and Pythagoras theorem. Taking the radius of an iron atom = r and the unit cell length = a, from the BCC structure in Figure 3.2c, the diagonal from one corner to the opposite corner across the center of the

Table 3.2 Crystal structure and atomic radius of some common metals at room temperature.

Metal	Crystal structure[a]	Atomic radius (nm)	Metal	Crystal structure[a]	Atomic radius (nm)
Aluminum	FCC	0.1431	Molybdenum	BCC	0.1363
Cadmium	CPH	0.1490	Nickel	FCC	0.1246
Chromium	BCC	0.1249	Platinum	FCC	0.1387
Cobalt	CPH	0.1253	Silver	FCC	0.1445
Copper	FCC	0.1278	Tantalum	BCC	0.1430
Gold	FCC	0.1442	Titanium	CPH	0.1445
Iron (α)	BCC	0.1241	Tungsten	BCC	0.1371
Lead	FCC	0.1750	Zinc	CPH	0.1332

[a] FCC, face-centered cubic; CPH, close-packed hexagonal; BCC, body-centered cubic.

cell $c = 4r$, and the diagonal across one face $b = a\sqrt{2}$. Using Pythagoras theorem, we get $c^2 = a^2 + b^2$, giving $a = 4r/\sqrt{3}$. Since $r = 0.1241$ nm, we get that $a = 0.286$.

c) The theoretical density D_t is the mass of the equivalent number of atoms in the unit cell divided by the volume of the unit cell. Thus

$$D_t = \frac{2\times(\text{mass of Fe atom})}{a^3} = \frac{2\left(\dfrac{55.85}{6.023\times10^{23}}\right)}{\left(0.286\times10^{-7}\right)^3} = 7.93\ \text{g/cm}^3$$

where the mass of the atom and the unit cell length were expressed in units of grams and centimeters, respectively.

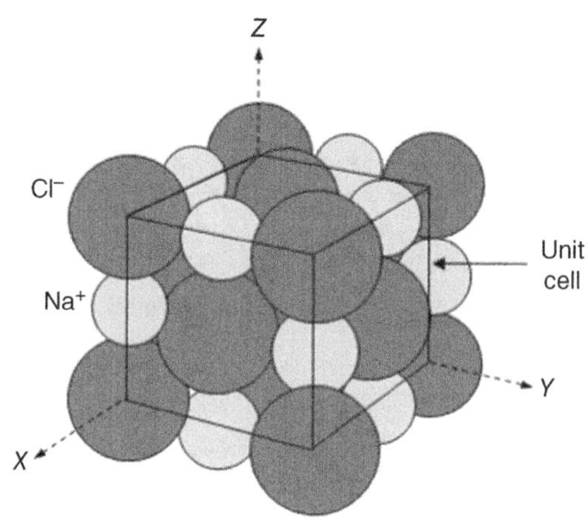

Figure 3.8 Arrangement of sodium ions (Na^+) and chlorine ions (Cl^-) in a sodium chloride unit cell.

3.3.2 Crystal Structure of Ceramics

The simplest ionic-bonded ceramics are compounds composed of ions of a metal and a non-metal that have discrete charges of opposite sign as in NaCl, for example (Chapter 2). Electrostatic bonding between the cations and anions provide the dominant contribution to the bonding energy. The bonding is non-directional and the ions tend to pack as densely as possible. One type of ion surrounds itself with as many ions of the opposite sign as possible, with the constraint that ions of the same sign must not touch, maximizing the electrostatic potential energy of attraction. Thus, the majority of ceramics with a high degree of ionic bonding have crystal structures based on FCC or CPH packing.

Although it is not used as a biomaterial, NaCl is often taken as an example of a ceramic material that shows a high degree of ionic bonding. In NaCl crystals, each type of ion is surrounded, that is, coordinated, by six ions of the opposite sign. (Figure 3.8). The unit cell is cubic and the structure can be considered as composed of two interpenetrating FCC lattices of the cations (Na^+) and anions (Cl^-).

Aluminum oxide (Al_2O_3), often referred to as alumina or α-alumina, is used as articulating bearings in hip implants (Chapter 1). Although alumina has a close-packed hexagonal (CPH) structure, the arrangement of the ions has to take into account the dissimilar cation to anion ratio (2 : 3) and the geometrical restriction that each type of ions must not touch. The structure can be viewed as composed of layers of O^{2-} ions stacked in a direction perpendicular to the basal plane between which two-thirds of the interstitial atomic sites are occupied by Al^{3+} ions and one-third is empty.

Zirconium oxide (ZrO_2), often referred to as zirconia, exists in three different crystalline forms called polymorphs (Chapter 7). The cubic form transforms to tetragonal below ~2370 °C that, in turn, transforms to a monoclinic structure below ~1240 °C. In addition to the change in crystal structure, the transformation to the monoclinic structure is accompanied by a volume expansion of ~4.7% that cannot be easily accommodated in the solid. When zirconia is cooled from its fabrication temperature, typically well above the tetragonal to monoclinic transition temperature, this volume change leads the creation of microcracks and, thus, to a weak material. Consequently, zirconia is not useful in engineering or biomedical applications. On the other hand, upon the addition of a few percent of yttrium oxide (yttria), Y_2O_3, or magnesium oxide (magnesia), MgO, for example, the cubic or tetragonal phase can be retained as a metastable phase on cooling to room temperature. When fabricated under well-controlled and reproducible conditions, yttria-stabilized zirconia (YSZ) with a tetragonal structure (~3 mol% Y_2O_3), is among the ceramics with the best mechanical properties. YSZ is shaped and used as dental restorations such as crowns and bridges and, in particulate form, it is incorporated into alumina in the production of zirconia-toughened alumina (ZTA) bearings for hip implants. These bearings have better resistance to cracking than alumina due to transformation of the YSZ particles to the monoclinic phase in the vicinity of an approaching crack (Chapter 7). As this transformation uses up some of the energy that would be otherwise available for crack propagation, growth of the crack is blunted.

Many of the strongest, hardest, and most refractory ceramics have structures with a high degree of covalent bonding. Several of these ceramics, such as carbon in the form of diamond, silicon carbide (SiC) and silicon nitride (Si_3N_4), for example, show a tetrahedral arrangement of the bonds due to sp^3 hybridization of the valence electron orbitals in atoms such as carbon and silicon (Chapter 2). As there are a fixed number of directional bonds in covalent bonding, the position and number of neighboring atoms in the structure are also fixed. Diamond is often considered prototypical of solids that show strong covalent bonds. It consists of a cubic unit cell in which the atoms are located at the corners of a tetrahedron due to sp^3 hybridization of the valence electron orbitals of the carbon atoms (Section 3.3.4). Silicon nitride is used in spinal fusion implants and it is being developed for use as femoral head bearings in hip implants. It has a hexagonal crystal structure in which each silicon atom is bonded to four nitrogen atoms due to sp^3 hybridization of the silicon atom. These SiN_4 tetrahedra are joined together by three nitrogen atoms in a trigonal planar arrangement due to sp^2 hybridization of the electron orbitals in nitrogen.

In silicon dioxide (SiO_2) commonly called silica, the four hybridized sp^3 orbitals in silicon leads to a strong preference for the formation of SiO_4 tetrahedra, in which the oxygen atoms are located at the corners of the tetrahedron in both the crystalline and amorphous states.

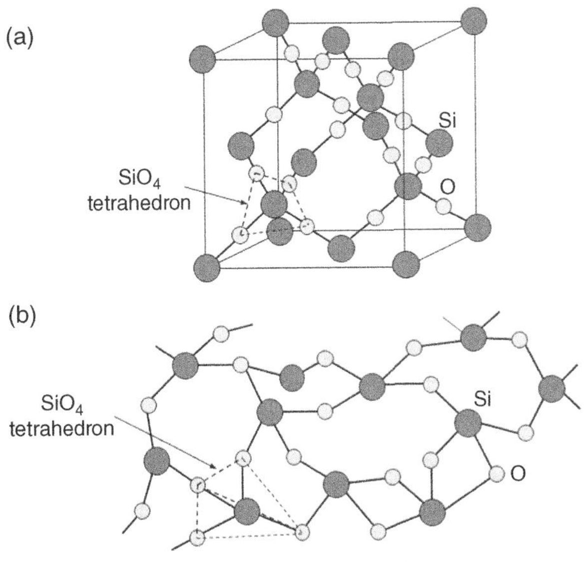

Figure 3.9 Illustration of (a) ordered arrangement of SiO_4 tetrahedra in crystalline silica (cristobalite) and (b) non-ordered arrangement in amorphous silica glass.

Crystalline silica, for example, is composed of an ordered repeating pattern of three-dimensional structures in which the oxygen atom at each corner of an SiO_4 tetrahedron is shared with that of a neighboring tetrahedron. Crystalline silica occurs in three different crystalline forms (polymorphs) called cristobalite, tridymite, and quartz. Figure 3.9a illustrates the arrangement of the SiO_4 tetrahedra in cristobalite.

Crystal Structure of Hydroxyapatite

Several types of synthetic calcium phosphate materials have been studied or developed for use in biomedical applications such as bone reconstruction, drug delivery devices and coatings for metal prostheses (Chapter 7). Implants composed of hydroxyapatite, with the formula $Ca_{10}(PO_4)_6(OH)_2$, β-tricalcium phosphate, $Ca_3(PO_4)_2$, and a two-phase mixture called biphasic calcium phosphate (BCP), composed of varying ratios of hydroxyapatite and β-tricalcium phosphate, for example, have been used for decades in healing bone defects. The primary building block of β-tricalcium phosphate and hydroxyapatite is the phosphate ion $(PO_4)^{3-}$ in which the phosphorus and oxygen atoms form a slightly distorted tetrahedron (Figure 3.10a). As the main inorganic (mineral) constituent of hard tissues such as bone and teeth is composed of a material that resembles hydroxyapatite, considerable attention has been devoted to the structure and properties of this material. Hydroxyapatite exists in a monoclinic or a hexagonal crystal structure. The unit cell in the hexagonal structure has dimensions $a = b = 0.9430$ nm and $c = 0.6891$ nm, and bond angles $\alpha = \beta = 90°$ and $\gamma = 120°$,

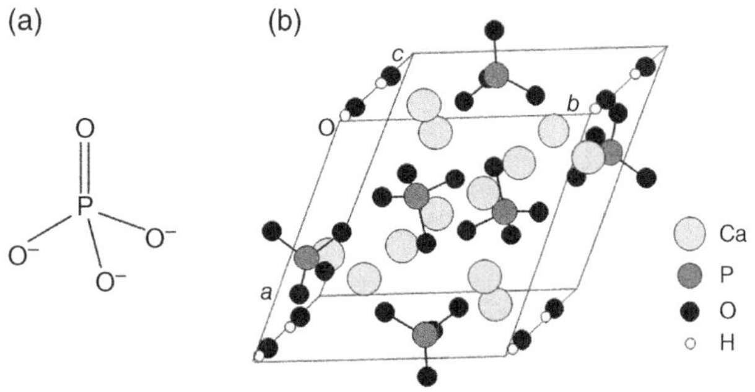

Figure 3.10 (a) Arrangement of atoms in a phosphate $(PO_4)^{3-}$ ion; (b) arrangement of atoms (ions) in unit cell of hydroxyapatite.

and it is composed of 6 $(PO_4)^{3-}$ ions surrounded by 10 calcium ions and with 2 hydroxyl $(OH)^-$ ions along the *c*-axis (Figure 3.10b).

The name hydroxyapatite refers to a crystalline material with the stoichiometric composition $Ca_{10}(PO_4)_6(OH)_2$. In comparison, the inorganic constituent of hard tissues such as bone and teeth is not pure hydroxyapatite. Instead, it has a composition that deviates from this stoichiometric composition to varying extent, depending on the living organism and the location of the hard tissue in the organism. The host ions in hydroxyapatite are substituted by small to trace amounts of other ions, such as $(CO_3)^{2-}$, $(HPO_4)^{2-}$, Mg^{2+}, Na^+, Sr^{2+}, K^+, and F^- (Section 3.4.1). This substituted composition, is one factor that contributes to the unique properties of the hydroxyapatite-like material of bone. As the properties of a material are dependent on its composition, there has been significant interest in producing synthetic hydroxyapatite-like materials with compositions that approximate the inorganic constituent of bone (Chapter 7).

3.3.3 Structure of Inorganic Glasses

The term glass commonly refers to inorganic materials that have an amorphous structure. Although the structure of a glass has short-range order arising from bonding of an atom with its immediate neighbors, the structure has no long-range order. Silica can be produced in both crystalline and amorphous states and a comparison of their structures provides a useful illustration of the difference between a crystalline material and a glass (Figure 3.9). Although silica glass is composed of SiO_4 tetrahedra with corner-sharing oxygen atoms as in crystalline silica, its three-dimensional structure is composed of a network with no long-range order.

A problem with silica glass is that it starts to soften only at high temperature (above ~1200 °C), making it difficult to form the glass into objects with desirable shapes. Consequently, metal oxides are commonly added to silica during the production process to obtain a glass that can be formed and shaped more easily. These metal oxides, referred to as network modifiers, add cations to the glass structure and break up a fraction of the Si–O–Si bonds. In quantities of a few percent to a few tens of percent by weight, the network modifiers often include sodium oxide (Na_2O) and calcium oxide (CaO), but a variety of network modifiers are used in many glasses to achieve the desired properties. Figure 3.11 illustrates the structure of a glass with one of the simplest compositions, formed by adding Na_2O to silica and referred to as a sodium–silicate glass. Overall, the positive charge of the sodium cations balance the negative charge of the oxygen atoms that are no longer bonded to two neighboring silicon atoms, called nonbridging oxygen atoms.

Silicate glasses, composed of the SiO_2 glass-forming network, are the most commonly used glasses, accounting for over ~95% of the tonnage in industrial and commercial applications. A common silicate glass composition is one that is used for windows of buildings, composed of ~70% SiO_2, ~15% Na_2O, and ~10% CaO by weight, plus minor amounts of other oxides. Borate glasses, composed of the B_2O_3 glass-forming network, and borosilicate glasses, composed of interspersed SiO_2 and B_2O_3 networks, also see considerable use commercially. Phosphate glasses, composed of the P_2O_5 glass-forming network, are used to a more limited extent.

Glasses, nondegradable or bioactive, find limited use as biomaterials, mainly to treat diseased or damaged tissues and organs where their compositional flexibility and ease of forming provide attractive benefits (Chapter 7). Radioactive glass microspheres, for example, are used to treat liver

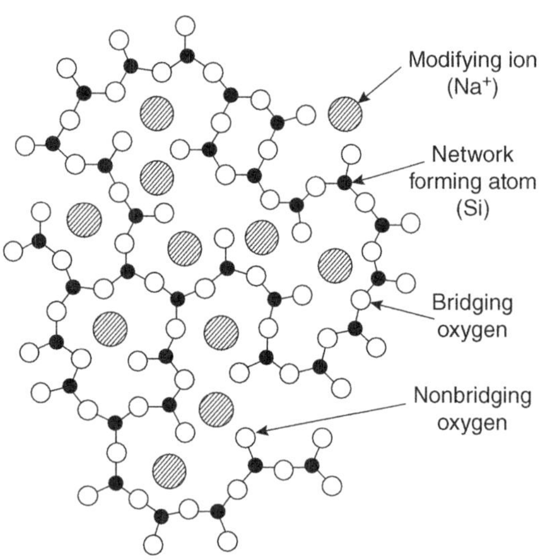

Figure 3.11 Schematic representation of the structure of a sodium silicate glass.

cancer. Bioactive glasses, such as the silicate glass designated 45S5, are used to a limited extent to heal bone defects while bioactive borate glass in a microfibrous form is used to heal skin wounds.

3.3.4 Structure of Carbon Materials

In the solid state, carbon can exist in several different structures, called allotropes, due to the different ways in which carbon atoms can bond together. Three allotropes, diamond and graphite, both of which are crystalline, and amorphous carbon have been well known for many years. Although they are not compounds, these three carbon materials are often classified as ceramics because they show several properties that are characteristic of covalent-bonded ceramics. Pyrolytic carbon is an important material in the manufacture of heart valves whereas diamond-like carbon (DLC) has been investigated as coatings for articulating bearings in hip and knee implants. Diamond and graphite are not used as biomaterials. On the other hand, more recently discovered allotropes of carbon such as fullerenes, carbon nanotubes and graphene, have been the subject of an enormous amount of investigation for use in technological applications and are now receiving considerable interest for potential biomedical applications.

Diamond is often considered prototypical of strongly bonded covalent solids. It consists of a cubic unit cell in which the atoms are located at the corners of a tetrahedron due to sp^3 hybridization of the valence electron orbitals of the carbon atoms (Figure 3.12). While the properties of diamond are very attractive due to its strong covalent bonds, they are more suitable to industrial applications. Diamond is also difficult and expensive to produce industrially.

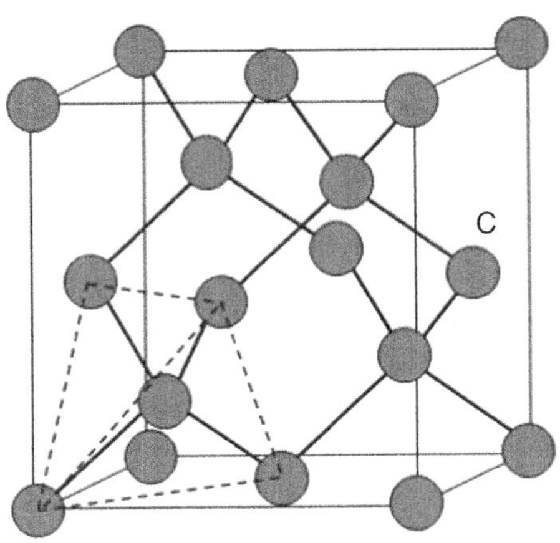

Figure 3.12 Tetrahedral arrangement of covalent bonded carbon atoms in diamond.

Diamond-like carbon, abbreviated DLC, is a term used to describe a variety of amorphous carbon materials that contain varying amounts of hydrogen. Commonly deposited as a film or coating, DLC has properties that are dependent on the method and carbonaceous precursors used in its production. A high degree of sp^3 bonding and a low amount of hydrogen enhance the diamond-like properties of these materials. Based on their attractive properties such as high hardness, high wear resistance, low coefficient of friction, chemical inertness and biocompatibility, DLC has been investigated for orthopedic applications, such as coatings on articulating metal bearings in hip and knee implants to reduce the amount of wear particles and to improve the corrosion resistance of the metal bearings (Dearnaley and Arps 2005). DLC has also elicited interest for use in cardiovascular applications, such as coatings on stents and heart valves.

DLC coatings for orthopedic applications typically have a high degree of sp^3 bonding (approximately 80–85%) and a low amount of hydrogen (less than 1 atomic percent). However, the use of hard coatings on articulating metal bearings has not proved successful in clinical applications due to their unpredictable performance in critical stress conditions. The coating can fracture and chip off when subjected to high local stresses due to wear particles between the articulating surfaces or by inadvertent contact with hard components such as the metal rim of the acetabular cup, resulting in catastrophic wear rates.

Although graphite is not used as a biomaterial, the basic building block of its structure, composed of a planar array of hexagonally arranged carbon atoms (Figure 3.13), is common to a variety of graphitic materials. In each plane, one s orbital and two p orbitals form three sp^2 orbitals that have a trigonal planar arrangement in which each atom is bonded by σ bonds to three neighboring atoms (Chapter 2). The remaining one electron per carbon atom is delocalized, residing in p orbitals that are perpendicular to the planar arrangement. Neighboring p orbitals overlap, forming π bonds, allowing these delocalized electrons to move more easily within each plane. Thus, each bond in the hexagonal arrangement has partial double bond character. Graphite has a three-dimensional structure that shows strong covalent bonding within each plane and weak intermolecular bonding between neighboring planes. These structural characteristics are responsible for the well-known properties of graphite, such as lubricity and electrical conductivity, which are suitable for engineering applications but not relevant for biomedical applications.

Pyrolytic carbon is used in the manufacture of mechanical heart valves (Chapter 24), where it is deposited as a thick coating (~25–500 μm) on graphite to produce valve leaflets (Figure 1.1e). In addition to adequate mechanical

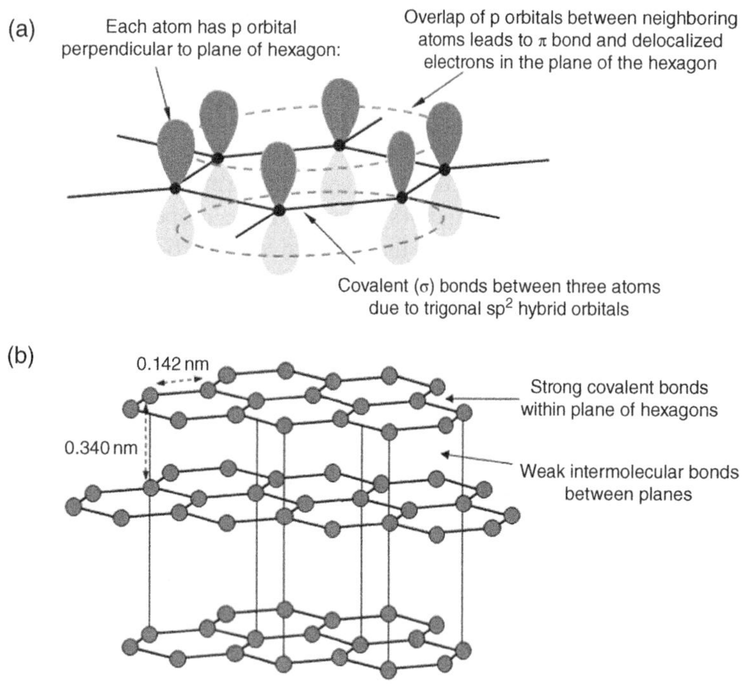

Figure 3.13 (a) Basic building block of graphite composed of a planar array of hexagonally arranged carbon atoms; (b) three-dimensional arrangement of planar arrays in graphite.

properties, such as strength, wear resistance, and durability, pyrolytic carbon is highly resistant to blood clotting and causes little damage to blood cells. This combination of properties makes pyrolytic carbon-coated graphite suitable for use as heart valve leaflets. Pyrolytic carbon belongs to a family of carbon materials called turbostratic carbons that have a similar structure to graphite. However, instead of the flat hexagonal arrays that are stacked and held together by weak interlayer bonding as in graphite, pyrolytic carbon, and other turbostratic carbons are composed of layers that are disordered, resulting in wrinkles or distortions within layers. These disordered layers give pyrolytic carbon improved durability and a smoother surface when compared to graphite.

Fullerenes, Graphenes, and Carbon Nanotubes

Fullerenes, graphenes, and carbon nanotubes share the same basic building block as graphite, a monolayer of hexagonally arranged carbon atoms. Graphene is the name given to a two-dimensional flat monolayer of carbon atoms arranged in a hexagonal lattice (Figure 3.14a). The stacking of these flat monolayers gives the three-dimensional structure of graphite. Graphene is a one-atom thick, two-dimensional crystal, often of size less than 1 mm (Geim and Novoselov 2007). The term graphenes is used in a more general sense to describe two-dimensional crystals composed of one, two or a few (3–10) monolayers that are each distinguishable by their electronic structure. Structures thicker than ~10 monolayers are typically described as thin films of graphite.

The structure of carbon nanotubes can be viewed as graphene rolled to form a tubular geometry. Carbon nanotubes may consist of single tubes (Figure 3.14b), called single-walled carbon nanotubes (SWNTs), or concentric tubes called multi-walled carbon nanotubes (MWNTs). SWNTs have a diameter from ~1 nm to a few tens of nanometers and lengths of hundreds of nanometers to a few millimeters (Iijima and Ichlhashi 1993). As the long-range periodicity of the atomic arrangement is retained along the axial direction of the tube, SWNTs may be viewed as one-dimensional crystals.

Fullerenes are large molecules with a structure that can be viewed as a monolayer composed of hexagonal and pentagonal arrangements of carbon atoms which has been used to form a cage-like geometry (Giacolone and Martin 2006). The most commonly investigated fullerene is the buckminsterfullerene, C_{60}, also called a buckyball, composed of 60 carbon atoms joined together to form 20 hexagons and 12 pentagons (Figure 3.14c). This arrangement of hexagons and pentagons, called a truncated icosahedron, has a pattern similar to a soccer ball. The carbon atoms in the pentagons are bonded by single bonds whereas the bonds in the hexagons consist of resonant double bonds. Although classified as a molecule, the C_{60} fullerene has a diameter

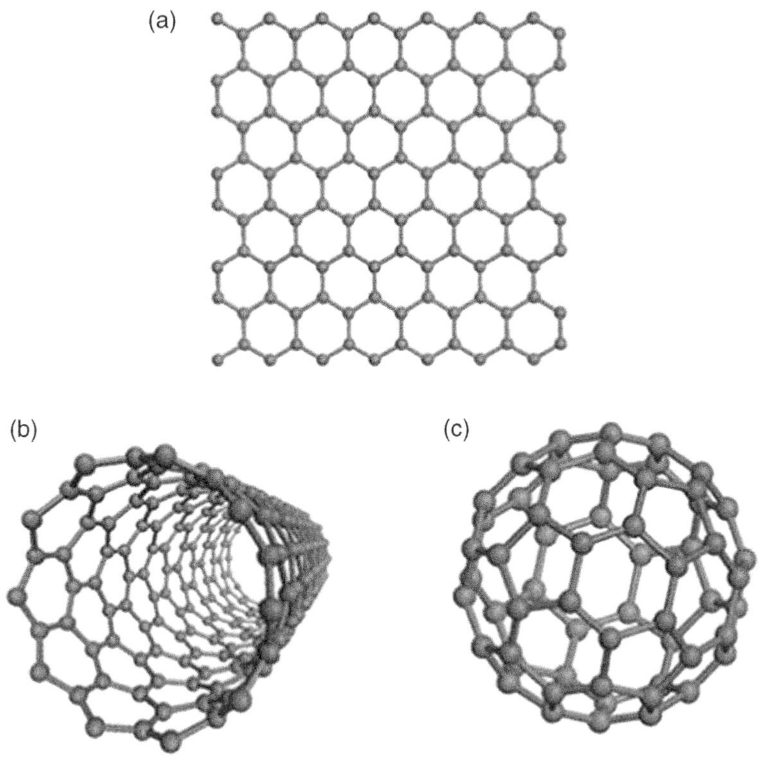

Figure 3.14 Arrangement of carbon atoms in (a) graphene, (b) single-walled carbon nanotube, and (c) buckminsterfullerene, C_{60}.

(0.7 nm) close to the lower limit of nanoparticles. When dispersed in a solvent, fullerenes often form aggregates due to weak van der Waals attractive forces, showing properties characteristic of nanoparticles rather than molecules.

The strong covalent (σ) bonds due to the sp^2 hybrid orbitals, delocalized electrons of the π bonds due to the overlapping p orbitals in the hexagonally arranged carbon atoms and small size endow graphenes, carbon nanotubes, and fullerenes with unique physico-chemical properties. These structures have received an enormous amount of scientific and technological interest since their discovery and are now receiving considerable interest for potential biomedical applications, such as drug delivery, gene delivery, biomedical imaging, biosensing, and tissue engineering (Eatemadi et al. 2014; Zhao et al. 2017).

As carbon is chemically inert and hydrophobic, a related area of investigation is the functionalization of graphenes, carbon nanotubes, and fullerenes with appropriate molecules for optimal use in these applications (Chapter 13). In common with nanostructured materials such as nanoparticles intended for use *in vivo*, the possible toxicity of graphenes, carbon nanotubes, and fullerenes has been widely studied. While the vast majority of studies have shown no serious adverse response from cells and tissues, some questions still remain about possible toxic effects.

3.3.5 Structure of Polymers

Polymers are composed of long molecules (macromolecules) in which hundreds or thousands of atoms are joined together by covalent bonds to form a chain. The chain backbone in the vast majority of polymers is composed of carbon atoms, as in polyethylene (PE), for example (Figure 2.10), but many polymers have a chain backbone that contains atoms other than carbon, such as silicon, or in addition to carbon, such as silicon, nitrogen, or oxygen. Macromolecules of commonly used degradable polymers such as polylactic acid (PLA), polyglycolic acid (PGA), and polycaprolactone (PCL), for example, contain carbon-oxygen bonds in addition to carbon–carbon bonds (Chapter 9).

Intermolecular forces resulting from van der Waals bonding, hydrogen bonding or a combination of these two types of physical bonding are present in all polymers. On the other hand, depending on the chemical composition of the polymer and how it is synthesized, neighboring chains can also be joined chemically at various points, that is, cross-linked by covalent bonds. The elasticity of synthetic rubbers used in engineering applications and elastin, a component of skin and other tissues, is due to the presence of chemical crosslinks between the molecules. Macromolecules in the vast majority of polymers are also more complex than the simple linear structure of PE

(Chapter 8). Instead of just hydrogen atoms bonded to carbon atoms in the chain backbone, the side groups can have a variety of composition, structure and size, and they can be bonded to carbon atoms of the chain backbone at regular or random intervals. These factors influence the way in which the macromolecules pack together in a solid.

In the majority of solid polymers, when the macromolecules pack together they do so in a random manner, giving an amorphous structure (Figure 3.15a). In other polymers, commonly those in which the macromolecules have a rather simple structure with little or no crosslinking, the chains can pack in an ordered pattern by folding backward and forward (Figure 3.15b). This regular repeating pattern of chain folding leads to the formation of crystalline regions. As long as the macromolecular structure allows this type of ordered arrangement, the formation of crystalline regions is possible because there is a driving force for it. This is because a crystal has a lower energy than an amorphous structure of the same composition due to the ordered and closer packing of the polymer chains. However, it is difficult for a macromolecule to pack completely into a crystalline pattern even for the simplest structure. Consequently, these polymers are composed of crystalline regions interspersed with amorphous regions (Figure 3.15b). The term semicrystalline has often been used to describe these polymers, in a sense to indicate that they are not totally composed of crystals. Generally, formation of crystalline regions is more favorable when the side groups are less bulky, are bonded at regular intervals to the chain backbone, or allow substantial intermolecular bonding such as hydrogen bonding to stabilize the ordered arrangement of the macromolecules.

Whether a polymer is amorphous or semicrystalline has significant consequences for its properties and applications. The crystalline regions typically have better mechanical properties than the amorphous regions, such as higher strength, higher elastic modulus and better wear resistance. Improvement in the percentage of crystalline regions in PE, for example, has resulted in reduced wear of ultrahigh molecular weight polyethylene (UHMWPE) bearings in hip and knee implants. In degradable polymers such as PLA and PGA, for example, the crystalline regions have a lower degradation rate than the amorphous regions. Consequently, amorphous copolymers of PLA and PGA, referred to as polylactic-co-glycolic acid (PLGA), are often preferred in drug delivery devices because they provide a more predictable degradation rate than semicrystalline PLA or PGA.

3.4 Defects in Crystalline Solids

A perfect crystal is an idealization. Imperfections, often called defects, are present at random positions in the structure. These defects play an important role because they control several physico-chemical properties of the solid. Metals, for example, would not show the attractive mechanical property of ductility when subjected to an applied stress were it not for the presence of a particular type of defect in the crystals called dislocations. Defects in crystals occur for two main reasons. Structurally, atoms in the pure material do not pack perfectly in a crystal due to thermal fluctuations during their processing or growth. Chemically, other atoms can be added accidentally or deliberately during production of the material, which disrupt the regular-repeating pattern of the host atoms. As this type of compositional modification can influence the defect structure to a far larger extent than purely packing irregularities in the pure material, it is a widely used approach to modify or improve the properties of metals and ceramics.

Defects in crystalline solids are commonly divided into three types, described as point defects, line defects, and planar defects. Three-dimensional defects, such as microcracks,

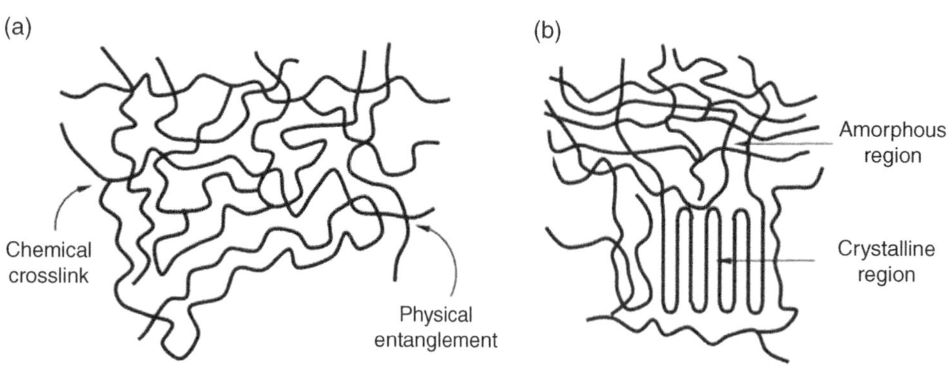

Figure 3.15 Schematic representation of (a) random arrangement of macromolecules in a polymer to give an amorphous structure and (b) ordered packing of macromolecule in crystalline regions of a semicrystalline polymer. For clarity, only the chain backbone of the macromolecules are shown.

pores, and fine particulate inclusions can be accidentally incorporated into a solid, whether crystalline or amorphous, during its production or use. These three-dimensional defects have a significant effect on the bulk properties of a solid, such as its strength and the way it fractures under an applied mechanical stress (Chapter 4).

3.4.1 Point Defects

Point defects are defects associated with one or a few atomic sites in the crystal. The major types of point defects are illustrated in Figure 3.16. A different atom, called a foreign atom, that occupies the site of a regular atom in the pure crystal, called a host atom, is said to be a substitutional defect. A vacant atomic site is called a vacancy. Foreign atoms and host atoms can occupy the interstitial sites between the regular atomic positions, giving foreign interstitial defects and self-interstitial defects, respectively.

The majority of metals are not used in their pure state because they often have inadequate properties for most applications. Instead, other metals are commonly added to them in various amounts, forming alloys with an improved combination of properties such as higher strength, fatigue resistance, and corrosion resistance. Atoms of the alloying metal occupy sites in the host metal structure, forming substitutional or interstitial defects. The alloying metal is said to form a solid solution with the host metal. When examined at a scale larger than the atomic scale, a solid solution is homogeneous. It can be visualized as a solid-state analog of a liquid solution such as a solution formed by dissolving common salt (sodium chloride) in water. The amount of one metal that can be alloyed with another metal to form a solid solution can vary from less than a fraction of a percent by weight (abbreviated wt%) to nearly 100 wt%, depending on the combination of metals. Copper, for

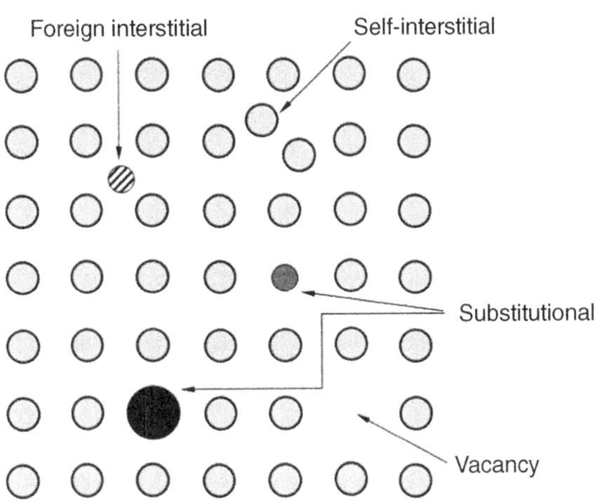

Figure 3.16 Schematic representation of the types of point defects in a crystal.

example, can dissolve more than 30 wt% zinc to form brass, forming a substitutional solid solution in which the zinc atoms occupy the regular atomic sites of copper. On the other hand, iron can dissolve no more than ~0.007 wt% carbon at room temperature to form an interstitial solid solution known as mild streel. Brass and mild steel, however, are not suitable for use as biomaterials on account of their low corrosion resistance *in vivo*.

A Ti alloy that is widely used as implants in orthopedic and dental surgery is Ti6Al4V, composed of ~6 wt% aluminum and ~4 wt% vanadium (V), with the remainder (~90 wt%) being Ti (Chapter 6). Although Ti6Al4V has an elastic modulus comparable to Ti, its superior strength, fatigue resistance, and corrosion resistance make it well suited for applications such as implants for total joint replacement, fracture fixation plates and dental implants. Certain stainless steels and cobalt–chromium (Co–Cr) alloys also find considerable use in orthopedic surgery.

Ceramic solid solutions are widely used in technological applications. One solid solution used in biomedical applications is YSZ, formed by adding ~3 mol% of yttria (Y_2O_3) to zirconia (ZrO_2). In this solid solution, the yttrium (Y) atoms predominantly occupy the regular zirconium (Zr) sites in the host ZrO_2 structure, resulting in the formation of substitutional defects. Vacant defect sites (vacancies) are also created in the process to compensate for the difference in valence between Y and Zr.

The inorganic constituent of bone and teeth is not pure hydroxyapatite, $Ca_{10}(PO_4)(OH)_2$, but a highly substituted hydroxyapatite-like material. This material has a structure that resembles that of hydroxyapatite but it contains approximately 4–8 wt% of $(CO_3)^{2-}$, often the main substituting species, smaller amounts of ions such as $(HPO_4)^{2-}$, Mg^{2+} and Na^+ (less than ~1 wt%), and trace amounts of ions such as Sr^{2+}, K^+, F^-, and Cl^- (Posner and Betts 1975). While the defect chemistry of hydroxyapatite is complex, $(CO_3)^{2-}$ and $(HPO_4)^{2-}$ ions typically substitute at the anionic $(PO_4)^{3-}$ sites in the crystal structure, whereas ions such Mg^{2+}, Na^+, Sr^{2+}, and K^+ substitute at the cationic Ca^{2+} sites, and F^- and Cl^- ions substitute for OH^- ions (Figure 3.17).

3.4.2 Line Defects: Dislocations

A dislocation is a discontinuity in atomic packing that is associated with several atomic sites along a line in a crystal. The interatomic bonds are highly distorted in the immediate vicinity of a dislocation (Figure 3.18), causing some degree of mechanical strain. Consequently, a crystal that contains dislocations has a higher energy than a similar but dislocation-free crystal due to the energy associated with the mechanical strain. The higher the number of

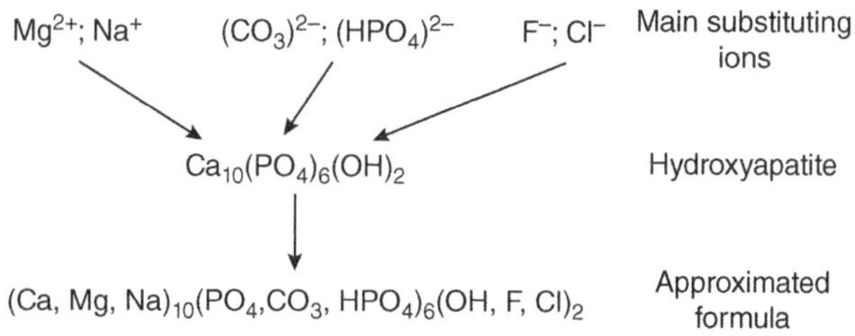

Figure 3.17 Major substituting ions and approximated formula of hydroxyapatite-like constituent of bone.

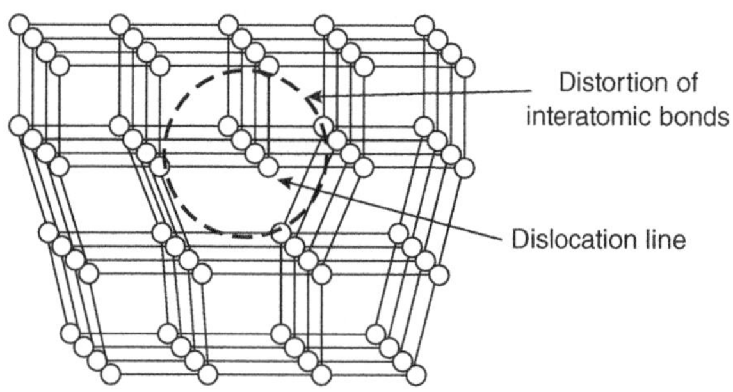

Figure 3.18 Schematic representation of a dislocation in a crystal and the distortion of the interatomic bonds in its immediate vicinity.

dislocations per unit volume, the higher the strain energy of the crystal. Because of the ease with which metal ions can break and reform bonds (Chapter 2), the mechanical properties of metals are strongly dependent on the presence of dislocations. The property of ductility, for example, that makes metallic biomaterials such as Ti6Al4V, stainless steel and Co–Cr alloys attractive for use as fracture fixation plates, hip, and knee implants and dental implants is due to the presence of dislocations in these materials.

Dislocations can become more numerous when a metal is deformed under appropriate mechanical stresses such as those during rolling or extrusion, thereby enhancing the strain energy of the crystals. Practically, deformation of a metal to increase its strain energy followed by thermal treatment of the highly strained material is one of the common methods used to control the microstructure of a metal and, thus, its mechanical properties (Chapter 6).

In comparison, dislocations are unimportant in the majority of ceramics at ordinary temperatures. A considerably higher amount of energy is required for the creation and migration of dislocations in ceramics due to their strong ionic or covalent bonding, and their composition composed of two or more dissimilar atoms. Consequently, the dislocation density (number of dislocations per unit volume) is low or negligible, and the role of dislocations can be neglected in the majority of ceramics, including those used as biomaterials such as aluminum oxide, hydroxyapatite, and β-tricalcium phosphate.

Types of Dislocations

Dislocations are divided into two basic types, called edge dislocations and screw dislocations (Figure 3.19) but they are often more complex in real crystals, composed of some mixture of these two basic types. In a simple sense, an edge dislocation can be visualized as making a cut in a perfect crystal and inserting an additional half-plane of atoms. The edge of the half-plane of atoms within the crystal is the dislocation line, that is, defects along a line of atomic sites. Instead of inserting an extra half-plane of atoms after making a cut in the crystal, we can displace one part of the crystal relative to the other part in a direction parallel to the bottom of the cut. This gives the other basic type of dislocation, called a screw dislocation because, if traced in the direction of the dislocation line, the planes of atoms form a helical surface or a screw.

Slip or Plastic Deformation Resulting from Dislocation Motion

When a crystal containing edge dislocations is subjected to a sufficiently high mechanical stress, one plane of atoms slides or is successively displaced over atoms in the adjacent plane in a direction perpendicular to the dislocation line. However, all the atoms in one plane do not slide simultaneously over those in the adjacent plane. Instead, the atomic bonds across the plane of displacement break

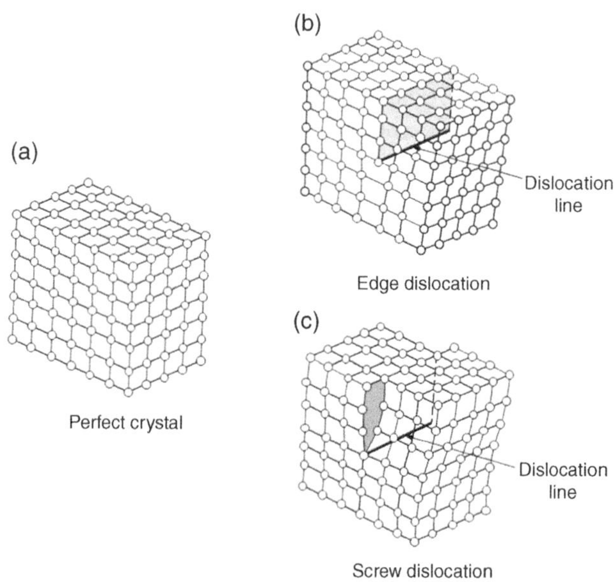

Figure 3.19 Schematic representation of part of a perfect crystal (a) and the arrangement of atoms near the two basic types of dislocations, edge dislocation (b) and screw dislocation (c).

and reform one at a time (Figure 3.20). Complete movement of the dislocation through the crystal results in permanent displacement of one plane of atoms over the other plane. This type of deformation is called plastic deformation or slip, because, unlike elastic deformation, the atoms do not return to their original positions upon release of the mechanical stress (Chapter 4). Although the geometry of a screw dislocation is more complicated, its properties are similar to those described for the edge dislocation.

If a perfect crystal is subjected to tensile loading, the stress required to simultaneously rupture all the bonds across a plane and cause fracture is defined as the ideal strength or theoretical strength of the crystal. Theoretical analysis shows that the ideal strength of a crystal is in the range approximately $E/3$ to $E/8$, where E is the elastic modulus in tensile loading called the Young's modulus (Chapter 4). Measurements show that when a metal is loaded in tension, the stress required to cause plastic deformation, called the yield stress, is considerably lower than the ideal strength, by a factor of approximately 10^2–10^6 depending on the purity and composition of the metal. This is because plastic deformation of a crystal by dislocations motion is far easier and requires far less mechanical stress or energy due to the sequential breaking and reforming of the atomic bonds. This ability to deform rather than fracture suddenly in a brittle manner at some higher stress is responsible for the attractive property of ductility shown by metals.

Another property of dislocation motion is that slip occurs along preferred planes and in preferred directions in these planes, and not just in any plane or direction (Figure 3.21). Of the possible slip planes and directions in a crystal, under a specific set of conditions, slip normally occurs in a plane in which the atoms have the closest packing and in a direction in this close-packed plane which has the closest spacing (that is, in a close-packed direction) (Table 3.3). This is

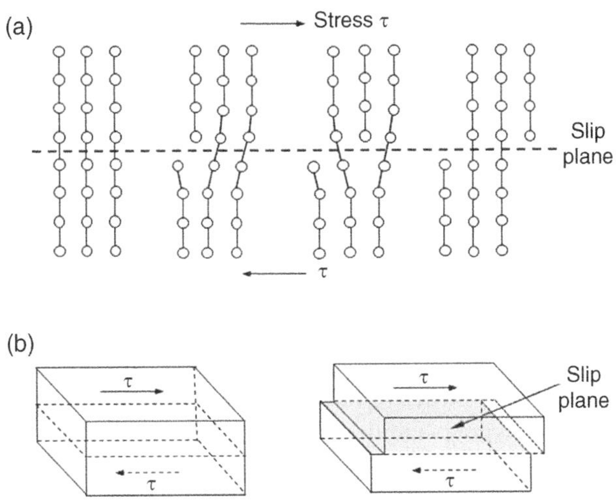

Figure 3.20 Schematic representation of the movement of an edge dislocation through a crystal. The atomic bonds at the center of the dislocation break and reform in a sequential manner, allowing the dislocation to move. Complete movement of the dislocation through the crystal results in slip of one part of the crystal over the other part.

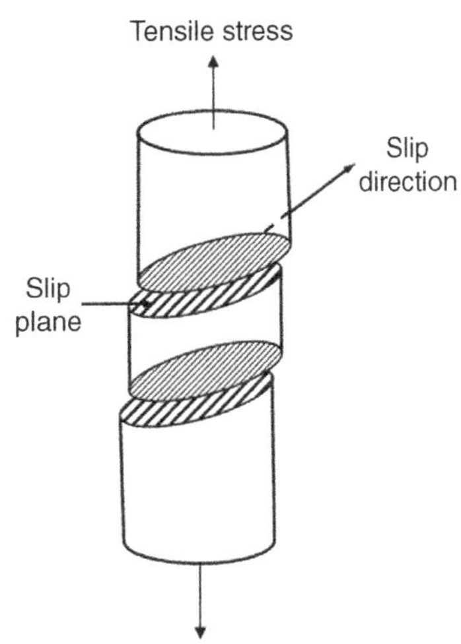

Figure 3.21 Schematic representation of slip in a metal that is subjected to a tensile stress. Slip occurs preferentially along closed-packed planes and in close-packed directions in these planes.

Table 3.3 Characteristic properties of the three common metallic structures.

Structure	N	CN	PD (%)	Slip planes/number	Slip directions/number	Slip system per unit cell
BCC	2	8	68	{110}	⟨111⟩	$6 \times 2 = 12$
				6	2	
FCC	4	12	74	{111}	⟨110⟩	$4 \times 3 = 12$
				4	3	
CPH	6	12	74	(0001)	$\langle 11\bar{2}0 \rangle$	$1 \times 3 = 3$
				1	3	

N, number of atoms per unit cell; CN, coordination number (number of atoms surrounding a given atom); PD, packing density.

because a lower amount of energy is required for the dislocation to move in a close-packed direction of a close-packed plane due to the smaller distance that an atom has to move from one atomic position to a neighboring position.

3.4.3 Planar Defects: Surfaces and Grain Boundaries

Unless formed under highly controlled and, often, expensive conditions, metals and ceramics do not consist of a single crystal. Instead, they are composed of a large number of small crystals, called grains, of size approximately a fraction of a micron to several tens of microns (Figure 3.22). The material is described as polycrystalline. At the two-dimensional interface where two grains meet, called the grain boundary, the atoms do not pack in an ordered crystalline arrangement because the atomic planes in adjacent grains have different orientations. Consequently, the grain boundary is generally considered a two-dimensional or planar defect in crystalline solids. However, it is not an infinitesimally thin plane. Electron microscopy has shown that the grain boundary in pure solids is 0.5–1.0 nm wide and, in this region, the atoms are more loosely packed when compared to the atoms within the crystal itself. The presence of grain boundaries have a strong influence on several properties of polycrystalline materials. As the atoms are more loosely packed, grain boundaries provide an easier path for atoms to migrate through the material. Grain boundaries also obstruct the motion of dislocations due to the looser atomic packing and the change in lattice orientation, a property that is used in controlling the strength and ductility of metals (Chapter 6).

The atoms at a free surface have a different atomic environment and, often, a different composition than those within the solid itself. Consequently, the surface of a material can also be considered a two-dimensional or planar defect. As the surface of a biomaterial has a strong influence on its performance *in vivo*, the surface characteristics of solids relevant to their use as biomaterials are discussed in detail in Chapter 5.

3.5 Microstructure of Biomaterials

The engineering properties of a material depends on its microstructure and, consequently, in the design and creation of biomaterials, we are concerned not just with the material itself but with its microstructure also. Microstructure refers to the structure of a material at a scale of approximately 0.1–100 μm, such as the nature of the phases present in the material, the quantity of each phase and its distribution. Unless specifically emphasized, structure at a nanoscale is often included within the realm of microstructure. A phase is defined as a region of material with uniform physical and chemical properties. As biomaterials are solids, we are essentially concerned with one or more solid phases, crystalline or amorphous, and, if present, porosity.

Controlling the microstructure of a biomaterial to achieve a desired combination of properties for a particular application is an important area of biomaterials science. Inadequate microstructural control often leads to unpredictable properties, which can lead to catastrophic failure and, for biomaterials used as implants for example, to serious complications for the patient. A large variety of methods are available to create biomaterials with the requisite shape and microstructure, but robotic methods, referred to as rapid prototyping, additive manufacturing or, more commonly, three-dimensional printing (3D printing), have been used increasingly in recent years. These robotic methods provide unprecedented control of the shape and microstructure

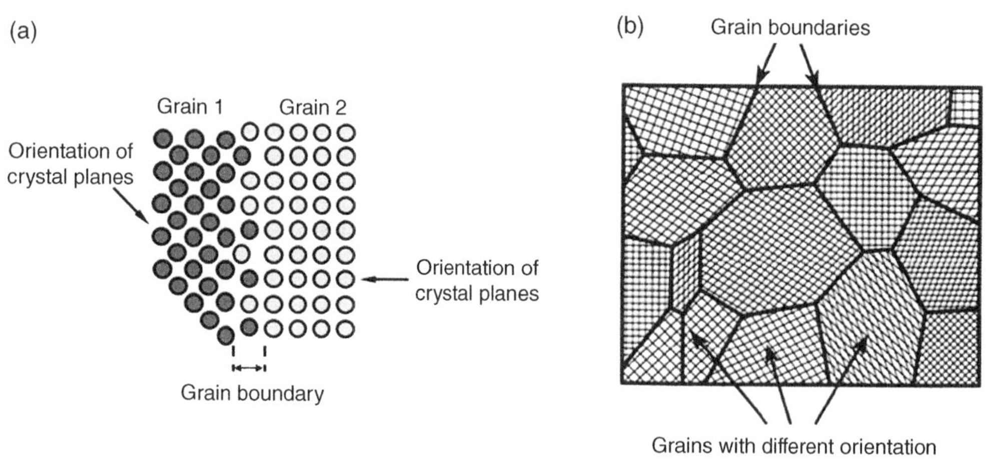

Figure 3.22 Illustration of (a) the boundary region between two grains and (b) part of a polycrystalline solid. The atomic planes in adjacent grains have different spatial orientation.

of a biomaterial, particularly when a high degree of complexity is required.

Microstructures are commonly examined in a microscope, such as an optical microscope or an electron microscope, depending on the scale of resolution required. Often, metals and ceramics are etched using a thermal or chemical treatment to reveal the grain boundaries or the phases within the solid. The microstructures of biomaterials are numerous, varying from simple to more complex and, thus, it is not possible to describe them all. Instead, we discuss microstructural features relevant to the design and creation of biomaterials in two broad categories, namely dense biomaterials and porous biomaterials.

3.5.1 Microstructure of Dense Biomaterials

The simplest type of microstructure is a dense solid, such as a solid containing little or no porosity, composed essentially of a single phase as, for example, a Ti dental implant, an Al_2O_3 femoral head in a hip implant or a polymethyl methacrylate (PMMA) intraocular lens. As they are polycrystalline materials, pure metals and ceramics such as Ti and Al_2O_3, respectively, are composed of grains (Figure 3.23a). Assuming the use of proper production methods, resulting in a homogeneous microstructure free from flaws, the properties of the pure polycrystalline material are controlled mainly by the size of its grains. Finer grains, for example, often lead to better mechanical properties. Microstructure is not relevant to a nonporous glass or a nonporous amorphous polymer such as PMMA, for example, because when produced using proper methods, each consists of a homogeneous single phase without grains or boundaries.

At a higher level of complexity, we can consider a dense solid composed of more than one phase. The Ti alloy, Ti6Al4V, for example, used as fracture fixation plates, in hip implants and as dental implants, is typically composed of two crystalline phases at ordinary temperatures, termed an alpha (α) phase and a beta (β) phase. Depending on the production method and subsequent heat treatment, this alloy can show a variety of microstructures. Figure 3.23b shows one such microstructure, composed of nearly parallel α lamellae within large β grains. ZTA femoral heads used in hip implants are composed of a matrix of Al_2O_3 grains and a distribution of YSZ grains (Figure 3.23c). The volume fraction of the ZrO_2 is often kept within a useful range, approximately 10–15%. For this composition, the optimal mechanical properties such as strength and resistance to fracture are achieved through production methods that give a fine grain size of the Al_2O_3 phase, less than a few micrometers, a finer grain size of the ZrO_2 phase and a homogeneous distribution of the ZrO_2 grains within the Al_2O_3 matrix.

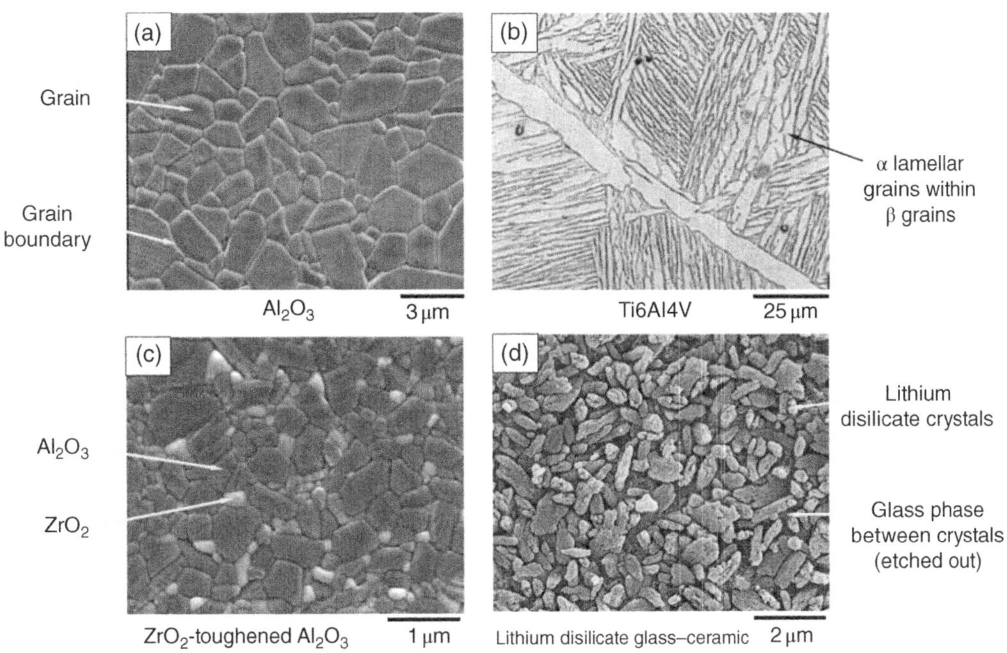

Figure 3.23 Examples of microstructures of dense biomaterials. (a) Al_2O_3 showing grains and grain boundaries; (b) Ti6Al4V, a two-phase (α + β) solid composed of α lamellae in β grains; (c) zirconia-toughened alumina (ZTA) composed of zirconia (ZrO_2) particles (grains) dispersed in a matrix of alumina (Al_2O_3) grains; (d) lithium disilicate glass-ceramic composed of elongated lithium disilicate ($Li_2Si_2O_5$) crystals in a glass matrix (removed by etching). The surface of the materials was etched thermally (a), (c), or chemically (b), (d) to better reveal the microstructure. (b) *Source* From Peters et al. (2003); (d) From Holand and Beall (2012).

Glass-ceramics, produced by controlled thermal treatment of a parent glass, show a microstructure composed of crystals within a glass matrix. A glass-ceramic typically has better properties, such as mechanical and thermal properties, than its parent glass. The volume fraction of the crystals in the majority of glass-ceramics is in the range 30–70% but some glass-ceramics have volume fractions outside this range. Lithium disilicate glass-ceramics are widely used for dental restorations such as crowns and bridges. While these materials have excellent resistance to chemical degradation, optimal mechanical properties are achieved through a controlled thermal treatment of the parent glass to achieve a fine crystal size, a high volume fraction of crystals and a homogeneous distribution of the crystals within the glass matrix (Figure 3.23d).

In a nonporous form, semicrystalline polymers, such as PE are composed of a mixture of amorphous and crystalline phases (Figure 3.15b). The crystalline regions are referred to as crystallites. They are not called grains because they have a different structure from the grains in polycrystalline metals and ceramics. The properties of semicrystalline polymers often improve with a higher volume fraction and a homogeneous distribution of crystallites as, for example, in UHMWPE used as articulating bearings in hip and knee implants.

As polymers are weaker than bone, one way to improve their mechanical properties is to reinforce them with a strong solid phase, commonly a ceramic biomaterial in the form of particles or fibers, to form a composite (Chapter 12). Composites composed of hydroxyapatite particles dispersed in a PE matrix have been created for use as implants in healing bone defects (Bonfield et al. 1981). The hydroxyapatite particles enhance the elastic modulus of the PE, bringing it closer to that of bone. They also provide an additional benefit by imparting some degree of bioactivity to the implant and, thus, enhance its potential to form a stronger interfacial bond and integrate with host bone.

3.5.2 Microstructure of Porous Biomaterials

Porous biomaterials are desirable in a variety of biomedical applications for two major reasons. First, as pores modify the properties of a solid, their incorporation into a biomaterial is often used as a strategy to bring its properties such as degradation rate, strength, and elastic modulus closer to those desired for a given application. Porosity can be used, for example, to modulate the degradation rate of a polymer in tissue engineering applications to some optimal value. Incorporation of pores into strong materials such a Ti and hydroxyapatite is often used to bring their strength and elastic modulus closer to those of bone when they are used in healing bone defects. This is because a large difference in mechanical properties between the implant and bone can lead to adverse cell and tissue responses, resulting in bone resorption, loosening of the implant, bone fracture or the need for revision surgery (that is, replacement of the implant in a second surgery) (Chapter 17).

Second, pores are often required to allow cells and nutrients to infiltrate an implant and create new tissue, resulting in healing of a tissue defect and integration of the implant with host tissue. When used in healing bone defects, for example, Ti implants often contain pores in order to achieve better bonding with bone. Lack of integration of an implant often results in unsatisfactory healing of a bone defect. In tissue engineering applications, pores are often incorporated into degradable biomaterials (called scaffolds) to allow cells to create new tissues or organs while the scaffold degrades away.

The design and creation of a porous implant is important because the properties and performance of the implant depend strongly on a variety of pore characteristics. Furthermore, each application will have its own optimal pore characteristics. The important pore characteristics that require consideration are:

- The volume fraction of the pores in the biomaterial, that is, the porosity of the biomaterial;
- The average size of the pores, their shape and whether they are oriented in a given direction or not;
- How the pores are distributed within the solid, either randomly or in some ordered manner;
- Whether the pores are isolated, that is, disconnected from the surface of the material and from one another, or interconnected, that is, they from a continuous channel through the solid.

An interconnected pore channel is desirable to allow cells and nutrients to migrate throughout an implant. For a random distribution of spherical pores of the same radius, the porosity should be at least ~16% for the pores to be interconnected. The porosity of implants used in healing bone defects and in tissue engineering, for example, should be well above this value, often greater than ~50%.

The majority of forming methods produce a random distribution of pores within the solid as well as pores that have a range of sizes (Figure 3.24a). On the other hand, the use of additive manufacturing (3D printing) techniques can provide pores with well-controlled sizes and with a predesigned distribution within the solid (Figure 3.24b). As different cell types have different sizes and may require different environments for creating new tissue, controlling the size of the pores within some narrow range is often desirable for optimal healing or regeneration of some tissues. Implants for healing bone defects, for example,

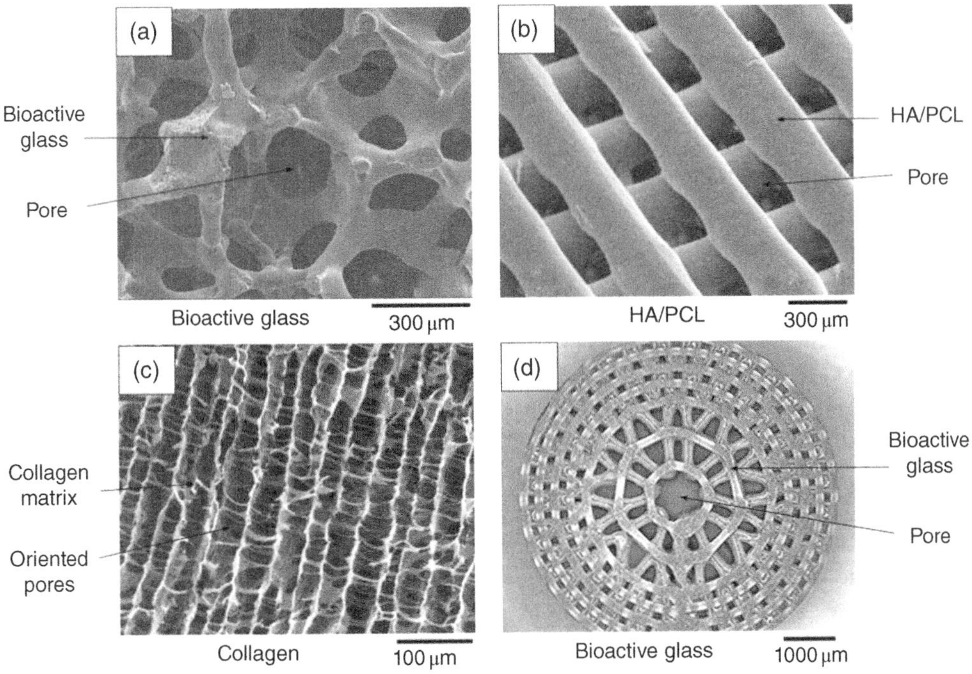

Figure 3.24 Examples of microstructures of porous biomaterials. (a) Bioactive glass with a microstructure approximating human trabecular bone; (b) composite composed of hydroxyapatite (HA) particles in a polycaprolactone (PCL) matrix. *Source:* From Russias et al. (2007); (c) oriented pores in collagen. *Source:* From Schoof et al. (2001); (d) bioactive glass with a gradient in pore size and porosity approximating human long bone of the limbs.

should have interconnected pores of size ~300–500 μm to provide an environment that is conducive to bone formation as well as angiogenesis, the formation of blood vessels that is essential for bone formation and growth (Karageorgiou and Kaplan 2005).

At another level of complexity, we can envisage microstructures in which the pores are aligned in a given direction, have two or more discretely different pore sizes, or have a controlled gradient in porosity along a given direction. Implants containing pores that are oriented in a given direction (Figure 3.24c) are particularly important for directional growth of cells in some tissue engineering applications such as nerve regeneration. Porous implants (scaffolds) with a more complex microstructure may be required for engineered regeneration of some tissues and organs. The liver, for example, contains five cell types and a complex network of blood vessels. Consequently, the use of scaffolds with a more complex microstructure may be relevant for the regeneration of a liver. The long bones of human limbs are composed of a less porous, stronger outer region of cortical bone and a more porous, weaker inner region of trabecular bone. These bones are also subjected to significant physiological stresses during normal activity. Synthetic implants composed of a gradient microstructure may be relevant to healing segmental defects in these long bones (Figure 3.24d) In addition to mimicking the bone structure, such implants may be better able to provide the requisite mechanical properties.

3.6 Special Topic: Lattice Planes and Directions

Different planes and directions in crystals can have a different packing of the atoms and, for crystals composed of compounds such as ceramics, they can also have a different composition of atoms or ions. Differences in atomic packing and composition can influence the properties of crystals in different planes and along different directions. Plastic deformation (slip) of metals, as noted earlier, occurs in planes and along directions that have the closest packing of the atoms and not just in any plane or along any direction. Consequently, in correlating properties of materials with their crystal structures, it is often necessary to identify specific atomic positions, planes, and directions in the crystal structure in a succinct manner.

Unit Cell Geometry

The atomic positions, lattice planes, and directions in a crystal are specified with reference to the unit cell of the structure, as described in Figure 3.5. Directions opposite to the x, y, and z axes are taken to be negative.

Lattice Positions

A point in the crystal lattice is denoted by its coordinates along the x, y, and z axes expressed as a ratio of the unit cell lengths a, b, and c in these directions (Figure 3.25). The origin of the unit cell is 0,0,0. Then, a point at the far corner of the unit cell, opposite the origin, with coordinates a,b,c is written 1,1,1 because each coordinate divided by the respective unit cell length is equal to 1. Similarly, 2,1/2,1 denotes a point in the lattice with coordinates $2a$, $b/2$, and c in the x, y, and z directions, respectively.

Lattice Planes

Any plane A′B′C′ can be defined by the intercepts OA′, OB′, and OC′ with the three principal axes of the crystal system (Figure 3.25). In the common notation, referred to as Miller indices,

- We take the reciprocals of the ratios of the intercepts to the corresponding unit cell dimensions
- Then the numbers are converted to the smallest integers

As an example, for a cubic crystal structure illustrated in Figure 3.25, the plane A′B′C′ has intercepts OA′ = $2a$, OB′ = $3a$, and OC′ = $3a$. The reciprocals of the ratios of the intercepts to the corresponding cell dimensions are respectively

$$\frac{1}{2a/a};\ \frac{1}{3a/a};\ \frac{1}{3a/a}$$

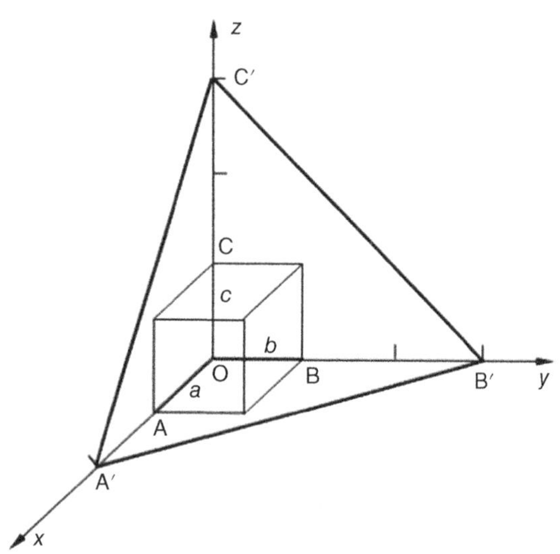

Figure 3.25 Diagram illustrating the specification of lattice planes in a crystal.

Converting these numbers to the smallest integers, we get 3, 2, and 2, respectively. These numbers are written in parentheses (322) and they are called the Miller indices of the plane. Thus, the plane A′B′C′ is referred to as the (322) plane. In general, a plane with Miller indices h, k, and l is written as (hkl).

A negative intercept of a plane on a crystal axis is written with a bar over the respective index. A plane that has intercepts of $2a$, $3a$, and $-3a$, respectively, on the x, y, and z axes, for example, has Miller indices of $(32\bar{2})$. A plane that is parallel to a principal axis of a crystal system is said to have an intercept of infinity (∞) on that axis and, thus, the reciprocal of the intercept is 0. In the cubic system, for example, the (100) plane intercepts the x axis at a and is parallel to the yz plane, that is, it intercepts the y and z axes at ∞.

Depending on the crystal system, two or more planes may belong to the same family of planes. In the cubic system, for example, the six sides of the unit cell has Miller indices of

$$(100)\quad(010)\quad(001)\quad(\bar{1}00)\quad(0\bar{1}0)\quad(00\bar{1})$$

These planes are equivalent except for our arbitrary choice of axis labels and directions, or, alternatively, on how we view the orientation of the unit cell in space. They have the same arrangement of atoms and, thus, the properties in these planes are identical. These planes are described collectively as {100} planes where the Miller indices are enclosed within curly brackets. In general, a family of equivalent planes with Miller indices h, k, and l is written $\{hkl\}$.

Lattice Directions

Any direction in a crystal lattice (Figure 3.26) is described by:

- A line through the origin O of the unit cell that is parallel to the given direction;
- Resolving the length of this line along the three principal axes x, y, and z and expressing these lengths as a fraction of the unit cell dimensions a, b, and c;
- Converting the ratios to the three smallest integers, u, v, and w, for example

The convention used to describe the direction is to write the three integers u, v, and w in square brackets, that is, $[uvw]$. The direction LM in Figure 3.26, for example, is found by drawing the line OE through the origin that is parallel to it. Resolving the length of OE along the x, y, and z axes and expressing them as a fraction of the unit cell dimensions, we get

$$\frac{OA}{a}, \frac{OB}{b}, \frac{OC}{c} \quad \text{or} \quad \frac{a}{a}, \frac{b}{b}, \frac{c}{c}$$

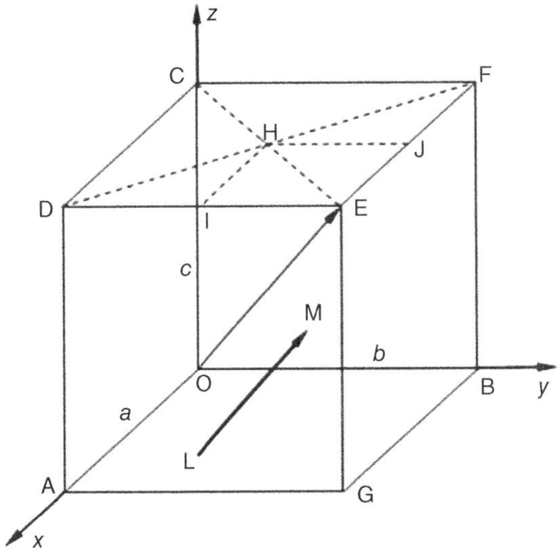

Figure 3.26 Diagram illustrating the specification of lattice directions in a crystal.

Thus, the direction LM is described as [111], as are directions parallel to LM or OE.

Similar to planes, we can have families of equivalent directions. In the cubic system, for example, the directions

$$[111] \quad [11\bar{1}] \quad [1\bar{1}1] \quad [\bar{1}11]$$

$$[\bar{1}\bar{1}\bar{1}] \quad [\bar{1}\bar{1}1] \quad [\bar{1}1\bar{1}] \quad [1\bar{1}\bar{1}]$$

are equivalent except for our arbitrary choice of axis labels and directions. They are denoted collectively as $\langle 111 \rangle$ directions where the integers are enclosed within angle brackets. In general, a family of equivalent directions described by the integers u, v, and w is written $\langle uvw \rangle$. Properties of a cubic crystal that depend on direction, such as the elastic modulus, for example, will be identical in these eight directions.

3.7 Concluding Remarks

In this chapter, we discussed how atoms pack in three dimensions to form solids, defects in crystalline solids and microstructure of materials, which, along with atomic bonding (Chapter 2), are the structural features that control the properties of a material.

- Packing of atoms in solids is divided into two broad classes, crystalline, and amorphous.
- The vast majority of metals and ceramics are not composed of a single crystal but, instead, are polycrystalline, composed of a large number of small crystals called grains. Atomic packing defects composed of point defects, line defects (dislocations) and grain boundaries exert a strong effect on the physical and mechanical properties of crystalline solids.
- Inorganic glasses are amorphous but when subjected to a controlled heat treatment, some glasses can be converted to glass-ceramics, which often have better properties than their parent glasses.
- The majority of polymers are amorphous but under appropriate conditions, some polymers develop a semicrystalline structure composed of crystalline and amorphous regions. Whether a polymer is amorphous or semicrystalline has a strong influence on their properties.
- We discussed more recently discovered allotropes of carbon, such as graphene, carbon nanotubes, and fullerenes, which are receiving considerable interest for use in biomedical applications due to their unique properties.
- As the engineering properties of materials are dependent on their microstructure, the ability to control the microstructure of a biomaterial is important for achieving a desired combination of properties for a particular application.

Problems

3.1 Aluminum has an atomic radius of 0.1431 nm and an FCC structure. Determine (i) the number of atoms in a unit cell, (ii) the unit cell length, and (iii) the density of aluminum. (Atomic weight of aluminum = 26.97; Avogadro number = 6.023×10^{23}.)

3.2 Determine the atomic packing fraction for the simple cubic, BCC and FCC structures composed of a single type of atom. Explain the steps in your calculation.

3.3 A metal has an atomic radius r and a CPH structure. Determine the unit cell lengths in terms of r and the atomic packing fraction. Explain the steps in your calculation.

3.4 Using the unit cell illustrated in Figure 3.10 and the unit cell lengths $a = b = 0.9430$ nm and $c = 0.6891$ nm, determine the density of hydroxyapatite.

3.5 The use of fluorinated water or toothpaste is known to improve the resistance of teeth to decay. Can this be accounted for in terms of the structure of hydroxyapatite. Explain.

3.6 Some drug delivery devices are composed of drug molecules incorporated within a degradable polymer. As the polymer degrades, these molecules are released at a controllable rate. In this application, amorphous polymers are often preferred over semicrystalline polymers. Explain why.

3.7 Explain why copper forms a substitutional solid solution with zinc whereas carbon forms an interstitial solid solution with iron.

3.8 Using a diagram, explain how dislocations reduce the amount of stress required to cause plastic deformation.

3.9 Explain the meaning and significance of microstructure.

3.10 What microstructural features would be desirable for (i) a hydroxyapatite implant for healing bone defects; (ii) a lithium disilicate glass-ceramic for use as a crown in dental restoration; (iii) a degradable polymer such as PLA reinforced with hydroxyapatite particles for healing bone defects? Explain.

3.11 Using the axes in Figure 3.25, sketch (i) two planes with Miller indices (122); (ii) a plane with Miller indices $(32\bar{2})$.

3.12 Determine the lattice directions of the lines OI, DE, FG, and HI in Figure 3.26.

3.13 Aluminum atoms (radius = 0.1431 nm) pack to form a FCC structure. Determine the number of atoms per square meter in the (100) and (111) planes of aluminum.

3.14 Using a unit cell, show the ions in the {100} and {111} planes of sodium chloride.

References

Bonfield, W., Grynpas, M.D., Tully, A.E. et al. (1981). Hydroxyapatite reinforced polyethylene – a mechanically compatible implant material for bone replacement. *Biomaterials* 2: 185–186.

Dearnaley, G. and Arps, J.H. (2005). Biomedical applications of diamond-like carbon (DLC) coatings: a review. *Surface and Coatings Technology* 200: 2518–2524.

Eatemadi, A., Daraee, H., Karimkhanloo, H. et al. (2014). Carbon nanotubes: properties, synthesis, purification, and medical applications. *Nanoscale Research Letters* 9: 393 (13 pages).

Geim, A.K. and Novoselov, K.S. (2007). The rise of graphene. *Nature Materials* 6: 183–191.

Giacolone, F. and Martin, N. (2006). Fullerene polymers: synthesis and properties. *Chemical Reviews* 106: 5136–5190.

Holand, W. and Beall, G.H. (2012). *Glass Ceramic Technology*. Wiley.

Iijima, S. and Ichlhashi, T. (1993). Single-shell carbon nanotubes of 1 nm diameter. *Nature* 363: 603–605.

Karageorgiou, V. and Kaplan, D. (2005). Porosity of 3D biomaterial scaffolds and osteogenesis. *Biomaterials* 26: 5474–5491.

Peters, M., Heptenmacher, J., Kumpfert, J., and Leyens, C. (2003). Structure and properties of titanium and titanium alloys. In: *Titanium and Titanium Alloys. Fundamental and Applications* (eds. C. Leyens and M. Peters), 1–36. Wiley-VCH.

Posner, A.S. and Betts, F. (1975). Synthetic amorphous calcium phosphate and its relation to bone mineral structure. *Accounts of Chemical Research* 8: 273–281.

Russias, J., Saiz, E., Deville, S. et al. (2007). Fabrication and in vitro characterization of three-dimensional organic/inorganic scaffolds by robocasting. *Journal of Biomedical Materials Research Part A* 83 (2): 434–445.

Schoof, H., Apel, J., Heschel, I., and Rau, G. (2001). Control of pore structure and size in freeze-dried collagen sponges. *Journal of Biomedical Materials Research* 58: 352–357.

Zhao, H., Ding, R., Zhao, X. et al. (2017). Graphene-based nanomaterials for drug and/or gene delivery, bioimaging, and tissue engineering. *Drug Discovery Today* 22: 1302–1317.

Further Reading

Callister, W.D. (2007). *Materials Science and Engineering: An Introduction*, 7the. New York: Wiley.

Dorozhkin, S.V. (2009). Calcium orthophosphates in nature, biology and medicine. *Materials* 2: 399–498.

Jones, D.R.H. and Ashby, M.F. (2018). *Engineering Materials I, 5th Edition, An Introduction to Properties, Applications and Design*. Oxford (UK): Butterworth-Heinemann.

LeGeros, R.Z. (2002). Properties of osteoconductive biomaterials: calcium phosphates. *Clinical Orthopaedics and Related Research* 395: 81–98.

4

Bulk Properties of Materials

4.1 Introduction

The atoms at the free surface of a solid have a different atomic environment and, often, a different composition than those within the solid itself. On this basis, we can divide the properties of materials into two broad categories:

- Bulk properties, that is, the properties of the three-dimensional solid itself
- Surface properties, the properties of the surface of the solid that interfaces with its environment, such as the biological environment *in vivo*.

Bulk properties include mechanical properties such as strength and elastic modulus, electrical properties, magnetic properties, and optical properties, commonly grouped together as physical properties. In comparison, surface properties include topography (surface roughness or smoothness), contact angle and surface charge. Together, the bulk properties and surface properties determine the performance of a biomaterial *in vivo*.

In this chapter, we will consider the physical properties of three-dimensional solids that are relevant to their use as biomaterials. Surface properties are discussed in the next chapter. To round off our treatment of the combination of properties relevant to the applications of biomaterials, chemical properties such as degradation and corrosion are discussed in Chapters 14 and 15 while biocompatibility phenomena are discussed in Chapters 16–22.

Mechanical properties are often among the most important physical properties because many biomaterials are subjected to loads (forces) at some point or during almost their entire application and catastrophic failure of an implant *in vivo* is a serious complication. Biomaterials should have the requisite mechanical properties to support physiological loads when used, for example, as a prosthetic hip and knee joint, or to withstand the pressure pulsations of blood flow through the arterial vessels when used as a stent. In addition, biomaterials are typically designed to have some desirable combination of properties that normally depend on the intended application and not just adequate mechanical properties (Chapter 1). The electrodes in a cardiac pacemaker, for example, have to deliver an electric charge to the heart muscle. Consequently, in addition to having adequate mechanical reliability, the electrode material should have a strong ability to conduct electrical signals, that is, a high electrical conductivity. Dental restorations such as crowns and bridges, designed to restore the function and appearance of tooth structure, should have not just the requisite mechanical properties to avoid fracture but a low thermal conductivity comparable to the tooth itself and the right color for esthetic appearance as well.

The physical properties of materials cover a very broad field and, consequently, we will discuss only the properties relevant to the applications of biomaterials in this chapter. Whereas mechanical properties form the major part of this discussion because of its aforementioned importance, other properties such as electrical, magnetic, thermal, and optical properties are discussed as well. In general, the following main issues will be considered:

- Response of a biomaterial to a physical stimulus, such as a mechanical force, an electric or magnetic field, or a change in temperature
- The origins of this response and its dependence on atomic bonding and microstructure of the biomaterial
- Relation among the physical properties, application, and performance of biomaterials.

4.2 Mechanical Properties of Materials

Biomaterials often have to withstand a complex combination of loads that may operate in more than one direction. An implant designed to heal large defects in the long bones

Materials for Biomedical Engineering: Fundamentals and Applications, First Edition. Mohamed N. Rahaman and Roger F. Brown.

Companion website: www.wiley.com/go/rahaman/biomedicalengineering

of the limbs, for example, should be capable of withstanding appropriate physiological loads not only in compression but also in bending (flexure) and twisting (torsion). The implant should also be capable of withstanding loads due to walking, running, and jumping which are composed predominantly of repeated cyclic loads and not just a constant load. Thus, an understanding of the mechanical response of biomaterials when subjected to these different loading modes is essential.

Different classes of materials, such as metals, ceramics, and polymers, show different characteristic responses when subjected to mechanical forces, such as the ductility of metals and the brittleness of ceramics. Unlike metals and ceramics, synthetic polymers and many natural materials show a property called viscoelasticity in which the mechanical response depends not just on the magnitude of the load but also on the time over which the load is applied. These different mechanical responses strongly influence the selection and design of biomaterials.

Our discussion of mechanical properties is confined to materials composed of a single type of solid. These properties can be modified or improved by combining two or more different solids to form composites, a topic considered in Chapter 12. While the general aspects of mechanical testing procedures are discussed when appropriate, these procedures are not covered in detail because standard procedures are well described by organizations such as the American Society for Testing and Materials (ASTM) and the International Organization for Standardization (ISO).

4.2.1 Mechanical Stress and Strain

The mechanical behavior of a material is commonly determined by taking nominally identical specimens and deforming them in a mechanical testing machine according to a standard procedure. This type of experiment commonly gives data for the applied load and the deformation of the specimen. Load is actually a force and these two terms are often used interchangeably. Force, which has the unit newton (N), is commonly converted to a stress by dividing it by the area of the specimen over which it acts. This means that stress has units of N/m^2, also called pascal (Pa). The stresses encountered in testing of materials are often quite high and, thus, units of MPa (10^6 Pa) or GPa (10^9 Pa) are often used.

The measured deformation of a specimen in response to a stress, such as its elongation, is normalized to the appropriate dimension of the specimen, such as its original length, thus converting it to a strain, expressed as a fraction or percent (that is, strain is unitless). Conversion of load and deformation to stress and strain, respectively, serves to normalize the data. This compensates for dimensional changes of the specimen during testing and the use of different specimen dimensions in different testing laboratories. A common way to show the mechanical response of a material is a diagram in which the measured stress (Y-axis) is plotted versus the measured strain (X-axis) to give what is called a stress versus strain curve, sometimes written stress–strain curve for short. However, in some experiments, it may also be required to show a plot of the applied force versus the deformation.

In mechanical testing, the load can be applied in a few different ways, giving data for the mechanical response of a material under these different loading modes. These loading modes include uniaxial tension and compression, shear, torsion, and bending.

Uniaxial Tension and Compression

In uniaxial tension, a common testing mode for metals and polymers, a specimen of uniform cross section is pulled in one direction along its length, that is, uniaxially (Figure 4.1a), whereas in uniaxial compression the specimen is compressed in one direction (Figure 4.1b). In tension, a load (force) F is applied at the larger ends of the specimen of circular or rectangular cross section, and the stress and strain are measured for the central portion, called the gage length. For a specimen of gage length l_o and uniform cross section A, the applied stress σ is defined as

$$\sigma = \frac{F}{A} \tag{4.1}$$

and the strain ε is defined as

$$\varepsilon = \frac{\delta}{l_o} = \frac{l - l_o}{l_o} \tag{4.2}$$

where, δ is the increase in the gage length from l_o to l.

As the specimen does not have to be gripped at its ends in compression testing, the use of a cylindrical or tetragonal specimen of uniform cross section, that is, without the larger ends, is easier and less expensive. The common convention in mechanics is that tensile stresses and strains are positive in sign whereas compressive stresses and strains are negative in sign.

Shear Deformation

Mechanical testing in shear is more difficult to perform when compared to tension or compression and, thus, it is of interest only when shear is a significant deformation mode in the intended application. For a specimen in the

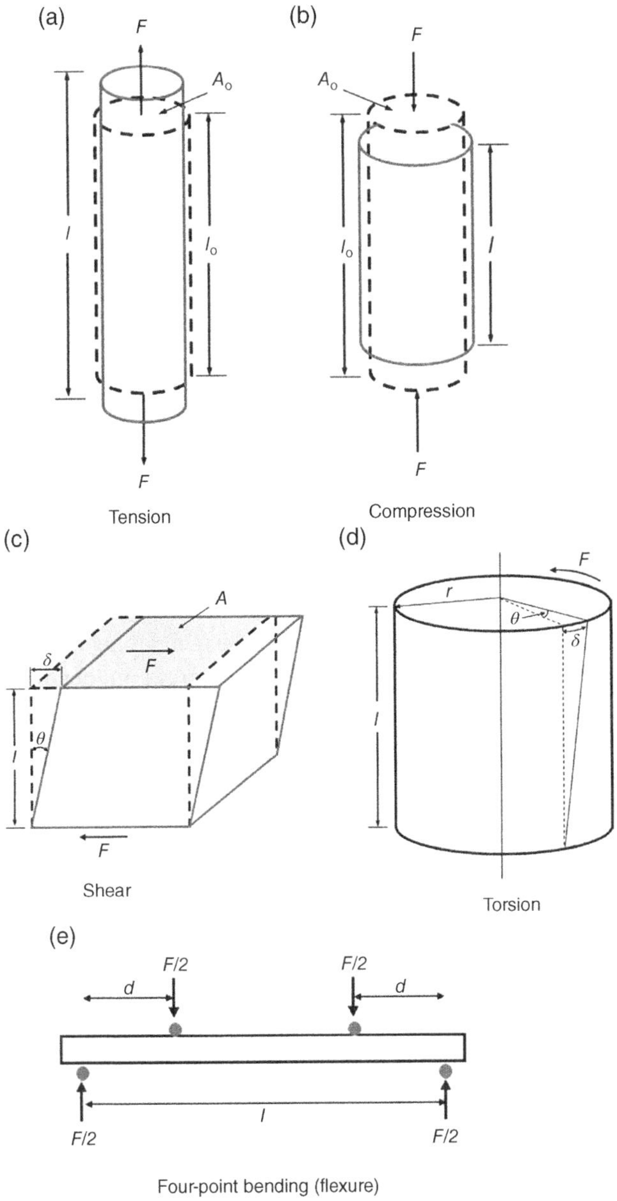

Figure 4.1 Different loading modes in mechanical testing of materials: (a) tension, (b) compression, (c) shear, (d) torsion, and (e) four-point bending (flexure).

shape of a cube of length l, for example, a force F is applied transversely over two opposite faces of the specimen of area A (Figure 4.1c). The horizontal faces of the specimen tend to slide relative to each other, with their edge lengths remaining the same, and the vertical faces tilt to accommodate this motion. The shear stress τ is defined as

$$\tau = \frac{F}{A} \tag{4.3}$$

where, A is the area over which the shear force acts, equal to l^2 for this example. The shear strain is defined as

$$\gamma = \frac{\delta}{l} \tag{4.4}$$

where, δ is the displacement of one horizontal face of the cube relative to the other. The shear strain is also given by

$$\gamma = \tan\theta \tag{4.5}$$

where, θ is the angle (in degrees) by which the one vertical face has tilted relative to the other. For small displacements (δ/l less than ~5%), the strain is also given to a good approximation as $\gamma = \theta$, where θ is now expressed in radians.

Torsional Deformation

In torsional deformation, a torque is applied to one end of a specimen to twist it relative to the other end and the angle of twist is measured. Although the geometry is different from that described in the previous section, torsional deformation results in shearing one part of the specimen over another and, thus, we can also define a shear stress and shear strain for this mode of deformation. A force F applied to one end of a specimen in the shape of a cylinder of length l and radius r to twist it in the circumferential direction relative to the other end (Figure 4.1d) creates a torque T given by

$$T = Fr \tag{4.6}$$

The angle of twist θ is the angle of rotation in the circumferential plane of one face relative to the other and the displacement δ of one face relative to the other is given by

$$\delta = r\theta \tag{4.7}$$

where, θ is the angle in radians. The maximum shear stress at the outer surface of the cylinder τ is given by

$$\tau = \frac{Tr}{J} \tag{4.8}$$

where, J is called the polar moment of inertia. For a cylinder of radius r, it is given by

$$J = \frac{\pi r^4}{2} \tag{4.9}$$

The shear strain is given by

$$\gamma = \frac{r\theta}{l} \tag{4.10}$$

where, θ is the angle of twist (in radians).

Flexural Deformation (Bending)

Because of their high hardness, most ceramics are difficult and expensive to machine into the shape required for tensile testing. Ceramic specimens are also more difficult to grip stably at their ends when compared to metals and polymers. The ease of testing in a flexural (bending) mode, coupled with the simple geometry and low cost of the specimens makes this test very popular for ceramics. There are two ways of performing flexural testing, referred to as three-point bending and four-point bending, but four-point bending is often preferred and commonly used in testing ceramics when design data are required.

In four-point bending, the specimen, commonly in the shape of a beam, is supported on two rollers, a distance l apart (called the outer span), the load (force) F is applied by means of two upper rollers that are symmetrically arranged, and the deflection at the center of the beam is measured (Figure 4.1e). In a common testing configuration, the distance d between the upper rollers (called the inner span) is equal to $l/2$. As the specimen is bent, the lower surface of the beam is in tension whereas the upper surface is in compression. Halfway across the thickness, called the neutral axis, the length of the beam is unchanged and, thus, the resultant stress is zero. The highest tensile stress σ is at the bottom surface of the beam and is constant over the inner span. It is given by

$$\sigma = \frac{3Fd}{h^2 w} \tag{4.11}$$

where, h is the thickness and w the width of the beam. The deflection of the center of the beam δ, relative to the outer support points, is given by

$$\delta = \frac{Fd}{4Eh^3 w}\left(3l^2 - 4d^2\right) \tag{4.12}$$

where, E is the Young's modulus.

In three-point bending, the load is applied by means of a single roller midway between the two lower rollers. The corresponding relations for the highest tensile stress σ (at the bottom surface below the upper roller) and deflection at the center of the beam are

$$\sigma = \frac{3Fl}{2h^2 w} \tag{4.13}$$

$$\delta = \frac{Fl^3}{4Eh^3 w} \tag{4.14}$$

4.2.2 Elastic Modulus

For small deformations, less than ~0.1% strain, the measured stress of most solids is proportional to the measured strain, in accordance with Hooke's law. In this region, the stress versus strain follows a straight line, and the slope of this straight line is called the elastic modulus. This type of behavior is sometimes described as perfectly elastic to distinguish it from other types of elastic behavior. If the specimen is unloaded, it returns to its original undeformed state along the same line, in much the same way as a spring. The elastic modulus in different loading modes is given different names to distinguish one from another. In tension or compression for example, it is called the Young's modulus, defined as

$$E = \frac{\sigma}{\varepsilon} \tag{4.15}$$

where, σ is the applied tensile or compressive stress and ε is the strain. For a perfectly elastic solid, E is the same in tension and compression. The Young's modulus is dependent on the interatomic force versus displacement curve (Chapter 2). It is one of the important mechanical properties used in design. Values for some solids used as biomaterials are given in Table 4.1. In shear and torsion, the elastic modulus is called the shear modulus or, less frequently, the rigidity modulus, whereas in bending (flexure), it is called the flexural modulus. The Young's modulus E and the shear modulus μ are related by the equation

$$\mu = \frac{E}{2(1+\nu)} \tag{4.16}$$

where, ν is the Poisson's ratio, equal in magnitude to the lateral strain ε_y divided by the tensile strain ε_x, given by

$$\nu = -\frac{\varepsilon_y}{\varepsilon_x} \tag{4.17}$$

The negative sign is included because most materials contract in the direction perpendicular to a tensile stress, that is, the strain in the lateral direction is opposite in sign to the strain in the tensile direction. Consequently, ν is positive for most materials. Values for the Poisson's ratio for the majority of materials are in the range ~0.25–0.40.

There are other forms of elastic behavior, although the stress versus strain curve is not a straight line or the unloading part of the curve does not follow the same path as the loading part. Polymers show elastic behavior in that they recover completely when the stress is removed but the recovery is slow and the stress–strain curve can be linear or nonlinear. Materials such as synthetic and natural rubbers, and the natural material elastin undergo large deformations and when unloaded, they return to their original shape. In spite of this, their stress–strain curve is nonlinear.

Table 4.1 Mechanical properties of selected materials used as biomaterials.

Material	Density (g/cm^3)	Young's modulus (GPa)	Yield strength (MPa)	Tensile strength (MPa)	Flexural strength (MPa)	Compressive strength (MPa)	Elongation to failure (%)	Fracture toughness ($MPa\,m^{1/2}$)	Vickers hardness (GPa)
CP[a] Ti (F67 Grade 1–4[b,c])	4.5	105	170–485	240–550			24–15	100–150	1.5–2
Ti6Al4V (F136 Grade 5[b])	4.4	115–120	795	860			10–15	50–100	3
Stainless steel (316L)	7.8	200	240–1000	550–1240			12–55	50–150	2–3
CoCr (F75[b]; as-cast/annealed)	8.3–8.5	210	448–517	655–889			10–20	50–100	3–4
CoCr (F799[b]; hot forged)	8.3–8.5	210	896–1200	1399–1586			5–15	50–100	3–4
Al_2O_3	3.98	350–400			360–400	2000–3000	<0.1	4–5	18–20
ZrO_2-toughened Al_2O_3 (ZTA)	4.40	300–350			700–1000	4000–4500	<0.1	6–10	16–20
YSZ	6.05	200–220			1000–1200		<0.1	6–12	12–13
Si_3N_4	3.2	290–310			600–1000		<0.1	6–10	14–15
Hydroxyapatite	3.16	90–120		80–200	100–200	500–1000	<0.2	~1.0	5–6
β-TCP	3.07	35–90			100–150	460–700			
UHMWPE	0.93–0.94	0.5–1.3	20–30	30–40			130–400	2–3	0.06–0.09
PMMA	1.2	2.5–3.2		60–80			4–5	1–1.5	0.1–0.2
Cortical bone (porosity 5–10%)	1.7–2.0	10–20		125–150(//) 50(⊥)	135–190	130–180	1.5–3	2–12	0.7–0.8(//) 0.5–0.6(⊥)
Trabecular bone (porosity 50–90%)		0.1–5		1–5	10–20	2–12		0.1–0.8	

These data correspond to those for the dense (pore-free) metals and polymers and for dense or nearly dense ceramics (<5% porosity) at room temperature (approximately 25 °C). The mechanical properties of metals are dependent on their microstructure and heat treatment whereas for ceramics, they are dependent on the amount of residual pores and grain size.
YSZ, yttria-stabilized zirconia (3 mol% Y_2O_3); ZTA, alumina matrix composite containing 15–20 vol% tetragonal ZrO_2 particles; UHMWPE, Ultrahigh molecular weight polyethylene; PMMA, polymethyl methacrylate.

[a] CP, commercial purity.

[b] ASTM designation.

[c] Strength increases but elongation to failure decreases from Grade 1 to Grade 4.

//, parallel to long direction.

⊥, perpendicular to long direction.

4.2.3 Mechanical Response of Materials

Materials show a variety of mechanical response that is strongly dependent on their interatomic bonding and structure. Following perfectly elastic deformation, ceramics (including glasses and glass-ceramics) typically fracture catastrophically in a brittle manner whereas metals (and their alloys) deform plastically prior to eventual failure. Unlike metals and ceramics, the mechanical response of polymers at a given temperature is strongly dependent on the time over which the stress is applied (or the rate at which the stress is increased). This time-dependent mechanical response of polymers is called viscoelasticity. Although large changes in temperature are not relevant to the application of biomaterials, it should be noted that unlike metals and ceramics, the mechanical response of polymers is also highly sensitive to the temperature of testing.

Elastic and Plastic Deformation

The measured stress–strain curves of solids show a variety of shapes but the schematic curve in Figure 4.2 can be used to illustrate the distinction between elastic and plastic deformation. For metals, perfectly elastic behavior (less than approximately 0.1% strain) is followed by plastic deformation that can continue to much larger strains until the specimen fails, typically by breaking into two pieces. The strain at which the specimen fails, called the elongation to failure, can be in the range ~1% to over 50% for many metals. In comparison, ceramics fail within their elastic region, at strains less than ~0.1%, and the material fractures catastrophically into two or more pieces, a failure mode called brittle fracture. The ability to deform plastically is commonly referred to as ductility and, thus, metals are said to be ductile whereas ceramics are brittle.

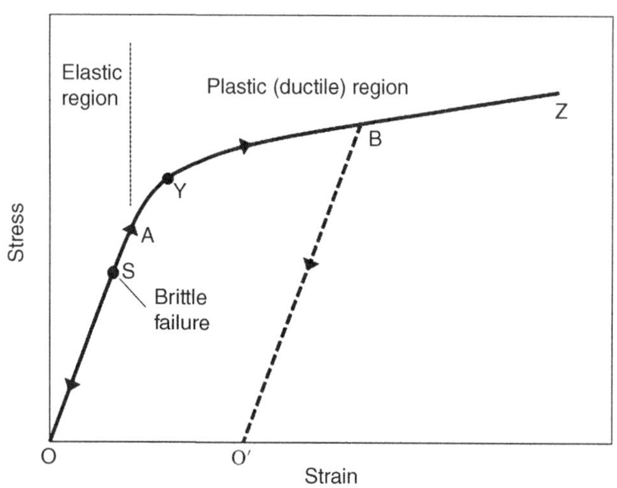

Figure 4.2 Schematic stress–strain curve to illustrate the distinction between elastic and plastic deformation of materials.

A key difference between elastic deformation and plastic deformation is that unlike elastic deformation, the strain or elongation due to plastic deformation is not recovered upon unloading the material. Upon loading a specimen from its unloaded dimension (point O in Figure 4.2) to point A in the elastic region, the specimen returns to its original dimension along the same line AO. Upon loading, atoms in the solid are displaced slightly from their equilibrium sites but do not take up new sites (Chapter 2). When the load is removed, the atoms return to their original equilibrium sites. In comparison, plastic deformation is irreversible. Upon loading a specimen to point B, the specimen follows the line BO′ upon unloading which is approximately parallel to AO. The small elastic portion of the strain is recovered but not the plastic portion. The atoms take up new sites relative to each other by movement of dislocations (Chapter 3). To summarize at this stage, the key features of elastic and plastic deformation are:

- Elastic deformation
  - Strains commonly small
  - Reversible: the material returns to its original unloaded dimension
  - Atoms are displaced slightly upon deformation but do not take up new sites
- Plastic deformation
  - Stains much larger than elastic deformation
  - Irreversible: the plastic strain is not recovered upon unloading
  - Atoms take up new sites relative to each other

Elastic Limit and Yield Point

The transition from elastic to plastic deformation is commonly called the yield point. Some stress–strain curves show a well-recognized bump in this transition region which is taken as the yield point (Figure 4.3a). On the other hand, it is often difficult to locate the yield point in some stress–strain curves because they change gradually in this region (Figure 4.3b). When difficult to locate, the yield point is often taken at what is called an offset strain, equal to 0.2% strain. A straight line is drawn from the strain axis at 0.2%, parallel to the line for elastic deformation, and where this line crosses the stress–strain curve is taken as the yield stress.

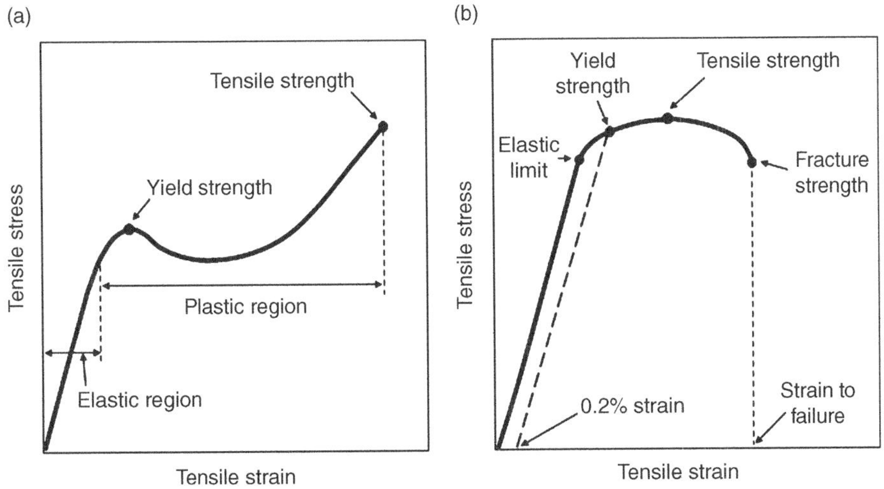

Figure 4.3 Schematic diagrams to illustrate two types of plastic deformation and yield behavior of ductile materials.

Definition and Determination of Strength

Various types of strength are used in the literature and, thus, it is useful to understand their definitions. As strength is determined from data obtained in various loading modes and is different in each loading mode, the specific type of strength should be stated when appropriate, such as tensile, compressive, shear, or flexural strength. In any loading mode, the strength is the highest stress that the material can support, sometimes referred to as the ultimate strength. The fracture strength (failure strength) is the stress at which the material fractures into two or more pieces. In tension or compression, strength is determined by taking the measured load and dividing it by the appropriate cross-sectional area of the specimen. In torsion, the measured torque and specimen radius at failure along with Eqs. (4.8) and (4.9) are used, whereas in flexure, the measured load and the initial dimensions of the beam-shaped specimen are substituted in Eq. (4.11) or Eq. (4.13). Commonly, several nominally identical specimens are tested and the strength is expressed as an average value plus or minus its standard deviation.

For ductile materials, such as metals, an important type of strength is the tensile yield strength, that is, the tensile stress at the yield point. Another type of strength that is often used is the tensile strength, sometimes called the ultimate tensile strength. Depending on the nature of the stress–strain curve, the tensile strength can be equal to or different from the fracture strength, that is, the strength at failure (Figure 4.3). If the fracture strength is the highest stress in the curve, for example, the tensile strength and the fracture strength are the same.

True Stress and Strain Versus Engineering (Nominal) Stress and Strain

Specimens of ductile metals can show a significant decrease in their cross-sectional area during tensile testing due to a large deformation in the ductile region. Consequently, for these specimens, stress can be defined in two ways:

- True stress σ, the load divided by the instantaneous cross-sectional area of the specimen which compensates for the reduction in area
- Engineering stress σ_n, sometimes called nominal stress, the load divided by the initial cross sectional area of the specimen which does not compensate for a change in the area.

Similar to stress, the strain can also be defined in two ways. For a specimen of length l that is deformed by an infinitesimally small value dl, the true strain ε is defined as the integral of all the infinitesimally small strains, that is

$$\varepsilon = \int_{l_0}^{l} \frac{\mathrm{d}l}{l} = \ln\left(\frac{l}{l_0}\right) \tag{4.18}$$

where, l_0 is the initial length of the specimen. The relation between the true strain and the engineering (nominal) strain ε_n is

$$\varepsilon = \ln\left(1 + \varepsilon_n\right) \tag{4.19}$$

The true stress σ, that is, the applied load divided by the instantaneous area, can also be found from the engineering or nominal stress σ_n and ε_n according to the equation

$$\sigma = \sigma_n\left(1+\varepsilon_n\right) \tag{4.20}$$

For small strains, such as strains in the elastic region of most solids but not in rubbers, a distinction between the true value of the stress and strain and their engineering (nominal) value is not relevant because their true value is almost equal to their engineering value. On the other hand, at large deformations in the ductile region, it is important to distinguish between true and engineering values for stress and strain.

Example 4.1 Tensile testing of a metal alloy using a cylindrical dog-bone specimen of gage length 60.00 mm and diameter 12.8 mm yielded the following data:

Force (N)	Length (mm)
0	60.00
15 442	60.06
30 883	60.12
46 325	60.18
61 766	60.24
77 208	60.30
92 649	60.36
105 517	60.42
107 834	60.48
109 506	60.54
110 150	60.60
110 664	60.66
108 992	60.72
106 547	60.78

Determine (a) the Young's modulus, (b) the elastic limit, and (c) the yield strength of the metal alloy.

Solution:
As the elongation at the highest force is small, only 1.3% of the original length, the engineering values of the stress and strain are almost identical to their true values and, thus, we can use the engineering values. While the data can be converted to stress and strain, it is not necessary in solving this problem as the area remains nearly constant during the deformation. It is often preferable to determine the requisite properties from a plot of force versus elongation but due to the simple nature of the data in this problem, we will determine them directly from an examination of the data.

a) The Young's modulus is the slope of the linear part of the stress versus strain curve. Examining the data we find that the force increases linearly with elongation up to a force of 92 649 N. The stress σ at this point is given by

$$\sigma = \frac{F}{A} = \frac{92649\,\mathrm{N}}{\pi\left(\frac{d}{2}\right)^2}$$

where $d = 12.8\times10^{-3}$ m is the diameter of the specimen. Substituting for d gives $\sigma = 720\times10^6$ Pa or 720 MPa. The strain ε at this point is the elongation divided by the original length, equal to $0.36 \div 60.00 = 0.006$ (or 0.6%). The Young's modulus is $720\times10^6\,\mathrm{Pa} \div 0.006 = 120\times10^9$ Pa or 120 GPa.

b) The stress–strain curve becomes nonlinear above a force of 92 649 N. Although the changeover from elastic to plastic deformation cannot be accurately determined, to a high approximation, the elastic limit is the stress or strain found in part (a), that is $\sigma = 720$ MPa or $\varepsilon = 0.006$.

c) Based on the slowly varying force above the elastic limit, this material does not show a clearly defined yield point. Thus, the yield stress σ_y will be taken as the stress at an offset strain of 0.002 or an offset elongation of 0.12 mm. Starting at this elongation, a line parallel to the linear portion of the force versus elongation data will intersect the force data at approximately 109 506 N. Dividing this force by the cross-sectional area of the specimen gives $\sigma_y = 851$ MPa.

It is left as an exercise to confirm these mechanical properties from a plot of force versus extension.

Viscoelasticity
When metals and ceramics are subjected to a constant stress below their elastic limit at temperatures relevant to biomedical applications and the majority of engineering applications, the resulting strain remains constant with time. In comparison, if a constant tensile stress is applied to a polymer specimen, the resulting strain does not remain constant but will increase slowly with time (Figure 4.4a). This slow extension with time is called creep. It is due to rearrangement of the mobile sections of the polymer chain in response to the stress. If the stress is removed, the polymer chains return slowly to their original conformation and the strain decreases slowly to zero. This strain recovery in polymers has its origins in the overall entropy of the polymer chains. Rearrangement of the polymer chains during deformation leads to a decrease in their entropy. Consequently, upon removal of the

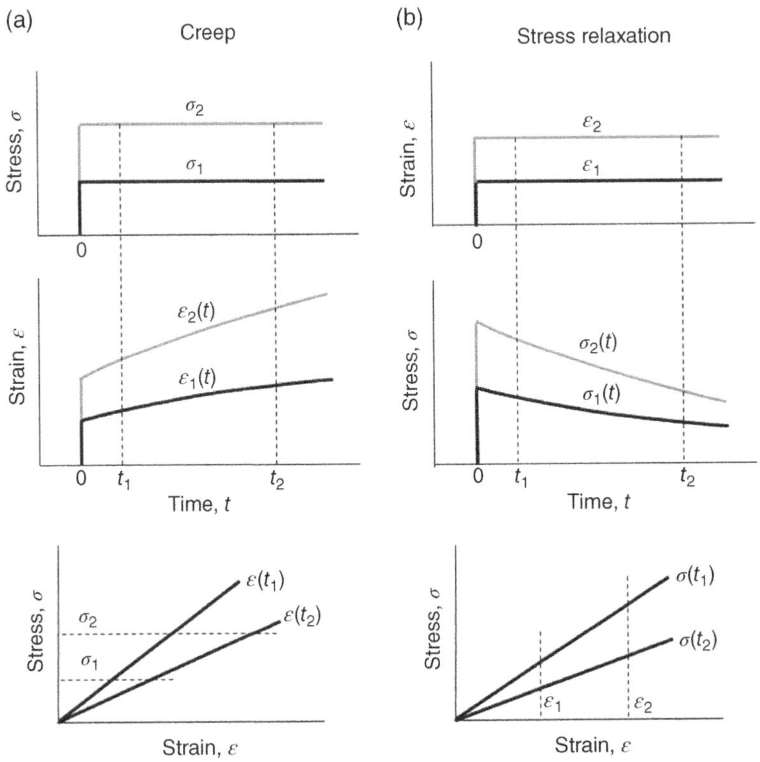

Figure 4.4 Linear viscoelastic behavior of polymers. (a) In creep, a constant stress σ applied at $t = 0$ leads to time-dependent strain $\varepsilon(t)$; at a given time, the strain increases linearly with the applied stress. (b) In stress relaxation, a constant strain ε applied at $t = 0$ leads to a time-dependent stress $\sigma(t)$; at a given time, the stress increases with increasing strain.

stress, there is a driving force for the chains to return to their higher entropy conformation. If a polymer specimen is extended to a certain constant strain, the tensile stress required to maintain this strain is not constant but decreases slowly with time (Figure 4.4b). This effect is called stress relaxation.

Creep and stress relaxation are manifestations of a general property of polymers called viscoelasticity. The overall response is a combination of two responses:

- An elastic response because, given sufficient time, the solid recovers completely upon removal of the stress
- A viscous response that results in creep or stress relaxation, in much the same way that a very thick syrup of high viscosity will deform slowly under an applied stress.

Polymers are viscoelastic over a wide range of temperature, starting from as low as approximately −100 to −200 °C. Consequently, in designing polymers for applications in which they are subjected to a mechanical stress, we must take into account that the strain is dependent not only on the magnitude of the stress but also on the time over which it is applied. The mechanical response of polymers is also strongly dependent on temperature. Although the majority of biomaterials are used within a narrow range of temperature, such as near room temperature or the physiological temperature (~37 °C), in general, the effect of temperature must also be taken into account when polymers are used in applications that cover a wider temperature range.

For small strains (less than ~0.5%), the strain at a specific time increases linearly with the applied stress and this type of behavior is called linear viscoelastic (Figure 4.4). On the other hand, at higher strains, the strain at a specific time increases faster with applied stress than predicted by extrapolation of the linear relationship, a behavior described as nonlinear viscoelastic. In the linear viscoelastic range, the relationship between applied stress, such as an applied tensile stress σ, and the time-dependent strain $\varepsilon(t)$ is

$$J_1(t) = \frac{\sigma}{\varepsilon(t)} \tag{4.21}$$

where $J_1(t)$ is the time-dependent creep modulus. This relation is similar to Eq. (4.15) for a perfectly elastic solid except that the elastic modulus and strain are now time-dependent.

The stress relaxation modulus $J_2(t)$ can be defined in a similar way, that is

$$J_2(t) = \frac{\sigma(t)}{\varepsilon} \tag{4.22}$$

The creep modulus $J_1(t)$ can be measured by subjecting a specimen to a constant load and measuring the tensile strain as a function of time. In the linear viscoelastic region, as $J_1(t)$ is independent of the applied stress σ, it can be determined by measuring $\varepsilon(t)$ for a single specific applied stress. Once $J_1(t)$ is known for a given time range, $\varepsilon(t)$ can be determined for any stress within this time range from Eq. (4.21). On the other hand, for nonlinear viscoelastic deformation, each value of the stress leads to a time-dependent strain that can be obtained only by experiment performed at that stress. However, it is not necessary to measure $\varepsilon(t)$ for every conceivable stress of interest. Interpolation between a few curves measured at appropriate stresses can often provide sufficiently precise data for design.

Mechanical models composed of various combinations of elastic springs and viscous dashpots are sometimes used to provide a phenomenological description of linear viscoelastic behavior. While these models do not provide a description of viscoelastic behavior at a molecular level, they are useful for fitting experimental data and, subsequently, for use in design to predict linear viscoelastic behavior under certain circumstances. One such model is the Zener model, also referred to as the standard linear solid. This model can be depicted in two equivalent ways, as a spring in series with a Kelvin model (Figure 4.5a) or a spring in parallel with a Maxwell model (Figure 4.5b). The springs account for the elastic contribution to the deformation whereas the dashpot accounts for the viscous contribution. Typically, the models are used to develop an equation for the creep modulus $J_1(t)$ or the stress relaxation modulus $J_2(t)$ (McCrum et al. 1997). Then, by fitting the model equation to the appropriate experimental data for creep or stress relaxation, model parameters relating to the elastic modulus of the springs, E_1 and E_2, and the viscosity of the dashpot η are determined. Once these parameters are determined from data for a specific polymer, the phenomenological equations can be used to predict linear viscoelastic behavior of the same polymer with reasonable accuracy under appropriate conditions.

4.2.4 Stress–Strain Behavior of Metals, Ceramics, and Polymers

Ceramics (including glasses and glass-ceramics) are brittle and, thus, they undergo only elastic deformation prior to fracture (Figure 4.6). The majority of ceramics used as biomaterials such as alumina, zirconia-toughened alumina (ZTA), yttria-stabilized zirconia (YSZ), and silicon nitride typically have a higher Young's modulus than most metals such as titanium and its alloys, stainless steel and cobalt–chromium alloys (Table 4.1). The strength of these ceramics, often determined in flexural testing, is also typically higher than the tensile strength of many metals and alloys. Consequently, the stress–strain curve consists of straight line and the stress at failure is often larger than the yield stress of many metals. Glasses typically have a lower Young's modulus and flexural strength than most ceramics used in biomedical and engineering applications. The key parameters obtained from the stress–strain curves for brittle materials are their Young's modulus, strength and strain

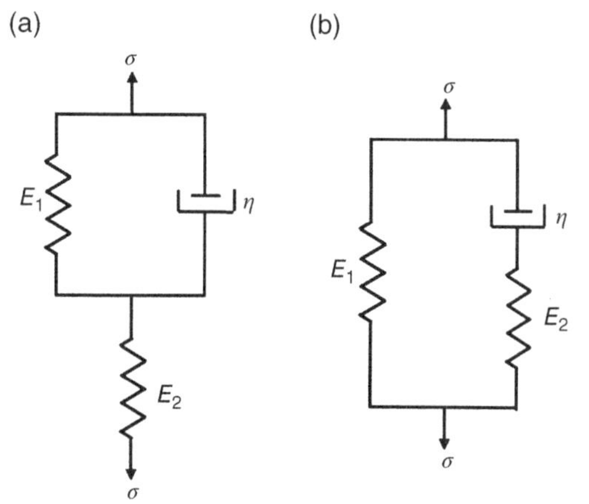

Figure 4.5 Two alternative versions of the Zener model, also called the standard linear solid, used to provide a phenomenological theory of linear viscoelastic behavior. The important parameters of the models are the elastic modulus of the two springs E_1 and E_2, and the viscosity of the dashpot η.

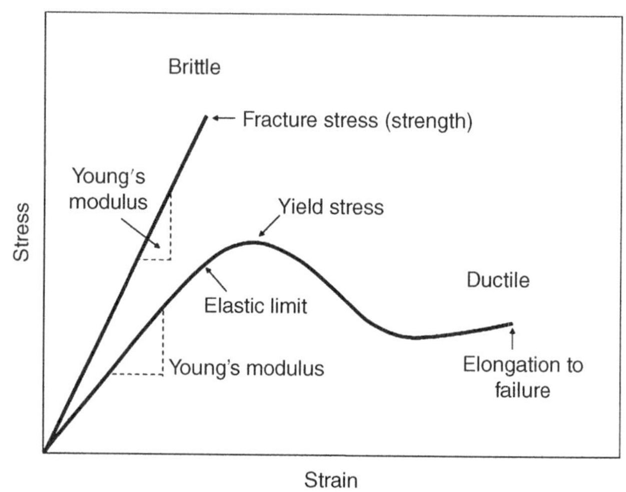

Figure 4.6 Schematic stress–strain curves to illustrate the characteristic response of brittle ceramics and ductile metals.

to failure in flexural loading. In comparison, metals show a stress–strain behavior characterized by an elastic deformation followed by plastic deformation, and the key parameters are their Young's modulus, yield strength, elastic limit, and elongation to failure.

At room temperature, polymers can show a brittle, ductile, or elastic response depending on their composition and structure (Figure 4.7). Some amorphous polymers such as polystyrene, for example, show a brittle response under some appropriate rate of loading, fracturing at tensile strains of less than approximately 0.5%. In comparison, semicrystalline polymers typically show a ductile response. Polyethylene, for example, shows a ductile response in which the strain to failure can be up to a few hundred percent. This difference in mechanical response between polystyrene and polyethylene is attributed to differences in their structure (see Section 8.2.6). Amorphous polymers with a crosslinked structure such as natural rubber show an elastic response over a very large strain (a few hundred percent for natural rubber) followed by failure, but the stress–strain curve is nonlinear. The strength and elastic modulus of polymers are far lower than those for ceramics and metals (Figure 1.5).

As viscoelastic materials, polymers show a mechanical response that is highly sensitive to the rate of loading (or the time over which the load is applied) and to temperature. A given polymer, for example, can show all three types of response illustrated in Figure 4.7 depending on the rate of loading and the temperature. High loading rate or low temperature typically leads to a brittle response whereas a low loading rate and higher temperature lead to an elastic response comparable to a rubber. At intermediate loading rate or temperature, a ductile response is commonly observed. Many natural materials, such as collagen, bone, cartilage, tendons, and ligaments show a viscoelastic response similar to that of synthetic polymers whereas elastin, a major component of skin, shows an elastic response similar to that of rubber (Chapter 10).

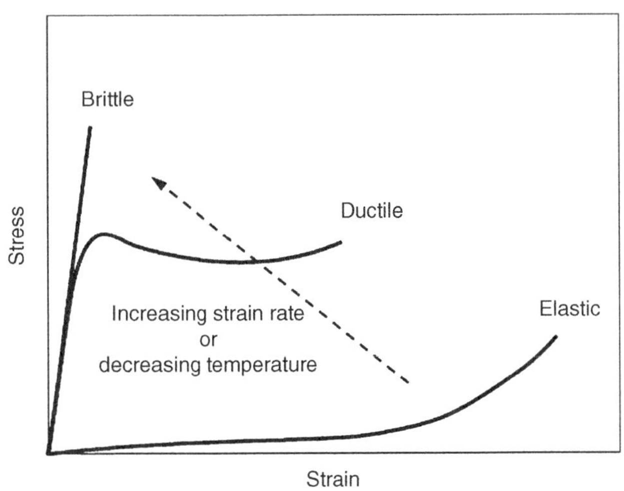

Figure 4.7 Schematic stress–strain curves to illustrate the characteristic response shown by different polymers near room temperature. A given polymer can show all three types of response depending on the rate of loading (or strain rate) and the temperature of the test.

4.2.5 Fracture of Materials

Fracture is the separation of a material into two or more pieces under an applied stress. Fracture can result from

- Rapid growth of a crack through the material, a process described as fast fracture but often referred to simply as fracture
- Slow crack growth, a process described as fatigue (Section 4.2.7).

Unless when necessary to distinguish between fast fracture and fatigue, we shall use the term facture to mean fast fracture. Fracture typically occurs by propagation of existing cracks in a material which suddenly become unstable when the material is subjected to a sufficiently high stress which may be constant or slowly varying. Often, the magnitude of this stress is equal to a significant fraction of its measured yield strength or fracture strength. In comparison, fatigue often occurs when a material is subjected to repeated stress cycles over a long period, the magnitude of the stress often being a small fraction of its measured strength.

Fracture is classified into two types, ductile fracture and brittle fracture, which depend on the mechanical response of the solid. Ductile fracture occurs in most metals (including their alloys) whereas ceramics (including glasses and glass-ceramics) show little or no overall plastic deformation but, instead, fail catastrophically in a brittle manner (Figure 4.6). At room temperature, some amorphous polymers, such as polystyrene for example, fracture in a brittle manner with little sign of ductility. In comparison, semicrystalline polymers and some amorphous polymers show ductile fracture to varying degrees at room temperature.

Fracture of materials is said to involve two successive steps, crack formation and crack propagation (growth), regardless of the fracture type. On the other hand, once a crack is formed, there is a marked difference in the mechanism of crack propagation and fracture between ductile and brittle fracture.

Crack Formation

Crack formation in metals subjected to a tensile stress, for example, often occurs by linking up of microvoids already present in the material, resulting in the formation of an elongated crack perpendicular to the loading direction.

Ceramics often contain residual pores and crack-like voids resulting mainly from their fabrication process (Chapter 7). Ceramics can also develop microcracks during cooling from their fabrication temperature due to differences in thermal expansion coefficient between adjacent grains when the grains have a non-cubic crystal structure. This difference in thermal expansion coefficient generates local stresses during cooling in much the same way that a stress develops upon cooling a sandwich structure composed of two well-bonded metal sheets that have different thermal expansion coefficients. If sufficiently high, these local stresses can generate microcracks in the ceramic.

Crack Propagation

As a basis for understanding the mechanism of crack propagation and fracture in ductile and brittle materials, we can examine the local stress near the tip of a crack (Figure 4.8). Theoretical analysis shows that the local stress σ_l ahead of a sharp crack of length c in an elastic solid due to an applied tensile stress σ in the Y direction is

$$\sigma_l = \sigma\left[1+\left(\frac{c}{2x}\right)^{1/2}\right] \tag{4.23}$$

where, x is the is the distance ahead of the crack tip in the X direction. The crack has the effect of amplifying the applied stress σ as the crack tip is approached. The closer the crack tip is approached, the higher the local stress σ_l. Depending on the values of σ and c, σ_l can become enormous near the crack tip, considerably larger than the yield stress of a ductile metal or the theoretical strength of a brittle solid.

In ductile metals, as the crack tip is approached, the local stress reaches the yield stress at some distance from the crack tip. Between this distance and the crack tip, plastic deformation occurs, turning the initially sharp crack into a blunt crack. As this process consumes a significant amount of energy, the energy available for the crack to propagate further is reduced considerably. This is equivalent to saying that the material has a high toughness (Section 4.2.6) or a high resistance to crack propagation through it. The stress at the blunted crack tip is, however, just sufficient to deform the material plastically and this, coupled with further linking up of microvoids within the metal, leads to crack growth and, eventually, to ductile failure.

In brittle materials such as ceramics, the stress at the crack tip can reach very high values because there is little or no plastic deformation to dissipate some of the energy available for crack propagation. When the local stress reaches the theoretical or ideal strength of the solid, equal to approximately $E/3$ to $E/8$, where E is the Young's modulus, it is sufficient to break the interatomic bonds and the crack propagates rapidly between two atomic planes, a process described as cleavage. Ceramics, glasses and glass-ceramics have low toughness or a low resistance to crack propagation when compared to ductile metals.

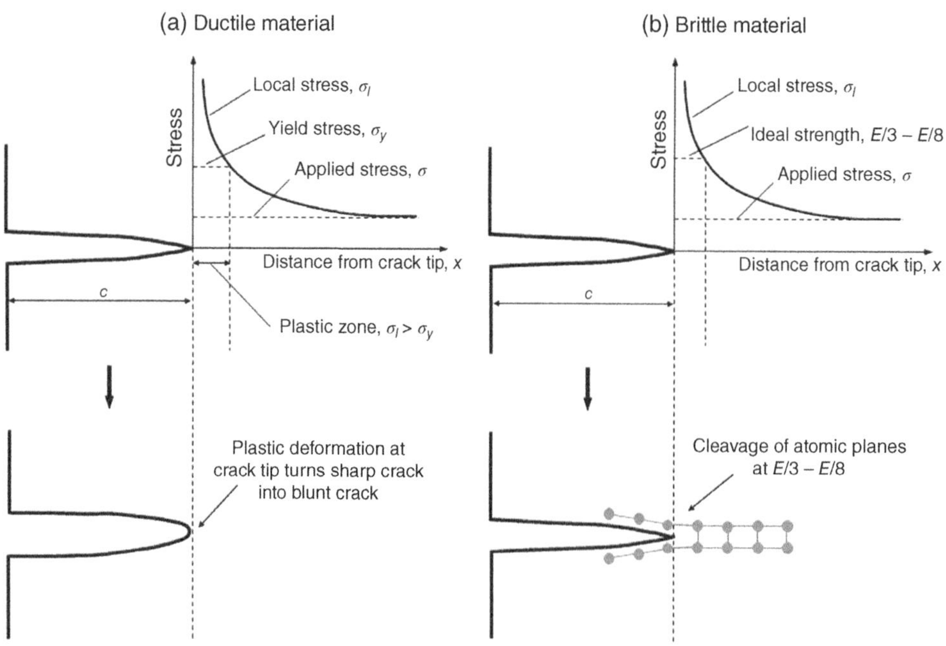

Figure 4.8 Local stress σ_l as a function of distance x from the tip of a sharp crack of length c in (a) a ductile material and (b) a brittle material that is subjected to an applied stress σ. If the local stress is sufficiently high, plastic flow near the crack tip of the ductile material leads to crack blunting whereas cleavage of the atomic planes can occur in brittle materials. E, Young's modulus.

As ceramics almost invariably contain microstructural flaws, their measured strength in tensile or flexural loading is much lower than their theoretical strength, often by a factor of 10^2 to 10^3 or more, due to the high amplification of the applied stress at the sharp tips of microcracks and pores. Another feature of ceramics is that their tensile or flexural strength is also much lower than their compressive strength, often by a factor of ~5–10 or more (Table 4.1). This is because cracks propagate more stably under compressive loads. Cracks have to twist out of their original orientation to propagate parallel to the direction of compression. Consequently, fracture does not occur by the rapid propagation of one crack (usually the largest crack) as in tensile loading. Instead, fracture in compression occurs by the slow extension of many cracks that eventually lead to crushing of the specimen.

Overall, because of the difference in crack propagation:

- Metals (and other ductile materials) have approximately the same measured strength in compression and in tension
- Ceramics (and other brittle materials) have a measured compressive strength that is much higher than their tensile (or flexural) strength

Consequently, proper design of structural ceramics is required to avoid their exposure to excessively high tensile stresses.

The fracture surface of ductile metals is often considerably rougher than that of ceramics or glasses due to the high degree of plastic deformation. Ceramics show a smoother fracture surface because crack propagation involves little or no plastic deformation but instead, involves cleavage of atomic planes. Another characteristic difference is that ductile fracture in metals occurs more slowly than brittle fracture of ceramics due to the energy absorbing process of plastic deformation during crack propagation.

Theoretical Analysis of Brittle Fracture

A criterion for fracture of brittle solids could be developed by determining the stress at a sharp crack tip and equating it to the theoretic strength. However, this is challenging because it requires detailed consideration of the atom arrangement at the crack tip. Instead, a widely recognized theory of fracture in brittle solids, called the Griffith theory, is based on energy considerations and assumes that the solid is a continuum, thus neglecting atomic considerations of the fracture process. An energy balance concept is assumed such that the limiting condition for crack growth occurs when the energy released by the elastic strain energy is equal to the energy required to create new crack surface. The analysis typically assumes a model composed of a thin wide plate containing a long thin elliptical-shaped crack of length $2c$ through it which is subjected to a tensile stress σ in the direction perpendicular to the length of the crack (Figure 4.9). For this model, the theory predicts that fracture will occur when the stress at the crack tip becomes equal to σ_f, called the fracture stress, given by

$$\sigma_f = \left(\frac{2E\gamma}{\pi c}\right)^{1/2} \tag{4.24}$$

where, E is the Young's modulus and γ is the surface energy per unit area of the solid. Equation (4.24), commonly called the Griffith equation, is often taken as a necessary condition for fracture to occur. As the surface energy of many ceramics falls within a narrow range, ~0.5–2.0 J/m^2, Eq. (4.24) can be seen to provide theoretical underpinning for the strong effect of microstructural flaws on their tensile strength.

Equation (4.24) is often written

$$\sigma_f = \left(\frac{EG_c}{\pi c}\right)^{1/2} \tag{4.25}$$

where, G_c is called the toughness, equal to 2γ (Section 4.2.6). Thus, we can say that brittle facture will occur when

$$\sigma(\pi c)^{1/2} = (EG_c)^{1/2} \tag{4.26}$$

According to Eq. (4.26), fracture will occur when, for a material subjected to a stress σ, a crack reaches a certain critical size c or, alternatively, when a material containing cracks of size c is subjected to some critical stress σ. The term on the left-hand-side of Eq. (4.26) occurs frequently in fracture mechanics and is often referred to as the stress

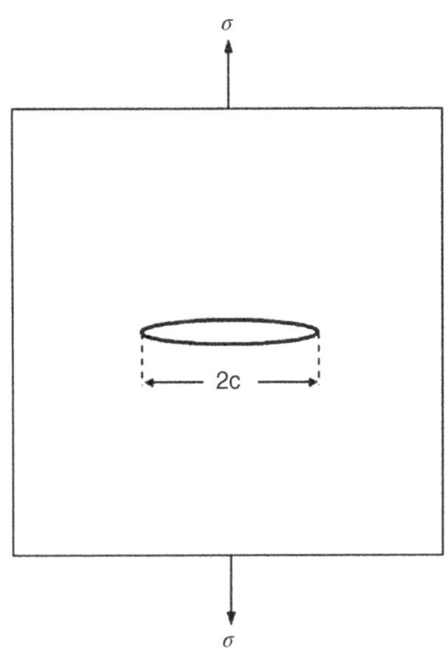

Figure 4.9 Geometrical model used in the Griffith theory of brittle fracture.

intensification factor K. Thus, we can also say that fracture will occur when $K = K_c$ where K_c is called the critical stress intensification factor or more commonly, the fracture toughness, given by

$$K_c = (EG_c)^{1/2} \tag{4.27}$$

Using Eqs. (4.26) and (4.27), the tensile strength is given by

$$\sigma_f = \frac{K_c}{(\pi c)^{1/2}} \tag{4.28}$$

Brittle materials do not contain just one crack but many cracks that differ in size. In tension (or flexure), fracture occurs typically by rapid propagation of the largest crack of length $2c$. On the other hand, fracture in compression typically occurs less rapidly by extension of many cracks. Thus, the fracture strength in compression is given as

$$\sigma_C = H\frac{K_c}{\bar{c}}$$

where, H is ~10 and $2\bar{c}$ is the average crack length.

4.2.6 Toughness and Fracture Toughness

Toughness refers to the ability of a material to withstand rapid propagation of a crack through it. The toughness G_c of a material is defined as the energy absorbed per unit area of crack (units J/m^2). A high G_c means that it is difficult for a crack to propagate through a material, as in pure ductile metals such as aluminum and copper, for example, which have G_c values in the range 100–1000 kJ/m^2. In comparison, brittle materials such as ceramics have low G_c values, in the range 0.01–0.1 kJ/m^2, and, thus, it is easy for cracks to propagate through them.

The toughness of a material is difficult to measure and, consequently, more easily measured parameters are used to provide a measure of toughness. One such parameter is the area under the stress–strain curve in a given loading mode such as tension or flexure (Figure 4.10). This area can be used to compare the relative toughness of specimens with a similar geometry but it is not equal to the toughness G_c of the material. As Figure 4.10 indicates, there is often no correlation between strength and toughness. The area under the elastic region of the stress–strain curve is often referred to as the resilience because it gives a measure of the energy recovered upon unloading a specimen.

A more widely used parameter of toughness is the fracture toughness K_c. It is determined experimentally using standard techniques by inserting a crack of length c into a specimen and loading it until fracture occurs. K_c (units MN/m$^{3/2}$or MPa m$^{1/2}$) is related to G_c by Eq. (4.27).

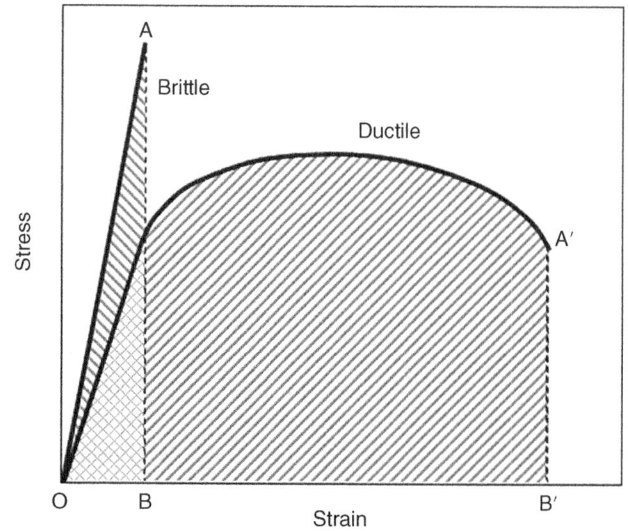

Figure 4.10 Area under the stress–strain curve used as measure of the relative toughness between materials of the same geometry.

Pure metals have K_c values in the range ~100–350 MN/m$^{3/2}$ (Table 4.1). The K_c values of the majority of ceramics and glasses are in the range ~0.5–5 MN/m$^{3/2}$. On the other hand, a few ceramics such as YSZ and silicon nitride (Si_3N_4) can have K_c values equal to ~10 MN/m$^{3/2}$. The higher fracture toughness of YSZ is due to the occurrence of a phase transformation in the region of the crack tip, which dissipates some of the energy available for crack propagation. This process is described as transformation toughening (Chapter 7). On the other hand, a unique fibrous structure of elongated grains coupled with an appropriate thin glass phase at the grain boundaries is responsible for the higher fracture toughness of Si_3N_4. A crack propagates along the weaker glass phase at the grain boundaries, making the crack path more tortuous when compared to a straighter path directly across the grains. This tortuous path consumes more energy than a straighter path. Polymers have G_c values between those for metals and ceramics but low K_c values because their Young's modulus is low.

4.2.7 Fatigue

The resistance of a material to fatigue, a process of slow crack growth under repeated stress cycles, is important in several biomedical and engineering applications. Implants used in total joint replacement or in repairing large defects in the long bones of the limbs, for example, are subjected not just to a constant or slowly varying stress but to repeated cyclic stresses as well during normal activities of walking, running, and jumping. Stents used to keep coronary arteries open have to withstand the pressure pulsations of blood flow through the arterial vessels.

Fatigue occurs by a process of slow crack growth under cyclic stresses that are often well below the strength of the material. Cracks formed during cyclic loading or accidentally present in the original material can grow slowly and, when they have propagated sufficiently far, failure occurs, often in a brittle manner, at stresses well below the strength of the material (Figure 4.11).

Fatigue behavior is commonly studied by subjecting specimens to cyclic loads often sinusoidal in nature, in the requisite loading mode such as tension, compression or bending. Specimens, commonly of a geometry similar to those used to measure the strength of the material (Section 4.2.1), are loaded for the requisite number of cycles of until they fail.

4.2.8 Hardness

Hardness provides a measure of the resistance of a material to penetration by a sharp object. Loading a material with a sharp indenter results in the creation of a permanent irreversible impression that is indicative of plastic deformation (Figure 4.12). This occurs even for brittle materials such as ceramics because the stresses due to a sharp indenter are sufficiently high to cause local plastic deformation. The hardness H is determined from the equation

$$H = \frac{F}{A} \tag{4.29}$$

where, F is the applied load (force) and A is the area of the residual deformation. While indenters of various geometry are available, the Vickers indenter consisting of a square-shaped pyramid usually made of diamond is often used. The area A used in Eq. (4.29) is the actual area of the residual indent, that is, the area between the four faces of the pyramid and the surface of the solid after the indenter has been removed. Thus, for the Vickers indenter

$$H = \frac{1.8544F}{(2a)^2} \tag{4.30}$$

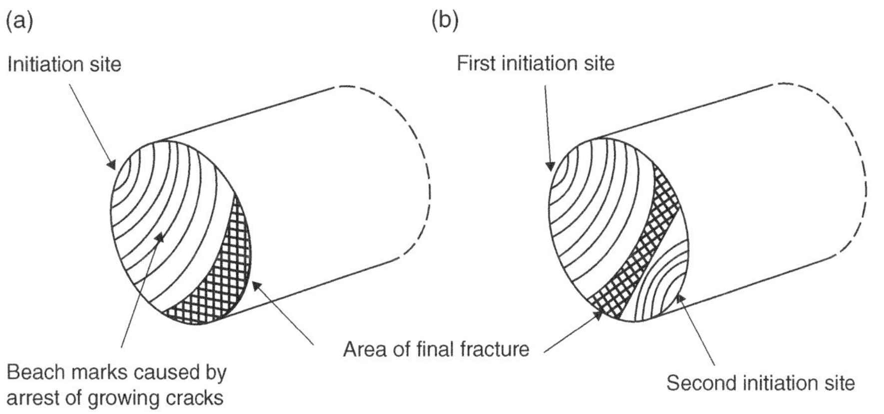

Figure 4.11 Schematic representation of the fracture surface of a ductile metal after fatigue failure in one-way bending (a) and two-way bending (b).

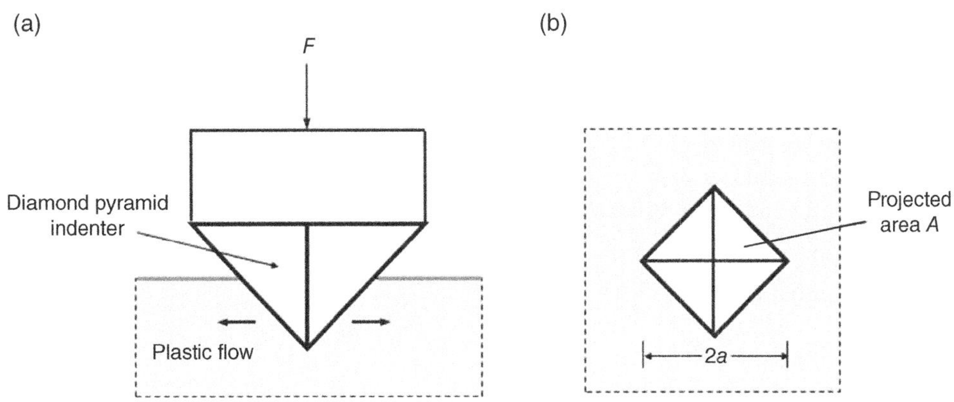

Figure 4.12 Geometry of hardness test using a Vickers indenter consisting of a square-shaped pyramid. (a) Side view of the indentation test. (b) View of the indent looking directly at the surface of the material.

where, $2a$ is the length of the diagonal of the indenter (Figure 4.12). The hardness determined from Eq. (4.30) is called the Vickers hardness. In another measure of hardness, sometimes called the true hardness, the projected area of indent is used in Eq. (4.29), giving

$$H = \frac{2F}{(2a)^2} \tag{4.31}$$

The true hardness is ~8% higher than the Vickers hardness and, thus, it is useful to state which hardness is being reported.

For some ductile metals, H is related to the yield strength σ_y by the equation

$$H = 3\sigma_y \tag{4.32}$$

but a correction factor is needed for other metals that become stronger with increasing plastic deformation (a process called work hardening or strain hardening). As seen from Eq. (4.32), measuring the hardness provides a simple and nondestructive test for determining the yield strength of a material. There is no need to make tensile specimens and the indenter is so small that several measurements can be performed on the same material sample without severely damaging it.

Hardness controls the resistance to abrasive wear between the articulating surfaces of two materials. The harder the material, the less prone it is to wear. Wear is particularly important, for example, in implants used for total joint replacement (Chapter 23). The production of excessive wear particles between the articulating surfaces in these implants, such as between the CoCr ball and ultra-high molecular weight polyethylene (UHMWPE) liner in a hip implant, leads to a severe foreign body response that invariably causes implant failure (Chapter 18).

4.3 Effect of Microstructure on Mechanical Properties

The microstructure of a material controls its mechanical properties and other physical properties (Chapter 3). Microstructural parameters that have a major influence on mechanical properties are

- Porosity, the volume fraction of pores in the material, as well as their distribution within the material and their shape
- Grain size, relevant to metals and ceramics
- Presence of microstructural flaws such as microcracks and pores, often introduced accidentally during fabrication of the material, particularly relevant to brittle materials.

4.3.1 Effect of Porosity

Porosity has a strong effect in reducing the elastic modulus (Figure 4.13) and the strength of materials. Introduction of porosity into metals and ceramics is a well-known approach to bring their high strength and elastic modulus closer to those of bone. A variety of equations, theoretical and empirical, has been proposed to account for the effect of porosity on elastic modulus and strength. A popular equation based on a combination of simplicity and accuracy is

$$E = E_0 \exp(-bP) \tag{4.33}$$

where, E is the Young's modulus, E_0 is the Young's modulus at zero porosity, P is the porosity and b is parameter that depends on the shape and distribution of the pores in the material. Commonly, b is determined empirically from a fit of the experimental data for E versus P. For example, a good fit of Eq. (4.33) to the data shown in Figure 4.13 for a titanium alloy is obtained using E_0 = 80 GPa and b = 4.7. Equation (4.33) often provides a good fit to data for P values up to ~25–40%. For higher porosity, one modification gives

$$E = E_0 \{1 - \exp[-b'(1-P)]\} \tag{4.34}$$

where, b' is a parameter determined from fitting the experimental data. Relationships similar in form to Eqs. (4.33) and (4.34) are often used to empirically describe the decrease in strength with porosity.

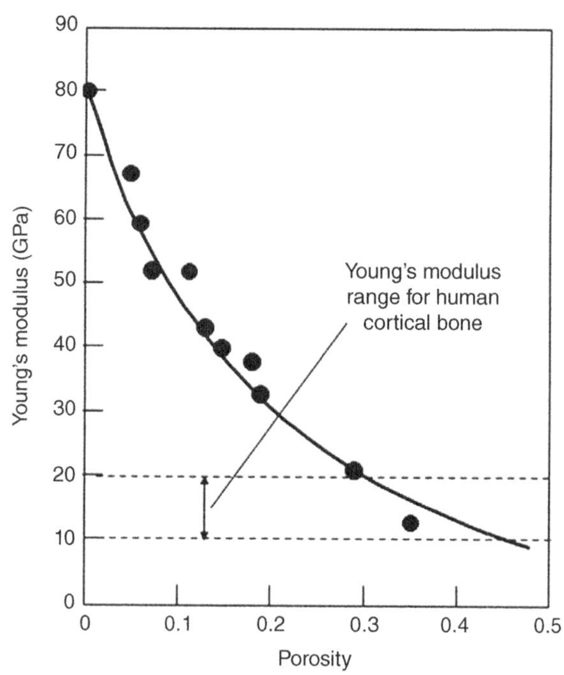

Figure 4.13 Data for the Young's modulus as a function of porosity for a titanium specimen. The data points are fitted by a smooth curve.

4.3.2 Effect of Grain Size

The yield stress of ductile metals is observed to increase with decreasing grain size (Figure 4.14), according to the well-known Hall–Petch equation, given by

$$\sigma_y = \sigma_0 + kG^{-1/2} \quad (4.35)$$

where, σ_y is the yield stress, σ_0 is the yield stress for the easiest slip system in a single crystal, and G is the grain size of the polycrystalline metal. Thus, reduction of grain size provides a powerful method to enhance the yield stress of metals (Chapter 6). Equation (4.35) can be derived theoretically by assuming that dislocations pile up at the grain boundaries that provide obstacles to dislocation motion. As the grain size decreases, the number of boundaries per unit volume increases, providing a larger number of obstacles.

The effect of grain size on the strength of brittle ceramics is more complicated than that for ductile metals. This is because the strength of brittle materials is strongly influenced by the presence of flaws such as microcracks, particularly those at their surface (Section 4.4.2). For ceramics composed of grains of size smaller than ~10–20 μm, the flaw size is often larger than the grain size and, consequently, the strength of these ceramics should be independent of the grain size. However, data for ceramics often do show an increase in average strength with decreasing grain size below 10–20 μm. Although there is a wide scatter in the data due to various degrees of surface finish of the specimens prior to testing, the average flexural strength of alumina, the most widely studied ceramic, shows an increase from ~400 MPa to ~600 MPa with a decrease in grain size from 10 to 1 μm (Wachtman et al. 2009).

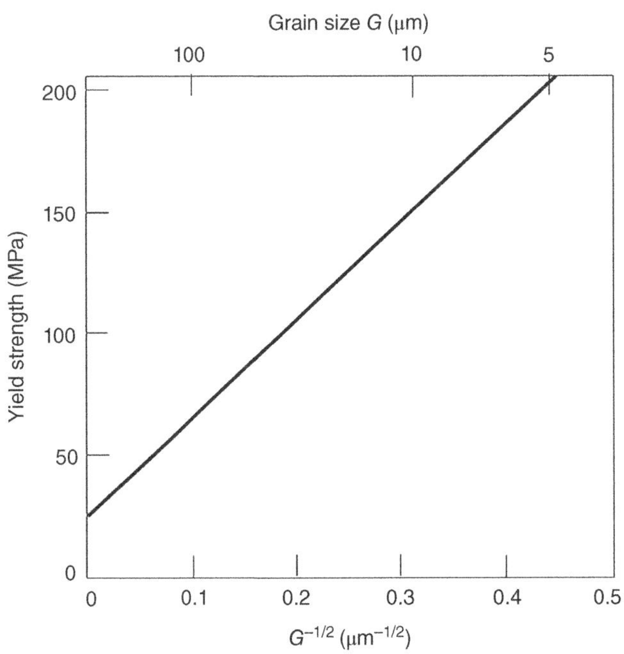

Figure 4.14 The influence of grain size on the yield strength of a 70Cu–30Zn brass alloy. *Source:* From Callister (2007) / with permission of John Wiley & Sons.

4.4 Designing with Ductile and Brittle Materials

An understanding of mechanical property principles is important in designing materials for use in biomedical applications and engineering applications as well. These principles are particularly important in applications where the biomaterial is subjected to a significant stress, such as fracture fixation plates, total joint replacement, healing of large defects in the long bones of the human limbs and dental restorations such as crowns and bridges.

4.4.1 Designing with Metals

A common requirement in the design of some biomaterials is that their shape and dimensions should remain constant. For ductile metals and alloys, this means that while plastic deformation might occur locally at a few specific points, such as the vicinity of a microvoid, it should not spread throughout the entire material. Consequently, the highest stress that the metal is subjected to should be less than its measured yield stress. As the fracture of ductile metals is considerably less sensitive to the presence of microstructural flaws such as pores and microcracks when compared to brittle ceramics, the average value of their measured yield stress is a useful design property.

4.4.2 Designing with Ceramics

The design requirements for ceramics are often more complex than those for metals. These requirements should be recognized because brittle fracture of an implant *in vivo* is a serious complication. In general, the measured strength of a number of nominally identical ceramic specimens often shows considerable scatter because the size, position, and orientation of flaws such as microcracks vary from one specimen to another (Section 4.2.5). This has important consequences for design. As failure at any flaw can lead to failure of the entire specimen, the strength that can be safely used in design is often far lower than the average measured strength. A specimen size effect also has to be taken into account because there is a higher probability of a strength-determining flaw present in a larger volume. Another factor is that the measured strength depends on the testing technique because each technique subjects a different volume of the specimen to a tensile stress. In order to account for all of these factors quantitatively, the design of brittle materials such as ceramics and glasses is often treated by a statistical method called Weibull statistics.

Weibull statistical analysis employs a weakest link assumption that failure of a single element of the specimen results in failure of the entire specimen. Instead of the more common distribution functions used to describe experimental data, such as the normal distribution, Weibull analysis assumes a function, called the Weibull distribution, to describe the measured strength of brittle materials. One form of the distribution, called the two-parameter distribution, is

$$P_f(\sigma) = 1 - \exp\left[-\left(\frac{\sigma}{\sigma_0}\right)^m\right] \tag{4.36}$$

where, $P_f(\sigma)$ is the cumulative probability of failure, that is, the probability that failure has occurred by a stress σ, m is the Weibull shape parameter, a number often called the Weibull modulus, which is an inverse measure of the width of the distribution, that is, a higher value of m corresponds to a narrower distribution, and σ_0 is the Weibull scale parameter, a measure of the center of the distribution, equivalent to the probability of failure occurring at or below a stress σ_0 is 0.63.

Weibull analysis of experimental data starts with measuring the strength of ~30 or more specimens in a given loading mode such as bending or compression using a standard procedure. Then the data are fitted, often using a least-squares method, by the Weibull distribution. Taking the natural logarithm twice of Eq. (4.36), we get

$$\ln\ln\left(\frac{1}{1-P_f}\right) = m\ln\left(\frac{\sigma}{\sigma_0}\right) \tag{4.37}$$

The Weibull parameters m and σ_0 are determined by fitting a straight line to a plot of the left-hand-side of Eq. (4.37) versus $\ln\sigma$ for the measured data. The general procedure for determining P_f from the data is to assign a rank to each value of the measured strength after sorting in an ascending order, by assigning a rank of 1 to the lowest value and n to the highest value. The failure probability assigned to the ith strength value is

$$P_f = \frac{i-0.5}{n} \tag{4.38}$$

As an example of fitting strength data to obtain the Weibull parameters, Figure 4.15 shows data for porous bioactive glass specimens with a uniform grid-like microstructure (similar to that shown in Figure 3.24b). The data were obtained by measuring the strength of 30 or more nominally identical specimens each, in their as-fabricated state, in compression and in bending (Liu et al. 2013). Data taken from the literature for hydroxyapatite and β-tricalcium phosphate specimens with a similar microstructure, tested in compression only, are shown for comparison. For the purpose of comparison, the data for the different materials are presented as a double logarithmic plot of $\ln[1/(1-P_f)$ versus σ, instead of a plot represented by Eq. (4.37).

These data illustrate features that are significant in the design of brittle materials for use in biomedical applications that are subjected to significant stresses, such as healing large defects in the long bones of the human limbs. At a specific applied stress, the bioactive glass specimens are more reliable in compression than in bending due to their higher m and σ_0 values in compression. This may be attributed to the higher volume of the specimen subjected to a tensile stress in bending. For example, the maximum compressive stress on the human femur in walking or running is estimated at less than ~10 MPa. Based on the data in Figure 4.15, the probability P_f that one of these bioactive glass specimens will fail under a compressive stress of 10 MPa is estimated at less than 10^{-6}, that is, less than one in a million specimens is predicted to fail. On the other hand, for the same stress (10 MPa) in bending, the probability of failure P_f is 0.3, or approximately 1 in 3 specimens will fail.

The data in Figure 4.15 also show that the Weibull parameters for β-tricalcium phosphate are considerably lower than those of hydroxyapatite that, in turn, are lower than those of the bioactive glass. Due to its low mechanical reliability, β-tricalcium phosphate should not be used

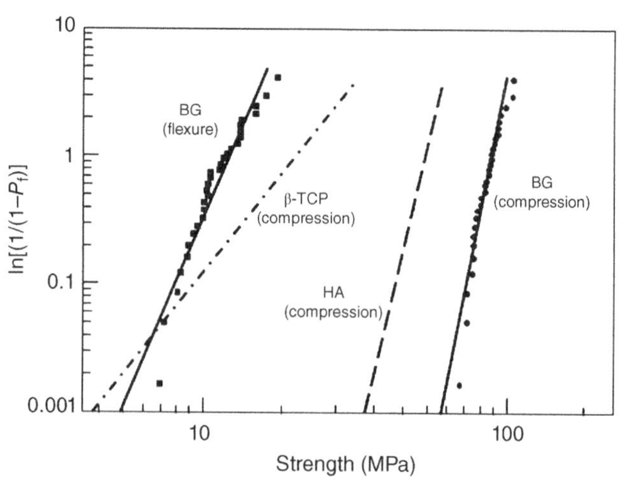

Figure 4.15 Weibull plots for porous bioactive glass (BG) specimens in compression and flexural loading. Data for hydroxyapatite (HA) and β-tricalcium phosphate (β-TCP) specimens with a similar microstructure are shown for comparison. *Source:* From Liu et al. (2013) / with permission of Elsevier.

in any application that is subjected to even a low compressive stress. In general, bioactive ceramics such as bioactive glass, hydroxyapatite and β-tricalcium phosphate degrade in the biological environment *in vivo* (Chapter 7) and lose a significant fraction of their strength with time, an important factor that is not represented by the data in Figure 4.15.

4.4.3 Designing with Polymers

In designing with polymers, the property of viscoelasticity must be taken into account. For applications in which a tensile load is present, for example, data for the tensile strength and creep modulus over the appropriate range of conditions such as stress, time scale, and temperature are relevant. If the polymer degrades *in vivo*, the effects of environmental conditions should also be taken into account. Designing to avoid yielding of the polymer is often straightforward once the time and temperature dependence of the yield strength is accounted for. In comparison, designing to avoid brittle fracture is more difficult due to the presence of flaws such as microcracks and pores (Section 4.2.5).

4.5 Electrical Properties

Electrical properties are important for biomaterials used in devices to deliver an electric current or an electrical signal. Electrically conducting metals such as platinum are used as electrodes in cardiac pacemakers and neural probes. On the other hand, electrically insulating materials such as polyurethane are used as coatings to isolate or insulate sensitive electronic devices from surrounding tissues and fluids. Whereas polymers are typically electrical insulators, several polymers have been synthesized recently which show a strong ability to conduct an electrical current. These so-called conducting polymers have attracted interest for use in biomedical applications such as biosensors, neural probes, tissue engineering, and drug delivery.

4.5.1 Electrical Conductivity of Materials

The ability of a material to transmit an electric current is quantified by its electrical conductivity or, less commonly, by its electrical resistivity which is the inverse of the electrical conductivity. The resistivity ρ of a material is independent of its geometry but, for a wire of length l and uniform cross sectional area A, it is related to the measured electrical resistance R by the equation

$$R = \frac{\rho l}{A} \tag{4.39}$$

As the unit of R is ohm (Ω), the unit of ρ is ohm-meter (Ω m), and the electrical conductivity, equal to $1/\rho$, has the unit $(\Omega\,\mathrm{m})^{-1}$, commonly written S/m, where S, the inverse of Ω, is the unit Siemens.

The electrical conductivity of materials covers an enormous range of ~28 orders of magnitude. Metals are good conductors, with a conductivity in the range ~10^4–0^8 S/m, whereas most ceramics and polymers are insulators with conductivity in the range ~10^{-20}–10^{-10} S/m. Materials with intermediate conductivity, ~10^{-6}–10^4 S/m, are called semiconductors (Figure 4.16).

Transmission of an electric current results from the motion of electrically charged particles in response to forces that act on them from an externally applied electric field. The high electrical conductivity of metals arises from the ease with which the sea of nearly free electrons that surround the metal cations move through the material (Chapter 2). Thus, metals are said to show electronic conductivity because electrons are essentially responsible for their ability to conduct an electric current. In comparison, a few ionic-bonded ceramics can show a limited capacity to conduct an electric current, which arises from the migration of ions, referred to as ionic conductivity, plus any contribution from electronic conductivity. While the total electrical conductivity due to the electronic and ionic contributions increases with temperature, ceramics are essentially insulators at temperatures relevant to biomedical applications.

Platinum alloys (conductivity ~10^7 S/m) are commonly used in cardiac pacemakers while silver alloys with a higher conductivity (~10^8 S/m) find use in some implantable defibrillators. On the other hand, polyurethane (conductivity ~10^{-12} S/m), is often used as a coating to isolate or insulate sensitive electronic devices from surrounding tissues and fluids.

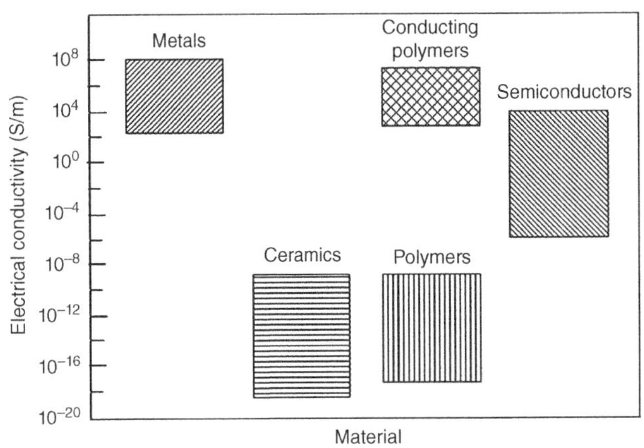

Figure 4.16 Bar chart showing the range of electrical conductivity for different types of materials at room temperature.

4.5.2 Electrical Conductivity of Conducting Polymers

Conducting polymers have attracted considerable interest in recent years because they can show an electrical conductivity as high as some metals (Figure 4.16). In addition to a high electrical conductivity, conducting polymers have an attractive combination of properties that make them suitable for use as biomaterials in applications such as biosensors, neural probes, tissue engineering and drug delivery (Guimard et al. 2007). These properties include ease of synthesis, flexibility of forming into desirable shapes, which are characteristic of polymers in general, and the potential for functionalizing their surface with appropriate molecules.

Polyacetylene has shown one of the highest electrical conductivities, up to ~10^7 S/m, whereas polypyrrole is the most widely studied conducting polymer. These and other conducting polymers typically show a characteristic structure composed of alternating single and double bonds in the polymer chain (Chapter 3). While conducting polymers can show an electrical conductivity as high as some metals, the source of their electrical conductivity is not the sea of almost free electrons that surround the cations in a metal. Instead, their high conductivity arises from a combination of factors that depend on the atomic bonding and structure of the polymer.

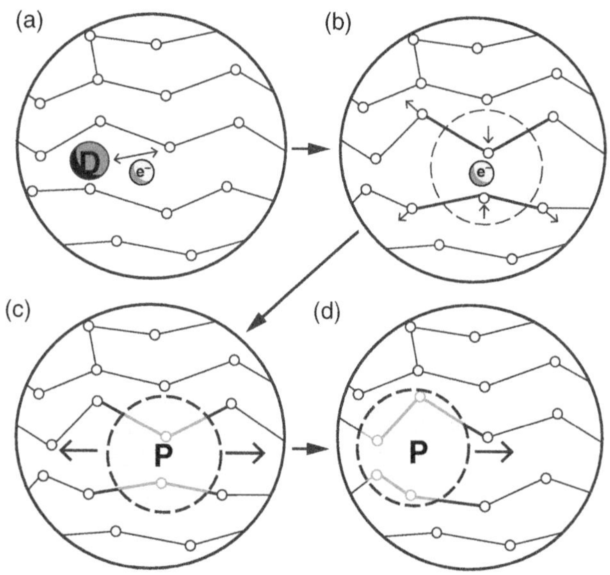

Figure 4.17 Simplified explanation of the electrical conductivity of conducting polymers. (a) A dopant D removes or adds an electron from/to the polymer chain, creating a delocalized charge. (b) It is energetically favorable to localize this charge and surround it with a local distortion of the crystal lattice. (c) A charge surrounded by a distortion is known as a polaron P (a radical ion associated with a lattice distortion). (d) The polaron can travel along the polymer chain, allowing it to conduct electricity. *Source:* From Balint et al. (2014).

A key factor is the alternating single and double bonds in the polymer chain backbone, allowing the π electrons of the double bond to be more easily delocalized and move more freely between the atoms of the chain (Figure 3.13a). However, this type of atomic bonding alone is not sufficient to endow the polymers with a high conductivity. Another key factor is the ability of these polymers to be doped with appropriate molecules that introduce a charge carrier into the system by removing electrons from, or adding electrons to, the polymer chain. While the mechanism is more complex, Figure 4.17 illustrates a simplified explanation of the electrical conductivity (Balint et al. 2014).

4.6 Magnetic Properties

Magnetic properties have not traditionally played an important role in the development of biomaterials but the interesting and improved properties observed for nanoparticles and nanostructured materials have generated considerable interest in the use of magnetic nanoparticles for biomedical applications. Magnetic nanoparticles are being considered for a variety of biomedical applications such as drug delivery, hyperthermia treatment of tumors and contrast agents for magnetic resonance imaging (Pankhurst et al. 2003).

Magnetism is commonly viewed as an attraction or repulsion between two materials. Whereas all materials are magnetic to some extent, the magnetism that we know from daily experience, such as iron filings or nails attracted to a bar magnet, is an important but special type of magnetic behavior called ferromagnetism. This type of magnetism is relevant to biomedical applications of magnetic nanoparticles and to engineering applications as well but there are weaker forms of magnetism called paramagnetism and diamagnetism. These different types of magnetism have their origins in the atomic structure of the constituent atoms or ions of the material. It is useful to understand how they arise in order to control or improve the magnetic properties of materials used as biomaterials.

4.6.1 Origins of Magnetic Response in Materials

The magnetic response of a material depends on its atomic structure and, in the case of some materials, on its temperature, although the effect of temperature is not normally relevant to the application of biomaterials. We can view a magnet as a dipole with a certain magnetic moment in much the same way as we view an electric dipole with an electric dipole moment (Chapter 2). The magnetic response of a material has its origins in the tiny magnetic dipole

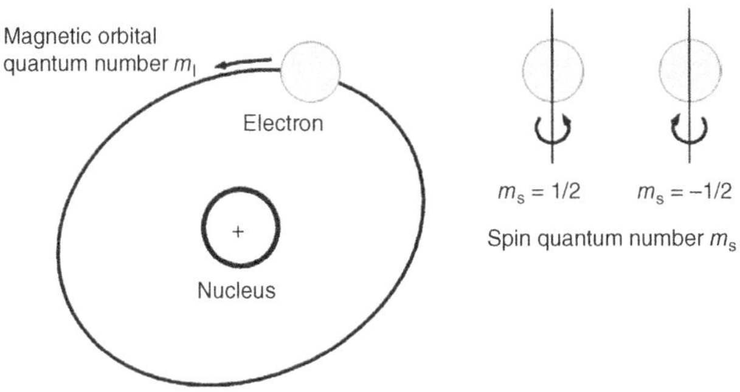

Figure 4.18 Schematic diagram illustrating the two contributions to the magnetic dipole moment of an electron in an atom due to its intrinsic spin and its orbital motion around the nucleus.

moments associated with the individual electrons of its constituent atoms or ions.

There are two contributions to the magnetic dipole moment of the electron (Figure 4.18). One is due to the intrinsic spin of the electron analogous to a charged body spinning about its own axis. The other is due to its orbital motion around the nucleus. Each contribution is quantized, that is, it depends on

- The spin quantum number m_s of the electron, which can have values of $+1/2$ (sometimes referred to as spin up) and $-1/2$ (spin down)
- The magnetic orbital quantum orbital quantum number m_l of the electron, which can have values from $-l$ to $+l$, where l is the angular momentum (azimuthal) quantum number.

The important quantity is the Bohr magneton μ_B, the magnetic moment of an unpaired electron, equal to 9.27×10^{-27} J/T. For each electron in an atom, for example, the intrinsic spin magnetic moment is $+\mu_B$ (spin up) or $-\mu_B$ (spin down), whereas the orbital magnetic moment is $m_l\mu_B$.

4.6.2 Meaning and Definition of Relevant Magnetic Properties

If a magnetic material is placed in an applied magnetic field of strength H, the individual atomic magnetic moments contribute to its overall response such that the internal field strength B in the material, also called the magnetic induction or magnetic flux density, is given by

$$B = \mu_o(H + M) \tag{4.40}$$

where, μ_o is a physical constant called the permeability of free space (equal to 1.257×10^{-6} H/m) and M is the magnetic moment per unit volume of the material, commonly referred to as its magnetization. B, H, and M are vectors but to simplify the discussion, we will not emphasize this point and just consider their magnitude. As the magnetic flux density in a vacuum B_0 is equal to $\mu_0 H$, the term $\mu_0 M$ represents the change in the flux density due to the presence of the material.

A few different parameters are used to classify the magnetic properties of magnetic materials. The magnetic susceptibility χ_m, defined by

$$M = \chi_m H \tag{4.41}$$

relates the magnetization induced in a material by an applied field H. As both M and H have the same units (A m), χ_m is unitless, that is, a number without units. The relative permeability μ_r is defined as

$$\mu_r = \frac{\mu}{\mu_o} \tag{4.42}$$

where, μ is the permeability of the material. The susceptibility or the relative permeability provides a measure of the degree to which a material can be magnetized by an applied field. They are related by

$$\chi_m = \mu_r - 1 \tag{4.43}$$

4.6.3 Diamagnetic and Paramagnetic Materials

For most atoms (or ions), the magnetic effects of the electrons, including their spin and orbital motions, cancel out and, consequently, the atom is not magnetic. This is true for atoms of the inert gasses and atoms of elements such as zinc that have completely filled electron shells. The Pauli exclusion principle requires that two electrons in any energy level of an atom must have opposite spins and, thus, the resultant spin magnetic moment of paired electrons is zero. Materials composed of atoms in which all the electrons are paired have no intrinsic magnetic moment and are described as diamagnetic. A weak magnetic dipole may be induced in these materials when they are placed in a magnetic field but the induced dipole points in a direction opposite to that of the field.

On the other hand, for other atoms or ions such as those of some transition elements, the magnetic effects of the electrons do not cancel, since one or more unpaired electrons are present, and the atom as a whole has a magnetic moment. In the absence of an applied magnetic field, the orientations of the atomic dipoles are random and, thus, the material as a whole has no net magnetic moment. However, when placed in an applied magnetic field, the atomic dipoles line up in the direction of the applied field. This type of magnetic behavior is referred to as paramagnetism.

Diamagnetic materials have χ_m values in the range approximately -10^{-6} to -10^{-3}, whereas paramagnetic materials have χ_m values in the range ~10^{-6} to 10^{-1}. The susceptibility of diamagnetic materials is negative whereas it is positive for paramagnetic materials. This is because in an applied magnetic field, the induced atomic dipoles in a diamagnetic material point in a direction opposite to the magnetic field, making the field within the material lower than that of a vacuum. On the other hand, the atomic dipoles present in a paramagnetic material line up in the direction of the applied field, in much the same way as a compass needle lines up in the earth's magnetic field. This has the effect of increasing the magnetization of the material.

4.6.4 Ferromagnetic Materials

Diamagnetism and paramagnetism are weak forms of magnetism and, consequently, materials that show these two types of behavior are often considered nonmagnetic. Certain materials that contain paramagnetic atoms or ions, however, do show a strong permanent magnetization even in the absence of a magnetic field. This type of response, called ferromagnetism, is shown by materials composed of atoms of certain transition elements such as iron and cobalt, and certain rare earth elements such as gadolinium. In ferromagnetic materials, a special type of coupling occurs between adjacent atoms, resulting in cooperative alignment of their magnetic moment throughout the crystal. This coupling is purely a quantum mechanical effect that cannot be adequately explained in terms of classical physics. It leads to a high net magnetic moment of the material even in the absence of a magnetic field. Ferromagnetism is a property not just of the individual atoms but is a result of the interaction of each atom with its neighbors in the crystal lattice of the solid. Although the magnetization can vary with the magnetic field, magnetic susceptibility χ_m values as high as 10^6 are possible for ferromagnetic materials.

Another property of ferromagnetism is that each grain in a polycrystalline material (or the whole crystal of a single crystalline material) is made up of a number of magnetic domains, tiny regions in which the atomic dipoles are essentially perfectly aligned. These domains are originally oriented in such a way that they cancel each other as far as their external magnetic effects are concerned (Figure 4.19a). However, when a ferromagnetic material is subjected to an applied magnetic field of gradually increasing strength, the domains that are favorably oriented grow at the expense of those that are not, and the material becomes highly magnetized (Figure 4.19b).

The maximum possible magnetization of a ferroelectric material, called the saturation magnetization M_{max},

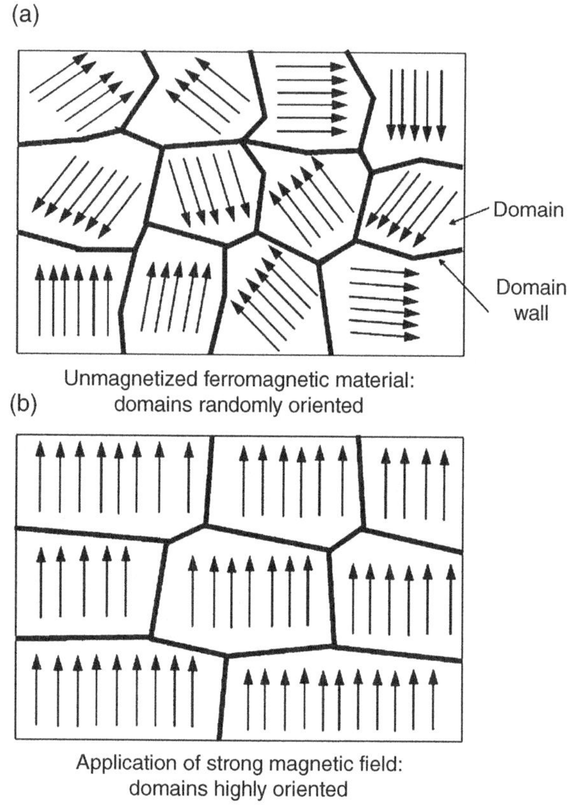

Figure 4.19 Schematic illustration of magnetic domains in a ferromagnetic material: (a) Randomly oriented domains in an unmagnetized material. (b) The domains become oriented upon application of a magnetic field, resulting in a highly magnetized material. Each arrow represents a huge number of atoms.

corresponds to the magnetization that would result if all the atomic dipoles of the material were completely aligned with the magnetic field. There is also a saturation magnetic flux density, B_{max} (Eq. (4.40)). For the transition metal atoms (those with a partially filled d electron subshell), the contribution from the orbital angular momentum is negligible and, thus, the magnetic moment of the atom is determined by the number of unpaired electrons. Consequently, in the transition metals where the 3d shells of the atoms are partially filled, we can determine the magnetic moment of the atoms by filling the 3d shell with aligned spins up to a maximum of five beyond which the spins must have an opposite or antiparallel alignment to the first five. This is required by the Pauli exclusion principle. In the iron (Fe) atom, for example, the net number of unpaired electrons is four and, thus, the net magnetic moment of the atom is $4\mu_B$, where μ_B is the Bohr magneton. Experimental magnetization curves of M (or B) versus H show a saturation magnetization that is lower than the ideal value due to microstructural factors and pinning of the domain walls by defects such as grain boundaries.

4.6.5 Ferrimagnetic Materials

Another type of magnetism, called ferrimagnetism, more common in ionic-bonded ceramics, refers to a type of ferromagnetism in which the magnetic moment of ions at one type of atomic sites is partly cancelled by antiparallel interactions with ions of another site. However, there remains a net magnetic moment of the material in the absence of a magnetic field. Ferrimagnetic ceramics have a lower saturation magnetization than ferromagnetic metals but their electrically insulating properties provide an advantage in some engineering applications where a low electrical conductivity is required.

The ferrimagnetic iron oxides Fe_3O_4, commonly referred to as magnetite, and ferric oxide, γ-Fe_2O_3, referred to as maghemite, are the most widely studied nanoparticles for hyperthermia treatment of tumors because of their favorable biocompatibility. Magnetite has a crystal structure that belongs to a class of compounds called spinels. The formula of normal spinels has the general formula AB_2O_4 (or $AO{\cdot}B_2O_3$), where A^{2+} and B^{3+} are divalent and trivalent ions, respectively. These normal spinels have a characteristic crystal structure based on a close packed face-centered cubic (FCC) lattice of O^{2-} ions, with the A^{2+} ions and B^{3+} ions occupying tetrahedral and octahedral interstitial sites within the lattice. Magnetite has an inverse spinel structure $B(AB)O_4$ in which the occupancy of the interstitial sites is slightly different. One-half of the B^{3+} ions in magnetite occupy tetrahedral sites (called *a* sites) whereas the A^{2+} ions and remaining half of B^{3+} ions occupy octahedral sites (called *b* sites) (Figure 4.20a). In magnetite, the O^{2-} ions are nonmagnetic due to all their electrons occupying filled shells. Consequently, we need to consider only the interactions between one-half of the Fe^{3+} ions (with five unpaired electrons in their 3d shell) on the *a* sites and the Fe^{2+} ions (four unpaired electrons in their 3d shell) and the other half of the Fe^{3+} ions on the *b* sites (Figure 4.20b). The antiparallel *a*–*b* interaction between the ions in the *a* and *b* sites dominates and, thus, the saturation magnetization per formula unit (Fe_3O_4) is $4\,\mu_B$.

Maghemite (γ-Fe_2O_3) has the same crystal structure as magnetite but it is often considered a ferrous ion (Fe^{2+}) deficient magnetite because there are no Fe^{2+} ions in the crystal structure when compared to magnetite. Although there is no clear agreement about the distribution of the cations in the maghemite crystal structure, it is often stated that the Fe^{3+} ions occupy both the tetrahedral *a* sites and octahedral *b* sites, with five Fe^{3+} ions and one vacancy in the *b* sites for every three Fe^{3+} ions in the *a* sites. This can be expressed by the formula $Fe[Fe_{5/3}\square_{1/3}]O_4$ where the symbol $\square$ represents a vacant site.

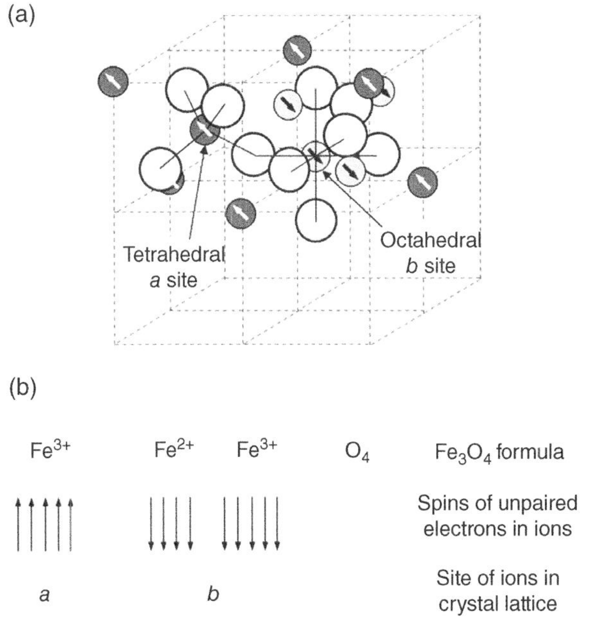

Figure 4.20 (a) Part of Fe_3O_4 crystal structure showing the tetrahedral *a* sites and octahedral *b* sites. (b) Illustration showing the arrangement of the electron magnetic moments of the Fe ions at the *a* and *b* sites in the Fe_3O_4 crystal structure.

4.6.6 Magnetization Curves and Hysteresis

Measured curves of magnetization *M* versus applied magnetic field *H* (or magnetic flux density *B* vs *H*) for diamagnetic and paramagnetic materials are straight lines and they retrace themselves as *H* is increased or decreased. In comparison, magnetization curves for multi-domain ferromagnetic and ferrimagnetic materials are not straight lines and they do not retrace themselves as *H* is increased or decreased (Figure 4.21). This lack of reversibility is a well-known phenomenon called hysteresis. Another feature of the magnetization curves for ferromagnetic and ferrimagnetic materials is that they show a remanent magnetization (point C or point D) even though there is no applied field. This is the phenomenon of permanent magnetism. Hysteresis is due to the presence of multiple magnetic domains within the material. The motion of the domain boundaries and the reorientation of the domain directions are not fully reversible. The domains do not return completely to their original arrangement but, instead, some alignment of the magnetic dipoles remains even when the magnetic field is completely removed.

4.6.7 Hyperthermia Treatment of Tumors using Magnetic Nanoparticles

Hysteresis in a time varying or alternating magnetic field leads to the generation of heat in ferromagnetic and

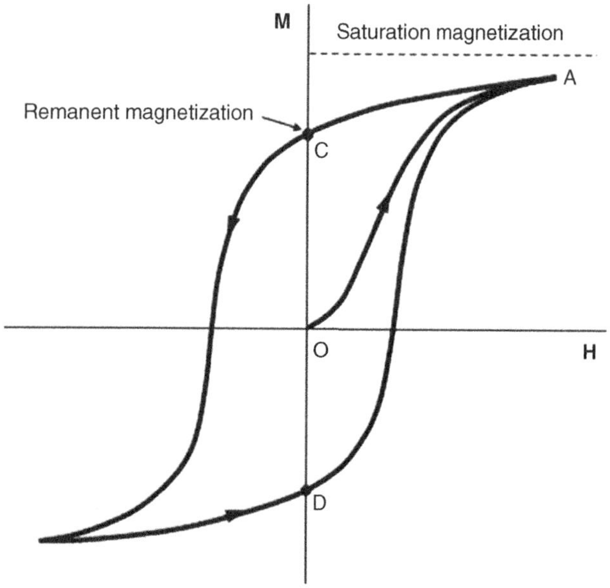

Figure 4.21 Magnetization curve for a ferromagnetic or ferrimagnetic material showing a hysteresis loop caused by domain motion.

ferrimagnetic materials. The amount of heat P generated per unit volume is given by

$$P = \mu_0 f \oint H \mathrm{d}m \tag{4.44}$$

where, μ_0 is the permeability of free space, f is the frequency of the alternating applied field, and the circular integral sign represents integration over the closed hysteresis loop, yielding the area of the loop.

The generation of heat in ferromagnetic and ferrimagnetic nanoparticles when they are subjected to an alternating magnetic field forms the basis of hyperthermia treatment of tumors. Nanoparticles of the ferrimagnetic oxide Fe_3O_4 or γ-Fe_2O_3 are most commonly used because of their favorable biocompatibility. The procedure involves dispersing nanoparticles in the target tumor and applying an alternating magnetic field of sufficient strength and frequency to generate heat in the particles. This heat is transferred to the surrounding tumor and if the temperature is maintained above ~42 °C for a sufficient period (~30 minutes or more), the tumor can be destroyed. This process is attractive because of the possibility of eradicating the intended tumor without causing significantly adverse effects to healthy tissue. However, there are constraints on the magnitude of H and f that healthy tissues can tolerate and questions about whether a sufficient number of nanoparticles can be delivered to the tumor to ensure the generation of sufficient heat over the required time to eradicate the tumor.

4.7 Thermal Properties

As the majority of biomedical applications occur at or near the body temperature, the thermal properties of biomaterials in these applications are of little consequence. On the other hand, thermal properties are important for biomaterials used in dental restorations such as crowns and bridges, for example, where they are required to match the thermal properties of the existing tooth structure. Thermal properties are also important for biomaterials whose processing or manufacturing process involves large changes in temperature. In the area of physical properties, the important thermal properties relevant to biomaterials are their thermal conductivity and coefficient of thermal expansion.

4.7.1 Thermal Conductivity

Whether a material can conduct heat or not is quantified by its thermal conductivity κ, defined by the equation

$$q = -\kappa \frac{\mathrm{d}T}{\mathrm{d}x} \tag{4.45}$$

where, q is the heat flux, that is, the amount of heat flowing across unit area per unit time, and $\mathrm{d}T/\mathrm{d}x$ is the temperature gradient down the material. Equation (4.45) is valid for the steady-state condition in which the heat flux is independent of time and the minus sign signifies that the heat flow is down the temperature gradient, that is, from high to low T. As q has units of W/m^2, the units of κ are W/m/K.

Transport of heat in solids occurs by flow of electrons and by vibration of atoms (or ions) in the crystal lattice. Thus, we can write

$$\kappa = \kappa_e + \kappa_l \tag{4.46}$$

where, κ_e and κ_l are the contributions to the thermal conductivity due to the electrons and lattice vibrations, respectively. Depending on the material, one or the other conduction mechanism often dominates. In pure metals with a high degree of metallic bonding, such as copper, thermal conduction occurs predominantly by the sea of nearly free electrons that surround the metal ions. Because of the large number of electrons and ease with which they flow, metals are good thermal conductors, that is, they have a high thermal conductivity. As the flow of electrons is responsible for both thermal and electrical conduction, metals are both good electrical conductors and good thermal conductors, a relationship quantified by the Weideman–Franz law given by

$$\frac{\kappa}{\sigma' T} = L \tag{4.47}$$

where, σ' is the electrical conductivity, T is the temperature and L is a constant equal to 2.44×10^{-8} WΩ/K^2.

Thermal conduction by lattice vibrations dominates in ceramics as these materials contain few free electrons. An atom vibrating about its equilibrium position due to its thermal energy can be thought of as a sphere supported by springs (Figure 4.22a). However, the atoms in a crystal lattice are bonded together and, thus, they do not vibrate independently of one another. Instead, the oscillations are connected which leads to a wave behavior (Figure 4.22b). Using the wave-particle duality of quantum mechanics, the waves associated with atomic oscillations are quantized and the quantum of energy or "particle" associated with these lattice vibrations is called a phonon. A phonon represents a quantum of atomic lattice vibration, analogous to a photon that is a quantum of electromagnetic radiation (light). Phonons can show properties such as scattering, for example, that are typical of particles.

As phonons are not as efficient as electrons in conducting heat energy due to scattering by lattice imperfections, ceramics typically have a lower thermal conductivity than metals (Figure 4.23). Glasses typically have a lower thermal conductivity than ceramics because phonon scattering becomes more effective with a more disordered atomic structure. Ceramics and glasses have a low thermal conductivity and a low electrical conductivity. On the other hand, diamond has a low electrical conductivity and a high thermal conductivity, higher than that of metals. The low electrical conductivity of diamond is due to its strong covalent bonding (no free electrons). In comparison, the high thermal conductivity is due to two factors, the highly ordered crystal structure and the strong atomic bonding

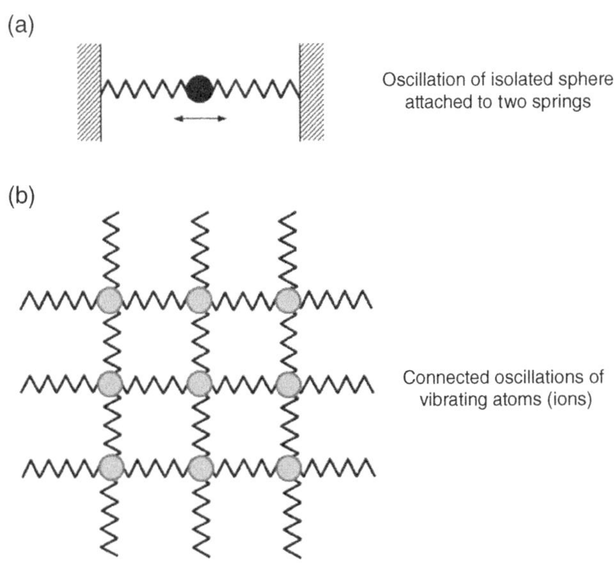

Figure 4.22 (a) Vibration of a sphere connected by springs. (b) The atomic vibrations in a crystal lattice are connected which leads to a wave behavior.

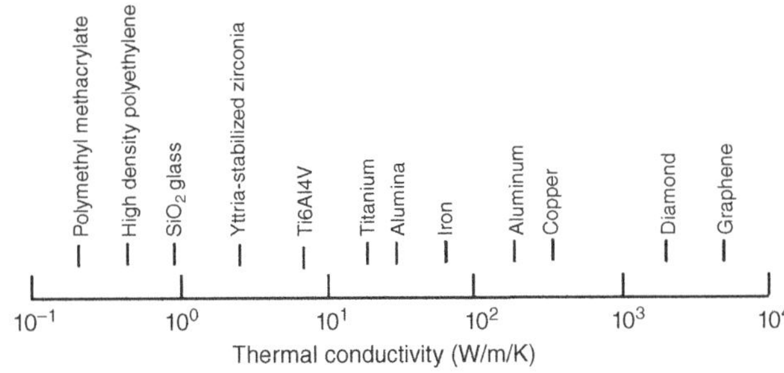

Figure 4.23 Chart showing the thermal conductivity values for a variety of materials.

that strongly limits the incorporation of impurities into its structure. These two factors severely limit phonon scattering in the crystal lattice.

In polymers, heat conduction occurs by vibration and rotation of the long chain molecules. Polymers have a low thermal conductivity, although, for a semicrystalline polymer, the conductivity increases with increasing volume fraction of crystalline regions because phonon scattering in the crystalline region is lower than that in an amorphous region of the same composition.

4.7.2 Thermal Expansion Coefficient

The expansion or contraction of a material upon heating or cooling is commonly quantified by its linear coefficient of thermal expansion α defined by the equation

$$\alpha = \left(\frac{\Delta l}{l_o}\right)\frac{1}{\Delta T} \tag{4.48}$$

where, Δl is the change in length of a specimen of length l_o due to a change in temperature ΔT. The thermal expansion coefficient has the unit $°C^{-1}$ (or K^{-1}) but is often expressed in units of $10^{-6}\ °C^{-1}$ because of its low value. The thermal expansion coefficient of one material relative to another is dependent on its interatomic bonding energy versus displacement curve (Section 2.2).

Although α varies slightly, depending on the temperature range of measurement, ceramics and glasses typically have low α values relative to other classes of materials, in the range $\sim 5 \times 10^{-6}$ to $\sim 15 \times 10^{-6}\ °C^{-1}$ over a temperature range of a few hundred degrees Celsius above room temperature. On the other hand, some glasses, such as fused silica glass, and a few glass-ceramics have values as low as $\sim 0.5 \times 10^{-6}\ °C^{-1}$. Metals have higher α values, in the range $\sim 10 \times 10^{-6}$ to $\sim 25 \times 10^{-6}\ °C^{-1}$ but tungsten and a few metal alloys, for example, have values lower than $\sim 5 \times 10^{-6}\ °C^{-1}$. Polymers have the highest values, typically $\sim 100 \times 10^{-6}\ °C^{-1}$ and over.

A difference in thermal expansion coefficient leads to the development of mechanical stresses between two adherent materials upon heating or cooling. If high enough, these stresses can lead to cracking and delamination of one material from the other, such as a coating deposited on a three-dimensional material. Metal implants, such as Ti6Al4V for example, have been coated with a bioactive material such as hydroxyapatite or bioactive glass, of thickness several tens of microns to a few hundred microns, to improve their bioactivity and osseointegration with host bone. These coatings are often formed on the metal implant at several hundred degrees Celsius. Ti6Al4V, for example, has a thermal expansion coefficient of ~9.5×10^{-6} °C^{-1}. Consequently, coating materials for Ti6Al4V should have a thermal expansion coefficient that is not significantly different from this value to minimize stresses in the coating and reduce the potential for cracking or delamination of the coating during cooling from the fabrication temperature. Bioactive glass-coated implants have seen little clinical application despite efforts to design glass compositions having thermal expansion coefficients that match those of implants such as titanium or Ti6Al4V (Peddi et al. 2008). In comparison, hydroxyapatite (thermal expansion coefficient of 10.0×10^{-6}–10.5×10^{-6} °C^{-1}) has been used for a few decades to coat dental and orthopedic implants, particularly those composed of titanium or Ti6Al4V. While considerable efforts have been devoted to creating adherent hydroxyapatite coatings using a variety of techniques, indications for the capacity of these coatings to enhance the long-term viability of the implants remain controversial (Le Guéhennec et al. 2007).

4.8 Optical Properties

Optical properties of biomaterials are relevant to their use in applications such as contact lenses and intraoptical lenses, and dental restorations for repairing existing tooth structure, particularly in the anterior of the mouth. Important optical properties for biomaterials are their degree of transparency, refractive index, and color. Light is an electromagnetic wave and, consequently, electrostatic interactions with its electric field component determine the optical properties of a material.

Materials are broadly classified as transparent, translucent, or opaque, depending on their ability to reflect, absorb, or transmit an incident light beam (Figure 4.24). A transparent material is one in which there is little reflection and absorption whereas in an opaque material, there is little or no transmission. In a translucent material, light is scattered within the material and, thus, is transmitted diffusely through the material. If a material is not perfectly transparent, the intensity of light I decreases exponentially with distance x through it, given by Lambert's law

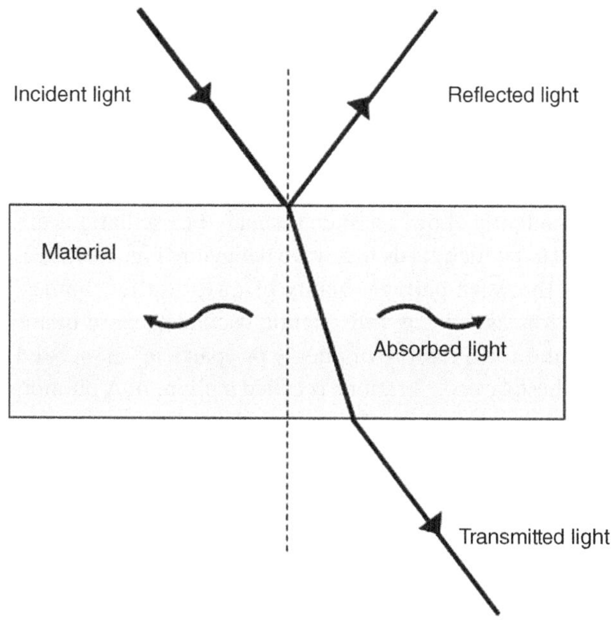

Figure 4.24 Reflection, transmission, and absorption of a light beam incident on the surface of a material.

$$I = I_0 \exp(-\alpha x) \tag{4.49}$$

where, I_0 is the intensity of the incident light beam and α is the absorption coefficient that can be obtained experimentally by plotting ln I versus x.

Absorption of light depends on the atomic bonding of the material. Ionic bonded solids such as ceramics show high absorption because of the interaction of their oppositely charged ions with the electric field component of light. Absorption occurs in the region of the electromagnetic radiation spectrum such as the infrared, for example, where these frequencies match the frequencies of the lattice vibrations (phonons). The absorption spectrum of polymers is dominated by absorption by their constituent macromolecules. Light is preferentially absorbed in regions where its frequency excites the various absorption modes in the polymer. Because of the high concentration of nearly free electrons, almost all the incident light over a wide frequency range of the electromagnetic radiation spectrum on a metal is absorbed within ~0.1 μm of its surface and reflected back. Thus, metals of thickness greater than this value are both opaque and reflective.

Refraction (bending) of a light beam occurs when it passes from one medium into another, such as from air into glass. The refractive index n of a material is a measure of the extent to which an incident light beam is bent when it passes from a vacuum into the material. It is defined by

$$n = \frac{c}{v} \tag{4.50}$$

where, c is the velocity of light in a vacuum and v is the velocity of light in the material. As v must be smaller than c, n has values larger than 1. Several glasses, for example have a refractive index of ~1.5. Refraction as well as reflection of light occurs at the interface between two materials (or media) with different refractive index n_1 and n_2. Assuming that the light is incident normally on the surface, the reflectivity R, defined as the ratio of the intensity of the reflected light I_R to the intensity of the incident light I_o, is given by

$$R = \left(\frac{n_2 - n_1}{n_2 + n_1}\right)^2 \tag{4.51}$$

Light that has entered a material can be reflected out due to scattering inside the material. If the amount of scattering is not too excessive, the material can appear translucent. Even the transmitted light can be scattered as it passes through the material, giving a transmitted beam that will spread out and appear blurred. Scattering occurs at the grain boundaries of polycrystalline materials, at fine pores within a material and at the interface between two phases of different refractive index. A dense homogeneous pore-free glass lacks these microstructural features and, thus, suffers from little scattering.

Small changes in composition can lead to large changes in the color of a material as, for example, the formation of glasses with a variety of colors by the addition of less than ~1% by weight of a transition metal oxide to a composition that would normally form a transparent glass. Changes in the electronic energy levels of the transition metal ions give rise to selective absorption of light at certain frequency ranges in the visible region of the electromagnetic spectrum, and, thus, to different colors.

Lithium disilicate glass-ceramics have become one of the most commonly used materials for dental restorations such as crowns and bridges (Chapter 7). The microstructure of these glass-ceramics consists of fine lithium disilicate crystals within a glass matrix (Figure 3.23d). In addition to an attractive combination of mechanical properties, thermal conductivity, and chemical stability in the oral environment, lithium disilicate glass-ceramic compositions have been designed to provide desirable optical properties such as color and fluorescence to match those of the existing tooth structure. This is achieved by adding small quantities of various transition metal oxides to the glass composition during its manufacturing process. In this way, the refractive index of the glass phase is made approximately equal to that of the crystals to reduce scattering of light at the enormous number of interfaces between the fine crystals and the glass matrix. A few of these metal ions and, if required, the use of additional metal ions, provide an additional benefit of esthetic appearance. They lead to the production of a wide range of desirable colors and fluorescence to match the existing tooth structure.

4.9 Concluding Remarks

In this chapter, we discussed how a three-dimensional material responds to an applied physical stimulus such as a mechanical stress, an electric field, a magnetic field or an electromagnetic field (light). Overall, the response has its origins in the atomic structure (atomic bonding) of the material.

- The physical properties of a material, such as strength, elastic modulus, electrical conductivity, magnetic susceptibility, thermal conductivity, and refractive index are controlled by its atomic structure and they can be modified by the microstructure of the material
- An understanding of the mechanisms by which materials respond to an applied physical stimulus and how the properties of materials depend on their atomic bonding and microstructure is critical in designing and creating biomaterials with a desirable combination of properties for the intended application
- Mechanical properties are among the most important physical properties for biomaterials that are subjected to mechanical stresses at some point or during their entire application. Consequently, designing biomaterials for mechanical reliability must take into account the inherent response of different types of materials to stresses.
- As the applications of biomaterials normally involve a combination of physical properties, an understanding of other physical properties in addition to, or instead of mechanical properties, is desirable.

Problems

4.1 The figure below shows general stress–strain curves for different types of materials at room temperature:

A Which curve best represents the mechanical response of (i) alumina, (ii) high density polyethylene, (iii) polystyrene, (iv) stainless steel, and (v) titanium.

B Which curve represents the toughest material?

C Which curve represents the material with the lowest Young's modulus?

Explain your answers.

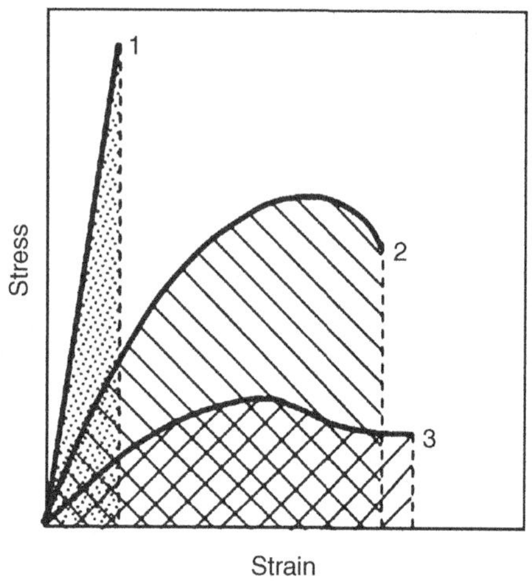

4.2 A cylindrical specimen of length 100 mm and diameter 10 mm is loaded in tension in a mechanical testing machine. Upon application of a force of 1000 N, the length increased to 100.5 mm. Determine the engineering (nominal) stress and strain in the specimen. If all the deformation occurred within the elastic region of its mechanical response, determine the Young's modulus of the material.

4.3 Determine the stress on the femoral bone of average diameter 2.5 cm in a human when it is subjected to a compressive force equal to the weight of a human of mass 90 kg (~200 pounds). How does this stress compare with the tensile strength of human cortical bone?

4.4 The following data were obtained in tensile testing of an aluminum alloy specimen of gage length 50.8 mm and diameter 12.8 mm:

Force (kN)	Length (mm)
0	50.80
8.9	50.85
17.8	50.90
35.6	51.00
44.5	51.05
53.4	51.16
57.8	51.31
62.3	52.07
71.2	53.3
69.4 (fracture)	54.2

A Plot the engineering stress–strain curve

B Determine the Young's modulus, yield strength (at an offset strain 0.2%), and the ultimate tensile strength

C Determine the engineering fracture strength and the true fracture strength, given that the diameter of the fractured specimen was 10.16 mm.

4.5 Define toughness and resilience. Draw a stress–strain curve for a ductile material and indicate how the toughness and resilience can be determined from it.

4.6 Explain why and how grain size influences the strength of metals. Give a relationship (name and equation) between strength and grain size, and define the terms in the equation.

4.7 Explain why a metal that has undergone mechanical fatigue often fails at stresses far smaller than those for a similar metal that has not. Is the fracture of a fatigued metal expected to be ductile or brittle in character?

4.8 Explain why a ceramic material such as Al_2O_3 commonly shows a compressive strength that is far higher than its flexural strength.

4.9 Discuss the most important properties that should be considered in designing metals, ceramics, and polymers for use as biomaterials in load-bearing applications *in vivo*.

4.10 Explain the differences between diamagnetism, paramagnetism, ferromagnetism, and ferrimagnetism, and how these differences influence the applications of biomaterials.

4.11 Distinguish between phonons and photons. Explain how phonons influence the thermal conductivity of materials.

4.12 Metals typically have high electrical and thermal conductivities. On the other hand, diamond has a high thermal conductivity but is an electrical insulator. Explain.

4.13 Determine the number of unpaired electrons in the following atoms or ions: Cr, Al^{3+}, Zn, Ni, O^{2-}, Co^{2+}.

4.14 Assuming that the magnetization of nickel results from its unpaired electrons only, calculate the saturation magnetization per kilogram of nickel which has a density of 8.9 g/cm^3 and an FCC structure of unit cell length 0.352 nm.

References

Balint, R., Cassidy, N.J., and Cartmell, S.H. (2014). Conductive polymers: toward a smart biomaterial for tissue engineering. *Acta Biomaterialia* 10: 2341–2353.

Callister, W.D. (2007). *Materials Science and Engineering: An Introduction*, 7e. New York: Wiley.

Le Guéhennec, L., Soueidan, A., Layrolle, P., and Amouriq, Y. (2007). Surface treatments of titanium dental implants for rapid osseointegration. *Dental Materials* 23: 844–854.

Guimard, N.K., Gomez, N., and Schmidt, C.E. (2007). Conducting polymers in biomedical engineering. *Progress in Polymer Science* 32: 876–921.

Liu, X., Rahaman, M.N., Hilmas, G.E., and Bal, B.S. (2013). Mechanical properties of bioactive glass (13-93) scaffolds fabricated by robotic deposition for structural bone repair. *Acta Biomaterialia* 9: 7025–7034.

McCrum, N.G., Buckley, C.P., and Bucknall, C.B. (1997). *Principles of Polymer Engineering*, 2e. New York: Oxford University Press Chapter 4.

Pankhurst, Q.A., Connolly, J., Jones, S.K., and Dobson, J. (2003). Applications of magnetic nanoparticles in biomedicine. *Journal of Physics D: Applied Physics* 36: R167–R181.

Peddi, L., Brow, R.K., and Brown, R.F. (2008). Bioactive borate glass coatings for titanium alloys. *Journal of Materials Science. Materials in Medicine* 19: 3145–3152.

Wachtman, J.B., Cannon, W.R., and Matthewson, M.J. (2009). *Mechanical Properties of Ceramics*, 2e. New York: Wiley Chapter 6.

Further Reading

Chiang, Y.M., Birnie, D.P., and Kingery, W.D. (1997). *Physical Ceramics*. New York: Wiley Chapter 1.

DRH, J. and Ashby, M.F. (2018). *Engineering Materials I: An Introduction to Properties, Applications and Design*, 5e. Oxford: Butterworth-Heinemann.

Ratner, B.D., Hofmann, A.S., Schoen, F.J., and Lemons, J.E. (eds.) (2013). *Biomaterials Science: An Introduction to Materials in Medicine*, 3e. New York: Elsevier Chapter 1.

5

Surface Properties of Materials

5.1 Introduction

Surface properties are important because, together with bulk properties (Chapter 4), they determine the performance of biomaterials *in vivo*. Upon implantation, it is the surface of the biomaterial that is first encountered by the physiological environment. The interaction between the surface and components of the physiological fluid, such as water molecules, inorganic ions, amino acids and proteins, strongly influences the response of cells and, thus, the biocompatibility of the implant. In the healing of bone defects using nondegradable implants, for example, if the surface properties and characteristics of the implant are not selected properly, cells can form fibrous tissue around the implant instead of new bone. Consequently, the ability of the implant to integrate with host bone is limited, jeopardizing a successful outcome of the procedure (Chapter 23).

The surface of a material consists of a thin layer with properties and characteristics that are different from those of its interior due to a variety of physical and chemical factors (Figure 5.1). Depending on the composition of the material, its production process and the environment to which it is exposed, the thickness of this surface layer can vary from a few nanometers to hundreds of nanometers. Due to their lower coordination with surrounding atoms, the outermost surface atoms have a higher energy than atoms in the interior and, consequently, they have a higher reactivity and mobility. Ions, molecules, and contaminants from the ambient environment can also adsorb physically or chemically onto the surface. Together, these factors commonly lead to a surface composition that is different from that of the interior.

Except for the noble metals such as gold, silver, and platinum, atoms at the surface of clean metals react with chemisorbed oxygen molecules to form a metal oxide surface layer upon exposure of the metal to a dry atmosphere containing oxygen, even for a short time. The surface oxide layer formed under such conditions is often very thin, ~2–3 nm for titanium (commercial purity) and stainless steel (316L) used as implants. Autoclaving and thermal sterilization, however, often lead to an increase in the thickness of the surface oxide layer, to ~5–10 nm for titanium, for example.

In an atmosphere containing moisture, such as common ambient conditions, all materials are typically covered with ~2–3 layers of adsorbed water molecules. In the case of most metals and ceramics, the outermost more reactive metal atoms react with adsorbed water molecules to form surface hydroxyl (OH) groups in order to increase their coordination. Thus, the surface of metals and ceramics is normally composed of chemically bonded OH groups on top of which are ~2–3 layers of physically adsorbed water molecules. The surface of polymers that have undergone some oxidation during their processing may contain a limited amount of OH groups but, generally, the surface of most polymers are covered with physically adsorbed H_2O molecules.

The aqueous physiological fluid *in vivo* contains a variety of components such as inorganic ions, amino acids, proteins, and substances released by cells. Upon implantation, the surface of a biomaterial becomes rapidly covered with positive or negative ions and water molecules, which subsequently influences its interaction with the other components of the physiological fluid. Depending on the scale of observation, the surface of a biomaterial can show a variety of structural features, such as a crystalline or amorphous structure, and topography (roughness or smoothness). These structural features can also influence the interaction of the surface with the physiological environment.

The inherent composition of the material itself and its exposure to the environment can, in some cases, result in surface characteristics and properties that are not optimal for the intended application. In these cases, a variety of techniques are available to modify the surface of the material in order to achieve a more desirable composition (Chapter 13) and topography (Chapter 21).

Materials for Biomedical Engineering: Fundamentals and Applications, First Edition. Mohamed N. Rahaman and Roger F. Brown.

Companion website: www.wiley.com/go/rahaman/biomedicalengineering

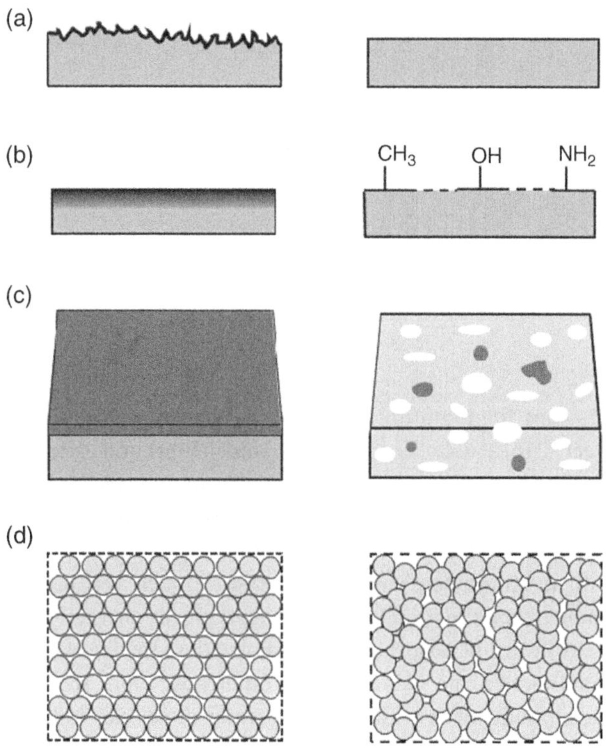

Figure 5.1 Various types of surface characteristics: (a) rough or smooth; (b) compositionally inhomogeneous with depth or composed of different functional groups; (c) chemically and structurally homogeneous or inhomogeneous; (d) crystalline or amorphous.

Overall, then, this chapter will discuss the following topics:

- Surface properties of biomaterials related to surface energy, composition, charge, and topography
- Physical and chemical principles underlying these surface properties and the techniques used to measure them.

Detailed descriptions of equipment and procedures are not covered because they are well provided elsewhere, such as in texts devoted more specifically to surface characterization or manuals provided by the equipment manufacturers.

5.2 Surface Energy

The outermost atoms at the surface of a material have a higher energy when compared to atoms in the interior due to their lower coordination and disrupted bonding (Figure 5.2). This excess energy is referred to as the surface energy, defined as the work required to create a unit area of surface. Surface energy (J/m^2) and surface tension (N/m) have different units but they are numerically equal. These two terms are often used interchangeably but, largely, surface energy has been more commonly associated with a solid whereas surface tension has been associated with a liquid. The difference in units stems from the way in which the terms are defined. Surface tension is defined mechanically in terms of a pressure difference Δp across a curved surface by the Laplace equation

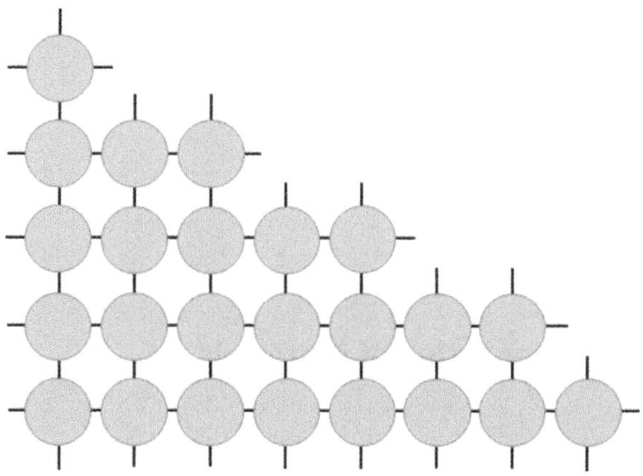

Figure 5.2 Illustration of lower coordination and disrupted bonding of outermost atoms at the surface of a crystalline material that gives rise to a surface energy.

$$\Delta p = \gamma_s \left(\frac{1}{r_1} + \frac{1}{r_2} \right) \tag{5.1}$$

where γ_s is the surface tension, and r_1 and r_2 are the principal radii of curvature of the surface. According to Eq. (5.1), the units of γ_s are N/m. On the other hand, surface energy γ_s is defined thermodynamically as the surface energy per unit area, giving units of J/m^2.

Unless a material is in an ideal vacuum, its surface will be in contact with another medium such as a vapor (gas) phase or a liquid phase, which will influence the bonding at the solid surface and, thus, its surface energy. Consequently, the surface energy γ_s of a solid commonly refers to the energy of the solid–vapor interface, that is, the solid surface in contact with the appropriate vapor (gas) phase such as air. It is often designated γ_{sv} to signify this, where the subscript *sv* refers to the solid–vapor interface. Similarly, the surface energy (surface tension) of a liquid is the interfacial energy of the liquid–vapor interface, designated γ_{lv}. At room temperature, γ_{sv} of many synthetic polymers are in the range ~20–50 mJ/m^2. In comparison, many metals and ceramics show much higher γ_{sv} values, in the range ~0.2–2 J/m^2, while a few metals show γ_{sv} values higher than 2 J/m^2. The low γ_{sv} for polymers is related to their weak van der Waals intermolecular bonds whereas the higher γ_{sv} for metals and ceramics is related to their strong interatomic bonding.

Surface energy has a significant influence on reactions that take place at the surface of a material. As there is a thermodynamical driving force to lower its energy, a material with a high surface energy will tend to encourage adsorption of substances from its surroundings if this will lead to a reduction in energy. In this sense, the surface energy of a solid has often been discussed in terms of the degree of contact between a drop of liquid and the solid surface (Figure 5.3). The change in the Gibbs free energy dG when the area A of the drop in contact with the solid increases by an infinitesimal amount dA is given by

$$dG = \gamma_{sl}dA - \gamma_{sv}dA + \gamma_{lv}dA\cos\theta \tag{5.2}$$

where γ_{sv}, γ_{sl}, and γ_{lv} are the specific energies (energy per unit area) of the solid–vapor, solid–liquid, and liquid–vapor interfaces, respectively, and θ is the contact angle between the liquid and the solid (the angle between the tangent to the liquid–vapor interface at the contact point with the solid surface). At equilibrium, $dG/dA = 0$ and Eq. (5.2) gives

$$\gamma_{sv} = \gamma_{sl} + \gamma_{lv}\cos\theta \tag{5.3}$$

Equation (5.3), sometimes referred to as the Young and Dupré equation, is what we would obtain by taking the horizontal components of the interfacial tensions in Figure 5.3. Roughly, good wetting of a solid by a liquid is said to be characterized by a low contact angle ($\theta < 90°$) and poor wetting by a high contact angle ($\theta > 90°$). Complete wetting and spreading of a liquid occurs when $\theta = 0°$ (Figure 5.4).

As the physiological fluid is aqueous in nature, the extent to which water will wet a biomaterial and spread over it has significant consequences for its interaction with the aqueous medium *in vivo*. A material that shows good wetting and spreading by water (low θ) is referred to as hydrophilic (literally, water-loving). If the solid has a higher surface energy than water, there is a thermodynamic driving force for wetting and spreading of the liquid in order to reduce the energy of the system. In comparison, a material that shows poor wetting by water (high θ) is referred to as hydrophobic (literally, water-hating). In this case, the material has a lower surface energy than water and, thus, wetting is thermodynamically unfavorable because it will lead to an increase in the energy of the system.

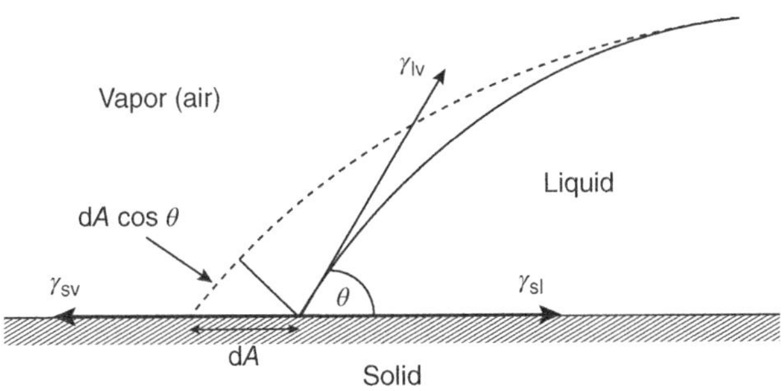

Figure 5.3 Contributions to the Gibbs free energy change due to change in area *dA* of a liquid drop on a solid.

5.2.1 Determination of Surface Energy of Materials

According to Eq. (5.3), the surface energy γ_{sv} is related to the surface energy of the liquid γ_{lv}, the solid–liquid interfacial energy γ_{sl} and the contact angle θ. Whereas γ_{lv} and θ can be easily measured, γ_{sv} and γ_{sl} cannot. Consequently, methods for estimating γ_{sv} often involve a combination of measuring γ_{lv} and θ, and estimating γ_{sl} using theoretical analyses, some of which can be fairly complex. While a variety of methods have been proposed, a useful method for polymers, based on its simplicity and accuracy, involves measuring θ for a single liquid of known surface energy (surface tension) γ_{lv} and using a theoretical expression for γ_{sl} (Girifalco and Good 1957):

$$\gamma_{sl} = \gamma_{sv} + \gamma_{lv} - 2\Phi(\gamma_{sv}\gamma_{lv})^{1/2} \tag{5.4}$$

where Φ is expressed in terms of V_s and V_l, the molar volume (molecular weight divided by the density) of the

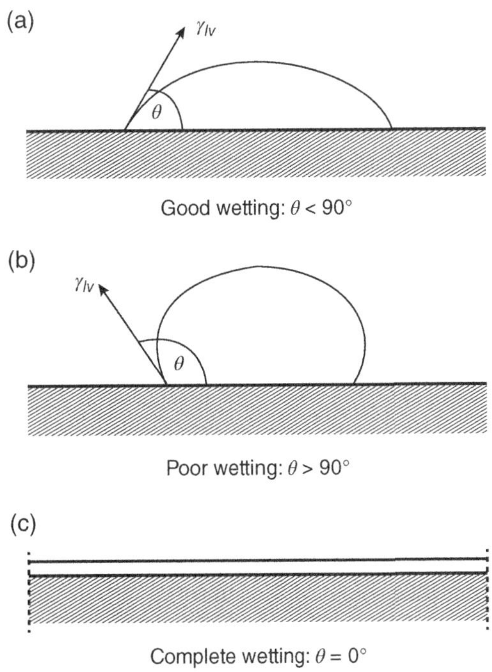

Figure 5.4 Wetting behavior between a liquid and a solid showing (a) good wetting, (b) poor wetting, and (c) complete wetting for a liquid of contact angle θ.

molecules or molecular segments of the solid and the liquid, respectively, given by

$$\Phi = \frac{4\left(V_s V_l\right)^{1/3}}{\left(V_s^{1/3} + V_l^{1/3}\right)^2} \tag{5.5}$$

For a polymer, the repeating unit in the polymer chain can be taken to approximate the molecular segment. Using Eqs. (5.3) and (5.4), we get

$$\gamma_{sv} \approx \gamma_{lv}\left[\frac{\left(1+\cos\theta\right)^2}{4\Phi^2}\right] \tag{5.6}$$

Example 5.1 The measured contact angle of a water droplet on polymethyl methacrylate (PMMA) is 70°. Determine the surface energy of PMMA, given that the densities of PMMA and water are 1.2 and 1.0 g/cm^3, respectively, and the surface tension of water is 73.0 mN/m.

Solution:
The molecular weight of water is 18.0 g/mol and the molecular weight of the repeating unit in PMMA is 100.1 g/mol, giving V_l = 18.0 cm^3, and V_s = 83.4 cm^3. Substituting in Eq. (5.5), we get $\Phi = 0.94$. Then, substituting the values for Φ, γ_{lv}, and θ in Eq. (5.6) gives $\gamma_{sv} = 37\,\text{mJ/m}^2$.

Another method applicable to materials of low surface energy, such as polymers, is referred to as the Zisman method. Fox and Zisman (1950, 1952) found that cosθ for a variety of polymers was approximately a monotonic function of γ_{lv}, that is

$$\cos\theta = a - b\gamma_{lv} \tag{5.7}$$

where a and b are constants. Eq. (5.7) can be expressed in the form

$$\cos\theta = 1 - c\left(\gamma_{lv} - \gamma_{cr}\right) \tag{5.8}$$

where c is a constant and γ_{cr} is a characteristic property of the material called the critical surface tension. The significance of γ_{cr} is that a liquid of surface tension less than or equal to γ_{cr} will wet and spread over the material. This follows from the limiting condition for spreading is $\theta = 0°$ and the fact that the contact angle cannot be less than 0°. The Zisman method involves measuring the contact angle θ for several liquids of known surface tension on a given polymer and plotting cosθ versus γ_{lv} (Figure 5.5). For many polymers, it is found that γ_{cr} is approximately equal to γ_{sv}, as exemplified by the data for PMMA in Figure 5.5. However, because of the time-consuming nature of the experiments and the less than rigorous theoretical justification, the Zisman method has found only limited application in recent years.

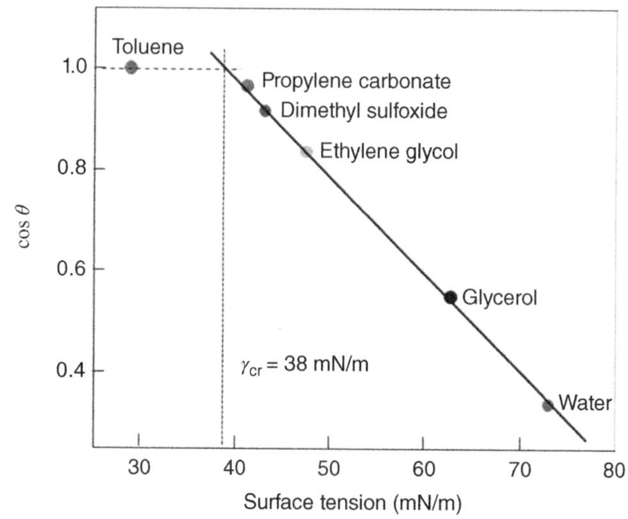

Figure 5.5 Zisman plot for polymethyl methacrylate (PMMA) using various liquids.

5.2.2 Measurement of Contact Angle

While a variety of methods can be used to measure the contact angle of a liquid on a solid (Ratner 2013), the sessile drop technique is easy to perform and, thus, finds considerable use. In this technique, a drop of the appropriate liquid, for example deionized water, is placed on a flat surface of the material according to a standard procedure and the contact angle is determined from images of the drop using automated equipment (Figure 5.3). The surface tension γ_{lv} of liquids used in contact angle measurements to determine γ_{cr} or to estimate γ_{sv} of a material (Eq. (5.6)) can be measured using simple techniques such as capillary rise of a liquid. However, this is often not necessary because the surface tension values of pure liquids at an appropriate experimental temperature such as room temperature are given in reference tables.

For a given liquid, the contact angle and, thus, the wettability of a material depends mainly on its surface chemistry but they can depend on the surface topography as well. Most metals and ceramics are typically more hydrophilic to varying extent due to the lower coordinated bonds at their surface whereas most polymers are more hydrophobic due to their organic surface groups. A rough surface that is wetted without any trapped air between a liquid droplet and the solid (Figure 5.6), referred to as homogeneous wetting, has an apparent contact angle θ_W given by (Marmur 2003):

$$\cos\theta_W = P\cos\theta \quad (5.9)$$

where θ is the contact angle for a smooth surface of the same composition and P is the roughness ratio defined as the true area of the rough surface relative to its nominal cross sectional area. In comparison, when wetting of the rough surface is inhomogeneous, that is, when air is trapped between the drop and the rough surface, the apparent contact angle θ_{CB} is given by

$$\cos\theta_{CB} = P_f f\cos\theta + f - 1 \quad (5.10)$$

where f is the fraction of the projected area of the solid surface that is wetted by the liquid and P_f is the roughness ratio of the wet area. Equations (5.9) and (5.10) are valid when the drop is sufficiently large relative to the roughness scale of the surface. The significance of these equations is that they predict how the roughness of a surface influences its contact angle with a liquid drop.

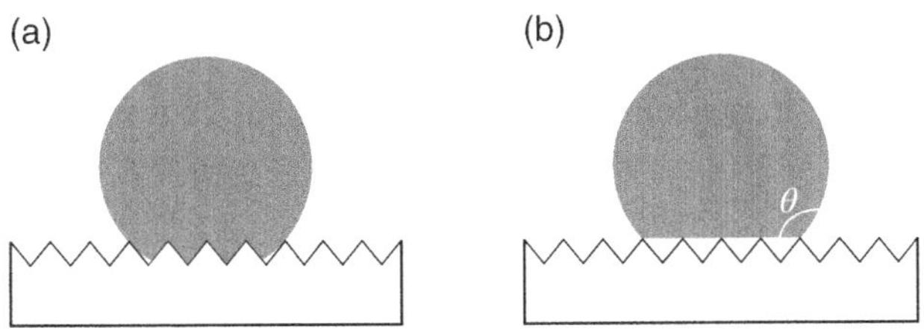

Figure 5.6 Homogeneous wetting (a) and heterogeneous wetting (b) of a rough surface illustrated for a hydrophobic liquid.

For homogeneous wetting, Eq. (5.9) predicts that a rough surface will decrease the contact angle of a hydrophilic material ($\theta < 90°$), that is, the hydrophilicity of the material will increase. In comparison, the contact angle of a hydrophobic material ($\theta > 90°$) will increase with surface roughness, that is, the hydrophobicity of the material will increase. For a greater amount of air trapped between the liquid and the rough surface, f decreases and, consequently, the apparent contact angle is higher. Subsequently, if the liquid slowly infiltrates the areas of trapped air, the contact angle is predicted to decrease with time, eventually becoming smaller than the contact angle of a smooth surface of the same composition. Thus, the measured contact angle of a rough surface can vary, depending on the time.

As an example, Figure 5.7 shows the contact angle of deionized water drops on machined surfaces of polyether ether ketone (PEEK) and titanium alloy (Ti6Al4V), and on a microrough surface of silicon nitride (Si_3N_4) in its as-fabricated state, three biomaterials used in spine repair (Bock et al. 2017). Due to its organic nature, PEEK has the highest contact angle of $86 \pm 4°$, whereas Ti6Al4V and Si_3N_4, composed of an oxidized surface layer, are more hydrophilic, having a contact angle of $71 \pm 5°$ and $66 \pm 9°$, respectively. Si_3N_4 showed a significant decrease in contact angle to $30 \pm 9°$ after 30 minutes, presumably resulting from the effect of heterogeneous wetting of its microrough surface. The surface of this as-fabricated Si_3N_4 is not composed of pure SiO_2 but, instead, a silicon oxynitride composition

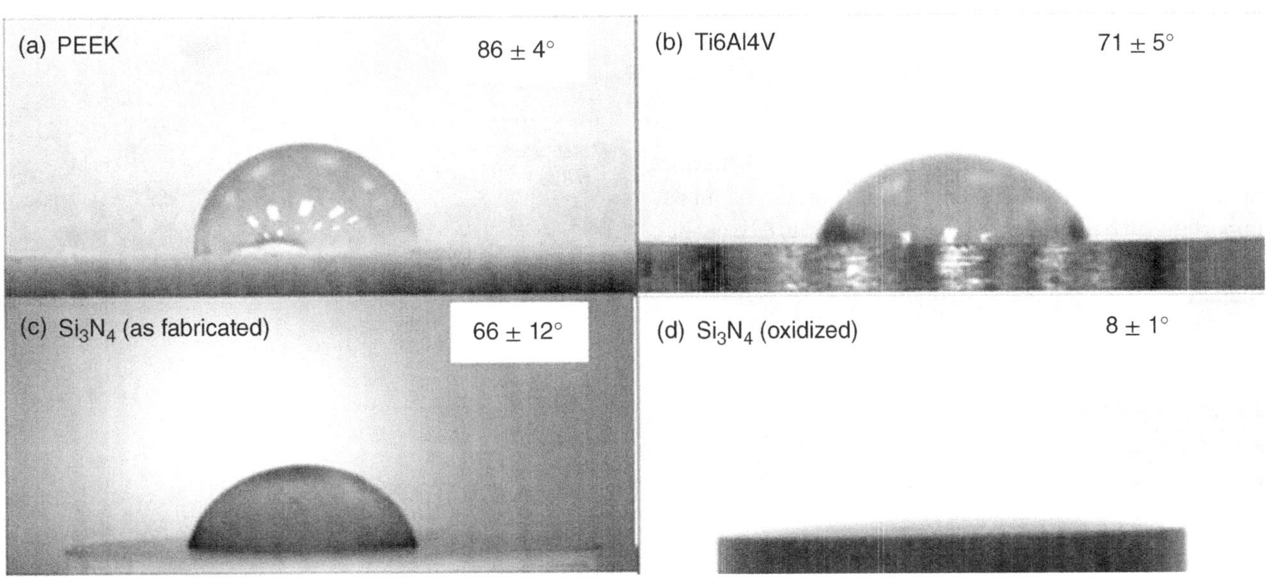

Figure 5.7 Images showing contact angle of a deionized water drop on (a) machined surface of polyether ether ketone (PEEK), (b) machined surface of Ti6Al4V, (c) surface of as-fabricated silicon nitride (Si_3N_4), and (d) surface of as-fabricated Si_3N_4 after an oxidation treatment. *Source:* From Bock et al. (2017).

composed of Si, O, and N atoms. Thermal treatment in an oxidizing atmosphere, in this case seven hours at 1070 °C in air, produced a thicker oxidized layer, the outermost region of which is composed of SiO_2. This resulted in a more hydrophilic surface of contact angle $8 \pm 1°$.

5.2.3 Effect of Surface Energy

Apart from its effect in controlling hydrophilicity or hydrophobicity, surface energy by itself has not played a major role in the applications of biomaterials. This is due to a variety of factors. One factor is that the true effect of surface energy is often difficult to separate from contributions from the effects of other surface properties such as surface charge and surface topology. Another factor is the difficulty in correlating surface energy with the interactions of biomaterials with the physiological environment. Often, surface energy is determined from experiments performed under certain ideal conditions that are different from practical conditions *in vitro* and *in vivo*. Prior to surface energy measurement, the biomaterial is often subjected to grinding and polishing, and thermal treatment to remove physically and chemically adsorbed water. These treatments can lead to a surface that is different from that of the implanted biomaterial due to adsorption of moisture and impurities such as hydrocarbons from the environment or from the container used for packaging. Standard liquids such as deionized water or, in some cases, phosphate buffered saline used experimentally are also different from the aqueous medium of the physiological environment which contains a variety of ions, small molecules, and proteins.

5.3 Surface Chemistry

The surface chemistry of a material influences its behavior in any given environment and, thus, forms the most important surface property of biomaterials. Surface chemistry influences several characteristics of a biomaterial, such as

- Chemical reactivity toward its environment as, for example, oxidation in air
- Composition of adsorbed ions and molecules
- Development of a surface electrostatic charge
- Mobility of side groups attached to polymer chains and their structural reorientation at the surface.

Apart from the noble metals such as gold, silver, and platinum, the surface of metals, exposed even for a short time to an environment containing oxygen (such as air) is typically composed of a thin layer of a metal oxide. This is due to the reaction of the surface atoms with chemically adsorbed oxygen molecules, which can be written

$$xM + \frac{y}{2}O_2 \rightarrow M_xO_y \tag{5.11}$$

The driving force for the reaction is the free energy change associated with formation of the metal oxide but the growth of the oxide layer is controlled by kinetic factors such as time and temperature. Similarly, ceramics such as silicon nitride (Si_3N_4) that are not composed of a metal oxide, often referred to as nonoxide ceramics, commonly develop a thin oxide surface layer when exposed to an environment containing oxygen.

In an environment containing moisture, such as common ambient conditions of relative humidity 40–50%, the surface of all materials is covered with two to three layers of physically adsorbed water molecules. The outermost, lower-coordinated metal atoms at a metal oxide surface react with the adsorbed water molecules to form surface hydroxyl (OH) groups in order to increase their coordination (Figure 5.8). On top of this hydroxylated surface, the water molecules are physically adsorbed by hydrogen bonds. Consequently, prior to implantation, most metals and ceramics used as biomaterials have a surface com-

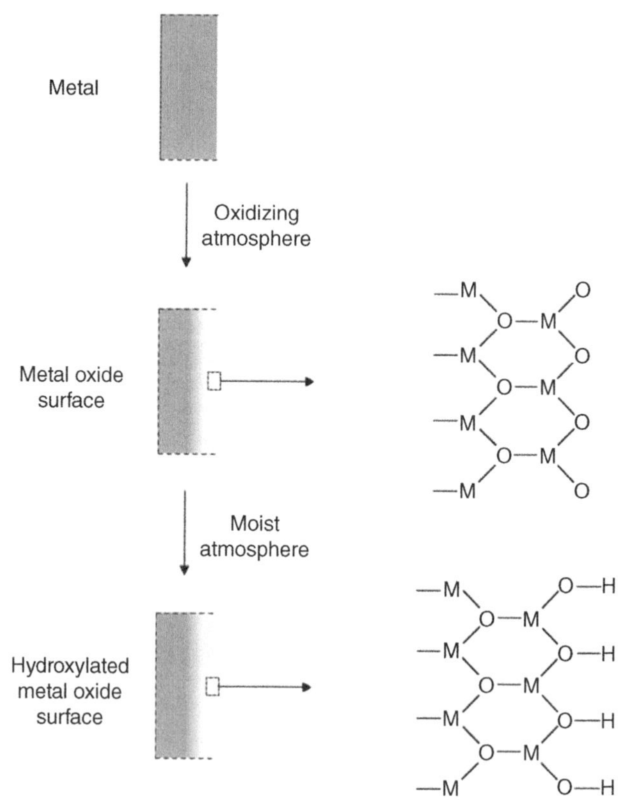

Figure 5.8 Illustration of the formation of an oxide surface layer on a clean metal exposed to an oxidizing atmosphere and a hydroxylated surface upon exposure to a moist atmosphere.

posed of OH groups attached to the outermost metal atoms on top of which are physically adsorbed H_2O molecules. The presence of these surface OH groups strongly influences the development of an electrostatic charge on the surface of the biomaterial in an aqueous medium *in vitro* and in the physiological medium *in vivo* (Section 5.4).

Depending on the nature of the surrounding medium, the polymer chains at the surface of some polymers can undergo structural reorientation of their side groups, which results in a change in surface composition. The thermodynamic driving force for this structural change is the reduction in energy due to a lower interfacial energy between the polymer surface and the medium. However, the rate at which this structural reorientation can occur is a kinetic problem, related to factors such as the flexibility of the chain backbone and the nature of the side groups. Typically, structural reorientation of the side groups occurs faster for polymers composed of single bonds between carbon atoms in the chain backbone and flexible, less bulky side groups. Surface structural reorientation in different polymers has been estimated to take place in time scales ranging from minutes to months.

The occurrence of surface structural reorientation has been studied for polyhydroxyethyl methacrylate (PHEMA), a biomaterial used for contact lenses, which has both hydrophilic and hydrophobic side groups (Figure 5.9). In air, the hydrophobic side groups are placed preferentially at the surface, resulting in a low surface energy. However, in an aqueous medium, the side groups can reverse their arrangement, presenting the hydrophilic (polar) side groups preferentially at the surface to interact with water molecules of the aqueous medium by hydrogen bonding and polar mechanisms.

5.3.1 Characterization of Surface Chemistry

In view of the significant role that surface chemistry plays in determining the performance of materials in biomedical applications and engineering applications as well, several techniques are available to study and characterize the surface chemistry of materials. The most common techniques are Auger electron spectroscopy (AES), X-ray photoelectron spectroscopy (XPS) also referred to as electron spectroscopy for chemical analysis (ESCA), and secondary ion mass spectroscopy (SIMS) (Table 5.1). AES, XPS, and SIMS are truly surface sensitive techniques in that they provide chemical information about the outermost layer (1–2 nm). They can also provide information from much deeper layers, commonly by sequential or simultaneous removal of surface layers by sputtering the surface with an ion beam and analyzing the sputtered surface, a mode of analysis referred to as depth profiling. A problem with these and most other surface analysis techniques is that they require an ultrahigh vacuum, necessitating the use of dry specimens that often differ in surface chemistry from that of biomaterials implanted *in vivo*. As noted earlier, in practice, biomaterials are often exposed to the ambient environment for a limited or prolonged period and, consequently, their surface is typically covered with physically and chemically adsorbed water and contaminants such as hydrocarbons.

Auger Electron Spectroscopy (AES)

AES is based on a two-step mechanism (Figure 5.10a). Collision of an incident electron beam with the surface of a material leads to the emission of an electron from an inner atomic orbital (binding energy = E_1) leaving a vacant

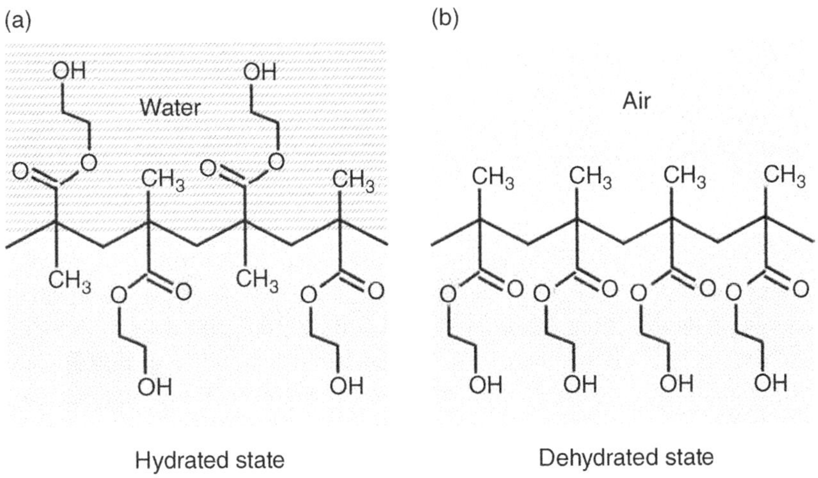

Figure 5.9 A model for reorientation of polymer surface functional groups during dehydration of polyhydroxyethyl methacrylate (PHEMA) hydrogels. The proportion of methyl and hydroxyethyl groups depicted at the polymer surface reflects qualitative changes in surface composition. *Source:* From Chen et al. (1999) / with permission of American Chemical Society.

Table 5.1 Common methods for the surface chemical analysis of materials.

Measurement parameter	Auger electron spectroscopy (AES)	X-ray photoelectron spectroscopy (XPS)	Secondary ion mass spectrometry (SIMS)
Incident particle	Electrons (1–20 keV)	X-rays (1.254 keV; 1.487 keV)	Ions (He^+, Ne^+, Ar^+) (100 eV–30 keV)
Emitted particle	Auger electrons (20–2000 eV)	Photoelectrons (20–2000 eV)	Sputtered ions (~10 eV)
Element range	>Li ($Z = 3$)	>Li ($Z = 3$)	>H ($Z = 1$)
Detection limit	10^{-3}	10^{-3}	10^{-6}–10^{-9}
Depth of analysis	2 nm	2 nm	1 nm
Lateral resolution	>20 nm	>150 μm	50 nm–10 mm

site that is soon filled by an electron from an outer orbital (binding energy = E_2). The energy released in this transition may appear as an X-ray photon or may be transferred to another electron in an outer orbital (binding energy = E_3) which is ejected with kinetic energy E_k given by

$$E_k = E_1 - E_2 - E_3 \tag{5.12}$$

The ejected electron is called an Auger electron. The electron ejected from the outer orbital moves through the solid and soon loses its energy through inelastic collisions with bound electrons. On the other hand, if the Auger electron is emitted sufficiently close to the surface of the material, it may escape and be detected using a spectrometer. The number of emitted electrons (or the derivative of the counting rate) is measured and plotted as a function of the kinetic energy of the electron. Since each type of atom has its own characteristic energy levels, the peaks in the observed Auger spectrum can be used to determine the elemental composition of the surface by comparison with standard Auger spectra for the elements.

X-ray Photoelectron Spectroscopy (XPS)

In XPS, the specimen is irradiated with a low-energy X-ray beam that leads to the emission of electrons from the inner atomic orbitals by the well-known Einstein photoelectric effect (Figure 5.10b). The kinetic energy of the emitted photoelectrons E_k is given by

$$E_k = h\nu - E_b - W \tag{5.13}$$

where h is the Planck constant, ν is the frequency of the incident X-ray photons, E_b is the binding energy of the photoelectron, and W is the work function of the spectrometer, a factor that corrects for the electrostatic environment in the instrument in which the photoelectron is produced and measured. By measuring E_k in a spectrometer of known work function, E_b can be determined from Eq. (5.13). The data are plotted as the number of emitted electrons versus their binding energy. As the binding energy of an electron is characteristic of the atom and the orbital from which it is emitted, the spectrum can be used to provide information on the surface composition by comparing with standard spectra.

Because the photoelectrons typically have a higher energy than the Auger electrons, XPS is not as surface sensitive as AES, that is, AES can probe a shallower depth than XPS. However, by varying the emission angle at which the photoelectrons are collected, a technique known as angle-resolved XPS, information from smaller and varying depths can be obtained. A benefit of XPS is that it can provide not just qualitative and quantitative information about the elemental surface composition but information about the chemical bonding (or oxidation state) of the surface atoms as well. As the incident beam is composed of X-rays, XPS suffers from fewer problems related to electrostatic charging of the specimen and damage to the sample surface when compared to AES that relies on the use of an incident electron beam. Consequently, XPS is often a preferred technique for surface characterization of ceramics that are typically electrically insulating and polymers that are typically insulating and have a low hardness as well. However, as they provide similar information about elemental composition, XPS and AES tend to be used in a complementary manner.

A common mode of using XPS is to perform a survey scan over a wide binding energy range (typically 0–1000 eV) to provide a qualitative analysis of the surface composition. The fractional concentration of the elements present at the surface can be determined from the area of the major peaks of each element using software. Then, information about the chemical bonding or oxidation state of the relevant atoms can be determined from higher resolution scans of the relevant peaks and measuring their chemical shift, that is, the change in their binding energy. A variation in the number of valence electrons or the types of bonds that they form results in a change in the binding energy of the innermost electrons and, thus, to a chemical shift.

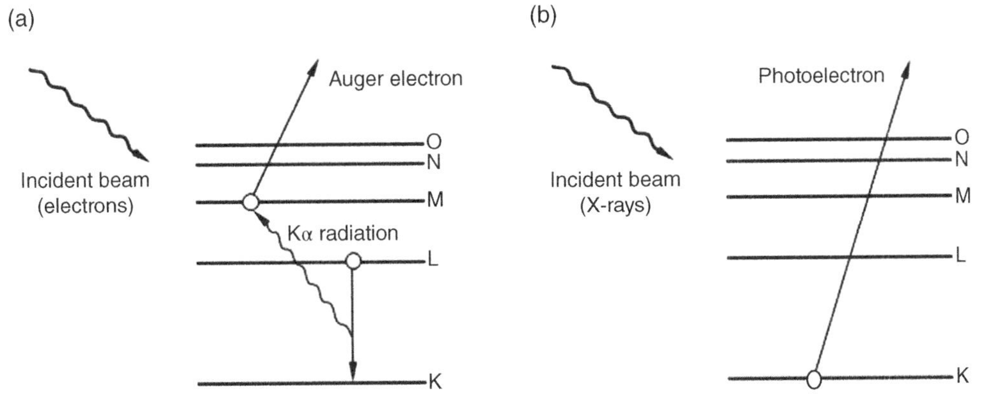

Figure 5.10 Interaction of incident beam (electrons or X-rays) with a solid, producing atomic excitation with (a) emission of electrons followed by de-excitation and emission of Auger electrons or (b) emission of photoelectrons.

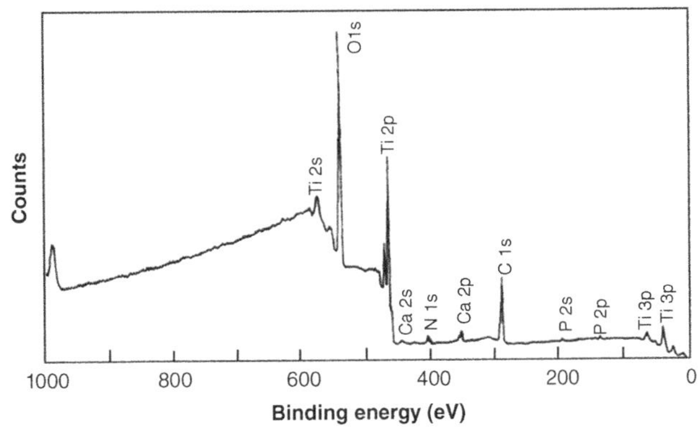

Figure 5.11 XPS survey spectrum for an autoclaved titanium dental implant. *Source:* From Lausmaa (1996) / with permission of Elsevier.

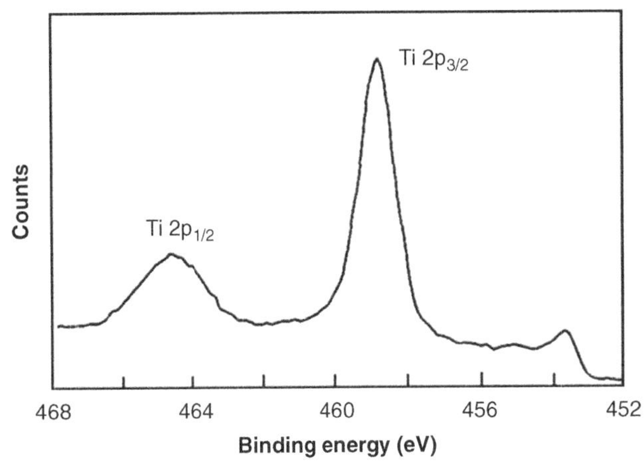

Figure 5.12 XPS high-resolution spectrum of the Ti 2p peak for a machined titanium implant. *Source:* From Lausmaa (1996) / with permission of Elsevier.

As an example, Figure 5.11 shows an XPS survey spectrum of commercial purity titanium that, prior to analysis, was subjected to treatments commonly used for titanium implants, such as machining, cleaning with various solvents, and steam autoclaving (at 135 °C for 20 minutes) (Lausmaa 1996). The major peaks corresponding to Ti and O reveal an oxidized surface. The minor peak corresponding to C is often due to contamination and, in many cases, to carbon present adventitiously in the spectrometer. Except for Ti and O, the peaks corresponding to C and the minor elements were found to disappear or to decrease significantly after sputtering off a few nanometers of the surface, indicating that these elements were present mainly as impurities on the surface. A peak at ~459 eV, which, from reference spectra, corresponds to Ti in the compound TiO_2, dominates the high-resolution spectrum of the Ti 2p peak (Figure 5.12). A smaller peak at ~454 eV corresponds to Ti in the underlying metal. Overall, then, the surface oxide layer on this titanium specimen corresponds to TiO_2.

The alloy Ti6Al4V also sees considerable use as a biomaterial. When subjected to the same treatments, its XPS spectrum is similar to that of commercial purity titanium but, in addition, it often shows a small amount of Al_2O_3. Typically, the concentration of Al in the surface oxide layer is approximately the same as that in the interior of the alloy.

Secondary Ion Mass Spectroscopy (SIMS)

SIMS consists of bombarding a surface with a primary beam of Ar, Ne, or He ions and analyzing the emitted ions and ion clusters in a mass spectrometer. As the emitted ions and ion clusters are characteristic of the surface, SIMS provides information about the chemical composition of the surface. Some information on the chemical bonding of the atoms can also be extracted by analyzing the composition of emitted ions and ion clusters.

SIMS can be used in two distinct modes of analysis. In static SIMS, an ion beam of low intensity is used so that analysis is confined to the outermost layers of the surface. In comparison, in dynamic SIMS, a high intensity beam is used to erode successive atomic layers at a relatively rapid rate. Static SIMS is more relevant to the application of biomaterials as it provides information about the outermost surface atoms. While SIMS has the advantages of high spatial resolution, high sensitivity for qualitative elemental analysis and the ability to provide a detailed analysis of the chemical composition of the surface, quantitative analysis of the surface composition is often difficult.

5.4 Surface Charge

As noted earlier, prior to implantation in the physiological environment, the surface of most polymers is normally covered with physically adsorbed water molecules. In comparison, most metals and ceramics have a surface composed of OH groups attached to the outermost metal atoms, on top of which are physically adsorbed H_2O molecules. The physiological fluid *in vivo*, on the other hand, can be approximated as an aqueous medium of homeostatic temperature 37.4 °C and pH 7.4, which contains a variety of ions, small molecules such as amino acids, macromolecules such as proteins, and substances released by cells. Upon implantation, the surface of a biomaterial acquires a positive or negative charge due to adsorption of ions or molecules from the aqueous medium or dissociation of certain surface functional groups, depending on the surface chemistry of the biomaterial.

5.4.1 Surface Charging Mechanisms

The surface of most metals and ceramics used as biomaterials, normally hydroxylated due to chemically adsorbed water molecules, acquire a surface charge typically by adsorption of hydrogen (H^+) ions, equivalent to hydronium (H_3O^+) ions, or hydroxyl (OH^-) ions (Figure 5.13). These ions are often referred to as the charge-determining ions for these materials. In an acidic medium (lower pH), preferential adsorption of H^+ ions leads to a positively charged surface, whereas in a basic medium (higher pH), a negatively charged surface is formed due to preferential adsorption of OH^- ions. At some intermediate pH, called the point of zero charge (PZC), there is a balance between adsorption of H^+ and OH^- ions, which leads to an electrically neutral surface. The PZC depends on the acidity or basicity of the surface composition. The more acidic oxides such as SiO_2 have a lower PZC whereas the more basic oxides such as MgO have a higher PZC.

The PZC can be measured from acid–base titrations but, often, it is easier to measure the zeta (ζ) potential corresponding to the electrostatic potential at a small distance from the surface (a few tenths of a nanometer). The pH at which the measured ζ potential is zero is referred to as the isoelectric point (IEP). Upon implantation in the physiological environment, then, a material whose IEP is lower than ~7.4, such as a more acidic metal oxide, will have a negative surface charge and electrostatic potential, whereas one having an IEP higher than ~7.4, such as a more basic metal oxide, will have a positive surface charge and potential.

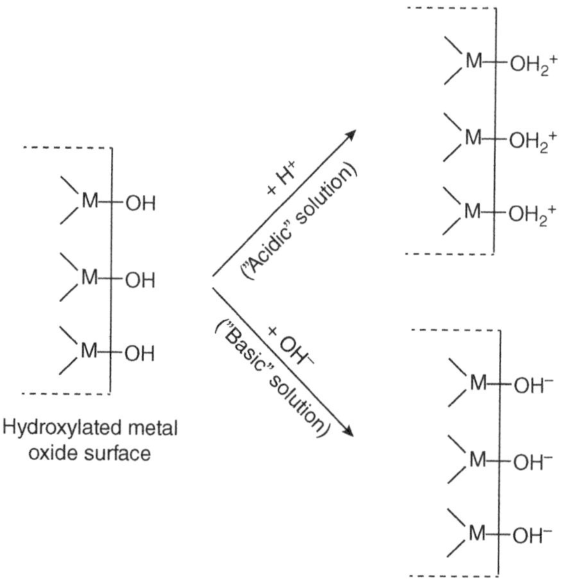

Figure 5.13 Production of surface charge on a hydroxylated metal oxide surface by adsorption of ions from an "acidic" or "basic" solution.

Some polymers that contain ionizable surface groups can acquire a charge by dissociation, such as dissociation of the H atom in the carboxyl (C=O)OH group, resulting in a negatively charged surface (Figure 5.14). In comparison, the presence of amine (NH_2) groups can lead to a positively charged surface due to adsorption of H^+ ions from an aqueous medium. For a given composition, the extent of dissociation or adsorption and, thus, the magnitude of the surface charge depends on the pH of the aqueous medium. The dissociation of H^+ from the (C=O)OH group, for example, often starts at a pH of ~3 and is essentially completed at a pH of ~7 to 8 when a large fraction of the H atoms at the surface has dissociated. In comparison, NH_2 groups are essentially neutral at pH higher than ~10 but the fraction at the surface that is protonated increases at lower pH.

The presence of ionizable functional groups at the surface of a material is not a prerequisite for the acquisition of a surface charge. Polymers devoid of such functional groups can acquire a surface charge in aqueous media by dispersion forces, that is, van der Waals forces of attraction between the material and ions present in the aqueous medium. These attractive forces are similar in nature to the van der Waals forces between molecules described in

Chapter 2. Ions in aqueous solutions are normally hydrated, that is, they are surrounded by an adsorbed layer of water molecules. Attraction between the atoms at the material's surface and the hydrated ions, positive or negative depending on the surface composition, leads to adsorption and, thus, to the production of a charged surface (Figure 5.15). The aqueous medium of the physiological environment, for example, contains a variety of ions such as H^+, Na^+, K^+, Mg^{2+}, Ca^{2+}, OH^-, Cl^-, HCO_3^-, HPO_4^{2-}, and SO_4^{2-}, and a variety of biomolecules (Chapter 14). The ensuing zeta potential of the biomaterial reflects the nature of the ions or molecules adsorbed at its surface.

Overall, then, in an aqueous medium, the surface of metals and ceramics acquire an electrostatic charge by adsorption of H^+ or OH^- ions (Figure 5.13) whereas polymers composed of certain surface functional groups can also acquire a surface charge by, for example, dissociation or adsorption of H^+ ions (Figure 5.14). Adsorption of water molecules and ions of the opposite charge, called counterions, leads to the formation of an ion atmosphere that is spread out in the solution. In colloid chemistry, this ion atmosphere is referred to as an electrical double layer (Figure 5.16). It consists of a rather tightly bound layer of counterions, referred to as the Stern layer, and a more diffuse layer in which the positive and negative ions migrate freely. The electrostatic potential at the boundary between the Stern layer and the more diffuse layer approximates the zeta potential. As noted earlier, polymers that do not possess appropriate surface functional groups acquire a surface charge by adsorption of hydrated ions from the solution (Figure 5.15).

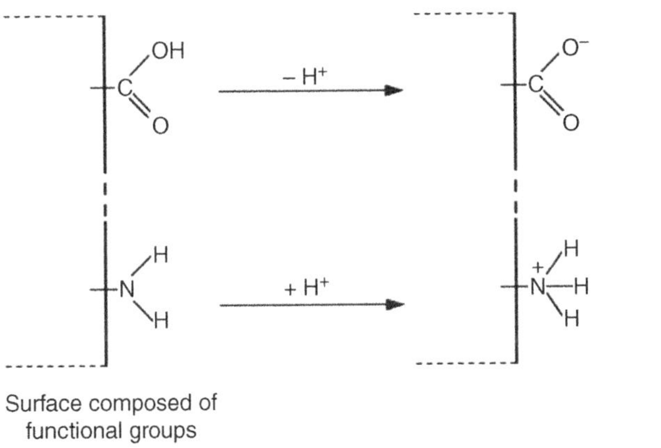

Figure 5.14 Production of negative or positive surface charge on surface composed of functional groups, as exemplified by the carboxyl and amine groups.

5.4.2 Measurement of Surface Charge and Potential

Zeta potential measurements are often used to characterize the surface charge and surface potential of fine particles in a suspension using techniques that track the motion of the particles when an electric field is applied between two electrodes in the suspension. Although zeta potential measurements of macroscopic solids are more difficult to perform, they are more relevant to biomaterials used in their macroscopic solid form. This is because the zeta potential of a macroscopic solid is typically different from that of fine particles of the same nominal composition. The zeta potential of macroscopic solids is commonly determined from measurements of their streaming potential. Flow of an aqueous liquid over the charged surface of a material leads to shearing and shifting of the ions adsorbed at the surface. This leads to an electrokinetic effect called the streaming potential. The surface charge can be determined from the measured surface potential and the use of theoretical equations, but this is often not necessary because the zeta

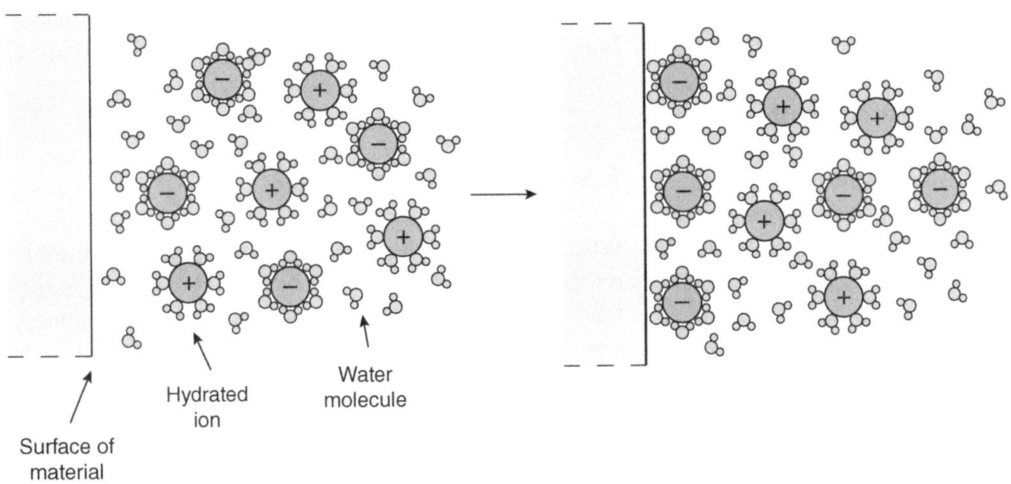

Figure 5.15 Production of surface charge on a surface devoid of functional groups by van der Waals attraction of ions in solution.

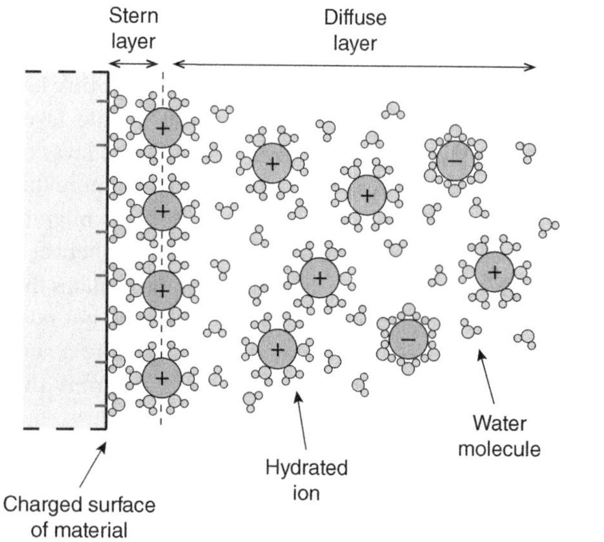

Figure 5.16 Illustration of the electrostatic charge distribution surrounding a solid surface upon introduction of the solid into an aqueous solution, shown for a solid that develops a negatively charged surface. The vertical dashed line corresponds approximately to the plane of the zeta potential.

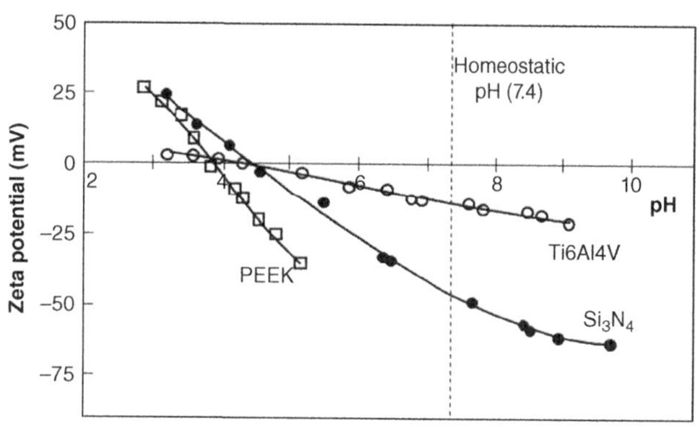

Figure 5.17 Zeta potential as a function of pH, as measured by the streaming potential method, for silicon nitride (Si_3N_4) as fabricated, and machined surfaces of Ti6Al4V and polyether ether ketone (PEEK). *Source:* From Bock et al.(2017) / with permission of John Wiley & Sons,

potential is commonly used as a measure of extent of the surface charge.

As an example, Figure 5.17 shows streaming potential data for the zeta potential as a function of pH for three biomaterials, PEEK, a titanium alloy (Ti6Al4V), and silicon nitride (Si_3N_4), the same materials described in Figure 5.7 (Bock et al. 2017). The pH of the aqueous liquid in these measurements was controlled using 0.1 M HCl solution at pH values in the range 3–5.5 and 0.1 M NaOH solution at pH between 5.5 and 10.0. In this medium, PEEK, Ti6Al4V, and Si_3N_4 have an IEP of 3.9, 4.4, and 4.5, respectively, and, at a pH of 7.4 (equal to the homeostatic pH of the physiological medium), a negative zeta potential of −50 (extrapolated by nonlinear regression), −15, and −45 mV, respectively. As PEEK has no ionizable functional groups, its negative zeta potential at the homeostatic pH is presumably due to preferential adsorption of negative ions present in the medium, such as Cl^- and OH^-. The IEP of Ti6Al4V is in the measured range for TiO_2 (~4–6) and, thus, the negative ζ potential at the homeostatic pH is most likely due to preferential adsorption of OH^- ions at TiOH groups (Figure 5.13). As the surface of Si_3N_4 is composed of both, SiOH and NH_2 groups, the IEP does not correspond to that (2–3) commonly observed for SiO_2. Instead, the higher IEP results from deprotonation and protonation reactions, respectively, at the SiOH and NH_2 groups. Presumably, deprotonation at SiOH dominates at pH values above ~4.5, giving a negative surface charge and potential.

5.4.3 Effect of Surface Charge

As the surface of a biomaterial rapidly develops an electrostatic surface charge and an associated electrical double layer upon implantation in the physiological environment, it is realistic to assume that this surface charge system should influence subsequent interactions with the physiological environment. Depending on their composition, the components of the physiological fluid such as ions, amino acids, and proteins can be positively charged, negatively charged, polar or nonpolar (Chapter 3). Consequently, we might expect varying degrees of electrostatic interaction between these components and the charged surface. This interaction should influence, for example, the type of ions and molecules adsorbed at the surface and the conformation of adsorbed proteins, which, in turn, should influence the response of cells. In practice, a correlation between the surface charge of a biomaterial and its interaction with the physiological environment has been difficult to establish. While this is due to a variety of reasons, a major factor is the difficulty in separating the true effect of surface charge from the contributions of other surface properties such as wettability (contact angle) and surface topography.

5.5 Surface Topography

When applied to materials, the term topography commonly refers to the roughness or smoothness of a material's surface. It has been well established that surface roughness can have beneficial effects on the response of certain cells *in vitro* and *in vivo* when compared to a smooth surface of the same material (Schwartz et al. 2008). As several biomaterials used in dental and orthopedic surgery are composed of metals, ceramics, and polymers that are neither degradable nor bioactive, surface roughness plays a key role in improving their interaction with cells and tissues *in vivo*. Nowadays, most dental implants composed of titanium or its alloy Ti6Al4V are designed with some degree of surface roughness to improve their integration with host bone, although there is some debate about the optimal nature of the surface roughness.

Surface topography of materials can be accidentally or deliberately introduced (Figure 5.18). Accidental features normally result from the fabrication process or from subsequent treatments such as abrasion, machining or grinding, and consist typically of marks, lines or groves. They are commonly random and variable over the surface, which presents difficulties in studying and interpreting their interaction with cells *in vitro* and *in vivo*. In comparison, deliberate topographical features, often composed of ordered spikes, grooves, or pores, can be created using a variety of controllable techniques, such as photolithography, electron beam lithography, or laser interference lithography. When this deliberately introduced topography has an ordered pattern, it is often referred to as surface texture. Although accidentally introduced topographical features are more common in biomaterials, much information on understanding cell response to topography has been achieved from deliberately introduced surface features.

The effect of topography on cell response has been shown to depend on the scale and geometry of the surface roughness, and on the type of cell as well. Consequently, an understanding of the interaction between surface topography and cell response is essential in creating implants with the desired surface topography. The scale of roughness that has the greatest effect on cell response is approximately equal to the cell dimension or smaller (Figure 5.19). This is because a surface will be perceived as approximately smooth by cells if the scale of roughness is larger than the cell diameter. Additionally, cells can respond differently to a "spiky" surface when compared to a smoothly undulating surface even though the peak to valley distance of the roughness is the same (Xu et al. 2017).

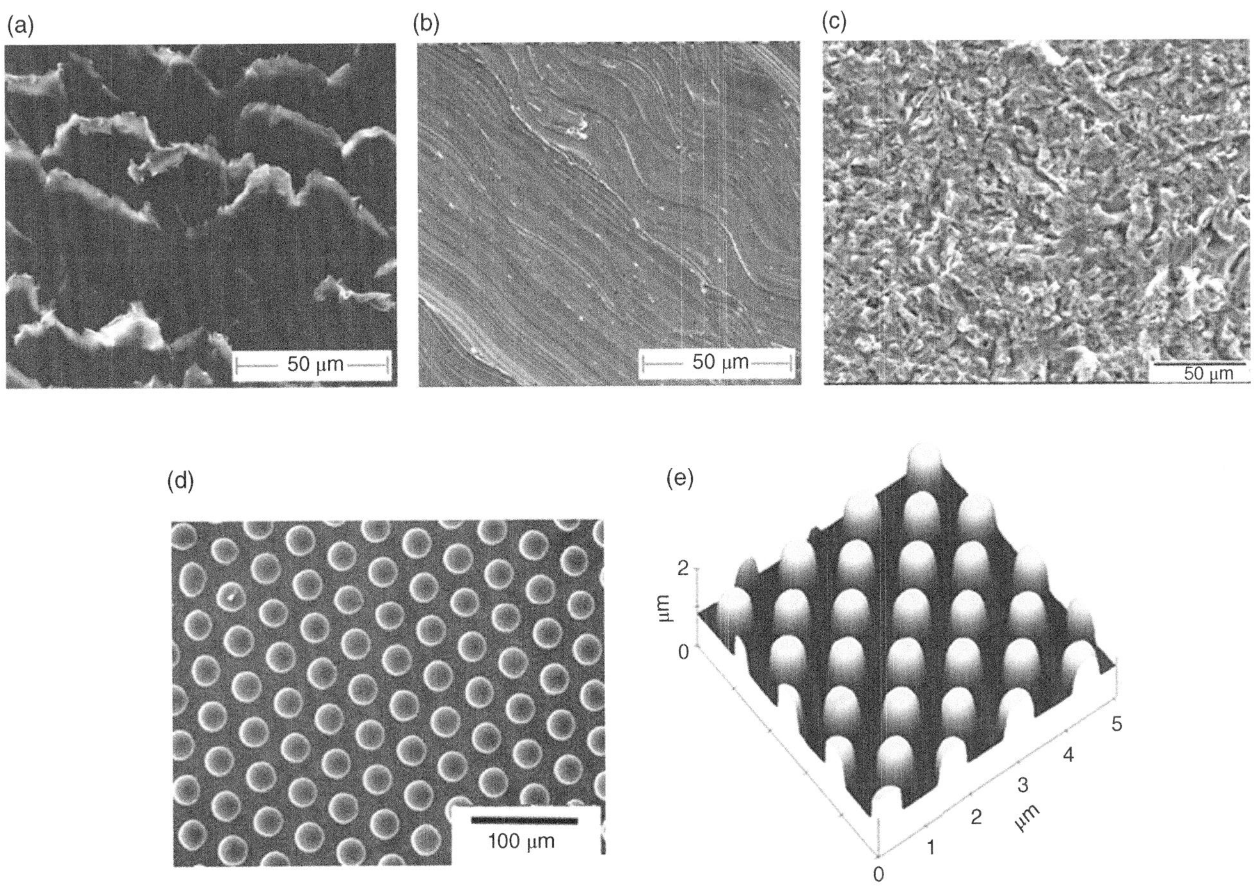

Figure 5.18 Examples of surface topography accidentally introduced (a, b) or deliberately introduced (c–e) in biomaterials. (a) Machined surface of polyether ether ketone (PEEK); (b) machined surface of Ti6Al4V; (c) sand-blasted surface of Ti6Al4V. (a–c) *Source:* From Bock et al. (2017). (d) Hemispherical depressions in titanium formed by photolithography. (e) Micro-pillars on polyurethane produced by lithography. *Source:* From Xu and Siedlecki (2012).

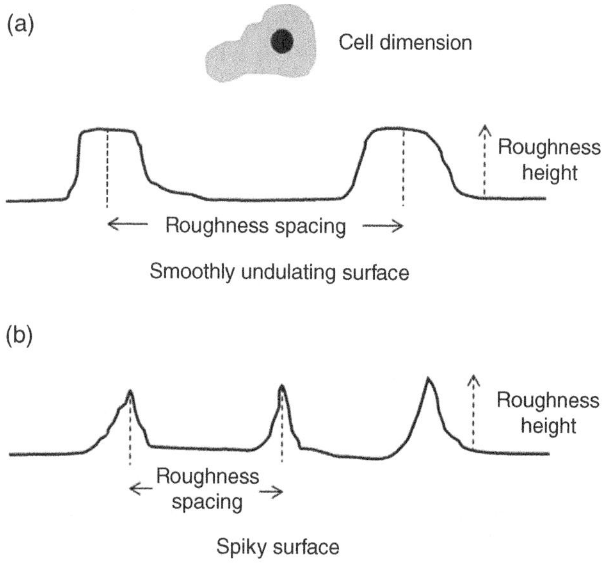

Figure 5.19 Schematic illustrating surface roughness parameters that can have a strong influence on cell response: scale of roughness spacing relative to the cell dimension and spikiness of roughness feature.

5.5.1 Surface Roughness Parameters

Surface roughness is commonly quantified in terms of the vertical height of the surface relative to a mean or reference line, defined such that the area between the roughness profile and the mean line is the same above and below the line. While a variety of parameters are used to quantitate the topography of materials, the average roughness R_a and the root mean square roughness R_q are most often used. R_a is the arithmetic mean of the average values of the vertical deviation from the mean line of the roughness profile whereas R_q is the square root of the arithmetic mean of the square of the vertical deviation from the mean line. As these two parameters are primarily concerned with the relative departure of the roughness profile in the vertical direction only, they do not provide information about the slopes, shapes, and sizes of the surface asperities or about their spacing. It is possible for surfaces of widely different profiles to have approximately the same R_a or R_q values. In general, these two parameters are more useful for characterizing surfaces of the same type that are produced by the same method.

5.5.2 Characterization of Surface Topography

Characterization of surface topography commonly starts with examining the surface of one or more representative specimens of the material in a scanning electron microscope (SEM) (Figure 5.18). Qualitative information about topographical features such as surface roughness at a microscale or nanoscale, the waviness and spacing of the surface roughness, and the presence of surface flaws can be obtained using this technique. Thereafter, other techniques, classified into two broad categories, depending on whether a component of the measuring instrument makes contact with the specimen surface (contact type) or not (noncontact type), can be used to obtain topographical data such as surface roughness parameters. As a contact-type technique has the potential for damaging a soft surface, applied loads on the contacting component (for example, a stylus) should not exceed the hardness of the surface. The major techniques used for biomaterials are profilometry and atomic force microscopy (AFM), both of which can be used in a contact or noncontact mode.

Scanning Electron Microscopy (SEM)

SEM is a widely used and versatile technique for characterizing the microstructure and topography of materials based on its simplicity, type of information obtained and its scale of resolution (down to a few nanometers). It relies on scanning a finely focused electron beam over the surface of a specimen and analyzing the interaction of the beam with the material. In general, an incident electron beam can produce a variety of emissions from a material, a few of which are used in the SEM (Figure 5.20). Depending on the thickness of the specimen and the energy of the incident electron beam, a certain fraction of the incident electrons will be scattered in the forward direction, used in transmission electron microscopy (TEM), a fraction will be absorbed and the remaining fraction will be scattered in the backward direction. As the capacity of atoms in the specimen to scatter the incident electrons is a function of their atomic number, these backscattered electrons are used in the SEM

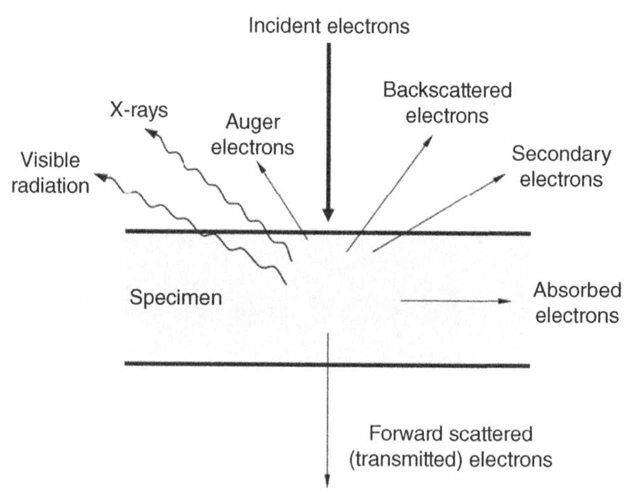

Figure 5.20 Emissions produced by the interaction of an electron beam with a solid specimen.

to provide compositional contrast between different (solid) phases at the surface.

The incident electron beam can also generate secondary effects. One type of effect is that the incident electrons can knock electrons out of their orbits around the atom. These electrons may have enough energy to escape from the specimen and become what are called secondary electrons if they are near the surface (within ~20–100 nm). These secondary electrons are used to produce an image of the microstructure (Figure 3.23) and topography (Figure 5.18) of the material. Another type of secondary effect occurs when an electron undergoes a transition from one energy level to fill a vacant site in a lower energy level of an excited atom, generating radiation in the form of X-rays or light. As the X-rays from each element will have a different energy (or wavelength), we can detect which element emitted them by measuring their energy in a spectrometer. This is the basis of energy-dispersive X-ray (EDS) analysis in the SEM.

EDS analysis provides rapid qualitative and quantitative information about the elemental composition of the specimen volume that is being analyzed by the electron beam, which, in the SEM is commonly an area of approximately 1 μm by 1 μm and a depth of ~1 μm. For qualitative analysis, the presence of an element is determined from a comparison of the energy values of the peaks in the spectrum with the energy values of known elements (Figure 5.21). Quantitative analysis, such as the percentage of each element in the specimen, is determined from the relative intensity of the peaks present in the spectrum. Semiquantitative analysis without standards, with an accuracy of approximately ±5%, can be quickly performed using software, but much greater accuracy is achieved using calibration standards of known compositions that contain the elements present in the specimen.

In common with most electron-beam techniques, conventional SEMs require the use of a high vacuum environment that is shared by the electron beam column and the specimen chamber. This necessitates the use of clean, dry specimens that often have a different surface chemistry from the actual biomaterials implanted *in vivo*. Electrically insulating materials such as ceramics and polymers must also be sputter-coated with a thin layer of a conducting material, typically a gold–palladium alloy or carbon, to reduce static charging at the specimen surface. In view of these limitations, the environmental scanning electron microscope (ESEM) has been developed which allows the examination of specimens under a variety of conditions more relevant to the use of biomaterials. The ESEM uses pressure-limiting apertures that separate the electron beam column from the specimen chamber, allowing the use of a variety of humidity levels, pressure, temperature, and ambient gas or liquid in the specimen chamber. Consequently, specimens that are electrically insulating or conducting, covered with adsorbed water molecules or impurities such as hydrocarbons, or prone to gaseous emission can be examined without cleaning or sputter-coating with a conducting layer.

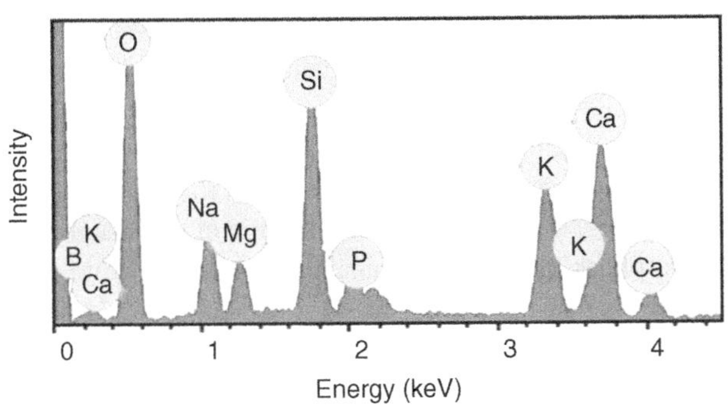

Figure 5.21 EDS spectrum of a borosilicate glass examined in the SEM, showing the elements (in addition to B) present in a surface layer of thickness ~1 μm.

Atomic Force Microscopy (AFM)

AFMs operate by measuring the tiny forces (less than ~1 nN) between atoms of a probe in the shape of a sharp tip and those of the specimen surface (Figure 5.22). These tiny forces cause a very small flexible cantilever to bend, which is used to sense the proximity of the tip to the surface and, thus, the topographical features of the surface. A variety of tip materials and shapes are available but commonly used tips include silicon nitride and single-crystal silicon in the shape of a pyramid several micrometers high and an end radius of ~20 nm or larger. To acquire a three-dimensional image, the tip mounted on a flexible cantilever is scanned with high precision over the surface and the relative lateral and vertical deflection of the tip due to its interaction with the surface is measured using an optical lever system. In some instruments, the tip is stationary and the specimen is moveable. The resolution of the topographical features depends on the sharpness of the tip and sensitivity of the cantilever system but modern AFMs can have a lateral resolution of less than 20–30 nm and a vertical resolution down to less than 0.1 nm. This makes AFM ideally suited for characterizing nanoscale topography, although it is also used for characterizing microscale features.

AFM imaging of a surface can be performed in three different scanning modes referred to as contact, noncontact, and tapping modes. In contact mode imaging, a tiny constant force is applied to the cantilever and the tip is brought into contact with the surface. Repulsive forces between the atoms of the tip and the surface (Figure 5.23), similar to those between atoms considered in Chapter 2, produce a

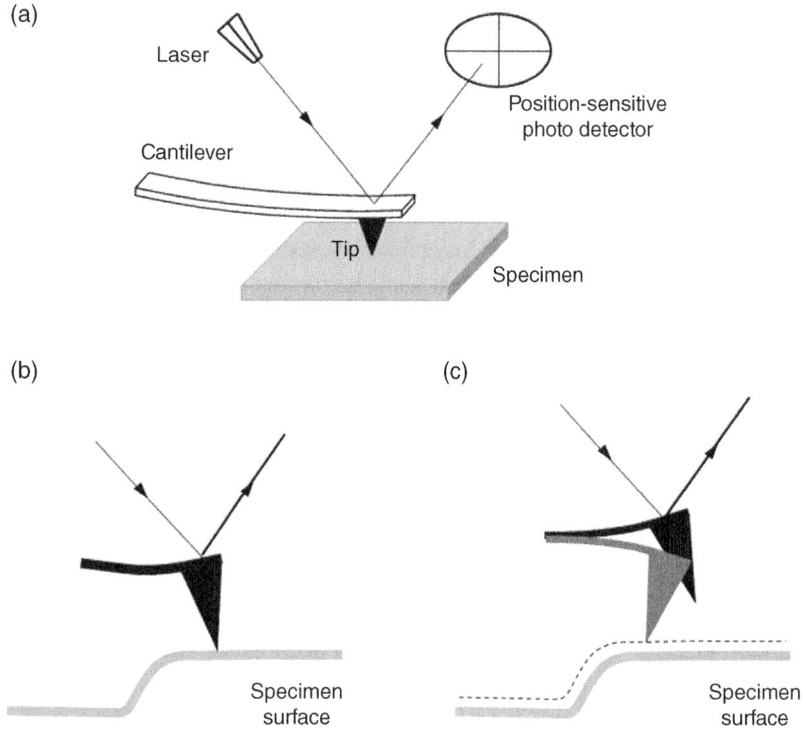

Figure 5.22 Schematic illustrating (a) the main components of the AFM technique, (b) imaging in the contact mode, and (c) imaging in the noncontact mode.

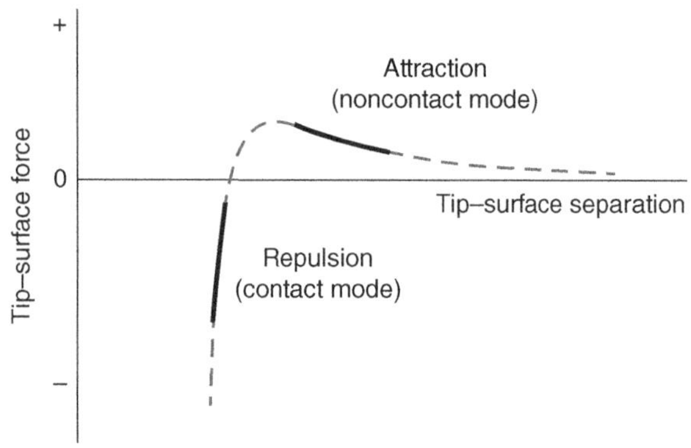

Figure 5.23 Schematic curve of force versus separation between the tip and specimen surface, showing the repulsive and attractive regions corresponding to contact mode and noncontact mode imaging.

deflection of the cantilever. This deflection is used in a feedback circuit to move the scanner up or down in the vertical direction in response to the topography by keeping the cantilever deflection constant. Contact mode imaging can damage the surface of soft samples and, consequently, it is commonly used for metals, ceramics, and hard polymers. In noncontact mode imaging, the cantilever is vibrated at its resonant frequency and at a constant amplitude using a piezoelectric device, and the tip is brought to within a few nanometers of the surface but not in contact with the surface. Weak van der Waals forces of attraction produce a deflection of the cantilever that is used to form an image. As the tip does not come into contact with the specimen surface, noncontact AFM is very suitable for soft polymers. Tapping mode imaging can be considered to be somewhat between contact and noncontact mode imaging. The cantilever is vibrated at its resonant frequency but with an amplitude lower than that in contact mode, and the tip slightly taps the surface of the sample. Tapping mode imaging gives a higher resolution than noncontact mode and can be used for a wide range of polymers.

Profilometry

Noncontact optical profilometry, based on the principle of interference between two light beams, is a widely used technique for characterizing the topography of materials. The reflected beams from two parallel plates placed normal to an incident beam interfere and result in the formation of fringes (lines) whose spacing is a function of the spacing of the two plates. If one of the plates is a reference plate and the other is a specimen surface whose roughness is to be measured, the fringe spacing can be related to the surface roughness (Figure 5.24). In white light interferometry, the incident beam is composed of all wavelengths in the visible spectrum, giving interference fringes of various colors in the spectrum. Profilometers based on white light interferometry

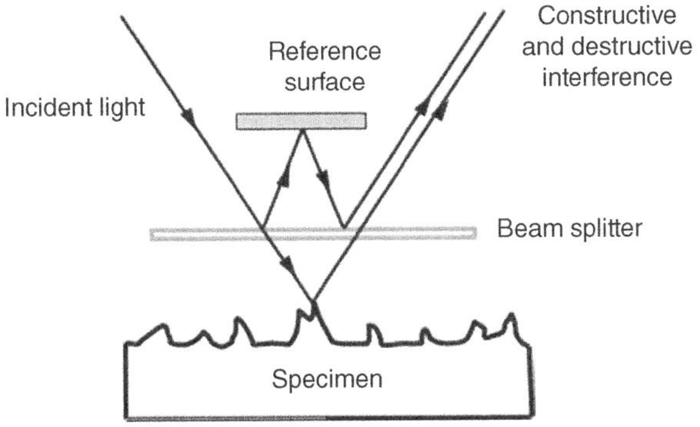

Figure 5.24 Schematic illustrating the principle of optical interferometry.

have a lateral resolution of ~0.5–1.0 μm and a vertical resolution of ~1 nm. Consequently, this technique has a lower resolution than AFM.

Figure 5.25 shows a comparison of SEM, contact mode AFM and white light interferometry images of the same nominal material, Si_3N_4 in its as-fabricated condition, a biomaterial used in spine repair. SEM shows a surface composed of elongated hexagonal grains, 0.2–2 μm in cross section, that protrude to various lengths, less than ~5 μm, in a random manner from the surface. Consequently, the roughness varies in scale from approximately a fraction of a micrometer to a few micrometers. Roughness parameters R_a and R_q obtained from AFM are 0.34 and 0.43 μm, respectively, values consistent with the SEM image. In comparison, R_a and R_q values obtained from white light interferometry are 0.64 and 0.83 μm, respectively. Whereas these values are almost twice those found from AFM, they are also consistent with the SEM image. One factor that could contribute to the difference between the AFM and interferometry values is that although the Si_3N_4 was fabricated using the same procedure, different regions of the material were imaged in the two techniques.

5.5.3 Effect of Surface Topography on Cell and Tissue Response

Although the mechanism of interaction is complex, beneficial effects of nanoscale to microscale surface topography on the response of cells *in vitro* and *in vivo* have been reported for several biomaterials, as discussed in Chapter 21. These effects have been well demonstrated for Ti and its alloy Ti6Al4V. Titanium and Ti6Al4V see considerable use as implants in dental and orthopedic surgery, and their microtopography can be varied using common surface treatments such as machining, grit blasting, and acid etching. When compared to smoother machined surfaces ($R_a \approx 0.2$ μm), microrough grit-blasted surfaces of Ti6Al4V ($R_a \approx 3$ μm) showed a better ability to stimulate the differentiation of osteoblastic cells *in vitro*

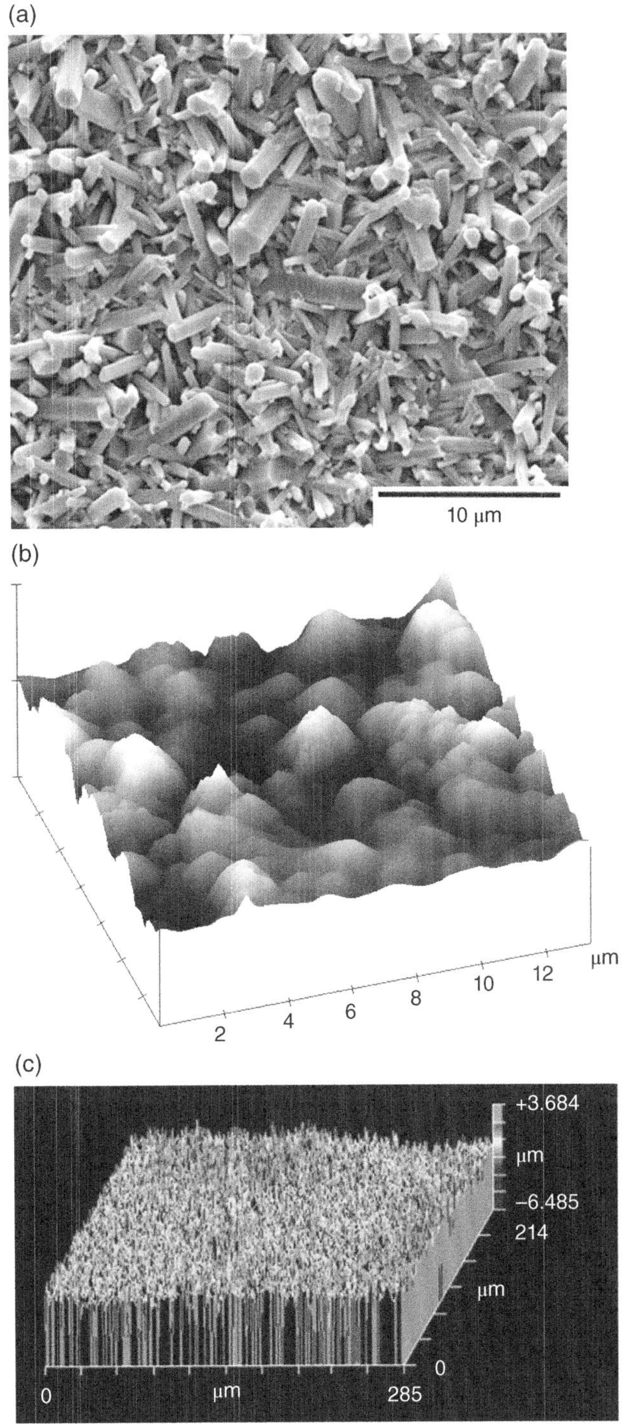

Figure 5.25 Topography of as-fabricated silicon nitride obtained by (a) SEM, (b) AFM, and (c) optical profilometry. *Source:* From Bock et al. (2017) / with permission of John Wiley & Sons.

(Schwartz et al. 2008). Additionally, when implanted in the vertebrae of sheep *in vivo*, implants with the microrough surface showed better bone formation around the implants and integration with host bone when compared to implants with the smoother machined surface.

The addition of microtopography, as noted earlier, is now a key design feature to improve outcomes associated with nondegradable biomaterials used as implants in medical and dental surgery. Adding nanoscale surface features to smooth (machined) surfaces using surface modification techniques has been reported to enhance the beneficial effects observed for microrough surfaces. For example, the formation of vertical TiO_2 nanotubes (outer and inner diameter ~100 and ~80 nm, respectively, and height ~250 nm) on commercial purity Ti surfaces, has been shown to improve bone formation around the implants and integration with host bone in rabbit tibiae *in vivo* when compared to machined titanium surfaces (Bjursten et al. 2010).

For the same material, different treatments such as grinding, polishing, and grit blasting can lead to changes in other surface characteristics, such as surface chemistry, surface energy, contact angle and wettability, and not just in the surface topography. Consequently, the effects of surface topography alone often cannot be separated statistically from the contributions of these other surface properties and characteristics. This creates an additional level of complexity in understanding the mechanism by which surface topography influences the response of cells and in designing the optimal surface topography for a specific application.

5.6 Concluding Remarks

Surface properties and characteristics, together with bulk properties, have a strong influence on the performance of biomaterials *in vivo*. In this chapter, we discussed important surface properties relevant to biomaterials, including surface chemistry, surface energy, surface topography, and surface charge.

- Surface chemistry is the most important surface property because it influences the behavior of a biomaterial in any given environment. We discussed a variety of techniques for characterizing surface chemistry of a material, but it should be noted that because most of these techniques require the use of an ultrahigh vacuum, the measured surface chemistry is often different from that of biomaterials implanted *in vivo*.
- Surface energy is difficult to measure for many materials and, consequently, it is often considered more simply in terms of the extent to which a liquid will wet and spread over a surface. The contact angle between a liquid (water) droplet and a solid surface is often taken as a measure of the wettability of the system. Overall, hydrophilic materials (low contact angle) show good wetting and spreading by water whereas hydrophobic materials (high contact angle) show poor wetting by water.
- When placed in an aqueous environment, such as the physiological fluid, biomaterials develop a surface charge that can influence subsequent adsorption of ions and molecules from the medium.
- Certain surface topographical features, such as certain roughness features, have been shown to have beneficial effects on the response of certain cells *in vitro* and *in vivo*, a topic that is discussed further in Chapter 21.

Problems

5.1 A dental implant composed of titanium (Ti), approximated as a cylinder of diameter 5.0 mm and length 8.0 mm, has a surface layer of thickness 3 nm composed of titanium dioxide (TiO_2). Determine the fractional volume of the TiO_2 surface layer, that is, the volume of the TiO_2 layer relative to the total volume of the implant.

5.2 Distinguish between the terms surface tension and surface energy.

5.3 Determine the contact angle of a droplet of deionized water on ultrahigh molecular weight polyethylene (UHMWPE), given that the densities of water and UHMWPE are 1.0 and 0.94 g/cm^3, respectively, the surface tension of water is 73.0 mN/m, and the surface energy of UHMWPE is 36 mJ/m^2.

5.4 The contact angle of a droplet of deionized water on a microrough surface of silicon nitride was observed to decrease from $66° \pm 9°$ to $30° \pm 9°$ in 30 minutes. Suggest an explanation for the decrease in contact angle.

5.5 Explain the significance of the critical surface tension in relation to biomaterials.

5.6 In an X-ray photoelectron spectrum of titanium, is the adventitious C 1s peak expected to be at a higher or lower binding energy than the O 1s peak? Is the Ti 2s peak expected to be at a higher or lower binding energy than the Ti 2p peak? Explain.

5.7 The composition of the oxide surface layer on silicon nitride (Si_3N_4) can change with depth. Briefly describe a method for studying the change in surface composition with depth and, in general terms, the expected change in surface composition.

5.8 Explain how the following biomaterials develop a surface charge when implanted in the physiological environment: (a) titanium, (b) silicon nitride, and (c) PEEK.

5.9 Briefly explain the type of surface topographical features that are expected to influence the response of cells.

References

Bjursten, L.M., Rasmusson, L., Oh, S. et al. (2010). Titanium dioxide nanotubes enhance bone bonding *in vivo*. *Journal of Biomedical Materials Research. Part A* 92: 1218–1224.

Bock, R.M., Jones, E.N., Ray, D.A. et al. (2017). Bacteriostatic behavior of surface modulated silicon nitride in comparison to polyetheretherketone and titanium. *Journal of Biomedical Materials Research. Part A* 105: 1521–1534.

Chen, Q., Zhang, D., Somorjai, G., and Bertozzi, C.R. (1999). Probing the surface structural arrangement of hydrogels by sum-frequency generation spectroscopy. *Journal of the American Chemical Society* 121: 446–447.

Fox, H.W. and Zisman, W.A. (1950). The spreading of liquids on low-energy surfaces. I. Polytetrafluoroethylene. *Journal of Colloid Science* 5: 514–531.

Fox, H.W. and Zisman, W.A. (1952). The spreading of liquids on low-energy surfaces. III. Hydrocarbon surfaces. *Journal of Colloid Science* 7: 428–442.

Girifalco, L.A. and Good, R.J. (1957). A theory for the estimation of surface and interfacial energies. I. Derivation and application to interfacial tension. *The Journal of Physical Chemistry* 9: 904–909.

Lausmaa, J. (1996). Surface spectroscopic characterization of titanium implant materials. *Journal of Electron Spectroscopy and Related Phenomena* 81: 343–361.

Marmur, A. (2003). Wetting on hydrophobic rough surfaces: to be heterogeneous or not to be. *Langmuir* 19: 8343–8348.

Ratner, B.D. (2013). Surface properties and surface characterization of biomaterials. In: *Biomaterials Science: An Introduction to Materials in Medicine*, 3e (eds. B.D. Ratner, A.S. Hoffman, F.J. Schoen and J.E. Lemons), 34–55. New York: Elsevier.

Schwartz, Z., Raz, P., Zhao, G. et al. (2008). Effect of micrometer-scale roughness of the surface of Ti6Al4V pedicle screws *in vitro* and *in vivo*. *The Journal of Bone and Joint Surgery. American Volume* 90: 2485–2498.

Xu, L.C. and Siedlecki, C.A. (2012). Submicron-textured biomaterial surface reduces staphylococcal bacterial adhesion and biofilm formation. *Acta Biomaterialia* 8: 72–81.

Xu, L.C., Wo, Y., Myerhoff, M.E., and Siedlecki, C.A. (2017). Inhibition of bacterial adhesion and biofilm formation by dual functional textured and nitric oxide releasing surfaces. *Acta Biomaterialia* 51: 53–65.

Further Readings

Gittens, R.A., Olivares-Navarrete, R., Schwartz, Z., and Boyan, B.D. (2014). Implant osseointegration and the role of microroughness and nanostructures: lessons for spine implants. *Acta Biomaterialia* 10: 3363–3371.

Schwartz, Z., Lohmann, C.H., Oefinger, J. et al. (1999). Implant surface characteristics modulate differentiation behavior of cells in the osteoblastic lineage. *Advances in Dental Research* 13: 38–48.

Part III

Classes of Materials Used as Biomaterials

6

Metallic Biomaterials

6.1 Introduction

Metals are commonly used as biomaterials when long-term mechanical durability must be guaranteed and when high electrical conductivity is a requirement. Pure metals have low mechanical strength and, except for the noble metals, they corrode in the physiological environment, releasing metal ions into the environment. Some metal ions are more toxic than others but above some threshold concentration, all metal ions are essentially toxic *in vivo*. Consequently, metals used as biomaterials have traditionally been selected from those that are passivated against corrosion by a surface oxide layer that is dense and strongly adherent to the underlying metal.

The requisite combination of durable mechanical properties and corrosion resistance limit the number of metals that can be used as implants *in vivo*. Typically, the majority of these metals were originally designed for use in engineering applications and subsequently introduced into biomedical applications. Table 6.1 provides a list of the majority of metals that are used as biomaterials and their areas of application. Of these, titanium, a titanium alloy referred to as Ti6Al4V, stainless steels, and cobalt–chromium alloys are the most commonly used.

Emerging developments in the use of metals as biomaterials include nanoparticles and degradable metals. Nanoparticles composed of gold, silver, or platinum are being designed and developed as devices for targeted drug delivery in treating cancer cells (Chapter 26) and diagnostic imaging of cells and tissues. Significant advances have been made in designing degradable metals for use as orthopedic and cardiovascular devices. Some such devices, composed typically of magnesium alloys, have already reached the stage of testing in human clinical trials. However, despite these advances, significant challenges remain in designing compositions to fulfill the demanding combination of properties, such as degradation rate, mechanical properties, and biocompatibility, required for these and other biomedical applications.

The properties of a material, as we have seen in previous chapters, are controlled by its composition and its structure at different length scales, such as crystal structure and microstructure. In metals, compositional and structural changes can be achieved mainly by:

- Crystallographic (polymorphic) transformations in metals that have more than one crystal structure
- Addition of alloying elements to the metals
- Thermal, mechanical, or thermomechanical treatments.

Consequently, an understanding of the principles and mechanisms of these processes is important in designing metals with an optimal combination of properties for the desired application. In addition, as many metals (alloys) are composed of two or more phases, a basic understanding of phase diagrams and the information that they provide is also useful.

This chapter is devoted to a discussion of the following topics:

- Principles and mechanism of structural changes in metals resulting from polymorphic transformations, addition of alloying elements, and thermal, mechanical, or thermomechanical treatments
- Basic understanding of phase diagrams and the information they provide
- Composition, structure, and properties of metals used as biomaterials, emphasizing the relationships between these characteristics.

6.2 Crystal Structure of Metals

At the smallest scale of controllable structure, atoms of pure metals pack together as tightly as possible to give simple crystal structures at room temperature, such as the face-centered cubic (FCC), close-packed hexagonal (CPH) or body-centered cubic (BCC) structure (Chapter 3). In general, the structure that a metal adopts at a given temperature

Materials for Biomedical Engineering: Fundamentals and Applications, First Edition. Mohamed N. Rahaman and Roger F. Brown.

Companion website: www.wiley.com/go/rahaman/biomedicalengineering

Table 6.1 Biomedical applications of metals.

Application	Metals
Orthopedic	
Fracture fixation: plates; screws; intramedullary nails	Ti; Ti6Al4V; stainless steel
Total joint replacement (hip and knee)	Ti; Ti6Al4V
Femoral stem and shell (hip); tibial insert (knee)	
Femoral head (hip); femoral bearing (knee)	Co–Cr alloy; stainless steel; oxidized Zr
Bone repair	Ti; Ti6Al4V; porous Ta
Spinal fusion	Ti; Ti6Al4V
Cardiovascular	
Stents	Co–Cr alloy; stainless steel; nickel–titanium (Nitinol)
Pacemakers	Ti6Al4V
Heart valves	Ti6Al4V
Electrodes	Platinum; silver
Neurological	
Electrodes in stimulators	Platinum; silver
Dental	
Implants	Ti; Ti6Al4V
Orthodontic: wires; files; reamers	Nickel–titanium (Nitinol)
Restorations (bridges; crowns)	Co–Cr alloy

is the one that has the minimum energy. When cooled extremely rapidly, metals can form an amorphous structure but this is not very relevant to biomaterials.

Depending on the temperature, some metals have more than one crystal structure, a property referred to as polymorphism. The most important examples of polymorphism relevant to metals used in biomedical applications and engineering applications as well are iron (Fe), titanium (Ti), and cobalt (Co) (Figure 6.1). Upon cooling, Ti changes from BCC, referred to as the β phase, to CPH, the α phase, at 882 °C, whereas Fe changes from BCC, the δ phase, to FCC, the γ phase (called austenite) at 1391 °C and back to BCC, the α phase (called ferrite) at 914 °C. Co changes from the α phase (FCC) to the ε phase (CPH) at 422 °C.

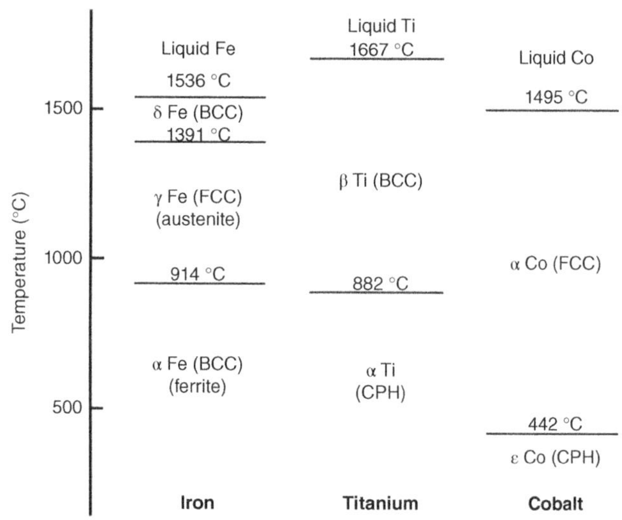

Figure 6.1 Polymorphic transformation in iron, titanium, and cobalt.

6.3 Polymorphic Transformation

The capacity of a metal to undergo a polymorphic transformation has important consequences for its structure and, thus, its properties. Polymorphism can be brought about not only by temperature changes but also by alloying, that is, the addition of other elements to the metal (Section 6.4).

Polymorphic transformations, such as those shown in Figure 6.1, are not accomplished instantaneously but, instead, they are completed over a certain temperature range as the material is cooled or heated at an appropriate rate. Upon cooling, the FCC to BCC transformation in pure Fe, for example, starts ~914 °C, reaches a maximum rate at ~700 °C and is essentially completed at ~500 °C. This trend is similar in nature to the phase transformation of water to ice. As the temperature decreases below 0 °C, a large number of fine ice crystals form throughout the water and, thereafter, solidification occurs at a rapid rate. Phase transformation in solids, such as the FCC to BCC transformation in Fe, occurs by a process of nucleation and growth, in much the same way as the solidification of ice crystals from water or the condensation of liquid droplets from a vapor. Tiny particles called nuclei form and those nuclei that reach a critical size grow to form the new phase.

6.3.1 Formation of Nuclei of Critical Size

Because it is easier to illustrate, we will discuss the formation of critical size nuclei in the solidification of a solid from a liquid, such as ice from water. However, similar principles apply for phase transformations in solids, such as the γ to α transformation in iron. We can imagine that upon cooling water to its freezing temperature, molecules will tend to arrange themselves to form ice crystals. When they do, they liberate an amount of energy proportional to their volume and to the Gibbs free energy change ΔG_n in going from the liquid to the solid state (a negative quantity below the freezing temperature). On the other hand, in forming each tiny ice particle, called a nucleus, we have created new solid–liquid interfacial area that requires energy to form. Consequently, the total Gibbs free energy change ΔG for the formation of an ice nucleus of radius a can be written as

$$\Delta G = 4\pi a^2 \gamma_i - (4/3)\pi a^3 \Delta G_n \tag{6.1}$$

where, γ_i is the energy of the solid–liquid interface. The two terms on the right hand side of Eq. (6.1) are sketched individually in Figure 6.2, along with ΔG for the system. ΔG reaches a maximum at some critical radius a^* found by putting $d(\Delta G)/da = 0$, giving

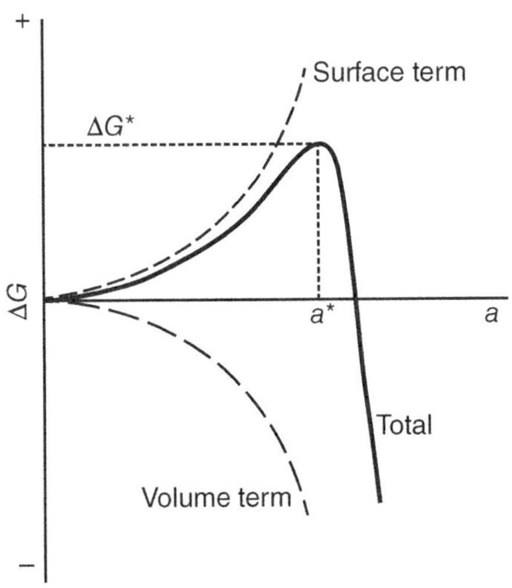

Figure 6.2 Free energy as a function of radius of nucleus. ΔG^* is the free energy required to create a critical nucleus of radius a^*.

$$a^* = \frac{2\gamma_i}{\Delta G_n} \tag{6.2}$$

Substituting this value of a^* for a in Eq. (6.1), the free energy change for the critical nucleus ΔG^* is

$$\Delta G^* = \frac{16\pi\gamma_i^3}{3\Delta G_n^2} \tag{6.3}$$

The Gibbs free energy of the system will start to decrease when a critical nucleus of radius a^* is exceeded. This represents the condition for the formation of stable nuclei. These nuclei can grow by addition of molecules, allowing the solidification process to continue.

6.3.2 Rate of Phase Transformation

The rate of transformation from one phase to another, such as the water to ice transformation or the γ to α transformation in Fe, depends on two factors:

- The thermodynamic driving force for the phase transformation, equal to the Gibbs free energy change ΔG_n for the transformation
- Kinetic factors, such as the speed at which each critical nucleus grows by the addition of more atoms (or molecules)

Using the definition of the Gibbs free energy, ΔG_n is given by

$$\Delta G_n = \Delta H - T\Delta S \tag{6.4}$$

where, ΔH and ΔS are respectively the change in enthalpy and entropy for the transformation and T is the temperature. As ΔH and ΔS are only weakly dependent on temperature, the variation of ΔG_n with degree of undercooling is due to the temperature T in the last term. Thus, if T_e is the equilibrium temperature for the transformation (0 °C for water or 914 °C for the γ to α transformation in Fe) and T is the actual temperature, because $\Delta G_n = 0$ at $T = T_e$, we get

$$\Delta G_n = \frac{\Delta H}{T_e}(T_e - T) \tag{6.5}$$

We can use Eqs. (6.3) and (6.5) to get an approximation of how the nucleation rate depends on the degree of undercooling, $(T_e - T)/T_e$. According to statistical thermodynamics, the number n of atoms (n assumed small) of energy ΔG in a system composed of a total number n_0 atoms is

$$n = n_0 \exp\left(\frac{-\Delta G}{kT}\right) \tag{6.6}$$

where, k is the Boltzmann constant and T is the temperature (in kelvin). We will assume that the rate ν at which stable nuclei are formed is proportional to the number n of critical nuclei of size a^* and to a rate constant K that describes how fast the critical nuclei grow by addition of atoms, that is

$$\nu \sim nK \tag{6.7}$$

Substituting for n from Eq. (6.6), where $\Delta G = \Delta G^*$ given by Eq. (6.3), we get

$$\nu \sim K \exp\left[\frac{-16\pi\gamma_i^3 T_e^2}{3kT(\Delta H)^2 (T_e - T)^2}\right] \tag{6.8}$$

The form of the rate constant K in Eq. (6.8) depends on the mechanism by which atoms move to cause structural changes.

In solids, structural changes can occur by two types of mechanism, referred to as diffusive and displacive. Displacive transformations are also referred to as martensitic, shear or diffusionless transformations. Most structural changes occur by a diffusive mechanism but the displacive mechanism is important in some materials, such as carbon steels widely used in engineering applications and nickel-titanium alloys used in orthopedic and dental applications and in engineering applications as well. Although pure Fe is not a useful material for biomedical and engineering applications, it is a prototype material for understanding diffusive and displacive transformation. Once diffusive and displacive transformations are understood in Fe, the mechanisms can be applied to other materials.

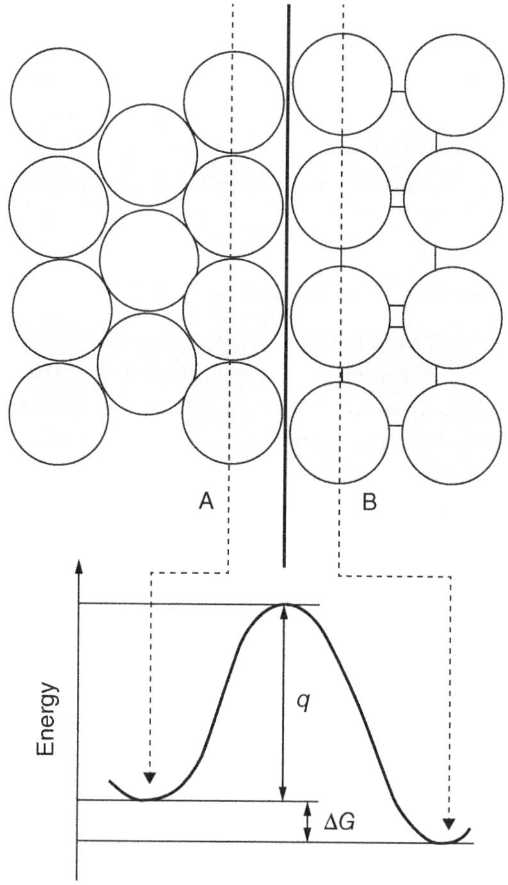

Figure 6.3 Activation energy q required for an atom to diffuse from A to B.

6.3.3 Diffusive Transformations

In a diffusive transformation, atoms move though the material, from one atomic site to another, by the process of diffusion. If Fe is cooled at a sufficiently slow rate to temperatures below 914 °C, atoms will diffuse across the interface from the γ-phase and join up with the growing α-phase grains formed from the critical size nuclei (Figure 6.3). In order to do so, atoms have to surmount an energy barrier q, referred to as the activation energy for diffusion, because they have to free themselves from the attractive forces of surrounding atoms in γ-phase. The rate at which the atoms diffuse across the γ to α interface in iron, that is, the rate constant K in Eq. (6.8), is given by

$$K = K_0 \exp\left(\frac{-q}{kT}\right) \tag{6.9}$$

where, k is the Boltzmann constant, T is the absolute temperature and K_0 is a constant for a given system that depends on the dimension of the atoms. This exponential type of dependence for the rate constant in Eq. (6.9) is characteristic of rate equations for physical and chemical processes and is often referred to as the Arrhenius equation.

Substituting for K in Eq. (6.8), we get that the rate of transformation is

$$\nu \sim K_0 \exp\left(\frac{-q}{kT}\right) \exp\left[\frac{-16\pi\gamma_i^3 T_e^2}{3kT(\Delta H)^2 (T_e - T)^2}\right] \tag{6.10}$$

We can now see how the rate of a diffusive transformation, such as the γ to α transformation in Fe, will change with temperature (Figure 6.4). On one hand, the nucleation rate would increase rapidly with undercooling below the equilibrium temperature, given by the $(T_e - T)$ term in the second exponential of Eq. (6.10). On the other hand, the diffusion step, that is, growth of the newly forming α phase, will become increasingly difficult and, eventually, will become nearly impossible (first exponential term in Eq. (6.10)) because the atoms do not have sufficient energy to overcome the activation energy barrier. At some temperature, ~700 °C for the γ to α transformation in Fe, the transformation rate is a maximum.

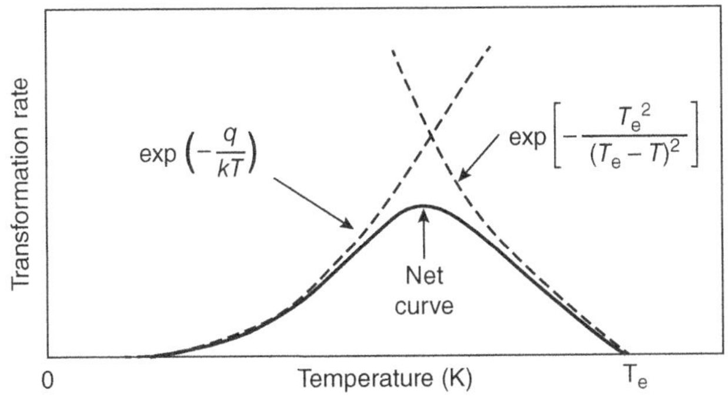

Figure 6.4 Illustration of the variation of the transformation rate with temperature for a diffusive transformation.

6.3.4 Displacive Transformations

If iron is cooled rapidly (quenched) from above 914 °C to room temperature, at a rate of several thousand degrees Celsius per second, there is a huge driving force for the γ to α phase transformation due the very large undercooling, T_e–T (Eq. (6.5)). The transformation is hindered kinetically because the atoms do not have sufficient time to diffuse and cause the nuclei to grow. Thus, we might expect the γ phase to survive almost indefinitely in some type of unstable or metastable state. However, this does not happen. As the temperature gets below ~550 °C, the driving force becomes so large that it can no longer be contained, and the FCC (γ) phase starts to transform by a displacive mechanism to a BCC phase called martensite.

Unlike diffusive transformations, displacive transformations occur almost immediately. Martensite grains nucleate at the grain boundaries of the γ phase and grow at speeds approaching the speed of sound in Fe to form plate-like or needle-like grains until they encounter the next grain boundary (Figure 6.5). Atomic bonds at the interface between the γ phase and the martensite phase, break and reform in such a way that the structure switches rapidly from the FCC structure of austenite to the BCC structure of martensite. However, the transformation uses up some of the driving force for the FCC to BCC transformation as strain energy. Consequently, fewer martensite grains will nucleate and grow, and the transformation will eventually stop, leaving residual unstable austenite if the temperature is held constant. Lowering the temperature increases the driving force due to greater undercooling, resulting in a larger amount of martensite. Typically, martensite starts to form at ~550 °C, called the martensite start temperature M_s, and it is not completed, giving 100% martensite, until the temperature reaches ~350 °C, called the martensite finish temperature M_f. The amount of martensite formed depends on the temperature only and not on the time.

In pure Fe, martensite has a BCC crystal structure that is identical to that of ferrite. However, because it is composed of finer plate-like grains, martensite has a yield strength and hardness that are approximately twice those for ferrite composed of coarser equiaxial grains. This is because finer grains provide a larger number of obstacles to dislocation motion (Section 4.3.2)

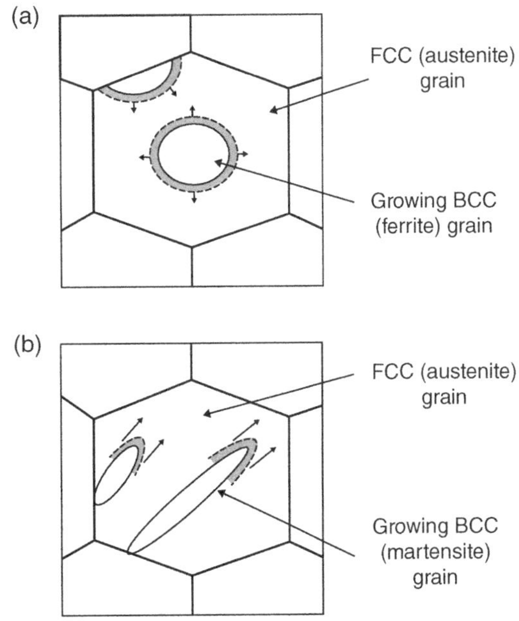

Figure 6.5 (a) Diffusive transformation and (b) displacive transformation in iron. In the diffusive transformation, ferrite (BCC) grains grow by diffusion of atoms to the interface with the austenite (FCC) grains, resulting in roughly equiaxial ferrite grains. In the displacive transformation, lens-like grains nucleate at the austenite grain boundaries and grow rapidly to give plate-like martensite (BCC) grains.

6.3.5 Time-Temperature-Transformation (TTT) Diagrams

It is common to use a chart, called a time-temperature-transformation (TTT) diagram to show the rates of phase transformations in a material succinctly, such as the transformations in Fe and its alloys, or the formation of crystals upon cooling a glass that is capable of crystallizing. Figure 6.6 shows the TTT diagram for the diffusive and displacive transformations in Fe. We see that the curves for the diffusive transformation have the shape approximating the letter C, one giving the time for the formation of 1% α (BCC) phase from γ (FCC) iron and the other for 99% formation of α phase. We can say that the 1% curve approximates the time required to form a detectable amount of the BCC phase whereas the 99% curve approximates the time required for completion of the transformation. Some TTT diagrams include curves for the formation of intermediate amounts of the BCC phase but they are omitted here.

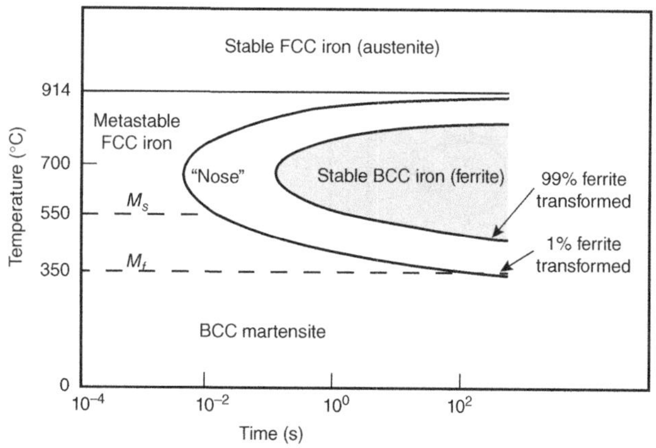

Figure 6.6 Time-temperature-transformation diagram for the FCC to BCC transformation in iron.

It is noticed that the shape of each C-curve reflects the shape of the overall curve for the transformation rate shown in Figure 6.4. As the transformation rate is essentially zero, the time for formation of 1% BCC iron is infinite above ~910 °C and below ~500 °C. The "nose" of the C-curve represents the fastest transformation rate and, thus, the time taken is a minimum. When compared to the 1% curve, the 99% curve is shifted to the right because it takes longer for more atoms to diffuse and form a larger amount of the BCC phase. Overall, then, the amount of (stable) BCC ferrite formed in the diffusive transformation is a function of time and temperature. In comparison, the amount of BCC martensite formed in the displacive transformation is a function of temperature only, as discussed in the previous section.

6.4 Alloys

The majority of metals are not used in their pure state but, instead, other elements are added to them in various amounts, forming alloys having an improved combination of properties (Chapter 3). Atoms of the alloying elements occupy sites randomly in the metal structure, forming substitutional or interstitial defects, and the alloying element is said to form a solid solution with the metal. The amount of one metal that can be alloyed with another metal to form a solid solution can vary from less than a fraction of a percent by weight (abbreviated wt%) to nearly 100 wt%, depending on the combination of metals.

The maximum amount of an alloying element that can be dissolved in a metal to form a solid solution is referred to as its solubility limit. At a concentration higher than the solubility limit, the excess alloying element will often separate out, typically as a chemical compound (that is, a new phase) at the grain boundaries or within the grains of the metal. Consequently, many alloys are composed of two phases. Formation of a new phase occurs because it leads to a lower energy of the overall system. For example, atoms of the alloying element lead to the creation of internal stresses due to strain in the crystal lattice. One mechanism to relieve these excessive stresses (or the strain energy) in the lattice is the formation of the new phase. A process based on this mechanism, known as precipitation hardening, is widely used for strengthening metals (Section 6.8).

Alloying is also widely used to control polymorphism and, thus, improve the properties of metals such as Ti and Fe, for example. As Figure 6.1 shows, cooling Ti (β phase) at a sufficiently slow rate from above ~882 °C to room temperature leads to the formation of the α phase. On the other hand, cooling the alloy Ti6Al4V formed by the addition of ~6 wt% Al and ~4 wt% V to Ti leads to the formation a two-phase alloy, sometimes referred to as an α–β alloy, which has much improved mechanical properties over Ti (Section 6.9.2). Aluminum improves the stability of the higher temperature α phase whereas vanadium stabilizes the β phase, leading to the two-phase microstructure. Stainless steels used in orthopedic surgery, such as the composition designated 316L, typically consist of the austenite phase and not the ferrite phase. This is because nickel (Ni) and other elements used as alloying elements stabilize the austenite phase by reducing the rate of the diffusive FCC to BCC transformation in Fe by orders of magnitude.

6.5 Shape (Morphology) of Phases

The phases present in an alloy can have a variety of three-dimensional shapes, often referred to by the term morphology. Using a pure metal as a starting point, its grain shape can often be approximated by a tetrakaidecahedron. One way of visualizing this grain shape, is to take a regular octahedron and truncate its eight corners, one-third the distance along each edge, giving a tetrakaidecahedron composed of fourteen faces. A section across the center of a tetrakaidecahedron gives a hexagon, often used in two-dimensional illustrations to approximate the shape of a grain.

In an alloy, taken as the α phase, in which a second phase β forms, Figure 6.7 shows some of the shapes that the β phase can have when interfacial energies between the two phases dominate. When the interfacial energy $\gamma_{\alpha\beta}$ is isotropic, that is, the same in all directions, the β phase can form as spherical single-crystal particles within the α grain to minimize the interfacial energy. This is because the spherical shape has the minimum surface area to volume ratio. On the other hand, if $\gamma_{\alpha\beta}$ is not isotropic but is lower

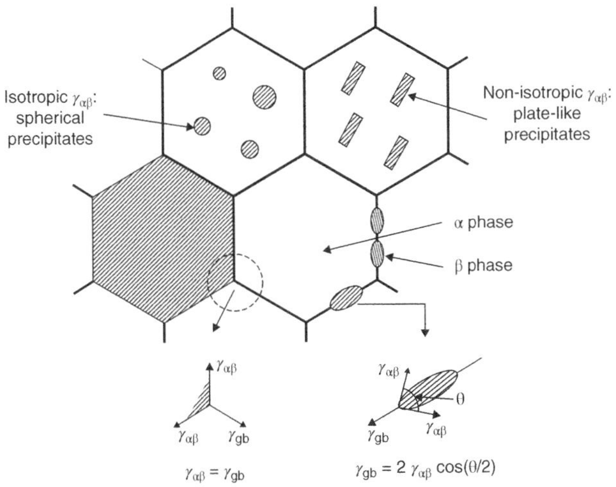

Figure 6.7 Types of phases that can be present in a metal composed of two phases.

for some planes, the β phase can from as plate-like particles within the α phase. Growth of the β phase particles occurs faster along the planes of lower interfacial energy, giving a lower overall interfacial energy for the same volume.

When formation of the β phase occurs at the grain boundaries of α phase, the β phase particles often adopt a lens-like shape, elliptical in cross section, which is controlled by the balance of interfacial tensions at the intersection of the β particles with the grain boundary of the α phase. This is given by

$$2\gamma_{\alpha\beta}\cos\left(\theta/2\right)=\gamma_{gb} \tag{6.11}$$

where, γ_{gb} is the interfacial energy between the α grains, often called the grain boundary energy, and θ is the angle between the interfacial tensions $\gamma_{\alpha\beta}$ at the intersection of the β particles with the α grain boundary. Formation of the β phase can also occur as complete grains between those of the α grains. Assuming that the β grains have a tetrakaidecahedron shape also, the balance of interfacial tensions requires that $\gamma_{\alpha\beta} = \gamma_{gb}$.

6.6 Phase Diagrams

Phase diagrams, such as the well-known H_2O phase diagram, show the phases or constituents present in a material or substance as a function of temperature and pressure. Phase diagrams of most practical metals and ceramics are available in reference publications. They have been constructed from experiments by identifying the phases present in a material or from calculations utilizing thermodynamic data for phases that might be present. Thus, whenever information about the phases present in a material is required, a useful first step is to examine its phase diagram. Phase diagrams are sometimes called phase equilibrium diagrams to emphasize that the diagrams show the phases present in a material under equilibrium conditions, that is, under conditions of temperature and time for the system to reach equilibrium. Often thermodynamic equilibrium is not achieved or not desired because the non-equilibrium phase composition, structure or morphology add useful properties to the material. However, the equilibrium constituents provide a useful guide from which nonequilibrium constituents can be inferred.

Phase Diagram Principles: The Fe-C Phase Diagram

Phase diagrams play an important role in designing metals having the requisite microstructure and, thus, properties. They vary from simple to complex, even for an alloy composed of two metals, referred to as a two-component system. One of the most widely recognized two-component phase diagram is that for the iron-carbon (Fe-C) system, particularly the Fe-rich part of the diagram composed of Fe and less than ~4 wt% C (Figure 6.8). This compositional range includes steels, often referred to as carbon steels, composed of less than ~1.5 wt% C, and cast irons composed of ~1.5–4 wt% C,

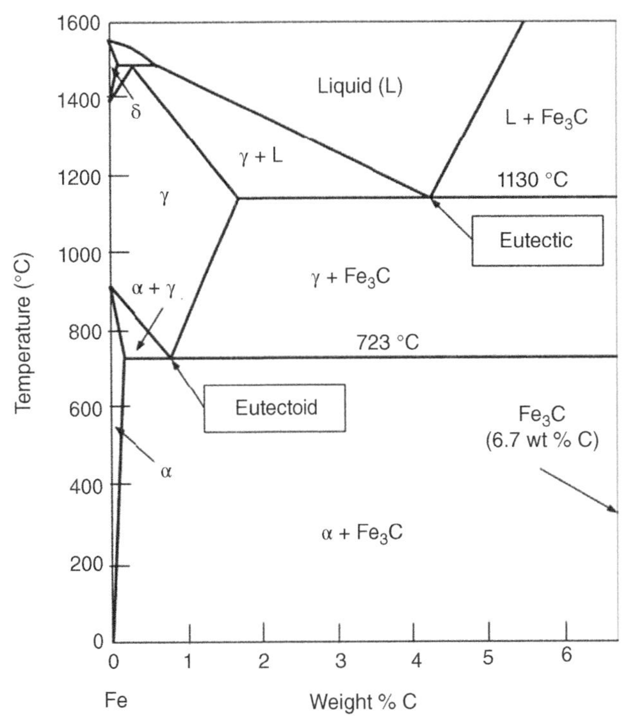

Figure 6.8 The iron-rich portion of the iron-carbon phase diagram.

which are used in a wide range of applications. Although these alloys are not used in biomedical applications mainly because of their low corrosion resistance, the Fe-C phase diagram illustrates many of the principles associated with phase diagrams and the type of information that can be obtained from examination of a phase diagram.

For two-component systems, the convention is to plot the temperature, typically in degrees Celsius, vertically (*Y*-axis) and the composition, typically in wt%, horizontally (*X*-axis), with the understanding that pressure is held constant (normally 1 atmosphere). The extreme left of the compositional axis corresponds to 100% Fe (equivalently 0% C) and as we go to the right, the amount of Fe decreases monotonically while the amount of C increases, with the extreme right corresponding to 100% C (0% Fe). To show the Fe-rich phases in more detail, Figure 6.8 shows only a part of the overall phase diagram, up to 6.7 wt% carbon, a composition corresponding to the compound Fe_3C, called iron carbide or cementite.

In the phase diagram, we observe a number of lines that delineate different phase fields, such as the fields labeled γ and $\alpha+\gamma$. A phase field is a combination of temperature and pressure for which all the phases present change their relative amounts but not their nature or structure. The α phase field represents a single solid solution composed of α iron (FCC) and carbon whereas the $\alpha+\gamma$ phase field represents a two-phase field composed of two solid solutions, the α solid solution and the γ solid solution composed of γ iron (FCC) and carbon. In comparison, the phase field labeled $\alpha+$ Fe_3C represents a material composed of the α solid solution containing less than ~0.035 wt% dissolved C and the compound Fe_3C. The compositions of most carbon steels lie in this two-phase field.

At any temperature, as we go to the right in the $\alpha+Fe_3C$ phase field, that is, with increasing carbon content, the amount of Fe_3C in the steel increases. The relative amount of each phase at any specific temperature and composition in a two-phase field can be determined from a construction known as the lever rule, which states that the relative amount of each phase is inversely proportional to its distance from the vertical line representing the composition. The phase diagram in Figure 6.8 shows two additional types of transformation, a eutectic and a eutectoid, which are present in the diagrams for many engineering alloys and are often mentioned in the literature. In a eutectic transformation, a liquid of a certain composition transforms to a solid of the same overall composition at the unique temperature indicated. A eutectoid transformation is much like the eutectic transformation but now a solid transforms to two other solids of the same overall composition.

Example 6.1 The eutectoid composition in carbon steels contains 0.80 wt% C and, at room temperature, the solubility limit of C in iron is 0.02 wt%. If a steel of the eutectoid composition is equilibrated in the γ phase field and cooled slowly to room temperature, determine the phases present in the steel and the amount of each phase.

Solution:

According to the Fe-C phase diagram (Figure 6.8), at room temperature the steel will consist of two phases, an α iron solid solution containing 0.02 wt% C and Fe_3C. The eutectoid composition is situated at a distance (in wt% C) of (0.80–0.02) from the α phase boundary and (6.7–0.80) from the Fe_3C boundary. Using the lever rule, the amounts of α phase relative to the amount of Fe_3C will be in the ratio 5.9/0.78. Thus, the amount of the α phase and Fe_3C will be 88 wt% and 12 wt%, respectively.

Composition–Structure–Property Relationships in Carbon Steels

Normalized carbon steels, that is, carbon steels cooled at a slow enough rate from temperatures in the γ phase field to room temperature, show a tensile strength and yield strength that increase with carbon content. This is due to the increasing amount of the Fe_3C phase in the steel that acts as a strengthening phase. The ductility, however, decreases more rapidly with increasing carbon content due to an increasing tendency for microcracks to form at the boundary between the α phase and Fe_3C phase. In comparison, if similar compositions in the γ phase field are cooled rapidly (quenched) to miss the nose of the TTT curve (Section 6.3.5), the γ phase will transform by a displacive mechanism to martensite, a supersaturated solid solution of carbon in iron. Martensite does not appear in the phase diagram (Figure 6.8) because it is not an equilibrium phase.

Whereas martensite in pure Fe has the same BCC structure as α iron (ferrite), the martensite in carbon steels does not. The displacive transformation happens so fast that the carbon atoms are frozen in place, remaining in their original lattice positions. As the solubility limit of C in the α phase (BCC iron) is only ~0.035 wt% under normal (equilibrium) conditions, martensite is supersaturated with C. In order to accommodate the excess C, the lattice deforms along one of the cube directions to form a body-centered tetragonal unit cell. This distortion creates significant lattice strain, making it more difficult for dislocations to move through the structure, thus enhancing the yield strength and hardness of martensite. As the distortion increases with the amount of C in solid solution, the hardness and yield strength also increase with carbon content. On the other hand, martensite is very brittle. The austenite to martensite transformation is also accompanied by an increase in volume that depends on the amount of C in the steel but is approximately 3–4 vol% for a low carbon steel.

Reheating martensite in a controlled manner to a temperature in the range 300–600 °C leads to a significant improvement in its toughness with only a moderate reduction in its yield strength and hardness. This reheating to an appropriate temperature, referred to as tempering, is a common treatment used in the production of many metals. Tempering provides sufficient thermal energy for the carbon atoms to diffuse out of the supersaturated solid solution and react with iron atoms to form fine closely spaced particles (precipitates) of Fe_3C. The lattice relaxes back to the undistorted BCC structure of α iron, resulting in an increase in ductility. On the other hand, the Fe_3C particles serve to maintain a high yield strength and hardness by a mechanism called precipitation hardening (Section 6.8.2). As exemplified for carbon steels, heat treatments have a strong influence on the structure and mechanical properties of metals and, consequently, they play an important role in the production of metals having a desirable combination of properties.

6.7 Production of Metals

The method used to produce a metal has a significant effect on its microstructure and, thus, on its mechanical properties. Consequently, some familiarity with the methods and techniques used to produce metals is useful for understanding the properties of metals used as biomaterials. The two major types of methods give rise to products that are often differentiated by the terms wrought metals and cast metals. Alternative methods such as powder metallurgy and chemical vapor deposition (CVD) are used to a limited extent. On the other hand, there is growing interest in the use of robotic methods referred to as additive manufacturing or three-dimensional (3D) printing.

6.7.1 Wrought Metal Products

The production of wrought metals involves two major steps

- Melting the metal and casting it into a mold to form a large object called an ingot or a billet, or continuously casting the molten metal into large sections
- Working (forming) the metal ingot into a standard section of the requisite shape using methods such as rolling, extrusion, wire drawing, or forging (Figure 6.9).

Forging involves pressing the ingot in a die to produce a desired shape, either by a single continuous compression or by multiple successive shocks. In rolling, the ingot is pressed through a set of parallel rollers, which may also impart a particular shape to the rolled metal. Extrusion involves pushing the ingot through a die or a shaped opening by means of a piston or a screw pushing a ram. Drawing is similar to extrusion, but the ingot is pulled through an orifice by means of a tensile force applied at the downstream end of the process. These forming methods are

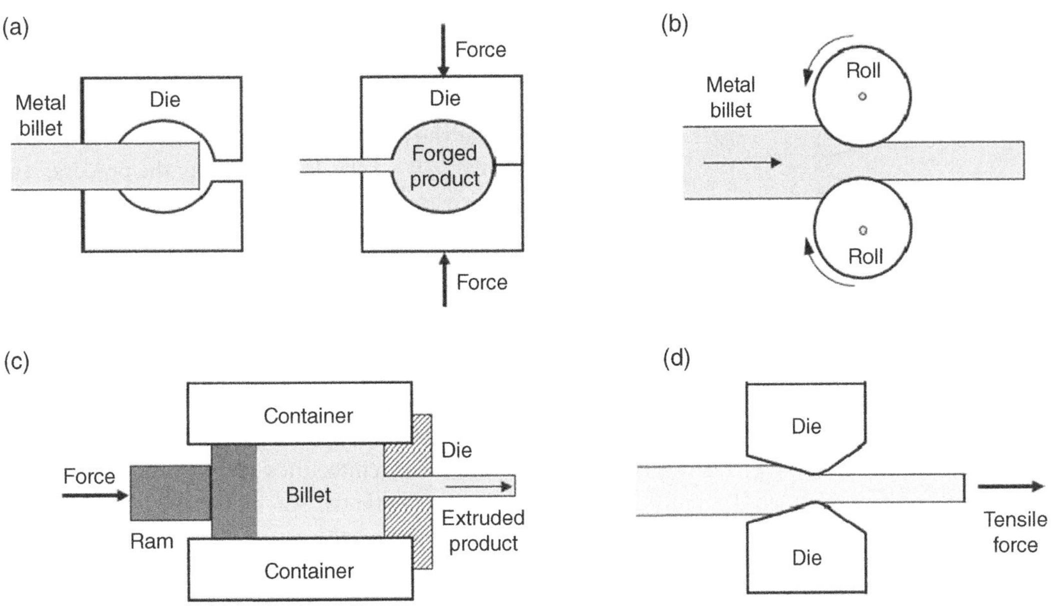

Figure 6.9 Common production methods for metals: (a) forging, (b) rolling, (c) extrusion, and (d) drawing.

often referred to as cold working when carried out at room temperature whereas they are often referred to as hot working when carried out at higher temperature.

As they cause permanent deformation of the ingot, these forming methods rely on plastic deformation of the metal by dislocation motion (Chapter 4). Consequently, the applied stresses should be greater than the yield stress, often a few times the yield stress or higher depending on the desired shape. The stress in cold working can be greatly reduced by using hot working due to the reduction in yield strength of the metal at higher temperature. Plastic deformation, particularly in cold working, has a significant effect on defect structure (dislocations) and microstructure and, consequently, it often forms a key step in techniques used to strengthen metals (Section 6.8).

6.7.2 Cast Metal Products

Cast metal products are formed by pouring the molten metal into a cool mold of the desired shape and allowing the metal to solidify. Solidification involves a process of nucleation and growth, similar in nature to that described for phase transformations (Section 6.3). This process often leads to a microstructure composed of large dendritic grains (or columnar grains) in the outer part of the product and more equiaxial grains in the interior. Nucleation starts preferentially at the surface in contact with the mold, due to faster cooling, followed by growth of the grains into the liquid, typically in a dendritic manner (Figure 6.10). The dendrites have a branched shape, approximating that of a tree. Upon completion of the solidification process, the metal consists of large dendritic grains and an interdendritic phase that fills in the region between the branches. However, the use of catalysts can lead to a more fine-grained equiaxial microstructure with better mechanical properties.

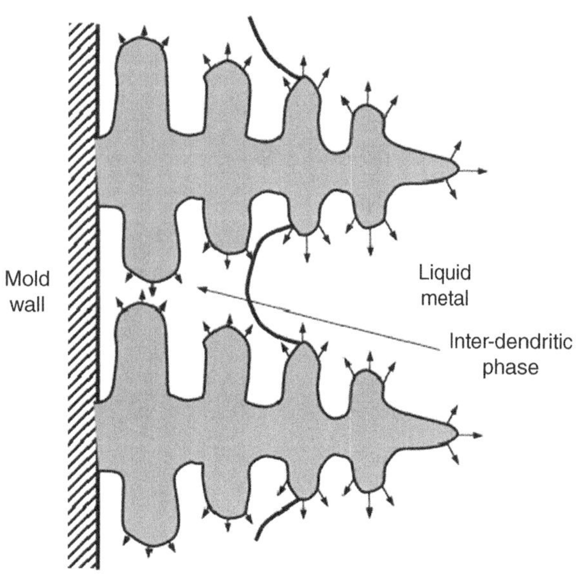

Figure 6.10 Illustration of growth of dendritic grains during casting of a molten metal in a mold.

6.7.3 Alternative Production Methods

In powder metallurgy, particles of size a few micrometers to a few tens of micrometers are compacted at room temperature to form a porous object of the desired shape. Then the porous object is heated to a suitable temperature, typically between 0.5 and 0.9 of the melting temperature of the metal to form a dense object. This heating process is referred to as sintering. Rapid atomic diffusion leads to bonding at the interface between the particles and transport of atoms into the pores, driven by a reduction in surface energy associated with the pores, resulting in a dense object with better mechanical properties. The powder metallurgy process can also be used to form metals with a range of porosities, from over 50% to nearly zero. Consequently, it forms a useful technique for creating porous metal implants, such as porous Ti implants for bone repair. A variation of this sintering process is hot isostatic pressing, in which a powder or as-formed porous object is subjected to a combination of heat and isostatic pressure. Hot isostatic pressing finds use in forming dense objects by the powder metallurgy route or for removing flaws, such as pores, from metals formed by the casting route. Sintering and hot isostatic pressing find limited use in the production of metals but they are the most common methods used to produce ceramics (Chapter 7).

CVD is a process by which reactive molecules in the gas phase are transported to a surface at a suitable temperature where they react chemically to form a solid phase. It is a well-established technique used to deposit all classes of materials but it is slow and expensive. Consequently, it is used mainly for the production of coatings, although it can be used to create 3D objects by prolonging the process. An example of the use of CVD is the production of porous tantalum (Ta) implants for use in bone repair (Section 6.9.5).

Additive manufacturing (3D printing) allows the creation of objects, layer by layer, by robotic deposition from a computer-aided design file. It is witnessing growing use for the creation of all classes of materials. The technique is well suited for the creation of dense implants having intricate external shapes and porous implants having well-controlled internal architectures that cannot be easily obtained by other methods. An important advantage of additive manufacturing is the ease with which the external geometry or internal architecture of the product can be changed, offering the ability to create implants with the dimensions and shape tailored to specific patient needs.

6.8 Mechanisms for Strengthening Metals

The majority of metals are not used in their pure state because they invariably have inadequate strength. Dislocations move with ease through the crystal structure, resulting in a low yield strength (Chapter 3). Consequently, metals need to be strengthened to provide better mechanical properties. The main mechanisms for strengthening (hardening) metals are

- Solid solution hardening
- Precipitation and dispersion hardening
- Work hardening, also referred to as strain hardening
- Grain size refinement

A crystal starts to yield when the force acting on a dislocation exceeds the intrinsic resistance of the crystal to dislocation motion due to breaking and reforming of atomic bonds. Most pure metals have a low yield strength because the lattice resistance to dislocation motion is low. Strengthening mechanisms enhance the yield strength by increasing the resistance to dislocation motion.

6.8.1 Solid Solution Hardening

Atoms of an alloying element that go into solid solution in a metal have a radius that is different from that of the metal and, thus, they generate strain in the crystal lattice. This occurs regardless of whether the atoms of the alloying element (the solute atoms) are smaller or larger than the atoms of the metal (the host atoms) because the host atoms surrounding a solute atom are pulled in or pushed out slightly from their equilibrium position. Stresses arising from lattice strain lead to an increase in the resistive force to dislocation motion, thereby enhancing the yield strength of the metal. The greater the concentration C of solute atoms (number of atoms per unit volume), the shorter their spacing in the lattice and, consequently, we might expect that the yield strength of the metal should increase. Indeed, the yield strength of several solid solution alloys increases as $C^{1/2}$ because the spacing of the solute atoms on a plane, including the slip plane, varies as $C^{-1/2}$. Stainless steels and many other single-phase alloys rely on this mechanism for their strength.

6.8.2 Precipitation and Dispersion Hardening

If an alloying element is dissolved in a metal at high temperature and the alloy is cooled, some of the alloying element may precipitate out as fine particles of a second phase

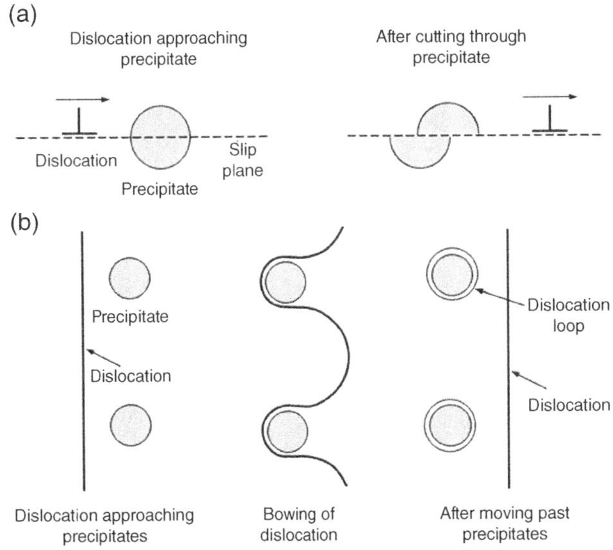

Figure 6.11 Obstruction of dislocation motion by precipitates. Dislocation can get past precipitates by (a) cutting through them or (b) by bowing around them, both of which require the expenditure of energy.

if its concentration is higher than its solubility limit at the relevant temperature, such as room temperature. In practice, however, the optimal process involves quenching and tempering, as discussed earlier for carbon steels (Section 6.6.2). Fine particles can also be incorporated into a metal in other ways, such as mixing them with the metal particles and forming a product by the powder metallurgy route. This process gives a dispersion of fine particles within the metal, thus the term dispersion hardening. Overall, both methods give a distribution of fine second-phase particles within the metal that can obstruct the motion of dislocations. Upon encountering these particles, dislocations can move past them by cutting through them or by bowing around them (Figure 6.11). Strong particles (greater resistance to cutting) that are closely spaced within the grains (greater resistance to bowing) provide the most effective strengthening.

6.8.3 Work Hardening

Metals work harden when they are subjected to cold working, such as rolling, extrusion, wire drawing, and forging. There is an increase in yield strength and hardness but this is accompanied by a decrease in ductility (elongation to failure) (Figure 6.12). Work hardening is an important method for strengthening many metals. The extent of cold working (CW) is expressed as a percentage reduction in area A or, for a product of uniform width, in thickness h:

$$\mathrm{CW} = \frac{A_0 - A}{A_0} \times 100 \tag{6.12}$$

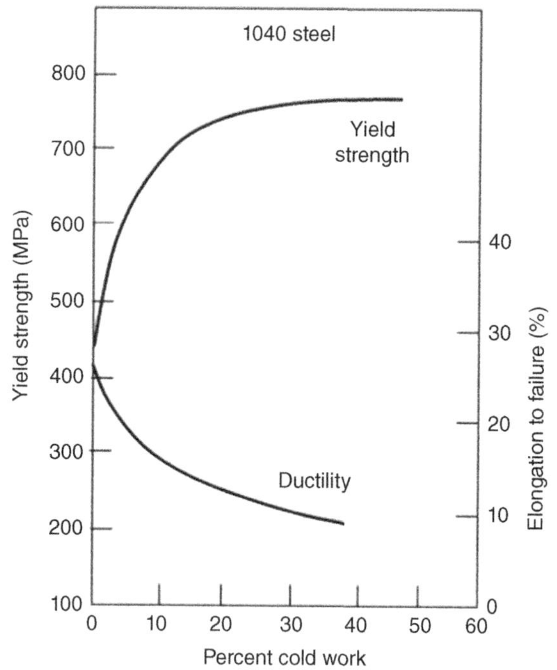

Figure 6.12 Yield strength and elongation to failure (ductility) as a function of amount of cold working for a 1040 steel. *Source:* Adapted from: Callister (2007).

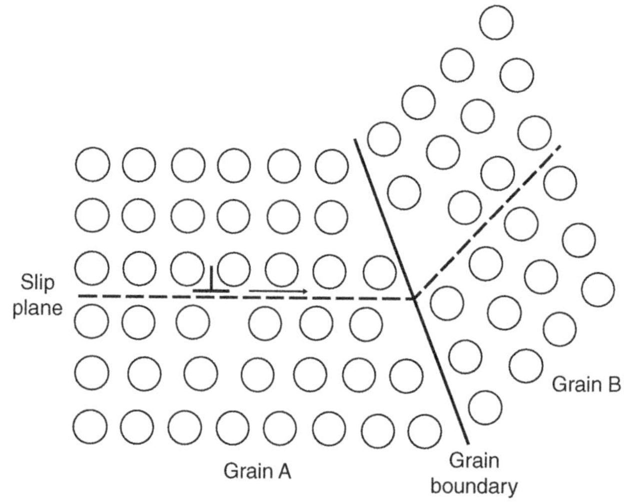

Figure 6.13 Motion of a dislocation as it encounters a grain boundary. The boundary provides a barrier to dislocation motion because slip planes are discontinuous across a boundary.

$$\text{CW} = \frac{h_0 - h}{h_0} \times 100 \tag{6.13}$$

where, A_0 and h_0 are the original area and thickness, respectively, of the cold-worked metal.

When metals deform during cold working, dislocations already present in the metal start to move, while new dislocations are created and they move as well. As these dislocations move along the slip planes, they interact and obstruct each other, leading to piling up of the dislocations and making it more difficult for them to move. Consequently, the yield strength and hardness increase but the metal becomes less ductile.

6.8.4 Grain Size Refinement

Grain size refinement, that is, reduction of grain size, provides a powerful method to enhance the yield strength of metals (Section 4.3.2). Dislocations cannot move past grain boundaries due to the discontinuity of the crystal planes at the grain boundary (Figure 6.13). They pile up at the grain boundaries, which limits further motion. As the grain size decreases, the number of boundaries per unit volume increases, providing a larger number of obstacles to dislocation motion. The relationship between the yield stress and grain size is described by the Hall–Petch equation (Eq. (4.34)).

Grain size refinement is often achieved by a combination of cold working and a controlled heat treatment. A metal that has not been cold-worked or one that has been annealed will typically have a low dislocation density (number of dislocations per unit area) (Figure 6.14a) and, thus, will have a low yield strength and high ductility. Upon appreciable cold working, the dislocation density increases considerably (Figure 6.14b), and the metal will work-harden and become less ductile. Work hardening will also lead to a significant increase in lattice strain energy because each dislocation, being a lattice defect, introduces strain into the lattice.

Subjecting a highly cold-worked metal to a controlled heat treatment, at a temperature in the range approximately one-third to one-half the melting temperature T_M of the metal, leads to the elimination of most of the dislocations and the generation of fine strain-free grains that have replaced the old highly strained grains. This controlled heat treatment is referred to as annealing. It consists typically of heating to a high enough temperature, holding for the desired time and cooling at a suitable rate to room temperature. Annealing leads to a reduction in yield strength and hardness but an increase in ductility due to removal of dislocations, resulting in the production of a suitable microstructure having the desired properties.

In annealing, cold-worked metals often undergo a two-step process called recovery and recrystallization. Upon heating the metal, recovery occurs first, at temperatures of ~$0.1\,T_M$. The additional thermal energy allows some dislocations to move to lower energy positions. However, while the dislocation density decreases slightly due to this rearrangement, the yield strength and ductility are almost unchanged (Figure 6.14c).

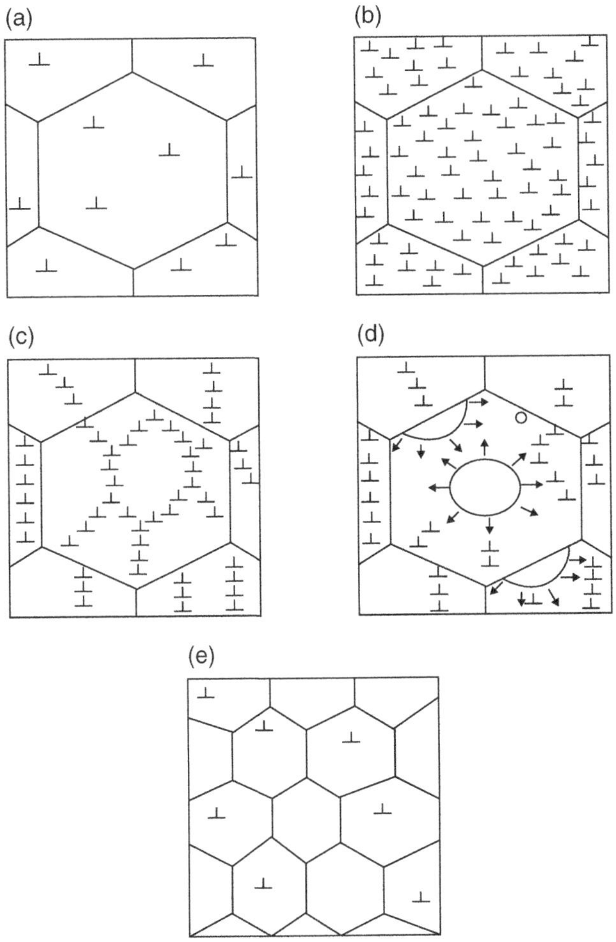

Figure 6.14 Sketch illustrating how the microstructure of a metal is changed by plastic working and annealing. (a) If the starting metal has already been annealed it will have a comparatively low dislocation density. (b) Plastic working greatly increases the dislocation density. (c) Annealing leads initially to recovery – dislocations move to low-energy positions. (d) During further annealing new grains nucleate and grow. (e) The fully recrystallized metal consists entirely of new undeformed grains. *Source:* From Ashby and Jones (2006) / with permission of Elsevier.

In comparison, major changes occur during recrystallization, upon heating the metal to higher temperatures, in the range ~0.3 to ~0.5 T_M. New grains nucleate and grow at the expense of the highly strained grains, driven by the large lattice energy associated with these strained grains, until the entire metal consists of almost strain-free grains (Figure 6.14d). The dislocation density returns to a level approximately similar to that prior to cold working. If the grain size of the annealed metal is similar to its original size prior to annealing, the yield strength and ductility will also return to their original values. On the other hand, if annealing leads to a finer grain size (Figure 6.14e), the yield strength will increase but the ductility will show a decrease. Thus, annealing is not just a process for removing work hardening but a powerful way to control the grain size of cold-worked metals. However, under nonoptimal conditions, if the recrystallization temperature is too high or the time at the recrystallization temperature is too long, grain growth of the strain-free grains can occur, driven by a reduction in grain boundary area. According to the Hall–Petch equation (Eq. (4.35)), this will lead to a decrease in yield strength.

6.9 Metals Used as Biomaterials

The vast majority of metals used as biomaterials are alloys, although a few metals such as gold (Au), silver (Ag), platinum (Pt), and Ta are used in their pure state. On the other hand, Ti finds considerable use both in an almost pure state, often referred to as commercial purity (CP) titanium, and as an alloy (Ti6Al4V) as well. This section is devoted mainly to a discussion of the composition, structure, and properties of metals most commonly used as biomaterials, such as stainless steels, Ti and Ti6Al4V, and cobalt–chromium (Co–Cr) alloys. These metals provide a wide range of mechanical properties (Table 4.1) and are attractive for use as biomaterials when long-term mechanical durability under a variety of physiological stress must be guaranteed. Metals such as nickel-titanium alloys, zirconium alloys, Ta and the noble metals (Au, Ag, and Pt) that witness more limited application are also discussed. In addition to providing a description of the composition, structure, and properties applications of these metals, emphasis is placed on the methods and mechanisms of controlling their structure in order to achieve desirable properties.

6.9.1 Stainless Steels

Steels are the most widely used class of metals in engineering applications. There are several types of steels, the simplest of which, in terms of composition, are the carbon steels composed of iron containing less than ~1.5 wt% carbon (Section 6.6.1). Although carbon steels are not used as biomaterials due mainly to their low corrosion resistance *in vivo*, they provide an extensive range of mechanical properties that are suitable for a wide range of engineering applications. It will be recalled that these carbon steels can be produced in a variety of crystal structures and microstructures depending on their composition and heat treatment.

Steels used as biomaterials belong to a class called stainless steels, metals that have an excellent range of mechanical properties in addition to high corrosion resistance. Stainless steels were invented around 1913 when Harry Brearly accidentally found that adding chromium (Cr) to low-carbon steels made them stain resistant. Modern stainless steels contain, in addition, Ni, molybdenum (Mo), manganese (Mn), and smaller amounts of other elements to produce not just corrosion resistance but a wide range of mechanical properties as well. The effects of these alloying elements and the heat treatments used in their production are largely responsible for the attractive combination of properties provided by stainless steels.

For a low carbon steel containing less than ~1 wt% C, the minimum amount of Cr required to provide corrosion resistance is ~15 wt%, but higher amounts may be required for applications in hostile environments. The chromium forms a dense Cr_2O_3 coating that strongly adheres to the surface of the underlying metal, which is responsible for the corrosion resistance of stainless steels.

Martensitic Stainless Steels

A steel of a composition composed of less than ~0.5–1 wt% C and ~15 wt% Cr lies in the γ phase field (austenite) of the iron-chromium phase diagram at temperatures in the region of ~1000–1200 °C. Quenching to room temperature gives the martensitic phase which can be tempered to give a fine dispersion of carbides. These martensitic steels have high strength and high hardness, but only moderate corrosion resistance. While they find use in a variety of applications such as knives, surgical scalpels and ball bearings, martensitic steels are not used as implanted biomaterials.

Ferritic Stainless Steels

Ferritic stainless steels typically contain a higher amount of Cr (~15–30 wt%). For a low carbon content (less than ~0.5–1 wt% C), these compositions lie outside the γ phase field in the iron-chromium phase diagram. These steels are composed of the α phase (ferrite) from above ~1200 °C down to below room temperature. They are used in engineering applications but not in biomedical applications.

Austenitic Stainless Steels

Addition of Ni (more than ~8 wt%) to a low carbon steel (less than ~0.5 wt% C) containing ~15 wt% Cr lowers the γ to α phase transformation for iron from ~914 °C to ~700 °C, while the Cr and Ni, together with Mn (less than ~2 wt%) slow the rate of the γ to α phase transformation by orders of magnitude. Additionally, the Cr and Ni lower the martensitic start temperature M_S to below 0 °C. Consequently, austenite can be cooled in air from above 700 °C to room temperature without transforming to the α phase or martensite. Unlike martensitic or ferritic steels, austenitic steels are nonmagnetic and have adequate strength, high toughness, and high corrosion resistance for biomedical applications.

Stainless steels used as implants are austenitic steels, with an ASTM (American Society for Testing and Materials) designation of 316L, composed of (in wt%) 60–65 Fe, 17–19 Cr, 13–15 Ni, 2–3 Mo, <2 Mn, <0.03 C, plus trace quantities of other elements often present in steels as impurities. Molybdenum enhances the corrosion resistance provided by the Cr, particularly in saline media whereas the carbon content is kept below 0.03 wt%, to reduce the potential for corrosion of the metal *in vivo*. With improper control, higher amounts of carbon can lead to the formation of carbides such as $Cr_{23}C_6$ that precipitate at the grain boundaries and grow. As the formula shows, a small amount of carbon can consume a considerably larger amount of chromium in the formation of $Cr_{23}C_6$, which depletes the amount of chromium in the grain boundary region of the steel. This can lead to intergranular corrosion (Chapter 14) and, eventually, to failure. Austenitic steels designated 316L have a range of suitable mechanical properties (Table 4.1) and corrosion resistance for use in a variety of orthopedic, cardiovascular, and dental applications (Table 6.1).

6.9.2 Titanium and Titanium Alloys

Titanium and its alloy Ti6Al4V have been widely used for decades in medical and dental applications (Table 6.1). These two metals also form the most widely used titanium compositions in engineering applications, with Ti6Al4V accounting for ~50% and Ti accounting for approximately another 20% of the total titanium usage. Although Ti6Al4V is among the most biocompatible metals, concerns have been sometimes expressed about the release and possible toxicity of vanadium ions *in vivo*. This has resulted in the development of various Ti alloys in which V has been replaced by one or more metals, such as Nb, Ta, and Zr, whose ions have shown a lower level of toxicity. One such alloy, Ti6Al7Nb, has a microstructure and properties similar to Ti6Al4V.

Titanium and Ti6Al4V have an attractive combination of properties for use as biomaterials, such as

- *Durable mechanical properties*: When subjected to a controlled heat treatment, Ti6Al4V can show strengths comparable to those of stainless steels and cobalt–chromium alloys (Table 4.1). However, titanium alloys stand out when the specific strength (that is, the strength-to-weight ratio) of materials is considered. Because of their lower density, the strength-to-weight ratio of titanium alloys (below ~500 °C) is superior to those of other high-performance alloys used in engineering applications, such as stainless steels and cobalt–chromium alloys. The Young's modulus of titanium and Ti6Al4V (100–120 GPa) is approximately half that of stainless steels and Co–Cr alloys but closer to that of human cortical bone. This low Young's modulus makes Ti and Ti6Al4V implants more suitable for use in bone repair because they provide the potential to reduce stress shielding. A drawback of Ti and Ti6Al4V is that they are weak in shear deformation and prone to sticking when articulating against themselves or other materials. Consequently, they are not suitable for use as articulating bearings in total joint replacement, for example.

- *Corrosion resistance*: Ti and Ti6Al4V have high corrosion resistance due to a strong adherent passivating oxide surface layer (Chapter 5).
- *Biocompatibility*: Ti and Ti6Al4V have been shown to have superior biocompatibility when compared to other metals used as implants. Consequently, they are often the implant composition of choice in applications where tissue or direct bone contact is a requirement, such as endosseous dental implants and porous uncemented orthopedic implants.

Similar to steels discussed in the previous section and other metals, the mechanical properties of titanium and titanium alloys are predominantly determined by two factors:

- Composition, such as the amount and type of alloying elements
- Structure at various length scales, such as crystal structure and microstructure, which, for a given composition, is controlled by thermal, mechanical or thermomechanical treatments used in production.

Changes in composition and structure can, to a lesser extent, influence the corrosion resistance and biocompatibility of titanium alloys. Similar to steels discussed in the previous section and the majority of other metals, phase diagrams provide a useful basis for understanding the constituent phases and microstructure of titanium alloys.

Phase Diagrams for Titanium Alloys

Titanium, and the majority of titanium alloys as well, crystallize in a CPH structure at lower temperature, referred to as α titanium, and in a BCC structure at higher temperature, referred to as β titanium (Figure 6.1). The transformation temperature from one structure to another, called the β-transus temperature, is ~882 °C for pure Ti. The existence of these two crystal structures and their transformation temperature form the basis for a large variety of microstructures and, thus, the large range of properties shown by Ti and its alloys.

Titanium used in biomedical and engineering applications is not pure. One composition, referred to as CP titanium, contains small amounts of elements in solid solution such as oxygen (~0.2–0.4 wt%) and impurities such as iron (~0.2–0.5 wt%) introduced during production. Upon cooling at a sufficiently slow rate from above the β-transus, α grains nucleate and grow within the β grains that often have a size as large as a few millimeters, depending on the cooling rate in the β phase field. Individual α grains nucleate at the boundaries of β grains and grow to form colonies (packets) of lamellar (plate-like) grains within each β grain in such a way that the lamellae in each colony have a common orientation (Figure 3.23). This orientation is dictated by the possible 12 slip systems in the BCC structure of the β phase (Chapter 3). As the α grains grow by diffusion, faster cooling leads to a fine lamellar structure whereas a coarse lamellar structure forms upon slower cooling. At high cooling rates from above the β-transus, the β phase (BCC) transforms completely by a displacive mechanism to the α (CPH) phase, resulting in a martensitic structure composed of metastable fine plate-like or acicular grains.

Alloying elements added to Ti are commonly classified according to their ability to stabilize the α or β phase (Figure 6.15). The α stabilizing elements extend the α phase field to higher temperature and lead to the development of a two-phase α + β field as well. Aluminum is the most important α stabilizer, while oxygen atoms, often in small concentrations that occupy interstitial sites in the lattice, are also an α stabilizer. Molybdenum (Mo), V and Nb are among the most important β stabilizers due to their high solubility in titanium. Depending on the phase(s) present, titanium alloys are often classified as α, β, or α + β alloys. CP titanium is often referred to as α alloy as it is composed of an α solid solution containing oxygen interstitials, but the most widely used titanium alloys are α + β alloys of which Ti6Al4V is the most important.

Microstructure and Properties of Titanium Alloys

The microstructure and, consequently, the mechanical properties of titanium alloys depend primarily on the size, volume fraction, and arrangement of their constituent phases. This microstructure is often generated by thermomechanical treatments, involving some combination of cold or hot working, annealing and aging which can sometimes be complex (Peters et al. 2003).

Commonly, α alloys, composed of a single phase, show moderate strength. They often find use when corrosion resistance and ductility are of primary importance and high strength is a secondary design consideration. The four CP titanium grades (Grade 1 to Grade 4) are essentially α alloys in which oxygen atoms are deliberately introduced interstitially in different concentrations (0.2–0.4 wt%). In going from Grade 1 to Grade 4, the strength increases but the ductility decreases (Table 4.1). A popular grade is Grade

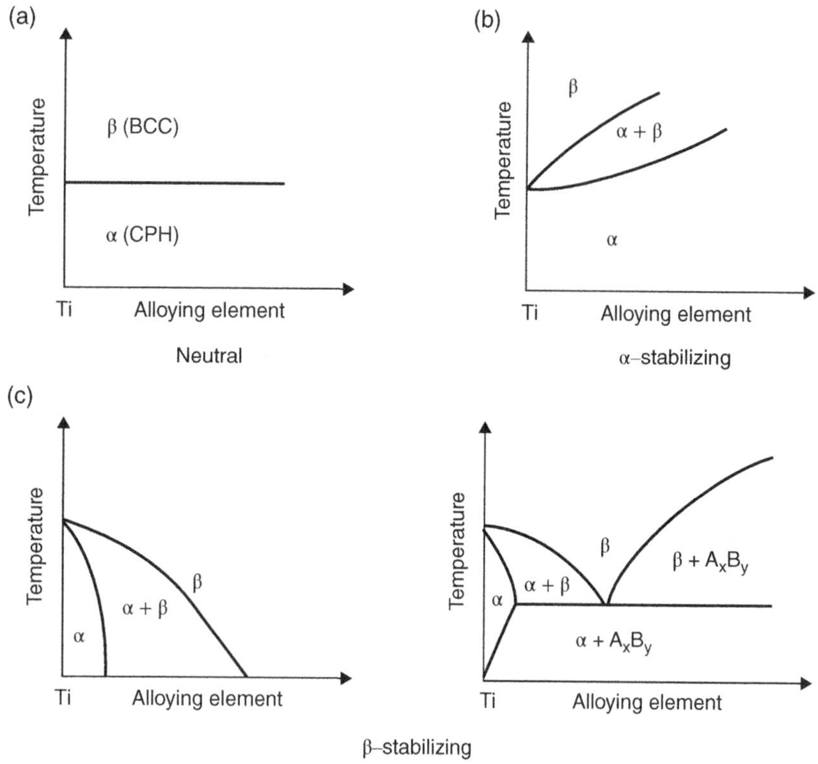

Figure 6.15 Influence of alloying elements on the phase diagram of titanium alloys, classified in terms of their effect on the β-transus temperature.

2, having a yield strength of ~275 MPa and an elongation to failure of ~20%, which finds use as dental implants.

Ti6Al4V, an α + β alloy, is the most commonly used titanium alloy. The presence of two phases means that its microstructure can be controlled over a wide range by suitable mechanical and thermal treatments, giving a range of attractive mechanical properties. Overall, the microstructure of Ti6Al4V is often classified into three broad types, two extreme types called equiaxial and lamellar, and a third type called bimodal, which can be considered to be a combination of these two extreme types. A key reference point is the β-transus temperature that separates the β phase from the α + β region (Figure 6.16).

Heat treatment in the β phase field and slow cooling below the β-transus temperature leads to nucleation of the α phase at the boundaries of the β grains and growth as lamellae into the β grains. A lamellar microstructure is formed, composed of colonies of α grains within each (prior) β grain, with the α grains in each colony having the same crystallographic orientation (Figure 6.17a). Rapid cooling leads to a finer-scale martensitic microstructure composed of fine plate-like or needle-like grains. Equiaxial microstructures are formed by deformation (cold working) to introduce lattice strain into the alloy and subsequent annealing in the α + β phase field (Figure 6.17b). The

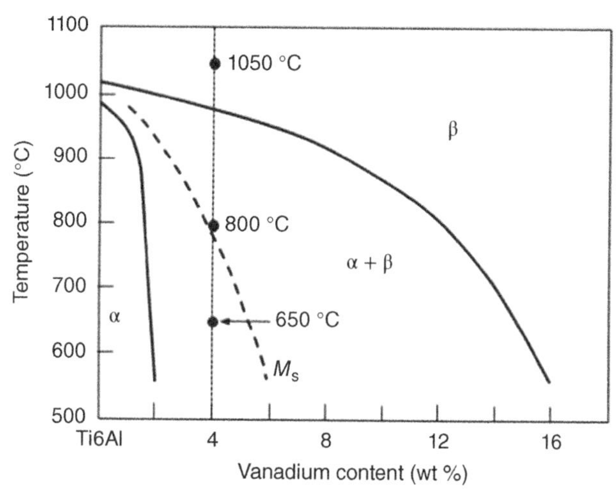

Figure 6.16 Schematic phase diagram for Ti6Al4V that shows the martensite start temperature M_s. *Source:* From Peters et al. (2003) / with permission of John Wiley & Sons.

recrystallization temperature and time at this temperature determines the volume fraction and size of the α grains. In comparison, heat treatment just below the β-transus temperature and slow cooling leads to a bimodal microstructure composed of equiaxial α grains in a colony-type

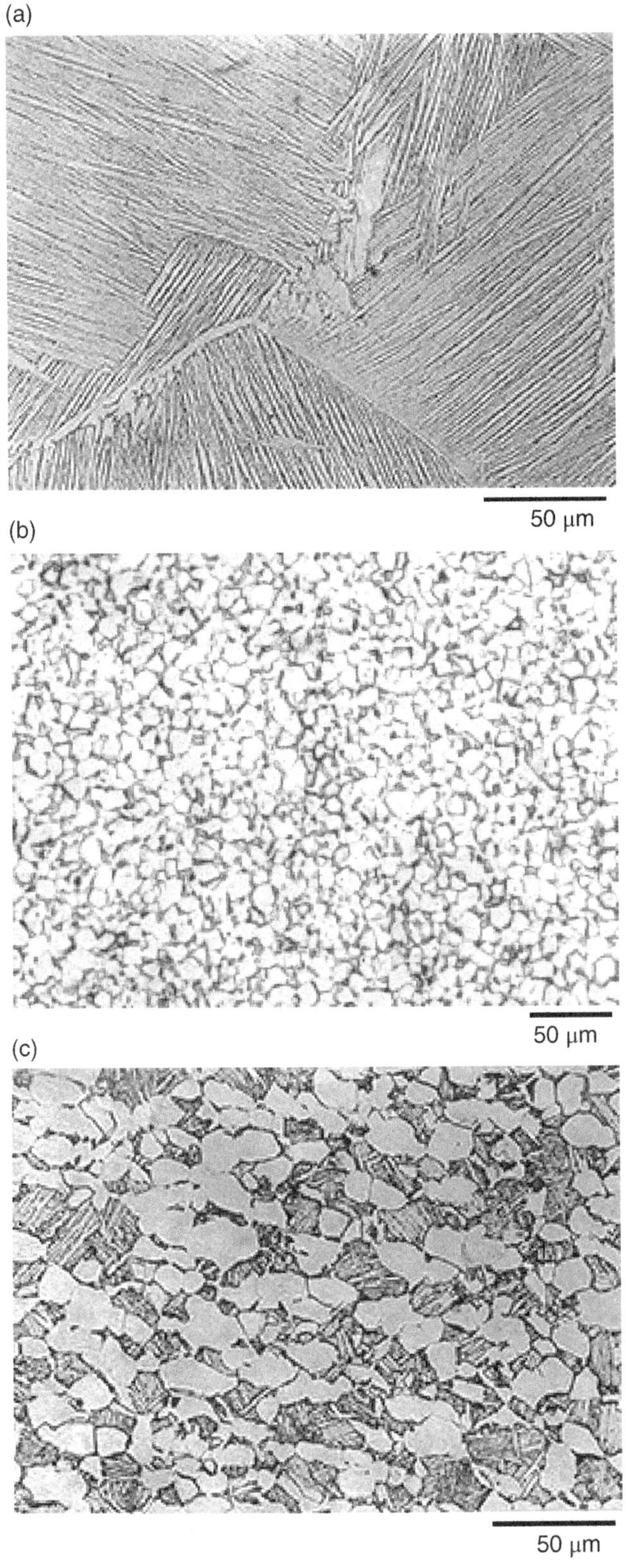

Figure 6.17 Images of Ti6Al4V microstructure produced under different thermal treatments after normalizing in the β phase region: (a) lamellar, (b) equiaxial, and (c) bimodal. *Source:* a, b: from Nalla et al. (2002); c: from Peters et al. (2003).

lamellar matrix of alternating α and β plate-like grains (Figure 6.17c).

Ti6Al4V can be produced with mechanical properties that are comparable to the strongest stainless steels (Table 4.1). Consequently, it is preferred to Ti when high strength is a requirement, such as femoral stems and acetabular shells of hip implants, tibial inserts of knee implants, and plates and screws in fracture fixation (Table 6.1). In comparison to the TiO_2 surface oxide layer on Ti, the surface of Ti6Al4V often contains, in addition, small amount of Al_2O_3, in which the Al to Ti content is approximately the same as that in the interior of the alloy (Chapter 5). This small difference in surface chemistry does not appear to cause any significant difference in biocompatibility between Ti and Ti6Al4V.

6.9.3 Cobalt-Based Alloys

Cobalt-based alloys, composed predominantly of Co and Cr, and varying amounts of other alloying elements such as Mo, W, and Ni, are often referred to generally as cobalt–chromium (Co–Cr) alloys. Developed for high-performance engineering applications because of their attractive combination of mechanical properties and corrosion resistance, particularly at high temperature, these alloys were introduced into biomedical applications over 70 years ago. They find use in several medical and dental applications (Table 6.1).

The corrosion resistance of Co–Cr alloys is due to a passivating surface oxide layer, composed essentially of Cr_2O_3, which results from the presence of a high amount of Cr in the alloy. This surface layer is similar in nature to the passivating Cr_2O_3 layer on stainless steels, although it often contains a small amount of cobalt oxide. When compared to stainless steels and Ti6Al4V described earlier, a distinguishing feature of Co–Cr alloys used in biomedical applications is the wide range of mechanical properties that they can provide. In addition to having a higher Young's modulus (stiffness) and hardness (wear resistance), some Co–Cr alloys can provide a better combination of strength, ductility, and fatigue resistance than stainless steels. These high performance alloys are particularly suitable for use as biomaterials when superior mechanical properties must be guaranteed, such as articulating bearings in hip and knee implants.

Four Co–Cr alloys that have found most use in biomedical applications are, using their ASTM designations, F75, F799, F90, and F562 (Table 6.2). F799 has a composition within the range of F75 and these two alloys, composed of Co, Cr, and Mo, plus small amounts various elements, are sometimes described as Co–Cr–Mo alloys. The high concentration of Cr (26–30 wt%) imparts high corrosion resistance to these two alloys, particularly in aqueous media containing chlorine ions. F90 is composed of a lower amount of Co and Cr, and Mo is replaced by a substantial amount of W and Ni. It is described as a Co–Cr–W–Ni alloy. F562 contains a still lower amount of Co, a high amount of Ni, approximately equal to its Co content, a substantial amount of Mo but no W. Essentially a cobalt-nickel alloy with substantial amounts of Cr and Mo, it can be described as a Co–Ni–Cr–Mo alloy.

F75 was one of the first Co–Cr alloys introduced into biomedical applications. F75 products are commonly formed by casting while the other three alloys are used mainly as wrought products. While the mechanical properties of cast F75 alloys are typically inferior to those of the wrought alloys, they are still attractive for several biomedical applications. On the other hand, F75 alloys have the advantage of lower production cost and capacity for forming into complex shapes due to the casting process. F799 products are formed by hot forging and can be considered wrought analogs of F75. It provides a Co–Cr–Mo product for use when more demanding mechanical properties are required.

Similar to steels and titanium alloys discussed earlier, and other metals as well, the mechanical properties of Co–Cr alloys are determined predominantly by their composition and their structure at various length scales. Co–Cr alloys have a range of compositions (Table 6.2) and their microstructures can be controlled over a wide range by a variety of thermal, mechanical, or thermomechanical treatments used in their production. These compositional and microstructural factors allow for the operation of some combination of the strengthening mechanisms in metals (Section 6.8), enabling the production of Co–Cr alloys having a range of attractive mechanical properties.

Cobalt–Chromium Phase Diagrams

Cobalt has a melting point of 1495 °C and, on cooling, it crystallizes in a FCC structure at higher temperature, referred to as the α phase, and in a CPH structure at lower temperature, the ε phase. If pure cobalt is cooled slowly, the transformation occurs at 422 °C, but it is sluggish due to a low Gibbs free energy change for the transformation (α to ε phase and vice versa). The existence of this transformation and the addition of alloying elements determine the phases present in Co–Cr alloys under a variety of conditions used in their production.

Addition of Cr to Co leads to a significant increase in the phase transformation temperature. Chromium can be said to stabilize the ε (CPH) structure because it leads to reduction in the α (FCC) phase region. According to one

Table 6.2 Composition and mechanical properties of four cobalt–chromium alloys[a] used as biomaterials.

Composition (wt%)	F75	F799	F90	F562
Co	58.9–68.9	58.0–59.0	45.5–56.2	29.0–38.0
Cr	27.0–30.0	26.0–30.0	19.0–21.0	19.0–21.0
Mo	5.0–7.0	5.0–7.0		9.0–10.5
W	<0.2		14.0–16.0	
Ni	<2.5	<1.0	9.0–11.0	33.0–37.0
Mn	<1.0	<1.0	<2.0	<1.0
Si	<1.0	<1.0	<0.4	<0.15
Fe	<0.75	<1.5	<3.0	<1.0
Ti	<0.1			<1.0
C	0.15–0.35	0.15–0.35	<0.15	<0.05
N	<0.25	<0.25		

Mechanical properties	F75	F799	F90		F562	
	As-cast/annealed	Hot forged	Annealed	44% Cold-worked	Hot forged	Cold-worked/aged
Young's modulus (GPa)	210	210	210	210	232	232
Yield strength (MPa)	448–515	896–1200	448–648	1606	965–1000	1500
Tensile strength (MPa)	655–889	1399–1586	951–1220	1896	1206	1795

[a] ASTM designation.

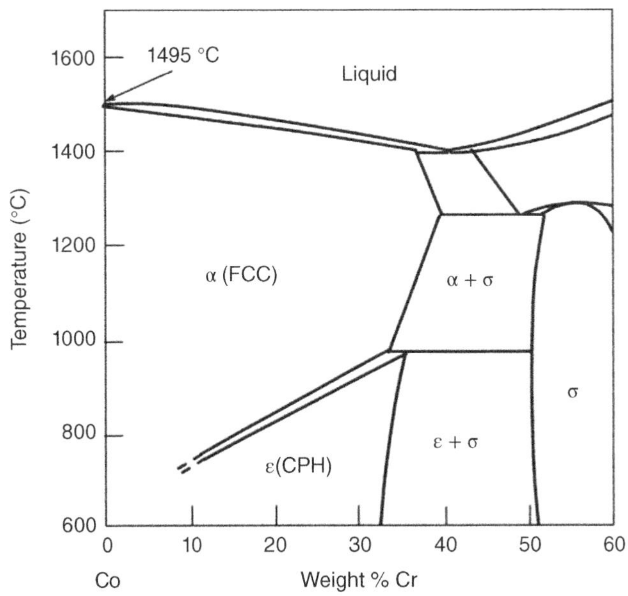

Figure 6.18 Portion of the cobalt–chromium phase diagram. *Source:* Adapted from Ishida and Nishizawa (1990).

version of the Co–Cr phase diagram (Ishida and Nishizawa, 1990), Co can dissolve up to ~35–40 wt% Cr at ~1400 °C, forming an α solid solution (Figure 6.18). The solubility of Cr in Co does not change considerably with decrease in temperature but, starting at ~950 °C, the α phase starts to undergo a martensitic-like transformation to the ε phase. The transformation temperature decreases with decreasing amount of Cr, down to ~700 °C at a Cr concentration of ~10 wt%.

Cobalt–Chromium–Molybdenum (Co–Cr–Mo) Phase Diagram

Co–Cr alloys used as biomaterials can show a variety of phases and phase distributions, depending on their composition and production method. As Co–Cr–Mo alloys approximating the F75 composition have been used for decades in engineering applications, the phase diagram for this system has received some attention. Overall, 5–7 wt% Mo (plus small concentrations of a variety of elements) is added to a Co–Cr alloy containing 26–30 wt% Cr to form the F75 composition (Table 6.2). The addition of ~6 wt% Mo to Co leads to a small reduction in the solubility limit of Cr in Co but it does not significantly affect the α to ε phase transformation temperature (Gupta 2005). Due to the high reactivity of Co, Cr, and Mo toward C, carbide precipitates form if the concentration of C in the alloy is sufficiently high (greater than ~0.15 wt%). Nitride precipitates can also form but they are less important. The carbide precipitates are typically composed of mixed carbides with the general formula $M_{23}C_6$, where M represents more than one of the three major elements Co, Cr, and Mo. Consequently, the equilibrium phases that can be present in Co–Cr–Mo alloys approximating the F75 composition are often

- The α (FCC) solid solution (above ~900 °C) or ε (CPH) solid solution (below ~900 °C) composed of a Co solid solution with Cr and Mo, which forms the major phase
- Carbide precipitates, primarily $M_{23}C_6$, present at the grain boundaries of the major phase, within the grains or a combination of both.

Microstructure and Mechanical Properties of Co–Cr–Mo Alloys

Phase diagrams, as we have seen, show the phases present in a material under equilibrium conditions but they provide a useful guide from which other nonequilibrium constituents can be inferred. Due to the sluggishness of the α to ε phase transformation, the major phase formed upon cooling from the α phase region to room temperature is not necessarily the ε phase. Instead, depending on the time and temperature of the thermal or thermomechanical treatment used in their production, Co–Cr–Mo alloys can show a major phase composed of the metastable α phase (FCC), the thermodynamically stable ε phase (CPH) or a combination of both phases at room temperature. Grains of the α phase commonly have a more equiaxial shape whereas the ε phase is commonly composed of martensitic plate-like grains.

As the FCC unit cell has a considerably larger number of independent slip systems than the CPH unit cell (Table 4.3), the α phase is more ductile than the ε phase. Consequently, the α phase provides better ductility whereas the ε phase provides better strength and hardness (wear resistance) but a lower ductility (elongation to failure). In general, the relative amount of these two phases in Co–Cr–Cr alloys is controlled through the selection of suitable alloying elements and forming conditions (thermal or thermomechanical treatments).

The amount of carbide precipitates, such as $M_{23}C_6$, their size and their distribution in the alloy depends primarily on the amount of C in the alloy and the production treatments. Fine precipitates at the grain boundaries are beneficial in controlling grain size and, when present within the grains, they impede dislocation motion. Consequently, for optimal strengthening, fine carbide precipitates both at the grain boundaries and within the grains might be desirable. On the other hand, excessive amounts of precipitates, particularly of large size, lead to a reduction in ductility. Overall, the more ductile Co–Cr–Mo alloys tend to have a low concentration of C (less than ~0.15 wt%).

- *Cast alloys*: Alloys approximating the F75 composition are commonly made by melting at 1400–1450 °C and casting into ceramic molds. In common with the casting of other metals, nucleation and growth starts at the mold wall and the solid phase tends to grow into the liquid typically in a dendritic manner (Figure 6.10). Upon completion of the solidification process, the alloy often consists of large dendritic grains (typically larger than ~100 μm) composed of the α phase and an interdendritic phase that fills in the region between the branches.

The solidification process is prone to micro-segregation of the alloy constituents. The "high" amount of C (~0.15–0.35 wt%) in cast F75 alloys leads to the formation of carbides such as $M_{23}C_6$ in both the dendritic grains and interdendritic region. Precipitates of the σ phase, an intermetallic compound having the formula CoCr and a tetragonal crystal structure, can also form in the interdendritic region. The presence of flaws such as small cavities is also a common feature of casting.

Cast F75 alloys are commonly annealed within the α phase field (~1200 °C) to homogenize their microstructure. Annealing serves to reduce dendritic and interdendritic carbide segregation by dissolution and re-precipitation as fine carbide particles within the grains and at the grain boundaries and to remove any σ phase that might be present. Strengthening mechanisms are primarily solid solution hardening due to the alloying elements and precipitation hardening due to carbide precipitates. However, large grains, excessive amounts of carbides at the grain boundaries and casting defects limit the mechanical properties of cast F75 alloys (Table 6.2).

- *Wrought alloys*: F799 products are formed by hot forging of cast ingots. Commonly, the cast ingot is annealed at ~1200 °C to homogenize the as-cast microstructure. Then it is subjected to several iterations of hot working (~800 °C) and "cold" working at lower temperature. This thermomechanical treatment leads to grain refinement and, thus, to a microstructure composed typically of a combination of α and ε grains having a nearly equiaxial shape and a size far smaller than those in cast F75 alloys.

Despite its low content (less than ~0.15 wt%), carbon plays an important role in controlling the microstructure and, thus, the mechanical properties of hot forged F799 alloys. First, hot working and cold working in the forging operation rely on the ability to deform the alloy plastically. Because of the low amount of carbon, the formation of an excessive amount of carbide precipitates at the grain boundaries is avoided, resulting in good ductility for forging. Second, carbon atoms go into solid solution interstitially and serve to enhance the stability of the more ductile α phase. Third, the thermomechanical treatment leads to a dispersion of fine carbide precipitates within the grains, enhancing precipitation strengthening.

Overall, the microstructure composed of a combination of ductile α (FCC) grains and harder martensitic ε (CPH) grains, fine grain size and fine intragranular carbide particles provides an attractive combination of strength and ductility. Despite having a composition approximating that of cast F75 alloys, hot-forged F799 alloys have a strength that is approximately two times higher (Table 6.2) and better fatigue resistance. On the other hand, hot forging is more expensive and has a limited capacity to form complex shapes.

Microstructure and Mechanical Properties of Co–Cr–W–Ni

Co–Cr–W–Ni alloys designated F90 can provide a better combination of strength and ductility (elongation to failure) when compared to cast Co–Cr–Mo (F75) alloys (Table 6.2). A combination of high strength and ductility is often desirable in most metals used as biomaterials but it is especially so in biomedical devices such as stents that are thin and have to undergo a large deformation upon deployment. Despite a substantial Ni content (~10 wt%), F90 alloys have sufficient biocompatibility for use *in vivo*. These alloys also have good forming and machining characteristics and, because of their good ductility, they can be hot-worked by, for example, forging at ~1200 °C, or cold-worked. Annealing of hot-worked alloys gives strengths comparable to cast F75 alloys whereas cold-worked alloys have strengths that are two to three times higher (Table 6.2). Both hot-worked and cold-worked alloys have far better ductility than cast F75 alloys.

Similar to our earlier discussion for other metals and alloys, the corrosion resistance and attractive mechanical properties of F90 alloys are dependent on their composition and on their structure at different length scales. However, it is often difficult to produce F90 alloys with optimal mechanical properties because of the complexity of their compositional and the thermomechanical treatments used in their forming. Similar to stainless steels, the high amount of Cr (~20 wt%) is primarily responsible for their corrosion resistance. When compared to F75 and F799 alloys, carbon precipitates are commonly not present in F90 alloys, either intergranularly or intragranularly. This is due primarily to the low amount of C in the alloys (less than 0.1 wt%) and a final annealing treatment at ~1200 °C used in their production. Consequently, unlike F75 alloys, carbide precipitates at the grain boundaries do not limit the ductility of F90 alloys.

Carbon and nickel atoms substitute interstitially and substitutionally, respectively, in the Co lattice and they are known to stabilize the more ductile α (FCC) phase (Betteridge 1982). Consequently, they provide solid solution strengthening and enhance ductility. As W has a considerably larger atomic radius (193 pm) than Co (152 pm), the substantial amount of W (~15 wt%) in F90 alloys introduces considerable strain in the Co lattice. The strain field associated with the W atoms can limit dislocation motion and, thus, increase the yield strength of these alloys.

Because of the thermomechanical treatments used in their production and a final annealing treatment at ~1200 °C, F90 alloys are commonly composed almost entirely of α-phase grains and often have a small grain size. This means that solid solution strengthening and grain size refinement are expected to be the primary mechanisms responsible for their strength and ductility.

Co–Ni–Cr–Mo Alloys

When compared to the three Co–Cr alloys presented so far, Co–Ni–Cr–Mo alloys designated F562 are characterized by a high Ni content (~35 wt%), approximately equal to their Co content. Thus, the F562 alloys might be better classified as Co–Ni (or Ni–Co) alloys rather than Co–Cr alloys. Like F90 alloys, F562 alloys provide an attractive combination of strength and ductility. However, while F562 alloys show good corrosion resistance, their high Ni content has generated concerns about potential adverse biological effects, such as metal ion sensitivity and allergy, resulting from Ni ions that can dissolve out from the alloys over time (Chapter 19). Consequently, the F562 alloys are used only in applications where their attractive mechanical properties such as high strength and fatigue resistance cannot be matched by other alloys.

6.9.4 Nickel-Titanium Metals and Alloys

Nickel, as we have seen, is an important alloying element in stainless steels and Co–Cr alloys used as biomaterials. However, in recent years, interest in the use of nickel-based alloys or alloys composed of a high Ni concentration has declined due to concerns about potential adverse biological effects arising from nickel ions released from these alloys. On the other hand, a nickel-titanium metal, referred to as Nitinol, has witnessed growing use in medical and dental applications because of its ability to show unique types of behavior referred to as shape memory and superelasticity.

Nitinol is composed of approximately equal atomic amounts of Ni and Ti, equivalent to ~55 wt% Ni and ~45 wt% Ti. Although it is often called an alloy, Nitinol is actually an intermetallic compound. Unlike alloys such as Ti6Al4V, steels and Co–Cr alloys in which the alloying metal atoms are incorporated randomly in the host metal lattice, Ni and Ti atoms in Nitinol occupy specific sites in the unit cell and show a regular repeating pattern in three dimensions. The shape memory effect in Nitinol was found in 1963 at the Naval Ordinance Laboratory (United States) and its name is derived from the symbols for nickel and titanium, and the first letters in this laboratory's name. Nitinol is not the only metal that shows shape memory and superelasticity but these effects, along with suitable strength, ductility, and corrosion resistance, endow Nitinol with an attractive combination of properties for practical applications. Consequently, Nitinol has been used in a variety of engineering applications.

Despite its high Ni concentration, Nitinol shows good biocompatibility. This is primarily due to its high Ti concentration that leads to the formation of a passivating TiO_2 surface layer that provides high corrosion resistance. This passivating oxide surface layer limits the amount of Ni ions (and Ti ions) that can dissolve out of Nitinol, in much the same way that a passivating oxide surface layer does for the metallic biomaterials described earlier. Based on its good biocompatibility and aforementioned properties, Nitinol has witnessed growing use in biomedical devices, such as stents and orthodontic wires, reamers and files (Table 6.1).

Shape Memory and Superelasticity of Nitinol

Upon cooling Nitinol below ~30–50 °C, it undergoes a displacive transformation from an austenitic (BCC) phase to a martensitic (monoclinic) phase (Figure 6.19). If Nitinol, composed of this martensitic phase, is deformed and then heated above the transformation temperature, it returns to its austenitic phase and shape even after significant deformation (up to ~10% strain). It is as if Nitinol has the ability to "memorize" its original pre-deformed shape under certain thermal conditions. This effect is termed shape memory. It involves both a temperature change and a mechanical stress.

Above the transformation temperature, deformation of Nitinol can induce its transformation from the austenitic phase to the martensitic phase. Even after being highly deformed (up to ~10% strain), Nitinol has the ability to recover its original pre-deformed shape almost spontaneously upon removing the stress. This is the superelastic effect. The material has the ability to deform reversibly to high strains and recover its original pre-deformed shape upon unloading (Figure 6.20). Unlike shape memory, superelasticity does not require thermal changes, that is, it is temperature independent, but it can be influenced by the actual temperature at which the test is performed.

It will be recalled that the elastic limit that defines the recoverable strain is typically less than ~0.2% for metals (Chapter 4). Deformation to higher strains occurs by dislocation motion, leading to unrecoverable plastic strain. The large recoverable strains in Nitinol are produced not by dislocation motion but, instead, they depend on different structural features specific to Nitinol.

Phase Transformation in Nitinol

The austenite to martensite transformation in steels (Section 6.3) forms a starting point for understanding the transformation and, thus, shape memory and superelasticity in Nitinol. Like carbon steels, the higher temperature phase in Nitinol is called austenite whereas lower temperature phase is called martensite (Figure 6.19). When Nitinol in the austenitic phase is cooled, transformation starts at the martensite start temperature (M_s) and ends at the martensite finish temperature (M_f). Typically, all the austenitic phase has transformed to the martensitic phase at M_f. Conversely, when Nitinol in the martensitic phase is heated, transformation to the austenitic phase stars at the austenite start temperature A_s and is completed at the austenite finish temperature A_f. The transformation curves upon cooling and heating do not coincide (Figure 6.19). They have a similar S-shape but one curve is shifted by ~20–30 °C relative to the other.

While the terms austenite and martensite were used originally to refer to the higher temperature phase and lower temperature phase in carbon steels (Section 6.6.2), they are now generalized to identify phases participating in similar types of displacive solid-state transformations. The use of these terms does not mean that the structural changes resulting from the transformation in Nitinol are the same as those in carbon steels.

As noted earlier, the austenitic phase in Nitinol has a BCC structure while the martensitic phase has a monoclinic structure. Whereas the austenite to martensite transformation in steels results in a highly strained lattice, a shape change and a volume change (~3–4 vol% for a low carbon steel), the transformation in Nitinol is accompanied by a shape change only (the volume change in Nitinol is −0.003 vol%). The BCC (austenite) to monoclinic (martensite) transformation is accommodated by a mechanism referred to as twinning. Instead of forming a single crystal in the transformation, the martensite forms two crystals (twins) in a symmetrical fashion (Figure 6.19). This twinned microstructure is a consequence of energy minimization in the austenite to martensite transformation in Nitinol.

Twinning is a reversible process and is the dominant accommodation process in the shape memory effect. Upon heating, the twinned martensitic structure returns to the BCC austenitic structure above A_f. Unlike martensite in steels, the twinned martensite in Nitinol is more deformable and, upon application of a stress, it deforms by motion of the twin boundaries and not by dislocation motion. The twins that are more favorably oriented to the applied stress grow at the expense of the others. If a sufficiently high stress is applied to Nitinol, it is possible to produce a structure composed of the most favorably oriented twins, a process referred to as de-twinning. Heating this deformed and

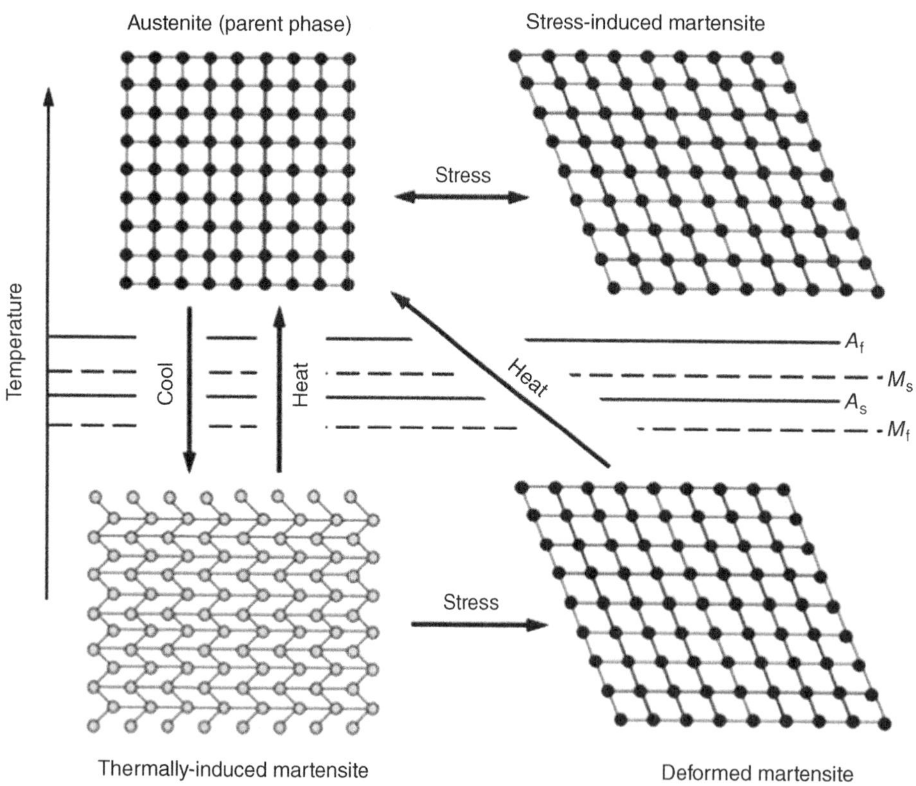

Figure 6.19 Schematic illustration of the shape memory and superelastic effects in Nitinol.

de-twinned martensite above A_f transforms Nitinol back to austenite in its original pre-deformed shape.

The austenite to martensite transformation in Nitinol can also be induced at temperatures above M_s by the application of a stress. As austenite is the thermodynamically stable phase above M_s, upon release of the stress, Nitinol returns to the austenite phase with its pre-deformed shape (Figure 6.19). This is the superelastic effect.

As the shape memory and superelastic effects occur within a narrow range of temperature dictated by the transformation temperatures, control of these transformation temperatures is crucial in the application of Nitinol. Several factors influence the transformation temperatures but the most important is the Nitinol composition, particularly the Ni concentration. For a Nitinol that has been normalized in the austenitic phase, M_s is ~50 °C for a composition composed of 50.0 atomic percent of Ni. A reduction in Ni content by ~1.0 atomic percent (to 49.0%) leads to a continuous increase in M_s by only ~25 °C, that is, increasing the Ti concentration in the metal has a weak effect on M_s. On the other hand, increasing the Ni content above the equiatomic composition leads to a steep reduction in M_s. An increase in Ni content by 1.0 atomic percent (to 51.0%), for example, leads to reduction in M_s by over 100 °C (to approximately −70 °C). Consequently, the most effective way to change M_s is to change the Ni content.

Use of Nitinol as a Biomaterial

An application that utilizes the shape memory effect of Nitinol is the stent (Figure 1.1g). After forming Nitinol into the requisite shape and dimensions at a temperature above M_s, the stent is cooled below M_f and deformed into a more compact shape for delivery into a patient's blood vessel by means of a catheter (Chapter 24). The temperature increase to the body temperature (~37 °C) and release from the catheter allows the stent to expand until it is pressing against the lumen of the blood vessel. As this shape memory effect depends on transformation of the deformed martensite phase to the austenite phase, A_f should be lower than the body temperature.

Another application of Nitinol is wire in orthodontic braces. This application utilizes the superelastic effect and, thus, A_f should be below the temperature in the mouth (~37 °C). Strained Nitinol (martensite phase) wire is inserted into brackets attached to the teeth and as the teeth move, the stress on the wire is gradually reduced as the wire transforms to austenite. Consequently, a continuously decreasing stress is maintained on the teeth, reducing the number of visits that a patient has to make to the orthodontist.

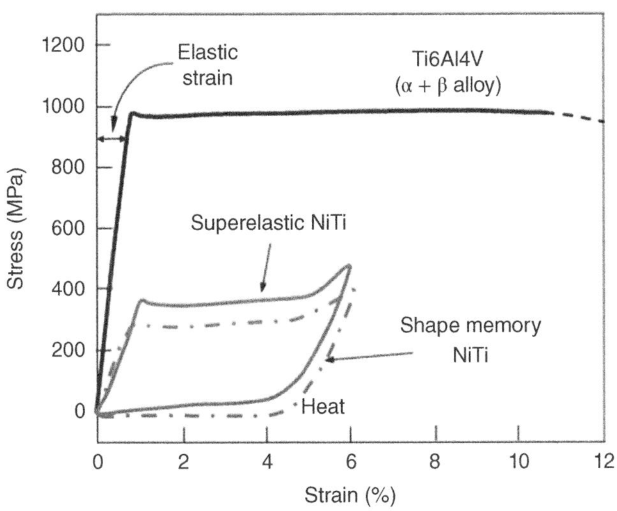

Figure 6.20 Illustration of stress versus strain response of superelastic and shape memory NiTi alloy and a Ti6Al4V alloy. Whereas the large strains are recovered upon release of the stress (superelastic NiTi) or heating the deformed shape memory NiTi alloy above the austenite finish temperature, only the small elastic strain is recovered in the Ti6Al4V alloy.

6.9.5 Tantalum

Tantalum has a high density (16.65 g/cm^3), nearly twice that of stainless steels and Co–Cr alloys used as biomaterials. Nonporous Ta implants would be very heavy and expensive as well because tantalum raw materials occur only to a limited extent naturally. On the other hand, in a highly porous form, Ta finds use in certain orthopedic applications to heal bone defects, particularly those resulting from revision hip and knee joint surgery. Tantalum used in these applications typically has a highly porous microstructure (70–80% porosity) of interconnected pores (~500 μm in size) that mimics the microstructure of dry human trabecular bone (Figure 6.21). This microstructure

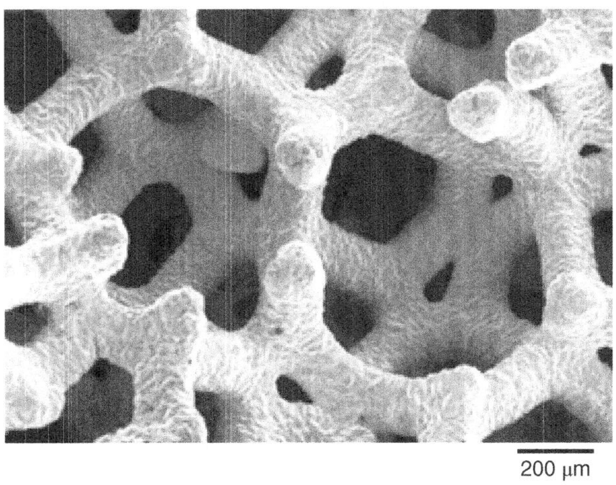

Figure 6.21 Microstructure of porous tantalum formed by a CVD process. *Source:* From Hacking et al. (2000).

is created using CVD that deposits a thick Ta coating (~50 μm) on a carbon foam having the requisite porous architecture. (The material produced in this way is known commercially as Trabecular Metal™.) As this process is expensive, porous Ta is used mainly in applications where it provides an advantageous combination of properties over other metals such as porous titanium.

In addition to good ductility, porous Ta has an elastic modulus of ~3 GPa and a compressive strength of 50–80 MPa, mechanical properties that are weaker than those of human cortical bone (Table 4.1). Consequently, porous Ta is more suitable for healing defects in bones that are not subjected to high physiological stresses. When implanted *in vivo*, porous Ta is conducive to rapid bone and soft tissue ingrowth and attachment (Hacking et al. 2000), attributed mainly to a combination of its attractive porous microstructure and micro-rough surface produced by the CVD process.

6.9.6 Zirconium Alloys

An alloy composed of Zr and ~2.5 wt%, Nb has witnessed limited use as bearings in total joint replacement, typically as the femoral head in prosthetic hips and the femoral bearing in prosthetic knees. For these applications, the alloy is subjected to an oxidizing treatment of its articulating surface, resulting in a surface layer (~5 μm thick) composed of zirconium dioxide (ZrO_2) (Figure 6.22). Unlike coatings deposited by several surface-coating techniques (Chapter 13), the interface between the ZrO_2 surface layer and the underlying metal does not show an abrupt change in composition. Instead, the interface has a graded composition, from ZrO_2 to Zr, which provides strong adhesion between the ZrO_2 surface layer and the underlying metal.

Oxidized zirconium bearings (known commercially as Oxinium™) are intended to provide the following benefits

- Low wear of the articulating bearing surface, due to a hard ceramic surface (ZrO_2)
- Reduction in the potential for catastrophic fracture of the bearing, when compared to an all-ceramic bearing (such as Al_2O_3), due to the bulk of the bearing being a ductile metal
- Reduction in the unpredictable failure of hard coatings on metal bearings caused by delamination and cracking, due to strong adhesion resulting from the compositionally graded interface.

Oxidized zirconium has shown a reduction in wear in both clean and abrasive conditions when compared to Co–Cr alloys (Ries et al. 2002). In comparison, oxidized zirconium femoral heads show higher wear than Al_2O_3 when articulating against commonly used ultrahigh molecular weight polyethylene (UHMWPE) acetabular cups, due mainly to the ZrO_2 surface having a lower hardness than Al_2O_3.

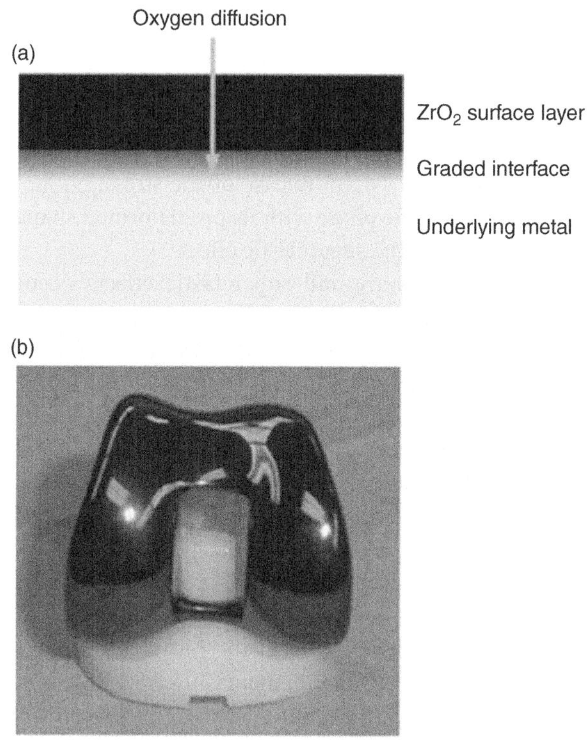

Figure 6.22 (a) Schematic of the surface modification of Zr alloy to produce a ZrO_2 bearing surface. (b) Oxinium™ bearing for total knee arthroplasty. *Source:* From Rahaman et al. (2007) / with permission of John Wiley & Sons.

6.9.7 Noble Metals

The noble metals Au, Ag, and Pt are ductile but, because of their low strength, they are used in applications where properties other than strength are of primary importance. For example, because of their high electrical conductivity, high ductility, and the high corrosion resistance, silver, platinum and certain platinum alloys are used as electrodes in medical devices. Gold is expensive and its low strength can be improved by alloying with elements such as Cu and Pt. Whereas Au and its alloys have witnessed use for many years in dental restorations, they have been almost completely replaced in these applications by strong, corrosion-resistant, and less expensive ceramics such as yttria-stabilized zirconia and glass-ceramics such as lithium-aluminum-silicate (Chapter 7).

Emerging applications of Au, Ag, and Pt in the last couple of decades have relied on these metals in the form of nanoparticles (less than ~50–100 nm) rather than macroscopic 3D objects. For example, Ag in the form of nanoparticles and coatings shows antibacterial properties under some conditions (Chapter 20). Gold nanoparticles have been widely studied for use as nanocarriers in the delivery of chemotherapy drugs and other therapeutic biomolecules to treat cancer (Chapter 26) and in diagnostic imaging as well.

6.10 Degradable Metals

Considerable effort has been devoted in the last few decades to the design and development of degradable metals as alternative biomaterials to mechanically durable, corrosion-resistant metals for use in applications that do not require the long-term presence of an implant *in vivo*. In these applications, it would be beneficial to rely on the attractive mechanical properties of metals coupled with the capacity to degrade and disappear completely in a safe manner after serving their function. Devices for use in orthopedic surgery, such as intramedullary rods and screws for bone fracture healing, and stents for use in cardiovascular surgery are the areas that have witnessed the most significant advances.

Durable corrosion-resistant metals such as Ti6Al4V, stainless steel, and Co–Cr alloys have the advantages of mechanical reliability under physiological loads, retention of shape and mechanical properties, and biocompatibility. Dissolution of metal ions into the physiological fluid is severely limited by the high corrosion resistance of these metals. On the other hand, devices composed of these metals remain in the body for the lifetime of the patient or until they are removed in a second (or revision) surgery, an operation that is not just costly but can often be a traumatic experience for the patient as well.

In some applications, such as bone fracture healing, the long-term presence of a device is not required and, in some situations, can have drawbacks. For example, most metallic implants used in orthopedic applications, even those with some moderate level of porosity, have a Young's modulus (or stiffness) that is much higher than that of bone (Table 4.1). This large difference in stiffness leads to stress shielding in which a large portion of the physiological stresses is supported by the metal implant. This can cause resorption of the bone surrounding the implant, weakening of the bone and, eventually, catastrophic failure of the weakened bone (Chapter 17). In cardiovascular surgery, the long-term presence of a durable metallic stent in an artery can cause accumulating adverse biological responses that can lead to restenosis. This necessitates revision surgery that, in some cases, can lead to death of the patient.

Degradable metals having an optimal combination of mechanical properties, degradation rate and biocompatibility can potentially circumvent or reduce some of the drawbacks of durable metals in some biomedical applications. In bone healing, for example, after the implant has served its function, it would be beneficial for it to degrade at an acceptable rate and eventually disappear or remodel into a bone-like tissue. This is more in line with the tissue engineering approach (Chapter 25). Degradable metal implants may be particularly attractive for pediatric patients where a temporary implant is more often desirable. In cardiovascular surgery, it would be beneficial for a stent to degrade away safely after healing a blocked artery, thus reducing the potential for further irritation of the endothelium of the artery and restenosis.

Significant advances have been made in designing and developing degradable metallic biomaterials, particularly in the last couple of decades. Orthopedic devices composed of magnesium alloys, such as fracture fixation rods and screws, and cardiovascular devices, such as stents, have been tested in animals and human clinical trials. However, despite these advances, major problems remain in creating degradable metals with optimal properties for use in these and other applications.

6.10.1 Designing Degradable Metals

A major problem in creating degradable metal implants is in designing alloy compositions with the requisite combination of properties for the desired application. An important property of the implant is its degradation rate because this has a strong influence on its performance *in vivo*, as related to the following factors:

- *Toxicity*: The degradation rate controls the release of metal ions from the implant into the physiological environment. Too rapid a degradation rate can lead to the release of ions at a rate that exceeds the rate at which these ions can be cleared from the body, which can lead to toxicity. Ions of some alloying elements, for example, are toxic even in low concentrations.
- *Reduction in mechanical properties over time*: If the degradation rate is too rapid, the mechanical properties of an implant can become inadequate to maintain its integrity and to support physiological stresses over the desired implantation time, which can lead to failure of the implant. The reduction in mechanical properties with implantation time is also difficult to predict because the degradation rate varies with implantation site.
- *Release of H_2 gas*: The degradation of metals in an aqueous liquid, such as the physiological fluid, often leads to the evolution of hydrogen (H_2) gas, which, in excessive amounts, can lead to cell and tissue damage. Additionally, concerns have been expressed about the release of excessive amounts of H_2 gas from degradable metal stents. This

can potentially lead to a temporary reduction in blood flow in the artery, resulting in serious complications.

- *Increase in pH*: The degradation process can lead to an increase in pH (alkalization) of the physiological medium, particularly near the surface of an implant, which can adversely influence the response of cells.

6.10.2 Degradable Magnesium Alloys

Magnesium alloys and, to a lesser extent, iron alloys have been most widely studied and developed for use as degradable metallic biomaterials. Magnesium ions are essential constituents of the human body. They have the fourth largest concentration (~1 mM) of all cations in human blood, and they are involved in several metabolic reactions and biological mechanisms. At concentrations below some threshold level, Mg ions can be safely cleared from the human body by normal biological processes (such as excretion through the urinary tract). On the other hand, CP Mg is limited by inferior mechanical properties and an undesirably high degradation rate. Consequently, Mg is alloyed with other elements to provide a better combination of properties for use as implants *in vivo*.

Degradation of Magnesium Alloys

As discussed in Chapter 14, degradation of metals in an aqueous environment is generally referred to by the term corrosion. Corrosion is an electrochemical process, which involves atoms releasing electrons within the metal and forming ions that dissolve into the aqueous environment, and conduction of the electrons to another region where they are consumed. Above a certain concentration determined by their solubility product in the aqueous medium, magnesium ions can form an insoluble hydroxide. The overall reaction for magnesium-based alloys can be written

$$Mg(s) + 2H_2O \leftrightarrows Mg(OH)_2(s) + H_2(g) \tag{6.14}$$

where, (s) and (g) are included in Eq. (6.14) to denote the solid and gaseous phase, respectively. In the presence of chlorine ions, present at a concentration of 100–110 mM in the physiological fluid, $Mg(OH)_2$ dissolves to form the soluble salt magnesium chloride, generating hydroxyl ions in the process:

$$Mg(OH)_2 + 2Cl^- \leftrightarrows Mg(Cl)_2 + 2OH^- \tag{6.15}$$

The degradation rate of magnesium alloys *in vivo* depends not only on the composition and microstructure of the alloy but also on the implantation site as well. Ions and molecules in the physiological fluid at these sites can strongly influence the degradation rate. In general, rapid degradation is an intrinsic response of magnesium alloys in the physiological fluid. It is accompanied by the evolution of H_2 gas and an increase in pH (alkalization). These two effects are most pronounced near the surface of the degrading alloy and, if severe enough, can lead to adverse biological effects.

In vitro, the degradation rate of a metal can be measured using an electrochemical method or an immersion method. The immersion method is easy to perform and does not require any sophisticated equipment. The mass W of a material of area A and density d is measured as a function of immersion time t in the appropriate liquid and the degradation rate $\dot{R}$ is determined using the equation

$$\dot{R} = \frac{\Delta W}{Adt} \tag{6.16}$$

where, $\Delta W = W_o - W$, and W_o is the initial mass (at $t = 0$). It is impractical to measure the weight loss *in vivo* but the loss in volume ΔV (equal to $\Delta W/d$ in Eq. (6.16)) can be determined as a function of t using microcomputed tomography (microCT). Then the degradation rate is determined using the equation

$$\dot{R} = \frac{\Delta V}{At} \tag{6.17}$$

The degradation rate of an implant is an important design property as it controls the rate at which it releases metal ions, the reduction in mechanical properties over time, the rate at which H_2 gas is evolved and the degree of alkalization. Consequently, considerable effort has been devoted to designing metals having an acceptable degradation rate for the desired application. By far the most common approach to controlling the degradation rate has been the use of alloying elements. In addition, a suitable coating composed of a degradable polymer, for example, can be used to reduce the degradation rate in the early period of healing when the degradation rate of the uncoated implant is fastest.

Example 6.2 An intramedullary nail (5 mm in diameter × 20 cm) composed of a Mg-Zn alloy (5 wt% Zn) degrades at a rate of 1.2 vol% per day. Determine the mass of magnesium and the volume of H_2 released per day, given that the density of the alloy is 1.80 g/cm^3.

Solution:

The volume V of the nail is $\pi(d/2)^2h$, where d is the diameter and h is the length of the nail, giving $V = \pi \times (0.5/2)^2$

$\times 20\,cm^3 = 3.93\,cm^3$. As the rod degrades at 1.2 vol% per day, the change in volume $\Delta V = 0.047\,cm^3$ per day. The change in mass ΔW of the rod per day is equal to the change in volume times the density of the rod, giving $\Delta W = 0.085$ g per day. As Mg makes up 95% of the mass of the rod, the mass of Mg ions released = 0.080 g per day or 80 mg per day.

According to Eq. (6.14), upon degradation, 1 mole (24.31 g) of Mg liberates 1 mole of H_2 gas. Thus 0.080 g of Mg releases 0.080/24.31 mole of H_2 gas per day. As the volume of 1 mole of gas at STP (standard temperature and pressure) is 22.4 liters, the volume of H_2 gas released per day is $(0.080/24.31) \times 22.4 \times 1000\,cm^3 = 73.7\,cm^3$.

Alloying Elements

Alloying elements, mainly from the following groups, have been studied to determine their effect on the degradation rate and mechanical properties of magnesium as a function of time:

- Ca, Sr, Zn, and Mn that are, in various concentrations, essential constituents of the human body and, thus, are expected to be nontoxic below some threshold concentration
- Zr, Si, and Sn that are commonly nontoxic at sufficiently low concentrations
- Al, a common alloying element in engineering alloys. However, concerns have been expressed about possible adverse biological effects of Al ions *in vivo*, such as neurotoxicity (Bondy 2010)
- Rare-earth elements (REs), such as Y and elements selected from the lanthanide series, but the toxicology and metabolic pathways associated with ions of these elements are not clear.

The degradation rate and mechanical properties of magnesium show a complex dependence on the composition and concentration of the alloying element even for single elements (Zheng et al. 2014). This dependence can be further complicated by the presence of trace impurities of various metals, such as Fe, Co, and Ni. Calcium has been one of the most common alloying elements. Calcium has a low solid solubility limit in Mg (1.3 wt%), presumably due to its ionic radius being much larger than that of Mg. At low concentration, below its solid solubility limit, Ca leads to solid solution strengthening of Mg whereas at higher concentration, it leads to a reduction in strength and ductility, and an enhancement in degradation rate, presumably due to the formation of intermetallic precipitates such as Mg_2Ca.

Zinc has a higher solid solubility limit (~6 wt%) than Ca in Mg, which can be rationalized in terms of its smaller ionic radius, comparable to that of Mg. Zinc provides solid solution strengthening that increases with its concentration. On the other hand, too high a Zn content is undesirable because it leads to a higher degradation rate due to the formation of large amounts of intermetallic precipitates such as $Mg_6Zn_3Ca_2$.

Yttrium has as a high solid solubility limit (~11 wt%) in Mg. At concentrations below this limit, Y leads to an improvement in mechanical properties, presumably by solid solution strengthening, and corrosion resistance, presumably due to little or no formation of precipitates. However, the toxicology and metabolic pathways associated with yttrium ions are not clear.

Trends determined from the use of single alloying elements have been incorporated into designing magnesium alloys composed of two or more alloying elements to provide a better combination of degradation rate and mechanical properties. One approach has been to use a low amount of Zn in the alloy, thereby reducing the tendency to form intermetallic precipitates (such as $Mg_6Zn_3Ca_2$) and compensate for the reduction in solid solution strengthening by addition of a second alloying element such as Y. Table 6.3 shows data for two alloys, denoted ZX50 and WZ21 according to their ASTM designation, that exemplify this approach. These alloys have been developed and evaluated *in vitro* and in animal models *in vivo* for potential use in bone healing. Alloy ZX50, composed of a higher Zn content and small amounts of Ca and Mn, degrades considerably faster *in vivo* than alloy AZ21 in which a large fraction of the Zn is replaced Y (Section 6.10.2.4). However, the addition of Y does not fully compensate for the reduction in strength due to the lower Zn content.

Another approach has been to use a low amount of Zn when compared to the ZX50 alloy but, instead of using a second alloying element such as Y, compensate for the reduction in solid solution strengthening by grain refinement. The effectiveness of this approach can be enhanced by improving the purity of the alloy, that is, by reducing the level of Mn and trace impurities. The ZX10 alloy in Table 6.3 is an example of this approach (Hofstetter et al. 2014). A reduction in Zn content (to ~1.0 wt%) leads to the formation of a homogeneous distribution of fine precipitates composed of the intermetallic M_2Ca, where M represents a combination of Mg and Zn. When compared to $Mg_6Zn_3Ca_2$ precipitates in ZX50, these precipitates inhibit grain growth during hot forming of the alloy and lower the degradation rate of the product. Consequently, the ZX10 alloy shows a higher yield strength due to its small grains (~2 μm) and a lower degradation rate.

Mechanical Properties

Wrought magnesium alloys show better mechanical properties than their cast counterparts but, overall, the mechanical properties of magnesium alloys (Table 6.3) are far inferior to those of nondegradable alloys such as Ti6Al4V, stainless steel, and Co–Cr alloys (Table 4.1). As fabricated, wrought (extruded) rods of alloys ZX50 and WZ21, for example, have an average yield strength of 210 MPa and 150 MPa, respectively, values that are only comparable to the compressive strength of human cortical bone. Consequently, the use of magnesium alloys to repair large defects in structural bone is challenging.

In addition to the mechanical properties of the as-fabricated metal, the reduction in mechanical properties over time is also a critical design parameter. The alloy should degrade at a rate such that its strength and stiffness are sufficient to maintain its integrity and ability to support physiological stresses during the healing process. Thereafter, a degradation rate to avoid toxicity and excessive evolution of H_2 gas is suitable.

According to some guidelines, degradable magnesium alloys should maintain their integrity for at least 12–18 months when used as implants for bone healing in adult patients, whereas for pediatric patients, a shorter duration may be acceptable. Degradable magnesium alloy stents used in clinical trials have been safely degraded within 4–5 months.

Use of Degradable Magnesium Alloys as Biomaterials

Magnesium alloys have been evaluated in animals and in human clinical trials, mainly as stents in cardiovascular surgery and orthopedic devices in fracture healing. Clinical trials of a drug-eluting magnesium alloy stent, composed of ~7 wt% rare earth elements and coated with a polylactic-co-glycolic acid (PLGA) layer containing the antiproliferative drug paclitaxel have shown promising results over an evaluation period of 12 months (Haude et al. 2013). These stents could provide alternatives to degradable polymer stents (Chapter 24).

In orthopedic applications, the use of degradable magnesium alloys to heal large defects in highly loaded bone is challenging due to their inferior mechanical properties. These alloys are also unsuitable for use in devices where durable mechanical properties must be guaranteed over a long duration, such as prosthetic hip and knees. In comparison, degradable magnesium alloys are more promising for use as load-sharing devices, such as screws and intramedullary rods in bone fracture healing, where they support a portion of the physiological load, or as devices that support little load. In these applications, degradable magnesium alloys compete with degradable polymers and their composites, although they have better strength and stiffness.

Some degradable magnesium alloys are showing promising performance in bone healing. Figure 6.23 shows images from an *in vivo* study in which extruded cylindrical pins (1.6 mm in diameter × 8 mm) of degradable magnesium alloys ZX50 and WZ21 (Table 6.3) were evaluated in a rat model (Kraus et al. 2012). The pins were implanted in the femurs of five-week old male Sprague–Dawley rats in a direction perpendicular to the long axis of the femurs. Whereas the degradation of ZX50 implants commenced soon after implantation at a rapid rate (~1.2% volume loss per day), generating considerable H_2 gas in the process, WZ21 implants degraded more slowly, maintained their integrity for more than four weeks and, thereafter, degraded more slowly (~0.5% volume loss per day). A larger amount

Table 6.3 Composition and mechanical properties of degradable magnesium alloys.

Alloy[a]	ZX50	WZ21	ZX10
Nominal composition (wt%)			
Mg	Balance	Balance	Balance
Zn	5	1	1
Ca	0.25	0.25	0.3
Mn	0.15	0.15	
Y		2	
Mechanical properties			
Yield strength (MPa)	210	150	240
Tensile strength (MPa)	295	250	255
Elongation to failure (%)	26	28	27

[a] ASTM designation.

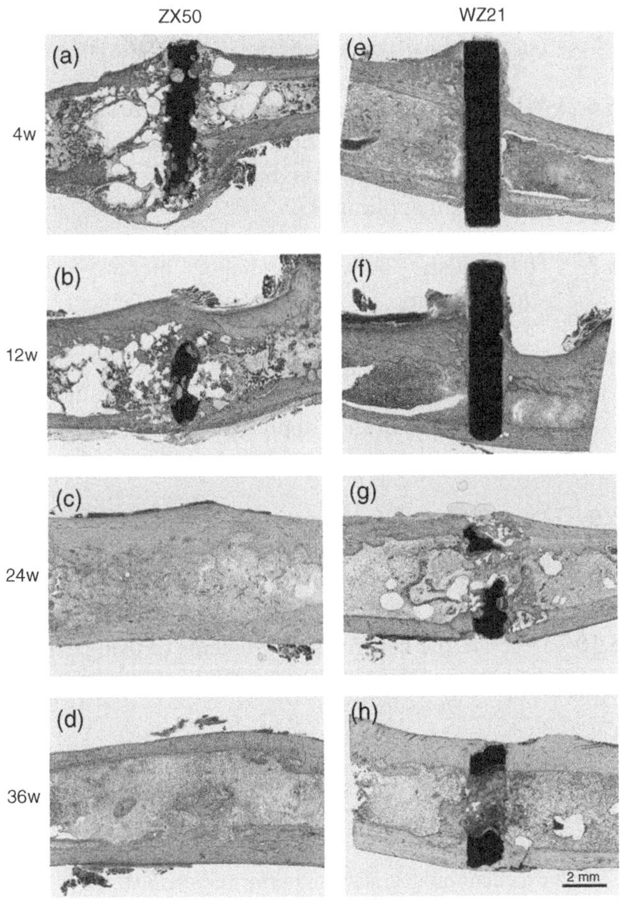

Figure 6.23 Optical images of Levai-Laxzko stained sections of rat femurs implanted with ZX50 (a–d) and WZ21 (e–h) pins at 4, 12, 24, and 36 weeks (w). *Source:* From Kraus et al. (2012).

of new bone was formed around the WZ21 implants, indicating their better osteoconductive capacity. Bone recovered following complete degradation of both implants, even for the ZX50 implants that produced a considerable amount of H_2 gas.

6.11 Concluding Remarks

Metals are commonly used as biomaterials when long-term mechanical durability must be guaranteed and, in some applications, when high electrical conductivity is a requirement. Overall, the majority of pure metals have weak mechanical properties and, except for the noble metals, they corrode in the physiological environment. Consequently, most metals used as biomaterials have traditionally been selected from a group composed of strong metals and alloys that are passivated by an adherent surface oxide layer. However, the last few decades have witnessed considerable efforts to design and develop degradable metal alloys for use in applications that optimally do not require the long-term presence of an implant *in vivo*.

- The properties of metals, including alloys, are determined by their composition, such as the type and concentration of alloying elements, and by their structure at different length scales, such as their crystal structure and microstructure
- Some metals, such as iron, titanium, and cobalt exist in more than one crystal structure, a phenomenon referred to as polymorphism. Alloying elements and thermal treatments dictate the actual crystal structure present at a given temperature. Transformation from one crystal structure to another occurs by a nucleation and growth process, which can occur by a slower diffusion mechanism that is dependent on temperature and time, or by a very rapid displacive mechanism that is dependent on temperature only.
- Polymorphic transformations, alloying elements and thermal, mechanical or thermo-mechanical treatments determine the structure of metals at different length scales and, thus, their properties
- Phase diagrams are a useful source of information for the solid phases present in a metal under equilibrium conditions but they also provide a useful guide to the phases that might be present under non-equilibrium conditions
- Metals used as biomaterials are predominantly titanium and its alloy Ti6Al4V, stainless steel and cobalt–chromium alloys that provide a wide range of superior mechanical properties, while nickel–titanium intermetallic compound, referred to as Nitinol, porous tantalum and a zirconium alloy find more limited use in certain selected applications
- Significant advances have been made in the development of degradable metals, particularly magnesium alloys, for use as temporary implants *in vivo*, but the design and creation of such biomaterials with an optimal combination of properties can be challenging.

Problems

6.1 The austenite to ferrite transformation temperature in iron is 914 °C but upon cooling, the transformation shows a maximum rate at ~700 °C. Explain.

6.2 Distinguish between diffusive and displacive transformations. In doing so, compare the characteristic features of these two transformations.

6.3 The eutectoid temperature of carbon steels is 723 °C and the eutectoid composition is composed of 0.80 wt% C. At this temperature, the solubility limit of C in iron is 0.035 wt%. Assuming that the phase boundaries in Figure 6.8 are straight lines, determine the equilibrium phases and their amounts present in a carbon steel composed of 0.4 wt% C at (i) 1000 °C, (ii) 750 °C, (iii) 600 °C, and (iv) 25 °C.

6.4 If you increase the amount of cold working prior to annealing, what happens to the number and size of the grains formed? Explain why this happens. What material property does this change?

6.5 Ti6Al4V is very biocompatible and, when compared to stainless steel and cobalt–chromium alloys, has a superior strength-to-weight ratio than stainless steel and cobalt–chromium alloys, and a Young's modulus closer to that of bone. Explain the advantages and drawbacks, if any, of replacing the modular Ti6Al4V stem and cobalt–chromium femoral head in common prosthetic hips with a single component (stem and ball) composed entirely of Ti6Al4V.

6.6 What is responsible for the corrosion resistance of stainless steels? Why is the carbon content of stainless steels limited to less than 0.03 wt% for orthopedic implant applications?

6.7 Distinguish between shape memory and superelasticity.

6.8 Explain the advantage of using Nitinol in (i) orthodontic braces and (ii) stents. Briefly explain how Nitinol works in each application and the desirable transformation temperatures.

6.9 Describe the advantages and drawbacks of using degradable magnesium alloys over durable metals such as Ti6Al4V in healing bone defects.

6.10 Describe the benefits and problems of using degradable magnesium alloys over Nitinol in stents.

References

Ashby, M.F. and Jones, D.R.H. (2006). *Engineering Materials 2: An Introduction to Microstructure, Processing and Design*, 3e. Oxford (UK): Butterworth-Heinemann.

Betteridge, W. (1982). *Cobalt and Its Alloys*. Hemel Hempstead, UK: Ellis Horwood Limited.

Bondy, S.C. (2010). The neurotoxicity of environmental aluminum is still an issue. *Neurotoxicology* 31: 575–581.

Callister, W.D. (2007). *Materials Science and Engineering: An Introduction*, 7e. New York: Wiley Chapter 10.

Gupta, K.P. (2005). The Co–Cr–Mo (cobalt-chromium-molybdenum) system. *Journal of Phase Equilibria and Diffusion* 26: 87–92.

Hacking, S.A., Bobyn, J.D., Toh, K.-K. et al. (2000). Fibrous tissue ingrowth and attachment to porous tantalum. *Journal of Biomedical Materials Research* 52: 631–638.

Haude, M., Erbel, R., Erne, P. et al. (2013). Safety and performance of the drug-eluting absorbable metal scaffold (DREAMS) in patients with de-novo coronary lesions: 12 month results of the prospective, multicenter, first-in-man BIOSOLVE-I trial. *Lancet* 381: 836–844.

Hofstetter, J., Becker, M., Martinelli, E. et al. (2014). High-strength low-alloy (HSLA) Mg–Zn–Ca alloys with excellent biodegradation performance. *JOM* 66: 566–572. https://doi.org/10.1007/s11837-014-0875-5.

Ishida, K. and Nishizawa, T. (1990). The Co–Cr (cobalt-chromium) system. *Bulletin of Alloy Phase Diagrams* 11: 357–358.

Kraus, T., Fischerauer, S.F., Hänzi, A.C. et al. (2012). Magnesium alloys for temporary implants in osteosynthesis: in vivo studies of their degradation and interaction with bone. *Acta Biomaterialia* 8: 1230–1238.

Nalla, R.K., Boyce, B.L., Campbell, J.P. et al. (2002). Influence of microstructure on high-cycle fatigue of Ti-6Al-4V: Bimodal vs. Lamellar structures. *Metallurgical and Materials Transactions A: Physical Metallurgy and Materials Science* 33: 899–918.

Peters, M., Hemptenmacher, J., Kumpfert, J., and Leyens, C. (2003). Structure and properties of titanium and titanium alloys *Titanium and Titanium Alloys: Fundamentals and Applications* (eds. C. Leyens and M. Peters). Weinheim, Germany: Wiley-VCH Verlag GmbH Chapter 1.

Rahaman, M.N., Yao, A., Bal, B.S. et al. (2007). Ceramics for prosthetic hip and knee joint replacement. *Journal of the American Ceramic Society* 90: 1965–1988.

Ries, M.D., Salehi, A., Widding, K., and Hunter, G. (2002). Polyethylene wear performance of oxidized zirconium and cobalt–chromium knee components under abrasive conditions. *The Journal of Bone and Joint Surgery. American Volume* 84: S129–S135.

Zheng, Y.F., Gu, X.N., and Witte, F. (2014). Biodegradable metals. *Materials Science and Engineering R* 77: 1–34.

Further Reading

Niinomi, M. (2002). Recent metallic materials for biomedical applications. *Metallurgical and Materials Transactions A: Physical Metallurgy and Materials Science* 33: 477–486.

7

Ceramic Biomaterials

7.1 Introduction

Ceramics are inorganic materials composed of compounds of metallic and nonmetallic elements (Chapter 1). Carbon is often grouped with ceramics because it shows several physico-chemical properties characteristic of ceramics, but it is composed of just one element. As illustrated in Figure 7.1, the term ceramics has traditionally been used to refer to the overall field of materials composed of inorganic compounds regardless of their structure (crystalline or amorphous), as well as to a specific type of material within this field, those composed of a crystalline structure. In this chapter, we will use the term in this traditional sense. Consequently, the types of biomaterials within the overall field of ceramics that we will discuss are:

- Ceramics, composed of a crystalline structure.
- Glasses, characterized by an amorphous structure composed of a three-dimensional network having no long-range atomic order and lacking detectable crystallinity.
- Glass-ceramics, a two-phase mixture composed of crystals in a glass matrix, typically formed by controlled crystallization of a parent glass.

Because of their strong interatomic bonding, the majority of ceramics are chemically unreactive, and they have an attractive combination of physical properties, which may include high compressive strength, high Young's modulus, high hardness and high wear resistance (Chapter 3). However, this strong bonding also makes ceramics brittle, that is, they have a low fracture toughness. This strong bonding also makes ceramics generally more difficult and expensive to manufacture than metals and polymers. Overall, ceramics find use in applications that take advantage of their attractive properties while minimizing the consequences of their brittleness.

The majority of ceramics used as biomaterials find applications in orthopedic, cranial, and maxillofacial surgery as implants in bone repair (Table 7.1). Chemically unreactive ceramics such as those based on alumina (Al_2O_3) and zirconia (ZrO_2), for example, are used as articulating bearings in hip implants. Calcium phosphates such as hydroxyapatite and beta-tricalcium phosphate (β-TCP), composed of the same ions as the main mineral constituent of human bone, find considerable use as bone graft substitutes. Calcium phosphate cements (CPCs) are formed into a paste and implanted into a bone defect, where they harden *in situ*, thus providing the benefits of ease of implantation and a better capacity to match the shape of the defect, particularly for complex shaped defects.

Pyrolytic carbon is widely used as coatings in heart valves whereas fullerenes, carbon nanotubes, and graphene are receiving considerable interest for potential biomedical applications (Section 3.3.4). The iron oxides Fe_2O_3 and Fe_3O_4 are magnetic and biocompatible (Chapter 4), and in the form of nanoparticles, they are receiving growing interest for use in treating tumors (Chapter 26).

Chemically unreactive glasses find little use as biomaterials, but microspheres of one glass composition containing the rare earth element yttrium, after irradiation to make them radioactive, have been used to treat liver cancer. Bioactive glasses degrade *in vivo*, convert to hydroxyapatite and release ions that stimulate cell responses, resulting in the formation of a direct bond at their interface with host bone and even soft tissues. Whereas bioactive glasses have not realized their early promise for use as implants to heal bone defects, a significant recent development has been the use of microfibrous forms of special bioactive glasses to heal soft tissue wounds, such as skin wounds.

Bioactive glass-ceramics are used to a limited extent as spinal fusion implants in orthopedic surgery. While several chemically unreactive glass-ceramics have been developed and commercialized for use as dental restorations, such as crowns and bridges, lithium disilicate compositions have emerged as the most commonly used glass-ceramics in dental restorations. Lithium disilicate glass-ceramics provide an attractive combination of

Materials for Biomedical Engineering: Fundamentals and Applications, First Edition. Mohamed N. Rahaman and Roger F. Brown.

Companion website: www.wiley.com/go/rahaman/biomedicalengineering

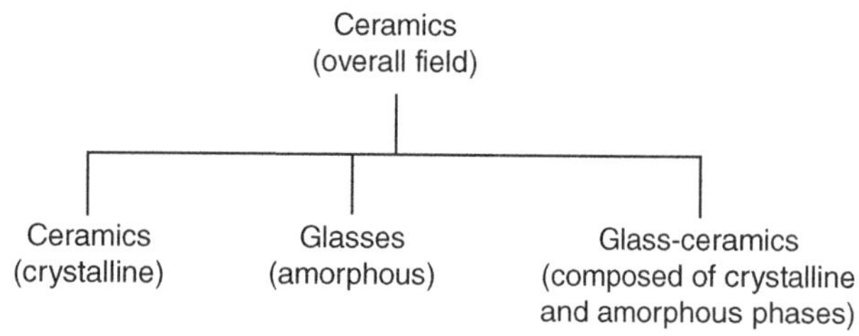

Figure 7.1 Types of materials in the overall field of ceramics.

Table 7.1 Biomedical applications of ceramics, glasses, and glass-ceramics.

Material	Applications
Ceramics	
Alumina (Al_2O_3); alumina matrix composite (AMC)	Bearings in total hip replacement
Calcium phosphates: hydroxyapatite; β-tricalcium phosphate (β-TCP); biphasic calcium phosphate (BCP); calcium phosphate cement (CPC)	Bone repair (bone graft substitute)
Pyrolytic carbon	Heart valve coating
Yttria-stabilized zirconia (YSZ)	Dental restorations (crowns; bridges)
Silicon nitride (Si_3N_4)	Spinal fusion
Calcium sulfate (plaster of Paris)	Antibiotic delivery (treatment of bone infection)
Glasses	
Radioactive glass (Y_2O_3–Al_2O_3–SiO_2) microspheres	Treatment of liver cancer
Bioactive glass	Treatment of skin wounds; bone repair
Glass-ceramics	
Lithium disilicate	Dental restorations (crowns; bridges)
Glass-ceramic A–W	Spinal fusion

properties, such as chemical stability in the oral environment, adequate mechanical properties, and optical properties such as color and fluorescence to match the existing tooth structure.

In this chapter, we will discuss the following issues relevant to the use of ceramics, glasses and glass-ceramics as biomaterials:

- Design and processing principles for ceramics.
- Principles and methods for producing ceramics, glasses, and glass-ceramics.
- Composition–structure–property relationships in ceramics, glasses, and glass-ceramics.

7.2 Design and Processing of Ceramics

Like metals, the properties and performance of ceramics are controlled by their composition (atomic bonding) and microstructure (Chapter 4). However, microstructure plays a far more dominant role in ceramics due to their lack of a plastic deformation mechanism. An important key to developing mechanically reliable ceramics is in controlling their microstructure.

Several ceramics have an attractive combination of physico-chemical properties such as chemical inertness, high compressive strength, high Young's modulus, high hardness, and high wear resistance, which make them desirable in many engineering applications and a few biomedical applications as well. On the other hand, ceramics are brittle and microstructural flaws such as pores and microcracks limit their strength, particularly in tensile and flexural loading. The common manufacturing method for ceramics, referred to as sintering, is a major source of these microstructural flaws. In general, this method also makes ceramics more difficult and expensive to produce when compared to the common methods used for metals and polymers. Overall, then, the production of mechanically reliable ceramics requires an understanding of proper design principles and proper manufacturing techniques.

7.2.1 Design Principles for Mechanically Reliable Ceramics

Design principles for ceramics, as noted in Chapter 4, are different from those for metals. Following elastic deformation, metals, as we have seen, yield by plastic deformation prior to failure. The yield strength is an important design property in metals. Metals also have high resistance to crack propagation, that is, high fracture toughness, due to plastic flow at the crack tip and energy absorption in the plastic zone. In comparison, ceramics fail in a catastrophic brittle manner at small strains within their elastic region of deformation. Although local plastic deformation can occur at the crack tip, it is very limited. Ceramics have low resistance to crack propagation (low fracture toughness) and their fracture strength is an important design parameter.

The strength of ceramics is strongly dependent on the mode of loading (compression, tension, or bending). In tensile loading, the stress at the tip of a microcrack is amplified and, thus, the tensile or flexural strength of a ceramic specimen is considerably lower than its compressive strength (Section 4.2.5). Thus, proper design is important to avoid the exposure of ceramics to high tensile or flexural stresses. Generally, the measured strength of a number of nominally identical ceramic specimens often shows considerable scatter due to the presence of microstructural flaws of varying size, position, and orientation. Consequently, the average strength is not the only important parameter in designing mechanically reliable ceramics. Instead, the spread in the measured strength (Weibull modulus) and the use of Weibull statistics to evaluate the probability of catastrophic failure under a given stress are also important.

Microstructural flaws present in ceramics vary in size from a fraction of the average grain size of the specimen to several times the average grain size (Figure 7.2). While they originate in a variety of ways, a common source is the sintering route that is used to produce almost all three-dimensional ceramics. For example, unless special techniques are used, it is difficult to produce a fully dense (pore-free) ceramic by sintering. Thus, pores of varying size and shape remain in the product and serve as flaws. Sintering is also performed at high temperature, and cooling the specimen to room temperature generally leads to the generation of internal stresses due to differences in thermal expansion coefficient between adjacent grains or regions of the specimen. If high enough, these stresses can also generate microcracks. After fabrication, surface abrasion due to machining or chemical attack by water vapor in the atmosphere can lead to the generation of new microcracks or growth of existing microcracks.

If the longest microcrack in a ceramic specimen is of length $2c$, the tensile strength σ_f of the specimen given by Eq. (4.28), that is

$$\sigma_f = \frac{K_C}{(\pi c)^{1/2}} \tag{7.1}$$

where K_C is the fracture toughness. We can see from Eq. (7.1) that one way to improve the strength of ceramics is to decrease the size of microstructural flaws. The most effective way of doing so is to use proper manufacturing methods. According to Eq. (7.1), another approach to increasing the strength of ceramics is to improve their fracture toughness, K_C. This is often achieved by reinforcing the ceramic with strong particles or fibers, for example, to form composites (Chapter 12). An example of this approach is the addition of ZrO_2 particles to Al_2O_3 to form ZrO_2 toughened Al_2O_3 composites that are now commonly used instead of Al_2O_3 as articulating bearings in hip implants. Another approach is to form a solid solution but this is only applicable to a few ceramics, the most notable example of which is Y_2O_3-stabilized ZrO_2 (Section 7.3.2).

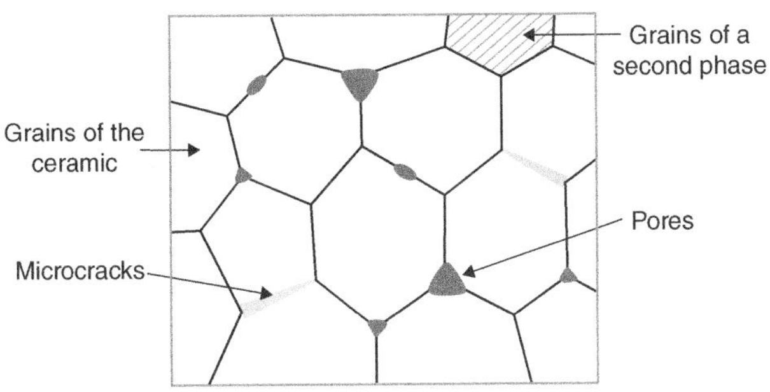

Figure 7.2 Microstructural flaws that may be present in ceramics.

Overall, then, in spite of their brittleness, ceramic products having sufficient mechanical reliability for a wide range of structural applications can be attained through:

- Proper design to ensure a sufficiently low probability of failure under the appropriate stresses.
- Use of proper manufacturing methods to achieve an optimal microstructure as free from strength-limiting flaws as possible.

7.2.2 Principles of Processing Ceramics

The strong ionic and covalent bonding in ceramics is responsible not just for their brittleness. It also limits the types of processing methods that can be used to produce useful ceramics. Because of their lack of ductility, useful ceramics cannot be produced by methods such as forging, rolling, extrusion, and drawing that are commonly used for wrought metals products (Chapter 6). Casting is also impractical because most ceramics have a high melting point. Even if it were practical to melt some ceramics, casting would lead to an undesirable microstructure composed of large grains and a considerable amount of flaws, giving weak and unreliable mechanical properties.

A process referred to as sintering or powder processing is used to produce the vast majority of ceramics. Sintering of ceramics is identical to the powder metallurgy process used to create some metal products (Chapter 6). As the production of ceramics having the requisite microstructure by sintering can often be challenging, an understanding of the basic principles and techniques of the sintering process is useful.

The Sintering Process

Figure 7.3a summarizes the major steps in the overall process of sintering. Typically, a mass of fine particles (a powder) having a particle size less than ~1 μm, often available commercially, is formed into a porous specimen of the desired shape using a variety of methods at room temperature. Pressing the particles in a die, for example, is a simple and common method but there is growing use of additive manufacturing (3D printing) methods, particularly when a complex external shape or well-controlled internal porosity

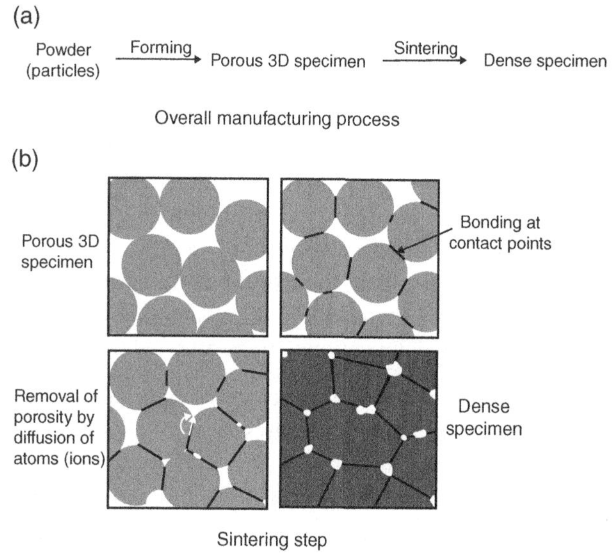

Figure 7.3 (a) Major processing steps in the production of ceramics by the sintering route. (b) Major microstructural changes that occur during the sintering step.

is required. Adhesion between touching particles during this consolidation step leads to the formation of a weak object (compressive strength of a few megapascals) having a porosity of ~35–50%. Then this object is heated (sintered) to produce a useful material having the desired microstructure.

In the sintering step, the as-formed object is heated to a high temperature, typically 0.5–0.9 of its melting temperature T_m to obtain a product having the requisite microstructure. Sintering is a complex process but the basic features can be illustrated by taking a pure single-phase ceramic such as Al_2O_3 (Figure 7.3b). Initially, the particles become joined together at their contact points, by diffusion of their constituent ions, forming necks that improve the strength of the porous object. As sintering proceeds further, diffusion leads to an increase in the diameter of the necks and transport of ions into the pores, resulting in removal of porosity and shrinkage of the specimen (that is, the specimen becomes denser). Ideally, all the pores should be removed, giving a fully dense specimen, but this rarely happens unless special techniques are used (Figure 3.23a). During all of this, the grains grow. Overall, the sintered product often contains a small amount of residual porosity (less than ~5%), composed mainly of small pores at the corners of the grains but a few large pores between the grains may also be present. The grains have varying sizes, characterized by an average size and a distribution of sizes approximating a log-normal distribution.

Principles Underlying Sintering

Sintering is dependent on certain thermodynamic and kinetic factors. Thermodynamically, for a physico-chemical process to occur there must be a reduction in the Gibbs free energy of the system. In sintering, this driving force arises from a reduction in surface free energy of the particles as the object goes from a porous state (high surface area) to a dense state (low surface area) (Figure 7.3b). For a mass M of particles (assumed spherical in shape) of size (diameter) $2a$, the change in surface free energy ΔE_s in going from an as-formed object of particles in point contact to a fully dense product is given by

$$\Delta E_s = \frac{3M\gamma_{sv}}{\rho a} \quad (7.2)$$

where γ_{sv} is the surface energy of the particles per unit area and ρ is the density of the material. Equation (7.2) shows that finer particles are better for sintering because they have a higher driving force for sintering.

The kinetics of the process are also important. Bonding between the particles and removal of porosity requires transport of atoms (ions), which occurs by diffusion. Whereas diffusion in ceramics is essentially nonexistent at room temperature, it increases upon heating in accordance with the Arrhenius equation (Eq. 6.9). Above ~$T_m/2$, it becomes sufficiently fast for sintering to take place at a reasonable rate. However, temperature is not the only important factor. In order to fill up the pores, atoms have to diffuse a certain characteristic distance, from within the particles (grains) to the surface of the pores. As this characteristic diffusion distance scales as the particle size, finer particles also enhance the rate of sintering.

Densification of the porous as-formed object is not the only process that occurs during sintering. In polycrystalline materials such as ceramics, the grains also grow because there is a driving force for them to do so. As the grains grow, there is a reduction in the overall area of the grain boundaries, which leads to a decrease in energy of the system. Grain growth occurs by diffusion but by a different mechanism from densification. Grain growth is problematic in that it increases the distance that ions have to diffuse to cause densification, thereby slowing down densification and reducing the capacity to achieve a sufficiently dense product. Large grains, particularly abnormally large grains, are also detrimental to the strength of ceramics. Consequently, control of grain growth is often essential for achieving a dense product having sufficiently high strength and mechanical reliability.

Sintering Process Optimization

Using an optimal sintering temperature, the sintering step can be completed within a reasonable time, usually within a few hours, depending on the size of the product. However, several complications can arise in this simplified picture of sintering described so far. For example, large pores in the as-formed specimen, of size several times the particle size of the starting powder, are difficult to remove during sintering and, thus, remain in the product. The presence of abnormally large particles or impurities in the starting powder often leads to the growth of abnormally large grains during sintering. As these large pores and abnormally large grains are particularly detrimental to the strength of the ceramic product (Eq. (7.1)), it is critical to use the right starting powders, forming methods and sintering conditions to eliminate them as far as possible.

For some ceramics, particularly those having a high degree of covalent bonding, sintering alone may be inadequate for attaining the desired microstructure, particularly when a low-porosity product is required. In such cases, applying a pressure during sintering as, for example, in hot isostatic pressing (Chapter 6) or adding a small amount (less than a few percent) of an appropriate oxide (or mixture of oxides) to the starting powder to aid the sintering process is often beneficial. A combination of sintering and hot isostatic pressing is also effective.

Overall, the capacity to produce a ceramic product having an optimal microstructure is enhanced by using:

- A sufficiently pure starting powder having a fine particle size (often less than ~1 μm).
- A suitable method to form the powder into a porous object having as homogeneous a microstructure as possible that is devoid of abnormally large pores and flaws.
- Appropriate sintering conditions (temperature, time, and atmosphere) to attain a product having the requisite microstructure. Commonly, this means that the solid phase of the product should be as dense as possible (close to zero porosity) and have as homogeneous a microstructure as possible, composed of fine grains and free from strength-limiting flaws.

The major types of ceramics relevant to biomaterials are classified in this chapter into three groups, chemically unreactive ceramics, calcium phosphates, and CPCs. However, other ceramics such as carbon materials (Section 3.3.4) and iron oxide nanoparticles (Section 4.6.5) are under development for use as biomaterials (Table 7.1).

7.3 Chemically Unreactive Ceramics

Other than carbon materials, chemically unreactive ceramics used as biomaterials are based on alumina (Al_2O_3), zirconia (ZrO_2), and silicon nitride (Si_3N_4). Like many ceramics, Al_2O_3, ZrO_2, and Si_3N_4 have strong interatomic bonding, although there are differences in the nature of their bonding. For example, Al_2O_3 and ZrO_2 have a high degree of ionic bonding whereas Si_3N_4 is bonded covalently. The strong atomic bonding in ceramics, as we have seen (Chapter 3), determine their intrinsic properties. Our primary interest in this section is to consider key factors relevant to the production of Al_2O_3, ZrO_2, and Si_3N_4 having an optimal microstructure and, thus optimal properties, for use in implant applications.

7.3.1 Alumina (Al_2O_3)

Al_2O_3 has been used since the 1970s as articulating bearings in prosthetic hips and has been the most widely studied ceramic for these applications. Improvements in material purity and processing methods over time have resulted in the production of bearings having high mechanical reliability and low rate of fracture *in vivo*. However, despite these improvements, the low fracture toughness of Al_2O_3 (~4–5 MPa m$^{1/2}$) increases the risk of catastrophic brittle fracture. Consequently, Al_2O_3 has been largely replaced as a bearing material in hip implants with ZrO_2-toughened Al_2O_3 composites that have a higher fracture toughness (~6–10 MPa m$^{1/2}$).

Al_2O_3 can exist in a few metastable forms called transition alumina but above ~1000 °C, they transform to the stable form referred to as α-Al_2O_3, which is the common form referred to simply as Al_2O_3. Al_2O_3 has a close packed hexagonal (CPH) crystal structure, strong atomic bonding (Chapter 3), and a high melting point (2053 °C).

Al_2O_3 products are manufactured using the sintering process (Section 7.2.2). Attainment of the requisite microstructure for optimal mechanical properties and reliability, as we have seen, is strongly dependent on the quality of the starting powder and the use of proper processing techniques. It is necessary to add ~0.25 wt% of MgO to the starting powder to reduce grain growth and, thus, enhance densification. An optimal temperature in the range 1300–1600 °C provides sufficiently rapid diffusion for densification while an atmosphere of O_2 or He gas improves the ability to achieve a fully dense product. For structural applications, such as bearings in hip implants, a fully dense product composed of fine grains (less than ~2–5 μm) is desirable (Figure 3.23a).

Dense fine-grained Al_2O_3 is biocompatible, and its high hardness and low coefficient of friction give low wear rates when used as bearings in hip implants. When articulating against polyethylene liners, Al_2O_3 femoral heads generate a lower amount of wear debris than commonly used cobalt-chromium alloys and even much lower wear rates when articulating against Al_2O_3 liners. Although Al_2O_3 has been largely replaced as a bearing material in hip implants with

ZrO_2-toughened Al_2O_3 composites, it provides a useful reference material for comparing other bearing materials.

7.3.2 Zirconia (ZrO_2)

ZrO_2-based ceramics, originally developed for engineering applications, have been introduced into orthopedic and dental applications. Unlike Al_2O_3, which is essentially a pure single-phase ceramic, the materials science of ZrO_2-based ceramics involves additional complications:

- ZrO_2 undergoes a crystallographic phase transformation upon cooling from its manufacturing (sintering) temperature, which makes pure ZrO_2 virtually useless as a structural engineering material.
- Useful ZrO_2 ceramics are typically stabilized ZrO_2, composed of ZrO_2 containing a small amount of a metal oxide that is used to stabilize its structure. The oxide stabilizer allows the ZrO_2 to retain its high temperature phase as a metastable phase upon cooling to room temperature, thus avoiding the deleterious phase transformation associated with pure ZrO_2.
- Transformation of the metastable phase to the stable phase when stabilized ZrO_2 is used at ordinary temperatures (such as room temperature) leads to mechanisms that enhance the fracture resistance (fracture toughness) of the material.
- Transformation toughening together with an optimal microstructure enables stabilized ZrO_2 to have a combination of high strength and fracture toughness that is among the highest for polycrystalline ceramics.

Overall, then, an understanding of crystallographic transformations, microstructure control during processing and mechanisms of transformation toughening are essential for creating ZrO_2 ceramics having optimal mechanical properties.

Crystallographic Transformation in ZrO_2 ZrO_2 exists in three polymorphs (crystal structures), cubic, tetragonal, and monoclinic. Transformation between these forms occurs by a displacive mechanism (Section 6.3). Cubic ZrO_2 is the highest temperature phase and, in its pure state, is stable between ~2370 °C and its melting point (~2680 °C) whereas the tetragonal phase transforms to the monoclinic phase at ~1240 °C.

Production of ZrO_2 specimens by the common sintering route typically requires a temperature of ~1400–1600 °C and, thus, pure ZrO_2 exists in the tetragonal phase at the sintering temperature. Upon cooling to room temperature, transformation to the monoclinic phase involves not only a change in crystal structure but also a volume increase of ~4.7%. This volume increase creates internal stresses that lead to the generation of microcracks within the material, making pure ZrO_2 weak and unsuitable for structural applications.

Upon the addition of a few percent of "stabilizers" such as MgO, CaO, Y_2O_3, or CeO_2 to ZrO_2, the cubic or tetragonal phase can be retained as a metastable phase upon cooling to room temperature. Stabilized ZrO_2 in the cubic phase is used in a variety of technological applications. On the other hand, for ZrO_2 stabilized in the tetragonal phase, the tetragonal to monoclinic transformation actually forms the basis for transformation toughening, a mechanism for enhancing the fracture toughness ZrO_2-based materials (section titled "Transformation Toughening Mechanism in Stabilized ZrO_2").

Zirconia Phase Diagrams As we have seen for metals (Chapter 6), phase diagrams provide useful information about the phases that might be present in a material at various temperatures and their concentration. Figure 7.4 shows the ZrO_2-rich portion of the ZrO_2–Y_2O_3 phase diagram. Although the phase boundaries depend on the composition of the stabilizing oxide, the principles are similar. The most common composition used for orthopedic implants and dental restorations contains 2–3 mol% Y_2O_3, referred to as yttria-stabilized tetragonal zirconia polycrystal (Y-TZP) but this is often shortened to yttria-stabilized zirconia (YSZ). At room temperature, this material is composed predominantly

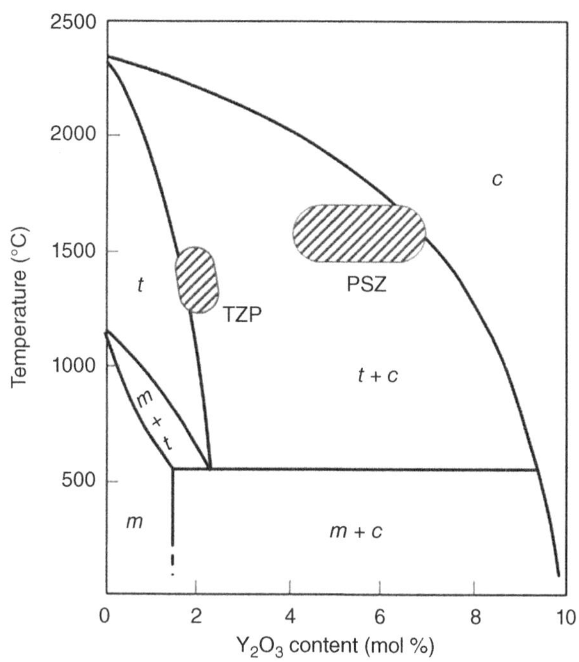

Figure 7.4 High zirconia portion of the zirconia–yttria phase diagram. The shaded regions show compositions and sintering temperatures for tetragonal zirconia polycrystals (TZP) and partially stabilized zirconia (PSZ). m, Monoclinic; t, tetragonal; c, cubic. *Source:* From Scott (1975) / with permission of Springer Nature.

of a single phase (that is, the metastable tetragonal phase). When produced in a fully dense microstructure of fine grains (0.1–1 μm) that is free from flaws (Figure 7.5a), Y-TZP has the highest strength and fracture toughness of all ZrO_2-based ceramics, which are also comparable to the highest values for polycrystalline ceramics.

Another material, referred to as partially stabilized zirconia (PSZ) contains a higher amount of stabilizer but this amount is insufficient to form the fully stabilized cubic phase (Figure 7.4). PSZ consists of a two-phase mixture of fine metastable tetragonal precipitates (often ~40–50% of the total volume of the material) within a matrix of large cubic grains (larger than ~50 μm). A composition composed of 8–10 mol% MgO, referred to as Mg-PSZ, has been developed for use as femoral bearings in knee implants for articulating against polyethylene tibial bearings. However, because the strength and fracture toughness of Mg-PSZ are inferior to those of Y-TZP, interest in using Mg-PSZ as bearings in hip and knee implants has remained low.

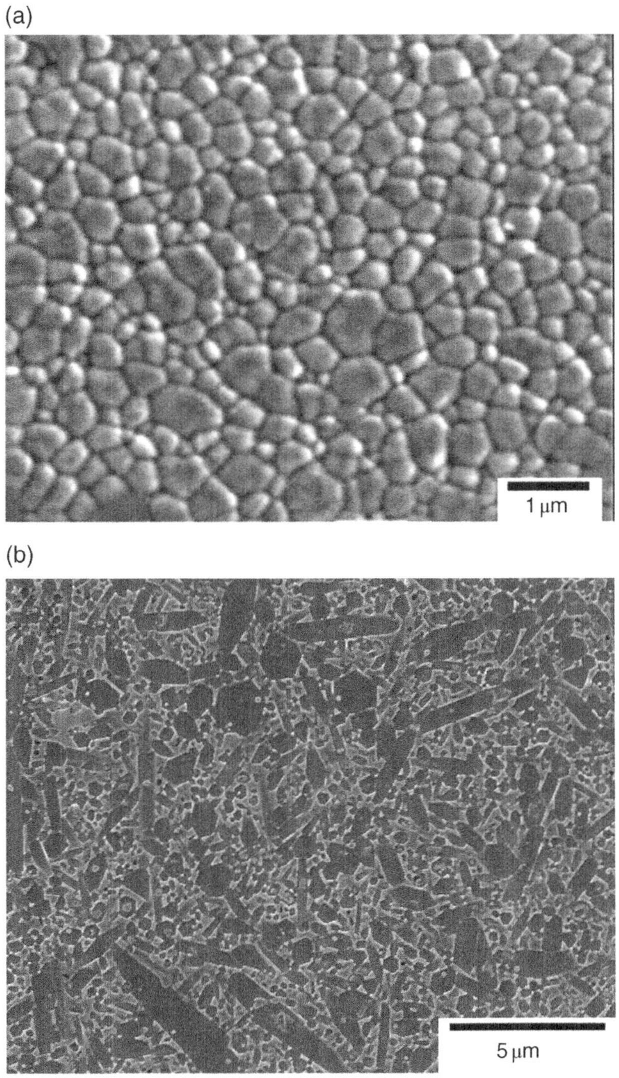

Figure 7.5 Scanning electron microscope images showing typical microstructure of (a) tetragonal zirconia polycrystals (Y-TZP) and (b) silicon nitride (Si_3N_4) used as biomaterials in orthopedic surgery. The glass phase in Si_3N_4 has been etched to better reveal the grains. *Source:* From McEntire et al. (2015).

Production of Stabilized ZrO_2 Creation of Y-TZP having an optimal combination of high strength and high fracture toughness by the commonly used sintering route is often challenging. It is dependent on having not only the right concentration of Y_2O_3 (in the range 2.0–3.0 mol%) but also on developing the right microstructure. Generally, this microstructure consists of a highly dense material (close to zero porosity) composed of fine grains and free from flaws. As the attainment of this microstructure is often challenging, an understanding of sintering principles and techniques is essential (Section 7.2.2). A fine grain size is important not just for improved strength but to avoid the tetragonal to monoclinic transformation upon cooling the sintered product to room temperature. It is difficult to avoid this phase transformation during cooling if the grain size were above a critical value. Y-TZP having optimum fracture toughness is obtained with ~2.0 mol% Y_2O_3 and a grain size of ~0.3 μm. With ~3.0 mol% Y_2O_3, a higher strength is obtained but the fracture toughness is lower. However, the sintering process is less demanding at this higher amount of Y_2O_3 because the critical grain size for avoiding the tetragonal to monoclinic transformation increases to ~1 μm.

Transformation Toughening Mechanism in Stabilized ZrO_2 The mechanism of transformation toughening in Y-TZP is illustrated in Figure 7.6. Toughening results from two major processes. First, the volume expansion that accompanies the tetragonal to monoclinic transformation produces a compressive stress that lowers the local tensile stress at the crack tip. Consequently, there is less amplification of the applied tensile stress at the crack tip. Second, energy is absorbed in the tetragonal to monoclinic transformation and,

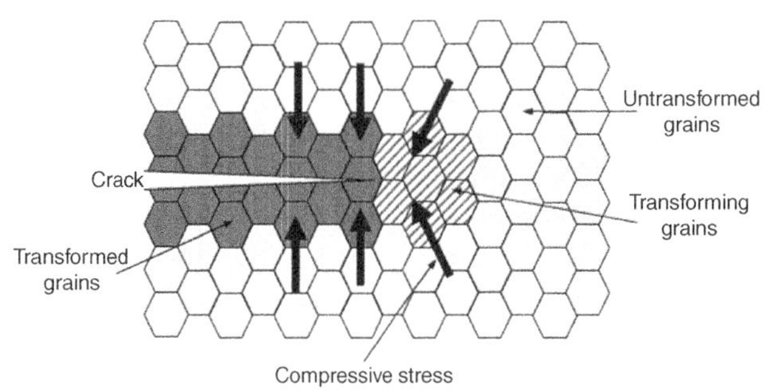

Figure 7.6 Illustration of transformation toughening mechanism in tetragonal zirconia polycrystals (Y-TZP).

thus the energy available for crack propagation is lowered. The fracture toughness of Y-TZP (6–10 MPa m$^{1/2}$) is small compared to that of most metals because the energy absorbing mechanisms in transformation toughening are not as effective as plastic flow in metals. However, the improvement in strength and fracture toughness brought about by transformation toughening is significant for ceramics.

Orthopedic Applications of Stabilized ZrO_2 Because of their higher strength and fracture toughness when compared to Al_2O_3, coupled with their low wear against polyethylene liners, Y-TZP femoral heads found considerable use upon their introduction in the 1980s. However, a change in manufacturing conditions around the year 2000 resulted in a significant increase in the incidence of catastrophic fracture *in vivo*, often only a few months after implantation. The cause of this higher failure rate was found to be an inherent limitation of Y-TZP for load-bearing applications *in vivo*, that is, its tendency to spontaneously transform to the monoclinic phase upon contact with an aqueous medium (Chevalier et al. 2009). This transformation leads to an increase in surface roughness and wear, weakening of the implant and, eventually, brittle facture. Consequently, despite their high initial strength and fracture toughness, Y-TZP femoral heads are no longer approved for use in the United States and the European Union.

7.3.3 Alumina–Zirconia (Al_2O_3–ZrO_2) Composites

Al_2O_3–ZrO_2 composites used in orthopedic applications, mainly as femoral heads and acetabular liners in hip implants, provide the high hardness and low wear rate of Al_2O_3, better mechanical reliability than Al_2O_3 due to their higher strength and fracture toughness, and they do not suffer from the inherent instability Y-TZP *in vivo*. Toughening in these composites results primarily from transformation of ZrO_2 particles dispersed within the Al_2O_3 matrix. These particles are often composed of Y_2O_3 stabilized ZrO_2 (YSZ) containing ~3 mol% Y_2O_3, but pure (unstabilized) ZrO_2 particles can also be used.

As the particles are constrained by the strong Al_2O_3 matrix of the product, they are prevented from transforming to the monoclinic phase upon cooling the sintered composite to room temperature. Consequently, it is not critically important whether the particles are composed of YSZ or pure ZrO_2. Except for the YSZ (or ZrO_2) particles at the surface of the implant, the Al_2O_3 matrix shields the Y-TZP particles from contact with the aqueous physiological fluid. Consequently, the inherent instability of monolithic Y-TZP is essentially avoided.

Two types of Al_2O_3–ZrO_2 composites have been used as articulating bearings in hip implants, one referred to zirconia toughened alumina (ZTA) and the other alumina matrix composites (AMCs). Toughening of these composites results from transformation of their metastable tetragonal ZrO_2 particles at the crack tip by the same mechanism described for monolithic Y-TZP (Figure 7.6). ZTA and AMC are similar in composition but because of a difference in the grain morphology of its Al_2O_3 matrix, AMC has better mechanical properties, particularly fracture toughness. ZTA was introduced around 2000, but has now been largely replaced by AMC, which is now the most commonly used ceramic bearing material in hip implants.

Zirconia Toughened Alumina (ZTA) ZTA is composed of a fully dense Al_2O_3 matrix containing ~15 vol% of ZrO_2 particles, commonly ZrO_2 stabilized with ~3 mol% Y_2O_3 (YSZ) (Figure 3.23c). It is manufactured by the sintering route from a mixture of Al_2O_3 and ZrO_2 powders. The particle size of the Al_2O_3 and ZrO_2 powders, the volume fraction of ZrO_2 and the sintering conditions are critical for creating ZTA having an optimal microstructure. Whereas both powders should have a fine particle size, the ZrO_2 particles should be finer. This is required for two main reasons. First, the ZrO_2 particles should ideally be present between the Al_2O_3 grains, that is, at the grain boundaries. Second, the ZrO_2 particles should have a size smaller than a critical value, to avoid transformation of the metastable tetragonal phase to the monoclinic phase upon cooling the sintered product to room temperature (Section 7.2.3.2). As the tetragonal to monoclinic transformation is accompanied by a volume expansion (~4.7%), ideally, the Al_2O_3 matrix should be fully dense and, thus, strong enough to constrain the ZrO_2 particles from expanding. In this way, the Al_2O_3 matrix helps to prevent the transformation of the ZrO_2 particles upon cooling the sintered product to room temperature.

The volume fraction of the ZrO_2 particles is also important. It should be high enough to provide sufficient toughening but not too high to cause significant deterioration of the composite due to the inherent instability of tetragonal ZrO_2 in an aqueous medium. This requires that the ZrO_2 volume faction should be less than ~16%, to prevent the ZrO_2 particles from forming a continuous interconnecting phase between the surface and the interior of the composite. Deterioration of the particles at the surface of the composite can occur, but the particles in the interior are shielded by the Al_2O_3 matrix.

Alumina Matrix Composite (AMC) AMC is essentially a special ZTA that contains, in addition to the ZrO_2 particles (~17 vol%), ~1.7 wt% of a mixture composed of Y_2O_3, Cr_2O_3, and SrO. In the sintering step, this mixture reacts with Al_2O_3, resulting in the formation of a microstructure composed of interlocking plate-like grains. Thus, the major difference between AMC and ZTA is in their microstructure, that is, the Al_2O_3 matrix of AMC consists of elongated grains whereas ZTA has more equiaxial grains.

Because of its microstructure of interlocking plate-like grains, AMC has higher strength and fracture toughness than ZTA (Table 4.1). As described for Si_3N_4 (Section 7.3.4), instead of propagating directly through the strong plate-like grains, an advancing crack is deflected at the weaker grain boundaries and has to go around the elongated grains to propagate further across the material. This makes the crack path more tortuous, which requires the expenditure of more energy, thus increasing the resistance to crack propagation. Overall, toughening of AMC results from the tetragonal to monoclinic transformation of ZrO_2 particles, as in ZTA, and from a crack deflection mechanism as well.

7.3.4 Silicon Nitride (Si_3N_4)

Si_3N_4 is a nonoxide ceramic that has been well established as a ceramic material for use in several structural applications, due to its attractive combination of properties such as high strength, elastic modulus, hardness, and wear resistance (Table 4.1). Si_3N_4 is also biocompatible. It has been used as implants in spinal fusion and is under development for use as bearings in hip and knee implants.

Structure and Chemical Stability of Si_3N_4 Si_3N_4 exists in two polymorphs, called α-Si_3N_4 and β-Si_3N_4, both of which have a hexagonal crystal structure but differ in the sequence in which the Si–N layers are arranged in the *c*-axis of the crystal. The β phase is more stable at high temperature, but the transformation from α to β, which occurs by a diffusion mechanism (Section 6.3), is slow unless it is accelerated by, for example, the presence of a liquid phase. Atomic diffusion and dislocation motion are essentially nonexistent in Si_3N_4 because of its strong covalent bonding. Si_3N_4 is chemically stable in a nitrogen atmosphere up to ~1500 °C but it starts to dissociate into Si and N atoms at higher temperature. As a non-oxide ceramic, Si_3N_4 is less stable in an oxidizing atmosphere (such as air) and it suffers from considerable oxidation to SiO_2 above ~1000 °C. These characteristics of Si_3N_4 strongly influence its processing, microstructure, and properties.

Microstructure and Properties of Si_3N_4 Si_3N_4 can be produced in a fully dense microstructure composed of interlocking, highly elongated (acicular) grains that are separated by a thin (less than ~2 nm) glass phase (Figure 7.4b). This unique microstructure provides Si_3N_4 products with a combination of high strength (flexural strength >1 GPa) and fracture toughness (up to ~10 MPa $m^{1/2}$) that are among the highest measured values for polycrystalline ceramics.

The high strength of Si_3N_4 is due to its interlocking arrangement of strong elongated grains that are bonded together by a thin glass phase at their grain boundary. Unlike Y-TZP, the high toughness of Si_3N_4 is due not to transformation toughening but to its unique microstructure. An advancing crack is deflected at the grain boundary and follows a tortuous path through the weaker glass phase at the grain boundary instead of cutting directly through the strong grains (Figure 7.7). This requires the expenditure of a larger amount energy, resulting in an increase in toughness. The glass phase should be sufficiently strong for high strength, but it should be weaker than the grains for enhanced fracture toughness.

Production of Si_3N_4 Like Y-TZP, production of Si_3N_4 having an optimal microstructure can be challenging. A major challenge is to create the right combination of elongated grain morphology and grain boundary phase. Typically, the use of fine α-Si_3N_4 powder (average particle size <0.5–1.0 μm) mixed with a combination of ~6 wt% fine Y_2O_3 powder and ~4 wt% fine Al_2O_3 powder, proper forming methods, and a combination of sintering and hot isostatic pressing (at ~1700 °C) are appropriate for producing a desirable microstructure (Figure 7.4b). Upon heating to the sintering temperature, these oxide additives react with a thin SiO_2-rich layer present on the surface of the Si_3N_4 particles to form an yttria–alumina–silica liquid phase that surrounds the Si_3N_4 particles. This liquid phase aids both densification and the formation of a microstructure composed of interlocking elongated grains. Upon cooling the product to room temperature, it solidifies to a glassy phase between the Si_3N_4 grains.

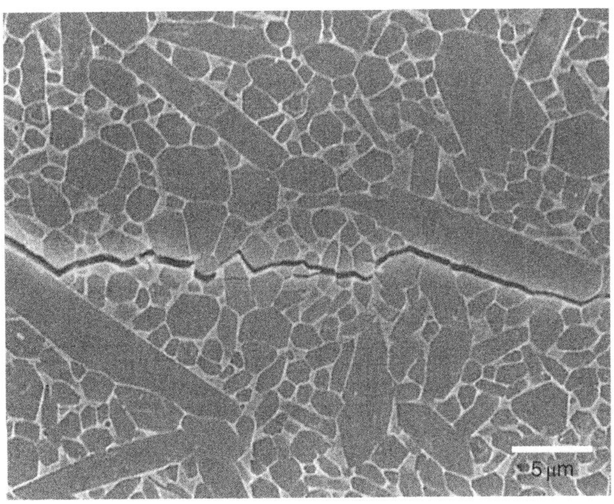

Figure 7.7 Illustration of crack deflection mechanism in silicon nitride (Si_3N_4). A crack is deflected at the grain boundary of the silicon nitride grains and follows a more tortuous path when compared to a crack that propagates directly through the grains.

Orthopedic Applications of Si_3N_4 The biocompatibility and mechanical properties of Si_3N_4 make it an attractive biomaterial for use as load-bearing implants in orthopedic surgery. Si_3N_4 femoral heads can be polished to give a smooth surface and they show low wear rates, comparable to ZTA, when articulating against polyethylene liners. However, Si_3N_4 femoral heads have not yet been approved for use in humans. On the other hand, Si_3N_4 implants have been used in spinal fusion, although to a lesser extent than Ti6Al4V and polyether ether ketone (PEEK). Si_3N_4 implants have been reported to show better integration with host bone than Ti6Al4V and PEEK, and antibacterial properties as well, which have been attributed to their surface chemistry and micro-rough surface in their as-fabricated state (Figure 5.25).

7.4 Calcium Phosphates

Calcium phosphates have been widely studied for decades for use as biomaterials. An attractive feature of these materials is that they are composed of the same ions as the mineral constituent of calcified tissues such as bone and teeth. Consequently, calcium phosphates used as biomaterials are generally highly biocompatible, particularly within bone.

In the form of dense or porous implants, particles, or granules, these ceramics have become increasingly important for use as bone graft substitutes in orthopedic, oral, and maxillofacial surgery.

The calcium phosphates used in these applications are calcium orthophosphates, composed of three major elements, calcium (oxidation state +2), phosphorus (oxygen state +5), and oxygen (oxidation state −2), and the orthophosphate group $(PO_4)^{3-}$. This group is different from the metaphosphate $(PO_3)^-$, pyrophosphate $(P_2O_7)^{4-}$ and polyphosphate $(PO_3)_n^{n-}$ groups, where n represents the number of units in the structure.

Calcium phosphates relevant to biomaterials cover a wide range of compositions but they can be divided into two groups depending on their production method (Table 7.2). Calcium phosphates in the first group are typically synthesized in the form of a powder at low temperature, such as room temperature, by precipitation or hydrolysis reactions in an aqueous solution. These calcium phosphates have traditionally not been used as bone graft substitutes but, instead, as starting materials in the

Table 7.2 Major properties of main calcium phosphate compounds relevant to biomaterials.

Material	Abbreviation	Chemical formula	Ca/P atomic ratio	Solubility at 25 °C: −log K_{sp}	pH stability range
Monocalcium phosphate monohydrate	MCPM	$Ca(H_2PO_4)_2{\cdot}H_2O$	0.5	1.1	0–2.0
Monocalcium phosphate	MCP	$Ca(H_2PO_4)_2$	0.5	1.1	[a]
Dicalcium phosphate dihydrate[1]	DCPD	$CaHPO_4.2H_2O$	1.0	6.6	2.0–6.0
Dicalcium phosphate[2]	DCPA	$CaHPO_4$	1.0	6.9	[a]
Octacalcium phosphate	OCP	$Ca_8H_2(PO_4)_6.5H_2O$	1.33	47	5.5–7.0
Amorphous calcium phosphate	ACP	$Ca_xH_y(PO_4)_z.nH_2O$; (n = 3–4.5; 15–20% H_2O)	1.2–2.2	[b]	~5–12
Calcium-deficient hydroxyapatite[3]	CDHA	$Ca_{10-x}(HPO_4)_x(PO_4)_{6-x}(OH)_{2-x}$; $(0 < x < 1)$	1.5–1.67	~85	6.5–9.5
α-Tricalcium phosphate	α-TCP	$Ca_3(PO_4)_2$	1.5	25.5	[c]
β-Tricalcium phosphate	β-TCP	$Ca_3(PO_4)_2$	1.5	29	[c]
Hydroxyapatite	HA or HAp	$Ca_{10}(PO_4)_6(OH)_2$	1.67	117	9.5–12
Oxyapatite	OXA	$Ca_{10}(PO_4)_6O$	1.67	~69	[c]
Fluorapatite	FA	$Ca_{10}(PO_4)_6F_2$	1.67	120	7–12
Tetracalcium phosphate[4]	TTCP	$Ca_4(PO_4)_2O$	2.0	38–44	[c]

[a] Stable above 100 °C.
[b] Cannot be measured accurately because of its variable composition.
[c] Cannot be precipitated from aqueous solution.
[1] Mineral brushite.
[2] Mineral monetite.
[3] Also referred to as precipitated hydroxyapatite (PHA).
[4] Mineral hilgenstockite.
Source: Compiled from Wang and Nancollas (2008); Dorozhkin (2011).

production of the second group at high temperature by the sintering process. Several calcium phosphate compounds in this first group are also relevant to CPCs (Section 7.5).

Calcium phosphates in the second group include hydroxyapatite, β-TCP and a composite composed of a two-phase mixture of hydroxyapatite and β-TCP, referred to as biphasic calcium phosphate (BCP). In the form of dense or porous implants or granules, these three calcium phosphates have been widely studied and applied as bone substitutes. They are biocompatible and bioactive, form a direct bond with host bone, and are degradable (resorbable) to varying extent.

Like the vast majority of ceramics, porous or dense three-dimensional implants or granules of hydroxyapatite, β-TCP and BCP are commonly produced by sintering. Powders, granules and three-dimensional specimens, porous or dense, for use as biomaterials are available commercially. Alternatively, considerable information is available in the literature for synthesizing powders of these materials and converting them into implants having the desired microstructure (LeGeros 1991).

7.4.1 Solubility of Calcium Phosphates

An important property of calcium phosphates relevant to biomaterials is their solubility in aqueous media. Solubility determines the direction of several chemical reactions involving these compounds, such as dissolution, precipitation, hydrolysis, and phase transformation. It also plays a major role in biological processes that involve the formation and resorption of bone and other calcified tissues, and in abnormal or pathological calcifications.

Solubility refers to the amount of solute that can be dissolved in a solvent to give a saturated solution, and it is often expressed in units of moles per liter. The solubility of calcium phosphates under standard conditions of temperature and pressure vary over several orders of magnitude, depending on their chemical composition and the pH of the solution. The solubility product, commonly written K_{sp}, a thermodynamic property, is the most fundamental way of defining the solubility of a solid. For example, for the dissolution reaction of hydroxyapatite, written as

$$\begin{aligned} &Ca_{10}(PO_4)_6(OH)_2(\text{solid}) \\ &\leftrightarrows 10Ca^{+} + 6(PO_4)^{3-} + 2(OH)^{-} \end{aligned} \tag{7.3}$$

the solubility product of the ions in solution is given by

$$K_{sp} = \left[Ca^{2+}\right]^{10}\left[(PO_4)^{3-}\right]^{6}\left[OH^{-}\right]^{2} \tag{7.4}$$

where the quantities in the square brackets represent the concentrations of the respective ions in a saturated solution. K_{sp} for calcium phosphate compounds relevant to biomaterials are given in Table 7.2. Because the composition of amorphous calcium phosphate (ACP) is variable, depending on the precipitation conditions, its K_{sp} cannot be measured accurately. However, the comparative solubility in an acidic buffer is

$$\begin{aligned} &\text{ACP} \gg \alpha\text{-TCP} \gg \beta\text{-TCP} > \text{CDHA} \\ &\gg \text{HA(HAp)} > \text{FA} \end{aligned} \tag{7.5}$$

Although K_{sp} does not directly give the amount of a solid that can dissolve under a particular set of conditions, it can be used to provide information about the solubility as a function of solution parameters such as pH and ionic concentration.

Calcium Phosphate Solubility Phase Diagrams

Calcium phosphates relevant to biomaterials cover a wide range of compositions and the capacity to form the preferred phase in a solution is complicated by the possibility of forming several other calcium phosphate phases. In general, the pH and ionic concentration of the solution have a strong effect in determining the formation of the preferred phase. This can be understood in terms of a solubility phase diagram in which ionic concentration and pH are plotted. Figure 7.8 shows one such solubility phase diagram for calcium phosphate phases at 37 °C, in which the calcium ion concentration $[Ca^{2+}]$ is plotted as a function of pH. Each line, called a solubility isotherm, defines the composition of a series of solutions that are saturated with respect to the same solid, such as hydroxyapatite, for example, for which Eqs (7.3) and 7.4 apply.

The isotherms in Figure 7.8 have a negative slope in the neutral and acidic regions (pH less than ~7), reflecting the fact that these calcium phosphate compounds become more soluble with decreasing pH. The steepness of the slope of the isotherm indicates how rapidly the solubility increases with decreasing pH. Isotherms for the more basic hydroxyapatite, α-TCP, β-TCP, and tetracalcium phosphate (TTCP) increase more steeply than the more acidic dicalcium phosphate dihydrate (DCPD) and its anhydrous form (DCPA). This is because for a given decrease in pH in this region, the solubility of a more basic compound increases more rapidly than that of a more acidic compound. In the alkaline region of the phase diagram, the calcium ion concentration increases with increasing pH.

The solubility phase diagram (Figure 7.8) is also useful for assessing the relative stability of the calcium phosphates

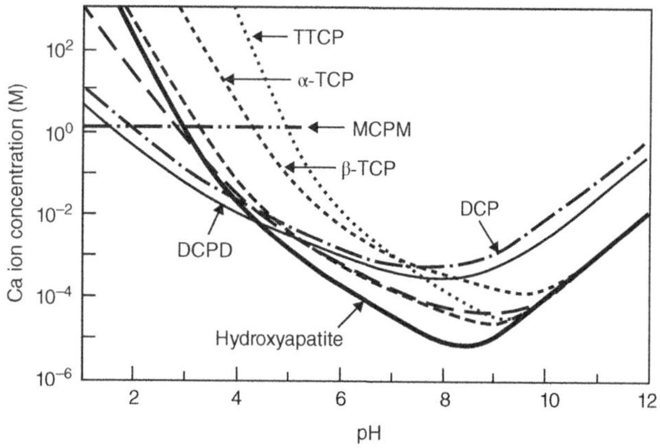

Figure 7.8 Solubility isotherms of various calcium phosphate compounds as a function of pH in water at 25 °C and ionic concentration of 0.1 M (molar). The solubility is expressed as the molar concentration of calcium (Ca) ions in solution. *Source:* From Ishikawa (2010).

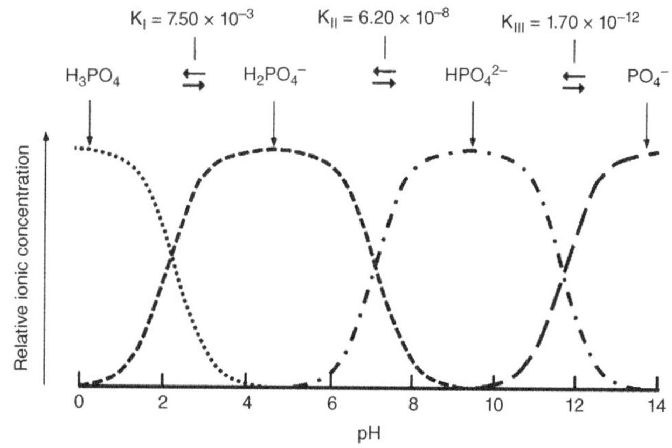

Figure 7.9 Variation of ionic concentrations in triprotic equilibrium for phosphoric acid solutions versus pH. Different pH alters the relative concentration of the deprotonated forms of phosphoric acid and, thus, both the chemical composition and the amount of calcium phosphate compound. K_I, K_{II}, and K_{III} are the rate constants for the equilibrium reactions. *Source:* From Lynn and Bonfield (2005) / with permission of American Chemical Society.

at different pH. For a given pH, a calcium phosphate whose isotherm lies below that of another calcium phosphate is less soluble (more stable) than the other. Hydroxyapatite is the most stable of all the calcium phosphates at pH greater than ~4 and, thus, should form preferentially. In comparison, DCPD should form preferentially below this pH as it is the most stable calcium phosphate. Similarly, TTCP is the most soluble calcium phosphate at pH lower than ~8, whereas above this pH, DCPD is the most soluble. However, in determining the likelihood of the formation of a preferred phase, kinetic factors can sometimes be important also. For example, the formation of octacalcium phosphate (OCP) or DCPD is faster than hydroxyapatite and it is possible for DCPD, OCP, or a combination of DCPD and OCP to form as a precursor phase and subsequently convert to hydroxyapatite.

The effect of pH on the formation of the various calcium phosphate phases is strongly related to the properties of phosphate-containing solutions. As orthophosphoric acid (H_3PO_4) is a triprotic acid (that is, composed of three acidic protons), variation in pH alters the relative concentrations of its three deprotonated forms (Figure 7.9). Consequently, the chemical composition and amount of calcium phosphate formed can vary with the pH of the solution.

7.4.2 Degradation of Calcium Phosphates

Degradation of calcium phosphate implants *in vivo* normally occurs by solution-mediated physico-chemical dissolution but certain calcium phosphates have been reported to degrade, in addition, by a cell-mediated process resulting from the activities of osteoclastic cells (Chapter 14). Although they cannot reproduce the more complex and dynamic conditions present *in vivo*, such as the presence of proteins, other molecules and cells, *in vitro* experiments provide a useful guide for predicting the relative degradation rate *in vivo*. Solution-mediated and cell-mediated degradation are strongly dependent on the physico-chemical properties of the material. For a given calcium phosphate compound, degradation by these two mechanisms is enhanced by fine particle size, high surface area and high porosity.

7.4.3 Hydroxyapatite

Pure hydroxyapatite has the composition $Ca_{10}(PO_4)_6(OH)_2$, sometimes referred to as the stoichiometric composition, which has a Ca to P (Ca:P) atomic ratio of 1.67. The formula of hydroxyapatite is sometimes written $Ca_5(PO_4)_3(OH)$, equivalent to half the number of ions present in a unit cell, but we will not use this formula. Of the calcium phosphate compounds relevant to biomaterials, hydroxyapatite is the most stable at the physiological pH and, in the dense form, it has the highest strength and Young's modulus. Although hydroxyapatite is widely used as a bone graft substitute, its use in healing structural bone defects is challenging because of its low fracture toughness (Table 4.1).

Structure and Composition of Hydroxyapatite Hydroxyapatite has a compositional and structural similarity to the mineral (inorganic) constituent of bone and other calcified tissues. However, the mineral constituent of bone, which we will refer to as biological hydroxyapatite, is composed of a substituted hydroxyapatite, that is, hydroxyapatite in which the calcium, phosphate and hydroxyl ions in the crystal structure are substituted to varying extent by a variety of ions (Chapter 3).

Hydroxyapatite exists in a monoclinic or hexagonal crystal structure but the hexagonal structure is more relevant to biological hydroxyapatite. The hexagonal structure has a high capacity for substitution of its constituent ions and, as biological hydroxyapatites are highly substituted, considerable effort have been devoted to studying the effects of ionic substitution on the physico-chemical and biological properties of hydroxyapatite. The most widely studied substituting ions have been K, Mg Sr, Zn, Cu, Si, CO_3, and F because they are present in minor or trace quantities in bone and other calcified tissues (LeGeros 1991). The effects of these ions generally depend on their composition. For example, substitution of $(CO_3)^{2-}$ for $(PO_4)^{3-}$, or Mg^{2+} or Sr^{2+} for Ca^{2+} in the hydroxyapatite structure leads to a decrease in crystal size and an increase in solubility, whereas substitution of F^- for OH^- leads to a decrease in solubility.

Production and Properties of Hydroxyapatite Dense or porous hydroxyapatite implants are typically produced by sintering the as-formed porous objects above ~1000 °C (Figure 7.10). Porous granules, a few hundred micrometers to a few millimeters in diameter, can be formed using a variety of methods, such as tumbling or vibrating the powder with a liquid granulating agent, or freezing droplets of a suspension of the fine powder. Then the granulated powder is heated above ~1000 °C to bond the particles together into a strong network while maintaining the requisite porosity. The use of sintering temperatures above ~1350 °C is often undesirable as they lead to decomposition to TTCP and α-TCP, via a dehydrated phase called oxyapatite, expressed by the reaction

$$\begin{aligned} &Ca_{10}(PO_4)_6(OH)_2 \rightarrow Ca_{10}(PO_4)_6O + H_2O \\ &\rightarrow Ca_4P_2O_9 + 2Ca_3(PO_4)_2 \end{aligned} \tag{7.6}$$

This decomposition can hinder sintering, although hydroxyapatite is generally reconstituted upon cooling to room temperature.

Hydroxyapatite has a very low solubility at the physiological pH (~7.4), which means that implants can remain *in vivo* for a considerable time (on the order of years). In some applications, it is desirable for the implant to degrade and be replaced by bone or to remodel into bone in a shorter time. Substitution by $(CO_3)^{2-}$, Mg^{2+}, or Sr^{2+} increases degradation by dissolution and by the action of osteoclastic cells, but the degradation rate is still much lower than the mineral constituent of bone. Addition of β-TCP to form BCP provides an effective route to increase the degradation rate of hydroxyapatite but at the expense of lowering its mechanical properties.

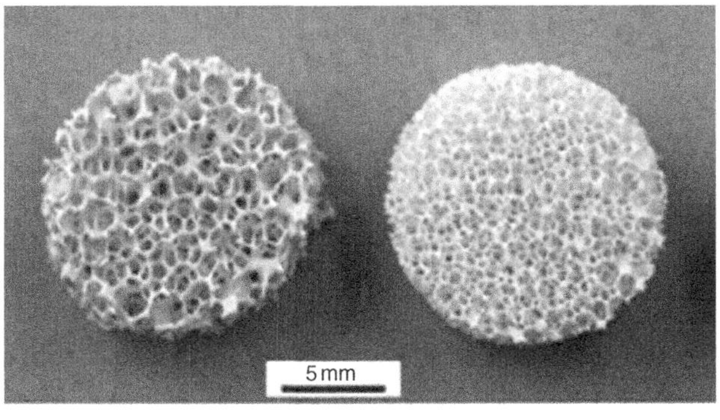

Figure 7.10 Hydroxyapatite discs of varying porosity and pore size produced by the sintering route.

7.4.4 Beta-Tricalcium Phosphate (β-TCP)

Tricalcium phosphate (TCP), $Ca_3(PO_4)_2$, has three polymorphs at atmospheric pressure. The β-phase (rhombohedral structure) is stable at room temperature, and transforms to the α-phase (monoclinic structure) at ~1125 °C, which can be retained on cooling to room temperature. The α-phase forms a continuous solid solution with the alpha'-phase (hexagonal structure) above ~1430 °C, but the alpha'-phase is not relevant to biomaterials. β-TCP is commonly used in bone healing whereas in the form of a powder, α-TCP is used in the preparation of CPCs (Section 7.5).

β-TCP has a much higher solubility than hydroxyapatite and, depending on the size of the implant, it can degrade fully *in vivo* within a period of weeks to months. When compared to hydroxyapatite, the effect of ionic substitution on its physico-chemical and biological properties has received much less attention. While substituting ions such as Mg, Zn, Sr, and Si, either singly or in binary combinations, have been studied, Mg ions have received the most attention. Magnesium ions occupy calcium vacancies in the β-TCP lattice and stabilize the structure. Doping with magnesium raises the transformation temperature to the α phase above ~1350 °C but reduces the solubility of β-TCP. The most common Mg-stabilized β-TCP is the naturally occurring mineral whitlockite, $Ca_{18}Mg_2(PO_4)_{12}(HPO_4)_2$, an analog of which can form locally in the physiological environment, typically under abnormal conditions and only when Mg ions are present.

The production of β-TCP in the form of dense or porous implants or granules follows procedures similar to those for hydroxyapatite. The sintering temperature is generally kept below ~1150 °C to avoid the formation of α-TCP. Doping the powder, typically with 0.5–1.0 wt% MgO, allows the use of higher sintering temperatures but at the expense of lowering the solubility of the product due to the ability of Mg to stabilize the β phase.

7.4.5 Biphasic Calcium Phosphate (BCP)

BCP is a two-phase mixture composed of varying ratios of hydroxyapatite and β-TCP. Depending on the ratio of hydroxyapatite to β-TCP, the overall degradation rate and mechanical properties of BCP can cover a wide range. Because of the higher solubility and lower strength of β-TCP, the degradation rate of BCP decreases with higher hydroxyapatite to β-TCP ratio while the strength increases. BCP powders can be synthesized by a variety of methods, such as heating calcium deficient hydroxyapatite (CDHA) above ~700 °C, hydrolysis of DCPD, or by direct precipitation from a solution composed of, for example, calcium nitrate $Ca(NO_3)_2$ and diammonium hydrogen phosphate $(NH_4)_2HPO_4$. Three-dimensional implants are normally produced by sintering above ~1000 °C.

7.4.6 Other Calcium Phosphates

Tetracalcium Phosphate (TTCP)

TTCP, $Ca_4(PO_4)_2O$, also referred to as the mineral hilgenstockite, has a monoclinic structure. It is not used as a biomaterial in bone healing but, in particulate form, it is a starting material in some CPCs (Section 7.5). Of the calcium phosphate compounds in Table 7.2, TTCP has the highest Ca:P ratio (2.0), is the most basic and has the highest solubility at pH less than ~7. In aqueous media, slow hydrolysis of TTCP can occur, leading to the formation of hydroxyapatite and calcium hydroxide, and a strongly alkaline solution, described by the equation

$$\begin{aligned} &Ca_4(PO_4)_2O + 3H_2O \\ &\rightarrow Ca_{10}(PO_4)_6(OH)_2 + 2Ca(OH)_2 \end{aligned} \tag{7.7}$$

However, TTCP can remain unreacted in water at room temperature for several weeks, presumably due to the initial formation of a thin hydroxyapatite layer on its surface that acts as a barrier to further reaction.

As TTCP reacts with water (Eq. 7.7), it cannot be synthesized in an aqueous medium. Instead, it is often synthesized by solid-state reactions in a dry atmosphere at temperatures above ~1350 °C. One such reaction is based on a mixture of DCP and calcium carbonate (CaCO3) powders having an overall Ca:P ratio of 2.0:

$$\begin{aligned} &2CaHPO_4 + 2CaCO_3 \\ &\rightarrow Ca_4(PO_4)_2O + 2CO_2 + H_2O \end{aligned} \tag{7.8}$$

After heating, the product should be cooled rapidly to avoid the formation of undesired phases such as hydroxyapatite, β-TCP and CaO.

Dicalcium Phosphate Dihydrate (DCPD)

DCPD, a hydrated calcium hydrogen phosphate with the composition $CaHPO_4.2H_2O$, is also known as the mineral brushite. DCPD has a monoclinic crystal structure in which parallel chains composed of alternating calcium and phosphate ions are interlayered with water molecules. An anhydrous calcium hydrogen phosphate, $CaHPO_4$, abbreviated DCPA, also known as the mineral monetite, has a lower solubility than DCPD due to the absence of water molecules in the crystal structure.

DCPD is the most stable of the relevant calcium phosphate compounds in aqueous solution at pH lower than ~4.0 (Figure 7.8). It can be a precursor phase in the formation of hydroxyapatite in aqueous media and is a major phase in some CPCs. Whereas DCPD does not exist in normal bone, it has been observed in abnormal or pathological calcifications such as dental calculus, urinary stones and soft tissue calcifications.

Octacalcium Phosphate (OCP)

OCP, $Ca_8(HPO_4)_2(PO_4)_4.5H_2O$, has a triclinic crystal structure that shows a remarkable similarity to the hexagonal crystal structure of hydroxyapatite. It has a layered structure composed of apatitic layers and hydrated layers. The apatitic layers have calcium and phosphate ions distributed in manner similar to that of hydroxyapatite while the hydrated layer consists of water molecules and less densely packed calcium and phosphate ions. Like DCPD and ACP, OCP can serve as a precursor phase in the formation of hydroxyapatite. It does not exist in normal bone but it has been observed in abnormal or pathological calcifications such as dental calculus, urinary stones, and soft tissue calcifications.

Calcium Deficient Hydroxyapatite (CDHA)

CDHA has a Ca:P atomic ratio smaller than the value (1.67) for pure hydroxyapatite. Its composition can be represented in a few different ways, one of which is $Ca_{10-x}(PO_4)_{6-x}(HPO_4)_x(OH)_{2-x}$. CDHA is commonly synthesized in a powdered form by precipitation or hydrolysis reactions. The degree of calcium deficiency depends on the synthesis method and conditions such as pH and temperature. Upon heating to temperatures above ~700 °C, CDHA gives a mixture of hydroxyapatite and β-TCP, as described by the equation

$$\begin{aligned} &Ca_{10-x}(PO_4)_{6-x}(HPO_4)_x(OH)_{2-x} \\ &\rightarrow (1-x)Ca_{10}(PO_4)_6(OH)_2 \\ &+ 3xCa_3(PO_4)_2 + H_2O + [(1-x)/2]O_2 \end{aligned} \tag{7.9}$$

This decomposition reaction provides one method for the formation of BCP. As Eq. (7.9) shows, the ratio of hydroxyapatite to β-TCP in the product is determined by the Ca:P ratio in the CDHA. For example, when $x = 0.5$, the CDHA powder has a Ca:P ratio of 1.585 and the molar ratio of β-TCP to hydroxyapatite in the product is 3.

Amorphous Calcium Phosphate (ACP)

Similar to many sparingly soluble inorganic compounds, the formation of hydroxyapatite by precipitation reactions between calcium and phosphate ions, both *in vitro* and *in vivo*, generally involves the formation of a precursor calcium phosphate phase that is chemically and structurally different from hydroxyapatite. One such phase is referred to as ACP, which is composed of aggregates of tiny particles and a structure possessing little long-range order. The composition of ACP formed by precipitation reactions, written in the general form $Ca_xH_y(PO_4)_z.nH_2O$, where $n = 3–4$, depends strongly on the pH of the aqueous medium. Ca:P atomic ratios of ACP increase with pH, from ~1.2 at pH ~6.5 to ~1.5 at pH ~12.0, and even as high as 2.2. At normal temperatures, such as the physiological temperature, ACP typically transforms rapidly in aqueous media to the more stable phase (hydroxyapatite) but the presence of certain ions and molecules can delay the transformation. Whereas ACP finds limited use as a biomaterial in bone healing, it provides a route for the production of BCP powders (by heating it above ~700 °C) and is a major phase in some CPCs.

7.4.7 Mechanical Properties of Calcium Phosphates

The mechanical properties of hydroxyapatite, β-TCP and BCP cover a wide range, depending on the composition and microstructure of the material. Similar to other ceramics, the strength of these calcium phosphate ceramics are dependent not just on their total porosity but on the size, geometry and distribution of the pores as well. For the same material produced by the same method, the strength and Young's modulus often show an exponential dependence on porosity (Eq. 4.33). Hydroxyapatite shows a significantly higher strength and elastic modulus than β-TCP when specimens of the same microstructure are compared (Figure 4.15).

Overall, the mechanical properties of BCP show a decreasing trend with increasing amount of β-TCP, although a few studies have shown a peak in strength at 10–40 wt% β-TCP (Wagoner Johnson and Herschler 2011). Hydroxyapatite, β-TCP and BCP show a low fracture toughness in their dense form (Table 4.1), which becomes lower still for porous specimens. These fracture toughness values are in the range reported for human trabecular bone but they are lower than the lowest values reported for cortical bone. Other mechanical properties such as the Weibull modulus and fatigue resistance (Chapter 4) are also relevant to bone repair applications but they have been rarely determined for these calcium phosphate bioceramics.

7.5 Calcium Phosphate Cement (CPC)

CPCs are obtained by mixing a powder of one or more calcium orthophosphate compounds described in the previous section with an aqueous solution to form a paste that sets (hardens) at normal temperatures within a desired time (~15 minutes) (Figure 7.11a). This overall process is similar to that for hydraulic cements, composed of calcium silicate compounds and other oxides, which are widely used in the construction industry. In fact, several advances in CPCs can be are attributed to an understanding of hydraulic cements. Although CPCs are composed of calcium phosphate compounds, they have a different microstructure and, thus, properties from sintered hydroxyapatite, β-TCP and BCP. Suggested for use as biomaterials in the early 1980s, CPCs have been the subject of considerable research and development. Several commercial formulations are now available for use in orthopedic and maxillofacial surgery to reconstruct non-load-bearing bone such as trabecular bone (Bohner 2010a).

CPCs offer the following benefits over rigid implants produced by sintering:

- Capacity to be molded into shape during implantation to match the shape of bone defects, particularly complex-shaped defects.
- Ease of formation involving processes near room temperature.
- Potential for implantation by minimally invasive surgery.

CPCs also have drawbacks. They are brittle and generally have lower strength and lower Weibull modulus when compared to sintered hydroxyapatite or BCP, which limit their use to non-load-bearing applications. Proper control of the CPC formulation is critically important because chemical reactions involving the initial cement mixture

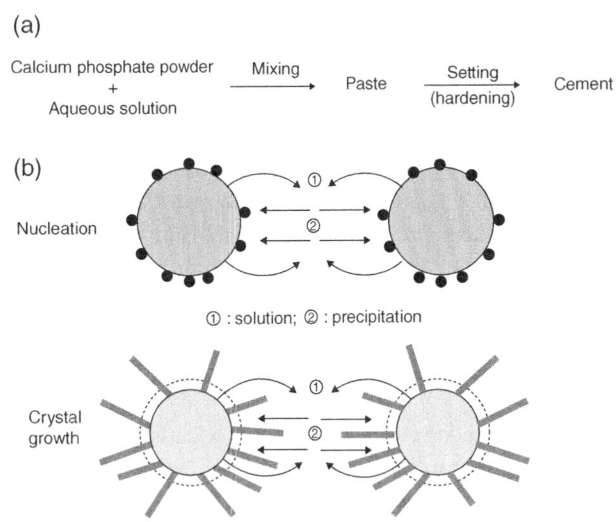

Figure 7.11 (a) Major steps in the setting (hardening) of calcium phosphate cements. (b) Illustration of setting of calcium phosphate cement by a solution–precipitation mechanism.

determine the properties of the resulting paste and the microstructure of the hardened cement as well.

7.5.1 CPC Chemistry

Many CPC formulations have been developed for use in clinical applications and they give rise to a wide variety of chemical reactions that can occur during their setting (hardening). However, the chemical reactions involved in the majority of these formulations can be divided into two major types. The first type involves a classic acid–base reaction between a more acidic compound and a more basic compound in an aqueous solution to form a more neutral compound, as for example the reaction

$$2CaHPO_4 + 2Ca_4(PO_4)_2O \rightarrow Ca_{10}(PO_4)_6(OH)_2 \quad (7.10)$$

In Eq. (7.10), the slightly acidic DCPA reacts with the basic TTCP to give hydroxyapatite (slightly basic). Another example of this type of acid–base reaction is between monocalcium phosphate monohydrate (MCPM) (acidic) and β-TCP (slightly basic) to give DCPD (slightly acid)

$$Ca(H_2PO_4)_2 \cdot H_2O + \beta\text{-}Ca_3(PO_4)_2 + 7H_2O \rightarrow 4CaHPO_4 \cdot 2H_2O \quad (7.11)$$

The second type of reaction can be described as a hydrolysis reaction of a calcium phosphate compound of lower chemical stability, giving a more stable compound, as exemplified by the hydrolysis of α-TCP:

$$3\alpha\text{-}Ca_3(PO_4)_2 + H_2O \rightarrow Ca_9(HPO_4)(PO_4)_5(OH) \quad (7.12)$$

The reaction described by Eq. (7.12) is one of the most commonly used in cement formulations. In this type of reaction, the reactant and the product have the same Ca:P atomic ratio and there is little change in the pH of the solution.

The product in Eq. (7.11) is DCPD (brushite) whereas Eqs 7.10 and 7.12 give hydroxyapatite or a CDHA as the product. Consequently, an alternative division of CPCs has been in terms of these two type of products, giving the terms brushite CPCs and hydroxyapatite (or apatite) CPCs.

7.5.2 CPC Setting (Hardening) Mechanism

Setting (hardening) of CPCs involves a classical physical chemistry mechanism, often referred to as solution–precipitation. The initial step involves dissolution from the surface of the reactant particles into the aqueous setting liquid (Figure 7.11b). Because the dissolution rate of a particle is inversely proportional to its radius, smaller particles lead to more rapid dissolution. The concentration of calcium and phosphate ions in the solution increases with time and, when it reaches a critical value above the saturation limit of the calcium phosphate product, nuclei start to form. This is the start of the precipitation step. Nuclei form preferentially on the surface of the reactant particles because they provide readily available nucleation sites. With continuing dissolution, calcium and phosphate ions generated in the solution diffuse to the nuclei and align on their surface, resulting in their growth to form crystals (particles) of the product. Eventually this dissolution–precipitation process slows and finally stops as the reactants are used up. The fractional amount of product formed as a function of time follows a classic S-shape curve.

The solubility phase diagram (Figure 7.8) determines the final composition of the CPC, that is, the product is generally composed of the most stable calcium phosphate phase in aqueous media. Brushite, as we have seen, is the most stable phase at pH lower than ~4 whereas hydroxyapatite is the most stable phase above this pH. Consequently, despite the many CPC formulations, these are only two major types of product formed, although the involvement of precursor phases such as ACP or OCP is possible (Section 7.2.4).

Hardening of CPCs and the final microstructure of the product are related to the solution–precipitation mechanism and the crystal structure of the calcium phosphate product. Because nucleation occurs preferentially at the surface of the reactant particles, crystals of the product tend to grow around the reactant particles, roughly in a radial direction, during solution–precipitation. These crystals grow in a plate-like (acicular) shape in an attempt to mimic the unit cell geometry of brushite (monoclinic) or hydroxyapatite (monoclinic or hexagonal). As the reaction proceeds and the crystals grow, the viscosity of the mixture increases and the cement progressively sets. Eventually, the crystals on one particle impinge on those of adjacent particles, forming a three-dimensional structure of interlocking crystals. When this occurs, the viscosity increases very steeply and the cement hardens into a solid mass. Typically, only a small amount of heat is released and the overall volume stays almost constant during the setting process.

7.5.3 Microstructure of CPCs

CPCs show a wide range of microstructures that depend on the formulation used in their production. However, Figure 7.12a provides an example to illustrate the major microstructural characteristics of CPCs. Overall, CPCs are composed of a porous structure of interlocking fine plate-like crystals. Whereas the grains in sintered ceramics (dense or porous) are bonded at their boundaries, as shown in Figure 7.12b for a porous BCP specimen, the crystals in

(a)

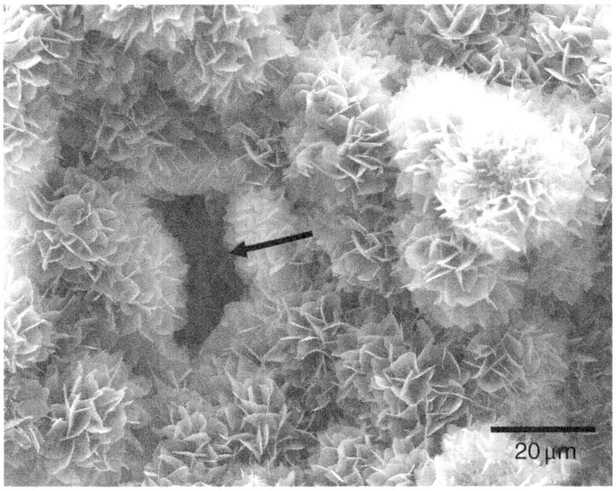

(b)

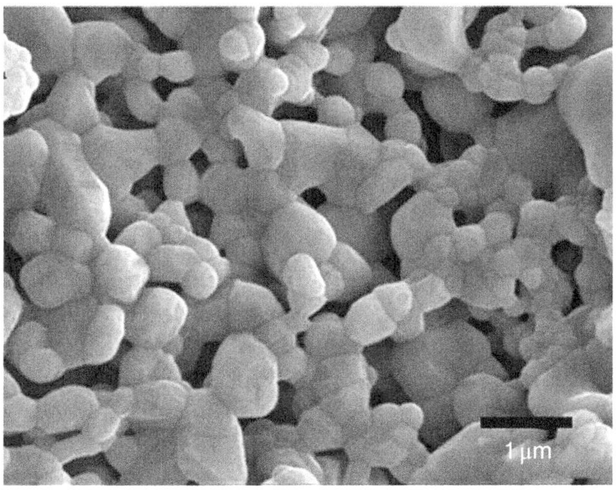

Figure 7.12 Comparison of typical calcium phosphate cement microstructure with that for a calcium phosphate material produced by sintering. (a) Scanning electron microscope (SEM) image of cement microstructure composed of fine pores between lath-like crystals and a small fraction of large pores (arrow). *Source:* From Bohner (2010b). (b) SEM image of porous biphasic calcium phosphate (BCP) microstructure composed of grains bonded at their boundaries. *Source:* From LeGeros (2008).

CPCs show predominantly physical interlocking. Crystals in CPCs generally show a smaller size (~1 μm) and a more plate-like (or acicular) morphology when compared to the larger, more equiaxial grains (less than a few micrometers) that are often present in sintered calcium phosphates.

Another common feature of CPC microstructures is their high porosity and fine pore size. Porosity can range from ~40 to ~80%, but several CPCs have a porosity of ~50%. The majority of the pores have a size less than ~1 μm and are present between the individual crystals. Because of their fine crystal size and pore size, CPCs generally have a much higher surface area (~10 m^2/g) than sintered calcium phosphates such as hydroxyapatite at the same total porosity. CPCs also contain a low percentage (less than ~10%) of large pores, of size larger than ~20–50 μm, resulting predominantly from non-optimal characteristics of the original reactant particles and inefficient mixing with the aqueous setting liquid.

7.5.4 Properties of CPCs

Hydroxyapatite and brushite CPCs are biocompatible, although inflammatory responses have been reported when large specimens of brushite CPCs are implanted *in vivo*. These inflammatory responses are due to the release of a large amount of phosphoric acid, which lowers the local pH, when a large mass of brushite is converted to hydroxyapatite *in vivo*, as described by the reaction

$$10CaHPO_4.2H_2O \rightarrow Ca_{10}(PO_4)_6(OH)_2 + 4H_3PO_4 + 18H_2O \quad (7.13)$$

CPCs show a wide range of solubility (degradation rate) in aqueous media because of their many formulations and microstructures. However, for a comparable microstructure, brushite cements have a higher solubility than hydroxyapatite cements (Figure 7.8). Because of their fine crystal size and high surface area, CPCs typically have a higher degradation rate than sintered calcium phosphates of the same composition.

The microstructure of sintered ceramics, as we have seen (Chapter 3), can be designed to give a wide range of pore sizes, such as pores of size ~300–500 μm that are optimal for enhancing bone ingrowth and integration with host bone. In comparison, because CPCs contain predominantly fine pores (less than ~1 μm), they generally degrade from their surface inward and, as they degrade, they are replaced by bone.

The strength of CPCs cover a wide range, due to their wide range of composition and microstructure. Compressive strengths have been reported in the range less than ~5 MPa (porosity ~80%) to ~100 MPa (porosity ~30%). For the same porosity, brushite CPCs generally show a lower strength than hydroxyapatite CPCs. Because of the weak structure of their physically interlocking crystals, for the same porosity, hydroxyapatite CPCs show a much lower strength than sintered hydroxyapatite. For example, for a porosity of ~50%, a hydroxyapatite CPC prepared by the reaction of Eq. (7.10) showed an average compressive strength of ~4 MPa, approximately 10 times lower than that of sintered hydroxyapatite (~40 MPa).

Like other brittle ceramics, the tensile and flexural strengths of CPCs are much weaker than their compressive strength. Because of their wide range of pore sizes, CPCs show a low Weibull modulus, that is, specimens produced by the same method show a wide spread in their strength. Consequently, CPCs have a high probability of catastrophic brittle failure when subjected to stresses that are much lower than their average strength (Chapter 4). In view of their low strength, low fracture toughness, low Weibull modulus and, thus, low mechanical reliability, CPCs are suitable for use in healing defects in non-load-bearing bones, such as trabecular bone.

In addition to the properties of the hardened cement, the properties of the paste formed by mixing the reactant particles and the aqueous setting liquid are critically important in clinical applications. Like the cement, the properties of the paste, such as setting time, cohesion, and injectability, are dependent on the cement formulation. The setting time should be slow enough to provide enough time for proper mixing and implantation but it should also be fast enough to prevent delay of the operation. Because the paste is implanted or injected into an aqueous physiological medium, it should have the capacity to harden and maintain its cohesiveness without disintegrating *in vivo*. The rheological properties of the paste are important for providing the desired flow behavior and for avoiding separation into its component phases (solid and liquid), particularly for injectable cements.

7.6 Calcium Sulfate

Calcium sulfate occurs widely as a soft mineral called gypsum, composed of calcium sulfate dihydrate, $CaSO_4.2H_2O$. Gypsum has the same crystal structure (monoclinic) as dicalcium phosphate dihydrate (DCPD) and a similarity in composition, except that the $(HPO_4)^{2-}$ is replaced by $(SO_4)^{2-}$, but it is ~10 times more soluble at neutral pH near room temperature. A well-known property of gypsum is its ability to lose and regain its water of crystallization. When heated above ~120 °C, gypsum loses water to form the hemihydrate $CaSO_4.½H_2O$, known as plaster of Paris (POP).

This hemihydrate has traditionally been given two different designations, α-hemihydrate and β-hemihydrate, depending on its method of preparation from gypsum. The α-hemihydrate has typically been prepared in a wet environment, such as by autoclaving, whereas the β-hemihydrate has been prepared by dry methods, such as calcination. Although they have the same crystal structure, these two hemihydrates show differences in their reactivity with water and in the strength of their hydration products after setting. These differences arise from a difference in crystal size due to the different production methods. Typically, the β-hemihydrate consists of a porous mass of particles that are themselves composed of fine crystals whereas the α-hemihydrate is composed of larger individual crystals. The α and β prefixes indicate a difference in production method and not a difference in crystal structure.

The reaction of POP with water can be written as

$$CaSO_4.½H_2O + 3/2H_2O \rightarrow CaSO_4.2H_2O \qquad (7.14)$$

POP is only slightly soluble in water at room temperature but the solubility of gypsum is lower still. At 20 °C, the solubility of the α and β hemihydrates are ~9 and ~7 g/l, respectively, whereas the solubility of gypsum is ~2 g/l. During the hydration reaction, gypsum forms by a solution–precipitation process, similar in nature to that described for CPCs (Section 7.5), giving a rigid mass of interlocking acicular crystals. For the same initial POP to water ratio, the α-hemihydrate forms a stronger plaster upon setting than the β-hemihydrate.

In the α-hemihydrate form, POP is used as to make molds for casting dental restorations. The use of POP (β-hemihydrate) in casts to heal fractured bones is well known, but other materials such as fiber-reinforced plastics are now also being used. POP has also been used as a carrier material for local antibiotic delivery in the treatment of infected bone and joints, although concerns have also been expressed about its possible cytotoxicity. Typically, an antibiotic (such as vancomycin) is mixed with an aqueous liquid (such as sodium chloride solution), which is added to the POP powder to form a paste. Then the paste is allowed to set in a mold to form the desired shape (such as spheres several millimeters in diameter). The calcium sulfate encapsulates the antibiotic and, as it degrades *in vivo*, it releases the antibiotic, giving reasonably predictable release rates. Because of its rapid degradation rate, POP gives fast antibiotic release rates, particularly over the first few days, and suffers from a rapid loss in strength in the process. Consequently, POP is more suitable for short-term antibiotic delivery in non-load-bearing applications.

7.7 Glasses

Glasses have an amorphous structure composed of a three-dimensional network showing no long-range atomic order and lacking detectable crystallinity (Chapter 3). A simple and common way of identifying a glass is the lack of identifiable peaks in its X-ray diffraction pattern. Glasses gradually soften upon heating and, at a sufficiently high temperature, they flow like a highly viscous liquid. Upon cooling from the molten state, glasses solidify gradually and their viscosity increases continuously with decreasing temperature. Unlike crystalline materials, glasses do not show a distinct phase transformation, from liquid to solid, at a well-defined melting temperature T_m. Instead, the transition from liquid to solid occurs at a glass transition temperature T_g that lies below T_m at which crystallization would take place if the glass were crystallizable.

7.7.1 Glass Transition Temperature (T_g)

At T_g, glasses show a change in physical properties, from those of a liquid to those of a solid. One characteristic change that is often used to illustrate the glass transition is the specific volume (volume per unit mass) as a function of temperature upon cooling (Figure 7.13). Above T_m, the material is a liquid, between T_m and T_g, it is said to be a supercooled liquid, and below T_g the material is a glass. Thus, T_g characterizes the temperature at which the transition from supercooled liquid to glass occurs. Unlike T_m, T_g depends on the thermal history of the glass, such as its cooling rate from the molten state. For example, a higher cooling rate gives a higher T_g.

T_g is one of the most important properties of a glass because the transition from liquid to glass is accompanied by a significant change in physical properties. For example, a glass shows mechanical properties such as compressive strength, Young's modulus, hardness and brittleness that are comparable to those of many ceramics. Although both the liquid and the glass have a disordered structure, the transition to a glass marks a state where structural rearrangements are no longer able to take place within a reasonable time scale. The potential energy barrier for molecular motion in the glass is large, such that the structure is essentially frozen-in. Consequently, the relaxation time for structural change is high. In comparison, the structure of the liquid can relax rapidly due to the low energy barrier for molecular motion, which is easily overcome by the thermal energy of the system.

Structural rearrangements that occur in the glass transition region lead to discontinuities in thermodynamic and physical properties such as enthalpy, thermal expansion coefficient and specific volume. These discontinuities allow the detection and measurement of T_g and, if the glass is also crystallizable, T_m. Although the specific volume is often used to illustrate the glass transition, standard techniques such as differential thermal analysis (DTA), differential scanning calorimetry (DSC) or dilatometry are more commonly used because they are easier to perform.

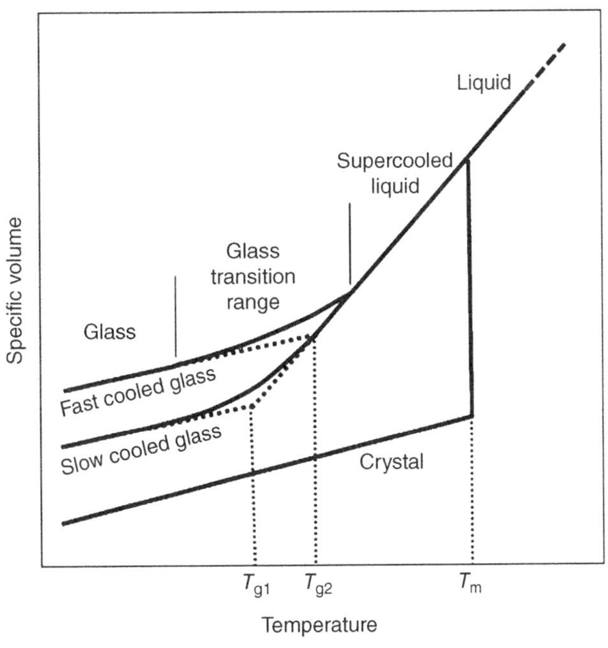

Figure 7.13 Specific volume (volume per unit mass) as a function of temperature for liquid, crystal and glass phases. The transformation from a supercooled liquid to a glass occurs at the glass transition temperature T_g that is lower than the melting temperature T_m of the crystalline phase if the glass were crystallizable. Both the specific volume of the glass and T_g depend on the cooling rate of the liquid.

7.7.2 Glass Viscosity

Another important property of a glass is its viscosity. Generally, the viscosity η of a substance is defined as the ratio of the applied shear stress τ to the strain rate $\dot{\gamma}$, that is

$$\eta = \frac{\tau}{\dot{\gamma}} \tag{7.15}$$

Consequently, the viscosity of a material provides a measure of its deformability. The viscosity of a glass is dependent on its composition and temperature. It has theoretical interest and practical interest as well, providing useful information related to melting conditions, working the glass into a desired shape, annealing the glass to relieve internal stresses, maximum use temperature and, if crystallizable, the rate of crystallization.

Figure 7.14 shows the viscosity versus temperature curves for three common types of inorganic glasses, a

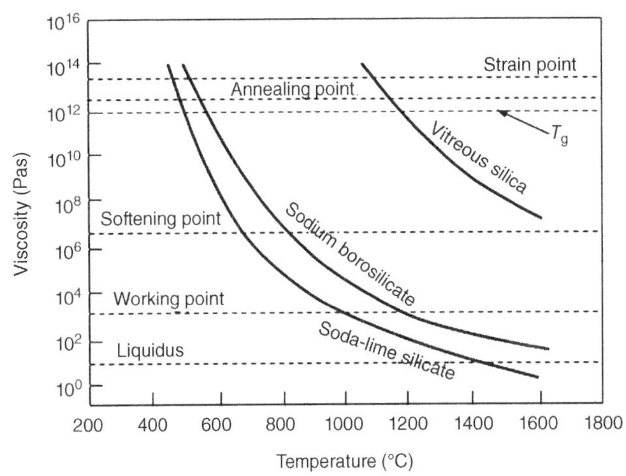

Figure 7.14 Viscosity of three silicate glasses as a function of temperature. Important viscosities relevant to glass manufacture are indicated.

soda-lime–silica glass that approximates the composition of glasses used for windows, a sodium borosilicate glass that approximates the composition of chemical glassware, and vitreous silica (100% SiO_2) used in high-temperature applications such as furnace tubes and crucibles (Table 7.3). Glasses are often characterized by the temperature at which various viscosities of practical interest are achieved. The melt used in casting a glass typically has a viscosity η of ~10^2–10^3 Pa s. In comparison, the viscosity of water at 25 °C is ~0.9 Pa s. By general agreement, T_g is defined as the temperature at which $\eta = 10^{12}$ Pa s. The working range corresponds to a region where the glass softens to a point such that it can be worked into a more desirable shape. When cooled rapidly from the melt to room temperature, a glass develops internal stresses due to temperature gradients within it. These stresses can lead to the generation of microcracks that lower the strength of the glass. Consequently, glasses used in structural applications are often annealed, typically at a viscosity of ~$10^{12.4}$ Pa s, corresponding to a temperature of ~10 °C below T_g, and cooled slowly to relieve internal stresses. The strain point corresponds to $\eta = 3\times10^{13}$ Pa s, above which structural relaxation times are so long that annealing cannot be achieved within a reasonable time. Well below T_g, the viscosity is considerably higher still.

7.7.3 Production of Glasses

Glasses are generally produced by heating a mixture of appropriate starting materials in an unreactive crucible to a high temperature to form a homogeneous melt. Then the melt is cooled sufficiently fast to avoid crystallization. For example, a soda-lime-silica glass would typically be produced from a starting mixture composed of the appropriate amounts of SiO_2 sand, sodium carbonate (soda) Na_2CO_3, and calcium carbonate (limestone) $CaCO_3$. The carbonates decompose to oxides prior to the formation of the melt and, at a sufficiently high temperature, the mixture reacts together, forming a melt. Products of the desired shape are formed by casting the melt or, after casting, by heating the glass to the working range and forming it by methods such as pressing, blowing or drawing. Bulk glasses are rigid but when formed into fine-diameter fibers, they are more pliable, particularly when the fiber diameter is smaller than a few micrometers.

Example 7.1 It is required to produce 100 g of a soda-lime–silica glass of composition (wt%) 75.0 SiO_2, 15.0 Na_2O, 10.0 CaO, using SiO_2, Na_2CO_3, and $CaCO_3$ as starting materials. Determine the amount of each starting material that should be used.

Solution:

Prior to formation of the melt, Na_2CO_3 and $CaCO_3$ decompose to their respective oxides according to the equations

$$Na_2CO_3 \rightarrow Na_2O + CO_2 \quad (7.16)$$

$$CaCO_3 \rightarrow CaO + CO_2 \quad (7.17)$$

Thus, we require the amount of $NaCO_3$ and $CaCO_3$ that will give 15.0 g Na_2O and 10 g CaO, respectively. Using the atomic weights of the elements, we find that the composition of the starting mixture should be 75.0 g SiO_2, 25.6 g Na_2CO_3, and 17.8 g $CaCO_3$.

7.8 Chemically Unreactive Glasses

Glasses used in industrial and commercial applications are generally unreactive at normal temperatures. As noted in

Table 7.3 Nominal composition (in wt%) of two common glasses and some bioactive glasses.

Glass	SiO_2	Na_2O	K_2O	MgO	CaO	SrO	Al_2O_3	B_2O_3	P_2O_5
Chemically durable									
Window	72.0	14.2	0.6	2.5	10.0	—	0.6	—	—
Chemical glassware	81.0	4.5	—	—	—	—	2.0	12.0	—
Bioactive									
45S5	45.0	24.5	—	—	24.5	—	—	—	6.0
S53P4	53.0	23.0	—	—	20.0	—	—	—	4.0
13–93	53.0	6.0	12.0	5.0	20.0	—	—	—	4.0
13–93B1	34.4	5.8	11.7	4.9	19.5	—	—	19.9	3.8
2B6Sr	15.6	5.4	10.9	1.2	17.8	9.0	—	36.0	4.1
13–93B3	—	5.5	11.1	4.6	18.5	—	—	56.6	3.7

Chapter 3, they are typically silicate, borate, borosilicate, or phosphate glasses but the vast majority (over ~95% by mass) are silicate glasses. Depending on their composition, glasses used as biomaterials can be chemically unreactive, degradable or bioactive when exposed to aqueous media such as the physiological fluid.

Chemically unreactive glasses see little use as biomaterials but one important application has been the use of microspheres composed of an yttria–alumina–silica glass (Y_2O_3–Al_2O_3–SiO_2) in radiotherapy treatment of liver cancer (Day 2012). This application takes advantage of the ability to form a chemically unreactive glass composition containing a high concentration (~35–55 wt% Y_2O_3) of the rare-earth element yttrium (stable ^{89}Y isotope) and the ease of forming the glass into microspheres of the requisite size (~25–40 μm) (Figure 7.15). After neutron activation in a nuclear reactor to convert the stable ^{89}Y isotope to the radioactive ^{90}Y isotope, the requisite amount of microspheres is injected by means of a catheter into the blood supply of the hepatic artery. Delivery of the microspheres to liver tumors is achieved by taking advantage of the hypervascularity of the tumors. Liver tumors larger than ~3 mm receive 80–100% of their blood supply from the hepatic artery whereas normal liver tissue receives its blood via the portal vein. The microspheres become embedded in the liver because they exceed the capillary size (average diameter 6–8 mm), thus embolizing the microvessels. Decay of the ^{90}Y radionuclide (half-life ~65 hours) to stable ^{90}Zr leads to the emission of β particles (electrons) that travel an average distance of ~2.5 mm (but no more than ~10 mm), allowing the microspheres to irradiate the tumors while sparing the healthy liver tissue. When compared to chemotherapy and external radiation treatment of tumors, these glass microspheres deliver radiation locally, allowing the use of an optimal dose without significant side effects.

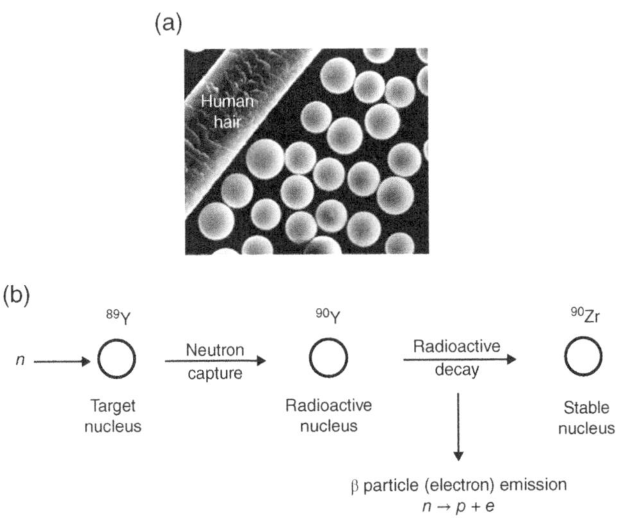

Figure 7.15 (a) Microspheres of Y_2O_3–Al_2O_3–SiO_2 glass (25–40 μm) used in radiotherapy treatment of liver cancer, compared to the diameter of a human hair. *Source:* Courtesy of Delbert E Day, Missouri University of Science and Technology. (b) Neutron activation and decay of yttrium nucleus, resulting in the emission of β particle for radiotherapy treatment.

7.9 Bioactive Glasses

The distinctive characteristics of bioactive glasses are the following (Figure 7.16):

- They react with aqueous solutions containing phosphate ions, such as the physiological fluid, and convert to a hydroxyapatite-like material that mimics the composition of bone mineral.
- Upon implantation, bioactive glasses form a direct bond to bone without any discernable intervening fibrous tissue between their surface and bone due to the hydroxyapatite-like material formed at their surface.
- During their reaction with aqueous solutions, bioactive glasses also release inorganic ions into the medium, some of which have therapeutic effects *in vivo*.

7.9.1 Bioactive Glass Composition

Bioactive glasses have the compositional flexibility of glasses, which allows the design of glasses that have a controllable reaction rate, from rapidly reacting to slowly

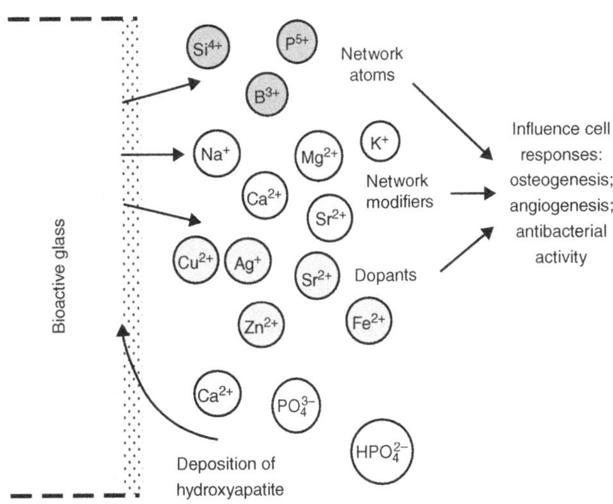

Figure 7.16 Illustration of distinctive characteristics of bioactive glasses when implanted *in vivo*: reaction in the physiological fluid, release of ions, and deposition of a hydroxyapatite-like material on the surface of the glass. Released ions influence cell responses whereas the hydroxyapatite layer leads to direct bonding with bone and soft tissues.

reacting glasses. This compositional flexibility can also be used to incorporate small amounts of elements such as Sr, Cu, and Zn that are known to stimulate specific cellular activity, such as osteogenic and angiogenic gene expression. As the glass reacts *in vivo,* these elements are released as ions or molecules at a biologically acceptable rate. Because of the ease of forming glasses, bioactive glasses can be created in a variety of forms, such as dense three-dimensional solids, granules, particles, fibers, and fibrous masses that are suitable for use in healing bone and soft tissue. Porous implants having optimal characteristics for use in bone healing are generally formed by the sintering route described for ceramics, after first grinding the glass to form particles.

Since ~1970 when the first bioactive glass, a silicate glass designated 45S5 and known commercially as Bioglass®, was reported, bioactive glasses have been the subject of considerable research and development. These efforts have resulted in the creation of several other bioactive glasses that include not just silicate glasses but borate, borosilicate and phosphate glasses as well. Table 7.3 gives the composition of a few bioactive glasses used clinically or under development for use in bone and soft tissue healing.

45S5 glass was originally created for use in healing bone defects but despite considerable efforts over nearly 50 years, it is used clinically only to a limited extent. Overall, bioactive glasses have not realized their early promise as biomaterials for use in bone healing. However, a significant recent development has been the creation of microfibrous forms of special bioactive glasses that have expanded the use of bioactive glasses to the healing of soft tissue wounds, such as skin wounds. These glasses are typically composed not of silicate compositions but of faster-reacting borate compositions, similar to that of the glass designated 13–93B3 in Table 7.3.

7.9.2 Mechanism of Conversion to Hydroxyapatite

A characteristic feature of bioactive glasses , as noted earlier, is their capacity to undergo specific reactions at their surface *in vivo*, resulting in the formation of a hydroxyapatite surface layer that is responsible for its capacity to bond directly to bone. More correctly, the product is a hydroxyapatite-like material because it differs slightly from the composition of hydroxyapatite, $Ca_{10}(PO_4)_6(OH)_2$. However, the term hydroxyapatite will sometimes be used in a general sense. 45S5 glass has been the most widely studied and applied bioactive glass, and it is often considered a prototype for understanding and designing bioactive glasses, particularly silicate bioactive glasses.

The nominal composition of 45S5 glass (in wt%) is

- 24.5 Na_2O, 24.5 CaO, 45.0 SiO_2, 6.0 P_2O_5

In the designation 45S5, the "45S" refers to the weight percent of SiO_2, whereas the "5" following the S refers to the approximate weight ratio of CaO to P_2O_5 (although the ratio is closer to 4). The compositional features that are responsible for the reactivity of 45S5 glass and its bioactivity as well are:

- Low SiO_2 content when compared to more chemically durable glasses such as window glass (Table 7.3).
- High concentration of network modifiers Na_2O and CaO that break up the SiO_2 network, giving a high fraction of nonbridging oxygen in the structure (Chapter 3).
- High CaO content (or CaO to P_2O_5 ratio).

Overall, these compositional features make the surface of 45S5 glass reactive in an aqueous medium and allow it to release a sufficiently high concentration of Ca^{2+} ions to form a hydroxyapatite layer at its surface.

The conversion of 45S5 glass to hydroxyapatite in an aqueous phosphate solution occurs in a series of stages:

- *Stage 1*: Upon exposure of the glass to an aqueous medium, rapid ion-exchange reactions between the glass network modifiers (Na^+ and Ca^{2+}) with H^+ (or H_3O^+) ions from the solution, leads to hydrolysis of the silica groups and the creation of silanol (Si–OH) groups on the glass surface, as described by the reaction

$$\begin{aligned}&-Si-O-Na^+ + H^+ + OH^- \\ &\quad \rightarrow -Si-O-H^+ + Na^+ + OH^-\end{aligned} \tag{7.18}$$

- *Stage 2*: Local increase in pH due to consumption of H^+ ions (Eq. 7.18) makes the SiO_2 glass network more susceptible to chemical attack by OH^- ions. This leads to dissolution of silica, in the form of silicic acid, $Si(OH)_4$, into the solution and the continued formation of Si–OH groups on the glass surface:

$$\begin{aligned}&-Si-O-Si- + H_2O \\ &\quad \rightarrow -Si-O-H + H-O-Si-\end{aligned} \tag{7.19}$$

- *Stage 3*: Condensation and polymerization of the silanol groups to form an amorphous SiO_2 layer (approximately 1–2 μm thick) on the surface of the glass that has been depleted in Na^+ and Ca^{2+}ions:

$$\begin{aligned}&-O-Si-O-H + H-O-Si-O- \\ &\quad \rightarrow -O-Si-O-Si-O- + H_2O\end{aligned} \tag{7.20}$$

- *Stage 4*: Dissolution of the glass continues, and migration of Ca^{2+} ions from the glass through the SiO_2 layer and their reaction with $(PO_4)^{3-}$ ions from the aqueous solution leads to the deposition of an ACP layer on the SiO_2 layer.

- *Stage 5*: As the glass continues to dissolve, the ACP layer thickens, incorporates $(OH)^-$ ions and $(CO_3)^{2-}$ ions normally present in the medium, and crystallizes as a carbonate-substituted hydroxyapatite, composed of a mesoporous structure of nanosize crystals.

The conversion mechanism of bioactive borate glasses follows a similar mechanism but without the formation of the SiO_2 layer. *In vivo*, following the initial formation of a carbonate-substituted hydroxyapatite surface layer (~100 μm thick in 12–24 hours), bonding between the implant surface and host bone starts to occur. The biological mechanisms of bonding involve adsorption of growth factors, followed by the arrival and attachment of stem cells that proliferate and differentiate into osteoblasts (bone-forming cells). Thereafter, osteoblasts create extracellular matrix (collagen) that mineralizes to form a nanocrystalline mineral on the surface of the implant, which leads to a firm bond with bone (Figure 7.17). The glass continues to react and convert to hydroxyapatite over time, at a rate that depends on its composition, until it is fully converted.

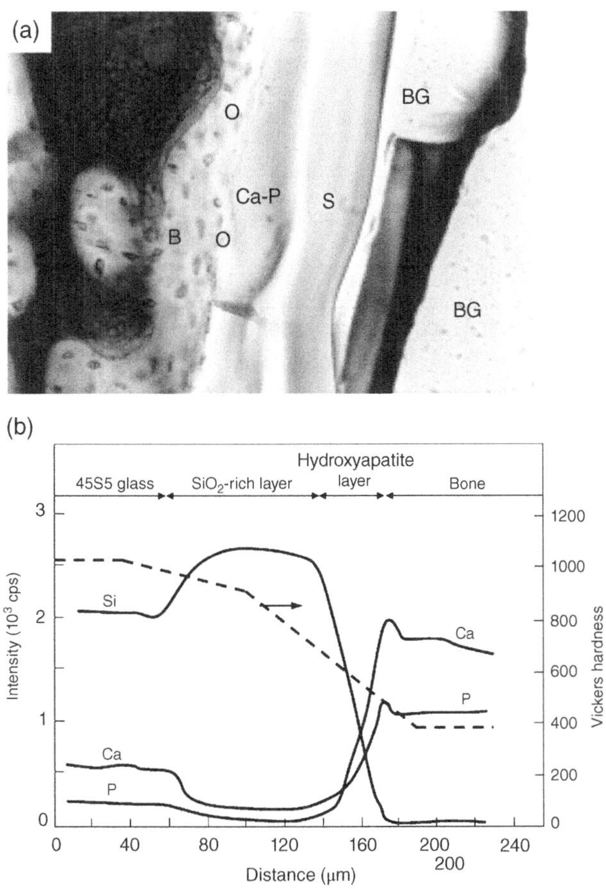

Figure 7.17 (a) Optical micrograph of a 45S5 bioactive glass implant (BG) bonded to rat bone (B) after one year, showing osteocytes or bone cells (O) in conjunction with a carbonate-substituted hydroxyapatite layer (Ca–P) formed on top of the silica layer (S). (b) Electron microprobe analysis across the implant-bone interface shown in (a). *Source:* From Hench (1998) / with permission of John Wiley & Sons.

7.9.3 Reactivity of Bioactive Glasses

Reactivity in aqueous solution containing phosphate ions is a critical property of bioactive glasses because it controls the rate at which they release ions into the medium and convert to hydroxyapatite, and how steeply their mechanical properties decrease with time.

It depends primarily on the glass composition but also on the properties of the aqueous medium (such as pH and presence of ions and molecules) and the surface area of the specimen. Traditionally, the effect of these parameters has been investigated *in vitro* because *in vivo* experiments are difficult or impractical to perform. Although the conditions of the environment are different, *in vitro* experiments provide a useful guide to the reactivity of bioactive glasses *in vivo*.

Typically, bioactive glass specimens in the form of particles of a given size (or surface area) or implants of the same microstructure are immersed in a standard solution, such as phosphate-buffered saline or simulated body fluid composed of the same ionic concentration as the physiological fluid. As a bioactive glass reacts and converts to hydroxyapatite, its mass (weight) decreases, which can be easily measured as a function of time, providing a useful method for evaluating the rate of reaction and conversion to hydroxyapatite. Changes in pH of the medium can also be measured, whereas the concentration of ions released into the medium can be determined using spectroscopic techniques. These types of kinetic data, when combined with standard structural and analytical techniques, such as X-ray diffraction, scanning electron microscopy, and Fourier transform infrared spectroscopy, provide a useful guide to the behavior of bioactive glasses *in vivo*.

Example 7.2 Particles (total mass = 10 g) of a reactive borate glass of composition (wt%) 20 Na_2O, 20 CaO, 60 B_2O_3 are immersed in 500 cm^3 of an aqueous solution containing orthophosphate $(PO_4)^{3-}$ ions to convert them completely to hydroxyapatite $Ca_{10}(PO_4)_6(OH)_2$. Determine (i) the minimum concentration of $(PO_4)^{3-}$ ions that should be present in the solution for complete conversion of the glass particles, and (ii) the weight loss of the particles upon complete conversion.

Solution

In the conversion process, we can assume that the Ca ions of the glass will react with the $(PO_4)^{3-}$ ions in the solution to form $Ca_{10}(PO_4)_6(OH)_2$ according to the overall equation

$$10Ca^{2+} + 6(PO_4)^{3-} + 2(OH)^- \rightarrow Ca_{10}(PO_4)_6(OH)_2 \quad (7.21)$$

a) According to Eq. (7.21), 401.0 g Ca^{2+} ions react with 6.0 moles $(PO_4)^{3-}$ ions to form 1005.0 g hydroxyapatite. The glass contains 2.0 g CaO or 1.43 g Ca^{2+}. This amount of Ca^{2+} ions requires $(1.43/401.0) \times 6.0 = 0.020$ moles of $(PO_4)^{3-}$ ions to react completely and form hydroxyapatite. As the volume of the aqueous solution is 500 cm^3 (= 0.5 l), the minimum concentration of $(PO_4)^{3-}$ ions in the solution should 40 mM (40 mmol/l).
b) Equation (7.21) shows that 401.0 g Ca^{2+} ions react to form 1005.0 g hydroxyapatite. Thus, when the glass is completely reacted, 1.43 g of Ca^{2+} ions in the glass forms 3.60 g hydroxyapatite. As 10 g of the glass reacts to form 3.6 g hydroxyapatite, the weight loss of the particles is 6.4 g or 64.0%. (The Na_2O and B_2O_3 dissolve out of the glass, forming Na^+ ions and $B(OH)_3$, respectively, which remain in the medium.)

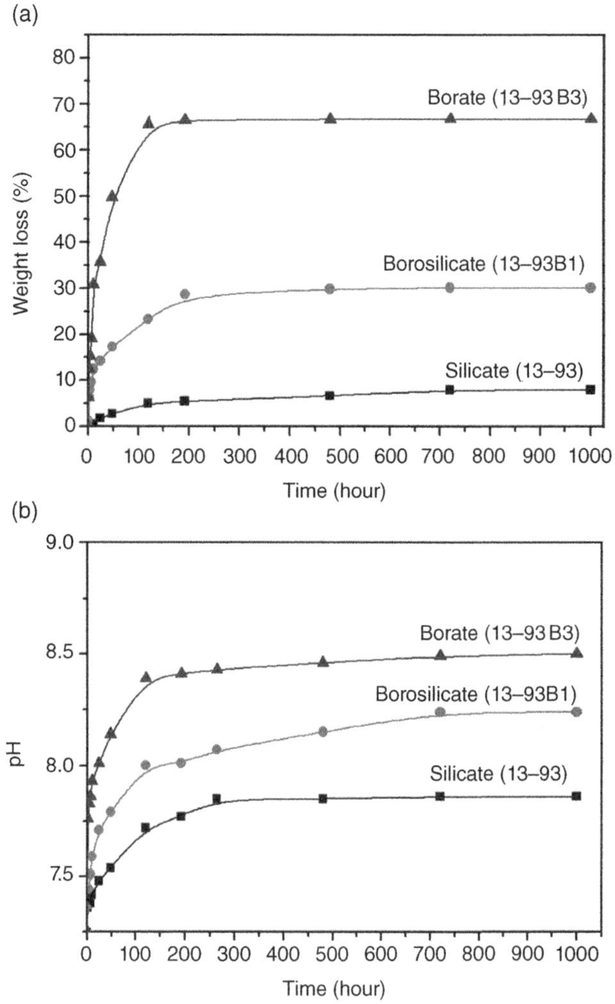

Figure 7.18 Effect of composition on (a) reactivity of bioactive glasses as measured by their weight loss as a function of time in an aqueous phosphate solution *in vitro* and (b) pH of the aqueous medium. The glass specimens, having the same microstructure (Figure 3.24a), are composed of a silicate (13–93), borosilicate (13–93B1) and borate composition (Table 7.3).

Figure 7.18a shows an example of the strong effect that the bioactive glass composition can have on its reactivity and conversion to hydroxyapatite. When compared to their silicate counterparts such as the glass designated 13–93 (Table 7.3), borate glasses such as 13–93B3 react and convert faster to hydroxyapatite (steeper initial slope), and they convert more completely as well (larger total weight loss). In comparison, borosilicate glasses, such as the glass designated 13–93B1 in which one-third of the molar concentration of SiO_2 in 13–93 glass is replaced by B_2O_3, show a behavior that falls between those for their borate and silicate counterparts. Overall, this general trend has also been observed *in vivo*, using structural and analytical techniques to examine specimens implanted for specific times in animal models.

The capacity of borate glasses to react faster than their silicate analogs can be rationalized in terms of differences in their atomic bonding. When compared to the fourfold coordination of Si atoms in the silicate (13–93) glass network, B atoms in the borate (13–93B3) glass network have a lower (threefold) coordination predominantly, which makes them more susceptible to hydrolytic attack. Boron atoms in 13–93B3 glass are also bonded to a larger fraction of nonbridging oxygen atoms, which also makes this glass more reactive.

The dissolution and conversion reactions of silicate and borate bioactive glasses generally lead to an increase in pH of the medium (Figure 7.18b). Because of the more dynamic fluid flow *in vivo*, the physiological liquid surrounding a bioactive glass implant is expected to show a lower average pH when compared to a stirred solution of fixed volume commonly used *in vitro*. However, the local pH immediately surrounding the implant is considerably higher than the average pH, which can enhance the potential for inflammatory responses of tissues (Chapter 18).

Overall, then, varying the composition of a bioactive glass provides an important way to control its reactivity *in vivo*.

7.9.4 Mechanical Properties of Bioactive Glasses

As bioactive glasses are brittle, their design for mechanical reliability follow principles similar to those described for ceramics (Section 7.2.1). Additionally, unlike nondegradable materials, bioactive glasses suffer a strong decrease in mechanical properties with time as they react and convert to hydroxyapatite *in vivo*. Consequently, we are concerned not just with the effect of composition and microstructure on the mechanical properties of bioactive glasses in their as-fabricated state but also with their reactivity in aqueous media as well.

As fabricated, bioactive borate glasses generally show a lower compressive strength than their silicate counterparts of the same microstructure. This can be attributed to a difference in atomic environment and to their higher reactivity that makes their surface more prone to degradation and the formation of flaws when exposed to moisture in the environment.

Three-dimensional implants for use in bone reconstruction should often have a porous microstructure to allow infiltration of bone and integration with host bone. Thus, the effects of pore characteristics such as the amount, size, and distribution of pores on the mechanical properties of bioactive glasses are relevant. Depending on these characteristics, bioactive glasses show a wide range of compressive strengths, from values comparable to that of human trabecular bone to values as high as that for cortical bone. For example, bioactive glass (13–93) specimens (porosity ~70%; pore size = 100–400 μm) having a microstructure mimicking dry human trabecular bone show an elastic response initially and a low Young's modulus (Figure 7.19). Thereafter, the stress versus deformation profile shows several peaks and valleys due to progressive breaking of the solid glass struts and compaction of the structure. In comparison, specimens (porosity ~50%; pore size = 60–80 μm) containing pores that are oriented in the direction of testing show a classic brittle response and much higher strength and Young's modulus. Specimens prepared by an additive manufacturing (3D printing) technique, having a gridlike microstructure of uniform pores (porosity ~50%, pore size ~300 μm) show the highest strength, comparable to that of human cortical bone.

Bioactive glasses, porous or dense, show a strong decrease in strength and Young's modulus when immersed in aqueous media that is nonlinear with time. A steep decrease occurs in the first few weeks after which the decrease is slower. For example, when immersed in simulated body fluid, specimens composed of bioactive silicate glass (13–93), showed a reduction in compressive strength of 50% from their as-fabricated value (from ~90 to ~45 MPa) within approximately two weeks (Liu et al. 2013). Thereafter, the decrease was less steep, to ~30 MPa at approximately six weeks. These specimens, of size 6 × 6 × 6 mm, had a gridlike microstructure consisting of pores of size (width) ~300 μm. The rate at which the strength and Young's modulus decrease is also a

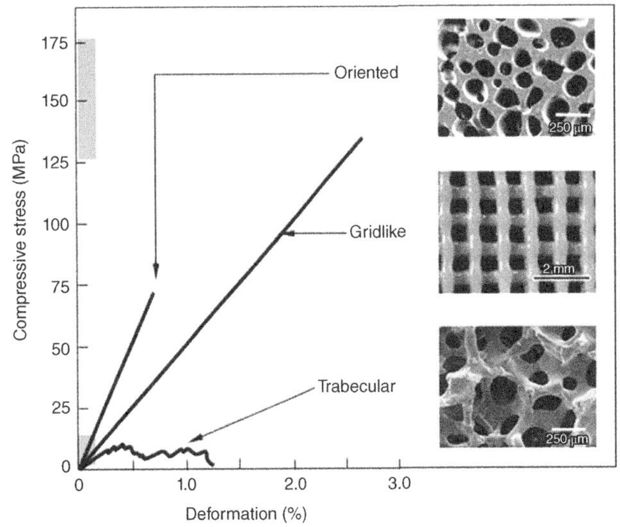

Figure 7.19 Effect of microstructure on the mechanical response of porous bioactive glass (13–93) specimens consisting of a trabecular microstructure mimicking dry human trabecular bone, a grid-like microstructure of orthogonal glass filaments, and a microstructure of oriented pores. The compressive strength range for cortical bone and trabecular bone is indicated by the upper and lower shaded region, respectively.

function of the glass composition. Faster-degrading bioactive borate glasses, for example, show a faster decrease with immersion time when compared to their silicate counterparts.

7.9.5 Release of Ions from Bioactive Glasses

An attractive benefit of bioactive glasses is their capacity to release inorganic ions as they react in an aqueous solution such as the physiological liquid (Figure 7.16). Certain ions such as copper, boron, strontium and zinc have therapeutic effects *in vivo*, such as the capacity to stimulate osteogenesis, angiogenesis and antibacterial activity. The compositional flexibility of bioactive glasses allows them to be doped with a wide variety of inorganic ions in varying concentration. As the glass degrades, these ions are released at a therapeutically desirable rate.

The release of inorganic ions from bioactive glasses has been receiving interest as an alternative approach to the use of growth factors in stimulating cell responses (Hoppe et al. 2011). It forms the basis for the use of bioactive glass microfibers in the treatment of soft tissue wounds (Chapter 24). As an example Figure 7.20 shows images of a felt-like material composed of bioactive borate glass microfibers (diameter = 0.4–1.2 μm) having a composition designated 13–93B3 (Table 7.3). Because of their fine diameter, the microfibers react rapidly in an aqueous phosphate solution. They are almost completely reacted within two weeks,

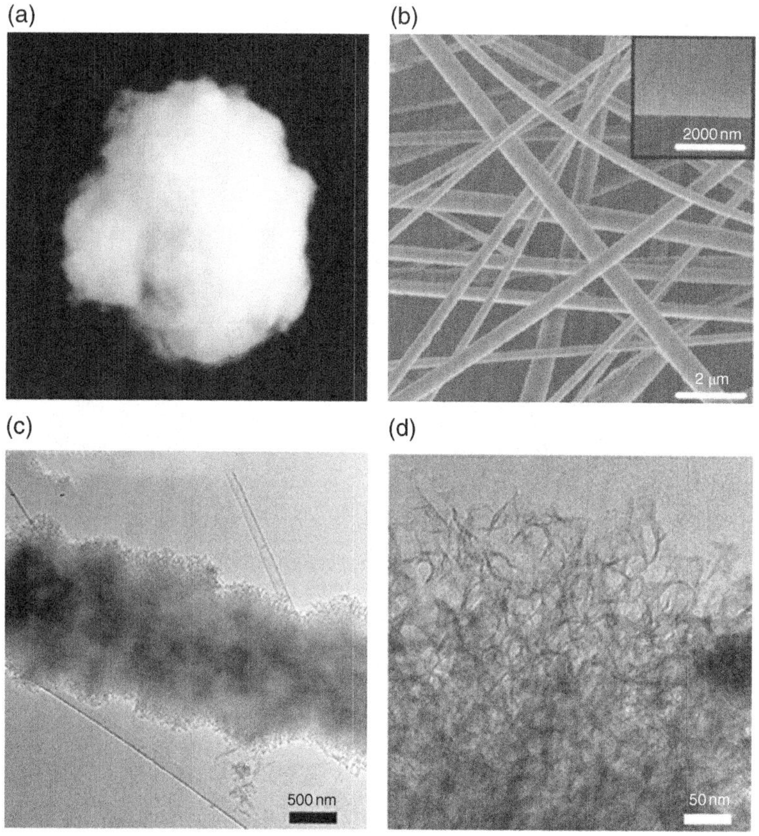

Figure 7.20 Bioactive borate glass (13–93B3) composed of microfibers used in treating soft tissue wounds. (a) Optical image of felt-like material composed of microfibers; (b) Scanning electron microscope (SEM) image showing individual microfibers; (c), (d) Transmission electron microscope (TEM) images of an individual microfiber after immersion in simulated body fluid for two weeks.

forming a nanofibrous hydroxyapatite-like material. When doped with Cu, the microfibers release Cu ions at a rate that is dependent on the concentration of Cu in the fibers (Figure 7.21).

7.9.6 Applications of Bioactive Glasses

Despite their several attractive characteristics, bioactive glasses have seen only limited clinical use in bone healing, and only for healing nonstructural bone. Bioactive glasses are brittle and, until recently, porous implants were created in forms that showed low strength. These low-strength implants compete with calcium phosphate implants and cements that have well-established biocompatibility. Recent efforts using additive manufacturing (3D printing) techniques have shown the capacity to produce bioactive glass implants having compressive strengths comparable to that of cortical bone. These implants have shown promising results in healing structural bone defects in small animal models, such as rodents (Chapter 23) but their low flexural and tensile strength and their brittleness remain areas of concern.

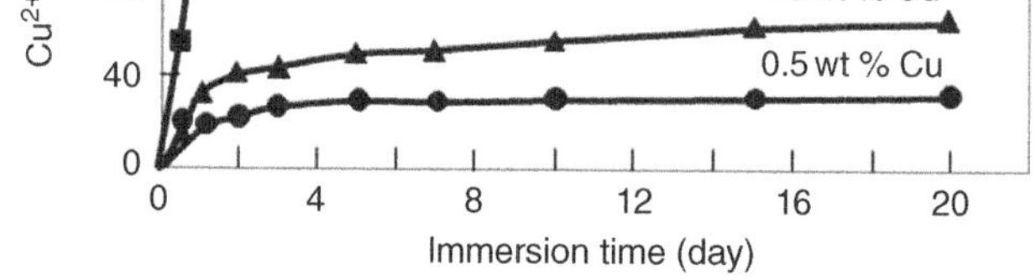

Figure 7.21 Concentration of copper (Cu^{2+}) ions released from copper-doped bioactive borate glass (13–93B3) microfibers into simulated body fluid as a function of immersion time. The microfibers were doped with 0.5, 1.0, and 3.0 wt% Cu.

A significant recent advance has been the use of bioactive glass in healing soft tissue wounds such as skin wounds (Chapter 24). The United States Food and Drug Administration has approved a microfibrous felt-like material composed of special compositions of a rapidly-reacting bioactive borate glass, similar to the 13–93B3 composition in Table 7.3, for use in healing skin wounds in humans. Application of these glasses in wound healing relies primarily on the rapid reactivity of their constituent microfibers and the release of specific ions that stimulate specific cell responses.

7.10 Glass-Ceramics

Glass-ceramics are composed of a glass matrix in which one or more crystalline phases have formed (Figure 3.23d). The crystals typically have a fine size, less than a few micrometers and often less than ~1 μm, and a volume fraction in the range 30–70%, but some glass-ceramics have a volume fraction outside this range. Glass-ceramics begin life as a glass that is subsequently converted to a glass-ceramic by a controlled heat treatment to nucleate and grow crystals within the glass matrix (Figure 7.22). A glass-ceramic is commonly shaped as a glass, which takes advantage of the ease of forming glasses, and thereafter converted to the glass-ceramic, often with little change in volume.

Composition and microstructure are the most important factors in designing glass-ceramics. Compositional design of the parent glass can often be complex because it determines several characteristics, such as the capacity to form a glass with the requisite workability, the nature of the crystallization process (crystallization at the glass surface or throughout the bulk), the crystalline phases that are formed, and the composition of the residual glass phase. The microstructure of a glass-ceramic controls its properties. In glass-ceramics, microstructure describes the quantity and distribution of the crystalline and glass phases within the glass-ceramic, and any porosity that might be present. It depends on the composition of the parent glass and can be modified by varying the thermal treatment used in converting the glass to the glass-ceramic.

A glass-ceramic generally has better physico-chemical properties than its parent glass such as higher strength, Young's modulus, fracture toughness, hardness, and chemical durability. Higher strength and fracture toughness are due to microstructural factors that are generally similar in nature to those for Si_3N_4 (Section 7.3.4). The glass phase bonds the strong crystals together, enhancing strength. Because it is weaker than the crystals, the glass phase provides a tortuous crack propagation path, enhancing fracture toughness as well. Glass-ceramic compositions can be designed to have a range of thermal expansion coefficients, from close to zero to over 10×10^{-6} °C^{-1}. A low thermal expansion coefficient, for example, in combination with high strength and high Young's modulus, makes certain glass-ceramics more resistant to thermal shock (shattering due to rapid temperature changes). Because of their attractive properties, glass-ceramics are widely used in a variety of commercial and technological applications, a widely encountered example being kitchen appliances and cookware. They find use as biomaterials as well, such as spinal fusion implants and dental restorations.

7.10.1 Production of Glass-Ceramics

Figure 7.23 illustrates a typical processing cycle for glass-ceramics. First, an appropriate mixture of raw materials is melted, cast into a warm mold, and cooled to room temperature to produce a glass. Then the glass is heated to its annealing temperature, held for the appropriate time (less than 15–30 minutes) to relieve internal stresses generated by the rapid casting. Alternatively, the melt is cooled directly to the annealing temperature and held for the appropriate time. Thereafter, a controlled heat treatment is used to convert the glass to a glass-ceramic. A common objective in this step is to produce a homogeneous distribution of fine

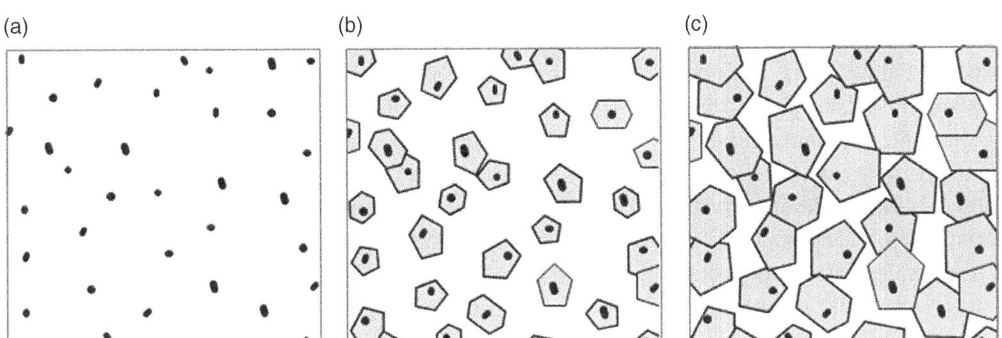

Figure 7.22 Illustration of the production of a glass-ceramic by controlled thermal treatment of a parent glass having an appropriate composition. The major processes in the conversion are nucleation (a) and crystal growth (b), (c).

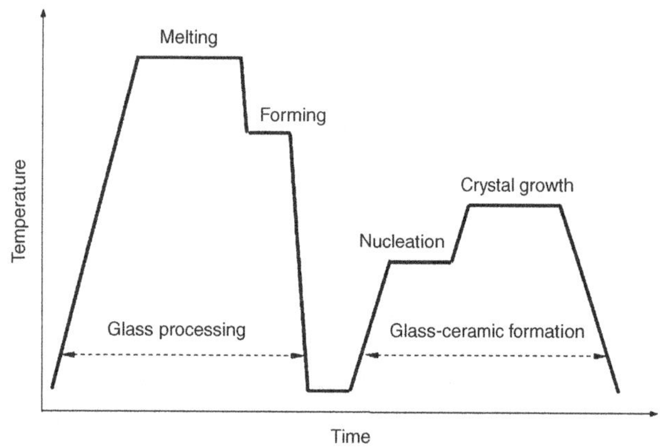

Figure 7.23 Typical processing cycle for the production of glass-ceramics.

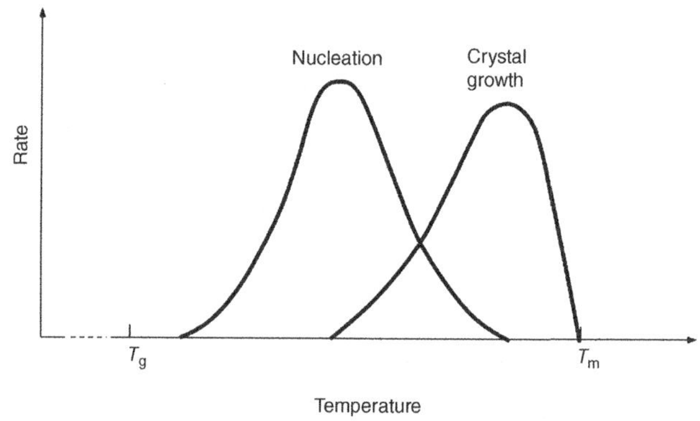

Figure 7.24 Illustration of relative nucleation and crystallization rates as a function of temperature upon heating a parent glass. The glass transition temperature T_g and melting temperature T_m are shown for reference.

crystals throughout the glass matrix (Figure 7.22) because this microstructure generally gives optimal mechanical properties. It is often easier to produce this microstructure by performing nucleation and crystal growth in separate steps.

The principles underlying nucleation and crystal growth in a glass are similar to those for the phase transformation of a metal (Section 6.3). In the nucleation step, the glass is heated for the appropriate time, often less than 15–30 minutes, at an optimal temperature (higher than T_g) determined, for example, from thermal analysis (DTA or DSC) experiments (Figure 7.24). Certain oxides such as P_2O_5, TiO_2, and ZrO_2, either singly or in combination of two or more, are often added during forming of the parent glass to serve as nucleating agents. They lead to the formation of a distribution of fine particles randomly throughout the glass that serve as preferred nucleation sites. Then the material is heated to a higher temperature and held for an appropriate time to grow the nuclei to form crystals and produce the desired glass-ceramic microstructure. Finally, the glass-ceramic is cooled to room temperature.

Alternative routes can be used to form glass-ceramics of the requisite shape but they find use only in special situations. In one method, after nucleation and crystallization of the bulk glass (as illustrated in Figure 7.22), the glass-ceramic is ground into particles and formed into a three-dimensional specimen by the sintering route used for ceramics (Section 7.2.2). As the particles are already composed of the glass-ceramic, there is little additional crystallization during the sintering step. Consequently, the three-dimensional specimen generally has a composition and volume fraction of crystals comparable to those of the particles. Another method is to grind the parent glass into particles and form the specimen by the sintering route. In this method, crystallization of the glass occurs during the sintering step. Both methods generally suffer from an inability to produce a fully dense glass-ceramic unless an external pressure is applied to the material during sintering.

Some glass-ceramics can be machined into the desired external shape using diamond-coated tools. Computer-aided design and manufacturing allows machining to be optimized and automated, which is becoming very popular, particularly for shaping dental restorations. Shaping can also be done by heating the glass-ceramic to an appropriate temperature to soften it and pressing it into a mold of the desired shape.

7.10.2 Bioactive Glass-Ceramics

Several glass-ceramics have been developed and commercialized for use in bone repair. A major objective of these efforts has been to provide implants having better mechanical properties than bioactive glasses, and bioactivity as well. As a crystalline phase has a lower reactivity than a glass of similar composition, bioactive glass-ceramics generally react more slowly than bioactive glasses, such as 45S5 glass. Nevertheless, they undergo similar types of surface reactions when implanted *in vivo*, which lead to a direct bond to bone and even soft tissues. As fabricated, bioactive glass-ceramics generally have a higher strength and fracture toughness than bioactive glasses and when placed in an aqueous phosphate solution, they retain a higher fraction of their strength as a function of time due to their lower reactivity. Consequently, bioactive glass-ceramics generally have better mechanical reliability than bioactive glasses over time in load-bearing applications but they also remain *in vivo* for a considerably longer time.

Created in the early 1980s, the glass-ceramic designated A–W, referred to commercially as Cerabone®, has been among the most widely studied and applied bioactive glass-ceramics in bone repair (Kokubo 1991). It is used mainly in spinal fusion surgery but to a more limited extent than titanium or

PEEK. The parent glass composition is designed to produce, upon thermal treatment, a microstructure composed of a high concentration of two crystalline phases, apatite, $Ca_{10}(PO_4)_6(O,F_2)$, which is essentially a fluorine-substituted hydroxyapatite, and wollastonite, $CaSiO_3$, within a glass matrix. Based on information from phase diagrams, a silicate glass of composition (in wt%) of 4.7 MgO, 44.7 CaO, 34.0 SiO_2, 16.5 P_2O_5, 0.5 CaF_2 is selected. When subjected to a controlled heat treatment, this glass produces a glass-ceramic composed of ~38 wt% apatite, ~34 wt% wollastonite, and a glass matrix (~28 wt%) of composition 16.6 MgO, 24.2 CaO, 59.2 SiO_2.

Bioactivity of glass-ceramic A–W results from dissolution of calcium ions from its constituent phases and deposition of a hydroxyapatite-like layer on the surface of the glass-ceramic by a mechanism similar to that described for 45S5 glass (Section 7.9.2). The glass-ceramic has an average flexural strength (~220 MPa) and a fracture toughness (~2.0 MPa $m^{1/2}$) that are far superior to those of the parent glass (flexural strength ~70 MPa; fracture toughness ~0.8 MPa $m^{1/2}$). Whereas the flexural strength of glass-ceramic A–W is higher than that of human cortical bone, its fracture toughness is still low, comparable to the lowest values for cortical bone (Table 4.1).

7.10.3 Chemically Unreactive Glass-Ceramics

Glass-ceramics have been widely used as dental restorations, such as crowns, bridges, onlays, inlays, veneers, and dentures to restore the function and appearance of the existing tooth structure. Materials used in these applications are required to have not just an optimal combination of several physico-chemical properties but also characteristics suitable for use by dental surgeons and technicians, and acceptability by patients as well. For example, dental restorations should have the following major properties:

- High chemical durability in the acidic and basic oral environment.
- High strength and adequate fracture toughness to avoid deformation and fracture under the forces of mastication.
- Low thermal conductivity to match that of teeth.
- Ability to be easily shaped and bonded to the existing tooth structure.
- The right color to match a patient's teeth.

7.10.4 Lithium Disilicate Glass-Ceramics

While several glass-ceramics have been developed and commercialized for use as dental restorations, glass-ceramics based on lithium disilicate have emerged as the most widely used. Development of these lithium disilicate glass-ceramics has relied on extensive fundamental research over the last few decades. Lithium disilicate crystals have the composition $Li_2Si_2O_5$ ($Li_2O{\cdot}2SiO_2$), equal to an SiO_2:Li_2O molar ratio = 2.0, or ~80 wt% SiO_2 to ~20 wt% Li_2O, whereas lithium disilicate glass-ceramics are composed of lithium disilicate crystals in a glass matrix. The microstructure of these glass-ceramics used in dental restorations such as crowns and bridges consist of a high volume fraction (over ~60%) of fine $Li_2Si_2O_5$ crystals having an elongated (lath-like) geometry that form an interlocking network within the glass matrix (Figure 3.23d). Attainment of this microstructure depends critically on the parent glass composition, use of a nucleating agent, and controlled thermal treatment of the parent glass.

- Parent glass compositions used for producing dental restorations (Table 7.4) are composed of an SiO_2:Li_2O molar ratio of ~2.2, that is, higher than the ratio of 2.0 for lithium disilicate crystals. This higher ratio enhances the capacity to form a glass and convert it to a glass-ceramic having the desired $Li_2Si_2O_5$ crystalline phase. Addition of K_2O is necessary to improve the workability of the glass whereas a major role of Al_2O_3 is to enhance the chemical durability of the glass-ceramic. The parent glass also contains ZrO_2.

Table 7.4 Composition of parent glass and properties of lithium disilicate ($Li_2Si_2O_5$) glass-ceramic formed by controlled crystallization of the parent glass.

Parent glass composition	Weight percent
SiO_2	~70
Li_2O	13–17
Al_2O_3	1–5
K_2O	2–5
P_2O_5	2–5
ZrO_2	<4
Dopants (selected from CeO_2, V_2O_5, Er_2O_3, Tb_4O_7, MnO_2, and Co_3O_4)	<~2
Microstructure	
Crystalline phase ($Li_2Si_2O_5$)	>60 vol%
Crystal morphology	Acicular crystals ~1–2 μm in length
Mechanical properties	
Flexural strength (three-point bending)	>400–450 MPa
Fracture toughness (Vickers indentation)	3–4 MPa $m^{1/2}$
Weibull modulus	10–12
Thermal properties	
Coefficient of thermal expansion	~10×10^{-6} °C^{-1}

- A nucleating agent, preferably P_2O_5, is critical for converting the parent glass to a glass-ceramic having a high volume fraction of the right crystalline phase ($Li_2Si_2O_5$) and the desired microstructure. Prior to crystallization of the glass, nanosize amorphous Li_3PO_4 particles form throughout the glass by a phase separation mechanism, which serve as sites for heterogeneous nucleation and growth of crystals. This type of nucleation and growth process is similar in nature to the way particles precipitating out of a salt solution nucleate and grow first on the inside surface of a dirty beaker.
- Nucleation and crystal growth: Conversion of the parent glass to a glass-ceramic having an optimal microstructure is often performed in two separate heating treatment steps. First, the parent glass is heated for ~15–30 minutes at ~650 °C. Crystals of $Li_2Si_2O_5$ and another phase, lithium metasilicate, $Li_2O.SiO_2$ (Li_2SiO_3) start to nucleate and grow on the $LiPO_4$ particles at ~500 °C, but the amount of Li_2SiO_3 is much higher than that of $Li_2Si_2O_5$ (Figure 7.25). The glass-ceramic formed in this step is often called a lithium metasilicate glass-ceramic because of its high amount of Li_2SiO_3 crystals. It has a lower strength and lower hardness than the lithium disilicate glass-ceramic composed of $Li_2Si_2O_5$ crystals and, thus, can be more easily machined into the desired shape. Thereafter, the lithium metasilicate glass-ceramic is heated at ~850 °C for the appropriate time (~1–2 hours). In this second step, the Li_2SiO_3 crystals dissolve into the glass matrix and form additional $Li_2Si_2O_5$ crystals, as described by the equation

$$Li_2SiO_3(\text{crystal}) + SiO_2(\text{glass}) \rightarrow Li_2Si_2O_5(\text{crystal}) \tag{7.22}$$

The $Li_2Si_2O_5$ crystals grow with time, forming an interlocking network of lath-like crystals. Figure 7.26 illustrates the overall mechanism of nucleation and crystal growth. The difference in microstructure between lithium metasilicate and lithium disilicate, which gives rise to their very different properties, particularly their mechanical properties, is shown in Figure 7.27.

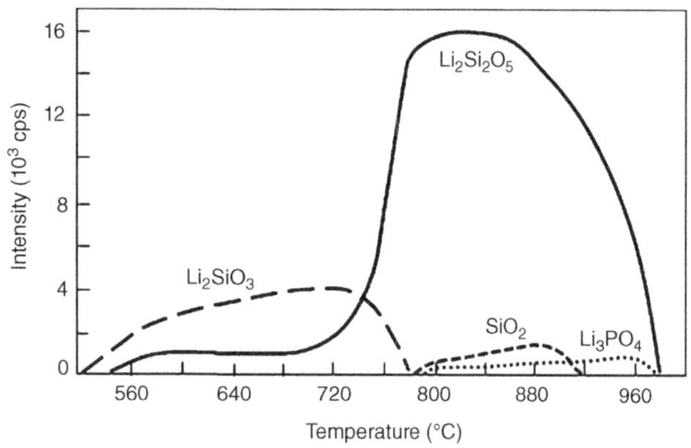

Figure 7.25 Formation of crystalline phases in lithium disilicate glass-ceramics as a function of temperature from a parent glass of composition similar to that given in Table 7.4. *Source:* From Höland and Beall (2012)/ with permission of John Wiley & Sons.

Properties of Lithium Disilicate Glass-Ceramics

Optical properties of $Li_2Si_2O_5$ glass-ceramic products are also critically important because dental restorations should match the translucency, color and fluorescence of a patient's teeth. Translucency is achieved by adjusting the refractive index of the glass matrix to match that of the $Li_2Si_2O_5$ crystals (~1.55), thereby reducing scattering of light at the numerous crystal-glass interfaces (Section 4.8). Translucency, color and fluorescence are varied over a wide range by adding small amounts of the right combination of various metal oxides to the parent glass during forming. These oxides are generally selected from those of certain elements that have unfilled *d* and *f* shells, such as CeO_2, V_2O_5, Er_2O_3, Tb_4O_7, MnO_2, and Co_3O_4.

$Li_2Si_2O_5$ glass-ceramics show a flexural strength (in three-point bending) of over 400–450 MPa, a fracture toughness of 3.0–4.0 MPa m$^{1/2}$, and a Weibull modulus of

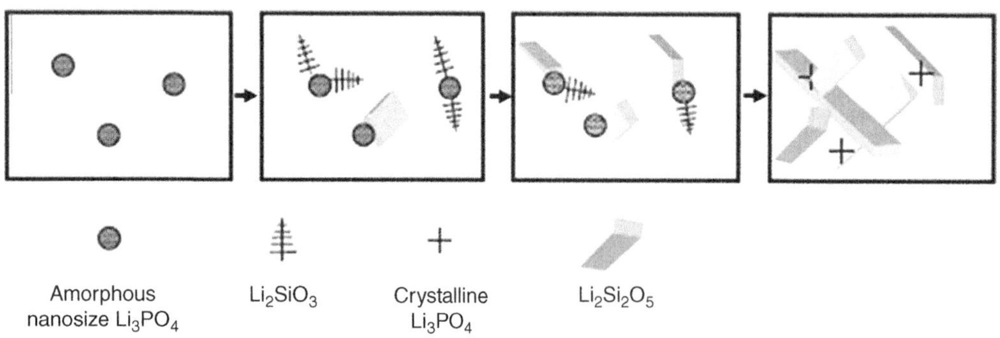

Figure 7.26 Illustration of the mechanism of phase formation in lithium disilicate glass-ceramics, from a parent glass of composition similar to that given in Table 7.4, using P_2O_5 as a nucleating agent. *Source:* From Höland and Beall (2012) / with permission of John Wiley & Sons.

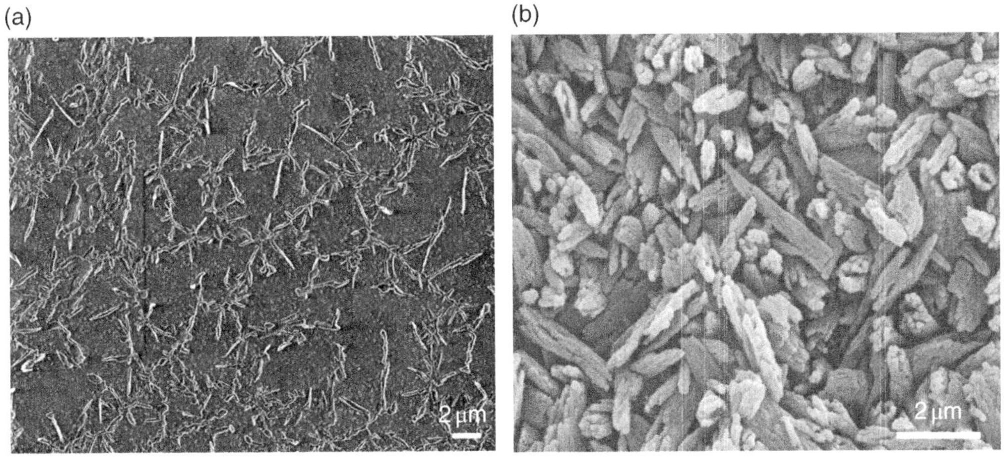

Figure 7.27 SEM images showing the difference in microstructure between lithium metasilicate (Li_2SiO_3) and lithium disilicate ($Li_2Si_2O_5$) glass-ceramics. (a) Formation of Li_2SiO_3 in an SiO_2–Li_2O–Al_2O_3–K_2O–P_2O_5 glass heated for two hours at 560 °C. The nucleating phase (Li_3PO_4) and lithium disilicate crystals were formed at a nanoscale dimension. (b) Precipitation of $Li_2Si_2O_5$ crystals in an SiO_2–Li_2O–Al_2O_3–K_2O–P_2O_5 glass heated for 20 minutes at 850 °C. The surfaces of the samples were etched for 30 s in 40% HF vapor. *Source:* From Höland et al. (2007).

10–12. They can be easily formed into complex shapes for dental restorations using two different methods. In one method, after heating to an appropriate temperature to soften the glass phase, the glass-ceramic is pressed into a mold of the requisite geometry. The other method relies on computer-aided design and manufacturing. Preferably, the material is machined as a Li_2SiO_3 glass-ceramic after a first heat treatment. Then it is converted to a $Li_2Si_2O_5$ dental restoration with little or no change in volume or shape, using a second heat treatment.

7.11 Concluding Remarks

In this chapter, we discussed a variety of ceramics, including glasses and glass-ceramics, which have been studied or developed for use as biomaterials in applications such as structural bone repair, bone substitutes, self-setting cements, dental restorations and even soft tissue healing. Overall, ceramics have an attractive combination of physico-chemical properties, but they are brittle and generally more difficult to manufacture when compared to metals and polymers.

- To reduce the risk of catastrophic (brittle) failure, the creation of reliable ceramics for load-bearing applications, such as structural bone repair, requires the use of proper design principles and proper manufacturing processes to produce an optimal microstructure that is free from strength-limiting flaws.
- Calcium phosphate compounds such as hydroxyapatite and β-TCP, composed of the same ions as the mineral constituent of bone, are highly biocompatible and bioactive as well, which make them attractive biomaterials for use as bone substitutes. We discussed factors that influence the solubility of calcium phosphate compounds, which is a critical factor in determining the direction of several chemical reactions involving these compounds. The solubility of calcium phosphate compounds also plays a major role in biological processes that involve the formation and resorption of bone and other calcified tissues.
- Because they set (harden) *in situ*, CPCs offer the capacity to be molded into shape during implantation and the potential for injection by minimally invasive surgery, in addition to the major benefits of other biocompatible calcium phosphate compounds.
- Bioactive glasses react *in vivo*, release ions into the medium, convert to a hydroxyapatite-like product and form a direct bond with bone and even soft tissues but, despite these attractive characteristics, they have seen limited use as biomaterials in bone repair.
- Some ions released from bioactive glasses during their reaction and conversion have been shown to have therapeutic effects, making bioactive glasses suitable for use as devices for controlled local delivery of such ions. This capacity to release therapeutic ions is central to the clinical use of microfibrous bioactive glass to heal soft tissue wounds.
- A glass-ceramic shows physico-chemical properties that are often significantly better than those of its parent glass. The ability to achieve these improved properties is largely dependent on the composition of the parent glass and on the microstructure of the glass-ceramic produced by controlled nucleation and crystallization of the glass during a heat treatment step.

Problems

7.1 From the following list of materials, select the most appropriate one for the questions below:

Al_2O_3;
$Ca_{10}(PO_4)_6(OH)_2$;
ZrO_2;
45S5 glass;
Si_3N_4;
$Ca_3(PO_4)_2$;
Y-TZP;
$Li_2Si_2O_5$ glass-ceramic;
AMC;
Calcium phosphate cement

A Femoral head in hip implants
B Dental restorations such as crowns and bridges
C Highest combination of strength and fracture toughness
D Capacity to bond directly to bone *in vivo*
E Readily resorbable *in vivo*
F Injectable by minimally invasive surgery
G Similar in composition to the mineral constituent of bone
H Three-dimensional implants not manufactured by sintering

7.2 Explain why the vast majority of ceramics are produced by sintering and not by the common methods used for metals and polymers.

7.3 Compare the surface area of 1 g of Al_2O_3 particles, assumed spherical in shape, when the size (diameter) of the particles is (i) 10 μm and (ii) 0.1 μm. If the surface free energy of Al_2O_3 is 1.0 J/m^2, determine the change in energy of the system when 1.0 g of Al_2O_3 particles of size 0.1 μm is sintered to form a fully dense specimen. (The density of Al_2O_3 is 3.986 g/cm^3).

7.4 The most common method for producing ceramics involves consolidating particles at room temperature to form a porous three-dimension specimen followed by sintering.
A Why are fine particles beneficial for sintering?
B What characteristics of the as-formed porous specimen are desirable for sintering? Explain.

7.5 What are the major compositional and microstructural differences between Al_2O_3 and AMC for use as femoral heads in hip implants? Why has AMC largely replaced Al_2O_3 in these applications?

7.6 Si_3N_4 shows high compressive and flexural strength, high hardness, and low friction and wear properties.
A This combination of properties makes Si_3N_4 a candidate for what type of biomedical applications?
B What is the ideal microstructure of Si_3N_4 for this application? Explain why?
C What are the possible drawbacks (if any) in using Si_3N_4 for this application?

7.7 Y-TZP and Si_3N_4 both show flexural strength and fracture toughness values that are among the highest for ceramics. What factors are responsible for this? Explain.

7.8 Explain the difference between the glass transition temperature of a glass and the melting temperature of a crystalline material.

7.9 The bioactive glass 45S5 (Table 7.3) is produced from a mixture of SiO_2, Na_2CO_3, $CaCO_3$ and $NaH_2PO_4.2H_2O$. Determine the amount of each material needed to produce 100 g of glass.

7.10 When immersed in an aqueous solution containing phosphate ions, nonporous particles of a bioactive glass of composition (in wt%) 20 Na_2O, 20 CaO, 60 B_2O_3 are completely converted to hydroxyapatite, $Ca_{10}(PO_4)_6(OH)_2$.
D Determine the fractional weight loss resulting from this conversion process.
E If the densities of the glass and hydroxyapatite in their fully dense state are 2.50 and 3.16 g/cm^3, respectively, and there is no change in the external volume of the particles upon conversion, determine the porosity of the hydroxyapatite product.

7.11 Briefly explain how a bioactive glass (such as 45S5 glass) works when it is used to heal a bone defect *in vivo*.

7.12 What is the difference between a glass and a glass-ceramic? Converting a glass to a glass-ceramic leads to a difference in several properties between the two. Name four of these properties and explain how you expect them to differ.

References

Bohner, M. (2010a). Design of ceramic-based cements and putties for bone graft substitution. *European Cells & Materials* 20: 1–12.

Bohner, M. (2010b). Resorbable biomaterials as bone graft substitutes. *Materials Today* 13: 24–30.

Chevalier, J., Gremillard, L., Virkar, A.V., and Clarke, D.R. (2009). The tetragonal-monoclinic transformation in zirconia: lessons learned and future trends. *Journal of the American Ceramic Society* 92: 1901–1920.

Day, D.E. (2012). Glasses for radiotherapy. In: *Bio-Glasses: An Introduction*, 1e (eds. J.R. Jones and A.G. Clare), 203–228. New York: Wiley, Chapter 13.

Dorozhkin, S.V. (2011). Self-setting calcium orthophosphate formulations: cements, concretes, pastes and putties. *International Journal of Materials and Chemistry* 1: 1–48.

Hench, L.L. (1998). Bioceramics. *Journal of the American Ceramic Society* 81: 1705–1728.

Höland, W. and Beall, G.H. (2012). *Glass-Ceramic Technology*. New York: Wiley.

Höland, W., Rheinberger, V., Apel, E., and Ritzberger, C. (2007). Principles and phenomena of bioengineering with glass-ceramics for dental restoration. *Journal of the European Ceramic Society* 27: 1521–1526.

Hoppe, A., Güldal, N.S., and Boccaccini, A.R. (2011). A review of the biological response to ionic dissolution products from bioactive glasses and glass-ceramics. *Biomaterials* 32: 2757–2574.

Ishikawa, K. (2010). Bone substitute fabrication based on dissolution–precipitation reactions. *Materials* 3: 1138–1155.

Kokubo, T. (1991). Bioactive glass ceramics: properties and applications. *Biomaterials* 12: 155–163.

LeGeros, R.Z. (1991). Monographs in oral sciences. In: *Calcium Phosphates in Oral Biology and Medicine*, vol. 15 (ed. H. Myers). Basel: Karger.

LeGeros, R.Z. (2008). Calcium phosphate-based osteoinductive materials. *Chemical Reviews* 108: 4742–4753.

Liu, X., Rahaman, M.N., Hilmas, G.E., and Bal, B.S. (2013). Mechanical properties of bioactive glass (13-93) scaffolds fabricated by robotic deposition for structural bone repair. *Acta Biomaterialia* 9: 7025–7034.

Lynn, A.K. and Bonfield, W. (2005). A novel method for the simultaneous, titrant-free control of pH and calcium phosphate mass yield. *Accounts of Chemical Research* 38: 202–207.

McEntire, B.J., Bal, B.S., Rahaman, M.N. et al. (2015). Ceramics and ceramic coatings in orthopaedics. *Journal of the European Ceramic Society* 35: 4327–4369.

Scott, H.G. (1975). Phase relationships in the zirconia–yttria system. *Journal of Materials Science* 10: 1527–1535.

Waggoner Johnson, A.J. and Herschler, B.A. (2011). A review of the mechanical behavior of CaP and CaP/polymer composites for applications in bone replacement and repair. *Acta Biomaterialia* 7: 16–30.

Wang, L. and Nancollas, G.H. (2008). Calcium orthophosphates: crystallization and dissolution. *Chemical Reviews* 108: 4628–4629.

Further Reading

Elliott, J.C. (1994). *Structure and Chemistry of the Apatites and Other Calcium Orthophosphates*. Amsterdam: Elsevier Science & Technology.

Liu, X., Rahaman, M.N., and Day, D.E. (2014). in vitro degradation and conversion of melt-derived microfibrous borate (13-93B3) bioactive glass doped with metal ions. *Journal of the American Ceramic Society* 97: 3501–3509.

Montazerian, M. and Zanotto, E.D. (2016). History and trends of bioactive glass-ceramics. *Journal of Biomedical Materials Research. Part A* 104: 1231–1249.

Rahaman, M.N., Day, D.E., Bal, B.S. et al. (2011). Bioactive glass in tissue engineering. *Acta Biomaterialia* 7: 2355–2373.

8

Synthetic Polymers I: Nondegradable Polymers

8.1 Introduction

Polymers are materials composed of long molecules (macromolecules) in which many small units (monomers) are joined end-to-end (Section 3.3.5). The chain backbone in the vast majority of polymers consists of carbon (C) atoms joined together by strong covalent bonds in a tetrahedral arrangement at each carbon atom, as in polyethylene (PE) for example (Figure 2.10). Consequently, most polymers are referred to as organic polymers. However, many polymers have a chain backbone that contains atoms other than C, such as silicon (Si), or in addition to C, such as Si, nitrogen (N), or oxygen (O). Polymers relevant to biomaterials are often broadly classified as synthetic or natural, depending on their origin, but each class can be further classified into subdivisions based on characteristics such as structure, composition, and properties.

This chapter and the following chapter are devoted to synthetic polymers, whereas natural polymers are the topic of Chapter 10. Synthetic polymers are vast in number but they can be classified into a few generic types. While some polymers may not fall within these generic types, they can be understood in terms of them. The types of synthetic polymers relevant to biomaterials are (Figure 8.1):

- *Thermoplastic polymers:* These polymers are generally composed of macromolecules consisting of linear chains that are not chemically crosslinked, although they may branch occasionally. Because the intermolecular forces between the chains are weak, these polymers soften on heating and can do so repeatedly upon heating and cooling. Consequently, thermoplastic polymers can be easily shaped by a variety of methods. Thermoplastic polymers can be amorphous, such as polymethyl methacrylate (PMMA) used in intraocular lenses, or semicrystalline, composed of a mixture of amorphous and crystalline regions, such as PE used as bearings in hip and knee implants.
- *Thermosetting polymers:* These polymers are composed of macromolecules that have been highly crosslinked during their production, often by mixing a low molecular weight resin with a hardener that react together and harden at room temperature or on heating. After hardening, the macromolecular network and, thus, the shape of the specimen, are set, giving the origin of the term thermoset (or thermosetting). Thermosetting polymers are generally amorphous, due to their high degree of chemical crosslinking. They do not soften on heating but, instead, generally maintain a high degree of their rigidity until they start to decompose. Consequently, thermosetting polymers, often referred to as resins, are difficult to reshape. Common examples of thermosetting polymers are epoxy resins used as adhesives or as matrices in glass fiber reinforced polymers.
- *Elastomers:* Often referred to as rubbers, elastomers are composed of nearly linear chains that have occasional chemical crosslinks. Secondary (intermolecular) bonds are negligible in these polymers. The crosslinks allow elastomers to be deformed to large strains and they provide the "memory" for them to return to their original shape upon unloading. Common examples are synthetic and natural rubbers used in car tires, and polydimethylsiloxane (PDMS), commonly referred to as silicone, used in a variety of industrial, cosmetic, and biomedical applications.
- *Hydrogels:* Hydrogels consist of water-filled hydrophilic polymers. These polymers are composed of highly crosslinked chains that, in the dehydrated state, contain large spaces between them. Upon immersion in water, they absorb varying amounts of water and expand, forming hydrogels that vary in their degree of hardness, from soft hydrogels to rigid hydrogels, such as the hydrogels used in contact lenses.

The majority of polymers used as biomaterials are composed of thermoplastic polymers and hydrogels, but a few elastomers, particularly PDMS, also find some use. In comparison, thermosetting polymers find little use.

Materials for Biomedical Engineering: Fundamentals and Applications, First Edition. Mohamed N. Rahaman and Roger F. Brown.

Companion website: www.wiley.com/go/rahaman/biomedicalengineering

Thermoplastic polymers and hydrogels can be nondegradable or degradable in an aqueous medium such as the physiological fluid.

When compared to metals and ceramics, synthetic polymers have a significant advantage in simplicity and ease of production, particularly when complex shapes are required. The mechanical properties of synthetic polymers can also be easily modified to make them more comparable to those of noncalcified tissues and nonstructural bone (Figure 1.5). This can reduce the potential for adverse biocompatibility effects due to mechanical forces (Chapter 17). Because of these advantages, synthetic polymers find considerable use as biomaterials in a variety of applications, such as medical devices, tissue repair, tissue engineering (or regenerative medicine), and drug delivery devices.

This chapter will cover nondegradable thermoplastic polymers, elastomers, and water-soluble thermoplastic polymers that are relevant to biomaterials. Table 8.1 presents a list of common polymers in these groups and their biomedical applications. Degradable thermoplastic polymers are the subject of the next chapter while hydrogels, nondegradable or degradable, are covered in Chapter 11. Overall, the following topics will be discussed in this chapter:

- Polymer science fundamentals that are applicable to polymers in general
- Common nondegradable thermoplastic, elastomeric, and water-soluble polymers used as biomaterials
- Relationships between composition, structure, and properties of these polymers.

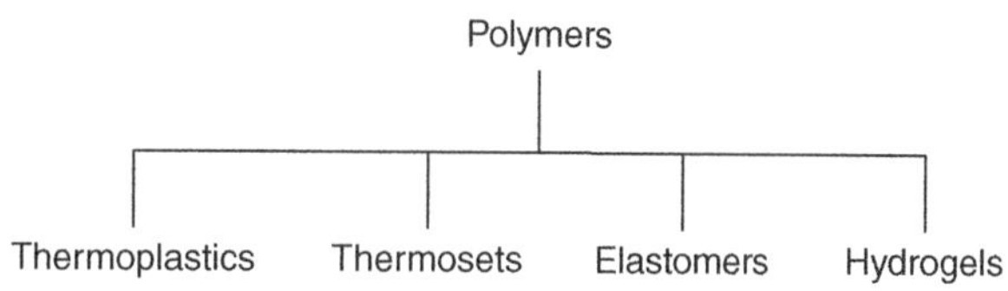

Figure 8.1 Types of synthetic polymers relevant to biomaterials.

8.2 Polymer Science Fundamentals

The physicochemical properties of polymers depend primarily on the chemical composition, structure, and shape (conformation) of their constituent molecules. Consequently, an understanding of these factors and their effect on the properties of polymers is important.

8.2.1 Copolymers

Some polymers, such as PE, for example, are composed of macromolecules consisting of a single type of monomer units (Figure 2.10). Because of this, they are sometimes referred to as homopolymers. However, many polymers, called copolymers, are composed of two or more different types of monomer units. For example, if ethylene and propylene are polymerized together, the resulting macromolecules will consist of both ethylene and propylene units in a ratio that mimics the concentration of these two monomers

Table 8.1 Common nondegradable synthetic polymers used as biomaterials and medical products.

Polymer	Applications
High-density polyethylene (HDPE)	Bearings in prosthetic joints
Low-density polyethylene (LDPE)	Syringes; bottles
Polypropylene (PP)	Nondegradable sutures; hernia repair; membrane oxygenators
Polytetrafluoroethylene (PTFE)	Vascular grafts; catheters; nondegradable sutures; hernia repair
Polyvinyl chloride (PVC)	Tubing; blood storage bags
Polymethyl methacrylate (PMMA)	Bone cement; intraocular lenses; hard contact lenses
Polyether ether ketone (PEEK)	Spinal fusion implants; craniofacial implants
Polyethylene terephthalate (PET)	Vascular grafts; nondegradable sutures; soft tissue (ligament) reconstruction
Nylon 6.6	Nondegradable sutures
Polydimethylsiloxane (PDMS)	Breast reconstruction; ear, chin, and nose reconstruction; finger joints
Polyurethane (PU)	Electrical insulation for pacemaker leads; outer layer of silicone breast implants; vascular grafts; catheters; wound dressings
Polyvinyl alcohol (PVA)	Contact lens; cell encapsulation
Polyethylene glycol (PEG); polyethylene oxide (PEO)	Surface modification of proteins and nanoparticles in drug delivery
Polyethylenimine (PEI)	Gene delivery

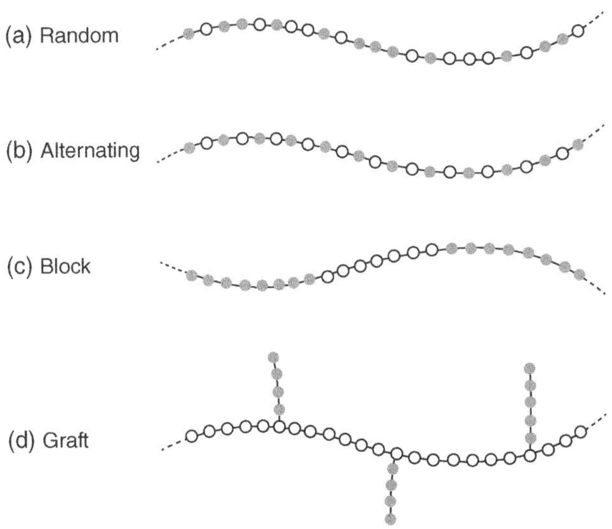

Figure 8.2 Arrangement of monomer units in (a) random copolymer, (b) alternating copolymer, (c) block copolymer, and (d) graft copolymer.

in the starting mixture. Copolymerization is a useful technique for modifying the physicochemical properties of a polymer, such as its strength, elastic modulus, toughness, and solubility. If the mixture of monomers is composed mainly of ethylene, the macromolecules will consist of ethylene sequences with a few propylene units at random points in the chain. The incorporation of the propylene units in the macromolecule will have the effect of modifying the properties of PE. Similarly, the incorporation of ethylene units into propylene sequences will modify the properties of polypropylene (PP).

For a copolymer composed of two different monomer units, A and B, the arrangement of the monomers can be random, or it can alternate in a regular A–B–A–B pattern. However, block or graft copolymers can be synthesized for specific purposes (Figure 8.2). Block copolymers contain segments of each monomer (A and B) in the chain backbone, whereas graft copolymers consist of segments of one monomer attached to the chain backbone composed of the other monomer.

8.2.2 Linear and Crosslinked Molecules

The macromolecules in some polymers are composed of linear chains because the monomers join end-to-end during synthesis of the polymer, although varying amounts of side branching are often present (Figure 8.3). In comparison, other polymers are composed of a network of crosslinked chains that are joined chemically at certain points. The chains in elastomers have few crosslinks whereas thermosetting polymers and hydrogels have a large number of crosslinks. Crosslinking provides another technique for modifying the physicochemical properties of a polymer. For example, when compared to the same polymer composed of linear chains, a high degree of crosslinking generally increases the stiffness of the polymer, prevents it from crystallizing and lowers its solubility.

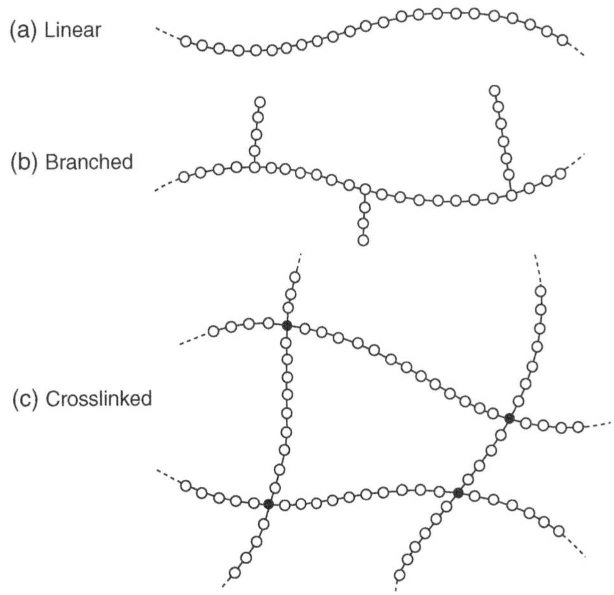

Figure 8.3 Arrangement of monomer units in polymers composed of (a) linear chains, (b) branched chains, and (c) crosslinked chains.

Crosslinked networks can be produced during the synthesis of a polymer, by mixing monomers of one polymer with a small fraction of a second monomer. Under the appropriate conditions, reaction between the two monomers will lead to a network that is bonded chemically at certain tie points. Crosslinked networks can also be formed by reactions involving an already synthesized polymer composed of linear chains as, for example, in PE and in vulcanization of rubber. One way to introduce crosslinks into PE, for example, is to irradiate it with an appropriate dose of gamma (γ) rays. Crosslinked PE, formed by irradiating ultrahigh molecular weight polyethylene (UHMWPE), is now commonly used as bearings in hip and knee implants because it provides significantly better wear resistance than conventional UHMWPE (Section 8.5.1.1).

8.2.3 Molecular Symmetry and Stereoregularity

Some polymers, as we have seen, can have an amorphous structure composed of a random arrangement of their constituent macromolecules or a semicrystalline structure composed of a mixture of amorphous regions and crystalline regions in which the macromolecules have an ordered arrangement (Section 3.3.5). Semicrystalline polymers generally have certain advantages over amorphous polymers in that they show an improvement in some mechanical properties such as strength, Young's modulus, toughness and hardness, as exemplified by PE used as bearings in hip and knee implants. As a crystalline phase generally has better chemical durability than an amorphous phase of similar composition, varying the ratio of crystalline to amorphous phase can be used to provide polymers having an optimal combination of strength and degradation rate, or just an optimal degradation rate alone. Copolymers of polylactic acid (PLA) and polyglycolic acid (PGA) used in degradable sutures and as drug delivery devices are well known examples of this approach (Chapter 9).

As the ability of a polymer to crystallize can be used in controlling many of its properties, a key question relates to what molecular characteristics distinguish crystallizable polymers from amorphous polymers. As we might expect from the regular repeating pattern of atoms in metals and ceramics, a critical characteristic is the regularity of the molecular structure. Irregularity in the macromolecular structure of a polymer prevents or severely limits its capacity to crystallize.

One form of regularity is the repeating sequence of the structural units, as in nylon that has an ABAB sequence of monomer units (Figure 2.16b). Random side branching provides a source of irregularity, as in PE for example. Irregularity in the size of the side branches or in their placement in the chain backbone inhibits the ability of the molecule to pack efficiently into a crystal.

Another type of irregularity, common in a type of polymers called vinyl polymers, arises from the way in which the side group is placed along the chain backbone. The molecular structure of vinyl polymers has the repeating unit CH_2–CHX, where X is CH_3 for PP, Cl for polyvinyl chloride (PVC), and C_6H_5 for polystyrene (PS), for example. Figure 8.4 illustrates three principal ways of placing the X group in a vinyl polymer, using PP as an example. There is no change in chemical composition but the placement of the X group gives three different patterns called stereoisomers. In one pattern, the X group falls on one side of the plane in which the carbon atoms lie, an arrangement described as isotactic, whereas in the second pattern, the X group alternates regularly from one side to the other, described as syndiotactic. In the third pattern, the X group is arranged randomly, described as atactic. In general, under the appropriate conditions, crystals are always formed from isotactic polymers, sometimes from syndiotactic polymers, and never from atactic polymers.

8.2.4 Molecular Weight

Molecular weight or atomic weight is actually a mass, but we shall continue to use the term weight because it is more common in the literature. For most inorganic materials,

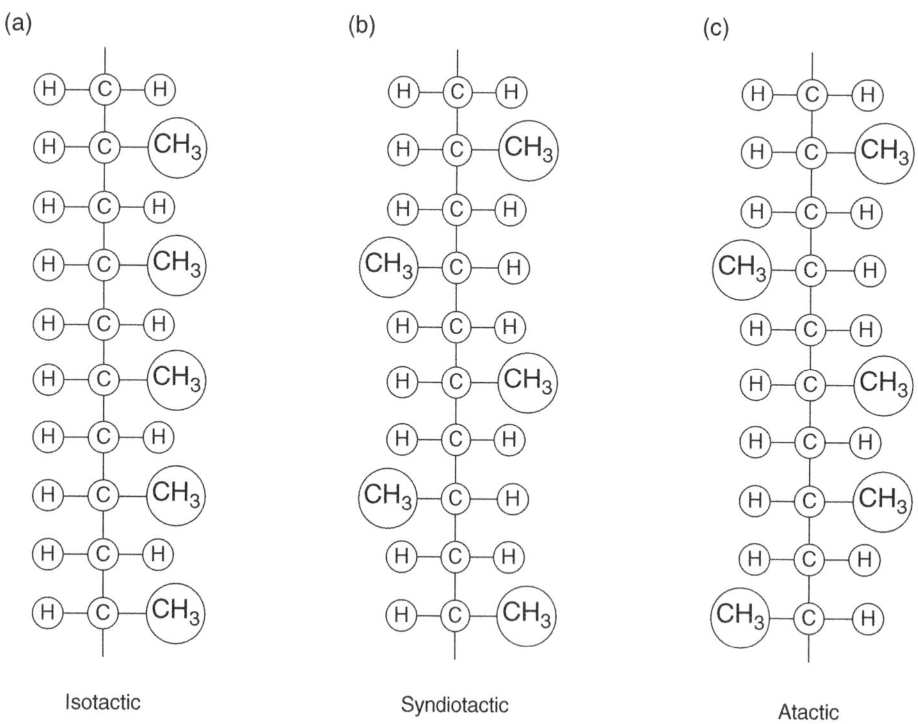

Figure 8.4 Stereoregularity of molecules in a vinyl polymer such as polypropylene.

such as sodium chloride (NaCl), for example, knowledge of their chemical composition or atomic structure yields their molecular weight. However, this is not so for polymers because they are composed of molecules of varying size or molecular weight. The size of a polymer molecule is described most directly by the number n of monomer units in the chain, called the degree of polymerization (DP). Of more practical value is the molecular weight, which is defined as

$$\text{Molecular weight} = \frac{\text{Weight of the molecule}}{\left(\frac{1}{12}\right)\text{the weight of a carbon 12 atom}} \tag{8.1}$$

As the molecular weight defined by Eq. (8.1) is the ratio of two weights, it is unitless. This ratio is often termed the relative molecular mass (RMM). In some disciplines, a unit is often assigned to the molecular weight defined by Eq. (8.1), called the dalton (abbreviated Da). As an example, taking a PE molecule having a degree of polymerization $n = 10^4$, the molecular weight M of the molecule is

$$M = n \times \text{molecular weight of ethylene} = 10^4 \times (2 \times 12.01 + 4 \times 1.01) = 280600\,\text{Da}$$

This molecular weight neglects the insignificant effect of the end groups, which is often typical for macromolecules. The RMM is simply 280 600.

In some disciplines, the RMM or the molecular weight expressed in daltons is often used. In this book, we will express the molecular weight in daltons. However, in the literature, molecular weight has been expressed in other units such as atomic mass unit (amu) or grams per mole (g/mol). Whereas the units are different, the magnitude is the same, that is, the molecular weight of the PE molecule in these units is 280 600 amu or 280 600 g/mol.

Polymerization reactions invariably lead to the synthesis of polymers having a distribution of macromolecular sizes and not just one size (Figure 8.5). As the molecular weight of a polymer influences several of its properties, such as viscosity and strength, it is common to specify it as an average of the various values in the distribution. Generally, the average molecular weight is based on

- The number of molecules in the distribution, giving a number average molecular weight ($\bar{M}_n$).
- The weight of the molecules, giving a weight average molecular weight ($\bar{M}_w$).

$\bar{M}_n$ is defined as

$$\bar{M}_n = \frac{\sum n_i M_i}{\sum n_i} \tag{8.2}$$

where n_i is the number of molecules in the ith fraction of the polymer of molecular weight M_i. As the weight (mass) w_i of the ith fraction is given by

$$w_i = n_i M_i \tag{8.3}$$

Equation (8.2) can also be expressed as

$$\bar{M}_n = \frac{\sum w_i}{\sum \frac{w_i}{M_i}} = \frac{W}{\sum \frac{w_i}{M_i}} \tag{8.4}$$

where W is the total weight of the various fractions.

$\bar{M}_w$ is defined as

$$\bar{M}_w = \frac{\sum w_i M_i}{\sum w_i} = \frac{\sum w_i M_i}{W} \tag{8.5}$$

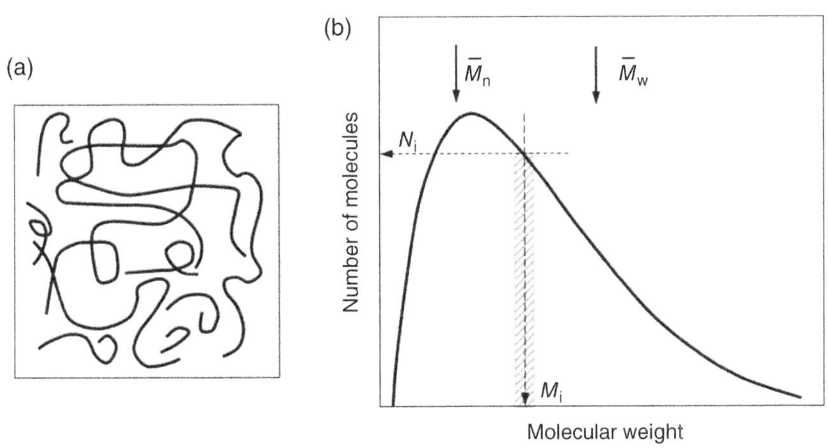

Figure 8.5 (a) Illustration of polymer composed of molecules having many different sizes. (b) Distribution of molecular weight in a polymer specimen. The distribution is thought of as a series of closely spaced fractions. The ith fraction is composed of a number N_i (or weight w_i) of molecules of molecular weight M_i.

$\bar{M}_w$ is sensitive to the presence of the high molecular weight molecules. Consequently, except for polymers in which all the molecules have the same molecular weight (that is, monodisperse systems), $\bar{M}_w$ is always larger than $\bar{M}_n$. The polydispersity, taken as a measure of the spread or broadness of the molecular weight distribution, is defined as

$$\text{Polydispersity} = \frac{\bar{M}_w}{\bar{M}_n} \tag{8.6}$$

A common method used to measure the molecular weight distribution of a polymer is gel permeation chromatography. A given mass of polymer is dissolved in a liquid to form a dilute solution, which has the effect of separating the molecules from one another. Then the solution is passed down a column of porous silica or polymer beads in which pores have a size comparable to that of the macromolecules. The molecules pass down the column at a rate determined by their size, such that the largest molecules pass down most rapidly. Measurement of the amount of polymer appearing at the bottom of the column as a function of time using optical or spectroscopic methods yields the distribution of molecular weights.

Example 8.1 Data for the molecular weight of a polystyrene specimen are as follows:

Molecular weight (kDa)	10	20	30	40	50
Weight (g)	0.10	0.20	0.30	0.30	0.10

Determine (a) the number average molecular weight, (b) the weight-average molecular weight, and (c) the polydispersity of the specimen.

Solution:

As the data give the weight w_i of the various fractions and not the number of molecules in each fraction, we can use Eq. (8.4) to find $\bar{M}_n$. A useful way to solve this problem is to tabulate the appropriate values w_i/M_i and w_iM_i for the various fractions:

M_i (10^4 Da)	w_i (g)	w_i/M_i (10^{-4}g/Da)	w_iM_i (10^4gDa)
1	0.1	0.1	0.1
2	0.2	0.1	0.4
3	0.3	0.1	0.9
4	0.3	0.075	1.2
5	0.1	0.02	0.5
	$\Sigma w_i =$ 1.0 g	$\Sigma w_i/M_i =$ 0.395×10^{-4} g/Da	$w_iM_i =$ 3.1×10^4 gDa

Using Eq. (8.4), Eq. (8.5) and the tabulated data, we get

a) $\bar{M}_n = 1/0.395 \times 10^4\,\text{Da} = 25.3\,\text{kDa}$
b) $\bar{M}_w = 31.0\,\text{kDa}$
c) Polydispersity $= \bar{M}_w/\bar{M}_n = 1.2$

8.2.5 Molecular Conformation

The conformation of a molecule refers to the overall geometric shape that the molecule takes up, whether in the molten state or solidified state of the polymer, or when dissolved in a liquid. Conformation should not be confused with the term configuration, which refers to the geometry in which the atoms are arranged in the molecule itself, based on their interatomic bonding. Whereas the conformation can be changed by mechanical forces or by temperature, for example, the configuration cannot be altered unless atomic bonds are broken and reformed, that is, by changing the composition of the molecule itself.

Molecular Conformation in Amorphous Polymers

To visualize the molecular conformation in an amorphous polymer (that is, a polymer that has not crystallized), it is useful to start by examining the molten polymer. The macromolecules in a molten polymer are entangled together, like a loose ball of string. They have a randomly coiled-up or crumpled conformation (Figure 8.6) due to rotation of the atomic bonds, such as the carbon–carbon (C–C) bonds in a PE molecule. Structural relaxation is rapid because segments of the chain backbone have sufficient thermal energy to allow them to change places rapidly. Consequently, at any instant of time, a macromolecule does not have a fixed conformation or size over time. As a macromolecule is composed of a large number of C–C bonds, the number of possible conformations is enormous.

While it is impossible to give precise information about the conformation of a macromolecule at any given instant of time, statistical models have been employed to provide information about the relative probability of a particular conformation. This information is generally adequate because the properties of the polymer, such as mechanical properties, depend on averaging over all its constituent molecules. One model assumes a freely jointed chain, that is, the real molecular chain is modeled by a hypothetical chain composed of rigid C–C bonds that can rotate freely without any restriction on the angle between successive bonds. Instead of a tetrahedral arrangement of the bonds at each C atom, the angle between successive bonds is selected randomly. Then, a description of the molecular conformation is given in terms of the distance r between the ends of the molecule. The model determines $\langle r^2 \rangle$, that

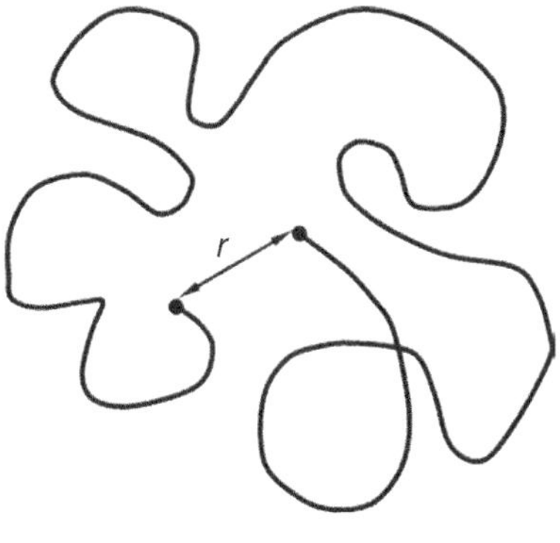

Figure 8.6 Illustration of the conformation of a molecular chain in a molten polymer at some instant of time.

is, the average of r^2, and $\langle r^2\rangle^{1/2}$, called the root mean square end-to-end distance, is taken as a description of the molecular conformation.

For a molecule composed on N segments, such as C–C bonds of length l, the root mean square end-to-end distance in the freely jointed chain model given as

$$\langle r^2\rangle^{1/2} = N^{1/2}l \tag{8.7}$$

As the fully extended length of the chain L is equal to Nl, the ratio $\langle r^2\rangle^{1/2}/L$ is

$$\frac{\langle r^2\rangle^{1/2}}{L} = \frac{1}{N^{1/2}} \tag{8.8}$$

Thus for $N = 10^6$, $\langle r^2\rangle^{1/2}$ is only 0.1% of L, showing that the molecule is highly coiled up or crumpled.

Cooling a molten polymer leads to a reduction in the ease of segmental motion and the polymer becomes stiffer, attaining a state comparable to a rubbery solid. Upon further cooling, segmental motion essentially stops. The conformation of the macromolecules is similar to that in the molten state but the structure is essentially frozen in, making the polymer hard and brittle, comparable to a glassy solid. The transition from rubbery solid to glassy solid is called the glass transition temperature T_g. The polymer below T_g is not a glass because this term is used to describe an inorganic glass, but it is said to be glassy or in the glassy state.

8.2.6 Glass Transition Temperature (T_g)

T_g of an amorphous polymer is just as important as T_g of an inorganic glass (Chapter 7) because it marks a significant change in physical properties. This change is particularly noticeable in the mechanical properties of the polymer, such as its strength and modulus (Section 8.4.1).

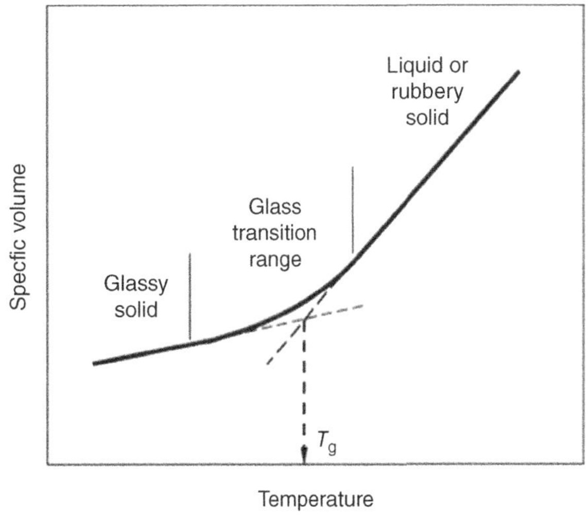

Figure 8.7 Schematic plot of the specific volume of a polymer as a function of temperature, showing the glass transition temperature T_g.

Methods used to measure T_g of an inorganic glass, such as differential scanning calorimetry, differential thermal analysis, and thermal expansion are also applicable to polymers. One way of obtaining T_g is from observations of the specific volume as a function of temperature on cooling (Figure 8.7). T_g of a polymer depends slightly on the cooling rate, being lower at slower cooling rate, but it is often determined at a cooling rate of 1 °C per minute. As illustrated in Figure 8.7, the coefficient of thermal expansion undergoes a marked reduction as the polymer goes through the glass transition region. Generally, the coefficient of linear thermal expansion of glassy polymers is ~10 times higher than that of metals, a factor that should be recognized in designing polymers for use at different temperatures.

The value of T_g depends on the mobility of the macromolecule and, thus, anything that hinders the rotational motion of the chain will lead to an increase in T_g. In general, the chain mobility depends primarily on the composition of the chain backbone and secondarily on the composition of the side groups attached to the chain backbone, but there is also a dependence on molecular weight.

- *Molecular weight:* For many polymers, T_g has been observed to depend on the number average molecular weight M_n according to an empirical equation known as the Fox–Flory equation:

$$T_g \approx T_{g,\infty} - \frac{K}{M_n} \tag{8.9}$$

where $T_{g,\infty}$ is the limiting value of T_g at high molecular weight, and K is a constant for a given polymer that is related to the free volume present in the polymer. T_g increases with M_n but, above ~10^4 Da, it shows little dependence on M_n. Because a higher free volume is associated with the ends of a chain, then for a given volume of material, polymers composed of shorter chains have a higher free volume and, thus, a lower hindrance to chain mobility.

- *Composition of the chain backbone:* Carbon single bonds (C–C bonds) in the chain backbone, for example, can rotate easily whereas double and triple bonds cannot (Chapter 2). Consequently, replacement of C–C bonds with a nearly equivalent number of carbon double bonds (C=C bonds) or carbon triple bonds (C≡C bonds) leads to an increase in T_g. Likewise, replacement of C–C bonds with paraphenylene groups (benzene rings linked to functional groups in the para position) leads to an increase in T_g, as seen by comparing polyethylene adipate ($T_g \sim -70\,°C$) with polyethylene terephthalate (PET) ($T_g \sim 69\,°C$) (Figure 8.8). On the other hand, replacing C–C bonds with carbon–oxygen (C–O) single bonds or with carbon–nitrogen (C–N) single bonds generally enhance the ability of the chain backbone to rotate and, thus, increase T_g. This is due to the lower number of single bonds that O and N can form when compared to C.
- *Composition of the side groups:* T_g is influenced by a variety of factors associated with the side groups attached to the chain backbone, such as their length, flexibility, and polarity. For flexible side groups, such as those composed of C–C bonds, T_g decreases as the length of the side group increases (Figure 8.9a). This is because the longer the side group, the more free volume it creates between the chains, making it easier for the chains to rotate. For less flexible side groups, more bulky side groups lead to an increase in T_g because they hinder rotational motion of the chain backbone (Figure 8.9b). More polar side groups lead to higher T_g due to the stronger intermolecular bonding that they introduce between the chains. For example, atactic PVC has a higher T_g (~81 °C) than atactic PP ($T_g \sim -20\,°C$). Chemical crosslinking reduces chain mobility and, thus, increases T_g.

When used at temperatures above T_g, polymers are in their rubbery state and, thus, are flexible and deformable as, for example, polyisoprene ($T_g \sim -70\,°C$), the main constituent of natural rubber, at room temperature. In comparison, polymers such as PMMA and PS are hard and glassy at room temperature because they have a T_g that is higher than room temperature (Table 8.2).

Low molecular weight compounds called plasticizers are often added to industrial polymers during their production to reduce their brittleness and improve their deformability. Plasticizers are typically low molecular weight organic substances such as ethylene glycol, low molecular weight polyethylene glycol (PEG), and dioctyl phthalate, but absorption of moisture from the atmosphere also has a plasticizing action in some polymers, such as polyvinyl alcohol (PVA) and nylon composed of molecules containing polar groups. Plasticizers generally get between the long chain molecules, causing the chains to pack less densely and reducing the intermolecular bonding between the chains. In this way, they lower T_g of the polymer but they also reduce its strength and modulus. A problem with some plasticizers is that they can migrate out of the polymer and, if toxic, can elicit adverse cell responses *in vivo*. This becomes particularly detrimental with polymers used as long-term implantable devices.

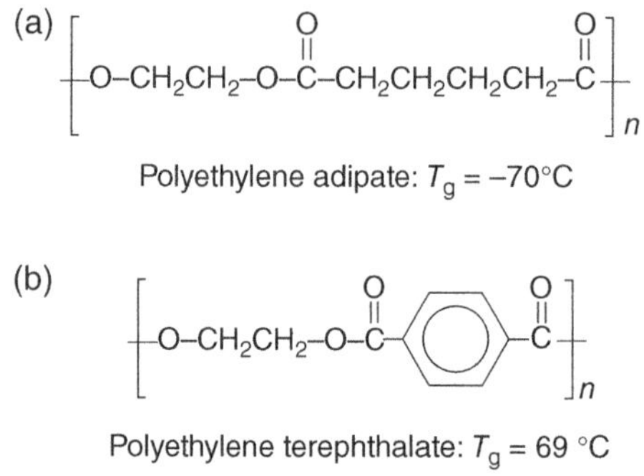

Figure 8.8 Effect of chain backbone composition on the glass transition temperature T_g. Polymers composed of less flexible units in the chain backbone generally have a higher T_g, as exemplified by comparing the structures of polyethylene adipate and polyethylene terephthalate,

8.2.7 Semicrystalline Polymers

Some polymers, as we have seen (Section 3.3.5), can crystallize, giving a two-phase structure composed of a mixture of amorphous and crystalline regions. These polymers can generally attain a crystallinity in the range 10–80% (by weight) and, as they cannot be completely crystallized, they are often referred to as semicrystalline polymers. The capacity of a polymer to crystallize depends on:

- The composition and structure of its constituent macromolecules, such as the stiffness and tacticity of the chain and the bulkiness and polarity of the side groups.
- Processing factors such as the cooling rate, isothermal crystallization temperature, orientation of the molecules and, for fibers, the drawing rate used in forming them.

Linear chains composed of a regular repeating arrangement of monomer units, a high degree of stereoregularity and strong polar interactions are more prone to crystallization whereas branched, crosslinked, or atactic chains inhibit crystallization.

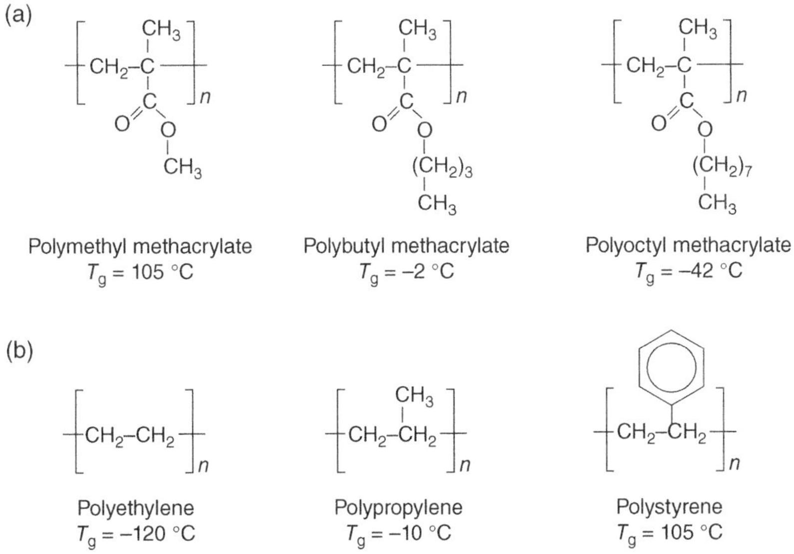

Figure 8.9 Effect of side chains on the glass transition temperature T_g. (a) Molecular structures of polymethyl methacrylate, polybutyl methacrylate, and polyoctyl methacrylate, illustrating that longer flexible side groups generally lead to lower T_g. (b) In comparison, as illustrated by the structures of polyethylene, polypropylene, and polystyrene, more bulky or less flexible side groups generally lead to higher T_g.

Table 8.2 Data for some common polymers available commercially for use as biomaterials.[a]

Polymer[b]	T_g (°C)	T_m (°C)	σ_T (MPa)	E (GPa)	ε_f (%)	α (10^{-6}/°C)	Density (g/cm^3)
HDPE	−125	120–135	20–30	0.8–1.2	300	90	0.92–0.96
PP	−10	170	30–35	1.5	50	110	0.90
PTFE	120	325	10	0.6	400	100	2.1–2.2
PVC	85		55	3.0	30	60	1.4
PMMA	105		72	3.2	2–5	70	1.2
PEEK	145	345	90–100	4	30–40	50	1.3
PET	75	260	55–75	2.5–3.0	50–150	70	1.3–1.4
Nylon 6.6	50–60	260	80	2.5–3.0	60	80	1.14
PDMS	−125		5–6	0.001–0.002	>150	90–95	0.97

[a] Data for three-dimensional isotropic polymers at or near room temperature. Mechanical properties vary depending on the actual microstructure and production conditions.
[b] Abbreviations of polymer names are given in Table 8.1. T_g = glass transition temperature; T_m = melting temperature; σ_T = tensile strength; E = Young's modulus (in tension); ε_f = strain to failure in tension; α = linear thermal expansion coefficient.

The model in Figure 3.15b describes the general features of a semicrystalline polymer microstructure, such as the presence of two phases that are held together by the macromolecules that pass from one region to another and back again many times. However, this model lacks critical details of the actual structure, which is considerably more complex at a finer scale. As described in the next section, semicrystalline polymers are composed of polyhedral objects, referred to as spherulites, which are themselves composed of a complex ordered aggregation of submicroscopic lamellar crystals.

Crystallization of Polymers

To get an idea of the crystallization process, we can start with the polymer in the molten state. The molecules, as noted earlier (Section 8.2.5), have a crumpled disordered conformation and they are highly mobile. Upon cooling to below the melting temperature T_m but above T_g, transformation from the molten state to a crystalline state occurs. In this temperature range, the chain segments have sufficient mobility to adopt the requisite conformation for crystallization. Overall, crystallization occurs by growth of spherulites composed of lamellar crystals (Figure 8.10).

Initially the spherulites have a spherical external shape. Growth of these spherulites occur radially by growth of their constituent lamellar crystals. As the spherulite grows into the melt, amorphous polymer is trapped between the crystals. Radial growth ceases when neighboring spherulites impinge, forming polyhedral objects.

Crystallization of polymers occurs by nucleation and crystal growth. Whereas the mechanisms are more complex, the underlying principles are similar to those described for phase transformations in metals (Section 6.3).

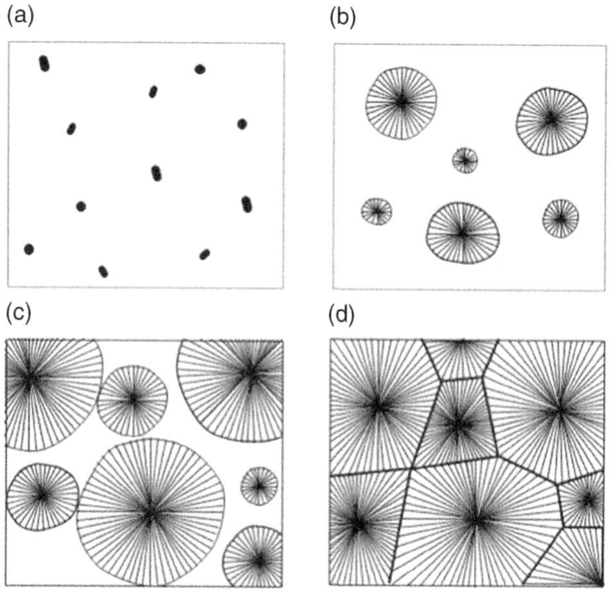

Figure 8.10 Illustration of stages in nucleation and crystal growth in polymers, leading to a spherulitic structure in semicrystalline polymers.

Nucleation Upon cooling the molten polymer to below T_m, segments of the chain tend to align and form tiny ordered domains at random points within the molten polymer. Disentanglement of chain segments from the melt and their accretion to the molecular smooth surface of these domains leads to the formation of stable nuclei. This type of nucleation is called homogeneous nucleation. Another type of nucleation, referred to as heterogeneous nucleation, can occur at sites such as impurities, additives, or production defects if they are present in the polymer. Heterogeneous nucleation is thermodynamically more favorable because it has a lower energy barrier to the formation of stable nuclei.

Crystal Growth Crystal growth occurs by disentanglement of chain segments from the melt and attachment to the crystal planes of the nuclei. Individual crystals grow from each nuclei in a radial direction to form spherulites. As shown in Figure 8.11, individual crystals develop a thin lamellar shape because accretion of the polymer segments occurs preferentially in the lateral plane whereas the thickness of the lamellae remains nearly constant (~20 nm).

Unlike metals and ceramics, high molecular weight polymers generally do not show a specific T_m but, instead, they show a small band of melting temperatures. This is because the crystals in a polymer do not all have the same size. Thicker crystals are formed at higher crystallization temperature whereas lower temperatures lead to the

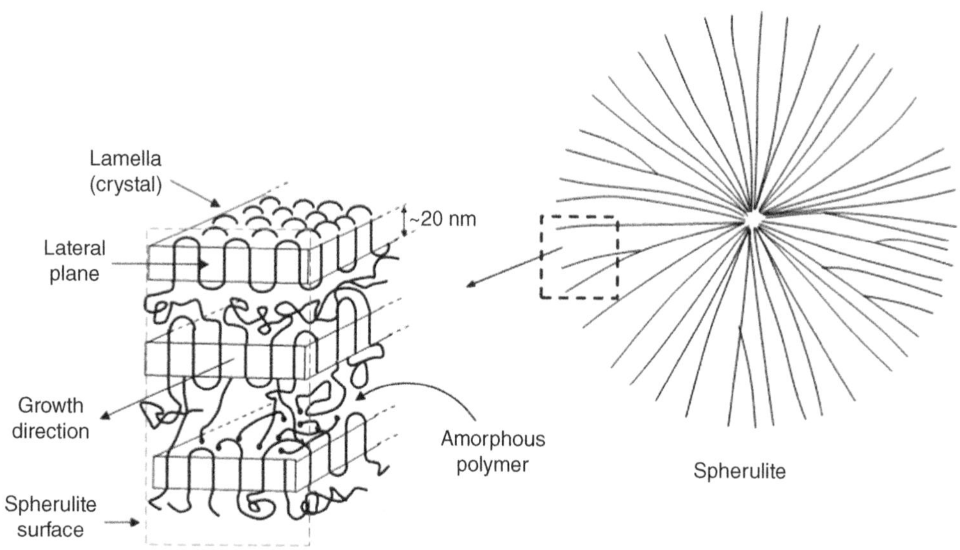

Figure 8.11 Illustration of the structure of a spherulite growing within a molten polymer. The lamellae of the spherulite grow in a radial direction by attachment of molecular segments to their lateral plane, with amorphous polymer left trapped between the lamellae.

formation of thinner crystals. Although the crystals are formed from the same polymer, they melt at different temperatures, giving a small range of melting temperatures rather than a specific melting temperature. Often, T_m is taken as the temperature at which the largest crystals melt, that is, the highest temperature in the small band of melting temperatures.

Semicrystalline polymers generally show two major transitions as they are heated or cooled, one at higher temperature (T_m) associated with crystalline phase, and the other at lower temperature (T_g) associated with the amorphous phase. However, minor transitions due to certain structural relaxations are also present in many polymers. The occurrence of these two major transitions has significant consequences for the mechanical properties of semicrystalline polymers (Section 8.4.1).

8.2.8 Molecular Orientation in Amorphous and Semicrystalline Polymers

Amorphous polymers, as we have seen, are generally composed of a random arrangement of coiled up or crumpled molecules (Section 8.2.6). However, tensile forces applied to a molten polymer during its production can cause the chains to align in the direction of the applied force (Figure 8.12). Cooling this molten polymer at a sufficiently rapid rate to room temperature, such that the molecules do not have enough time to relax to their random conformation, will lead to the formation of a solid polymer having a frozen-in molecular orientation. Plastic deformation of a solid polymer in tension at temperatures above T_g can also lead to molecular orientation. One consequence of molecular orientation is that the polymer can develop anisotropic physical properties, particularly anisotropic mechanical properties (Section 8.4.4). This is particularly useful in the formation of strong fibers.

Semicrystalline polymers can also develop molecular orientation, by extending the molten polymer and crystallization of the oriented melt, or by plastic deformation of the solid polymer in tension above T_g of the amorphous phase. In fact, many commodities in everyday use are composed of uniaxially oriented semicrystalline polymers such as nylon, polyester, and PP, particularly in the form of fibers. The structure of an oriented semicrystalline polymer is quite different from the spherulitic structure of the unoriented polymer, as illustrated in Figure 8.13. Orientation in semicrystalline polymers develops by a different and more complex mechanism when compared to amorphous

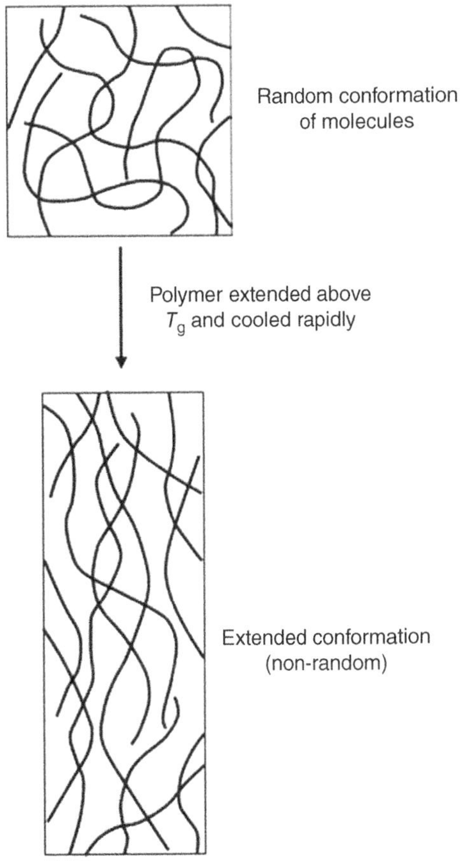

Figure 8.12 Molecular orientation in amorphous polymers. Extending an amorphous polymer at temperatures above its glass transition temperature followed by rapid cooling to the glassy state leads to molecular orientation that remains when the stress is removed.

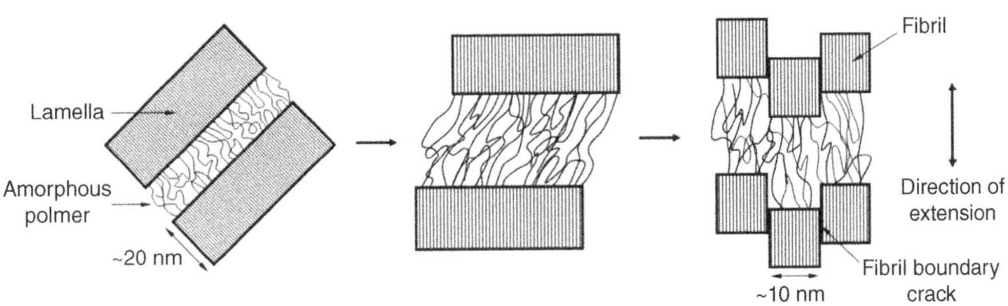

Figure 8.13 Stages of molecular orientation in semicrystalline polymers. Extending a semicrystalline polymer leads to molecular orientation in both the crystalline and amorphous regions.

polymers. Initially, the chains in the amorphous regions extend and the individual lamellar crystals within the spherulites slide past each other. Then the lamellae themselves become oriented such that the chain folds are aligned in the direction of the applied force. Blocks within individual lamellae are only weakly bonded to one another and, at higher deformation, may separate from one another, although they remain connected by tie chains in the direction of orientation.

8.3 Production of Polymers

Many polymers of sufficient purity are available commercially in the form of flakes, pellets, filaments or rods that are subsequently converted into the requisite three-dimensional shape and internal architecture using a variety of common production methods. However, several applications, particularly tissue engineering and drug delivery, may require the creation of polymers having physicochemical properties that are different from those of commercial polymers. Thus, an understanding of polymer synthesis and production methods is often useful.

8.3.1 Polymer Synthesis

Methods used to synthesize polymers from their monomers fall into two broad categories, called addition polymerization and condensation polymerization.

- *Addition polymerization:* In this method, polymers are synthesized by the addition of unsaturated monomers, such as monomers containing a double bond, to the growing chain. For example, the synthesis of PMMA from its monomer methyl methacrylate, a liquid at room temperature (boiling point = 101 °C), can be represented by the overall reaction illustrated in Figure 8.14a. The reaction occurs by a free radical mechanism, commonly initiated by the addition of a small quantity of a photoinitiator such as dibenzoyl peroxide (Figure 8.14b).
- *Condensation polymerization:* In this type of polymerization, a small molecule, usually H_2O, is released in each step of the reaction. An example is the reaction between ethylene glycol and terephthalic acid to form PET, a polymer belonging to a group called polyesters (Figure 8.15). As exemplified by PET, some condensation polymers have an alternating A–B–A–B pattern. This type of regularity enhances the capacity of the polymer to crystallize

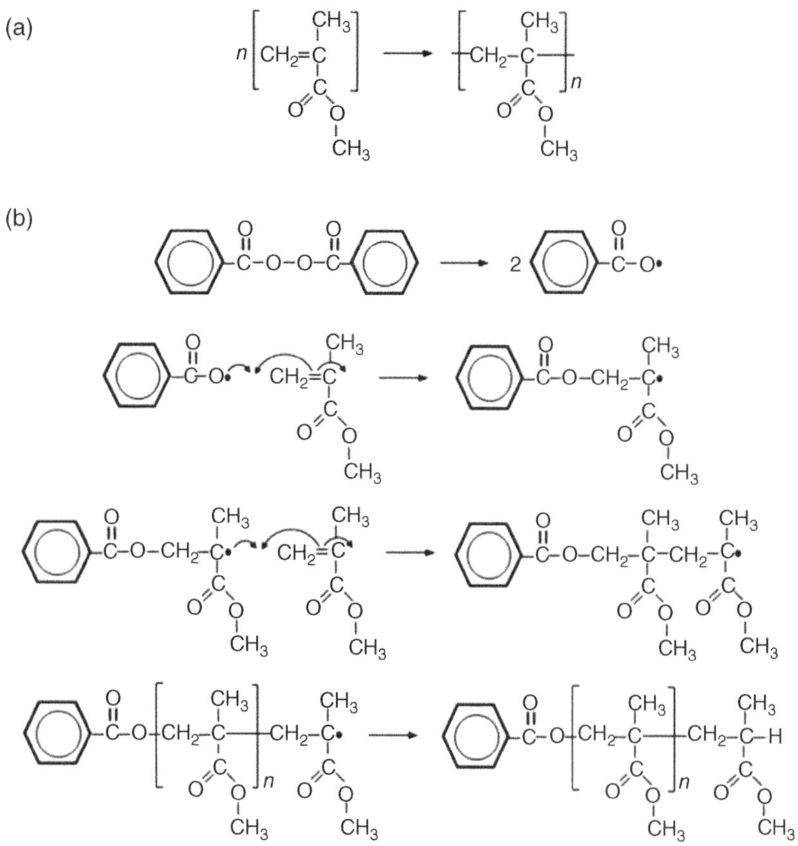

Figure 8.14 (a) Representation of the synthesis of polymethyl methacrylate (PMMA) from methyl methacrylate (MMA) by an addition reaction. (b) Illustration of the synthesis of PMMA from MMA by a free radical mechanism using dibenzoyl peroxide as a catalyst.

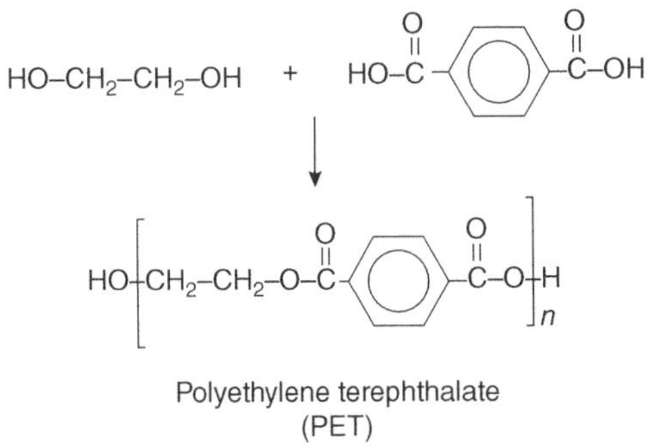

Figure 8.15 Representation of the synthesis of polyethylene terephthalate (PET) by an addition reaction between ethylene glycol and terephthalic acid.

and form a semicrystalline polymer upon cooling from the molten state. Block copolymers and graft copolymers can also be synthesized. In general, the chemical composition of the monomer unit and the number of monomer units (degree of polymerization) of each block will influence the physicochemical properties of the polymer.

8.3.2 Production Methods

An important advantage in the application of polymers, as noted earlier, is their simplicity and ease of production, particularly when complex shapes are required. Generally, the polymer in the form of pellets or rods is softened by heating to appropriate temperature, often less than only a few hundred degrees Celsius, deformed into the desired geometry and cooled to room temperature. The most common forming methods, include extrusion, injection molding, and blow molding. Additive manufacturing (3D printing) methods are now seeing considerable use because of the ease with which they can produce specimens having intricate external shape and internal architecture tailored to specific patient needs.

8.4 Mechanical Properties of Polymers

The mechanical properties of polymers generally show the following characteristic features, described earlier in the book (Chapter 4):

- Polymers, amorphous or semicrystalline, are inherently much weaker than most metals and ceramics (Figure 1.5). This makes their use in structural bone repair challenging, but they are well suited for healing nonstructural bone and noncalcified tissues.
- Polymers show a viscoelastic response, that is, their response to an applied mechanical stress depends not just on the magnitude of the stress or strain but also on the time over which the load is applied (or the rate of loading) (Section 4.2.4). Viscoelasticity is an important property in designing polymers for applications in which they are subjected to a mechanical stress over time.
- The mechanical response of polymers depend strongly on temperature, even temperatures not far above or below room temperature
- Because of the sensitivity of its mechanical response to temperature and rate of loading, a given polymer can show all three major types of mechanical response (brittle, ductile, and elastic response) depending on the temperature or rate of loading. Overall, a decrease in temperature or an increase in strain rate leads to an increase in strength and modulus but to a decrease in ductility (Figure 4.7).

8.4.1 Effect of Temperature

An important property in the design of polymers for a specific application is how their mechanical properties such as strength and modulus change with temperature. Because polymers are viscoelastic, the effective time of the measurement should also be specified but in this discussion, we shall assume some effective time such as ~1 s. Important reference temperatures are T_g for amorphous polymers and a combination of T_g and T_m for semicrystalline polymers because they demarcate a drastic change in mechanical properties. As an example, Figure 8.16 shows the shear

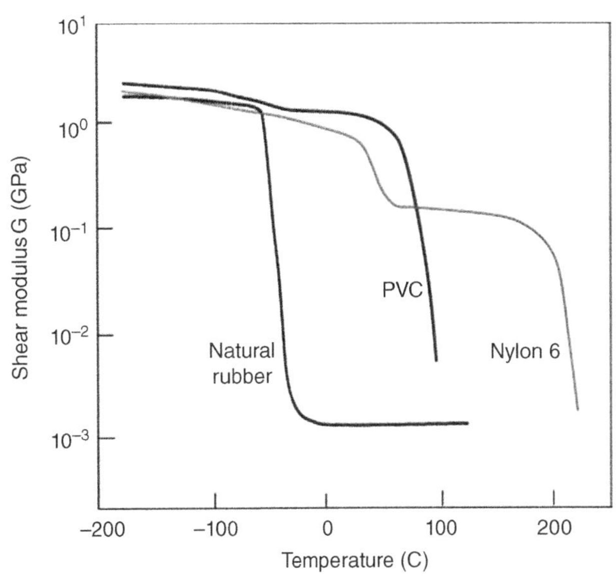

Figure 8.16 Illustration of the shear modulus as a function of temperature for three types of polymers: natural rubber, an amorphous polymer (polyvinyl chloride, PVC) and a semicrystalline polymer (nylon 6).

modulus G of three polymers, natural rubber, PVC, and nylon 6. All three curves show a characteristic trend composed of one or more regions in which G decreases slightly with increasing temperature, along with one or more regions in which there is a sharp decline. A small decrease in G with increasing temperature is due to thermal expansion as the molecules move further apart with increasing temperature. On the other hand, a steep decline is associated with a major transition, such as the glass transition or melting.

At very low temperatures (below approximately −100 °C), G for all three polymers lies within a narrow range (~1–5 GPa), even for natural rubber. Upon heating, G for natural rubber shows a steep decline in the region −50 to −25 °C (glass to rubber transition) and levels off at ~1 MPa at higher temperatures because the molecules are crosslinked. PVC, an amorphous polymer, shows a small drop in G at approximately −40 °C due to a minor structural relaxation but a major decline at ~75 °C (glass to rubber transition). In comparison, nylon 6, a semicrystalline polymer of crystallinity ~50 wt%, shows a significant decline in G at ~50 °C (glass to rubber transition of the amorphous phase) and a larger decline at ~200 °C (melting of the crystalline phase).

The major transition temperatures (T_g and T_m) determine the temperature range in which these polymers can be usefully applied. Natural rubber is used above its T_g when it is in its rubbery state which, together with its crosslinked molecules, gives rise to its well-known elasticity (that is, high extensibility and recovery after mechanical deformation). On the other hand, while PVC is a brittle glassy polymer below the minor structural relaxation at approximately −40 °C, it develops an increasing degree of ductility above this temperature. Consequently, PVC is used in a range from above −40 °C to below its T_g. Nylon 6 is brittle below approximately −30 °C due to the brittleness of its glassy amorphous phase but, with increasing temperature, it starts to show an increasing degree of ductility as the ductility of the amorphous phase improves. Consequently, the application of nylon 6 can cover a range from above −30 °C to below T_m (~200 °C). In this temperature range, the strength and modulus of nylon 6 decreases with temperature, particularly in the region of T_g, but its ductility increases.

For polymers used as implants, the physiological temperature (~37 °C) relative to T_g or T_m is an important reference point because it determines the nature of their mechanical response *in vivo*. For example, PMMA, an amorphous polymer, has a T_g of ~105 °C and, at room temperature, it shows limited ductility (elongation to failure ~2–5%). In comparison, while the T_g of the amorphous phase in high molecular weight PE, a semicrystalline polymer, is the subject of some controversy (Section 8.5.1), it is well below room temperature. Consequently, at room temperature, the amorphous phase is in a rubbery state, making PE ductile.

8.4.2 Effect of Crystallinity

The crystallinity of a polymer is commonly defined in terms of a mass fraction (percent), that is, the mass of the crystalline phase divided by the total mass of the specimen, but it is commonly stated as a weight percent (wt%). In general, some mechanical properties such as strength, modulus, hardness, and wear resistance of semicrystalline polymers increase with increasing crystallinity. This trend can be understood in terms of better mechanical properties of the crystalline phase dueto better packing and stronger intermolecular bonding of the molecules. In comparison, for the same polymer, the ductility (strain to failure) generally decreases with increasing crystallinity.

8.4.3 Effect of Molecular Weight

For many amorphous polymers, the measured tensile strength has been found to increase as their molecular weight increases but above ~20 kDa, there is little dependence on molecular weight. The increase in tensile strength σ with number average molecular weight M_n generally follows an expression similar in form to Eq. (8.9), that is

$$\sigma = \sigma_\infty - \frac{A}{M_n} \qquad (8.10)$$

where σ_∞ is the limiting tensile strength at very high M_n and A is a constant for a given polymer. Lower molecular weight polymers have a lower degree of chain entanglement, along with a higher chain mobility due to the higher free volume associated with shorter chains. Consequently, they have a lower capacity to resist an applied stress.

Higher molecular weight generally leads to better mechanical properties of semicrystalline polymers. Larger segments of the chains are able to form crystals, leading to larger crystals and a higher crystallinity. The longer chains also enter and leave a larger number of crystalline and amorphous regions, providing better linkage of these regions.

8.4.4 Effect of Molecular Orientation

Molecular orientation provides greatly improved mechanical properties in the direction of orientation when compared to the isotropic polymer. This is particularly effective for semicrystalline polymers such as nylon, PET, and PP that find considerable use as fibers in a range of products. For example, the measured tensile strength and Young's modulus of an isotropic PET specimen are in the range ~55–75 MPa and 2.5–3.0 GPa, respectively, whereas the tensile strength and Young's modulus of oriented fibers produced from the same PET are in the range 170–180 MPa and 14–15 GPa, respectively.

8.5 Thermoplastic Polymers

This section will cover common thermoplastic polymers relevant to biomaterials, which are nondegradable (chemically stable) in an aqueous environment. The use of the term nondegradable presents some difficulties because all polymers undergo some degree of degradation. However, we will use the term in a restricted sense to mean that the polymer shows no measurable degradation in an aqueous medium over a time scale appropriate to a specific application of interest. In general, the chemical composition of a polymer is the primary factor that determines whether it is chemically stable in an aqueous medium or not, but factors such as crystallinity and chemical crosslinking can modify its degradation rate.

Several chemically stable thermoplastic polymers find considerable use in consumer and engineering applications and some of them have been used for decades as biomaterials, mainly as implants or devices designed for long-term implantation. These polymers are often divided into various types depending on the presence of certain chemical groups within their constituent molecules or in the monomers used for their synthesis. Generally, strong interatomic bonds in the chain backbone, such as C–C bonds, and lack of significant polarity of the molecule are essential characteristics for chemical stability in aqueous media.

8.5.1 Polyolefins

Polyolefins are based on olefin monomers, that is, monomers composed of unsaturated aliphatic hydrocarbons having a C=C bond. They are easily formed by addition polymerization (Section 8.3.1). Composed of C–C bonds in the chain backbone, these polymers are strong and chemically stable. Two polyolefins, PE and PP, are among the most widely used thermoplastics in consumer products and they find use as biomaterials as well.

Polyethylene (PE)

The simplest of all polymers, PE is composed of molecules in which the repeating monomer unit is derived from the ethylene ($H_2C{=}CH_2$) molecule (Figure 2.10). PE used in commercial, industrial and biomedical applications is typically a semicrystalline polymer of high molecular weight (>10–100 kDa). PE crystals have a density of ~1.01 g/cm^3 and a T_m of ~135 °C. Amorphous PE has a lower density, ~0.85 g/cm^3 but its T_g has been a subject of some controversy. It is difficult to produce completely amorphous PE due to the high tendency of its linear PE molecules to crystallize. Studies on LDPE have shown two glass transitions, one at approximately −80 °C and the other at approximately −120 °C, but the transition at approximately −120 °C appears to be the main transition. Regardless of which of these two temperatures is taken as T_g, the amorphous phase in PE is in a rubbery state at room temperature. Consequently, semicrystalline PE shows good ductility due to the rubbery response of its amorphous phase.

Differences in crystallinity have given rise to various types of PE that are generally classified in terms of their density. The major types are low-density polyethylene (LDPE) and high-density polyethylene (HDPE). LDPE is composed of branched chains, ~3 per 100 C atoms in the chain backbone, whereas HDPE is composed of linear chains. LDPE has a lower amount of crystalline phase (~35–55 wt%) than HDPE (~50–85 wt %), a lower density (~0.90–0.94 g/cm^3) than HDPE (~0.92–0.96 g/cm^3), and a lower strength and modulus. LDPE is used in a variety of commercial and industrial applications, and in medical products such as syringes and bottles, but not in high performance applications. Acetabular liners in hip implants and tibial inserts in knee implants are composed of a type of HDPE commonly referred to UHMWPE.

Ultrahigh Molecular Weight Polyethylene (UHMWPE) UHMWPE has a molecular weight of at least 10^6 Da, typically in the range $2\text{–}6 \times 10^6$ Da, and a crystallinity of ~50 wt% (Table 8.3). Used for decades in hip and knee implants, UHMWPE has played a key role in the success of these prostheses, despite limitations related to its long-term biocompatibility and mechanical reliability. A critical limitation of UHMWPE, when used as articulating bearings in hip and knee implants, is its wear resistance. Cyclic compressive and tensile stresses due to normal activities of a patient activate wear mechanisms in UHMWPE bearings that generate wear particles. These particles elicit adverse cell responses near the implant, leading to osteolysis (bone resorption) and, eventually, to aseptic loosening of the implant (Chapter 18).

Because of the adverse effects related to the generation of wear particles, considerable efforts have been made in the last few decades to improve the wear resistance of UHMWPE. These efforts have shown that crosslinking is

Table 8.3 Effect of crosslinking and crystallinity on mechanical properties of ultrahigh molecular weight polyethylene (UHMWPE)[a].

Processing/property	UHMWPE	XPE[b]	HP-XPE[c]
Crosslinking (γ ray intensity)		50 kGy	50 kGy
High pressure treatment			240 °C; 500 MPa
Crystallinity (wt%)	50	46	68
Lamellae thickness (nm)	28	23	52
Tensile yield strength (MPa)	21.3 ± 0.2	19.0 ± 0.2	20.6 ± 0.4
Young's modulus (GPa)	0.50 ± 0.01	0.33 ± 0.08	0.70 ± 0.05
Vickers microhardness (MPa)	66 ± 4	57 ± 3	73 ± 2
Fatigue threshold stress intensity ($MPam^{1/2}$)	1.41	0.92	1.08

[a] Molecular weight = $5.5 \times 10^6 - 6.0 \times 10^6$ Da.
[b] Crosslinked UHMWPE.
[c] High pressure treated XPE.
Source: Data taken from Simis et al. (2006).

particularly effective in significantly improving the wear resistance of UHMWPE bearings but it can have an adverse effect on certain mechanical properties, particularly fatigue resistance. However, this problem has now been largely alleviated by modifications to the production process.

Crosslinked UHMWPE A simple and effective way to crosslink PE is to irradiate it with γ rays of a suitable intensity (~50 kGy). Irradiation leads to the creation of reactive free radicals (unpaired electrons) at random C atoms in the chain backbone, which react to form crosslinks. Crosslinking occurs predominantly in the amorphous phase of semicrystalline PE, reducing the mobility of the molecules and, thus, improving the modulus of the amorphous phase and of the polymer as well.

While crosslinking leads to an improvement in the Young's modulus and wear resistance of UHMWPE, it is also accompanied by a deterioration in other mechanical properties, such as tensile strength, ductility, and fatigue resistance. Reduction in fatigue resistance over a long-term is a critical concern because UHMWPE bearings, particularly in knee implants, are subjected to repeated cyclic stresses during normal activities of a patient. As the amplitude of these cyclic stresses can sometimes reach values that are a significant fraction of the yield stress of the polymer, there is a risk that fatigue failure might occur.

Although γ radiation is simple and effective for crosslinking UHMWPE, the free radicals generated in the process generally lead to oxidative degradation of UHMWPE bearings *in vitro* and *in vivo* over time through a series of chemical reactions (Chapter 15). This causes a deterioration in the mechanical properties of the bearings, such as their tensile strength, fracture toughness, ductility, and fatigue resistance. Consequently, several methods have been developed to annihilate free radicals from crosslinked UHMWPE while retaining its improved wear resistance.

One method has been to incorporate a natural antioxidant such as vitamin E into the polymer. More commonly used methods are based on melting and recrystallizing the irradiated polymer, or subjecting the irradiated polymer to a series of annealing steps below T_m of the crystalline phase. These methods have the effect of modifying the structural characteristics of the irradiated polymer, such as its crystallinity and size of the lamellar crystals, which, in turn, influence the mechanical properties of the product. They require precise control to produce an optimal combination of mechanical properties while retaining the improved wear resistance of the crosslinked polymer.

Because a higher crystallinity generally leads to improved mechanical properties, one goal has been to improve the crystallinity of crosslinked UHMWPE to compensate for the deterioration in mechanical properties resulting from crosslinking. Annealing below T_m allows the irradiated polymer to retain a high degree of its crystallinity but free radicals are not fully annihilated. In comparison, heating above T_m leads to complete annihilation of free radicals but the crystals may not fully reform upon annealing below T_m, leading to a decrease in crystallinity. However, a subsequent thermal treatment above T_m under high pressure has been shown to produce an increase in crystallinity, giving a product that has the improved wear resistance due to crosslinking but without a significant deterioration in fatigue resistance (Table 8.3).

Example 8.2 Determine the percent crystallinity (by weight) of a semicrystalline PE of density 0.94 g/cm^3, given that the densities of amorphous PE and PE crystals are respectively 0.85 and 1.01 g/cm^3.

Solution:
Although there are a few ways of solving this problem, we will use a method that involves simple principles related to mass, volume, and density. Consider unit mass, that is, 1 g of semicrystalline PE of total volume V and density d. Let V_a and m_a be the volume and mass, respectively, of the amorphous phase, and V_c and m_c the volume and mass, respectively of the crystalline phase. Units of g, cm^3, and g/cm^3 are used for mass, volume and density, respectively.

As we are considering 1 g of PE, we can write

$$m_a + m_c = 1 \tag{8.11}$$

The total volume is

$$V = V_a + V_c \tag{8.12}$$

and, as volume is simply mass ÷ density, Eq. (8.12) can also be written

$$1/d = m_a/d_a + m_c/d_c \tag{8.13}$$

As $m_a = 1 - m_c$, we get

$$1/d = (1 - m_c)/d_a + m_c/d_c \tag{8.14}$$

Substituting the density values into Eq. (8.14), we get $m_c = 0.60$, that is, the percent crystallinity is 60%.

Polypropylene (PP)

PP is produced by the addition polymerization of propylene. Although included here as a polyolefin, it can also be classified as a vinyl polymer based on its chemical structure (Figure 8.4). Due to the CH_3 side group in its configuration, PP can exist in three different stereoregular forms, isotactic, syndiotactic, and atactic (Section 8.2.3). However, it is of interest only when polymerized in the isotactic form. Isotactic PP is a semicrystalline polymer that has physicochemical properties that are generally similar to those of HDPE, although it has a higher Young's modulus and yield strength. The most widespread use of PP in biomedical engineering is in the form of surgical sutures. Because of their useful mechanical properties and resistance to degradation, PP sutures have good long-term performance. PP has also been used as a mesh for tissue repair such as hernia repair and in urological reconstruction procedures, but its use in these applications has been controversial.

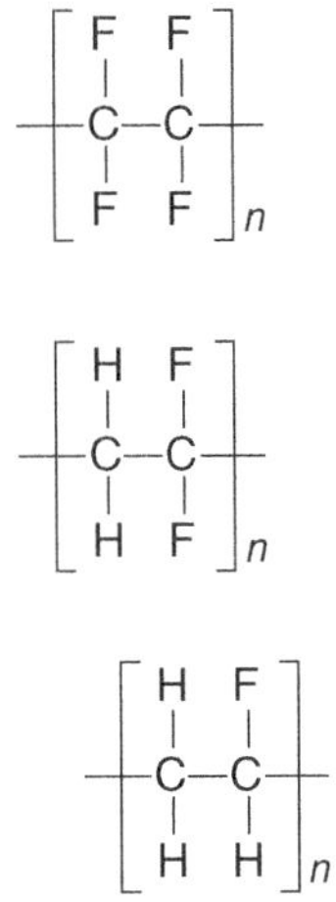

Polytetrafluoroethylene (PTEE)

Polyvinylidene fluoride (PVDF)

Polyvinyl fluoride (PVF)

Figure 8.17 Chemical structures of three fluorinated hydrocarbon polymers: polytetrafluoroethylene (PTFE), polyvinylidene fluoride (PVDF), and polyvinyl fluoride (PVF).

8.5.2 Fluorinated Hydrocarbon Polymers

Fluorinated hydrocarbon polymers are composed of PE molecules in which F replaces one or more H atoms in a monomer unit (Figure 8.17). Fluorinated hydrocarbons have some interesting physicochemical characteristics. The C–F bond, for example, is one of the strongest bonds in organic chemistry. Overlap of the electron orbitals between these two small atoms leads to a strong covalent bond. There is a large electronegativity difference between the two atoms (2.5 and 4.0 for C and F, respectively, on the Pauling electronegativity scale), which gives a strong polarity to the C–F bond (Chapter 2). However, when these bonds exist in a symmetrical manner, as in the monomer unit of polytetrafluoroethylene (PTFE), the molecule can be nonpolar.

PTFE, well known commercially as Teflon®, is the most widely used fluorocarbon polymer industrially. It is used as a biomaterial as well, in a variety of applications, most widely as implantable vascular grafts and as catheters, and to a more limited extent as sutures and as meshes for soft tissue repair.

Polytetrafluoroethylene (PTFE)

PTFE is very stable chemically and is nonpolar due to the strong C–F bonds and their symmetric configuration in the monomer units. It has a surface free energy (~20 mJ/m^2) that is among the lowest for all polymers, a high contact angle (~110°) with water, and is hydrophobic, that is, wetting by aqueous liquids is difficult. The low coefficient of friction of PTFE is well known.

PTFE is a semicrystalline polymer, having a high T_m of ~325 °C and a T_g of ~120 °C. Commercially available PTFE has a molecular weight of ~10^6 Da and a high degree of crystallinity (~60 wt%). PTFE is not very strong, having a tensile strength and Young's modulus of ~10 MPa and ~0.6 GPa, respectively, at room temperature. However, it has good ductility, showing a strain to failure of over ~400% at room temperature.

Although PTFE is a thermoplastic polymer, it is difficult to form into three-dimensional specimens directly by the common polymer production methods such as extrusion and injection molding because of its unsuitable rheological properties. Instead, a sintering process, similar to that described for ceramics (Chapter 7) is used to produce three-dimensional specimens. Commercially available PTFE particles, typically having a high degree of crystallinity (>90%), are pressed under pressures as high as ~35 MPa to form a porous specimen. Alternatively, the PTFE particles are mixed with a liquid such as mineral spirits to form a paste that is extruded to form the desired shape and dried. Then the porous specimen is sintered under a controlled temperature profile to ~350 °C (that is above T_m) and cooled at a controlled rate to room temperature, giving a fully dense specimen of crystallinity ~60–70 wt%.

Vascular grafts for blood vessel replacement and meshes for tissue repair are composed of an expanded form of PTFE that has a porous microstructure (Figure 8.18). First, an unsintered specimen of the desired geometry, such as a thin sheet or a tube, is formed by extruding a paste of the PTFE particles, followed by drying. Then the specimen is expanded using appropriate mechanical stresses and subjected to a controlled thermal treatment at temperatures up to ~350 °C.

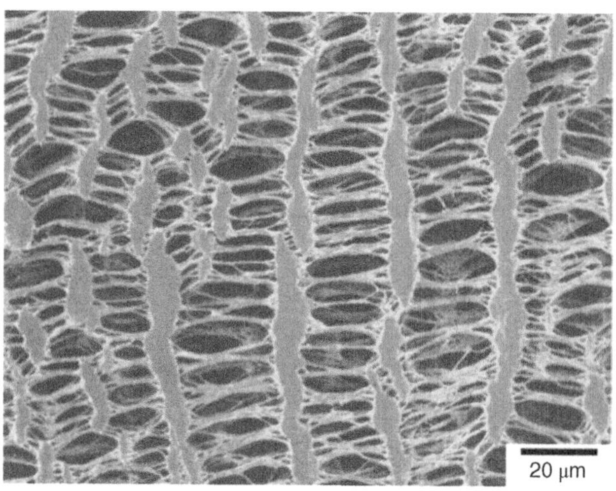

Figure 8.18 Scanning electron microscope image of the microstructure of a porous expanded polytetrafluoroethylene (PTFE) specimen used as a vascular graft in blood vessel repair. *Source:* From Begovac et al. (2003).

8.5.3 Vinyl Polymers

The molecular structure of vinyl polymers, as we have seen, consists of PE molecules in which an atom or group (denoted as X) replaces one H atom in the monomer unit (Figure 8.4a). Common vinyl polymers are PP (X = CH_3), PVC (X = Cl), PS (X = C_6H_5), and PVA (X = OH). PP can also be considered as a polyolefin and was discussed as such in Section 8.1. Whereas PVA is soluble in aqueous media due to its polar OH group, PP, PVC, and PS are chemically stable (insoluble) in aqueous media and in some other liquids such as alcohols.

PVC is commonly produced in the atactic form and, thus, it is amorphous. As its *Tg* is ~85 °C, PVC is in the glassy state at room temperature, which makes the polymer hard and brittle. Low molecular weight plasticizers are commonly used to reduce its T_g and make the polymer more deformable. While PVC finds use in both the unplasticized and plasticized state, it is more frequently used in its plasticized state. Like PVC, PS is often produced in the atactic (amorphous) form. PS has a T_g of ~105 °C and, thus, is hard and brittle at room temperature.

PVC and PS, along with PE, PP, and PET make up a considerable fraction (over ~85% by volume) of the worldwide consumption of plastics and are used in an enormous range of domestic and industrial applications. PVC and PS are not used in implantable devices but they find considerable use in the healthcare industry for a variety of applications such as tubing, packaging, and containers due to their chemical stability in aqueous media, optical transparency, low cost, and ease of sterilization.

8.5.4 Acrylic Polymers

Acrylic polymers are formed by the addition polymerization of monomers based on acrylic acid or methacrylic acid. The most commonly used acrylic polymer in biomedical applications is PMMA, which has been widely used for decades as intraocular lenses (Figure 1h) and as bone cement to stabilize total joint prostheses. It is also used as hard contact lenses. PMMA is composed of atactic macromolecules and it has a T_g of ~105 °C. Consequently, it is an amorphous, hard and brittle polymer at room temperature. PMMA is also highly transparent and chemically unreactive in aqueous media, although it can absorb up to ~1–2 wt% water. It can be easily formed into three-dimensional products at temperatures not far above room temperature using common polymer production methods.

Acrylic acid, methacrylic acid, and polymers formed from them, such as polyacrylic acid (PAA) and polymethacrylic acid (PMAA), are soluble in aqueous (polar) solvents because the H atom in their carboxylic acid (C=O)

OH group can dissociate, giving negatively charged oxygen atoms. In comparison, methyl methacrylate and PMMA are insoluble in aqueous solvents, although they are soluble in solvents that have a similarity in functional group such as acetone, or in nonpolar solvents such as toluene. The ester (OCH_3) group that replaces the OH group in methacrylic acid or PMAA has little polarity, making these substances very hydrophobic.

As shown in Figure 8.14, PMMA is commonly synthesized by addition polymerization of methyl methacrylate, via a free radical mechanism initiated commonly by a substance such as dibenzoyl peroxide. Polymerization can be achieved by heating the system to a few tens of degrees above room temperature. Alternatively, an accelerator such as *N,N*-dimethyl-*p*-toluidine can be added to achieve polymerization and hardening *in situ*, a process that is employed in formation of PMMA bone cement. The accelerator enhances the rate at which dibenzoyl peroxide decomposes to form free radicals.

PMMA Bone Cement

Polymerization of methyl methacrylate, a liquid at room temperature (boiling point = 101 °C), leads to a shrinkage of ~20 vol% and to a temperature increase of as high as a few tens of degrees Celsius that can lead to cell and tissue damage *in vivo*. To reduce these effects, bone cement is commonly formed from a mixture of PMMA powder and methyl methacrylate, in a ratio of two to three parts of PMMA to one part methyl methacrylate. In commercially available formulations, the powder is mixed with ~1–2 wt% dibenzoyl peroxide and a radiopacifier such ~4–8 wt% $BaSO_4$ particles (~1–2 μm) that is used to make the cement visible in X-ray imaging. The monomer contains ~1–2 wt % *N,N*-dimethyl-*p*-toluidine and a stabilizer (less than ~0.01 wt%) such as hydroquinone. Mixing is often performed manually but methods such as vacuum mixing have been introduced to provide more homogeneous mixing and better consistency.

PMMA bone cement has been used successfully for decades in orthopedic surgery, but concerns have been expressed occasionally about its potential for causing adverse biological effects *in vivo*. The temperature increase arising from the polymerization reaction can cause cell and tissue damage. Release of toxic methyl methacrylate monomers from the bone cement over time can also occur because the *in situ* polymerization reaction generally does not go to completion, leaving residual monomers in the cement.

In some bone cements, an antibiotic such as gentamycin or tobramycin is included to provide local antibiotic delivery in the treatment of bone infection. Typically, ~10 wt% of the antibiotic is added to the cement mixture to provide a desirable elution rate at a sustainable therapeutic level. However, PMMA is not degradable and can provide a surface upon which secondary bacterial infection can occur. Consequently, PMMA must often be removed upon completion of the antibiotic treatment prior to reconstruction of the infected bone. Inclusion of the antibiotic can also lead to a reduction in the strength of the cement, which can be an important factor in treating structural bone.

8.5.5 Polyaryletherketones

Polyaryletherketones are a family of high-performance thermoplastics composed of macromolecules in which the chain backbone contains the benzene ring linked by O atoms characteristic of an ether group and by carbonyl (C=O) groups characteristic of a ketone group (Figure 8.19). The presence of the aromatic benzene ring in the chain backbone makes the polymer stable against chemical attack by many liquids and against thermal and oxidative degradation. As the benzene rings are bonded by the O and C=O bridges in the para position (that is, opposite sides of the ring), they also contribute to the stiffness of the chain. The ratio and sequence of the ether O and ketone C=O influences the physicochemical properties of the polymer.

Polyaryletherketones generally have high T_m, high tensile strength, high Young's modulus, good ductility, and good retention of their mechanical properties to temperatures as high as ~250 °C. Three-dimensional specimens are produced by common methods used for polymers. One polyaryletherketone, referred to as polyether ether ketone (PEEK) (Figure 8.19), has been used as implants in spinal fusion surgery and has been developed for use as craniofacial and dental implants.

Polyether Ether Ketone (PEEK)

Commercially available PEEK is a semicrystalline polymer having a molecular weight of ~10^5 Da, a crystallinity of ~35 wt%, T_m ~345 °C and T_g ~145 °C. Because of its high T_g, the amorphous phase is in a glassy state at the physiological temperature. The tensile yield strength and Young's modulus of PEEK at room temperature are ~140 MPa and ~4 GPa, respectively, values that are comparable to those at the lower end of the range for human

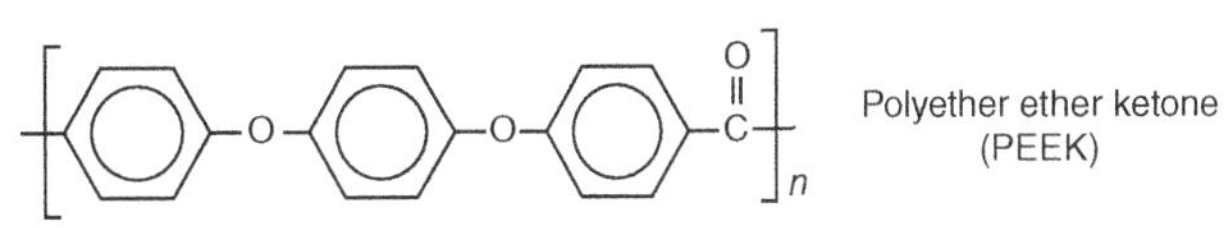

Figure 8.19 Chemical structure of polyether ether ketone (PEEK).

cortical bone. PEEK also has good ductility (strain to failure ~30–40%). Despite the presence of the O and C=O bridges in the chain backbone, PEEK is chemically stable *in vivo* due to the presence of the benzene rings in the chain. PEEK has sufficient biocompatibility for long-term implantation but it has poor wettability by aqueous media based on the high contact angle (~90°) of water droplets on its surface.

Overall, PEEK is an attractive biomaterial for several bone repair applications because of its superior mechanical properties, chemical stability, and biocompatibility. Reinforcing PEEK with short or continuous fibers leads to a significant improvement in its strength and Young's modulus. Such composites are being studied for highly loaded applications such as bearings and stems in joint prostheses.

8.5.6 Polycarbonate, Polyethersulfone, and Polysulfone

Polycarbonate (PC), polyether sulfone (PES), and polysulfone (PSF) generally refer to the polymers having the chemical structures shown in Figure 8.20, but these names can also be applied to any polymer based respectively on the carbonate, ethersulfone, and sulfone linkage. Like the polyaryletherketones (Section 8.5.5), these polymers are composed of macromolecules in which the benzene rings in the chain backbone are linked by functional groups in the para position. Consequently, they generally share several characteristic properties of polyaryletherketones, such as those described for PEEK.

8.5.7 Polyesters

Polyesters are polymers composed of macromolecules that contain the ester –(C=O)–O– repeating group in their main chain. They are commonly formed via a condensation reaction involving a dicarboxylic acid and a diol. A well-known example is the reaction between terephthalic acid and ethylene glycol to form PET (Figure 8.15). The term polyester is also often used to refer specifically to PET, the most commonly used thermoplastic polyester. In the form of fabrics (clothing) or fibers, PET is commonly known by its trade name Dacron®.

PET is widely used in the clothing industry, and it is an important polymer in consumer products such as bottles and in engineering applications. It can be easily formed into three-dimensional products by the common methods used for thermoplastic polymers. In the form of monofilaments or multifilament yarns, PET is used as nondegradable sutures, vascular grafts to repair blood vessels and implants to reconstruct soft tissues such as ligaments.

Polyethylene Terephthalate (PET)

PET has a T_m of ~260 °C and a T_g of ~75 °C. Cooling the polymer rapidly from the molten state to below T_g leads to an amorphous polymer that is weak and transparent. On the other hand, a semicrystalline opaque polymer of higher strength is formed upon cooling the molten polymer at an appropriately slow rate or heating the rapidly cooled amorphous polymer to temperatures above T_g but below T_m.

The mechanical properties of PET depend on its crystallinity. Commercially available PET in the form of isotropic rods or blocks have a molecular weight of ~2×10^4 Da, crystallinity of ~30–50 wt%, a tensile strength of ~55–75 MPa, a Young's modulus of 2.5–3.0 GPa, and an elongation to failure (a measure of ductility) of ~50–150%. In the form of fibers, PET can have a tensile strength and Young's modulus of 170–180 MPa and ~14–15 GPa, respectively.

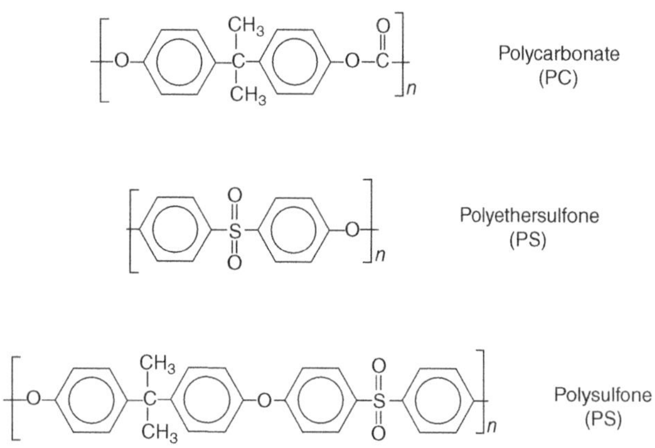

Figure 8.20 Chemical structures of polycarbonate (PC), polyether sulfone (PES), and polysulfone (PSF).

8.5.8 Polyamides

Polyamides (PAs) are composed of macromolecules in which the amide bond –NH–(C=O)– links the repeating units together. This bond is the same as the amide bond in proteins that links amino acid residues together (Chapter 2). Commercial polyamides are commonly synthesized by condensation reactions involving an amine and a carboxylic acid as, for example, the reaction between hexamethylene diamine and adipic acid to form the alternating block polymer commonly known as nylon 6.6 (Figure 8.21). A well-known type of commercial polyamides, nylons have a semicrystalline structure and they are widely used as engineering thermoplastics. Nylons are also excellent polymers for making fibers. Nylons see little use in biomedical

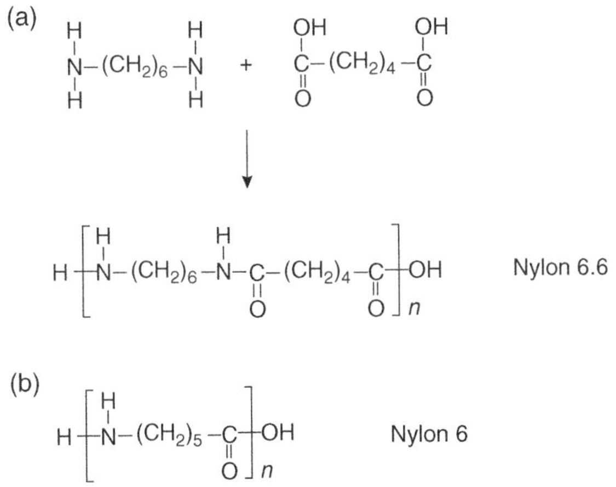

Figure 8.21 (a) Representation of the synthesis of nylon 6.6 by a condensation reaction between hexamethylene diamine and adipic acid. (b) Chemical structure of nylon 6.

applications but one type, nylon 6.6, finds limited use as nondegradable sutures.

There are six main types of nylon, referred to as nylon 6.6, nylon 6.10, nylon 6.12, nylon 6, nylon 11, and nylon 12, depending on the number of C atoms in each repeating monomeric unit in the chain backbone (Figure 8.21). The polar N–H and C=O groups in the nylon molecules, which lead to strong intermolecular (hydrogen) bonding (Figure 2.16b), and the linear structure of the chains provide these polymers with a strong capacity to crystallize. Thus, nylons are semicrystalline polymers possessing attractive mechanical properties. For example, nylon 6.6 has a T_m of ~260 °C and, in the dry state, a T_g of ~50–60 °C. Commercially available isotropic nylon 6.6 has a crystallinity of 35–50 wt% and, in the dry state, has a tensile yield strength of ~80 MPa, a Young's modulus of 2.5–3.0 GPa, and an elongation to failure of over 50%. These mechanical values are generally comparable to those of isotropic PET.

The polarity of the amide bond in the nylon molecules also causes these polymers to absorb water. Typically, the capacity to absorb water decreases as the CH_2 content of the molecules increases. Nylon 6.6, which has the highest capacity to absorb water, can absorb up to ~2.5 wt% water at room temperature in a moist atmosphere of relative humidity ~50%. As the H_2O molecules act as a plasticizer, exposure of nylon 6.6 to a moist environment can lead to a significant reduction in its mechanical properties. For example, when exposed to a moist atmosphere of relative humidity ~50% at room temperature, the Young's modulus of commercial nylon 6.6 can decrease from 2.5–3.0 GPa to ~1.2 GPa, while the tensile yield strength can decrease from ~80 MPa to ~60 MPa. On the other hand, the elongation to failure can increase from ~50% to ~300%.

8.6 Elastomeric Polymers

Elastomers, often referred to as rubbers, are amorphous polymers that are above their T_g. They show the following characteristic properties:

- High extensibility, often over several hundred percent, even under low mechanical stress
- Complete recovery upon removal of the mechanical stress.

These two properties have their origin in entropy changes of the material. Stretching an elastomer leads to a decrease in entropy because the number of conformations available to the stretched molecules is reduced. This reduction in entropy provides a strong driving force for recovery of the elastomer upon removal of the stress. This can be understood in terms of the Boltzmann equation for the entropy S of a system, given by

$$S = k \ln W \tag{8.15}$$

where k is the Boltzmann constant, and W is the number of microstates available to the system. For an elastomer, we can approximate W as the number of conformations available to a molecule. Stretching leads to a decrease in entropy and, upon removal of the mechanical stress, the system returns to its undeformed state of higher entropy, in accordance with the second law of thermodynamics.

Ideally, elastomers are composed of flexible molecular chains that are crosslinked by chemical bonds. The crosslinks provide the memory for the polymer to recover after removal of the mechanical stress. Because the polymer is above its T_g, structural relaxation of the macromolecules between the crosslinks is rapid, that is, these molecules can change their conformation rapidly. On the other hand, many polymers show a mechanical response comparable to that of chemically crosslinked elastomers even without chemical crosslinks. These polymers are generally composed of large chains (molecular weight greater than ~10^4–10^5 Da) that are highly entangled. The entanglements pin together the molecules, providing what can be described as physical crosslinks. Rubbery behavior can also arise if the mobile molecular chains are pinned by entering and leaving crystalline or glassy regions.

Elastomers are widely used industrially. Common engineering elastomers include natural rubber composed mainly of chemically crosslinked polyisoprene molecules and synthetic rubbers composed of chemically crosslinked molecules such as polyisoprene, polybutadiene, or styrene-butadiene molecules. One elastomer, polydimethylsiloxane (PDMS), is the most widely used polymer in a group of elastomers referred to as silicones. It is used in a variety of cosmetic and industrial applications. PDMS is widely used in biomedical applications as well, particularly when elastic deformability is a critical design property.

8.6.1 Polydimethylsiloxane (PDMS)

Silicones (also referred to as polysiloxanes) are composed of macromolecules in which the chain backbone consists of silicon–oxygen (Si–O) bonds, referred to as siloxane bonds. The most important silicone, PDMS, consists of methyl groups attached to each Si atom (Figure 8.22) but other groups such as phenyl, vinyl, or trifluoropropyl groups can substitute for the methyl group. Polymerization reactions allow the synthesis of PDMS composed of linear chains or branched chains that can be crosslinked into three-dimensional networks by a variety of chemical reactions. Overall, the following two factors are central to understanding the physicochemical properties of PDMS:

- Chemical composition of its constituent macromolecules, consisting of organic methyl groups attached to an inorganic Si–O chain backbone.
- Chemical structure of these macromolecules consisting of linear, branched, or crosslinked chains.

The capacity to control the molecular structure over a wide range allows formulation of the polymer into products such as fluids, adhesives, resins and elastomers, endowing PDMS with a unique combination of properties.

Production of PDMS Elastomers

Conversion of PDMS into an elastomer composed of crosslinked chains is achievable using a variety of addition or condensation reactions. Silicone adhesives are commonly composed of macromolecules formulated to have sufficiently reactive end groups. The adhesive is in a liquid-like state and, thus, it can be stored in sealed cartridges. Upon opening of the cartridge, hydrolysis of the reactive end groups by moisture from the atmosphere, followed by condensation reactions, leads to crosslinking of the chains and to the formation of a sufficiently strong adhesive. The reaction occurs at room temperature but an organometallic tin compound is often included in the adhesive formulation to catalyze the reaction.

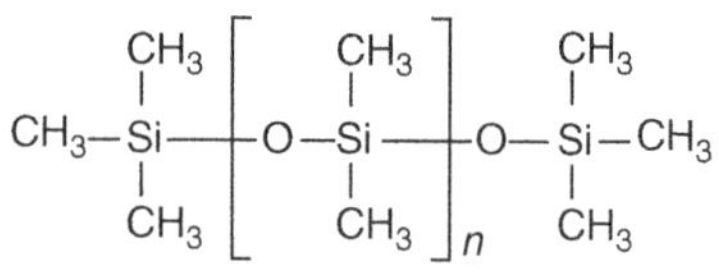

Figure 8.22 Chemical structure of polydimethylsiloxane (PDMS).

Most PDMS products formed by casting or molding are formed by addition reactions involving a two-part mixture. One part consists of PDMS composed of molecules containing reactive vinyl ($H_2C{=}CH_2$) end groups, and a catalyst composed of an organometallic platinum complex. The other part consists of a low molecular weight compound (called an oligomer) in which the small molecules contain Si–H end groups. Addition reactions involving the vinyl end groups of the PDMS molecules and the end groups of the oligomer leads to crosslinking of the chains. This reaction releases no by-products and involves no measureable shrinkage, thus allowing the formation of products having accurate dimensions.

Physicochemical Properties of PDMS Elastomers

The low energy barrier for rotation around the Si–O bonds in the chain backbone and the small size of the CH_3 groups attached to the Si atoms are mainly responsible for the low T_g (~−125 °C) of PDMS. Relaxation of chain segments between the crosslinks is rapid at room temperature, which give PDMS its elastomeric mechanical properties. Because of the difference in electronegativity between Si (1.8 on the Pauling scale) and O (3.5), the Si–O bond has a high degree of polarity. On the other hand, the organic CH_3 groups attached to the Si atoms shield the chain backbone, limiting intermolecular bonding between the chains. The strong Si–O bonds (bond energy ~443 kJ/mol) in the chain backbone, similar to those in inorganic silicate glasses (Chapter 7), together with the organic CH_3 side groups make PDMS very stable against hydrolytic attack in aqueous media.

PDMS has a low surface tension (20–24 mN/m) because it presents an organic surface due to the CH_3 side groups attached to the chain. This accounts for the ability of PDMS adhesives to wet most solid surfaces. PDMS is hydrophobic, has a large contact angle with water droplets (105–110°), and is not wetted by water. Because of its crosslinked structure and CH_3 side groups, PDMS has a higher free volume than many organic polymers and, thus, is highly permeable to the migration of gases such as O_2 and N_2, and of water vapor through it.

The mechanical properties of PDMS elastomers depend on the temperature (called the curing temperature) used to form the material from its reactants. When cured at room temperature (~25 °C), commercial PDMS elastomers have shown a tensile strength of ~5.0 MPa and a Young's modulus of ~1.3 MPa. These values increase to ~6.0 and ~2.0 MPa, respectively, for a curing temperature of ~100 °C. The strain to failure is greater than ~150% and ~100%, respectively, for curing at room temperature and at 100 °C. For the same curing temperature, the strength and Young's modulus of PDMS are much higher in compression. For example, when cured at room temperature, the same commercial PDMS showed a compressive strength of ~50 MPa and Young's modulus ~190 MPa.

Applications of PDMS

The high gas (oxygen) permeability of PDMS elastomers is partly responsible for their use as gas permeable membranes in artificial respirators and as contact lenses. As it can be easily fabricated to provide a range of desirable mechanical properties, PDMS is widely used in the creation of microfluidic devices, composed of well-controlled micro-patterned surfaces or channels. These devices have been applied in several areas of biomedical research and development, such as cell culture, regenerative medicine, and medical diagnostics. For example, microfluidic devices provide unique substrates to study the influence of microenvironments on cellular responses, an understanding of which is vital for the creation of engineered tissues.

PDMS implants have good stability *in vivo* and they are biocompatible, although they elicit a mild inflammatory response. Commonly referred to as silicone, PDMS has been widely used as implants for breast reconstruction or enlargement, although its use in this application has been associated with controversy and litigation. Silicone breast implants take advantage of the capacity to convert PDMS into various forms, such as gels and elastomers. For example, breast implants typically consist of a PDMS elastomer shell that encapsulates a PDMS gel to achieve the desired deformability and shape retention.

8.7 Special Topic: Polyurethanes

Polyurethanes (PUs) are polymers composed of macromolecules in which the repeating units are linked together by the urethane group –NH–(C=O)–O–, also referred to as the carbamate group. PUs can be formulated to provide a range of physicochemical properties, from rigid thermoplastics to soft elastomers, which can be degradable or nondegradable. Thus, PUs cannot be appropriately classified under any one of these individual polymer types.

An understanding of the molecular chemistry and molecular structure of PUs is central to understanding how their physicochemical properties arise. PUs can be synthesized by several routes but a commonly used method involves a condensation reaction between a diisocyanate and a diol, such as the reaction between hexamethylene diisocyanate and adipic acid (Figure 8.23). This reaction gives an alternating copolymer in which the monomer units denoted A and B have a simple A–B–A–B repeating pattern. Except for a difference in the linking group (urethane versus amide), the molecular pattern of this alternating polyurethane copolymer is similar to that of nylon 6.6 (Figure 8.21a). Thus, we might expect that this PU would have mechanical properties that are not significantly different from those of nylon 6.6. However, most commercial PUs are not alternating copolymers but, instead, are block copolymers in which the molecules are composed of alternating segments of either A or B (Figure 8.2).

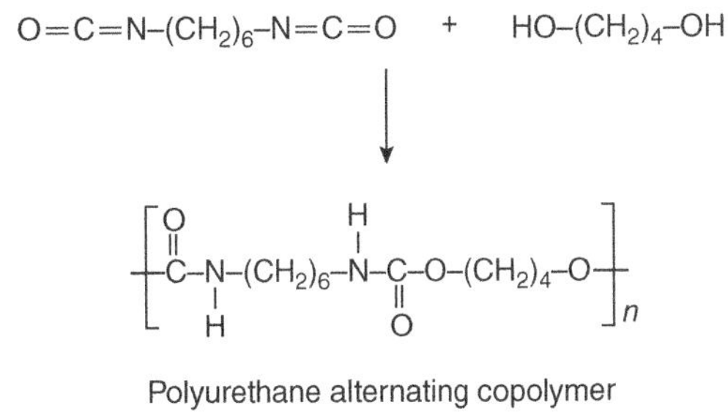

Figure 8.23 Illustration of the synthesis of a polyurethane alternating copolymer by a condensation reaction between hexamethylene diisocyanate and adipic acid.

PU block copolymers are designed such that one of the segments, referred to as the hard segment, is composed of hard (stiff) repeating units and, thus, generally has a semicrystalline or glassy structure. The other segment, the soft segment, is composed of flexible monomeric units and is generally in a rubbery state. In some PUs, the soft phase can be designed to contain a crystallizable segment, which provides an additional component to improve their load-bearing capacity.

Proper selection of the monomer units and polymer synthesis conditions is a key to the production of PUs having the desired physicochemical properties. Control of these properties, particularly mechanical properties, can often be challenging because they depend on several factors, such as

- Monomer composition of the hard and soft segments
- Molecular weight of the hard and soft segments
- Overall molecular weight of the macromolecule composed of hard and soft segments
- Ratio of hard segment content to soft segment content
- Degree of ordering of the hard and soft segment molecules, such as their degree of crystallization or orientation.

8.7.1 Production of Polyurethanes

As illustrated in Figure 8.24, PU block copolymers are often synthesized using a two-step process. The first step involves reacting a diisocyanate with a polyol, an organic compound containing multiple hydroxyl groups, to form a prepolymer. Then the prepolymer is reacted with a chain extender to form high molecular weight PU molecules.

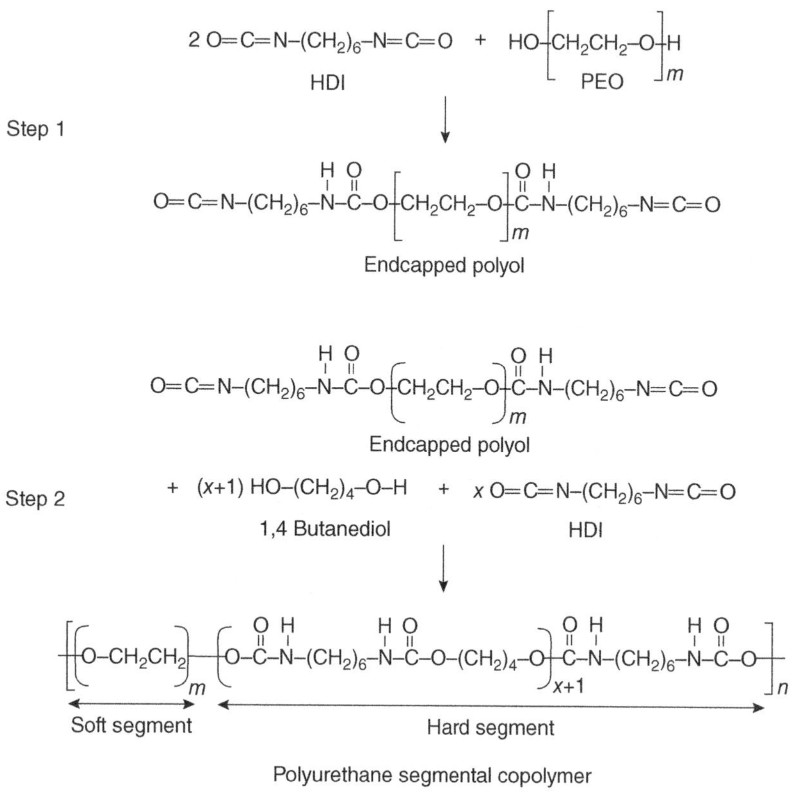

Figure 8.24 Synthesis of a polyurethane segmental block copolymer by a two-step process involving hexamethylene diisocyanate (HDI), polyethylene oxide (PEO), and 1,4 butanediol.

Isocyanates used in synthesizing PUs have two or more isocyanate (–N=C=O) end groups in each molecule. They can be aromatic or aliphatic, although aromatic diisocyanates are more often used, particularly in the production of nondegradable PUs. Commonly used isocyanates are methylene diphenyl diisocyanate (MDI), toluene diisocyanate (TDI), and hexamethylene diisocyanate (HDI), which make up the hard segments (Figure 8.25). Polyether or polyester diols generally make up the soft segments. These diols generally have a low molecular weight (~1–2 kDa). At room temperature, they are well above their T_g and, thus, they are in a rubbery state. Low molecular weight polyethylene oxide (PEO), also referred to as PEG, for example, has a T_g of ~ −55 to −60 °C. The rubbery response of these soft segments plays a strong role in modifying the mechanical response of the PU. Monomers containing more than two functional groups can be used to produce crosslinked structures.

Commonly, molecules of aliphatic diols or diamines, containing 2–6 carbon atoms, such as 1,4-butanediol (BDO), are used as chain extenders to synthesize PU molecules of a high enough molecular weight. The hard segment and the chain extender together make up the hard phase content of the polymer, often expressed as a fraction (percent) of the total mass of the polymer.

8.7.2 Structure–Property Relations in Polyurethanes

PUs synthesize from a given combination of diisocyanate and diol have a microstructure that depends on the molecular weight of the hard and soft segments of the macromolecule and the fraction of hard phase in the polymer. This microstructure, in turn, determines the mechanical behavior of the PU. To illustrate the relationship between microstructure and mechanical properties, we can take a polymer consisting of hard segments composed of MDI, soft segments composed of a polyether diol such as polytetramethylene oxide (PTMO) of molecular weight ~1–2 kDa and T_g ~ −80 °C, and a chain extender such as 1,4-butanediol.

A low fraction of hard segments (less than ~20 wt%) generally leads to an overall amorphous structure composed of interspersed hard and soft segments, which gives low tensile strength, low Young's modulus, and high extensibility typical of an elastomer (Figure 8.26). As the hard fraction increases, a sufficient amount of the hard segments nucleate and crystallize, forming isolated small crystalline regions. These crystals reinforce the soft matrix and physically crosslink the mobile molecules of the amorphous phase, giving a mechanical response comparable to that of a crosslinked rubber. A further increase in

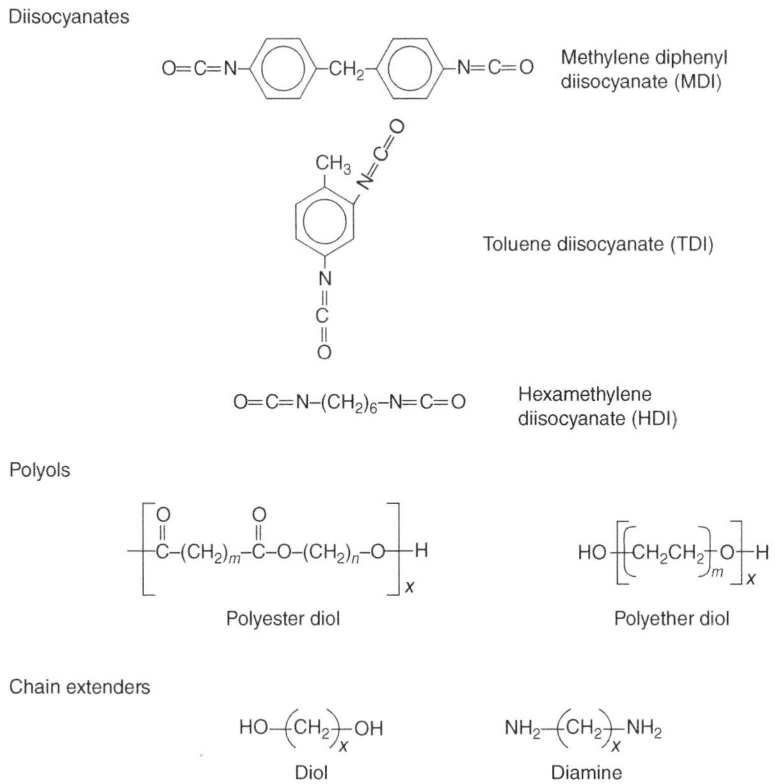

Figure 8.25 Chemical structures of common diisocyanates, polyols, and chain extenders used in the synthesis of polyurethane segmental block copolymers.

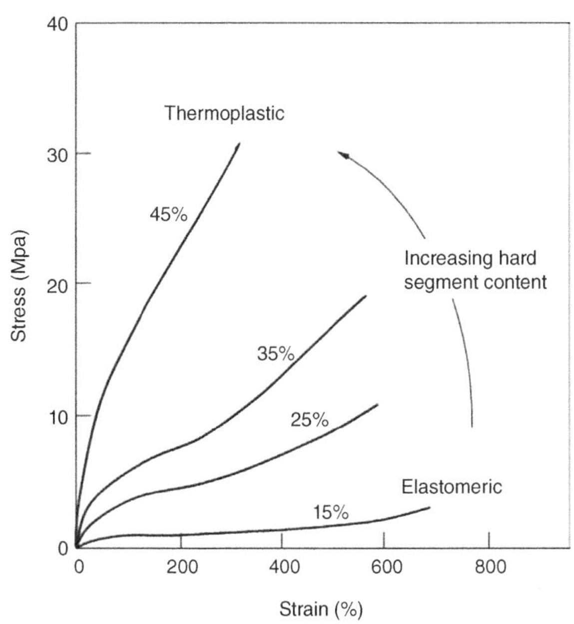

Figure 8.26 Schematic representation of the effect of hard segment content (15–45 wt%) on the tensile mechanical response of a polyurethane segmental block copolymer consisting of hard segments composed of methylene diisocyanate, soft segments composed of polytetramethylene oxide (PTMO) and a chain extender composed of 1,4-butanediol (BDO). *Source:* Adapted from Abouzahr et al. (1982).

the hard segment fraction (greater than ~35–40 wt%) leads to the formation of longer hard segments, which, in turn, leads to crystalline regions of larger size that may eventually touch. The polymer shows a mechanical response comparable to that of a thermoplastic, that is, it shows high tensile strength and high Young's modulus but a reduced extensibility.

8.7.3 Chemical Stability of Polyurethanes *in vivo*

PUs originally used in biomedical applications were considered to have sufficient chemical stability in the aqueous physiological medium *in vivo*, even though the urethane group –NH–(C=O)–O– was known to have a low susceptibility to hydrolytic attack. However, the long-term stability of these PUs came under considerable scrutiny following their failure in critical applications such as coatings on pacemaker leads and outer layers of silicone breast implants. Studies showed that these PUs were prone to degradation not just by hydrolytic mechanisms but by enzymatic, oxidation, and cellular mechanisms as well (Chapter 15). A better understanding of PU degradation mechanisms derived from these studies led to the design of a variety of compositions, some of which have improved chemical stability for long-term use *in vivo* whereas others have a controllable degradation rate for use in short-term applications such as tissue engineering.

For long-term implantation, PUs are generally synthesized using aromatic diisocyanates such as MDI or TDI for the hard segments, and polyether diols that have a higher number of CH_2 groups than PEO in their repeating units because of their resistance to hydrolytic and oxidative attack. In comparison, the toxicity of their degradation products is of critical importance for degradable PUs designed for use as implantable devices. Because of concerns that aromatic diisocyanates might lead to the formation of carcinogenic degradation products, aliphatic diisocyanates such as HDI or lysine diisocyanates are preferred for the hard segments. PLA, PGA, and polycaprolactone (PCL) have been investigated for use as the soft segments because they degrade to form nontoxic products (Chapter 9).

8.7.4 Biomedical Applications of Polyurethanes

PUs are used in a variety commercial and industrial application such as adhesives, coatings, rigid and flexible foams, and textile fibers, and they find considerable use in biomedical applications as well. PUs have good compatibility with blood and tissues, are good electrical insulators and, we have seen, they can be formed into products having a wide range of physicochemical properties. Nondegradable PUs find considerable use in a variety of biomedical applications such as electrical insulating layers on pacemaker leads, outer layers on breast implants, catheters, vascular grafts, and wound dressings. Degradable PUs are being developed for use as scaffolds in tissue engineering and as devices for drug delivery.

8.8 Water-soluble Polymers

Water-soluble polymers have a measurable solubility in an aqueous medium, and the dissolved molecules have a chemical composition and molecular weight that are similar to those of the solid polymer. This can be visualized as similar to the dissolution of sugar in water, where the sugar molecules dissolve but do not change their composition. Water-soluble polymers by themselves find limited use in biomedical applications, mainly in drug and gene delivery. However, in the form of hydrogels composed of crosslinked macromolecules (Chapter 11), they find wider use in a variety of applications, particularly drug delivery and tissue engineering.

As H_2O molecules have a high degree of polarity, water-soluble polymers are generally composed of monomer units that contain one or more of the following:

- Chemical groups that have a high degree of polarity
- Atoms having one or more pairs of unshared electrons in their outer (valence) shell
- Acidic or basic groups that can be deprotonated or protonated to give positively or negatively charged atoms.

The structures of common water-soluble polymers relevant to biomaterials are illustrated in Figure 8.27.

Molecules of PEG, also known as PEO, have been used to modify the surface of a variety of pharmaceutical agents, ranging from proteins to nanoparticles, a process sometimes referred to as PEGylation (Chapter 26). In addition to modifying the surface chemistry of these agents, PEGylation makes them larger. Together, this chemical and physical modification is used to improve the effectiveness of the pharmaceutical agents.

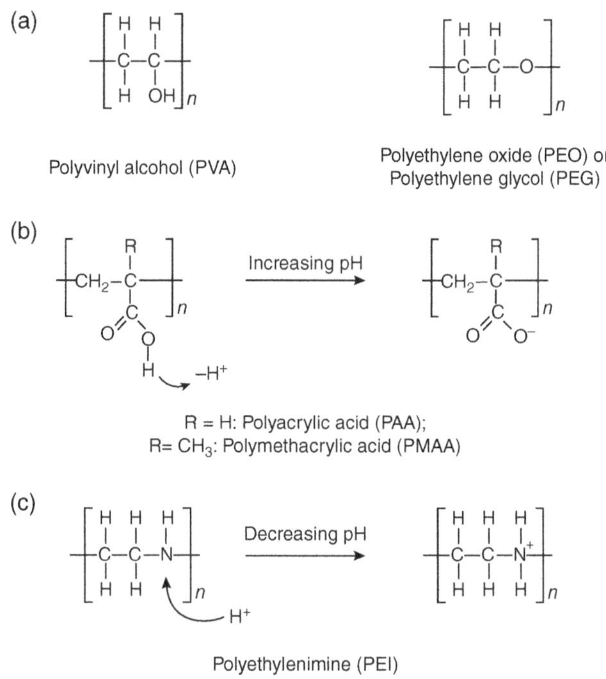

Figure 8.27 Chemical structures of common water-soluble polymers. (a) Polyvinyl alcohol (PVA) and polyethylene oxide (PEO) composed of polar molecules. (b) Polyacrylic acid (PAA) and polymethacrylic acid (PMAA) composed of monomer units in which the hydroxyl (OH) group becomes increasingly deprotonated with increasing pH, resulting in negatively charged (anionic) molecules. (c) Polyethylenimine (PEI) composed of monomer units in which the nitrogen (N) atom becomes increasingly protonated with decreasing pH, resulting in positively charged (cationic) molecules.

The hydroxyl group in PAA or PMAA molecules starts to become deprotonated at pH ~3 and, at the pH of the physiological fluid (~7.4), PAA and PMAA molecules acquire a high negative charge, forming anionic molecules. In comparison, molecules of polyethylenimine (PEI) start to become protonated at pH ~10 and the degree of protonation increases with decreasing pH, forming cationic molecules at the physiological pH. PEI has been used as a delivery system for negatively charged nucleic acids, although they may be associated with some degree of cytotoxicity (Chapter 26).

8.9 Concluding Remarks

In this chapter, we considered several physical and chemical principles that are fundamental to solid polymers. Then common types of synthetic nondegradable polymers relevant to biomaterials were discussed in three main groups, referred to as nondegradable thermoplastics, elastomers and water-soluble polymers. Polyurethanes, polymers that are designed to provide a range of behavior, were described as a special topic.

- When compared to metals and ceramics, synthetic polymers have a significant advantage in simplicity and ease of production, and their mechanical properties can be easily modified to make them more comparable to those of noncalcified tissues and nonstructural bone.
- Although very different from metals and ceramics, composition and structure still form the basis for understanding the physicochemical properties of solid polymers and for designing polymers having desirable properties.
- Molecular conformation and molecular mobility, which are dependent on chemical composition, along with molecular weight determine the mechanical properties of solid polymers.
- Thermoplastic polymers are amorphous or semicrystalline. The glass transition temperature T_g and the melting temperature T_m, which are themselves dependent on the composition of the polymer, determine the temperature range up to which an amorphous polymer and a semicrystalline polymer, respectively, can be usefully applied.
- Nondegradable thermoplastic polymers have played a key role in the success of long-term implantable devices and they continue to see considerable use in these applications. In the next chapter, we shall discuss degradable synthetic polymers, which also find widespread use as biomaterials.

Problems

8.1 Distinguish between linear, branched, and crosslinked polymer chains.

8.2 Distinguish between thermoplastic and thermosetting polymers.

8.3 Give the chemical formulae of the polymers formed by an addition reaction from the following monomers:

A $H_2C{=}CHF$
B $H_2C{=}C(CH_3)_2$
C $H_2C{=}CH(OH)$

In each case, give the molecular weight of a polymer having a degree of polymerization of 1000. Which of these monomers is capable of forming isotactic or syndiotactic sequences?

8.4 The repeating unit in PMMA is shown in Figure 8.9. Determine the molecular weight of a PMMA molecule composed of 1000 repeating units.

8.5 A polymer is composed of 5 chains having a degree of polymerization (DP) = 100, 5 chains having a DP = 150, and 5 chains having a DP = 200. What is the average DP of the polymer? If the molecular weight of each repeating monomer unit in the chain is 100 Da, determine the number average molecular weight, weight average molecular weight and polydispersity of the polymer.

8.6 Data for the molecular weight (MW) of a polystyrene (PS) specimen are as follows:

MW (kDa)	15	27	39	56	78	104	120	153
Weight (g)	0.10	0.18	0.25	0.17	0.12	0.08	0.06	0.04

Determine (a) the number average molecular weight, (b) the weight average molecular weight, and (c) the polydispersity of the specimen.

8.7 Discuss how crosslinking a polymer composed of linear chains influences the following properties: (a) ability to crystallize, (b) mechanic properties, and (c) solubility.

8.8 A linear polyethylene (PE) molecule has a molecular weight of 10^5 Da.

A Determine the length of the fully extended chain, assuming no distortion of the bond angles or bond lengths. (Length of C-C single bond = 0.154 nm)

B Determine the root mean square end-to-end distance based on a freely jointed chain model and the ratio of the root mean square end-to-end distance to the fully extended chain length.

8.9 Explain what is meant by the term glass transition temperature of a polymer. Is the glass transition temperature the same as the melting temperature? If not, explain the difference.

8.10 Polymers show a wide range of glass transition temperatures. Briefly discuss how the molecular structure of a polymer influences its glass transition temperature.

8.11 Briefly discuss the nucleation and crystallization process in a polymer such as polyethylene composed of linear molecules.

8.12 The specific volumes of the crystalline phase v_c and the amorphous phase v_a in PE are $v_c = 0.99\,cm^3/g$ and $v_a = 1.17\,cm^3/g$. Determine the crystallinity (in weight percent) of a PE specimen that has a specific volume of $1.04\,cm^3/g$.

8.13 Does the molecular weight of a polymer influence (a) its glass transition temperature and (b) its melting temperature? Explain.

8.14 Is the glass transition temperature of an amorphous polymer different from that of a semicrystalline polymer of the same composition? Explain.

8.15 In the same figure, sketch the Young's modulus as a function of temperature for an amorphous polymer and a semicrystalline polymer that have the same chemical composition. Indicate the main regions in each modulus.

8.16 Polyethylene liners in hip implants are now commonly made from crosslinked UHMWPE instead of conventional UHMWPE. Explain why.

8.17 Starting from UHMWPE, briefly describe a method for producing crosslinked UHMWPE having mechanical properties suitable for use as bearings in hip implants.

8.18 State three conditions that are necessary for a polymer to show elastomeric behavior.

8.19 The chain backbone in PDMS is composed of polar Si–O bonds, similar to those in an inorganic silicate glass such as window glass. However, PDMS is hydrophobic and shows elastomeric behavior whereas silicate glass is hydrophilic, hard, and glassy. Explain.

8.20 Polyurethanes show a wide range of mechanical behavior. Discuss how the molecular structure of polyurethanes influences their mechanical behavior at room temperature.

References

Abouzahr, S., Wilkes, G.L., and Ophir, Z. (1982). Structure-property behavior of segmented polyether-MDI-butanediol based urethanes: effect of composition ratio. *Polymer* 23: 1077–1086.

Begovac, P.C., Thompson, R.C., Fisher, J.L. et al. (2003). Improvements in GORE-TEX vascular graft performance by Carmeda bioactive surface heparin immobilization. *European Journal of Vascular and Endovascular Surgery* 25: 432–437.

Simis, K.S., Bistolfi, A., Bellare, A., and Pruitt, L.A. (2006). The combined effect of crosslinking and high crystallinity on the microstructural and mechanical properties of ultrahigh molecular weight polyethylene. *Biomaterials* 27: 1688–1694.

Further Reading

Billmeyer, F.W. (1984). *Textbook of Polymer Science*, 3e. New York: Wiley.

Johnston, I.D., McCluskey, D.K., Tan, C.K.L., and Tracey, M.C. (2014). Mechanical characterization of bulk Sylgard 184 for microfluidics and microengineering. *Journal of Micromechanics and Microengineering* 035017 (7pp): 24.

Korley, L.T.J., Pate, B.D., Thomas, E.L., and Hammond, P.T. (2006). Effect of soft and hard segment ordering on the morphology and mechanical behavior of semicrystalline segmented polyurethanes. *Polymer* 47: 3073–3082.

Kurtz, S.M., Villarraga, M.L., Herr, M.P. et al. (2002). Thermomechanical behavior of virgin and highly crosslinked ultra-high molecular weight polyethylene used in total joint replacements. *Biomaterials* 23: 3681–3697.

Kurtz, S.M. and Devine, J.N. (2007). PEEK biomaterials in trauma, orthopedic, and spinal implants. *Biomaterials* 28: 4845–4869.

McCrum, N.G., Buckley, C.P., and Bucknall, C.B. (1988). *Principles of Polymer Engineering*. Oxford (UK): Oxford University Press.

Ratner, B.D., Hofmann, A.S., Schoen, F.J., and Lemons, J.E. (eds.) (2013; Chapter 1.2.2). *Biomaterials Science: An Introduction to Materials in Medicine*, 3e, 64–111. New York: Elsevier.

Santerre, J.P., Woodhouse, K., Laroche, G., and Labow, R.S. (2005). Understanding the biodegradation of polyurethanes: from classical implants to tissue engineering materials. *Biomaterials* 26: 7457–7470.

9

Synthetic Polymers II: Degradable Polymers

9.1 Introduction

Nondegradable synthetic polymers, as we have seen (Chapter 8), have played a key role in the success of long-term implantable devices, and they continue to see considerable use in these applications. In comparison, degradable polymers provide significant advantages in short-term applications that require only the temporary presence of an implant. Synthetic degradable polymers generally have structural and physical properties such as molecular weight, molecular conformation, and mechanical properties that are generally comparable initially to those of nondegradable polymers. However, they are normally composed of molecular chains containing chemical bonds that are susceptible to chemical degradation by hydrolysis, enzymatic reactions, or a combination of the two. Upon implantation, scission of these bonds leads to the formation of small molecules (oligomers) and monomers that are soluble in an aqueous environment. Consequently, the polymer suffers a reduction in molecular weight, mass, and mechanical properties with time. Ideally, after serving its function, the polymer should degrade away completely in the physiological environment without causing toxic effects.

Synthetic degradable polymers based on the ester bond were developed for use as sutures and bone fracture fixation devices nearly 50 years ago. Since then, several types of degradable polymers have been synthesized, providing a wide range of degradation rates and physicochemical properties for use in a variety of biomedical applications (Table 9.1). Advances in the last few decades, particularly in areas such as drug delivery and tissue engineering (or regenerative medicine), have relied considerably on the use of degradable polymers.

In this chapter, we will cover synthetic degradable polymers that are relevant to biomaterials. Hydrogels and natural polymers are discussed in subsequent chapters. The following topical areas will be covered:

- Polymer degradation principles
- Common degradable polymers used as biomaterials
- How the composition and structure of degradable polymers influence their properties.

9.2 Degradation of Polymers

Degradation of polymers refers to a chemical process in which the macromolecular chain backbone of the polymer undergoes scission, forming smaller molecules and, eventually, monomer units (Figure 9.1). There are four major modes of polymer degradation, referred to as photodegradation, thermal degradation, mechanical degradation, and chemical degradation. Photodegradation can occur during exposure of the polymer to radiation, such as ultraviolet (UV) light or gamma (γ) radiation as, for example, in the use of UV light to reduce bacterial contamination during processing or the use of γ radiation for sterilization. Polymers subjected to elevated-temperature processing and to mechanical stresses during processing or during use can undergo thermal and mechanical degradation. However, the most important degradation mode for polymers relevant to biomaterials is generally chemical degradation.

Degradable polymers used as biomaterials can undergo chemical degradation by two principal reactions, by hydrolysis (that is, chemical attack by water molecules) and by enzyme-mediated hydrolysis. When the appropriate enzymes are present, natural polymers such as proteins and polysaccharides (Chapter 10) degrade predominantly by enzymatic-mediated hydrolysis. In comparison, hydrolysis is the most important degradation mechanism for most synthetic polymers. The various polymer degradation mechanisms are discussed in more detail in Chapter 15. In this chapter, we will focus on the degradation of synthetic polymers by hydrolysis.

Degradable synthetic polymers are insoluble in water but their molecular chain backbone contains chemical bonds that are susceptible to hydrolysis. Upon exposure to an aqueous medium, hydrolytic scission of these bonds occurs at

Materials for Biomedical Engineering: Fundamentals and Applications, First Edition. Mohamed N. Rahaman and Roger F. Brown.

Companion website: www.wiley.com/go/rahaman/biomedicalengineering

Table 9.1 Degradable synthetic polymers used or investigated for use as biomaterials and medical products.

Polymer	Applications
Poly(α-hydroxy esters): polyglycolic acid (PGA), polylactic acid (PLA), and polylactic-co-glycolic acid (PLGA)	Sutures; orthopedic devices (fracture fixation); drug delivery; hormone delivery; skin replacement materials; stents; staples; injectable fillers; barrier membranes; dura mater substitutes; scaffolds for tissue engineering
Polycaprolactone (PCL)	Long-term drug delivery; implantable contraceptive drug devices; orthopedic devices; staples
Polyanhydrides	Drug delivery
Poly(ortho esters)	Drug delivery; stents
Polydioxanone	Sutures; fracture fixation devices for nonstructural bone; wound clips
Polyhydroxyalkanoates: poly-3-hydroxybutyrate (P3HB), poly-3-hydroxyvalerate (P3HV), and their copolymers P(3HB-3HV)	Long-term drug delivery; orthopedic applications; stents; artificial skin; surgical patching materials for congenital heart defects; sutures
Poly(propylene fumarate) (PPF)	Orthopedic applications; scaffolds for tissue engineering
Polyketals: poly(cyclohexane-1,4-diyl acetone dimethyl ketal) (PCADK)	Drug delivery
Poly(polyol sebacates): poly(glycerol sebacate) (PGS)	Soft tissue repair and regeneration: cardiac muscle, vascular tissue engineering, cartilage, nerve conduits, and retina; drug delivery
Polycarbonates: poly(trimethylene carbonate) (PTMC) and copolymers	Sutures; orthopedic applications
Tyrosine-derived polycarbonates	Orthopedic applications: fracture fixation devices for nonstructural bone

Source: Adapted from Treiser et al. (2013).

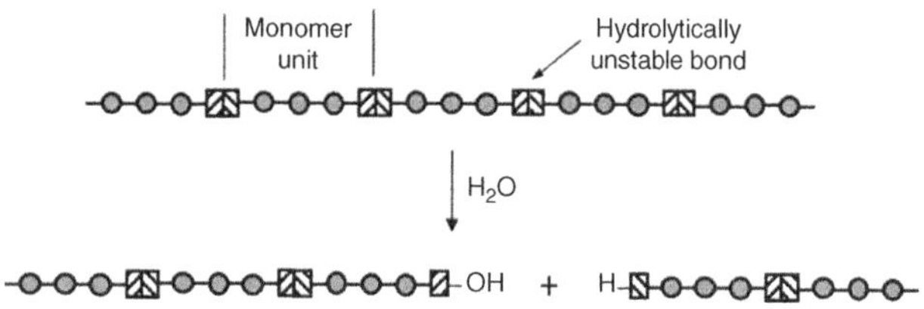

Figure 9.1 Schematic illustration of the cleavage of a hydrolytically unstable bond in a macromolecular chain backbone by a water molecule.

random points in the chain backbone, resulting eventually in the formation of small molecules (oligomers) or monomers that are soluble in the aqueous environment. Consequently, degradation leads to a reduction in the average molecular weight of the polymer, along with a reduction in its strength. Dissolution of the oligomeric or monomeric degradation products into the aqueous medium leads to erosion of the polymer, that is, to a decrease in its mass.

Changes in molecular weight, mass, and strength have been used as parameters to characterize polymer degradation as a function of time, but they do not all follow the same kinetics. For example, some polymers show a large decrease in molecular weight and tensile strength prior to any observable decrease in mass (Section 9.5). As degradation is a chemical process that involves chain scission, the most important measure of degradation is the reduction in molecular weight of the polymer as a function of time. In comparison, a reduction in mass as a function of time is the most important measure of erosion because it involves transfer of mass (oligomers or monomers) from the polymer to the aqueous environment. Strength provides a measure of the capacity of the degrading polymer to support a mechanical stress.

9.3 Erosion of Degradable Polymers

Degradation of synthetic polymers, as noted earlier, is a chemical process of molecular chain scission by hydrolysis. Degradation is accompanied by a physical process, referred to as erosion, in which the oligomeric or monomeric degradation products dissolve in the aqueous environment, leading to a change in mass, geometry, microstructure, and integrity of the specimen. Like degradation, erosion is a crucial factor in the design and use of degradable polymers for biomedical applications. For example, erosion can influence how drugs are released from a device, how the strength of bone fracture plates and screws decrease with time, and the function of scaffolds in engineering tissues and organs.

Erosion is often described in terms of two extreme types, referred to as surface erosion and bulk erosion, although several degradable polymers show some combination of the two. Surface-eroding polymers lose material from their

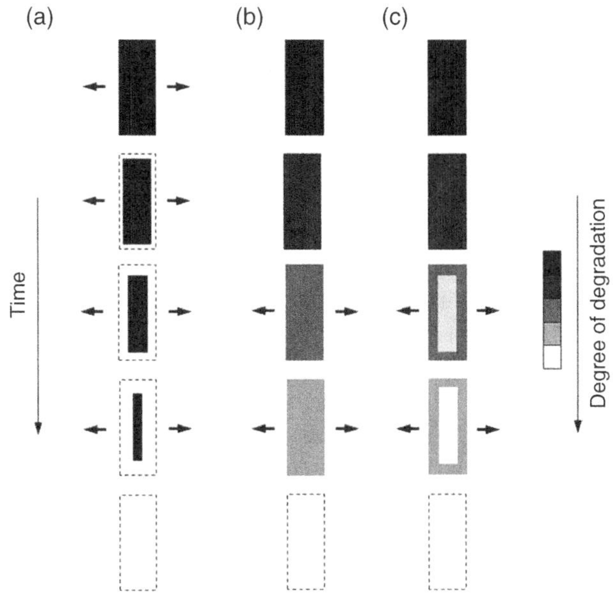

Figure 9.2 Schematic illustration of (a) surface erosion, (b) bulk erosion, and (c) bulk erosion with autocatalysis. Arrows represent dissolution of low molecular weight degradation products from the specimen into the aqueous environment.

surface only, becoming smaller with time, but they generally keep their original shape like, for example, a bar of soap (Figure 9.2a). In comparison, bulk erosion occurs throughout the volume of the specimen (Figure 9.2b). The specimen generally maintains its mass, external dimensions, and integrity over some period of the degradation process, after which it suffers a rapid loss in mass and, sometimes, in integrity. A cube of adherent sugar crystals placed in water provides an approximation of this type of erosion. Whether a polymer erodes by surface erosion or bulk erosion has important consequences for its use in a variety of biomedical applications (Section 9.6). For example, the type of erosion influences the capacity to control the release of a drug from a device, or the decrease in strength of a fracture fixation device or a suture with time.

9.4 Characterization of Degradation and Erosion

The most important measure of degradation, as noted earlier, is the reduction in molecular weight as a function of time, whereas a reduction in mass as a function of time is the most important measure of erosion. Reduction in tensile strength provides a measure of the capacity of a specimen to maintain its mechanical integrity over time and is dependent on the degradation and erosion profile of the specimen.

The degradation, erosion, and strength of a polymer *in vitro* are commonly determined by immersing appropriate specimens of a specific geometry, such as a thin disc or thin rectangular sheet, in an aqueous solution, such as a phosphate-buffered solution, at ~37 °C and pH ~7.4 (the physiological temperature and pH). At appropriate times, the requisite number of specimens is removed from the solution and dried in a vacuum. Then their molecular weight, mass and, if required, other properties such as tensile strength, Young's modulus, water absorption, and crystallinity, are measured using standard methods.

9.5 Factors Controlling Polymer Degradation

Biomedical applications of degradable polymers depend critically on their degradation and erosion. Consequently, an understanding of polymer degradation and erosion is critical in selecting polymers for specific applications and in designing polymers having optimal properties and performance for specific applications. Several factors influence polymer degradation, such as the type of hydrolytically unstable bond in the molecular chain backbone, the pH of the aqueous medium, copolymer composition, conformation of the molecular chains, molecular weight of the polymer, and the hydrophobicity (or hydrophilicity) of the monomer units in the chain. Physical and chemical changes that accompany degradation, such as pH changes within the interior of a polymer specimen and crystallization of oligomeric or monomeric degradation products can also have a significant effect on degradation.

9.5.1 Chemical Structure

The type of chemical bond in the molecular chain backbone has the greatest effect on polymer degradation because degradation proceeds by chain scission due to hydrolytic attack of these bonds. Consequently, there have been several approaches to classify degradable polymers based on the susceptibility of the chemical bonds in their chain backbone to hydrolytic cleavage. Figure 9.3 presents a list of some different types of chemical bonds present in degradable polymers that are susceptible to hydrolytic attack. The reactivity of these bonds has been determined as a reaction rate constant λ from the hydrolysis kinetics of specific polymers or by extrapolation of data from low molecular weight compounds having the same functional groups, typically in a buffer solution at ~37 °C and pH ~7.4. The rate constants cover a wide range, approximately 10 orders of magnitude, depending on the type of bond. Anhydride bonds, for example, are considerably more reactive than ester bonds, whereas amide bonds have a low reactivity.

Polymers that contain bonds having a higher reactivity (higher rate constant λ) will generally degrade faster under similar conditions. However, the rate constants in Figure 9.3 should be viewed with a high degree of caution because they can change considerably (by up to a few orders of magnitude) in response to various factors. These factors include changes in the chemical groups in the immediate neighborhood of the hydrolytically unstable bond, the pH of the aqueous medium, and the molecular weight and crystallinity of the polymer. For example, hydrolysis of the ester bond in the poly(α-hydroxy esters) such as polylactic acid (PLA) and polyglycolic acid (PGA), is catalyzed by acids or bases and, consequently, changes in pH can lead to large changes in their degradation (Section 9.8.1). Because of the presence of methyl (CH_3) side groups adjacent to the ester bonds in its molecular chain, PLA presents a more hydrophobic surface to an aqueous environment when compared to PGA. This reduces penetration (diffusion) of water molecules into the interior of PLA, reducing access of the ester bonds to water molecules that are required for hydrolytic cleavage. Consequently, PLA degrades more slowly than PGA even though they contain the same hydrolytically unstable ester bond in their chain backbone.

Polymer	Chemical structure	Degradation rate constant λ (1/s)
Polyanhydride	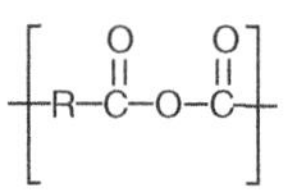	2×10^{-3}
Polyketal	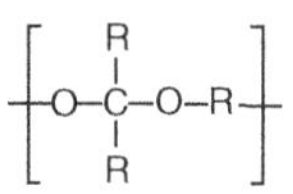	6×10^{-5}
Polyorthoester	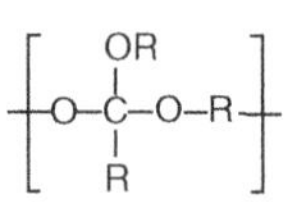	5×10^{-5}
Polylactic acid	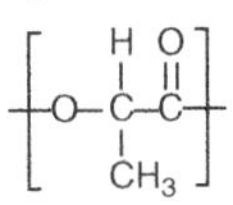	1×10^{-7}
Polycaprolactone	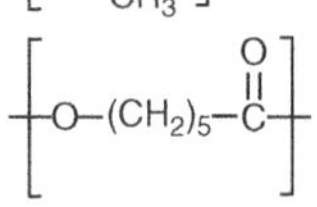	8×10^{-8}
Polyacetal	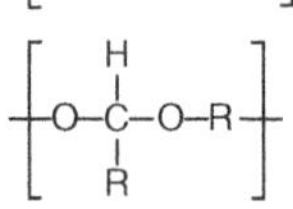	3×10^{-8}
Polyamide	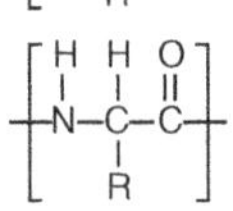	3×10^{-13}

Figure 9.3 Chemical structure and degradation rate constants for various degradable polymers. *Source:* von Burkersroda et al. (2002); Park et al. (1993).

9.5.2 pH

Changes in pH of the aqueous environment can have a strong effect in modifying the degradation of some polymers, by as much as a few orders of magnitude. This is important because the pH of various fluids and tissues in the human body can vary over a significant range, which can influence the performance of devices used as sutures or in drug delivery. For example, the pH of the gastric fluid in the stomach can reach values as low as ~1–3, the pH of the skin is ~4–6, and the pH of infected tissue can reach values as low as ~4–5. On the other hand, the ability to control the pH within certain degradable polymers, such as poly(ortho esters), has been used to control their degradation and, thus, improve their performance (Section 9.8.4).

The effect of pH on degradation has been investigated for a variety of degradable polymers. Poly(α-hydoxy esters) such as PLA, PGA, and their copolymer polylactic-co-glycolic acid (PLGA) provide a classic example of the significant effect that pH can have on the degradation of some polymers (Section 9.8.1). Hydrolysis of the ester bond in these polymers is catalyzed (accelerated) by acidic or basic pH, that is, by an increase in the hydronium (H_3O^+) or hydroxyl (OH^-) ion concentration in the aqueous environment when compared to a neutral pH (Figure 9.4).

As an example of the effect of pH on degradation, Figure 9.5 shows data for the molecular weight and mass as a function of time for PLGA, an amorphous copolymer (molecular weight 47 kDa) synthesized from equimolar amounts of D- and L-lactic acid and glycolic acid. A change in the pH of the aqueous medium from 7.4 to >13.0 leads to a dramatic change in degradation (reduction in molecular weight) and erosion (reduction in mass). Whereas the molecular weight decreased over a period of 14–21 days at pH = 7.4, it remained approximately constant at pH > 13. The mass remained nearly constant over 14–21 days at pH = 7.4 but disappeared almost completely within ~14–21 hours at pH > 13.

PLA, PGA, and PLGA also show a pH-dependent degradation phenomenon, referred to as autocatalysis (Figure 9.2c). Degradation of theses polymers leads to the generation of soluble acidic monomers (lactic acid, glycolic acid, or a combination of the two, depending on the polymer composition) (see Figure 9.9) that themselves cause a reduction in the local pH within the specimen, accelerating degradation in the interior of the specimen. This can cause inhomogeneous degradation in which the interior region of the specimen degrades considerably faster than the surface region. Inhomogeneous degradation is often undesirable, particularly in drug delivery devices, in which predictable degradation is critical.

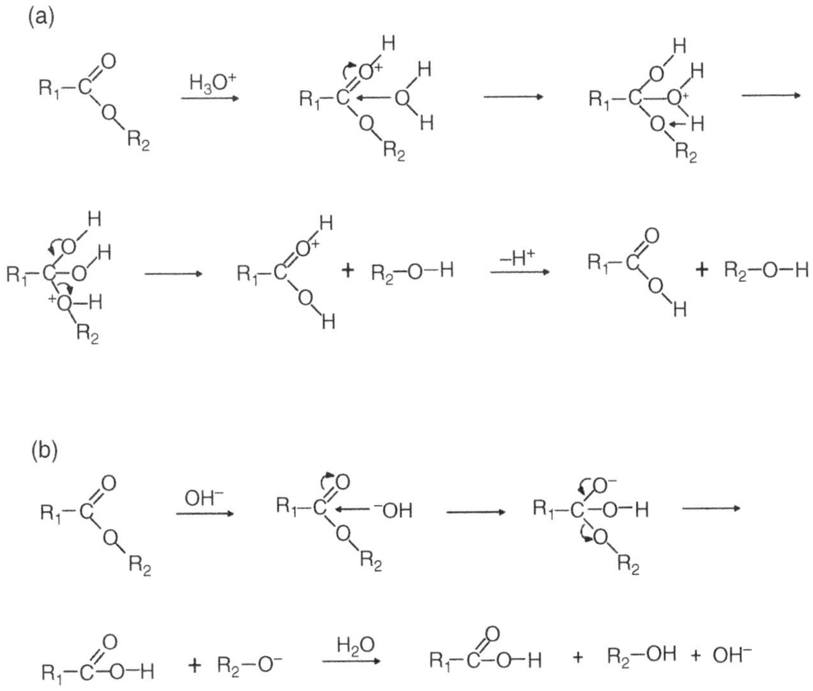

Figure 9.4 Simplified illustration of (a) acid-catalyzed and (b) base-catalyzed ester hydrolysis mechanism. Because of the polarity of the carbonyl bond, hydrogen H^+ (or hydronium H_3O^+) ions attack at the O atom (slight negative charge or two lone pair electrons) (electrophilic mechanism) whereas hydroxyl HO^- ions attack at the carbon atom (slight positive charge) (nucleophilic mechanism).

On the other hand, changes in the internal pH has been used as a way to control the degradation of a few polymers and improve their performance, particularly in drug delivery devices. One way to control the internal pH of a polymer is to physically add an acidic or basic substance (sometimes referred to as an excipient) to it. Dissolution of the excipient generates acidic or basic ions that modify the internal pH of the specimen during hydrolysis. Poly(ortho esters) provide a good example of this approach (Section 9.8.4). The hydrolytically unstable bond in these polymers degrade faster at acidic pH values when compared to neutral pH, but they are more stable at basic pH. By using acidic or basic excipients, the degradation of poly(ortho esters) can be varied in a controlled manner. On the other hand, the incorporation of drugs and other substances that contain acidic or basic functional groups into some degradable polymer devices requires serious consideration. These substances can modify the internal pH of the device and, thus, influence the degradation of the polymer, making the drug release rate unpredictable.

9.5.3 Copolymerization

Copolymerization, as we have seen for nondegradable polymers (Section 8.2.1), is a useful technique for modifying the physicochemical properties of a polymer, such as its strength, elastic modulus, crystallinity, and glass transition temperature. It is also an important technique for modifying the degradation of some degradable polymers used as biomaterials, particularly poly(α-hydroxy esters) and polyanhydrides.

Poly(α-hydroxy esters) again form a classic system for illustrating the effects of copolymerization on degradation. PLA can exist as three stereoisomers, referred to as poly(D-lactic acid) (D-PLA) and poly(L-lactic acid) (L-PLA), which are semicrystalline polymers having a crystallinity of ~40 wt%, and poly(D,L-lactic acid) (D,L-PLA), an amorphous polymer (Section 9.8.1). The incorporation of glycolic acid units into macromolecules of the D or L isomer leads to a copolymer, PLGA, which has a lower crystallinity and a more rapid degradation rate than D-PLA or L-PLA. PLGA formed by the incorporation of lactic acid units into PGA molecules shows a similar effect. However, the reduction in crystallinity and the increase in degradation are not linear with increasing amount of lactic acid or glycolic acid units in PLGA. When the amount of lactic acid or glycolic acid units in PLGA becomes higher than ~25–30 mol%, the copolymer becomes amorphous and shows a considerably faster degradation than semicrystalline PLA or PGA.

Overall, copolymerization is a commonly used technique to prepare degradable polymers having a wide range of physicochemical properties and varying degradation in

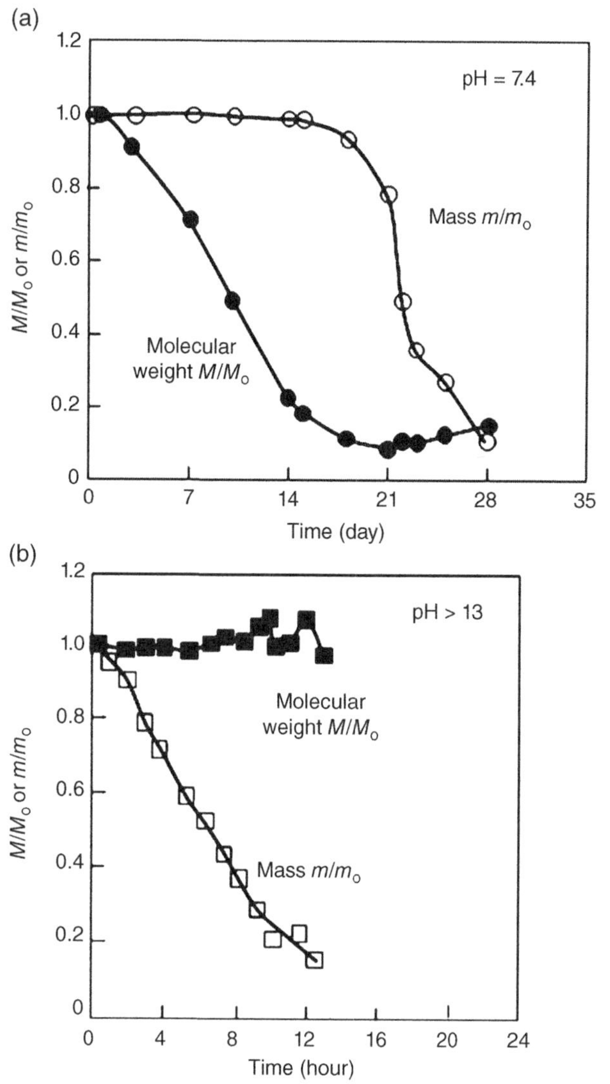

Figure 9.5 Effect of pH on degradation (molecular weight M relative to initial molecular weight M_o) and erosion (mass m relative to initial mass m_o) for polylactic-co-glycolic acid (PLGA) of initial molecular weight 47 kDa, composed of equimolar amounts of D,L-lactic acid and glycolic acid. Data are shown as a function of time in an aqueous medium at (a) pH = 7.4 and (b) pH > 13. *Source:* Göpferich (1997); von Burkersroda et al. (2002).

order to obtain an optimal combination of properties for specific applications.

9.5.4 Crystallinity

The conformation of the macromolecular chains in a polymer has a strong effect on its degradation. Polymer crystals, as we have seen from Chapter 8, are composed of an ordered arrangement of molecular chains whereas amorphous polymers are composed of a random arrangement of chains. Generally, the crystalline phase of a polymer degrades more slowly than the amorphous phase of the same composition. This can be understood in terms of the better packing and stronger intermolecular bonding of the molecular chains in the crystals, which lowers the access of their hydrolytically unstable bonds to water molecules. PLA provides a good example to illustrate the effects of crystallinity. As noted earlier, PLA can be have a semicrystalline structure or an amorphous structure. The semicrystalline polymer, D-PLA or L-PLA, generally has a crystallinity of ~40 wt% and shows a degradation rate that is several times slower than the amorphous polymer D,L-PLA.

Degradation of a semicrystalline polymer proceeds not only more slowly than its amorphous counterpart but in a less homogeneous manner as well. Semicrystalline polymers, have a spherulitic structure in which the spherulites are composed of lamellar crystals held together by amorphous polymer trapped between them (Chapter 8). Because of its faster degradation rate, the amorphous phase erodes faster, leaving behind a significant fraction of the crystalline phase, which can eventually crumble into individual lamellae, due to lack of cohesion. This type of inhomogeneous degradation is undesirable in some drug delivery applications where a more homogeneous degradation is critically required to provide more predictable degradation.

9.5.5 Molecular Weight

Molecular weight can influence degradation in a direct way as well as in an indirect way. Molecular weight directly influences degradation through the length of the polymer chain. For polymers of the same composition, a larger number of bonds has to be cleaved by hydrolysis to generate a given fractional molecular weight, such as 50% of its initial molecular weight, for example. Consequently, the degradation rate decreases with increasing molecular weight. As an example, the degradation of PLGA composed of equimolar amounts of D,L-lactic acid and glycolic acid of molecular weight 47 kDa is approximately two times slower than the same polymer of molecular weigh 8 kDa (Figure 9.6). Overall, the molecular weight versus time curve for degradation shifts to longer times with increasing molecular weight.

Molecular weight may have an indirect effect on degradation through its influence on the glass transition temperature T_g of the polymer. As noted in Chapter 8, for molecular weights below ~20 kDa, T_g decreases with decreasing molecular weight (Eq. (8.9)). However, this indirect effect of molecular weight is relevant only when the reduction in T_g is large enough to bring the polymer from the glassy state to a rubbery state relative to the experimental test temperature (such as 37 °C). This is because a polymer in its rubbery state will degrade faster than the

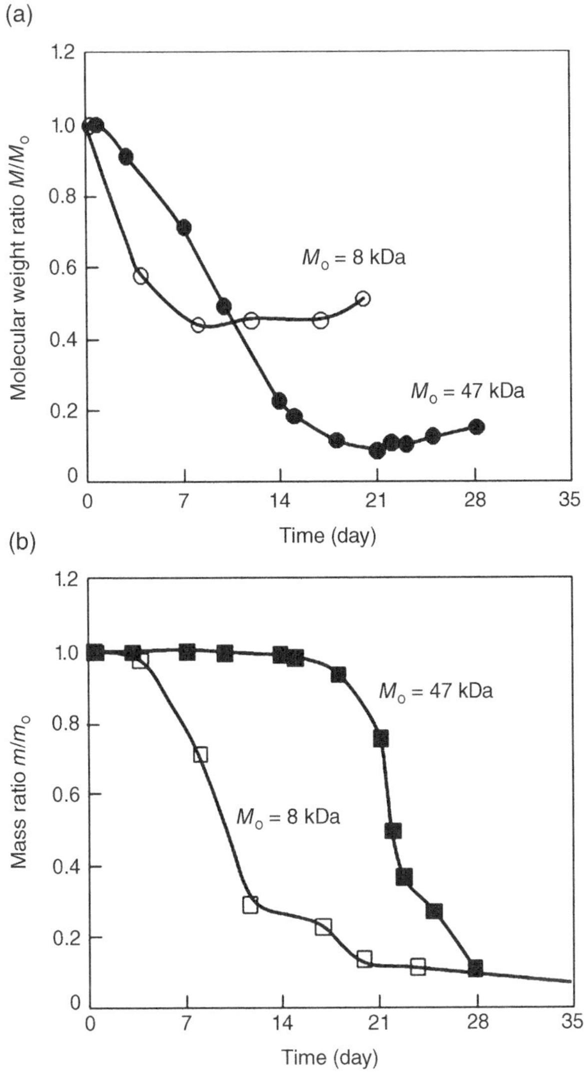

Figure 9.6 Effect of molecular weight on (a) degradation (molecular weight M relative to initial molecular weight M_o) and (b) erosion (mass m relative to initial mass m_o) for polylactic-co-glycolic acid (PLGA) composed of equimolar amounts of D,L-lactic acid and glycolic acid. Data are shown as a function of time in an aqueous medium at pH = 7.4. *Source:* Data replotted from Göpferich (1997).

same polymer in its glassy state due to the higher mobility of its chain segments and its higher free volume, which makes its hydrolytically unstable bonds more accessible to H_2O molecules.

As an example, a decrease in molecular weight from 41 to 17 kDa has been reported to lower the T_g of D,L-PLA in its dry state from 47.0 to 41.7 °C. Upon immersion in water for two days, the T_g values of the higher and lower molecular weight polymers were lowered to 39.4 and 34.2 °C, respectively, attributed to the absorbed H_2O molecules serving as a plasticizer between the chains. Following this two-day immersion, the T_g of the lower molecular weight polymer is below 37 °C whereas the higher molecular weight polymer has a T_g higher than 37 °C. The faster degradation observed for the lower molecular weight D,L-PLA at 37 °C has been attributed to its rubbery nature when compared to the glassy nature of the higher molecular weight polymer.

9.5.6 Water Uptake

Cleavage of the hydrolytically unstable bonds in the chain backbone of a polymer by H_2O molecules can be considered a bimolecular reaction. The reaction rate depends on the product of the concentrations of both reactants and, thus, it can be raised by increasing the concentration of either reactant. Water uptake into the interior of a degradable polymer depends on the hydrophobicity (or hydrophilicity) of its constituent molecules. Polymers that possess a higher degree of hydrophobicity have a lower tendency to take up water and, thus, generally degrade more slowly than similar types of polymers that are less hydrophobic. For example, for approximately the same molecular weight and crystallinity, L-PLA degrades more slowly than PGA. This can be attributed mainly to the methyl (CH_3) side groups attached to the monomer units of its molecular chain. These CH_3 groups present a more hydrophobic surface to an aqueous environment, limiting the penetration of H_2O molecules into the polymer. Another way to modify the hydrophobicity of a polymer is by copolymerization, by introducing hydrophilic monomers into the chain backbone of the polymer.

9.6 Factors Controlling Polymer Erosion

Erosion, particularly the type of erosion (surface or bulk erosion), is a critical process in determining the performance of degradable polymers in a variety of biomedical applications (Section 9.3). Two factors are central to whether a degradable polymer shows surface or bulk erosion:

- The degradation rate of the polymer by hydrolysis
- The rate of penetration (diffusion) of water molecules into the interior of the polymer.

If hydrolysis is slow when compared to diffusion, the entire volume of the polymer is susceptible to erosion, and the polymer shows bulk erosion. On the other hand, if hydrolysis is faster when compared to diffusion, erosion is a surface phenomenon because water molecules are consumed mainly at the surface in the hydrolysis reaction. In line with this view, only the fast-degrading polymers such as polyanhydrides and poly(ortho esters) have generally been reported to be surface-eroding polymers. However, a combination of experimental studies and theoretical modeling has provided an improved understanding of erosion when compared to this simple view.

9.6.1 Bulk Erosion

Degradation starts prior to erosion. For several bulk-eroding polymers such as D,L-PLA and PLGA, the first stage of degradation generally follows first-order kinetics, that is, the rate of change of a reactant is proportional to its concentration. Thus for the degradation of polymers, as determined by the reduction in molecular weight M, we can write

$$\frac{dM}{dt} = -\lambda M \tag{9.1}$$

where λ is a reaction rate constant and t is time. Integration of Eq. (9.1) leads to

$$M = M_0 \exp(-\lambda t) \tag{9.2}$$

where M_0 is the initial molecular weight (at $t = 0$). According to Eq. (9.2), the molecular weight decreases exponentially with time. An example of bulk degradation kinetics is shown in Figure 9.6a for two specimens of D,L-PLA, an amorphous polymer, of molecular weight 8 and 47 kDa.

In comparison, erosion, as measured by a reduction in mass, shows a more complicated behavior. After a period of no significant mass loss, erosion sets in suddenly, leading to a rapid decrease in the mass of the specimen. All bulk-eroding polymers show this trend but on a different time scale. For example, lower molecular weight specimens of D,L-PLA show a nearly constant mass over a few days whereas higher molecular weight specimens of the same polymer shows a nearly constant mass over a few weeks (Figure 9.6b). D,L-PLA specimens of still higher molecular weight can maintain their mass over a few months.

Studies have shown that the rapid mass loss starts to occur when the average molecular weight of the polymer reaches a certain value, as exemplified by the data in Figure 9.6. This value, which we can call a threshold value, depends on the composition of the polymer but it is often observed to be around a few thousand daltons. However, achieving a threshold molecular weight is not the only factor that triggers the start of rapid mass loss. Hydrolytic degradation is a random process and a degrading polymer, even in the period of nearly constant mass, contains molecules of molecular weight smaller than this threshold value. On the other hand, these low molecular weight molecules cannot migrate out into the aqueous environment if they have no connection to the surface of the polymer.

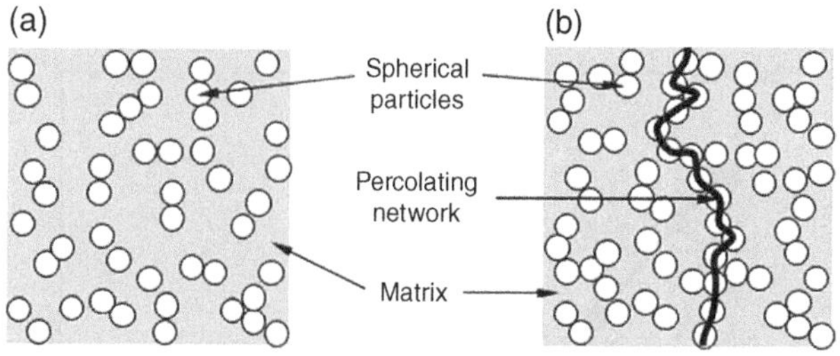

Figure 9.7 Illustration of percolation in a material composed of spherical particles in a homogeneous matrix. Below a certain volume fraction (the percolation threshold), the particles are isolated from one another (a). In comparison, above the percolation threshold, a path of interconnecting particles (solid line) crosses the entire matrix (b). Approximating the lower molecular weight polymers as spherical particles, rapid mass loss will occur when they form a percolating network.

One model assumes that another important factor governing the start of the rapid mass loss is that degradation should proceed to a certain stage when the lower molecular weight degradation products form a percolating (interconnected) network throughout the specimen (Figure 9.7). This is rather like water percolating through ground coffee in a coffee maker. The low molecular weight molecules form a continuous channel extending from the surface all way across the specimen. To do so, these molecules have to attain a certain fraction of the overall polymer volume. When this stage is reached, these low molecular weight degradation products can start to migrate out of the specimen into the aqueous environment, and the mass of the specimen starts to decrease rapidly.

9.6.2 Surface Erosion

In surface-eroding polymers, the planar surface of a polymer in the form of a thin disc or sheet moves uniformly inward with time. Consequently, the mass of the specimen decreases linearly with time and the molecular weight of the eroding polymer remains approximately constant (see Figure 9.5b). Surface erosion follows zero-order kinetics, that is, the mass m per unit area decreases linearly with time t:

$$\frac{dm}{dt} = -c \tag{9.3}$$

where c is a constant. Upon integration, Eq. (9.3) yields

$$m = m_0 - ct \tag{9.4}$$

where m_0 is the initial mass (at $t = 0$).

A common view based on a variety of studies has been that only fast-degrading polymers such as polyanhydrides and poly(ortho esters) show surface erosion. However, a combination of experimental studies and theoretical modeling has shown that this view holds only under certain conditions. Depending on the thickness of the specimen and the conditions of the aqueous medium (such as pH), an otherwise bulk-eroding polymer can become a surface-eroding polymer whereas an otherwise surface-eroding polymer can become a bulk-eroding polymer.

One model assumes that an important parameter is the ratio ε of the time taken for water molecules to diffuse an average distance $\langle x \rangle$ into the polymer and the average time for bonds to degrade over the same distance $\langle x \rangle$. For $\varepsilon > 1$, degradation is faster than water diffusion, and thus, surface erosion dominates. In comparison, bulk erosion dominates when $\varepsilon < 1$. One prediction of this model is that the value of ε and, thus, the erosion mechanism, can change depending on two main parameters:

- λ / D, the ratio of reaction rate constant λ of the hydrolytically unstable bond to the diffusion coefficient D of water molecules into the polymer
- The thickness L of the specimen.

For example, a bulk-eroding polymer can become surface-eroding when L is increased sufficiently or the ratio λ / D is high. Using data from the literature for λ (Figure 9.3) and for D, a critical thickness L_c can be determined such that when $L > L_c$, surface erosion dominates whereas bulk erosion dominates when $L < L_c$.

Figure 9.8 shows L_c for polymers containing a variety of hydrolytically unstable bonds. Although these values can vary widely for a polymer containing a given bond due to the factors discussed in Section 9.4, nevertheless they provide a useful indication of the large difference in erosion behavior among the various types of polymers. An important finding is that fast-degrading polymers such as polyanhydrides and poly(ortho esters), generally viewed as surface eroding, can become bulk eroding if their thickness were small enough. In comparison, poly(α-hydroxy esters) such as PLGA, generally viewed as bulk eroding, can become surface eroding if they are sufficiently thick.

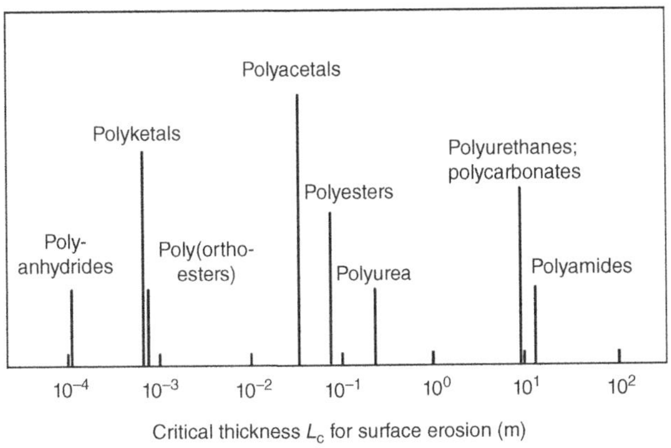

Figure 9.8 Critical thickness L_c that a polymer has to exceed in order to undergo surface erosion. *Source:* Von Burkersroda et al. (2002) / with permission of Elsevier.

Based on our discussion of bulk erosion and surface erosion, changes in molecular weight, mass, and tensile strength are summarized as follows for bulk-eroding and surface-eroding polymers having the geometry of a thin disc or sheet, immersed in an aqueous solution at a constant temperature and pH (such as 37 °C and pH ~7.4):

- Bulk-eroding polymers
  - Molecular weight M decreases approximately exponentially with time (Eq. (9.2))
  - Mass remains approximately constant for a certain period of time that depends on the molecular weight but decreases rapidly with time thereafter
  - Tensile strength decreases with time and can often reach a low value prior to any significant loss in mass.
- Surface-eroding polymers
  - Molecular weight remains approximately constant with time
  - Mass decreases approximately linearly with time
  - Tensile strength of the specimen remains approximately constant with time, although its load-bearing capacity decreases due to a reduction in cross-sectional area.

Overall, several degradable polymers are not ideally bulk-eroding or surface-eroding and, thus, will deviate from this behavior depending on the relative contribution of the two types of erosion.

9.7 Design Criteria for Degradable Polymers

Whereas degradable polymers can provide significant advantages in short-term applications that require only the temporary presence of an implant, their design can often be more challenging when compared to nondegradable polymers used as permanently implantable devices. Degradable polymers are often required to provide not just a mechanical function but a biological function or a drug-release function as well. Consequently, degradable polymers used as implantable biomaterials should generally have a combination of desirable characteristics and properties, as follows.

- *Biocompatibility*: The use of degradable polymers often requires a heightened consideration of biocompatibility phenomena when compared to nondegradable implants. Ideally, the polymer should have a chemical composition that does not elicit a sustained inflammatory or toxic response upon implantation. Another critical requirement is that degradation products of the polymer should be nontoxic and be capable of elimination from the body by various catabolic pathways and physiological clearance mechanisms. Additionally, the polymer should not contain toxic contaminants such as residual stabilizers and polymerization initiators.

- *Degradation and erosion*: A critical requirement is that the polymer should have predictable degradation and erosion kinetics over the requisite lifetime of the intended application. For example, too fast a degradation will lead to a rapid loss in strength of the implant, making it unable to satisfy its mechanical function. The mechanism of erosion is also critical in some applications. For example, in some drug delivery applications it is required for the polymer to erode by surface erosion and release the drug at a constant rate over a prolonged period.
- *Structural characteristics*: Whether a degradable polymer is semicrystalline or completely amorphous is critical in several applications. The capacity of a polymer to crystallize depends on its chemical structure (Chapter 8). In applications where the strength of the device needs to be sustained above a certain level for a desired period, such as sutures and bone fracture fixation devices, a semicrystalline degradable polymer is often necessary. On the other hand, drug delivery devices generally require a homogeneous dispersion of the drug within the device and homogeneous release of the drug. In such applications, a completely amorphous polymer is often preferred over a semicrystalline polymer.
- *Mechanical properties*: Unlike nondegradable polymers, whose mechanical properties are generally stable and predictable over time, the mechanical properties of degradable polymers are less predictable, being dependent on their degradation and erosion. A degradable polymer should have suitable mechanical properties over the desired period of the application when it is required to serve a mechanical function.
- *Processability*: A degradable polymer should be capable of being processed into the requisite geometry and microstructure, such as microspheres and three-dimensional shapes, solid or porous, using common polymer processing techniques.

9.8 Types of Degradable Polymers Relevant to Biomaterials

A variety of degradable polymers has been synthesized and investigated for use as biomaterials, particularly for applications in orthopedic surgery, drug delivery, and tissue engineering (Table 9.1). However, only the following polymers have been used in devices that have so far gained approval from the US Food and Drug Administration (FDA):

- Poly(α-hydroxy esters) containing glycolic acid
- Poly(α-hydroxy esters) containing lactic acid
- Polyesters containing caprolactone
- Polyesters containing dioxanone
- Polyanhydrides containing sebacic acid
- Photocrosslinkable polymers composed of lactic acid and polyethylene glycol
- Poly(trimethylene carbonate) and its copolymers
- Tyrosine-derived polyarylates.

Degradable polymers are often classified into various types depending on the hydrolytically unstable bonds present in their molecular chain backbone. The main types are described in the following sections. Additionally, polyurethanes, which can be degradable or nondegradable depending on their composition, are described in Chapter 8.

9.8.1 Poly(α-hydroxy Esters)

Poly(α-hydroxy esters) composed of PGA, PLA, and their copolymer PLGA (Figure 9.9) were developed for use as biomaterials starting nearly 50 years ago. They remain among the most widely investigated and widely used degradable synthetic polymers in biomedical applications (Table 9.1). Poly(α-hydroxy esters) are polymers composed of monomer units joined by the ester (C=O)O bond. They are synthesized from polymerization reactions involving molecules having a hydroxyl (OH) group attached to the carbon atom adjacent to a carboxylic acid (C=O)OH group (the α carbon atom). For example, PGA and PLA can be synthesized by condensation reactions involving the α-hydroxy acids glycolic acid and lactic acid, respectively. However, the attainment of a sufficiently high molecular weight by condensation reactions involving these molecules is difficult. Consequently, the preferred method to synthesize high molecular weight PGA and PLA involves ring-opening polymerization of their respective cyclic diesters called glycolide and lactide. Because PGA and PLA are commonly synthesized from glycolide and lactide, respectively, they are sometimes referred to as polyglycolide and polylactide. We shall use the chemical names glycolic acid and lactic acid but in synthesis reactions, they can be substituted by glycolide and lactide, respectively.

PLA and PLGA are soluble in common organic solvents such as trichloromethane (chloroform) ($CHCl_3$) and methylene chloride (dichloromethane) (CH_2Cl_2), a property that is used in certain techniques to convert them into various geometries, such thin sheets, discs, or microspheres. Because of its high degree of hydrophilicity, PGA is insoluble in these common organic solvents but it is soluble in highly fluorinated solvents such as hexafluoroisopropanol. Three-dimensional specimens of PGA, PLA, and PLGA having a variety of shapes and internal architecture can be formed using common polymer processing techniques and additive manufacturing (3D printing) techniques as well.

PGA, PLA, and PLGA are insoluble in aqueous media but they undergo hydrolytic degradation at varying rates to form soluble oligomers or monomers, depending on factors such as their composition, molecular weight, crystallinity, and pH of the aqueous environment.

Polyglycolic Acid (PGA)

PGA is the simplest linear aliphatic polyester, having a molecular structure composed of monomer units containing one nonpolar methylene group and a more polar ester group (Figure 9.9a). These compositional and structural characteristics determine the physicochemical properties of PGA. In general, PGA shows a high degree of crystallinity (~50 wt%) and a high degree of hydrophilicity. PGA of high molecular weight (>50–100 kDa) has a melting temperature T_m of ~225 °C and a glass transition temperature T_g of ~35 °C. The density of PGA is 1.5–1.7 g/cm^3, which is high for a polymeric material. Because of its high crystallinity, PGA is one of the strongest degradable polymers (Table 9.2), having a tensile strength of ~70–100 MPa and a Young's modulus of ~6–7 GPa at room temperature. Its elongation to failure is ~20–30%. Oriented forms of PGA, such as fibers, have a significantly higher strength and Young's modulus in the direction of orientation.

PGA was used in the development of the first synthetic degradable sutures, and PGA sutures have been available commercially for nearly 50 years (under the trade name Dexon™). Because of their hydrophilic nature, PGA sutures degrade significantly and lose a large fraction of their strength within two to four weeks of implantation. PGA also finds use as devices in bone fracture fixation, such as pins and screws.

Polylactic Acid (PLA)

PLA has a chemical structure similar to that of PGA, except for the methyl (CH_3) side group that replaces an H atom at the α carbon atom (Figure 9.9b). This seemingly small change in composition makes a large difference in the physicochemical characteristics and properties of PLA when compared to PGA. PLA is more hydrophobic than PGA, which has the effect of significantly lowering its water uptake when placed in an aqueous solution. Thus, access of water molecules to the hydrolytically unstable ester bonds is reduced, making the degradation of PLA slower than that of PGA.

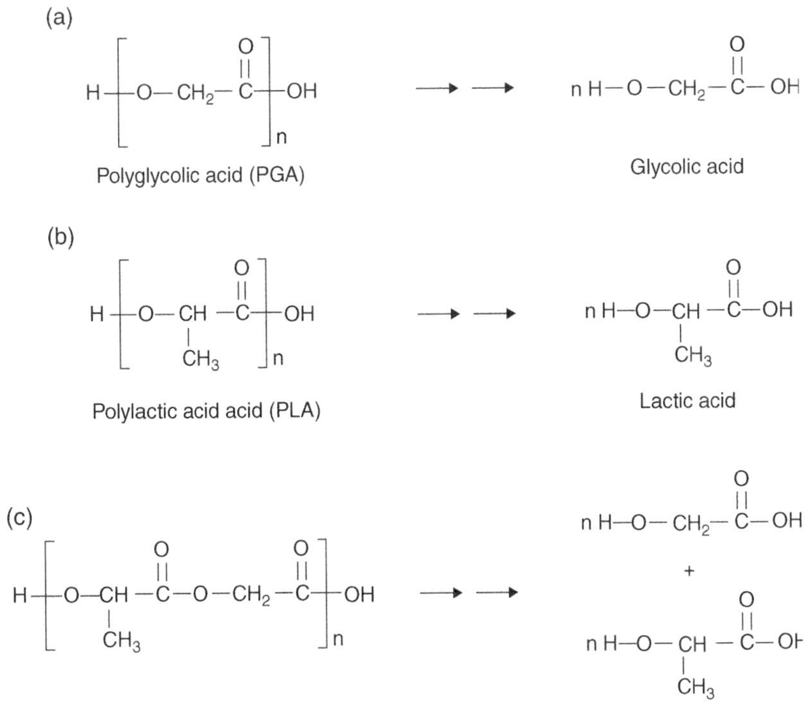

Figure 9.9 Chemical structure of the most widely used poly(α-hydroxy esters) and their monomeric degradation products: (a) polyglycolic acid (PGA), (b) polylactic acid (PLA), and (c) polylactic-co-glycolic acid (PLGA), shown for a copolymer composed of equimolar amounts of PLA and PGA.

Unlike PGA that is generally semicrystalline, PLA can be produced in a semicrystalline form or a completely amorphous form, which is again attributable to the CH_3 side group bonded to the α carbon atom of its monomer units. Lactic acid (or lactide) is a chiral molecule, existing in two stereoisomeric forms that are nonsuperimposable images of each other, called D-lactic acid and L-lactic acid (Figure 9.10). Because of this, PLA can exist in three stereoisomeric forms. This is similar in nature to the stereoisomerism discussed in Chapter 8 for vinyl polymers. D-PLA and L-PLA, synthesized respectively from D-lactic acid and L-lactic acid, are two stereoregular polymers. In comparison, the third stereoisomer, D,L-PLA, synthesized from a mixture of equimolar amounts of D-lactic acid and L-lactic acid, is composed of macromolecules in which the two types of monomer are randomly arranged in the chain backbone. Thus, D-PLA and L-PLA are semicrystalline polymers, whereas D,L-PLA is always amorphous.

Among these PLA polymers, L-PLA and D,L-PLA, along with their copolymer PLGA, have been the most widely investigated and widely used degradable synthetic polymers in biomedical applications. L-PLA is often preferred over D-PLA for use in implantable devices because its hydrolysis yields L-lactic acid, the naturally occurring stereoisomer of lactic acid, which can be eliminated from the body by common biological mechanisms. Consequently, we will focus on L-PLA, D,L-PLA, and their copolymers.

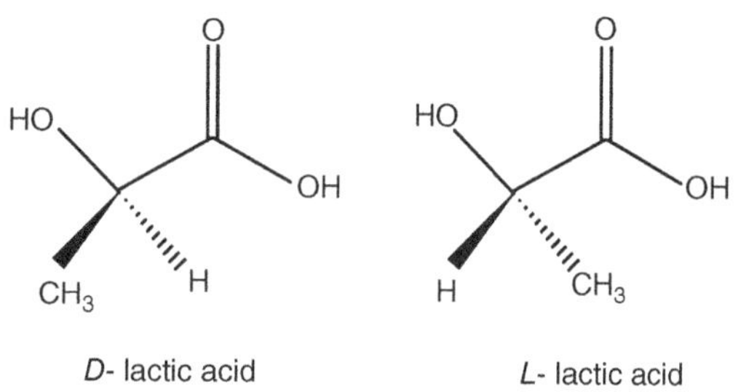

Figure 9.10 Chemical structure of D-lactic acid and L-lactic acid.

High molecular weight L-PLA (>50–100 kDa) is a semicrystalline polymer having a T_m of ~175 °C and a T_g of ~65 °C. It has a crystallinity of ~35 wt.% and a density of 1.2–1.3 g/cm^3, which are lower than the corresponding values for PGA. The room temperature tensile strength (~50–70 MPa) and Young's modulus (~3–4 GPa) of L-PLA are also lower than those for PGA (Table 9.2), which may be attributed largely to its lower crystallinity. However, the strength and modulus of L-PLA are still among the highest values for degradable polymers investigated for use as biomaterials. The elongation to failure of L-PLA (~5–10%), is also lower than that of PGA, which may be attributed to its higher T_g (Chapter 8). In general, L-PLA is a weaker, more hydrophobic and more slowly degrading polymer when compared to PGA. Consequently, it finds use in applications similar to those for PGA, such as sutures and fracture fixation devices, particularly when a more slowly degrading device is required.

D,L-PLA, as noted earlier, is a completely amorphous polymer. It has a T_g of ~55–60 °C, and its density (1.2–1.3 g/cm^3) is similar to that of L-PLA. Because of its amorphous structure, D,L-PLA of high molecular weight (>50–100 kDa) has a tensile strength (~25–35 MPa) and a Young's modulus (~1.5–2.5 GPa) at room temperature that are significantly lower than those of L-PLA, but its elongation to failure (~3–10%) is comparable to that of L-PLA (Table 9.2). D,L-PLA has the same chemical composition as L-PLA and, thus, it is also hydrophobic and degrades slowly. However, because of its amorphous structure, D,L-PLA degrades faster and more homogeneously when compared to L-PLA. On the other hand, it degrades more slowly than PGA, which, although semicrystalline, is less hydrophobic. D,L-PLA has

Table 9.2 Physical properties of poly(α-hydroxy esters) and polycaprolactone at room temperature: approximate values for polymers of high molecular weight (>50–100 kDa) typically used as biomaterials.

Property	PGA	L-PLA	D,L-PLA	PLGA	PCL
Crystallinity (wt.%)	50	35	Amorphous	Amorphous	35–50
T_m (°C)	225	175			60
T_g (°C)	35	65	55–60	55–60	−60
Density (g/cm^3)	1.5–1.7	1.2–1.3	1.2–1.3	1.2–1.3	1.1
Tensile strength (MPa)	70–100	50–70	25–35	25–35	20–30
Young's modulus (GPa)	6–7	3–4	1.5–2.5	1.5–2.5	0.2–0.3
Elongation to failure (%)	20–30	3–10	3–10	3–10	300–500

PGA, polyglycolic acid; L-PLA, poly(L-lactic acid); D,L-PLA, poly(D,L-lactic acid); PLGA, polylactic-co-glycolic acid; data shown for copolymer composed of equimolar amounts of lactic acid and glycolic acid; PCL, polycaprolactone.

been investigated for use in drug delivery devices and as scaffolds for tissue engineering.

Poly(Lactic-co-Glycolic Acid) (PLGA)

PLGA has been the most widely investigated and widely used degradable synthetic polymer for biomedical applications such as sutures, drug-delivery devices, and scaffolds for tissue engineering. This can be attributed mainly to the ease of synthesizing the copolymer and the wide range of physicochemical properties that can be attained by varying the molar ratio of lactic acid to glycolic acid in the copolymer. For example, sutures composed of ~90 wt.% glycolic acid and ~10 wt.% lactic acid are available commercially (under the trade name Vicryl®).

PLGA can be synthesized by copolymerization reactions involving glycolic acid (glycolide) and lactic acid (lactide), typically either L-lactic acid or equimolar mixtures of L-lactic acid and D-lactic acid. The type of lactic acid starting material influences the structure of the copolymer and, thus, its physicochemical properties. Using L-lactic acid, as the amount of glycolic acid is increased, the crystallinity of the PLGA decreases from a starting value of ~35–40 wt.% for semicrystalline L-PLA (Figure 9.11). This reduction in crystallinity occurs because the segments composed of lactic acid monomers in the chain backbone decrease in length with increasing amount of glycolic acid, which limits the size of the crystals that they can form. When the amount of glycolic acid reaches ~30 mol%, the PLGA becomes completely amorphous and it remains completely amorphous until ~70 mol% glycolic acid. With higher amounts of glycolic acid, the segments composed of glycolic acid monomers in the chain backbone become large enough for them to form crystals. Consequently, the crystallinity increases to ~50 wt.%, the value for semicrystalline PGA.

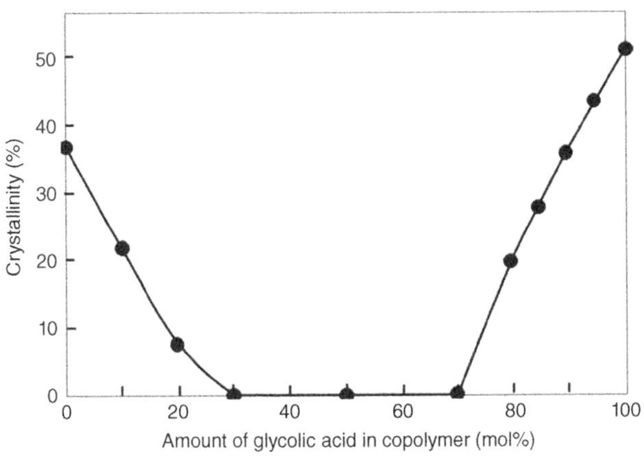

Figure 9.11 Crystallinity of polylactic-co-glycolic acid (PLGA), synthesized from L-lactic acid and glycolic acid, as a function of composition. *Source:* Gilding and Reed (1979) / with permission of Elsevier.

When equimolar amounts of D-lactic acid and L-lactic acid are used instead of just L-lactic acid, the PLGA is completely amorphous from 0 to ~70 mol% glycolic acid in the copolymer. This is because the lactic acid segments in the chain backbone have a random arrangement of D-lactic acid and L-lactic acid monomers and, thus, lack the capacity to form crystals.

Physicochemical Properties of PLGA

The physicochemical properties of PLGA depend mainly on its chemical composition (that is, the molar ratio of lactic acid to glycolic acid) and its crystallinity. Starting with the homopolymer L-PLA, as the molar ratio of glycolic acid to L-lactic acid increases from zero, the tensile strength and Young's modulus of the PLGA decreases due mainly to a reduction in its crystallinity. This decrease in strength and modulus continues until ~30 mol% glycolic acid in the copolymer, when the PLGA becomes amorphous. The tensile strength and Young's modulus of the copolymer at this composition are comparable to those of D,L-PLA, and they do not show any significant changes in the amorphous compositional range of PLGA (~30–70 mol% glycolic acid) (Table 9.2). Above ~70 mol% glycolic acid, the PLGA shows an increase in tensile strength and Young's modulus, up to the values for the homopolymer PGA.

PLGA synthesized from equimolar amounts of D-lactic acid and L-lactic acid instead of just L-lactic acid, as noted earlier, is amorphous in the range 0 to ~70 mol% glycolic acid. Its tensile strength and Young's modulus in this compositional range are comparable to those of D,L-PLA. Overall, the tensile strength and Young's modulus of PLGA can be varied over a wide range, from those for the semicrystalline L-PLA (or PGA), to approximately those for amorphous D,L-PLA by controlling the copolymer composition.

Water uptake by a degradable polymer has a strong effect on its degradation (Section 9.5.6). PLGA provides a good example of the effect of crystallinity and hydrophobicity on water uptake by degradable polymers. Figure 9.12 shows the water uptake (measured as a percentage of the initial dry mass of the specimen) by thin sheets of PLGA (weight average molecular weight 60–90 kDa) as a function of composition after constant weight attainment (after three to four days) in a phosphate buffered solution at 37 °C and pH ~7. Water uptake is low for L-PLA and for PLGA of low glycolic acid content because of their crystallinity and hydrophobicity. With increasing glycolic acid content (~10–30 mol%), water uptake increases rapidly due to further reduction in crystallinity and hydrophobicity. Then the water uptake increases more slowly with glycolic acid content in the range (~30–70 mol%). As the polymers are amorphous in this compositional range, the increase in water uptake is due primarily to an increase in hydrophilicity resulting from the increasing amount of glycolic acid monomers in the copolymer. Thereafter, water uptake decreases slowly due to two opposing effects, increasing crystallinity and increase in hydrophilicity.

Degradation of PLGA occurs by random hydrolytic scission of the ester bonds in its molecular chain, producing soluble oligomers or monomers composed of lactic acid and glycolic acid. Erosion at the physiological pH (~7.4) generally occurs by a bulk mechanism unless the specimen is sufficiently thick (Figure 9.8). One measure of the capacity of polymers to degrade is in terms of the time $t_{1/2}$, called the half-life, for their molecular weight to decrease to one-half its initial value. For example, a longer $t_{1/2}$ value represents a slower degradation rate. For hydrolytic degradation approximating first order kinetics, $t_{1/2}$ is inversely proportional to the reaction rate constant λ (Eq. (9.2)).

Although the degradation of PLGA has been widely studied, the use of measured $t_{1/2}$ values to compare the degradation of PLGA compositions presents some problems due to a variety of factors, such as the use of specimens having varying molecular weight and crystallinity. Consequently, we will illustrate the variation of $t_{1/2}$ in an approximate manner for specimens of high molecular weight (>50–100 kDa) prepared by a similar method (Figure 9.13). In general, the variation of $t_{1/2}$ as a function of PLGA composition will show the combined effects of crystallinity (Figure 9.11) and water uptake (hydrophobicity or hydrophilicity) (Figure 9.12).

Taking $t_{1/2}$ for semicrystalline L-PLA as a starting point, as the amount of glycolic acid in the copolymer increases (0 to ~30 mol% glycolic acid), $t_{1/2}$ decreases steeply due to the reduction in crystallinity and in hydrophobicity. In this compositional region, $t_{1/2}$ for completely amorphous PLGA (synthesized using equimolar concentrations of D- and L-lactic acid instead of just L-lactic acid) decreases less

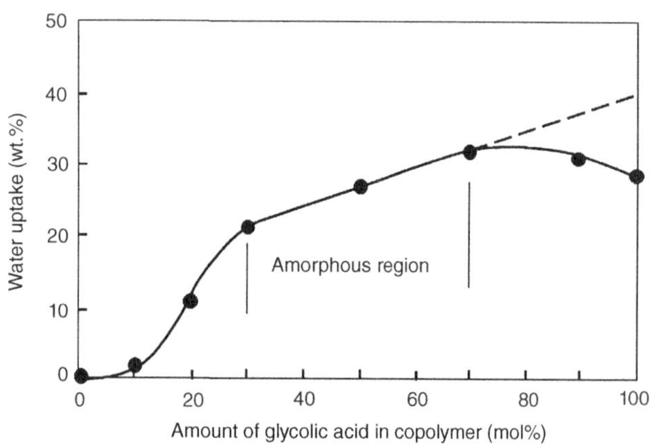

Figure 9.12 Water uptake of polylactic-co-glycolic acid (PLGA), synthesized from L-lactic acid and glycolic acid, as a function of composition. *Source:* Gilding and Reed (1979) / with permission of Elsevier.

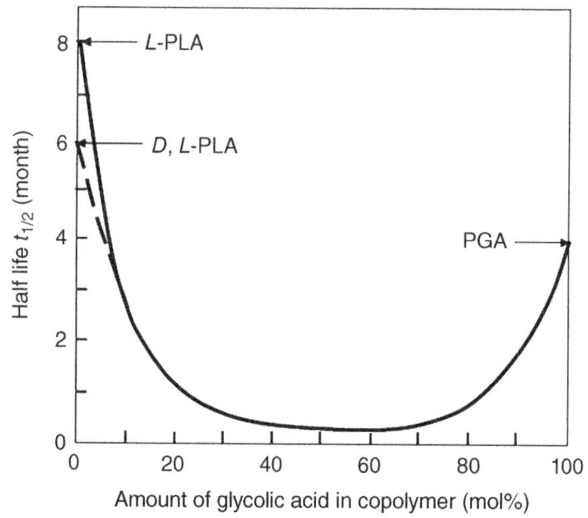

Figure 9.13 Approximate half-life $t_{1/2}$ (time taken for the molecular weight to decrease to half its initial value) for high molecular weight (~50–100 kDa) poly(L-lactic acid) (L-PLA), poly(D,L-lactic acid) (D,L-PLA), and polyglycolic acid (PGA), and copolymers synthesized from L-lactic and glycolic acid (solid line). The dashed line represents $t_{1/2}$ for PLGA synthesized using an equimolar mixture of D-lactic acid and L-lactic acid instead of just L-lactic acid.

steeply due to its lack of crystallinity (dashed line). Then, for completely amorphous compositions (~30–70 mol% glycolic acid), $t_{1/2}$ shows little change. Finally, above ~70 mol% glycolic acid, $t_{1/2}$ increases significantly due primarily to an increase in crystallinity.

9.8.2 Polycaprolactone

Polycaprolactone (PCL) is composed of macromolecules that contain the same hydrolytically unstable ester (C=O)O bond as the poly(α-hydroxy esters), but each monomer unit in the chain backbone contains, in addition, five methylene (CH_2) groups (Figure 9.14). PCL is commonly synthesized by ring-opening polymerization reactions of the monomer ε-caprolactone (often referred to simply as caprolactone), a cyclic ester consisting of a seven-membered ring. PCL has good solubility in common solvents such as chloroform and methylene chloride, and it can be formed into a variety of shapes and internal architecture using common polymer processing methods and additive manufacturing (3D printing) techniques. When compared to the poly(α-hydroxy esters), PCL degrades by the same mechanism (random hydrolytic scission of ester bonds) but it shows significant differences in its degradation rate and other physicochemical properties due to the different chemical structure of its monomer units.

PCL of high molecular weight (~50–100 kDa) commonly used as biomaterials is a semicrystalline polymer of crystallinity ~50 wt% but its crystallinity decreases with higher molecular weight, to ~30–35 wt% for a molecular weight of ~200 kDa. A decrease in crystallinity with increasing molecular weight is unusual for polymers. For PCL, it has been explained in terms of a higher degree of entanglement of its molecular chains and an increase in its viscosity, which hinders the ability of its chain segments to migrate and align with the surface of a growing crystal. PCL has a density of ~1.1 g/cm^3, a melting temperature T_m of ~60 °C, and a glass transition temperature T_g of approximately −60 °C (Table 9.2). This means that the amorphous phase in PCL is rubbery at the body temperature (~37 °C). Thus, despite its semicrystalline structure, PCL is a weak polymer, having a tensile strength of 20–30 MPa and Young's modulus of 0.2–0.3 GPa at room temperature, but it has a high elongation to failure in tension (300–500%). Consequently, PCL is generally unsuitable for use in load-bearing applications, such as implants for healing of structural bone, unless it is reinforced by strong continuous fibers.

Degradation of PCL occurs by random scission of the ester bonds in its molecular chain backbone, resulting in the production of soluble oligomers and monomers composed of 6-hydroxyhexaonic acid. Because hydrolysis occurs by the same mechanism, like the poly(α-hydroxy esters), degradation of PCL is prone to autocatalysis.

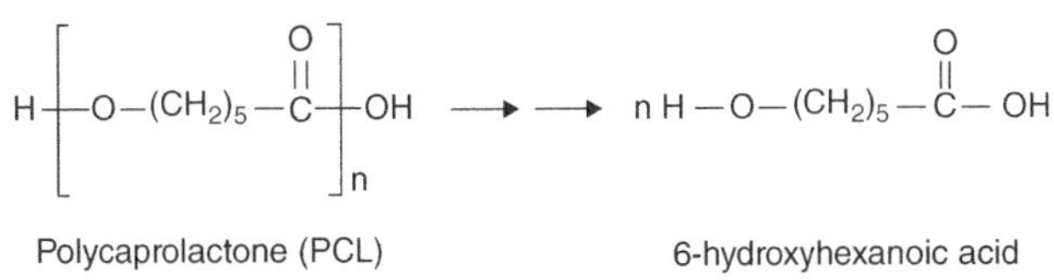

Figure 9.14 Chemical structures of polycaprolactone (PCL) and its monomeric degradation product composed of 6-hexanoic acid.

Erosion occurs predominantly by bulk erosion unless the specimen cross section is thick enough (Figure 9.8).

PCL is known to degrade slowly. The chemical structure of PCL is similar to that of PGA, except that the monomers in its chain backbone contain five methylene (CH_2) groups instead of one. An increase in organic content of the chain backbone of a polymer generally leads to higher degree of hydrophobicity and, thus, slower hydrolytic degradation. PCL also degrades more slowly than D,L-PLA. For specimens of comparable molecular weight (>50–100 kDa), the reaction rate constant λ for the degradation of PCL (~3.0×10^{-3}/day) is approximately three times lower than that for D,L-PLA (~8.4×10^{-3}/day) at 37 °C.

Overall, PCL is a weak, slowly degrading polymer that has good formability and good biocompatibility *in vivo*. It has been approved by the FDA for use as coating on one type of PGA sutures, and it has been investigated for use in devices for long-term drug delivery and more widely for use as scaffolds in tissue engineering. Copolymerization has been investigated to improve the degradation and modify the physicochemical properties of PCL. For example, block copolymers composed of caprolactone and glycolic acid are available commercially for use as sutures (under the trade name Monocryl®) (Chapter 24).

9.8.3 Polyanhydrides

Polyanhydrides are composed of macromolecules in which the monomer units of the chain backbone are joined by the anhydride (C=O)–O–(C=O) bond (Figure 9.15), one of the most hydrolytically unstable bonds in degradable polymers investigated for use in biomedical applications. Polyanhydrides provide a good example in the development of degradable polymers that show surface erosion

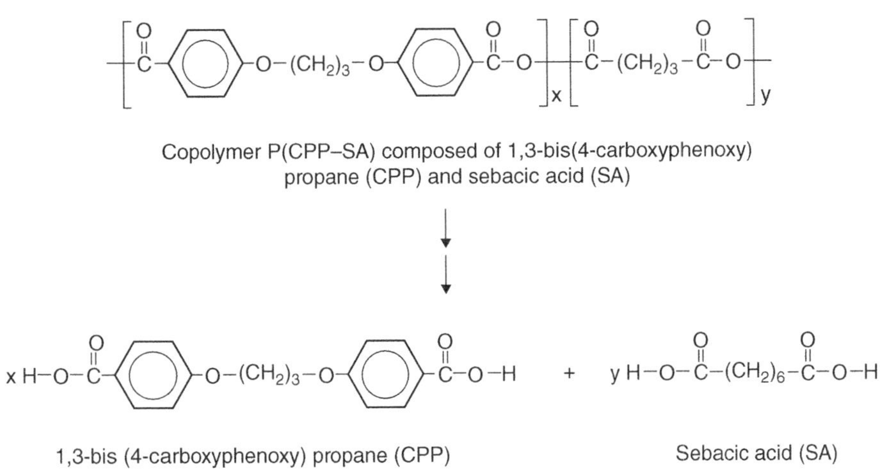

Figure 9.15 Chemical structures of a polyanhydride synthesized from 1,3-bis(4-carboxyphenoxy) propane (CPP) and sebacic acid (SA) and its monomeric degradation product composed of CPP and SA.

rather than bulk erosion, thus providing predictable erosion kinetics and better control of device properties, particularly drug release. For example, a drug incorporated homogeneously in a specimen, which has a low diffusivity through the specimen, will be released at a uniform rate over time as the specimen erodes at a constant rate by surface erosion (Eq. (9.3)). In fact, polyanhydrides were designed and developed to erode by surface erosion specifically for drug delivery applications.

Whether a degradable polymer erodes by bulk erosion or surface erosion, depends on the relative rates of hydrolytic degradation and penetration (diffusion) of water molecules into the polymer (Section 9.6). Because of the rapid hydrolysis of the anhydride bond, water molecules are essentially used up in hydrolytic scission at the surface of the specimen and, thus, they have little tendency to penetrate into the interior. Additionally, as degradation is controlled by access of the anhydride bonds to water molecules, the rate of hydrolytic degradation and, thus, the rate of surface erosion can be controlled by the hydrophobicity of the polymer.

Compositional Effects on Degradation and Erosion

The composition of polyanhydrides is a critical factor in determining their degradation and erosion rates. As we have seen, an attractive way to modify the physicochemical properties of polymers, nondegradable or degradable, is through copolymerization. By employing different monomer ratios, a wide range of properties can be obtained.

A prototype in the development of polyanhydrides for use in drug delivery, is a copolymer synthesized from 1,3-bis(4-carboxyphenoxy) propane (CPP) and sebacic acid (SA) monomers (Figure 9.15), a formulation that is used clinically in drug delivery devices. CPP, an aromatic anhydride, is more hydrophobic due to its phenylene groups whereas SA, an aliphatic anhydride, is more hydrophilic. By varying the ratio of CPP to SA in the chain backbone, copolymers having a range of hydrophobicity (hydrophilicity) and, thus, a range of erosion rates can be synthesized. Depending on the copolymer composition, erosion times can vary from several days to over several months.

As an example, Figure 9.16 shows the erosion profiles of specimens (weight average molecular weight ~6–15 kDa) composed of the homopolymer PCPP and copolymers P(CPP-SA) composed of various molar ratios of CPP and SA. The more hydrophobic polymers PCPP and P(CPP-SA) 85:15 (composed of 85 mol% CPP and 15 mol% SA) show approximately constant erosion kinetics over a few months and, by extrapolation, will erode completely in a few years. In comparison, the more hydrophilic copolymers P(CPP-SA) 45:55 and P(CPP-SA) 21:79, respectively show erosion rates that initially are ~60 and ~100 times faster than PCPP.

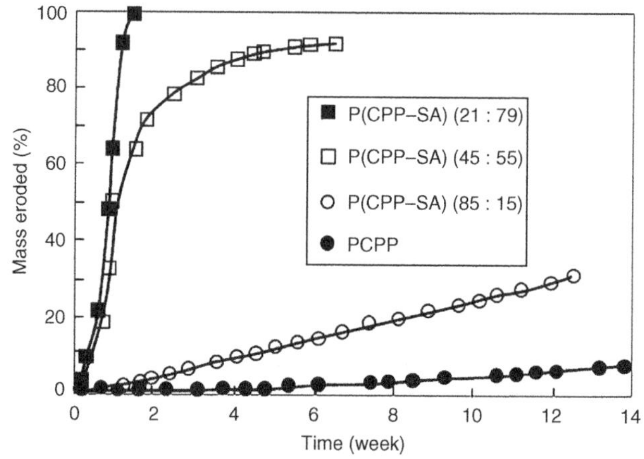

Figure 9.16 Erosion kinetics (mass eroded relative to the initial mass) for the homopolymer PCPP, synthesized from 1,3-bis(4-carboxyphenoxy) propane (CPP), and copolymers P(CPP-SA), synthesized from varying molar ratios of CPP and sebacic acid (SA), in a phosphate buffer solution at 37 °C and pH = 7.4. *Source:* Leong et al. (1985) / with permission of John Wiley & Sons.

However, P(CPP-SA) 45:55 and P(CPP-SA) 21:79 tend to disintegrate after ~60 wt% erosion, which has been attributed to a contribution from bulk erosion at longer times.

In addition to the CPP-SA molar ratio, the composition of the aromatic anhydride provides another way to modify the erosion rate of the copolymer. For example, a change in the number of methylene (CH_2) groups in the backbone of the aromatic anhydride molecule from one to six leads to reduction in its erosion rate by ~10^3 times. Erosion rates of P(CPP-SA) decrease slowly with more acidic pH but they increase steeply with increasing pH above ~7–8 due to catalytic hydrolysis by hydroxyl (OH^-) ions.

Biocompatibility of Polyanhydrides

Polyanhydrides composed of P(CPP-SA) are biocompatible and they cause little inflammatory response *in vivo*. Degradation leads to the generation of nontoxic molecules composed of diacid monomer units in the molecular chain backbone (Figure 9.15). The SA portion of P(CPP-SA) is metabolized by β-oxidation with the CO_2 byproduct eliminated by exhalation. In comparison, the CPP portion is metabolized more slowly, with the resulting byproduct eliminated mainly in the urine. The FDA has approved devices composed of P(CPP-SA) for the delivery of chemotherapy drugs to treat brain tumor.

Structure of Polyanhydrides

Polyanhydrides composed of P(CPP-SA) are semicrystalline polymers whose chemical structure and crystallinity depend on their composition. Polyanhydrides form block copolymers or random copolymers depending on their

CPP : SA molar ratio (see Figure 8.2). Copolymers composed of a high concentration of one monomer (generally > ~75 mol% of CPP or SA) are generally block copolymers in which the number of monomers in each block reflects the relative molar concentration of each monomer. For example, the number of SA monomers in each block decreases from ~15 to ~2 as the CPP : SA molar ratio increases from 5:95 to 50:50. In this compositional range, the number of CPP monomers in each block increases from ~1 to ~2. It is expected to increase more steeply as the CPP : SA ratio increases further, particularly above a CPP : SA molar ratio of 75:25, as found for other anhydride systems. In comparison, copolymers having a CPP : SA molar ratio in the region ~75:25 to ~25:75 are generally random copolymers. Whether a block copolymer or a random copolymer is formed depends on the degree of randomness of the polymerization reaction. For example, for higher CPP : SA molar ratios, CPP molecules have a higher probability of reacting with themselves than with SA molecules.

The copolymer composition also influences the crystallinity of the polymer (Figure 9.17). Homopolymers composed of CPP or SA monomeric units have a crystallinity of ~60–65 wt% but copolymerization leads to a strong reduction in crystallinity. For example, for copolymers of high molecular (>50–100 kDa), the reduction in crystallinity is steep in the region 0 to ~25 mol% CPP (or SA). In the region ~25 to ~75 mol% CPP (or SA), the crystallinity is approximately constant at ~10–15 wt%. The crystallinity of P(CPP-SA) shows a trend that is similar to that of poly(α-hydroxy esters) composed of L-PLA and PGA (Figure 9.11), except that P(CPP-SA) copolymers are not

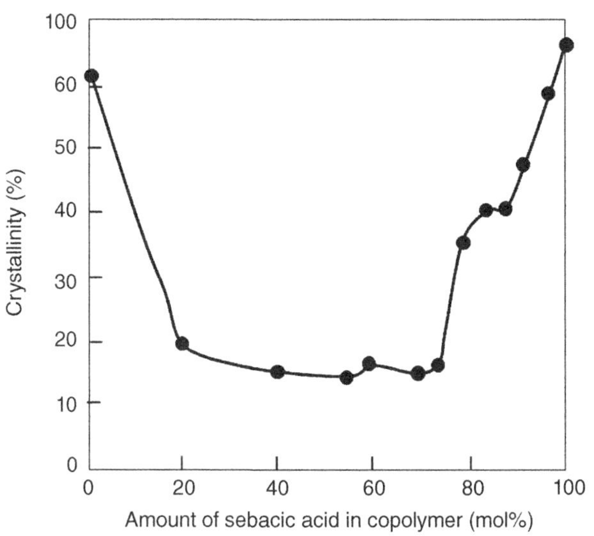

Figure 9.17 Percent crystallinity of copolymer P(CCP-SA) composed of varying molar ratios of 1,3-bis(4-carboxyphenoxy) propane (CPP) and sebacic acid (SA). *Source:* Mathiowitz et al. (1990) / used with permission of American Chemical Society.

completely amorphous in the region ~25 to ~75 mol% CPP (or SA).

Physicochemical Properties of Polyanhydrides

The solubility of P(CPP-SA) in common organic solvents such as chloroform and methylene chloride is dependent on the copolymer composition. Whereas copolymers containing up to ~60 mol% CPP are highly soluble in these common organic solvents, those containing over ~70 mol% CPP are only sparingly soluble. Overall, however, P(CPP-SA) copolymers can be easily formed into a variety of shapes, such as sheets and discs, using common polymer processing techniques.

Polyanhydrides, as noted earlier, have been designed and developed specifically for drug delivery applications. Consequently, their mechanical properties are not a primary design criterion. In general, polyanhydrides composed of P(CPP-SA) are weak, making them unsuitable even for low-load applications. In order to provide surface-eroding polymers having better mechanical properties, other polyanhydrides have been synthesized and investigated. One approach has been to copolymerize anhydrides with imides, that is, molecules containing the –(C=O)–NR–(C=O)– bond, where R is H, an alkyl group or an aryl group. This approach takes advantage of the higher strength and stiffness of imide segments in the chain backbone. Another approach has been the development of crosslinked polyanhydrides synthesized from anhydride monomers such as CPP and SA that are terminated with methacrylate groups.

Because the anhydride bond is very susceptible to hydrolysis, it is necessary to store polyanhydrides in a moisture-free environment at temperatures below 0 °C. In organic solutions or in the solid state, P(CPP-SA) copolymers tend to undergo depolymerization reactions into lower molecular weight molecules at temperatures near or above room temperature.

9.8.4 Poly(Ortho Esters)

Poly(ortho esters) are composed of macromolecules containing the ortho ester $-C(OR)_3$ group in which three alkoxy OR groups are attached to one carbon atom, where R represents one or more type of organic group. Poly(ortho esters) have been under development for use in biomedical applications for nearly 50 years and during that time, several compositions have been studied and characterized. However, interest in most of these compositions has waned over time and only one composition (Figure 9.18) is currently under development for biomedical applications.

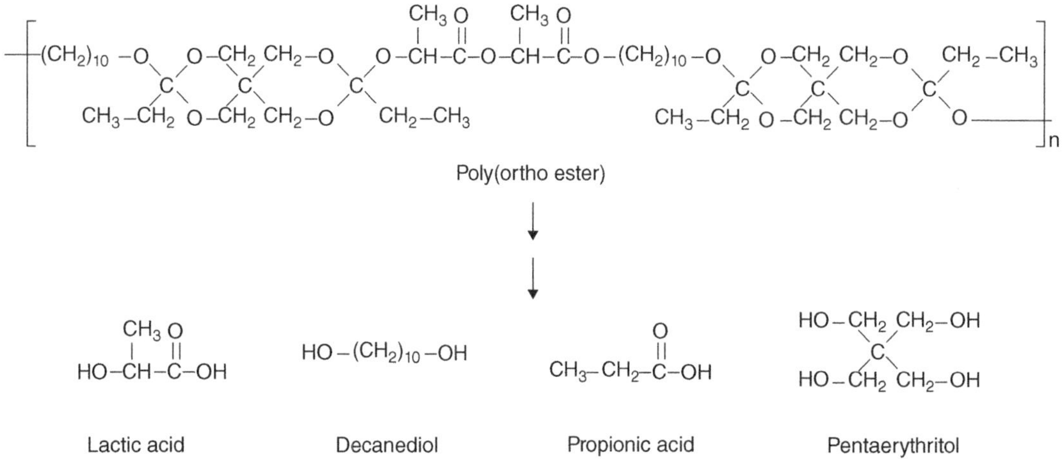

Figure 9.18 Chemical structures of a poly(ortho ester) under investigation for biomedical applications, and its monomeric degradation products.

Like the polyanhydrides, poly(ortho esters) are another type of degradable synthetic polymer that have been developed to show surface erosion and, thus, to provide more predictable erosion kinetics. The early poly(ortho ester) compositions suffered from various drawbacks, such as very slow degradation and erosion rates, autocatalytic hydrolysis, and difficulty in synthesizing reproducible polymer compositions. The polymers were extremely hydrophobic and, despite the high susceptibility of the ortho ester linkage to hydrolysis, they were very stable in the aqueous physiological medium, making them unsuitable for drug release applications. Autocatalytic degradation occurred because hydrolytic scission of the ortho ester linkage is acid-catalyzed and degradation generated acidic molecules that lowered the pH within the polymer. This made the polymer degradation and erosion rates unpredictable. Although excipients were subsequently incorporated into some poly(ortho esters) to reduce autocatalytic hydrolysis or to control their degradation, their use was only marginally successful.

The major drawbacks of the early poly(ortho esters) have been largely eliminated in the current composition shown in Figure 9.18. First, segments of more hydrophilic molecules such as lactic acid (or glycolic acid) are incorporated into the poly(ortho ester) chain backbone. These chain segments readily hydrolyze, generating their monomeric acids that catalyze hydrolysis of the ortho ester linkages. By controlling the concentration of these hydrophilic segments in the chain, the rate of surface erosion can be well controlled. Second, the method used to synthesize the polymer is easier, and it allows reproducible control of the polymer composition and its molecular weight to a sufficiently high value (~60–80 kDa). Third, whereas the early poly(ortho esters) were generally weak, the present composition can be designed to provide a more desirable range of strength and modulus. This is achieved by modifying the glass transition temperature T_g of the poly(ortho ester) by varying the amount of diol in the chain backbone and the composition of the diol, that is, the number of methylene (CH_2) groups in the diol.

Hydrolysis of the poly(ortho ester) of composition shown in Figure 9.18 proceeds in three consecutive steps. In the first step, hydrolysis of the lactic acid segments leads to the generation of free lactic acid that catalyzes the hydrolysis of the ortho ester bonds. Further hydrolysis proceeds in two steps to generate a diol, followed by pentaerythritol and propionic acid. Overall, this poly(ortho ester) shows good biocompatibility in animal models *in vivo* and it is being investigated for potential applications in drug delivery, treatment of postoperative pain and osteoarthritis, and ophthalmic diseases.

9.8.5 Polydioxanone

Poly(*p*-dioxanone), often referred to simply as polydioxanone, is composed of macromolecules that contain both an ester (C=O)O bond and an ether R_1–O–R_2 bond in their repeating monomer units, where R_1 and R_2 are alkyl or aryl groups (Figure 9.19). Whereas polydioxanone monofilament sutures (designated PDS® or PDS II®) have been used clinically since the early 1980s, polydioxanone and its copolymers have only recently elicited significant interest for use in other biomedical applications, such as devices for drug and gene delivery, and scaffolds for tissue engineering.

Polydioxanone is a semicrystalline polymer (crystallinity ~55 wt%), having a melting temperature T_m of ~110 °C and a glass transition temperature T_g of −5 to −10 °C. The properties of polydioxanone have been characterized more commonly in the form of commercially available sutures or fibers. For the isotropic polymer, a comparison of its chemical structure with those of PGA and PCL can provide a useful guide to its physicochemical properties. For example, all three polymers contain the hydrolytically unstable ester bond in their chain backbone and, thus, they degrade by hydrolytic scission of the ester bond. The monomeric units in polydioxanone contain the $-CH_2-CH_2-O-CH_2$ group whereas those in PGA and PCL contain a less hydrophobic $-CH_2-$ group and a more hydrophobic $-(CH_2)_5-$ group, respectively. Thus, for comparable molecular weights, this difference in chemical structure would cause polydioxanone to degrade at a rate that is slower than PGA but faster than PCL.

9.8.6 Polyhydroxyalkanoates

Polyhydroxyalkanoate (PHA) is a general term used to describe a polymer synthesized from a hydroxyalkanoic acid, that is, an acid having the general structure HO–R–(C=O)OH, where R is an alkyl group (Figure 9.20). The functional group in these polymers which is susceptible to hydrolysis is the ester (C=O)O group. Whereas synthetic polymers such as PLA and PCL that contain the ester functional group are also PHAs, they are generally not considered as such. Instead, when applied to biomaterials, the term refers to a specific group of polymers derived from microorganisms such as bacteria. However, synthetic routes can also be used to produce PHAs. Because of their chemical similarity to the synthetic polyesters, PHAs are included in this chapter, rather than in the next chapter devoted to natural polymers.

PHAs can be homopolymers or copolymers. Poly-3-hydroxybutyrate (P3HB) is the simplest and most widely investigated PHA, but poly-4-hydroxybutyrate (P4HB) and copolymers of P3HB and poly-3-hydroxyvalerate (P3HV), abbreviated P(3HB-3HV), have also been investigated for use as biomaterials (Figure 9.20).

P3HB is produced naturally by a wide variety of bacteria that store it as a ready source of carbon and energy, in a manner similar to the storage of fat in humans. A widely used industrial process involves fermentation in a reactor using an appropriate bacterial strain and controlled conditions consisting of a carbon source (such as glucose), nitrogen, phosphorus, sulfur, trace elements, water, and oxygen. If the bacteria are fed, in addition, a controlled amount of an organic acid, such as propionic acid $CH_3-CH_2-C(=O)-OH$ during the fermentation process, a predictable amount of 3-hydroxyvalerate (3HV) monomeric units are incorporated randomly in the chain backbone to form the copolymer P(3HB-3HV). The polymer is extracted from the cells using a solvent, such as an aqueous solvent. Using this process, large quantities of copolymers having a range of 3-hydroxybutyrate (3HB) to 3HV molar ratios and, thus, a range of physicochemical properties are produced industrially.

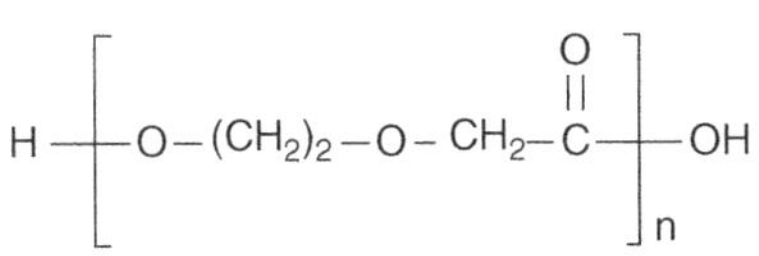

Polydioxanone

Figure 9.19 Chemical structure of polydioxanone.

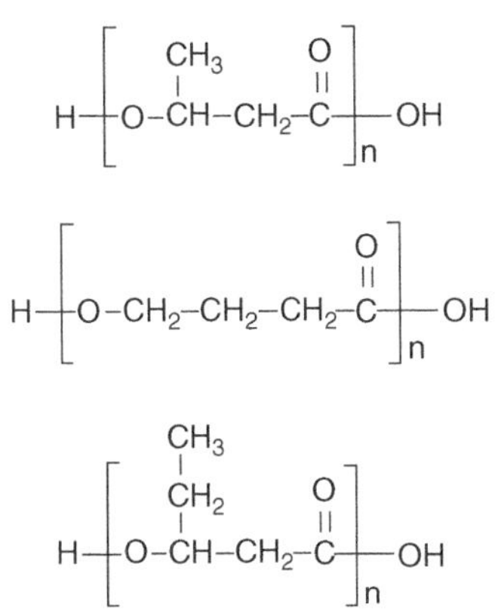

Poly-3-hydroxybutyrate (P3HB)

Poly-4-hydroxybutyrate (P4HB)

Poly-3-hydroxyvalerate (P3HV)

Figure 9.20 Chemical structures of the polyhydroxyalkanoates poly-3-hydroxybutyrate (P3HB), poly-4-hydroxybutyrate (P4HB), and poly-3-hydroxyvalerate (P3HV).

Structure and Properties of PHAs

P3HB and P(3HB-3HV) produced by a fermentation process described in the previous section typically have a weight-average molecular weight of over ~500 kDa. Like polypropylene (PP) molecules for example (Chapter 8), P3HB molecules can exist in different stereoisomeric forms. However, P3HB produced naturally exists in the isotactic form, has a high degree of crystallinity (up to ~80 wt%), a melting temperature T_m of ~175 °C and a glass transition temperature T_g of ~5 °C (Table 9.3). It has a tensile strength of ~40 MPa and a Young's modulus of ~3.5 GPa, but despite the amorphous phase being in the rubbery state at room temperature, it shows little ductility (an elongation to failure ~5%) due to its high crystallinity. P(3HB-3HV) shows a strong reduction in crystallinity, melting temperature T_m, tensile strength, and Young's modulus with increasing molar concentration of 3HV in the copolymer, but a significant increase in ductility (elongation to failure).

P3HB is strongly hydrophobic and, thus, it degrades very slowly in aqueous media. For example, the time $t_{1/2}$ for a 50% reduction in the molecular weight of P3HB

Table 9.3 Typical properties of poly-3-hydroxybutyrate (P3HB) and its copolymers containing 10 and 20 mol% 3-hydroxyvalerate (3HV).

	Composition (mol% 3HV)		
Property	**0**	**10**	**20**
Crystallinity	80	60	35
Melting temperature T_m (°C)	177	140	130
Tensile strength (MPa)	40	25	20
Flexural modulus (GPa)	3.5	1.2	0.8
Elongation to failure (%)	8	20	50

Source: Luzier (1992).

(from an initial value of >500 kDa) is over one year in a phosphate buffered solution at 37 °C and pH = 7.4. Erosion, as measured by a reduction in mass, is negligible after over one year under similar conditions. This slow degradation along with its low ductility limits the potential of P3HB for use in biomedical applications. Although P(3HB-3HV) shows an increase in degradation and erosion with increasing 3HV molar ratio, they are still too slow for some biomedical applications. Copolymerization of 3HB with 4-hydroxybutyrate (4HB) or with less hydrophobic molecules such as glycolic acid, lactic acid or dioxanone, and physically mixing (blending) P3HB with synthetic atactic P(3HB), or with D,L-PLA, PCL, or polydioxanone, have also been studied to enhance the degradation of P3HB and P(3HB-3HV).

Biocompatibility of PHAs

Like many naturally derived polymers, P3HB shows good biocompatibility *in vivo*. Degradation of P3HB leads to the generation of 3-hydroxybutyric acid, a common metabolite in human blood, which is partly responsible for its low toxicity *in vivo*. However, the presence of residual protein and other substances in P3HB produced by some fermentation procedures can lead to an inflammatory response. In comparison, P(3HB-3HV) has shown a greater tendency to elicit an inflammatory response, which increases with the molar concentration of 3HV in the copolymer. This has been attributed to leeching of 3HV out of the polymer.

Overall, P3HB, P4HB, and P(3HB-3HV) are very stable under normal storage conditions, such as low-humidity air at room temperature, and they can be easily formed into the desired shape using common polymer processing techniques. They show good biocompatibility *in vivo* and, despite their slow degradation, these and other PHAs are receiving growing interest for a variety of biomedical applications.

9.8.7 Poly(Propylene Fumarate)

Fumarate polymers are based on fumaric acid HO(C=O)–CH=CH–(C=O)OH. The most widely investigated fumarate polymer for use in biomedical applications is poly(propylene fumarate) (PPF), composed of linear chains in which the monomers contain one carbon double bond (C=C) bond and two ester (C=O)O bonds (Figure 9.21). As synthesized, PPF has a consistency from that of a high-viscosity liquid to that of a putty, which can make its handling somewhat cumbersome. On the other hand, PPF has two main characteristics that make it attractive for use in biomedical applications:

- The unsaturated C=C bonds in the chain backbone allow the molecules to be chemically crosslinked into a three-dimensional network, converting the viscous polymer into a rigid solid.
- Degradation of PPF by hydrolytic scission of the ester bonds in the chain backbone produces nontoxic products that can be eliminated from the body by common biological processes.

The physicochemical properties of crosslinked PPF are strongly dependent on the properties of its constituent macromolecules, such as their molecular weight, and on its crosslinking as well, such as the degree of crosslinking and the type of crosslinking agent used. Consequently, by varying the molecular weight of as-synthesized PPF and its crosslinking, PPF-based polymers having a range of controllable properties can be synthesized for use in a variety of biomedical applications.

Several methods have been developed to synthesize PPF, but one method used to obtain the requisite purity for biomedical applications involves reacting diethyl fumarate

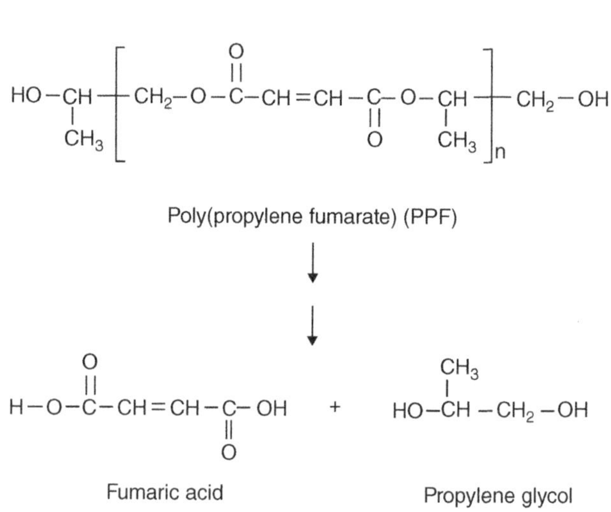

Figure 9.21 Chemical structures of poly(propylene fumarate) (PPF) and its monomeric degradation products.

and propylene glycol. This method gives controllable molecular weights, in the number average molecular weight (M_n) range ~0.5 to ~10 kDa and a polydispersity of ~1.5–3. The glass transition temperature T_g of PPF increases with M_n in accordance with Eq. (8.9), from approximately −30 to ~30 °C as M_n increases from ~1 to ~5 kDa. Thereafter, T_g is approximately constant with molecular weight. Thus, depending on its molecular weight, the viscosity of PPF varies from that of a liquid-like material such as ketchup to that of a putty at the body temperature (~37 °C). Crosslinking through the C=C bond in the chain backbone, using *N*-vinylpyrrolidone (NVP) or poly(propylene fumarate)-diacrylate (PPF-DA) as a crosslinking agent, and an initiator such as benzoyl peroxide, converts PPF to a rigid solid form. Consequently, PPF can be used as an injectable paste that sets (hardens) *in situ* or as a solid implant, porous, or dense.

The mechanical properties of crosslinked PPF also vary widely, depending on the molecular weight of its constituent molecules and their degree of crosslinking, and the crosslinking agent used, but they are generally low. For example, PPF with an M_n of ~2.5–5 kDa crosslinked with NVP and benzoyl peroxide has shown compressive strengths of ~4–12 MPa and Young's moduli of ~0.10–0.25 GPa. These values are within the range for human trabecular bone (Table 4.1). Significant improvement in mechanical properties can be achieved by reinforcing PPF with a variety of particulate materials such as β-tricalcium phosphate (β-TCP) and carbon nanotubes.

Because of the presence of the ester bonds in its molecular chain backbone, PPF is susceptible to degradation by hydrolysis, giving fumaric acid and propylene glycol as degradation products. Fumaric acid is a naturally occurring metabolite in the citric acid cycle (also known as the tricarboxylic acid cycle), a common metabolic pathway for the *in vivo* oxidative metabolism of acetyl fragments and other intermediates derived from carbohydrates, fatty acids, and amino acids. In comparison, propylene glycol is a commonly used diluent (diluting agent) in drug formulations. Consequently, the degradation products of PPF or PPF crosslinked with PPF-DA are nontoxic. However, concerns have been expressed about the formation of nondegradable links in the macromolecular network when PPF is crosslinked with NVP. When immersed in a phosphate buffered solution at ~37 °C and pH ~7.4, crosslinked PPF have shown only small changes in mass, dimensions, compressive yield strength, and Young's modulus over 52 weeks, which indicate slow degradation and erosion.

Overall, because of their nontoxicity, injectability, and their capacity to show a range of physicochemical properties, PPF-based polymers have been investigated for a number of biomedical applications, such as implants for orthopedic surgery, scaffolds for tissue engineering, delivery devices for biomolecules, and vehicles for cell transplantation.

9.8.8 Polyacetals and Polyketals

Polyacetals are composed of macromolecules in which the monomer units in the chain backbone contain the acetal O–C(R_1R_2)–O–R_3 linkage, where R_1 is an organic group (such as CH_3), R_2 is H, and R_3 is an organic group. The monomers in polyketals contain the same functional group except that R_1, R_2, and R_3 are all organic groups (Figure 9.22). Acetals and ketals are derived from aldehydes and ketones, respectively, and ketals are generally considered as a subset of acetals. Introduced within the last 10–15 years as a new family of degradable polymers for use in drug delivery, polyketals, and copolymers of ketals and acetals in the form of microparticles and nanoparticles are attracting interest as drug delivery devices. The characteristics of ketals and their copolymers that make them attractive for these applications include:

- Hydrolysis of ketal (acetal) linkages in the chain backbone generates neutral degradation products such as alcohols and ketones (aldehydes), that is, products that are neither acidic nor basic. Consequently, when compared to the poly(α-hydroxy esters) such as D,L-PLA and PLGA, these degradation products cause little change in the pH of the local aqueous environment.
- Whereas polyketals degrade slowly at the physiological pH such as the pH (~7.4) of blood, they degrade more rapidly with decreasing pH, such as the acidic pH of macrophage phagosomes (Chapter 18). Consequently,

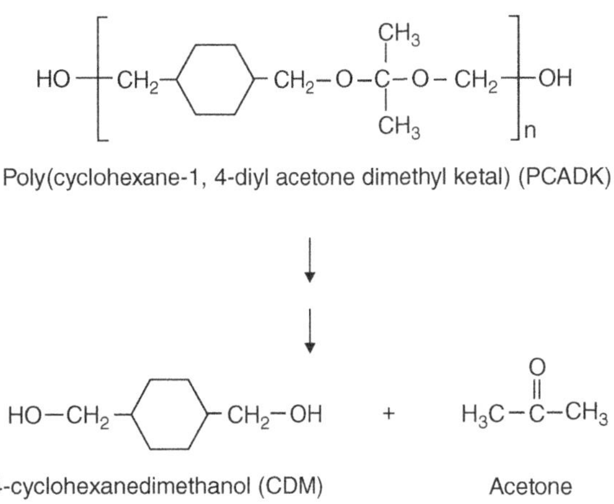

Figure 9.22 Poly(cyclohexane-1,4-diyl acetone dimethyl ketal) (PCADK) and acid hydrolysis of PCADK into 1,4-cyclohexanedimethanol (CDM) and acetone.

polyketal particles can be effective in delivering drugs to a variety of cells.

- Hydrolytic degradation can be varied over a wide range by modifying the ketal composition or by copolymerization with acetals
- Polyketals and their copolymers can be easily synthesized, and they can be converted into nanoparticles and microparticles by common techniques such as emulsion techniques.

One of the most widely studied polyketals for drug delivery has been poly(cyclohexane-1,4-diyl acetone dimethyl ketal) (PCADK), synthesized from 1,4 cyclohexanedimethanol (CDM) and 2,2-dimethoxypropane (DMP) (Figure 9.22). Hydrolysis of PCADK generates CDM and acetone. An example of PCADK microspheres formed by an emulsion technique is shown in Figure 9.23a. A reduction in pH leads to a rapid increase in hydrolysis of PCADK, as the ketal linkages are susceptible to acid catalysis. For example, whereas the molecules in PCADK particles are hardly hydrolyzed at pH = 7.4, the time $t_{1/2}$ for hydrolysis of 50% of the ketal linkages in similar particles is ~24 days at pH = 4.5, while the molecules are completely hydrolyzed in ~2 days at pH = 1.0 (Figure 9.23b).

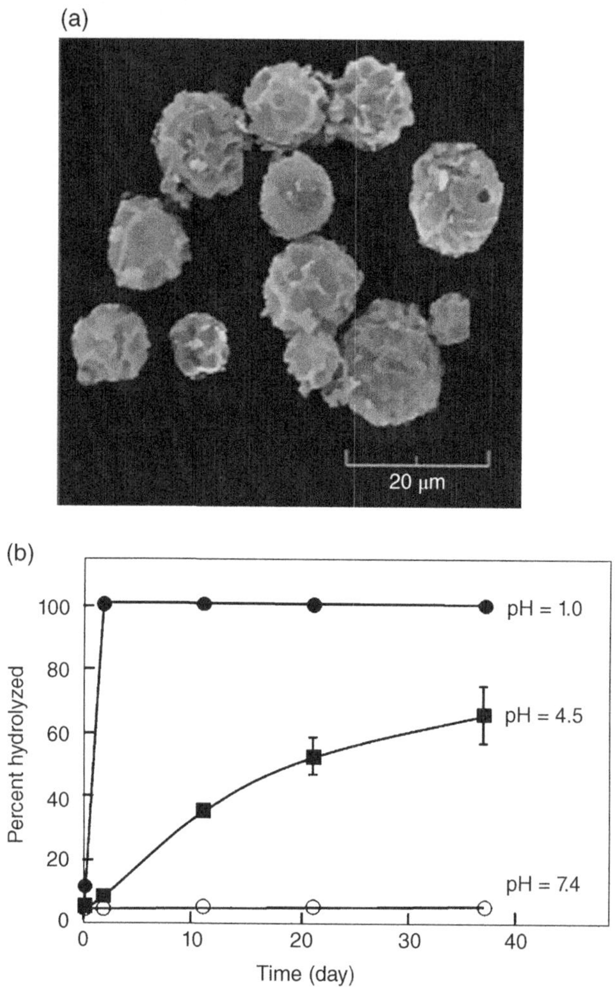

Figure 9.23 (a) Scanning electron microscope image of poly(cyclohexane-1,4-diyl acetone dimethyl ketal) (PCADK) microparticles. (b) The hydrolysis of PCADK is pH-sensitive. Hydrolysis kinetics of PCADK (ground powder) at pH 1.0, 4.5, and 7.4 at 37 °C. Half-lives of PCADK are 24.1 days (pH 4.5) and over four years (pH 7.4). PCADK was completely hydrolyzed in two days at pH 1.0. *Source:* Lee et al. (2007) / with permission of American Chemical Society.

9.8.9 Poly(polyol sebacate)

Poly(polyol sebacate) is a family of degradable polymers synthesized by condensation reactions involving a polyol (an organic compound containing more than two hydroxyl groups but no other functional groups) and SA. The first poly(polyol sebacate) introduced for use in biomedical applications was poly(glycerol sebacate) (PGS), synthesized from glycerol and SA (Figure 9.24). Introduced 15–20 years ago, PGS has been the most widely investigated poly(polyol sebacate) polymer, but other compositions in this family have also been receiving interest more recently for a variety of biomedical applications. PGS provides a useful system for illustrating the design of a new degradable polymer for a particular type of biomedical application.

Design and Synthesis of Poly(Glycerol Sebacate) (PGS)

PGS was originally designed specifically for use in soft tissue repair and regeneration. Overall, this meant designing a suitable degradable polymer that has the characteristics of an elastomer (Section 8.6), rather than a thermoplastic typical of several degradable polymers that we have

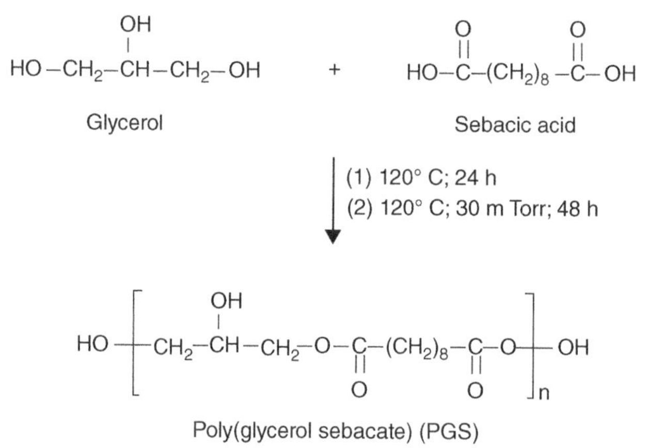

Figure 9.24 Synthesis of poly(glycerol sebacate) (PGS) from glycerol and sebacic acid.

described in this chapter. The main criteria used in designing the polymer are as follows:

- Degradation should occur by hydrolysis rather than by enzymatic reactions to minimize differences in degradation rate between individuals
- The hydrolytically unstable bond in the molecular chain backbone should consist of the ester (C=O)O bond because it can be incorporated by a variety of established and versatile techniques
- Molecular chains in the polymer should have a low degree of crosslinking to reproduce the structure of well-known elastomers such as natural and synthetic rubbers (Chapter 8). In comparison, a high degree of crosslinking would lead to a stiff and brittle polymer, typical of thermosetting polymers
- Both the chain backbone and the crosslinks in the macromolecular network should contain the same hydrolytically unstable (ester) bond to reduce the potential for heterogeneous degradation of the network in an aqueous environment
- Monomers used in synthesizing the polymer should be nontoxic. However, at least one of the monomers should have at least three functional groups to provide for crosslinking (rather than forming just linear chains) and at least one should contain hydroxyl (OH) groups to form hydrogen bonds to stabilize the molecular structure.

Glycerol and SA, the monomers used in synthesizing PGS, satisfy the aforementioned design criteria, and they are also inexpensive. These monomers are known to be nontoxic, and they have been used in devices approved by the FDA for medical applications. Whereas several methods have been developed to synthesize PGS, the most common is a two-step condensation reaction that was used to synthesize the original PGS introduced for biomedical applications. Equimolar quantities of glycerol and SA are reacted (typically at 120 °C for ~24 hours in argon) to form a polymer composed of uncrosslinked chains, sometimes referred to as a prepolymer. Then the gas pressure is lowered and the polymer is held (cured) for 24–96 hours to form a three-dimensional macromolecular network. The degree of crosslinking depends on the duration of this step.

Properties and Applications of PGS

PGS synthesized from equimolar quantities of glycerol and SA (Figure 9.24) is amorphous and has a glass transition temperature T_g of ~6 °C. It shows the typical elastomeric response of high extensibility and recovery under low stresses. Depending on its degree of crosslinking, PGS shows a tensile strength of 0.2–0.5 MPa, a Young's modulus of 0.04–0.12 MPa, and a tensile strain to failure of ~300–350%. A wider range of mechanical properties can be obtained by varying the composition of the polymer (molar ratio of glycerol to SA), and the curing time and temperature.

When compared to the poly(α-hydroxy esters) such as D,L-PLA and PLGA that typically show bulk erosion, erosion of PGS in an aqueous medium occurs predominantly by surface erosion. For example, when implanted subcutaneously in female Sprague–Dawley rats, PGS synthesized by the method depicted in Figure 9.24 showed a linear decrease in mass to ~20% of its original mass after seven weeks, and a linear decrease in tensile strength to ~60% of its original value (Figure 9.25). Over the same period, water uptake increased linearly to ~15 wt%, without any distortion in the external shape of the specimens. In comparison, PLGA (molecular weight ~18 kDa), synthesized from equimolar amounts of D- and L-lactic acid and glycolic acid, shows little loss in mass over the first two weeks but a rapid loss to ~40% of its original value over the next week. These PLGA specimens also showed a rapid loss in strength

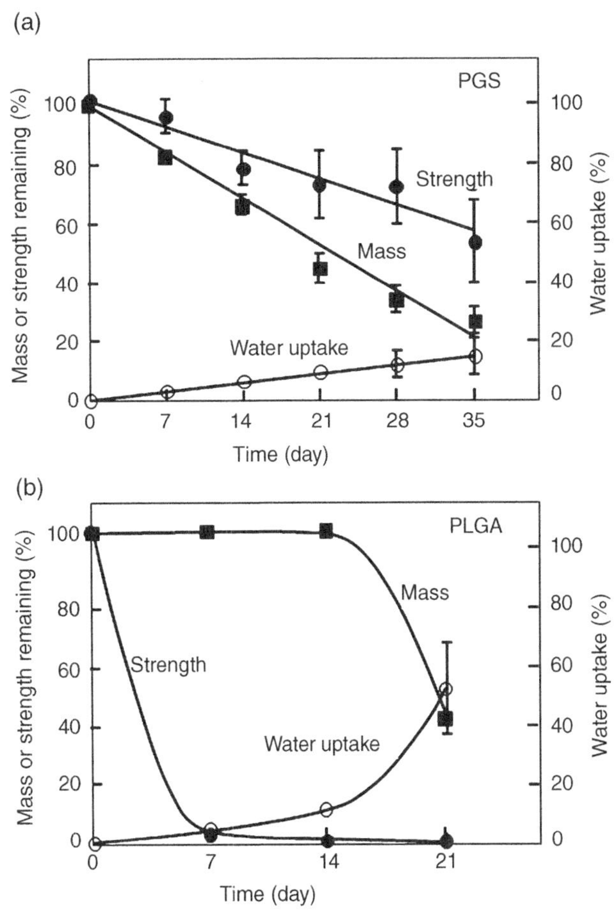

Figure 9.25 Comparison of the changes in mass, compressive strength, and water uptake as a function of time for specimens of (a) poly(glycerol sebacate) (PGS) and (b) polylactic-co-glycolic acid (PLGA) implanted subcutaneously in a rat model. Data are shown as a mean with standard deviation. *Source:* Wang et al. (2003) / with permission of John Wiley & Sons.

during the first week to only a few percent of their original value, a considerably higher water uptake (~50 wt.% after three weeks) and considerable distortion in shape.

PGS and several other poly(polyol sebacate) polymers have good biocompatibility. Overall, these polymers have an attractive combination of physicochemical properties for a variety of biomedical applications focused primarily on soft tissue repair and regeneration, such as cardiac muscle, vascular tissue engineering, cartilage, nerve conduits and retina, and for drug delivery as well.

9.8.10 Polycarbonates

The term polycarbonate (PC) generally refers to the engineering thermoplastic having the specific chemical structure shown in Figure 8.20 but it can be applied to any polymer based on the carbonate O–C(=O)O linkage. PC used in engineering applications is a strong glassy polymer (T_g ~ 160 °C). It has properties such as transparency, strength, Young's modulus, and chemical and thermal stability comparable to polymethyl methacrylate (PMMA) used in intraocular lenses but much better toughness. The PC-based materials that have attracted interest for biomedical applications are PTMC and its copolymers, and polymers derived from the amino acid tyrosine.

Poly(Trimethylene Carbonate) (PTMC) and Its Copolymers

The chemical structure of PTMC is shown in Figure 9.26. When compared to PC, PTMC is a rubbery polymer at room temperature due to the flexible methylene CH_2 groups in its chain backbone. PTMC has a glass transition temperature T_g of approximately −17 °C and a Young's modulus of ~5 MPa. The carbonate linkage is hydrolytically stable and, thus, PTMC is nondegradable in aqueous media *in vitro*, such as a phosphate buffered solution at 37 °C and pH ~7.4. On the other hand, PTMC shows rapid surface erosion *in vivo*, which is attributed to degradation by enzymatic reactions. Surface erosion, as we have seen, is desirable for achieving properties that are more predictable over time, but enzymatic reactions generally lead to a variability in the degradation rate between individuals. In order to provide a wider range of mechanical properties and more predictable degradation by hydrolysis rather than by enzymatic reactions, copolymers of trimethylene carbonate (TMC) with, for example, D,L lactic acid or ε-caprolactone have been synthesized and evaluated.

Tyrosine-Derived Polycarbonates

Polymers derived from amino acids are another type of degradable polycarbonate-based polymers which have been investigated for biomedical applications. These polymers are often referred to as pseudo-poly(amino acids) because they have a chemical structure related to natural proteins but the amino acids in the chain backbone are linked by carbonate bonds rather than by amide bonds (Section 2.10). Although a variety of pseudo-poly(amino acids) has been synthesized, those derived from the amino acid L-tyrosine have been the most widely investigated for use in biomedical applications (Figure 9.27).

The chemical structure of tyrosine-derived PCs show two interesting characteristics that have a strong influence on their physicochemical properties. First, the chemical structure of the chain backbone can be maintained while independently altering the length of the alkyl ester side group, which can strongly influence properties such as the glass transition temperature T_g, strength, modulus, surface free energy, and degradation rate. Second, the structure contains three bonds, that is, carbonate, ester, and amide bonds, which are susceptible to scission. The carbonate and ester bonds are susceptible to hydrolysis but the amide bond is stable in aqueous media that do not contain enzymes. Consequently, the relative hydrolysis rate of the carbonate and ester bonds determines the overall degradation pathway and the degradation products *in vitro*.

High molecular weight tyrosine-derived PCs (weight average molecular weight ~120–450 kDa) having the alkyl-ester side groups shown in Figure 9.27 have an amorphous structure. With increasing length of the alkyl (R) group, from ethyl CH_2CH_3 to octyl $(CH_2)_7CH_3$, these PCs generally show a decrease in several physicochemical properties. For example, the glass transition temperature T_g decreases from ~100 to ~50 °C, the tensile strength decreases from ~70 to ~50 MPa, while the Young's modulus decreases from ~1.5 to ~1.2 GPa. On the other hand, the ductility (elongation

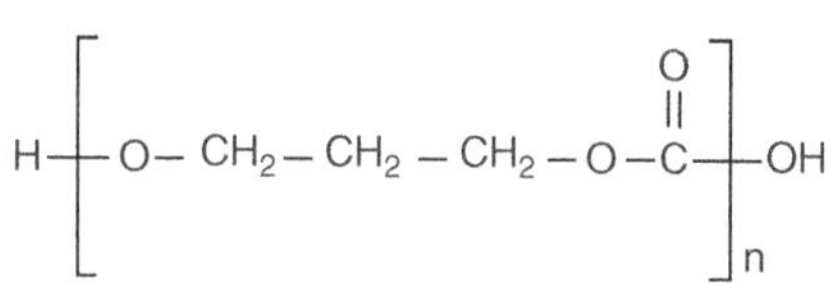

Figure 9.26 Chemical structure of poly(1,3-trimethylene carbonate).

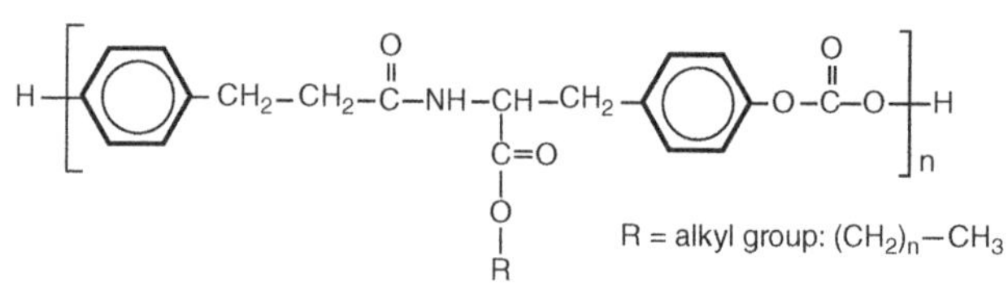

Figure 9.27 Chemical structure of tyrosine-derived polycarbonate.

to failure) increases from ~3 to >400%. These polymers degrade very slowly in a phosphate buffered solution at 37 °C and pH ~7.4, by random scission of the carbonate linkages, accompanied by a small amount of cleavage of the ester bonds. The time $t_{1/2}$ for 50% reduction in molecular weight for the polymer having R = C_2H_5 is ~200 days and there is little loss in mass over this period.

Tyrosine-derived PCs have good thermal stability and they can be formed into the desired shape by common polymer processing methods. They have been investigated for biomedical applications that require a slowly degrading polymer having good mechanical properties, such as pins, screws, and plates for stabilizing fractures in nonstructural bone.

9.9 Concluding Remarks

In this chapter, we discussed a variety of factors that influence the degradation and erosion of synthetic polymers in an aqueous environment. Then we discussed several degradable polymers that have been investigated or developed for use in biomedical applications, emphasizing the relationship between their structure and physicochemical properties. Interest in degradable polymers has increased considerably in the last few decades, particularly for applications in drug delivery and tissue engineering (or regenerative medicine). Overall, an understanding of the factors that influence degradation and erosion is critical in designing degradable synthetic polymers having the requisite properties and performance.

- When used for polymeric biomaterials, the term degradation refers to a chemical process in which certain bonds present in the chain backbone of the polymer undergo scission, typically at random points in the chain, resulting eventually in the production of small molecules (oligomers) or monomers that dissolve in an aqueous environment
- Degradation of synthetic polymers in an aqueous medium *in vitro*, such as a phosphate buffered solution, occurs by hydrolysis, although degradation by enzymatic reactions can make a contribution *in vivo*
- The major factors that influence the degradation of synthetic polymers are the nature of the hydrolytically unstable bond in their molecular chain backbone, the pH of the aqueous medium, compositional modification by copolymerization, the polymer molecular weight and crystallinity, and the capacity of the polymer to absorb water
- Degradation is accompanied by erosion, a process in which the oligomeric or monomeric degradation products dissolve into the aqueous environment, which leads to a reduction in mass of the polymer, and to changes in other characteristics such as the geometry, microstructure, and integrity of the specimen
- Erosion is generally described in terms of two extreme types referred to as bulk erosion and surface erosion. Whether a polymer degrades by bulk erosion or surface erosion depends generally on the rate of degradation relative to the rate at which water molecules diffuse into the polymer. Overall, one type of erosion often dominates in some polymers, whereas other polymers can show a combination of the two types.
- Both the degradation rate and the type of erosion (bulk or surface) have a strong effect on the properties and performance of a degradable polymer
- We discussed the chemical structure, degradation, and erosion of a variety of degradable synthetic polymers, grouped in terms of the chemical structure of their hydrolytically unstable bonds, focusing on the relationship between their composition, structure, and properties. These polymers provide a wide range of degradation rates and physicochemical properties for a variety of biomedical applications, particularly drug delivery, and tissue engineering.

Problems

9.1 Explain the meaning of the terms degradation and erosion when applied to degradable polymers intended for use as biomaterials.

9.2 How are degradation and erosion of polymers characterized? Explain.

9.3 The degradation of a polymer is sometimes characterized in terms of its half-life. Explain what the term half-life of a degradable polymer means and how it can be measured.

9.4 Explain the main differences between bulk erosion and surface erosion of a degradable polymer.

9.5 Upon immersion in an aqueous medium such as a phosphate buffered solution, some degradable synthetic polymers show no measurable change in mass for a long period followed by a rapid change in mass. In comparison, other degradable synthetic polymers show a linear decrease in mass over time. Explain why.

9.6 Why are surface-eroding polymers often preferred for use in some drug delivery devices when compared to bulk-eroding polymers?

9.7 Which erosion process (bulk erosion or surface erosion) would be preferred for a degradable polymer intended for use as a load-bearing implant? Explain.

9.8 Draw the chemical structures of the repeating unit in PLA and PGA and use them to explain differences in their hydrolytic degradation.

9.9 (a) Why is D,L-PLA preferred over L-PLA for use in some drug delivery devices? Explain.

(b) Why is PLGA preferred over L-PLA for use in some drug delivery devices? Explain.

9.10 The degradation of poly(α-hydroxy esters) such as PGA, PLA, and PLGA is said to be susceptible to autocatalysis. Explain what is meant by autocatalytic degradation and how it influences the degradation of these polymers.

9.11 Although the monomer units in the chain backbone of PCL contains the same hydrolytically unstable bond as PGA, PCL degrades much more slowly than PGA of a comparable molecular weight. Explain.

9.12 (a) Describe the relationship between the structure and properties of the polyanhydride P(CCP-SA) composed of CCP and SA. (b) These properties make P(CCP-SA) particularly suitable for which biomedical applications?

9.13 (a) Describe the relationship between the structure and properties of PCADK.

(b) Briefly describe a biomedical application for which these properties make PDADK particularly attractive.

9.14 (a) Describe the relationship between the structure and properties of PGS.

(b) Compare the properties of PGS with those of a poly(α-hydroxy ester) such as PLGA composed of equimolar amounts of L-lactic acid and glycolic acid.

References

Gilding, D.K. and Reed, A.M. (1979). Biodegradable polymers for use in surgery - polyglycolic acid/polylactic acid homopolymers and copolymers: 1. *Polymer* 20: 1459–1464.

Göpferich, A. (1997). Polymer bulk erosion. *Macromolecules* 30: 2598–2604.

Lee, S., Yang, S.C., Hefferman, S.J. et al. (2007). Polyketal microparticles: a new delivery vehicle for superoxide dismutase. *Bioconjugate Chemistry* 18: 4–7.

Leong, K.W., Brott, B.C., and Langer, R.S. (1985). Bioerodible polyanhydrides as drug-carrier matrices, I: characterization, degradation, and release kinetics. *Journal of Biomedical Materials Research* 19: 941–955.

Luzier, W.D. (1992). Materials derived from biomass/biodegradable materials. *Proceedings of the National Academy of Sciences of the United States of America* 89: 839–842.

Mathiowitz, E., Ron, E., Mathiowitz, G. et al. (1990). Morphological characterization of bioerodible polymers. 1. Crystallinity of polyanhydride polymers. *Macromolecules* 23: 3212–3218.

Park, H., Park, K., and Shalaby, W.S.W. (1993). *Biodegradable Hydrogels for Drug Delivery*. Boca Raton, FL: Taylor and Francis.

Treiser, M., Abramson, S., Langer, R., and Kohn, J. (2013; Chapter 1.2.6). Degradable and resorbable biomaterials. In: *Biomaterials Science: An Introduction to Materials in Medicine*, 3e (eds. B.D. Ratner, A.S. Hofmann, F.J. Schoen and J.E. Lemons), 179–195. New York: Elsevier.

von Burkersroda, F., Schedl, L., and Göpferich, A. (2002). Why degradable polymers undergo surface erosion or bulk erosion. *Biomaterials* 23: 4221–4231.

Wang, Y., Kim, Y.M., and Langer, R. (2003). *In vivo* degradation characteristics of poly(glycerol sebacate). *Journal of Biomedical Materials Research. Part A* 66: 192–197.

Further Readings

Bruggeman, J.P., de Bruin, B.-J., Bettinger, C.J., and Langer, R. (2008). Biodegradable poly(polyol sebacate) polymers. *Biomaterials* 29: 4726–4735.

Chen, Q., Liang, S., and Thouas, G.A. (2013). Elastomeric biomaterials for tissue engineering. *Progress in Polymer Science* 38: 584–671.

Fisher, J.P., Dean, D., and Mikos, A.G. (2002). Photocrosslinking characteristics and mechanical properties of diethyl fumarate/poly(propylene fumarate) biomaterials. *Biomaterials* 23: 4333–4343.

Goonoo, N., Jeetah, R., Bhaw-Luximon, S., and Jhurry, D. (2015). Polydioxanone-based biomaterials for tissue engineering and drug/gene delivery. *European Journal of Pharmaceutics and Biopharmaceutics* 97: 371–391.

Göpferich, A. (1996). Mechanisms of polymer degradation and erosion. *Biomaterials* 17: 103–114.

Heller, J., Barr, J., Ng, S.Y. et al. (2002). Poly(ortho esters): synthesis, characterization, properties and uses. *Advanced Drug Delivery Reviews* 54: 1015–1039.

Kohn, J., Welsh, W.J., and Knight, D. (2007). A new approach to the rational discovery of polymeric biomaterials. *Biomaterials* 28: 4171–4177.

Kumar, N., Langer, R.S., and Domb, A.J. (2002). Polyanhydrides: an overview. *Advanced Drug Delivery Reviews* 54: 889–910.

Pêgo, A.P., Poot, A.A., Grijpma, D.W., and Feijen, J. (2003). Biodegradable elastomeric scaffolds for soft tissue engineering. *Journal of Controlled Release* 87: 69–79.

Reed, A.M. and Gilding, D.K. (1981). Biodegradable polymers for use in surgery - polyglycolic acid/polylactic acid homopolymers and copolymers: 2. *in vitro* degradation. *Polymer* 22: 494–498.

Tangpasuthadol, V., Pendharkar, S.M., and Kohn, J. (2000). Hydrolytic degradation of tyrosine-derived polycarbonates, a class of new biomaterials. Part I: study of model compounds. *Biomaterials* 21: 2371–2378.

Tangpasuthadol, V., Pendharkar, S.M., Peterson, R.C., and Kohn, J. (2000). Hydrolytic degradation of tyrosine-derived polycarbonates, a class of new biomaterials. Part II: 3-year study of polymeric devices. *Biomaterials* 21: 2379–2387.

Ullery, B.D., Nair, L.S., and Laurencin, C.T. (2011). Biomedical applications of biodegradable polymers. *Journal of Polymer Science. Part B, Polymer Physics* 49: 832–864.

Wang, Y., Ameer, G.A., Sheppard, B.J., and Langer, R. (2002). A tough biodegradable elastomer. *Nature Biotechnology* 20: 602–606.

Wang, S., Lu, L., and Yaszemski, M.J. (2006). Bone tissue-engineering material poly(propylene fumarate): correlation between molecular weight, chain dimensions, and physical properties. *Biomacromolecules* 7: 1976–1982.

Yang, S.C., Bhide, M., Crispe, I.N. et al. (2008). Polyketal copolymers: a new acid-sensitive delivery vehicle for treating acute inflammatory diseases. *Bioconjugate Chemistry* 19: 1164–1169.

10

Natural Polymers

10.1 Introduction

Natural polymers are polymeric materials made by living organisms. They are obtained from human and other vertebrate sources, invertebrate organisms, and plants. Examples are collagen isolated from mammalian tissues such as skin and bone, silk from silkworms, chitin extracted from the shells of crustaceans, alginate produced by seaweed, hyaluronic acid extracted from rooster combs or from genetically engineered bacteria, and cellulose extracted from plants or produced by certain types of bacteria. Unlike synthetic polymers, natural polymers relevant to biomaterials generally have some degree of chemical and structural similarity to components of the extracellular matrix (ECM) of tissues and organs in living organisms.

The ECM is composed mainly of structural and other specialized proteins, and proteoglycans consisting of a protein core to which long linear chains of disaccharide units called glycosaminoglycans (GAGs) are attached (Chapter 16). Natural polymers relevant to biomaterials are often subdivided into protein-based polymers and polysaccharide-based polymers. Protein-based polymers are composed of macromolecules (proteins or polypeptides) in which amino acid residues are linked together in the chain backbone by amide (peptide) bonds (Chapter 2). In comparison, polysaccharide-based polymers are carbohydrates composed of long chains of monosaccharide units linked together by glycosidic bonds. Some polysaccharides relevant to biomaterials, such as hyaluronic acid, are GAGs, composed of long linear chains of disaccharide units. A third type of natural polymers, the polyhydroxyalkanoates (PHAs), commonly obtained from microorganisms such as bacteria but which can also be produced by synthetic routes, were described in Chapter 9.

The attractiveness of natural polymers as biomaterials stems largely from their chemical and structural similarity to macromolecules present in the ECM. When compared to synthetic polymers, natural polymers often have a better capacity to interact favorably with cells and tissues *in vivo*, while adverse tissue responses due to mechanical incompatibility and the risk of toxicity are often lower. However, this attractiveness of natural polymers is offset by a few disadvantages. Biomaterials composed of natural polymers, particularly in three-dimensional isotropic forms such as hydrogels and foams, generally show weak mechanical properties. Consequently, they are often limited to nonstructural applications where their mechanical properties are not a primary consideration. Some natural polymers, particularly collagen, also show low thermal stability, being susceptible to decomposition or structural changes upon heating to temperatures not far above the physiological temperature. This limits the techniques and conditions that can be used for their processing. Additionally, natural polymers have a more complex composition and structure when compared to synthetic polymers, and they have an inherent variability from species to species.

Collagen, the most abundant protein in mammals, has been used in one form or another as a biomaterial for many years, and it remains one of the most widely used natural polymers in biomedical applications. In comparison, while other natural polymers currently find limited use as biomaterials in commercial products, they have been the subject of considerable research and development efforts for a variety of biomedical applications, particularly for drug delivery and tissue engineering. Like synthetic polymers, natural polymers are generally applied in the form of solids, porous or dense, such as coatings, thin sheets, particles, foams, fibers and tubes, and in the form of hydrogels (water-filled polymers) as well.

In this chapter, we are mainly concerned with solid biomaterials, porous or dense, composed of natural polymers. Although discussed when appropriate in this chapter, hydrogels, natural and synthetic, form the topic of Chapter 11. Overall, the following areas relevant to the design, selection, and application of natural polymers for

Materials for Biomedical Engineering: Fundamentals and Applications, First Edition. Mohamed N. Rahaman and Roger F. Brown.

Companion website: www.wiley.com/go/rahaman/biomedicalengineering

use as implants and devices in biomedical applications will be discussed:

- Composition and structure of natural polymers
- Processability of natural polymers into desirable shapes and internal architectures
- Relationship between the composition, structure, and properties of natural polymers.

10.2 General Properties and Characteristics of Natural Polymers

Like all biomaterials, natural polymers have certain attractive properties and characteristics for use in biomedical applications but these advantages are offset by a few disadvantages. In this section, major properties and characteristics of natural polymers related to their use as biomaterials are discussed in general terms.

- *Biocompatibility*: As natural polymers are generally obtained from living species, they should not suffer from many of the inherent risks of toxicity that may arise from synthetic polymers unless toxic substances are introduced during their processing. Because they generally have a chemical and structural similarity to some ECM components, natural polymers have a better capacity to interact more favorably with cells and tissues *in vivo* when compared to synthetic polymers. Unlike synthetic polymers that often elicit an inflammatory response from tissues *in vivo*, natural polymers generally elicit little inflammatory response. Polysaccharide-based polymers generally do not elicit an immune response because they lack the requisite amino acid sequences to trigger an immune response. In comparison, whereas most protein-based polymers generally elicit an immune response, some highly conserved proteins, such as collagen, have the potential to elicit little or no immune response depending on their source.
- *Mechanical properties*: Natural polymers generally have weak mechanical properties, particularly in isotropic three-dimensional forms such as foams (sponges) and hydrogels. On the other hand, in the form of fibers or thin sheets, where their chains have a high degree of orientation, some natural polymers can have considerable strength and stiffness. For example, single fibers of some protein-based polymers, such as collagen and silk, have high tensile strength and Young's modulus, comparable to those of some common synthetic fibers. Consequently, except for fibrous forms of collagen and silk, natural polymers are suited mainly for nonstructural applications where their mechanical properties are not a primary consideration.
- *Stability in the physiological environment*: In comparison with synthetic degradable polymers that degrade by hydrolysis (Chapter 9), natural polymers that degrade *in vivo* do so predominantly by an enzyme-mediated hydrolysis mechanism. Consequently, the degradation of these natural polymers *in vivo* may be less controllable when compared to synthetic polymers and may vary between individuals. Commonly, chemical modification of natural polymers by, for example, crosslinking their molecular chains or by modification of their functional groups, is vital in controlling their degradation and other physicochemical properties.
- *Processability*: Natural polymers are generally susceptible to changes in their composition or native structure upon heating to temperatures not far above the physiological temperature (~37 °C). For example, collagen undergoes significant structural disruption, referred to as denaturation, when heated to only ~10 °C above the physiological temperature. Consequently, several natural polymers need to be processed under mild conditions, such as around room temperature or below, which may limit the techniques that can be used to form them into practical biomaterials.

10.3 Protein-Based Natural Polymers

Protein-based natural polymers are composed of macromolecules, referred to as proteins or polypeptides, in which amino acid residues are linked together in the chain backbone by amide (peptide) bonds (Chapter 2). The polypeptide chain, as we have seen, can take up a variety of three-dimensional shapes (conformations), but two broad classes of structures, termed fibrous and globular, are often distinguished (Figure 2.22), giving rise to:

- Fibrous proteins in which the polypeptide chain has a more elongated or filamentous three-dimensional structure
- Globular proteins in which the polypeptide chain is more compactly folded, like a ball of string having an irregular surface.

Fibrous proteins form the main structural framework and support for connective tissues such as bone, skin, tendons, ligaments, and blood vessels. They provide an attractive source for biomaterials for use in a variety of biomedical applications. In comparison, globular proteins generally serve functional roles such as catalytic and molecular recognition functions.

10.3.1 Collagen

Approximately 28 different molecular forms (types) of collagen have been identified in vertebrates but the most recognizable types in mammals are the fibrous collagens type I, II, III, and XI and the nonfibrous collagen type IV. Collagen type I, together with some type III, is present in tissues such as bone, tendons, ligaments, and skin, whereas type II and type XI are the dominant proteins in cartilage. The nonfibrous collagen type IV, composed of molecules that form antiparallel sheet-like structures (Chapter 2), is present in the basement membranes of surface tissues such as those found in the epidermis, the eye lens, and the linings of blood vessels and kidney tubules.

Collagen type I is the most abundant type of fibrous protein, accounting for ~25% of the dry protein mass in mammals. It occurs in a variety of hierarchical structural forms, such as aligned fibrous networks in bone, tendons, and ligaments, alignable networks in skin and intestinal walls, and a combination of aligned and alignable networks in the intervertebral disc. Our discussion of collagen will focus mainly on type I because it is the most commonly used collagen in biomedical devices and implants. However, other collagens, such as type II and type III, are being investigated for use in biomedical applications, particularly in tissue engineering. The atomic structure and bonding in proteins, discussed in Chapter 2, provide essential background material for the following discussion of collagen-based biomaterials. Occasionally, there may be some overlap with this information discussed in Chapter 2 when it is necessary for clarity.

Structure of Fibrous Collagen (Type I)

Collagen type I used as a source for producing biomaterials is prepared not only from animal tissues such as bovine, porcine and equine skin, and tendons, but also from the bone, skin, and tendons of human donor cadavers. It has a hierarchical structure, that is, a structure that can be distinguished and considered at different length scales, from a molecular scale (such as a single molecule) to a macroscopic scale (such as a tissue). A simplified version of the hierarchical structure of collagen is illustrated in Figure 10.1. At the molecular scale, individual polypeptide chains, referred to as protocollagen chains are synthesized from amino acids within fibroblasts. These protocollagen chains have a short non-helical peptide (called a telopeptide) and additional globular peptides (called propeptides) at each end of the chain (the N-terminus and the C-terminus). Then each protocollagen chain, composed of ~10^3 amino acid residues and molecular weight ~95 kDa, combines with two others to form a hydrogen-bonded, triple-stranded helical molecule, referred to as a procollagen triple helix, ~1.5 nm in diameter and ~300 nm long, of molecular weight ~300 kDa. Cleavage of the globular propeptide end groups by enzymes leads to the formation of triple-helical structures, referred to as collagen molecules. Then these collagen molecules self-assemble to form fibrils, in which they are aligned in parallel in a quarter-stagger arrangement. The short non-helical telopeptides at the ends of the collagen molecules are essential in the enzymatic formation of natural crosslinks between the collagen molecules to stabilize the fibrillar structure.

Each fibril has a diameter in the range ~50–500 nm and a length of up to a few millimeters, depending on the tissue and its stage of development. Studies have shown that these fibrils are composed of smaller fibrillar structures called microfibrils, of diameter ~5 nm, consisting of five collagen molecules in cross-section, that have a straight or helical arrangement within the fibrils. It was once thought that collagen fibrils are formed by a process in which procollagen molecules are secreted by cells and, after cleavage of their propeptide end groups, they self-assemble rapidly in the extracellular space. However, it is now generally accepted that formation of collagen fibrils does not occur in a single self-assembling step extracellularly but, instead, by a combination of both intracellular and extracellular stages.

At the next level of the hierarchical structure, multiple fibrils aggregate to form a collagen fiber, aided by crosslinking macromolecules such as proteoglycans, of diameter up to ~20 μm and length up to several centimeters. Collagen fibers are not present in all collagen-based structures but in those in which they are present, such as tendons, skin and arteries, the arrangement of the collagen fibers strongly influences the mechanical properties of the tissues. In tendons, for example, collagen fibrils align themselves almost parallel to the long axis of the fiber and numerous collagen fibers, in turn, align themselves into bundles that are almost parallel to the long axis of the tendon. This structure is mainly responsible for the high tensile strength and Young's modulus of tendons. In comparison, although the structure of skin varies widely between different species, the fibers are generally arranged in a more wavy pattern in which they have different orientations, which gives skin the capacity to stretch in different directions when subjected even to low stresses.

When viewed in a transmission electron microscope, collagen fibrils show a characteristic banded structure having a periodicity of ~67 nm (sometimes termed the d-period), which is due to the regularly staggered packing of the individual collagen molecules in each fibril. This staggered packing is a direct result of chemical crosslinking between adjacent collagen molecules, which is important for enhancing the adhesion strength of the fibrils.

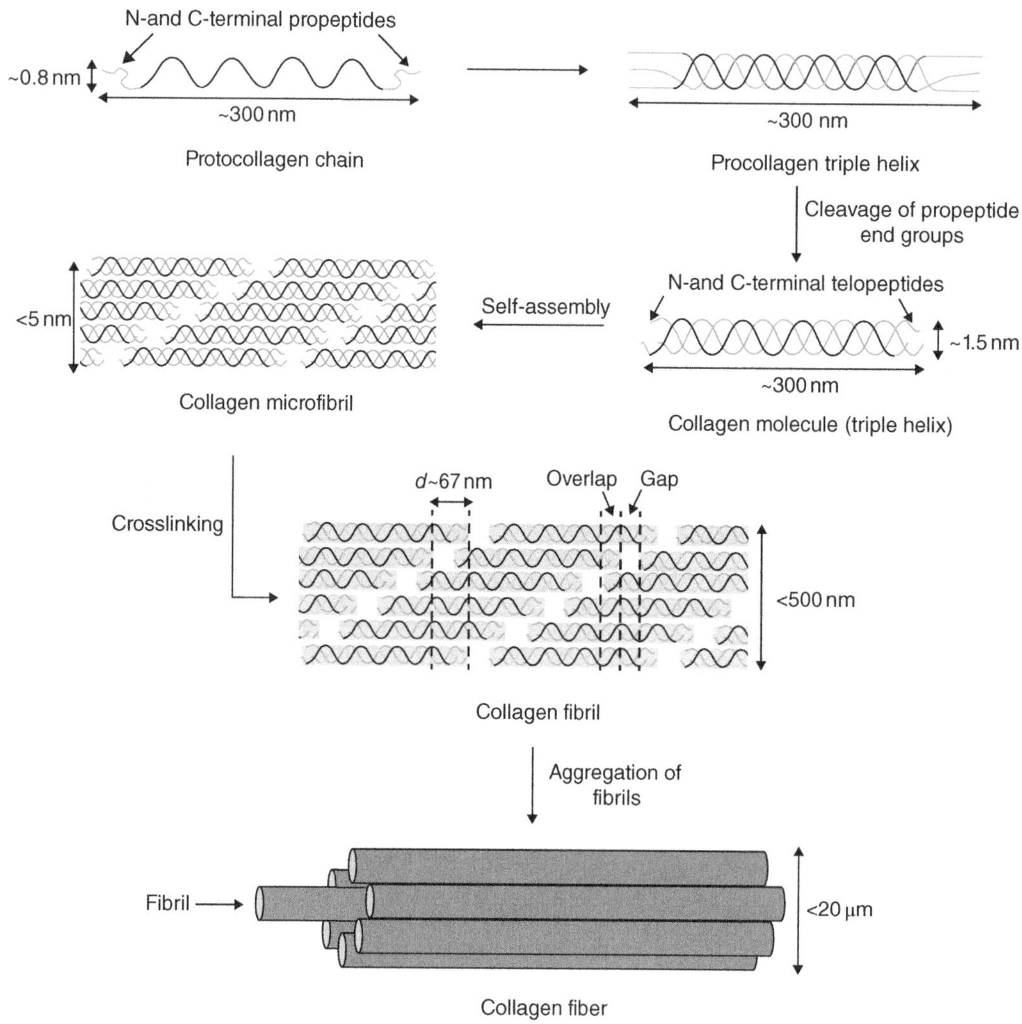

Figure 10.1 Illustration of the hierarchical structure of collagen (type I).

Overall, the major levels of collagen hierarchical structure are the molecule, fibril, fiber, and tissue. In relation to the use of these terms for collagen, fibril should not be confused with fiber because they define structures at different length scales.

Preparation of Collagen for Use as Biomaterials

Collagen (type I) available commercially for use in the creation of biomaterials is prepared mainly from animal tissues such as bovine, porcine, and equine skin and tendons. Collagen can also be prepared by recombinant technologies, which is more consistent and reproducible when compared to the batch-to-batch variability of collagen prepared from animal and human tissues. However, the process suffers from low yield and is more expensive.

Preparation of collagen from animal tissues involves a series of chemical, physical, and mechanical steps that may involve chemical treatment to remove non-collagenous material, cutting the tissue into small strips, washing in acidic solution, mechanical grinding in the wet or dry state, and various purification steps. Collagen is available in two forms, referred to as insoluble collagen and soluble collagen. Insoluble collagen is essentially a comminuted form of collagen fibers, composed of particles in the dry state (a few hundred micrometers in size) or in the wet state, dispersed in an aqueous medium as a suspension. Thus, insoluble collagen consists of particles that retain the fibrillar structure of native collagen.

In comparison, soluble collagen consists of triple-helical collagen molecules dispersed in an aqueous medium. Preparation of soluble collagen commonly includes chemical treatment of purified animal tissue in an acidic solution and treatment with a noncollagenase enzyme such as pepsin, which destroys the crosslinks between the collagen molecules in the fibrils and removes the telopeptides at the ends of the collagen molecules. Then additional steps, such as precipitation and filtration, are used to purify the collagen. Collagen molecules in cold neutral salt solutions maintain their stability for up to several months. However, upon warming the solution to the physiological temperature (~37 °C), the collagen molecules reassemble to form fibrils that precipitate to form a rigid gel-like material.

Production of Collagen Biomaterials

Collagen implants and devices used in biomedical applications cover a wide range of forms and characteristics, but they may be divided into two main groups:

- Tissue-based implants and devices, such as bioprosthetic heart valves and blood vessels, composed of treated mammalian tissues in essentially their natural form that have undergone varying degrees of processing.
- Implants and devices, in forms such as thin sheets, particles, and three-dimensional solids (porous or dense), which have been produced from purified collagen (soluble or insoluble) extracted from animal or human tissues.

We will focus on the latter group because of their relevance to designing and creating biomaterials. For the same geometry, biomaterials formed from insoluble collagen generally have a higher strength than those formed from soluble collagen. Consequently, biomaterials in the form of particles, thin sheets, foams, fibers, and tubes are commonly formed from insoluble collagen because their capacity to support a mechanical stress is often a relevant design parameter. In comparison, soluble collagen is better suited for forming coatings, although it may also be used to form thin sheets. After forming, the biomaterial is often crosslinked, primarily to control its degradation *in vivo* and improve its mechanical properties. If required, crosslinking may also be used to lower the immunogenicity to a certain extent.

Collagen is susceptible to denaturation (disruption of its native structure) when processed at temperatures not far above the physiological temperature. Consequently, the techniques and conditions used for processing collagen products are more limited when compared to those used for synthetic polymers. For example, two common techniques used for forming synthetic polymers, injection molding and extrusion of the thermally softened polymer, are generally unsuitable for collagen-based biomaterials. The commonly used forming methods are casting a solution of collagen molecules or a suspension of collagen particles, freeze-drying a collagen suspension, and extruding a paste-like mixture, generally performed below the collagen denaturation temperature.

- Casting of a solution of collagen molecules (soluble collagen) or a suspension particles (insoluble collagen) is a common method for preparing collagen coatings used, for example, to improve the biological response of some implants and to prepare thin sheets for use in applications such as hernia repair, temporary skin replacement, and dental applications. The concentration of the solution or suspension is generally adjusted to control the thickness of the coating or thin sheet and drying is achieved by convection or in a vacuum.
- Freeze-drying of insoluble collagen suspensions is a common method for forming porous devices, such as particles for drug delivery and foam-like structures (sponges) for wound healing, and tissue engineering applications (Figure 10.2). The diameter and orientation of the pores in these foams depend on the concentration of the collagen suspension, the rate of freezing, and the directionality of freezing. For example, unidirectional

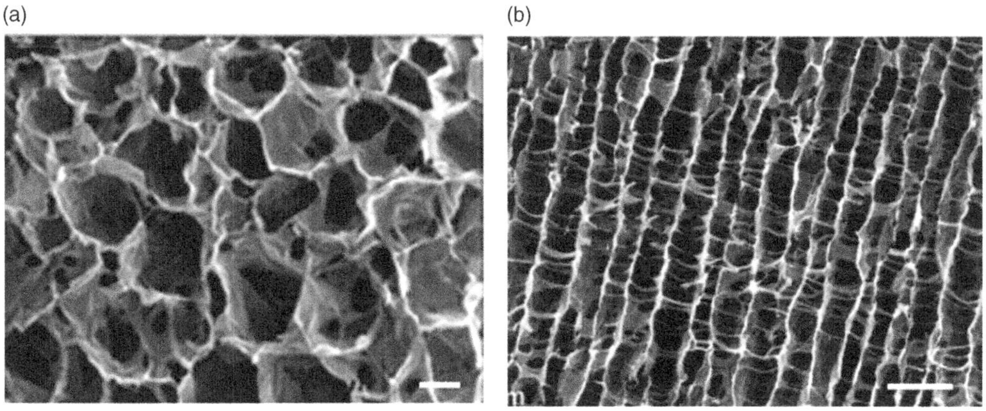

Figure 10.2 Scanning electron microscope images of collagen sponge microstructures. (a) Cross section of sponge formed by isotropic freezing of collagen suspension followed by freeze-drying. *Source:* From Chan et al. (2016); (b) Cross section of sponge formed by unidirectional freezing of collagen suspension followed by freeze-drying. *Source:* From Schoof et al. (2001). (Bar = 100 μm).

freezing yields pores oriented in the direction of freezing whereas a random pore arrangement is obtained from isotropic freezing. In general, the size (diameter) of the pores, generally in the range ~20–200 μm, increases with a decrease in the concentration of the suspension and in the rate of freezing.

- Extrusion of a highly viscous mass of insoluble collagen is a common method for forming fibers for use in wound closure for example, filamentous structures for tendon and ligament repair, and hollow tubes for use in nerve guides, for example.

Crosslinking of Collagen Biomaterials

Crosslinking is a widely used and effective means to control the degradation of collagen biomaterials, improve their mechanical properties (tensile strength and Young's modulus) and, to some extent, reduce their immunogenicity (Chapter 19). After their formation into the desired shape, collagen-based biomaterials may be crosslinked by chemical or physical methods, or a combination of both methods. Ideally, the crosslinking procedure should be effective without introducing toxic substances that can subsequently be eluted from the implant or device.

Chemical Crosslinking A variety of crosslinking agents such as bifunctional aldehydes, isocyanates, and carbodiimides may be used to chemically crosslink collagen biomaterials, but glutaraldehyde has been the most common. With optimal conditions, such as pH and temperature of the medium, and concentration of the crosslinking reagent, the effectiveness of a crosslinking reagent depends on the number of amino acid residues with which it can react, the chemical stability of the crosslink formed, and the bond strength of the crosslink.

Crosslinking collagen biomaterials with glutaraldehyde typically involves soaking the biomaterial for up to several hours in a solution composed of the requisite concentration of glutaraldehyde (typically less than 0.5–1.0 wt%) in a phosphate buffered solution (pH ~7.4) at room temperature. Crosslinks are formed by the reaction between free amine (NH_2) groups of lysine and hydroxylysine amino acid residues of adjacent polypeptide chains and aldehyde groups of glutaraldehyde to form covalent bonds. The first reaction that takes place involves the formation of an intermediate compound, called a Schiff base (Figure 10.3), which is considered the central intermediate compound from which a variety of subsequent reactions may occur to form a crosslink. However, these subsequent reactions and, thus, the overall crosslinking mechanism, are the subject of controversy because the intermediate reaction products are often unstable in an aqueous medium and, thus, difficult to isolate.

In the crosslinking reaction, the number of free amine groups of the collagen decreases rapidly with time whereas its denaturation temperature increases. For example, after one hour, the primary amine groups in collagen suspensions prepared from sheep dermis has been found to decrease from ~34 per 1000 amino acids to ~10 per 1000 amino acids. The denaturation temperature increased from ~56 to ~77 °C but leveled off and became approximately independent of glutaraldehyde concentration above ~0.5 wt% (Figure 10.4). A longer reaction time (24 hours)

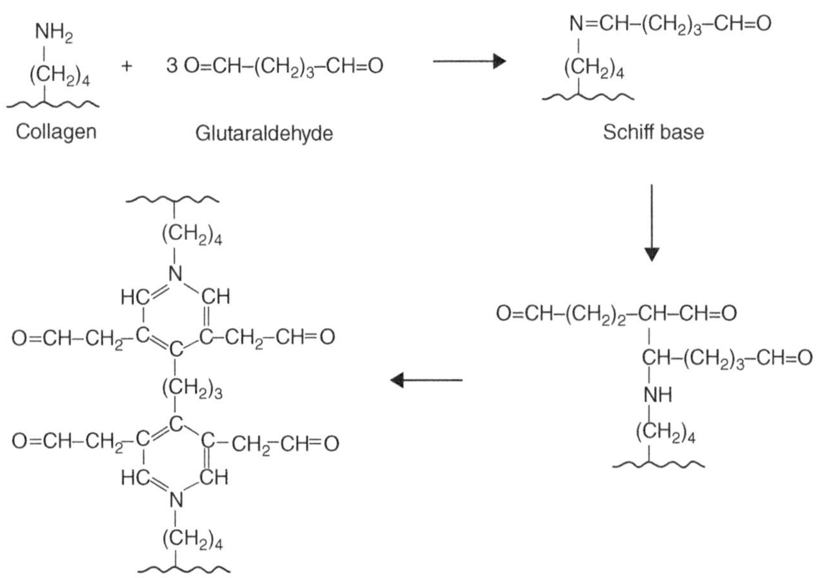

Figure 10.3 Reaction of glutaraldehyde with the free amino groups of lysine (or hydroxylysine) to form covalent crosslinks between collagen molecules.

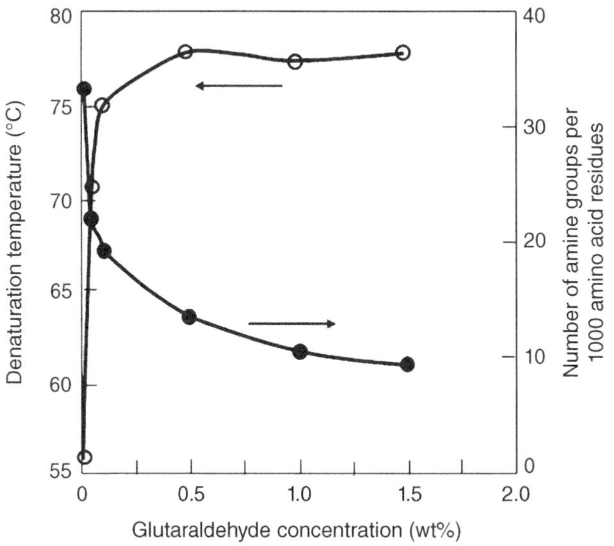

Figure 10.4 Denaturation temperature and primary amine group content for dermal sheep collagen crosslinked with glutaraldehyde for one hour at room temperature and pH = 7.4 as a function of glutaraldehyde concentration. *Source:* From Olde Damink et al. (1995) / with permission of Springer Nature.

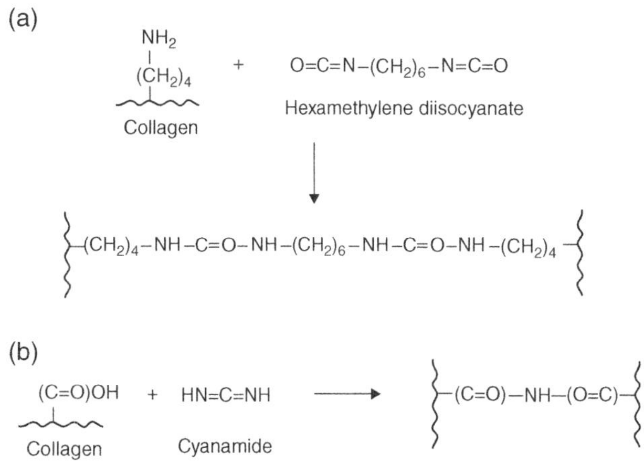

Figure 10.5 Reaction of (a) hexamethylene diisocyanate with the free amine groups and (b) cyanamide with carboxyl groups to form covalent crosslinks between collagen molecules.

produced a further decrease in the number of free amine groups (to ~5 per 1000 amino acids) but the denaturation temperature remained essentially constant. This indicates that longer reaction times may lead to the incorporation of glutaraldehyde molecules into the collagen biomaterial but does not lead to the formation of additional crosslinks. On average, approximately three glutaraldehyde molecules react with an amine group to form crosslinks at these longer reaction times. One possible reaction mechanism involves formation of a Schiff base, followed by condensation reactions to form a heterocyclic compound that subsequently forms cyclic pyridinium ring crosslinks (Figure 10.3).

Concerns have been expressed occasionally about undesirable side effects associated with the use of glutaraldehyde as a crosslinking agent, particularly cytotoxicity due to elution of glutaraldehyde molecules from the implant during its degradation. However, despite its potential for introducing toxic effects, at the levels relevant to crosslinking, glutaraldehyde is generally well tolerated *in vivo*. Overall, glutaraldehyde is very effective in forming stable crosslinks in collagen biomaterials and remains the most commonly used crosslinking agent for these biomaterials.

Alternative agents such as certain diisocyanates and carbodiimides, which do not produce toxic degradation products, have been used to crosslink collagen biomaterials. Diisocyanates, such as hexamethylene diisocyanate, react with amino groups of adjacent polypeptide chains, typically at room temperature and neutral pH (Figure 10.5a). In comparison, carbodiimides such as cyanamide HN=C=NH, lead to formation of short crosslinks by reacting with the carboxyl (C=O)OH groups of adjacent polypeptide chains (Figure 10.5b).

Physical Crosslinking The most common physical techniques for crosslinking collagen biomaterials are radiation treatment using ultraviolet (UV) light or γ rays, and dehydrothermal treatment. UV irradiation is a rapid and easily controlled method for crosslinking collagen-based biomaterials, particularly those in the form of coatings and thin sheets. Typically, the biomaterial is placed under a short wave UV lamp (wavelength ~254 nm) for the appropriate time (a few hours). UV radiation has been reported to produce radicals (unpaired electrons) located predominantly on the aromatic nuclei of the tyrosine and phenylalanine amino acid residues of the chain, which react to form crosslinks. Crosslinking with UV light does not introduce any toxic substances but exposure for more than ~five hours may lead to fragmentation of at least a portion of the collagen molecule and subsequent denaturation.

Dehydrothermal treatment is a type of drying process that is performed typically in a vacuum at ~110 °C for up to a few days. In a solution of soluble collagen or a suspension of insoluble collagen, collagen molecules or particles are separated by water molecules, which prevents them from adhering. However, upon dehydrothermal treatment, removal of water molecules leads to the formation of crosslinks due to condensation reactions between carboxyl (C=O)OH and hydroxyl OH groups (to form an ester bond) and between (C=O) groups and amine NH_2 groups (to form an amide bond) on adjacent polypeptide chains. This leads to a reduction in free (C=O)OH and NH_2 groups. Amino acid residues that may be involved in these reactions are those of aspartic acid, glutamic acid, serine, threonine, arginine, and lysine (Figure 2.17). Like UV irradiation, prolonged dehydrothermal treatment may lead to fragmentation of at least a portion the collagen molecule and subsequent denaturation.

Effectiveness of Crosslinking Techniques The effectiveness of four crosslinking techniques has been evaluated from the effects they have on the properties of the crosslinked collagen biomaterials, such as their resistance to degradation and swelling, and the improvement in their mechanical properties. The techniques evaluated were UV irradiation, dehydrothermal treatment, exposure to glutaraldehyde, exposure to cyanamide. Overall, the more effective the crosslinking technique, the greater the resistance to degradation and swelling and the greater the improvement in mechanical properties.

While all four crosslinking techniques lead to an increase in the degradation resistance of the specimens when compared to the uncrosslinked specimens, a significant result is the much higher degradation resistance of the glutaraldehyde-crosslinked specimens. Enhancement of the elastic modulus and the swelling ratio of the specimens, determined as the ratio of the total volume (solid + water) to the solid volume of the specimens, provide additional measures for the effectiveness of the glutaraldehyde crosslinking. A combination of dehydrothermal treatment and exposure to cyanamide is as effective as glutaraldehyde, although it involves a more elaborate and time-consuming process.

Properties of Collagen Biomaterials

In addition to their capacity to be formed into a variety of shapes and architectures (Section 10.3.1.3), the major characteristics of collagen biomaterials are their biocompatibility, chemical stability in the physiological environment and mechanical properties. As collagen biomaterials are susceptible to denaturation when subjected to even modest temperature changes, their thermal stability is also relevant.

Biocompatibility of Collagen Biomaterials As collagen (type I) biomaterials are commonly produced from purified collagen that has been extracted from animal species, they should not suffer from the inherent risk of toxicity that may arise from synthetic polymers unless toxic substances are introduced during their processing. Collagen biomaterials generally elicit a limited immune response but this is often well tolerated due to the conservation of the collagen structure and amino acid sequence across human and animal species. Crosslinking, used primarily to control the degradation and mechanical properties of collagen biomaterials, and enzymatic removal of non-helical telopeptides from the ends of the collagen molecules provide common techniques to suppress any immunogenicity of collagen. On the other hand, collagen does have the potential to be immunogenic and a small number of patients (less than a few percent) do show an immune response to collagen, particularly injectable forms of collagen.

Degradation of Collagen Biomaterials Degradation of collagen (type 1) biomaterials *in vivo* occurs by hydrolysis mediated by enzymes called collagenases, a group of enzymes that belong to a family of proteolytic enzymes known as matrix metalloproteinases. These enzymes cleave peptide bonds within the polypeptide chains, producing degradation products composed of peptides and amino acids. Degradation of collagen-based implants and devices takes place over a period of a few months to several months, depending on the degree of crosslinking, effectiveness of the crosslinking agent, and the architecture of the implant or device.

Thermal Stability of Collagen Collagen undergoes denaturation, that is, disruption of its native structure, and shrinks upon heating to elevated temperatures. Differential scanning calorimetry (DSC) is the most accurate technique for determining the denaturation temperature T_D of collagen, but other methods, such as measuring changes in viscosity and shrinkage, may also be used. DSC provides the capacity to determine T_D for not just the biomaterial but the collagen molecule, fibril and fiber as well, thermodynamic properties such as the enthalpy of denaturation, and the effects of various treatments such as hydration and crosslinking on T_D.

Collagen molecules in solution (soluble collagen) denature into individual polypeptide chains upon warming to ~35–40 °C, for collagen prepared from bovine skin, a common source of collagen used in the creation of biomaterials (Figure 10.6). In comparison, hydrated fibrillar collagen (insoluble collagen) does not denature until ~60–65 °C due to the restricted mobility imposed by the bonding between adjacent collagen molecules in the fibrils. Crosslinking with, for example, glutaraldehyde further stabilizes the fibrillar structure and further reduces the mobility of the collagen molecules, leading to an increase in T_D to ~85–95 °C. This increase in T_D correlates with a decrease in the free NH_2 groups of lysine and hydroxylysine residues of the polypeptide chains that react with glutaraldehyde (Figure 10.4). Dehydration of uncrosslinked collagen also leads to an increase in T_D because the formation of crosslinks leads to a reduction in the mobility of the collagen molecules. For example, fully dehydrated collagen shows a T_D of ~220 °C.

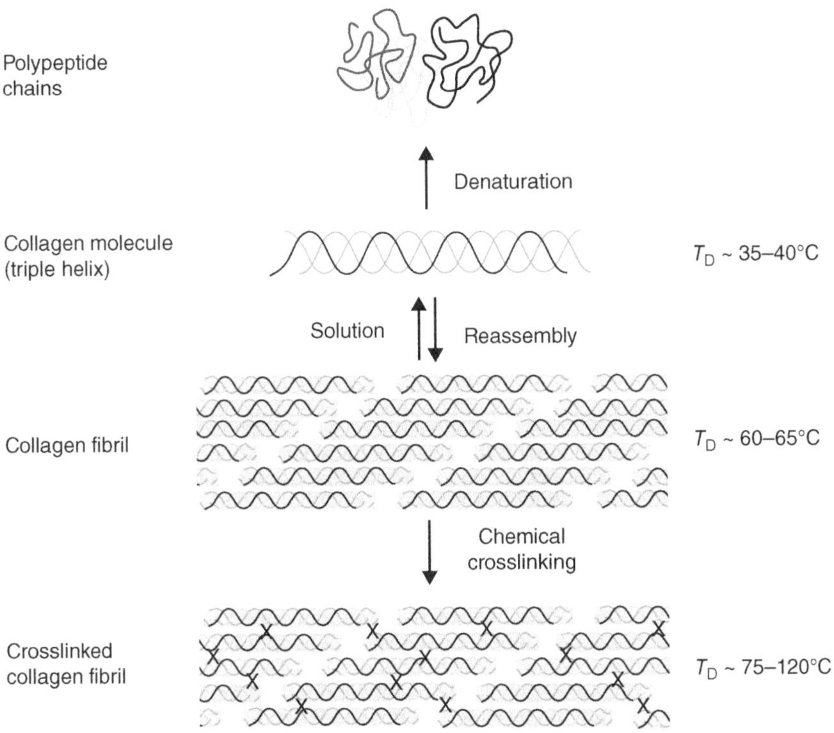

Figure 10.6 Schematic of different collagen structures and their denaturation temperature T_D.

Mechanical Properties of Collagen The mechanical properties of fibrillar collagen have been studied and modeled at different length scales of its hierarchical structure, that is, at the molecular scale (response of an individual triple-helical collagen molecule), at the fibrillar scale (response of an individual fibril), at the microscale (response of an individual fiber), and at the macroscale (response of a collagenous tissue). Although there are large variations in the results for the same hierarchical level, depending on the model or technique used, a general outcome is that the Young's modulus decreases as the length scale of the hierarchical structure increases (Table 10.1). This trend is generally attributed to the activation of different deformation mechanisms. For example, inter-fibrillar sliding is a dominant mechanism at the smaller structural scale, whereas straightening and orientation of fibrils and fibers, along with inter-fibrillar and inter-fiber sliding are the dominant mechanisms at the macroscale of a tissue.

At the molecular scale, several experimental and modeling techniques lead to an estimate in the range ~2–7 GPa for the Young's modulus of a collagen molecule. For wet fibrils, which are relevant to *in vivo* conditions, a Young's modulus of ~0.2–1.0 GPa has been estimated in some studies. Fibers extracted from rat-tail tendon and tested in a hydrated state have shown a linear modulus in the range ~50–250 MPa depending on their length, but crosslinking can lead to a value of ~1 GPa, comparable to the Young's modulus of a fibril.

Table 10.1 Comparison of the Young's modulus of collagen at multiple hierarchical levels.

Hierarchical level	Young's modulus (GPa)
Molecule	~2–7
Fibril (wet)	~0.2–1
Fiber	
Non-crosslinked (rat tail tendon)	0.05–0.25
Crosslinked (rat tail tendon)	~1
Tissue	
Human tendon	~1
Human skin (longitudinal body direction)	~0.01–0.02

At the macroscale level, fibrils and fibers in tissues are arranged in a vast variety of structures, resulting in a wide range of mechanical properties. However, the mechanical response (stress versus strain curve) of collagenous tissues often show linear and nonlinear regions, followed by region of failure (Figure 10.7). Attempts have been made to correlate each region with deformation mechanisms at different structural levels of the material. For example, an initial region of low but increasing slope, sometimes called a toe region, has been associated with straightening and alignment of the fibril and fiber arrangement. Then the slope becomes approximately linear, attributed to slipping of the fibrils and stretching of the fibers. Finally, the crosslinks and fibers begin to fail due to delamination of the fibrils, leading to failure.

The slope of the linear region of the stress versus strain curve in Figure 10.7 is generally taken as a measure of the stiffness of the collagenous tissue and, thus, it is often reported as a modulus. However, this modulus does not correspond to the classical definition of the Young's modulus (Chapter 4) and, consequently, it is sometimes referred to as the linear modulus or the tangent modulus. Like synthetic polymers, collagenous tissues and biomaterials are viscoelastic, and their mechanical properties such as strength and modulus are dependent on the strain rate (or time) of the test. However, tissues have often been treated as nonlinear elastic materials, which neglects the viscous component of their response.

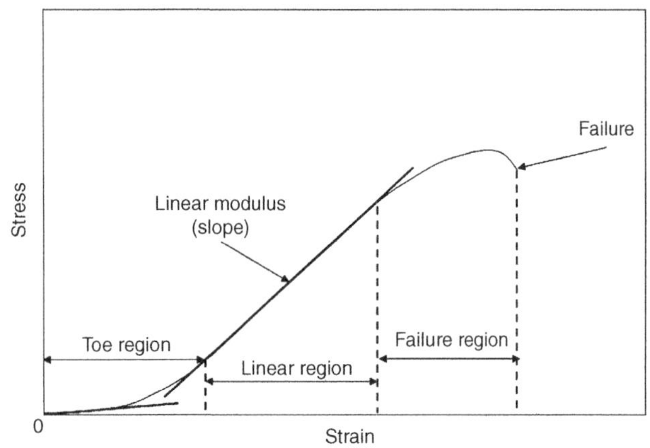

Figure 10.7 Illustration of the mechanical response of collagenous tissues in tension. The response can generally be separated into three regions, designated toe, linear, and failure. The linear modulus is often taken as a measure of the elastic modulus in tension.

The fibrils and fibers in tendons are arranged in a slightly wavy pattern parallel to the length of the tendon. Because of the limited amount of straightening and alignment of the fibrils and fibers upon application of a stress, the mechanical response of tendons shows a small toe region. This is followed by an approximately linear region of steep slope, followed by failure (Figure 10.8a). Human tendons, for example, show a high linear modulus (up to ~1 GPa) and a high tensile strength (up to ~200 MPa), but a strain to failure of less than ~10–15%. In comparison, while the fibrils and fibers in skin have a variety of arrangements, depending on the species, they are often arranged in a highly curved pattern in which they have different orientations. This arrangement gives a much larger toe region and a lower linear modulus (Figure 10.8b). Measurements of the linear modulus of human skin *in vivo* present significant difficulties but they have yielded values of ~0.4–0.8 MPa using a technique based on torsional deformation to as high as ~5–20 MPa using a tensile deformation technique.

Collagen biomaterials for use in biomedical applications, as noted earlier, are commonly produced from insoluble collagen, and they are designed to have a desired three-dimensional shape and architecture, such as a thin sheet, fiber or foam. While factors such as alignment of the fibrillar structure, crosslinking and microstructure (volume fraction, size, and orientation of pores) have a strong effect, the mechanical properties of these collagen biomaterials are generally weak. Crosslinked single fibers (diameter ~125 μm in the hydrated state) formed by extrusion

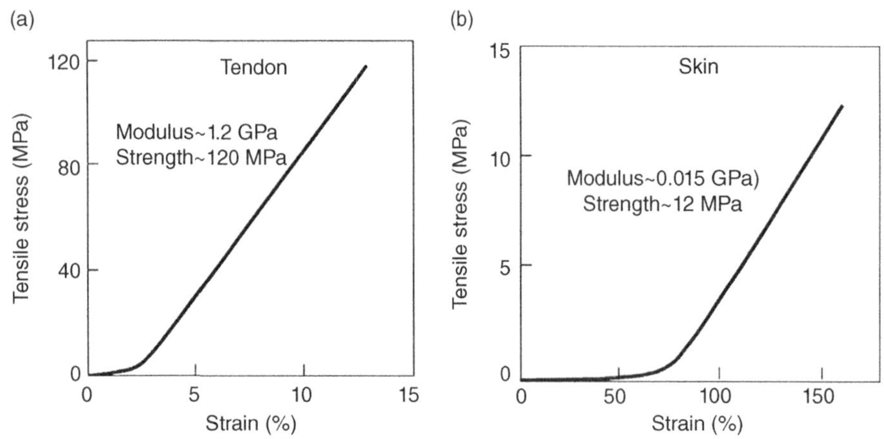

Figure 10.8 Typical tensile stress versus strain curves of collagen structures for (a) human tendon and (b) human skin (longitudinal body direction). The vast variation in the mechanical response shows the importance of the geometric arrangement of the collagen fibrils in the structures.

have been reported to have a linear modulus of ~360 MPa and a tensile strength of ~36 MPa, while crosslinked bundles composed of 10 similar fibers showed a liner modulus of ~260 MPa and a tensile strength of ~20 MPa. In comparison, collagen foams have shown a linear modulus in the range ~10–50 kPa and a tensile strength in the range ~5–20 kPa.

Applications of Collagen Biomaterials

Collagen is one of the most widely used natural polymers in biomedical applications. In the form of particles, thin sheets, fibers, foams or three-dimensional solids (porous or dense), collagen (type I) biomaterials, are used in a variety of biomedical applications, such as wound and burn dressings, drug delivery, and tissue engineering. Overall, these biomaterials make use of the advantages of collagen, such as its biocompatibility, degradation, and capacity to interact in a favorable manner with the ECM. In an injectable form, collagen is used as a filler for tissues such as the vocal chords and the urethra, and for cosmetic effects. An important property of this injectable form is that collagen molecules in a cold solution can reform into fibrils upon warming to the body temperature (Figure 10.6). Composites of collagen, commonly with hydroxyapatite or β-tricalcium phosphate (β-TCP) particles (Chapter 12), are also used in healing and regenerating nonstructural bone. The hydroxyapatite or β-TCP particles enhance the stiffness and bioactivity of the collagen matrix. Blends (mixtures) of collagen with and other polymers, such as polysaccharides, for example, have also attracted interest for use in tissue engineering applications.

10.3.2 Gelatin

Gelatin is a denatured form of collagen, obtained by partial hydrolysis of collagen in which the triple-helical chains of collagen are separated, giving individual polypeptide chains that have molecular weights in the range ~20–250 kDa (Figure 10.9). As it is a denatured form of collagen, gelatin is biocompatible *in vivo*. Gelatin is degraded by hydrolysis mediated by enzymes called proteases that cleave the polypeptide chains into peptides and amino acids. Degradation of gelatin can be controlled by chemical crosslinking, using agents such as glutaraldehyde, in a manner similar to that of collagen. Commercial gelatin is often produced in a particulate form, of size ranging from fine particles to granules. It is soluble in warm water, forming a sol (that is, a solution of polypeptide chains), which

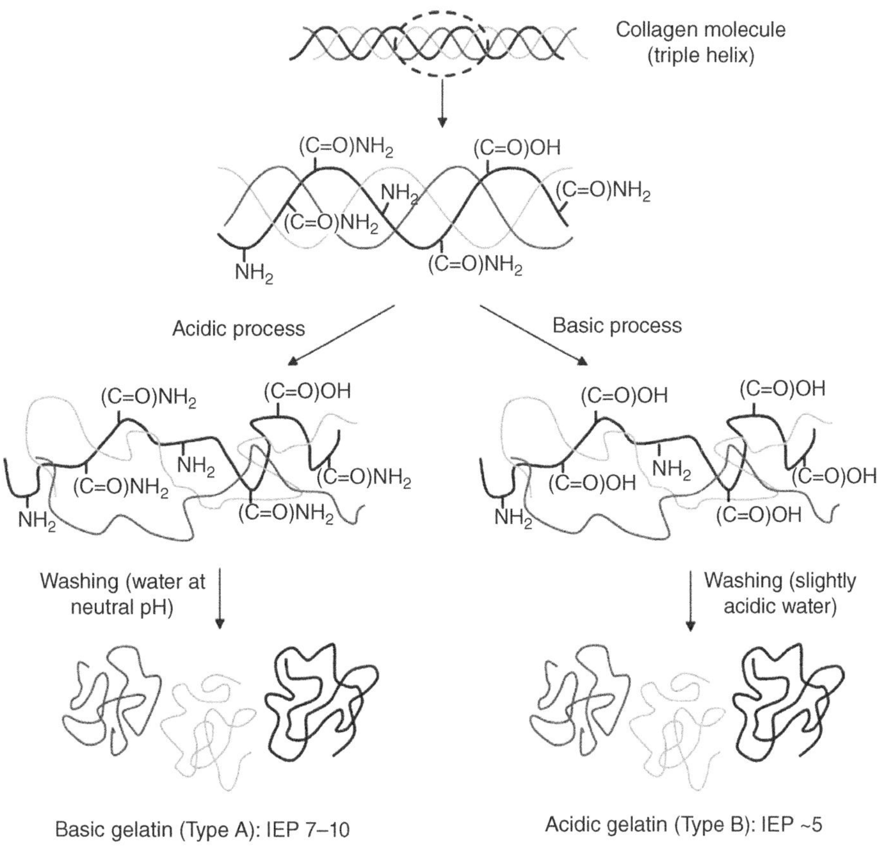

Figure 10.9 Schematic showing method for extracting gelatin from collagenous tissues by an acidic and a basic process.

can be converted into a gel by cooling below ~30 °C or processed by common techniques into microspheres or blocks (porous or dense). The major biomedical applications of gelatin are in the form of microspheres and porous blocks for use as a carrier for controlled delivery of biomolecules, such as antibiotics to cure infected tissues and growth factors in tissue engineering. Porous blocks (sponges) also provide an effective scaffold for seeding cells, such as human bone marrow-derived mesenchymal stem cells for creating engineered soft tissues.

Gelatin is commonly prepared from animal tissues rich in collagen type I, such as bovine and porcine skin. It is prepared by two separate methods that include treatment in acidic conditions (giving type A gelatin) or in basic conditions (type B gelatin). In aqueous solution, gelatin polypeptides are amphoteric in character. This is due primarily to the presence of amine NH_2 and carboxyl (C=O)OH groups in some amino acid residues and at the ends of the chains, which can be protonated and deprotonated, respectively (Figure 5.14). However, type A and type B polypeptides have small differences in composition, such as the ratio of (C=O)OH to NH_2 groups, which alter their electrostatic characteristics and, thus, their interaction with electrostatically charged molecules.

In aqueous media, the pH at which particles or polymer molecules have a net electrostatic charge of zero is termed the isoelectric point (IEP). At pH values less than the IEP, the particles or molecules have a net positive charge whereas they have a net negative charge at pH values greater than the IEP (Chapter 5). Whereas acidic conditions used in preparing gelatin has little effect on the NH_2 groups of polypeptides, alkaline conditions lead to hydrolysis of the (C=O)NH_2 groups in glutamine and asparagine, converting them to (C=O)OH groups. Consequently, gelatin prepared under acidic and basic conditions differ in electrostatic behavior when dissolved in an aqueous medium. Whereas type A gelatin chains have an IEP (~7–10) comparable to that of collagen molecules, type B chains have an IEP of ~5 due to its higher proportion of (C=O)OH groups. This has significant consequences for the interaction between gelatin polypeptides and positive or negative charged molecules when using gelatin as a carrier in drug delivery. For example, under physiological conditions (pH ~ 7.4), type B gelatin (IEP ~5) is better suited as a carrier for basic therapeutic molecules whereas type A collagen (IEP ~7–10) is better suited for acidic molecules.

Gelatin sols can be easily formed into solid microspheres by common techniques such as emulsion techniques, or into gels by cooling. Gels can subsequently be converted into blocks, dense or porous, by direct freeze-drying or by a two-step process in which the gel is frozen to solidify the aqueous liquid to ice followed by subliming the ice in a freeze dryer. After crosslinking, the particles or blocks may be impregnated with an aqueous solution of biomolecules. Electrostatic interaction between charged biomolecules and gelatin polypeptides leads to stable immobilization of the biomolecules within the gelatin matrix. Release of the biomolecules is controlled by the degradation of the gelatin, which can be controlled by its degree of crosslinking.

10.3.3 Silk

Silk has been used as threads and cloths for a few thousand years, whereas silk fibers have been used as sutures for a few centuries. In the last few decades, silk has been attracting interest as a biomaterial for use in other biomedical applications, particularly in tissue engineering. Silks are proteins synthesized in the form of fibers by certain arthropods such as silkworms and spiders. Depending on their source, silks vary widely in composition, structure, and mechanical properties, which are related to their specific natural function. Silk fibers derived from the cocoons of the domesticated silkworm *Bombyx mori* (*B. mori*) and from the webs of spiders such as *Nephila clavipes* and *Araneus diadematus* have been the most widely studied and characterized silks relevant to biomaterials. Synthetic silks produced by recombinant methods have also been attracting interest. However, purified silk derived from *B. mori* will form the major focus of the following discussion because it is the most popular silk relevant to biomaterials.

Composition and Structure of Silkworm Silk

As with other materials, the composition and structure of silk control its physicochemical properties. Purified silk derived from the silkworm *B. mori* consists of single continuous fibers of diameter ~10–25 μm. Like collagen fibers, these fibers have a hierarchical structure, from the molecular scale to the microscale (Figure 10.10), but their structure at different length scales are different from those of collagen. This hierarchical structure, as we have seen for collagen, controls the physicochemical properties of silk, particularly its mechanical properties.

The primary structure of silk fibroin, the protein that constitutes almost all of the silk fiber, dominates the hierarchical structure at a molecular scale. Silk fibroin from *B. mori* consists of two distinct parts, a high molecular weight chain (~391 kDa), often referred to as H-fibroin and a chain of much lower molecular weight (~26 kDa), referred to as L-fibroin. A disulfide (covalent) bond at the carboxyl (C) terminus of H-fibroin links H-fibroin and L-fibroin together. H-fibroin from *B. mori* is composed primarily of three amino acids, ~46% glycine (G), ~30% alanine (A) and ~12% serine (S), plus ~5% tyrosine (Y), ~2% valine (V), plus

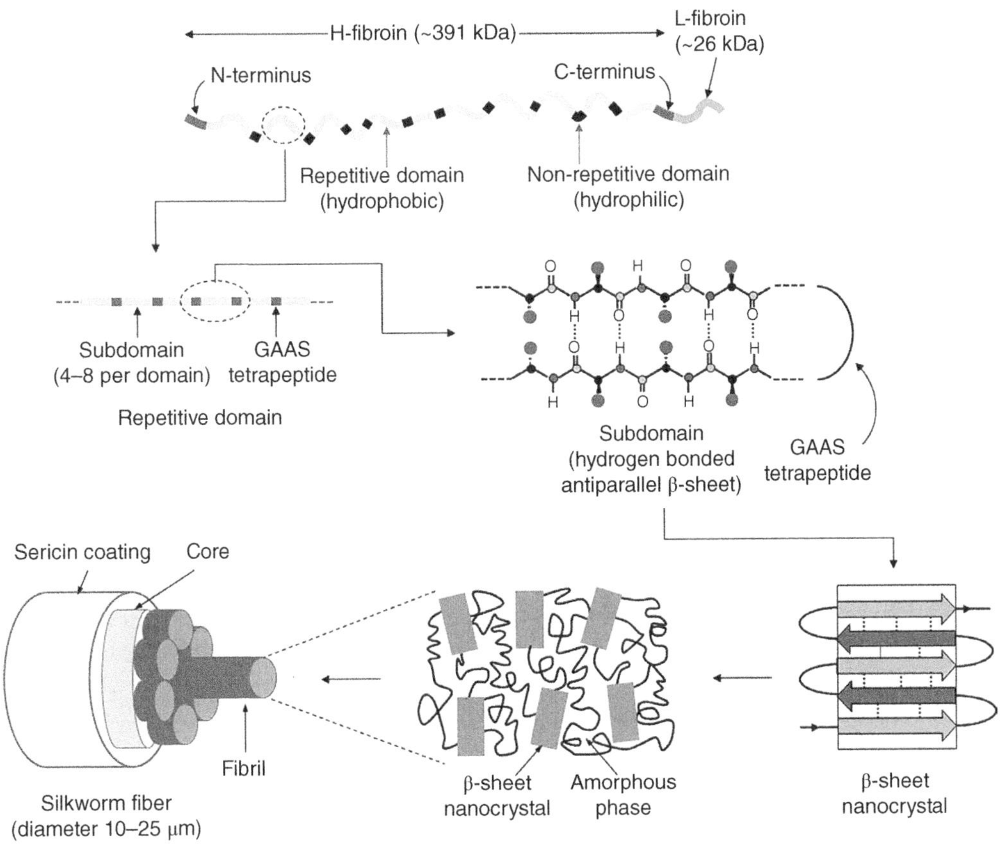

Figure 10.10 Illustration of the hierarchical structure of silkworm silk fiber.

~5% of other amino acids (based on the total number of amino acid residues). In comparison, L-fibroin, has a very different composition consisting of a variety of amino acids, such as A, D (aspartic acid), G, E (glutamic acid), S, V, I (isoleucine), L (leucine), present in varying amounts in the range 7–15%, plus lower amounts of several other amino acids. Generally, L-fibroin plays only a marginal role in the properties of silkworm silk fibers.

H-fibroin consists of a total of 5263 amino acid residues, consisting of 12 longer regions, referred to as repetitive domains or "crystalline" domains, which are connected to one another by 11 short segments (non-repetitive domains or "amorphous" domains), plus short chain segments at its N and C terminals. Each of the 12 repetitive domain, composed mainly of the amino acid residues G, A, and S, is made up of subdomains consisting of ~70 residues. These subdomains have a repetitive sequence that begins with a repeating hexapeptide arrangement composed predominantly of GAGAGS, and terminates with a GAAS tetrapeptide. In comparison, the short segments connecting the repetitive domains consist of 42–44 residues each. They are composed of a nearly identical sequence of amino acids, which contains a segment composed of 25 amino acid residues arranged in a non-repetitive manner. This non-repetitive sequence contains some amino acids that are not present in the repetitive domains.

The next scale of the *B. mori* silk structure is dominated by the secondary protein structure. Some parts of the repetitive domains in H-fibroin, particularly those composed of the smaller amino acid residues G, A, and S, organize themselves into antiparallel β-sheet structures stabilized mainly by hydrogen bonds (Figure 2.21), forming small crystals (nanocrystals) having dimensions of a few nanometers. In comparison, other parts of the repetitive domains and the non-repetitive domains have a less ordered structure, forming disordered chain segments (amorphous regions) that connect the nanocrystals together (Figure 10.10). Although these regions are considered amorphous, as determined by a lack of coherent X-ray diffraction, they may consist not just of randomly arranged chain segments but α-helical structures and less ordered β-sheets as well.

At the next level, the nanocrystals and amorphous regions of a single polypeptide arrange themselves to form a fibril of diameter ~30 nm, in which disordered chain segments of the amorphous regions connect the nanocrystals. At the microscale, these fibrils are arranged in an approximately parallel manner to form a fiber of diameter ~10–25 μm.

The β-sheets making up the nanocrystals are arranged approximately parallel to the axis of the fibrils and, thus, approximately parallel to the fiber axis. The amount of crystalline phase in the fibers is ~60%, with the remaining 40% composed of the amorphous phase. Overall, fibers from *B. mori* may be considered a semicrystalline polymer consisting of an amorphous phase composed of a disordered chain segments of amino acid residues, which is reinforced with nanocrystals composed of β-sheets.

Production of Silk Biomaterials

Purified silk fibers can be converted into implants and devices having a variety of shapes and architectures using standard procedures. Braided, twisted, knotted, or nonwoven structures are formed directly from the purified fibers themselves. On the other hand, structures such as films, microspheres, sponges, nanofibrous mats, and gels are formed from appropriate suspensions of silk fibers using techniques such as casting, emulsion techniques, freeze-drying, and additive manufacturing (3D printing).

Properties and Applications of Silk Biomaterials

A well-known characteristic of silkworm and spider silk fibers is their strong mechanical properties. For example, single fibers from *B. mori* have a Young's modulus of ~10 GPa, a tensile strength of ~300–700 MPa, a strain to failure of ~20%, and a toughness of ~70–80 MJ/m^3 (determined from the area under the stress versus strain curve). The tensile strength and Young's modulus of single *B. mori* fibers are significantly higher than the values for single bovine collagen fibers and comparable to the values for some commercial synthetic fibers such as propylene (Table 10.2). These attractive mechanical properties of *B. mori* fibers may be surprising when it is considered that the β-sheets in their constituent nanocrystals are held together by hydrogen bonds, one of the weakest chemical bonds. A simulation has shown that the size and orientation of the β-sheet nanocrystals have a strong effect on the mechanical properties of *B. mori* fibers. The dimensions of the nanocrystals and the nearly parallel orientation of the β-sheets to the fiber axis have been shown to favor shear deformation of the β-sheet structures (rather than bending), which makes the hydrogen bonds more effective in resisting deformation of the fiber.

The mechanical response (tensile stress versus strain) of *B. mori* single fibers generally shows a region of steep slope (less than ~2% strain) followed by a more extended region of shallower slope (up to ~20–25%) (Figure 10.11). Upon application of a tensile stress, deformation occurs primarily by stretching of the disordered chains in the amorphous regions and a considerable fraction of the stress is transferred to the (strong) β-sheet nanocrystals, giving the stress versus strain curve a steep slope. As the stress increases further, some chains in the amorphous regions are ruptured and the randomly arranged chains slide past each other and past the nanocrystals. A deformation process that approximates yield characterizes this stage, and the slope of the stress versus strain curve becomes less steep. Further increase in the stress leads to larger extension of

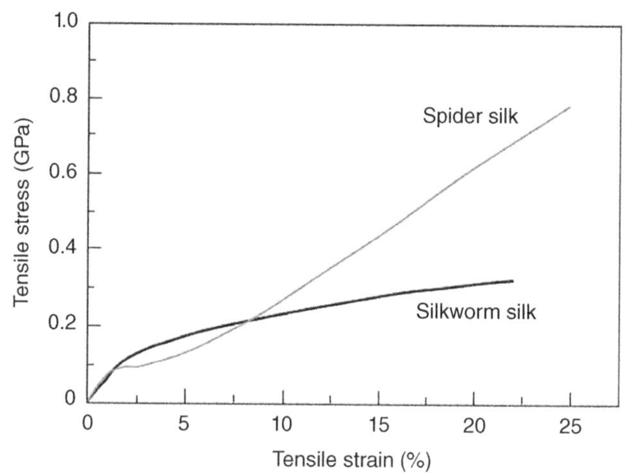

Figure 10.11 Tensile stress versus tensile strain curves for silkworm (*Bombix mori*) silk and spider (*Nephila edulis*) fibers. *Source:* From Sirichaisit et al. (2003) / with permission of American Chemical Society.

Table 10.2 Approximate mechanical properties of some protein-based polymers.

Material	Young's modulus (GPa)	Tensile strength (MPa)	Strain to failure (%)
Collagen (mammalian tendon)	1.2	120	12–15
Elastin (bovine ligament)	0.001	2	150
Silkworm (*B. mori*) silk fiber	10	300–700	20–25
Spider (*Nephila edulis*) fiber	10	1000	20–25
Polypropylene fiber	3–5	500–700	20–25
Kevlar fiber	100–150	3500	3–4

For comparison, the properties of two synthetic fibers, polypropylene and Kevlar, are shown.

the chains, with most of the stress being taken up by the β-sheet crystals. When the stress can no longer be supported, the β-sheet crystals rupture, resulting in failure.

Thin sheets formed from solutions of silk fibers by casting, for example, are weaker than single *B. mori* fibers because their hierarchical structure deviates from that of single fibers. Three-dimensional specimens such as sponges and porous implants created by additive manufacturing (3D printing) techniques are weakened even further because of the presence of substantial porosity.

Biomaterials produced from purified *B. mori* fibers generally show acceptable biocompatibility, although silk sutures have been known to elicit a greater inflammatory response than some synthetic polymers. Although, silk fibroin is cleaved by proteases such as chymotrypsin, particularly in the amorphous regions to form peptides, silk biomaterials degrade slowly. For example, silk sutures lose ~50% of their tensile strength over ~6 months and lose almost all of their strength in ~12 months. Because of this slow degradation, silk sutures are classified in the medical device industry as nondegradable.

Although silk sutures are now used less commonly, the last few decades have seen a heightened interest in silk as a biomaterial for use in a variety of biomedical applications, particularly tissue engineering. This recent interest in silk has been largely driven by its attractive combination of properties and characteristics. Silk biomaterials produced from purified *B. mori* fibers are generally biocompatible and, in the form of fibers (braided or twisted) or thin sheets, they have attractive mechanical properties. The good thermal stability of silk allows it to be formed by a variety of techniques that are not often applicable to some natural polymers. Because of the diversity of the side-group chemistry in its polypeptide chains, silk biomaterials can be modified by attaching growth factors and adhesion molecules to enhance their interaction with cells and tissues. Synthetic forms of silk are also attracting interest because they can be genetically engineered to control their composition and, thus, their degradation. In various forms, silk biomaterials produced from *B. mori* have been evaluated for use in the tissue engineering of a variety of tissues and organs, such as ligaments, blood vessels, nerves, skin, bone, and cartilage.

10.3.4 Elastin

The protein elastin is an important component in tissues of the ECM that have to possess a significant degree of elasticity for their physiological function. In vertebrate tissues, elastin is found in elastic fibers, which may comprise a small (2–4%) but important fraction of the dry weight of the tissue, as in the skin, or much larger amounts, as in the lungs (~30%), large arteries (~50%), and ligaments (~70%). Elastin has a chemical composition that is rich in glycine (~33%), proline (~10–13%) and hydrophobic amino acids constituting ~44% of the total amino acid residues. It is secreted into the extracellular space as a polypeptide precursor called tropoelastin (molecular weight ~72 kDa) where it becomes highly crosslinked by hydrophobic molecules into a network through the activity of the copper-requiring enzyme lysyl oxidase. Although elastin contains ~4% lysine, most of the lysine is present in the crosslinks. The structure of elastin is difficult to determine with a high degree of preciseness, but a common view is that elastin in tissues is composed largely of a crosslinked network of random chains.

These compositional and structural characteristics define the physicochemical properties of elastin. Overall, elastin is hydrophobic, and it is insoluble in aqueous media at the physiological pH (~7.4) and temperature (~37 °C). Upon application of a stress, it shows the classic deformation characteristics of a rubber (Section 8.6). Its high extensibility and recovery after removal of the stress have their origins in entropy changes of the crosslinked network (Figure 10.12). When compared to collagen fibers (Table 10.2), elastin has a much lower Young's modulus

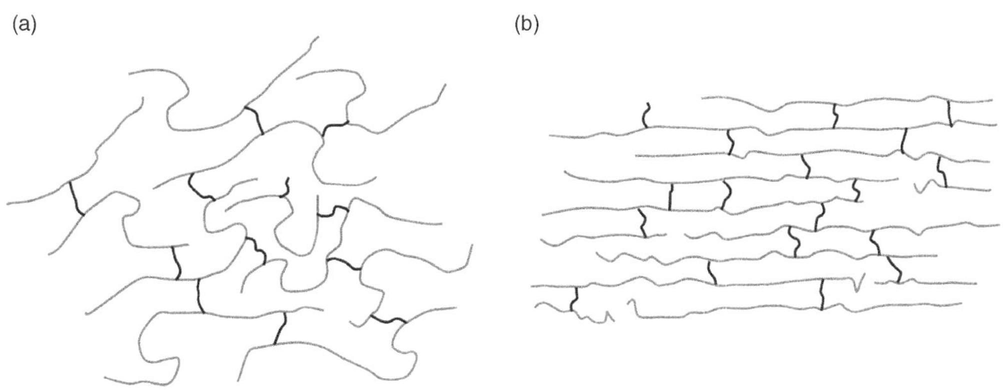

Figure 10.12 Illustration of elastin in (a) relaxed and (b) stretched states.

(~1.0 MPa) and tensile strength (~1 MPa), but a much higher strain to failure (~150%).

Elastin-based biomaterials can be created and used in a few different forms. For example, decellularized tissue specimens containing elastin are created by methods that remove the cells but maintain the three-dimensional architecture of the tissue. Another method involves extracting elastin from animal tissues, commonly bovine and equine ligaments. Soluble and insoluble forms of elastin are obtained by suitable chemical treatments, which can be converted into biomaterials having a variety of shapes and architectures by common processing techniques, as described earlier for collagen. These elastin-based biomaterials have been investigated for use in a variety of biomedical applications, particularly skin substitutes and vascular grafts.

Elastin-like polypeptides (ELPs) synthesized by recombinant methods have been attracting significant interest for use as biomaterials because they possess several attractive characteristics. These polypeptides can be synthesized with a composition consisting of pentapeptide repeating units present in the hydrophobic domains of human tropoelastin and with precise polypeptide length (molecular weight) and polydispersity. They can also be synthesized in the form of di-block copolymers to control their physicochemical properties, such as their hydrophobic to hydrophilic nature. Additionally, biomolecules can be incorporated into them to enhance their biological interaction with tissues *in vivo*.

ELPs synthesized by recombinant methods show a thermally responsive behavior, referred to as thermally induced phase transition or as coacervation. At low temperatures, ELPs are soluble in aqueous media. However, as the temperature of the solution is raised, they become insoluble and aggregate at a certain critical temperature, termed the inverse transition temperature T_t. This phenomenon is similar to the lower critical solution temperature of aqueous polymer solutions in which hydrogen bonds play a significant role (see Section 11.10.2). The process is reversible so that, upon cooling below T_t, the system returns to a solution. Additionally, T_t can be controlled in the range ~0–100 °C to a precision of a few degrees Celsius by controlling the composition of the polypeptide.

ELPs are generally biocompatible and degradable *in vivo*. The thermally induced phase-transition of ELPs can be exploited to create biomaterials in a variety of forms, such as hydrogels, nanoparticles, sponges, and nanoporous devices, which are also suitable for encapsulating cells and therapeutic drugs. The capacity to control T_t is particularly useful in encapsulating and releasing drugs. Overall, elastin-like biomaterials have attractive characteristics that make them suitable for a variety of biomedical applications, particularly drug delivery and tissue engineering.

10.3.5 Fibrin

Fibrin has been used clinically for decades in the form of a glue or sealant to join soft tissues together when sutures are unsuitable, and as a hemostatic dressing to stop bleeding. In the last few decades, fibrin has attracted interest for use in variety of other biomedical applications, particularly as scaffolds for stem cell differentiation and to deliver cells and growth factors in tissue engineering. The formation of fibrin for use as a biomaterial generally aims to recapitulate the process by which fibrin is formed naturally *in vivo*.

In vivo, fibrin is formed as an insoluble fibrous network in the blood coagulation cascade through the hydrolysis of fibrinogen (a soluble protein in blood), a process catalyzed by the protease thrombin that is activated in response to an injury (Chapter 16). Figure 10.13 illustrates major steps in the formation of fibrin. Fibrinogen molecules are elongated structures (~45 nm long), consisting of three pairs of polypeptide chains, called Aα, Bβ, and γ, including the fibrinopeptides FpA and FpB. The amino N-termini of all six chains are joined in the central region E, which is connected to the end domains by α-helical coiled coils. The Aα chains fold back to form a fourth strand of the coiled coil and connect to the αC domains, which bind to the central region.

Fibrin polymerization starts when thrombin cleaves fibrinopeptides FpA and FpB from the central domain of fibrinogen to form a fibrin monomer. This exposes peptide sequences at the N-termini of the α and β chains, termed knob A and knob B, respectively, which are then available to interact with the complementary holes *a* and *b*, located at the C-termini of the γ and β chains, respectively. These interactions occur between two adjacent fibrin monomers to form non-covalent bonds between the E region of one molecule and the D region of another molecule, resulting in the formation of half-staggered dimer. Addition of other molecules leads to the formation of double-stranded structures referred to as protofibrils, which are stabilized by specific protein–protein interactions. Protofibrils extend to a length of ~600–800 nm and then they assemble laterally into fibers by specific weak protein-protein bonds and by interactions with chlorine ions. Then, the fibers branch, leading to the formation of a three-dimensional fibrin network. Most of the branching sites are trifunctional, that is, they consist of junctions between three fibers.

Fibrin glues used as biomaterials aim to recapitulate the *in vivo* process by mixing separate preparations containing fibrinogen and thrombin, just prior to application at the surgical site, as in cardiopulmonary bypass surgery, or at the appropriate time for tissue engineering applications. Blood components used to form fibrin glues may be derived from autogenous blood or allogenic blood, but the latter is commonly used in commercial products. They are obtained using

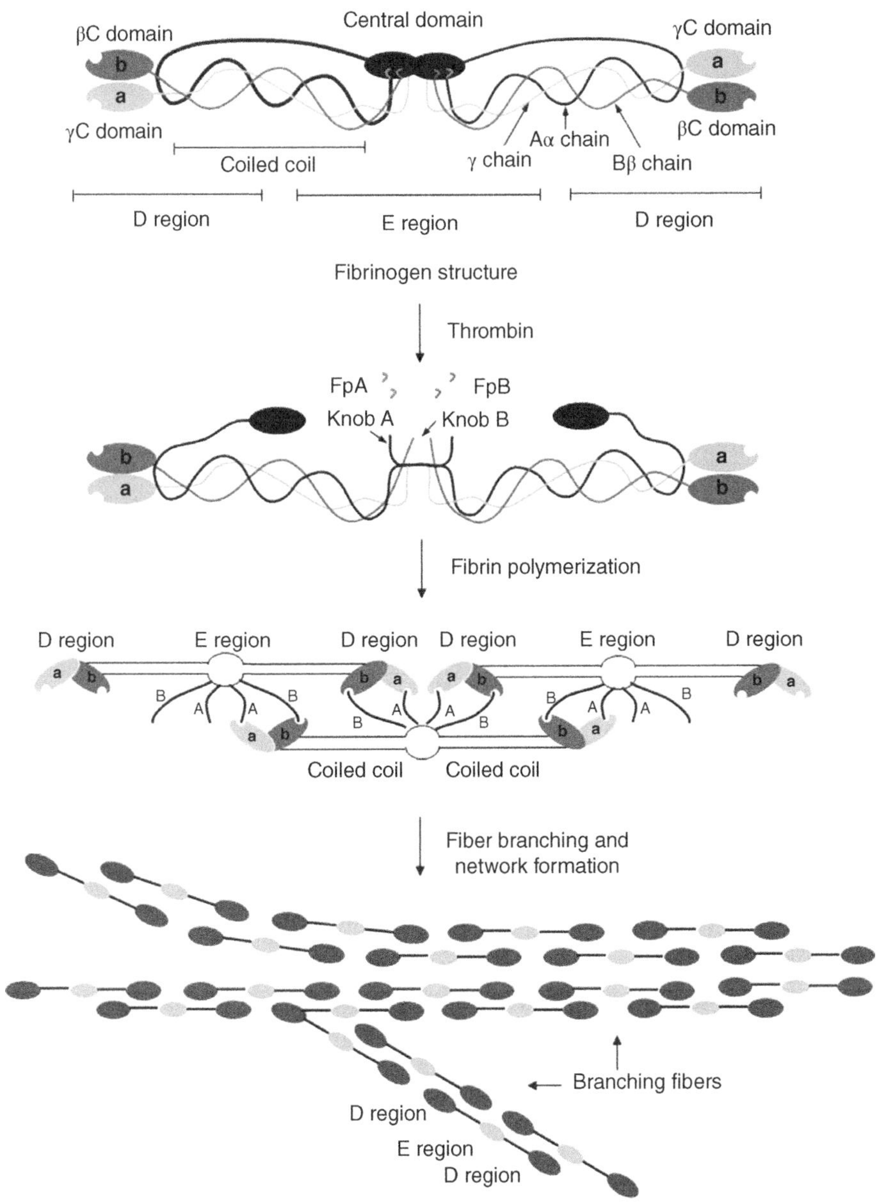

Figure 10.13 Illustration of major steps in the formation of fibrin.

methods that often include a variety of steps such as purification, cryoprecipitation, centrifugation, and filtration.

Fibrin biomaterials used in tissue engineering applications are typically in the form of hydrogels (Chapter 11), in which the solid phase consists of a fibrous three-dimensional network (Figure 10.14). Fibrin hydrogels are easy to produce, and they have desirable features such as bioactivity and biocompatibility due to the role of fibrin as a coagulation protein *in vivo*. However, they also have a few disadvantages. Fibrin hydrogels are weak, and they are highly susceptible to degradation (termed fibrinolysis) by cleavage at specific sites, mediated predominantly by the protease plasmin. Several tissue-engineering applications require the use of scaffolds having an optimal combination of strength, porosity, and degradation rate for the intended application. These desirable properties may be achieved through modification of the fibrin composition and structure at a molecular level using appropriate polymerization techniques, such as synthesis conditions, copolymerization, crosslinking, and functionalization by the addition of biologically active molecules (Chapter 13).

10.3.6 Laminin

Laminin is a high molecular weight (~400–900 kDa) protein of the ECM, which forms a major component of

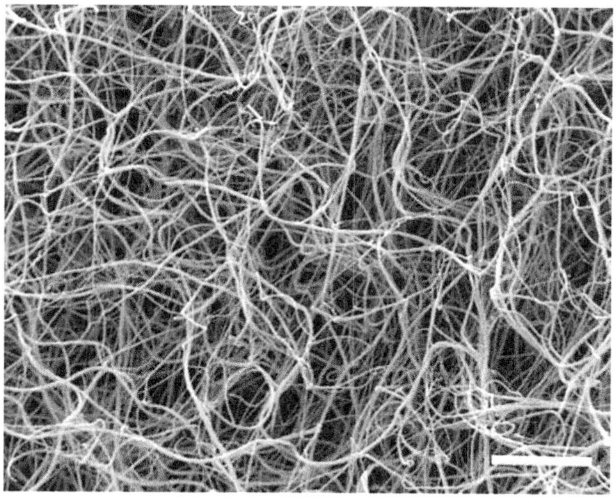

Figure 10.14 Scanning electron microscope image of the structure of fibrin formed by polymerizing 1 mg per liter human fibrinogen with 1 unit (U) of thrombin per milliliter in an aqueous solution containing 150 millimoles NaCl at pH = 7.4. (Bar = 10 μm). *Source:* Courtesy of Paul A. Janmey, University of Pennsylvania.

basement membranes. These membranes take the form of a dense layer beneath many types of cell sheets, particularly the epithelium and endothelium, as in skin and the lining of blood vessels, for example. Although laminin will not form a three-dimensional biomaterial, there is significant interest in using it as a surface coating on biomaterials to enhance their interaction with cells in tissue regeneration. The rationale is that cells are typically in contact with a basement membrane or some other component of the ECM *in vivo*. Consequently, a laminin coating may influence the biomaterial-cell interaction, such as cell adhesion and cell signaling pathways (Chapter 16).

The attachment of laminin molecules to the surface of degradable polymeric biomaterials forms an important strategy in neural tissue engineering. Whereas several methods may be used to attach laminin molecules to the surface of these biomaterials, methods that rely on covalent bonding provide more stable attachment (Chapter 13). A variety of laminin-coated biomaterials such as the degradable synthetic poly(α-hydroxy esters) poly(L-lactic acid) (L-PLA) and polylactic-*co*-glycolic acid (PLGA), and polysaccharides such hyaluronic acid, chitosan and alginate, have shown the capacity to promote nerve regeneration.

10.4 Polysaccharide-Based Polymers

Polysaccharide-based polymers are the main group of natural alternative polymers to protein-based polymers. Polysaccharides are generally defined as carbohydrates

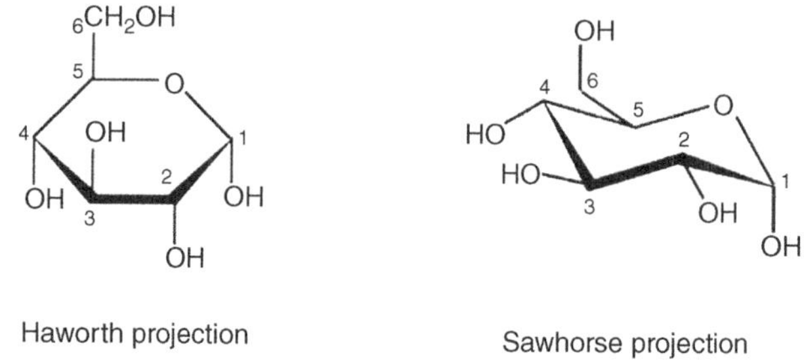

Figure 10.15 Two common ways to depict the chemical structure of a monosaccharide, illustrated for the α-glucose structure. (The C atoms in the ring structure and H atoms at the C atoms are omitted for clarity.)

composed of long linear chains of monosaccharide units that are linked together by glycosidic bonds. The glycosidic bond is generally defined as a bond between one carbohydrate molecule and another molecule that may or may not be a carbohydrate molecule. Carbohydrates are polysaccharides composed of C, H, and O atoms only. For example, whereas sucrose (table sugar) is a disaccharide composed of two monosaccharides (glucose and fructose), the carbohydrates cellulose and starch are polysaccharides composed of many monosaccharide (glucose) units linked together.

Polysaccharide-based polymers relevant to biomaterials are generally produced from animals or plants (including algae). They are composed of long linear (unbranched) chains of high molecular weight (~10–10^4 kDa), consisting of monomer units that normally have a cyclic ring-like structure made up of five C atoms and one O atom. Two common ways to illustrate the cyclic structure of these monomer units are shown in Figure 10.15 for α-glucose. One way, sometimes referred to as the Haworth projection, is to show the structure as a planar hexagonal arrangement, with the side groups above or below the plane. The other way, termed the sawhorse projection, gives a view of the structure down a particular C–C bond. Conventionally, the C atoms are numbered consecutively, starting with the C atom bonded to the O atom in the ring that is also bonded to OH and H.

Hyaluronic acid, alginates, and chitosan have been the most widely studied polysaccharides relevant to biomaterials. They are composed of long linear chains that contain functional side groups, such as polar hydroxyl (OH) groups together with carboxyl (C=O)OH or amine NH_2 groups. As these (C=O)OH and NH_2 groups may become deprotonated and protonated, respectively, in aqueous media depending on the pH, these polysaccharides show certain characteristic features such as:

- *Hydrophilicity*: The molecular chains in these polysaccharides are strongly hydrophilic and have a high capacity to bind with water molecules

- *Electrostatic charge*: Depending on their composition and the pH of the aqueous medium, the molecular chains in these polysaccharides acquire a negative or positive electrostatic charge. Under pH conditions when a large fraction of the functional groups is deprotonated or protonated, these molecular chains may acquire a high negative or positive charge.
- *Solubility*: Molecular chains of these polysaccharides are very soluble in aqueous media over a certain range of pH. On the other hand, under other pH conditions when they are effectively neutral electrostatically, the chains may form a weak insoluble gel due to intermolecular hydrogen bonding.

Because of their strong hydrophilicity and their high solubility under certain pH conditions, hyaluronic acid, alginates and chitosan chains are normally chemically modified to convert them into practical biomaterials. Two common types of chemical modification used for these polysaccharides are crosslinking of their chains and altering the composition of the side groups in the chains. These modifications are generally used for two main purposes:

- To modify physicochemical properties, such as solubility, degradation, and mechanical properties, in order to achieve an optimal combination of properties for a specific application.
- To enhance interaction with proteins and cells *in vivo* because polysaccharides may discourage protein adsorption and cell adhesion due to their strong hydrophilicity (Chapter 16).
- Blends (mixtures) composed of one polysaccharide with a variety of other polysaccharides or other polymers have also received attention but they do not involve chemical modification of the chains.

Hyaluronic acid, alginates, and chitosan can be converted into a variety of forms, such as thin sheets, fibers and fibrous meshes, microspheres, hydrogels, and foams, for a variety of biomedical applications. As discussed in Chapter 11, hydrogels have received considerable interest, particularly for drug delivery and tissue engineering applications. Attractive features of these hydrogels are their capacity to retain a high amount of water (up ~30 wt%) and the ease with which cells and biomolecules (such as growth factors) may be incorporated into them under reasonably mild conditions, particularly in the case of alginate hydrogels. Biomaterials composed of hyaluronic acid and chitosan degrade *in vivo* to form nontoxic molecules, but the degradation of alginate biomaterials is slow. They are biocompatible *in vivo*, eliciting only a limited inflammatory response. Overall, these three polysaccharides provide an attractive set of alternative biomaterials to protein-based polymers for use in a variety of biomedical applications, particularly to deliver cells and biomolecules in healing and regenerating soft tissues.

10.4.1 Hyaluronic Acid

Hyaluronic acid, often abbreviated HA, is also referred to as hyaluronan, which is the generally accepted chemical name for hyaluronic acid and salts of hyaluronate, the conjugate base of hyaluronic acid formed by dissociation of the hydrogen atom in its (C=O)OH groups. Whereas alginates and chitosan are not present in mammalian tissues, HA is an essential GAG in the ECM of mammalian tissues. It is composed of long linear chains (molecular weight ~10^3–10^4 kDa) of disaccharide repeating units composed of D-glucuronic acid and *N*-acetyl-D-glucosamine (Figure 10.16).

The attractiveness of HA as a biomaterial, particularly in tissue engineering applications, stems from two major features. First, HA is an essential component of the ECM of mammals, and it contributes to the properties and function of several tissues of the body. This suggests that it may be possible to recapitulate these properties and function to some extent within HA biomaterials. HA is present in cartilage, synovial fluid, skin, vertebral discs, bone, urinary tract, heart valves, eyes, and various other soft tissues. As an example, the presence of HA is largely responsible for cartilage having the requisite mechanical properties to withstand the complex stresses in joints, whereas its presence in the synovial fluid provides the major source of lubrication for the smooth functioning of joints.

Second, the composition and structure of HA can be modified by a variety of well-known chemical techniques, which can be used to create HA biomaterials having a wide range of physicochemical properties without altering the structure of the molecular chain backbone. For example, esterification of the carboxyl (C=O)OH groups in the chains with hydrophobic groups and chemical crosslinking of the chains can be used to synthesize HA biomaterials

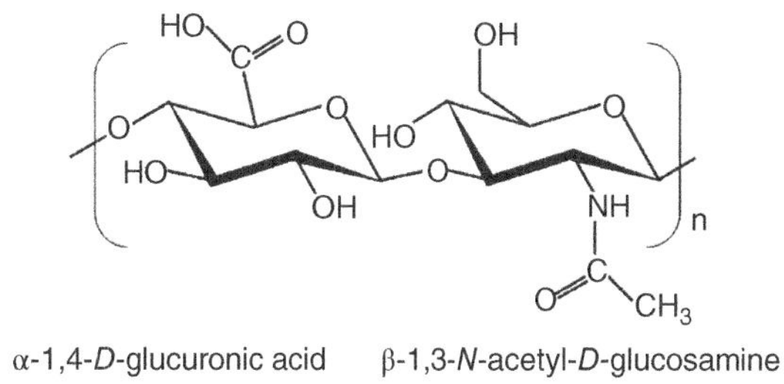

Figure 10.16 Chemical structure of the repeating unit in hyaluronic acid, composed of disaccharide units of α-1,4-D-glucuronic acid and β-1,3-*N*-acetyl-D-glucosamine linked by alternating β-1,4 and β-1,3 glycosidic bonds.

having properties such as solubility, degradation, and stiffness that are different from those for unmodified HA.

Commercial HA has been commonly extracted from rooster combs but it is now also produced on a large scale by genetically engineered bacteria. Because of their high degree of hydrophilicity, HA molecules are very soluble in water. Dissociation of the H atom of the (C=O)OH group in the molecules starts at pH ~3.0, forming (C=O)O$^-$ groups, and the fraction of H atoms dissociated increases with pH such that the molecules are almost fully dissociated at pH ~7–8. Under these conditions, the molecular chains exist as expanded coils due to electrostatic repulsion between the negatively charged (C=O)O$^-$ groups. On the other hand, if the pH is lowered to ~2.5, the sol (solution of molecules) may form a gel because of the absence of an electrostatic repulsion and the presence of an attraction between the chains due to intermolecular hydrogen bonding. Further reduction in pH may convert the gel back to a sol, attributed to protonation of the NH groups in the molecules, which again leads to a strong electrostatic repulsion between the chains.

Chemical Modification of Hyaluronic Acid

HA biomaterials, as noted earlier, are commonly modified chemically to control their physicochemical properties and to enhance their interaction with cells and tissues *in vivo*. The chemical groups in the repeating units of the chain that could be used for chemical modification reactions are the C-2 and C-3 OH groups and the C-6 (C=O)OH group of glucuronic acid, and the C-4 and C-6 OH groups and the C-2 NH group of N-acetylglucosamine.

Acetylation or, more generally, esterification of hydroxyl groups is one of the most important and common reactions for modifying the physicochemical properties of polysaccharides. For example, reactions involving the (C=O)OH group and the OH group of an alcohol ROH, where R is an organic group, have been used to synthesize HA derivatives having different degrees of esterification. In these reactions, different fractions of the more hydrophilic (C=O)OH groups may be replaced by the more hydrophobic benzyl ester (C=O)OR. One such reaction is between an ammonium or sodium hyaluronate solution and benzyl alcohol, where R is the phenyl group (Figure 10.17a). Biomaterials created in this way belong to a family referred to commercially as HYAFF®, which, in various forms such as thin sheets, membranes, microspheres, gauzes, and sponges, are used in a number of tissue engineering applications. As an example of the effect of esterification, tablets of HYAFF 11, in which ~100% of the (C=O)OH groups have been esterified with benzyl alcohol, showed no observable signs of degradation even after implantation for 90 days in rat subcutaneous sites. In comparison, similar implants of HYAFF 11 p75, fractional esterification ~75%, were almost completely degraded within 30 days.

Covalent crosslinking is a common technique used in creating HA biomaterials, commonly in the form of hydrogels or foams, particularly for use in drug delivery and tissue engineering applications. Crosslinking may be achieved using a variety of reactions. For example, one type of reaction does not involve a separate crosslinking agent. Instead, crosslinking is achieved directly between the chains through the use of a base-catalyzed reaction involving the formation of an ester bond between the (C=O)OH and OH groups of the HA chains themselves. Another type of crosslinking reaction utilizes water-soluble agents such as glutaraldehyde and carbodiimide, described earlier for crosslinking collagen (Section 10.3.1). This type of reaction may be used to crosslink other polysaccharides as well.

In a third type of reaction, HA molecules are modified by reacting them with glycidyl methacrylate or methacrylic anhydride. Then they are crosslinked using a photochemical technique such as exposure to UV radiation. For example, sodium hyaluronate molecules in solution are reacted with an excess of methacrylic anhydride to form methacrylated HA molecules (Figure 10.17b). Then a solution composed of purified methacrylated HA and an initiator in phosphate buffered saline is irradiated with UV light to form a crosslinked hydrogel, by a reaction involving the carbon double bond of the modified side group.

Properties and Applications of Hyaluronic Acid Biomaterials

As may be expected from its presence in several tissues, HA having the requisite purity for use in biomedical applications is biocompatible *in vivo*. Degradation of its long molecular chains *in vivo* occurs by hydrolysis catalyzed by a family of enzymes called hyaluronidases, which leads to the formation of smaller nontoxic molecules. HA chains are strongly hydrophilic and, depending on the pH, they may possess a high degree of negative charges, making them capable of interacting strongly with water molecules and positive ions (such as Na^+). Aqueous HA sols, composed of a solution of high molecular weight uncrosslinked chains or a mixture of uncrosslinked and lightly crosslinked chains, have attractive viscoelastic and rheological properties. These properties, plus the well-known function of HA as a major source of lubrication in joints, account for the wide application of aqueous HA sols in cosmetics and in visco-supplementation in joint diseases (such as injections into the joint spaces to treat osteoarthritis).

In the form of hydrogels or sponges, HA biomaterials are widely used in tissue engineering applications, particularly to deliver cells or biomolecules such as growth factors, and as scaffolds to regenerate soft tissues. One advantage of HA

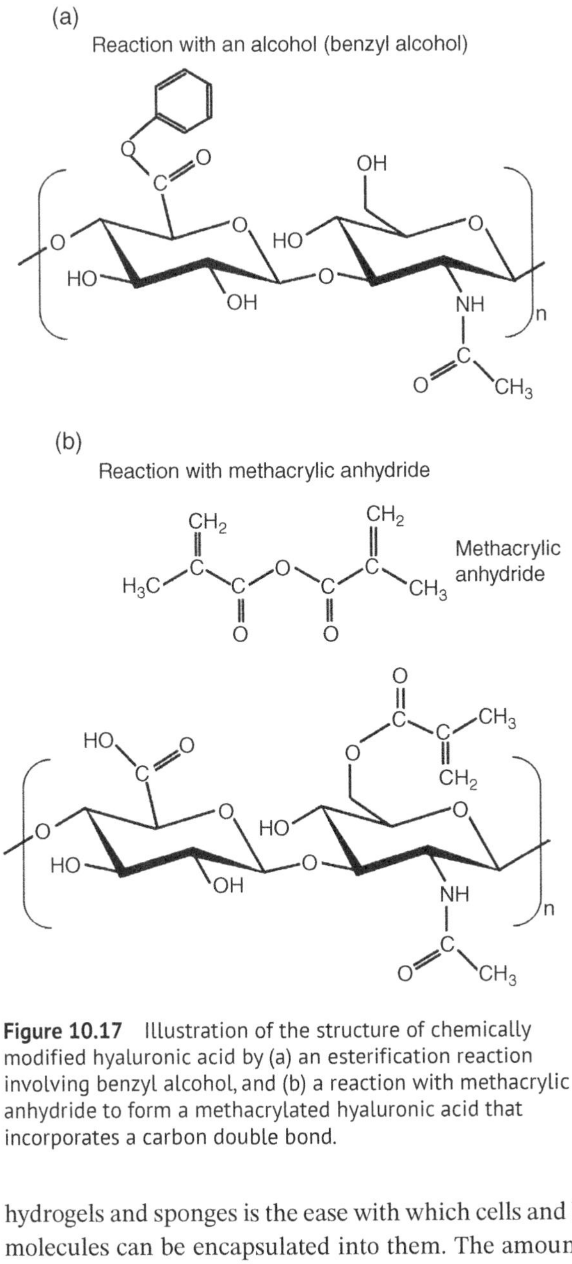

Figure 10.17 Illustration of the structure of chemically modified hyaluronic acid by (a) an esterification reaction involving benzyl alcohol, and (b) a reaction with methacrylic anhydride to form a methacrylated hyaluronic acid that incorporates a carbon double bond.

hydrogels and sponges is the ease with which cells and biomolecules can be encapsulated into them. The amount of swelling and the degradation of HA hydrogels and sponges decrease with their degree of crosslinking. Although the strength and stiffness of these gels and sponges increase with their degree of crosslinking, they are generally low. Consequently, HA hydrogels and sponges are suitable for engineering soft tissues, where it has received significant attention as a scaffold for engineering cartilage.

10.4.2 Sulfated Polysaccharides

A family of sulfated polysaccharides that includes chondroitin sulfate, dermatan sulfate, keratan sulfate, heparin and heparan sulfate have also received interest for use in biomedical applications. Figure 10.18 shows the major disaccharide repeating unit in the chains of these polysaccharides, along with the minor disaccharide units for heparin and heparan sulfate. Overall, the structures of these sulfated polysaccharides have a high degree of variability and, thus, are complex because the sulfonyl hydroxide $S(=O)_2OH$ groups can be bonded to the chains at different positions.

Chondroitin sulfate and heparin have been the most widely studied and applied sulfated polysaccharides in this family of sulfated polysaccharides. Chondroitin sulfate consists of linear chains composed mainly of alternating monosaccharides D-glucuronic acid and *N*-acetyl-D-galactosamine (Figure 10.18). Sulfation of *N*-acetyl-D-glactosamine normally occurs at the OH groups of the C-4 or C-6 atoms, giving two types of chondroitin sulfate, referred to respectively as chondroitin 4-sulfate (or chondroitin sulfate type A), and chondroitin 6-sulfate (or chondroitin sulfate type C). However, sulfation may occur at one or both positions and, thus, disaccharide units having a different number and position of the sulfate groups may occur in the polysaccharide chains. The total number of monosaccharide units in the chains vary, depending on the tissue, from ~50 to ~100, giving a molecular weight in the range ~10–30 kDa, with an average of ~20 kDa. Because of the presence of $S(=O)_2OH$ and carboxyl (C=O)OH groups, chondroitin sulfate chains possess a high negative charge at pH values above 3.0.

Chondroitin sulfate may be viewed as approximating a sulfated form of hyaluronic acid. Like hyaluronic acid, chondroitin sulfate is a major component of the ECM and it contributes to the properties and function of several tissues, particularly cartilage. Chondroitin sulfate is produced commercially by enzymatic digestion of bovine or marine animal tissues, such as bovine nasal septum and trachea, and shark cartilage. In combination with glucosamine, it is widely marketed as a nutritional supplement in the treatment of joint diseases. Like hyaluronic acid, chondroitin sulfate chains are very soluble in aqueous media. Consequently, it is generally crosslinked or blended (mixed) with other polymers (such as collagen, chitosan, hyaluronic acid, polyvinyl alcohol, or polylactic-co-glycolic acid) to form useful biomaterials, typically in the form of hydrogels or foams. These biomaterials have been investigated mainly for tissue engineering applications, to deliver cells and biomolecules (such as growth factors) and as scaffolds for cartilage regeneration.

Heparin consists of linear chains composed of repeating units of pyranosyluronic acid and 2-amino-2-deoxyglucopyranose (D-glucosamine) linked by glycosidic bonds between their C-1 and C-4 atoms (Figure 10.18). Like all polysaccharides, heparin consists of a polydisperse mixture of macromolecular chains. Heparin in the GAGs of

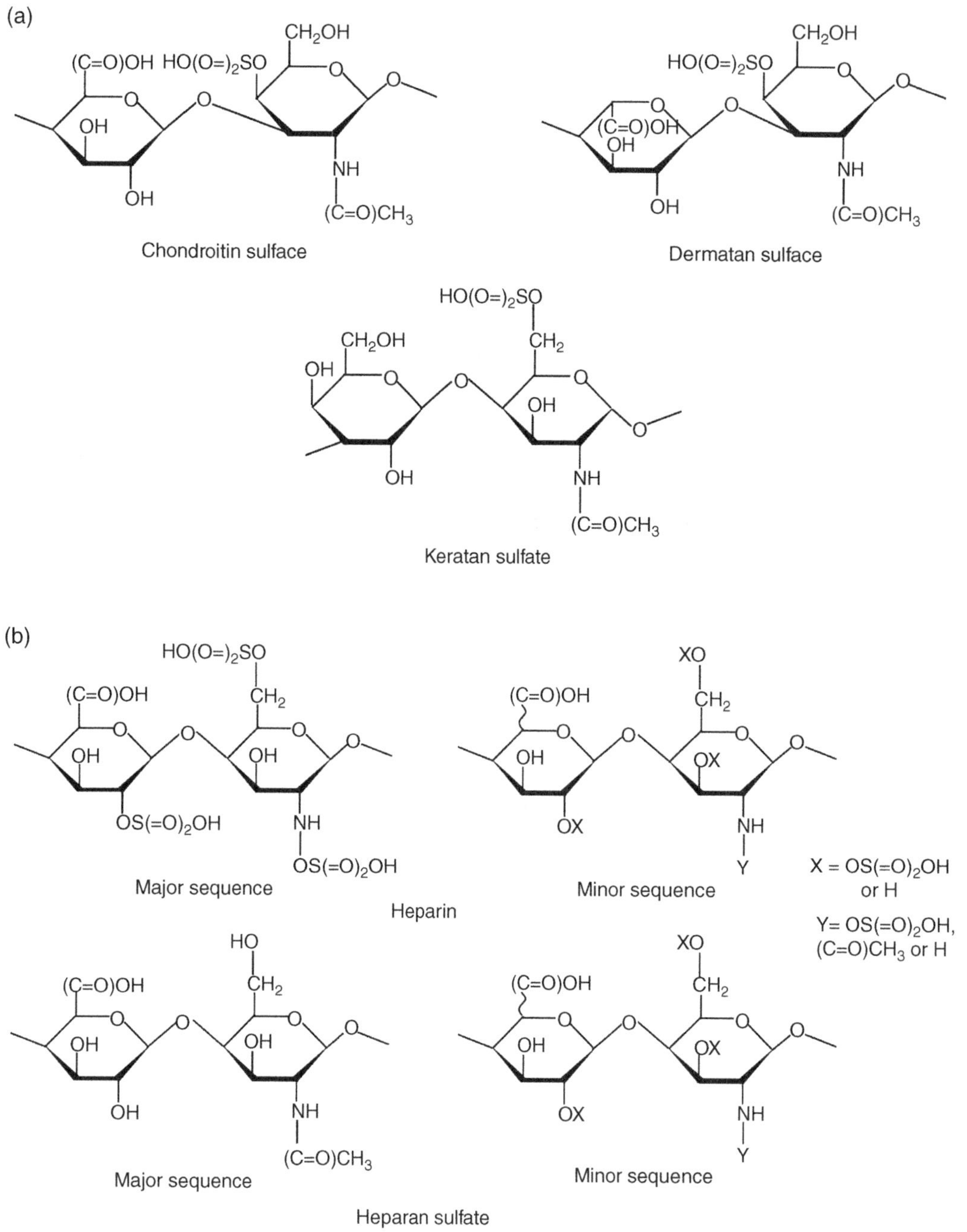

Figure 10.18 Chemical structure of the repeating unit in chondroitin sulfate, dermatan sulfate, keratan sulfate, heparin and heparan sulfate.

tissues has a molecular weight in the range ~5–40 kDa, with an average of ~12 kDa. The major repeating unit of heparin chains consists of a fully sulfated (tri-sulfated) disaccharide but a number of additional disaccharides are present, which makes the structure of heparin complex. Because of the high content of $S(=O)_2OH$ and (C=O)OH groups in its chains, fully sulfated heparin has the highest charge density of any known biological macromolecule.

Heparin is produced commercially by extraction from animal tissues, such as porcine intestine and bovine lung. It has been used for decades as an anticoagulant drug. Heparin has also been used for decades as a coating on biomaterials to reduce their tendency to form a clot when in contact with blood (Chapter 13). More recently, immobilization of heparin on or within biomaterials has been investigated in the delivery of protein-based drugs such as growth factors.

10.4.3 Alginates

Alginates are extracted from marine algae in their wild state because cultivation is too expensive. The supply of alginates is considerable because of the abundance of algae in the marine environment. However, there is some variability in the quality of the extracted alginate, such as composition, sequence of monomer units, and molecular weight, due to its dependence on the algae species and their source. Production of alginates from algae involves a variety of steps. Overall, algae are washed with a dilute acid (such as HCl) to extract their alginate as insoluble alginic acid, which is treated in a solution of sodium carbonate or sodium hydroxide to form soluble sodium alginate (the sodium salt of alginic acid). Then precipitation, purification, drying, and grinding steps lead to a product composed of sodium alginate in the form of a powder.

Structurally, alginates may be viewed as block copolymers consisting of long linear chains composed of sequences of β-D-mannuronic acid (M blocks) and α-L-guluronic acid (G blocks) interspersed with MG blocks composed of alternating M and G units (Figure 10.19). β-D-mannuronic acid and α-L-guluronic acid are stereoisomers, differing only in the placement of the (C=O)OH group in the monomers. Commercially available sodium alginate typically has a molecular weight in the range ~50–400 kDa. Two factors strongly influence the physicochemical properties of alginates:

- The ionic (or nonionic) state of the (C=O)OH groups of the chains
- The ratio of M to G monomers and their sequential distribution in the chains.

Alginic acid, with its (C=O)OH groups in their undissociated form, is insoluble in water, whereas sodium alginate,

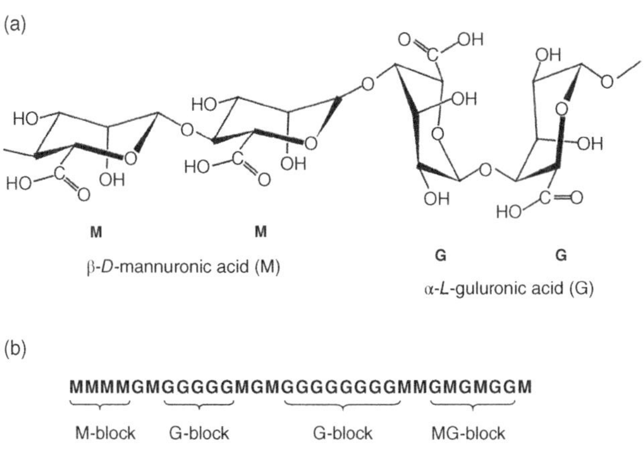

Figure 10.19 Representative alginate structure showing (a) chain conformation and (b) block distribution. *Source:* From Pawar and Edgar (2012) / with permission of Elsevier.

with its $(C=O)O^-$ groups is very soluble. On the other hand, Ca^{2+} ions chelate $(C=O)O^-$ groups between adjacent chains, leading to the formation of insoluble calcium alginate. Chain segments composed of MG blocks are more flexible than M or G blocks, whereas M blocks are less flexible than G blocks. G blocks composed of more than 10 monomer units form stable crosslinked junctions with divalent Ca^{2+} ions, providing the capacity to form gels.

Alginates can be easily created in forms such as thin sheets, microspheres, fibers and fibrous meshes, hydrogels, and foams by techniques commonly used for natural polymers, but hydrogels are receiving the most interest for tissue engineering applications. An attractive feature of alginates is the ease with which alginate solutions (typically sodium alginate) can be converted to a hydrogel (typically calcium alginate) under mild conditions. This makes alginate hydrogels particularly attractive as biomaterials for encapsulating and delivering cells and biomolecules (such as growth factors) in healing and regenerating soft tissues.

Alginate Hydrogels

Alginic acid solutions will form a hydrogel when the pH of the solution is lowered in a controlled manner to below ~3.0, that is, when the (C=O)OH groups in the chains are undissociated. Hydrogen bonds between functional groups of the chains, such as OH and (C=O)OH, lead to the formation of a three-dimensional network. However, the low pH required for gelation is often a limitation. A far more useful method utilizes the capacity of alginates to chelate with divalent cations in solution, allowing the formation of hydrogels under milder conditions. Chelation is driven by electrostatic attraction between the negatively charged $(C=O)O^-$ groups of G blocks and the divalent cations, which lead to the formation of ionic bonds between segments of the chains (Figure 10.20). MG blocks may also participate but they lead to weaker crosslinks. Thus, alginates with a higher fraction of G blocks yield stiffer hydrogels. Alginates have a high affinity for chelating with divalent ions such as Ba^{2+}, Sr^{2+}, and Ca^{2+}, but not with Mg^{2+}.

While the affinity decreases in the order $Ba^{2+} > Sr^{2+} > Ca^{2+}$, that is, with decreasing radius of these ions, Ca^{2+} ions are the most commonly used cations used to induce alginate gel formation. A simple way to form calcium alginate hydrogels is to add a sodium alginate solution in a controlled manner to calcium chloride ($CaCl_2$) solution.

Chemical Modification of Alginates

A variety of chemical reactions has been studied to synthesize alginate derivatives in order to modify the physicochemical properties of alginates and to enhance their interactions with cells. These chemical modifications

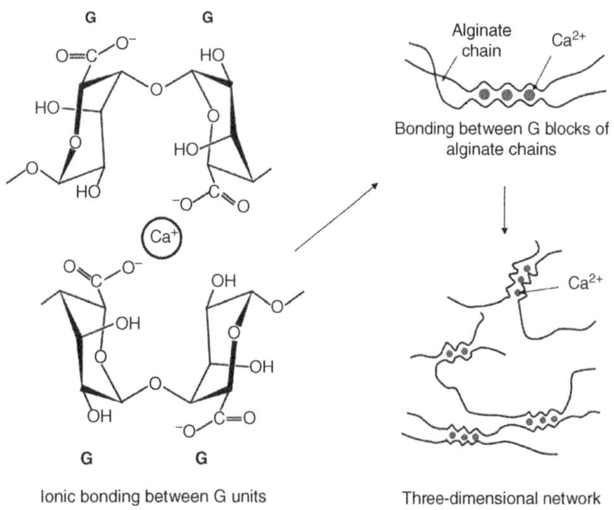

Figure 10.20 Illustration of the formation of a three-dimensional network of alginate chains in aqueous media by calcium (Ca^{+}) ions.

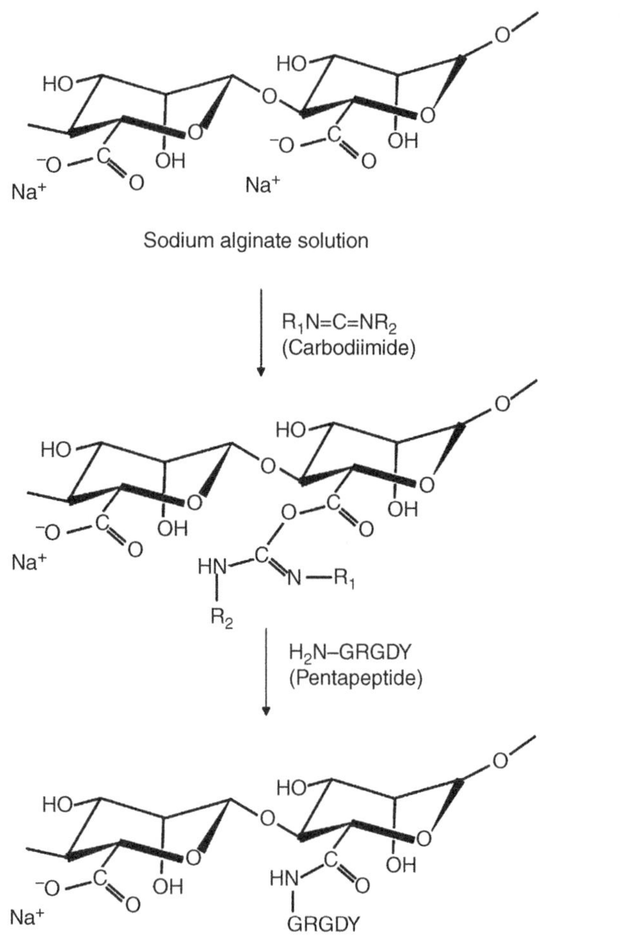

Figure 10.21 Illustration of a possible reaction scheme for the covalent bonding of a pentapeptide GRGDY to alginate chains in a sodium alginate solution, mediated by a water-soluble carbodiimide.

generally involve the C-2 and C-3 OH groups and the C-6 (C=O)OH group in the monosaccharide units. They generally involve aqueous solutions but modifications in organic media have also been developed.

Acetylation or esterification of hydroxyl groups, as noted earlier for hyaluronic acid, is one of the most important and common reactions for reducing the hydrophilicity of polysaccharides. A variety of reactions may be used to synthesize less hydrophilic alginate derivatives. For example, acid-catalyzed acetylation reactions involving alginic acid and an alcohol ROH have been used to synthesize alginic acetate in which the carboxyl (C=O)OH or hydroxyl OH groups in the chains are partially or fully substituted by ester group (C=O)OR or OR groups, where R is an organic group (Figure 10.17a). Another type of reaction, involving sodium alginate solution and long-chain alkyl amines, such as octyl amine $C_8H_{17}NH_2$, has been used to synthesize alginate derivatives in which the $(C{=}O)O^-$ groups are substituted by (C=O)NHR, where R is an alkyl group such, as C_8H_{17}.

Biomaterials composed of alginate molecules crosslinked by ionic bonds, such as calcium alginate hydrogels, lose their mechanical properties over time, often attributed to a diffusion of Ca^{2+} ions from the crosslinks into the surrounding medium. More stable covalent crosslinks may be introduced into alginate hydrogels using a variety of approaches, such as the use of bifunctional carbodiimide molecules, as described for collagen (Figure 10.6b) and the incorporation of methacrylate molecules into the chains that can be crosslinked ptotochemically using UV light, as described for hyaluronic acid (Figure 10.17b).

Covalent bonding of specific cell-adhesion molecules to alginates is a common type of chemical modification that is used to improve the interaction of alginates with a variety of cells (Chapter 13). Peptides containing the RGD sequence of fibronectin, a cell-adhesion molecule having a broad activity, are commonly used in these reactions, where the single letter amino acid abbreviations are used for arginine (R), glycine (G), and aspartic acid (D). In one type of reaction illustrated in Figure 10.21, the pentapeptide GRGDY, where Y is the abbreviation for tyrosine, is attached to sodium alginate molecules or hydrogels in aqueous media. This reaction utilizes a water-soluble carbodiimide such as 1-ethyl-(dimethylaminopropyl) carbodiimide (EDC), which reacts with the dissociated (C=O)OH groups of the alginate chains, thus providing a suitable intermediate for forming an amide bond with the pentapeptide. The reaction normally requires the use of a co-reactant, *N*-hydroxysulfosuccinimide (sulfo-NHS), to stabilize this intermediate against a competing hydrolysis reaction (Section 13.11.3).

Properties and Applications of Alginate Biomaterials

Alginate biomaterials are generally biocompatible *in vivo*. Calcium alginate hydrogels commonly used in a variety of biomedical applications degrade slowly in the physiological environment due to cation exchange of the divalent Ca^{2+} ions with monovalent ions (such as Na^{+}) and their diffusion out of the network. This leads to the generation of soluble molecular chains but further degradation to form smaller molecules is difficult because cleavage of the glycosidic bonds in the chains requires highly acid or basic conditions, or the presence of the enzyme alginase, which is lacking in mammals. As these long chains can be larger than the threshold size for renal clearance by the kidneys, it may be difficult to remove them completely from the body. Consequently, a few techniques have been investigated to modify the degradation of alginate hydrogels, such as the introduction of hydrolytically unstable chemical groups into the chains via oxidation reactions and varying the molecular weight distribution of the oxidized alginates.

The physicochemical properties of alginate biomaterials, as we have seen, depend on the ratio of M to G monomers and their sequential distribution in the molecular chains, and on the type and degree of crosslinking of the chains. For example, a higher fraction of G to M blocks leads to stiffer calcium alginate hydrogels, whereas covalently crosslinked hydrogels are stiffer and more resistant to degradation when compared to ionically crosslinked hydrogels.

Alginate biomaterials have been widely investigated for use in a variety of biomedical applications. In the form of freeze-dried sheets and nonwoven fibrous meshes, they have been used as wound dressings because of their ability to absorb a considerable amount of exudate from the wounds. However, they have received most attention in the form of hydrogels and freeze-dried foams for use in tissue engineering applications to deliver cells and biomolecules in healing and regenerating soft tissues and organs. These applications take advantage of the ease of synthesizing calcium alginate hydrogels and of incorporating cells and biomolecules into them. Calcium alginate gels have been investigated for regenerating cartilage and other soft tissues and organs such as muscle, nerve, pancreas, and liver.

10.4.4 Chitosan

Chitin, derived from the exoskeletons of crustaceans such as crabs and shrimps, is the second most abundant natural polymer in the world after cellulose. Structurally, chitin consists of linear chains composed of repeating units of *N*-acetyl-D-glucosamine linked by glycosidic bonds between the C-1 and C-4 atoms (Figure 10.22a). Chitosan is derived from chitin, being composed of chains with repeating units of *N*-D-glucosamine, in which the repeating units of chitin are fully deacetylated, that is, the acetamide $NH(C{=}O)CH_3$ group in all the repeating units has been replaced by NH_2 (Figure 10.22b). However, the polymer rarely consists of pure chitin or pure chitosan but, instead, it consists of a copolymer composed of these two monomers. Generally, chitin refers to the polymer in which the degree of acetylation exceeds ~50%, that is, more than 50% of the acetamide $NH(C{=}O)CH_3$ groups remain in the chains. In comparison, chitosan refers to the polymer in which the degree of

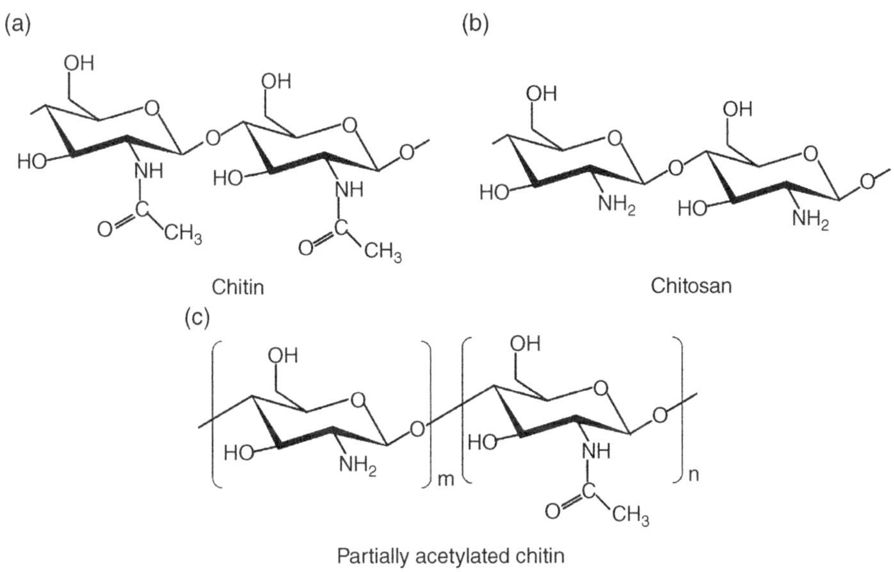

Figure 10.22 Chemical structure of the repeating unit in (a) chitin, (b) chitosan and (c) partially acetylated (deacetylated) chitin. The percentage of n (or m) units in the chain is known as the degree of acetylation (or deacetylation).

acetylation is less than ~50%, that is, amine NH_2 groups have replaced more than 50% of the acetamide groups in the chains. Alternatively, the term degree of deacetylation, the inverse of the degree of acetylation, is used to signify the fraction of the $NH(C{=}O)CH_3$ groups in the chains that have been replaced by NH_2 groups. The polymers may consist of random copolymers or block copolymers (Figure 10.22c).

Chitosan has been attracting increasing interest as a biomaterial, particularly for use in drug delivery and tissue engineering applications. It is prepared from chitin by deacetylation reactions under basic conditions or by enzyme-catalyzed hydrolysis. The quality of chitosan can vary, depending on the source and the preparation process, but commercially available chitosan generally has a degree of acetylation in the range ~5–30% and a molecular weight of ~50–1000 kDa.

The OH and NH_2 groups in the molecular chains and the degree of acetylation strongly influence the physicochemical properties of chitosan. Like hyaluronic acid and alginates such as sodium alginate, chitosan is strongly hydrophilic. However, unlike hyaluronic acid and sodium alginate that are anionic polysaccharides, chitosan is a cationic polysaccharide. The NH_2 group in the monomer units of the chitosan chains have a pK_a of ~6.5 and, thus, as the pH of the aqueous medium decreases below ~6.5, the chitosan chains develop a high positive charge due to protonation of the N atoms of the NH_2 group. In comparison, above this pH, there is essentially no protonation of the NH_2 group and the chains are effectively neutral electrostatically. Consequently, chitosan becomes soluble only when the pH falls below ~6.5, because the chains have an expanded conformation due to electrostatic repulsion between them, which prevents them from aggregating. In comparison, chitosan is insoluble above pH ~6.5 and may form a hydrogel due to linking of the chains by hydrogen bonds.

Chemical Modification of Chitosan

Because of their cationic nature, chitosan chains can bond ionically with various types of negatively charged biomolecules or with deoxyribose nucleic acid (DNA) to form complexes, which can provide the opportunity to enhance the biological activity of chitosan or to protect molecules such as DNA from inactivation or degradation. For example, chitosan molecules can bond ionically with anionic GAGs such as heparin (Figure 10.23a) to form a complex, which can enhance the capacity of chitosan to bond with heparin-binding proteins such as basic fibroblast growth factor (bFGF), a protein that plays an active role in the growth of various tissues. However, as we have seen for hyaluronic acid and alginates, chemical modification provides an alternative and more powerful technique to

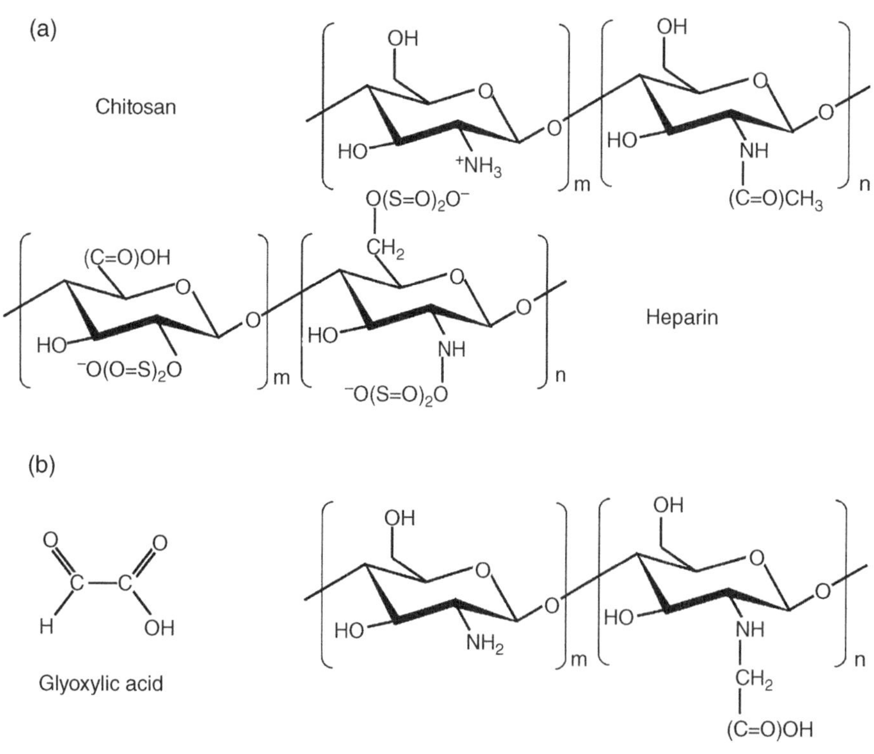

Figure 10.23 Illustration of the chemical modification of chitosan. (a) Ionic bonding between chitosan and heparin, and (b) chitosan derivative formed by reaction with glyoxylic acid.

enhance the physicochemical properties and biological activity of chitosan.

The abundant OH and NH_2 groups in the molecular chains of chitosan provide sites for a variety of reactions, making chitosan very amenable to chemical modification. In general, the majority of these modifications focus mainly on improving the solubility or biological activity of chitosan biomaterials. The reactions may be grouped into specific reactions involving the NH_2 group at the C-2 position and nonspecific reactions involving OH groups at the C-2 and C-6 positions.

Chitosan, as noted earlier, is soluble in aqueous media only when the pH is less than ~6.5, which makes it unsuitable for some biomedical applications. A common derivative of chitosan that provides better solubility over a wider pH range is N-carboxymethylchitosan. One preparation method involves dissolving chitosan is a solution of acetic acid and treating the solution with glyoxylic acid (also known as oxoacetic acid) H(C=O)(C=O)OH. The NH_2 group is modified to $NHCH_2(C{=}O)OH$ (Figure 10.23b). Depending on the starting chitosan material (degree of acetylation and molecular weight, for example) and the amount of glyoxylic acid used in the reaction, chitosan derivatives containing a combination of amine NH_2, acetamide $NH(C{=}O)CH_3$, and $NHCH_2(C{=}O)OH$ groups may be synthesized.

The cationic nature of chitosan, as noted earlier, enables pH-dependent ionic bonding with anionic GAGs such as heparin, enabling heparin-binding proteins to bond to it. Consequently, an assortment of biologically active molecules, such as growth factors, may be attached indirectly to chitosan biomaterials through heparin-bonded intermediates. However, chitosan itself still lacks the ability to support cell adhesion directly because it does not contain chemical groups that are recognized by cell receptors, thus limiting cell adhesion. To address this limitation, peptides containing the RGD sequence of fibronectin may be bonded covalently to chitosan molecules or hydrogels using methods such as water-soluble carbodiimides (as described for alginates) or photochemical methods.

Chitosan chains in aqueous media can be crosslinked to form hydrogels by a variety of physical or chemical methods, which utilize hydrogen bonding, ionic bonding or covalent bonding using crosslinking agents. For example, chitosan-based hydrogels may be formed through ionic bonding between cationic chitosan chains and tripolyphosphate $(P_3O_{10})^{5-}$ anions by adding a chitosan solution in a controlled manner to a solution of sodium tripolyphosphate. Covalently bonded hydrogels may be formed by using water-soluble crosslinking agents such as glutaraldehyde or isocyanates, as described for the crosslinking of collagen.

Properties and Applications of Chitosan

Chitosan is biocompatible, eliciting only a minimal inflammatory response with little or no fibrous encapsulation. It degrades *in vivo* by enzyme-catalyzed hydrolysis to give nontoxic oligomers. Lysozyme is the primary enzyme responsible for hydrolysis, which operates by attacking the acetylated residues in the chain. Consequently, degradation times for chitosan, in the range of weeks to months, are strongly dependent on the degree of acetylation. Compositions having a lower degree of acetylation (less than ~15%) degrade more slowly when compared to those of higher degree of acetylation. Chitosan shows intrinsic antimicrobial activity against a broad spectrum of bacteria, which has been attributed to the interaction of its cationic amino acid groups with anionic groups on the bacterial cell wall to suppress bacterial activity or to disrupt mass transport across the cell wall to accelerate the death of bacteria.

Chitosan solutions can be converted into a variety of forms, such as thin sheets, particles (such as nanoparticles, microspheres, or beads), hydrogels, tubes, nonwoven fibrous mats, and foams, using common methods that are similar to those mentioned for hyaluronic acid and alginates. In common with several natural polymers, chitosan biomaterials generally have weak mechanical properties and, consequently, they are suitable mainly for nonstructural applications. Fibrous mats and sponges are widely used in healing soft tissue wounds, particularly in dermal wound healing, which takes advantage of certain intrinsic properties of chitosan to promote wound healing.

Chitosan hydrogels and foams have attracted considerable interest for use in tissue engineering applications, particularly to deliver cells and biomolecules in healing and regenerating soft tissues and organs. Chitosan molecules have the capacity to bond to native anionic GAGs, which makes chitosan-based hydrogels particularly attractive for cartilage tissue engineering. Chitosan, chitosan derivatives, and blends of chitosan with other polymers have also been widely studied for other tissue engineering applications such as skin, nerve, liver, and blood vessels.

10.4.5 Agarose

Agarose is a polysaccharide extracted in the form of a powder from red algae. Structurally, agarose consists of long chains (molecular weight ~100 kDa) composed of alternating units of D-galactose and 3,6-anhydro-L-galactopyranose (Figure 10.24). Agarose is biocompatible but it is not degradable. The major use of agarose in biomedical applications has been in the form of hydrogels for tissue engineering of cartilage, but agarose hydrogels have also received attention in engineering other tissues and organs, such as the intervertebral disc and the pancreas. The use of agarose

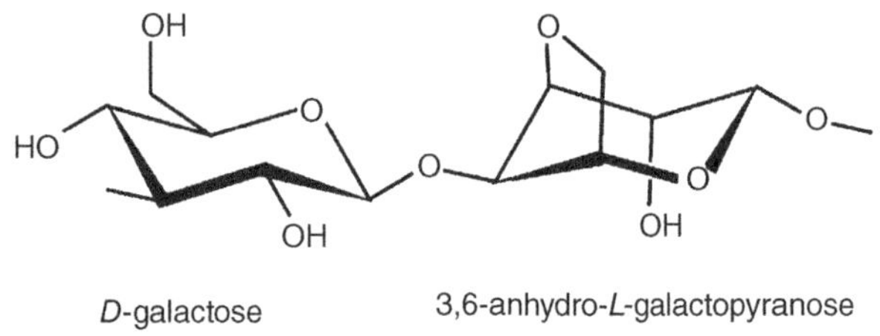

Figure 10.24 Chemical structure of the repeating unit in agarose.

hydrogels in cartilage tissue engineering stems largely from their capacity to maintain the chondrogenic phenotype of cells in culture over a long period to form new cartilage. In fact, a hydrogel composed of agarose and alginate is used as a carrier for chondrocytes in a clinical treatment to heal cartilage defects, termed autologous chondrocyte transfer (Chapter 25).

Individual agarose chains adopt an α-helical structure, and two chains combine to form a double helix stabilized by hydrogen bonds. This double-helical structure is largely responsible for the capacity of agarose solutions to show the property of thermally reversible gelation. Agarose is soluble in warm aqueous media but on cooling below ~40 °C, its forms a hydrogel. A key step in the formation of the hydrogel is the aggregation of the double helical chains, which is stabilized by hydrogen bonds. Thus, agarose forms hydrogels by physical crosslinking, unlike polysaccharides such as hyaluronic acid and alginates, which generally form hydrogels by chemical crosslinking due to covalent or ionic bonds.

Agarose hydrogels are generally stiffer than several other polysaccharide hydrogels because of their double helical chain structure. However, the capacity to modify the stiffness of physically crosslinked agarose hydrogels by, for example, changing the concentration of the agarose solution, is limited. For example, changing the concentration of agarose solutions over a practical range from ~1 to ~4 wt% leads to an increase in the elastic modulus of agarose hydrogels from ~13 to ~78 kPa. In comparison, chemical modification of agarose can lead to a wider range of mechanical properties. This can provide a better capacity to control cell function because the physical characteristics of the environment, such as stiffness, strongly influence cell response.

One type of chemical modification involves the precise oxidation of the primary alcohol group of the D-galactose units in the chains to a carboxyl group, providing a family of biomaterials termed carboxylated agarose. By controlling the number of carboxylated D-galactose units in the chain, the elastic modulus of the chemically modified agarose hydrogels may be varied by over 3–4 orders of magnitude, from ~6 to ~36 kPa. Another advantage of these carboxylated agarose hydrogels is that they provide the opportunity to covalently bond peptides containing the RGD sequence of fibronectin to their chains to improve their interaction with a variety of cells.

10.4.6 Cellulose

Cellulose is considered the most abundant organic polymer in the world. It is found in all plant matter, such as wood, cotton, and hemp, and it is the main structural component of the cell wall in green plants and many types of algae. The structure of cellulose consists of long linear chains of β-D-glucose units that are bonded by 1,4 glycosidic bonds, that is, bonds between the C-1 atom of one monomer and the C-4 atom of another monomer. As shown in Figure 10.25, glucose exists in two stereoisomeric forms, referred to as α-D-glucose and β-D-glucose, which differ only in the direction in which the H and OH atoms at the C-1 atom point. Cellulose is composed of β-glucose chains whereas starch is composed of α-glucose chains.

Like collagen, cellulose has a hierarchical structure. In fact, the fibrillar structure of cellulose bears a certain degree of similarity with that of collagen. Every other glucose monomer in the cellulose chain is rotated ~180° about its axis, making the repeating unit in the chain a disaccharide, sometimes referred to as cellobiose (Figure 10.26). Intermolecular and intramolecular hydrogen bonding due to the multiple OH groups in cellobiose holds the chains tightly together, side-by-side, to form an elementary structure, referred to as a nanofibril but also by other names (Figure 10.27). Depending on the source of the plant matter, these nanofibrils may have a diameter ~3–20 nm and length ~0.1–1 μm, composed of ~30–40 chains. Because of the capacity of the chains to form ordered arrangements, these nanofibrils are composed of crystalline regions interspersed with amorphous regions. Several nanofibrils assemble to form a microfibril of approximately the same length as the nanofibrils but of larger diameter (over ~50–500 nm). These microfibrils, in turn, assemble to form indi-

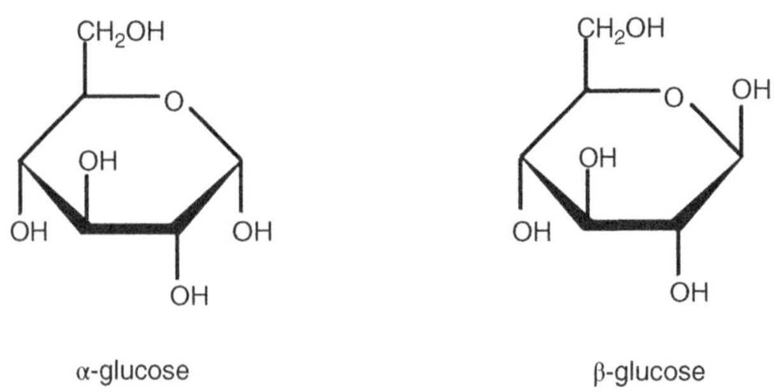

Figure 10.25 Chemical structure of α-glucose and β-glucose.

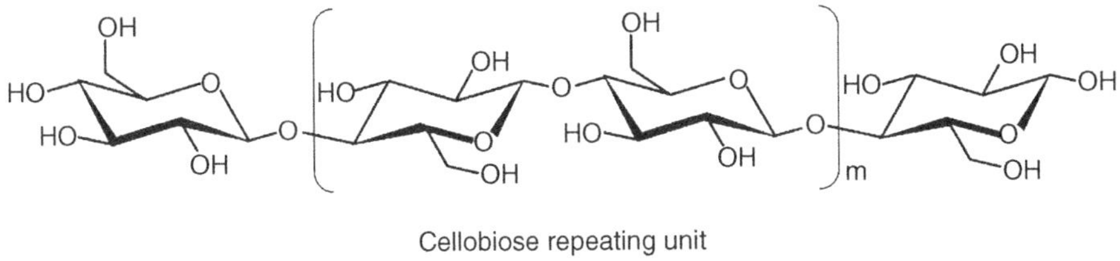

Figure 10.26 Chemical structure of cellulose.

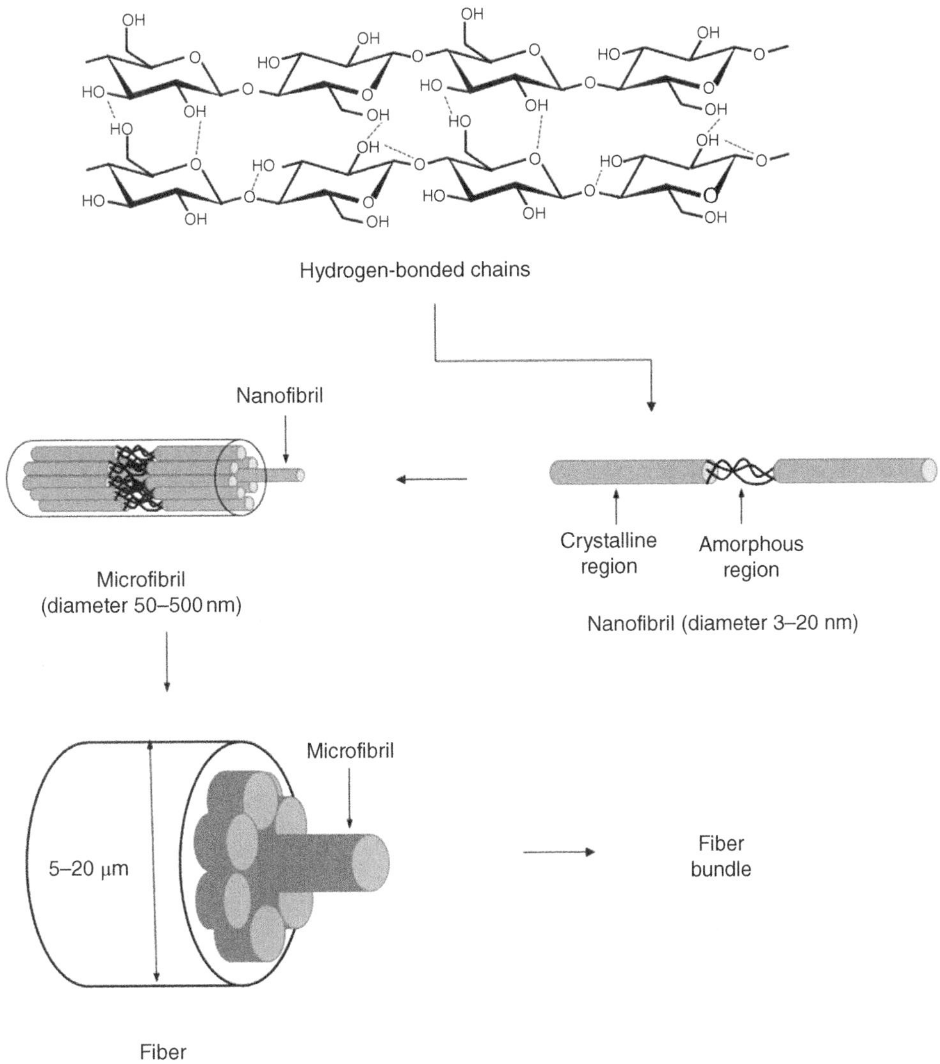

Figure 10.27 Hierarchical structure of cellulose fiber.

vidual fibers of diameter ~5–20 μm and length ~1–3 mm. Depending on its source, cellulose may have a crystallinity of ~50–90 wt%. Bundles of cellulose fibers make up the cell walls of plant-based matter.

Because of its semicrystalline structure, cellulose has attractive mechanical properties, particularly in the form of fibers of high crystallinity, such as hemp fibers. Despite its strongly hydrophilic nature, cellulose is insoluble in aqueous media, which may be partly attributable to the compact arrangement of its chains and its high degree of crystallinity. Cellulose is also nondegradable *in vivo* because the physiological environment lacks the requisite enzymes. Cellulose is generally considered to be biocompatible. However, because it cannot be degraded by mammalian tissues, it is possible that cellulose particles could elicit a foreign body response (Chapter 18).

Cellulose does not show significant deformation prior to melting and this, coupled with its aforementioned insolubility, makes it difficult to form cellulose into desirable shapes for use as biomaterials. Consequently, it is often necessary to modify cellulose chemically to improve its processability. A common type of chemical modification is the conversion of cellulose into cellulose acetate by an esterification reaction (Section 10.4.1). Cellulose modified in this way can be produced in forms such as thin sheets, fibers, tubes, and porous scaffolds by techniques such as casting, molding and additive manufacturing (3D printing) techniques. Chemical modification also has the effect of modifying the physicochemical properties of cellulose, such as its capacity to degrade.

Cellulose has been used in a variety of biomedical applications such as cotton gauze for hemostatic wound dressings and membranes for renal dialysis. More recently, cellulose-based biomaterials have received limited interest for use in tissue engineering applications, such as bone, cartilage and cardiac tissue engineering, and in drug delivery. However, as described in the next section, cellulose produced by species of bacteria has been attracting considerably greater interest for use in biomedical applications.

10.4.7 Bacterial (Microbial) Cellulose

Bacterial cellulose, also referred to as microbial cellulose, is synthesized by a variety of bacteria, but the most widely studied is cellulose synthesized by *Acetobacter*, particularly the species *Acetobacter xylinium*, a gram-negative bacterium. Overall, in the synthesis process by *Acetobacter xylinium*, the bacteria use a saccharide such as sucrose and smaller amounts of inorganic nutrients in a culture medium to form β-D-glucose chains within the cell (Figure 10.28). Then these chains are secreted through pores in the cell membrane in the form of sub-fibrils of diameter 2–4 nm composed of ~6 parallel chains, which self-assemble to form microfibrils of diameter ~5–10 nm. Because of the parallel arrangement of the chains, these microfibrils have a high degree of crystallinity, consisting of crystalline regions interspersed with amorphous regions. Microfibrils, in turn, assemble to form bundles (ribbons) of width ~50–100 nm by ~4 nm in cross section, which aggregate to form a three-dimensional gelatinous structure, called a pellicle, on the surface of the culture medium. As bacterial motion is random, the three-dimensional structure of the pellicle consists of randomly oriented nanofibrous ribbons. However, the arrangement of the ribbons is anisotropic, that is, their random arrangement in the plane parallel to the interface with the culture medium is different from the random arrangement in the plane perpendicular to the interface.

Bacterial cellulose has the same composition as plant cellulose but it its three-dimensional structure composed of a highly crystalline nanofibrillar network (Figure 10.29) is very different from the coarser and more compact structure of plant cellulose. Consequently, the physicochemical properties of bacterial cellulose are different from those of plant cellulose. For example, bacterial cellulose has a much higher surface area and the capacity to hold a considerably higher amount of water due to its fine pores, high porosity and nanofibrillar structure. Additionally,

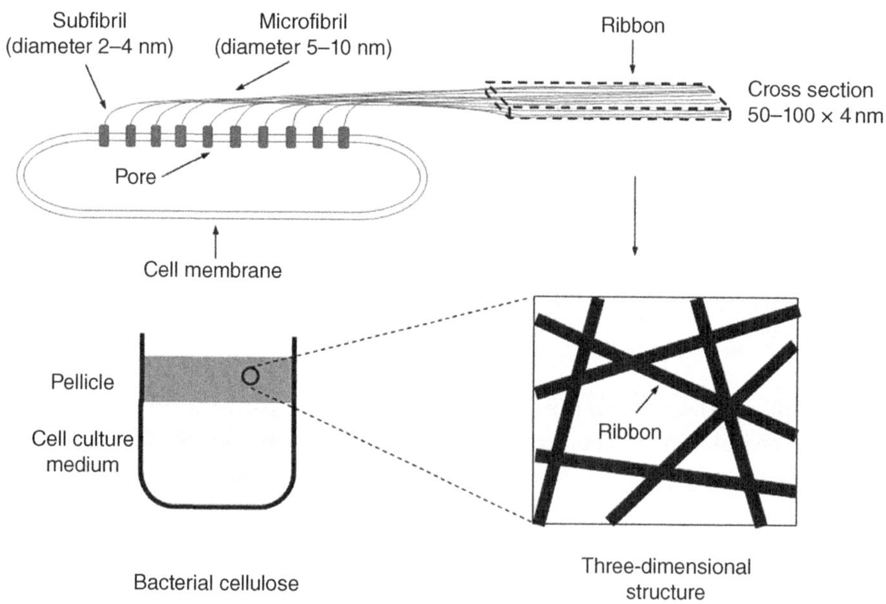

Figure 10.28 Illustration of the hierarchical structure of bacterial cellulose.

despite their high porosity, bacterial cellulose pellicles synthesized in the form of flat sheets under static culture conditions can show considerable mechanical strength due to their high degree of intermolecular and intramolecular hydrogen bonding and the high crystallinity of their nanofibrillar network.

Although plant cellulose is plentiful, it requires intensive processing to extract it in a pure state suitable for biomedical applications from the hemicelluloses, lignin and various proteins present in the cell wall of plants. In comparison, whereas large-scale fermentation has been difficult and expensive, bacterial cellulose is generally synthesized in a pure state, free from contaminant molecules. Often, only a simple washing step is required to remove residual cells. However, like other biomaterials synthesized in microbial processes, a concern is the potential presence of endotoxins remaining as residues from the bacterial cell wall.

Bacterial cellulose may be formed into sheets of varying thickness (pellicles) under static culture conditions and into particulates of varying size and shape under agitated culture. A drawback of bacterial cellulose as a biomaterial, particularly for tissue engineering applications, is its fine pores, which limit cell migration into its interior (Figure 10.29). However, the introduction of a more desirable pore size into bacterial cellulose has been achieved using a variety of methods, such as the use of a pore-forming agent (porogen), a well-known method for creating porous biomaterials (Chapter 25).

Implantation in animal models has shown the biocompatibility of bacterial cellulose. Like plant cellulose, bacterial cellulose is nondegradable *in vivo*. Consequently, a variety of chemical techniques, some similar to those described for other polysaccharides, have been studied to modify the structure of bacterial cellulose in order to modify its physicochemical properties and to improve its interaction with cells. There has been considerable interest in the last two to three decades in the biomedical applications of bacterial cellulose. In the form of sheets, bacterial cellulose has been used clinically in wound healing, an application that takes advantage of its network structure, capacity to hold a considerable amount of water and mechanical properties. Bacterial cellulose has also been investigated as a biomaterial for a variety of tissue engineering applications, such as blood vessels and bone.

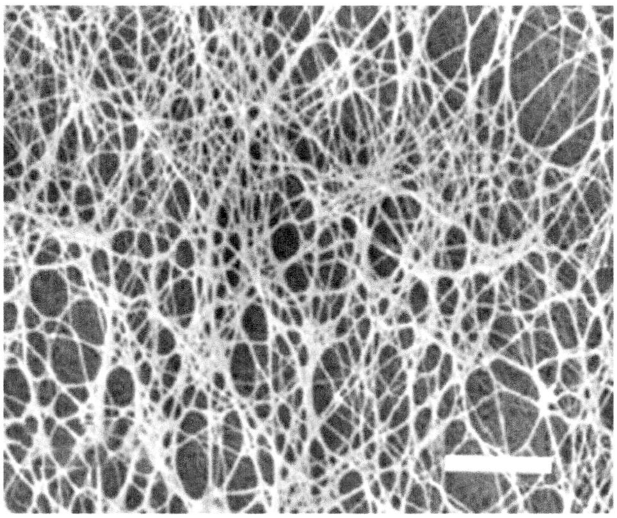

Figure 10.29 Scanning electron microscope image of the surface of a bacterial cellulose pellicle. (Bar = 4 μm). *Source:* From Svennson et al. (2005).

10.5 Concluding Remarks

Following the discussion of nondegradable and degradable synthetic polymers in the last two chapters, this chapter focused on a discussion of several natural polymers relevant to biomaterials. We broadly discussed major characteristics of natural polymers related to their use as biomaterials, pointing out features that distinguish natural polymers from synthetic polymers, and their advantages and drawbacks as well. Then we discussed a variety of natural polymers, grouped in terms of whether they are protein-based or polysaccharide-based, emphasizing relationships between their composition, structure, and properties

- Because they are obtained from living species and have some degree of compositional and structural similarity with components of the ECM, natural polymers are generally biocompatible and nontoxic, and they have the capacity to interact more favorably with cells and tissues *in vivo* when compared to synthetic polymers.
- In porous three-dimensional forms, natural polymers generally show weak mechanical properties, but oriented fibers and thin sheets of some polymers may show considerable strength and stiffness.
- Natural polymers may suffer from some variability in composition depending on their source. Unlike synthetic polymers that degrade in the physiological medium predominantly by hydrolysis, natural polymers degrade predominantly by enzyme-catalyzed hydrolysis, which may make their degradation less controllable due to variability between individuals.
- The techniques and conditions used for processing natural polymers into biomaterials are often more limited when compared to those for synthetic polymers because they generally have a lower thermal stability.
- As we have discussed for other types of materials, the design and selection of natural polymers for use as

biomaterials require an understanding of their composition and their structure at different length scales.

- Chemical modification of natural polymers by, for example, crosslinking and modification of their functional groups, and blending (mixing) natural polymers with other polymers (natural or synthetic) are often vital in converting them into biomaterials having an optimal combination of properties for specific biomedical applications.
- Collagen is the most widely used natural polymer in biomedical applications, but other natural polymers have been receiving considerable interest in the last few decades for use in a variety of biomedical applications, particularly drug delivery and tissue engineering.

Problems

10.1 Explain the difference between protein-based natural polymers and polysaccharide-based natural polymers.

10.2 Describe the hierarchical structure of collagen (type I). How do the structures at different length scales influence the Young's modulus of collagen?

10.3 Although collagen (type I) is the major structural protein in both tendons and skin, these two tissues show a considerable difference in mechanical properties such as tensile strength, Young's modulus and elongation to failure. Explain why.

10.4 Explain what the term denaturation of collagen means. What factors influence the denaturation of collagen?

10.5 Explain why collagen-based biomaterials are often used in a crosslinked form. Describe three methods for chemically crosslinking collagen and, for each method, the mechanism of crosslinking.

10.6 Like collagen fibers, silkworm silk fibers have a hierarchical structure. With the aid of illustrations, compare the hierarchical structures of collagen and silkworm silk fibers at different length scales. Explain how these structures produce differences in properties between collagen fibers and silkworm silk fibers.

10.7 Single silkworm silk fibers show considerable tensile strength, Young's modulus, elongation to failure and toughness. Explain what factors are responsible for this attractive combination of properties.

10.8 Compare the structures of collagen (type I) and elastin. How do these structures influence the mechanical response of collagen and elastin?

10.9 Explain what changes in elastin are largely responsible for age-related wrinkles in the skin of humans.

10.10 Draw the repeating unit in hyaluronic acid chains. What are the major functions of hyaluronic acid in the body?

10.11 Alginates relevant to biomaterials are alginic acid, sodium alginate, and calcium alginate. Explain the difference between these three alginates.

10.12 Sodium alginate is soluble in water whereas calcium alginate forms an insoluble gel. Explain why.

10.13 Briefly describe a method for forming calcium alginate hydrogels. What factors influence the properties of calcium alginate hydrogels? Explain.

10.14 Draw the chemical structure of the repeating unit in chitin and in chitosan. Explain the meaning of the terms degree of acetylation and degree of deacetylation when used in relation to chitin and chitosan.

10.15 Explain how the degree of acetylation influences the properties of chitosan.

10.16 Glucose is soluble in water but plant-based cellulose, composed of chains of β-glucose units, is insoluble in water. Explain why.

10.17 With the aid of illustrations, compare the hierarchical structures of plant-based cellulose and collagen (type I) at various length scales. Explain how these structures produce differences in the properties of cellulose and collagen.

10.18 Is the structure of bacterial cellulose different from that of plant-based cellulose? If different, describe the major differences and what effects they have on the properties of bacterial cellulose when compared to plant-based cellulose.

References

Chan, E.C., Kuo, S.-M., Kong, A.M. et al. (2016). Three dimensional collagen scaffold promotes intrinsic vascularization for tissue engineering applications. *PLoS One* 11: e0149799.

Schoof, H., Apel, J., Heschel, I. et al. (2001). Control of pore structure and size in freeze-dried collagen sponges. *Journal of Biomedical Materials Research* 58: 352–357.

Sirichaisit, J., Brookes, V.L., Young, R.J. et al. (2003). Analysis of structure/property relationships in silkworm (*Bombyx mori*) and spider dragline (*Nephila edulis*) silks using Raman spectroscopy. *Biomacromolecules* 4: 387–394.

Svennson, A., Niclasson, E., Harrah, T. et al. (2005). Bacterial cellulose as a potential scaffold for tissue engineering of cartilage. *Biomaterials* 26: 419–431.

Olde Damink, L.H.H., Dijkstra, P.J., van Luyn, M.J.A. et al. (1995). Glutaraldehyde as a crosslinking agent for collagen-based biomaterials. *Journal of Materials Science. Materials in Medicine* 6: 460–472.

Pawar, S.N. and Edgar, K.J. (2012). Alginate derivatization: a review of chemistry, properties and applications. *Biomaterials* 33: 3279–3305.

Further Reading

Brown, A.C. and Barker, T.H. (2014). Fibrin-based biomaterials: modulation of macroscopic properties through rational design at the molecular level. *Acta Biomaterialia* 10: 1502–1514.

Buehler, M.J. (2006). Nature designs tough collagen: explaining the nanostructure of collagen fibrils. *Proceedings of the National Academy of Sciences of the United States of America* 103: 12285–12290.

Buehler, M.J. (2006). Atomistic and continuum modeling of mechanical properties of collagen: elasticity, fracture, and self-assembly. *Journal of Materials Research* 21: 1947–1961.

Chilcoti, A., Christensen, T., and MacKay, J.A. (2006). Stimulus responsive elastin biopolymers: application in medicine and biotechnology. *Current Opinion in Chemical Biology* 10: 652–657.

Daamen, W.F., Veerkamp, J.H., van Hest, J.C.M., and van Kuppevelt, T.H. (2007). Elastin as a biomaterial for tissue engineering. *Biomaterials* 28: 4378–4398.

Forget, A., Christensen, J., Lüdeke, S. et al. (2013). Polysaccharide hydrogels with tunable stiffness and provasculogenic properties via α-helix to β-sheet switch in secondary structure. *Proceedings of the National Academy of Sciences of the United States of America* 110: 12887–12892.

Gentleman, E., Lay, A.N., Dickerson, D.A. et al. (2003). Mechanical characterization of collagen fibers and scaffolds for tissue engineering. *Biomaterials* 24: 3805–3813.

Keten, S., Xu, Z., Ihle, B., and Buehler, M.J. (2010). Nanoconfinement controls stiffness, strength and mechanical toughness of β-sheet crystals in silk. *Nature Materials* 9 (4): 359–367.

Lynn, A.K., Yannas, L.V., and Bonfield, W. (2004). Antigenicity and immunogenicity of collagen. *Journal of Biomedial Materials Research Part B Applied Biomaterials* 71: 343–354.

Ramshaw, J.A.M., Werkmeister, J.A., and Glattauer, V. (1996). Collagen-based biomaterials. *Biotechnology & Genetic Engineering Reviews* 13: 335–382.

Rinaudo, M. (2008). Main properties and current applications of some polysaccharides as biomaterials. *Polymer International* 57: 397–430.

Rockwood, D.N., Preda, R.C., Yücel, T. et al. (2011). Materials fabrication from *Bombyx mori* silk fibroin. *Nature Protocols* 6: 1612–1631.

Sherman, V.R., Yang, W., and Meyers, M.A. (2015). The materials science of collagen. *Journal of the Mechanical Behavior of Biomedical Materials* 52: 22–50.

Shoulders, M.D. and Raines, R.T. (2009). Collagen structure and stability. *Annual Review of Biochemistry* 78: 929–958.

Vespari, C. and Kaplan, D.L. (2007). Silk as a biomaterial. *Progress in Polymer Science* 32: 991–1007.

Weadock, K., Olson, R.M., and Silver, F.H. (1983). Evaluation of collagen crosslinking techniques. *Biomaterials, Medical Devices, and Artificial Organs* 11: 283–316. 1984.

Young, S., Wong, M., Tabata, Y., and Mikos, A.G. (2005). Gelatin as a delivery vehicle for the controlled release of bioactive molecules. *Journal of Controlled Release* 109: 256–274.

11

Hydrogels

11.1 Introduction

Hydrogels are water-filled polymers, composed of a crosslinked three-dimensional (3D) network of hydrophilic polymer chains that, in the dehydrated state, contain spaces (pores) of varying size between them (Figure 11.1). Upon immersion in an aqueous medium, dehydrated hydrogels absorb varying amounts of water and swell. However, because of their crosslinked structure, they do not dissolve in the aqueous medium. Depending on their composition and 3D network structure, hydrogels vary from being soft and pliable to extremely rigid. Hydrogels can be created from synthetic polymers, natural polymers or a combination of synthetic and natural polymers whose chains are crosslinked by covalent or noncovalent bonds to give a network having different structural stability. Depending on their composition, hydrogels may be chemically stable (nondegradable) or they may degrade and eventually dissolve in an aqueous environment such as the physiological fluid.

Hydrogels based on poly(hydroxyethyl methacrylate) (PHEMA), created in the early 1960s, are often credited with being the first synthetic polymers designed specifically for use in biomedical applications (as contact lenses). Prior to that time, synthetic polymers used as biomaterials were generally adopted from polymers developed for use in engineering applications (Chapter 8). Since the creation of PHEMA, hydrogels have increased tremendously in variety, and they have attracted considerable interest in the last few decades for several biomedical applications, particularly drug delivery and tissue engineering (Table 11.1). Along with degradable synthetic polymers (Chapter 9), hydrogels have played a significant role in advances made in these areas over the last few decades.

Although hydrogels are generally more challenging to sterilize when compared to other types of materials because of their sensitivity to heat and the presence of water within their pores, they have a combination of properties and characteristics that make them attractive biomaterials for use in a variety of biomedical applications. For example, hydrogels are normally easy to synthesize in a wide range of compositions and structures under mild conditions, which makes it easy to control their physicochemical properties and for incorporating cells and biomolecules. Hydrogels also have a structural similarity to components of the extracellular matrix (ECM) and their mechanical properties can be varied to match those of several soft tissues, which makes them attractive biomaterials for use in soft tissue healing and regeneration. Some hydrogels contract or swell in response to environmental factors (such as changes in pH or temperature of the aqueous medium), which provides additional advantages in their use as injectable implants and controlled drug delivery devices.

In this chapter, we will discuss the following topics relevant to the use of hydrogels in biomedical applications:

- Principles underlying the creation, structure, and properties of hydrogels
- Types of hydrogels and their characteristic features
- Relationship between the composition, structure, and properties of hydrogels.

11.2 Characteristics of Hydrogels

As a background to the topics covered in this chapter, some properties and characteristics of hydrogels are discussed in broad terms in this section.

- *Ease of synthesis and chemical modification.* Hydrogels having a wide range of compositions and 3D network structures can be easily created and chemically modified under mild conditions. This provides the capacity to control their physicochemical properties over a wide range, to incorporate biomolecules and cells without significant risk of deactivating them, and to enhance their interactions with cells and tissues.
- *3D network structure.* Hydrogels, as noted earlier, are composed of a 3D network of crosslinked chains in which the

Materials for Biomedical Engineering: Fundamentals and Applications, First Edition. Mohamed N. Rahaman and Roger F. Brown.

Companion website: www.wiley.com/go/rahaman/biomedicalengineering

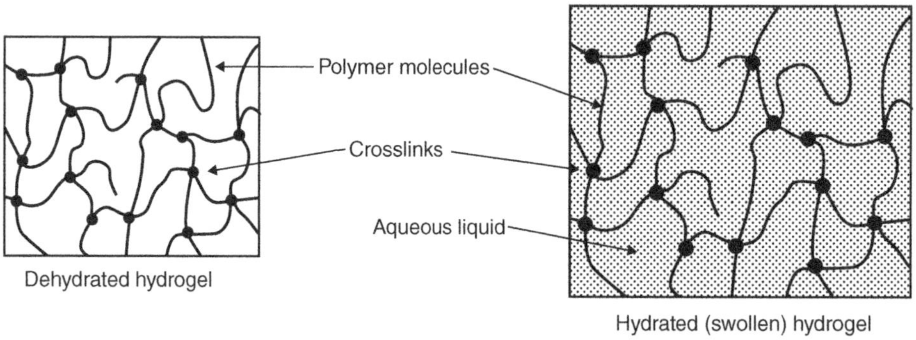

Figure 11.1 Illustration of a hydrogel in a dehydrated state and in a hydrated (swollen) state.

Table 11.1 Examples of polymers used to create hydrogels for biomedical applications.

Applications	Hydrogel polymer
Blood-compatible hydrogels	Polyethylene glycol (PEG); polyethylene oxide (PEO); polyvinyl alcohol (PVA); poly(*N*-vinyl pyrrolidone) (PVP); polyhydroxyethyl methacrylate (PHEMA); polyacrylamide (PAAm)
Contact lenses	Polyhydroxyethyl methacrylate (PHEMA); poly(HEMA-*co*-methyl methacrylate) P(HEMA-*co*-MMA); poly(HEMA-*co*-*N*-vinyl pyrrolidone) P(HEMA-*co*-VP); poly(HEMA-*co*-methacrylic acid) P(HEMA-*co*-MAA)
Drug delivery and tissue engineering	
Degradable natural polymers	Collagen; gelatin; hyaluronic acid; chondroitin sulfate; chitosan; polylysine; dextran
Nondegradable natural polymers	Alginate; agarose
Degradable synthetic polymers	Polylactic acid (PLA); polyglycolic acid (PGA); polylactic-*co*-glycolic acid (PLGA); PLA–PEG block copolymer (PLA-*b*-PEG-*b*-PLA); PGA-PEG block copolymer (PGA-*b*-PEG-*b*-PGA); poly(1,3-bis(4-carboxyphenoxy) propane-*co*-sebacic acid P(CPP-SA)
Nondegradable synthetic polymers	
Neutral	PHEMA; PVA; PVP
pH-sensitive	Polyacrylic acid (PAA); polymethacrylic acid (PMAA); PAAm
Temperature-sensitive	Poly(*N*-isopropyl acrylamide) (PNIPAAm); poly(NIPAAm-*co*-AA) P(NIPAAm-*co*-AA); PEO-*b*-PPO-*b*-PEO

spaces within the network are filled with an aqueous liquid. This structure, which is dependent on the composition, volume fraction, and degree of crosslinking of the chains, provides not just efficient transport of oxygen and nutrients through the hydrogel but also the capacity for controlled release of biomolecules.

- *Capacity to absorb water.* Hydrogels can absorb varying amounts of water, from ~5 to 10% of their dry mass to several times their dry mass, depending on their composition and degree of crosslinking, and on environmental factors such as pH, temperature and ionic concentration of the aqueous medium. Overall, the amount of absorbed water strongly influences the mechanical and transport properties of the hydrogel.
- *Weak mechanical properties.* Hydrogels normally have weak mechanical properties but these properties can be controlled over a wide range by controlling the capacity of the hydrogel to absorb water. For example, loosely crosslinked hydrogels, which absorb a considerable amount of water and undergo a high degree of swelling, normally show an elastic modulus comparable to that of a soft rubber. In comparison, hydrogels that absorb a small amount of water and, thus, undergo a small degree of swelling can be very rigid.
- *Similarity to soft mammalian tissues.* Hydrogels generally have a structural similarity to the ECM of soft mammalian tissues, which may be broadly approximated as aqueous gels composed of protein and polysaccharide networks. Some hydrogels such as collagen and hyaluronic acid also have a compositional similarity to the ECM components. This similarity to the ECM components, together with their mechanical properties, make hydrogels attractive biomaterials for use in tissue engineering to regenerate soft tissues and organs.

- *Capacity to respond to environmental factors*. Depending on their chemical structure, some hydrogels have the capacity to swell and contract reversibly in response to modest changes in their environment, such as temperature or pH. These environmentally responsive hydrogels can provide additional advantages in applications such as drug delivery and tissue engineering.

11.3 Types of Hydrogels

Like polymers, hydrogels used as biomaterials may be classified into several types, based on the characteristics or properties being emphasized. Overall, as we have seen for polymers, one broad classification is in terms of their origin, that is, whether they are created from natural or synthetic polymers, although some hydrogels are composed of a combination of natural and synthetic polymers. Some common polymers used to create hydrogels relevant to biomaterials are given in Table 11.1.

Another broad classification of hydrogels has been in terms of the type of crosslinks between their chains, which controls the stability of their 3D network structure. Crosslinking can be achieved by covalent, ionic, or secondary bonds (hydrogen or van der Waals bonds), or by physical mechanisms such as crystallization or simple entanglement of the chains. Networks formed by ionic bonds, secondary bonds, and physical mechanisms are generally reversible and may be disrupted by changing the physical conditions of the aqueous environment, such as its pH, temperature, or ionic strength. These hydrogels are normally referred to as physical hydrogels. In comparison, when the molecular chains are crosslinked by covalent bonds, the formation of the 3D network is not reversible, and these hydrogels are referred to as chemical hydrogels. Overall, chemical hydrogels have better network stability than physical hydrogels.

Classifications may sometimes emphasize other characteristics or properties. For example, based on their chemical structure, hydrogels may be composed of homopolymers, copolymers, or an interpenetrating network of two or more different polymers. They may be anionic, cationic, amphiphilic or neutral, depending on the electrostatic charging of their molecular chains in aqueous media. Hydrogels may also be classified in terms of whether they degrade or not in an aqueous medium, or whether they are environmentally responsive, such as pH or temperature responsive. Depending on the size of the spaces (pores) within their 3D network, hydrogels may be classified as macroporous, microporous, or nonporous.

11.4 Creation of Hydrogels

Hydrogels have been created using a wide variety of routes, which lead to the formation of a 3D network by primary bonds (covalent or ionic), secondary bonds (hydrogen or van der Waals), or by physical mechanisms (such as crystallization or simple physical entanglement).

11.4.1 Chemical Hydrogels

Chemical hydrogels, as noted earlier, are composed of a covalently bonded 3D network. Two common routes for creating chemical hydrogels are

- Crosslinking by a free radical mechanism
- Crosslinking by chemical reaction of complementary functional groups present in the chains.

Crosslinking by Free Radical Mechanism

This method involves polymerization of monomers, oligomers, macromolecules, or their derivatives that contain an unsaturated bond, such as a carbon double bond (C=C), commonly with the aid of a small amount of a photoinitiator and an accelerator. The reaction mechanism is similar to that described for the addition polymerization of polymethyl methacrylate (PMMA) (Figure 8.14b). A well-known example is the polymerization (and crosslinking) of hydroxyethyl methacrylate (HEMA) (Figure 11.2), in the presence or absence of water, using varying amounts of a crosslinking agent such as ethylene glycol diacrylate, to form PHEMA. Following polymerization, swelling in distilled water or a saline solution up to its thermodynamic equilibrium (attainment of constant mass) usually at ~37 °C,

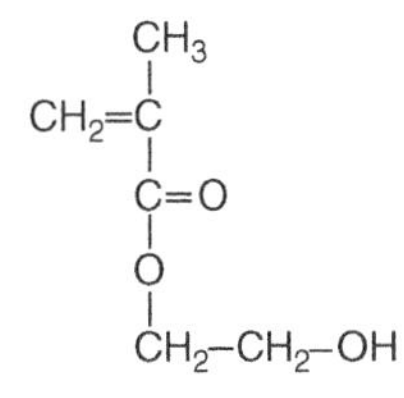

Hydroxyethyl methacrylate (HEMA)

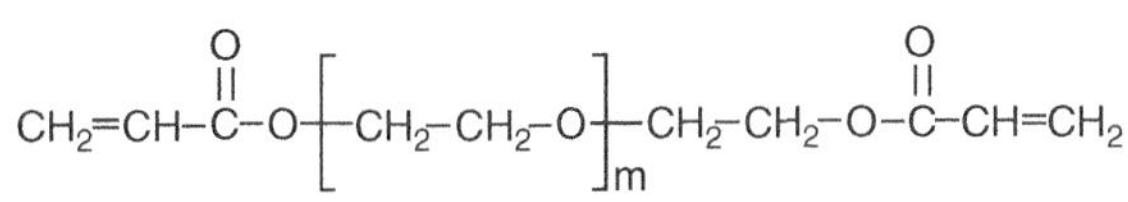

Polyethylene glycol diacrylate (PEGDA)

Figure 11.2 Example of the chemical structure of a monomer (hydroxyethyl methacrylate) (HEMA) and a macromolecule (polyethylene glycol diacrylate) (PEGDA) used in the creation of hydrogels by free radical mechanism.

leads to the formation of hydrogels used for contact lenses and drug delivery.

Covalently crosslinked hydrogels can also be synthesized by radical polymerization of water-soluble polymers. In the case of chemically saturated chains, such as polyvinyl alcohol (PVA), the chains in the aqueous medium are irradiated by high energy γ-rays or electron beams, which create free radicals and induce crosslinking. A far more widely used method, involves irradiation of water-soluble chains that contain unsaturated bonds (such as C=C bonds) using less-energetic radiation such as ultraviolet (UV) rays. This route has been used to synthesize a variety of hydrogels, both synthetic and natural. A well-known example is the creation of polyethylene glycol (PEG) hydrogels by crosslinking macromolecules of polyethylene glycol diacrylate (PEGDA) (Figure 11.2) using a photochemical method such as exposure to UV radiation.

Crosslinking by Chemical Reactions Between Functional Groups

Water-soluble polymers, as noted in previous chapters devoted to polymers, are composed of molecular chains that contain polar or ionizable functional groups, such as OH, (C=O)OH, or NH_2 groups. These functional groups are also valuable in crosslinking reactions using bifunctional water-soluble agents, such as glutaraldehyde, diisocyanates, and carbodiimides, as discussed for some natural polymers in Chapter 10. This route may be used for synthetic polymers as well. For example, one method that has been used to synthesize PVA hydrogels is crosslinking PVA molecules in an aqueous medium using glutaraldehyde. However, this reaction normally requires moderately harsh conditions, such as low pH, reaction times of ~10 minutes at 80 °C to ~50 minutes at 30 °C, and the use of methanol to quench the reaction (Figure 11.3).

11.4.2 Physical Hydrogels

Physical hydrogels are composed of a 3D network crosslinked by ionic bonds, secondary bonds, or physical mechanisms. A benefit of these crosslinking techniques is that they normally do not involve the use of crosslinking agents, which avoids the potential risk of toxicity associated with some agents.

Crosslinking by Ionic Bonds

Crosslinking by ionic bonds is normally achieved in an aqueous medium, between chains possessing negative or positive electrostatic charges and either ions or molecular chains having charges of the opposite sign (Figure 11.4). Calcium alginate is a well-known example of a hydrogel created by ionic crosslinking between negatively charged alginate chains and positively charged calcium (Ca^{2+}) ions (Chapter 10). Gelling is reversible in that the 3D network can be disrupted by the addition of a chelating agent such as sodium carbonate (Na_2CO_3), due to extraction of the Ca^{2+} ions from the hydrogel by carbonate $(CO_3)^{2-}$ ions.

Hydrogels created by ionic bonding between chains of opposite charge are sometimes referred to as coacervate hydrogels. An example is the hydrogel created from negatively charged alginate chains and positively charged polylysine chains. The (C=O)OH groups in alginic acid have a pK_a of ~3.5 and, thus, alginate chains become increasingly negatively charged as the pH of the aqueous medium increases above ~3.0 due to deprotonation of its (C=O)OH groups. In comparison, the NH_2 groups in polylysine have a pK_a of ~9–10 and, thus, polylysine chains become increasingly positive charged as the pH decreases below ~10 due to protonation of its NH_2 groups. Consequently, near the physiological pH (~7.4), the two types of chains have opposite charges, which leads to crosslinking by ionic bonding and formation

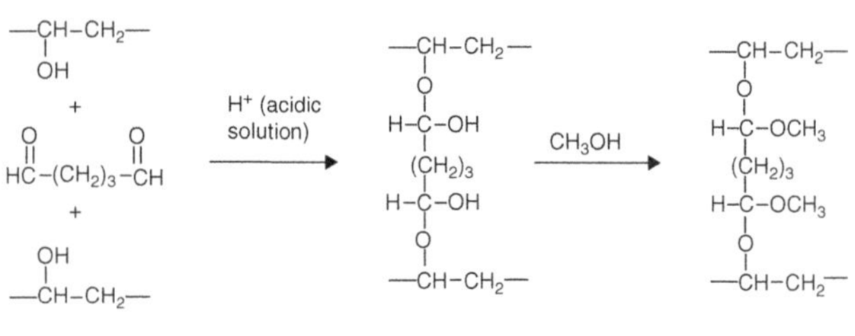

Figure 11.3 Schematic of chemical reaction between polyvinyl alcohol (PVA) chains and glutaraldehyde to create PVA hydrogels by covalent crosslinking.

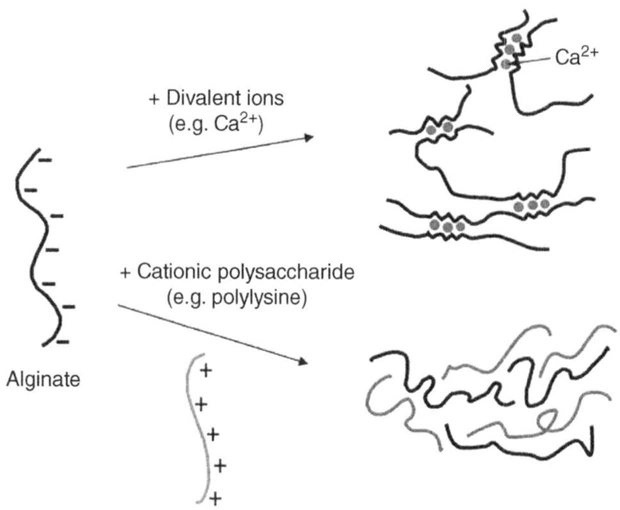

Figure 11.4 Illustration of alginate hydrogels created by ionic crosslinking between negatively charged alginate chains and divalent ions such as calcium (Ca^{2+}) and between chains of opposite charge such as alginate and polylysine. Electrostatic charges have been omitted from the hydrogel network for clarity.

of a 3D network. The network is disrupted at pH above ~10 when the polylysine chains become uncharged.

Crosslinking by Secondary Bonds

Crosslinking by secondary bonds, such as hydrogen bonds or van der Waals bonds can also lead to the creation of hydrogels. A well-known example of gelling by hydrogen bonding is the creation of gelatin hydrogels, by cooling a warm solution of gelatin to low temperature (Chapter 10). Hydrogen bonding between different types of molecular chains can also lead to the formation of a hydrogel as, for example, the complexation between polyacrylic acid (PAA) chains and polyacrylamide (PAAm) chains in aqueous media at low temperature (less than ~30 °C) and pH of ~3.0 (Figure 11.5).

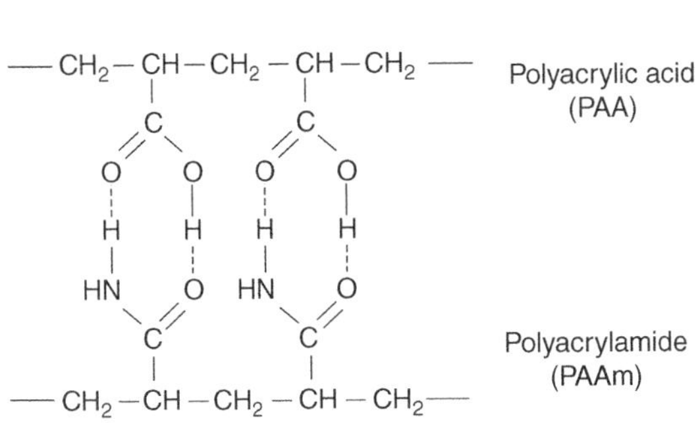

Figure 11.5 Illustration of hydrogen bonded structure between polyacrylic acid (PAA) and polyacrylamide (PAAm) chains in an aqueous medium at low pH (~3.0) and temperatures below ~30 °C.

van der Waals bonding involving associative interactions between hydrophobic groups can also lead to the creation of some hydrogels. A widely studied example consists of triblock copolymers composed of polyethylene oxide (PEO) and polypropylene oxide (PPO), abbreviated PEO-*b*-PPO--*b*-PEO, which are referred to commercially as Pluronics. PEO-*b*-PPO-*b*-PEO copolymers are amphiphilic, that is, they are composed of chains having hydrophilic blocks (PEO) and hydrophobic blocks (PPO) (Figure 11.6). They are soluble in water at low temperature but upon warming, moderately concentrated solutions (~20 wt%) of some Pluronics, such as the Pluronic designated F127, form a hydrogel above ~25 °C. The mechanism of gelling involves two processes:

- Dehydration of the hydrophobic blocks of the chains, that is, the water molecules migrate away from between these blocks into the surrounding medium
- Cooperative association of the dehydrated hydrophobic groups by van der Waals attraction to form spherical micellar structures.

In Pluronic F127, these micellar structures can go on to adopt an ordered cubic arrangement. Hydrogels composed of PPO-*b*-PEO-*b*-PPO copolymers can be created by a similar mechanism.

Another widely studied example of gelling by a similar mechanism is poly(*N*-isopropyl acrylamide), abbreviated PNIPAAm (Figure 11.7), whose chains contain hydrophilic

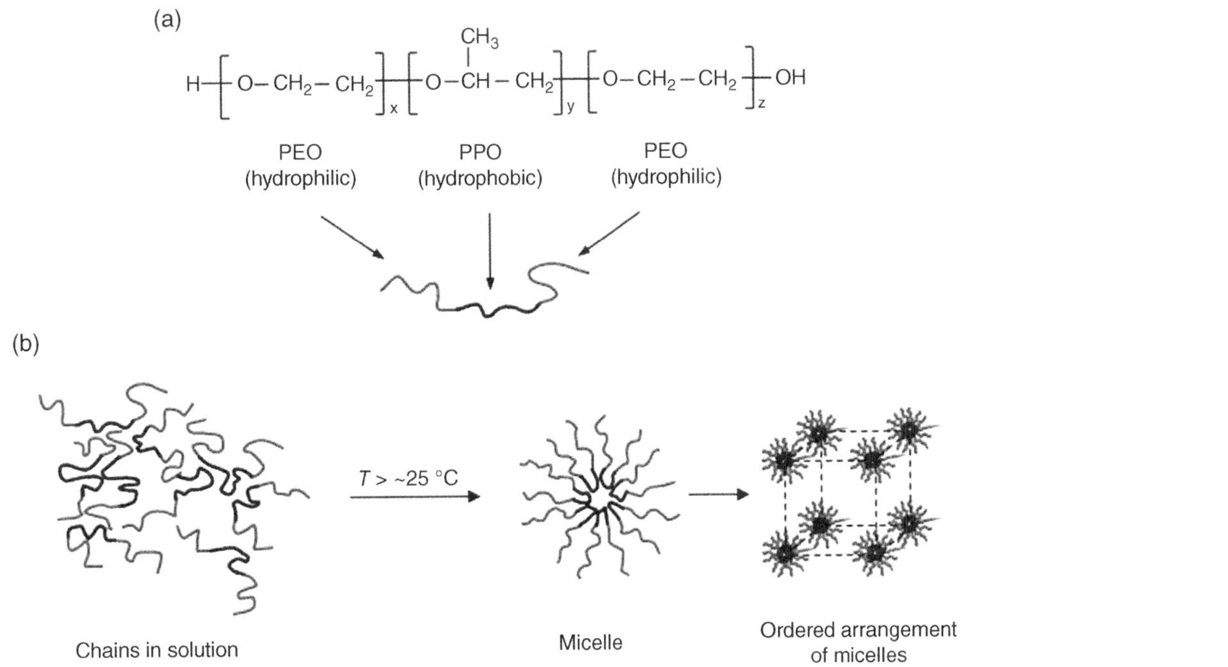

Figure 11.6 (a) Chemical structure of polymers (referred to as Pluronics) consisting of block copolymers composed of hydrophilic polyethylene oxide (PEO) and hydrophobic polypropylene oxide (PPO); (b) illustration of the mechanism of gelling upon warming a solution of the Pluronic designated F127 above ~25 °C.

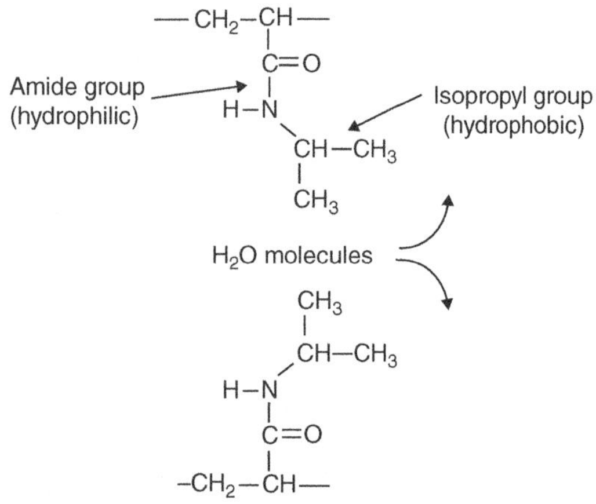

Figure 11.7 Illustration of the gelling mechanism upon warming a solution of poly(*N*-isopropyl acrylamide) (PNIPAAm) above ~32 °C.

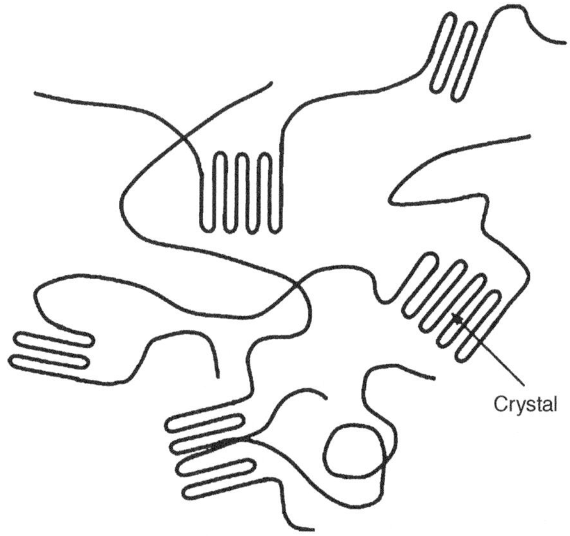

Figure 11.8 Illustration of gelling in polyvinyl alcohol (PVA) by a crystallization mechanism. The crystals form crosslinks that stabilize the three-dimensional network.

(amide) and hydrophobic (isopropyl) groups. In a solution at low temperature, hydrogen bonding between water molecules and the hydrophilic (amide) groups lead to an ordered layer of water molecules that surround the hydrophobic (isopropyl) groups. However, upon warming the solution above ~32 °C, the hydrogen bonds are disrupted, which leads to dehydration of the hydrophobic groups. Gelling occurs by cooperative association of these hydrophobic groups by van der Waals attraction.

Pluronic 127 and PNIPAAm are also widely studied examples of temperature responsive hydrogels. The relationship between their gelling mechanism and temperature responsive behavior are discussed further in Section 11.10.2.

Crosslinking by Physical Mechanisms

Some crosslinked hydrogels can be created by physical mechanisms, such as crystallization of chain segments, formation of helical coil (coiled-coil) chain arrangements, or physical entanglements, which link the chains into a 3D network. As described for semicrystalline polymers in Chapter 8, linear chains composed of a regular repeating arrangement of monomer units, a high degree of stereoregularity and strong polar interactions are more susceptible to crystallization. For example, as illustrated in Figure 11.8, crystallization provides one technique for the creation of PVA hydrogels. Certain proteins can form helical coil (coiled-coil) arrangements, which are stabilized by hydrogen bonds (Chapter 2), and when bonded to other chains, can link these chains into a 3D network. Although it can lead to the creation of hydrogels, physical entanglements find little practical use in the creation of hydrogels for biomedical applications because the stability of their network structure is weak.

11.5 Characterization of Sol to Gel Transition

A significant and abrupt change in several properties accompanies the sol to gel transition and, consequently, changes in one or more of these properties may be used to determine the transition. While several measurements may be used, the most common are:

- *Swelling ratio.* A large amount of water is expelled from the spaces between the chains as they become crosslinked and, thus, the volume of the gel becomes significantly smaller. Consequently, the water content of the gel, defined in terms of a swelling ratio given by

$$\text{Swelling ratio} = \frac{\text{Mass of swollen gel} - \text{mass of dry gel}}{\text{Mass of dry gel}} \tag{11.1}$$

provides an easily measurable parameter to characterize the creation of a hydrogel.

- *Optical transmission.* Generally, a sol (solution of macromolecules) is optically clear whereas a hydrogel becomes cloudy or opaque due to the aggregation of the chains, which leads to a less homogeneous system consisting of two phases (crosslinked chains and liquid). Thus, the creation of a hydrogel may be determined by measuring the optical transmittance of the sample.
- *Rheological behavior.* Like solid polymers, hydrogels are viscoelastic (Chapter 8). Thus, subjecting a hydrogel to a repetitive sinusoidal stress (or strain) in a rheometer (or dynamic mechanical analyzer) leads to elastic and viscous

responses. The measured elastic modulus undergoes a significant increase during gelling, which provides another facile method to determine the creation of a hydrogel.

11.6 Swelling Behavior of Hydrogels

The 3D network structure of a hydrogel controls its properties, such as its mechanical, transport, surface, and optical properties. Consequently, an understanding of the swelling behavior of hydrogels is important in their design and application. The important parameters of the 3D structure are (Figure 11.9):

- Volume fraction of the polymer phase in the swollen state of the hydrogel, denoted $v_{2,s}$, which provides a measure of the amount of liquid absorbed and retained within the hydrogel
- Molecular weight of the chains between two neighboring crosslinks ($\bar{M}_c$), which is a measure of the degree of crosslinking of the network
- Mesh size (ξ), also referred to as the correlation length, which is the distance between two adjacent crosslinks. The mesh size is particularly important in controlling the transport properties of the hydrogel. Approximating the network structure of the hydrogel as a 3D sieve, the mesh size provides a measure of the maximum size (diameter) of a solute molecule that is capable of diffusing through the hydrogel. In general, the mesh size of most hydrogels is in the nanoscale range (less than 50–100 nm).

These parameters, which are related to one another, can be determined using theoretical models or experimental techniques. Because of the random nature of the chain conformation and crosslinking, $\bar{M}_c$ and ξ can each be determined only as an average value for the hydrogel.

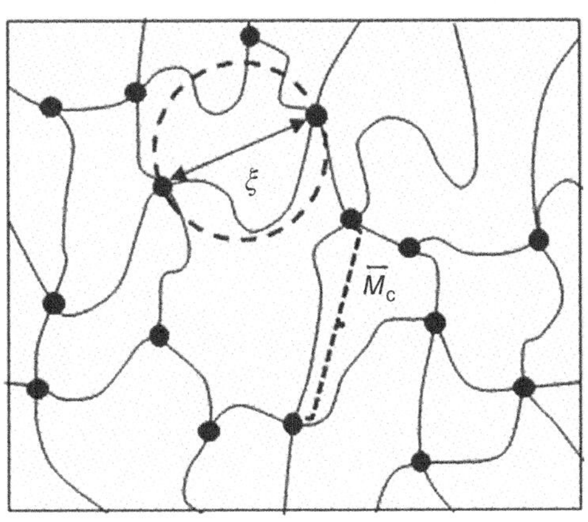

Figure 11.9 Illustration of the mesh size ξ and molecular weight between neighboring crosslinks $\bar{M}_c$ for a crosslinked polymer network.

Qualitatively, upon immersion of a non-solvated hydrogel in water, water molecules will diffuse into the interior of the gel to cover the surface of the hydrophilic polymer chains and reduce its interfacial energy, in much the same way as water wets and climbs up the walls of a glass capillary tube. This generates a tensile hydrostatic stress, which causes the hydrogel network to swell. On the other hand, this swelling causes the chain segments between the crosslinks to stretch, which leads to an elastic compressive hydrostatic stress on the network. This compressive stress is similar in nature to that of a stretched rubber band that attempts to return to its original (relaxed) state. When the tensile and compressive stresses balance, swelling stops and the hydrogel reaches its equilibrium swelling.

The degree of swelling Q of a hydrogel is commonly defined as the ratio of the volume of the swollen hydrogel (V_s) to the volume of the polymer phase (V_p). As Q is equal to $1/v_{2,s}$, we can write

$$v_{2,s} = \frac{1}{Q} = \frac{\text{volume of polymer}}{\text{volume of swollen hydrogel}} = \frac{V_p}{V_s} \quad (11.2)$$

The swelling of hydrogels has been analyzed theoretically, generating equations from which changes in other important parameters, such as $\bar{M}_c$ and ξ, may be predicted.

11.6.1 Theory of Swelling

The most widely used theory to explain the swelling of polymer networks is the Flory–Rehner theory, which is a thermodynamic theory for describing the equilibrium swelling of the network. The theory assumes an electrostatically neutral network (that is, a network possessing no ionic charges) and a chain conformation between the crosslinks corresponding to that of idealized rubber (Section 8.6). As the theory is based on thermodynamic considerations, it predicts the equilibrium swelling of the network. Again, upon immersion of a non-solvated hydrogel in water, water molecules will diffuse into the hydrogel. At equilibrium, the change in energy due to mixing of the chains with the water molecules balances the change in elastic energy due to swelling of the network (Figure 11.10).

The Gibbs free energy change ΔG in the swelling process can be written

$$\Delta G = \Delta G_{mix} + \Delta G_{el} \quad (11.3)$$

where ΔG_{mix} is the Gibbs free energy change due to mixing of the chains with the water molecules, and ΔG_{el} is the Gibbs free energy change due to the elastic stretching of the network. Differentiating Eq. (11.3) with respect to the

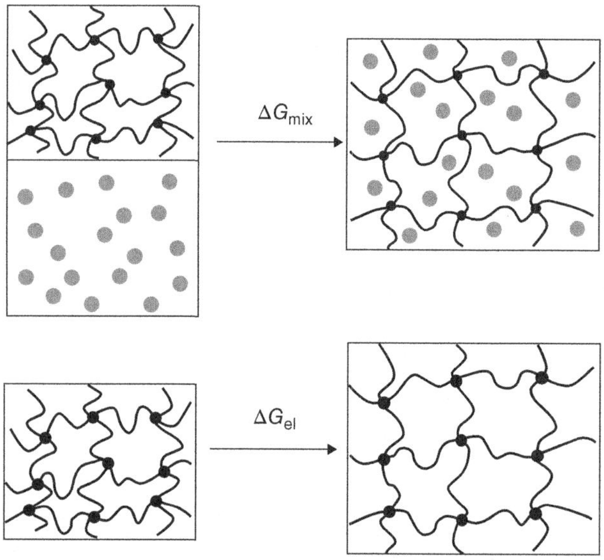

Figure 11.10 Illustration of the contributions to the overall Gibbs free energy change upon immersing a non-solvated hydrogel in water. ΔG_{mix} is the Gibbs free energy change due to mixing of the polymer chains with water molecules, while ΔG_{el} is the change due to swelling of the network.

number of moles of water molecules at constant temperature and pressure gives

$$\mu - \mu_o = \Delta\mu_{mix} + \Delta\mu_{el} \tag{11.4}$$

where μ is the chemical potential of the water molecules in the hydrogel, μ_o is the chemical potential of the molecules in the water surrounding the hydrogel, and $\Delta\mu_{mix}$ and $\Delta\mu_{el}$ are the chemical potential changes due to mixing and elastic deformation, respectively. The condition for equilibrium is that $\mu = \mu_o$, that is, the chemical potential of the water molecules within the hydrogel is equal to that in the surrounding medium. Thus,

$$\Delta\mu_{mix} + \Delta\mu_{el} = 0 \tag{11.5}$$

The chemical potential difference $\Delta\mu_{mix}$ can be determined from the thermodynamics of mixing between the polymer chains and the liquid (the Flory–Huggins solution theory). It consists of a term due to the change in entropy ΔS plus a term due to the change in enthalpy ΔH upon mixing. Overall, $\Delta\mu_{mix}$ is given by (Peppas and Merrill 1977):

$$\Delta\mu_{mix} = RT\left[\ln\left(1 - v_{2,s}\right) + v_{2,s} + \chi_1 v_{2,s}^2\right] \tag{11.6}$$

where R is the gas constant, T is the absolute temperature, and the parameter χ_1, known as the Flory–Huggins interaction parameter, takes into account the degree of resistance of the chains to mixing with the water molecules. χ_1 can have positive or negative values of magnitude less than 1. For example, hydrophilic polymers such as PEG and PVA, composed of an electrostatically neutral network, have a weak resistance to mixing with water, and χ_1 is ~0.5 at room temperature.

The chemical potential change $\Delta\mu_{el}$ has been determined from the theory of rubber elasticity. As discussed in Chapter 8, stretching of a rubber reduces the number of possible conformations that its chains can take up, which leads to a reduction in its entropy ΔS. Thus, ΔG_{el} is positive. Overall, $\Delta\mu_{el}$ is given by

$$\Delta\mu_{el} = \frac{RTV_1}{\bar{v}\bar{M}_c}\left(1 - \frac{2\bar{M}_c}{\bar{M}_n}\right)\left(v_{2,s}^{1/3} - \frac{v_{2,s}}{2}\right) \tag{11.7}$$

where $\bar{M}_n$ is the number-average molecular weight of the polymer chains prepared under the same conditions but without the crosslinks, $\bar{M}_c$ is the number-average molecular weight between the crosslinks, $\bar{v}$ is the specific volume of the bulk polymer (that is, the volume per unit mass, equal to the reciprocal of its density), and V_1 is the molar volume of water.

Using Eq. (11.4) to Eq. (11.7), we get

$$\frac{1}{\bar{M}_c} = \frac{2}{\bar{M}_n} - \frac{\left(\frac{\bar{v}}{V_1}\right)\left[\ln\left(1 - v_{2,s}\right) + v_{2,s} + \chi_1 v_{2,s}^2\right]}{\left(v_{2,s}^{1/3} - \frac{v_{2,s}}{2}\right)} \tag{11.8}$$

In Eq. (11.8), expressing $\bar{M}_n$ in units of g/mol, $\bar{v}$ in cm^3/g, and V_1 in cm^3/mol gives $\bar{M}_c$ in units of g/mol, which, as noted in Chapter 8, has the same numerical value if it were expressed in the unit dalton (Da). For the case of a network where the crosslinks are introduced into a polymer whose chains are already in solution, the presence of the liquid molecules alters $\Delta\mu_{el}$, which must now take into account the volume fraction of polymer chains present during the crosslinking. This leads to a small modification of Eq. (11.8), which becomes

$$\frac{1}{\bar{M}_c} = \frac{2}{\bar{M}_n} - \frac{\left(\frac{\bar{v}}{V_1}\right)\left[\ln\left(1 - v_{2,s}\right) + v_{2,s} + \chi_1 v_{2,s}^2\right]}{v_{2,r}\left[\left(\frac{v_{2,s}}{v_{2,r}}\right)^{1/3} - \left(\frac{v_{2,s}}{v_{2,r}}\right)\right]} \tag{11.9}$$

where $v_{2,r}$ is referred to as polymer volume fraction in the relaxed state, defined by

$$v_{2,r} = \frac{V_p}{V_r} \tag{11.10}$$

where V_r is the volume of the hydrogel after crosslinking (relaxed state) but prior to swelling and equilibration when immersed in more pure liquid.

The presence of ionic groups in the polymer network makes the analysis of swelling more complicated. A contribution ΔG_{ion} must be added to the total Gibbs free energy change (Eq. (11.3)) due to the interaction of the charged groups with the liquid molecules. Equation (11.4) will now contain an additional contribution $\Delta\mu_{\text{ion}}$ due to these ionic interactions, given by

$$\Delta\mu_{\text{ion}} = V_1 RT \sum_j \left(c_j^{\text{gel}} - c_j^{\text{soln}}\right) \tag{11.11}$$

where the summation is over all the ions j of concentration c_j^{gel} and c_j^{soln}, respectively, in the hydrogel and the solution. At equilibrium

$$\Delta\mu_{\text{mix}} + \Delta\mu_{\text{el}} + \Delta\mu_{\text{ion}} = 0 \tag{11.12}$$

For anionic hydrogels, such as PAA hydrogels prepared in the presence of water, the equilibrium swelling equation is (Brannon-Peppas and Peppas 1991):

$$\begin{aligned} &\frac{V_1}{4IM_{\text{r}}}\left(\frac{v_{2,\text{s}}}{\bar{v}}\right)^2\left(\frac{K_{\text{a}}}{10^{-\text{pH}} + K_{\text{a}}}\right)^2 \\ &= \left[\ln(1 - v_{2,\text{s}}) + v_{2,\text{s}} + \chi_1 v_{2,\text{s}}^2\right] \\ &+ \left(\frac{V_1}{\bar{v}\bar{M}_{\text{c}}}\right)\left(1 - \frac{2\bar{M}_{\text{c}}}{\bar{M}_{\text{n}}}\right) v_{2,\text{r}} \left[\left(\frac{v_{2,\text{s}}}{v_{2,\text{r}}}\right)^{\frac{1}{3}} - \left(\frac{v_{2,\text{s}}}{2v_{2,\text{r}}}\right)\right] \end{aligned} \tag{11.13}$$

where I is the ionic strength of the aqueous medium, K_{a} is the dissociation constant of the acid group of the hydrogel chain, and M_{r} is the molecular weight of the repeating monomer unit of the chain. Equation (11.13) gives a prediction for the relationship between the equilibrium swelling and the various parameters of the hydrogel and the aqueous medium. An equation of the same form as Equation (11.13) is obtained for cationic hydrogels but the term on the left-hand-side is modified to account for the difference in the acid–base reaction.

$\bar{M}_{\text{c}}$ can be determined using Eq. (11.9) or (11.13) once the other parameters in the equation are known. Another parameter that has been often used to characterize the network structure of hydrogels is the crosslink density ρ, which is related to $\bar{M}_{\text{c}}$ by

$$\rho = \frac{1}{\bar{v}\bar{M}_{\text{c}}} \tag{11.14}$$

Approximating the 3D network of a crosslinked hydrogel to that of an ideal rubber, Flory (1953) used the theory of rubber elasticity to develop an alternative equation for $\bar{M}_{\text{c}}$ in terms of the mechanical deformation behavior of the network. When an initially unswollen crosslinked network is swollen such that the ratio of the volume of the polymer phase to the volume of the swollen hydrogel is $v_{2,\text{s}}$, application of a shear stress τ to the swollen unstretched specimen leads to an extension ratio α given by

$$\frac{\tau}{\alpha - \dfrac{1}{\alpha^2}} = \frac{RT}{\bar{v}\bar{M}_{\text{c}}}\left(1 - \frac{2\bar{M}_{\text{c}}}{\bar{M}_{\text{n}}}\right) v_{2,\text{s}}^{1/3} \tag{11.15}$$

where R is the gas constant, T is the absolute temperature, and τ is expressed as the shear force per unit cross sectional area of the swollen unstretched specimen. The extension ratio α is the ratio of the stretched specimen length (L) to the unstretched length (L_{o}). For small extensions, where $\Delta L = L - L_{\text{o}}$ is less than ~$0.1L_{\text{o}}$, the extension ratio is related to the engineering strain ε by the equation

$$\alpha = \frac{L}{L_{\text{o}}} = \frac{L_{\text{o}} + \Delta L}{L_{\text{o}}} = 1 + \frac{\Delta L}{L_{\text{o}}} = 1 + \varepsilon \tag{11.16}$$

Within the limits of applicability of Eq. (11.15) (ε less than ~0.5), a plot τ versus $(\alpha - 1/\alpha^2)$ is approximately a straight line of slope equal to the shear modulus μ, given by

$$\mu = \frac{RT}{\bar{v}\bar{M}_{\text{c}}}\left(1 - \frac{2\bar{M}_{\text{c}}}{\bar{M}_{\text{n}}}\right) v_{2,\text{s}}^{1/3} \tag{11.17}$$

When the polymer is initially in solution (prior to crosslinking) at a concentration $C_{2,\text{r}}$ g/cm^3, the appropriate form of Eq. (11.15) is

$$\frac{\tau}{\alpha - \dfrac{1}{\alpha^2}} = \frac{RTC_{2,\text{r}}}{\bar{M}_{\text{c}}}\left(1 - \frac{2\bar{M}_{\text{c}}}{\bar{M}_{\text{n}}}\right)\left(\frac{v_{2,\text{s}}}{v_{2,\text{r}}}\right)^{1/3} \tag{11.18}$$

in which case, the shear modulus is

$$\mu = \frac{RTC_{2,\text{r}}}{\bar{M}_{\text{c}}}\left(1 - \frac{2\bar{M}_{\text{c}}}{\bar{M}_{\text{n}}}\right)\left(\frac{v_{2,\text{s}}}{v_{2,\text{r}}}\right)^{1/3} \tag{11.19}$$

The mesh size ξ of the network can be determined from the equation

$$\xi = \beta \left\langle r_{\text{o}}^2 \right\rangle^{1/2} \tag{11.20}$$

where β is the elongation ratio of the polymer chains in any direction and $\left\langle r_{\text{o}}^2 \right\rangle^{1/2}$ is the root mean square end-to-end distance of the chain segments between two neighboring crosslinks in the relaxed (unperturbed) network (Section 8.2.6). For a network that undergoes isotropic swelling (that is, it expands equally in all directions), β is related to $v_{2,\text{s}}$ by

$$\beta = v_{2,\text{s}}^{-1/3} \tag{11.21}$$

For a freely jointed chain, $\left\langle r_{\text{o}}^2 \right\rangle^{1/2}$ is given by Eq. (8.7). However, a correction is required because the chain segments in a hydrogel are not free chains but, instead, are restricted to some degree due to the presence of the crosslinks. One modification is

$$\langle r_o^2 \rangle^{1/2} = (C_n N)^{1/2} l \tag{11.22}$$

where C_n is termed the Flory characteristic ratio, N is the number of monomer units between crosslinks, and l is the bond length of the monomer units (such as the C–C bond length). The value of C_n depends on the polymer. For example, for PVA and PHEMA, C_n has been determined as ~9.0 and ~7.0, respectively. As N can be determined as

$$N = \frac{2\bar{M}_c}{M_r} \tag{11.23}$$

where M_r is the molecular weight of the monomer units in the chain segments, using Eq. (11.21) to Eq. (11.23), the mesh size is

$$\xi = v_{2,s}^{-1/3} \left(\frac{2C_n \bar{M}_c}{M_r} \right)^{1/2} l \tag{11.24}$$

The swelling relationships covered in this section, as noted earlier, describe the equilibrium swelling of hydrogels. They do not give information about the rate of swelling, which requires a kinetic analysis (De et al. 2002). For example, the rate of swelling decreases with increasing thickness of the hydrogel because the average diffusion distance of the water becomes longer. Additionally, for polyelectrolyte hydrogels (anionic or cationic) that possess electrostatic charges, the rate of swelling depends on the pH, ionic strength, and the type of ions in the swelling medium.

11.6.2 Determination of Swelling Parameters

Experiments for determining the swelling parameters of a hydrogel would normally start with measuring V_p, V_r, and V_s using standard techniques. When the hydrogel has a regular geometry, which is not often the case, its volume may be determined from its external dimensions. A more common method is to use the Archimedes method, which involves measuring the mass of the hydrogel in air and in a nonpolar liquid such as heptane. The difference in the two masses gives the mass of liquid displaced, which is the volume of liquid displaced times its density. Using a liquid of known density, the volume of the liquid displaced and, thus, the volume of the hydrogel can be determined. Prior to measuring V_s, the crosslinked hydrogel is immersed in an aqueous liquid, such as distilled water, at the physiological temperature (~37 °C) until thermodynamic equilibrium (that is, constant weight) is attained. On the other hand, V_p is measured after drying the crosslinked hydrogel, from the mass of the dried hydrogel and the density of the bulk (dense) polymer. Then $v_{2,s}$ and $v_{2,r}$ can be determined using Eqs. (11.2) and (11.10), respectively.

$\bar{M}_c$ can be determined from Eq. (11.8), (11.9), or (11.13), depending on the way in which crosslinking has been performed, and when $\bar{M}_n$ and M_r for the polymer used in creating the hydrogel, the bulk density of the polymer (equal to $1/\bar{v}$), V_1 (~18 cm^3/mol for water), and χ_1 are known. An alternative way to determine $\bar{M}_c$ is from the initial linear elastic region of tensile tests and using Eq. (11.15) or (11.18). For loosely crosslinked PVA hydrogels, experiments show good agreement between $\bar{M}_c$ values determined from swelling experiments and from tensile tests. On the other hand, significant deviations have been found for highly crosslinked PVA hydrogels, which have been attributed to stress-induced crystallization of the chains. The mesh size ξ of the swollen network can be determined from Eq. (11.24) when C_n and l are also known.

As an example, Figure 11.11 shows data for PVA hydrogels created from a polymer of $\bar{M}_n$ ~ 52.8 kDa and crosslinked using varying amounts of glutaraldehyde (moles of glutaraldehyde per mole of repeating unit of PVA). With increasing glutaraldehyde content, that is, increasing degree of crosslinking, $v_{2,s}$ increases, while $\bar{M}_c = 1/(\bar{v}\rho)$ decreases.

For PVA hydrogels chemically crosslinked using γ-radiation or glutaraldehyde, the following empirical correlation has been found to hold from experimental data for ξ and $v_{2,s}$:

$$\xi = k_1 + k_2 v_{2,s}^n \tag{11.25}$$

where k_1 and k_2 are constants for the specific hydrogel system and n is an exponent that depends on $v_{2,s}$. For $v_{2,s}$ below ~0.1, n is −1/2, whereas for $v_{2,s}$ above ~0.1, n is −1. Because

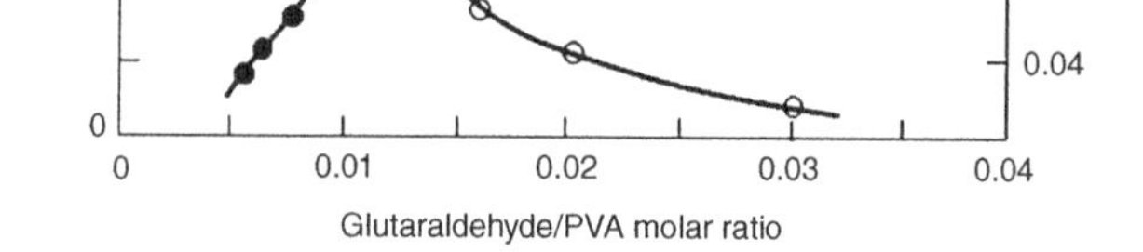

Figure 11.11 Dependence of the molecular weight between crosslinks $\bar{M}_c$ and the volume fraction of polymer in the swollen hydrogel $v_{2,s}$ for polyvinyl alcohol (PVA) hydrogels as a function of degree of crosslinking, defined as the molar ratio of glutaraldehyde per mole of repeating unit of PVA. *Source:* Peppas and Benner (1980) / with permission of Elsevier.

Eq. (11.25) does not involve the polymer parameters present in Eq. (11.24), it may provide an easier method to determine ξ. However, it is empirical and is normally valid only for the hydrogel for which k_1, k_2, and n have been determined.

11.7 Mechanical Properties of Hydrogels

The mechanical properties of hydrogels are important in their selection and design for specific applications. For example, hydrogels find considerable use as scaffolds in tissue engineering, where their 3D network structure is responsible for holding cells in place and for providing diffusional access to nutrients and biomolecules as well (Chapter 25). Another important function of the scaffold is to deliver the requisite molecular or mechanical signals to cells in order for them to create the desired tissue. It is well known that the elastic modulus of the scaffold has a strong effect in determining the fate of the encapsulated cells, that is, their capacity to differentiate into the desired phenotype.

Overall, hydrogels show weak mechanical properties when compared to most solid polymers but their mechanical properties cover a wide range depending on factors such as the composition of the polymer chains, degree of crosslinking, and environmental factors. For example, hydrogels may show an initial elastic modulus E_0 (slope of the initial region of the tensile stress versus strain curve) in the range of few kilopascals, typical of a soft hydrogel to over several hundred kilopascals, typical of a stiff hydrogel. While the mechanical response of lightly crosslinked hydrogels may be understood using the theory of rubber elasticity, it should be remembered that hydrogels, like all polymers, are viscoelastic (Chapter 8). Consequently, their mechanical properties at a given temperature depend not just on the magnitude of the applied stress (or strain) but on time (or the rate of testing) as well.

Equation (11.17) or (11.19) gives the shear modulus μ of a swollen hydrogel, depending on whether the hydrogel is crosslinked in the absence or presence of a liquid. The initial elastic modulus E_0 is related to μ through the Poisson's ratio (Eq. (4.16)), which, for a given hydrogel, is approximately constant (~0.4–0.5). Thus, Eq. (11.17) or (11.19) predicts that, for a given hydrogel, E_0 increases with decreasing $\bar{M}_c$ (or equivalently, with increasing crosslink density ρ) and with increasing $v_{2,s}$. Experimental data for lightly crosslinked hydrogels generally show good correlation with these predictions.

Overall, then, varying the polymer composition by copolymerization, for example, to modify the stiffness of the chain backbone and the crosslink density of the network provides an effective way to control the mechanical properties of hydrogels.

Common methods for measuring the mechanical properties of hydrogels involve tensile testing and indentation techniques. Tensile testing is typically performed using dog-bone-shaped specimens of standard dimensions, which are completely immersed in the desired aqueous environment in which they are in equilibrium, and applying the load at a constant strain rate (Chapter 4). Indentation techniques provide an alternative method to measure the elastic modulus when it is not practical to form the hydrogel into the requisite shape for tensile testing.

11.8 Transport Properties of Hydrogels

An important property of hydrogels, particularly for drug delivery and tissue engineering applications, is their ability to control or restrict diffusion of biomolecules such as therapeutic drug molecules or growth factors through them. Approximating these biomolecules as solute molecules of spherical shape, their transport through a hydrogel depends strongly on the size of the water-filled spaces delineated by the crosslinked polymer chains (Figure 11.12a). Consequently, factors that alter the size of these spaces will affect the diffusion of a solute through the hydrogel. Additional factors that influence solute diffusion are the mobility of the chains between the crosslinks and interactions between the solute and the chains, such as electrostatic interactions.

At a basic level, we will neglect solute–chain interactions and consider solute transport in a system composed of hydrophilic solute molecules within a hydrogel having a homogeneous crosslinked network of mobile chains. Qualitatively, for this type of system, the solute diffusion through the hydrogel will decrease as the size of the solute increases, the crosslink density of the hydrogel increases, and the volume fraction of water within the hydrogel decreases.

A variety of models has been developed to explain and predict solute diffusion in hydrogels, but the most popular model in drug delivery is based on a free volume theory originally put forward to explain the diffusion of solutes in a liquid. In this model, the solute diffuses by jumping into neighboring voids formed in the liquid by redistribution of the liquid molecules due to random fluctuations. In the case of a hydrogel, in addition to finding a sufficient free volume within the liquid, the solute must also find a large enough opening between the crosslinked polymer chains (Figure 11.12b). Consequently, the diffusion of the solute through the hydrogel may be taken to be proportional to

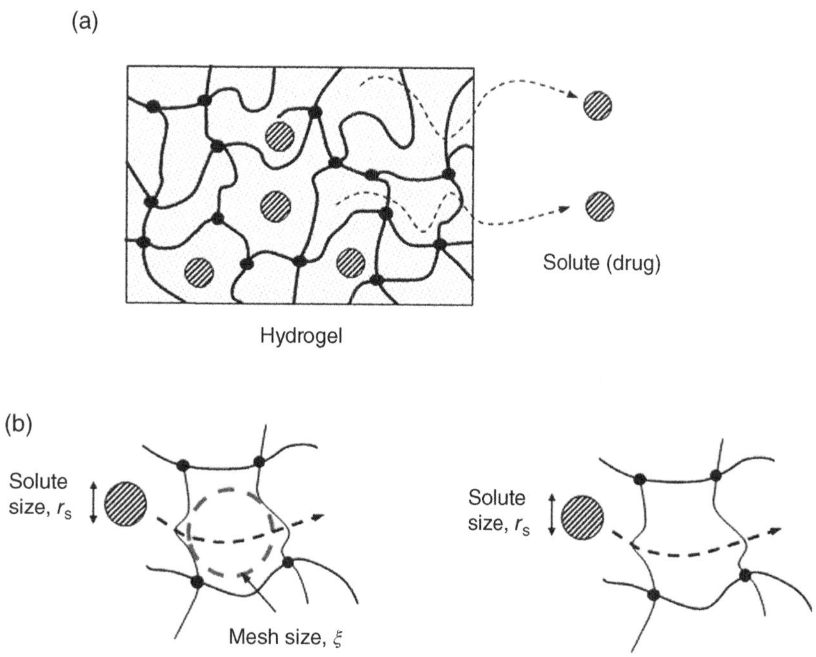

Figure 11.12 (a) Illustration of solute (drug) release through a hydrogel; (b) diffusion depends on the solute finding a sufficiently large space between the crosslinked chains and within the volume of the hydrogel.

the product of the probability P_V of finding a sufficient free volume within the hydrogel volume and the probability P_O of finding a sufficient opening between the chains. For a solute molecule of size (diameter) r_s, one model gives P_O as

$$P_O = \frac{\xi - r_s}{\xi} \tag{11.26}$$

where ξ is the mesh size of the hydrogel. Modification of a theory for solute diffusion in a liquid gives

$$P_V \sim \exp\left(-\frac{1}{Q-1}\right) \tag{11.27}$$

where Q is the degree of swelling given by Eq. (11.2). Substituting for Q, we get

$$P_V \sim \exp\left(-\frac{v_{2,s}}{1-v_{2,s}}\right) \tag{11.28}$$

One model gives the diffusion coefficient D_g of the solute molecule in the hydrogel relative to its diffusion coefficient D_o in the pure liquid as (Lustig and Peppas 1988):

$$\frac{D_g}{D_o} = \left(1 - \frac{r_s}{\xi}\right)\exp\left[-k_3 r_s^2\left(\frac{v_{2,s}}{1-v_{2,s}}\right)\right] \tag{11.29}$$

where k_3 is a constant for the specific hydrogel system.

Equation (11.29) predicts that the diffusion of the solute molecule in the hydrogel, relative to that in the pure liquid, decreases with increasing size of the solute and increasing volume fraction of the polymer in the hydrogel and with decreasing mesh size of the crosslinked network.

11.9 Surface Properties of Hydrogels

The surface of a hydrogel has a complex structure, containing numerous dangling chains that have only one end attached to the 3D network (Figure 11.13). These dangling chains have a higher mobility when compared to the crosslinked chains in the bulk of the hydrogel. They behave in a manner comparable to that of polymer brushes, that is, an assembly of polymer chains, each tethered at one end to a surface or interface and extending into the surrounding medium. This structure, along with the hydrophilicity of the chains, strongly influences the surface properties of a hydrogel.

Depending on the nature of the surrounding medium, the mobile chains at the surface of a hydrogen may undergo structural reorientation of their side groups driven by a reduction in interfacial energy between the hydrogel surface and the surrounding medium (Figure 5.9). Another characteristic of hydrogels is that strongly hydrophilic synthetic hydrogels generally show a lower capacity to support protein adsorption and cell adhesion to their surface when

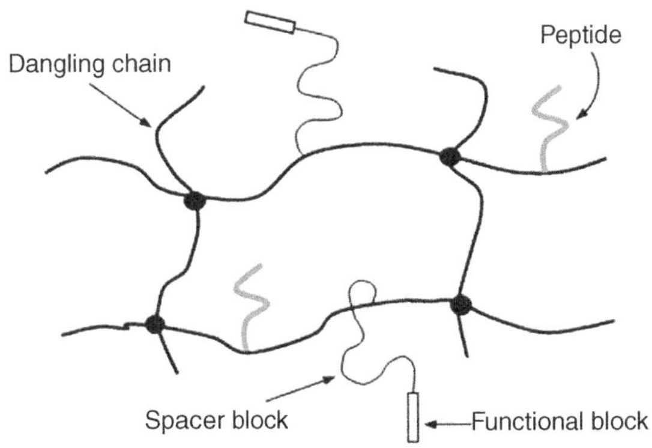

Figure 11.13 Illustration of types of structural features that may be present at the surface of a hydrogel, consisting of dangling chains, cell recognition sites (peptide), and functional groups tethered to the surface by spacer molecules.

compared to more hydrophobic hydrogels (Chapter 16). One way to improve the capacity of hydrophilic hydrogels to support cell adhesion is through surface modification, by attaching cell-adhesive peptides to their chains, such as peptides having the arginine-glycine-aspartic acid (RGD) sequence. This type of surface modification is covered later along with other surface modification techniques in Chapter 13, but when appropriate, it will be described for specific hydrogels in this chapter.

11.10 Environmentally Responsive Hydrogels

Environmentally responsive hydrogels consist of a polymer network that shows a significant change in properties in response to a modest change in one or more physical factors, such as the pH, temperature, or ionic strength of the aqueous environment. Some hydrogels can also show a response to an external physical stimulus, such as ultrasound, or an applied electric, magnetic, or electromagnetic field (Chapter 26). Hydrogels that respond to environmental factors or physical stimuli are also referred to as stimuli responsive hydrogels or smart hydrogels. Here, we shall consider the basic principles related to the response of hydrogels to changes in the aqueous environment because these changes are normally more relevant to the performance of many hydrogels used in biomedical applications.

11.10.1 pH Responsive Hydrogels

pH responsive hydrogels undergo significant changes in their 3D structure and, thus, in their properties such as swelling, transmittance, or elastic modulus, in response to a modest pH change of their aqueous environment. These hydrogels normally contain a high fraction of ionizable functional groups in their chains. Polyelectrolyte hydrogels, composed of chains in which the repeating unit contains one or more ionizable functional groups, are a widely studied type of such pH responsive hydrogels. For example, the monomer unit of PAA contains the carboxyl (C=O)OH group in which the H atom dissociates above a certain pH to give a negatively charged (C=O)O$^-$ group (Figure 11.14a). The deprotonation reaction and its equilibrium constant K_a are given by the following equations:

$$R-(C=O)OH \leftrightarrows R-(C=O)O^- + H^+ \tag{11.30}$$

$$K_a = \frac{[R-(C=O)O^-][H^+]}{[R-(C=O)OH]} \tag{11.31}$$

$$pK_a = -\log K_a \tag{11.32}$$

$$pH = pK_a + \log\frac{[R-(C=O)O^-]}{[R-(C=O)OH]} \tag{11.33}$$

where the square brackets represent the concentration of their respective species. The pK_a of PAA, as noted earlier, is ~4.5, which means that a significant fraction of the (C=O)OH groups in the chains is dissociated near or above a pH of ~4.5. The degree of ionization i, defined by

$$i = \frac{[R-(C=O)O^-]}{[R-(C=O)O^-]+[R-(C=O)OH]} \tag{11.34}$$

is related to the pH, equal to $-\log_{10}[H^+]$, by

$$pH = pK_a + \log_{10}\frac{i}{1-i} \tag{11.35}$$

The dependence of i with pH for PAA is shown in Figure 11.14a. At pH less than ~3–4 when i is ~0, there is almost no dissociation of the (C=O)OH groups. The chains are electrostatically neutral and the hydrogel has a compact structure due to hydrogen bonding between neighboring (undissociated) OH groups. In comparison, above pH ~5–6 when i is ~1, the (C=O)OH groups are almost fully dissociated and the network swells considerably due to the repulsion between the negatively charged (C=O)O$^-$ groups. Swelling and de-swelling in response to a change in pH is controlled by kinetic factors and does not occur very rapidly. Additionally, the kinetics of swelling and de-swelling are normally different, which means that the swelling and de-swelling curves normally do not coincide.

Cationic hydrogels, composed of chains in which the monomer units contain basic groups that can become protonated below a certain pH are also pH responsive. The swelling and de-swelling behavior in response to pH changes is similar in nature to that of PAA but it is reversed

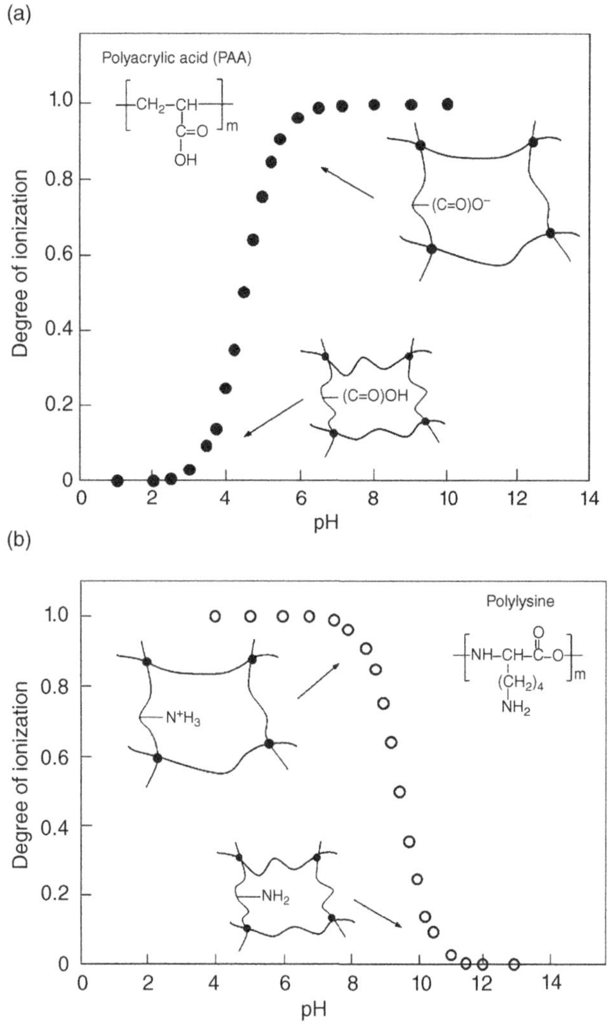

Figure 11.14 Degree of ionization of (a) the (C=O)OH group in polyacrylic acid (PAA) and (b) the NH_2 group in polylysine as a function of pH in water. The swollen and unswollen hydrogel network for the ionized and neutral chains are illustrated.

(Figure 11.14b). For example, the deprotonation–protonation reaction of the NH_2 groups in the monomer units of polylysine has a pK_a of ~9–10. Consequently, polylysine hydrogels expand considerably when the pH of the medium is lowered below ~9–10.

Polyelectrolyte hydrogels may respond not only to pH changes but also to changes in the ionic strength I of the swelling medium defined as

$$I = \frac{1}{2} \sum_{j}^{\text{All ions}} z_j^2 c_j \tag{11.36}$$

where z_j is the charge of the ion j of concentration c_j in the medium. For an anionic hydrogel such as PAA immersed in a solution of sufficiently high pH containing sodium (Na^+) and chlorine (Cl^-) ions), such as phosphate buffered

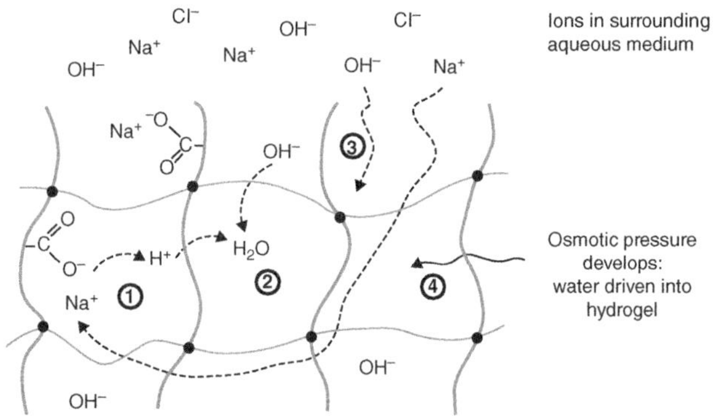

Figure 11.15 Illustration of the mechanism of swelling for a polyelectrolyte hydrogel such as polyacrylic acid (PAA) in a solution such as phosphate buffered saline (PBS) containing various ions.

saline (PBS), the mechanism of swelling is illustrated in Figure 11.15. First, the H atoms of the (C=O)OH groups dissociate and combine with OH^- ions from the solution to form H_2O molecules. Charge balance is maintained by cations (Na^+) and hydroxyl (OH^-) ions entering the hydrogel, which makes the concentration of Na^+ ions within the hydrogel higher than that in the solution. This generates an osmotic pressure that pushes the chains apart, causing water to be driven into the hydrogel, which leads to swelling. For polyelectrolyte hydrogels such as PAA whose chains develop a high negative charge, the osmotic pressure is normally much greater than the hydrostatic pressure due to mixing between the chains and water. Consequently, an equilibrium swelling occurs when the osmotic pressure balances the compressive pressure due to the elastic retractive force of the chains.

An understanding of swelling and de-swelling behavior is vital in the design and selection of pH responsive hydrogels for specific applications, particularly for drug delivery. One approach is to take advantage of the predictive capabilities of models, which can be correlated with experimental data and used subsequently in hydrogel design and selection. For example, when the surrounding medium and the aqueous medium within the hydrogel have comparable ionic strength, Eq. (11.13) predicts that the equilibrium swelling of anionic hydrogels depends on the pK_a of their ionizable functional groups and the pH and ionic strength of the surrounding medium. Changing the pK_a by using hydrogels that have a different ionizable functional group causes the swelling curve to shift along the pH axis with little change in the shape of the curve (Figure 11.16a). This shows that the relationship between the pK_a of the polymer and the pH of the swelling medium determines the equilibrium swelling of the hydrogel. On the other hand, for a given hydrogel, an increase in the ionic strength of the swelling medium has the effect of reducing the equilibrium swelling without significantly altering the pH over which swelling takes place (Figure 11.16b). Kinetic models, involving numerical simulations, predict that an anionic hydrogel will swell faster with increasing ionic strength of the swelling medium and with higher diffusivity of the appropriate ions into the hydrogel (Figure 11.17), in general accordance with experimental data.

11.10.2 Temperature Responsive Hydrogels

Temperature responsive hydrogels, also referred to as thermoreversible hydrogels, show significant changes in their network structure and, thus, in their properties when subjected to modest temperature changes. Like the case of pH responsive hydrogels, this temperature-dependent response is reversible but the kinetics of swelling and deswelling normally do not coincide. Temperature responsive hydrogels are attractive biomaterials for a variety of biomedical applications, especially when they undergo these reversible changes near room temperature.

Temperature responsive hydrogels are normally physical hydrogels that rely on cooperative bonding interactions involving secondary (weak) bonds, such as hydrogen bonds and van der Waals bonds, for their network stability. They show two general types of behavior (Figure 11.18):

- Hydrogels composed of polymers that show a lower critical solution temperature (LCST), that is, polymers that phase separate from their (single-phase) solution to form a solid phase (polymer network) and solvent when the temperature is increased above their LCST. Widely studied examples are the triblock copolymer PEO-*b*-PPO-*b*-PEO and the homopolymer PNIPAAm.
- Hydrogels composed of polymers that show an upper critical solution temperature (UCST), which is the more common type of temperature responsive behavior. Solutions of these polymers phase separate when the temperature is lowered below their UCST. Examples are gelatin (Chapter 10) and polymers composed of an interpenetrating network of PAA and PAAm chains.

A basic requirement in designing and selecting temperature responsive hydrogels for specific biomedical applications is to understand the causes of these two types of behavior. Thermodynamically, for a physical or chemical change to occur, there must be a decrease in the total Gibbs free energy for the system. The standard Gibbs free energy change ΔG^o for a system is

$$\Delta G^o = \Delta H^o - T\Delta S^o \qquad (11.37)$$

where ΔH^o and ΔS^o are the standard enthalpy and entropy change, respectively, and T is the absolute temperature. For a variety of PEO-*b*-PPO-*b*-PEO copolymers referred to as Pluronics, ΔH^o has a positive value (~200–300 kJ/mol). Consequently, for gelling to occur upon increasing the temperature there must be an increase in entropy (that is, ΔS must be positive). This increase in entropy arises from dehydration of the hydrophobic PPO chain segments. The ordered water molecules surrounding the PPO chain segments take up a disordered arrangement in the liquid surrounding the micellar structures (Figure 11.6). This dehydration allows the hydrophobic PPO chain segments to interact in a cooperative manner by van der Waals attraction, which leads to gelling. A similar type of mechanism is responsible for the gelling of PNIPAAm (Figure 11.7).

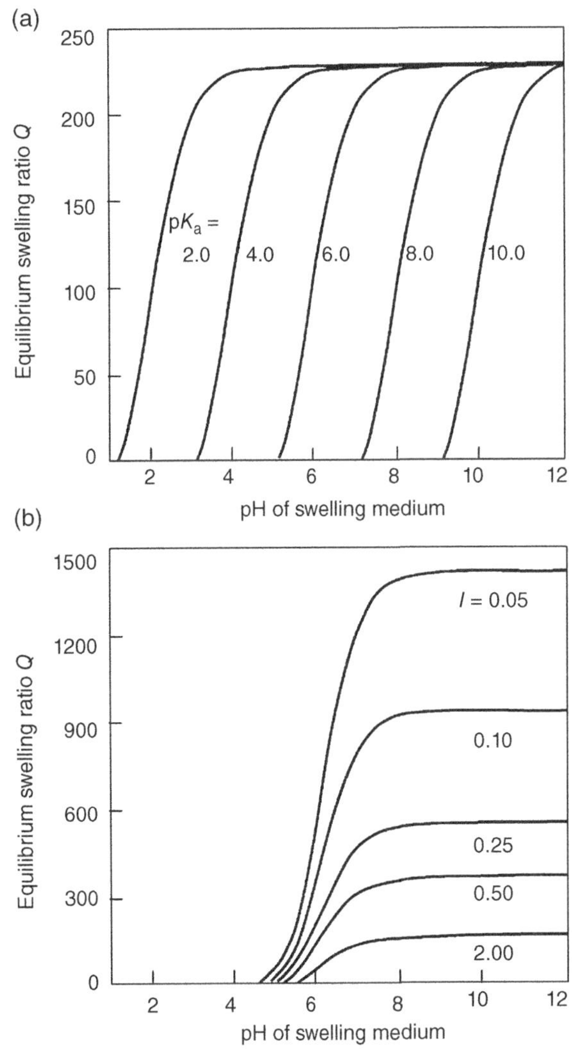

Figure 11.16 Theoretical predictions for the equilibrium swelling Q as a function of the pH of the swelling medium for (a) anionic hydrogels of different pK_a and (b) an anionic hydrogel immersed in swelling medium of different ionic strength. *Source:* Brannon-Peppas and Peppas (1991) / with permission of Elsevier.

In comparison, a solution of PAA and PAAm show a UCST. At low pH (~3.0) and at temperatures below ~30 °C, PAA and PAAm chains form complexes, which leads to gelling, due to hydrogen bonding between the chains (Figure 11.5). Upon warming above ~30 °C, these bonds are disrupted due to an increase in thermal energy of the hydrogen-bonding groups, and the hydrogel reverts to a solution. In these hydrogels, gelation is enthalpy driven, that is, cooling the solution leads to a decrease in enthalpy due to the formation of hydrogen bonds.

Overall, then, for LCST polymers, the temperature-dependent gelling is driven by an increase in entropy whereas for UCST polymers, gelling is driven by a decrease in enthalpy.

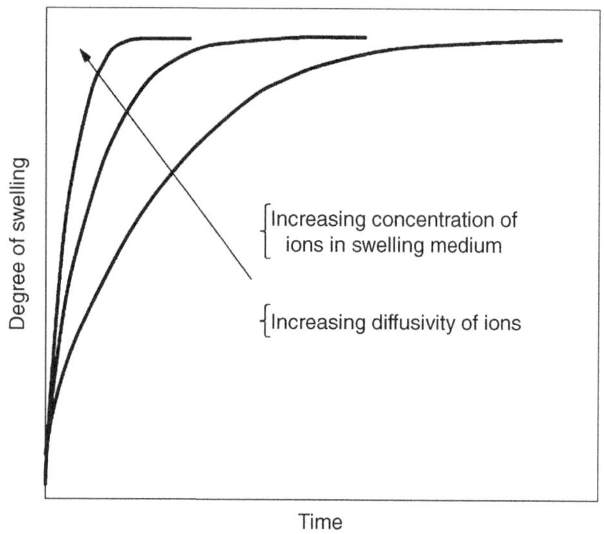

Figure 11.17 Schematic illustration of the trend in the kinetics of swelling with changes in the concentration and diffusivity of ions in the swelling medium.

11.11 Synthetic Hydrogels

Synthetic hydrogels have been attracting considerable interest for biomedical applications in the last few decades, due to a large extent on the ability to control their chemical composition at a molecular scale and, thus, control their physicochemical and biological properties. A list of some synthetic polymers that have been widely used to create hydrogels for biomedical applications is given Table 10.1.

11.11.1 Polyethylene Glycol and Polyethylene Oxide

PEG and PEO are composed of chains having the same ($-CH_2-CH_2-O-$) repeating unit but slightly different end groups (Figure 11.19). This is because they are normally synthesized by different chemical reactions, that is, ring opening polymerization of ethylene oxide for PEO and condensation of ethylene glycol for PEG. Because of this difference in synthesis technique, PEG normally has a lower molecular weight (less than ~20 kDa) than PEO (greater than ~20 kDa).

PEG hydrogels by themselves or in combination with other molecules have been one of the most widely studied hydrogels for a variety of biomedical applications, such as medical devices, drug delivery, and tissue engineering. PEG is biocompatible, strongly hydrophilic, and strongly resistant to adsorption of proteins and cells. Although PEG is soluble in aqueous media, it is not degradable and it

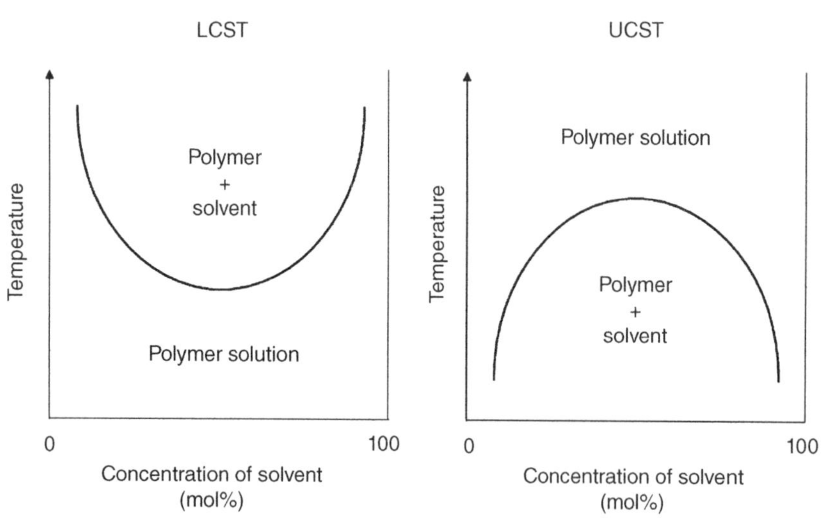

Figure 11.18 Schematic illustration of typical phase diagrams for polymers showing a lower critical solution temperature (LCST) or an upper critical solution temperature (UCST).

lacks intrinsic biological activity, which accounts for one of its major uses as covalently bonded or adsorbed molecules on surfaces to inhibit protein adsorption. PEG molecules have also been attached to proteins and to nanoparticle surfaces to modify their biological properties, a technique often referred to as PEGylation (Chapter 13).

Despite being nondegradable and lacking intrinsic biological activity, PEG has a chemical structure that makes it a versatile polymer for use as hydrogels in biomedical applications. For example, PEG hydrogels can be easily synthesized by a variety of chemical techniques under mild conditions. PEG can also be synthesized in the form of not only linear chains but also branched chains (such as four-arm chains), which expands the range of network structures in which its hydrogels can be created. Because of the reactive hydroxyl (OH) end groups in their chains, PEG hydrogels are very amenable to chemical modification or functionalization to control their chemical or biological activity. Additionally, by taking advantage of these OH or functionalized end groups, PEG molecules can be combined with other molecules (synthetic or natural) to create hydrogels having a range of physicochemical and biological properties.

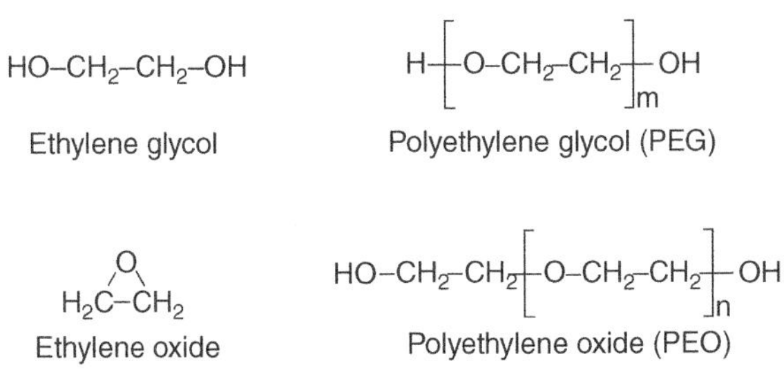

Figure 11.19 Chemical structures of polyethylene glycol (PEG) and polyethylene oxide (PEO).

Although a variety of techniques can be used to create PEG hydrogels, the most common involves free radical polymerization using photochemical methods such as irradiation with UV light. This technique uses PEG composed of molecules that contain crosslinkable end groups such as the acrylate group, as illustrated for PEGDA (Figure 11.2b).

Chemical Modification of PEG Hydrogel

The major modifications of PEG hydrogels consist of two types:

- Attachment of appropriate molecules such as peptides to PEG chains to enhance their biological activity toward cells
- Copolymerization of PEG molecules with other molecules to impart controllable degradation to the hydrogels.

Attachment of various peptides to biomaterials is a widely used technique to improve their interaction with cells. A commonly used peptide is RGD (using the one-letter abbreviation for the amino acids). For example, as illustrated in Figure 11.20, the pentapeptide YRGDS may be bonded directly to an acrylate group or indirectly through a PEG spacer molecule using a reaction with an *N*-hydroxysuccinimidyl-activated ester (Dern and Hubbell 1998). Then the mono-acrylated pentapeptide is covalently bonded to PEGDA molecules in PBS by irradiation with UV light (Figure 11.2). As an example, PEG modified with an optimal amount of RGD has been found to significantly enhance the osteogenic differentiation of bone marrow-derived mesenchymal stem cells *in vitro*.

A common modification of PEG, particularly for tissue engineering applications, is copolymerization with other molecules to enhance and control its degradation. For

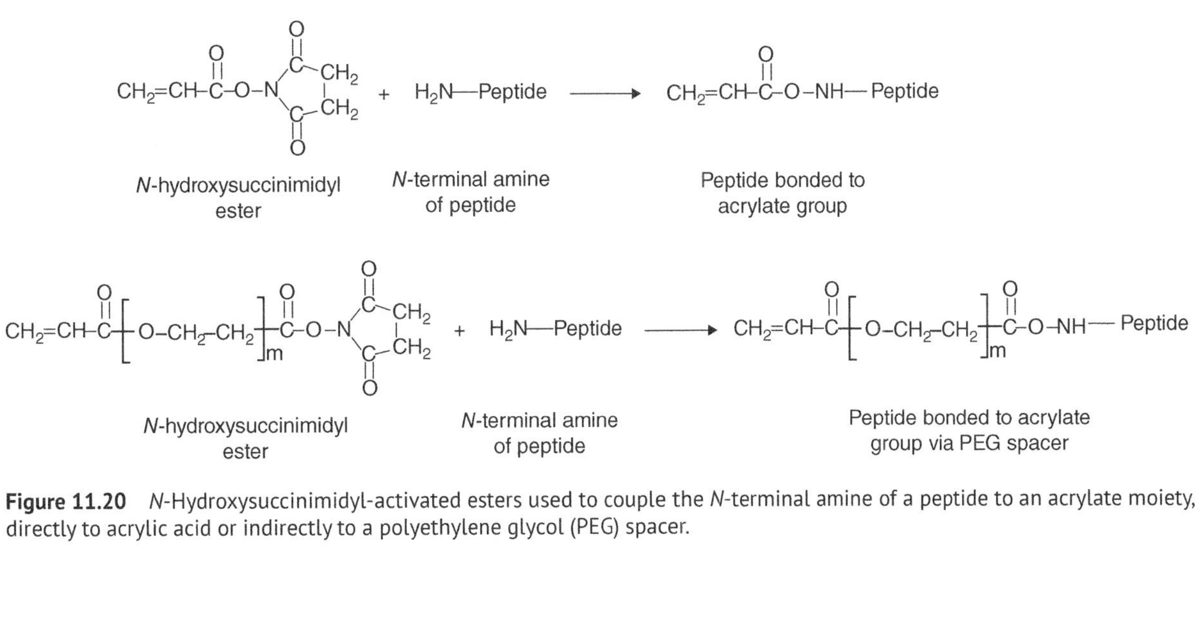

Figure 11.20 *N*-Hydroxysuccinimidyl-activated esters used to couple the *N*-terminal amine of a peptide to an acrylate moiety, either directly to acrylic acid or indirectly to a polyethylene glycol (PEG) spacer.

example, copolymerization with a degradable synthetic polymer such as an α-hydroxy ester (Section 9.8.1) can be used to create PEG-based hydrogels that degrade by hydrolysis. Although PEG hydrogels are insoluble and nondegradable, individual PEG chains of molecular weight less than ~10–15 kDa dissolve in the physiological fluid and can be eliminated from the body by excretion through the kidneys. Additionally, copolymerization with certain peptides can be used to create hydrogels that degrade by hydrolysis mediated by specific enzymes such as collagenase.

An example of a hydrolytically degradable hydrogel is a triblock PLA-*b*-PEG-*b*-PLA copolymer composed of a central PEG block (molecular weight less than ~5 kDa) extended with oligomers of lactic acid and terminated with acrylate groups (Figure 11.21). Alternatively, oligomers of glycolic acid may be used instead of lactic acid oligomers. These water soluble molecules can be crosslinked by irradiation with visible light (wavelength ~420 nm) to create hydrogels of varying composition and structure, depending on the molecular weight of the PEG and PLA chain segments in the copolymer. Degradation of these PLA-*b*-PEG-*b*-PLA hydrogels are controllable over a wide range (less than one day to several weeks) by varying the molecular weight of the PEG (between ~1 and ~10 kDa). Lower PEG molecular weight leads to slower degradation due to the increase in crosslink density.

Erosion of these PLA-*b*-PEG-*b*-PLA hydrogels in phosphate-buffered saline at the physiological pH (~7.4) and temperature (~37 °C) shows an approximately linear decrease in mass as a function of time over most of the erosion process (Figure 11.22). This erosion profile is very different from that of (pure) PLA composed of linear chains, which typically shows bulk erosion, that is, little mass loss over a considerable period of degradation followed by a rapid decrease (Section 9.6).

A hydrogel susceptible to enzyme-mediated hydrolysis may consist of a triblock B-*b*-PEG-*b*-B copolymer composed of a central PEG block (~5 kDa) extended with short peptide chains (the B blocks) that are the targeted cleavage sites for a specific enzyme. These copolymer chains are normally capped at each end with acrylate groups for creating a 3D crosslinked hydrogel network by photopolymerization. For example, a B block susceptible to cleavage by collagenase may consist of the tetrapeptide APGL (using

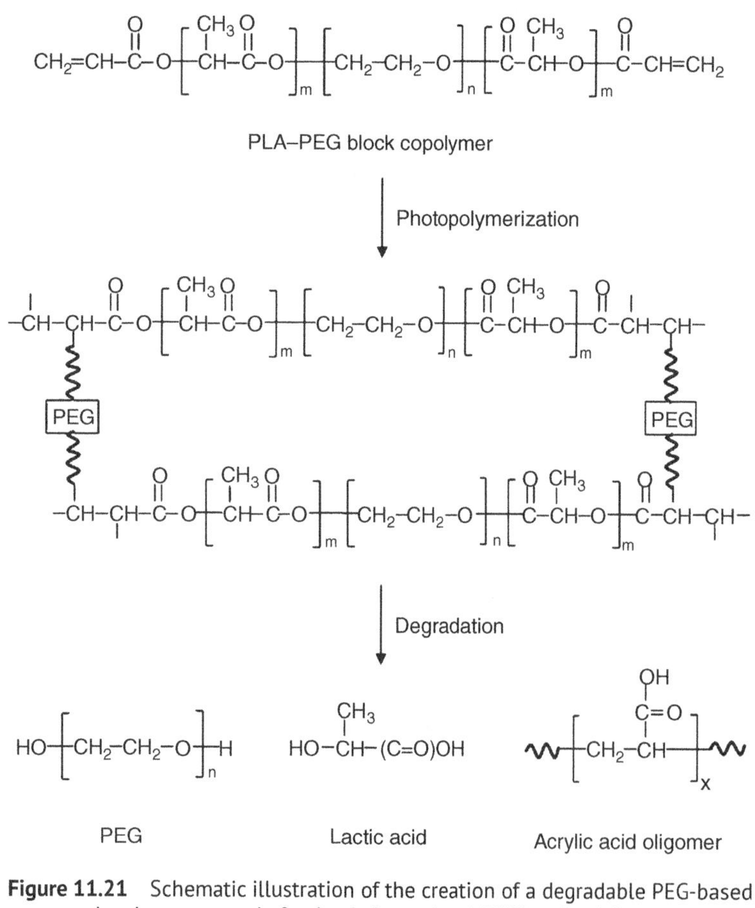

Figure 11.21 Schematic illustration of the creation of a degradable PEG-based hydrogel by photopolymerization of PLA-*b*-PEG-*b*-PLA macromolecules composed of polyethylene glycol (PEG) and polylactic acid (PLA), and its degradation products. *Source:* Sawhney et al. (1993) / with permission of American Chemical Society.

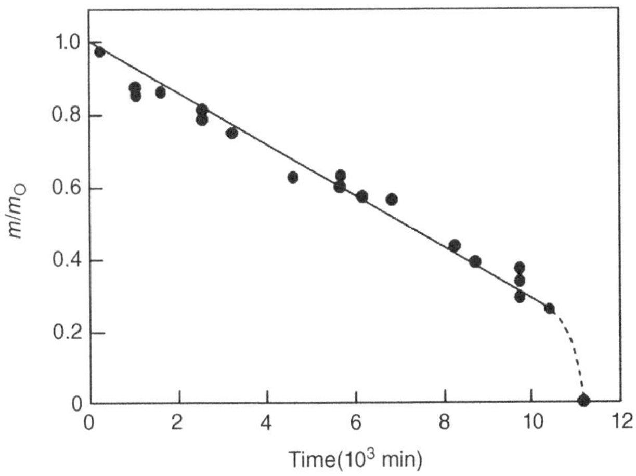

Figure 11.22 Measured erosion (mass m relative to initial mass m_0) as a function of time for a hydrogel created from a 50 wt.% solution of a PLA-*b*-PEG-*b*-PLA copolymer composed of polyglycolic acid (PEG) (molecular weight = 4.6 kDa) and polylactic acid (PLA) (five monomer units). *Source:* Metters et al. (2000) / with permission of Elsevier.

the one-letter abbreviation for the amino acids). When placed in an aqueous solution composed of HEPES buffered saline containing collagenase, these hydrogels swell and show an erosion rate that depends on the collagenase concentration in the solution. In comparison, no erosion occurs in a solution without collagenase or in a solution containing the enzyme plasmin but no collagenase, which shows the specificity of the degradation process.

As noted earlier, triblock copolymers composed of PEO-*b*-PPO-*b*-PEO known commercially as Pluronics have been widely studied because of their capacity to show temperature responsive gelling. They are available in various chain lengths (molecular weights) and compositions consisting of a different number of PEO and PPO monomer units in their chains. In moderately concentrated solutions (~20 wt%), the widely studied and applied Pluronic designated F127, composed of chains having a composition of approximately $PEO_{100}PPO_{65}PEO_{100}$, shows an LCST of ~25 °C. Despite being nondegradable, the capacity of these Pluronic F127 solutions to form a gel at the physiological temperature is an attractive feature in drug delivery and tissue engineering applications. Therapeutic drugs or biomolecules can be easily incorporated into the solution, which gels upon implantation at the physiological temperature.

11.11.2 Polyvinyl Alcohol

PVA, normally synthesized by hydrolysis of polyvinyl acetate, is typically composed of a copolymer of PVA and polyvinyl acetate due to the difficulty of completely replacing the acetate groups. Typically, commercially available PVA consists of over ~98% PVA. Although very hydrophilic, PVA has a low solubility in water at room temperature. Consequently, mixtures of water and PVA (molecular weight 50–100 kDa) normally need to be heated to ~60–80 °C to achieve a solution having an appropriate PVA concentration (~10–15 wt%) prior to creating the hydrogel. PVA hydrogels may be formed by covalent crosslinking using glutaraldehyde as a crosslinking agent (Figure 11.3) or by irradiation with an electron beam or with γ-rays.

Another method involves physical crosslinking by crystallization (Figure 11.8). Upon standing at room temperature, PVA solutions gradually undergo gelling by a process in which chain segments form small crystalline regions that link the chains into a 3D network. Subjecting PVA solutions to repeated freezing and thawing cycles leads to the creation of hydrogels having better mechanical properties. The molecular weight of the PVA, the concentration of the PVA solution, and the number of freeze–thaw cycles control the structure of the hydrogel (such as the degree of crystallinity and crystal size) and, thus, its physicochemical properties. Overall, a PVA molecular weight of ~50–100 kDa, a PVA concentration of ~10–15 wt%), and a higher number of freeze–thaw cycles (such as 5–10) lead to the creation of hydrogels having a combination of good mechanical properties and good network stability over a longer-term period (approximately six months). As this method does not involve a crosslinking agent, the potential risk of introducing toxicity into the hydrogels is avoided. Generally, PVA hydrogels created by this freeze–thaw method show improved mechanical properties when compared to PVA hydrogels created by other crosslinking techniques. These hydrogels find use as contact lenses and as implants for tissue repair.

Because of the presence of hydroxyl groups in their chains, PVA hydrogels can be chemically modified by a variety of reactions to control their physicochemical and biological properties. For example, using techniques similar to those described for PEG, methacrylate groups can be added to PVA chains to create photocrosslinkable hydrogels. Attachment of RGD peptides to the chains can be used to enhance the interaction of PVA hydrogels with cells and tissues, while attachment of PEG molecules can be used to inhibit protein and cell adsorption to PVA hydrogels. Another type of chemical modification reaction involves copolymerization of PVA molecules with PAA molecules to create pH-sensitive and temperature-sensitive hydrogels.

Although PVA hydrogels are not degradable in the physiological environment, their good mechanical properties are an advantage in some biomedical applications. For example, in addition to contact lens applications noted earlier, PVA hydrogels have been used as the lining of artificial hearts and as devices for drug delivery. They have also been studied for applications in soft tissue healing, artificial articular cartilage repair, and intervertebral disc replacement.

11.11.3 Polyhydroxyethyl Methacrylate

PHEMA, as noted earlier, was the first synthetic hydrogel created specifically for use in biomedical applications. The monomer unit in PHEMA chains (Figure 11.2) has a chemical structure similar to that of PMMA (Figure 8.9a), except that the methyl (CH_3) group in the ester side group of PMMA is replaced by the CH_2–CH_2–OH group. Because of this difference in chemical structure, PHEMA has a lower glass transition temperature (T_g ~ 87 °C) than PMMA (~105 °C), which may be attributed to a small increase in free volume between its chains due to the longer side group. However, the most significant effect, when compared to PMMA, is an improvement in hydrophilicity due to the polar OH group in its hydroxyethyl side group.

PHEMA hydrogels are normally lightly crosslinked by free radical polymerization (Section 11.4.1). Because of their hydrophilicity, they absorb water and swell, reaching an equilibrium swelling at ~40 wt% of absorbed water. This capacity to absorb water and swell endows PHEMA hydrogels with a combination of properties that are suitable for contact lens applications. PHEMA hydrogels have an optical transparency comparable to PMMA but are mechanically softer, are wetted by an aqueous film of tears, have good oxygen permeation and good chemical stability (insoluble and nondegradable), and can be manufactured at low cost. Copolymerization of HEMA with molecules such as ethylene glycol diacrylate, methacrylic acid, or vinylpyrrolidone has been used to create PHEMA-based hydrogels that have a wider range of physicochemical properties. In addition to their use as contact lenses, these hydrogels have been used as drug delivery devices as well.

11.11.4 Polyacrylic Acid and Polymethacrylic Acid

The chemical structure of polymethacrylic acid (PMAA) differs from that of PAA (Figure 11.14a) only in that the monomer unit of PMAA has a methyl (CH_3) side group bonded to the C atom to which the carboxyl (C=O)OH side group is bonded instead of H. As this difference in side group does not produce a significant difference in the physicochemical properties between PAA and PMAA, the difference between PAA and PMAA hydrogels will not be emphasized. Instead, we will focus mainly on PAA and include PMAA when appropriate.

PAA and PMAA are both strongly hydrophilic glassy polymers, having a T_g of ~105 and ~128 °C, respectively. When dissolved in water, the carboxyl (C=O)OH groups in their chains can become deprotonated to varying extent depending on the pH, giving negatively charged chains (Figure 11.14a). Because of this capacity of their chains to develop a high negative charge in water, PAA and PMAA are often referred to as anionic polyelectrolytes. PAA and PMAA have several uses industrially, particularly as a dispersant, and they have been used in biomedical applications as well, particularly as devices for controlled drug delivery.

PAA hydrogels, as noted earlier, show pH responsive behavior. They swell and collapse reversibly as the pH of the aqueous environment is cycled above and below ~3.5 to ~6 (Figure 11.14a), although swelling and collapse do not occur instantaneously or at the same rate. In acidic solutions, PAA chains can associate with PEG chains to form hydrogels in which crosslinking occurs mainly by hydrogen bonds between the ether O atom of PEG and the H atom of the (C=O)OH group of PAA. In comparison, van der Waals bonding involving hydrophobic CH_3 groups may also play a role when PMAA associates with PEG. Hydrogen bonding occurs at low pH (less than ~3.0) when the (C=O)OH groups are in their undissociated state. When these groups start to become deprotonated at higher pH, the tendency for hydrogen bonding between the PAA and PEG chains is eliminated because the negatively charged PAA chains repel each other.

Instead of using free chains in solution, covalent bonding of the chains to form a graft copolymer provides better control of the complexation process and, thus, better control of the physicochemical properties of the hydrogel. Hydrogels created from PMAA-g-PEG copolymers, composed of PEG chains grafted on to PMAA chains, show pH dependent gelling. They undergo swelling and contraction as the pH is cycled above and below ~4 to ~7. These pH responsive hydrogels provide attractive biomaterials for a variety of biomedical applications, particularly as devices for drug delivery (Section 11.13.1).

11.11.5 Poly(*N*-isopropyl acrylamide)

Hydrogels created from PNIPAAm (Figure 11.7), as noted earlier, have been one of the most widely investigated and best-known examples of synthetic hydrogels that show temperature responsive behavior. One advantage of PNIPAAm is that it has an LCST of ~32 °C, which is a convenient temperature for biomedical applications because it is below the physiological temperature (~37 °C). Another advantage is the ease in varying its temperature responsive behavior by modifying its chemical structure to change the relative number of hydrophilic and hydrophobic groups in its chains.

One way to change the hydrophilic to hydrophobic balance in PNIPAAm is to change the composition of the N-substituted groups, giving a family of N-alkyl-substituted PAAm (Figure 11.23a). Generally, an increase in the number of carbon atoms on the alkyl group or in the number of alkyl groups leads to a decrease in the LCST. However, when the alkyl group is longer than the propyl group, van der Walls interactions between the hydrophobic groups dominate, which makes the PNIPAAm-based polymers insoluble in aqueous media at room temperature. In comparison, the presence of hydroxyl (OH) groups on the N-substituted group enhances hydrogen bonding with water molecules, leading to an increase in the LCST.

Another way to modify the temperature responsive behavior of PNIPAAm is by copolymerization with other molecules. This provides a route to not only alter the LCST but also incorporate additional functionalities as well, such as the incorporation of electrostatic charges into the chains or the capacity of the chains to degrade. For example, copolymerization with acidic or basic monomers introduces anionic or cationic features into the chains (Figure 11.23b), which allows the copolymers to show a combination of pH and temperature responsive behavior. Generally, the LCST increases with increasing degree of electrostatic charge of the chains. This may be attributed to an increase in repulsion between the chains that reduces the potential for van der Waals interactions between the hydrophobic groups of the chains.

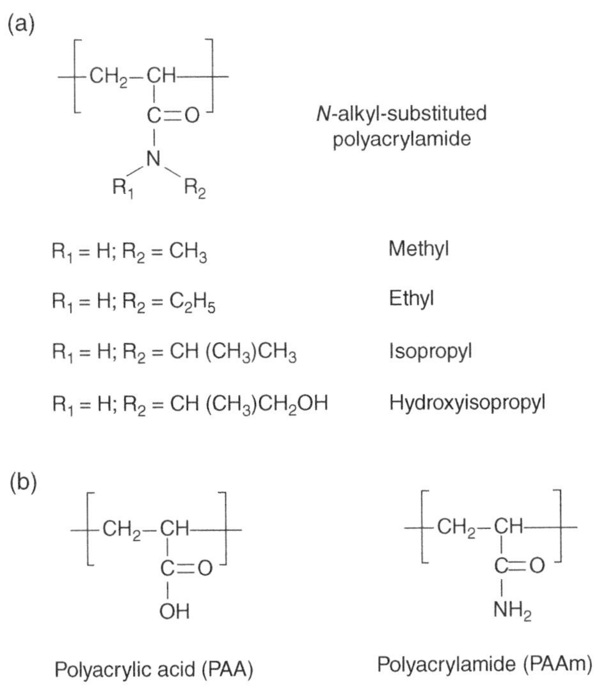

Figure 11.23 (a) Chemical structure of the monomer unit in *N*-alkyl-substituted polyacrylamides; (b) structures of two monomers (B) that may be incorporated within random P(NIPAAm-*co*-B) copolymers.

As an example, P(NIPAAm-*co*-AA) hydrogels, composed of *N*-isopropyl acrylamide (NIPAAm) and acrylic acid (AA) copolymers, lightly crosslinked to stabilize their network structure (Figure 11.24a), show an increase in LCST in PBS when compared to PNIPAAm hydrogels (Figure 11.24b). This is because the AA monomer develops a negative charge in PBS which hinders the dehydration of the chains, reduces the contraction of the hydrogel network, and offsets the van der Waals interaction between the hydrophobic groups. Additionally, P(NIPAAm-*co*-AA) hydrogels show a broader transition and a slower rate of de-swelling when compared to PNIPAAm hydrogels. This may be advantageous in some applications, such as when rapid de-swelling and shrinkage in a tissue defect *in vivo* are undesirable.

Temperature responsive PNIPAAm copolymers may consist of random or block copolymers. PNIPAAm chains can also be grafted on to other hydrogels or they can be conjugated with other molecules to impart temperature responsiveness to them or to modify their properties. Overall, PNIPAAm hydrogels provide a versatile set of biomaterials for use in a variety of biomedical application, particularly drug delivery and tissue engineering, based on the capacity to control their LCST, extent of de-swelling and rate of de-swelling.

11.12 Natural Hydrogels

The chemical structure, physicochemical properties and applications of some common natural polymers, both protein-based and polysaccharide-based, were described in Chapter 10, along with techniques to crosslink their chains into a 3D network. Overall, the crosslinking mechanisms for these hydrogels, involving covalent or noncovalent bonding, are generally similar to those described in Section 11.4.

Some natural polymers show pH responsive or temperature responsive behavior. For example, gelatin is a well-known protein-based polymer that shows temperature responsive behavior. Gelatin, approximately, has a UCST of room temperature and, thus, warm gelatin solutions convert to a gel upon cooling to low temperature. Strongly hydrophilic polysaccharide-based polymers such as hyaluronic acid and chitosan show pH responsive gelation. For example, hyaluronic acid is an anionic polysaccharide. The carboxyl (C=O)OH group in the repeating units of its chains has a pK_a of ~3.5, which means that a high fraction of these groups become deprotonated near or above a pH of ~3.5,

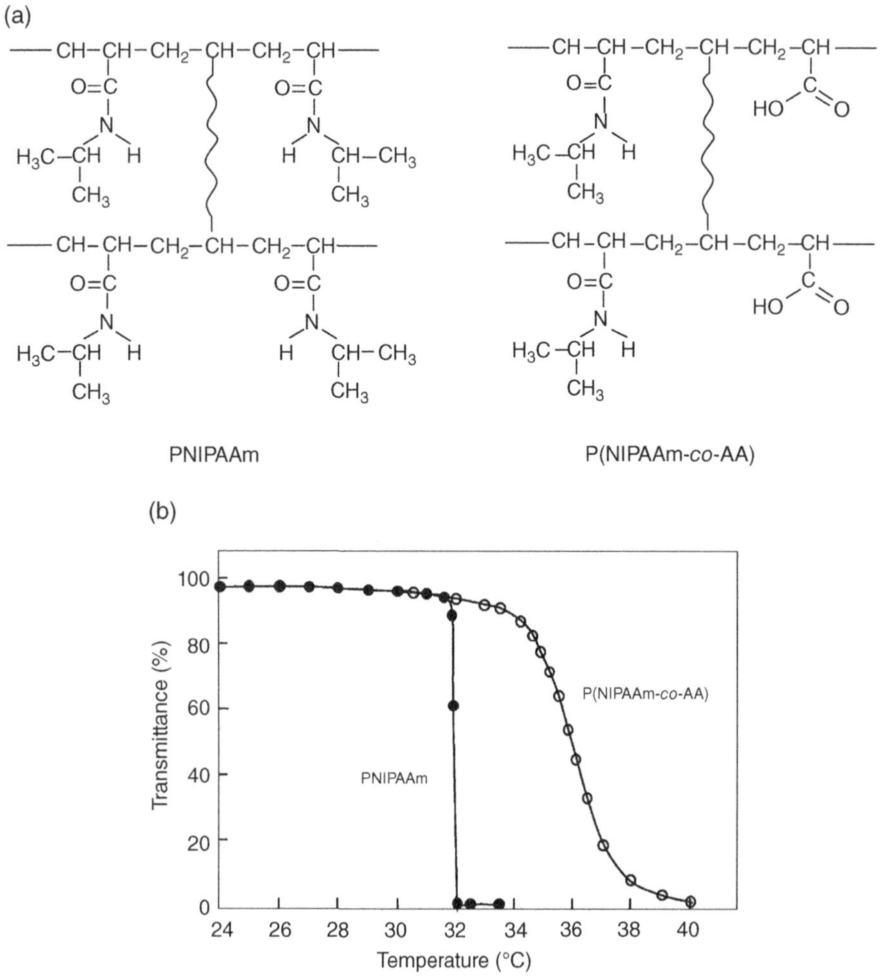

Figure 11.24 (a) Chemical structure of lightly crosslinked hydrogels composed of poly(*N*-isopropyl acrylamide) (PNIPAAm) and a copolymer P(NIPAAm-*co*-AA) composed of *N*-isopropyl acrylamide (NIPAAm) and acrylic acid (AA); (b) transmittance as a function of temperature for PNIPAAm and P(NIPAAm-*co*-AA) hydrogels. *Source:* Stile et al. (1999) / with permission of American Chemical Society.

giving the chains a high negative charge. Thus, hyaluronic acid is soluble in aqueous media above this pH value (due to repulsion between the negatively charged chains) but the solution undergoes pH reversible gelling as the pH is lowered to ~2.5–3.0 (due to hydrogen bonding between the chains). However, this low pH required for gelling is often a disadvantage in biomedical applications and, consequently, hyaluronic acid gels are commonly created by covalent crosslinking using water-soluble agents (Chapter 10).

Another pH responsive natural polymer is the cationic polysaccharide chitosan. The amino (NH_2) group in the repeating unit of chitosan chains has a pK_a of ~6.5, which means that a significant fraction of these groups is protonated at pH near or lower than ~6.5, giving the chains a high positive charge. Chitosan is soluble in aqueous media at pH lower than ~6.5 (due to repulsion between the chains) but as the pH is increased above this value, it forms a hydrogel (due to hydrogen bonding between the chains). This capacity to gel near the physiological pH (~7.4) makes chitosan attractive for use as an injectable biomaterial in drug delivery and tissue engineering applications.

Chitosan solutions can show not only pH dependent gelling but also temperature-dependent gelling by adding a solution of a polyol salt such as disodium glycerol phosphate to them (Figure 11.25). Such chitosan–glycerol phosphate solutions of the appropriate viscosity and a pH near the physiological value can be prepared and held below room temperature for incorporating cells or biomolecules. Then, because gelling occurs near the physiological temperature, they convert to a gel and set *in situ* when implanted *in vivo*. Figure 11.26 shows the temperature-dependent gelling profile, determined by measuring the elastic modulus for a chitosan–glycerol phosphate solution of pH ~7.2. This solution shows an LCST of approximately the physiological temperature and temperature reversible gelling in which the de-swelling curve has a different

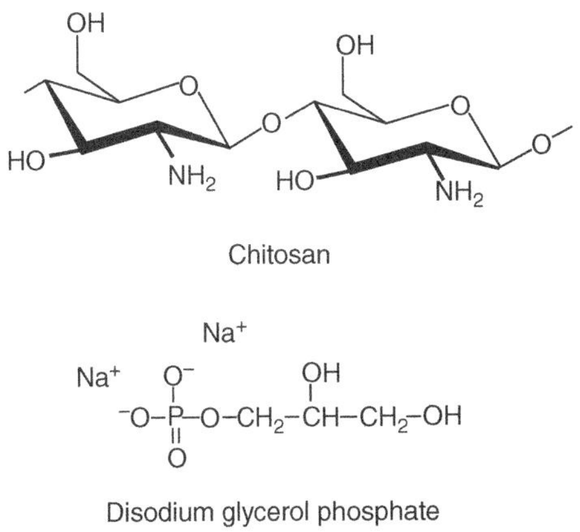

Figure 11.25 Chemical structures of chitosan and disodium glycerol phosphate.

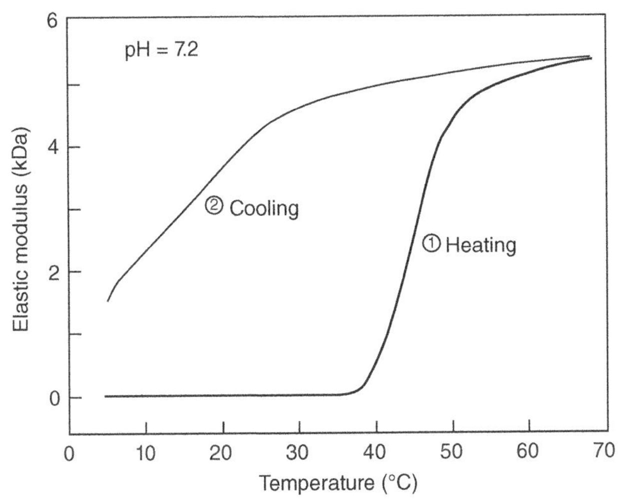

Figure 11.26 Elastic modulus as a function of temperature for chitosan–glycerol phosphate solution at pH = 7.2, showing temperature-dependent reversible gelling. *Source:* Chenite et al. (2000) / with permission of Elsevier.

profile from the swelling curve. The gelling mechanism for these chitosan–glycerol phosphate solutions is not clear but two contributions have been suggested. First, the negatively charged divalent glycerol phosphate ions may lead to ionic crosslinking of the positively charged chitosan chains. Second, the hydroxyl groups in the glycerol phosphate molecules may influence the hydration and dehydration of the chitosan chains, which may contribute to temperature-dependent gelling by a mechanism similar in nature to that of LCST polymers.

11.13 Applications of Hydrogels

Hydrogels, both synthetic and natural, as noted earlier, have been studied and used as biomaterials for a variety of biomedical applications but drug delivery and tissue engineering have been two areas of significant interest in the last few decades (Table 11.1). While several chapters later in this book are devoted to the applications of biomaterials, this section focuses specifically on features of hydrogels that make them attractive biomaterials for use in these two areas. It is noted that hydrogels used in these and other biomedical applications should have the requisite sterility and that the sterilization of hydrogels can normally be more challenging when compared to other types of biomaterials, such as metals and ceramics. In general, the two major approaches for sterilizing biomaterials are

- The use of aseptic processing of presterilized components
- Terminal sterilization, that is, sterilization of the fabricated biomaterial.

Because of their sensitivity to heat and radiation, and the presence of water within their structure, most hydrogels may have to be prepared by aseptic processing of presterilized components. In comparison, terminal sterilization is safer but it is often more challenging. The conditions employed in dry heat sterilization and autoclaving make them unsuitable for hydrogels. Ethylene oxide treatment has been the preferred method for sterilizing hydrogels but the use of γ-rays has been increasing. A less widely used method is hydrogen peroxide (HO–OH) sterilization. While these terminal sterilization methods are effective, they also suffer from drawbacks. One drawback is the creation of structural and chemical changes in the polymer chains, such as the formation free radicals. These reactive species can lead to the formation of additional crosslinks between the chains if they react together, which can adversely influence the properties and performance of the hydrogel.

11.13.1 Drug Delivery

Hydrogels generally provide an environment in which protein-based drugs and biomolecules are normally stable because of their hydrophilic nature and the lack of organic solvents during their creation. Additionally, when compared to some solid degradable polymers (Chapter 9), hydrogels generally provide easier outward transport of acidic or basic degradation products due to the water-filled spaces within their 3D network structure. On the other hand, this network structure may also lead to easier release of incorporated drugs and biomolecules, making it difficult to achieve long-term release.

As discussed in Section 11.10, some hydrogels have the capacity to shrink and expand in response to environmental changes, such as changes in pH, temperature, and ionic concentration. These hydrogels may provide drug release by two alternative strategies, termed expansion-release and squeeze-release, as illustrated in Figure 11.27 for a pH responsive hydrogel. In expansion release, drugs are released as the hydrogel expands, which provides larger spaces for the drug to diffuse out of the hydrogel, whereas in squeeze release, the shrinking gel forces out the drug. In each strategy, intermittent drug release, referred to as on–off release, may be achieved due to expansion and contraction of the hydrogel in response not only to environmental changes but also to external physical stimuli such as magnetic or electrical fields, sounds waves, or light (Chapter 26).

11.13.2 Cell Encapsulation and Immunoisolation

Hydrogels, as we have seen, have some degree of similarity to the ECM components, can be normally created under mild conditions, and provide easy transport of oxygen, nutrients, and biomolecules due to their 3D structure. Consequently, they provide attractive matrices for encapsulation and immunoisolation of cells. In immunoisolation, allogeneic or xenogeneic cells, once encapsulated within the hydrogel, are protected, completely or in part, from host immune-mediated destruction, and they are able to deliver specific therapeutic proteins to the desired anatomical site over an extended period. The hydrogel also defines a specific physical space for the cells, preventing them from migrating into the surrounding host tissue. This provides for both sustained viability of the encapsulated cells and their delivery to the anatomical site of interest. Figure 11.28 illustrates a method for encapsulating cells into PEG hydrogels by free radical polymerization of PEGDA using UV light (Section 11.10.1).

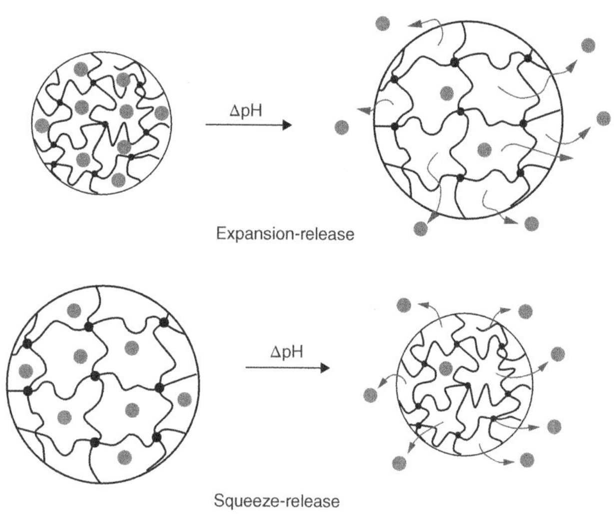

Figure 11.27 Illustration of drug release from polyelectrolyte hydrogels by expansion-release and squeeze-release.

11.13.3 Scaffolds for Tissue Engineering

The characteristics of hydrogels that make them attractive biomaterials for drug delivery and cell encapsulation and immunoisolation are also relevant to their application in tissue engineering as well (Chapter 25). As noted earlier, hydrogels are also easy to modify chemically to enhance their interaction with cells and tissues, and their mechanical properties may be controlled to make them comparable to those of soft tissues. Overall, these characteristics make hydrogels attractive biomaterials for use as scaffolds in tissue engineering to heal and regenerate soft tissues. In a common approach (Figure 11.29), cells are incorporated into an appropriate solution of polymer molecules, which can be easily converted to a hydrogel by, for example, irradiation with UV light. Then the cell-seeded scaffold is cultivated in a bioreactor to create the desired tissue prior to implantation or implanted directly into the tissue defect.

Temperature responsive hydrogels provide additional advantages for use as scaffolds in tissue engineering. For example, hydrogels composed of PEO-*b*-PPO-*b*-PEO

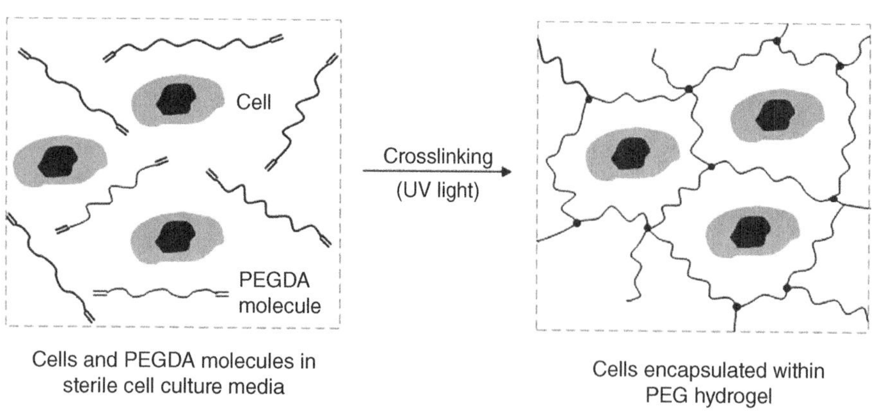

Figure 11.28 Illustration of a method for encapsulating cells in a polyethylene glycol (PEG) hydrogel.

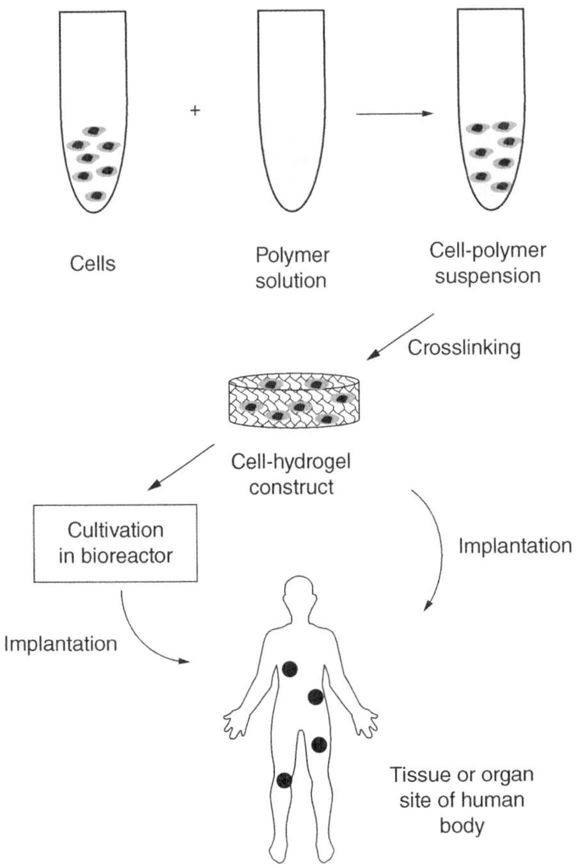

Figure 11.29 Illustration of a common approach in which hydrogels are used as scaffolds in tissue engineering.

block copolymers (Pluronics) and PNIPAAm-based polymers that show an LCST near room temperature provide the capacity for injection by minimal invasive surgery and setting *in situ*. Additionally, hydrogels that show a reasonable LCST or UCST allow for creating scaffolds having a well-defined external geometry and internal architecture by techniques such as 3D printing.

11.14 Concluding Remarks

In this chapter, we discussed key principles underlying the creation, structure, and properties of hydrogels. These principles provide a background in designing and selecting hydrogels for specific biomedical applications. We discussed hydrogels created from several polymers and relationships between their composition, structure, and physicochemical properties. Interest in hydrogels has grown considerably in the last few decades, particularly for drug delivery and tissue engineering applications.

- Hydrogels may be classified into several types, using criteria such as whether they are created from synthetic or natural polymers but an important classification is in terms of the type of crosslinks between their molecular chains because this has a strong effect on the structural stability of their 3D network
- Chemical hydrogels, crosslinked by covalent bonds, are generally the most stable and their crosslinking is not reversible, whereas the crosslinking in physical hydrogels (non-covalent crosslinks) is reversible
- As the physicochemical properties of hydrogels depend on their chemical composition and 3D network structure, an understanding of the factors that control these characteristics are important
- Understanding the factors that control the behavior of environmentally responsive polymers is important because hydrogels created from these polymers may provide additional advantages in some applications.

Problems

11.1 Explain the classification of hydrogels into chemical hydrogels and physical hydrogels.

11.2 Explain the factors that control the equilibrium swelling of a hydrogel composed of polyethylene glycol. Are these factors the same for a hydrogel composed polyacrylic acid? Explain.

11.3 After drying, the mass of a hydrogel composed of polyvinyl alcohol (PVA) is 0.06 g. Upon immersion in water, the hydrogel reaches an equilibrium swelling volume of 1 cm^3. If the density of PVA is 1.19 g/cm^3, determine the volume fraction of the polymer in the swollen hydrogel.

11.4 A hydrogel composed of polyvinyl alcohol (PVA) has a polymer volume fraction $v_{2,s} = 0.05$ and a molecular weight between crosslinks $\bar{M}_c = 5.0$ kDa. Given that the Flory characteristic ratio $C_n = 9.0$, determine the mesh size ξ of the hydrogel network.

11.5 Discuss the major factors that control the diffusion of solutes such as biomolecules through a hydrogel.

11.6 The carboxyl (C=O)OH group in the repeating unit of hyaluronic acid chains has a $pK_a = 3.5$. Determine and plot the degree of ionization of hyaluronic acid chains as a function of pH between pH = 2.5 and pH = 7.0.

11.7 The amino (NH_2) group in the repeating unit of chitosan chains has a pK_a of ≈6.5. Determine and plot the degree of ionization of the chitosan chains as a function of pH between pH = 3.0 and pH = 9.0.

11.8 A lightly crosslinked polyacrylic acid (PAA) hydrogel is placed in pure water (pH = 7.0) or in phosphate buffered saline (PBS) (pH = 7.4). Is the swelling behavior expected to be different in these two liquids? Explain.

11.9 Explain the mechanisms by which some polymers show environmentally responsive behavior.

11.10 Explain the meaning of the terms lower critical solution temperature (LCST) and upper critical solution temperature (UCST) as applied to polymers.

11.11 Explain why some hydrogels created from some temperature responsive polymers may show a lower critical solution temperature (LCST) whereas others may show an upper critical solution temperature (UCST).

11.12 Describe in general terms one method each for creating chemical hydrogels and physical hydrogels of polyvinyl alcohol (PVA), and compare the advantages and disadvantages of the hydrogel created by each method.

11.13 The Pluronic designated F127 is a block copolymer of polyethylene oxide (PEO) and polypropylene oxide (PPO), composed of chains having a composition of approximately $PEO_{100}PPO_{65}PEO_{100}$. Explain the mechanism by which solutions of Pluronic F127 convert to a hydrogel.

11.14 Poly(*N*-isopropyl acrylamide) (PNIPAAm) is a temperature responsive polymer that has a lower critical solution temperature (LCST) of ~32 °C. Explain the gelling mechanism of PNIPAAm. Can the LCST of PNIPAAm be altered? If so, explain how.

11.15 Discuss any advantages that pH responsive hydrogels may provide over conventional hydrogels for use as devices in drug delivery.

11.16 Discuss any advantages that temperature responsive hydrogels may provide over conventional hydrogels for use as biomaterials in tissue repair and regeneration.

11.17 Briefly describe the characteristics of hydrogels that make them attractive biomaterials for use as scaffolds in tissue engineering.

References

Brannon-Peppas, L. and Peppas, N.A. (1991). Equilibrium swelling behavior of pH-sensitive hydrogels. *Chemical Engineering Science* 46: 715–722.

Chenite, A., Chaput, C., Wang, D. et al. (2000). Novel injectable neutral solutions of chitosan form biodegradable hydrogels *in situ*. *Biomaterials* 21: 2155–2161.

De, S.K., Aluru, N.R., Johnson, B. et al. (2002). Equilibrium swelling and kinetics of pH-responsive hydrogels: models, experiments and simulations. *Journal of Microelectromechanical Systems* 11: 544–555.

Dern, D.L. and Hubbell, J.A. (1998). Incorporation of adhesion peptides into non-adhesive hydrogels useful for tissue resurfacing. *Journal of Biomedical Materials Research* 39: 266–276.

Flory, P.J. (1953). *Principles of Polymer Chemistry*. New York: Cornell University Press.

Lustig, S.R. and Peppas, N.A. (1988). Solute diffusion in swollen membranes. IX. Scaling laws for diffusion in gels. *Journal of Applied Polymer Science* 36: 735–747.

Metters, A.T., Anseth, K.S., and Bowman, C.N. (2000). Fundamental studies of a novel, biodegradable PEG-*b*-PLA hydrogel. *Polymer* 41: 3993–4004.

Peppas, N.A. and Benner, R.A. Jr. (1980). Proposed method of intracordal injection and gelation of poly(vinyl alcohol) solution in vocal cords: polymer considerations. *Biomaterials* 1: 158–162.

Peppas, N.A. and Merrill, E.W. (1977). Crosslinked polyvinyl alcohol hydrogels as swollen elastic networks. *Journal of Applied Polymer Science* 21: 1763–1770.

Sawhney, A.S., Pathak, A.P., and Hubbell, J.A. (1993). Bioerodible hydrogels based on photopolymerized polyethylene glycol-*co*-poly(α-hydroxy acid) diacrylate macromers. *Macromolecules* 26: 581–587.

Stile, R.A., Burghardt, W.R., and Healy, K.A. (1999). Synthesis and characterization of injectable poly(*N*-isopropyl acrylamide)-based hydrogels that support tissue formation *in vitro*. *Macromolecules* 32: 7370–7379.

Further Readings

Amsden, B. (1998). Solute diffusion within hydrogels. Mechanisms and models. *Macromolecules* 31: 8382–8395.

Bell, C.L. and Peppas, N.A. (1996). Modulation of drug permeation through interpolymer complexed hydrogels for drug delivery applications. *Journal of Controlled Release* 39: 201–207.

Canal, T. and Peppas, N.A. (1989). Correlation between mesh size and equilibrium degree of swelling of polymeric networks. *Journal of Biomedical Materials Research* 23: 1183–1193.

Hassan, C.M. and Peppas, N.A. (2000). Structure and morphology of freeze/thawed PVA hydrogels. *Macromolecules* 33: 2472–2479.

Hennink, W.E. and van Nostrum, C.F. (2002). Novel crosslinking methods to design hydrogels. *Advanced Drug Delivery Reviews* 54: 13–36.

Hoffman, A.S. (2002). Hydrogels for biomedical applications. *Advanced Drug Delivery Reviews* 54: 3–12.

Kanjickal, D., Lopina, S., Evancho-Chapman, M.M. et al. (2008). Effect of sterilization on poly(ethylene glycol) hydrogels. *Journal of Biomedical Materials Research. Part A* 87: 608–617.

Liu, R., Fraylich, M., and Saunders, B.R. (2009). Thermoresponsive copolymers: from fundamental studies to applications. *Colloid and Polymer Science* 287: 627–643.

Peppas, N.A., Huang, Y., Torres-Lugo, M. et al. (2000). Physicochemical foundations and structural design of hydrogels in biology and medicine. *Annual Review of Biomedical Engineering* 2: 9–29.

12

Composite Biomaterials

12.1 Introduction

Composites are normally used to modify the bulk properties of a material. A composite is a material composed of two or more physically distinct constituent materials or phases that are separated by a distinguishable interface. There are various types of composites but the most common consists of two phases, a matrix phase that is normally continuous and a dispersed phase such as fibers or particles. The two constituent phases typically have different properties and the normal objective in creating composites is to enhance the matrix properties by including the dispersed phase.

For biomedical applications, the constituent materials in the composite should each have good biocompatibility *in vivo*. A major objective is normally to enhance the mechanical properties or the biological activity of the matrix phase, or a combination of both. Overall, except for dental composites used as fillings, composites have not found considerable use as biomaterials. This is largely because any improvement in properties is often offset by an increased difficulty in producing composites having the requisite shape and architecture for use as implants.

This chapter will discuss the principles and factors that control the physicochemical properties of composites, particularly their mechanical properties.

12.2 Types of Composites

The most common type of composites consists of two phases, a matrix phase that is normally continuous and a dispersed phase, also referred to as a reinforcing phase. These two-phase composites may be further classified into various types depending on the characteristics of the matrix and reinforcing materials. A broad classification has been in terms of the type of reinforcing phase because this has a strong influence on the properties of the composite, giving (Figure 12.1):

- Fiber composites, in which the reinforcing phase consists of short fibers (called short-fiber composites) or continuous fibers (continuous fiber composites), which are normally aligned in a specific direction but may be randomly arranged in some composites.
- Particulate composites, in which the reinforcing phase, consisting of irregular-shaped, spherical or plate-like particles, is dispersed randomly within the matrix.

Another type of composite, referred to as lamellar composites, consists of thin alternating layers or sheets of two or more different materials, which are bonded together to create a thicker lamellar structure. Examples of this type of composite are adult lamellar bone, abalone shell (nacre) and, at a macroscopic level, plywood.

Other criteria may be used to classify composites. For example, composites such as bone and wood whose constituent phases have their origins in nature may be described as natural composites when a distinction from synthetic composites is appropriate. Depending on their structural scale, composites may be described as nanocomposites, microcomposites, or macrocomposites. In nanocomposites, at least one constituent phase has one dimension or more in the nanoscale range (less than ~50–100 nm). Microcomposites, which are more common in engineering and biomedical applications, have their phases dispersed at a microscopic scale (a fraction of a micron to several microns). In comparison, macrocomposites such as metal-reinforced concrete have their constituent phases dispersed at a macroscopic scale.

12.3 Mechanical Properties of Composites

Mechanical properties are normally important for most types of composites and, thus, in order to design composites having the requisite mechanical properties, it is useful to understand basic concepts underlying the mechanisms of strengthening composites. Generally, polymers show

Materials for Biomedical Engineering: Fundamentals and Applications, First Edition. Mohamed N. Rahaman and Roger F. Brown.

Companion website: www.wiley.com/go/rahaman/biomedicalengineering

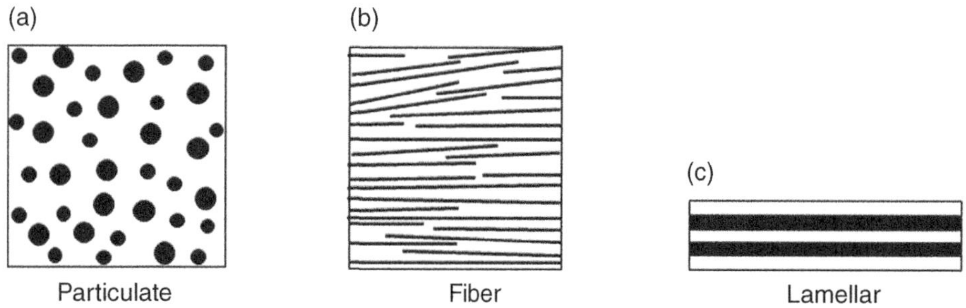

Figure 12.1 Idealized illustrations of common types of composites: (a) particulate composite, (b) fiber composite, and (c) lamellar composite.

weak mechanical properties (Chapter 4) and, thus, reinforcing them with strong fibers or particles (such as carbon or glass fiber-reinforced composites used in engineering) provides an effective approach to strengthen them. The capacity to improve their mechanical properties depends not only on the properties of the reinforcing phase, such as its strength and elastic modulus, but also on the strength of adhesion between the two phases. For example, strong interfacial adhesion between the matrix and the reinforcing phase is normally essential for improving the mechanical properties of polymer matrix composites (Section 12.3.1).

In comparison, most ceramics generally show high strength and high Young's modulus but low fracture toughness (Chapter 4) and, thus, a major objective in creating ceramic matrix composites is normally to enhance the fracture toughness of the ceramic material. One approach, as noted in Chapter 7, is to use a strong reinforcing phase consisting of particles, short fibers or continuous fibers that is weakly bonded to the matrix to achieve toughening predominantly by a crack deflection mechanism. Another approach is to activate a transformation toughening mechanism. Metals and their alloys used as biomaterials generally have attractive mechanical properties (Chapter 6) and, consequently, the creation of metal matrix composites for use in biomedical applications becomes unnecessary.

The remainder of this discussion on the mechanical properties of composites will focus on polymer matrix composites because of their greater relevance and potential for use in biomedical applications. Overall, the mechanical properties of these composites depend on the following factors:

- Amount (volume fraction) of the reinforcing phase
- Size of the reinforcing phase as, for example, coarse versus fine particles, or long (continuous) fibers versus short fibers
- Shape of the reinforcing phase, such as particles, platelets, or fibers
- Strength of adhesion between the matrix and the reinforcing phase.

12.3.1 Mechanical Properties of Fiber Composites

Although fiber composites see little use in biomedical applications when compared to particulate composites, analysis of their mechanical properties forms a useful starting point because of the simplicity of the models. For example, the Young's modulus of a fiber composite in which the fibers are aligned in a given direction can be estimated in a simple way. If a tensile stress σ is applied to the composite to deform it in a direction parallel to the fibers (Figure 12.2a), both the fibers and the matrix carry the stress. Thus, balancing the forces across a plane perpendicular to the applied stress gives

$$\sigma = A_f\sigma_f + (1 - A_f)\sigma_m \tag{12.1}$$

where A_f is the fractional area of the fibers in the plane, and σ_f and σ_m are the stresses in the fibers and in the matrix, respectively. As the volume fraction of each phase is equal to its area fraction in the plane, we get

$$\sigma = V_f\sigma_f + (1 - V_f)\sigma_m \tag{12.2}$$

where V_f is the volume fraction of fibers in the composite, given by

$$V_f + V_m = 1 \tag{12.3}$$

where V_m is the volume fraction of the matrix in the composite. The strain ε in the fibers and the matrix is the same, and from Eq. (12.2) we get

$$\sigma = E_f V_f \varepsilon + E_m (1 - V_f)\varepsilon \tag{12.4}$$

where E_f and E_m are Young's modulus of the fibers and of the matrix, respectively. As the Young's modulus of the composite $E_{c\|}$ is equal to σ/ε, Eq. (12.4) gives

$$E_{c\|} = E_f V_f + E_m (1 - V_f) \tag{12.5}$$

Equation (12.5) is an example of a rule of mixtures, that is, the Young's modulus of the composite is a weighted sum of the Young's modulus of the fibers and of the matrix.

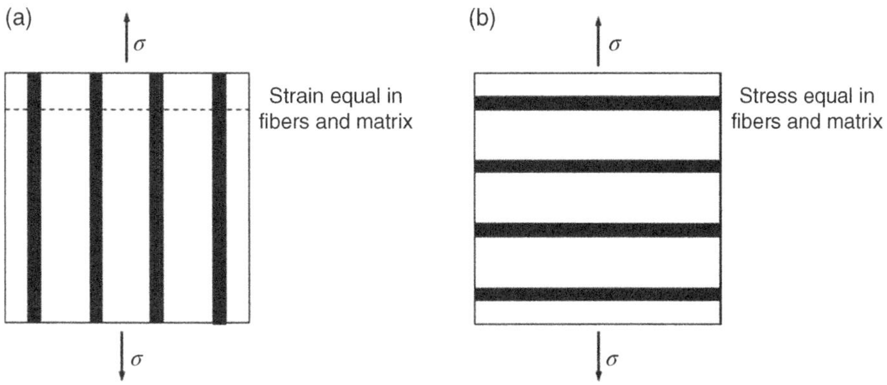

Figure 12.2 Illustration of fiber-reinforced composite loaded in the direction (a) parallel to the fibers and (b) perpendicular to the fibers.

In comparison, if the stress σ is applied in a direction perpendicular to the fibers (Figure 12.2b), it can be assumed that the stresses in the matrix and in the fibers are now the same. Consequently, the total strain ε in the composite is a weighted sum of the strain in the fibers ε_f and in the matrix ε_m, given by

$$\varepsilon = V_f\varepsilon_f + (1 - V_f)\varepsilon_m \tag{12.6}$$

As the Young's modulus $E_{c\perp}$ is simply the stress divided by the strain, Eq. (12.6) gives

$$\frac{\sigma}{E_{c\perp}} = V_f\frac{\sigma}{E_f} + (1 - V_f)\frac{\sigma}{E_m} \tag{12.7}$$

Rearranging Eq. (12.7) gives

$$E_{c\perp} = \frac{1}{\left\{\dfrac{V_f}{E_f} + \dfrac{(1 - V_f)}{E_m}\right\}} \tag{12.8}$$

Figure 12.3 shows plots of $E_{c\parallel}$ and $E_{c\perp}$ as a function of V_f. At any given V_f, $E_{c\parallel}$ gives an upper estimate for the modulus of the fiber reinforced composite, that is, the modulus cannot be greater than this value. In comparison, $E_{c\perp}$ gives a lower estimate, that is, the modulus cannot be lower than this value. Figure 12.3 also shows that there is considerable anisotropy in the mechanical response of the composite. The composite is stiffer in the direction parallel to the fibers and not as stiff in the direction perpendicular to the fibers.

Adding aligned fibers to a polymer matrix may initially lead to a reduction in tensile strength because the fibers are too few to support the applied stress. However, above a critical volume fraction, the tensile strength σ_{TS} of the composite may follow a rule of mixtures:

$$\sigma_{TS} = V_f\sigma_f^f + (1 - V_f)\sigma_y^m \tag{12.9}$$

where σ_f^f and σ_y^m are fracture strength of the fibers and the yield strength of the matrix, respectively. According to Eq. (12.9), the tensile strength of the composite increases linearly with increasing fiber volume fraction.

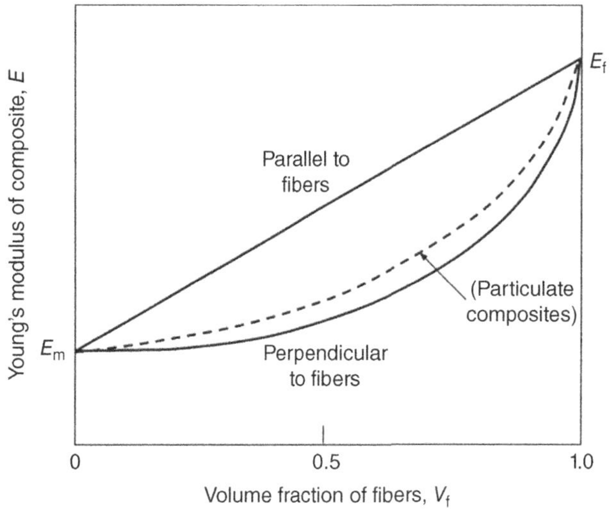

Figure 12.3 Young's modulus of fiber-reinforced composite in the directions parallel and perpendicular to the fibers as a function of the volume fraction of the fibers. E_m and E_f are the Young's modulus of the matrix and fibers, respectively. The predictions of a model for a particulate composite reinforced with a random arrangement of spherical particles of the same size are shown for comparison.

12.3.2 Mechanical Properties of Particulate Composites

Analysis of the Young's modulus for particulate composites is considerably more difficult when compared to aligned fiber composites. However, for polymer matrices reinforced with nearly spherical particles, theoretical and empirical models show that the Young's modulus follows a curve that is between the upper and lower bounds for fiber composites but only slightly above the lower bound (Figure 12.3). Unlike fiber composites, particulate composites show isotropic mechanical behavior unless the reinforcing particles have an elongated or plate-like shape and are aligned in a specific direction. Consequently, for practical volume fractions of nearly spherical reinforcing particles, the composite shows only a modest improvement in strength when compared to the polymer alone.

Above a certain size (typically ~100 nm), particle size has little effect on the Young's modulus of the composite. In comparison, below this threshold size, theoretical models predict an increase in Young's modulus of the composite with decreasing particle size (Figure 12.4). This threshold particle size cannot be predicted because it depends on the properties of the matrix and the particles. Overall, experimental data have shown a strong dependence on particle size below ~20–30 nm, although the effect is not as strong as the model predictions in Figure 12.4.

The Young's modulus of particulate composites is not affected significantly by the adhesive strength of the interface because it is determined at low strain and the deformation is normally insufficient to cause separation of the particle–polymer interface. On the other hand, the strength of the composite depends strongly on the adhesive strength of the particle-polymer interface. This is because the strength depends on the effectiveness of stress transfer between the polymer matrix and the particles. Consequently, composites having a strongly bonded interface normally have a higher strength when compared to similar composites having a weakly bonded interface. Overall, then, a strongly bonded interface is essential for creating particulate composites having optimal mechanical properties.

The strength of a given composite also depends on the particle size and particle volume fraction but factors such as the interfacial adhesive strength and the homogeneity of the particle dispersion within the matrix may complicate the observed effect. Overall, for a given volume fraction of particles, the tensile strength of particulate composites has been observed to increase with decreasing particle size, particularly for nanoparticles (less than ~50–100 nm) (Figure 12.5). This is normally attributed to an increase in surface area with decreasing particle size, which provides more effective stress transfer.

Some studies have shown a decrease in tensile strength with increasing particle volume fraction (Figure 12.5), whereas other studies have shown an increase. This difference in behavior may be related to differences in the interfacial adhesive strength. For example, adding particles to a polymer matrix may lead to a decrease in tensile strength when the interfacial adhesive strength is low because of inefficient stress transfer between the matrix and the particles. In comparison, for a well-bonded interface, the addition of particles to the polymer matrix may lead to an increase in strength, particularly for nanoparticles.

Coupling agents such as organofunctional alkoxysilanes (Section 13.10.1) are widely used to improve the interfacial bonding between the reinforcing phase and the polymer matrix in the production of industrial composites. However, in creating polymer composites for use as implants, the biocompatibility of any coupling agent used should be

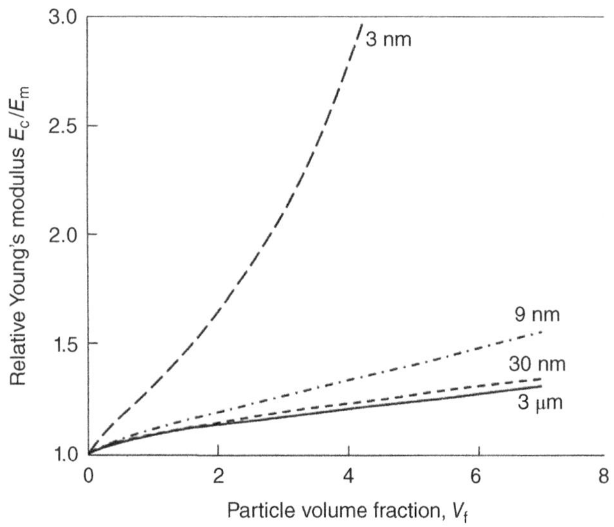

Figure 12.4 Predictions for the effect of particle size on the Young's modulus of a particulate composite E_c, relative to the Young's modulus of the matrix phase E_m, as a function of the volume fraction of particles V_f. The particles are assumed to be spherical and to have a Young's modulus E_f equal to $40E_m$. *Source:* Adapted from Ji et al. (2002).

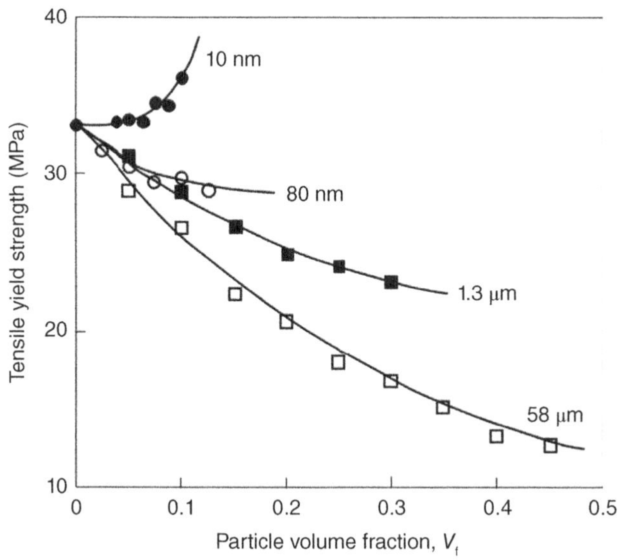

Figure 12.5 Effect of average particle size on the tensile yield strength of a polypropylene matrix reinforced with calcium carbonate ($CaCO_3$) particles. *Source:* Adapted from Puzanszky and Vörös (1993).

guaranteed and the effect of the coupling agent on the degradation of the particles and the matrix should be considered.

It has been assumed so far that the reinforcing particles are nearly spherical in shape and are dispersed homogeneously within the polymer matrix. Factors such as the shape of the particles and the homogeneity of their dispersion in the matrix may strongly modify the effect of particle volume fraction on the strength and modulus of the composite. For example, the tensile strength of some composites show an increase with increasing particle volume (or weight) fraction to a peak value, followed by a decrease (Figure 12.6). While a variety of factors may be responsible for this behavior, particle agglomeration is often a major factor in some composites. When the particles agglomerate, they touch and normally adhere weakly, which reduces their surface area for adhesion to the matrix. This effect is more common for a reinforcing phase consisting of fine particles (such as nanoparticles) because the tendency for agglomeration increases with decreasing particle size. Overall, regardless of particle size, homogeneous dispersion of the particles within the matrix, such that each particle is surrounded by an appropriate thickness of the matrix, is essential for creating composites having optimal mechanical properties.

Improvement in the mechanical properties of polymers may also be achieved by using elongated particles, that is, particles having a length to diameter (or thickness) ratio (called the aspect ratio) much greater than 1, when compared to the use of nearly spherical particles (aspect ratio ~1). A well-known example is the use of short glass fibers to improve the properties of engineering polymers when compared to the use of nearly spherical or crushed glass particles. The use of nanoparticles of high aspect ratio, such as carbon nanotubes (Section 3.3.4), layered silicates such as clays, or silica (SiO_2) can lead to much better improvement in mechanical properties.

A feature of these composites reinforced with high aspect ratio nanoparticles is that improvement in mechanical properties may be achieved with low particle volume fraction, as shown in Figure 12.7 for nylon 6 reinforced with multi-walled carbon nanotubes. However, caution should be exercised in using such high aspect ratio nanoparticles as a reinforcing phase because they have been associated with adverse biocompatibility effects *in vivo*, such as toxicity and significant inflammatory responses (Chapter 18).

Homogeneous dispersion and strong interfacial bonding are still essential requirements for the creation of composites having optimal mechanical properties, which may be promoted by an appropriate surface treatment of the nanoparticles. For example, the dispersion and interfacial bonding of carbon nanotubes in some polymers may be enhanced by treating them in a strong acidic solution, such as concentrated nitric acid. This treatment introduces functional groups such as carboxyl (C=O)OH, carbonyl C=O, and hydroxyl OH groups at the particle surface, which can interact with polar groups present in some polymers (for example, nylon).

12.4 Biomedical Applications of Composites

Except for composite resins used as dental restorations (Chapter 23), composites do not find considerable use as biomaterials. Ceramic composites find a limited amount of

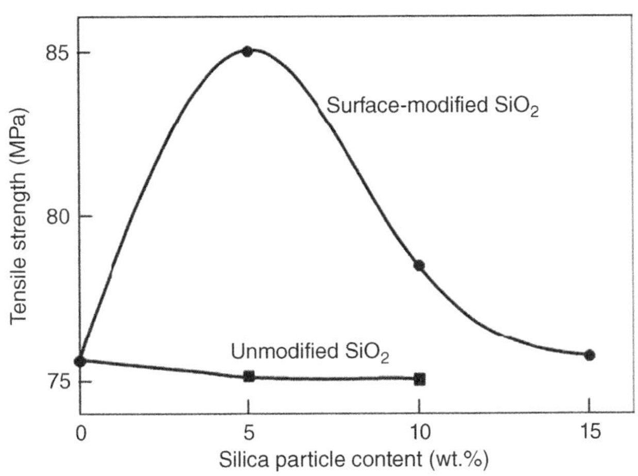

Figure 12.6 Tensile strength of particulate composites composed of nylon 6 reinforced with surface-modified and unmodified silica (SiO_2) particles. *Source:* Ou et al. (1998) / with permission of John Wiley & Sons.

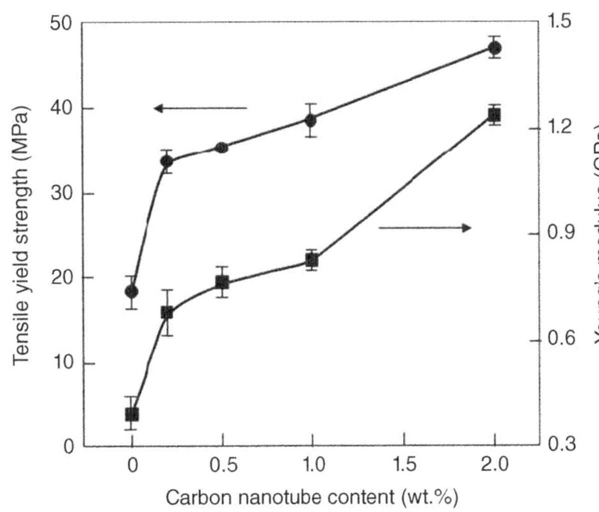

Figure 12.7 Tensile yield stress and Young's modulus as a function of particle content for nylon 6 reinforced with carbon nanotubes. *Source:* Liu et al. (2004) / with permission of Elsevier.

Example 12.1

A dense composite consists of 70 wt.% hydroxyapatite dispersed in a polyethylene (PE) matrix. Hydroxyapatite has a density of 3.16 g/cm^3 and a Young's modulus of 80.0 GPa, whereas PE has a density of 0.94 g/cm^3 and a Young's modulus of 1.2 GPa. Determine the Young's modulus of the composite based on Eqs. (12.5) and (12.8) for a model composite containing aligned continuous fibers.

Solution:

The fraction of hydroxyapatite in the composite must be converted into a volume fraction in order to use Eqs. (12.5) and (12.8). Taking an arbitrary composite mass such as 100 g, the masses of hydroxyapatite and PE in the composite are 70.0 and 30.0 g, respectively. As volume is simply mass divided by density, the volumes of hydroxyapatite and PE are 22.0 and 32.0 cm^3, respectively. Thus, the volume fraction of hydroxyapatite and PE are 0.41 and 0.59, respectively. Using Eq. (12.5), we get $E_{c\parallel} = 33.6$ GPa and using Eq. (12.8), $E_{c\perp} = 2.0$ GPa. As noted earlier, the Young's modulus of particulate composites normally follows a curve slightly above that for $E_{c\perp}$ for fiber composites. Consequently, an estimate of the Young's modulus of the particulate composite is ~2 MPa. This represents an increase of ~70% over the value for the unreinforced PE.

use as biomaterials in biomedical applications, mainly in orthopedic surgery. For example, one application is the use of alumina matrix composite (AMC) or zirconia toughened alumina (ZTA) as the articulating bearings of some hip implants to provide a higher wear resistance when compared to the more commonly used bearing couple consisting of Co-Cr and UHMWPE. Another application is the use of biphasic calcium phosphate (BCP) as implants to heal bone defects. BCP, as noted in Chapter 7, consists of a two-phase mixture of hydroxyapatite and beta-tricalcium phosphate (β-TCP), where a major objective is to achieve an optimal combination of mechanical properties (such as Young's modulus) and degradation rate by controlling the ratio of hydroxyapatite to β-TCP.

Polymer matrix composites, as noted earlier, have been more widely studied and applied when compared to ceramic composites. Fiber composites are more efficient in the way that the reinforcing phase contributes to the mechanical properties (Section 12.3), but they are normally very expensive to produce in the requisite shapes and architectures for use as implants. In comparison, whereas particulate composites show a more modest improvement in mechanical properties, they have the advantages of ease and economy of production because conventional polymer processing methods can be used to create them. Consequently, polymer composites relevant to biomedical applications are normally particulate composites.

Addition of hydroxyapatite particles to high-density polyethylene (HDPE) can lead to the creation of dense composites having a Young's modulus close to that of human cortical bone (Figure 12.8a). Because of the bioactivity of the hydroxyapatite particles, the composites can also show better biological activity. For example, while few cells adhere to the surface of HDPE specimens *in vitro*, the composites show a better capacity to support the attach-

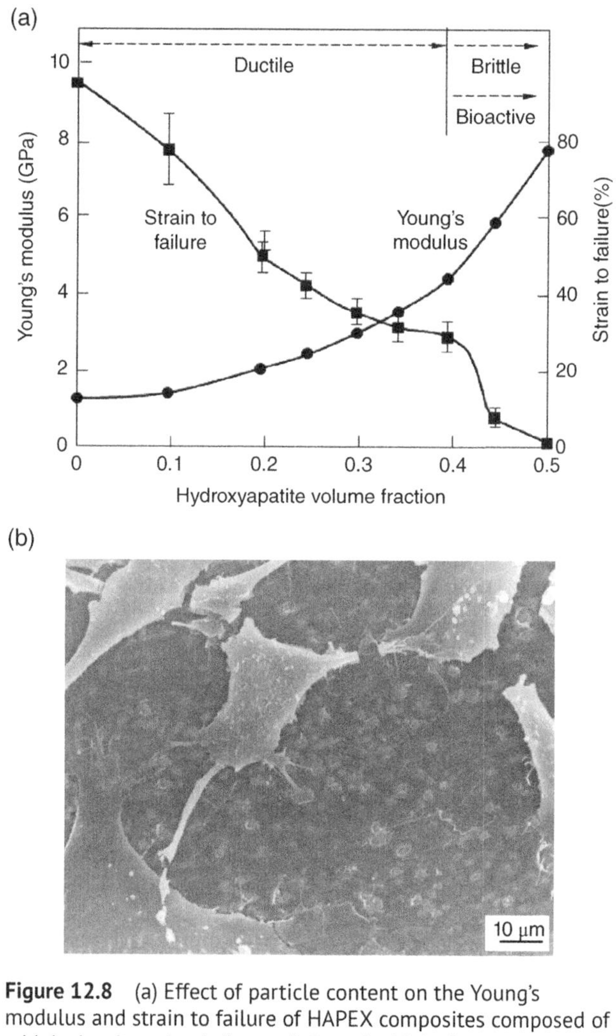

Figure 12.8 (a) Effect of particle content on the Young's modulus and strain to failure of HAPEX composites composed of a high-density polyethylene reinforced with hydroxyapatite particles. *Source:* Adapted from Hench (1998). (b) Scanning electron microscope image showing adherence and spreading of human osteoblast-like cells on HAPEX after 24 hours of culture. *Source:* From Huang et al. (1997).

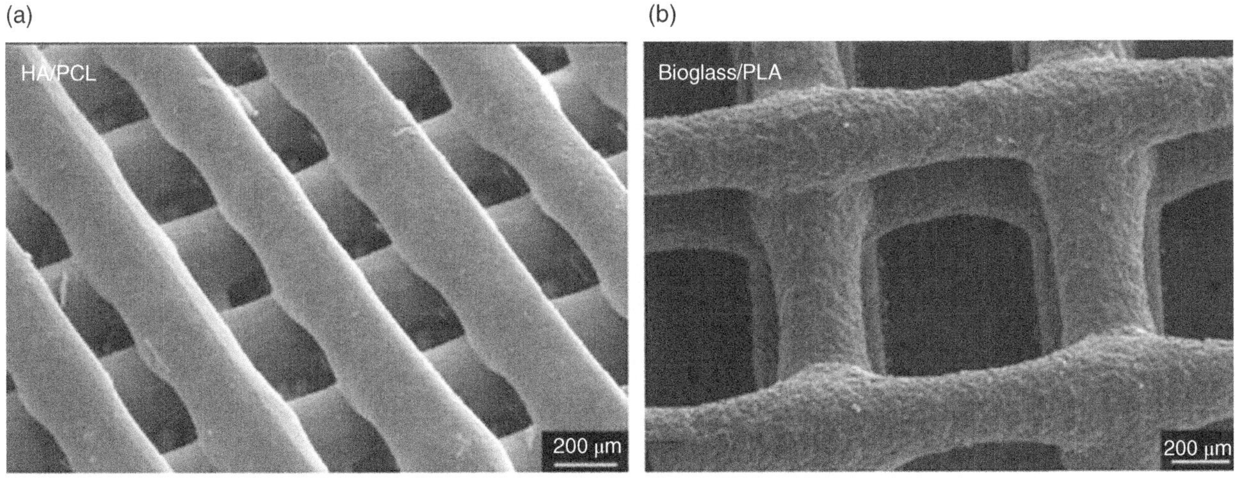

Figure 12.9 Scanning electron microscope images of porous composites composed of a degradable matrix reinforced with bioactive particles created by additive manufacturing (3D) printing. (a) Polycaprolactone (PCL) reinforced with hydroxyapatite (HA) (70 wt%); (b) polylactic acid (PLA) reinforced with bioactive glass 6P53B particles (70 wt%). *Source:* From Russias et al. (2007).

ment and proliferation of human osteoblast-like cells (Figure 12.8b). These composites, known commercially as HAPEX®, have been used to repair bone defects.

Dense composites consisting of a degradable polymer matrix such as polylactic acid (PLA), polyglycolic acid (PGA), or polycaprolactone (PCL) reinforced with bioactive particles, such as hydroxyapatite or bioactive glass, have been studied and applied as biomaterials in bone repair, such as screws in bone fracture fixation (Figure 1.1j). These composites show some desirable characteristics, such as the capacity of their matrix phase to degrade and be replaced by bone, improved Young's modulus when compared to the polymer alone, and surface bioactivity imparted by the particles. Bioactive glass particles can, in addition, degrade, and convert to hydroxyapatite, releasing various inorganic ions in the process. Depending on their composition, these ions may stimulate osteogenesis and angiogenesis *in vivo* to enhance bone regeneration or they may eradicate bone infection.

Porous composites composed of a degradable polymer matrix reinforced with bioactive particles have also been studied for use as scaffolds in the tissue engineering approach to regenerate bone (Chapter 25). Scaffolds having the requisite shape and porous architecture can be easily created using additive manufacturing (3D printing) or conventional processing techniques (Figure 12.9). However, because they normally contain a high amount of porosity (typically over 50%) to allow cells to infiltrate into them, these scaffolds have a much lower strength and Young's modulus when compared to dense composites of the same composition. Consequently, they are suited for regenerating nonstructural bone, that is, bone that is not subjected to significant physiological stress.

12.5 Concluding Remarks

This chapter discussed the use of composites to improve the physicochemical properties of biomaterials, particularly the mechanical properties and bioactivity of polymeric biomaterials. Although composites find limited use as biomaterials, one area of application has been the creation of implants for use in orthopedic surgery, where polymers are reinforced with strong bioactive particles such as hydroxyapatite to improve their mechanical properties and bioactivity. It is essential to understand the various factors that control the mechanical properties of these particulate composites in order to design composites having optimal mechanical properties.

Problems

12.1 Explain why a strong interface is normally required in enhancing the mechanical properties of polymer matrix composites whereas a weak interface is required for ceramic matrix composites.

12.2 Reinforcing a polymer with strong particles often leads to an increase in Young's modulus whereas the tensile strength may or may not show an increase. Explain.

12.3 Estimate the Young's modulus of a composite composed of a PLA matrix (density = 1.24 g/cm^3; Young's modulus = 2.7 GPa) and 70 wt.% hydroxyapatite particles (density = 3.14 g/cm^3; Young's modulus = 80 GPa), assuming strong interfacial bonding between the two phases. State any other assumptions that you make. Would the Young's modulus of the composite be different if the PLA matrix was replaced with a PCL matrix (density = 1.14 g/cm^3; Young's modulus = 0.4 GPa) but the hydroxyapatite content of the composite remained at 70 wt.%? Explain.

References

Hench, L.L. (1998). Bioceramics. *Journal of the American Ceramic Society* 81: 1705–1728.

Huang, J., Di Silvio, L., Tanner, K.E. et al. (1997). *In vitro* mechanical and biological assessment of hydroxyapatite-reinforced polyethylene composite. *Journal of Materials Science. Materials in Medicine* 8: 775–779.

Ji, X.L., Jing, K.L., Jiang, W. et al. (2002). Tensile modulus of polymer nanocomposites. *Polymer Engineering and Science* 42: 983–993.

Liu, T., Phang, I.Y., Shen, L. et al. (2004). Morphology and mechanical properties of multiwalled carbon nanotubes reinforced nylon-6 composites. *Macromolecules* 37: 7214–7222.

Ou, Y., Yang, F., and Yu, Z.-Z. (1998). A new conception on the toughness of nylon 6/silica nanocomposite prepared via *in situ* polymerization. *Journal of Polymer Science Part B: Polymer Physics* 36: 789–795.

Puzanszky, B. and Vörös, G. (1993). Mechanism of interfacial interactions in particulate filled polymers. *Composite Interfaces* 1: 411–427.

Russias, J., Saiz, E., Deville, S. et al. (2007). Fabrication and *in vitro* characterization of three-dimensional organic/inorganic scaffolds by robocasting. *Journal of Biomedical Materials Research. Part A* 83: 434–445.

Further Readings

Fu, S.-Y., Feng, X.-Q., Lauke, B., and Mai, Y.-W. (2008). Effects of particle size, particle/matrix adhesion and particle loading on the mechanical properties of particulate-polymer composites. *Composites B* 39: 933–961.

Tjong, S.C. (2006). Structural and mechanical properties of polymer nanocomposites. *Materials Science and Engineering R* 53: 73–197.

13

Surface Modification and Biological Functionalization of Biomaterials

13.1 Introduction

Surface modification is normally used to modify the surface characteristics and properties of a material without changing its bulk properties. A process related to surface modification is biological functionalization, which generally involves immobilizing biomolecules such as peptides, proteins, enzymes, antibodies, or drugs on or within a material to confer a desired biological functionality to the material.

Surface modification has become increasingly important for improving the performance of biomaterials *in vivo*, through control of their surface characteristics and properties, such as their surface chemistry, surface structure, and surface physical properties covered in Chapter 5. Upon implantation of a biomaterial, the physiological environment first encounters the surface of the biomaterial. Consequently, the interaction of the surface with components of the physiological fluid (such as water molecules, inorganic ions, and proteins) strongly influences the response of cells and, thus, the biocompatibility of the implant. A major objective of many surface modification processes, therefore, is to improve the interaction of the biomaterial's surface with the physiological environment.

A wide variety of techniques is available to modify the surface of biomaterials, providing the capacity to control their surface characteristics and properties over a wide range. These techniques generally involve chemically or physically altering the atoms or molecules at the existing surface without forming a separate coating or coating the surface with a layer of material of different composition. The modified surface layer is normally thin and, consequently, it has little effect on the bulk properties of the biomaterial.

Biological functionalization is often necessary because many biomaterials, particularly synthetic biomaterials, normally do not possess the requisite characteristics and properties to elicit the desired biological response *in vivo*. An example is immobilizing the polysaccharide heparin on the surface of biomaterials, which has been used for several decades to reduce the thrombogenecity of a variety of medical devices. In the last few decades, immobilization of peptides or growth factors on or within biomaterials has become an attractive approach to regulate cell response and to regenerate tissues and organs.

This chapter will discuss the following topics relevant to surface modification and biological functionalization of biomaterials:

- Principles underlying common techniques used for surface modification and biological functionalization.
- Properties and applications of surface modified and biologically functionalized biomaterials.

13.2 Surface Modification

The composition and structure of a biomaterial, as noted in Chapter 4, controls its bulk properties such as its mechanical and other physicochemical properties and, thus, its primary function. In comparison, the surface properties of a biomaterial, such as its surface chemistry, topography, and physical properties strongly influence its interaction with the physiological environment (Chapter 5). Often, the surface properties of the as-fabricated biomaterial may not be optimal for its intended application. In such cases, a variety of techniques is available to modify the surface of the biomaterial in order to improve its properties. Overall, as illustrated by the examples in Figure 13.1), surface modification is normally associated with a thin layer that affects only the outermost surface atoms or molecules of the biomaterial. Consequently, it normally has little effect on the bulk properties of the biomaterial and, thus, its main function.

For biomaterials used as implants, a major objective of surface modification is to modify the biological response of the biomaterial, such as compatibility with blood, protein adsorption or cell adhesion and growth. Surface modification may also be used to improve the lubricity, corrosion resistance, or wear resistance of a biomaterial,

Materials for Biomedical Engineering: Fundamentals and Applications, First Edition. Mohamed N. Rahaman and Roger F. Brown.

Companion website: www.wiley.com/go/rahaman/biomedicalengineering

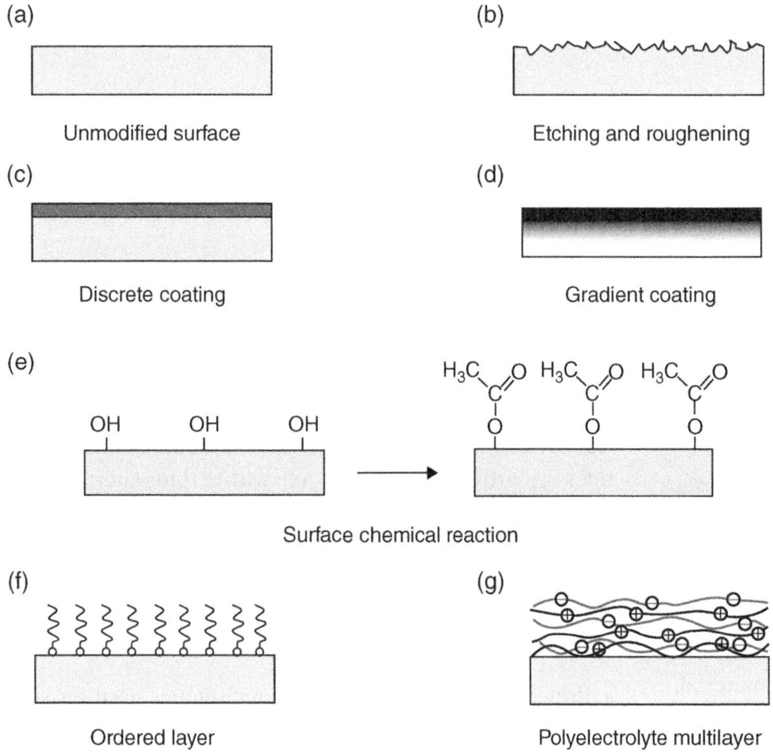

Figure 13.1 Examples of various types of surface modification.

or to modify its surface transport properties or its surface electrical properties.

Ideally, the modified surface layer should have the minimum thickness required for it to perform its function and the requisite uniformity and stability. It should also be resistant to delamination, that is, resistant to separation from the surface of the biomaterial. Consequently, when possible, primary bonds, particularly covalent bonds, should bond the modified surface layer to the biomaterial. This may sometimes require modifying the surface to incorporate suitable functional groups or depositing a suitable primer on the surface prior to the actual surface modification step to aid strong interfacial bonding.

Surface analysis techniques are normally required to characterize the chemistry, structure, and morphology of the modified surface accurately (Chapter 5), to ensure that the desired coating has been formed and that it is chemically pure, devoid of any contamination. Additionally, the surface modification process used to form the coating should be economically feasible.

13.3 Surface Modification Methods

A wide variety of techniques is available to modify the surface of biomaterials (Ratner and Hoffman 2013).

Table 13.1 Examples of chemical and physicochemical methods used to modify the surface of biomaterials.

Type of surface modification	Examples
Covalently bonded coating	Plasma treatment
	Chemical vapor deposition
	Radiation grafting
	Self-assembled monolayers
	Solution-based chemical reactions
	Physical vapor deposition
Physically adsorbed (non-covalent) coating	Langmuir–Blodgett films
	Layer-by-layer deposition
	Solution-based coating
Surface modification without a coating	Plasma treatment
	Ion beam implantation

These techniques may be classified into the following groups, depending on the type of modification produced (Table 13.1):

- Methods to modify the physical characteristics of the surface without changing the surface chemical composition.

- Chemical or physicochemical methods to change the composition of the surface without the formation of a discrete coating.
- Methods to deposit a coating on the material by covalent or noncovalent bonding.

The term coating has often been used interchangeably with the term film, which has often been classified into two types, thin film and thick film, to give a qualitative indication of the film thickness. Although there is no clear agreement about the definition of these terms, thin film is often associated with a thickness in the range of a fraction of a nanometer to ~1 μm, whereas thick film is associated with thickness of ~1 to ~100 μm. Overall, we shall use the term coating and film interchangeably and give an indication of their thickness when appropriate.

A variety of physical techniques can be used to introduce surface topographical features in materials (Section 5.5). Techniques such as abrasion, machining, or grinding introduce topographical features that are commonly random and variable over the surface. In comparison deliberate topographical features, often composed of ordered spikes, grooves, or pores, can be created using a variety of controllable techniques, such as photolithography, electron beam lithography, or laser interference. The following sections present a discussion of three broad categories of physicochemical techniques (Table 13.1) for modifying the surface properties and surface characteristics of biomaterials. A few common techniques in each category are covered to illustrate their basic principles and methods.

13.4 Plasma Processes

Plasma treatment, also referred to as glow discharge or gas discharge treatment, is one of the most widely used techniques to modify the surface of materials. Plasma treatment may be used for all three types of surface modification, that is, to modify the physical characteristics of the surface without changing its composition, to modify the atoms or molecules at the surface without the formation of a separate coating, or to deposit a coating on the surface. Overall, the technique involves the generation of a plasma composed of an atomically and molecularly dissociated gas containing positive ions, negative ions, free radicals, electrons, atoms, molecules, and photons, into which the material is placed (Figure 13.2).

13.4.1 Plasma Treatment Principles

Production of a plasma requires the separation of electrons from atoms or molecules of a gas, that is, ionization of the gas atoms or molecules. This occurs when the atoms or molecules gain enough energy from an external excitation source or from collisions with one another. Generally, the plasma is ignited by applying a high voltage electrostatic potential between two electrodes that generates an electric field through the gas. Initially, a few free electrons already present in the gas accelerate toward the anode. If the electric field is strong enough, such an electron can gain more than enough energy to liberate another electron from a gas molecule (or atom), that is, ionize the gas molecule, upon collision. These two electrons travel toward the anode and gain sufficient energy from the electric field to ionize other molecules, and so on, effectively causing a chain reaction. Impaction of ions on the anode frees additional electrons that contribute to the generation of a stable plasma. For a given gas, the voltage at which breakdown of the atoms or molecules occurs is a function of the gas pressure and the gap length between the electrodes.

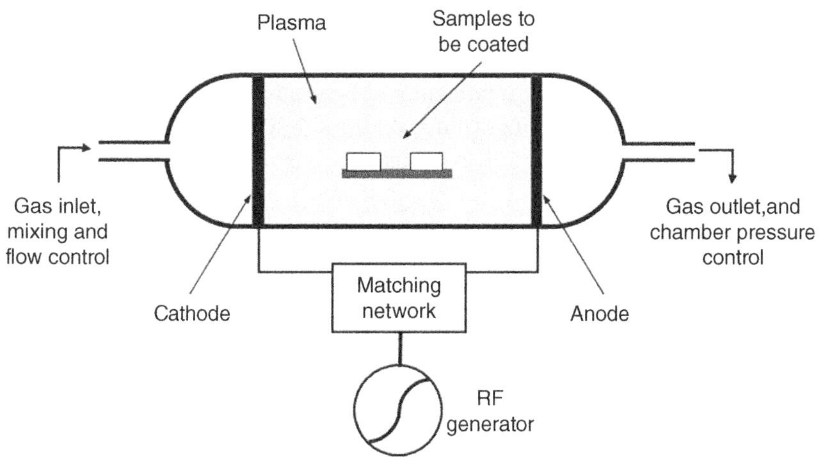

Figure 13.2 Schematic of the main components of a plasma treatment system.

A variety of sources may be used to generate a plasma but the radiofrequency glow discharge (rfGD) source finds widespread use in surface modification equipment because of its capacity to produce a sufficiently large volume of stable plasma. In this rfGD method, the plasma is typically generated in a chamber at a pressure of ~0.01–1 torr and a temperature up to ~60 °C, with an electron energy of ~1–10 eV. These conditions are mild enough to allow plasmas to be used for surface modification of most polymers.

Depending on the conditions, the plasma may cause ablation or etching of the surface, leading to the removal of material from the surface, or it may lead to deposition of a coating on the material. These are competing processes. No deposition will take place when ablation (or etching) is much faster than deposition but in many cases, a combination of ablation and deposition takes place. In ablation, a negative voltage is applied to the material, which causes positive ions to be accelerated toward the material. These ions have insufficient energy to penetrate the surface but a large fraction of their energy is transferred to the surface atoms by elastic and inelastic collisions. Some of these surface atoms will acquire sufficient energy and escape from the surface into the vacuum chamber. After the first layer of atoms has been removed, the underlying layer will be exposed and gradually removed by ablation or etching. As this process is confined to the topmost layers of the material, it hardly alters the physical and chemical properties of the material.

Several mechanisms have been proposed for the deposition process. One mechanism involves the creation of free radicals and other reactive species on the surface of the material by the reactive gaseous environment, which react with and polymerize molecules from the gas phase. In another mechanism, small molecules from the gas phase may combine to form higher molecular weight units or particles that settle or precipitate onto the surface. Overall, it is likely that deposition occurs by some combination of these two mechanisms.

13.4.2 Advantages and Drawbacks of Plasma Treatment

Plasma processes such as rfGD have several advantages for use in surface modification of biomaterials, particularly for depositing a coating, such as:

- The technology is mature and well understood, which facilitates easy deposition of coatings having controlled surface characteristics and properties.
- Coatings conform to the geometry of the material and may be deposited on almost any clean surface of all classes of materials (polymers, metals, and ceramics).
- Normally, the coatings are dense, free from pinholes and adherent on the material's surface.
- The techniques provide the capacity for preparing coatings having a unique chemistry, which normally cannot be achieved by conventional chemical methods.
- When removed from the reactor, the coatings are normally sterile and contain little leachable substances.

On the other hand, like other surface modification techniques, plasma processes have a few disadvantages. For example, the equipment is generally expensive because it requires a vacuum chamber and equipment to control the gas pressure within the chamber. Another disadvantage is that the chemical composition within the reactor can often be ill defined, making it difficult to achieve the desired coating composition. Additionally, plasma processes cannot be reliably used for coating the interior surface of tubular materials or the surface of elongated pores in a biomaterial.

13.4.3 Applications of Plasma Treatment

Plasma treatment is a versatile technique for surface modification of biomaterials, particularly polymeric biomaterials. Treatment in plasmas composed of various types of gases can be used to produce a wide range of surface characteristics and properties. For example, an inert-gas plasma such as argon can be used for cleaning, ablation, and etching, leading to modification of the physical characteristics of the surface without altering the chemical composition of the polymer.

In comparison, treatment in a plasma composed of other gases such N_2, NH_3, O_2, F, CO_2, and H_2O (water vapor) can be used to modify the surface composition of the biomaterial without depositing a separate coating. For example, treatment in an H_2/NH_3 plasma can be used to introduce amine (NH_2) groups at the surface of a polymer whereas hydroxyl (OH) groups can be introduced using an H_2/H_2O plasma.

Plasma treatment can be used to deposit discrete polymer coatings (films) on the surface of a material by a process referred to as plasma polymerization. This process involves converting monomers or low molecular weight molecules in the gaseous state into polymers with the aid of the energetic species present in the plasma, such as ions, electrons, and radicals. Examples of gases that can be used in the process are hydrocarbons such as methane, ethane, or ethylene, fluorine-containing gases such as HF or F_2, and silicon-containing gases such as silane (SiH_4). Plasma polymerization is different from conventional polymerization processes such as addition and condensation polymerization (Chapter 8). Instead of the addition of monomers to a growing chain, a large fraction of the monomers forms free radicals within the plasma. These radicals react together and form a polymer coating that has a highly crosslinked structure. Normally, the polymer has a chemical composition and physicochemical properties that are different from those of conventionally synthesized polymers even if the same monomers were used. Because of its highly crosslinked structure, the polymer film can sometimes have improved durability and improved resistance to aging and oxidation.

The mechanism of plasma polymerization is a subject of some controversy. While a variety of mechanisms have been proposed, the most commonly cited is a free radical mechanism. Overall, this mechanism involves activation of monomers to create radicals, combination of these radicals to form molecules, and activation of these molecules that also combine with radicals. The chemical structure produced by this mechanism does not consist of regular repeating monomer units found in conventionally synthesized polymer. Instead, it is more complex, consisting of crosslinked, fragmented, and rearranged units formed from the monomers.

13.5 Chemical Vapor Deposition

Vapor deposition techniques are divided into two types, chemical vapor deposition (CVD) and physical vapor deposition (PVD). As their names imply, PVD employs physical techniques to deposit a film, whereas CVD employs chemical reactions. In CVD, reactive molecules in the gas phase are transported to a surface at which they react chemically and form a solid film. CVD is a well-established and widely used technique for depositing coatings of inorganic materials (metals, ceramics, and semiconductors), particularly for applications in the electronics industry. More recently, there has been a growing interest in using CVD to deposit polymer films for a variety of engineering and biomedical applications. Although it is also a well-established technique, PVD finds much less use when compared to CVD, and it is used almost exclusively for surface modification of inorganic materials.

13.5.1 Chemical Vapor Deposition of Inorganic Films

As noted earlier, CVD has been widely used to deposit inorganic films. As an example, one reaction to form silicon (Si) films uses a mixture of silicon tetrachloride ($SiCl_4$) and hydrogen (H_2) gases, as expressed by the equation

$$SiCl_4(g) + H_2(g) \rightarrow Si(s) + 4HCl(g) \qquad (13.1)$$

Normally, the solid (substrate) to be coated must be heated to a sufficiently high temperature for the reaction to proceed because inorganic films are often formed by endothermic reactions. For example, formation of a Si coating by the reaction of Eq. (13.1) requires heating the substrate to ~500–800 °C. Although some CVD processes are carried out at atmospheric pressure (~100 kPa), most are performed in a vacuum chamber at a much lower pressure, generally ~1–15 kPa. The lower pressure leads to higher diffusivity of the gas molecules to the substrate and to better uniformity of the film. Overall, the temperature of the substrate, the flow rate of the gas through the reactor, and the pressure in the reaction chamber control the rate of deposition and the structure (crystalline, amorphous, or semicrystalline structure) of the film.

A variation of the CVD technique is plasma-enhanced CVD (PECVD), in which the reactant gases are activated by passing them through a plasma or past a heated filament at high temperature before contacting the substrate. A well-known example has been the production of diamond films (or, more correctly, diamond-like films). Diamond is well-known for having several attractive properties such as high hardness, low friction, and high corrosion resistance. However, diamond films have been difficult to produce by other methods because they require very high temperatures and pressures. In comparison, PECVD allows the deposition of diamond films at much lower temperatures and pressures. Although the deposition process is complex, the basic reaction involves the pyrolysis of a carbon-containing precursor gas, such as methane:

$$CH_4(g) \rightarrow C(\text{diamond}) + 2H_2(g) \qquad (13.2)$$

One process involves passing a mixture of CH_4 and H_2 through a plasma or past a heated filament (at ~2000 °C) before deposition of the film on a substrate at ~800–1000 °C.

Diamond-like coatings on cobalt–chromium (Co–Cr) femoral heads have been studied to reduce friction when articulating against UHMWPE acetabular liners in hip implants in order to reduce wear of the liners. However, hard coatings on metal femoral heads, deposited by CVD or by other methods, have not shown the ability to provide any significant benefits when compared to the uncoated femoral heads.

13.5.2 Chemical Vapor Deposition of Polymer Films

Building on the technology for depositing inorganic films and on the principles of polymer chemistry, CVD techniques have been developed more recently to deposit polymer films. CVD polymerization has shown good versatility for depositing films composed of a wide variety of polymers and copolymers. In CVD polymerization, monomers in the gas phase are converted directly to solid polymer films. The polymer film formed in the process normally has a chemical structure that is identical to the polymer formed by conventional free radical polymerization (Chapter 8).

The process starts with one or more monomers, which are the constituents of the polymer film to be created, and an initiator such as *tert*-butyl peroxide, which is used to accelerate the polymerization reaction at the surface of the substrate (Figure 13.3a). These starting materials are vaporized by heating or by reduction of the pressure in their containers, and introduced at a controlled flow rate into a vacuum chamber containing the material to be coated, which is maintained at a pressure of ~0.1–1 torr. Inside the chamber, the initiator vapor is thermally activated to form free radicals by filament wires heated to ~200–400 °C, which are suspended a few centimeters above the substrate. Then adsorption and reaction of these radicals and monomers to form a polymer film take place on the surface of the substrate, kept at a lower temperature (~20–50 °C) to promote adsorption. Any volatile products formed in the reaction are normally desorbed from the surface and diffuse away with the gas flow. An alternative way to activate the initiator within the chamber is to use a photochemical method, such as ultraviolet (UV) light. Some monomers may be activated photochemically, in which case there is no need to use an initiator.

Principles and Mechanisms of CVD Polymerization

CVD polymerization using an initiator involves three steps (Figure 13.3b):

- Thermal decomposition of the initiator molecules to form free radicals.
- Diffusion and adsorption of the initiator radicals and monomers to the surface of the substrate.
- Polymerization of the monomers on the substrate surface by initiation, propagation, and termination reactions to form a continuous film.

As any of these individual steps may control the overall rate of deposition, an understanding of which step is rate limiting is essential for optimal deposition of the film. For example, if the reaction at the substrate surface is rate controlling, raising the temperature of the surface while keeping the other variables constant will increase the deposition rate.

Two important parameters in CVD polymerization are the partial pressure of the monomer p_m in the chamber and its saturation vapor pressure p_s (that is, its vapor pressure when it is in equilibrium with the liquid monomer) at the temperature of the substrate used for deposition. The saturation ratio S, equal to p_m/p_s, determines the following:

- Concentration of the monomer adsorbed on the surface.
- Film growth rate.
- Number-average molecular weight of the polymer film.
- Degree to which the film conforms to the surface topography.

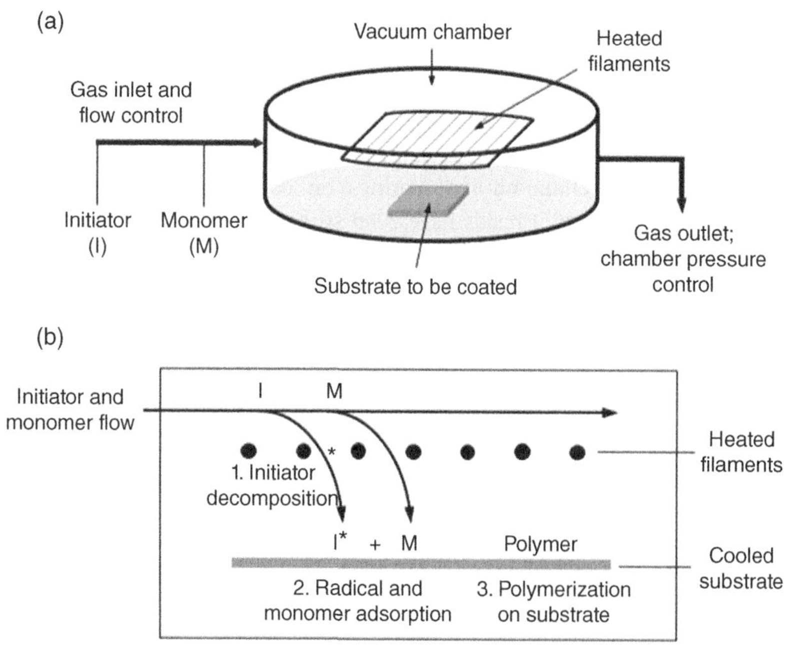

Figure 13.3 (a) Schematic of the main components of a reactor for chemical vapor deposition (CVD) of polymer films. (b) Steps in the CVD polymerization of polymer films. (b) *Source:* Lau and Gleason (2006) / with permission of American Chemical Society.

When $S = 1$, the monomer is in equilibrium with its liquid. As the presence of a liquid in the vacuum chamber is undesirable, CVD polymerization is normally carried out at values of S below 1, such as ~0.2–0.8.

For a series of alkyl acrylate monomers, of composition $CH_2{=}CH(C{=}O)OR$, where $R = C_nH_{2n+1}$ (n = 2–6), polymerized under the same conditions (gas flow rate, p_m, and temperature of the heating filaments and substrate), the film deposition rate and polymer molecular weight were found to increase strongly with increasing values of p_m/p_s (Figure 13.4). For example, the film deposition rate increased by over 300 times in going from ethyl acrylate to hexyl acrylate polymerization. As $1/p_s$ is a measure of the amount of monomer adsorbed on the surface of the substrate, these observations indicate that it is the monomer concentration (p_m/p_s) at the substrate surface that influences the deposition rate and molecular weight. For a given monomer (ethyl acrylate), the film deposition rate increased with decreasing substrate temperature (17–29 °C) and with increasing filament temperature (285–360 °C). However, the activation energy for film deposition (79 ± 5 kJ/mol) was almost independent of the filament temperature, indicating that the same process controlled film deposition at these filament temperatures. Overall, these results indicate that the adsorption of the monomer on the substrate may be the most important step that controls the CVD polymerization of these alkyl acrylates.

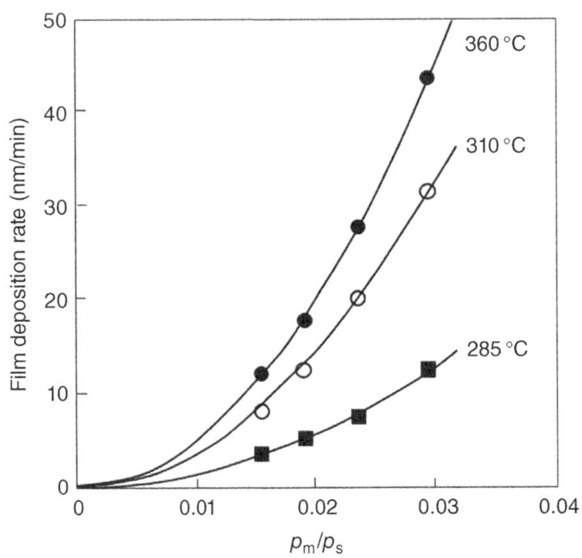

Figure 13.4 Film deposition rate plotted as function of the ratio of the partial pressure of the monomer p_m in the reactor chamber to the saturation vapor pressure p_s of the monomer, for ethyl methacrylate at three different filament temperatures. *Source:* Lau and Gleason (2006) / with permission of American Chemical Society.

Advantages and Drawbacks of CVD Polymerization

CVD polymerization is essentially similar to free radical polymerization in the bulk, such as the polymerization of liquid methyl methacrylate described in Chapter 8, but without the liquid. However, an important distinction is that CVD polymerization separates the initiator decomposition temperature (to form radicals) from the polymerization reaction temperature. This may allow better flexibility and control of the film deposition and, thus, provide the potential for better control of the film characteristics and properties. Overall CVD polymerization may provide several advantages, such as:

- Ability to deposit films on a wide variety of materials, including fragile, heat-sensitive, and soluble substrates due to the low substrate temperatures used in the process and the absence of a liquid phase.
- Ability to deposit dense pinhole-free films having a controlled thickness over a wide range, from <10 nm to >200 μm, generally at an efficient rate of ~1 μm per minute or greater.
- When compared to film deposition methods such as casting from a solution, CVD polymerization allows the deposition of insoluble polymers, while the absence of a liquid avoids the need for drying, impurities present in a solution, and residual liquid due to incomplete drying.
- The isotropic diffusion of the vapor phase reactants to the substrate surface and the low probability of the monomer molecules sticking to the substrate enable the deposition of conformal films of controlled thickness, which is essential for substrates that have microscale or nanoscale topography.
- Because functional side-groups present in the monomers are retained in the polymer, films having specific surface functions may be deposited through the selection of appropriate monomers, such as the ability to bond to specific biomolecules or particles, or to show pH responsive behavior.
- As one or more monomers may be introduced into the CVD reactor at a controllable flow rate, it is possible to deposit films composed of copolymers or having a gradient in composition.
- Although films are normally adherent to the substrate, their durability may be improved by strategies such as crosslinking, using monomers having multiple propagation sites, or by the formation of covalent interfacial bonds by activating the substrate surface prior to deposition.

A disadvantage of CVD polymerization is that the monomers need to have a high enough vapor pressure to achieve a reasonable and steady flow rate in the CVD reactor, which limits the use of monomers and initiators having a high molecular weight or polarity. CVD polymerization equipment is expensive for laboratory-scale experimentation. On the other hand, the process is efficient and economical at an industrial scale.

13.6 Physical Techniques for Surface Modification

Physical techniques for surface modification are normally used with inorganic materials (metals and ceramics) and generally involve the interaction of atoms, ions, or small molecules with the surface of the material. In PVD, material is ablated as atoms or molecules from a solid target of the desired chemical composition and deposited as atomic layers on the material to be coated (Figure 13.5). Ablation may be achieved by bombardment with an energetic ion beam produced by a plasma (Section 13.4) or by other sources, such as a laser beam. This technique deposits a film of the same composition as the target. It cannot be used to deposit polymer films because the polymer target will be degraded by the ablation step but polymer substrates may be coated if the process temperature is sufficiently low.

Ion beam implantation involves converting atoms into ions in a plasma, which are focused into a beam, accelerated to energies ranging from ~10 eV to ~10^3 keV, and directed at the surface of the material. Upon impact, an ion of sufficiently high energy has a high probability of entering the surface, transferring much of its energy to a localized surface zone and triggering a cascade of events, such as the creation of interstitials and vacancies in the crystal structure. At the same time, some atoms may be sputtered off the surface due to the intensity of the ion beam. Ion beam implantation does not create a separate coating. Instead, the ions are embedded individually within the structure of the material, with a distribution similar to that produced by a diffusion process, although the temperature for ion implantation is much lower. The degree of surface modification depends on the conditions and on the type of ions, but it is normally less than a few tenths of a micron.

Ion beam implantation has been used largely with metals, particularly steels. Although ions from most atoms in the periodic table can be used, small ions such as those of nitrogen, carbon, and boron have been commonly used with metals, mainly to improve their hardness, wear resistance, corrosion resistance, or fatigue resistance. Examples of metallic biomaterials that been modified by ion beam implantation are titanium and its alloy Ti6Al4V, and stainless steel (316L).

13.7 Parylene Coating

Parylene is a generic name for a series of polymers based on *para*-xylene. As Figure 13.6 illustrates, deposition of parylene coatings involves a few separate steps. Overall, monomers of the solid dimer di-*para*-xylylene are sublimed at ~120–160 °C, pyrolyzed at ~600–650 °C to crack the dimer vapor into a monomer vapor, which polymerize and form a polymer (parylene) film on the surface of a material in a low pressure reactor near room temperature. The polymerization mechanism may be viewed as involving some elements of plasma polymerization and conventional radical

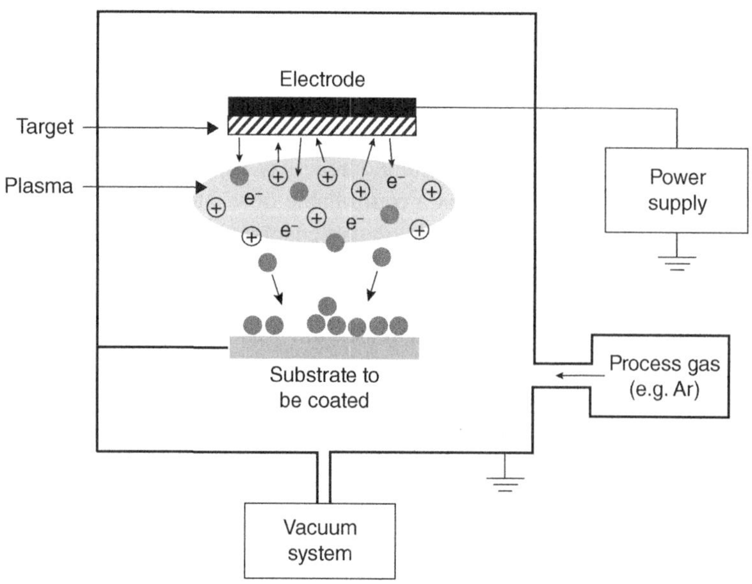

Figure 13.5 Schematic of the main components of a physical vapor deposition (PVD) system.

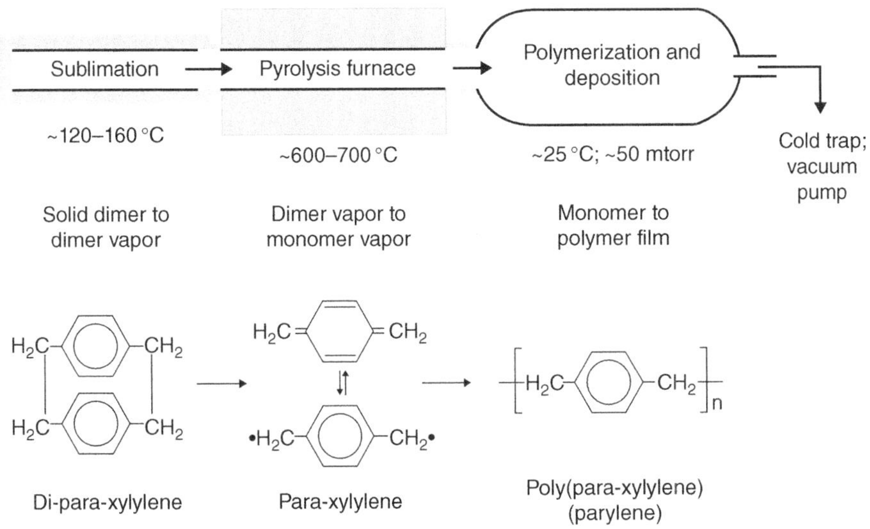

Figure 13.6 Schematic of the equipment and process steps in the deposition of parylene coatings.

polymerization, giving a polymer whose chain backbone consists almost entirely of carbon–carbon bonds.

Parylene coatings are normally smooth, strongly insulating electrically, have a high resistance to moisture permeation, and are not vulnerable to hydrolytic degradation in the physiological environment. Based on these characteristics, parylene has been used to coat a variety of biomedical devices, such as pressure sensors and ultrasound transducers, cardiac assist devices, stents, orthopedic devices and bone growth stimulators, catheters, needles and epidural probes, and cannulae.

13.8 Radiation Grafting

Radiation grafting involves irradiating a material with a beam of high-energy radiation, such as an electron beam, γ rays, or UV light, which breaks chemical bonds at the surface, forming reactive species such as free radicals, peroxides or other reactive species on the surface. Then these reactive species react with suitable monomers by a free radical polymerization mechanism to form a graft copolymer on the surface of the material (Figure 13.7). One type of radiation grafting involves placing the material in a monomer solution and irradiating the system with an electron beam or with γ rays. This method creates reactive species in both the surface of the material and the monomer. Another method involves first irradiating the material to create reactive species and then exposing it to a monomer. Alternatively, the surface may be oxidized by treating it with ozone or placing the material in a hydrogen peroxide solution and irradiating the system with UV radiation to create reactive peroxide species on its surface. Then the material is exposed to the monomer to form the coating.

The grafted layer normally has a high molecular weight and a thickness greater than ~1 μm, which is well bonded to the surface by covalent bonds. Homopolymer or copolymer coatings having a wide variety of compositions may be prepared by this method because of the availability of a wide range of monomers that polymerize by a free radical mechanism. As many synthetic polymers used as biomaterials are hydrophobic, radiation grafting has often been used to graft hydrophilic polymers to their surface to improve their hydrophilicity, and thus, to modify their interaction with the physiological environment. For example, polyurethane (PU) and polycaprolactone (PCL) grafted with polyhydroxyethyl methacrylate (PHEMA), polymethacrylic acid (PMAA), or polyacrylamide (PAAm) have shown improved hydrophilicity and endothelial cell adhesion.

13.9 Chemical Reactions

Certain chemical reactions involving a variety of reagents, such as a gas or a reagent solution, modify atoms or molecules at the surface of the biomaterial without the formation of a separate coating. They may be nonspecific, in which case they leave a distribution of different functional groups at the surface, or specific, changing only one functional group at the surface with few side reactions.

An example of a nonspecific reaction is the treatment of carbon nanotubes in a strong acidic solution, such as nitric acid or sulfuric acid, to introduce surface functional groups composed of hydroxyl OH, carboxyl (C=O)OH and carbonyl C=O groups at the surface. As noted in Chapter 12, these functional groups promote homogeneous dispersion and strong interfacial bonding in polymer

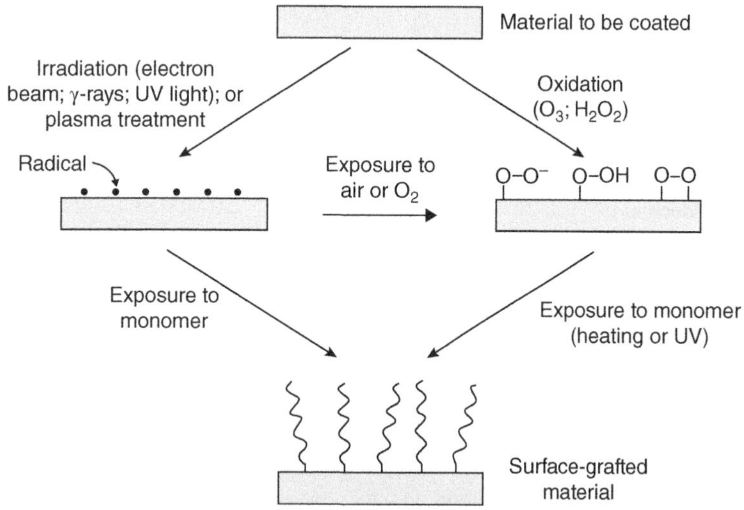

Figure 13.7 Illustration of method of grafting a polymer coating to the surface of a material.

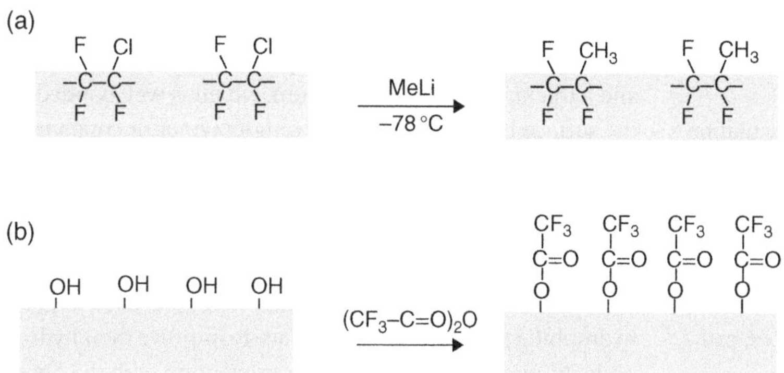

Figure 13.8 Examples of specific chemical reactions to modify the surface of a material.

composites reinforced with carbon nanotubes. Another example of a nonspecific surface reaction is plasma treatment of a biomaterial to introduce functional groups such as OH and NH_2 groups, which may be used subsequently in other surface modification reactions (Section 13.4.3).

Examples of specific surface reactions are shown in Figure 13.8. For example, treatment of polychlorotrifluoroethylene (PCTFE) in a solution of methyl lithium in ether at −78 °C leads to the replacement of the chlorine atom in the monomer unit by the methyl (CH_3) group. Another example is the treatment of a hydroxylated surface with reagents composed of a variety of functional groups to incorporate a different functional group, as illustrated for the reagent trifluoroacetic anhydride.

13.10 Solution Processing of Coatings

Solution or liquid-based coating methods are generally simple to perform and inexpensive because they do not require sophisticated equipment. One of the simplest methods is dip coating in which the material to be coated is dipped into a solution of the coating material (such as a polymer solution) and removed at a steady rate. Then the coated material is dried, leaving a solid coating on its surface. The solvent used in preparing the coating solution should be nontoxic and proper drying is necessary to avoid residual solvent in the coating. Examples of methods for solution-based deposition of coatings (films) are discussed in the following sections.

13.10.1 Silanization

Silanization is a surface modification method in which the surface of a material is coated with a layer of organofunctional alkoxysilane molecules. These molecules are normally composed of a silicon atom bonded to an alkyl group R and to three other groups that are normally alkoxy groups but may be other groups or atoms, such as the Cl atom (Figure 13.9). They are often referred to as *N*-alkyl silanes

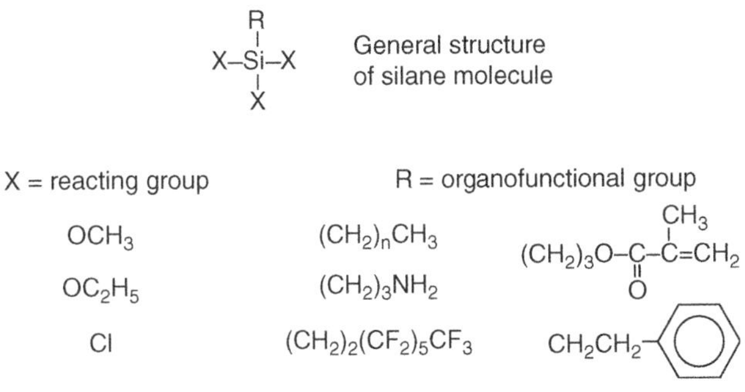

Figure 13.9 Chemical structure of silanes and examples of reacting group X and organofunctional group R.

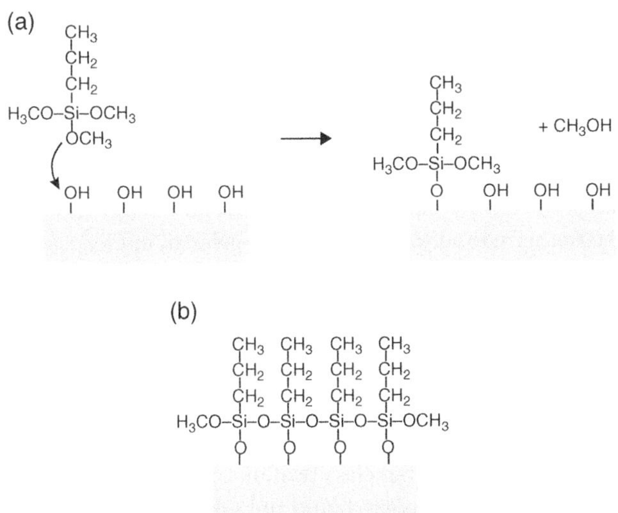

Figure 13.10 (a) Illustration of the reaction between a silane molecule and a hydroxyl OH group at the surface of a material to from a covalent bond. (b) Example of a three-dimensional network of silane molecules formed on the surface of a material.

or simply silanes. In the coating process, the material to be coated is placed in a solution of the silane for an appropriate time. Reaction between the alkoxysilane groups and appropriate functional groups on the surface of the material (such as hydroxyl OH groups) leads to the formation of a covalent bond that connects the pendant alkyl chain to the surface. Coatings composed of a single layer or a polymerized three-dimensional network can be created (Figure 13.10).

Inorganic materials such as ceramics, glasses, and metal oxide surfaces exposed to a moist environment normally contain an abundance of surface OH groups and, thus, silanization provides a simple and useful method to modify their surface. However, because the Si–O bond between the coating and the surface of the material is susceptible to hydrolysis, silanization can be easily reversed in the presence of water. Overall, silanization of inorganic materials serves two major functions:

- Reduction of the surface hydrophilicity, which may be desirable in some *in vitro* cell culture experiments.
- Improvement in the adhesion between the surface and an organic substance such as a protein or a synthetic polymer due to van der Waals bonding.

13.10.2 Langmuir–Blodgett Films

Langmuir–Blodgett films are ordered thin films created on the surface of a material by successive deposition of molecular monolayers from the surface of a liquid such as water that has a high surface tension (Figure 13.11). The films possess certain distinctive characteristics, such as controllable thickness, from a monolayer to multiple layers, high surface uniformity, and a high degree of orientation order. These characteristics have generated considerable interest for the use of Langmuir–Blodgett films in various applications, particularly in electronic devices and sensors.

An apparatus called a Langmuir trough is normally used to deposit Langmuir–Blodgett films on a solid. The main parts of this apparatus are a container of water (or another liquid) on which the monolayer-forming material is spread, barriers to control the surface area of the monolayer, and devices to measure the pressure applied to the barriers to control the monolayer surface area. Fatty acids such as stearic acid, $C_{17}H_{35}(C{=}O)OH$, have been a widely studied type of monolayer-forming material. They consist of amphiphilic molecules composed of a linear saturated alkyl chain, C_nH_{2n+1}, called the tail, which is hydrophobic, terminated at one end by a carboxyl (C=O)OH group, called the head or head group, which is hydrophilic.

Principles of Langmuir–Blodgett Film Formation

When a solution of an appropriate fatty acid (such as stearic acid) in a water-immiscible solvent (such as chloroform) is

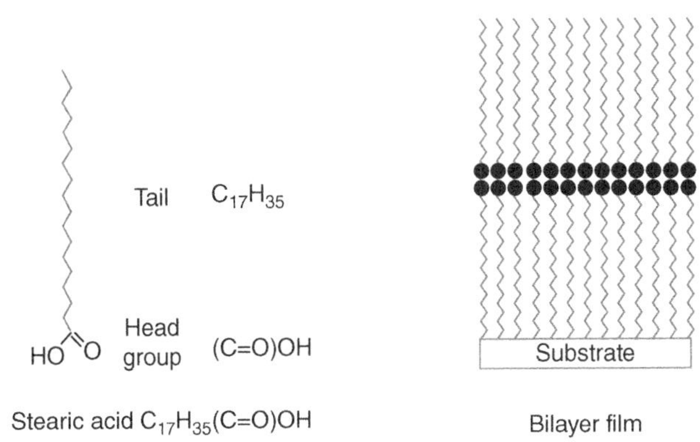

Figure 13.11 Chemical structure of stearic acid and an illustration of a bilayer Langmuir–Blodgett film on the surface of a hydrophobic material.

placed on a water surface, the solution spreads rapidly to cover the surface (Figure 13.12a). As the chloroform evaporates, a monolayer is formed on the water surface, dictated by the amphiphilic character of the stearic acid molecules, that is, the hydrophilic head groups are immersed in the water surface whereas the hydrophobic tails remain outside. When the distance between the stearic acid molecules is large, their interactions are small, and the molecules may be regarded as forming a two-dimensional gas. In comparison, upon reducing the surface area available to the molecules by moving the barriers, the molecules start to exert a repulsion on each other. The two-dimension pressure in the plane of the water surface is normally referred to as the surface pressure. For a planar liquid surface on which the monolayer is formed, the surface pressure p is given by

$$p = \gamma_o - \gamma \tag{13.3}$$

where γ_o is the surface tension of the liquid without the stearic acid molecules and γ is the surface tension with the molecules present. Thus, the maximum possible value of p for a monolayer on a water surface at room temperature is 73 mN/m (the surface tension of water). However, much lower values are normally measured in experiments.

A commonly used indicator of the monolayer properties is the surface pressure p as a function of the area A of the water surface available to each molecule, measured at a constant temperature, referred to as a surface pressure–area isotherm or, simply, an isotherm (Figure 13.12b). This isotherm normally shows a few distinct regions. As the surface area available to the stearic acid molecules is decreased from an initial high value, there is a gradual increase in surface pressure until an approximately horizontal region is reached. This region corresponds to the lifting of the hydrophobic tails of the molecules away from the water surface. It is characterized by weak interactions between the hydrophobic tails and the water surface and a low surface pressure, typically less than ~1 mN/m, which is often not observed in most equipment. The highly expanded monolayer in this region has been interpreted as a two-dimensional gas.

Then there is a transition to a steeply sloping linear region in which the compressibility of the molecules, defined as $-(1/A)(dA/dp)$, is approximately constant, where A is the area per molecule. This higher compressibility linear region has often been interpreted as a liquid. When the area decreases to ~20 nm^2 per molecule, there is an abrupt increase in the slope. This is attributed to a phase change due to a transition to a solid-like arrangement composed of a two-dimensional array of molecules. Although the compressibility in this region is also constant, it is lower by a factor of ~10. Extrapolation of this second linear region of the isotherm to zero surface pressure gives an intercept (A_o) equal to the area per stearic acid molecule that would be expected for the hypothetical state of an uncompressed close-packed layer. The observed value A_o = 0.23 nm^2 per molecule is close to that occupied by stearic acid molecules in single crystals, which supports the interpretation of the compact film as a two-dimensional solid.

Overall, the forgoing classification of the isotherm into regions interpreted as gas, liquid and solid is simplistic. The shape of the isotherm depends on the monolayer-forming material and the temperature, and different transitions

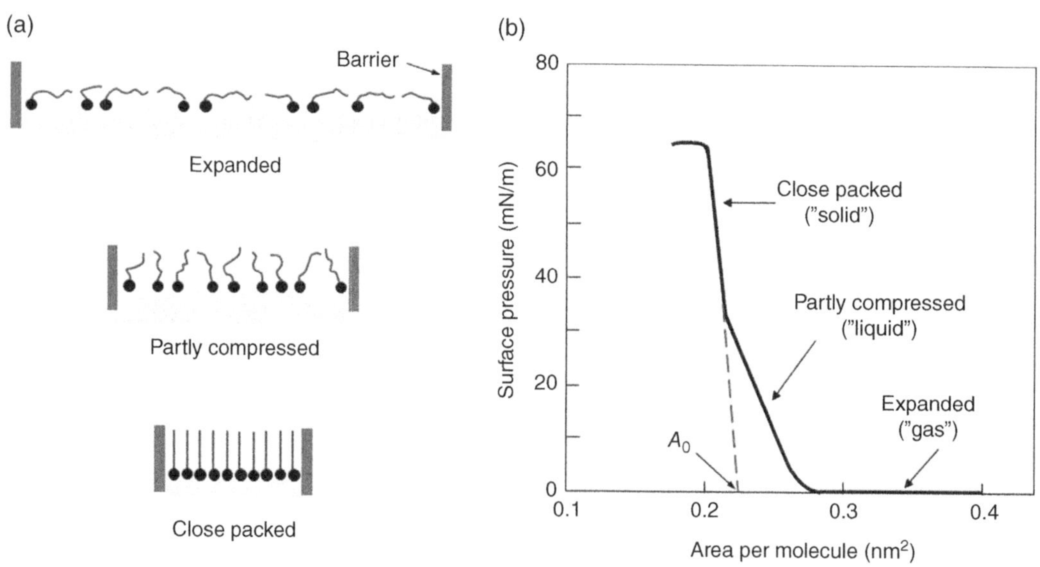

Figure 13.12 (a) Monolayer of stearic acid molecules on the surface of a water subphase illustrating the arrangement of the molecules in an expanded, partly compressed and close packed state; (b) surface pressure versus area isotherm for stearic acid on an acidified water subphase at room temperature, showing the regions of the isotherm corresponding to the arrangements in (a).

can be observed if the material or the temperature is changed.

Film Deposition Techniques

While fatty acids of the formula $C_nH_{2n+1}(C{=}O)OH$, as noted earlier, have been a widely studied type of monolayer-forming material, not all fatty acids can form stable films. The carboxyl (C=O)OH head group may be regarded as providing a constant amount of hydrophilicity to the molecule. On the other hand, the length of the alkyl group, that is, the value of n in the formula, determines the amount of hydrophobicity. Overall, then, the length of the alkyl group determines the hydrophilic to hydrophobic balance of the molecules. Fatty acids of n greater than ~12 can be spread on a water surface but their stability varies considerably with chain length. For example, myristic acid ($n = 13$) tends to dissolve slowly and, thus, is unsuitable for forming stable films. In comparison, stearic acid ($n = 17$) and arachidic acid ($n = 19$), which have a better hydrophobic to hydrophilic balance are more stable. They have been widely used in Langmuir–Blodgett film studies, although monolayers have also been formed from acids with longer alkyl groups, such as $n = 35$.

In the most commonly used method, shown for deposition on one side of a hydrophilic surface in Figure 13.13, the substrate is lowered through the compressed monolayer into the subphase (such as water) and withdrawn, keeping the surface pressure constant. The value of the surface pressure for optimal deposition depends on the composition of the monolayer molecules and is normally determined by trial-and-error experimentation. However, pressures less than ~10 mN/m are often too low whereas problems such as film collapse or film rigidity are often encountered above ~40 mN/m. Upon insertion, the water wets the hydrophilic substrate and the meniscus turns up to give a low contact angle. There is no deposition at this stage but, upon withdrawal, the meniscus wipes the surface of the material and leaves a monolayer in which the hydrophilic carboxyl (C=O)OH groups adhere to the surface of the substrate. A water film separates the surface of the material from the monolayer and must be removed, by evaporation for example, in order to achieve good adhesion of the monolayer to the surface.

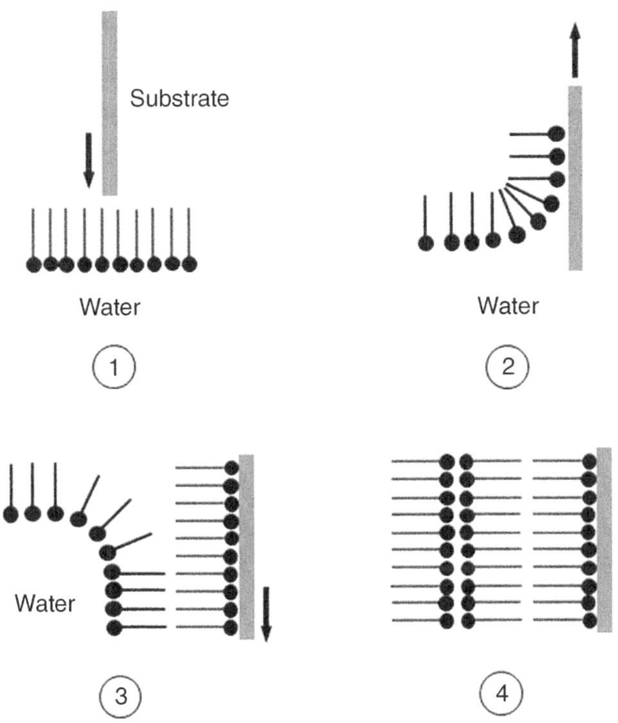

Figure 13.13 Deposition of Langmuir–Blodgett multilayer film on one side of a hydrophilic substrate.

Subsequently, a monolayer is deposited upon each traversal (insertion and withdrawal) of the substrate, giving a head-to-head and tail-to-tail stacking arrangement referred to as Y type deposition. This example of deposition on a hydrophilic surface leads to the creation of a film having an odd number of monolayers. In comparison, for a hydrophobic surface, a monolayer is deposited as the substrate is inserted into the subphase and upon withdrawal, giving a Y type film containing an even number of monolayers. Y type deposition is the most common type of deposition, but deposition modes in which deposition occurs only upon insertion or withdrawal, have also been reported.

Properties and Applications of Langmuir–Blodgett Films

Langmuir–Blodgett films have a highly ordered structure and can have a wide range of controllable thickness, from a monolayer to multiple layers, as noted earlier. Although the forgoing discussion has been based on the use of fatty acids composed of a linear saturated alkyl tail and a carboxyl (C=O)OH head group, monolayer-forming molecules having a variety of compositions can also be used. For example, fatty acids composed of modified tails, such as tails containing a double bond, or tails composed of a combination of hydrophobic and hydrophilic blocks, such as polystyrene-*b*-polyethylene oxide tails, have been widely studied. Additionally, the carboxyl head group may be modified to an ester (C=O)OR group, where R is an alkyl group such as C_2H_5, or to an amide $(C{=}O)ONH_2$ group.

Langmuir–Blodgett films are bonded to the substrate by secondary bonds (hydrogen or van der Waals bonds) and, thus, they do not adhere strongly to the substrate. However, the stability of the films can be improved by crosslinking or internally polymerizing the molecules, normally by using molecules that contain a double bond in their alkyl tail.

Langmuir–Blodgett films are useful scientifically as tools to study the behavior of monolayers at the air–water interface related to their transition from a gas to a liquid and to a solid phase (Figure 13.12). They have received considerable attention for potential use in a variety of electronic applications, such as sensors and electroluminescent displays. In biomedical applications, Langmuir–Blodgett films have been receiving interest for use in biosensors. As they have a structural resemblance to the membrane that surrounds cells, bilayer Langmuir–Blodgett films have attracted considerable interest in the creation of highly specific biosensor devices. At a simplified level, the cell membrane typically consists of amphiphilic macromolecules called phospholipids in a bilayer arrangement, with globular proteins inserted within the bilayer (Chapter 16). In addition to holding the cell together, the membrane provides sites for contact and recognition of other cells and for transmitting chemical and mechanical signals. The ordered structure of a bilayer Langmuir–Blodgett film provides a matrix in which biologically specific binding agents may be dispersed in creating highly specific biosensor devices.

13.10.3 Self-Assembled Monolayers

Self-assembled monolayers (SAMs) are thin films that form as highly ordered two-dimensional structures on specific solid or liquid surfaces by adsorption of molecules from a liquid or a gas. However, SAMs relevant to biomaterials are normally formed by adsorption of short-chain molecules from a liquid onto a solid surface. These molecules are generally amphiphilic, consisting of two essential parts, a functional anchoring group at one end that is capable of reacting with a specific functional group on the surface of the material, and a short molecular chain, such as an alkyl chain (Figure 13.14). Two processes dictate the formation of these small-molecule SAMs:

- An exothermic reaction between the anchoring group of the molecule and a specific functional group at the surface of the material to form a covalent bond. This provides a strong driving force to fill every possible functional site on the surface and displace any contaminants that may be present.
- Van der Waals interaction of the alkyl chains. Once the molecules have filled all the surface sites, they are close enough to one another for van der Waals forces to maintain their ordered arrangement.

Although it is not a prerequisite for forming SAMs, the molecules may also contain a functional head group capable of allowing the coated surface to interact with specific types of molecules in the environment. In this case, the head group of the molecules becomes the functional group of the coated material. The thickness of a SAM is dependent on the length of the alkyl chain but it is typically in the range of ~1–3 nm. Common examples of SAM films on materials are:

- *N*-alkyl silanes on hydroxylated surfaces, such as silica, glass, and many other inorganic materials (Section 13.10.1).
- Alkyl thiols (or dithiols) on certain metals such as gold and silver. Alkyl thiols have the general formula R–SH, where R is an alkyl group such as $CH_3(CH_2)_n$, whereas dithiols have the general formula $HS–(CH_2)_n–SH$.
- Amines ($R–NH_2$) or alcohols (R–OH) on platinum.
- Carboxylic acids R–(C=O)OH on aluminum oxide and silver.

Figure 13.15 illustrates the formation of a thiol coating on a gold surface.

SAMs formed from small molecules have some similarity to Langmuir–Blodgett films (Section 13.10.2), such as ease of formation and the ordered two-dimensional structure of the films. Additionally, the molecular chains may consist not only of alkyl chains but also of chains having other

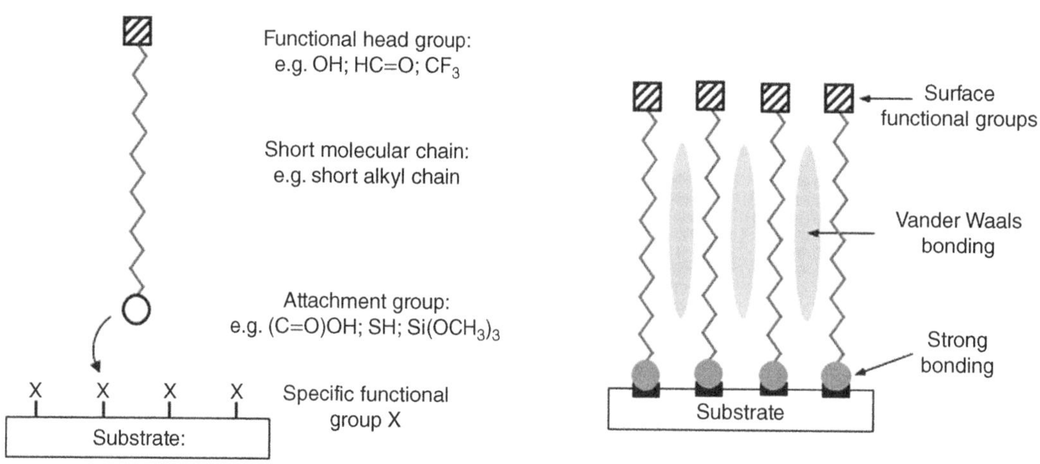

Figure 13.14 General characteristics of short-chain molecules and substrate for the creation of self-assembled monolayers (SAMs).

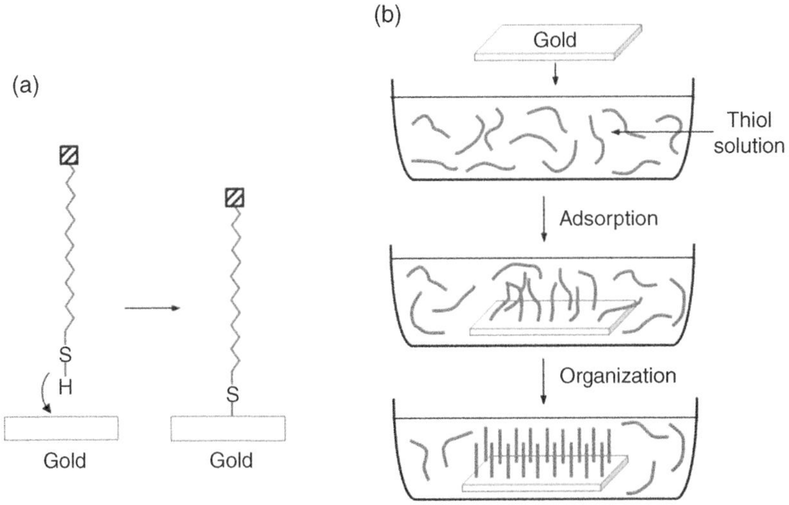

Figure 13.15 (a) Schematic of reaction of a thiol (S–H) terminated chain with the surface of a gold substrate. (b) Illustration of the creation of a self-assembled monolayer of thiol terminated molecules on the surface of a gold substrate.

chemical structures, such as copolymer structures or proteins. On the other hand, SAMs have better stability than Langmuir–Blodgett films because they are covalently bonded to the surface, and they provide the potential for changing the exposed functional group of the biomaterial that interfaces with the surrounding environment.

SAMs may also be created from molecules composed of macromolecular chains instead of short chains. These SAMs, sometimes referred to as polymeric SAMs to distinguish them from small-molecule SAMs, are being studied for a variety of electronic and biomedical applications. Like small-molecule SAMs, polymeric SAMs are chemisorbed to the surface of the material. However, they are structurally different. Instead of having an ordered two-dimensional structure, polymeric SAMs have a structure in which the conformation of the chains and, thus, the thickness of the film, depend on the way in which they are chemisorbed to the surface (Figure 13.16). For example, chains composed of surface-reactive monomer units tend to be chemisorbed with their chain axis extended parallel to the surface. In comparison, chains lacking reactive monomer units tend to extend in a direction perpendicular to the surface, forming a monolayer often referred to as a polymer brush. However, unlike small-molecule SAMs, the chains are not fully extended but, instead, adopt a coiled conformation. This is because the increase in entropy due to the large number of possible chain conformations offsets the decrease in energy arising from a uniform arrangement of the extended chains.

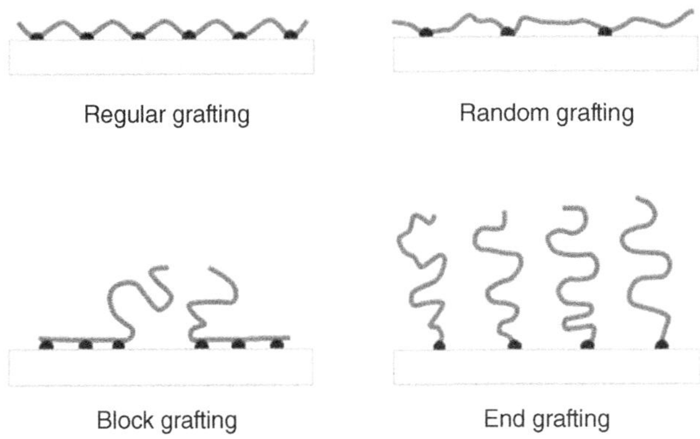

Figure 13.16 Illustration of various structures of self-assembled monolayers (SAMs) created from long-chain molecules.

13.10.4 Layer-by-Layer Deposition

In layer-by-layer (LbL) deposition, films of nanoscale thickness are formed by consecutively depositing individual macromolecular layers on a solid. A wide variety of macromolecules can be deposited on almost any type of surface but the most common LbL process involves the use of alternating molecular layers of opposite charge. Although ionic bonding between oppositely charged molecules is a well-known phenomenon, LbL deposition has attracted considerable interest since it was shown a few decades ago that it can be used to create stable multilayer films on a solid surface. Overall, LbL deposition can be used not only as a surface modification technique but also in the creation of thin film devices.

Figure 13.17 illustrates the first two steps in the deposition of anionic and cationic polyelectrolyte layers on a positively charged surface. As with most ionic systems, the pH and ionic strength of the polyelectrolyte solutions strongly influence the film deposition. Upon deposition of each layer, the surface charge is inverted, that is, it changes to the opposite sign. The chains within the layers are flexible

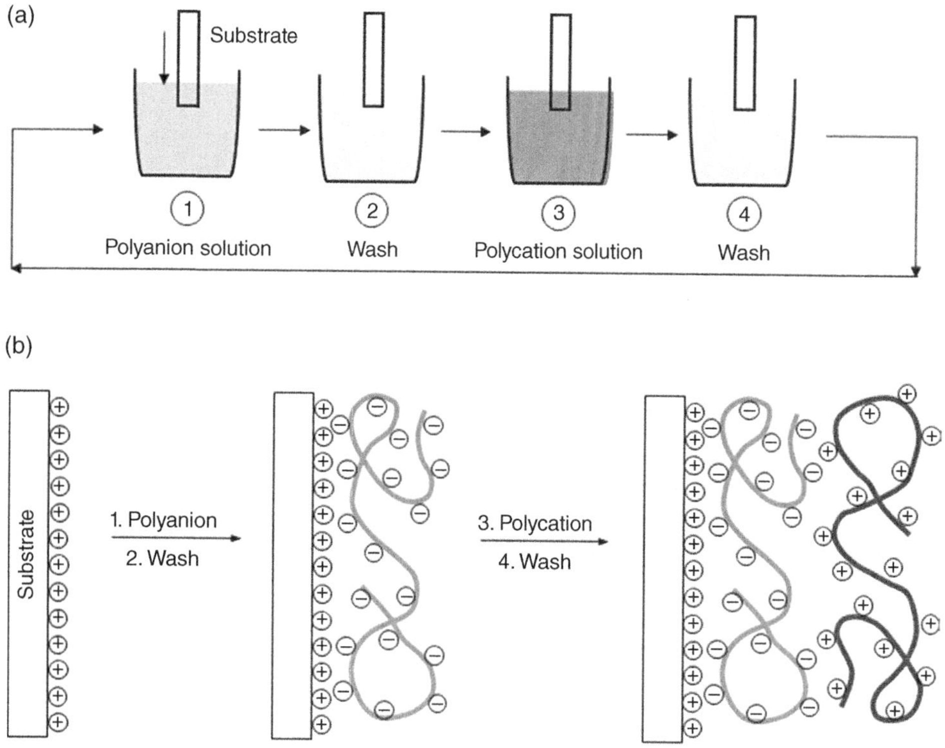

Figure 13.17 (a) Schematic of the layer-by-layer (LbL) film deposition process. (b) Simplified illustration of the molecular structure of a bilayer film deposited on the surface of a positively charged surface.

and do not have any ordered arrangement. LbL deposition has several attractive characteristics, such as

- Simplicity, involving adsorption form solution.
- Ability to control the thickness of the film over a wide range, from thin films to thick films, simply through the number of layers deposited.
- Good bonding of the layers due to the large number of ionic bonding sites in the polyelectrolyte chains.
- Ability to deposit conformal films on solids, dense or porous, of almost any type and geometry by a variety of macromolecules.

LbL deposition can be used not only with water-soluble macromolecules that have charges on them but also with water-soluble macromolecular pairs that have hydrogen-bonding capacity. Additionally, nanoparticles or biomolecules may be incorporated within the layers. Overall, LbL-deposited films may be viewed as multilayered nanocomposites. They have been studied for a variety of biomedical applications, such as devices for delivering growth factors to stimulate cell response and diffusion barriers to control drug release. Films of the appropriate composition may also be used to prevent adverse biological responses, such as bacterial buildup or thrombogenicity, which often plague the performance of biomedical devices such as stents. One biomedical product that has already been commercialized consists of multilayered coatings on contact lenses.

13.11 Biological Functionalization of Biomaterials

Biological functionalization generally refers to a process of immobilizing biomolecules such as peptides, proteins, enzymes, antibodies, or drugs on or within a biomaterial to confer a desired biological functionality to it. It may also involve combining cells with biomaterials, particularly when the surface of these biomaterials contains cell-adhesive peptides or proteins. Biological functionalization is often necessary because many biomaterials, particularly synthetic biomaterials, normally do not possess the requisite characteristics and properties to elicit the desired biological response *in vivo*. For example, surface modification with the polysaccharide heparin has been used for decades to reduce the thrombogenecity of biomaterials (Section 13.11.4). In the last few decades, immobilization of peptides or growth factors on or within polymeric biomaterials has become a very attractive approach to regulate cell response and to regenerate tissues and organs.

Biological functionality can be added to biomaterials of all classes and shapes, once they possess the requisite surface functional groups or can be treated by any of the aforementioned surface modification techniques to create the requisite functional groups. However, the methods are associated mainly with polymers because of their widespread use as biomaterials. Table 13.2 summarizes several types of biomolecules that may be immobilized on or within biomaterials.

13.11.1 Immobilization Methods

Methods to immobilize biomolecules on or within a biomaterial may be divided into the following broad categories:

- Physical immobilization, which includes immobilization by ionic or van der Waals bonds, entrapping or encapsulating the biomolecules within the biomaterial, or simply dispersing the biomolecules within the biomaterial by mixing.
- Chemical immobilization of the biomolecules on the surface of the biomaterial by covalent bonds.

Immobilization may be short-term or long-term based on the time scale of the application. Physical methods are normally associated with short-term immobilization but large biomolecules may also be immobilized over a long duration by these methods. Additionally, a degradable biomaterial may release physically immobilized or covalently bonded biomolecules as it degrades.

Table 13.2 Examples of biomolecules that may be immobilized on or within polymeric biomaterials.

Type of biomolecules	Examples
Peptides or proteins	Cell adhesive peptides
	Enzymes
	Antibodies
	Antigens
Saccharides	Polysaccharides
	Carbohydrates
	Oligosaccharides
Lipids	Fatty acids
	Phospholipids
Conjugates or mixtures of peptides, proteins, saccharides and lipids	
Drugs	Antithrombogenic agents
	Peptide or protein drugs
	Antibiotics
	Anticancer agents
Ligands	Hormone receptors
	Cell surface receptor ligands
Nucleic acids and nucleotides	Single or double stranded deoxyribonucleic acid (DNA) or ribonucleic acid (RNA)

13.11.2 Physical Immobilization

A simple method for immobilizing or encapsulating biomolecules within a biomaterial is to disperse the biomolecules in a solution or a suspension of the biomaterial prior to solidification or gelling. This method has been widely used to immobilize biomolecules such as growth factors and other drugs within polymers and hydrogels. The elution of the biomolecule depends on the composition of the polymer (or hydrogel) and its characteristics such as molecular weight and geometry, as discussed in previous chapters on degradable synthetic polymers, natural polymers and hydrogels.

Physical adsorption is another simple method, which normally involves dipping the biomaterial into a solution of the biomolecules for an appropriate time prior to implantation. The biomolecules are retained at the surface of the biomaterial by ionic, hydrogen or van der Waals bonding, depending on the presence or absence of electrostatic changes on the biomolecules and on the biomaterial. Upon implantation, the biomolecules are released by desorption. Because of its simplicity, this method has been widely used to immobilize biomolecules such as growth factors on the surface of biomaterials, particularly inorganic biomaterials such hydroxyapatite, β-TCP and bioactive glass. Several studies have shown the capacity of the released biomolecules to stimulate cell and tissue response. For example, adsorption of bone morphogenetic protein-2 (BMP2) has been observed to enhance the capacity of hydroxyapatite or bioactive glass implants to stimulate bone repair and regeneration. However, it should be noted that the release of BMP2 at concentrations much higher than the safe physiological concentration has been associated with undesirable biological effects, such as induced bone formation in ectopic (abnormal) sites and possible growth of cancerous tumors.

Overall, methods based on immobilization of biomolecules by physical adsorption suffer from drawbacks. A major drawback is the lack of control over the retention of the biomolecules and their delivery. Release of physically adsorbed biomolecules from a biomaterial normally shows a characteristic profile, consisting of a rapid initial release (often called a burst release) in which the biomaterial releases a significant fraction of the adsorbed biomolecules within the first few hours, followed by a more gradual release. Consequently, the concentration of the released biomolecules may initially be much higher than the physiologically safe concentration. Thereafter, the gradual release may not deliver the biomolecules at a sufficiently high rate to trigger optimal cell and tissue response. Another factor is that release of the biomolecules depends on the physiological environment, which varies from site to site and from patient to patient. Freely diffusible growth factors released from the biomaterial are also susceptible to rapid loss of bioactivity. In order to ensure the desired cell or tissue response, one approach has been to use large quantities of biomolecules, much higher than the safe physiological amount (sometimes referred to as supra-physiological doses). This is not only expensive but it may also cause undesirable effects, as noted earlier for BMP2.

Immobilization of biomolecules within coatings deposited by the LbL technique (Section 13.10.4) is a more recently developed strategy. Studies have shown promising potential of this method for controlled release of biomolecules to stimulate cell and tissue response. For example, repeated deposition of a four-layer film composed of a poly(β-amino ester), polyacrylic acid (PAA), the osteogenic growth factor BMP2, and PAA has been used to form a coating (~2 μm thick) on a porous composite composed of a PCL matrix and β-TCP particles (Figure 13.18a). BMP2 has an isoelectric point of ~8.5 and, thus, has a positive charge under acidic and physiological pH conditions. Under similar conditions, the poly(β-amino ester) develops as positive charge due to protonation of its amino group, whereas PAA develops a negative charge at pH values higher than ~3.0–4.0. This coating released BMP2 in a controlled manner over a two-week period in phosphate-buffered saline (PBS) at 37 °C *in vitro* (Figure 13.18b), which retained its biological activity and stimulated the differentiation of pre-osteoblastic MC3T3-E1 cells into an osteogenic lineage. When implanted into intramuscular sites in rodents *in vivo*, the coated scaffolds showed the capacity to induce bone formation.

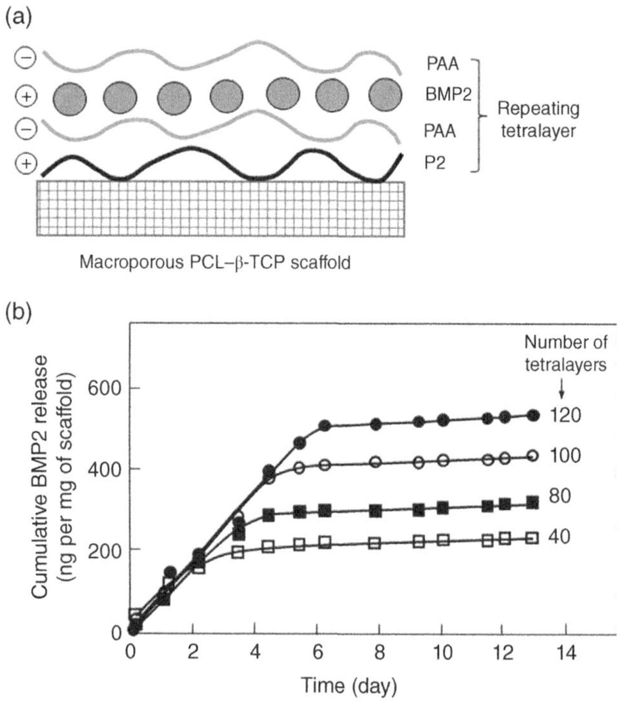

Figure 13.18 (a) Illustration of repeating tetralayer on the surface of a macroporous scaffold composed of polycaprolactone (PCL) reinforced with β-tricalcium phosphate (β-TCP). The tetralayer consists of individual layers of poly(β-amino ester) (P2), polyacrylic acid (PAA), bone morphogenetic protein-2 (BMP2), and PAA. (b) Release of BMP2 as a function of time from coatings consisting of different numbers of tetralayers in phosphate-buffered saline (PBS) at 37 °C. *Source:* Shah et al. (2011) / with permission of Elsevier.

Immobilization of growth factors within LbL layers has also been shown to have the capacity to simultaneously deliver controlled amounts of two or more growth factors, such as a combination of an osteogenic growth factor such as BMP2 and an angiogenic growth factor such as vascular endothelial growth factor (VEGF).

13.11.3 Chemical Immobilization

Chemical immobilization of biomolecules by covalent bonds has the potential to avoid or reduce the aforementioned drawbacks associated with physical immobilization, although the procedures are generally more complex. When compared to physical immobilization, covalent immobilization may provide the following advantages:

- The amount and distribution of the biomolecules at the surface of the biomaterial can be better controlled because the biomolecules are bonded to the surface by specific chemical reactions.
- Release of the biomolecules from the surface can be better controlled because desorption occurs by specific hydrolytic or enzymatic cleavage reactions.
- Biomolecules delivered to cells or tissues generally have improved stability and persistence, which avoids the need for using high doses of biomolecules.
- Two types of biomolecules are presented to cells or tissues, that is, soluble (released) and bound biomolecules, which may perform different functions and, thus, provide the potential for enhancing the biological significance of the chemically immobilized biomolecules.

A requirement for covalent immobilization is that the surface of the biomaterial and the biomolecules should possess the appropriate functional groups for the desired chemical reaction. Common functional groups for the biomaterial and the biomolecules are carboxyl (C=O)OH, amine NH_2 or hydroxyl OH groups. If the biomaterial does not possess the appropriate functional group, one of the surface modification techniques described earlier in this chapter may be used to introduce the functional group. The biomolecules may be attached directly to the surface or by means of a spacer molecule. These spacer molecules are normally bifunctional, that is, they contain appropriate reactive groups at each end to form covalent bonds with the biomolecule at one end and with the biomaterial at the other end. The use of spacer arms may provide additional functionality to the biomolecule, such as greater steric freedom to interact with cells or tissues or the capacity to undergo degradation by a specific reaction and release the biomolecule. Additionally, more than one type of biomolecule may be bonded to the biomaterial to provide enhanced biological functionality.

While a variety of methods has been used to covalently immobilize biomolecules to the surface of biomaterials, only a few of the common methods are described in the following sections. These methods are generally applicable to the immobilization of a variety of biomolecules that contain the appropriate functional groups but we shall focus mainly on the immobilization of adhesive peptides and growth factors because of their strong relevance to tissue engineering applications.

Immobilization Using Carbodiimide Chemistry

A common method to immobilize biomolecules on a variety of biomaterials involves the use of a water-soluble carbodiimide agent. Carbodiimides molecules have the general formula R_1–N=C=N–R_1 where R_1 and R_2 are chemical groups, a commonly used example being 1-ethyl-(dimethylaminopropyl) carbodiimide (EDC). Carbodiimides couple carboxyl (C=O)OH groups to amine NH_2 groups, forming stable amide bonds. Consequently, the reaction requires the presence of carboxyl or amine groups on the biomolecule and on the surface of the biomaterials. For example, the amine group may be provided by the N-terminus of a polypeptide (or peptide) or by lysine residues within a polypeptide, whereas the C-terminus or residues such as aspartic acid or glutamic acid within the polypeptide chain may provide the carboxyl group.

A common reaction involving carbodiimide chemistry is the attachment of adhesive peptides to biomaterials, an example of which is the bonding of the cell-adhesive peptide RGD to alginates described in Chapter 10. This reaction is also commonly used to immobilize growth factors on biomaterials. The overall reaction scheme is illustrated in Figure 13.19 for attaching a growth factor containing carboxyl groups to the surface of a biomaterial containing

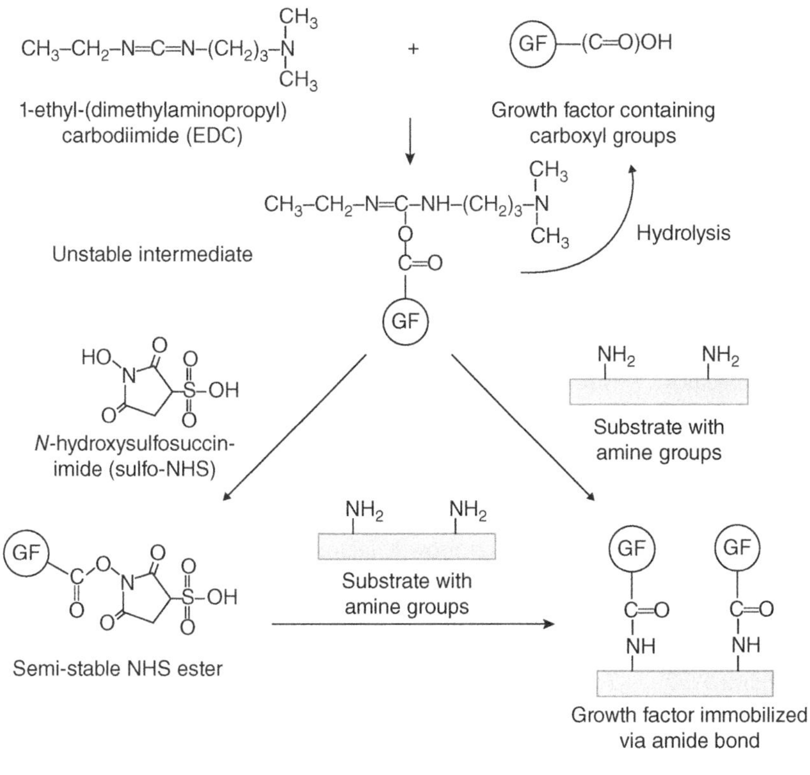

Figure 13.19 Reaction scheme for immobilization of growth factor (GF) containing carboxyl (C=O)OH groups on to the surface of a substrate containing amine (NH_2) groups via an amide bond.

amine groups. As noted for the RGD-alginate system, the reaction normally requires the use of a co-reactant, *N*-hydroxysulfosuccinimide (sulfo-NHS), to stabilize an intermediate compound against a competing hydrolysis reaction.

Carbodiimide coupling is a popular immobilization method because it is simple and inexpensive. Additionally, because it requires mild reaction conditions, the risk of reducing the biological activity of biomolecules such as growth factors is diminished. Examples are the immobilization of BMP2 on collagen, and on polyethylene terephthalate (PET) functionalized with (C=O)OH groups, VEGF on collagen, epidermal growth factor (EGF) on amine functionalized polymers composed of PCL and block copolymers of PCL with polyethylene glycol (PEG), and nerve growth factor (NGF) on amine functionalized PCL–PEG block copolymers.

On the other hand, carbodiimide coupling suffers from a few drawbacks. For example, carbodiimide chemistry is not able to differentiate between terminal reactive groups and reactive groups within the growth factor structure, which may lead to adverse biological effects. Depending on the location of the reactive amine or carboxyl groups relative to the cell surface receptor sites of the growth factor, recognition of the growth factor by its receptor may be impeded, which will lower the growth factor bioactivity (Chapter 16). Additionally, the presence of multiple amine or carboxyl groups in the growth factor structure can lead to bonding to the biomaterial at several points along its molecular chain. This can prevent the growth factor from adopting its optimal conformation for interacting with its cell surface receptor or hinder the receptor from reaching the growth factor binding site.

Immobilization Using Photochemical Methods

Another common method for immobilizing biomolecules on biomaterials involves the use of a photochemical method. This method relies on the biomaterial and the biomolecules possessing the requisite functional group, such as a carbon double bond (C=C), to take part in a free radical reaction upon exposure to electromagnetic radiation, typically UV light. The biomolecule can be immobilized directly on the biomaterial or via a spacer molecule. A well-known example is the bonding of the cell-adhesive peptide RGD to PEG hydrogels to enhance their interaction with cells and tissues (Chapter 11). The peptide YRGDS may be bonded directly to an acrylate group or indirectly through a PEG spacer molecule using a reaction between the N-terminal amine group of the peptide and an *N*-hydroxysuccinimidyl-activated ester (Figure 11.20). Then this mono-acrylated peptide is bonded to acrylate-containing PEG by free radical polymerization using UV light. The peptide may be immobilized by combining the monoacrylated molecules with PEG diacrylate (PEGDA) molecules prior to UV exposure or by adding to an already crosslinked polymer followed by exposure to UV light. A similar reaction can be used to immobilize growth factors using their N-terminal amine or lysine residues of the growth factor (Figure 13.20).

Photochemical immobilization is simple, inexpensive, and effective. It has been used to immobilize a variety of growth factors such as VEGF, platelet-derived growth factor (PDGF), basic fibroblast growth factor (bFGF) and transforming growth factor-β (TFGβ) on biomaterials, particularly PEG. However, the method also suffers from drawbacks. First, because it depends on acrylate reaction with another acrylate, the types of biomaterials that can be used is limited. Second, the acrylation reaction cannot distinguish between different amine groups in the growth factor, which may lead to a loss in bioactivity, as described for the carbodiimide coupling method.

13.11.4 Heparin Modification of Biomaterials

Heparin and structurally related heparan sulfate are sulfated polysaccharides that belong to a family of glycosaminoglycans in the extracellular matrix (ECM) (Chapter 10). They have a chemical structure composed of linear chains of disaccharide units of varying degrees of sulfation and acetylation (Figure 10.18). Heparin was discovered nearly a century ago and it has been used as an anticoagulant drug for several decades. Heparin binds to antithrombin, a protein in the blood stream that prevents blood from clotting too rapidly through a process of blocking the blood clotting mechanism by inactivating the major clotting protein thrombin.

Heparin modification of biomaterials has been used for decades to reduce their thrombogenecity, that is, the tendency of the biomaterial in contact with blood to produce a thrombus (clot). Heparin also binds to many biomolecules in the ECM, such as enzymes and other proteins, which has a strong effect in regulating several biological processes. In the last few decades, heparin-modified biomaterials have been utilized in the delivery of protein-based drugs such as growth factors for a variety of biomedical applications.

Heparin Modification of Biomaterials to Reduce Thrombogenicity

Heparin can be immobilized on the surface of biomaterials using a wide variety of methods that produce not only physical immobilization by ionic, hydrogen or van der Waals bonding but also chemical immobilization by covalent

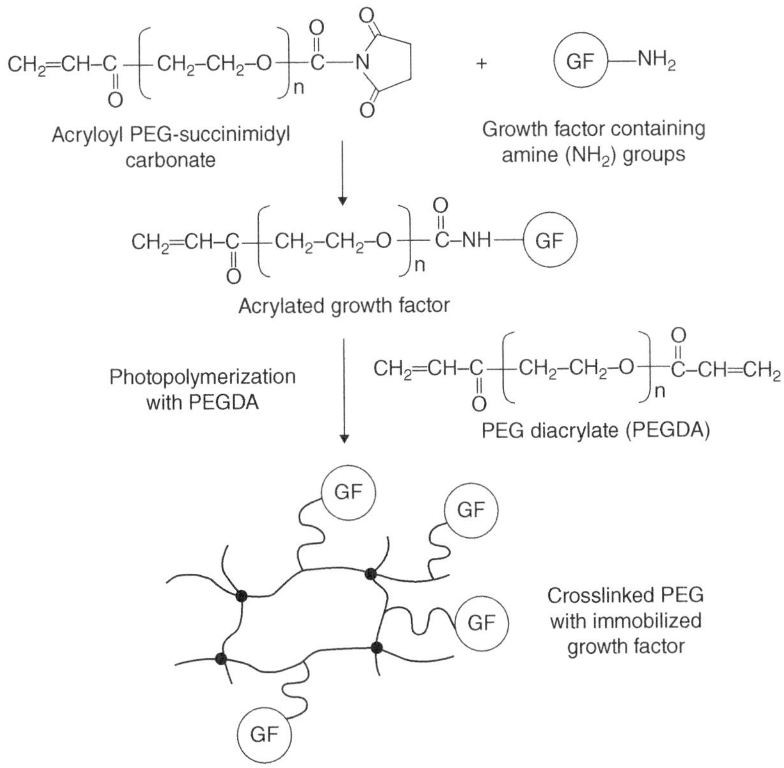

Figure 13.20 Reaction scheme for immobilizing a growth factor (GF) containing amine (NH_2) groups on polyethylene glycol (PEG) chains.

bonding (Figure 13.21). Immobilization by ionic bonding takes advantage of electrostatic attraction between the strong negative charge of heparin chains under acidic or physiological pH conditions and a positively charged surface. If the surface does not possess a positive charge, an appropriate technique from those described in this chapter can be used to introduce positively charged groups at the surface. For example, ionic bonding of heparin to some synthetic polymers has been achieved by introducing quaternary ammonium groups at the polymer surface.

Because of the abundance of reactive functional groups such as sulfonyl hydroxide $S(=O)_2OH$, carboxyl (C=O)OH and hydroxyl OH groups in its chains (Figure 10.18), heparin can be easily immobilized on biomaterials by covalent bonds using a variety of techniques. The heparin molecules can be bonded directly to the surface or by means of a spacer molecule. A common method involves the use of carbodiimide chemistry (Section 13.11.3). These methods normally lead to bonding of the heparin chains at multiple sites to the surface.

Methods have also been developed for covalently bonding the heparin chains at only one end to the surface of the biomaterial, which better preserve the bioactive sites by which heparin binds to antithrombin (Figure 13.22). One method involves treating heparin with nitrous acid, which

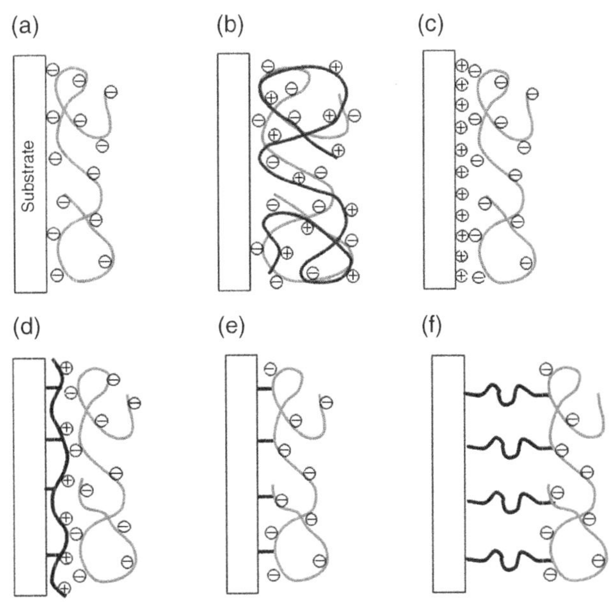

Figure 13.21 Examples of methods for immobilizing heparin molecules at the surface of biomaterials: (a) physical adsorption by secondary bonds (hydrogen bonds or van der Waals bonds); (b) complexation with a positively charged polymer and physical adsorption; (c) ionic bonding to a positively charged surface; (d) ionic bonding to a positively charged polymer covalently bonded to the surface; (e) covalent bonding directly to the surface; (f) covalent bonding to the surface via a bifunctional spacer molecule.

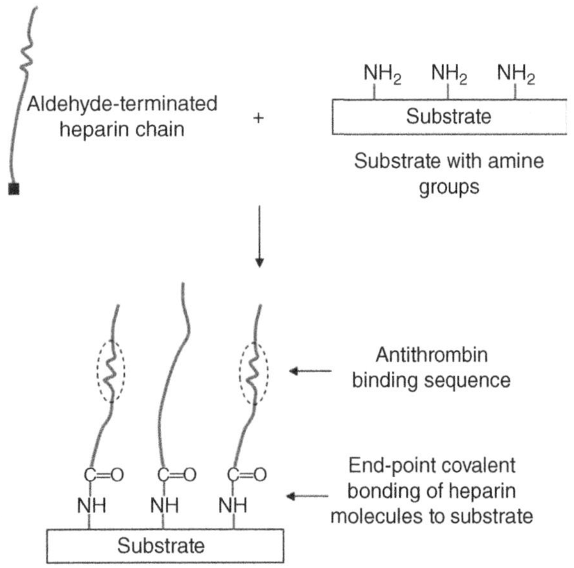

Figure 13.22 Illustration of end-point covalent bonding of aldehyde-terminated heparin molecules to the surface of a biomaterial (substrate) containing amine (NH_2) groups on its surface.

leads to partial depolymerization and the creation of chains of lower molecular weight that contain a single reactive aldehyde group at one end of the chain. Then the depolymerized heparin chains are covalently bonded to the surface of the biomaterial by a reaction between their aldehyde group and amine groups at the surface. If amine groups are not originally present at the surface of the biomaterial, they may be introduced by an appropriate surface modification method. This type of end-point immobilization has been used to create stable heparin coatings on biomaterials for use in a variety of applications, such as expanded polytetrafluoroethylene (PTFE) and Dacron® used as vascular grafts (Chapter 8).

Heparin Modification of Biomaterials for Drug Delivery

The affinity of heparin for protein-based drugs such as growth factors, as noted earlier, has been utilized in the creation of heparin-modified biomaterials for drug delivery. These types of drug delivery systems, sometimes referred to as affinity drug delivery systems, can potentially provide advantages over other systems. For example, they utilize specific non-covalent interactions to stabilize growth factors and immobilize them within a biomaterial matrix, thus protecting their biological activity and slowing their diffusion (migration) from the matrix. Additionally, interactions between affinity delivery systems and growth factors can potentially mimic those that occur naturally with native ECM proteoglycans.

The primary and secondary structure of heparin molecules play a strong role in their interaction with growth factors (and other proteins). Heparin chains, as noted earlier, are linear unbranched chains that develop a strong negative charge under acidic or physiological pH conditions due to dissociation of H atoms from the sulfonyl hydroxide $S(=O)_2OH$ and carboxyl (C=O)OH groups present in their repeating disaccharide units. These chains adopt a helical conformation in which the $S(=O)_2O^-$ and $(C=O)O^-$ groups are presented at defined intervals and in defined orientations along the chain backbone to promote interactions with growth factors. Unlike proteins, heparin is not known to adopt or fold into any specific tertiary structure.

Heparin interaction with growth factors occurs largely by electrostatic attraction (ionic bonding) between small regions of positive charge on the growth factor (arising from positively charged amino acid residues such as lysine or arginine) and $S(=O)_2O^-$ groups of heparin. However, in some cases, there is a significant contribution from hydrogen bonding due to polar amino acid residues (such as asparagine or glutamine) and, to a lesser extent, from van der Waals bonding, which can enhance this electrostatic interaction. Heparin-binding sites in growth factors are normally situated on the external surface of their chains. Binding to heparin induces a conformational change in the growth factor chain, which imparts some degree of specificity to the interaction. Examples of common growth factors that bind to heparin are bFGF, VEGF, members of the TGF family such as BMP2 and TGFβ, PDGF, and EGF.

Immobilization of Heparin on Biomaterials for Drug Delivery

Heparin can be immobilized on or within biomaterials by covalent or non-covalent bonds in the delivery of growth factors, using methods described earlier in this section. As an example, carbodiimide chemistry using EDC has been used to immobilize heparin to the chains of a crosslinked collagen matrix for the delivery of bFGF, by coupling the (C=O)OH groups of heparin to NH_2 groups of collagen, as illustrated in Figure 13.20. Then the heparinized collagen specimens, in the form of thin discs, are incubated in a solution of bFGF in PBS to bind the bFGF molecules to heparin. Release of bFGF from the heparinized collagen *in vitro* was initially slower when compared to similar collagen specimens without heparin, but after 10 days, the cumulative amount of bFGF released from the two types of specimens was comparable. Delivery of bFGF from heparinized collagen specimens significantly enhanced endothelial cell proliferation at 6 and 10 days *in vitro* and vascularization over 3 weeks in rat subcutaneous sites *in vivo*, when compared to similar collagen specimens without heparin. The end-point immobilization method for covalently bonding heparin to biomaterials for reduction of thrombogenicity (Figure 13.22) has also been used to immobilize heparin on various biomaterials for growth factor delivery.

Methods utilizing primarily ionic bonding rather than covalent bonding have also been developed to immobilize heparin on or within biomaterials for growth factor delivery. One approach utilized a three-component delivery system composed of a fibrin matrix, a bi-domain peptide and heparin. A key element of this system is the bi-domain peptide, composed of an N-terminal domain to bond covalently to fibrin and a C-terminal domain to bond ionically to heparin (Figure 13.23). In turn, the heparin serves to immobilize the growth factor and to control its release from the fibrin. This system showed the ability to deliver bFGF in a controlled manner, to enhance neurite extension from dorsal root ganglia *in vitro*, and to enhance peripheral nerve regeneration in a rat sciatic nerve model *in vivo*. This approach involving immobilization of heparin by ionic bonding has been used with other biomaterials for delivering growth factors.

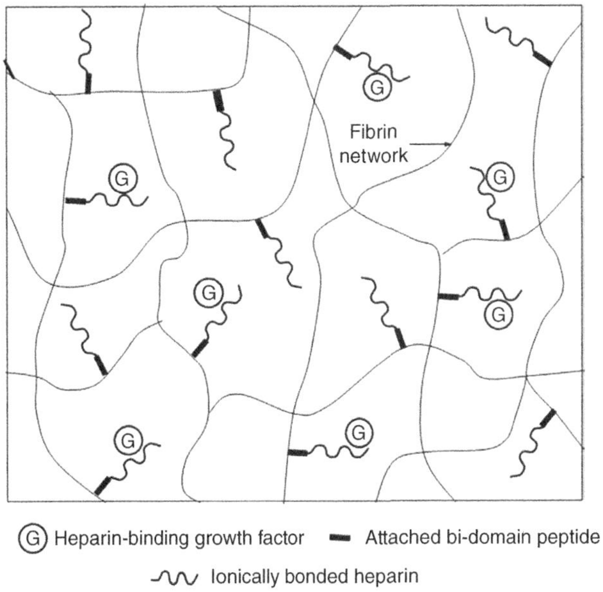

Figure 13.23 Illustration of a drug delivery system composed of a fibrin matrix, heparin and a heparin-binding growth factor. A bi-domain peptide is covalently bonded to the fibrin matrix, while heparin is ionically bonded to the heparin-binding domain of the peptide. The growth factor is immobilized to the heparin. *Source:* Adapted from Sakiyama-Elbert and Hubbell (2000).

13.12 Concluding Remarks

This chapter discussed two approaches for modifying the surface of biomaterials in order to improve their interaction with the physiological environment.

- First, a discussion of surface modification of biomaterials was presented. The normal objective of surface modification is to alter the surface characteristics and properties of biomaterials in order to enhance their biological response such as blood compatibility, protein adsorption, or cell adhesion and growth. A wide variety of techniques, which rely on modifying the outermost atoms of the surface or overcoating the surface with a thin layer, was covered. Emphasis was placed on the principles of the techniques and their capacity to alter the surface characteristics of biomaterials.
- Second, biological functionalization was discussed as an approach to confer a desired biological function to biomaterials by immobilizing biomolecules on or within them. Immobilization of peptides or growth factors forms an attractive approach to regulate cell response and to regenerate tissues and organs. Heparin modification of biomaterials was discussed in detail as an example of biological functionalization to reduce thrombogenicity and to deliver growth factors for tissue regeneration.

Problems

13.1 Compare the formation of polymer films on a biomaterial from ethyl acrylate $CH_2{=}CH(C{=}O)OC_2H_5$ monomers using plasma treatment and CVD polymerization. What are the advantages and disadvantages of each method? Do the films formed by these two methods have the same chemical structure? Explain.

13.2 Langmuir–Blodgett films of stearic acid deposited on a hydrophilic surface normally consist of an odd number of layers whereas films deposited on a hydrophobic surface normally consist of an even number of layers. With the aid of appropriate illustrations, explain why.

13.3 What are self-assembled monolayers (SAMs)? Explain what conditions are necessary for the formation of SAMs.

13.4 The surface of a biomaterial contains hydroxyl groups but it is required to modify the surface to contain carboxyl groups instead. Explain in general terms how this can be done by forming stable SAMs, giving general information about the chemical agents and methods.

13.5 Explain any advantages and drawbacks that SAMs may have over Langmuir–Blodgett films.

13.6 Compare the advantages and drawbacks of LbL deposition and the Langmuir–Blodgett technique for depositing thin films on biomaterials.

13.7 Describe in general terms a method by which the LbL technique may be used to deposit films on biomaterials for the delivery of biomolecules such as growth factors.

13.8 Discuss the significance of end-point immobilization of heparin to polymers (that is, covalent bonding at one end of the heparin molecule to the polymer) for reducing the thrombogenicity of the polymer.

13.9 Discuss the benefits of heparin modification to biomaterials in delivering growth factors.

References

Lau, K.K.S. and Gleason, K.K. (2006). Initiated chemical vapor deposition (iCVD) of poly(alkyl acrylates): an experimental study. *Macromolecules* 39: 3688–3694.

Ratner, B.D. and Hoffman, A.S. (2013). Physicochemical surface modification of materials used in medicine. In: *Biomaterials Science: An Introduction to Materials in Medicine*, 3rde (eds. B.D. Ratner, A.S. Hoffman, F.J. Schoen and J.E. Lemons), 259–276. New York: Elsevier.

Sakiyama-Elbert, S.E. and Hubbell, J.A. (2000). Development of fibrin derivatives for controlled release of heparin-binding growth factors. *Journal of Controlled Release* 65: 389–402.

Shah, N.J., Macdonald, M.L., Beben, Y.M. et al. (2011). Tunable dual growth factor delivery from polyelectrolyte multilayer films. *Biomaterials* 32: 6183–6193.

Further Reading

Chu, P.K., Chen, J.Y., Wang, L.P., and Huang, N. (2002). Plasma-surface modification of biomaterials. *Materials Science and Engineering R* 36: 143–206.

Decher, G. (1997). Fuzzy nanoassemblies: toward layered polymeric multicomposites. *Science* 277: 1232–1237.

Hann, R.A. (1990). Molecular structure and monolayer properties. In: *Langmuir-Blodgett Films* (ed. G. Edwards), 17–92. New York; Chapter 2: Springer Science + Business Media LLC.

Hoffman, A.S. and Hubbell, J.A. (2013). Surface immobilized biomolecules. In: *Biomaterials Science: An Introduction to Materials in Medicine*, 3rde (eds. B.D. Ratner, A.S. Hoffman, F.J. Schoen and J.E. Lemons), 339–349. New York: Elsevier.

Macdonald, M.L., Samuel, R.E., Shah, N.J. et al. (2011). Tissue integration of growth factor-eluting layer-by-layer polyelectrolyte multilayer coated implants. *Biomaterials* 32: 1446–1453.

Masters, K. (2011). Covalent growth factor immobilization strategies for tissue repair and regeneration. *Macromolecular Bioscience* 11: 1149–1163.

Park, J.Y. and Advincula, J.C. (2011). Nanostructuring polymers, colloids, and nanomaterials at the air–water interface through Langmuir and Langmuir–Blodgett techniques. *Soft Matter* 7: 9829–9843.

Sakiyama-Elbert, S.E. (2014). Incorporation of heparin into biomaterials. *Acta Biomaterialia* 10: 1581–1587.

Tenhaeff, W.E. and Gleason, K.K. (2008). Initiated and oxidative chemical vapor deposition of polymeric thin films: iCVD and oCVD. *Advanced Functional Materials* 18: 979–992.

Part IV

Degradation of Biomaterials in the Physiological Environment

14

Degradation of Metallic and Ceramic Biomaterials

14.1 Introduction

Biomaterials, as we have seen from previous chapters, are normally designed to be intentionally degradable, that is, to have a controllable degradation rate, or to be resistant to degradation in the aqueous physiological environment. In this book, we have used the term nondegradable to describe biomaterials that are designed to be resistant to degradation. However, no material is completely resistant to degradation. As defined in Chapter 1, the term nondegradable (or resistant to degradation) means that the degradation rate is sufficiently slow such that it does not affect the performance of the biomaterial over the lifetime of its application.

Regardless of whether it is designed to be degradable or resistant to degradation, an implanted biomaterial should show adequate biocompatibility. This means that for long-term (permanently) implantable devices, the biomaterial should not degrade in the physiological environment and it should not elicit any adverse tissue response. In comparison, biomaterials designed to be intentionally degradable for use in short-term implantable devices should have a physiologically tolerable degradation rate and their degradation products should be soluble, nontoxic, and capable of being eliminated from the body by various catabolic pathways and physiological clearance mechanisms.

When compared to an isotonic solution (that is, an aqueous solution having the same ionic concentration as the general physiological fluid) at pH ~7.4 and 37 °C, the specific physiological environment into which biomaterials are implanted is considerably more corrosive, even to those biomaterials designed to be resistant to degradation. The general physiological environment contains a variety of inorganic ions and biomolecules, but the action of cells, particularly inflammatory cells, can change the local chemistry of the implantation site and produce strong oxidizing agents, all of which can strongly enhance its corrosive nature.

Although normally slow, degradation can adversely affect the properties and performance of biomaterials designed to be resistant to degradation, particularly those biomaterials designed for use in long-term implantable devices. For example, degradation over time, particularly in conjunction with physiological stress, can lead to a reduction in the mechanical properties of the biomaterial, loss of its integrity and, eventually, failure of the implanted device, which can have disastrous consequences *in vivo*. Often of much greater significance is the release of degradation products from the implanted biomaterial, which can have a significant and controlling effect on host response, that is, the response of cells and tissues. For example, the release of certain ions from metallic implants, such as nickel, cobalt, and chromium ions, even in very small concentrations, can lead to adverse biological effects such as toxicity, metal ion hypersensitivity (Chapter 19), and carcinogenesis, while particulate degradation products can elicit an adverse foreign body response (Chapter 18). Overall, then, it is important to understand the mechanisms of degradation and how they are influenced by the physiological environment in order to avoid or reduce any adverse biological effects of degradation.

Degradation of metals in an aqueous environment is generally referred to as corrosion. Corrosion is an electrochemical process, which involves atoms releasing electrons within the metal and forming ions that dissolve into the aqueous environment, and conduction of the electrons to another region where they are consumed. Although some degradable metal alloys such as magnesium alloys are being studied for use in biomedical applications, metals used as biomaterials have normally been designed to be resistant to corrosion. As various aspects in the design of degradable metals were discussed in Chapter 6, this chapter will focus mainly on the corrosion of metallic biomaterials designed to be resistant to degradation *in vivo*.

Ceramics used as implanted devices are designed to be nondegradable, degradable, or bioactive (Chapter 7). Overall, nondegradable ceramics such as alumina (Al_2O_3) designed for use in long-term implantable devices have a greater resistance to degradation when compared to the common metallic biomaterials because they are already in

Materials for Biomedical Engineering: Fundamentals and Applications, First Edition. Mohamed N. Rahaman and Roger F. Brown.

Companion website: www.wiley.com/go/rahaman/biomedicalengineering

an oxidized state and generally have stronger atomic bonds. On the other hand, a few bioceramics, particularly the calcium phosphate bioceramics, are designed to have some controllable degradation rate in the physiological environment. These calcium phosphate ceramics, such as β-tricalcium phosphate (β-TCP), degrade mainly by dissolution to form nontoxic calcium and phosphate ions that are resorbed into the body. However, additional mechanisms can contribute to their degradation, such as disintegration into particles, which can lead to adverse tissue responses. As the physicochemical factors that influence the solubility of these calcium phosphate bioceramics were discussed in Chapter 7, the focus of this chapter is mainly on these additional mechanisms that may contribute to their overall degradation *in vivo*.

Overall, this chapter will focus mainly on the following topics:

- Principles of metal corrosion
- Effect of the physiological environment on the corrosion of metallic biomaterials designed to be corrosion-resistant in long-term implantable devices
- Types of metal corrosion and methods for reducing or avoiding them
- Effect of the physiological environment on the degradation of calcium phosphate bioceramics.

14.2 Corrosion of Metals

Metals have been used for decades as long-term implantable devices in a variety of biomedical applications (Chapter 6), and the mechanisms by which they degrade and interact with the physiological environment are well known and understood. However, knowledge of these mechanisms and the way in which the corrosive physiological environment influences them is important in avoiding adverse biological effects when new metallic biomaterials are introduced into new or existing applications.

14.2.1 Principles of Metal Corrosion

Corrosion of metals in an aqueous medium, as noted earlier, is an electrochemical process involving release of electrons from atoms in the metal, forming ions that dissolve into the medium. The chemical potential (or, more correctly, the electrochemical potential) μ_i of an ion i in solution that experiences an electric potential E is given by

$$\mu_i = \mu_i^o + RT \ln a_i + n_i FE \quad (14.1)$$

where a_i is the activity of the ion i of charge n_i in the electric field, R is the gas constant, T is the absolute temperature, F is the Faraday constant, and μ_i^o is the chemical potential of the ion i at standard state ($a_i = 1$). Now let us consider the case of a pure metal M in contact with an aqueous solution of its ions M^{n+} (Figure 14.1). This system forms a half-cell, in which the reaction

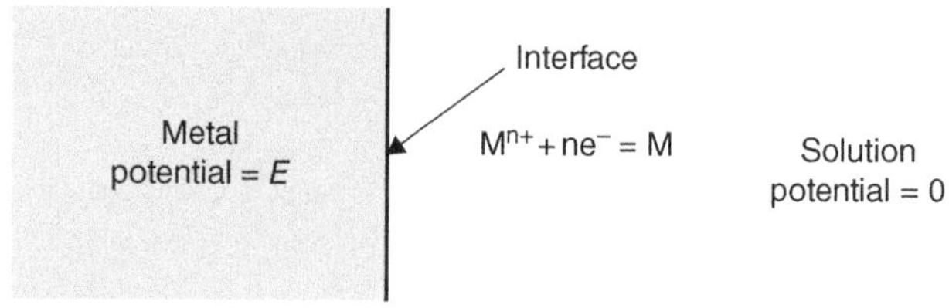

Figure 14.1 Schematic of a metal in contact with a solution of its ions.

$$M^{n+} + ne^- \rightarrow M \quad (14.2)$$

can occur at the interface between the metal and the solution. This will generate an electrostatic potential difference between the metal and the solution, even when the reaction given by Eq. (14.2) is in equilibrium.

The equilibrium potential difference between the metal and the solution can be calculated, and since we are interested in a difference in potential, we can take the potential in the solution further from a short distance from the metal surface as zero. As metals are good conductors, they tend to have uniform potentials, and we can assume that the metal has a uniform potential E right up to its interface with the solution. Following Eq. (14.1), the electrochemical potential of an ion at the interface is

$$\mu_i = \mu_i^o + RT \ln a_i^s + n_i FE \quad (14.3)$$

where a_i^s is the activity of the ion at the interface, which is different from a_i, the activity in the solution. As the potential in the bulk of the solution is taken to be zero, the electrochemical potential of an ion some distance from the metal surface is

$$\mu_i = \mu_i^o + RT \ln a_i \quad (14.4)$$

At equilibrium, the electrochemical potential of an ion at the interface is equal to that in the bulk of the solution. Thus, equating μ_i in Eqs. (14.3) and (14.4), we get

$$n_i FE = RT \ln a_i - RT \ln a_i^s \quad (14.5)$$

The activity a_i^s can be determined from the reaction given by Eq. (14.2) at equilibrium, using the equation for the equilibrium constant of the reaction, that is:

$$\frac{a_M}{a_{M^{n+}}^s (a_{e^-})^n} = K_a = \exp\left(-\frac{\Delta G_o}{RT}\right) \quad (14.6)$$

where the quantities on the left-hand side of Eq. (14.6) are the activities of the respective species in Eq. (14.2), and K_a is the equilibrium constant and ΔG_o is the Gibbs free energy change, respectively, for the reaction.

As the metal is pure, its activity a_M is equal to 1. The activity of the electrons is that for the electrons in the metal at the metal–solution interface. The principle of electroneutrality requires that the number of electrons must balance the charges on the nuclei of the metal atoms. This means that the activity of the electrons is constant. Thus, from Eq. (14.6), the activity of the ions at the interface is

$$a^{s}_{M^{n+}} = C\exp\left(\frac{\Delta G_o}{RT}\right) \tag{14.7}$$

where C is a constant. Substituting in Eq. (14.5), we get

$$n_i FE = RT\ln a_i - RT\ln C - \Delta G_o \tag{14.8}$$

Putting $E = E_o$ when $a_i = 1$ in Eq. (14.8) gives

$$n_i FE_o = -RT\ln C - \Delta G_o \tag{14.9}$$

and thus

$$E = E_o + \frac{RT}{n_i F}\ln a_i \tag{14.10}$$

Equation (14.10) is known as the Nernst equation, an equation that finds considerable use in metal corrosion studies. The potential difference E_o between the metal and the solution that contains unit activity of metal ions is called the standard electrode potential of the metal. This standard state is normally taken as a one-molar (1 M) solution of the relevant electrochemical species (such as the relevant metal ions) at 25 °C. As can be seen from an examination of Eq. (14.10), a_i is actually the activity of a solution relative to the standard state (1 M) and, thus, has units of mole per liter (M).

Example 14.1 Determine the electrode potential (with respect to the standard hydrogen electrode) for a copper electrode in a solution containing 2.0 g of Cu^{2+} ions per liter at 25 °C.

Solution:
Since the atomic weight of copper is 63.55 g, the concentration of Cu^{2+} ions in the solution is (2.0/63.55) = 0.03 M. Assuming that the activity of the Cu^{2+} ions is equal to their concentration, using the Nernst equation (Eq. (14.10)), we get

$$E = 0.34 + \frac{R(298)}{2F}\ln 0.03$$

where the standard electrode potential E_o for Cu is 0.34 V. Using values for the gas constant R = 8.314 J/K and the Faraday constant $F = 9.6485\times10^4$ C/mol, we get E = 0.295 V.

Measuring the standard electrode potential between a metal and a solution presents considerable difficulties and, consequently, a standard hydrogen electrode is normally used as a reference to which all electrode potentials are referred (Figure 14.2). The standard hydrogen electrode has been assigned a potential of zero, which means that the standard electrode potentials are not absolute values but, instead, they are values relative to that of the standard hydrogen electrode. This standard (reference) electrode consists of a platinum foil immersed in a solution of unit activity of H^+ ions (pH = 0) and into which H_2 gas at 1 atm is bubbled so that the bubbles contact the platinum electrode, bringing about the reaction

$$H_2 \rightarrow 2H^+ + 2e^- \tag{14.11}$$

The platinum serves to catalyze the reaction in Eq. (14.11) and to permit an electrical connection to a measuring device such as a voltmeter.

Standard electrode potentials for several pure metals are given in Table 14.1. These values conform to the generally accepted convention in metal corrosion, but some older textbooks may use a different convention in which the

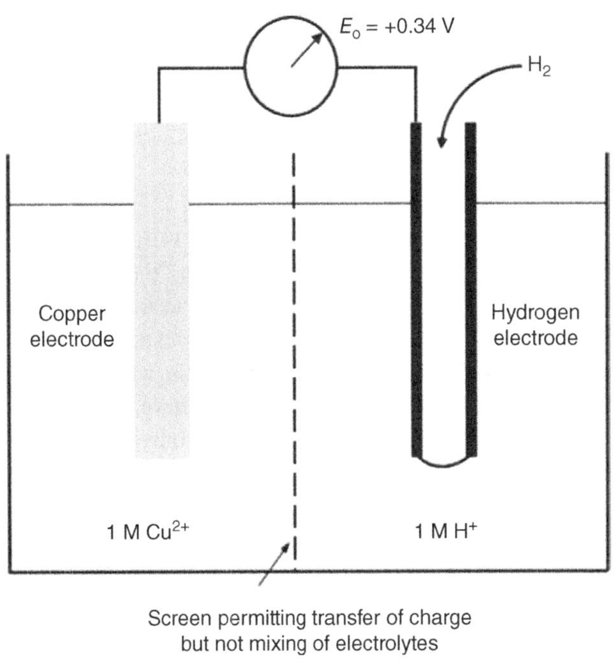

Figure 14.2 Potential difference between a copper electrode and a standard hydrogen electrode. Because copper produces a higher electrode potential than the hydrogen electrode, copper is the cathode and hydrogen the anode.

Table 14.1 Standard electrode potentials (E_o) at 25 °C and 1 atm.[a]

Half-cell reaction	E_o (V)
$Au^3 + 3e^- = Au$	1.50
$O_2 + 4H^+ + 4e^- = 2H_2O$	1.23
$Hg^{2+} + 2e^- = Hg$	0.80
$Ag^+ + e^- = Ag$	0.80
$Cu^+ + e^- = Cu$	0.52
$O_2 + 2H_2O + 4e^- = 4OH^-$	0.40
$Cu^{2+} + 2e^- = Cu$	0.34
$H^+ + e^- = ½H_2$	0.00
$Pb^{2+} + 2e^- = Pb$	−0.13
$Sn^{2+} + 2e^- = Sn$	−0.14
$Ni^{2+} + 2e^- = Ni$	−0.23
$Co^{2+} + 2e^- = Co$	−0.27
$Cd^{2+} + 2e^- = Cd$	−0.40
$Fe^{2+} + 2e^- = Fe$	−0.44
$Cr^3 + 3e^- = Cr$	−0.74
$Zn^{2+} + 2e^- = Zn$	−0.76
$Cr^{2+} + 2e^- = Cr$	−0.91
$Mn^{2+} + 2e^- = Mn$	−1.05
$Al^3 + 3e^- = Al$	−1.66
$Ti^{2+} + 2e^- = Ti$	−1.75
$Mg^{2+} + 2e^- = Mg$	−2.38
$Na^+ + e^- = Na$	−2.71
$Ca^{2+} + 2e^- = Ca$	−2.84
$K^+ + e^- = K$	−2.92
$Li^+ + e^- = Li$	−3.01

[a] The standard solution is 1 M of the appropriate ion.

values are the negative of those in Table 14.1. These electrode potentials provide a measure of the tendency of a pure metal to corrode, that is, for atoms to lose electrons and form ions that dissolve into the aqueous medium. For example, the alkali and alkali earth metals, which release their outer-shell electrons rather readily, have a high electrode potential at one end of the scale. In comparison, gold, platinum, and silver that release their electrons less readily are at the opposite end of the scale.

We can now go on to examine what happens when two different metals are placed in the same solution and connected by an external conductor such as a metal wire. As an example, Figure 14.3 shows a copper (Cu) electrode and a zinc (Zn) electrode in a solution containing Cu and Zn ions at unit activity. In this case, the electrode potentials will be the standard values, and thus the potential of Cu minus the potential of Zn = (+0.34)−(−0.76) V = +1.1 V. Thus, an electrical current will flow from the electrode having the more positive potential (Cu) to the electrode of lower potential. This corresponds to a flow of electrons in the opposite direction. Electrochemical reactions take place at the interface of each metal with the solution. The Zn electrode will generate electrons, represented by equation

$$Zn \rightarrow Zn^{2+} + 2e^- \quad (14.12)$$

These electrons will flow through the external conductor, react with Cu ions in the solution, and lead to the deposition of Cu atoms at the Cu–solution interface:

$$Cu^{2+} + 2e^- \rightarrow Cu \quad (14.13)$$

With electrons being removed from the Zn electrode, the reaction described by Eq. (14.12) goes to the right. Consequently, more Zn ions go into solution, and the Zn electrode is corroded. Meanwhile, the electrons supplied by this reaction react with Cu ions in the solution, leading to deposition of more Cu atoms on the Cu electrode (Eq. (14.13)).

The system in Figure 14.3 is an example of a galvanic cell (or battery), that is, a cell that produces electricity. The Cu electrode is the cathode (positive electrode), whereas the Zn electrode forms the anode (negative electrode). Reduction of an ion, as achieved electrochemically at a cathode, corresponds to adding an electron, whereas oxidation at an anode corresponds to the removal of an electron from an atom or ion. Overall, then:

- A pure metal having a higher standard electrode potential has a lower tendency to corrode in an aqueous

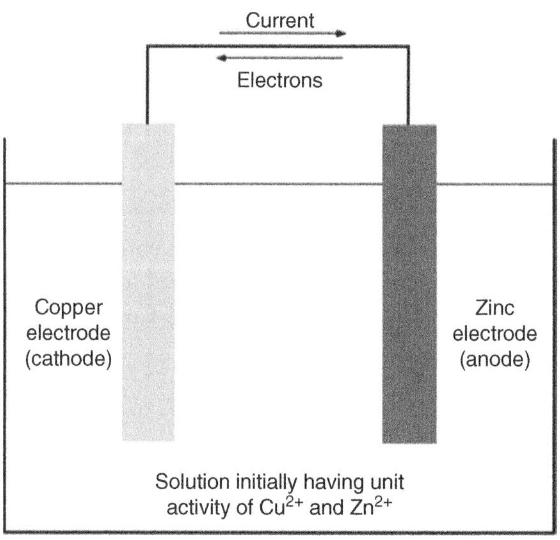

Figure 14.3 Schematic of two electrically connected dissimilar metals in contact with a solution of their ions, forming a galvanic cell. Because copper has the higher standard electrode potential, it forms the cathode whereas zinc of lower electrode potential is the anode.

solution when compared to a metal of lower electrode potential

- The metal having the higher standard electrode potential forms the cathode of a galvanic cell, whereas the metal of lower electrode potential forms the anode
- Corrosion occurs by the coupling of two reactions, an oxidation reaction at the anode (removal of an electrons from an atom or ion) (Eq. (14.12)), and a reduction reaction at the cathode (adding an electron to an ion) (Eq. (14.13))
- Corrosion occurs at only one electrode (the anode) of a galvanic cell.

Cathode Reactions

The oxidation reaction, exemplified by Eq. (14.12), and written in general for a pure metal M as

$$M \rightarrow M^{n+} + ne^- \quad (14.14)$$

is the prime anode reaction, and as we have seen in the previous section, an electron-consuming (cathodic) reaction always accompanies this electron-generating (anodic) reaction. A variety of cathodic reactions is possible and the precise reaction that dominates depends on the conditions present in the electrolyte solution, such as the pH and the availability of oxygen. In addition to the reaction given by Eq. (14.2) (sometimes described as an electroplating reaction), the major cathodic reactions are

Hydrogen evolution (acidic solution)

$$2H^+ + 2e^- \rightarrow H_2 \quad (14.15)$$

Hydrogen evolution (neutral or basic solution)

$$2H_2O + 2e^- \rightarrow 2OH^- + H_2 \quad (14.16)$$

Oxygen reduction (acidic solution)

$$O_2 + 4H^+ + 4e^- \rightarrow 2H_2O \quad (14.17)$$

Oxygen reduction (neutral or basic solution)

$$O_2 + 2H_2O + 4e^- \rightarrow 4OH^- \quad (14.18)$$

Each of these reactions consumes electrons. Equations (14.15) and (14.16) lead to the evolution of H_2 gas and are encountered mainly in anaerobic conditions. In comparison, Eqs. (14.17) and (14.18) both require the presence of oxygen, with Eq. (14.17) dominating at low pH whereas Eq. (14.18) dominating at basic or neutral pH.

Example of Metal Corrosion: Iron

As an example of wet corrosion of metals, let us consider the corrosion of iron (Fe) in aerated (oxygen-enriched) water, which is well-known as the rusting of iron. Pure Fe has a standard electrode potential of −0.44 V and, thus, there is a reasonable tendency for Fe atoms to release electrons, forming Fe^{2+} ions (or Fe^{3+}) ions that pass into solution (the anodic reaction) (Figure 14.4). The electrons are conducted through the metal to a place where the cathodic reaction (Eq. (14.18)) can take place to consume them. This reaction generates hydroxyl (OH^-) ions that react with the Fe^{2+} ions to form a hydrated iron oxide $Fe(OH)_2$ (also written $FeO{\cdot}H_2O$). This reaction product (rust) is insoluble in water and, instead of forming on the surface of the Fe, precipitates in the water. Rusting will occur in air even if there is only a small amount of adsorbed moisture from the air.

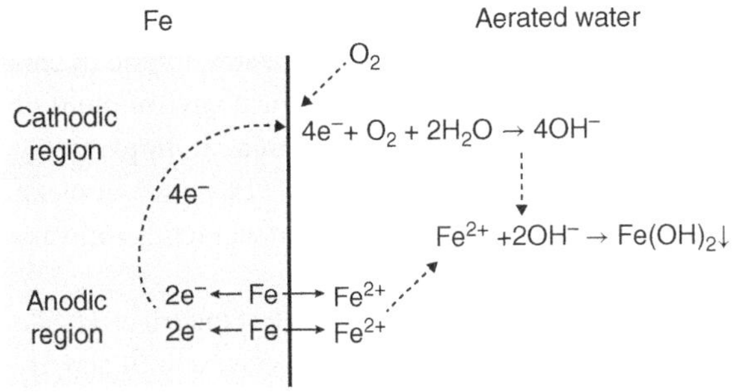

Figure 14.4 Illustration of rust formation. The oxidation reaction in the anodic region produces Fe^{2+} ions and electrons (e^-), while combination of these electrons with O_2 and H_2O in the cathodic region forms OH^- ions, which combine with Fe^{2+} ions to form a hydrated iron oxide $Fe(OH)_2$ precipitate.

14.2.2 Rate of Corrosion

The standard electrode potentials in Table 14.1 are useful as a guide to the tendency of a metal to corrode in an aqueous environment and as a basis for understanding the principles of corrosion. Normally, the rates of wet corrosion found in practice bear little relationship to these electrode potentials, assuming that the metal is prone to corrosion in the first place, and they may often be the reverse of the order expected from the standard electrode potentials. For example, titanium (Ti) has a standard electrode potential of −1.75 V, but in clean water or in the physiological environment, commercial purity Ti used in biomedical applications is one of the most corrosion-resistant metals (Chapter 6). This is because Ti is passivated against corrosion by a thin (~5–10 nm) protective layer of titanium dioxide (TiO_2) that forms almost immediately on the surface of Ti upon its exposure to an oxidizing medium such as air. This TiO_2 surface layer is dense (impervious) and strongly adherent, making Ti strongly resistant to corrosion. Another example is aluminum (Al), which is passivated against aqueous corrosion in clean water in a pH range of ~5–8 by a thin hydrated Al_2O_3

surface layer. On the other hand, this oxide layer is susceptible to attack by (Cl^-) ions, making Al susceptible to corrosion in media such as the physiological environment or seawater. Consequently, Al is not used in implantable devices despite its excellent corrosion resistance in clean water. Overall, the corrosion rates of most metals can vary widely in practical environments.

Another point to remember about the standard electrode potentials is that they are for solutions having unit activity (1 M concentration) of the specific metal ions. Such a concentration is high and is not regularly encountered in practice. It also makes it difficult for the metal to dissolve because its ions will suffer a large repulsion from its ions already present in the solution. Metals can corrode more easily in dilute solutions and, according to Eq. (14.10), a more dilute solution will lower the electrode potential from the standard value.

14.2.3 Pourbaix Diagrams

Pourbaix diagrams see considerable use in the corrosion of metals, well known to be a significant and costly problem in engineering and industrial applications. Pourbaix diagrams consist of plots of electrical potential E versus pH that map out regions of possible stability and corrosion of aqueous electrochemical systems (normally at 25 °C). They are calculated using the Nernst equation (Eq. (14.10)) and thermodynamic data such as the Gibbs free energy change for possible reactions in the system. As such, Pourbaix diagrams show the possible phases present at equilibrium, and they do not take into account the kinetics of the process. The role of Pourbaix diagrams in electrochemical corrosion may be viewed as being analogous to that of phase diagrams for metals and ceramics, which show the various phases of a system stable thermodynamically under appropriate conditions of temperature and composition (see, for example, Figure 6.8).

Figure 14.5 shows an example of a Pourbaix diagram for iron (Fe). Although Fe is not a useful biomaterial, the diagram illustrates basic features of many Pourbaix diagrams, in addition to Fe being a widely studied system for which reliable thermodynamic data are available. The two sloping dashed lines in the diagram have special significance in that they represent the stability limits for water. The upper dashed line represents the oxidation of water, according to

$$H_2O \rightarrow 2H^+ + \tfrac{1}{2}O_2 + 2e^- \qquad (14.19)$$

while at the bottom dashed line, reduction of water proceeds according to

$$2H_2O + 2e^- \rightarrow 2OH^- + H_2 \qquad (14.20)$$

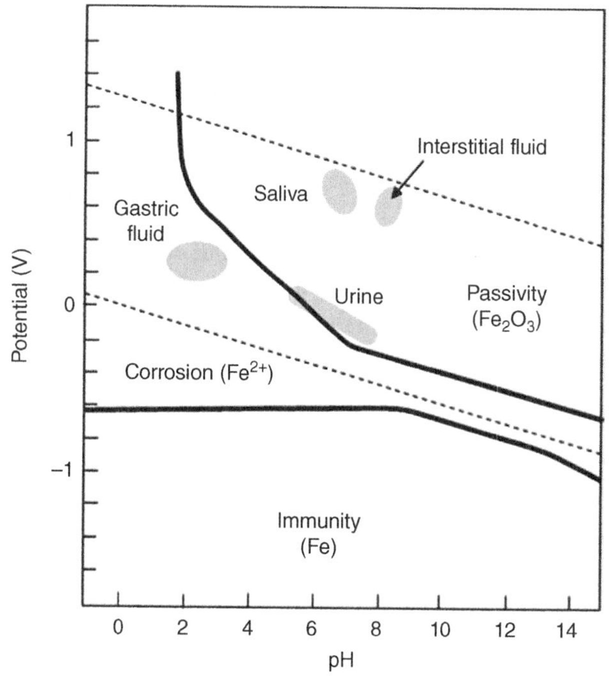

Figure 14.5 Example of a Pourbaix diagram for iron (Fe), assuming a passivating layer composed of Fe_2O_3, showing regions of immunity, corrosion, and passivation. Areas corresponding to some physiological environments, such as saliva, interstitial fluid, gastric fluid, and urine, are also shown. *Source:* Pourbaix (1984) / with permission of Elsevier.

Thermodynamically, the main part of the diagram for aqueous corrosion lies between these two dashed lines. The E value in a solution would be at the upper line if the solution were in equilibrium with pure O_2 at 1 atm, and at the lower line if the solution were in equilibrium with pure H_2 at 1 atm.

The principal regions in Figure 14.5 are the possible regions where Fe is immune to corrosion, susceptible to corrosion, or passivated against corrosion, based on certain assumptions. The voltage E is calculated using the Nernst equation for the equilibrium reaction

$$Fe \rightleftarrows Fe^{2+} + 2e^- \qquad (14.21)$$

giving the expression

$$E = E_o + \frac{RT}{2F}\ln\left[Fe^{2+}\right] \qquad (14.22)$$

where E_o is the standard electrode potential for Fe (−0.44 V) and the activity of Fe^{2+} is taken as its concentration in the solution. Various types of reactions are possible in this system but an example of one that involves H^+ ions or pH is the precipitation of a hydrous oxide from an aqueous solution containing Fe^{2+}:

$$Fe^{2+} + 2H_2O \rightarrow Fe(OH)_2 + 2H^+ \qquad (14.23)$$

As the activities of H_2O and $Fe(OH)_2$ are unity, the free energy change for this reaction is

$$\Delta G_o = -RT \ln K_a = \frac{\left[H^+\right]^2}{\left[Fe^{2+}\right]} \tag{14.24}$$

By calculating ΔG_o from thermodynamic data for the reaction in Eq. (14.23), E versus pH can be determined for this system.

Construction of the Pourbaix diagram requires additional assumptions. For example, the region of corrosion is normally defined as the region where the metal ion concentration $[M^{n+}]$ in the solution is $>10^{-6}$ M, while the region of immunity, where the metal is thermodynamically immune from corrosion, is defined as the region where $[M^{n+}]$ is $<10^{-6}$ M. In the region of passivity, the metal is protected from corrosion by a stable oxide or hydroxide surface layer. However, corrosion is possible in this region if this protective surface layer is disrupted or damaged.

As the foregoing discussion shows, construction of a Pourbaix diagram requires a variety of assumptions, depends on the availability of the appropriate thermodynamic data for the specific system, and relies on the ability to identify the correct phases that may be present. Additionally, the diagrams represent equilibrium conditions and do not consider the kinetics of corrosion. Consequently, Pourbaix diagrams should be used only as a guide.

Despite their limitations, Pourbaix diagrams are useful in visualizing the potential for a metal to corrode under a variety of conditions in a single map. For example, as shown in Figure 14.5, different areas of the physiological environment have different average pH values. Thus, an important point to note is that a metal that is passive or immune to corrosion and performs well in one region of the body may undergo an unacceptable level of corrosion in another region.

14.2.4 Types of Electrochemical Corrosion

Galvanic corrosion may be categorized into various types, depending on the factors that cause corrosion, such as differences in composition, mechanical stress, and electrolyte concentration. It is important to recognize how these factors cause corrosion in order to avoid or reduce their effects. Corrosion occurs because one part of the electrochemical system acts as an anode while another part acts as a cathode. As we have seen, only the anode is corroded, and only when it has some type of electrical contact with the cathode. If only the anode were present, corrosion would quickly stop because buildup of negative charge in the anode would lead to electrostatic attraction of the positive ions, preventing further release into the aqueous environment.

Compositional Effects

As noted earlier, galvanic corrosion occurs when two dissimilar metals (alloys) are present in an electrolyte, once there is some means of electrical connection between them. The metal of lower standard electrical potential acts as the anode and corrodes. Corrosion occurs in a uniform manner, with the thickness of the metal decreasing over its entire surface area. Thus, the use of dissimilar metals in the same device, such as dissimilar metals for the screws and the plate of an orthopedic fracture fixation device, should be avoided.

It is not necessary for the metals to be macroscopic monolithic components for galvanic corrosion to occur. Galvanic corrosion can also occur in a single metal if it contains different microstructural features, such as phases of different composition and structure. In this case, corrosion normally occurs selectively or locally at a microscopic scale. The term microgalvanic corrosion has sometimes been used to distinguish this type of corrosion from that of more uniform corrosion of a macroscopic monolithic metal. A well-known example of a system susceptible to microgalvanic corrosion is a carbon steel composed of cementite and pearlite phases (Chapter 6). The ferrite forms the anode and corrodes while the perlite serves as the cathode (Figure 14.6a).

Stainless steels provide another well-known example of microgalvanic corrosion. When the stainless steel contains more than ~0.05 wt.% C, chromium carbide ($Cr_{23}C_6$) particles may precipitate at the grain boundaries under certain processing heat treatments. This leads to a depletion of Cr in the grain boundary region, which serves as the anode of a microgalvanic cell and corrodes (Figure 14.6b), resulting in selective corrosion at the grain boundaries. The release of Cr ions and other ions, such as Ni ions, as noted earlier, would present a serious carcinogenic hazard. Consequently, austenitic stainless steels used as biomaterials, such as the steel designated 316L, contain <0.03 wt.% C to avoid precipitation of $Cr_{23}C_6$ and, thus, to avoid this type of selective corrosion (Chapter 6).

Corrosion can also occur at breaks in a passivating oxide layer, caused by abrasion for example (Figure 14.6c), or at precipitates in certain alloys (Figure 14.6d). These two types of selective corrosion are referred to as pitting corrosion because as the term indicates, they lead to the formation of tiny pits at the surface of the metal.

Mechanical Stress Effects

Mechanical stress can lead to selective or localized corrosion, instead of uniform corrosion of the metal. For example, atoms at the grain boundaries of a metal have a higher energy when compared to atoms within the grains, which is equivalent to these atoms experiencing a higher mechanical

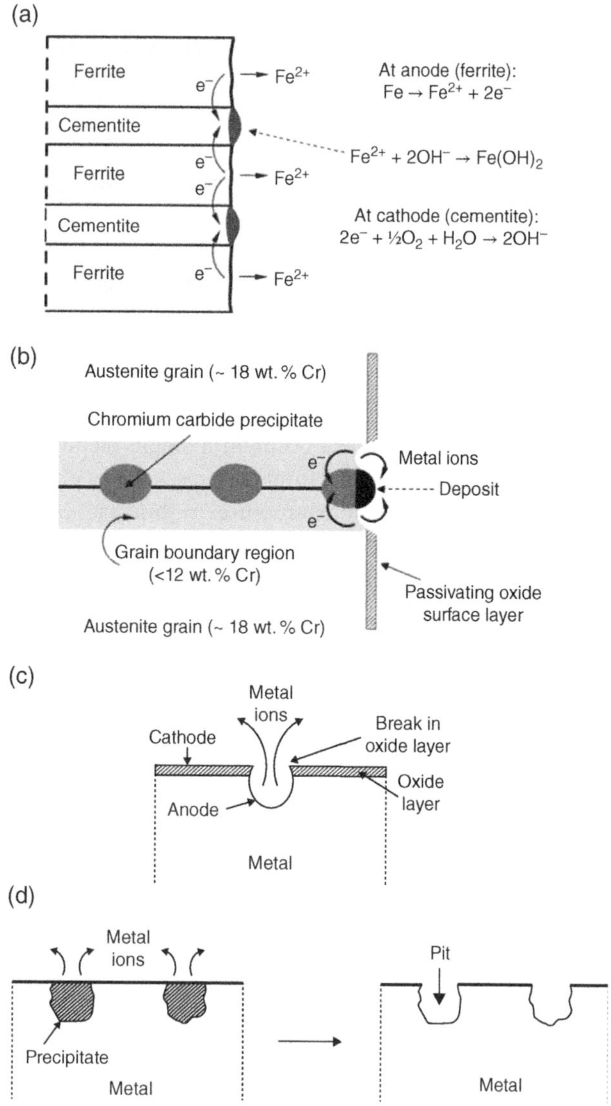

Figure 14.6 Examples of galvanic corrosion at a microscale due to compositional differences. (a) In carbon steels, ferrite is anodic to cementite; (b) precipitation of chromium carbide at the grain boundary of austenitic stainless steels depletes the chromium in the grain boundary region, making it anodic; (c) pitting corrosion due to a break in the passivating surface oxide layer; (d) pitting corrosion due to precipitated compounds in certain alloys.

stress (even without the application of an external stress). This leads to selective corrosion at the grain boundaries because they form the anode an electrochemical system (Figure 14.7a). Another example relates to local internal stresses introduced, for example, by cold working at the bend of an otherwise annealed wire, which cause the bent region to act as an anode (Figure 14.7b).

A type of corrosion that may be considered within this group is fretting corrosion, which occurs when two contacting surfaces undergo repeated cyclic motion of small amplitude (~10–100 μm) in an aqueous medium. Devices such as metal fracture fixation plates attached to bone by screws are susceptible to this type of corrosion because the screws cannot be tightened completely. Artificial hip joints are another example of devices that may suffer from fretting corrosion. The femoral head (Co–Cr or stainless steel) in these devices is attached to the metal stem (Ti6Al4V) not by direct bonding but by mechanical impact. Consequently, they may undergo repeated cyclic motion during normal physiological activity of a patient, which can lead to fretting corrosion.

Selective corrosion, caused by grain boundary attack or by pitting corrosion, for example, together with an applied tensile or flexural stress is particularly serious because it may lead to failure more rapidly when compared to what might be inferred from uniform corrosion of the overall metal. Crack-like defects may grow steadily under a stress intensity K much lower than the critical stress intensity K_C, which can lead eventually to unexpected failure under applied stresses that are much lower than the yield stress of a metal (Chapter 4). Corrosion may also increase the rate at which fatigue cracks already present in a metal grow. The crack growth rate is often much larger than what would be expected from the sum of corrosion and fatigue acting alone.

Concentration Effects

According to the Nernst equation (Eq. (14.10)), an electrode in a dilute electrolyte containing its ions is anodic with respect to a similar electrode in a concentrated electrolyte. The electrode in the dilute solution will corrode more readily and serve as the anode, whereas the electrode in the concentrated electrolyte will serve as the cathode and is protected from corrosion. Consequently, corrosion is enhanced where the electrolyte concentration is lower. Corrosion caused by changes in electrolyte concentration is not common in the physiological environment because fluid flow tends to equalize differences in concentration.

In comparison, corrosion caused by differences in oxygen concentration is normally more important in the physiological environment. Surprisingly, the most marked corrosion occurs in the part of the system in which the oxygen concentration is lower and not where it is higher. This may be rationalized in terms of the reactions that occur at the cathode surface where electrons are consumed, specifically Eq. (14.17) or (14.18) because each equation involves O_2. Since the cathodic reaction consumes electrons, more electrons must be supplied by adjacent areas that have a lower O_2 concentration. Thus, the area of lower O_2 concentration serves as the anode. Overall, then, corrosion may be accelerated in areas of a metal that have limited accessibility to O_2, such as the tips

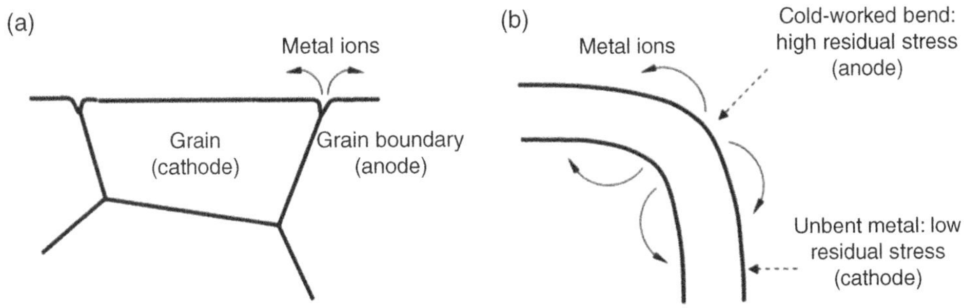

Figure 14.7 Examples of mechanical stress effects that can lead to corrosion. (a) As the atoms at the grain boundary have a higher energy when compared to those within the interior of the grain, the grain boundary is anodic and corrodes; (b) cold working required to bend a metal bar introduces high residual stresses at the bend, which is anodic and corrodes.

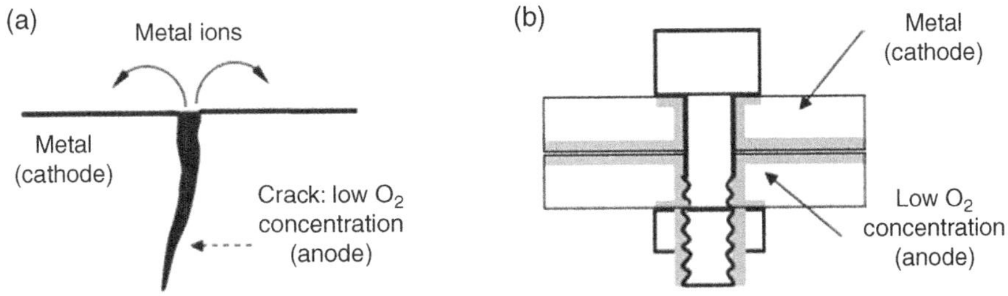

Figure 14.8 Examples of crevice corrosion due to low oxygen concentration. (a) The oxygen concentration within a crack is lower when compared to the surface of a metal, making the crack anodic; (b) fixation of one metal plate to another metal or to bone by means of screws has various regions of low oxygen content, which make them anodic.

of cracks or crevices (Figure 14.8). This type of corrosion is often referred to as crevice corrosion.

14.3 Corrosion of Metal Implants in the Physiological Environment

The physiological environment into which biomaterials are implanted, as noted earlier, is considerably more corrosive than an isotonic solution at pH ~7.4 and 37 °C, even to those biomaterials designed to be resistant to degradation. Although the corrosion of long-term implantable metallic devices is very slow, the chronic low-level release of certain ions is known to cause a variety of adverse systemic effects, such as carcinogenic, immunogenic, and toxic effects. The ions of Cr, Ni, and Co, metals that are components of stainless steels and cobalt–chromium (Co–Cr) alloys, are extremely carcinogenic. It is therefore important to understand how the corrosion of metallic implants is influenced by the physiological environment in order to minimize these adverse biological effects.

The physiological environment consists of an aqueous medium containing a variety of inorganic ions and biomolecules, and a variable concentration of dissolved O_2 gas. Additionally, implantation of a biomaterial can lead to significant changes in local pH. These characteristics can modify the equilibrium of the electrochemical reactions that occur in a system, strongly influencing the rate of corrosion and, thus, the stability of a metallic implant *in vivo*. Plasma, the liquid portion of the blood is composed of ~90% by volume of water containing a variety of inorganic ions, ~8% by weight of high molecular weight proteins and lipids, and less than ~2% by weight of lower molecular weight biomolecules such as fatty acids. Chlorine (Cl^-), bicarbonate (HCO_3^-), phosphate (monohydrogen phosphate HPO_4^{2-} and dihydrogen phosphate $H_2PO_4^-$), and sulfate (SO_4^{2-}) ions are the most abundant anions present in plasma, while the most abundant cations are the alkali (Na^+ and K^+) and alkaline earth (Mg^{2+} and Ca^{2+}) ions (Table 14.2). The common biomolecules present in plasma are shown in Table 14.3. The interstitial fluid, the fluid surrounding all cells, has approximately the same ionic composition as plasma but only approximately one-sixth the amount of proteins. The interstitial fluid and plasma together make up the extracellular fluid.

The major anion constituent in the physiological environment, Cl^- ions, is corrosive to most metals. Based on its significantly higher concentration of Na^+ and Cl^- ions, the physiological environment approximates ionically a mild form of seawater whose corrosive effect on metals is well known. Cl^- ions also have a strong tendency to attack the passivating surface oxide layer on some metals in aqueous

Table 14.2 Ionic concentration of plasma.

Ion	Concentration (mM)
Cl^-	100–110
HCO_3^-	20–30
$HPO_4^{2-} + H_2PO_4^-$	1–2
SO_4^{2-}	0.5
Na^+	130–150
K^+	3.5–5.0
Ca^{2+}	2–3
Mg^{2+}	0.7–2.0

Table 14.3 Organic and protein constituents of plasma.

Constituent	Concentration (g/l)
Organic	
Amino acids	0.05–0.10
Cholesterol	1.0–1.3
Fatty acids	2.5–4.0
Glucose	0.8–1.0
Total lipids	4.5–6.0
Phospholipid	2.0–3.0
Urea	0.02–0.04
Protein	
Albumin	40–55
Fibrinogen	2–4
Fibronectin	0.02–0.03
Immunoglobulin G	6–16
Macroglobulin	1.5–4.5
Transferrin	2.0–3.5

media. A widely studied system is Al, which is passivated against corrosion in clean water by a thin layer (~5 nm) of Al_2O_3 (or a hydrated aluminum oxide). In aqueous media containing Cl^- ions, this oxide layer becomes unstable and degrades locally, causing local breakdown of the layer. This leads to pitting corrosion on surfaces and to crevice corrosion in occluded sites. Consequently, as noted earlier, Al is not used as a biomaterial for applications *in vivo*, despite its strong corrosion resistance in clean water. The mechanism by which Cl^- ions attack the surface oxide layer on Al is the subject of some controversy, but it is well established that Cl^- ions adsorb to the oxide surface and can become incorporated into the entire layer. As Al_2O_3 has an isoelectric point of ~8–9, adsorption at the physiological pH is promoted by electrostatic attraction between Cl^- ions and the positively charged Al_2O_3 surface. While Cl^- ions might not attack all metals (alloys) by the same mechanism, their potential to destabilize the passivating oxide layer on other metals should be recognized.

The partial pressure of O_2 in blood varies from ~100 mmHg for arterial blood to ~40 mmHg for venous blood, compared to a value of ~160 mmHg in atmospheric air. The thin oxide surface layer that passivates some metals against corrosion may become damaged under certain conditions due to abrasion or wear, for example. As re-passivation of this oxide layer relies on the availability of O_2, this lower O_2 partial pressure can make re-passivation more difficult.

Although the physiological environment has a reasonably constant temperature (~37 °C) and its pH is well buffered at ~7.4, there are differences in regional pH (Figure 14.5). Additionally, upon implantation of a device, inflammatory responses of tissues can lead to a significant reduction in local pH to as low as ~4–5. Bacterial infection may lead to an even wider pH change near the implant surface, from ~4 to ~9. Although they may occur over a short term (days to weeks), these changes in pH can modify electrochemical reactions and, thus, lead to local corrosion.

Proteins and cells in the physiological environment can influence the corrosion of metals to varying extent, which is normally dependent on the composition of the metal (alloy), while the mechanisms by which they influence corrosion are sometimes controversial. Cations have been shown to bond to proteins and it has been suggested that they are transported away from the implant surface. For example, because albumin, the main protein in blood (Table 14.3), has an isoelectric point of ~5.0, it has a negative charge at the physiological pH and, thus, can bond to cations by electrostatic attraction. Transportation of cations away from the implant surface will drive the anode reaction (Eq. (14.14)) to the right, causing further corrosion of the metal.

Although iron is not used as a biomaterial, the Pourbaix diagram in Figure 14.5 shows that the stability of the passivating oxide on a metal depends on the electrode potential and the pH. Proteins can affect the electrode potential due to their electron-carrying capability, whereas bacterial cells can change the local pH by generating acidic metabolic products, which can shift the stability region of the passivated metal.

Another effect of the physiological medium is that proteins and cells can adsorb onto the surface of the metal, which can limit the diffusion of O_2 to the surface. This lower availability of O_2, as noted earlier, can make re-passivation of a damaged surface oxide layer on the metal more difficult. In such a case, the metal below adsorbed

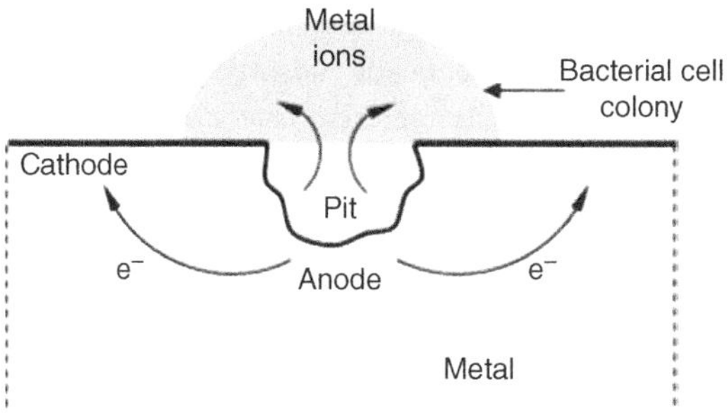

Figure 14.9 Illustration of the formation of a pit below a colony of bacterial cells due to a low oxygen concentration below the colony.

proteins or cells may undergo corrosion, as illustrated in Figure 14.9 for an adsorbed colony of bacterial cells.

14.3.1 Minimizing Metal Implant Corrosion *in vivo*

Corrosion of an implanted metal device designed to be resistant to corrosion, as noted earlier, can lead to serious adverse responses *in vivo*. As no metal is completely resistant to corrosion, it is worth recognizing some basic steps that can be taken to reduce the risk of unsafe corrosion:

- Select an appropriate metal (alloy) for the intended application, typically a noble metal or a passivated metal
- Minimize the number of metals and components in the implant design
- Recognize regional differences in the physiological environment and the effects that the physiological environment can have on corrosion
- Minimize the batches and types of processing that any one metal undergoes in producing the implant
- Minimize the existence of pits and crevices at the implant surface
- Avoid the transfer of a different metal from surgical tools to the implant or to surrounding tissue.

In some engineering and industrial applications, metals are protected from corrosion by various means of active intervention. One method is to connect a sacrificial metal of lower standard electrode potential to the metal. The sacrificial material becomes the anode of a galvanic cell (and corrodes) whereas the metal forms the cathode (and is protected). An example is the protection of underground steel pipes using a sacrificial magnesium-based alloy. However, this type of active corrosion control is essentially impractical for biomaterials used as implantable devices. The method is cumbersome but, more importantly, degradation of the sacrificial anode can lead to the release of toxic metal ions into the physiological environment or the release of metal ions at a physiologically intolerable rate.

14.4 Degradation of Ceramics

Ceramics used as biomaterials may be resistant to degradation (nondegradable), degradable, or bioactive (Chapter 7). Nondegradable oxide ceramics used as biomaterials, such as alumina (Al_2O_3) and zirconia (ZrO_2) are already in their oxidized state and are bonded by strong interatomic bonds. Consequently, they have a higher resistance to degradation in an aqueous environment when compared to the common metallic biomaterials designed to be corrosion resistant. On the other hand, although the nonoxide ceramic Si_3N_4 used in spinal implant devices is also bonded by strong interatomic bonds, it is chemically unstable in an oxidizing atmosphere. Consequently, the surface of Si_3N_4 is covered with a thin layer (~3–5 nm) of amorphous SiO_2 (or, more correctly, silicon oxynitride), which forms rapidly in an oxidizing atmosphere and prevents further oxidation (Chapter 7). This oxide layer dissolves very slowly in water at the physiological pH, releasing Si as soluble silicic acid $Si(OH)_4$, but is replenished by the hydrolysis of Si_3N_4 in an aqueous environment. The grain boundaries in Si_3N_4, normally consisting of a silicate glass, can also dissolve very slowly, releasing Si as $Si(OH)_4$, and hydrated ions, such as Al^{3+} and Y^{3+}.

Calcium phosphate bioceramics such as hydroxyapatite, $Ca_{10}(PO_4)_6(OH)_2$, and β-TCP, $Ca_3(PO_4)_2$, calcium phosphate cements, and calcium sulfate (plaster of Paris), $CaSO_4{\cdot}\frac{1}{2}H_2O$, degrade in an aqueous environment mainly by dissolution into their constituent cations and anions. As noted in Chapter 7, dissolution of calcium phosphate bioceramics is influenced by a variety of physicochemical factors, such as their Ca to P atomic ratio, particle size and porosity, and the pH of the aqueous medium.

Because calcium phosphate bioceramics normally have a polycrystalline structure, they are also susceptible to degradation by disintegration into particles due to faster dissolution of their grain boundaries. Additionally, *in vitro* studies have shown that certain calcium phosphate bioceramics, particularly carbonate-substituted hydroxyapatite, may be susceptible to degradation by cell-mediated resorption.

14.4.1 Degradation by Dissolution and Disintegration

As atoms (ions) at the grain boundary have a higher energy than those within the bulk of the grain, the grain boundaries of degradable polycrystalline ceramics can dissolve faster when compared to regions away from the grain boundary. Consequently, degradation of bioceramics by dissolution may be accompanied by disintegration and the release of particles consisting of one or more grains, particularly when the bioceramic has a fine-grained microstructure. Release of such particles from an implant *in vivo* may initiate inflammatory responses in the surrounding tissue (Chapter 21). Degradation by a combination of dissolution and disintegration has been observed *in vitro* and *in vivo* for certain calcium phosphate bioceramics, such as β-TCP (Figure 14.10). For the same calcium phosphate bioceramic, the *in vivo* degradation patterns generally match the *in vitro* pattern, which indicates that the disintegration is caused by dissolution, and not by cells or biomolecules in the physiological environment.

14.4.2 Cell-Mediated Degradation

Several *in vitro* studies have shown that certain calcium phosphate bioceramics may be susceptible to cell-mediated degradation by a process of resorption by osteoclastic cells. Osteoclasts are large cells having more than one nucleus that are responsible for bone resorption *in vivo* (Chapters 18 and 21). Shallow pits beneath adsorbed osteoclasts (Figure 14.11) normally characterize osteoclastic resorption. Overall, osteoclastic resorption of bioceramics is controversial. It has been observed for only a few calcium phosphate bioceramics and mainly in studies performed *in vitro*. Even for the same calcium phosphate composition, osteoclastic resorption has been observed in some studies but not in others. Because osteoclastic resorption will normally occur in combination with degradation by dissolution and disintegration mechanisms, it is normally difficult to distinguish its effects from those caused by dissolution and disintegration.

Most studies on osteoclastic cell-mediated degradation have focused on calcium phosphate bioceramics because of their widespread use as biomaterials, particularly in bone healing. In one study, degradation of disc-shaped specimens composed of sintered or compacted powders of hydroxyapatite, carbonate-substituted hydroxyapatite, α-TCP, β-TCP, tetracalcium phosphate (TTCP), dicalcium phosphate dihydrate (DCPD), and octacalcium phosphate

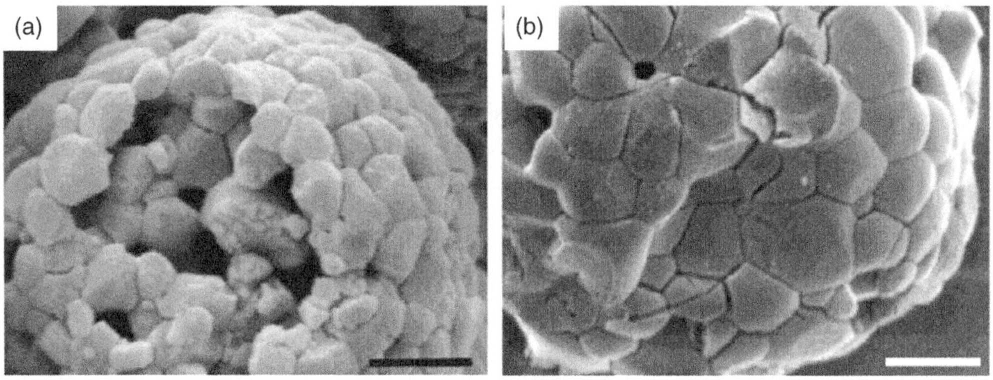

Figure 14.10 Scanning electron microscope images of (a) hydroxyapatite microsphere after immersion for 16 hours in an acetate buffer of pH = 4.0 at 37 °C, and (b) a β-tricalcium phosphate microsphere implanted for four weeks in a mouse peritoneal cavity (bar = 2 μm). *Source:* Koerten and van der Meulen (1999).

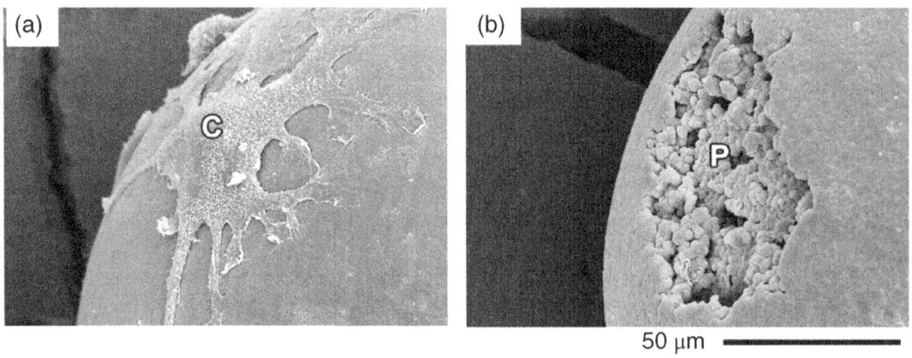

Figure 14.11 Scanning electron microscope images of carbonate-substituted hydroxyapatite microspheres (~10 wt.% carbonate ions) after incubation for one week with TIB-71 cells with added RANKL, showing (a) a cell (C) adsorbed on the microsphere surface, and (b) a pitted region (P) on the microsphere surface after rinsing off the cells. *Source:* Adapted from Xiao et al. (2016).

(OCP) by osteoclast-enriched rabbit bone marrow cells was studied *in vitro*. After 48 hours of culture, only the carbonate-substituted hydroxyapatite specimens showed osteoclastic degradation patterns. In another study, coatings of amorphous calcium phosphate (ACP), carbonate-substituted hydroxyapatite, and OCP were deposited on Ti6Al4V discs by chemical precipitation techniques. Then they were evaluated *in vitro* for their response to osteoclast-enriched mouse bone marrow cells. After seven days of culture, only the carbonate-substituted hydroxyapatite clearly showed osteoclastic resorption patterns.

Overall, *in vitro* studies have shown significant evidence for osteoclastic resorption of hydroxyapatite when it contains some amount of carbonate ion (CO_3^{2-}) substitution and has a fine-grained microstructure. These characteristics may not be surprising when it is considered that the main mineral constituent of natural bone consists of nano-size hydroxyapatite crystals that contain ~8 wt% CO_3^{2-} ions. While the role that the substituted CO_3^{2-} ions plays in the osteoclastic resorption of synthetic hydroxyapatite is unclear, it is noted that CO_3^{2-} substitution increases the solubility of hydroxyapatite (Chapter 7).

Coarse-grained calcium phosphate ceramics are normally less susceptible to osteoclastic resorption when compared to a similar material having a fine-grained microstructure. As solubility decreases with increasing grain size, this may be related to a lower solubility. Osteoclastic resorption has been shown to be dependent on the solubility of certain calcium phosphate ceramics. In one study, osteoclastic resorption of calcium phosphate ceramics having different solubility, composed of hydroxyapatite, β-TCP, and biphasic calcium phosphate (BCP) with a hydroxyapatite to β-TCP ratio of 1 : 3 and 3 : 1 (by weight), was studied *in vitro*. As noted in Chapter 7, the solubility of BCP increases with increasing β-TCP to hydroxyapatite ratio. The most extensive resorption was found for the BCP with the 1 : 3 ratio of hydroxyapatite to β-TCP. In comparison, β-TCP, with the highest solubility, showed little resorption, while no resorption was found for the other two calcium phosphate compositions with a low solubility. The reason why the high solubility of β-TCP should make it less susceptible to osteoclastic resorption is unclear, but it has been suggested that the high concentration of released Ca^{2+} ions could inhibit osteoclastic activity. Overall, these results suggest that there is an optimal solubility for osteoclastic resorption of these calcium phosphate ceramics.

14.5 Concluding Remarks

This chapter discussed the corrosion of metals, the degradation of calcium phosphate bioceramics, and the influence of the physiological environment on degradation of these materials *in vivo*. Because of the presence of various inorganic ions, biomolecules, and cells, the physiological environment can strongly modify the corrosion of metals and the degradation of ceramics when compared to an aqueous isotonic solution at the normal physiological pH and temperature.

- We discussed the principles of metal corrosion and their use in understanding the corrosion of metals and in limiting the potential for adverse corrosion effects in biomedical devices designed for long-term implantation
- Corrosion of metals designed for use in long-term implantable biomedical devices, although slow, can have serious consequences *in vivo*, such as failure of the implant or the release of toxic metal ions
- Degradable bioceramics used in implantable devices, consisting typically of polycrystalline calcium phosphate materials, degrade mainly by dissolution, which releases calcium and phosphate ions that are resorbed into the body, and by disintegration due to preferential dissolution of their grain boundaries, which releases particles into the physiological environment.

Problems

14.1 Explain the meaning and significance of the term standard electrode potential of a metal.

14.2 State the Nernst equation and define the terms in the equation. Explain the significance of the Nernst equation in metal corrosion.

14.3 Determine the electrode potential (with respect to the standard hydrogen electrode) for a zinc electrode in a solution containing 2.0 g of Zn^{2+} ions per liter at 25 °C. (The standard electrode potential of zinc is -0.76 V)

14.4 A standard cell is made in which the anode reaction is

$$Fe \rightarrow Fe^{2+} + 2e^- \qquad E_o = -0.44\,V$$

and the cathode reaction is

$$4OH^- \rightarrow O_2 + 2H_2O + 4e^- \quad E_o = 0.40\,V$$

What potential is established between the two electrodes? Explain your answer.

14.5 The standard electrode potentials of the following four metals are Ti = −1.75 V, Al = −1.66 V, Cr = −0.91 V, Au = 1.50 V. Based on these electrochemical potentials

A Which metal would be the most anodic?

B Which metal would corrode the most?

C What is the maximum voltage one can obtain from these metals in a standard electrochemical solution of the appropriate ions?

14.6 The standard electrode potentials of Ti and Al are −1.75 V and −1.66 V, respectively. Despite both metals having approximately similar high negative standard electrode potentials, Ti is used in implantable devices whereas Al is not. Explain why.

14.7 What is a Pourbaix diagram? Explain what type of information in depicted in a Pourbaix diagram and the usefulness of the information.

14.8 Give the types of corrosion that could occur in each of the following devices and, in each case, explain why each type of corrosion could occur:

A A low-carbon (<0.03 wt%) stainless steel plate used to repair a bone fracture, which is held in position by a combination of stainless steel screws of the same composition and cobalt–chromium screws.

B A niobium (Nb) femoral head on a titanium alloy (Ti6Al4V) stem in an artificial hip implant in which micro-motion between the femoral head and the stem can occur.

C A bone fracture fixation plate composed of a stainless steel having a carbon content of 0.1 wt%.

14.9 Discuss the factors that influence the degradation of β-TCP granules in the physiological environment.

14.10 Explain what is meant by the term osteoclastic resorption, and discuss its applicability to the degradation of calcium phosphate bioceramics.

References

Koerten, H.K. and van der Meulen, J. (1999). Degradation of calcium phosphate ceramics. *Journal of Biomedical Materials Research* 44: 78–86.

Pourbaix, M. (1984). Electrochemical corrosion of metallic biomaterials. *Biomaterials* 5: 122–134.

Xiao, W., Bal, B.S., and Rahaman, M.N. (2016). Preparation of resorbable carbonate-substituted hollow hydroxyapatite microspheres and their evaluation in osseous defects *in vivo*. *Materials Science and Engineering: C* 60: 324–332.

Further Readings

Doi, Y., Iwanaga, H., Shibutani, T. et al. (1999). Osteoclastic responses to various calcium phosphates in cell cultures. *Journal of Biomedical Materials Research. Part A* 47: 424–433.

Evans, J.W. and De Jonghe, L.C. (2016). *The Production and Processing of Inorganic Materials*, Chapter 9. Switzerland: Springer International Publishers.

Leeuwenburgh, S., Layrolle, P., Barrère, F. et al. (2001). Osteoclastic resorption of biomimetic calcium phosphate coatings *in vitro*. *Journal of Biomedical Materials Research. Part A* 56: 208–215.

Williams, D.F. and Williams, R.L. (2013). Degradative effects of the biological environment on metals and ceramics. In: *Biomaterials Science: An Introduction to Materials in Medicine*, 3e (eds. B.D. Ratner, A.S. Hoffman, F.J. Schoen and J.E. Lemons), 728–738. New York: Elsevier.

15

Degradation of Polymeric Biomaterials

15.1 Introduction

Chapter 14 presented a discussion of the mechanisms and processes that influence the degradation of metallic and ceramic biomaterials in the physiological environment. In this chapter, we will discuss similar effects for polymeric biomaterials. As the general introductory remarks of Chapter 14 (Section 14.1) are also relevant to polymeric biomaterials, a review of those remarks forms a useful background to this chapter. When compared to metallic and ceramic biomaterials, the influence of the physiological environment on the degradation of polymeric biomaterials is equally important. However, as noted in earlier chapters devoted to polymeric biomaterials, the use of polymers as implantable devices is more widespread and their chemical composition cover a wide range, from those designed to be intentionally degradable to those resistant to degradation. Consequently, the influence of the physiological environment on their degradation can often be more complex.

The main degradation mechanism for synthetic polymeric biomaterials is hydrolysis whereas natural polymers degrade mainly by enzyme-mediated hydrolysis. Hydrolytic degradation (including enzyme-mediated hydrolysis) involves the scission of chemical bonds in the polymer chain backbone by reactions involving water molecules, which yields oligomers and monomers that are soluble in the aqueous environment (Figure 9.1).

In the case of degradable polymers used in implantable devices, their degradation products should be formed at a physiologically tolerable rate, be nontoxic and be capable of elimination from the body by various catabolic pathways and physiological clearance mechanisms. In comparison, polymeric biomaterials designed to be resistant to degradation generally have a chemical structure that is stable against hydrolysis (Chapter 8). However, hydrolysis is not the only mechanism by which synthetic polymers can degrade *in vivo*. Even those polymers designed to be resistant to hydrolysis can degrade in the physiological environment to varying extent under certain conditions.

The physiological environment into which biomaterials are implanted, as noted in Chapter 14, is complex and more corrosive when compared to an isotonic solution at pH ~7 and ~37 °C. Implantation of a biomaterial can lead to a variety of cellular responses, such as lowering of the local pH, secretion of certain enzymes, and production of strongly oxidizing agents, all of which can strongly enhance the corrosive nature of the environment surrounding the implant. For example, some hydrolysis reactions are sensitive to changes in pH (Chapter 9). While natural polymers are particularly susceptible to enzyme-mediated hydrolysis, certain enzymes produced by cells in response to an implanted biomaterial can also catalyze the hydrolytic degradation of some synthetic polymers. Additionally, highly reactive species such as free radicals and hydrogen peroxide (H_2O_2) produced by these cells can attack and break susceptible chemical bonds in the polymer chain by an oxidative mechanism, regardless of whether the polymer is susceptible to hydrolysis or not.

Overall, an important question in the use of synthetic polymers as implanted biomaterials is their chemical stability in the physiological environment and the role of ions, biomolecules, cells, and cellular products present in the physiological environment. However, a clear answer to this question is often difficult because of the complexity of the mechanisms and the overlapping effects of different mechanisms.

Accelerated degradation in the physiological environment can have serious consequences for the use of polymers in implantable devices, particularly those designed for long-term (permanent) implantation. Accelerated degradation of polymers intentionally designed to degrade by hydrolysis, such as the aliphatic esters (Chapter 9) used as sutures and other surgical devices, can drastically modify their physicochemical properties, and, thus, adversely influence their performance *in vivo*. Degradation of hydrolytically stable polymers designed for use in long-term implantable devices, such as polyurethane (PU) insulation on pacemaker leads, polyethylene terephthalate (PET)

Materials for Biomedical Engineering: Fundamentals and Applications, First Edition. Mohamed N. Rahaman and Roger F. Brown.

Companion website: www.wiley.com/go/rahaman/biomedicalengineering

artificial arteries, and ultra high molecular weight polyethylene (UHMWPE) liners in artificial hip and knee implants, can have serious consequences *in vivo*. For example, oxidative degradation over time can cause UHMWPE liners in hip and knee implants to become brittle, thereby increasing the potential for catastrophic failure of these devices *in vivo*.

As the principles and mechanisms of hydrolytic degradation of polymeric biomaterials designed to be degradable in an aqueous medium were presented in Chapter 9, knowledge of those principles and mechanisms will be assumed in this chapter. Overall, this chapter will focus on the following topics related to the stability of polymeric biomaterials in the physiological environment:

- Pathways for hydrolytic degradation
- Principles and mechanisms of oxidative degradation
- Influence of the physiological environment on hydrolytic and oxidative degradation.

15.2 Hydrolytic Degradation

Hydrolytic degradation, as noted earlier, involves scission of chemical bonds in the polymer chain backbone by reactions involving water molecules, which normally leads to formation of soluble oligomers and monomers. This degradation reaction may be viewed as the reverse of condensation reactions used in the synthesis of certain polymers (Section 8.3). Although hydrolytic scission of bonds in the chain backbone is the main degradation pathway for solid polymers, there are other pathways for converting certain insoluble polymers into soluble molecules by hydrolysis.

15.2.1 Hydrolytic Degradation Pathways

Overall, there are three major pathways by which an insoluble polymer network can be degraded by hydrolysis reactions to form molecules that are soluble in the surrounding aqueous medium. One pathway involves scission of chemical bonds in the crosslinks between macromolecules in a three-dimensional macromolecular network structure (Figure 15.1a). The individual molecules or molecular chains produced in this process will dissolve in the aqueous medium once they have the appropriate chemical structure. This is the main pathway by which a hydrogel, normally insoluble because of its three-dimensional structure, is degraded into smaller soluble molecules (Chapter 11). An example is the hydrolytic degradation of hydrogels composed of crosslinked triblock PLA-*b*-PEG-*b*-PLA copolymer chains (Figure 11.21). It should be noted that while PEG molecules are stable against hydrolysis, they are soluble in an aqueous environment and can be eliminated from the body when their molecular weight is less than ~10–15 kDa.

Another pathway is hydrolysis of the hydrophobic side group in the polymer chains of an insoluble polymer to form soluble polymer chains containing polar or ionizable side groups (Figure 15.1b). This pathway does not necessarily lead to a reduction in the macromolecular chain length. An example is the conversion of the hydrophobic methyl acrylate (C=O)OCH_3 side group in polymethyl cyanoacrylate (a surgical adhesive), to a carboxyl (C=O)OH group. Under the appropriate pH conditions, the H atom in the carboxyl group easily dissociates, making the chain very hydrophilic and, thus, soluble in aqueous media.

The third pathway, which is the main degradation pathway for solid polymers used as biomaterials, involves scission of hydrolytically susceptible bonds in the polymer chain backbone, producing degradation products composed of soluble oligomers and monomers (Figure 15.1c). Hydrolytic degradation leads to erosion of the polymer from the surface inward (surface erosion), within its bulk (bulk erosion), or by combination of surface and bulk erosion (Chapter 9). Figure 9.3 shows the chemical structures of some common functional groups in a polymer chain backbone that are susceptible to hydrolytic attack. In comparison, the chemical structures of the chain backbone in some common polymeric biomaterials that do not have hydrolysable chemical bonds in their molecular chains are

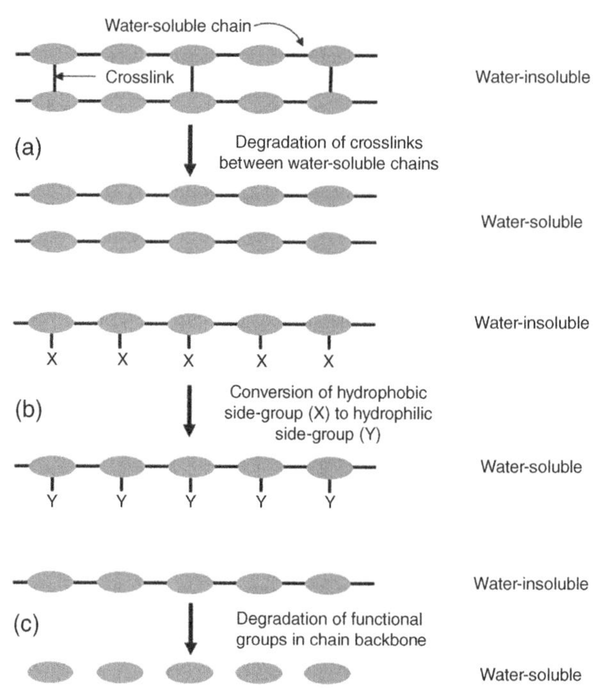

Figure 15.1 Main pathways for converting an insoluble polymer network into soluble molecules by hydrolysis.

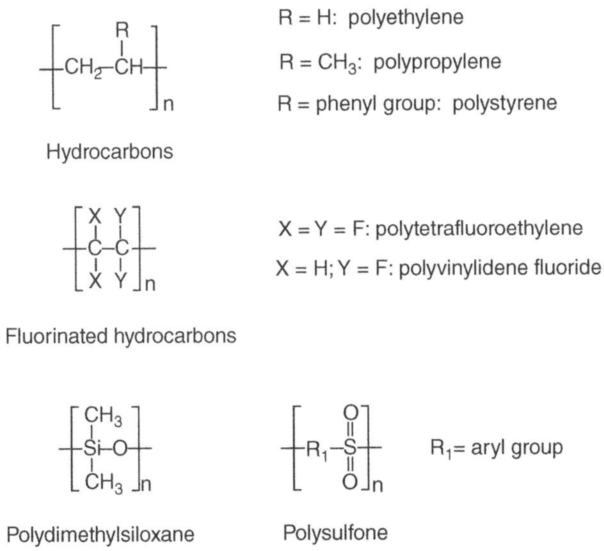

Figure 15.2 Examples of chemical groups in polymer chains that are resistant to hydrolysis.

shown in Figure 15.2. Typically, the monomer units in many of these polymers are bonded by strong covalent bonds, such as carbon–carbon (C–C) bonds.

15.2.2 Role of the Physiological Environment

In an isotonic aqueous medium at pH ~7.4 and ~37 °C, polymers that contain hydrolysable bonds in their chain backbone undergo degradation at various rates, depending on their chemical structure (Chapter 9). However, as noted earlier, the physiological environment is complex, containing a variety of inorganic ions (Table 14.2), biomolecules (Table 14.3), and cells. Consequently, it is necessary to consider the degradation of polymeric biomaterials in light of this more complex environment. Overall, the degradation process can be quite complicated.

15.2.3 Effect of Local pH Changes

The local pH of the physiological environment near an implanted biomaterial can decrease from the normal physiological value (~7.4) to as low as 4–5 during acute inflammation or infection (Chapter 20). *In vitro* studies have shown that acidic or basic pH can accelerate the hydrolytic degradation of aliphatic polyesters such as polyglycolic acid (PGA), polylactic acid (PLA), and their copolymer (polylactic-*co*-glycolic acid (PLGA)) by an acid-catalyzed or a base-catalyzed mechanism (Chapter 9). The degradation of these polymers is normally lowest at a certain pH, which depends on the composition of the polyester, and increases with a decrease or an increase from this pH value. However, the increase in degradation rate with decreasing pH is often less steep when compared to that for increasing pH. For example, the tensile strength of PGA fibers, used as a measure of their degradation rate, normally shows no significant difference at pH values in the range ~5.0 to ~7.5, but a significant decrease as the pH increases to ~10.0.

Like the aforementioned aliphatic polyesters, aromatic polyesters such as PET, known commonly as Dacron®, are also susceptible to hydrolytic degradation because of the presence of the ester functional group in their chain backbone. However, they degrade considerably more slowly because of their higher degree of hydrophobicity due to the benzene ring in their monomer units (Figure 8.15). Bulk erosion is unlikely and although surface erosion can occur slowly, PET is sufficiently stable against hydrolytic degradation at the normal physiological pH and temperature for it to be used in long-term implantable devices such as vascular grafts. For example, implantation in subcutaneous sites in the backs of rabbits and dogs has shown that the time for complete degradation of PET fibers is 30 ± 7 years, with a 50% reduction in their initial tensile strength at 10 ± 2 years. In comparison, when implanted in infected subcutaneous sites in dogs, where the local pH can drop to values as low as ~5, similar PET fibers have been found to degrade significantly faster and lose their tensile strength almost completely within a few months.

Evaluation of PET prostheses retrieved after 1–10 years from uninfected sites in the abdominal wall of humans showed a time for 50% reduction in initial strength comparable to that for uninfected dogs (10 ± 2 years) along with predominantly surface erosion. On the other hand, PET prostheses retrieved from similar but infected sites in another group of patients, characterized by a pH of ~5–6, showed significantly faster degradation and predominantly bulk erosion. These prostheses showed a 50% reduction of their initial strength after only four to seven months.

Based on their molecular size, many enzymes, such as those produced by bacterial cells, might be expected to influence polymer degradation by surface erosion because they may be too large to migrate into the interior of a solid polymer such as PET. Although bacterial cells were found on the surface of the retrieved prostheses, the ability of enzymes to influence hydrolytic degradation *in vivo* has been questioned (Section 15.3). Thus, the faster degradation by bulk erosion of PET in the infected sites has been attributed to acid-catalyzed hydrolysis. Overall, significant local pH changes, which can occur in some implantation sites, particularly infected sites, has been associated with a strong modification of the degradation rate of some polymers, such as those containing the hydrolytically susceptible ester group in their chain backbone.

15.2.4 Effect of Inorganic Ions

When compared to an isotonic liquid, the extracellular fluid contains a variety of cations and anions, of which Na^+ and Cl^- and, to a lesser extent, HCO_3^-, are the major ionic constituents (Table 14.2). When compared to deionized water alone, high concentrations (~1–4 M) of NaCl and $MgCl_2$ in deionized water have been found to retard the hydrolytic degradation of commercial PGA sutures (designated Dexon®), as determined from tensile strength measurements. However, these concentrations are considerably higher than those of the respective ions in the interstitial fluid. Another study has shown that when compared to water, phosphate buffers containing 100 mM to 1.0 M phosphate ions strongly accelerate the degradation of similar PGA sutures, as measured from their mass as a function of time (Figure 15.3). Again, these ionic concentrations are considerably higher than the normal phosphate ion concentration of the interstitial fluid (1–2 mM).

It is possible that cations can coordinate to the polar sites (such as the carbonyl C=O group) in the chains of polyesters, making certain sites in the chains more electronegative and, thus, promote hydrolysis. Experiments have shown no significant effect of Cu^{2+}, Co^{2+}, Ni^{2+}, Zn^{2+}, and Ag^+ on the hydrolytic degradation of PGA, PLA, PLGA, and polycaprolactone (PCL), but Fe^{3+} showed the ability to enhance the degradation of PCL. Overall, cations at concentrations comparable to those in the physiological environment have not shown a significant capacity to influence the hydrolytic degradation of these aliphatic polyesters used in implantable devices.

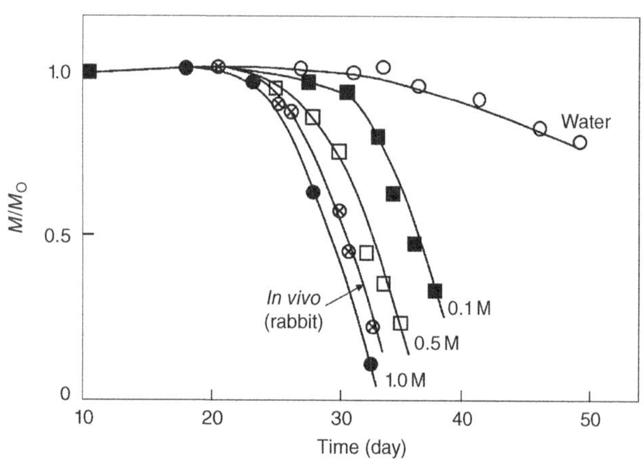

Figure 15.3 Mass of PGA sutures (designated Dexon®), relative to their initial mass, as a function of immersion time in water and in a phosphate buffer containing 0.1, 0.5, and 1.0 M phosphate ions at 37 °C, and as a function of implantation time in rabbit subcutaneous sites *in vivo*. *Source:* Zaikov (1985) / with permission of Taylor and Francis.

15.2.5 Effect of Bacteria

As discussed in the next section, while enzymes have been normally associated with causing the degradation of natural polymers, certain enzymes have also been found to accelerate the degradation of a wide variety of synthetic polymers *in vitro*. Consequently, the question arose as to whether bacteria might also affect the degradation of these synthetic polymers through the hydrolytic enzymes that they release. In one study (Williams 1980), sutures composed of PGA or catgut (a natural material) were incubated *in vitro* for three weeks in appropriate broths containing the bacterium *Streptococcus mitis* (*S. mitis*), *Streptococcus epidermis* (*S. epidermis*), or *Escherichia coli* (*E. coli*). Then they were tested for their tensile strength, used as a measure of degradation. The PGA sutures were found to degrade slower in broths containing the bacteria than in broths without bacteria. The same trend was seen even when the bacterial cultures were acidified to reproduce the change in pH caused by the metabolic products of the bacteria. In comparison, the catgut sutures showed no measureable difference in degradation when incubated in broths with or without bacteria.

When implanted for one week in rat subcutaneous sites inoculated with *S. epidermis*, PGA sutures again degraded slower than in similar sites that were not inoculated. In comparison, the degradation of catgut sutures showed a dependence on the bacterial concentration. While there was no difference in degradation between the two sites at lower bacterial counts, at higher concentrations, the catgut sutures showed a reverse trend to that of the PGA sutures.

Although the reasons are not clear, these results show that bacterial enzymes do not accelerate the degradation of PGA sutures under the conditions used in this study. They also show a difference in degradation behavior between (synthetic) PGA and (natural) catgut sutures.

15.3 Enzyme-Catalyzed Hydrolysis

Enzymes are normally globular proteins that serve as biological catalysts, that is, they accelerate chemical reactions in living organisms and, like most catalysts, do not undergo any permanent chemical change. A characteristic feature of enzymes is that they are usually specific as to the substrate (molecules) to which they bind and, then, the chemical reactions that they catalyze. On the other hand, some enzymes have broad specificity and act on a range of different physiologically relevant substrates. While the number of enzymes present in the human body is enormous, two classes are most relevant to the degradation of polymeric biomaterials in the aqueous physiological environment:

- Hydrolases (hydrolytic enzymes), such as proteases (also referred to as peptidases or proteinases), esterases, lipases, and phosphatases that catalyze hydrolysis reactions
- Oxidoreductases that catalyze the transfer of electrons from one molecule (the reductant) to another molecule (the oxidant or electron acceptor) (Section 15.4).

Enzymes are known to cause the degradation of some polymers but they have usually been associated with natural polymers, as exemplified by the degradation of collagen by collagenases and of hyaluronic acid by hyaluronidases (Chapter 10). On the other hand, *in vivo* experiments several decades ago showed that some polymers containing hydrolysable groups in their chain backbone, but normally stable against hydrolysis over a long period, degraded considerably faster than would be expected from thermal oxidation. Consequently, it was suggested that enzymes might play a role in the hydrolytic degradation of synthetic polymers. Since that time, a considerable number of studies has shown that enzymes can accelerate the hydrolytic degradation of a variety of synthetic polymers *in vitro*. These polymers cover a wide range, from aliphatic polyesters such as PGA and PLA that are designed to be intentionally degradable, to polymers that are designed to be very stable against hydrolysis, such as PET, nylon 6, and polyether urethane (PEU).

Overall, then, *in vitro* studies have shown that the mechanisms exist for enzymes to influence the degradation of polymers that contain hydrolysable functional groups in their chain backbone. On the other hand, whether this actually occurs *in vivo* has been a difficult question to answer. Although different enzymes have been shown to be present around an implant, their role in influencing the degradation of the implant has been the subject of controversy.

15.3.1 Principles of Enzyme-Catalyzed Hydrolysis

Enzymes are typically proteins that catalyze many biological processes by providing a reaction path of lower activation energy. The structure of an enzyme reflects two tendencies, stability and function, which are often opposed. To minimize their free energy, enzymes normally adopt a globular three-dimensional shape in which nonpolar side groups are buried within the interior while polar or ionizable molecules gather on the outside of the protein chain (Chapter 2). On the other hand, enzyme function arises from conformational flexibility of active (binding) sites distributed at certain positions on the chain where charged or polar groups are sequestered from water. Studies have generally shown that the active site of the enzyme may have a higher conformational flexibility when compared to the enzyme chain as a whole. At a basic level, an appropriate labile functional group in the chain backbone of a substrate (such as a polymer) induces a conformational change in the enzyme active site, which brings these two groups into proximity (Figure 15.4). This enhances the probability that the reaction could occur, thereby increasing the reaction rate.

The interaction of a substrate functional group with the enzyme active site is a dynamic one, which requires some degree of flexibility of the enzyme active site. Some

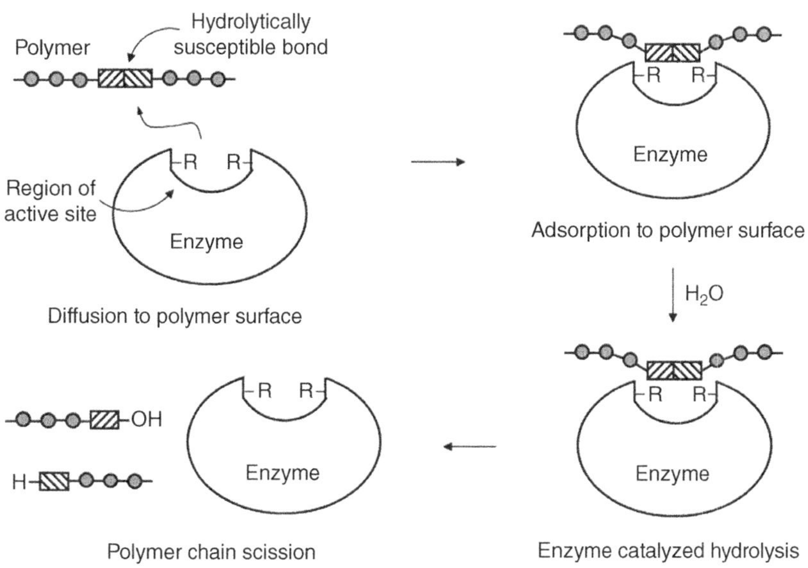

Figure 15.4 Schematic illustration of a model for enzyme-catalyzed hydrolysis. R represents essential protein residues for the enzymatic reaction.

enzymes, as noted earlier, show a high degree of specificity, which means that the active site of the enzyme binds and reacts with a single type of functional group in the polymer chain. This may imply that the active site has some degree of order or may have a high degree of conformational rigidity. In comparison, some enzymes that have broad specificity and act on a range of different physiologically relevant substrates may have less ordered active sites or more conformational flexibility. Overall, some degree of conformational flexibility of the binding site is required for the function of an enzyme.

Enzymatic degradation of a polymeric substrate typically involves four potential steps, any one of which can be rate limiting:

- Diffusion of the enzyme from the solution to the surface of the polymer
- Enzyme adsorption and the formation of an enzyme-substrate complex
- Catalysis of the hydrolysis reaction
- Diffusion of the degradation products from the polymer surface into the solution.

Enzyme-accelerated hydrolysis of polymers may be influenced by a variety of factors, such as

- The physicochemical properties of the polymer, for example, its chemical structure, molecular weight, crystallinity, and surface topography
- The inherent characteristics of the enzyme, such as its composition, concentration, stability, activity, and conformation
- The pH of the medium.

Chemical modification of the polymer, such as copolymerization of the chain backbone, altering the chemical composition of the side groups, or crosslinking can also influence the degradation rate. These chemical changes can alter the ability of the polymer (substrate) to induce the requisite conformational change in the active site of the enzyme to achieve optimal binding and accelerated hydrolysis.

The capacity of common synthetic polymers used as biomaterials to degrade in an aqueous medium has been discussed in previous chapters. These polymers range from aliphatic polyesters designed intentionally to degrade in short-term implantable devices to those designed to be strongly resistant to hydrolytic degradation, such as aromatic polyesters and PUs designed for use in long-term implantable devices. Examples of these polymers are used in the following sections to highlight various aspects related to the role of enzymes in hydrolytic degradation. However, it should be noted that there is no single mechanism of degradation and that the degradation process is dependent not only on the polymer but also on the conditions within the implanted site.

15.3.2 Role of Enzymes in Hydrolytic Degradation *in vitro*

Approximately four decades ago, the role of enzymes in the hydrolytic degradation of PGA was investigated *in vitro* using 15 enzymes consisting of several types, such as proteases, esterases, and phosphatases (Williams and Mort 1977). PGA surgical sutures (designated Dexon®) were incubated at 37 °C for up to 4.5 weeks in solutions composed of each enzyme in an appropriate buffer, and the tensile strength of the sutures was used as a measure of degradation. When compared to controls immersed in the buffer alone, the enzymes acid phosphatase, papain, pepsin peptidase, pronase, protease K, and trypsin showed no measurable effect on the degradation of the sutures over a three-week period.

In comparison, the enzymes carboxypeptidase A, clostridiopeptidase A, bromelain, esterase, leucine aminopeptidase, α-chymotrypsin, and ficin accelerated the rate of degradation. Enzymes such as esterase and leucine aminopeptidase showed a particularly strong effect on the degradation. Except for carboxypeptidase A and clostridiopeptidase A, all of these enzymes that enhanced the degradation of PGA have esterase activity, which may indicate that enzymes will tend to influence the degradation of polymers having a structural resemblance to the substrates upon which they normally act. However, this is not generally so, as found for these PGA sutures, for which a few enzymes possessing no esterase activity accelerated hydrolysis, whereas a few that possessed this activity had no effect on degradation.

Studies have shown that the susceptibility of PLA to enzyme-catalyzed hydrolysis depends not only on the specific enzyme but also on the stereoisomerism and crystallinity of PLA. One specific enzyme, proteinase K, has a particularly significant effect on PLA degradation. Proteinase K preferentially degrades L-lactic acid units in PLA chains but not the stereoisomeric D-lactic acid units. Figure 15.5a further illustrates the influence of stereoisomerism on the enzymatic degradation of PLA. Despite containing a small amount of crystallinity (~6 wt%), L-PLA specimens composed of 100% L-lactic acid units degrade at faster rate (as measured from their weight loss) when compared to amorphous specimens composed of 50% L-lactic acid units and 50% D-lactic acid units. This is compatible with the preferential degradation of L-lactic acid units by proteinase K. The effect of crystallinity on the enzyme-catalyzed degradation of L-PLA is shown in Figure 15.5b. While a crystallinity of less than ~25 wt% has little effect, the degradation rate decreases rapidly as the crystallinity increases above this value. Overall, the strong effect of

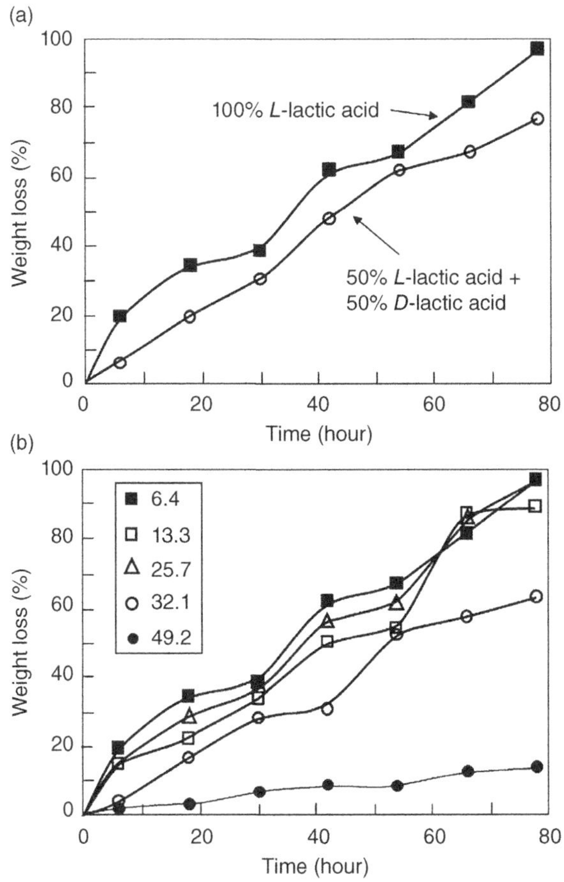

Figure 15.5 Weight loss versus time for PLA films during enzymatic degradation in a Tris/HCl buffer solution containing proteinase K at 37 °C. (a) Films composed of 100% L-lactic acid and of 50% L-lactic acid + 50% D-lactic acid. (b) Films composed of L-PLA having different crystallinity, shown in wt%. *Source:* Li and McCarthy (1999) / with permission of American Chemical Society.

proteinase K in catalyzing the hydrolytic degradation of L-PLA can be seen from a comparison of its degradation rate with that in the control medium (Tris HCl) without the enzyme, where the weight loss was negligible in all cases (<0.4 wt% after 78 hours).

Aliphatic esters such as PGA and L-PLA are synthesized from naturally occurring monomers (glycolic acid and L-lactic acid), while PCL can be considered to be synthesized from analogs of naturally occurring monomers (ε-caprolactone). Consequently, the capacity of some hydrolases to catalyze the hydrolytic degradation of these esters *in vitro* may not be surprising. However, their degradation illustrates various factors that can influence enzyme-catalyzed hydrolysis of polymers in general.

PUs are used in a variety of long-term implantable devices such as insulation on pacemaker leads and, thus, their resistance to degradation over a long-term period is important. The relationship between the chemical structure of PUs used in these biomedical applications and their physicochemical properties were discussed in Chapter 8, but a few features are summarized here to provide some relevant background. PUs are block copolymers composed of alternating segments of hard (stiff) repeating units, which have a semicrystalline or glassy structure at the physiological temperature, and soft segments composed of flexible repeating units, which is generally in the rubbery state (Figure 8.24). The hard segments are normally composed of a diisocyanate, such as toluene diisocyanate (TDI) and, above a certain fraction of the overall chain, they form small crystalline regions that crosslink the more mobile chains of the amorphous phase. The soft segments are composed of a polyester, giving a polyester urethane (abbreviated PESU) or of a polyether, such as polytetramethylene oxide (PTMO), giving a PEU.

The urethane group –NH–(C=O)–O– is known to have a low susceptibility to hydrolysis. Because the ether bond is more stable against hydrolysis than the ester bond, for the same hard segment composition and ratio of hard to soft segments, PEUs generally have a higher resistance to hydrolysis when compared to PESUs. Additionally, for the same soft segment composition, PEUs having a higher ratio of hard to soft segments generally show a higher resistance to hydrolysis. This can be rationalized in terms of the higher fraction of crystalline regions in which the chains have a high degree of hydrogen bonding.

In vitro experiments using single enzymes in an appropriate buffer have shown that certain enzymes catalyze the hydrolytic degradation of PUs. Cholesterol esterase (CE), an enzyme present in increasing quantities as monocytes differentiate into macrophages during the chronic inflammatory tissue response to an implant (Chapter 18), has a particularly strong ability to catalyze the hydrolysis of both PESUs and PEUs.

The CE-catalyzed hydrolysis is dependent on the hard segment content, as shown in Figure 15.6 for PEUs composed of TDI, PTMO, and ethylene diamine (ED) used as a chain extender, incubated in a solution composed of CE in phosphate buffer (0.05 M) at pH = 7.0 and 37 °C. As the ratio of hard segments to soft segments (TDI : PTMO) increases, the degradation, as measured by the release of radiolabeled (^{14}C) TDI, decreases. In comparison, similar specimens incubated in the buffer without CE all show low and similar levels of radiolabeled TDI release. Because of their highly ordered structure with a high degree of hydrogen bonding between the chains, the crystalline regions may limit the interaction between the hydrolysable groups of the polymer and the binding sites of the enzyme, particularly if the crystals form preferentially at the surface of the polymer. Consequently, degradation in the enzyme solutions will decrease with increasing ratio of hard to soft segments.

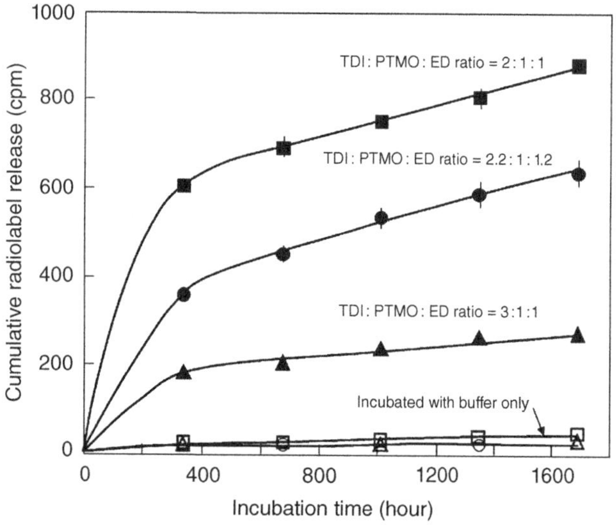

Figure 15.6 Effect of hard segment content on the degradation of polyether urethanes (PEUs) incubated in phosphate buffer containing cholesterol esterase (filled data points) and in phosphate buffer alone (unfilled points) at pH = 7.0 and 37 °C. The PEU chains consisted of toluene diisocyanate (TDI) (hard segment), polytetramethylene oxide (PTMO) (soft segment), and ethylene diamine (ED) (chain extender), and release of radiolabeled (^{14}C) TDI (counts per minute, abbreviated cpm) was used as a measure of degradation. *Source:* Santerre and Labow (1997) / with permission of John Wiley & Sons.

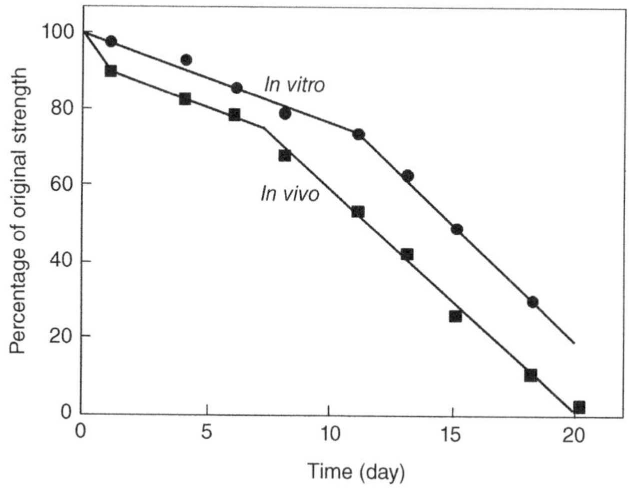

Figure 15.7 Degradation of polyglycolic acid (PGA) sutures determined from their reduction in tensile strength, as a function of implantation time in rat subcutaneous sites *in vivo* and as a function of immersion time in Tris buffer at pH = 7.4 and 37 °C *in vitro*. *Source:* Williams (1979) / with permission of ASTM International.

15.3.3 Role of Enzymes in Hydrolytic Degradation *in vivo*

The degradation of PGA sutures has been studied *in vivo*, by implanting specimens subcutaneously in rats and monitoring the reduction in their tensile strength as a function of time (Williams 1979). Similar specimens immersed in Tris buffer (pH = 7.4) served as controls. Although the strength of the implanted specimens was lower than that of the controls at any given time interval, this lower strength was due almost entirely to a rapid loss that occurred in the first two days only (Figure 15.7). Thus, except for the first two days, these PGA sutures degraded at the same rate and by the same hydrolysis mechanism regardless of the two different environments. This initial period of rapid degradation corresponds to the acute inflammatory stage of tissue response to an implanted biomaterial (Chapter 18) but the factors responsible for this rapid degradation are not clear.

Tissue response to PLA has been studied by monitoring the activity patterns of a variety of enzymes surrounding L-PLA hollow fibers implanted subcutaneously in rats for up to three weeks (Schakenraad and Hardonk 1990). Special attention was paid to enzyme patterns characteristic of polymorphonuclear granulocytes (PMNs), fibroblasts, and macrophages, which are involved in inflammatory responses and foreign body reactions to implanted biomaterials (Chapter 18). One day after implantation, an influx of neutrophilic and eosinophilic granulocytes into the implantation site was observed, coinciding with the activity of alkaline phosphatase (indicative of neutrophils) and β-glucuronidase (indicative of eosinophils). From day 3 onwards, macrophages containing ATPase (an enzyme that catalyzes the conversion of adenosine triphosphate, ATP, to adenosine diphosphate, ADP), acid phosphatase, and esterase were observed.

In comparison, from day 7 onward, lactate dehydrogenase (LDH), the reversible enzyme in the metabolism of lactic acid, along with its co-enzyme nicotinamide adenine dinucleotide (NAD^+), was observed in macrophages and fibroblasts. This enzyme showed a slightly higher activity when compared to that surrounding nondegradable polytetrafluoroethylene (PTFE) vascular grafts of a similar external diameter (~1 mm). The observed enzyme pattern was compatible with a normal wound-healing reaction, while the slightly higher level of LDH was attributed to the release of lactic acid resulting from hydrolytic degradation of PLA in an aqueous medium. Overall, while various enzymes were observed in the surrounding tissue, their role in catalyzing the hydrolytic degradation of the L-PLA fibers could not be clearly determined.

15.4 Oxidative Degradation

Phagocytic cells such as macrophages and neutrophils are very potent in defending the human body against invading microorganisms (Chapter 18). On the other hand, it is well established that they produce highly reactive molecules such as superoxide radicals (anions), hydrogen peroxide, and hydroxyl radicals. Consequently, the question arises as to whether these reactive molecules can also be hostile to polymers implanted into tissues.

Although studies have shown the presence of phagocytes around implanted biomaterials, there is no direct evidence for the involvement of radicals produced from macrophages and neutrophils in the degradation of polymeric implants. However, *in vitro* experiments have shown that such radicals can accelerate the degradation of polymeric biomaterials, while polymeric biomaterials retrieved after long-term implantation in animals and humans have shown certain characteristics, such as surface compositional changes, which are similar to those caused by oxidation in air or oxygen.

Overall, oxidative degradation by radicals produced by phagocytic cells in response to an implanted polymer may provide an additional degradation mechanism *in vivo*. However, it should be much slower than hydrolytic degradation of polymers that contain hydrolysable bonds in their chain backbone. Consequently, oxidative degradation should be more relevant to long-term implantable devices, such as PU insulation on pacemaker leads, polyester and PU vascular grafts, and even polyethylene (PE) liners in artificial hip and knee implants. Although the degradation may be slow, it can have serious consequences for these types of devices.

15.4.1 Principles of Oxidative Degradation

Although the reaction sequence is not clear, oxidative degradation of polymers *in vivo*, as noted earlier, shows features that are similar to their oxidation in air or oxygen, a process often referred to as autoxidation because it proceeds by a chain reaction. Autoxidation is normally considered to occur in three steps, called initiation, propagation, and termination (Figure 15.8). Initiation usually occurs when a radical abstracts an H atom from the polymer chain, forming a carbon-centered radical. In autoxidation, the radicals that abstract the H atom from the carbon chain may be produced by external sources such as ultraviolet (UV) light or γ-rays, or by certain transition metals if present as impurities in the polymer. The propagation step starts with this carbon-centered radical reacting with O_2, forming a peroxy radical that abstracts an H atom from a neighboring chain to form a hydroperoxide and a carbon-centered radical. Various chain branching reactions may occur, generating more radicals, such as dissociation of the hydroperoxide to form a peroxy radical and a hydrogen radical. Overall, propagation is a chain reaction in which a single reaction can lead to further reaction in a considerable number of other molecules. Chain reactions of propagation and chain branching are responsible for the degradation of the polymer. Termination occurs by coupling or disproportionation of propagating radicals, forming degradation products, some of which contain alcohol hydroxyl (OH) and carbonyl (C=O) groups.

Figure 15.9 shows examples of sites favorable for initial oxidative attack, allowing abstraction of an atom (or ion). The chemical structure of the polymer chain has a strong effect on oxidative degradation. Overall, the ease of H abstraction in the initiation step depends on the type of C–H bond present in the polymer chain. For the polyolefins such as linear PE, branched PE, and polypropylene (PP) (Figure 15.10), the bond between the H and C atoms at the point where chain branching occurs has a lower dissociation energy when compared to the bond in the methylene groups along the chain. Consequently, these bonds are more likely sites for the initiation of autoxidation. For example, PP is more susceptible to oxidative degradation than branched PE, which, in turn is more susceptible to oxidative degradation than linear PE.

Although it may not hold in every case, in general, whenever branched groups are present which lower the dissociation energy of carbon–hydrogen bonds, the polymer will have a greater susceptibility to oxidative degradation. In the case of polymers such as polyesters and polyethers that contain O atoms in addition to C atoms in their chain backbone, the H atom at the α-carbon site (the C atom adjacent to the carbonyl (C=O) or ether O groups) are more easily abstracted when compared to H atoms in methylene groups along the chain.

The physical characteristics of a polymer can also influence its oxidative degradation rate. For example, when the crystalline regions of a polymer (for example PE) has an ordered compact arrangement of chains such that O_2 molecules cannot diffuse into them, the oxidation rate of the polymer will decrease with increasing crystallinity. In this case, oxidative degradation will take place preferentially in the amorphous phase of the polymer.

15.4.2 Production of Radicals and Reactive Species *in vivo*

Phagocytic cells, as noted earlier, produce reactive molecules such as superoxide radicals (anions), hydrogen peroxide, and hydroxyl radicals. Figure 15.11 shows the electronic structures of these reactive molecules, along with those for the O_2 molecule and the hydroxyl (HO^-) anion, often written OH^-. Although the electronic structure of the O_2 molecule is normally depicted as containing a double bond (two shared pairs of electrons), only one

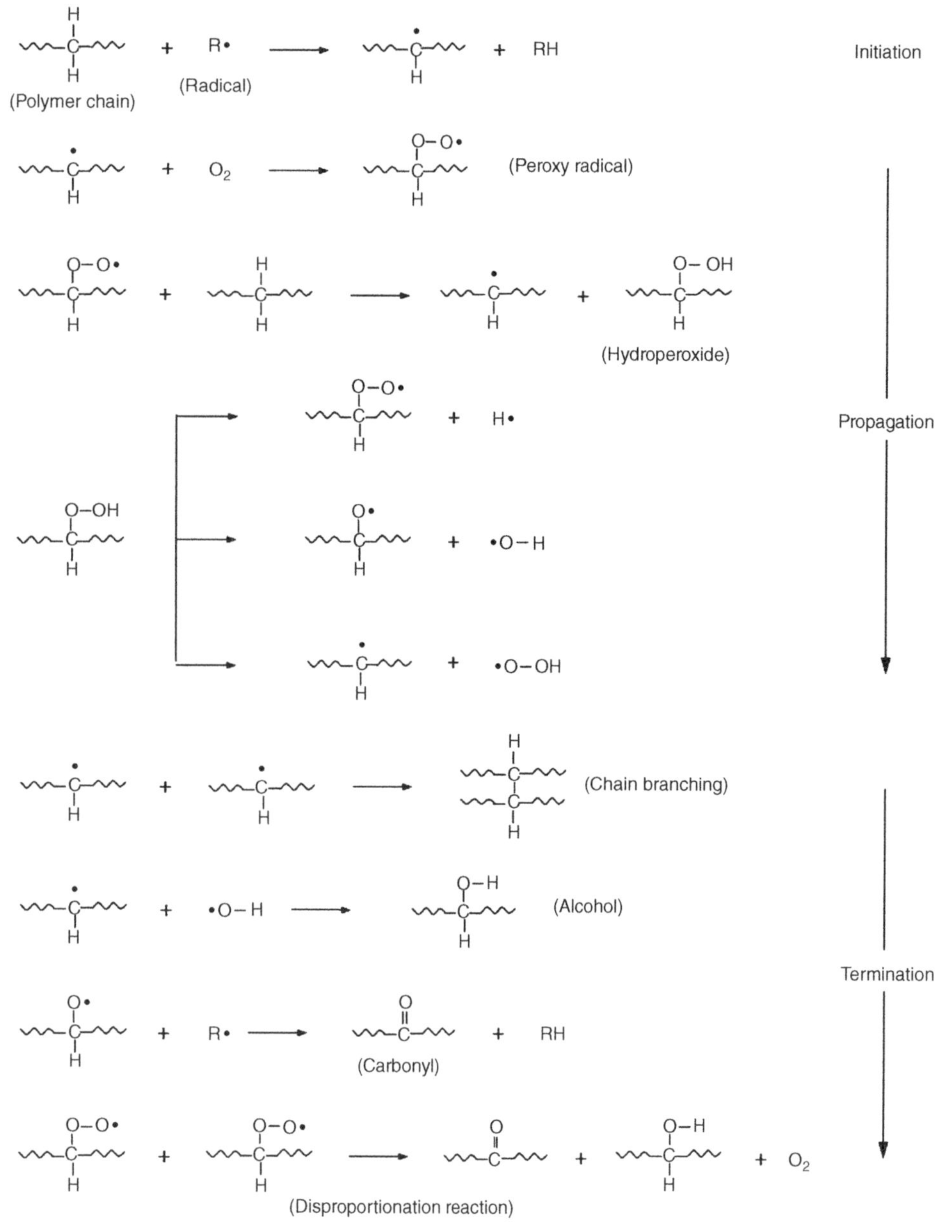

Figure 15.8 Possible initiation, propagation, and termination reactions in the autoxidative degradation of polymers, shown for a polymer chain containing the methylene (CH_2) group.

shared pair is formed, leaving two unshared electrons that are involved in a special type of coordinate bond. Adding one electron to the O_2 molecule produces the superoxide molecule, written $O_2^{-}\bullet$, which is called the superoxide anion or the superoxide radical because it has a negative charge and an unpaired electron (•). The electronic structure of the hydroxyl radical HO• may be viewed as removing an electron from the outer orbital of the OH^- anion, leaving an unpaired electron.

A relevant question relates to how these reactive molecules are produced *in vivo*. Although certain enzymes can release hydrogen peroxide (H_2O_2), most H_2O_2, superoxide radicals and other oxygen radicals are produced by neutrophils and macrophages when they are appropriately stimulated, such as by the presence of infectious microbes (Chapter 18). This process is accompanied by a marked increase in oxygen uptake and an increased utilization of glucose in a coordinated sequence of reactions commonly known as the respiratory burst, a key innate defense mechanism of neutrophils and macrophages. The primary reaction of the respiratory burst is the one-electron reduction of O_2 to the superoxide radical $O_2^{-}\bullet$ catalyzed by the enzyme NADPH oxidase (or NADH oxidase), summarized by the reaction

$$\mathrm{NADPH} + 2\mathrm{O_2} \underset{\text{oxidase}}{\overset{\text{NADPH}}{\rightarrow}} \mathrm{NADP^+} + \mathrm{H^+} + 2\mathrm{O_2}^{-}\bullet \quad (15.1)$$

In general, the superoxide radical can function either as an oxidant, in which case it gains an electron, forming O_2^{2-}, or as a reductant, in which case it loses an electron and is oxidized to O_2.

Two superoxide radicals can react in a manner such that one is reduced and the other oxidized, resulting in the formation of H_2O_2 and O_2, called a dismutation reaction, which is catalyzed by the enzyme superoxide dismutase (SOD):

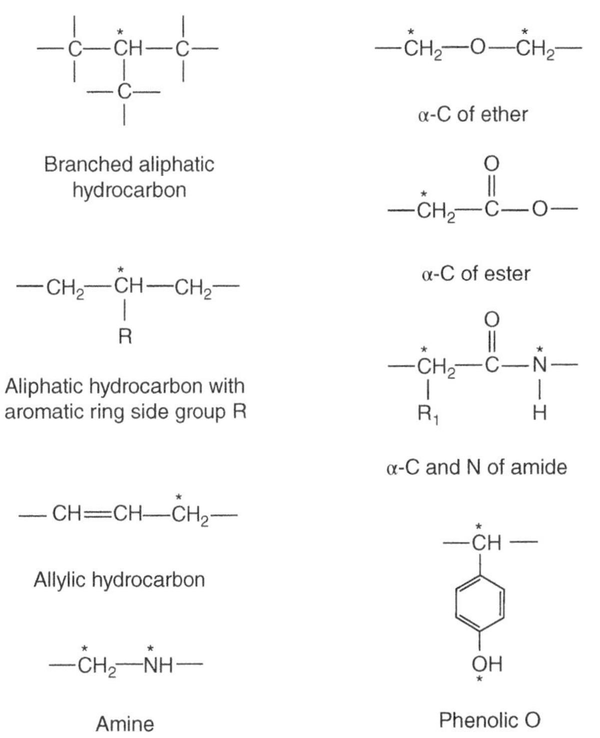

Figure 15.9 Examples of sites (*) favorable for initial oxidative attack by abstraction of an atom such as H.

$$\mathrm{O_2}^{-}\bullet + \mathrm{O_2}^{-}\bullet + 2\mathrm{H^+} \overset{\text{SOD}}{\rightarrow} \mathrm{H_2O_2} + \mathrm{O_2} \quad (15.2)$$

However, as the superoxide radical has a $pK_a = 4.8$, the rate constant for the reaction in Eq. (15.2) is small at the physiological pH (~7.4) due to repulsion of the negatively charged superoxide radicals. The preferred dismutation reaction involves a reaction between superoxide radicals and perhydroxy (protonated superoxide) radicals:

$$\mathrm{O_2}^{-}\bullet + \mathrm{HO_2}\bullet + \mathrm{H^+} \overset{\text{SOD}}{\rightarrow} \mathrm{H_2O_2} + \mathrm{O_2} \quad (15.3)$$

This reaction is rapid at acidic pH values needed to form the protonated superoxide radical and becomes slower as the pH increases. As the pH beneath an activated macrophage adhering to a surface has been reported to be ~5 or lower, production of H_2O_2 should occur at a rapid rate. The pK_a of H_2O_2 is high (~10.5) and, thus, uncharged H_2O_2 is the dominant form at acidic and neutral pH values.

Although they may possess some intrinsic toxic capacity, H_2O_2 and $O_2^{-}\bullet$ are generally considered harmless because they react with biomolecules at low rates and specific enzymes exist to remove them. On the other hand, they may be converted to more potent oxidants such as hydroxyl

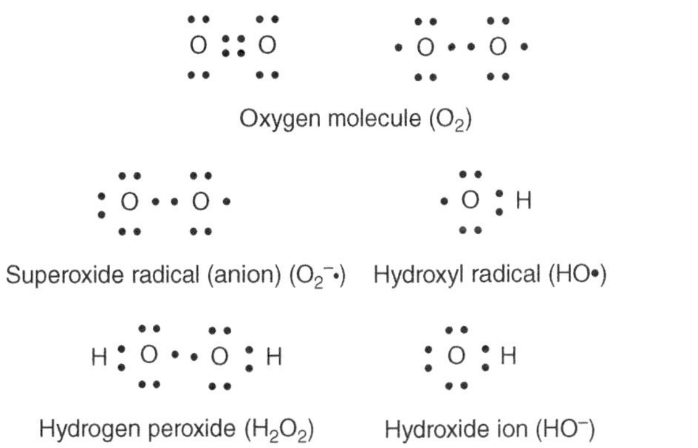

Figure 15.11 Electronic structures of the reactive molecules superoxide radical (anion), hydroxyl radical, and hydrogen peroxide. The electronic structures of the oxygen molecule and the hydroxide ion are shown for comparison.

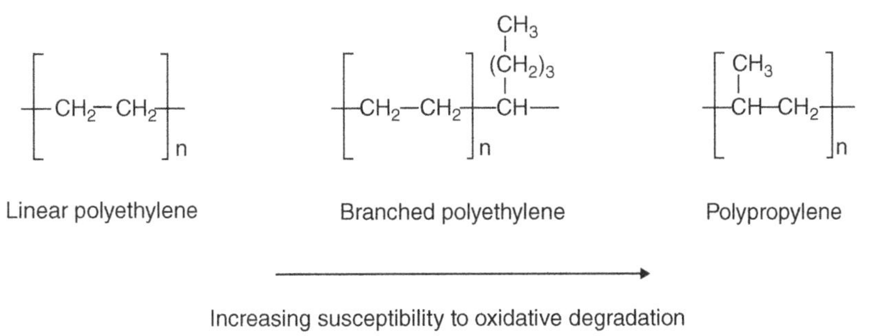

Figure 15.10 Illustration of the susceptibility to autoxidation in the polyolefins linear polyethylene (PE), branched PE, and polypropylene (PP) due to increased chain branching.

HO• radicals, which can lead to degradation by an autoxidation reaction similar to that in Figure 15.8. The most prominent reactions for the formation of these hydroxyl radicals are those catalyzed by the presence of certain metal ions (such as Fe^{2+} or Co^{2+}):

$$O_2^{-}\bullet + M^{n+} \rightarrow O_2 + M^{(n-1)+} \quad (15.4)$$

$$H_2O_2 + M^{(n-1)+} \rightarrow HO\bullet + HO^{-} + M^{n+} \quad (15.5)$$

giving the net reaction

$$O_2^{-}\bullet + H_2O_2 \rightarrow O_2 + HO\bullet + HO^{-} \quad (15.6)$$

where M is a metal ion of charge n or $(n-1)$. The reaction in Eq. (15.6) is often called the Haber–Weiss reaction. On the other hand, when M = Fe and $n = 2$, the reaction described by Eq. (15.5) is often called the Fenton reaction. Overall, the production of the reactive molecules $O_2^{-}\bullet$, H_2O_2, and HO• by phagocytic cells is summarized in Figure 15.12.

Potent radicals or oxidants may also be formed by other reactions. For example, in the presence of the enzyme myeloperoxidase (MOP), H_2O_2 can be converted to hypochlorous acid HOCl:

$$HOOH + H^{+} + Cl^{-} \xrightarrow{MOP} HOCl + H_2O \quad (15.7)$$

and, in the presence of metal ions, HOCl can be converted into HO• radicals according to the equation

$$HOCl + M^{(n-1)+} \rightarrow HO\bullet + Cl^{-} + M^{n+} \quad (15.8)$$

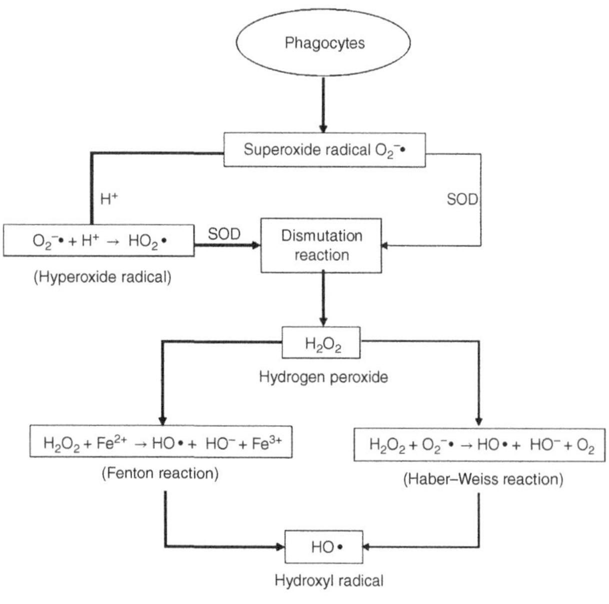

Figure 15.12 Schematic of the production of hydroxyl radicals in the body from phagocytic cells. Bolder lines represent the most prominent reaction and the enzyme superoxide dismutase is abbreviated SOD.

15.4.3 Role of Radicals and Reactive Species in Degradation

Polymeric biomaterials retrieved after long-term implantation in animals or humans, as noted earlier, show some characteristics that are similar to those of autoxidation. Consequently, except for a difference in reactive species, oxidative degradation *in vivo* should generally follow the three-step autoxidation process (Figure 15.8). H_2O_2 produced by phagocytic cells (and enzymes), as we have seen in Section 15.4.1, is converted to more reactive HO• radicals. Because of their high reactivity, most of these HO• radicals will undergo reaction in their immediate vicinity, producing secondary radicals. These secondary radicals and, if possible, a small fraction of the HO• radicals will initiate autoxidation. Then the propagation step begins, in which more H_2O_2 is released and converted to HO• radicals, which, in turn, are converted into secondary radicals, leading to a chain reaction. Termination occurs when two radicals react together.

Figure 15.13 shows one proposed mechanism for the oxidative degradation of a polymer *in vivo*. The highly reactive radicals such as HO• will initiate autoxidation at the surface of the implant, attacking the surface molecules before they have a chance to diffuse into the bulk of the polymer. If certain metal ions (such as Fe^{2+}) are present at the surface of the polymer, they may catalyze the production of HO• radicals, which will also initiate autoxidation at the implant surface. Depending on their lifetime and the accessibility of the polymer to them, the less reactive secondary radicals may diffuse into the interior of the polymer. However, the extent to which they can cause autoxidation is limited because of their lower reactivity. Consequently, degradation by this route occurs essentially at the surface of the implant.

In comparison, because of their lower reactivity, H_2O_2 molecules can diffuse away from their formation sites. If the polymer has good accessibility to H_2O_2 and contains transition metal ions (such as Fe^{2+}), introduced as impurities or catalysts during processing for example, the H_2O_2

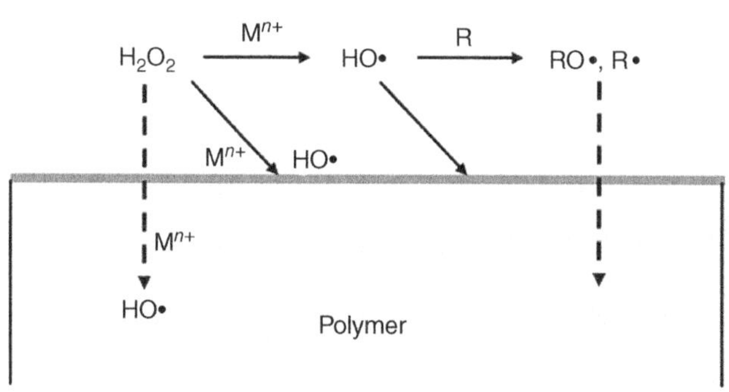

Figure 15.13 Illustration of free radicals attacking an implanted polymer. M^{n+} represents metal ions of charge n+, such as Fe^{2+} for example. *Source:* Williams and Zhong (1991) / with permission of John Wiley & Sons.

can be converted into HO• in the interior of the polymer (Eq. (15.5)). In this case autoxidation can occur within the bulk of the polymer. On the other hand, if the polymer has poor accessibility to H_2O_2, due to a high amount of crystallinity for example, autoxidation will be confined to the surface of the polymer.

15.4.4 Oxidative Degradation of Polymeric Biomaterials

In vitro studies on five different degradable suture materials containing hydrolysable bonds in their chain backbone (Figure 15.14) have shown that superoxide radicals (O_2^-•) can accelerate their degradation over a wide range, depending on their chemical structure, the superoxide concentration, and the degradation time. Because of the reactivity of the superoxide radical in aqueous solutions (Eq. (15.2)), degradation of these sutures has been studied in an aprotic solvent consisting of various concentrations (0.5–10 mM) of potassium superoxide $K^+O_2^-$• dissolved in a mixture of an ether (18-crown-6 ether) and tetrahydrofuran (THF). In these studies, the percentage retention of the tensile breaking force, equal to the tensile force at failure relative to the tensile failure force of the original suture, was used as a measure of degradation.

The tensile breaking force of these five sutures showed a characteristic two-stage pattern consisting of a rapid reduction in the first two hours followed by a less rapid reduction thereafter (Figure 15.15). The rapid reduction in the first stage depended strongly on the chemical structure of the polymer, whereas the reduction in the second stage was approximately independent of the chemical structure. In comparison, when immersed in a saline

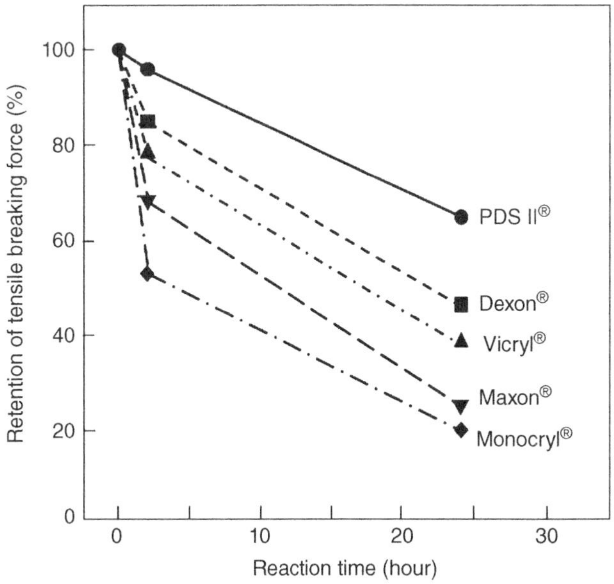

Figure 15.15 Percentage retention of tensile breaking force as a function of reaction time for five commercial suture materials in an aprotic solvent containing 0.005 M superoxide ions at 25 °C. *Source:* Lee and Chu (2000) / with permission of John Wiley & Sons.

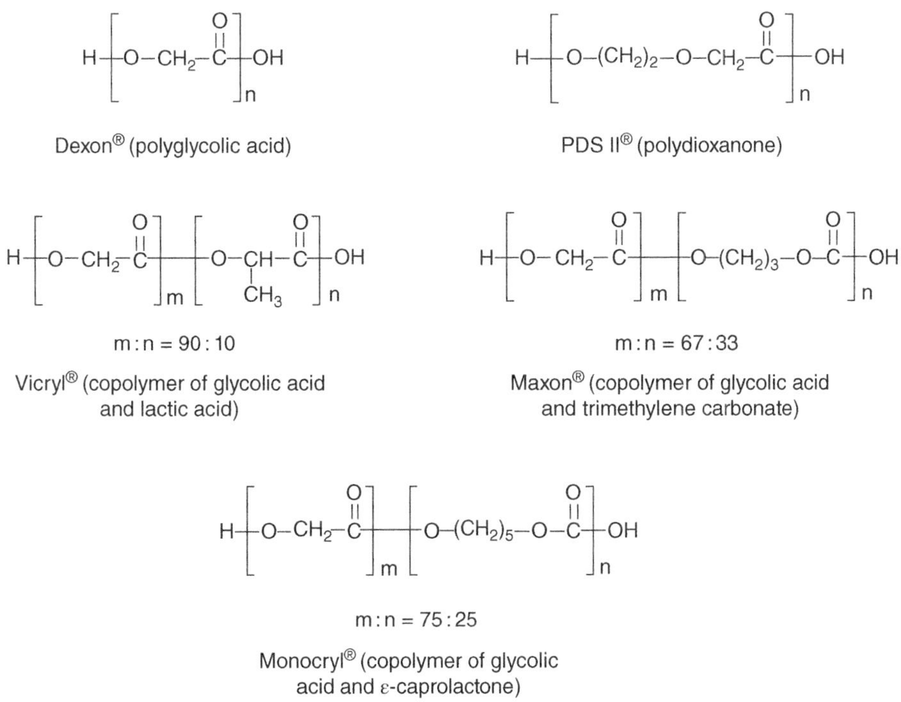

Figure 15.14 Chemical structure of five suture materials.

buffer at the same temperature (25 °C), these sutures did not show any measureable reduction in tensile breaking force even after several days. Varying the superoxide concentration (0.5–5 mM) did not change the two-stage pattern but the reduction in breaking force increased strongly with increasing superoxide concentration. At higher superoxide concentration (10 mM), the extent of degradation is such that the sutures lost their integrity within the immersion period (24 hours).

In the superoxide concentration range 0.5–5 mM, Monocryl® showed the largest reduction in tensile force, followed by Maxon®, Vicryl®, Dexon®, and PDS II®. The influence of the chemical structure on the susceptibility to superoxide attack is not clear. However, a few factors may be noted. The chemical structures in Figure 15.14 show that the polymers least susceptible to superoxide attack, Dexon and PDS II, are homopolymers, whereas the others are copolymers. On the other hand, like Dexon, the chemical structures of the three polymers most susceptible to superoxide attack all have the ester (C=O)O group in their chain backbone. Wetting of the sutures by the immersion medium, which is necessary for degradation, may also be a factor. For example, Monocryl sutures, most susceptible to attack, were found to have better wetting characteristics in the immersion medium.

Examination of the suture surface morphology after immersion in the superoxide solution using scanning electron microscopy (SEM) showed varying degrees of degradation. Monocryl, the suture most susceptible to attack, showed a considerable range of degradation that depended on the superoxide concentration (Figure 15.16).

Overall, these experiments using an aprotic solution show that superoxide radicals can have a strong effect on the degradation of some degradable sutures. On the other hand, superoxide radicals are very reactive in aqueous media and, at the pH beneath an activated macrophage adhering to an implant surface (~5 or less), they are rapidly converted to H_2O_2 molecules, which, in turn, are converted to hydroxyl radicals (HO•) (Section 15.4.2).

In vitro studies have shown that HO• radicals can play a significant role in the degradation of some polymers. These radicals can be easily generated by the reaction of aqueous H_2O_2 solutions with iron(II) sulfate, $FeSO_4$, or cobalt(II) nitrate, $Co(NO_3)_2$, in deionized water (Eq. (15.5)). In one study, cylindrical PCL specimens were incubated in solutions of H_2O_2 containing $FeSO_4$ or $Co(NO_3)_2$ at 37 °C for up to 30 weeks (Ali et al. 1993). The formation of HO• in these solutions was confirmed using electron spin resonance. Specimens immersed in deionized water or in deionized water containing H_2O_2 alone under the same conditions

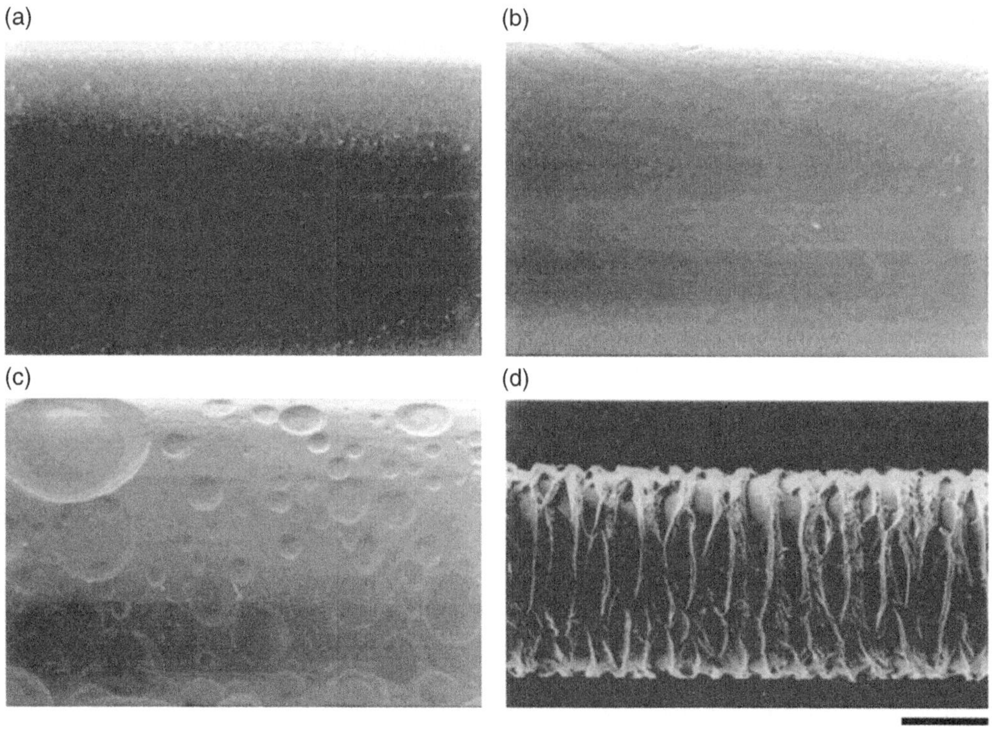

Figure 15.16 Scanning electron microscope images of the surface of Monocryl suture after reaction for 24 hours at 25 °C in an aprotic solvent containing varying concentrations of superoxide ions: (a) 0, (b) 0.0005 M, (c) 0.0025 M, and (d) 0.005 M (Bar = 100 μm). *Source:* Lee and Chu (2000).

served as controls. The number-average molecular weight M_n (or the weight-average molecular weight M_w), used as measure of degradation (Figure 15.17a), showed a gradual decrease with time in water or in H_2O_2 solutions, which can be attributed to hydrolytic degradation of the ester groups in the PCL chain. In comparison, specimens immersed in the H_2O_2 solutions containing $FeSO_4$ or $Co(NO_3)_2$ showed a sharp decrease in M_n (or M_w) with time.

Additionally, PCL specimens exposed to water containing varying amounts of H_2O_2 (1–6%) or to water containing two different amounts of metal ions (10 ppm and 0.1 M) showed an M_n dependence on time that was similar to those exposed to water. Overall, these results indicate that degradation in the H_2O_2 solutions containing $FeSO_4$ or $Co(NO_3)_2$ is due to hydrolytic degradation of the ester groups and radical-induced oxidative degradation. The sharp decrease in M_n (or M_w) in H_2O_2 solutions containing $FeSO_4$ or $Co(NO_3)_2$ is accompanied by an increase in polydispersity M_w/M_n (Figure 15.17b), which indicates that the radical-induced degradation is affecting nearly all the chains of PCL exposed to these solutions.

Another feature of the PCL specimens immersed in the H_2O_2 solutions containing $FeSO_4$ or $Co(NO_3)_2$ is that they showed a significantly higher increase in crystallinity when compared to those immersed in water or in H_2O_2 solutions. While the crystallinity of the specimens incubated in water or in H_2O_2 solutions increased from their initial value of ~57% to ~61–63% after the 30-week immersion period, those immersed in H_2O_2 solutions containing $FeSO_4$ or $Co(NO_3)_2$ showed a significantly higher increase to 80–87%. Overall, this increase in crystallinity is attributed to crystallization of chain segments that have acquired a greater degree of mobility due to chain scission in the amorphous phase of the polymer during the 30-week immersion at 37 °C. The low glass transition temperature of the PCL (approximately −60 °C) facilitates this crystallization because the amorphous phase of the polymer is in a rubbery state at 37 °C, which means that chains in the amorphous phase already have some degree of mobility.

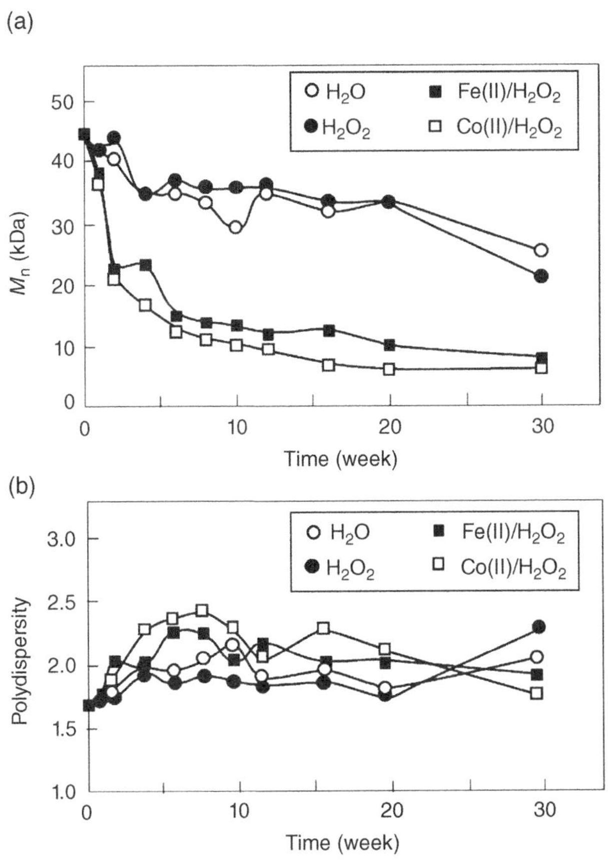

Figure 15.17 (a) Number-average molecular weight M_n and (b) polydispersity of polycaprolactone (PCL) cylindrical specimens immersed in deionized water (H_2O) (control), in deionized water containing hydrogen peroxide (H_2O_2), and in aqueous solutions of H_2O_2 with either iron(II) sulfate (Fe(II)/H_2O_2) or cobalt(II) nitrate (Co(II)/H_2O_2). *Source:* Ali et al. (1993) / with permission of Elsevier.

15.5 Other Types of Degradation

Under certain conditions, polymeric biomaterials may degrade *in vivo* by routes other than the mechanisms described so far. These routes are generally most relevant to biomaterials designed for use in long-term implantable devices and, often, their effects can be prevented or reduced by proper design and processing of the implant. A widely studied system has been PEUs used for electrical insulation of pacemaker leads. PEUs are strongly resistant to hydrolytic degradation and generally perform well for over several years in these devices. However, some devices have shown varying degrees of surface cracking after implantation times of months to a few years. The most common types of degradation have been cracks on the outer surface of the PEU insulation in contact with the tissue and on the inner surface in contact with the pacemaker lead. These two types of degradation have been attributed to the effects of residual stress in the PEU and metal-ion induced oxidative degradation, respectively.

15.5.1 Stress Cracking

Stress cracking normally arises from the interaction between residual stress that may be present in the implanted polymer (particularly within its surface layer) and the physiological environment. Stress cracking has also been referred to as environmental stress cracking but no distinction is made between these two terms in this discussion. The effects of stress cracking normally take the form of shallow cracks within the surface region of the polymer. The extent of stress cracking in PEUs, such as the frequency and depth of the cracks, is often found to be directly related to the amount of residual stress in the implanted polymer and to its ether (soft segment) content.

Phagocytic cells have been shown to play a direct role in stress cracking. For example, subcutaneous implantation of PEU thin sheets in female Sprague-Dawley rats has shown the presence of adherent foreign body giant cells on approximately 20–25% of the film surface five weeks post-implantation. Localized surface cracking was found directly under the adherent foreign body giant cells but not under adjacent areas that were devoid of such cells.

Overall, stress cracking of PEUs *in vivo* is consistent with a proposed two-step mechanism:

Step 1: Oxidative degradation of the PEU surface by phagocytic cells (Section 15.4), which likely leads to the initiation of surface damage

Step 2: The interaction between residual stress that may be present in the PEU and the physiological environment. This leads to crack propagation by enhancing the depth and width of the surface cracks without causing major chemical changes in the bulk of the polymer.

Overall, stress cracking in PEUs may be reduced by

- Proper polymer processing techniques to avoid residual stresses in the polymer, particularly at its surface
- Proper design of the chemical structure of the polymer by reducing its ether content.

15.5.2 Metal Ion-Induced Oxidative Degradation

As noted earlier, crack-like damage has also been observed in some cases on the inner surface of PEU insulation in contact with pacemaker leads. Figure 15.18 shows an often-cited example of this type of damage. While this type of damage is essentially specific to PEU insulation leads, it is important to understand its cause because of the serious consequences it can have, such as failure of the device. Overall, this type of damage has been attributed to metal-ion induced oxidation of the PEU surface in contact with the pacemaker lead. While it shows similarities to stress cracking, the mechanism of metal-ion induced oxidative degradation is different. Studies have shown that metal-ion induced degradation of PEUs is associated with two processes:

- Corrosion of the pacemaker lead to form metallic ions
- Oxidative degradation of the PEU by metal ions or radicals produced by reactions involving these metal ions.

Corrosion of metals, as discussed in Chapter 14, is an electrochemical process involving the combination of an anodic reaction (Eq. (14.14)), which produces metal ions, and a cathodic reaction, such as an appropriate reaction from those given by Eqs. (14.15)–(14.18). While a variety of reactions has been proposed for the initiation of oxidative degradation of PEUs by metal ions, one such reaction is the conversion of H_2O_2 by metal ions such as Fe^{2+} and Co^{2+} into highly reactive HO• radicals. After initiation, a chain reaction begins, propagating autoxidation (Section 15.4.1). If sufficiently high residual tensile stresses are present in the surface layer of the polymer, their interaction with oxidative degradation can lead to cracking in order to relieve the internal stresses.

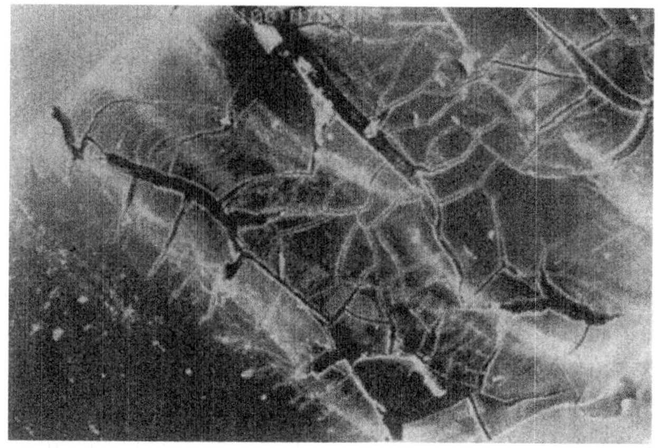

Figure 15.18 Crack pattern on inner surface of polyether urethane (PEU) insulation in contact with pacemaker lead, attributed to brittle fracture caused by metal-ion induced oxidative degradation. *Source:* Coury et al. (1996).

The extent of metal-ion induced degradation generally increases with the polyether content of the PEU, which is not surprising in light of the fact that the H atom attached to the C atom adjacent to the ether O atom is susceptible to abstraction by radicals (Figure 15.9). Consequently, one approach that has been used to reduce the susceptibility of PEUs to metal-ion induced oxidative degradation involves modifying their chemical structure by, for example, reducing the polyether content, or replacing the polyether segment with another polymer, such as a polycarbonate. Overall, however, metal-ion induced degradation involves a complex interaction between the PEU, the device, and the biological environment. For example, approaches that rely on modifying the chemical structure of the polymer are applicable only if they maintain the functionality and biocompatibility of the device.

15.5.3 Oxidative Degradation Induced by the External Environment

Degradation by UV radiation, referred to as photodegradation (or photo-oxidation), is the most relevant means of degradation by an external source for implanted polymers, but the circumstances under which it occurs are rather limited. The main source of UV radiation for biomaterials is sunlight, which has a UV wavelength range of ~300 to ~400 nm in its spectrum. Adsorption of photons by a molecule leads to the transition of an electron from its ground state to its first excited state. The energy change ΔE in the transition is given by

$$\Delta E = \frac{hc}{\lambda} \tag{15.9}$$

where h is the Planck constant, c is the velocity of light, and λ is the wavelength of the radiation. For photons of wavelength ~300 to ~400 nm, ΔE is in the ~300 to ~400 kJ/mol. This energy change, particularly for photons at the shorter wavelength end of the range, is comparable to the energy required for the dissociation of certain covalent bonds, such as the carbon–carbon bond and carbon–oxygen bond.

A prerequisite for photodegradation is that the polymer molecules should have the capacity to absorb the electromagnetic radiation. Many synthetic polymers, such as the polyolefins PE and PP, do not absorb radiation in the UV range of sunlight and, thus, are stable against photodegradation. On the other hand, some impurities, such as additives and metal catalysts used in processing, and certain functional groups, such as ketones and peroxides formed during processing, can significantly improve the capacity to absorb radiation, which will cause a greater amount of degradation. Overall, photodegradation will follow the three steps of initiation, propagation, and termination as described for oxidative degradation (Section 15.4).

Another factor that influences photodegradation is the time of exposure to the UV radiation, particularly the low wavelength range of sunlight, but for most implanted polymers, this time will be small. Photo-oxidative degradation has been implicated in the surface damage of PP sutures used for fixation of intraocular lenses and in the undesirable color and physical property changes of some polymeric maxillofacial prostheses, but a contribution from other means of degradation cannot be ruled out.

15.6 Concluding Remarks

In this chapter, we discussed the influence of the physiological environment on the degradation of polymeric biomaterials. Hydrolysis is the major degradation mechanism for synthetic polymers that contain hydrolytically susceptible functional groups in their chain backbone, whereas natural polymers degrade mainly by enzyme-mediated hydrolysis. However, implantation of a biomaterial can lead to a variety of cellular responses, which can strongly enhance the corrosive nature of the physiological environment surrounding the implant.

- *In vitro* studies have shown that certain enzymes can accelerate the degradation of polymers whose chains contain hydrolytically susceptible functional groups, which indicates that mechanisms exist for enzymes to influence the degradation of polymers *in vivo*. On the other hand, while enzymes are present around an implant *in vivo*, their role in influencing the degradation of the implant is unclear.
- Reactive molecules, such as superoxide radicals and hydroxyl radicals, which are produced by phagocytic cells in response to an implanted biomaterial, have been found to enhance the degradation of polymers *in vitro*. Additionally, the degradation characteristics of implanted polymers often match those for the autoxidation of polymers in an oxidizing atmosphere. However, direct evidence for the occurrence of autoxidation in implanted polymers is limited.
- We discussed other degradation mechanisms such as stress cracking and metal-ion induced oxidation, which are mainly relevant to PUs used in long-term implantable devices.

Problems

15.1 Briefly describe the possible degradation mechanisms for polyglycolic acid (PGA) sutures *in vivo*.

15.2 Implantation of PET fibers in uninfected subcutaneous sites in dogs showed a 50% reduction in tensile strength at 10 ± 2 years. In comparison, when implanted in infected subcutaneous sites in dogs, similar PET fibers lost their tensile strength almost completely within a few months. Discuss the factors that could cause this rapid deterioration in strength.

15.3 Tubular grafts of polyethylene terephthalate (PET) and polytetrafluoroethylene (PTFE) are used to repair a damaged section of an artery (blood vessel).

A Discuss and compare the degradation mechanisms of these two grafts.

B For each graft, would the degradation of the outer surface be different from the inner surface? Explain.

15.4 Polyether urethanes (PEUs) are normally preferred over polyester urethanes (PESUs) for use in long-term implantable devices, such as the insulation on pacemaker leads. Explain why.

15.5 Explain what is meant by the term autoxidation. What is the relevance of autoxidation in the degradation of polymeric biomaterials *in vivo*?

15.6 Briefly discuss how reactive molecules such as superoxide radicals, hydrogen peroxide, and hydroxyl radicals are produced *in vivo*.

15.7 Although they are designed to be resistant to hydrolytic degradation, some polyether urethane (PEU) specimens degrade slowly *in vivo*. Explain what mechanism(s) may cause this slow degradation.

15.8 Polyether urethane (PEU) insulation on some pacemaker leads show degradation characterized by crack-like damage on their outer and inner surfaces after long-term implantation.

A Is the mechanism of degradation on the outer and inner surfaces expected to be the same or different? Explain.

B Discuss the mechanism of degradation on each surface, whether the same or different.

References

Ali, S.A.M., Zhong, S.-P., Doherty, P.J., and Williams, D.F. (1993). Mechanism of polymer degradation in implantable devices. I. Polycaprolactone. *Biomaterials* 14: 648–656.

Coury, A.J. et al. (1996). *Biomaterials Science: An Introduction to Materials in Medicine* (eds. B.D. Ratner, A.S. Hoffman, F.J. Schoen and J.E. Lemons), 257. Elsevier.

Lee, K.-H. and Chu, C.C. (2000). The role of superoxide ions in the degradation of synthetic absorbable sutures. *Journal of Biomedical Materials Research* 49: 25–35.

Li, S. and McCarthy, S. (1999). Effect of crystallinity and stereochemistry on the enzymatic degradation of poly(lactide)s. *Macromolecules* 32: 4454–4456.

Santerre, J.P. and Labow, R.S. (1997). The effect of hard tissue segment size on the hydrolytic stability of polyether-urea-urethanes when exposed to cholesterol esterase. *Journal of Biomedical Materials Research* 36: 223–232.

Schakenraad, J.M. and Hardonk, M.J. (1990). Enzymatic activity toward poly(l-lactic acid) implants. *Journal of Biomedical Materials Research* 24: 529–545.

Williams, D.F. and Mort, E. (1977). Enzyme-accelerated hydrolysis of polyglycolic acid. *Journal of Bioengineering* 1: 231–238.

Williams, D.F. (1979). Some observations on the role of cellular enzymes in the *in vivo* degradation of polymers. In: *Corrosion and Degradation of Implant Materials*. ASTM STP 684 (eds. B.C. Syrett and A. Acharya), 61–75. American Society for Testing and Materials.

Williams, D.F. (1980). The effect of bacteria on absorbable sutures. *Journal of Biomedical Materials Research* 14: 329–338.

Williams, D.F. and Zhong, S.P. (1991). Are free radicals involved in biodegradation of implanted polymers? *Advanced Materials* 3: 623–626.

Zaikov, G.E. (1985). Quantitative aspects of polymer degradation in the living body. *Journal of Macromolecular Science, Part C* 25: 551–597.

Further Readings

Badwey, J.A. and Karnovsky, M.L. (1980). Active oxygen species and the functions of phagocytic leukocytes. *Annual Review of Biochemistry* 49: 695–726.

Coury, A.J. (2013). Chemical and biochemical degradation of polymers intended to be biostable. In: *Biomaterials Science: An Introduction to Materials in Medicine*, 3e (eds. B.D. Ratner, A.S. Hoffman, F.J. Schoen and J.E. Lemons), 696–715. New York: Elsevier.

Gumargalieva, K.Z., Moiseev, Y.V., Daurova, T.T., and Voronkova, O.S. (1982). Effect of infections on the degradation of polyethylene terephthalate implants. *Biomaterials* 3: 177–180.

Zhao, Q., Topham, N., Anderson, J.M. et al. (1991). Foreign-body giant cells and polyurethane biostability: *in vivo* correlation of cell adhesion and surface cracking. *Journal of Biomedical Materials Research* 25: 173–181.

Part V

Biocompatibility Phenomena

16

Biocompatibility Fundamentals

16.1 Introduction

This chapter forms the first chapter of the fifth part of the book devoted to a consideration of biocompatibility phenomena. Biocompatibility, as defined in Chapter 1, refers to the ability of a material to perform with an appropriate host response in a specific application. It is the most critical factor that determines the success of an implanted biomaterial *in vivo*. Biocompatibility is not simply a property of the biomaterial alone but, instead, a property of the interaction between the biomaterial and the tissue into which it is implanted, that is, it is a property of the biomaterial–tissue system.

In general, the ability of a biomaterial to achieve an acceptable host response depends on several factors of the biomaterial–tissue system, such as the composition and properties of the biomaterial itself and the specific tissue into which it is implanted, and the duration of the implantation as well. For example, a biomaterial that is initially biocompatible may later degrade, releasing degradation products such as particulate debris, highly acidic molecules, or certain metal ions that trigger adverse biocompatibility phenomena, such as an inflammatory tissue response (Chapter 18) or an immune hypersensitivity response (Chapter 19).

No biomaterial implanted *in vivo* can be said to be completely biocompatible. Additionally, the biocompatibility phenomena associated with a given biomaterial are complex and can vary from situation to situation. Consequently, the biocompatibility of a biomaterial needs to be considered in the context of the specific application. In general, because failure to achieve an acceptable host response can sometimes have severe consequences, a major goal of biomaterials science is to identify ways to minimize adverse tissue responses.

In this chapter, examples of biocompatibility phenomena associated with well-known implanted devices are first considered to introduce various types of biomaterial–tissue interactions and the consequences of failure to achieve an acceptable host response. This is followed by a discussion of the various properties and characteristics of the relevant cells, biomolecules, extracellular components, and tissues that interact with an implanted biomaterial and affect its biocompatibility. This basic information is designed to provide a background for the biocompatibility phenomena discussed in subsequent chapters.

Overall, then, the major topics that will be discussed in this chapter are:

- Examples of biomaterial–tissue interactions relevant to biocompatibility and the consequences of failure to achieve an acceptable host response.
- Structure and function of various cellular organelles and the extracellular components in which the cells reside.
- Structural characteristics of the various tissues of the body and the way in which they originate.
- Components of blood and the interactions between blood and biomaterials that lead to coagulation.

16.2 Biocompatibility Phenomena with Implanted Devices

The biomaterial–tissue interactions associated with the prosthetic hip implant, the stent, and the intraocular lens (IOL), three widely implanted devices, are considered in the context of introducing examples of various biocompatibility phenomena that can occur with implanted devices.

Total hip arthroplasty is a very effective procedure in orthopedic surgery, which has served to improve the quality of life for millions of patients (Chapter 1). However, the prosthetic hip implant is not the ideal solution to a degenerated hip joint. It has a finite lifetime, determined largely by inflammatory responses of the surrounding tissue to wear particles generated by the articulating surfaces of the device (Figure 21.28). This leads, in turn, to osteolysis and premature failure of the prosthetic joint by aseptic loosening (Chapter 21). This failure, if it occurred suddenly, can

Materials for Biomedical Engineering: Fundamentals and Applications, First Edition. Mohamed N. Rahaman and Roger F. Brown.

Companion website: www.wiley.com/go/rahaman/biomedicalengineering

be disastrous for the patient and clinician. More commonly, however, a major consequence of this failure is that it necessitates revision (repeat) surgery to replace a failed implant, a procedure that is associated with significant morbidity and risks. One effective approach to reducing the adverse tissue response has been the steady improvement in the wear resistance of the articulating bearing couples (Chapter 23).

Stents are used when a blood vessel such as a coronary artery becomes blocked (a process referred to as stenosis) due to the accumulation of plaque, a condition that, in many cases, can lead to a heart attack (Chapter 24). Stents are tubular in shape, normally composed of a durable metal such as stainless steel (316L) or a cobalt–chromium alloy (Figure 1.1g), which are inserted into the artery to compress the plaque and to remain at the site to keep the artery open. A common problem with such metal stents is that they can irritate the lining of the vessel and provoke fibrous scarring, which can cause the vessel to become blocked again (restenosis). Consequently, the procedure may have to be repeated at regular intervals. In order to reduce the problem of restenosis, some modern stent designs incorporate a drug-eluting function by coating the surface of the stent with a thin coating of a polymer in which an appropriate drug is dispersed.

The IOL (Figure 1h) is widely used to replace the lens of the eye and to restore vision in patients suffering from age-related lens deterioration such as the formation of cataracts. However, some IOLs, particularly those composed of hydrophobic polymers, are susceptible to anterior capsule opacification and adherence of the iris to the surface of the IOL (iris posterior synechiae) caused by adhesion of inflammatory cells and lens epithelial cells to the IOL (Chapter 24). One approach to address this problem is to reduce the potential for protein adsorption, the leading cause of cell adhesion and related complications by, for example, covalent bonding of a hydrophilic polymer to the surface of the IOL.

16.2.1 Consequences of Failed Biocompatibility

There are several potential consequences of failure to achieve an acceptable host response to implanted biomaterials, varying from mild to severe and, occasionally, even to the death of the patient, as exemplified by the following examples.

Fully Resolvable Complications

Some complications resulting from biocompatibility failure, such as allergy, inflammation, hyperplasia, hypoplasia/resorption, and foreign body reactions can often be fully resolved. An example of this type of resolvable complication is metal ion-induced hypersensitivity, a disorder that affects ~10–15% of the general population. Metals in contact with the physiological fluid can undergo electrochemical corrosion (Chapter 14), resulting in the release of metal ions that form a complex with endogenous proteins. These complexes can activate the immune system and cause a delayed hypersensitivity response. Delayed hypersensitivity is a complication associated with implants containing nickel, cobalt, and chromium. Overall, a significant fraction of total hip implant failures is caused by metal ion hypersensitivity (Chapter 20).

Partially Resolvable Complications

Although serious, certain outcomes associated with implanted biomaterials are partially resolvable, such as wear debris-induced osteolysis associated with hip implants, coronary stent restenosis, and biofilm-related infections. Osteolysis is a direct outcome of the multinucleated giant cell foreign body response, a serious inflammatory reaction triggered by wear debris (Chapter 18). Restenosis, as noted earlier, is caused largely by irritation-induced growth of fibrous tissue within the stent. Hydrophobic implants present favorable surfaces for adsorption of proteins, which, in turn, become susceptible to biofilm formation by bacteria introduced during surgical implantation. Because the bacteria embedded within and protected by a biofilm layer are extremely resistant to host immune defenses, severe infection often occurs at the implant site.

Unresolvable Pathologies Associated with Failed Biocompatibility

Biocompatibility failure is occasionally linked to serious, unresolvable conditions such as tumorigenesis, seriously painful or crippling neural damage, and autoimmune diseases. For example, there have been rare cases of osteosarcoma (cancer of the bone) and soft tissue tumor (fibrous histiocytoma) associated with metallic orthopedic prostheses, while ~1% of all dental implant procedures carried out each year result in permanent nerve injuries. Serious questions have also been raised about the safety of polydimethylsiloxane (PDMS) (silicone) breast implants. There are concerns that the encapsulated PDMS gel within the implant might escape and trigger inflammatory responses, which can lead to autoimmune diseases such as rheumatoid arthritis and systemic lupus erythematosus whereby the immune system of the body mistakenly attacks healthy tissue.

Mortality Linked with Failed Biocompatibility

The most serious consequences of biocompatibility failures are those that result in organ failure and mortality. An important example is thromboembolic complication brought on by thrombus formation on an artificial heart valve (Chapter 24), which can cause the valve to malfunction and, thus, impede blood flow. Subsequently, there is a high risk of a fatal cerebral stroke or a fatal myocardial infarction if part of the clot is released into the blood stream and becomes lodged in a small blood vessel in the brain. An example of organ failure associated with biocompatibility failure is right ventricular failure after implantation of a left ventricular assist device in patients with chronic congestive heart disease. Although these fatal occurrences are relatively rare, they point to the need for achieving as high a level of biocompatibility as possible.

16.2.2 Basic Pattern of Biocompatibility Processes

Implantation of a biomaterial *in vivo* triggers a complex sequence of interrelated events, which may be divided into various stages (Figure 16.1). The initial stage of these events involves the introduction of the biomaterial into the body by surgical implantation, injection, or other methods of delivery, which brings the biomaterial into contact with blood, living tissue, and interstitial fluid. The second stage, called the causative events stage, always includes the adsorption of proteins to the biomaterial surface, followed by adhesion of cells to the protein-covered surface. In some cases, there will also be causative events resulting from adverse mechanical biocompatibility phenomena due to differences in mechanical properties between the implant and the surrounding host tissue (Chapter 17). For example, fracture fixation plates composed of durable metals such as stainless steel or titanium used to stabilize fractures in the long bones of the limbs often lead to stress shielding of the adjacent bone, resulting often in bone resorption and, eventually, to bone fracture (Chapter 23). Another causative event is the release of ions or molecules due to degradation of an implant.

A progression stage of host responses follows the causative events stage, which is characterized by a sequence of tissue events that make up the inflammatory response. Additional events of the progression stage may include normal wound healing processes such as clot formation and fibrous ingrowth with the formation of scar tissue. The resolution stage, the final stage of the implant response, involves the restoration of normal tissue function, which is also the last step of the wound healing process. This last step of healing at the implant site includes the resorption and remodeling of scar tissue. It may also involve the complete degradation and removal of a degradable implant, bonding to and incorporation of host tissue, or, in many cases, satisfactory fibrous encapsulation to permit full resumption of host tissue activities. A more detailed description of these inflammatory responses is presented in Chapter 18.

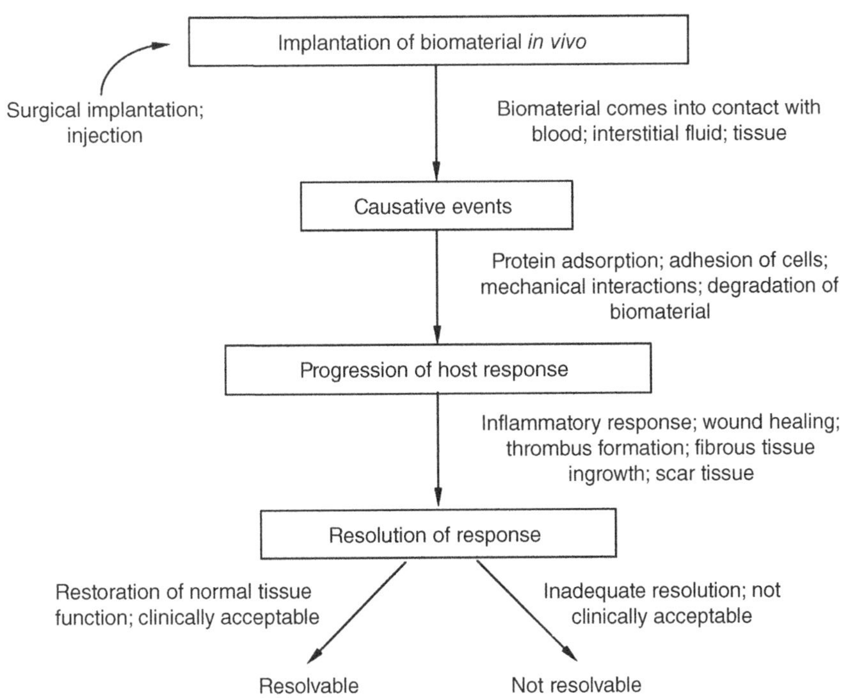

Figure 16.1 Basic stages of biocompatibility phenomena upon implantation of biomaterials *in vivo* and events associated with them. *Source:* Adapted from Williams (2014).

16.3 Protein and Cell Interactions with Biomaterial Surfaces

A major determinant of implant biocompatibility is the adsorption of proteins to the surface of the implant, which is a prerequisite for cell attachment and successful integration of an implant with host tissue. Thus, an understanding of protein adsorption and the interactions of cells with a protein-coated biomaterial surface forms a useful basis for understanding biocompatibility phenomena.

16.3.1 Protein Adsorption onto Biomaterials

As noted in Chapter 5, upon introduction of a biomaterial into the aqueous physiological environment, its surface immediately acquires an electrostatic charge by adsorption of ions from the environment. Then a layer of biomolecules, composed mainly of proteins, rapidly coats the surface by preferential adsorption from the physiological fluid. It is the surface covered with a layer of adsorbed biomolecules, and not the bare surface of the biomaterial, which initially interacts with cells in the physiological environment. In general, it is observed that hydrophobic surfaces adsorb a greater amount of protein when compared to hydrophilic surfaces.

The forces involved in protein adsorption include electrostatic attraction, hydrogen bonding, and van der Waals attraction. In order for protein adsorption to occur, some dehydration of the implant surface must occur, that is, water molecules must be displaced from the surface of the implant. This will be thermodynamically more feasible for hydrophobic surfaces than for hydrophilic surfaces because of the weaker interaction between water molecules and a hydrophobic surface. Adsorption of a protein to an implant surface leads to changes in the secondary structure of the adsorbed protein in which the macromolecule assumes a slightly flattened shape to allow greater interaction of the nonpolar domains of the macromolecule with the hydrophobic surface (Roach et al. 2005). Overall, the driving force for protein adsorption is the decrease in the Gibbs free energy of the system. Water molecules displaced from the surface of the implant take up a more disordered arrangement in the liquid, whereas the protein molecule adopts a different conformation from that in the physiological fluid to allow different segments of the chain to contact the surface and bond with it by the aforementioned forces of attraction.

In general, faster diffusing proteins, often those of smaller size (molecular weight), or proteins with a high concentration in the medium adsorb preferentially to the implant surface in the initial stage of adsorption. Later, these proteins are displaced by larger, more slowly diffusing proteins or by proteins that have a higher affinity for the implant surface, which are capable of being more stably adsorbed to the implant surface. The kinetics of this dynamic displacement process often give rise to a characteristic adsorption curve, in which the amount of protein adsorbed increases with time to a maximum, followed by a decrease to an approximately constant (equilibrium) value (Figure 16.2). Protein adsorption and desorption is a dynamic process and, in this equilibrium region, the amount of protein adsorbed approximately balances the amount desorbed. The adsorption phenomenon depicted in Figure 16.2 is often referred to as the Vroman effect (Vroman et al. 1980).

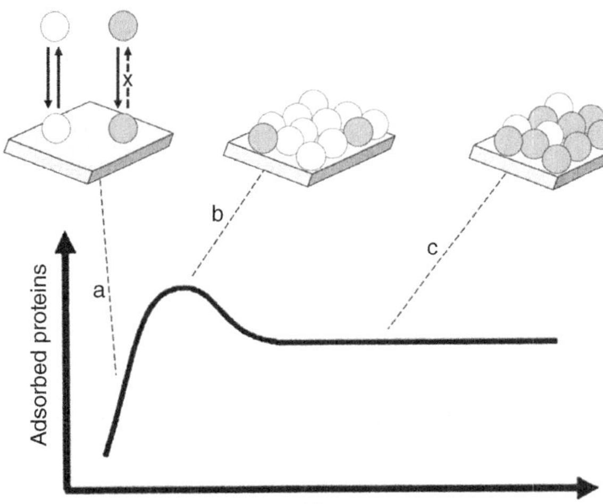

Figure 16.2 The Vroman effect describes the competitive adsorption of molecules onto the surface of an implant, consisting (a) of faster diffusing molecules or molecules of higher concentration in the medium, and more slowly diffusing molecules or molecules of higher surface affinity onto the surface of a material. (b) The faster diffusing molecules or molecules of higher concentration initially adsorb preferentially onto the surface, but they are subsequently displaced by (c) more slowly diffusing molecules or molecules with a higher affinity for the surface. *Source:* From Vroman L, et al. *Blood* 55, 156–159, 1980.

In a medium containing a single type of protein, the amount of protein adsorbed with time will generally follow the trend depicted in Figure 16.2. On the other hand, in a medium containing different proteins with a varying concentration, there will also be a change in the composition of the adsorbed protein layer with time. For example, in blood, the dominant proteins such as albumin and fibrinogen will initially adsorb in greater amounts, but they will later be displaced by vitronectin and fibronectin that have a higher affinity for the surface.

16.3.2 Cell–Biomaterial Interactions

The interaction between a cell and the surface of an implanted biomaterial has sometimes been rationalized in terms of the predictions of the DLVO theory (after Derjaguin, Landau, Verwey, and Overbeek) for the interaction between two colloidal particles in an aqueous ionic solution. Consequently, it is worth examining the basic features of the DLVO theory to understand how it may be related to the interaction between a cell and the surface of an implanted biomaterial.

The basic DLVO theory relates to the interaction of two identical colloidal particles, whose surface is surrounded by an electrical double layer of charges composed of a more compact surface charge and a more diffuse layer of charge extending into the medium (Figure 5.16). There are two components to the interaction, an attraction due to van der Walls forces and a repulsion due to repulsion between the electrical double layers when they overlap (Figure 16.3). At large separation distances, the interaction forces between the particles are negligible. However, as the particles approach closer due to random thermal motion, particularly when the electrical double layers start to overlap, the magnitude of the attractive and repulsive forces increase rapidly with decreasing distance of separation between the particles. Whether the particles can come close enough to adhere or remain dispersed depends on the resultant of the attractive and repulsive forces. For example, if the van der Waals attraction can overcome the repulsion due to the electrical double layers, the particles can approach close enough and adhere together. On the other hand, a sufficiently strong repulsion will prevent the particles from coming close enough to adhere.

Figure 16.3 is a schematic illustration showing one example of the resultant curve for the interaction between

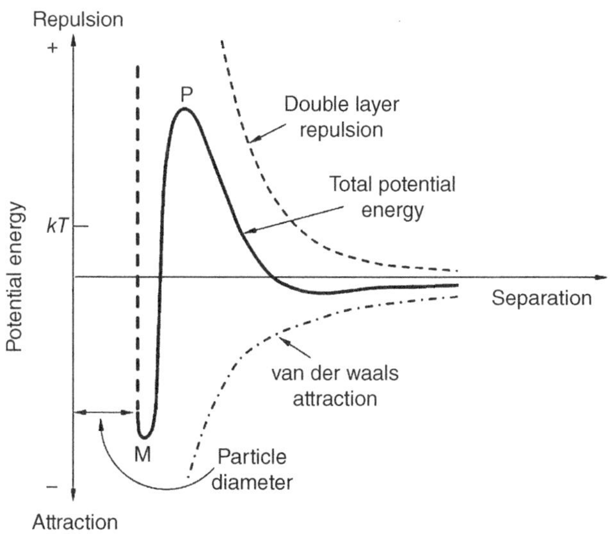

Figure 16.3 Schematic illustration of the potential energy of interaction between two colloidal particles as a function of their separation distance (center-to-center distance) in an aqueous medium based on the DLVO theory. The shape of the potential energy curve can vary widely depending on the conditions.

two colloidal particles of equal diameter, plotted in terms of the potential energy of the interaction versus the separation distance (center-to-center distance) between the two particles, along with the separate contributions from the attractive and repulsive interactions. In general, the shape of the resultant curve can vary widely from system to system, or with a change in conditions, depending on the magnitude of the attractive and repulsive components. The resultant curve in Figure 16.3 shows a peak (P), which can be visualized as an energy barrier that determines whether the particles can come close enough to adhere or remain dispersed. It is useful to consider the height of P relative to the thermal energy (kT) of the colloidal particle, where k is the Boltzmann constant and T is the absolute temperature (in kelvin). If the height of P is much larger than kT, the particles will not have sufficient energy to surmount the energy barrier and, thus will remain dispersed. On the other hand, if the height of P is a few kT or smaller, the particles will have a high probability of surmounting the energy barrier and come close enough to fall into the deep minimum M. In this case, the particles will adhere because their thermal energy is insufficient to allow them to escape from each other.

From the foregoing discussion, it can be said that the DLVO theory provides only an approximation to the forces of interaction between a cell and the surface of a biomaterial. Upon implantation, the surface of the biomaterial will acquire a surface charge due to adsorption of ions from the aqueous physiological medium and an associated diffuse double layer of charges. Subsequent adsorption of proteins from the medium will modify these charges but the way in which the charges are modified will depend on the surface composition of the biomaterial. On the other hand, mammalian cells have a negative surface charge due to phospholipid groups in their cell membrane and are surrounded by mobile ions. In general, attachment of cells to the surface of a biomaterial is promoted when the energy barrier of interaction is low enough to allow cells to approach the surface close enough for adhesion to take place.

As noted earlier, cells are not attached directly to the biomaterial surface but, instead, they are attached nonspecifically to the protein layer coating the biomaterial surface (Figure 16.4). The changes that occur in the secondary structure of proteins adsorbed onto a hydrophobic surface alters the functionality of the adsorbed protein, which, in turn affects cell attachment. As an example, experiments with cultured cells have shown that cells attach more strongly to protein-coated hydrophilic tissue culture dishes than to protein-coated hydrophobic bacteriological dishes. These dishes were composed of polystyrene and had the same size and shape, but they differed in that the tissue culture dishes had been treated in a glow discharge plasma

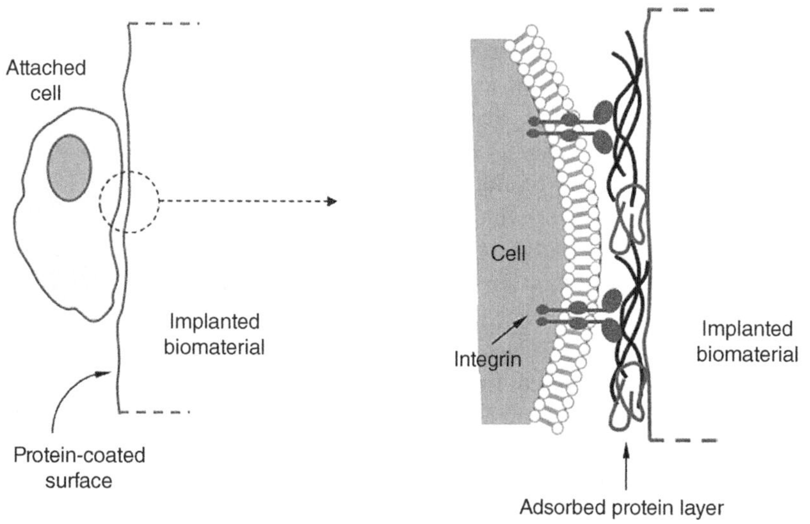

Figure 16.4 Illustration showing that cell attachment to a biomaterial surface occurs by adsorption to the protein-coated biomaterial rather than by adsorption to the actual surface of the biomaterial.

(Chapter 13), which oxidized the polymer chains at the surface, making the surface more hydrophilic.

Additional evidence of this phenomenon has been obtained from assays of the binding of radiolabeled anti-immunoglobulin G (anti-IgG) to tissue culture dishes and bacteriological dishes. These dishes were previously treated with anti-fibronectin, followed by incubation in physiological saline containing 4 μg/ml of fibronectin. The amount of radiolabeled anti-IgG bound to the tissue culture dishes was more than 10 times the amount bound to the bacteriological dishes. These results indicate that the fibronectin adsorbed onto the hydrophilic tissue culture dishes remained in a biological active conformation when compared to the fibronectin adsorbed onto hydrophobic bacteriological dishes where most of the fibronectin is distorted into a biologically inactive conformation (Grinnell and Feld 1982).

The attachment of cells to the surface of an implanted biomaterial triggers a series of host tissue interactions, which are initiated at the cell surface. In order to understand these interactions, it is important to first consider how cells communicate with the extracellular environment and the basic structure of cells and their many organelles.

16.4 Cells and Organelles

As an overview of the cells of the human body, it is useful to consider the number and sizes of cells that make up the body. The total number of cells in the body can only be estimated and the various estimates that have been made vary widely, depending on the methods used. One recent estimate based on the various organs and the many different cell sizes present gives the total number as ~30–40 $\times 10^{12}$. Approximately 80% of these cells are circulating red blood cells (referred to as erythrocytes). There are also ~200 different types of cells, which vary greatly in size and shape. At the low end of the size range are the circulating erythrocytes, each of which has a volume of about 10^{-10} cm^3, whereas the largest cells are the motor nerve cells, many of which have a length greater than ~1 m.

Figure 16.5 shows an illustration of a typical eukaryotic animal cell, which has dimensions of ~20 × 30 μm. The cell is enclosed by a plasma membrane and contains a nucleus and several different specialized parts called organelles which are membrane-bound structures specialized for specific functions. Many cells of the human body show this morphology and size. One of these cells is the parenchymal cell of the liver having a volume of ~3×10^{-9} cm^3, which is ~30 times greater than the volume of an erythrocyte.

16.4.1 Plasma Membrane

The outer layer and selective barrier that encloses the cell is referred to as the plasma membrane (or cell membrane) because it surrounds the cell plasma, also called the cytoplasm. The most widely accepted structural model of the plasma membrane, as well as other cellular membranes, is the fluid mosaic model (Singer and Nicholson 1972). This model (Figure 16.6) depicts the plasma membrane as a dynamic, fluid-like structure consisting mainly of two continuous layers of lipid molecules. The membrane lipids include phospholipids, the principal lipid component molecules, plus cholesterol and glycolipids. Discontinuously interspersed among the lipid layers are many different types of proteins. Some of the membrane proteins are

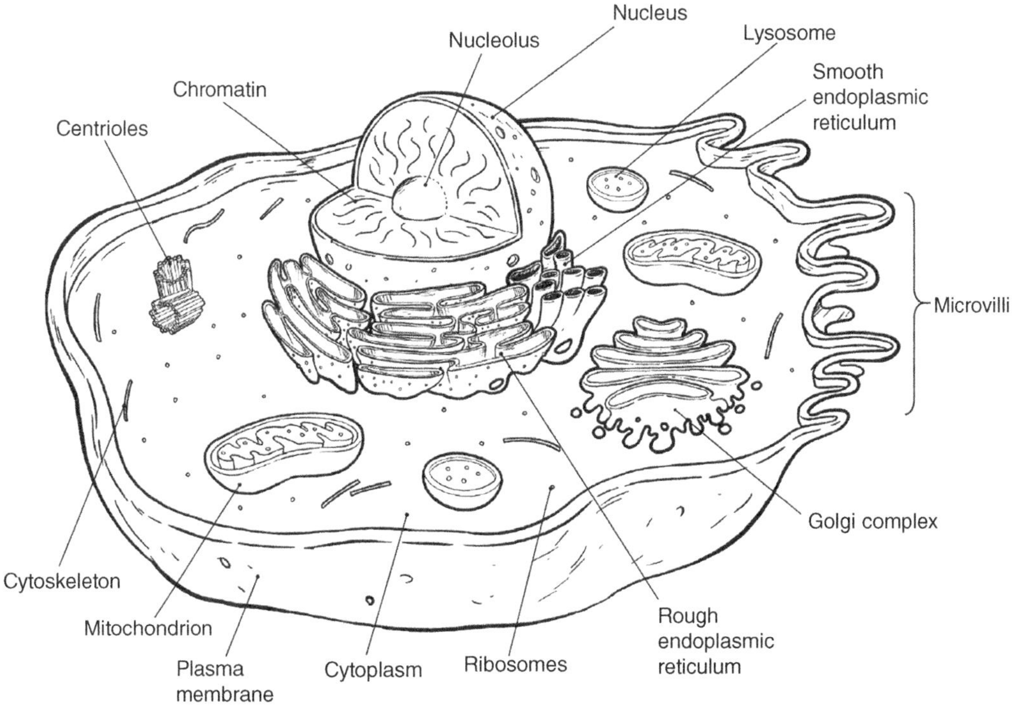

Figure 16.5 Illustration of the structure of a typical eukaryotic animal cell with multiple internal organelles. The relative sizes of the organelles are intentionally distorted to allow the visualization of the organelle architectural features. The relative size of the nucleus is decreased whereas the relative size of all other organelles is greatly increased. *Source:* Courtesy of Tim van de Vall.

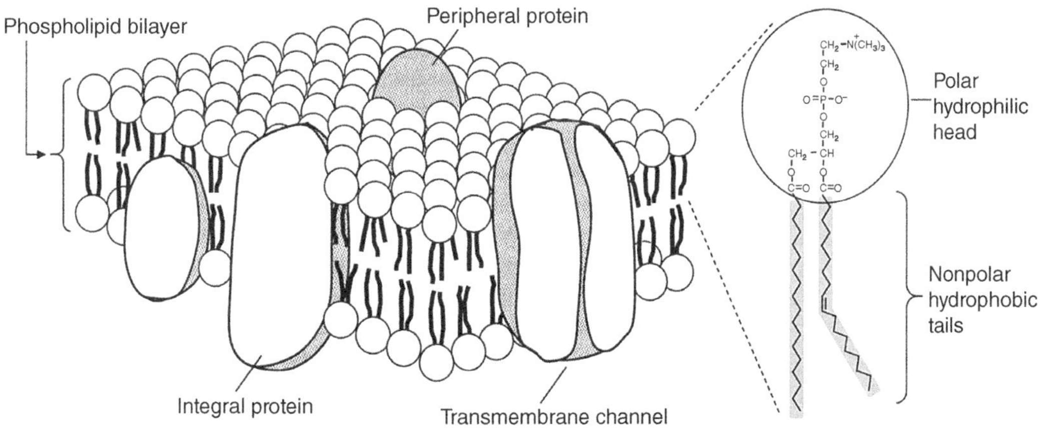

Figure 16.6 The fluid mosaic model of cellular membranes consisting mainly of a lipid bilayer structure plus multiple peripheral and integral proteins. Included is an enlarged view showing the components of phospholipid, the principal component of the membrane bilayer.

loosely associated peripheral proteins whereas others are deeply embedded integral proteins. The different membrane proteins are of various functional types that particularly include structural support proteins, hormone receptors, and enzymes.

An important property of the membrane phospholipids is that they are amphiphilic molecules, which have a chemical structure consisting of a hydrophilic polar group (e.g. phosphocholine) at one end and two nonpolar fatty acid moieties at the opposite end. The phospholipids are homogeneously aligned with the fatty acids of one layer oriented toward the fatty acids of the opposing layer as illustrated in Figure 16.6. This orientation of the phospholipid molecules results in a highly nonpolar, hydrophobic zone within the interior of the lipid bilayer that greatly enhances the stability of the lipid bilayer.

Another important functional property of the lipid bilayer that comprises the bulk of the plasma membrane is that fluid movement of the lipid and protein molecular components of the membrane are largely restricted to lateral diffusion within one of the two layers of the membrane bilayer. There is a very low probability of movement of a phospholipid molecule from one layer to the other layer of the membrane. This type of movement would require reorientation of the phospholipid molecule in a manner that would cause the hydrophilic polar group to pass through the hydrophobic interior, a process that is strongly resisted by electrostatic repulsion.

Table 16.1 shows the different biochemical compositions of three different types of cellular membranes. The measurably higher amount of carbohydrate in erythrocyte membranes is a manifestation of the carbohydrate portions of the glycoprotein type A or type B surface antigens, markers of the major blood types. In comparison, the high lipid content of the membranes of the myelin sheath is related to the role of this sheath as an insulating layer necessary for conduction of impulses along the axons of peripheral neurons. The high protein content of the inner mitochondrial membrane reflects the high content of enzymes that catalyze the many respiratory metabolic processes characteristic of mitochondria (Section 16.4.4).

Table 16.1 Biochemical compositions (in weight percent) of three different types of cellular membranes.

	Biochemical class		
Cell or organelle	**Protein**	**Lipid**	**Carbohydrate**
Erythrocyte (red blood cell)	49	43	8
Myelin membrane	18	79	3
Inner mitochondrial membrane	80	20	0

16.4.2 Cell Nucleus

The cells of the human body are eukaryotic, a term that denotes the presence of a membrane bound nucleus, the largest and most prominent organelle within the cell. The nucleus is commonly referred to as the cell command center because it contains most of the genes responsible for directly or indirectly controlling all cellular processes. The two nuclear events by which control of cellular processes is brought about are replication and transcription. Replication is a multistep, complex sequence by which double-stranded deoxyribonucleic acid (DNA) directs the synthesis of a duplicate double-stranded copy. Figure 16.7 shows some of the key events of replication, including partial separation of the two antiparallel parent strands, followed by pairing of complementary DNA bases, and elongation of nascent DNA chains. The pairing of complementary DNA bases involves pairing of thymine with adenine and guanine with cytosine. Elongation of the DNA chain is catalyzed by the enzyme DNA polymerase which moves in a 5′ to 3′ direction with respect to the direction of

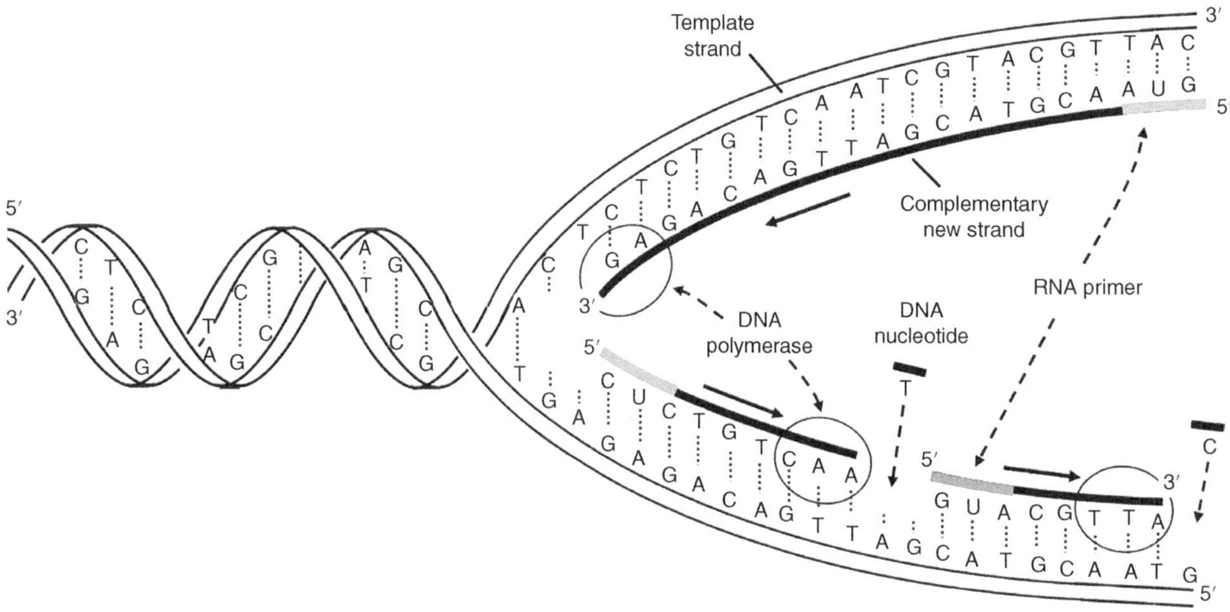

Figure 16.7 The major steps of DNA replication. DNA polymerase catalyzes formation of a continuously elongating copy along one parent strand and intermediate length fragments along the anti-parallel complementary strand. Short segments of RNA primer are required for initiation of the 5′ to 3′ polymerization process.

orientation of the deoxyribose phosphate backbone. The initiation of polymerization requires the presence of short ribonucleic acid (RNA) fragments that serve as primers to which deoxyribonucleotides can be attached and then lengthened by the polymerase enzyme. As shown in Figure 16.7, the polymerization of deoxyribonucleotides results in the formation of a continuously elongating copy along one parent strand while along the antiparallel strand, intermediate length fragments are formed and subsequently joined together by DNA ligase.

Figure 16.8 depicts transcription, the initial phase of gene expression whereby a DNA template strand directs the formation of a complementary single strand of RNA. The process of transcription begins with the partial separation of the two DNA strands allowing free ribonucleotide molecules to undergo complementary base pairing along the DNA template strand followed by polymerization. The complementary base pairing process during transcription is the same as that during replication with the exception that the RNA base uracil pairs with the DNA base adenine. The polymerization of ribonucleotides to form a continuous RNA chain is catalyzed by the enzyme RNA polymerase which, like the enzyme DNA polymerase, moves in a 5′ to 3′ direction. The transcription of a gene sequence is initiated when the polymerase enzyme attaches to a specific DNA region known as a promoter, which is located upstream from the site where transcription begins. Transcription generates three different types of RNA that include messenger RNA (mRNA), ribosomal RNA (rRNA), and transfer RNA (tRNA). Gene directed synthesis of mRNA is a prelude to the formation of the protein encoded by that specific gene. The ribosomes on which the protein synthesis process occurs are globular structures consisting of one large subunit and one small subunit, each composed of rRNA. Each of the 20 different amino acids used to form a protein chain is recognized and transported to the ribosome by a unique tRNA molecule.

Upon completion of mRNA transcription, the transcript exits to the cytoplasm where it directs translation, the second half of gene expression whereby a sequence of triplets of mRNA bases is translated to a corresponding sequence of amino acids of a newly formed peptide chain. Each triplet of mRNA bases is referred to as a codon and the reading of these codons is represented schematically in Figure 16.9. It is noted that there are 64 possible combinations of mRNA bases (codons), which constitute the genetic code that is common to all living organisms.

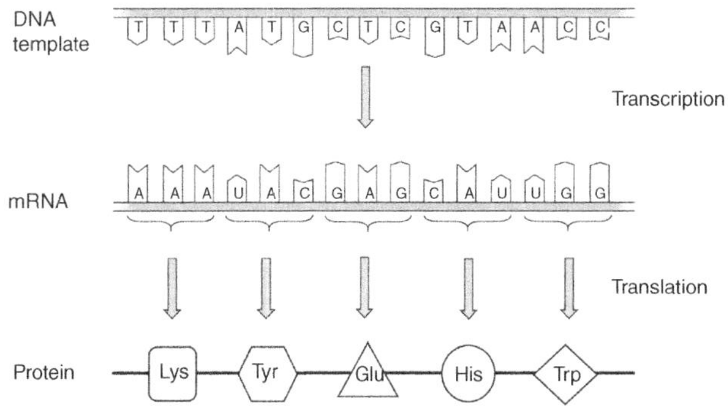

Figure 16.9 The two-part process of gene expression. Part I involves transcription of the coding stand of a gene to a complementary mRNA molecule. Part II involves translation of the mRNA to a peptide chain. As depicted, there is a triplet code in which a specific sequence of three RNA bases codes for a specific amino acid. Each specific sequence of three RNA bases is known as a codon.

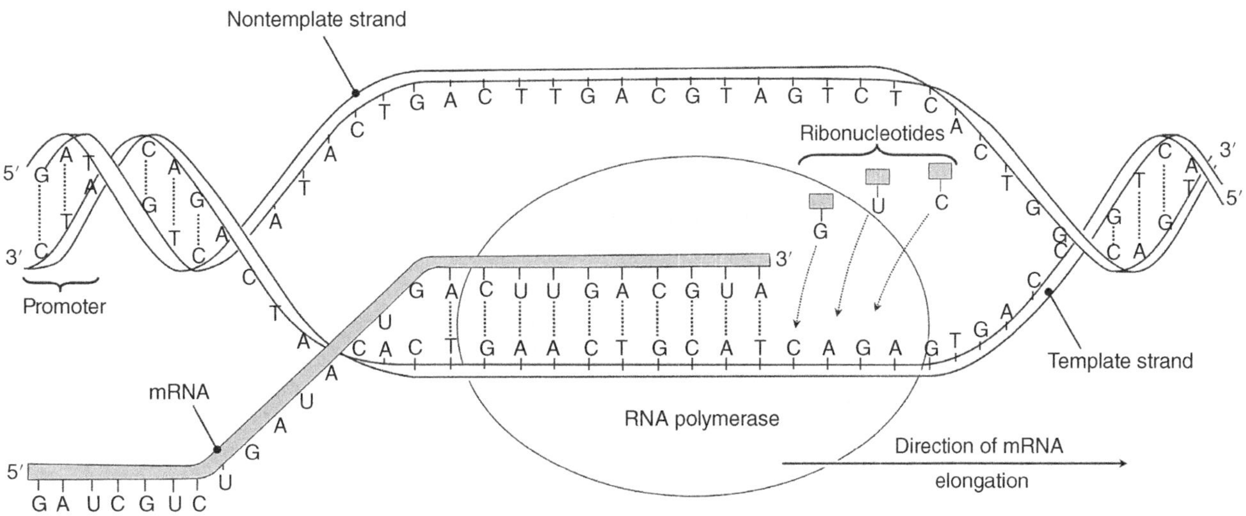

Figure 16.8 RNA transcription involving base pairing along a single DNA template strand followed by polymerization catalyzed by RNA polymerase moving in the 5′ to 3′ direction.

16.4.3 Ribosomes, Endoplasmic Reticulum, and the Golgi Apparatus

Ribosomes, the structures responsible for peptide synthesis (or translation), are not true organelles but they are included here because of their functional relationship to two organelles discussed in this subsection. Some ribosomes are free within the cytoplasm while others are described as bound ribosomes because they are attached to the endoplasmic reticulum (ER) (an organelle described in the next paragraph). The ribosomes of human cells consist of two subunits, a small 40S subunit and a large 60S subunit (Figure 16.10). Present on the surface of the ribosome are three separate groove-shaped sites located mostly on the large 60S subunit and partly on the small 40S subunit. One site is referred to as the peptidyl (or P) site, a second site is the aminoacyl (or A) site, and a third site is the exit (or E) site.

Initiation, the first stage of translation, begins by the formation of an initiation complex that first involves the attachment of an mRNA molecule to the 40S subunit in a manner such that the mRNA start codon (AUG) aligns with the P site (Figure 16.10a). The initiation complex is enlarged by attachment of a tRNA molecule carrying methionine (Met), the amino acid that is always the first amino acid in all newly forming peptide chains. Bonding of the Met tRNA occurs by base pairing between the Met tRNA anticodon (UAC) and the AUG start codon. Formation of the initiation complex is completed by attachment of the large 60S subunit.

The second stage of translation, the elongation stage, commences as a charged tRNA molecule carrying the amino acid specified by the next codon of the mRNA binds to the A site of the large subunit. The two amino acids are then joined by the formation of a peptide linkage catalyzed by peptidyl transferase, resulting in a dipeptide momentarily located in the A site. The peptidyl transferase (or ribozyme), is an integral part of the large ribosomal subunit and one of the very few examples of a nonprotein enzyme. The elongation of the peptide continues as the ribosome moves forward one codon unit along the mRNA, movement that occurs in a 5′ to 3′ direction. As the ribosome advances one codon unit, the peptide bound tRNA is relocated from the A site to the P site while the deacylated tRNA is simultaneously relocated from the P site to the E site. A charged tRNA carrying the amino acid specified by the third codon then attaches to the vacant A site, attachment that causes the deacylated tRNA to be ejected due to allosteric interaction between the A site and the E site (Figure 16.10b). The peptidyl transferase immediately catalyzes formation of the second peptide linkage resulting in elongation to the nascent peptide chain. The elongation continues until the next to last mRNA codon. One of three codons (UAA, UAG, or UGA) will be the last codon, which will signal termination, the final stage of translation. Following the completion of translation, the Met is usually removed during posttranslational processing of the newly formed peptide.

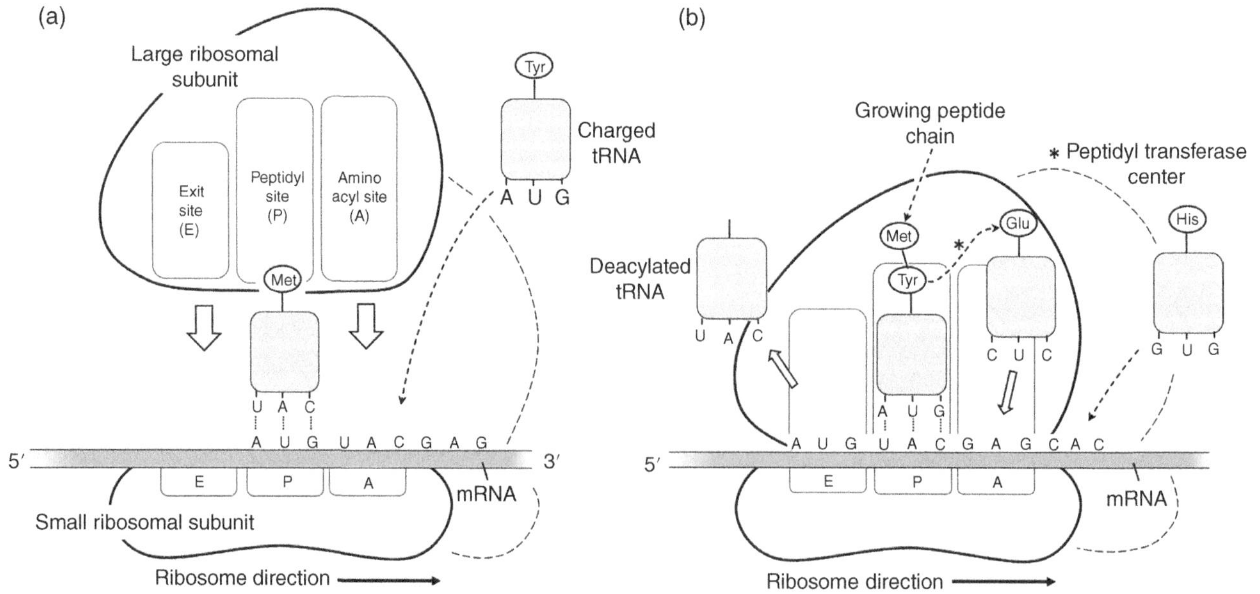

Figure 16.10 Translation of the sequence of mRNA codons to a sequence of amino acids on the ribosome surface. Translation begins by (a) formation of the initiation complex started by coupling mRNA with the small ribosomal subunit followed by bonding of Met tRNA and then the large ribosomal subunit. (b) The elongation stage particularly involves activity of peptidyl transferase and movement of ribosome along the mRNA one codon unit at a time.

The ER is a network of tubes and flattened sacs that includes the smooth ER and the rough ER, the latter covered with ribosomes (Figure 16.11). The smooth ER is a network of tubular channels whose main functions include the synthesis of lipids including steroid hormones such as the corticosteroids secreted by the adrenal cortex plus the detoxification of chemicals such as environmental pesticides. The tubular channels of the sarcoplasmic reticulum (modified smooth ER of skeletal muscle fibers) store calcium ions that are subsequently released to trigger contraction. In comparison, the rough ER serves as a site for the synthesis and subsequent processing of peptides destined for release and utilization outside the cell. A prime example of a product of the rough ER is the protein collagen. As synthesis proceeds to the point at which the elongating peptide reaches the mRNA stop codon, the full-length collagen polypeptide is released from the ribosome and enters the lumen of the rough ER.

The collagen polypeptide, which at this point is referred to as a pre-propeptide, then undergoes one or more modifications during a coordinated series of posttranslational processing steps involving the rough ER and a second organelle known as the Golgi complex, also referred to as the Golgi apparatus (Figure 16.11). Modifications within the rough ER include disulfide bond formation, initial glycosylation and polypeptide folding. The processed proteins, at this point referred to as propeptides, are then packaged into vesicles and transported to the Golgi complex where further processing occurs that particularly includes additional glycosylation plus sulfation of some tyrosine residues. Some of the final completed proteins are packaged in Golgi vesicles and subsequently released from the cell by exocytosis. The Golgi complex also packages degradative enzymes into vesicles that become lysosomes. These lysosomes are organelles that function in the intracellular digestion of particulate material taken into the cell by phagocytic endocytosis (Chapter 21).

The steps of posttranslational processing of polypeptides have been characterized by numerous investigations involving the incorporation of radioactively labeled amino acids followed by autoradiographic analyses. Visualization by autoradiography consistently reveals radioactive amino

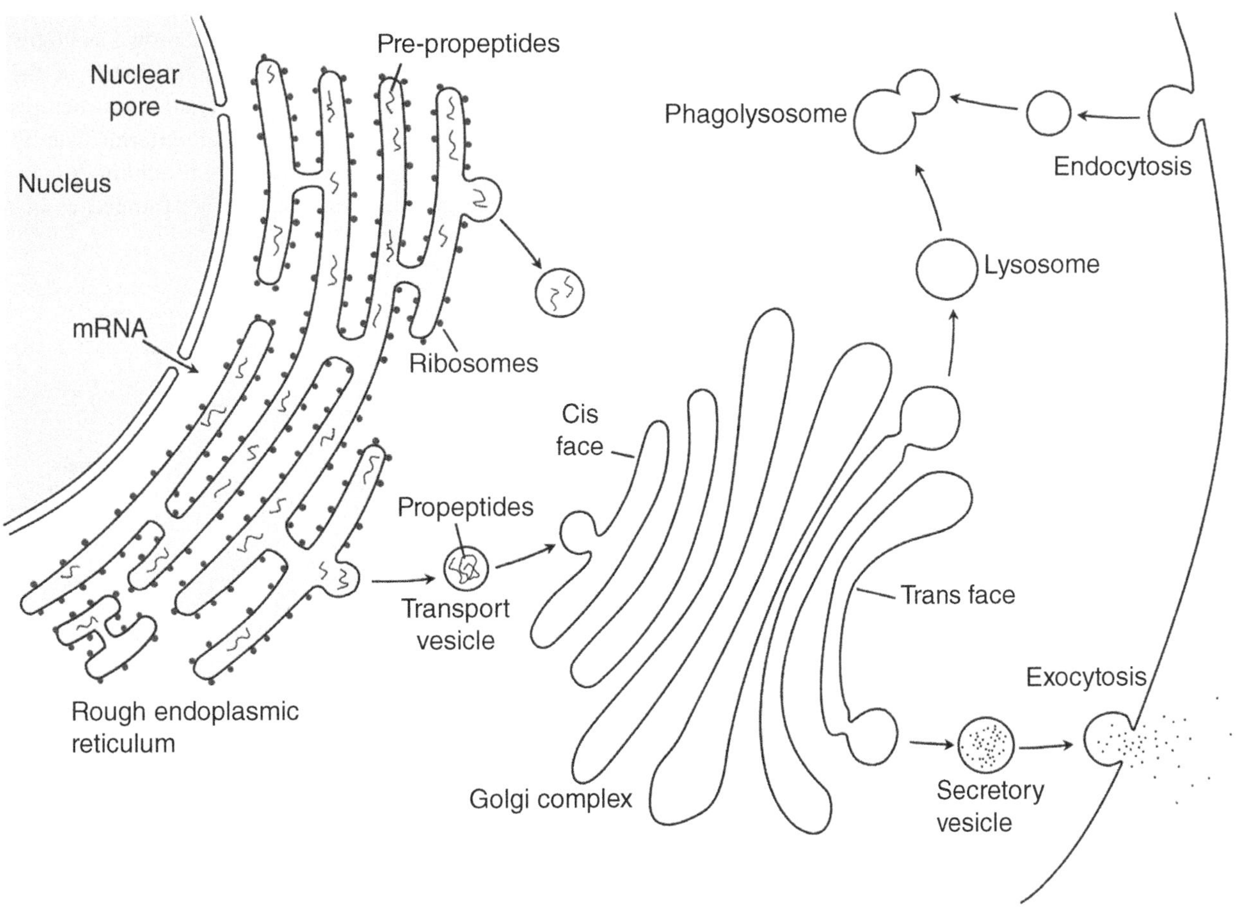

Figure 16.11 Movement of peptides from the rough endoplasmic reticulum to the Golgi complex for additional posttranslational processing and packaging into secretory vesicles for subsequent exocytosis from the cell.

acids appearing first over the rough ER. Radioactive granules next appear over vesicles adjacent to the rough ER, followed by the appearance of granules over nearby vesicles. The radioactively labeled material is then detected within the Golgi complex and nearby secretory vesicles, while the last event is the appearance of radioactively labeled material outside the cell. Thus, the sequence of posttranslational processing (Figure 16.11) proceeds as follows: rough ER → Golgi complex → secretory vesicles → cell exterior.

16.4.4 Mitochondria

Mitochondria are relatively large organelles that are the main sites of aerobic metabolism. These organelles have a smooth outer membrane plus an inner membrane that has numerous folds referred to as cristae (Figure 16.12a). As shown in Table 16.1, the inner mitochondrial membrane consists of about 80 wt% protein, a composition that reflects this membrane's rich enzyme content and high aerobic metabolic activity. Mitochondria possess a circular chromosome with a small subset of the cell's genetic information. These organelles also contain ribosomes that are the sites for the synthesis of proteins encoded by the circular chromosome. The matrix region within the interior of the mitochondrion is the location of the citric acid cycle, the key central metabolic pathway that oxidizes multiple nutrients to generate high amounts of the reduced form of nicotinamide dinucleotide ($NADH_2$), a cofactor central to cellular metabolism. Embedded within the inner mitochondrial membrane are the components of the electron transport chain (Figure 16.12b), the pathway that transports electrons from $NADH_2$ to molecular oxygen. Simultaneous with the electron removal, the electron transport chain moves hydrogen ions (protons) into the intermembrane space resulting in a large hydrogen gradient. Located on the surface of the inner mitochondrial membrane are adenosine triphosphate (ATP) synthase particles, each possessing a handle that extends through the membrane. The large hydrogen gradient provides the force to push H^+ ions through a channel within the ATP synthase which, as a result, functions as a pump that runs in reverse. The operation of this pump, referred to as the "chemiosmotic pump," drives the phosphorylation of adenosine diphosphate (ADP) to ATP, the cellular energy currency that in turn drives most cellular processes.

16.4.5 Cytoskeleton

Mammalian cells possess an internal cytoskeleton that maintains the distinctive shape of each cell type and supports cellular appendages such as cilia and flagella as well as the temporary membrane protrusions known as filopodia and lamellipodia. The principal components of the cytoskeleton are three different filamentous structures called microfilaments, microtubules, and intermediate filaments (Figure 16.13). Microfilaments, the thinnest component of the cytoskeleton, are double-stranded chains

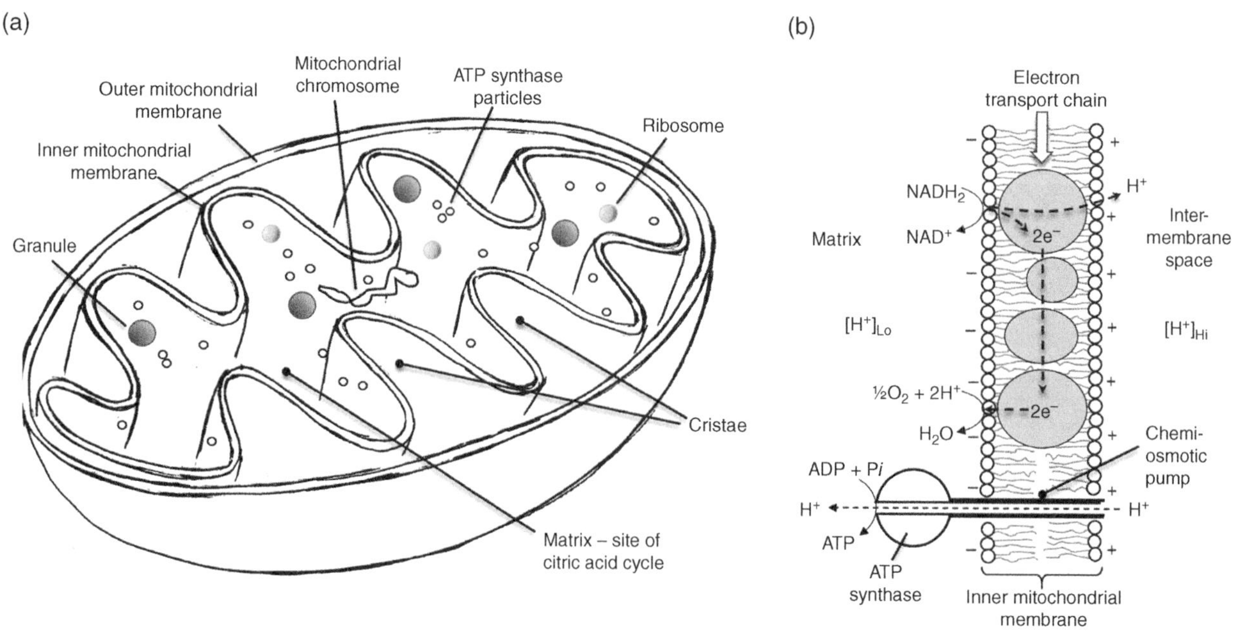

Figure 16.12 Functional architecture of the mitochondrion showing (a) outer mitochondrial membrane and inner mitochondrial membrane with cristae folds. The magnified representation of (b) the inner mitochondrial membrane with the electron transport chain and the chemiosmotic pump that drives the phosphorylation of ADP to ATP by ATP synthase particles.

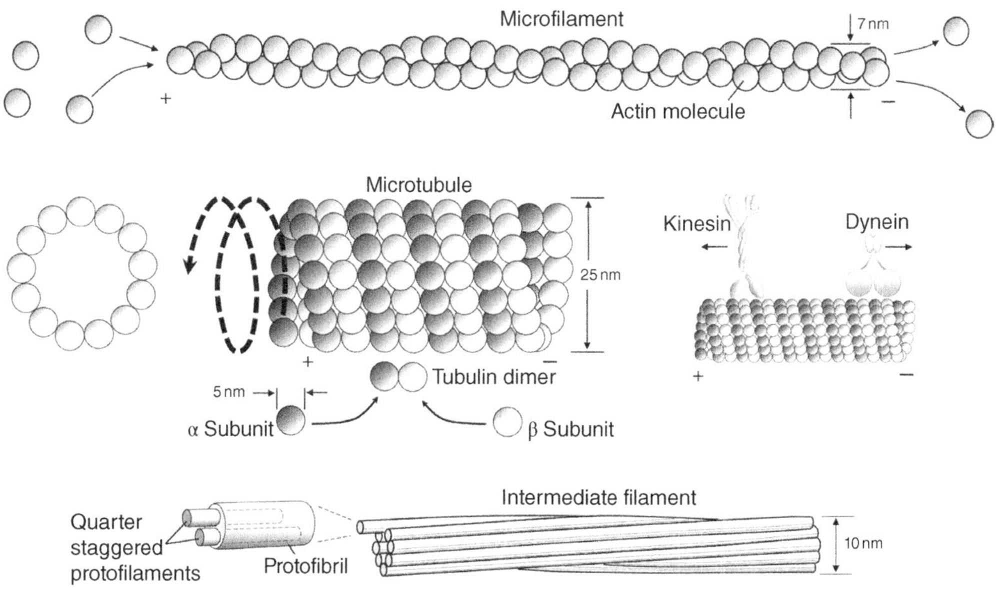

Figure 16.13 Illustration of the microstructure of microfilaments, microtubules, and intermediate filaments, the three types of filamentous components that comprise the cytoskeleton of eukaryotic cells. These dynamic structures are characterized by a lengthening (+) end and a shortening (−) end. Kinesin and dynein, two of the three types of polypeptide molecular motors, are shown with a short microtubule segment.

(~7 nm wide) of the protein actin, the most abundant intracellular protein. Microtubules, the thickest structural component of the cytoskeleton, have a width of ~25 nm, and are composed of α-tubulin and β-tubulin heterodimers. Intermediate filaments, ~10 nm in width, are composed of more than 50 different types of filamentous proteins, such as vimentin, desmin, multiple types of keratin, lamins, and other intermediate filament proteins.

All three of these components of the cytoskeleton are very dynamic and adaptable structures that continuously undergo assembly, disassembly, and reassembly as needed for functioning in cell movement, migration, division, tissue repair, and other processes. The manner in which microfilaments and microtubules form and undergo directional elongation is initiated by nucleation of actin monomers and tubulin heterodimers, respectively. The elongation of microfilaments and microtubules occurs by addition of actin or tubule subunits at the (+) end of each of the two structures while simultaneous removal of subunits occurs at the (−) end (Figure 16.13), normally resulting in no net increase in length. The assembly and disassembly of intermediate filaments occurs by a different mechanism. The different types of proteins that are the building blocks of intermediate filaments consist of soluble subunits that initially organize as two parallel helical chains collectively referred to as a protofilament. One protofilament then associates with a second protofilament in an antiparallel, quarter-staggered alignment to form a tetrameric assembly referred to as a protofibril. The protofibrils, in turn, organize into an array (~10 nm wide) that is the intermediate filament. Unlike microfilaments and microtubules, intermediate filaments do not exhibit polarity due to the antiparallel orientation of the paired protofilaments.

Microfilaments have multiple functions in the cell, which include serving as supportive cytoskeletal components that attach the cell to extracellular anchor sites referred to as focal adhesions (Figure 16.14). Investigations involving the use of fluorescent-labeled antibodies against actin and other protein constituents of the cytoskeleton reveal a characteristic architecture of cytoskeletal actin microfilaments radiating from the focal adhesion anchor sites. Cell attachment and anchoring to the extracellular matrix (ECM) (or to an implant) is essential for the continuance of almost all differentiated functions of mammalian cells. The only occasion when mammalian cells release most of their microfilament attachments to the ECM is during the final phase of mitosis immediately preceding the division of a cell into two new daughter cells. Microfilaments are also the principal cytoskeletal element that supports and maintains the shape of microvilli (microscopic cellular protrusions). In addition, microfilaments are involved in cytokinesis (the process of separation of daughter cells, the final event of mitosis), endocytosis and exocytosis, muscle contraction, and cell motility.

Microtubules also have multiple functions in the cells. Multiple organelles derive from microtubules including cilia, flagella, centrioles, and basal bodies. Microtubules form the spindle fibers of the mitotic spindle fiber apparatus.

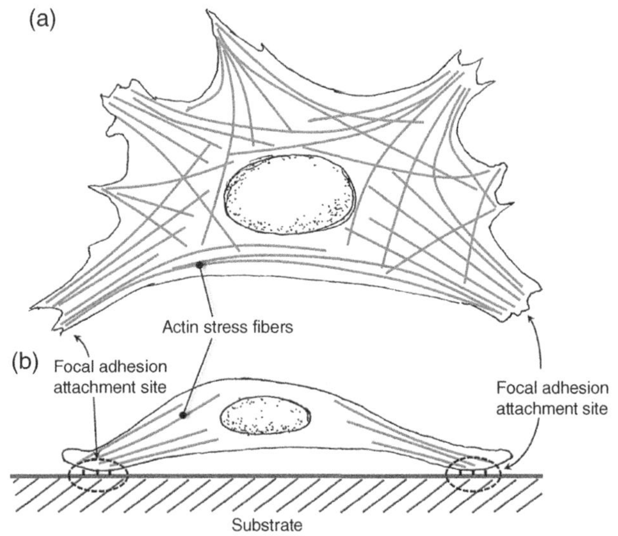

Figure 16.14 Illustration of a typical pattern of cytoskeletal actin microfilaments that radiate from focal adhesions, the sites of cell attachment to a substrate. (a) Superior view showing a spread, well attached cell with actin stress fibers radiating from the points of attachment; (b) lateral view showing components of focal adhesion anchoring the cell to the substrate.

Fully replicated chromosomes attach to individual spindle fibers followed by alignment during metaphase (second phase of mitosis) along the equatorial plate at the center of the spindle fiber apparatus. After separation of daughter chromatids in anaphase (third phase of mitosis) the microtubules undergo directional disassembly and reassembly in a manner that pulls and segregates the daughter chromatids to opposite poles in preparation for separation of the two new daughter cells. Microtubules also act as an axonal conduit along which vessels containing neurotransmitters are moved from the neuron cell body to the axon ending. Intermediate filaments are produced in abundance in epidermal cells to form structures such as hair, fingernails and horns. Intermediate filaments are also the principal supportive element of the two varieties of desmosomes: spot desmosomes and belt desmosomes.

A major mechanism by which cell movement and vesicle transport occurs involves the action of three groups of molecular motors that particularly associate with microfilaments and microtubules. Two types of molecular motors, kinesin, and dynein, are represented as part of Figure 16.13. Kinesin and dynein are both composed of two heavy polypeptide chains wound about each other in a long α-helical configuration plus two light chains in a globular configuration. One example of function involves kinesin driven movement of neurotransmitter containing vesicles along the microtubules in the axons of motor neuron proceeding from the (−) end, located in the cell body, to the (+) end at the axon ending where the vesicles are stored for subsequent release as the final event of an action potential and diffusion across the synaptic gap. In contrast to the $(-)\rightarrow(+)$ direction of kinesin movement, the molecular motor dynein moves in the $(+)\rightarrow(-)$ direction along microfilaments and microtubules. The third type of molecular motor is myosin, the protein that interacts with actin microfilaments to drive the contraction of skeletal muscle.

16.4.6 Cell Contacts and Membrane Receptors

Cells develop contacts with other cells via three different types of cell-to-cell junctions that include tight junctions, gap junctions and desmosomes (Figure 16.15). Tight junctions are complexes of three groups of transmembrane proteins that include occludin, claudins, and junction adhesion molecules. Occludin is a protein of molecular weight ~60 kDa, with two extracellular loops, whereas claudins are smaller proteins (molecular weight 18–27 kDa) that also possess two extracellular loops. The junction adhesion molecules are glycosylated proteins within the immunoglobulin superfamily. Tight junctions serve to join adjacent epithelial cells in a manner that effectively precludes diffusion of polar materials through the extracellular space between the cells. This type of cell-to-cell junction is of critical importance in areas such as the endothelial cells that form the blood–brain barrier.

Gap junctions consist of complexes of six proteins referred to as connexins that form portals for the passage of ions and other substances between adjacent cells. A major example of gap junctions is the type of portals between cardiac muscle cells of the ventricles. The passage of calcium ions via these myocardial portals is responsible for the synchronization of left ventricular contraction that allows effective propulsion of blood through the systemic circulation. Desmosomes are present in two forms, spot desmosomes and belt desmosomes. Spot desmosomes provide a rivet-like reinforcement for cell-to-cell attachments subjected to high amount of lateral stress such as in the lining of the bladder. The filamentous proteins known as cadherins account for the tensile strength of spot desmosomes and the linkage of those structures with intermediate filaments within the cell interior. Belt desmosomes are composed of the same molecular constituents organized in a continuous belt-like array.

The various interactions of cells with extracellular entities are mediated by protein cell surface receptor molecules, two of the most important of which are cadherins and integrins. Cadherins, which are so-named due to their dependence on calcium for function, are monomeric proteins that provide part of the attachment of cells to extracellular focal adhesions (Figure 16.14). The integrins are

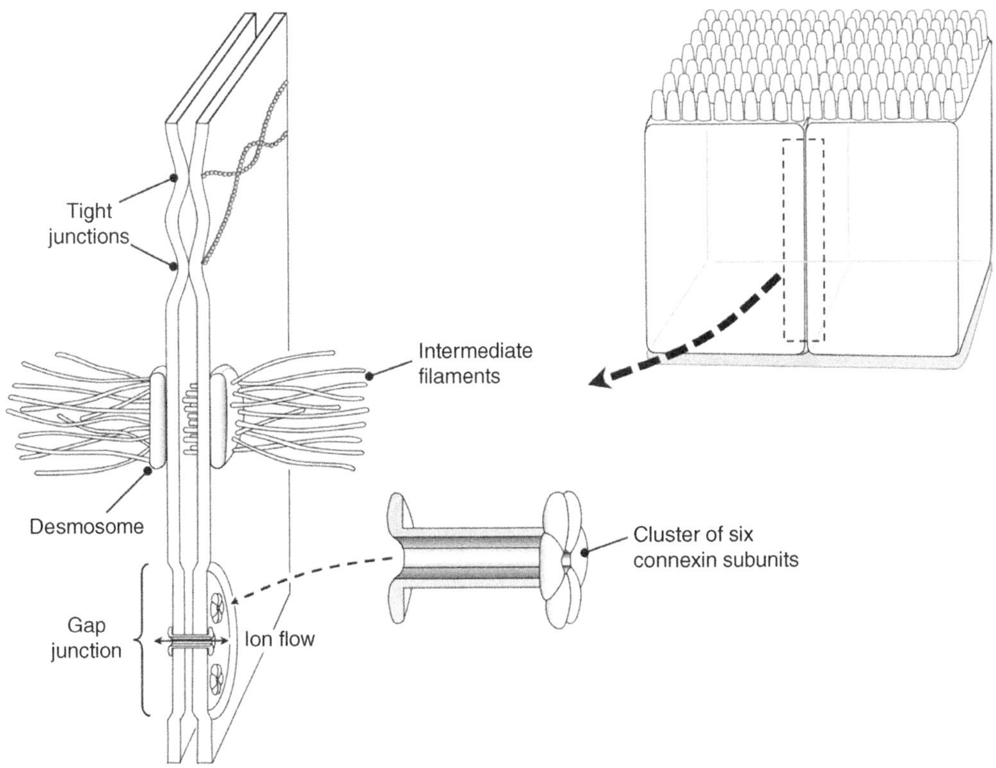

Figure 16.15 Illustration of types of cell-to-cell junctions including tight junctions, desmosomes and gap junctions found predominantly in epithelial tissues (example shown) and muscle tissues.

heterodimeric transmembrane proteins that have particular importance in cell attachment to components of the ECM (Section 16.5).

16.5 Extracellular Matrix and Tissues

16.5.1 Components of the Extracellular Matrix

The cells of the various tissues of the body are embedded in an ECM composed of polysaccharides and fibrous proteins. Much of the ECM consists of a glue-like mix of amino-sugars called glycosaminoglycans (GAGs), which are space-filling hydrogels that provide flexible support. Two examples of the five different types of GAG are hyaluronic acid (Figure 10.16), the GAG that makes up the bulk of hyaline cartilage, and keratan sulfate (Figure 10.18), the highly hydrated material that functions as a weight-bearing cushion in joints. The main protein of the ECM is collagen, the most abundant protein in the body. As discussed in Chapter 10, there are over 20 subtypes of collagen, each of which consists of three polypeptide chains, one $\alpha1$ helix and two $\alpha2$ helices, which assemble within the cell to form a triple-helical procollagen molecule (Figure 10.1). After secretion from the cell, the splayed ends of the procollagens are removed by extracellular proteinases to form collagen molecules. These collagen molecules assemble spontaneously into collagen microfibrils, which, in turn assemble into collagen fibers.

In addition to filling the space between cells, the ECM helps bind similar cells together to form tissues. The ECM is responsible for the distinguishing physical characteristics of some tissues such as the hardness of bone, the flexibility of cartilage, and the tensile strength of tendons. The ECM also directly influences cell migration, cell proliferation and differentiation, and many other cellular functions.

16.5.2 Attachment Factors

A number of different proteins function as attachment factors that bind cells to components of the ECM. Some specific protein attachment factors are fibronectin, chondronectin, osteonectin, and laminin. Fibronectin is particularly associated with fibroblasts and connective tissue. Chondronectin is the attachment factor for cartilage, while laminin is the principal attachment factor for epithelial tissues.

Fibronectin has been well characterized with regard to its chemical composition, structure, and function. The

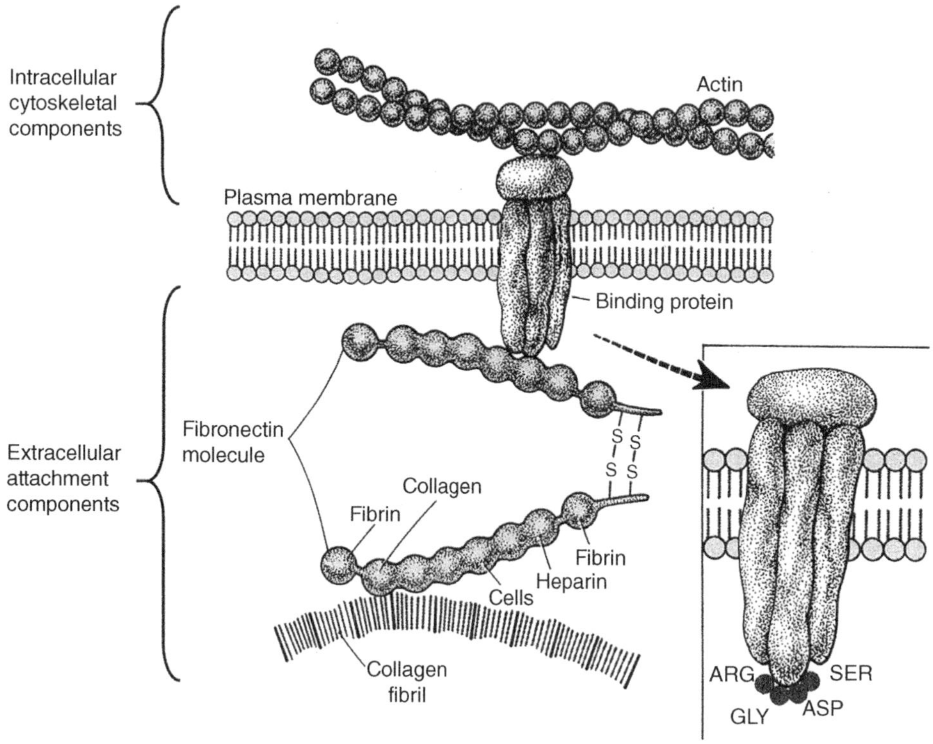

Figure 16.16 Mechanism of cell attachment to the extracellular matrix mediated by fibronectin interaction with integrin on one side and collagen on the other side. *Source:* From Carlson (1988) / with permission of McGraw-Hill.

fibronectin attachment factor consists of two peptide subunits, one of which bonds to collagen while the other bonds to the cell surface receptor protein integrin. The portion of fibronectin that is recognized by integrins to effect cell attachment is the tripeptide sequence arginine-glycine-aspartate (RGD). By functioning in this manner, fibronectin acts to mediate cell attachment to the ECM (Figure 16.16). The interactions of cells with the ECM that are mediated by fibronectin and other attachment factors have major effects on the proliferation and differentiation of cells making up the various tissues.

16.5.3 Cell Adhesion Molecules

In addition to the extracellular protein attachment factors, various transmembrane molecules function both in the adhesion of cells to the extracellular components and in cell-to-cell adhesion. These transmembrane cell adhesion molecules include integrins, immunoglobulin G (IgG) proteins, selectins, and cadherins.

Integrins are a family of transmembrane glycoproteins composed of two heterodimeric subunits referred to as α and β. The integrins form a reversible, low affinity binding between cytoskeletal actin filaments and the ECM. This reversible binding allows cells to detach easily and move to a new location. The integrins also function as receptors for both "inside-out" (intracellular to extracellular) signaling and "outside-in" (extracellular to intracellular) signaling. During the signaling activation, the extracellular portion of the integrin molecule undergoes a dramatic change of conformation, changing from an inactive configuration resembling an inverted V to a fully extended active form (Figure 16.17). Ligand binding to the active form causes the tail portion of the β subunit to shift laterally, the event thought to trigger "outside-in" signaling (Chapter 17).

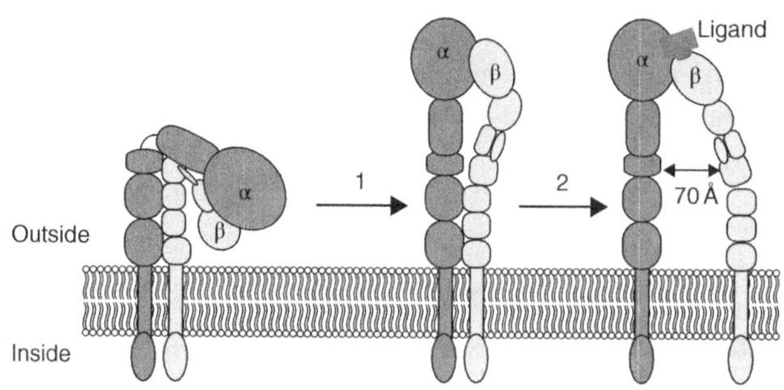

Figure 16.17 Ligand binding to the active form of integrin causes the tail portion of the β subunit to shift laterally, the event thought to trigger "outside-in" signaling. *Source:* From *Integrins, Cadherins, Signaling and the Cytoskeleton*. Curator: Larry H. Bernstein, MD, FCAP, 10/30/2014 / with permission of Pharmaceutical Intelligence.

Part of the IgG superfamily of surface glycoproteins includes five intercellular adhesion molecules (ICAMs) referred to as ICAM-1 to ICAM-5. The ICAMs are present on the surface of endothelial cells and immune cells. ICAM-1 functions as a receptor on the endothelial cell surface that facilitates the movement of leukocytes across the wall of blood vessels to extracellular sites of infection. The expression of ICAM-1 increases dramatically in endothelial cells at sites of inflammation and infection.

The selectins are a family of transmembrane adhesion molecules that consist of a single glycoprotein chain. An especially important member of this family is P-selectin found in endothelial cells. The endothelial cells of normal noninflamed tissues maintain P-selectin in storage vesicles referred to as Weibel–Palade bodies. The release of inflammatory cytokines causes P-selectin to translocate to the endothelial cell surface where it is used in transient, low affinity binding of neutrophils and other leukocytes. This low affinity binding of neutrophils is an important part of a process known as extravasation (Chapter 18), in which these cells pass through the vessel wall to the extracellular space where they perform multiple functions to defend tissues from invasive microbes.

As noted earlier, the name cadherin reflects the dependence of this family of transmembrane glycoproteins on calcium ions to function as cell adhesion molecules. There are about 30 types of cadherins, most of which are found in the brain where there is a vast number of very specific cell-to-cell contacts. The cadherin cell-to-cell contacts are described as homophilic because they involve a cadherin on one cell linking to a cadherin of an adjacent cell.

16.5.4 Molecular and Physical Factors in Cell Attachment

Attachments between a cell and an implant can vary in contact distance and strength. The principal type of biological contact of a cell with a substrate is the focal adhesion, an attachment mediated by the transmembrane protein integrin and the attachment factor fibronectin (Figure 16.16). Various physical factors can influence the attachment of cells to an implant, one of which is the topography of the implant surface (Chapter 21). Cells tend to adhere more easily to surfaces having certain roughness features, porosity or texture than to smooth surfaces.

16.5.5 Tissue Types and Origins

There are four basic types of tissues in the human body, which include the epithelium, connective tissue, muscle, and nerve. Examples of these four tissues, together with some of the distinguishing morphological characteristics of each tissue type, are illustrated in Figure 16.18. All epithelial tissues possess a free apical surface, tight packing of

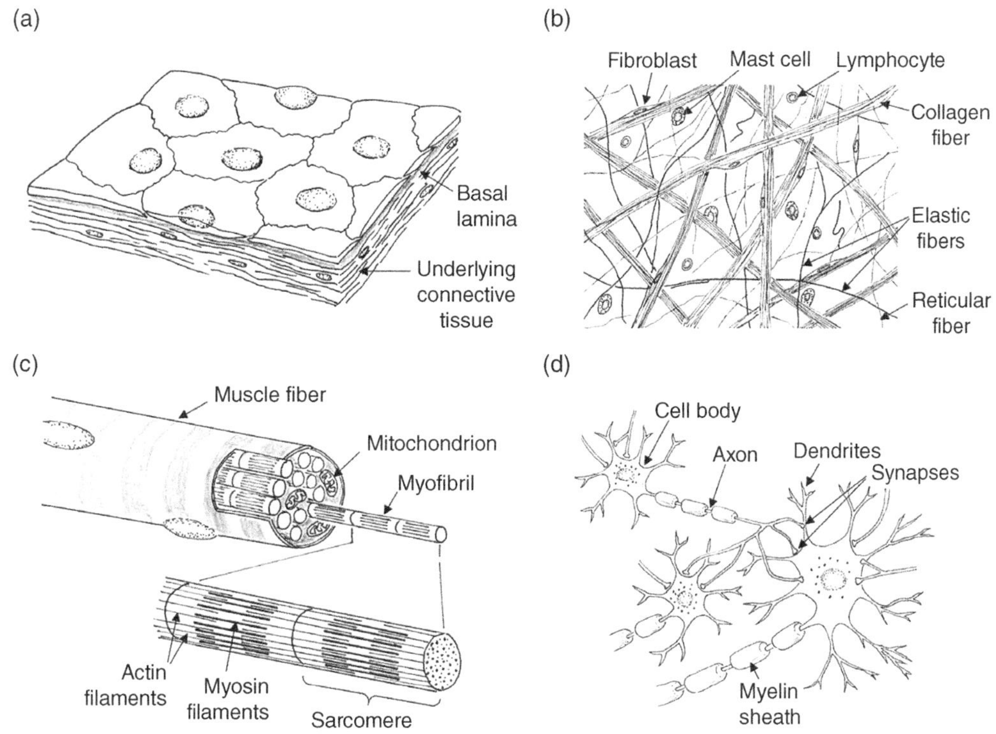

Figure 16.18 Illustration of the structure of the four basic types of mammalian tissues: (a) epithelium; (b) connective tissue; (c) muscle tissue; (d) nerve tissue.

component cells, absence of blood vessels within the tissue, and an underlying basal lamina that attaches the tissue to the underlying connective tissue. In comparison, all connective tissues are characterized by loosely packed cells lying within an abundance of extracellular material. The epithelial and connective tissues are referred to as non-excitable tissues because they are unable to conduct impulses, whereas muscle tissues and nerve tissues are classified as excitable tissues because of their ability to conduct impulses. Skeletal muscle tissue, the muscle tissue shown in Figure 16.18, is commonly referred to as voluntary because it is under conscious voluntary control, whereas cardiac and smooth muscle tissues are involuntary because they are not under conscious control. Also shown in Figure 16.18 is a multipolar neuron characteristic of the nerve tissue that conducts motor nerve impulses to skeletal muscle tissues.

Epithelium

Epithelial tissues function in protection, absorption, and secretion. There are two categories of epithelium, based on cell shape and the number of cell layers. The cell shape subcategories include squamous, cuboidal, and columnar. The number of cell layer subcategories include simple epithelium consisting of one cell layer only and stratified epithelium consisting of multiple layers. The epidermal layer of the skin is an example of stratified squamous epithelium. Much of the very important protective function of the epidermis is attributable to the cell-to-cell adhesion that holds the cells tightly to each other. Sitting beneath all epithelial tissues is the basal lamina, the layer that functions as a glue-like adhesive that bonds the epidermis to the underlying dermis.

Connective Tissues

Connective tissues are the most abundant tissue of the body. Their principal anatomical characteristics include a rich vasculature (present in most but not all connective tissues), an extensive ECM and the absence of a free surface. They function to provide structural support and to supply metabolic needs. The types of connective tissues include what is known as connective tissues proper, which include loose connective tissue (also known as areolar tissue), dense connective tissue such as tendon and ligament, three types of cartilage (hyaline cartilage, fibrocartilage, and elastic cartilage), bone (compact and trabecular), and blood. The nature of the ECM accounts for the most recognizable physical characteristics of connective tissues such as the hardness of bone, which is due to an extensive hydroxyapatite mineral matrix, or the flexibility of cartilage, which is mainly attributable to the highly flexible hydrated hydrogel matrix.

Embryo Layers and Tissue Origins

During the early stages of development, the human embryo transitions from zygote to morula to blastula and then to a gastrula (Figure 25.4). The three basic layers of the body, the ectoderm, mesoderm, and endoderm, form during the gastrula stage of development. It is from these three layers of the embryo that all of the various tissues and organs derive (Figure 16.19). The ectoderm gives rise to the epidermis, nervous tissues, and many glands, the mesoderm gives rise to the three types of muscle tissue, all connective tissues, dermis, many epithelia tissues, and the gonadal organs, and the endoderm gives rise to liver, pancreas, lungs, and the lining of the gastrointestinal tract.

Tissues, Growth Factors and Cytokines

There are numerous chemical factors that control the development and activities of tissues, among which are various growth factors and cytokines that induce tissue-specific responses. As their name implies, growth factors have a positive, growth promoting effect while cytokines are considered neutral since they have both negative and positive effects on growth. Cytokines and growth factors are peptide secretions that exert effects on cells and tissues by autocrine, paracrine and endocrine signaling mechanisms. Autocrine and paracrine secretions both function within a short distance from the source of the secretions. Cells that secrete autocrines and those affected by autocrines are of the same cell type whereas cells affected by paracrines are of a different cell type than those that secrete paracrines. Growth factors and cytokines that control tissues by endocrine signaling are transported via the blood to distance sites to exert control of the target tissue. The common manner by which growth factors and cytokines exert control of target tissues involves initial binding with a specific cell surface receptor. The interaction with a receptor leads to the activation of one or more signal transduction pathways, each a cascade of intracellular biochemical reactions that culminate in a cell response. Protein phosphorylation and dephosphorylation are key steps in most of the many different signal transduction pathways.

The first growth factor discovered was nerve growth factor (NGF), followed a decade later by the epidermal growth factors (EGFs), a family of multiple growth factors that are the most well characterized of the various mammalian growth factors. Continuing research resulted in the discovery of several additional growth factors such as fibroblast growth factor (FGF), platelet-derived growth factor (PDGF), transforming growth factor-β (TGFβ), vascular endothelial growth factor (VEGF), and several other growth factors. The discoveries also included many different cytokines such as the large family of interleukins, the chemotactic cytokines, referred to as chemokines, interferons, and other

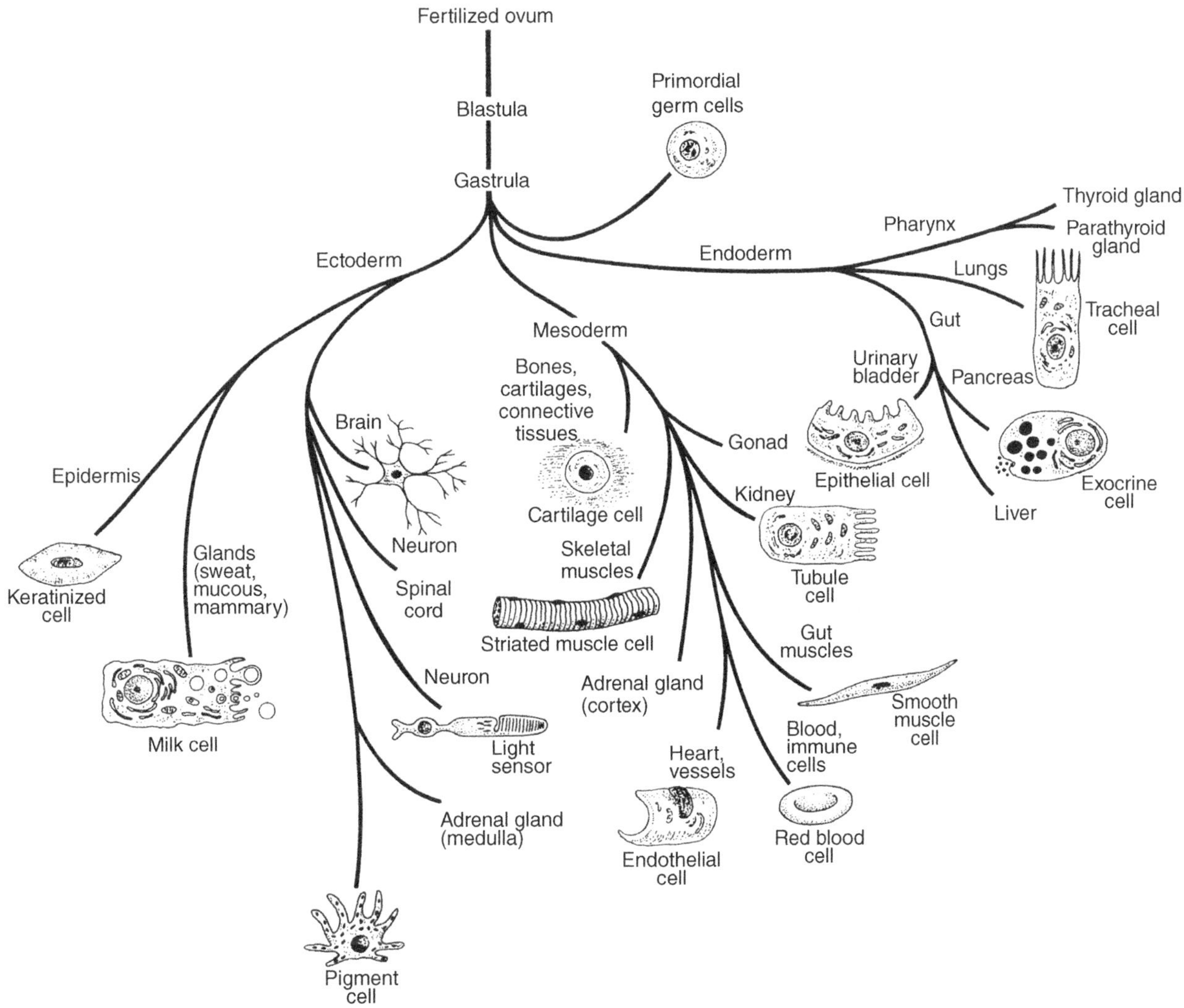

Figure 16.19 The families of cell types that derive from the three basic body layers (ectoderm, mesoderm, and endoderm) in developing mammalian embryos. *Source:* From Hopson and Wessells (1990) / with permission of McGraw-Hill.

cytokines. Table 16.2 provides a selected list of particularly important growth factors and cytokines together with the main sources, target tissues, and their principal effects. Most of these growth factors and cytokines have important roles in wound healing and the modulation of tissue responses to implants.

16.6 Plasma and Blood Cells

The average, healthy adult has a total of ~5 liters of blood. Approximately 55–60% of this volume is plasma, the pale amber fluid that contains many electrolytes and buffer systems that are essential for maintenance of the correct osmotic balance and pH of the interstitial fluid bathing the cells of the body. Present in the plasma are three major groups of proteins that include albumin, globulins, and fibrinogen. Albumin, the most abundant of the three, determines the colloidal osmotic pressure of plasma, while the globulins are particularly involved in immune defense functions. Fibrinogen is the precursor of fibrin, the filamentous protein that forms the supportive mesh of a clot.

16.6.1 Erythrocytes

Each milliliter of blood contains ~5×10^9 red blood cells (erythrocytes) and ~10^7 white blood cells (leukocytes). The circulating mature erythrocyte lacks a nucleus and all other organelles and, as a result, performs only anaerobic metabolism despite an enormous oxygen content. The erythrocyte is shaped as a smooth biconcave disk (Figure 16.20), a shape that greatly facilitates passage through capillaries and the release of oxygen needed to support aerobic metabolism in the cells. The mature erythrocyte is essentially a sack with a

Table 16.2 Selected growth factors and cytokines: source, target tissues, and principal effects.

Growth factor (GF) family	Source	Target	Principal effects
Epidermal GF: EGF; TGFα	Platelets; macrophages; fibroblasts	Keratinocytes; endothelial cells	Reepithelialization of wounds; angiogenesis
Fibroblast GF: FGF2; FGF7; FGF10	Keratinocytes; fibroblasts; myocytes	Multiple mesodermal tissues	Mitogenic; wound healing; angiogenesis
Vascular endothelial GF: VEGF-A; VEGF-B; VEGF-C	Monocytes; fibroblasts; keratinocytes; myocytes	Vascular endothelium	Vasculogenesis; angiogenesis
Platelet-derived GF: PDGF-AA; PDGF-AB; PDGF-BB; PDGF-CC	Platelets; macrophages; fibroblasts; keratinocytes	Mesoderm-derived tissues	Wound healing; reepithelialization
Transforming GFβ: TGFβ1; TGFβ2; TGFβ3	Bone; lung; kidney; placenta	Mesoderm-derived tissues	Wound healing; ECM production
Bone morphogenetic proteins: BMP2; BMP6; BMP9	Mesodermal tissues	Osteoprogenitor mesenchymal tissue	Osteogenesis; kidney and neural development
Erythropoietin (EPO)	Proximal convoluted tubules (kidneys)	Active bone marrow	Promotes erythrocyte formation
Cytokines			
Interleukins: IL-1; IL-2; IL-6; IL-10	Neutrophils; macrophages; keratinocytes	Cells of hemopoietic origin	Immune defense responses; neutrophil recruitment
Tumor necrosis factor-α (TNFα)	Activated macrophages	Systemic inflammation	Acute inflammatory response
Chemokines (chemotactic)	Multiple leukocytes and other cells	Circulating immune defense cells	Recruit leukocytes to sites of infection
Interferons: IFNα, IFNβ	Marrow-derived dendritic cells	Virus infected cells	Interfere with viral replication

Source: Adapted in part from Barrientos et al. (2008).

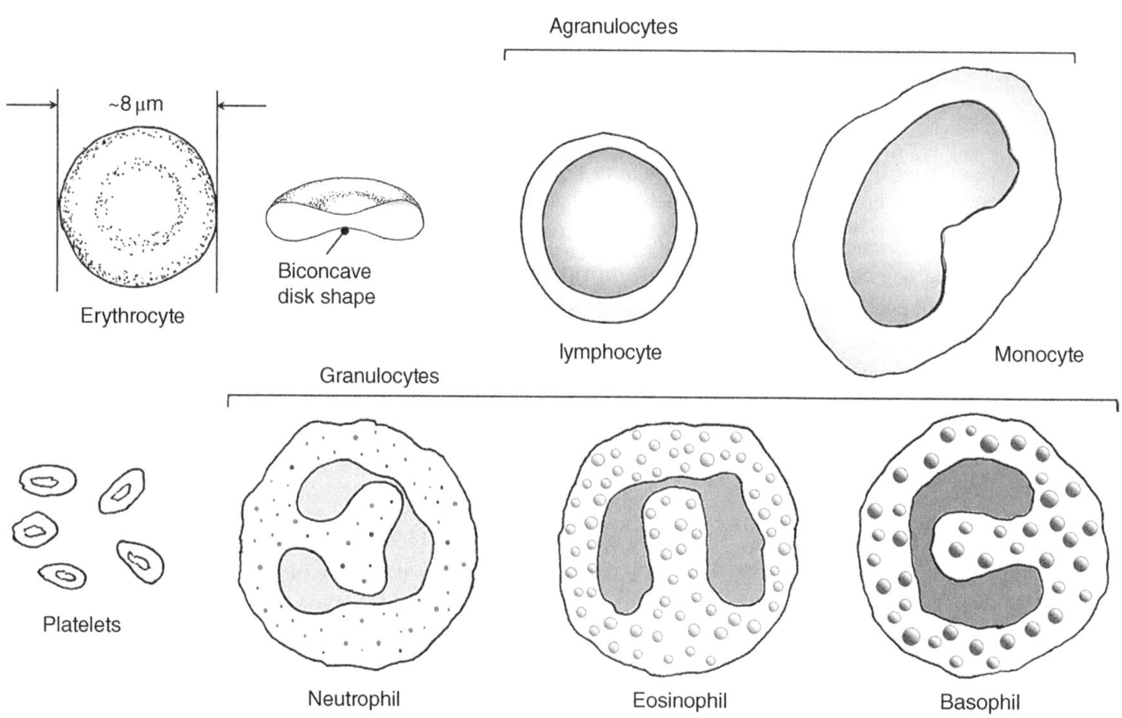

Figure 16.20 Cells of circulating blood including erythrocytes, agranulocytes, granulocytes, and platelets. The agranulocytes include lymphocytes and monocytes. Granulocytes are neutrophils, eosinophils, and basophils. Cells are depicted relative to the size of a mature erythrocyte.

high hemoglobin content that constitutes ~34% of the wet weight of the cell. There is a maximum of 15 g of hemoglobin per 100 ml of blood (in males), which provides the enormous oxygen transport capacity of erythrocytes. This high transport capacity allows a maximum of ~20 ml O_2 per 100 ml blood (in males), a key factor that permits the high metabolic rate of mammals.

In addition to hemoglobin, erythrocytes also contain the enzyme carbonic anhydrase, which catalyzes a very rapid reversible reaction between CO_2 and water to produce H_2CO_3. This important reaction allows blood to carry large amounts of CO_2 in the form of HCO_3^-, a component of the bicarbonate buffer system, a major pH buffering system of the blood. The hemoglobin content of erythrocytes provides the blood with much greater buffering capacity. Although the buffering effect of proteins tends to occur quite slowly, the buffering capacity of hemoglobin occurs quickly and effectively via cooperative interaction with the bicarbonate buffer system. Erythrocytes remain functional for ~120 days before they are broken down in the liver. Much of the product from hemoglobin degradation is recycled and utilized in the synthesis of bile.

16.6.2 Leukocytes

There are five different types of leukocytes with the characteristic cell morphologies depicted in Figure 16.20. Three of the five types are collectively referred to as granulocytes due to the presence of numerous granules in their cytoplasm. Among these three types of granulocytes are neutrophils, basophils, and eosinophils. The two remaining types, collectively referred to as agranulocytes, include monocytes and lymphocytes. The most abundant type of leukocyte is the neutrophil (also referred to as a polymorphonuclear granulocyte or PMN), a cell that functions as a first responder to provide protection in acute infections. Inflamed tissues release chemotactic factors that attract neutrophils causing these cells to squeeze between capillary endothelial cells by the process of extravasation (Chapter 18). After exiting the capillary, the neutrophils migrate through the ECM to the site of infection where they perform phagocytosis and release free radicals to eradicate bacterial invaders (Chapter 18).

The least abundant leukocyte is the basophil, an important cell in the inflammatory process. Upon activation by tissue injury, the basophil releases the vasodilator histamine that increases blood flow through the inflamed tissue. The basophil also releases the anticoagulant heparin, which prevents local blood clotting, thereby maintaining continued blood flow through the affected tissue. The basophils also give rise to mast cells, which are long-lived resident tissue cells that function both in adaptive immunity and in innate immunity, particularly the acute phases of allergic reactions.

Eosinophils account for slightly more than 2% of circulating leukocytes. One role of eosinophils is to seek out parasitic invaders which they then destroy via the release of oxygen radicals and hydrolytic enzymes. Another important protective function of eosinophils is to destroy antibody-allergen complexes. The latter activity serves to modulate and limit the extent of the inflammatory reaction.

Approximately 5% of the circulating leukocytes are monocytes, the largest type of leukocyte. Monocytes play multiple roles that are especially important when infections continue beyond the acute stage. Like neutrophils, circulating monocytes also respond to chemotactic factors released from inflamed tissues at sites of infection or tissue injury. The monocytes move between capillary endothelial cells and, upon passage into the extravascular tissue space, undergo conversion to tissue macrophages, cells of primary importance in combating chronic infections. The protective actions of macrophages are particularly attributable to their voracious ingestion of bacteria and cell debris by phagocytosis, plus their release of highly reactive oxygen radicals that are lethal to most bacterial invaders. Macrophages have additional protective functions, which will be discussed in Chapters 18 and 19.

Approximately 30% of the circulating leukocytes are lymphocytes, major cells of the immune system, consisting of two main groups. These include B cells that protect the body via the production and release of antigen-specific antibodies and T cells that are responsible for cell-based immunity (Chapter 19).

In addition to the leukocytes, blood contains platelets, also referred to as thrombocytes, the smallest type of blood cells. Platelets are not true cells. Instead, they are actually the remains of large multinucleated cells referred to as megakaryocytes that, upon maturation, fragment to form minute platelets. Although platelets have several functions, they are especially important for the activation of protein factors involved in coagulation (Section 16.7). The circulating platelets have a discoid shape of diameter 2–3 μm. Upon encountering sites of tissue injury or where very small vessels are cut, platelets undergo a change in morphology from a nonactivated state to a partially activated state and then to a fully activated state. Partially activated and fully activated platelets are characterized by numerous slender membrane projections (Figure 16.21) that cause the platelets to agglutinate, resulting in the formation of a platelet plug that quickly seals the vessel and temporarily halts bleeding. The platelet plug also becomes a site favorable for formation of a blood clot, the primary function of platelets. Another important role of platelets is the release of growth factors that promote the formation of

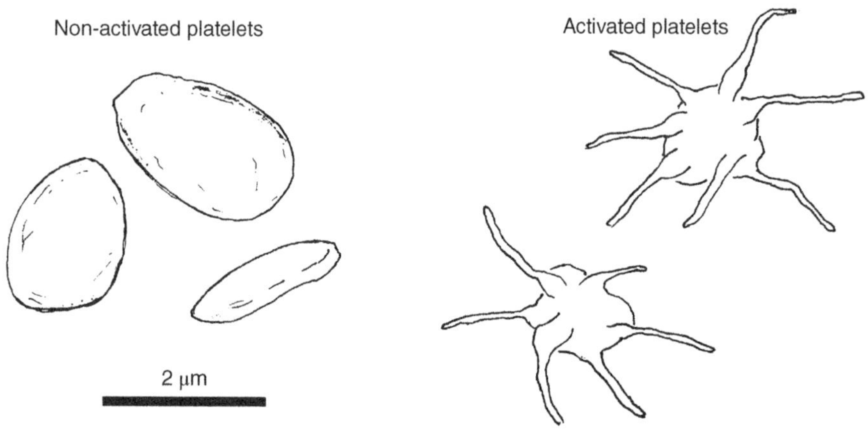

Figure 16.21 Illustration comparing discoid morphology of circulating nonactivated platelets compared with noncirculating activated platelets. The slender projections of the activated state promote platelet aggregation and formation of the platelet plug that temporarily halts blood loss from small vessels.

vascular endothelial cells and tissue repair. In addition, platelets form prostaglandins that trigger many localized tissue responses.

16.7 Platelet Adhesion to Biomaterial Surfaces

Some biomaterials are designed to promote coagulation. An example is collagen sponge (Figure 10.2), often used by surgeons to stanch blood flow from small vessels accidentally or intentionally cut during surgery. Contact with collagen directly activates plasma factor XII and subsequent steps of the coagulation process that lead to clot formation. In comparison, other materials such as polytetrafluoroethylene (PTFE) provide a very smooth surface that discourages platelet adherence and clot formation, which is a major reason for the common use of PTFE in intravascular catheters and vascular grafts.

Present in plasma is the protein plasminogen, which, upon activation, becomes the fibrinolytic enzyme plasmin. Injured tissues and endothelium release a separate protein plasminogen activator that converts plasminogen to plasmin. By virtue of its fibrinolytic activity, plasmin destroys the fibrin mesh of emboli, small clots that form in blood. This causes dissolution of the emboli, thereby preventing the blockage of blood flow in a critical area that could trigger an event such as a stroke or myocardial infarction.

16.8 Platelets and the Coagulation Process

It is clear that the process of coagulation is essential for the maintenance of life. Failure to form a clot capable of stanching the flow of blood from a severed vessel can lead to excessive blood loss, which could be fatal. However, coagulation must also be tightly controlled to prevent blood from clotting too rapidly and cause blockage of blood flow in a vital area such as the coronary or cerebral arteries, an event that would trigger myocardial infarction or a stroke.

Over the past several decades an extensive number of investigations, performed mostly with static *in vitro* systems, have identified and characterized several proteins that are involved in the coagulation process as components of two separate pathways, referred to as the intrinsic pathway and the extrinsic pathway. The intrinsic pathway includes four proteins of the circulating blood, three of which are inactive serine proteases plus one nonenzymatic protein referred to as antihemophilic factor or factor VIII. These four blood proteins are commonly identified by Roman numerals as factors XII, XI, IX, and VIII with a lower case "a" added to indicate the active form of each of the four factors as XIIa, XIa, IXa, and VIIIa, respectively. Upon activation, each protease catalyzes the activation of the next protease in the sequence. The extrinsic pathway includes a membrane protein referred to as tissue factor (TF) and factor VIIa, another serine protease in the circulating blood. TF is located throughout the body with particularly high concentrations in the tunica adventitia of blood vessels. Factor VIIa is unique in that it is present in the circulating blood in the active form.

Investigations using confocal microscopy and high-speed intravital microscopy have allowed the characterization of *in vivo* processes of coagulation in animal models. An important outcome of these investigations is the development of the cell-based model of coagulation (Smith 2009). This model describes coagulation as the result of a complex series of events that occur on the surface of TF-bearing

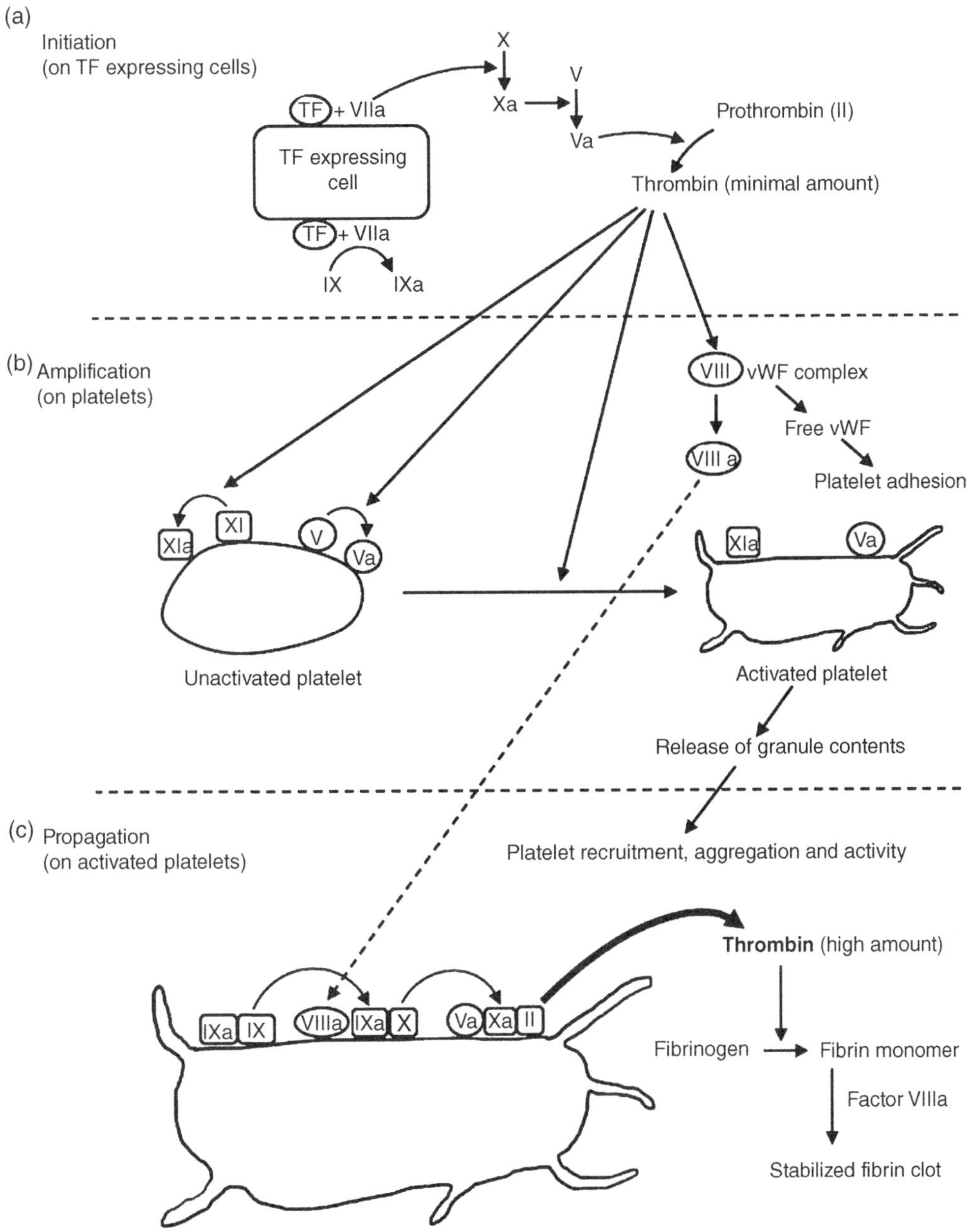

Figure 16.22 The cell-based model of coagulation consisting of three successive stages ((a) initiation stage, (b) amplification stage, and (c) propagation stage) that occur on the surface of tissue factor (TF)-bearing cells and platelets.

cells and platelets during three successive phases, referred to as the initiation stage, the amplification stage and the propagation stage (Figure 16.22).

During the initiation stage, tissue injury causes factor VIIa to spill from the circulation and come into contact with TF-bearing cells resulting in the formation of a TF/VIIa complex that generates low amounts of activated factor IX and factors X, IXa, and Xa. The small quantity of Xa produced triggers the formation of a minimal amount of thrombin, which, in turn, triggers the events of the amplification stage that occurs on the surface of the platelets (the basis of the cell-based model). The events of the amplification stage include the hydrolysis of the factor VIII/von Willibrand factor complex, activation of factor V to Va, activation of factor XI to XIa, and activation of platelets. The platelet activation triggered by the binding of thrombin to the surface receptors causes the platelets to undergo a dramatic change to a morphology with membrane projections (Section 16.5.2), which greatly enhances platelet clumping and aggregation.

Another important consequence of activation is the shuttling of phosphatidylserine from the internal layer of the phospholipid bilayer to the external layer to form a procoagulant membrane surface. In the propagation stage, factor IXa combines with factor VIIIa on activated platelets to form a IXa/VIIIa complex that activates factor X to Xa. Factor Xa then combines with factor Va to form a Xa/Va complex (also referred to as tenase complex). This tenase complex catalyzes the immediate formation of a high amount of thrombin that rapidly cleaves fibrinogen to fibrin. The soluble fibrin molecules spontaneously polymerize to fibrin threads. Factor VIII crosslinks the fibrin threads to greatly strengthen and stabilize the fibrin clot.

In addition to hemostasis, the fibrin clot functions in other roles that are critically important for healing. Especially important in this regard are the platelets entrapped within the gelatinous clot. These tiny cells release several growth factors and cytokines, such as PDGF, VEGF, TGFα, TGFβ1, TNFα, IL-1, and other cytokines that stimulate growth processes essential to tissue repair.

16.9 Cell Types and Their Roles in Biocompatibility Phenomena

As discussed throughout this chapter, the body is composed of many different tissues and cell types. The cell types that compose these tissues perform several different functions that are essential to biocompatibility phenomena and the various responses (positive or negative) to biomaterial implants. Table 16.3 summarizes these different cell types and their roles in biocompatibility phenomena.

Table 16.3 Summary of cell types involved in biocompatibility phenomena and their roles.

Cell type	Principal functions in biocompatibility phenomena
Platelets	Have key role in coagulation essential for maintenance of life; source of growth factors essential for wound repair
Neutrophils	Initial response to attack infectious microbes; secrete cytokines to recruit other leukocytes; release matrix degrading proteinases
Monocytes	Respond to inflammatory chemotactic signals and give rise to tissue macrophages; secrete cytokines for recruitment of leukocytes
Macrophages	M1 phenotype macrophages combat chronic infections and remove apoptotic cell debris; M2 phenotype macrophages active in wound repair and tissue reconstruction
Lymphocytes: two types	Responsible for adaptive defense of body and defense to implants
B-lymphocytes	Secrete antigen-specific antibodies
T-lymphocytes	Cell-mediated immunity; hyperactive response to some implants
Fibroblasts	Form collagen fibers for tissue support and repair of wounds; produce extracellular matrix that interacts with biomaterials
Keratinocytes	Maintains protective barrier; responsible for regrowth of epidermis at implant wound sites
Mesenchymal cells: multiple locations	Progenitor cells that differentiate to specific tissues to maintain function
Bone cells: three types	
Osteoblasts	Responsible for formation of new bone tissue
Osteocytes	Mature cell that monitors load and signals adaptive responses
Osteoclasts	Degrades bone matrix remodeling and implant accommodation
Endothelial cells	Source of angiogenic buds that revascularize tissues damaged during implantation of biomaterials
Endocrine cells: multiple types and locations	Release of hormones for systemic delivery to coordinate tissues; transmit signals for tissue response to implanted biomaterials
Neurons: three types	Required for tissue coordination and homeostasis
Afferent neurons	Sensory input for subsequent biocompatibility response
Interneurons	Process information and make decision for appropriate response
Efferent neurons	Send signals to effectors for biocompatibility response as needed

16.10 Concluding Remarks

In this chapter, we discussed basic aspects relevant to the interaction of implanted biomaterials with cells and tissues, which should provide a background for the discussion of biocompatibility phenomena in subsequent chapters.

- Biocompatibility is the most critical factor that determines the success of an implanted biomaterial *in vivo*. Failure to achieve an acceptable host response can lead to consequences ranging in severity from modest and resolvable to unresolvable, such as organ failure or mortality.
- We discussed various properties and characteristics of the relevant cells, biomolecules, extracellular components and tissues that interact with an implanted biomaterial and affect its biocompatibility.

Problems

16.1 Briefly explain why biocompatibility should be assessed for an implanted material and host tissue as an interacting system.

16.2 Describe one example of a biomaterial that is biocompatible when implanted in one tissue but not with other tissues and explain the biologic basis for this tissue-specific biocompatibility.

16.3 Briefly explain why protein adsorption is a requirement for the biocompatibility of all biomaterials.

16.4 Explain the physical mechanism by which proteins adsorb to a greater extent to hydrophobic surfaces compared to hydrophilic surfaces.

16.5 Explain the physical basis of the observation that fibronectin adsorbed onto biomaterials with a hydrophobic surface often shows less cell binding activity compared to fibronectin adsorbed onto biomaterials with a hydrophilic surface.

16.6 Indicate whether the phospholipids or the membrane proteins are mainly responsible for the biochemical and physical properties of the plasma membrane, and explain the basis of your answer.

16.7 Briefly explain why red blood cells have a much shorter life span in the body than liver parenchymal cells.

16.8 Describe three examples of how the ECM largely accounts for the characteristics of different tissues of the body, including which extracellular component(s) is (are) the main determinant of those properties.

References

Barrientos, S., Stojadinovic, O., Golinko, M.S. et al. (2008). Growth factors and cytokines in wound healing. *Wound Repair and Regeneration* 16: 585.

Carlson, B.M. (1988). *Patten's Foundations of Embryology*, 5the. McGraw-Hill.

Grinnell, F. and Feld, M.K. (1982). Fibronectin adsorption on hydrophilic and hydrophobic surfaces detected by antibody binding and analyzed during cell adhesion in serum containing medium. *The Journal of Biological Chemistry* 257: 4888–4893.

Hopson, J.L. and Wessells, N.K. (1990). *Essentials of Biology*. McGraw-Hill.

Roach, P., Farrar, D., and Perry, C.C. (2005). Interpretation of protein adsorption: surface-induced conformational changes. *Journal of the American Chemical Society* 127: 8168–8173.

Singer, S.J. and Nicolson, G.L. (1972). The fluid mosaic model of the structure of cell membranes. *Science* 175 (4023): 720–731.

Smith, S.A. (2009). The cell-based model of coagulation. *Journal of Veterinary Emergency and Critical Care* 19: 3–10.

Vroman, L., Adams, A.L., Fischer, G.C., and Munoz, P.C. (1980). Interaction of high molecular weight kininogen, factor XII, and fibrinogen in plasma at interfaces. *Blood* 55: 156–159.

Williams, D. (2014). *Essential Biomaterials Science*, 143. Cambridge University Press.

Further Reading

Williams, D. (2008). On the mechanisms of biocompatibility. *Biomaterials* 29: 2941–2953.

17

Mechanical Factors in Biocompatibility Phenomena

17.1 Introduction

Cells and tissues of the body are continuously being subjected to a variety of mechanical forces, such as forces due to pulsating blood flow and to normal physiological activity, as well as external forces. Particularly within connective tissues, these forces play an important role in homeostasis, that is, the maintenance of an almost constant internal environment. Cells are able to sense mechanical stresses (or mechanical strains) within and translate them into some type of adaptive response, a process referred to as mechanotransduction. Because most biomaterials have mechanical properties that are different or even, in many cases, vastly different from the tissues into which they are implanted, they will cause some perturbation of the normal mechanical stress (or strain) system within the tissue. This means that the mere presence of an implanted biomaterial will generally trigger an adaptive response by cells, independent of any additional effects that the biomaterial can have.

In this chapter, we shall discuss biocompatibility phenomena caused by the perturbation of normal mechanical forces in tissues or by the application of mechanical forces to cells and tissues. Subsequent chapters in this part of the book will discuss additional biocompatibility phenomena, such as those arising from direct chemical interaction between an implanted biomaterial and tissues.

Mechanotransduction is generally considered to occur in four stages, involving transmission of force to the cell surface, force transduction through the cell membrane, propagation of signals within the cytoskeleton, and cellular response. Several pathways connect the initial causative event of force transmission to the end effect of cellular response. They are closely associated with the manner in which cells are linked to the extracellular matrix (ECM) via connections that are also linked to the cytoskeleton. Because of this linkage, perturbation of mechanical forces within the ECM or application of mechanical forces is transmitted to the cell interior, triggering a chain of intracellular biochemical events and ending with an adaptive cellular response.

The resulting events may be a specific cellular or tissue-level response, such as a change in cell phenotype, shape or motility, cell destruction, production of ECM, tissue resorption, or fibrosis. Because these events can have important consequences for the use of biomaterials in healing hard and soft tissues, a consideration of mechanical stress-induced biocompatibility phenomena should be an important component in biomaterials design.

In this chapter, we shall discuss the following topics related to biocompatibility phenomena induced by mechanical forces:

- The stages and mechanisms of mechanotransduction
- Examples of mechanical stress-induced biocompatibility phenomena associated with implantable biomaterials.

17.2 Stages and Mechanisms of Mechanotransduction

Mechanotransduction, as noted earlier, is generally considered to occur in four stages, which are shown in Figure 17.1. The first stage involves the application or perturbation of mechanical forces, which are transmitted to the system involving both the biomaterial and the host. The second stage, force transduction, involves the sensory action of the cell that transduces the mechanical stimulus to a chemical signal. This signal is transmitted within the cell in the third stage. Then the cell responds to the transduced signal (fourth stage), resulting in specific cellular or tissue-level events. The pathways that connect these events with the initial causative event are discussed in the following sections.

17.2.1 Force Transduction Pathways

There appear to be at least two basic types of mechanosensors that, in response to external forces, initiate the intracellular cascade of biochemical signals that bring about cellular responses to a mechanical force. One type, with an activation

Materials for Biomedical Engineering: Fundamentals and Applications, First Edition. Mohamed N. Rahaman and Roger F. Brown.

Companion website: www.wiley.com/go/rahaman/biomedicalengineering

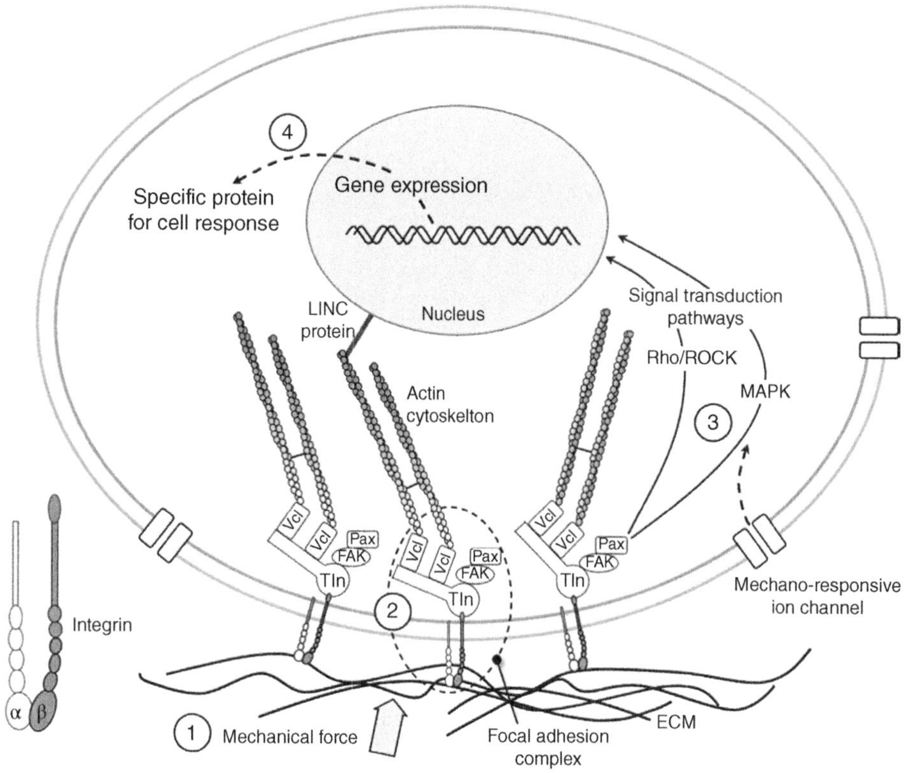

Figure 17.1 The four stages of mechanotransduction: (1) force transmission with perturbation of ECM; (2) force transduction via integrin-based focal adhesion (FA) complex or mechano-responsive ion channels; (3) signal propagation via signal transduction pathways; and (4) cell response via gene expression and other cellular mechanisms.

processing duration of seconds to minutes, consists of a focal adhesion with its associated complex of proteins. The other type of mechanosensor is a mechano-sensitive ion channel, a sensor type that can become activated within milliseconds, which enables more rapid cellular responses.

Focal Adhesions

An introduction to transmembrane focal adhesions, the large multi-protein complexes that mechanically link the ECM with the cytoskeleton, was presented in Chapter 16. Here we address the molecular composition of focal adhesions as well as the manner in which they transmit external force perturbations through all parts of the cell, including the nucleus. Focal adhesions are highly dynamic structures whose size and number vary with the magnitude of the mechanical force. The heterodimeric molecule integrin forms much of the extracellular portion of the focal adhesion. The α and β subunits of the heterodimer project outside the plasma membrane where they bind with fibronectin, laminin, and other tissue-specific adhesion proteins of the ECM (Figure 17.1). There are several types of α and β chains, which account for the ~24 different integrin molecules that are produced and associated with various tissues and organ locations.

The intracellular portion of the focal adhesion is a cluster of multiple proteins, the most important of which include focal adhesion kinase (FAK), talin, vinculin (Vcl), paxillin (Pax), and α-actinin. There are other proteins of the focal adhesion complex but they are not considered in this basic description. FAK is a tyrosine kinase, which, in response to mechanical stimuli transmitted inward through the cluster of focal adhesion proteins, indirectly activates an intracellular signal transduction cascade. It is also responsible for the initial organization and recruitment of focal adhesion complexes in response to a mechanical force. FAK performs this role by initially recruiting talin, a structural protein that forms the integrin–actin connection.

The molecular structure of talin includes a globular head plus a long rod-like domain. The head portion of talin binds to the tail of the integrin β subunit, while the rod portion has two sites to which actin binds, plus additional binding sites for Vcl, a major component of the focal adhesion core. In response to a mechanical force, Vcl is recruited to, and strengthens nascent focal adhesions. α-Actinin has several binding sites that serve to crosslink and strengthen actin filaments. Pax is a signal transduction adaptor protein that requires phosphorylation by FAK for activation.

Immediately upon activation, Pax triggers a signal transduction cascade.

Mechano-Sensitive Ion Channels

Mechano-sensitive ion channels are one of three types of protein-lined channels that control ion passage through the cell membrane. The other two types of channels are voltage-gated channels and ligand-gated channels. The mechano-sensitive ion channels, which appear restricted to transmembrane passage of cations, are widely distributed among all the tissue types of the body and they play a crucial role in many physiological functions. In their normal rested, non-stressed state, these ion channels are in a closed conformation that blocks cation movement across the plasma membrane.

There are several different types and categories of mechano-sensitive ion channels but two of the particularly important categories are described briefly. One category is the large family of so-called transient receptor potential (TRP) ion channels that function as sensors for a wide variety of physical and chemical stimuli. The second category includes the more recently discovered Piezo 1 and Piezo 2 ion channels, which are larger and more complex than the TRP channels. Like the TRP channels, the Piezo channels also function as sensors for a wide variety of physical and chemical stimuli.

The force-activated opening of TRP and Piezo channels results in a transient influx of Ca^{2+} ions, which trigger the intracellular events associated with these two types of mechano-sensitive ion channels. Force-induced influx of Ca^{2+} ions via the TRPV4 ion channel, an ultra-rapid TRP channel, has been found to occur within ~4 ms. The Ca^{2+} ions, which enter during the transient opening of mechano-activated channels, function as a second messenger in various signaling transduction mechanisms that elicit many different cellular responses. In many cases, small GTPases, enzymes that catalyze the hydrolysis of guanosine triphosphate (GTP) to guanosine diphosphate (GDP), mediate the calcium ion-stimulated cellular processes. It is noted that the actual amount of calcium that enters the cell during the transient opening of these channels is kept low to limit potential injury from Ca^{2+}, a divalent ion that is highly toxic when it exists freely in the intracellular fluid compartment.

17.2.2 Signal Transduction Pathways and Other Mechanisms

Results from more than two decades of laboratory investigations have shown the involvement of multiple signal transduction pathways in the transduction of cues from external mechanical forces to biochemical signals, which then trigger various cellular responses. The generalized cell responses triggered by mechanical stress often includes activation of mitogen-activated protein kinases (MAPKs), the Rho/Rho-associated protein kinase (ROCK) family of GTPases, and other signaling pathways as represented in Figure 17.1. The mechanism of transduction of external mechanical forces inward to activate the signal transduction pathways involves changes in the conformation of the tail portion of the integrin molecule (Figure 16.17). In this sense, the integrin heterodimer is a key to this type of "outside-in" signaling process.

MAPK and Rho/ROCK Signal Transduction Pathways

The MAPK signal pathways regulate multiple cellular processes that include cell proliferation, specific gene expression, differentiation, inflammatory responses, apoptosis, and other events (Figure 17.1). Because many MAPKs are involved in signaling events initiated by mitogenic growth factors, the phrase mitogen-activated is included in the name of these pathways. All MAPKs are protein kinases that catalyze phosphorylation of the hydroxyl group of a serine or threonine moiety of the next kinase enzyme within a cascade of enzyme reactions involving sequential phosphorylation of the remaining kinases of the cascade. The initially discovered MAPK is the protein serine/threonine kinase referred to as extracellular signal-regulated kinase 1 (ERK1). Another example of a MAPK signal pathway is the c-Jun NH2-terminal kinase (JNK) subfamily of serine/threonine protein kinases, which are known to affect nearly 100 target proteins that promote a wide variety of cellular responses. It may also be noted that the JNK pathway is responsive to mechanical stress stimuli in addition to purely mitogenic stimuli. As an example, JNK is involved in promoting the hypertrophy of skeletal muscle that follows the mechanical stress of repeated bouts of high-intensity eccentric contractions.

Rho/ROCK signaling pathways consist of a family of small GTPases responsive to a variety of stimuli that include chemical stimuli (hormones, growth factors, cytokines, and Ca^{2+}) and mechanical stimuli. The Rho label is used to distinguish a group of GTPases within the Ras GTPase superfamily, a name derived from the protein coded by the *ras* gene, a human cancer oncogene. ROCK is an acronym for Rho-associated protein kinase, a component of the pathway. The Rho/ROCK signaling pathways are particularly activated by mechanical stresses involved in regulation of the remodeling of all three types of cytoskeletal microfilaments: actin microfilament, intermediate filaments, and microtubules. The remodeling of the cytoskeleton that occurs causes changes in cell shape and controls cell movement and migration. These types of cytoskeletal changes are part of the remodeling that allows embryonic cells, for example, to undergo morphogenesis and mature cells to withstand increased tensile stress.

LINC Proteins

Forces transmitted from focal adhesions to the cytoskeletal filaments are also transmitted to the nucleus via the protein complex LINC (linker of nucleoskeleton and cytoskeleton) that physically links the cytoskeleton to the nucleus. The principal LINC proteins identified so far include proteins known as nesprins and a group of proteins known as SUN-domain proteins associated with the inner nuclear membrane. The transfer of mechanical force from the cytoskeleton to the nucleus via LINC proteins directly affects specific gene expression, thereby partially accounting for nuclear mechanotransduction. Another part of the coupling of mechanical signals to the nucleus involves soluble proteins that shuttle from the plasma membrane to the nucleus. One example of this type of shuttle protein is zyxin, a component of the intracellular portion of the focal adhesion complex. Dynamic stretching causes zyxin to detach from the focal adhesion complex, allowing it to diffuse to the nucleus where it affects specific gene expression. One example of a zyxin-mediated effect is aberrant stretch-induced gene expression in endothelial cells leading to endothelial dysfunction.

17.2.3 Mechanisms of Cellular Response

The cellular mechanisms and responses to mechanical forces described so far in this chapter have included a generalized description of basic cellular processes such as cell migration, gene expression, proliferation, differentiation, and apoptosis. Here we address some examples of specific cellular and tissue adaptive responses that occur in response to various mechanical stresses that impinge on cells and tissues. Many of the adaptive responses that occur are initiated by "outside-in" signaling events. There are also adaptive responses that are controlled by "inside-out" signaling processes.

Focal Adhesion Strengthening

An important example of an "inside-out" signaling event is the strengthening of focal adhesions to withstand increased tensile stress. One characteristic of focal adhesions, as noted earlier, is their highly dynamic structure that undergoes continuous changes in response to both external and internal forces. The load-induced enlargement and strengthening of focal adhesions to withstand a directional force involves recruitment of integrins, especially $\alpha5\beta1$ integrins. It has been shown recently that fibroblasts are capable of commencing the process of recruiting fibronectin-attached $\alpha5\beta1$ integrins in less than ~0.5 s after application of a mechanical load. The mechanical load-induced recruitment process begins by the transition of $\alpha5\beta1$ integrins from the inactive state to the active state capable of fibronectin attachment. Then the active, fibronectin-attached $\alpha5\beta1$ integrins begin to cluster in order to strengthen the focal adhesion complex (Figure 17.2). A measurable increase in fibroblast attachment strength is detectable within ~5 s of applying a mechanical load.

The recruitment and strengthening of focal adhesions occur as part of the process by which cells adjust their alignment to accommodate and withstand directional forces such as uniaxial and shear forces. The adjustment to withstand directional forces also involves reorganization and directional orientation of cytoskeletal filaments to form what are often referred to as stress fibers (Figure 16.14). Directional orientation of cytoskeletal filaments has been

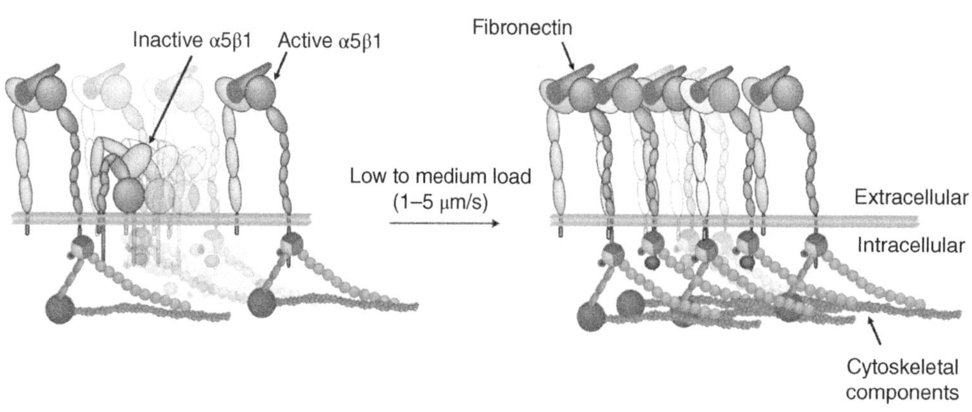

Figure 17.2 Illustration depicting the process by which fibronectin-bound $\alpha5\beta1$ integrins respond to external mechanical load. The integrin mediated sensing of load and signaling for reinforcement of focal adhesions occurs in less than a second. *Source:* From Strohmeyer et al. (2017) / with permission of Springer Nature.

well documented in numerous investigations with cell cultures subjected to various types of cyclic stretching. The culturing systems used in these investigations have included flexible two-dimensional and three-dimensional cell support systems. An intriguing observation that has emerged from some studies is that the cellular axis and cytoskeletal filaments of cells maintained on flexible two-dimensional substrates orient in a direction perpendicular to the direction of uniaxial cyclic stretching. In comparison, the cell axis and cytoskeleton of cells embedded in three-dimension gels orient in a direction parallel to the direction of uniaxial cyclic stretching. The molecular basis of these different patterns of cytoskeletal orientation observed for two-dimensional and three-dimensional substrates remains unclear.

Cell Migration

Cell migration is a process of fundamental importance for embryogenesis and morphogenesis. Cell migration is also the underlying basis of re-epithelialization of a wound created during the surgical implantation of a biomaterial. The cell migration process involves a highly coordinated series of events that include dynamic changes in focal adhesions to modulate anchoring to the extracellular substrate or matrix. The movement of mesenchymal cells cultured *in vitro* is often studied as a model of cell migration. Coordinated movement *in vitro* begins as the cultured cell extends a leading edge forward. The advancing edge is typically fan-shaped, as represented in Figure 17.3, and involves ruffling of the edge as well as extension of multiple lamellipodia. Adjacent to this fan-shaped leading edge is a dynamic zone in which immature focal adhesions undergo assembly. This is followed by focal adhesion maturation and mechanical coupling with contractile actin stress fibers, which enable the contractile fibers of the cytoskeleton to pull the cell body forward. The retracting, trailing portion of a migrating cell in culture typically has a tapered shape and includes a zone of focal adhesion disassembly. The assembly, maturation and disassembly of focal adhesions are particularly mediated by integrins and the force-sensitive protein Vcl.

Extracellular Matrix Adjustments to Maintain Homeostasis

The extracellular cues to which cells respond include matrix stiffness, a parameter that affects many cellular processes. Matrix stiffness affects mechanical force signals transmitted via the various mechanotransduction mechanisms that control cellular homeostasis and behavior. In this context, adjustments of matrix stiffness by continuously balancing the relative amounts of soft and stiff components of the ECM are important for the maintenance of tissue homeostasis. Maintenance of an appropriate balance of soft and stiff components is especially important in soft connective tissues.

Fibroblasts have primary responsibility for the degradation and synthesis of most of the extracellular constituents of soft connective tissue. The various components of the

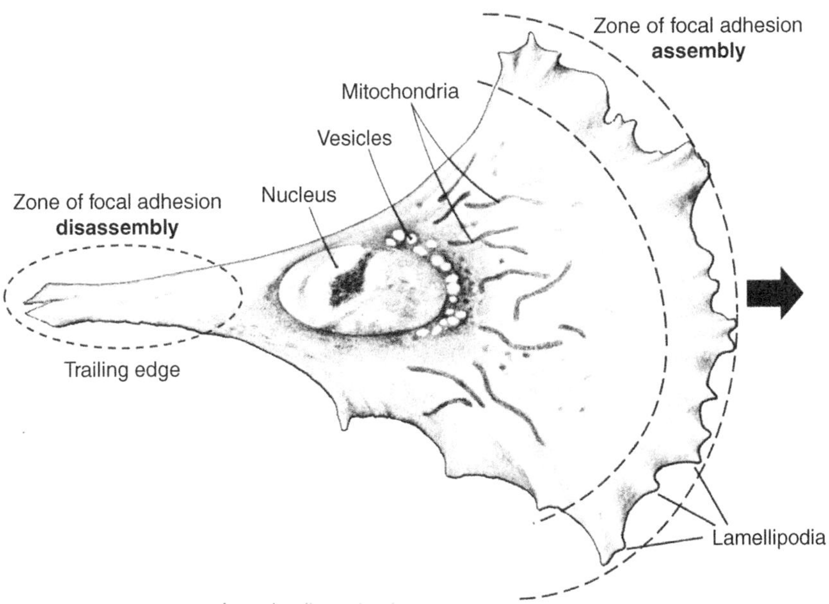

Figure 17.3 The process of cell migration occurs by dynamic assembly of focal adhesions at the leading, fan-shaped edge of the cell and focal adhesion disassembly at the trailing edge. Also involved, but not shown, are continuous rearrangements of actin and myosin cytoskeletal filaments that pull the cell forward. *Source:* Carlson (1988) / with permission of McGraw-Hill.

ECM of soft connective tissues produced by fibroblasts include collagen and elastic fibers, glycosaminoglycans and related proteoglycans (Chapter 10). Increasing the relative amount of glycosaminoglycans enhances matrix softness, whereas increasing the relative amount of collagen enhances matrix stiffness. In addition, fibroblasts also secrete matrix metalloproteinases (MMPs) that can degrade collagen and other ECM constituents, actions that reduce the stiffness of the tissue. When soft connective tissues are subjected to high tensile stresses, fibroblasts often differentiate to myofibroblasts, cells capable of contraction and of synthesizing high amounts of collagen. Development of this phenotype is often associated with pathological fibrotic conditions.

Fibroblasts are very sensitive to external mechanical stresses and to the stiffness of the ECM. Constant loading triggers signaling patterns in fibroblasts that lead to continuance of an appropriately compliant matrix resulting in the maintenance of homeostasis. In comparison, the absence of loading results in an unstressed ECM and loss of signaling which leads to a type of apoptosis known as anoikis (Figure 17.4) (Humphrey et al. 2014). Homeostasis is maintained during conditions of increased mechanical stress accompanied by signaling that promotes ECM remodeling to achieve an appropriate compliance. Aberrant signaling during states of overload often causes excessive collagen deposition, resulting in matrix stiffening and the aforementioned condition of fibrosis.

Gene Expression Responses

There is abundant evidence of changes in gene expression in response to mechanical forces. Although the mechanism of force-induced changes in gene expression is unclear, it has been speculated that a mechanical force transmitted to the nucleus might cause changes in the conformation of chromatin, which could directly influence transcription. Here, examples of some effects of mechanical force on gene expression in a few specific cells and tissues are briefly discussed.

Osteoblasts The application of mechanical stress to the MC3T3-E1 established line of osteoblasts has been shown to enhance expression of a transcription factor known as cox-2. This finding is of particular significance because the cox-2 transcription factor is known to be important in bone formation. It has also been observed that exposure of osteoblasts to shear forces due to fluid flow promotes the expression of epidermal growth factor receptor (EGFR), which allows the osteoblasts to be more responsive to epidermal growth factor (EGF), resulting in enhanced proliferation.

Endothelium Fluid shear forces have been shown to influence the expression of gene products that affect the synthesis of nitric oxide (NO) in endothelial cells. This finding is physiologically relevant because NO is the principal factor for endogenous regulation of arteriole

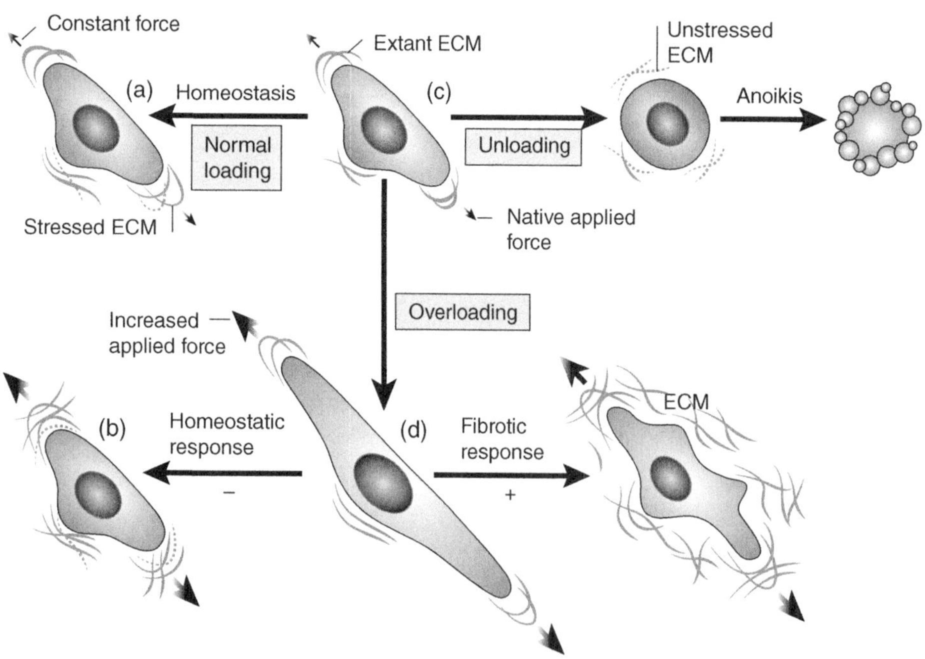

Figure 17.4 Depiction of cell-ECM interactions to (a) maintain homeostasis during loading or (c) anoikis type apoptosis during unloading. (b) Appropriate adjustment for compliance during overloading allows maintenance of homeostasis; (d) fibrotic stiffening results when aberrant overload signaling occurs. *Source:* From Humphrey et al. (2014) / with permission of Springer Nature.

diameter and blood flow. NO also has a beneficial anti-apoptotic effect in endothelial cells. One mechanism by which fluid shear forces influence NO production in endothelial cells involves the expression of genes that code for specific microribonucleic acids (miRNAs), which are a class of noncoding ribonucleic acids (RNAs) that can regulate protein synthesis by a post-translational mechanism. In one study, prolonged exposure of human umbilical vein endothelial cells (HUVECs) to unidirectional shear stresses was found to produce a five times increase in the expression of a type of miRNA known as miR-21. Correlated with this large increase in miR-21was a subsequent increase in phosphorylation of endothelial nitric oxide synthase (eNOS), an enzyme that catalyzes the production of NO in HUVECs. Based on these observations, it was proposed that the resulting NO production could have an anti-hypertensive effect and could also modulate endothelial apoptosis.

Tendon Tendon cells subjected to cyclic stretching for durations of 1–7 days have been observed to show increased expression of connexin-43 messenger ribonucleic acid (mRNA). The physiological relevance of this observation is that connexin-43 is one of the multiple connexin proteins that form gap junctions. The increased assembly of gap junctions would contribute to enhanced communication between tendon cells for organization to accommodate increased mechanical load.

Fibroblasts When fibroblasts encounter a boundary between materials of different stiffness, they preferentially migrate toward the material of greater stiffness. Matrix stiffness also controls gene expression in mesenchymal stem cells (MSCs) to direct their differentiation toward the fibroblast phenotype.

17.3 Mechanical Stress-Induced Biocompatibility Phenomena

Mechanical stresses affect the development and physiological function of tissues in several ways, while the presence of an implanted biomaterial often influences the manner in which a tissue responds to mechanical stresses. This section presents examples of tissue-specific responses and adaptations to various mechanical stresses and how they are modulated by implanted biomaterials.

17.3.1 Implantable Devices in Bone Healing

There is extensive evidence that mechanical stresses are essential for the maintenance of bone homeostasis and health, and that the absence of mechanical stresses, due to prolonged bed confinement, for example, leads to bone resorption and osteolytic weakening. It has been known since about 1995 that the mechanosensory processes of detecting mechanical stresses are associated with canaliculi, the system of intercommunicating micro-channels responsible for much of the porosity of compact bone (Duncan and Turner 1995). The application of a bending force to a long bone leads to a compressive stress on the concave side of the bone and to a tensile stress on the convex side, resulting in a stress gradient that drives fluid flow through the canaliculi (Figure 17.5a). Located at the surface of osteocytes, the mature bone cells that populate the lacuno-canalicular cavities of compact bone, are numerous mechanoreceptors sensitive to fluid flow (Figure 17.5b). These flow-sensitive cells transmit osteogenic signals to osteoblasts or degradative signals to osteoclasts (Figure 17.5c). The osteogenic signals particularly include the Wnt signal pathway (a pathway known to be associated with osteogenesis), as well as MAPK pathways and the Rho/ROCK pathway. The most important osteoclastogenic signal released by osteocytes is the signal protein RANKL, a potent stimulator of osteoclast activity. Osteocytes also release sclerostin, a protein that causes direct inhibition of osteoblasts, which tips the balance in favor of osteoclast activity.

One of the factors that contributes to the frequent problem of bone resorption in fractures stabilized by metallic fracture fixation plates is the large difference in the Young's modulus (or stiffness) between the metallic plates and host bone (Table 4.1). Because of their much higher Young's modulus, fracture fixation plates support almost all of the physiological stresses. Consequently, the surrounding bone is shielded from mechanical stresses, an undesirable phenomenon that impairs bone healing and contributes to bone resorption.

A variety of approaches has been used to design orthopedic devices that provide a better match to the mechanical properties of bone in order to reduce stress shielding. One approach has been to incorporate porosity into the metallic devices, which has the effect of lowering their Young's modulus (Figure 4.13). However, the incorporation of porosity into some orthopedic devices often complicates the fabrication process and may reduce their mechanical durability. As polymers are normally weaker than human cortical bone, another approach has been to reinforce synthetic polymers (degradable or nondegradable) with particles of a strong ceramic phase to create composites with better mechanical properties (Chapter 12). Hydroxyapatite particles are a commonly used ceramic phase, because normally they not only enhance the Young's modulus of the polymer matrix but also improve its bioactivity and, thus, its ability to directly bond to adjacent bone. However, such composite

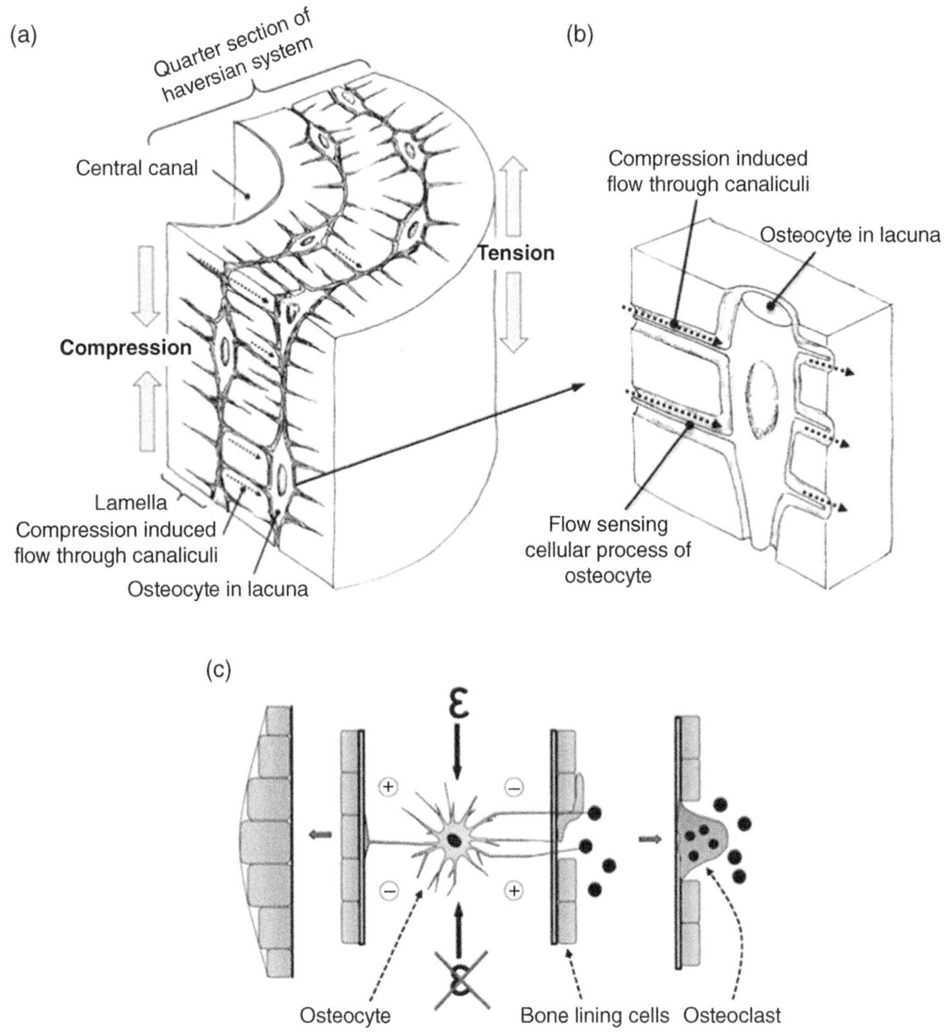

Figure 17.5 Bone mechanotransduction process associated with fluid flow through canaliculi, the system of intercommunicating microchannels. (a) Bending forces cause compressive stress on one side of the bone and tensile stress on the opposite side producing a pressure gradient that (b) drives fluid flow through the canaliculi. (c) Mature osteocytes that populate the lacuno-canalicular cavities are fluid flow sensitive mechanotransducers that transmit osteogenic signals to osteoblasts or degradative signals to osteoclasts. *Source:* (c): From Bonewald (2007) / with permission of John Wiley & Sons.

devices, such as hydroxyapatite-reinforced polyglycolic acid (PGA) screws, for example, are often suitable for stabilizing small fractures only. Approaches based on the design of new metal compositions, such as degradable metals have received significant attention recently (Chapter 6). Devices composed of degradable metals can potentially provide the requisite mechanical properties during fracture healing but thereafter degrade away. Consequently, the period over which stress shielding can occur is limited. However, as noted in Chapter 6, major problems remain in creating degradable metals with optimal properties for use as orthopedic devices.

While stress shielding is an important adverse phenomenon, the effectiveness of a device in bone healing depends on a combination of properties. Despite their high Young's modulus, commonly used devices composed of metals such as titanium and titanium alloys, cobalt-chromium alloys, and stainless steel (316L) have shown their effectiveness over many years because they provide a useful combination of properties (Chapter 23).

17.3.2 Implantable Devices in the Cardiovascular System

The ventricular chambers, heart valves, systemic blood vessels and other components of the cardiovascular system are continuously subjected to mechanical stresses, such as shear stresses and cyclic tensile stresses. Consequently, implantable devices designed for repairing tissues in the cardiovascular system should be capable of withstanding and functioning under these types of stresses. This section discusses two examples of biocompatibility phenomena related to the use of implantable devices for the repair of arteries and heart tissue.

Vascular Grafts

Arteries in the human body can become narrowed (or occluded) due to atherosclerosis, the formation of plaque or deposits on the luminal (inner) surface of the arteries, resulting in a reduction or a blocking of normal oxygenated blood flow. As discussed in Chapter 24, a common solution is to replace or circumvent the narrowed or blocked part of the artery using a device called a vascular graft, composed of a biological material, such as an autologous saphenous vein, or a synthetic material, such as Dacron® or expanded polytetrafluoroethylene (ePTFE). Synthetic grafts are successfully and widely used for repairing large-diameter arteries (greater than ~6–7 mm in diameter) but they have shown limited success with small-diameter arteries (diameter less than ~6–7 mm). In comparison, biological grafts are more effective for small diameter arteries. However, all grafts deteriorate at varying rates with time.

An important property of vascular grafts is their ability to remain patent (unoccluded) over a long-term duration. One factor that strongly influences graft patency, particularly for synthetic grafts, is the mismatch in mechanical properties between the graft and the healthy artery. Studies have shown a direct correlation between the compliance of the graft, that is, its ability to dilate when subjected to a pressure, and the duration over which the graft remains patent (Chapter 24). Efficient transmission of a pressure pulse through the peripheral vessels requires adequate compliance of the vessel wall. In general, stiff grafts lead to a higher degree of compliance mismatch when compared to grafts that are more compliant, and the mismatch is greater at the anastomoses (where the graft is sutured to the healthy artery). Differences in compliance lead to hemodynamic changes such as turbulent flow and shear stresses in the direction of the blood flow, eventually resulting in occlusion of the vascular graft, particularly at the distal anastomosis, the primary mechanism of which is known as intimal hyperplasia.

The intimal hyperplasia that develops is an outcome of the dysfunctional state of the vascular endothelium injured by turbulence resulting from the compliance mismatch. The endothelial cells of normal healthy arteries maintain vascular tissue homeostasis by releasing NO, prostacyclin and other factors that control and modulate the activities of smooth muscle cells in the tunica media, the contractile middle layer of the arterial wall. These regulatory factors released by the endothelium restrict vascular smooth muscle cells to a differentiated state in which they divide infrequently and express proteins such as smooth muscle α-actin essential for vessel contraction and tone. Dysfunctional endothelial cells release lower amounts of the factors that inhibit proliferation of vascular smooth muscle cells and increased amounts of factors such as basic fibroblast growth factor (bFGF) that promote proliferation of the smooth muscle cells. Because of this shift toward release of stimulatory agents, the vascular smooth muscle cells transition to a more proliferative phenotype, which leads to their accumulation in the tunica media.

The endothelial dysfunctional events linked with compliance mismatch also include release of TGFβ1, IL-6, and TNFα, cytokines that promote activation of matrix metalloproteases. These proteolytic enzymes catalyze remodeling of the ECM and degradation of the basal lamina, the layer that separates the tunica media from the tunica intima, the innermost layer of the vessel wall. The resulting changes of the ECM and basal lamina allow the immature smooth muscle cells to migrate into the tunica intima leading to formation of a mass of neointima or intimal hyperplasia. A secondary factor that contributes to obstruction and failure of synthetic vascular grafts in smaller diameter arteries (diameter less than ~6–7 mm) is thrombosis. This is an outcome of the tendency of platelets to adhere to the surface of vascular grafts fabricated from synthetic materials such as Dacron® and ePTFE.

One approach to the development of vascular grafts with improved hemodynamic properties has been the creation of synthetic grafts, such as polyurethane (PU) grafts, which provide a better match to the mechanical compliance of natural arteries when compared to Dacron® and ePTFE grafts. However, PU grafts have been plagued by limitations related to inadequate chemical stability in the physiological environment and to a loss in mechanical compliance following implantation (Chapter 24). Another approach, which is receiving significant attention, is application of tissue engineering principles to the regeneration of functional vascular grafts. Tissue engineering approaches to the regeneration of functional tissues and organs are discussed in Chapter 25.

Cardiac Remodeling

A myocardial infarction (MI) event (heart attack) is followed within days by cardiac fibrosis and remodeling of the left ventricle, the main pumping chamber of the heart. Although fibrosis initially serves to prevent rupture of the weakened, necrotic portion of the ventricular wall, continued fibrosis with formation of excess fibrotic tissue causes the ventricle to enlarge and dilate, which leads to a serious reduction of pumping efficiency. The excess fibrosis that occurs, an example of an aberrant mechanotransduction response, is an outcome of stretch-induced release of

TGFβ1 and other pro-inflammatory cytokines. This realization of the underlying basis of the stretch-induced enlargement of the infarcted ventricle has led to a new strategy for reducing the pathological remodeling, which involves injection of a hydrogel directly into the infarcted tissue.

The rationale for this approach is that the injected hydrogel could function as a filler that would expand and stiffen the infarcted myocardial wall, which would decrease mechanical stress in the surrounding non-infarcted myocardium and thereby reduce stretch-induced release of inflammatory cytokines. While several natural polymers such as fibrin, collagen, hyaluronic acid, and alginate gel are being studied for use as injectable fillers, the results of one study suggest that hydrogels formed from sericin might be potentially effective for early intervention and treatment of infarcted myocardial tissue (Song et al. 2016). Sericin is the protein that surrounds silk fibers derived from the cocoons of the domesticated silkworm *Bombyx mori* (see Figure 10.10). Sericin hydrogels were prepared by mixing an aqueous solution of sericin and a cross-linking agent (genipin). MI was induced in C57BL/6 mice by surgical occlusion of the left anterior coronary artery and, within minutes, the center of the infarcted ventricle was injected with saline (designated the MI group) or the sericin hydrogel (the S/G group). Sham control animals underwent the same surgery without occlusion of the coronary artery.

The effect of the hydrogel injection on cardiac remodeling and physiology was assessed 6 weeks after MI. Two-dimensional echocardiography was performed to measure left ventricular ejection fraction (EF) and fractional shortening (FS), two key indicators of ventricular function. Then the animals were euthanized, and their hearts were excised, fixed in formalin and embedded in paraffin. Histological analyses showed a greater myocardial wall thickness (Figure 17.6a) and lower total infarct area (Figure 17.6b) in the left ventricles of the animals treated with the sericin hydrogel (S/G group) when compared to the MI group. In addition, the pre-euthanasia echocardiography showed a significant increase in left ventricular EF and FS for the animals treated with the sericin hydrogel when compared to the MI group (Figure 17.6c, d). These results indicate that the injection of a hydrogel filler appears to be effective in limiting pathological cardiac remodeling and in preserving ventricular pumping function in a rodent model. This approach has also been shown to be effective in studies performed with large animals, such as sheep and pigs and, thus, might be a potentially valuable method for early intervention to limit pathological cardiac remodeling in human patients.

17.3.3 Soft Tissue Healing

Mechanical stress-induced fibrosis is particularly problematic in soft tissues. The occasional occurrence of stress-induced soft tissue fibrosis is a manifestation of aberrant responses that principally involve fibroblasts and the related cell type known as a myofibroblast. As noted earlier, fibroblasts are highly sensitive to external mechanical stresses and to the stiffness of the ECM (Section 17.2.3.3). In response to extracellular mechanical stress cues, fibroblasts can adjust the composition of the ECM (such as the relative amounts of stiff collagen fibers and soft glycosaminoglycans) and, thereby, alter the stiffness of soft tissues. Conditions of overload and high mechanical stress trigger the aberrant signaling that can result in soft tissue fibrosis and formation of scar tissue (Figure 17.4). In addition, the release of TGFβ1 is one of the factors that appear particularly important in promoting the events that culminate in

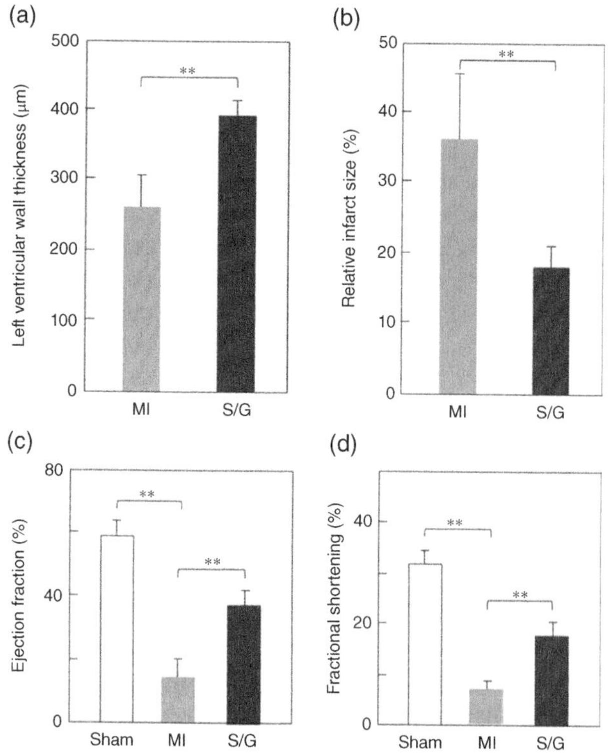

Figure 17.6 Effects of sericin hydrogel injection on cardiac remodeling and functional recovery of infarcted hearts in rodents. Animals that underwent surgically induced infarction received injection of saline (MI group) or sericin hydrogel (S/G group). Sham control animals underwent the same surgery without occlusion of the coronary artery. Assessment of left ventricular remodeling was performed at six weeks post-surgery by quantification of (a) the left ventricular (LV) wall thickness and (b) relative infarct size. Recovery of ventricular function was assessed by echocardiography measurement of (c) left ventricular ejection fraction (EF) and (d) fractional shortening (FS). (**$p < 0.01$). *Source:* From Song et al. (2016) / with permission of Elsevier.

fibrosis. Another factor is the release of MMPs that can promote tissue fibrosis by causing the breakdown of the basal lamina.

Reduction of Peritoneal Fibrosis and Adhesions

Recognition of the role of TGFβ1and MMPs as causative factors in fibrosis has led to the development of several approaches for the reduction of mechanical stress-induced release of TGFβ1, MMPs and related pro-inflammatory factors. One approach that has been studied for blocking MMP release involves intraperitoneal application of a membrane fabricated from electrospun fibers (~0.5–2 μm in diameter) composed of polycaprolactone (PCL) functionalized with a fragment of the β1-chain of laminin, which is known to block expression of the enzyme matrix metalloproteinase-2 (MMP2). The potential effectiveness of the functionalized PCL membrane in reducing peritoneal fibrosis was tested in a mouse model (Horejs et al. 2017). Gelatin electrophoresis zymographs showed that the functionalized PCL membranes were effective in blocking TGFβ-driven release of MMP2 (Figure 17.7). In addition, histological analyses showed that explanted peritoneal tissues incubated with the functionalized PCL membranes were almost completely free of fibrotic tissue formation. Overall, the results of this study indicate that this type of biomaterial and treatment approach could potentially be effective for reducing peritoneal fibrosis, which is a common complication of abdominal surgical procedures in human patients.

Control of Scar Formation and Contractures

The hypertrophic scars that form in burn patients, which cause inelasticity of the skin, disfigurement and loss of joint mobility, are directly attributable to the activity of myofibroblasts and their effects on the ECM components of the dermis. The ECM of normal dermis has randomly oriented collagen fibers, a property that accounts for much of the elasticity of normal skin. After a burn injury, the myofibroblasts produce a matrix with aligned, densely packed collagen fibers that cause the skin to become stiff. This stiffening of the skin triggers additional differentiation of fibroblasts to myofibroblasts, further intensifying the problem. To address this problem, polymer scaffolds have been created with an appropriate combination of elastomeric and tensile response and durability in an approach to blunt much of the matrix stiffening that occurs after a burn injury. Two such scaffolds have been prepared using the electrospinning technique (Chapter 25), one composed of a copolymer of poly(L-lactic acid) (L-PLA) and PCL, and the other composed of a PU to which collagen type 1 molecules were covalently bonded. The effectiveness of these scaffolds was evaluated by implantation in the wound bed of mice with second-degree burns. Both scaffolds were sufficiently durable in that they remained in the wound bed for 30 days, a duration that matches the remodeling phase of wound healing (Chapter 24). In addition, both scaffolds reduced the extent of scar formation and maintained greater elasticity of the skin. The use of such scaffolds could potentially be effective in reducing hypertrophic scar in human burn patients.

Overall, there is need for advances in biomaterials and treatments procedures to reduce mechanical stress-induced soft tissue fibrosis and fibrotic complications. The magnitude of the need is brought into focus by the high number of mortalities in the western world attributable to organ or tissue fibrosis, which is estimated to be as high as ~45% of total mortalities.

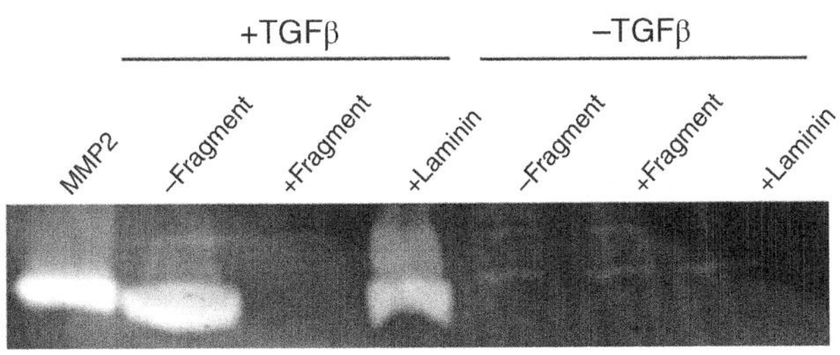

Figure 17.7 Electrophoretic analysis of effectiveness of electrospun PCL membranes functionalized with laminin β1-fragment for blocking TGFβ-driven fibrosis of peritoneal explants from C57BL6N mice. Representative gelatin zymographs of conditioned media from explanted peritoneal membrane tissues show functionalized PCL membranes (+fragment) were effective in completely blocking TGFβ-driven release of the enzyme MMP2 (matrix metalloproteinase-2) which causes extracellular matrix destruction associated with fibrosis. *Source:* From Horejs et al. (2017).

17.3.4 Stem Cells in Tissue Engineering

There is extensive evidence that cues from extracellular mechanical stresses and from substrate stiffness influence the development of stem cells. Recognition of the importance of these cues has led to their incorporation into various tissue engineering approaches for the regeneration of functional tissues and organs. As discussed in Chapter 25, biomaterials, referred to as scaffolds, often in the form of porous solids, hydrogels, or decellularized mammalian tissue, are an essential component of the tissue engineering approach. As noted, the scaffold should provide a suitable microenvironment for cells to reside, proliferate and differentiate into the desired phenotype, and should be capable of providing the right mechanical and molecular signals for optimal cell proliferation, differentiation and gene expression.

Hydrogels find considerable use as scaffolds in tissue engineering, particularly for the engineering of soft tissues, based on a number of factors, particularly their ability to approximate the aqueous gel-like nature of the ECM. Additionally, hydrogels can vary from being soft and pliable to extremely rigid, providing a range of stiffness and viscoelastic properties for optimal mechanical signaling. In this section, we will discuss a few examples of differential cellular responses to scaffolds with adjustable stiffness, as well as how those different cellular responses can be utilized to control the differentiation of stem cells to produce tissue of a desired type.

Scaffolds with Adjustable Stiffness for Engineering of Tissues

Hydrogels with adjustable stiffness have been used in several studies for guided tissue engineering with MSCs. An example is the guided differentiation of human MSCs toward an osteogenic phenotype or an adipogenic phenotype by an extended duration of culturing on stiff or soft hydrogels, respectively (Guvendiren and Burdick 2012). Soft hydrogels were formed from a solution of methacrylated hyaluronic acid (MeHA) by partially crosslinking the MeHA molecules using dithiothreitol (DTT). Prior to this crosslinking step, the MeHA molecules were functionalized with an adhesive peptide (arginine-glycine-aspartic acid, RGD). The soft hydrogels, which were formed on glass slides, had a Young's modulus of ~3 kPa and a thickness of ~250 μm. Subsequently, these soft hydrogels were subjected to rapid dynamic stiffening in a second crosslinking step, giving hydrogels with a Young's modulus of ~30 kPa.

In experiments with cells, MSCs were seeded onto the soft MeHA hydrogel substrates and incubated in complete growth medium for 24 hours. Then the growth medium was replaced by differentiation medium composed of equal parts adipogenic and osteogenic induction media, and incubation was continued for up to 14 days. At specific times (1, 3, and 7 days) during this incubation, the soft hydrogels were stiffened by introducing a photoinitiator and further crosslinking the molecules by exposure to ultraviolet light (UV) for 2 minutes (Figure 17.8a). This short exposure to UV light had a negligible effect on MSC survival or differentiation. At the conclusion of the 14-day incubation (with the induction media refreshed every 3 days), the cultures were histochemically stained for alkaline phosphatase activity and the presence of lipid droplets as indicators of osteogenic or adipogenic differentiation, respectively. The results showed that a longer duration on the stiff hydrogel promoted differentiation of MSCs toward the osteogenic phenotype whereas longer duration on the soft hydrogel promoted differentiation toward the adipogenic phenotype (Figure 17.8b).

Cyclic Stretching in Stem Cell Engineering of Tissues

Scaffolds in some tissue engineering applications are designed to undergo cyclic mechanical stretching as part of the *in vitro* protocol. This procedure addresses the

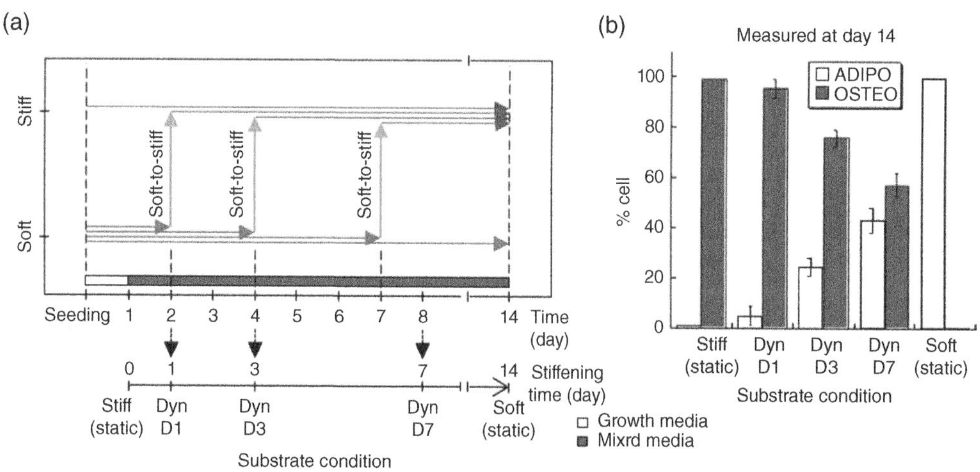

Figure 17.8 Effect of substrate stiffening for guided tissue engineering of human mesenchymal stem cells (MSCs) cultured on hydrogels formed from methacrylated hyaluronic acid (MeHA) functionalized with the adhesive peptide RGD. (a) Representation of time line for cell seeding and subsequent ultraviolet UV light-activated stiffening of MeHA hydrogel substrates. Cell seeded hydrogels were initially cultured 1 day in standard growth media followed by 14 days in adipogenic/osteogenic) differentiation media (mixed media). Substrate condition is indicated as static or dynamic (Dyn) for gels stiffened at day 1, day 3 or day 7 (D1, D3 or D7, respectively). (b) Percentage of MSCs stained positive for alkaline phosphatase, indicator of osteogenic phenotype (osteo), or presence oil red O stained lipid droplets, indicator of adipogenic phenotype (adipo), as a function of substrate stiffening time. *Source:* From Guvendiren and Burdick (2012) / with permission of Springer Nature.

importance of cyclic stretching not only to obtain physiologically functional mature tissues but also for differentiation of the progenitor cells from which those tissues develop. Both two-dimensional and three-dimensional scaffolds have been used in such cyclic stretching procedures.

An example of a two-dimensional scaffold designed for cyclic stretching experiments is a system devised for a tissue engineering procedure with human patellar tendon fibroblasts (HPTFs). This system involved the culturing of HPTFs on polydimethylsiloxane (PDMS) substrates with a microcontact-printed surface pattern of ridges and grooves, each having a width of 10 μm. After coating the surface of the substrates with an agent to promote cell attachment, HPTFs were seeded on the substrates, cultured for 48 hours, and placed in a cyclic stretching apparatus. The stretching protocol involved either 4% or 8% cyclic stretching at 0.5 Hz for four hours, followed by a four-hour incubation without stretching. Evaluation of the molecular response to the stretching protocol was achieved by measuring the levels of mRNA expression for collagen type I, collagen type III, and TGFβ1. It was found that the level of collagen type I mRNA expression and TGFβ1 mRNA expression in HPTFs subjected to the 8% cyclic stretching increased by ~22% and ~25%, respectively, relative to the expression level in control HPTFs without cyclic stretching. These results suggest that the exposure to cyclic stretching promoted increased expression of tendon fibroblast phenotype traits, an indication that cyclic stretching on two-dimensional substrates could be an effective strategy for some types of guided tissue engineering.

As noted earlier, three-dimensional scaffolds have also been used in cyclic stretching procedures. An example is an investigation undertaken to determine whether the use of cyclic stretching during the culturing of human bronchial smooth muscle cells on three-dimensional collagen hydrogels could promote the expression of genes for smooth muscle contractile proteins (Asano et al. 2018). The rationale for this approach was that because cells of the lungs and bronchi are continuously subjected to cyclic stretching and shear stresses during breathing, it would be reasonable to anticipate that mechanical forces could also have a positive effect on the development of pulmonary tissues. The hydrogel system used in this investigation was prepared from a suspension composed of bronchial smooth muscle cells, complete growth medium and collagen type I (2 mg/cm^3). After being placed in silicone chambers (10 mm × 5 mm × 5 mm) with sponge anchors at opposite ends of their length, the suspension was gelled. Following incubation for 24 hours to initiate cell growth, the cell-seeded hydrogels were mounted in a computer-controlled cyclic stretching apparatus and a uniaxial strain of 12% was applied to them at a frequency of 0.5 Hz for 48 hours. Cell-seeded hydrogels incubated for the same duration under static conditions were used as controls. Then the constructs were analyzed to determine cell orientation and gene expression for four different smooth muscle proteins.

Two of the gene expression assays were for the genes ACTA2 and MYH11 that code for the contractile proteins α-smooth muscle actin and myosin heavy chain 11, respectively. The results showed that the expression of the ACTA2 and MYH11 genes were respectively ~three times and ~six times greater in the hydrogels subjected to cyclic stretching when compared to the controls (Figure 17.9a, b). It was also observed that the bronchial smooth muscle cells tended to orient along the long axis of the stretching. Overall, several studies have shown that three-dimensional culture systems combined with cyclic stretching can provide an effective strategy for guided *in vitro* engineering of various tissues.

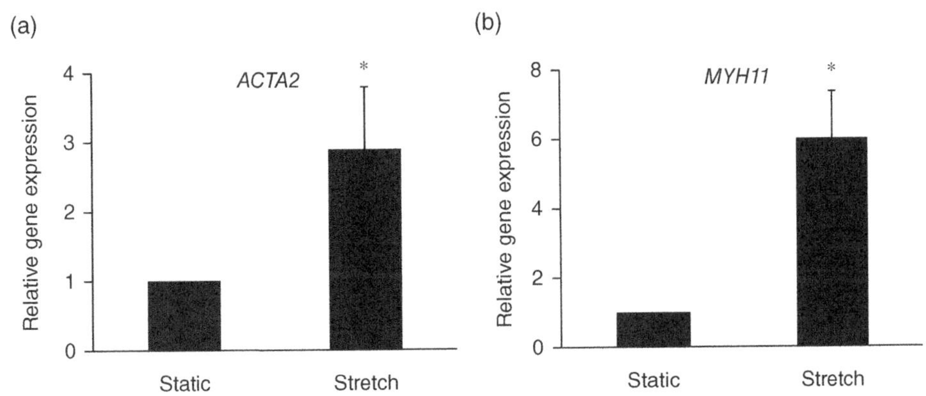

Figure 17.9 Effects of cyclic uniaxial stretching on expression of contractile protein genes in human bronchial smooth muscle cells cultured within three-dimensional collagen gel constructs. The gene expression assays were for (a) the α-smooth muscle actin gene *(ACTA2)*, and (b) the myosin heavy chain 11 gene (*MYH11*). (*$p < 0.05$). *Source:* From Asano et al. (2018).

17.4 Outcomes of Transduction of Extracellular Stresses and Responses

The principal theme of this chapter has been a consideration of the various mechanisms by which mechanical stresses acting on cells and tissues of the body are detected and converted through a mechanotransduction process to biochemical signals that bring about specific cellular responses. As discussed, the central mechanism of the mechanotransduction process involves mechanosensing of external mechanical stresses via focal adhesion and ion channel complexes and transduction through cytoskeletal components and signaling pathways leading to specific cellular responses. Biomaterials, as we have seen, can influence the mechanotransduction process. Implanted biomaterials with appropriate mechano-responsive properties allow normal transduction of extracellular mechanical stress while those with inappropriate mechano-responsive properties impede transduction, often resulting in pathological outcomes. Figure 17.10 summarizes the mechanosensing of extracellular mechanical stresses and transduction through cytoskeletal components and signaling pathways leading to specific cellular responses.

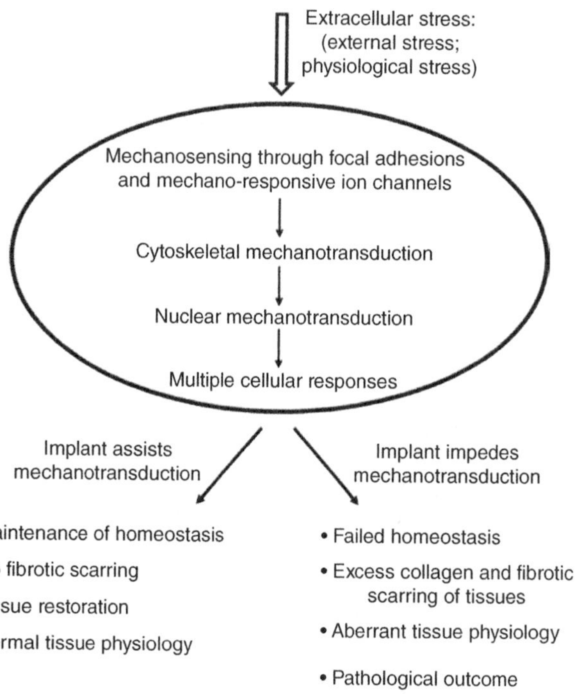

Figure 17.10 Representation of mechanosensing of extracellular stresses via focal adhesion/ion channel complex and transduction through cytoskeletal components and signaling pathways leading to specific cellular responses.

17.5 Concluding Remarks

In this chapter, we discussed biocompatibility phenomena caused by mechanical forces. Cells are able to sense mechanical stresses (or strains) within the ECM and translate them into some type of adaptive response, a process referred to as mechanotransduction. We discussed the stages and pathways of mechanical transduction.

- Because implanted biomaterials generally have mechanical properties that are different from the tissues into which they are implanted, the presence of an implanted biomaterial will generally trigger an adaptive response by cells, independent of any additional effects that the biomaterial can have, such as a direct chemical interaction with the tissue
- Cell responses triggered by mechanotransduction mechanisms include, among others, changes in cell shape, cell movement, cell proliferation, specific gene expression, differentiation, inflammatory responses, and apoptosis
- The resulting events, which can be at a cellular or tissue-level response, can have important consequences for the use of biomaterials in repairing or regenerating hard and soft tissues. Consequently, mechanical stress-induced biocompatibility phenomena should be an important consideration in biomaterials design.

Problems

17.1 External mechanical stresses affect the macromolecular components of focal adhesions in a manner that result in an "outside-in" signal that initiates various mechanotransduction mechanisms. What is the key component of the focal adhesion that is responsible for this "outside-in" signal and what change takes place with this component that produces the signal?

17.2 Until approximately 50 years ago, it was common practice to have patients remain in bed for three or more days following an appendectomy surgical procedure. Since that time, it has been standard practice to have appendectomy patients resume limited activities such as modest walking within less than a day after surgery. Briefly explain how mechanical stresses due to physiological activity could be beneficial to recovery. Related to this, explain how three or more days of bed rest could be detrimental to recovery.

17.3 Explain how changes in the relative amount of glycosaminoglycans affect the mechanical properties of soft tissues. In comparison, explain how changes in the relative amount of collagen affect tissue mechanical properties.

17.4 In occasional cases, mechanical stretching of healing surgical incisions results in the formation of a thick surgical scar. Assume that as a biomedical engineer, you are assigned the task of developing a new biomaterial or biomedical device to minimize thickening of surgical scar tissue. Describe the new biomaterial or biomedical device with an explanation of its mechanical properties that account for the reduction of scar thickness.

17.5 Most bioprosthetic heart valves (BHVs) are fabricated from porcine aortic valves treated with glutaraldehyde, a chemical agent that crosslinks collagen but does not preserve glycosaminoglycans (GAGs), which are lost during the treatment process. Explain two potential improvements of BHV performance that might occur by using a crosslinking agent that does not cause loss of GAGs.

17.6 Assume that you have the responsibility of selecting the type of biomaterial to be used in the surgical repair of a fractured tibia in a young patient. The orthopedic surgical procedure requires realigning and stabilizing the fractured parts of the bone using a fracture fixation plate of dimensions ~8 cm long by ~2 cm wide secured with screws. Indicate a specific type of material you would select for this procedure and explain the rationale for your selection based on the mechanical properties of the material and the bone to which it is attached.

References

Asano, S., Ito, S., Morosawa, M. et al. (2018). Cyclic stretch enhances reorientation and differentiation of 3-D culture model of human airway smooth muscle. *Biochemistry and Biophysics Reports* 16: 32–38.

Bonewald, L.F. (2007). Osteocytes as dynamic multifunctional cells. *Annals of the New York Academy of Sciences* 1116: 281–290.

Carlson, B.M. (1988). *Patten's Foundations of Embryology*, 5e. McGraw-Hill.

Duncan, R.L. and Turner, C.H. (1995). Mechanotransduction and the functional response of bone to mechanical strain. *Calcified Tissue International* 57: 344–358.

Guvendiren, M. and Burdick, J.A. (2012). Stiffening hydrogels to probe short- and long-term cellular responses to dynamic mechanics. *Nature Communications* 3: 792.

Horejs, C.M., St-Pierre, J.P., Ojala, J.R.M. et al. (2017). Preventing tissue fibrosis by local biomaterials interfacing of specific cryptic extracellular matrix information. *Nature Communications* 8: 15509.

Humphrey, J.D., Dufresne, E.R., and Schwartz, M.A. (2014). Mechanotransduction and extracellular matrix homeostasis. *Nature Reviews. Molecular Cell Biology* 12: 802–812.

Song, Y., Zhang, C., Zhang, J. et al. (2016). An injectable silk sericin hydrogel promotes cardiac functional recovery after ischemic myocardial infarction. *Acta Biomaterialia* 41: 210–223.

Strohmeyer, N., Bharadwaj M., Costerll M. et al. (2017). Fibronectin-bound α5β1 integrins sense load and signal to reinforce adhesion in less than a second. *Nature Materials* 16: 1262–1270.

Further Reading

Henry, J.J.D., Yu, J., Wang, A. et al. (2017). Engineering the mechanical and biological properties of nanofibrous vascular grafts for *in situ* vascular tissue engineering. *Biofabrication* 9: 035007.

18

Inflammatory Reactions to Biomaterials

18.1 Introduction

The implantation of a biomaterial into the body will invariably and unavoidably involve some degree of wounding or tissue injury as a consequence of surgical penetration of the skin. When tissue injury occurs, whether caused by bacterial infection, surgical trauma, or any other phenomena, a complex sequence of host tissue inflammatory reactions will be initiated. These inflammatory reactions are normal protective responses that are physiological manifestations of the second of the body's three lines of defense against invading pathogens and foreign materials. Of these three lines of defense, the first two are nonspecific innate defenses, which act rapidly against anything identified as not belonging to the body.

The first line of defense consists of skin and mucous membranes that, together, form an external physical barrier which is very effective in walling out most bacteria and viruses and small particles of foreign matter. Any pathogens and foreign matter that may occasionally penetrate this outer barrier will be met by the second line of defense provided by the innate immune system, which is a system of internal, nonspecific defense mechanisms. Components of the second line of defense include the sequence of inflammatory reactions and the phagocytic leucocytes. Working with these is a family of antimicrobial plasma proteins known as the complement system.

The third and final line of defense is the adaptive (acquired) immune system, which will be further addressed in the next chapter. It will be seen there that, unlike the innate immune system, the adaptive system acquires the ability to defend against specific pathogens. Also delayed until the next chapter is a discussion of complement proteins. This is because the complement system functions along with both the innate and adaptive immune systems.

Any biomaterial implanted in the body, regardless of its chemical composition or structure, will elicit tissue inflammation. The extent and duration of inflammatory reactions are directly related to the level of the biocompatibility of the implanted biomaterial. In general, the biocompatibility of an implanted biomaterial is described in terms of the acute and chronic inflammatory responses. Biomaterials with a high level of biocompatibility may elicit only an acute inflammatory reaction that can normally be resolved within a week. In comparison, biomaterials with a lower level of biocompatibility will elicit more severe chronic inflammatory reactions, which, in some cases, may persist for several months or longer.

Within a short duration after surgical implantation of a biomaterial, the wound tissue around the implant will release numerous cytokines and chemokines (Table 16.2) that attract defender cells already present in the tissue, as well as several types of circulating leukocytes. The resident defender cells include mast cells and tissue macrophages. The leukocytes attracted to the implant site include neutrophils, monocytes, eosinophils, and lymphocytes. After passing through the capillary wall and migrating to the wound site, the leukocytes engage in multiple tasks essential to wound healing and the restoration of normal tissue function. The temporal sequence of inflammatory reactions to an implanted biomaterial has been described in terms of seven reactions (Anderson 2001):

Implant interaction with plasma proteins
Provisional matrix formation
Acute inflammation
Chronic inflammation
Formation of granulation tissue
Foreign body reaction
Fibrosis and fibrous encapsulation

Complete degradation or removal of the material provoking the inflammatory reactions is followed by a resolution phase to restore normal tissue function.

This chapter will discuss implant-induced inflammatory responses and the extent of those responses. The topics to be discussed will cover the following main areas:

- The temporal sequence of inflammatory responses to implanted biomaterials.

Materials for Biomedical Engineering: Fundamentals and Applications, First Edition. Mohamed N. Rahaman and Roger F. Brown.

Companion website: www.wiley.com/go/rahaman/biomedicalengineering

- The relationship between the severity of the inflammatory response and the level of biocompatibility of the implanted biomaterial.
- Potential approaches to modulate inflammatory responses to implanted biomaterials.

18.2 Implant Interaction with Plasma Proteins

A sequence of inflammatory reactions starts almost immediately after the implantation of a biomaterial. The surface of the biomaterial is rapidly coated with an adsorbed layer of proteins from the surrounding interstitial fluid and plasma. The proteins in this adsorbed layer include albumin, immunoglobulin G, fibrinogen, complement system proteins, and several other proteins. The adsorption of fibrinogen has special significance because it leads to two immediate outcomes. One outcome is the activation and enzymatic conversion of fibrinogen to a loose fibrin mesh, which becomes the main component of a provisional matrix that fills the wound site. The second outcome involves fibrinogen's action as a chemotactic factor that attracts mast cells and macrophages from the surrounding tissue to the implant site. These mast cells attracted to the implantation site immediately commence releasing the physiologically active contents of their granules, such as the vasodilator histamine and the cytokine tumor necrosis factor-alpha (TNFα). The released histamine triggers the dilation of local capillaries, which increases their permeability, resulting in fluid leakage and tissue swelling. In combination, the tissue swelling and the formation of a fibrin-based provisional matrix has the effect of "walling-off" the affected wound area. This walling-off of the wound area has pathophysiological significance in that it limits the spread of infectious pathogens that may have breached the outer protective barrier, thereby providing part of a second line of host defense.

18.3 Formation of Provisional Matrix

The formation of a fibrin-based provisional matrix is a very early event in the inflammatory response to infection or to the implantation of a biomaterial (Figure 18.1). The newly formed fibrin mesh provides a supportive scaffold for the adhesion and migration of the multiple types of leucocytes, fibroblasts, endothelial progenitor cells, and other cells which participate in the defense and repair of the wound site. However, the provisional matrix is more than just a passive support scaffold. It also functions as a dynamic microenvironment that actively contributes to tissue repair, while undergoing continuing changes in composition as wound healing proceeds. Much of this contribution is due to the provisional matrix being a source of various cytokines and chemotactic substances, which attract the multiple types of cells that migrate into the implantation site. Two examples of these cytokines are platelet-derived growth

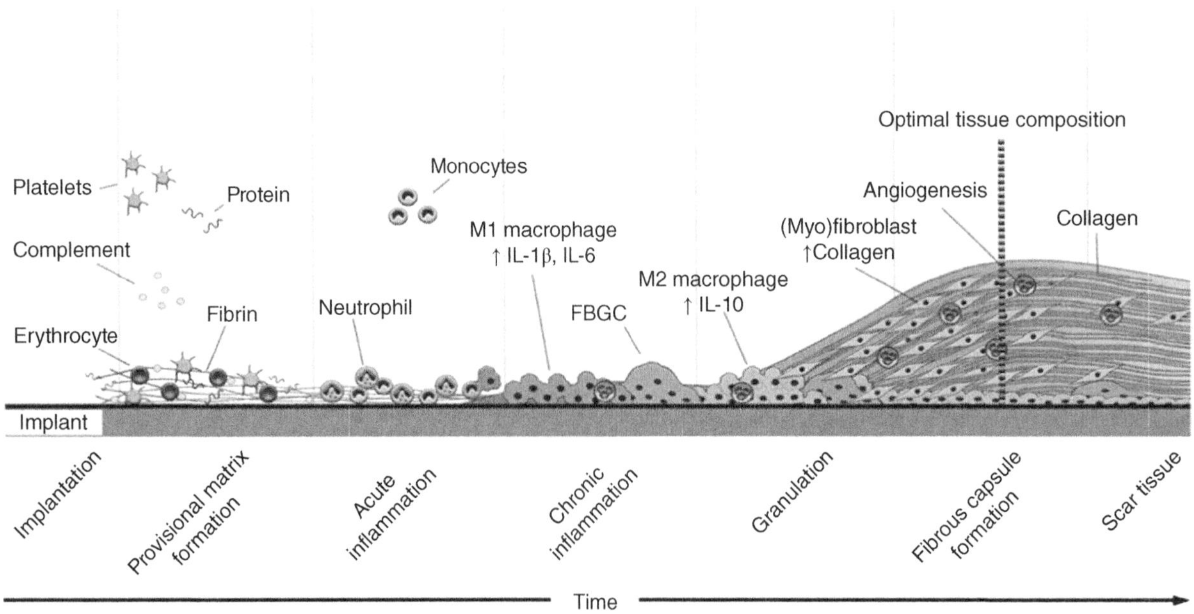

Figure 18.1 Depiction of sequence of events of tissue inflammatory response to an implant. The sequence begins with formation of a provisional matrix and culminates with fibrous encapsulation and eventual resolution. *Source:* Geelhoed et al. (2017).

factor (PDGF) and transforming growth factor-β (TGFβ), which are released by platelets. These PDGF and TGFβ cytokines contribute to further recruitment of fibroblasts. In addition, several other cytokines are then released by leukocytes that migrate into the provisional matrix.

Resident macrophages that migrate into the provisional matrix of the wound release the pro-inflammatory cytokines interleukin-1 (IL-1) and TNFα, which induce endothelial cells of the local microvasculature to express P-selectins, a group of cellular adhesion molecules. Circulating leukocytes then move toward the wound site due to the action of chemotactic factors and the presence of P-selectins on local endothelial cells. There is evidence that the most important effectors of the early recruitment of circulating leukocytes, to the site of an infection or an implanted biomaterial, are tissue mast cells. The chemotactic substance released by mast cells, which promotes the recruitment of neutrophils and monocytes to the implant wound site, is histamine, the same agent that also causes vasodilation.

The provisional matrix continues to change as the early events of wound healing proceed. Because of the incomplete nature of the developing microvascular network of the matrix, the microenvironment develops a slightly reduced oxygen tension. The significance of this reduced oxygen tension is that it promotes the retention of neutrophils, the short-lived population of leukocytes that are particularly associated with the acute phase of inflammation.

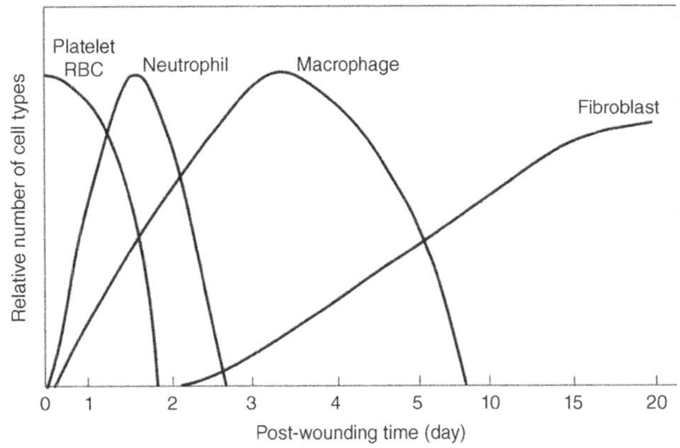

Figure 18.2 Relative number of cells as a function of post-wounding time, illustrating the time scale of migration of various cell types into tissue undergoing inflammatory response to an implant. *Source:* Velnar et al. (2016).

18.4 Acute Inflammation and Neutrophils

The start of the acute phase of inflammation is marked by the arrival of a large number of neutrophils that function as the primary defender cell at the wound site. Subsequently, other cell types arrive and participate in wound repair (Figure 18.2). Although neutrophils make up about two-thirds of the circulating leukocytes, they normally remain in the circulating blood for only 4–8 hours. The action of cytokines, chemokines, and the activated microvascular endothelium lengthens the life span of neutrophils to 4–5 days, enabling them to accumulate at the wound site to combat infection and to initiate wound healing.

18.4.1 Neutrophil Activation and Extravasation

Before the neutrophils can engage in their protective functions at the wound site, they must migrate across the blood vessel wall by the process of extravasation as represented in Figure 18.3. The portion of the microvasculature where neutrophil extravasation occurs is mainly in the venules where the flowing shear stress is lowest. The circulating neutrophils, which are in a nonactivated resting state, move mainly along a path at the center of vessels where the laminar blood flow is highest. Because of the localized inflammation and the histamine-induced dilation of the microvasculature, the velocity of blood flow at the wound site is much reduced, causing the neutrophils to "marginate" or move toward the venule periphery. At the same time, localized inflammatory signals are released, which cause the activation of the microvascular endothelium and partial activation of the nearby neutrophils. The marginated, partially activated neutrophils experience brief, loose binding with P-selectin molecules on the exterior of endothelial cells. This transient binding further slows the movement of the neutrophils, causing them to begin to roll along the vessel wall. The slowly rolling neutrophils engage in higher affinity binding with endothelial integrins and intercellular adhesion molecules (ICAMs). These tightly bound neutrophils then undergo cytoskeletal changes that allow them to squeeze between endothelial cells and extravasate into the interstitial space. Upon entering the interstitial space, the neutrophils migrate toward the wound site, the source of the chemotactic substances.

An example of neutrophil migration and accumulation at the sites of biomaterial implantation in BALB/c mice is shown in Figure 18.4 (Zhou et al., 2012). Two-photon microscopy and time-lapse imaging were used to monitor *in vivo* neutrophil migration and accumulation at sites of inflammation in the lower back of the animals. The inflammation was triggered by the subcutaneous implantation of polylactic acid (PLA) microparticles 24 hours earlier. Fluorescent detection of the neutrophil accumulation was achieved by means of a fluorescent complex consisting of a

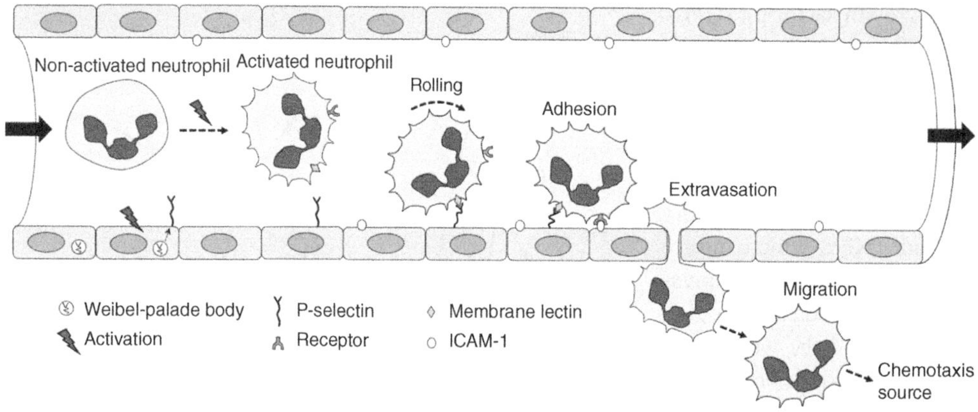

Figure 18.3 Process of neutrophil extravasation from microcirculation. Activated neutrophils rolling along vessel wall adhere to P-selectin on endothelial cell surface, then squeeze between endothelial cells, and emerge to migrate toward site of tissue injury.

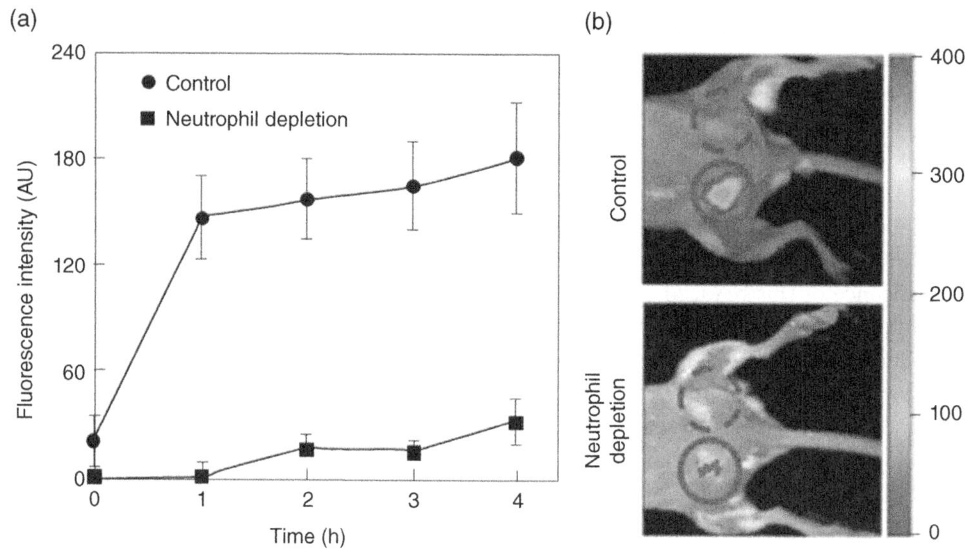

Figure 18.4 Fluorescent detection of neutrophil accumulation at sites of inflammatory response to 24 hours prior implantation of PLA microcapsules compared to sham sites injected with saline. (a) Fluorescence intensity (in arbitrary units, AU) as a function of time after intraperitoneal administration of a neutrophil-specific peptide receptor/fluorescent nanoprobe complex to control and neutrophil depleted mice. (b) *in vivo* image at 4 hours postinjection comparing fluorescence intensities in control and neutrophil depleted animals. *Source:* Zhou et al. (2012).

neutrophil-specific formyl receptor peptide (FRP) conjugated with a fluorescent compound. The fluorescent-labeled FRP was administered by intravenous injection to two groups of mice: a normal (control) group, and a group of neutrophil-depleted animals. The neutrophil depleted state was an outcome of pre-treatment of the animals by intraperitoneal injection of neutrophil neutralizing antibody. Intravenous administration of the fluorescent, neutrophil-binding FRP was followed by a relatively rapid fluorescent signal accumulation at the sites of PLA microparticle injection in the control animals (Figure 18.4a). In comparison, very little accumulation of the fluorescent signal was detected at the PLA injection sites in the neutrophil depleted animals. Figure 18.4b shows the high level of fluorescent signal observed at the PLA injection sites in the control group versus the much lower level of fluorescent signal observed in the neutrophil-depleted group. Collectively, these results provide evidence for the rapid accumulation of neutrophils at sites of inflammatory reaction triggered by an implanted biomaterial.

The main functions performed by neutrophils upon reaching the wound site include the secretion of cytokines/chemokines that recruit additional leukocytes, the formation of reactive oxygen species (ROS), the degranulation or exocytosis of membrane-bound secretory granules, the destruction of pathogenic microbes by phagocytosis and the formation of special structures known as neutrophil extracellular traps (NETs). Neutrophil-derived cytokines/

chemokines that are effective in the recruitment of additional leukocytes include interleukin-8 (IL-8), monocyte chemotactic protein-1, and macrophage inflammatory protein-1α.

The change in morphology as the neutrophil undergoes transformation from the resting, nonactivated state to the activated state is shown in Figure 18.5. As shown, the neutrophil has a dimpled but relatively smooth exterior surface in the resting, nonactivated state (Figure 18.5a). In comparison, the exterior surface of the neutrophil in the activated state has numerous membrane ruffles (Figure 18.5b). The wrinkling or ruffling of the exterior surface is thought to be a morphological change that enhances phagocytic activity.

18.4.2 Formation of Reactive Oxygen Species

The recruitment and activation of neutrophils triggers the "respiratory burst," a process of increased oxygen metabolism resulting in the formation of ROS that have varying levels of bactericidal activity. The reactions and mechanisms by which the anion radical superoxide, $O_2^{-}\bullet$, and other ROS are formed were presented in Chapter 15. Here, we will reiterate a few brief highlights of ROS formation that pertain to neutrophil function. The reactive oxidants generated by phagocytic cells, as mentioned in Chapter 15, include $O_2^{-}\bullet$, H_2O_2, $HO\bullet$, and HOCl. Of these four oxidants, HOCl is known to have the most potent bactericidal effect. The importance of HOCl for neutrophil function is reflected in the cell's high content of myeloperoxidase, the enzyme that catalyzes HOCl formation.

18.4.3 Phagocytosis by Neutrophils

Another major protective function of neutrophils is phagocytosis. Upon coming into contact with plasma and interstitial fluid, invading microbes and implanted biomaterial acquire a coating of adsorbed proteins. The adsorbed proteins include immunoglobulin G (IgG) and complement C3, two protein groups of special significance because they function as opsonins that mark invading microbes and foreign particles for phagocytosis by neutrophils and other phagocytic cells. The phagocytic process begins as Fc receptors mediate neutrophil attachment to opsonins on the foreign particle, followed by dynamic movement of the plasma membrane, which leads to the formation of pseudopod extensions that surround and envelope the microbe or particle. When the envelopment is complete, the microbe or particle is internalized within a membranous vesicle referred to as a phagosome. The phagosome subsequently fuses with intracellular lysosomes to form a phagolysosome. Endocytosis and the mechanisms of formation of phagosomes together with a figure depicting the process are discussed in Chapter 21.

Two mechanisms are then employed by the phagolysosome to destroy the ingested microbe or foreign particle. One mechanism involves the enzymatic digestion by lysosomal proteases that include lysozyme and elastase, two enzymes that have broad killing effects on a variety of bacteria. The second mechanism involves the generation of bactericidal ROS. The phagosome membrane contains activated NADPH oxidase, the enzyme complex that catalyzes formation of superoxide, the metabolic precursor of H_2O_2, HOCl, and other bactericidal oxidants.

18.4.4 Neutrophil Extracellular Traps (NETs)

A recently discovered mechanism by which neutrophils can destroy invasive pathogenic microbes involves the formation of web-like structures referred to as NETs. The principal components of these large extracellular structures

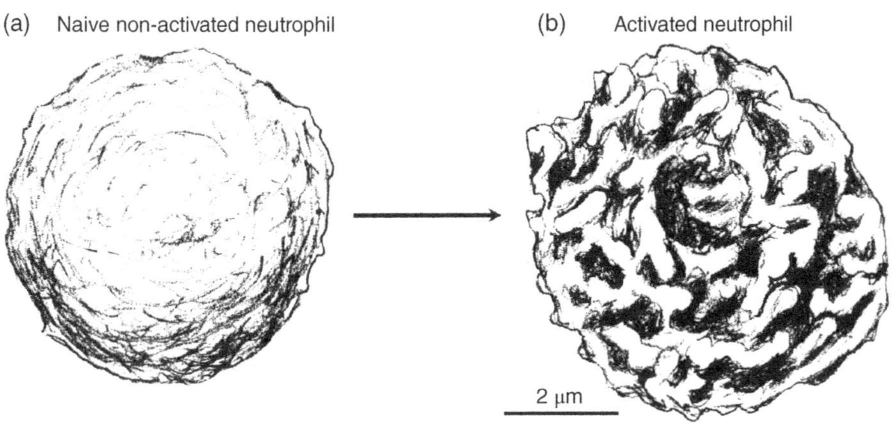

Figure 18.5 Illustration comparing the morphology of (a) nonactivated neutrophil and (b) activated neutrophil showing the wrinkling or ruffling of the exterior surface that enhances phagocytic activity.

include DNA, histones, and proteins such as neutrophil elastase, one of the main contents of neutrophilic granules. The formation of NETs provides an additional mechanism for neutrophil destruction of pathogens by extracellular entrapment followed by proteolytic degradation plus bactericidal action of histones. The presence of several types of gram-positive and gram-negative bacteria has been found to lead to the formation of NETs.

In vivo imaging procedures have revealed evidence of at least two basic mechanisms of NETosis or NET formation by activated neutrophils (Figures 18.6). One mechanism is a suicidal NETosis that involves extensive morphological change followed by cell rupture, death, and rapid release of most of the cellular content. A second mechanism, referred to as vital NETosis, occurs slowly, involving only partial release of cellular content. This occurs by nuclear envelope blebbing and vesicle exocytosis with the cell remaining alive following NET formation. Results of *in vivo* imaging procedures also reveal that vital NETosis occurs in response to bacterial infections while suicidal NETosis is observed in response to stimuli by chemical agents such as phorbol 12-myristate 13-acetate, a potent agent of neutrophil activation. Additional investigations have shown that NETosis particularly occurs when local microbe numbers are high. This observation has been interpreted as an indication that NETs might become especially important when local microbe numbers exceed the phagocytic capacity of neutrophils.

The formation of NETs is also dramatically increased following implantation of some types of sterile biomaterials. It was recently observed that intraperitoneal injection of sterile capsules of various sizes and compositions into the peritoneal cavity of mice leads to a 30–500 times increase in the number of neutrophils and NETs in the peritoneal exudates withdrawn from the animals 2 weeks after microcapsule injection (Jhunjhunwala et al., 2015). The capsules used in the investigation had sizes in the range 300 μm to 2 mm and compositions that included both degradable and nondegradable materials. The degradable capsules were composed of alginate and polylactic-*co*-glycolic acid (PLGA), while the nondegradable microcapsules were composed of glass, polystyrene (PS), and polymethyl methacrylate (PMMA). It is noteworthy that although neutrophil infiltration of tissues is typically observed during the few days that comprise the acute phase of inflammation, the high numbers of neutrophils in the peritoneal exudates were observed 2 weeks after injection of the nondegradable microcapsules, a duration well beyond the acute response. Related to this, significantly lower numbers of neutrophils were found 2 weeks after injection of the microcapsules composed of a degradable biomaterial.

The results of a recent *in vitro* investigation indicated that neutrophil adhesion and formation of NETs might play a role in osseointegration of titanium implants with roughened surfaces (Vitkov et al., 2015). The samples in the investigation were medical grade titanium plates with a surface that had been acid-etched and sandblasted with large grit to give surface roughness features of different length scales (referred to as an SLA surface). These samples were coated with albumin or a combination of albumin and acetylsalicylic acid, or left uncoated. Then they were covered with fresh capillary blood from human volunteers, and incubated for 4 hours at 37 °C in a humid 5% CO_2 atmosphere. Additional blood samples were incubated on uncoated or poly-D-lysine-coated glass coverslips. Analyses at the conclusion of the incubation showed that

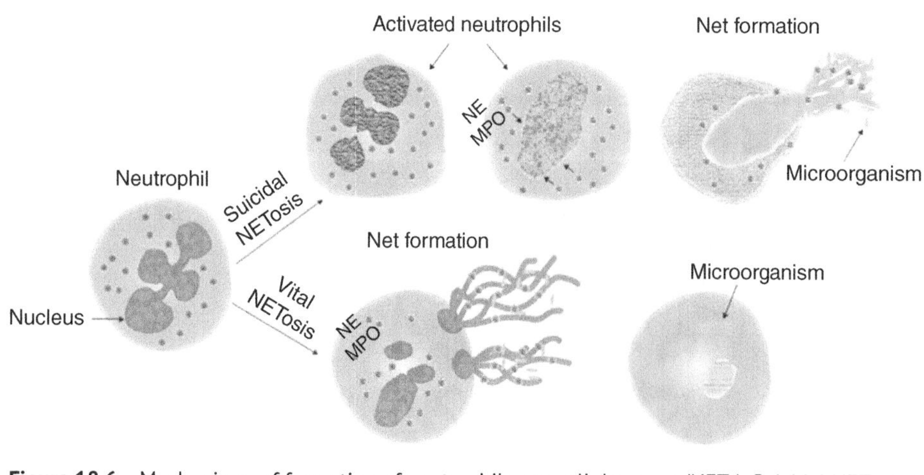

Figure 18.6 Mechanisms of formation of neutrophil extracellular traps (NETs). Suicidal NETosis involves extensive morphological change followed by cell rupture, death, and rapid release of most of the cellular content. Vital NETosis involves slow, partial release of cellular content by nuclear envelope blebbing and vesicle exocytosis with the cell remaining alive following NET formation. NE, neutrophil elastase; MPO, myeloperoxidase. *Source:* Alasmari (2020).

the SLA surfaces caused rapid adherence of human neutrophils followed by NETosis. In addition, the coating (albumin or a combination of albumin and acetylsalicylic acid) had no significant effect on the inflammatory response to the SLA surfaces. In comparison, NETosis did not occur on poly-D-lysine-coated glass surfaces. These results indicate that NETosis during the initial inflammatory response might play a role in the osseointegration of titanium implants with SLA surfaces.

18.4.5 Neutrophil Apoptosis

Upon completion of their tasks in acute inflammation, neutrophils normally undergo apoptosis, a state of programmed cell death, and clearance from the tissue. Apoptotic neutrophils secrete additional factors that contribute to the attraction of macrophages to engulf and digest the dying neutrophils at the wound site. As the clearance of neutrophils proceeds, inflammation and swelling subsides allowing the tissue to begin a gradual return to a state of normal homeostasis.

Although neutrophils are essential for combatting infection, errant neutrophils are capable of causing extensive damage to normal tissues. Neutrophil-generated ROS can cause extensive oxidative damage that is linked with rheumatoid arthritis, atherosclerosis, and other pathological conditions. In addition, the release of the lysosomal proteases elastase and cathepsin G by errant neutrophils can cause extensive degradation of extracellular matrix components, a mode of degradation that is thought to be a contributing factor to emphysema and acute respiratory distress syndrome.

18.5 Chronic Inflammation and Macrophages

Chronic inflammation is marked by a state of longer duration inflammation that may include tissue necrosis and fibrosis. The most important cells in chronic inflammation are macrophages. The name macrophage means "big eater" from Greek origin and refers to the cell's capacity for phagocytosis which surpasses that of neutrophils. Macrophages are versatile cells capable of multiple roles. They are also relatively long-lived cells that can remain at the wound site for months.

18.5.1 Macrophage Differentiation and Recruitment to Implant Surfaces

The macrophages present in tissues derive from circulating monocytes. In response to chemotactic signals released from injured tissues, including tissues injured by the implantation of a biomaterial, monocytes adhere to P-selectins of activated endothelium and complete the process of extravasation via essentially the same steps described for neutrophils (Section 18.4.1). Upon entering the interstitial space, monocytes undergo differentiation, mainly to one of two macrophage phenotypes, depending on the signals present in the local microenvironment. Exposure to interferon-γ (IFNγ) or lipopolysaccharide (LPS) triggers the differentiation of monocytes to the M1 macrophage phenotype that secretes the pro-inflammatory mediators IL-1β, TNFα, and IL-8, whereas exposure to the cytokines IL-4 and IL-10 induces differentiation of monocytes to the anti-inflammatory M2 macrophage phenotype.

Most types of biomaterials implanted into the body will lead to the robust recruitment and adhesion of macrophages to the implant surface. The time course of macrophage adhesion to an implant typically occurs over days rather than hours as occurs with neutrophils (Figure 18.4). An example of the time course of macrophage adhesion is represented by the results of implantation of cellulose acetate membranes in the peritoneal cavity of mice (Figure 18.7). At multiple intervals postimplantation, the implanted membranes were recovered and processed for immunohistological staining with the macrophage-specific antibody Mac 3. The staining showed detectable macrophage adhesion to the membranes at 12 hours postimplantation, followed by a steady increase in macrophage adhesion, reaching a peak at 8 days postimplantation. Also detectable at 8 days postimplantation was the fusion of macrophages to form foreign body giant cells as well as the initial formation of an avascular fibrotic capsule, which indicated the occurrence of more severe phases of the inflammatory response, discussed further in Sections 18.6 and 18.8.

The factors that promote macrophage adhesion to implanted biomaterials and frequent pathological outcomes have been investigated in numerous studies. These studies show that one of the more important factors in the promotion of macrophage adhesion is the initial adsorption of fibrinogen. For example, in a recent clinical investigation, uncoated and protein-coated discs (12 mm in diameter) composed of polyethylene terephthalate (PET) were implanted into subcutaneous sites in the upper arm of human volunteers (Zdolsek et al., 2007). The protein coating was formed by soaking the discs for 12 hours in either autologous fibrinogen-containing plasma or autologous fibrinogen-free serum. Each volunteer received a plasma-coated disc in one arm and a serum-coated disc in the other arm. After 24 hours, the discs were retrieved, rinsed in saline, and placed in a Triton X-100 solution to extract indicator phagocyte enzymes for quantitative analyses. The measurements of enzyme activity included

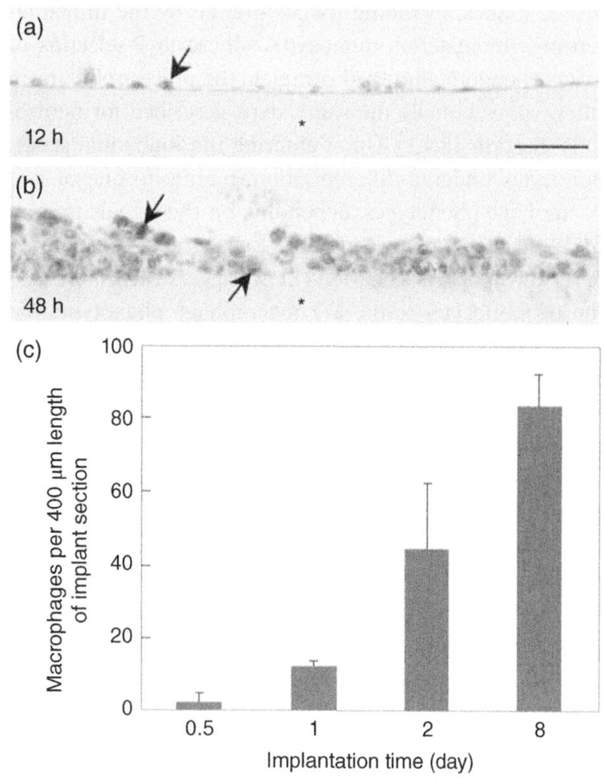

Figure 18.7 Macrophage adhesion and accumulation on cellulose acetate membranes implanted in the peritoneal cavity of mice. Images of immunohistological staining with the macrophage-specific antibody Mac 3 show macrophages on membranes (arrow) recovered from mice after (a) 12 hours and (b) 48 hours. (c) Quantitation of accumulated macrophages on membranes of length 400 μm at various times after intraperitoneal implantation. *Source:* Jay et al. (2010) / with permission of John Wiley & Sons.

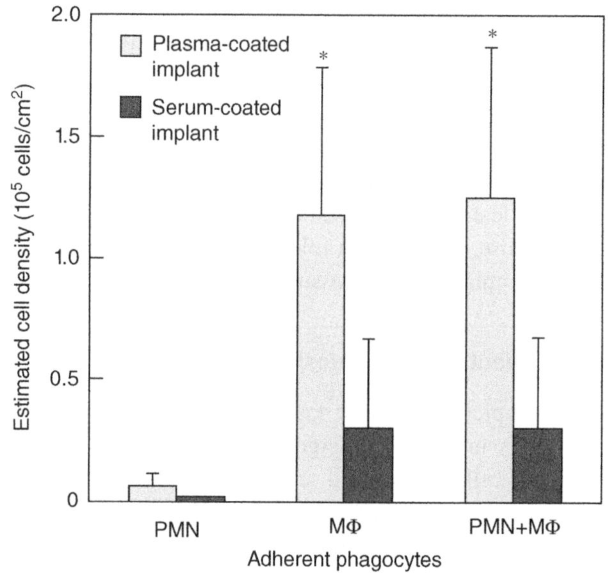

Figure 18.8 Assessment of adhesion of polymorphonuclear neutrophils (PMN) and monocytes/macrophages (MØ) to polyethylene terephthalate discs implanted in subcutaneous pouches in human volunteers. Prior to implantation, the discs were soaked for 12 hours in either autologous fibrinogen-containing plasma or autologous fibrinogen-free serum. Numbers of adherent PMN and MØ on each disc were estimated from measurements of activities of two cell-specific, indicator enzymes. The average numbers of adherent PMN and MØ on fibrinogen-free serum-coated implants were four- to five-fold lower than those on the fibrinogen-containing plasma-coated implants. * $p < 0.05$. *Source:* Zdolsek et al. (2007).

quantitation of myeloperoxidase and non-specific esterase, enzymes specific to polymorphonuclear neutrophils (PMN) and monocytes/macrophages (MØ), respectively. Then the measured activities of these two indicator enzymes were used to estimate the number of adherent PMN and MØ on each implant. These estimates showed that the average numbers of adherent PMN and MØ on the fibrinogen-free serum-coated implants was ~4–5 times lower than those on the fibrinogen-containing plasma-coated implants (Figure 18.8). Overall, these results strongly suggest that fibrinogen adsorption is a major factor in implant-mediated inflammatory responses.

The central role of fibrinogen adsorption in the adhesion of macrophages to implanted biomaterials, and subsequent pathological outcomes, has focused attention on strategies for blocking fibrinogen adsorption. A common procedure for reducing surface adsorption of a specific protein is to pretreat the biomaterial in bovine serum albumin (BSA), a protein that is widely used as a nonspecific blocking agent. However, BSA pre-treatment has not been successful in preventing fibrinogen adsorption to several types of biomaterials, due to the displacement of the adsorbed BSA by fibrinogen.

One approach that may be effective for preventing the displacement of BSA by fibrinogen is first pretreating the surface of the biomaterial to increase its hydrophobicity. An example of the use of this approach has been reported in a recent investigation involving stainless steel (316L) (Wood et al., 2018). Pre-treatment of the stainless steel samples with the detergent sodium dodecyl sulfate (SDS), followed by placement in a BSA blocking solution was found to be effective in preventing BSA displacement by fibrinogen, resulting in a significantly lower level of fibrinogen adsorption to the stainless steel samples (Figure 18.9). It has been suggested that this two-step approach might potentially be effective in reducing the extent of the inflammatory response triggered by the implantation of some biomaterials.

18.5.2 Phagocytosis by M1 Macrophages

M1 macrophages appear in the early stages of inflammation, generate oxidative radicals, and have vigorous phagocytic activity. In apparent preparation for phagocytosis, the

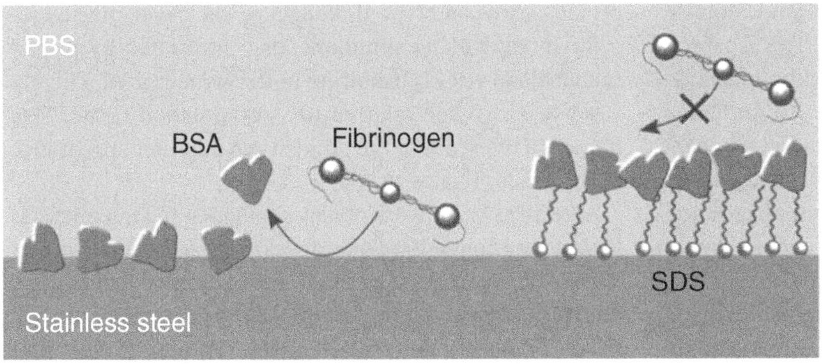

Figure 18.9 Depiction of pretreatment of stainless steel with sodium dodecyl sulfate (SDS) for increased surface hydrophobicity to lessen fibrinogen displacement of adsorbed bovine serum albumin (BSA). The SDS pretreatment followed by soaking in BSA blocking agent would more effectively prevent the adsorption of fibrinogen which attracts inflammatory cells to implants. *Source:* Wood et al. (2018) / with permission of American Chemical Society.

M1 macrophages enlarge and express various transmembrane receptor proteins that mediate phagocytosis. One type of receptor of special importance to macrophage function is the mannose receptor, a receptor that recognizes the mannose repeat units common to many viral and bacterial pathogens. Two other important receptor proteins are Fc receptors and C3 receptors that mediate recognition and binding of target particles with adsorbed immunoglobulin (IgG) or complement C3 proteins, respectively. Macrophages also respond rapidly to implanted biomaterials via recognition of and binding to IgG, complement C3, and other proteins adsorbed onto the implant surface. Complement C3, bound to the surface of an invasive microbe, functions as a molecular marker that identifies the microbe as a target for phagocytosis by M1 macrophages and is an example of opsonization, a process discussed in Chapter 19.

The driving force for phagocytic internalization of the target particle involves remodeling of the actin cytoskeleton to form pseudopodial extensions and a phagocytic cup that surround the target (Chapter 21). The edges of the phagocytic cup fuse to completely enclose and internalize the particle within a phagosome of the M1 macrophage. Then the phagosome fuses with intracellular lysosomes to form a phagolysosome. The various proteolytic lysosomal enzymes incorporated into the phagolysosome are effective in digesting most ingested microbes. Another phagocytic function of M1 macrophages is the clearance of late apoptotic neutrophils. The phagocytosis of apoptotic neutrophils is dependent on the ability of M1 macrophages to recognize one or more ligands found on the apoptotic neutrophils that are not present on healthy cells, which include phosphatidylserine, thrombospondin, and various sugar moieties.

In additional to bactericidal effects, the lysosomal enzymes of M1 macrophages are capable of degrading some types of foreign particles, including small particulate debris from degradable materials. For example, biomaterials composed of the degradable poly(α-hydroxy esters) L-PLA and PGA are subject to hydrolysis by macrophage lysosomal enzymes.

18.5.3 Generation of Oxidative Radicals by M1 Macrophages

Another major function of M1 macrophages involves generation of ROS that exert a strong oxidative killing effect against invasive microorganisms. M1 macrophages produce high amounts of oxygen radicals via the NADPH oxidase system that functions in essentially the same manner as those in neutrophils. The M1 macrophages also generate large quantities of highly reactive nitric oxide (NO) radicals by a metabolic pathway that involves the oxidation of L-arginine to NO and L-citrulline (Chapter 15). The L-arginine to L-citrulline pathway of M1 macrophages is catalyzed by the inducible high-output isoform of nitric oxide synthase (iNOS). Induction of transcriptional expression of iNOS in M1 macrophages occurs in response to inflammatory cytokines and microbial products, thereby limiting the NO production to those occasions when protection against infection is needed and avoiding damage to normal tissues.

18.5.4 Anti-inflammatory Activities of M2 Macrophages

When inflammatory stimuli at the wound site begin to subside, there is a change in the tissue microenvironment that leads to the transition from predominantly M1 macrophages to M2 macrophages, a transition critical for the resolution of inflammation. The M2 macrophages secrete cytokines and perform processes associated with the proliferation stage of wound healing and tissue repair. As part of this activity, the M2 macrophages hydrolyze and remove the old collagen scaffold of damaged tissues. This is a preliminary step needed to permit infiltrating fibroblasts to begin rebuilding a new collagen scaffold. An indirect, related role of M2 macrophages is the formation of polyamine and proline precursors that the fibroblasts then utilize in the biosynthesis of collagen. M2 macrophages are also sources of the cytokine vascular endothelial growth factor (VEGF) which promotes angiogenesis that is an essential part of wound healing. In addition, M2 macrophages secrete other factors including TGFβ, a growth factor that is anti-inflammatory and promotes the process of re-epithelialization. Another anti-inflammatory cytokine released by M2 macrophages is IL-10.

It is known that macrophages are functionally plastic and do not remain rigidly fixed in either the M1 or the M2 phenotype category (Ortega-Gomez et al., 2013). The presence of apoptotic neutrophils promotes transformation of M1 macrophages to M2 macrophages. It appears that in response to subtle changes in the tissue microenvironment, macrophages can transiently adopt various functional phenotypes as needed for tissue repair or for maintenance of normal tissue homeostasis.

Some of the additional important activities of macrophages include immune surveillance and antigen presentation. These topics will be addressed in the next chapter.

18.6 Granulation Tissue

As the sequence of inflammatory reactions continues, dermal fibroblasts migrating from nearby skin become much more numerous at the wound site, and this increasing population of fibroblasts produces an extracellular matrix of proteoglycans and type III collagen. The developing extracellular network is simultaneously infiltrated by microscopic blood vessels, resulting in the formation of a highly vascularized tissue referred to as granulation tissue. This name is a reference to the pinkish-red granular appearance of the tissue (Figure 18.10a).

An expansive network of new capillaries and microvessels within the granulation tissue develops by angiogenic sprouting of capillary tubes from existing nearby microvasculature. As the microvasculature forms and expands, the tissue histology becomes highly vascularized (Figure 18.10b). This rich vascularization allows the microenvironment of the wound tissue to become highly aerobic, a change that accelerates anabolic metabolism and growth and is also conducive to the completion of neutrophil apoptosis.

Fibroblasts that synthesize the extracellular matrix of granulation tissue initially form a loose network of type III collagen fibers that serves as a temporary scaffold to support the ingrowth of various cell types, which migrate into the site to contribute to the tissue repair process. The temporary scaffold of type III collagen fibers is subsequently remodeled and replaced by a much stronger, better-organized network of type I collagen, the collagen type also responsible for scar tissue.

The presence of granulation tissue is usually a positive sign and an indication that the injured tissue is undergoing rebuilding and beginning to return to normal. In some cases, however, granulation tissue becomes a pathological state that may persist for an extended period of time, as sometimes occurs with continuing tissue injury due to implant movement or strength mismatch with host tissue. Excessive production of granulation tissue, which becomes visible as "proud flesh," is detrimental to healing because it inhibits both wound contraction and re-epithelialization of the wound site.

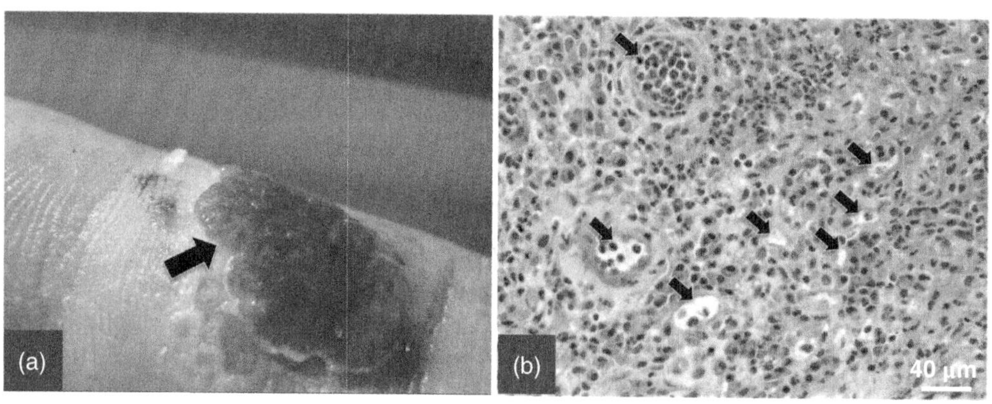

Figure 18.10 (a) Granulation tissue (arrow) at a surface wound showing characteristic reddish, granular appearance. (b) Histological analysis shows an abundance of various sized vessels (indicated by arrows) typical of granulation tissue. (a) *Source:* Public domain image from Wikipedia.org. (b) *Source:* Sheikh et al. (2015).

18.7 Foreign Body Response

Macrophages are recruited to migrate to any tissue site containing foreign material, regardless of whether the foreign material might be a cluster of invading microbes, a splinter lodged beneath the skin, or an implanted biomaterial. The responding macrophages quickly bind to, and begin phagocytizing, foreign particulate matter. Very often the macrophages phagocytize particulate matter that is either too large for effective digestion (larger than ~10 μm in diameter) or chemically resistant to enzymatic hydrolysis (such as metal shards, ceramic debris, and nondegradable polymer fragments). A foreign body response occurs when macrophages internalize particles that cannot be digested. They become "frustrated" and proceed through a sequence of steps (Figure 18.11) in which frustrated macrophages fuse with each other to form much larger multinucleated cells referred to as foreign body giant cells (FBGCs). An example of the progressive formation and accumulation of FBGCs in a tissue containing foreign particulate material is shown in Figure 18.12. The presence of FBGCs is the most distinguishing feature of the foreign body response, which can have damaging effects on both the host tissue and the implanted biomaterial.

The adverse effects caused by FBGCs during the foreign body response are an outcome of the activity of inflammasomes, large multi-protein complexes within the cytoplasm of macrophages and neutrophils. Inflammasomes are important components of the innate defense system, and are responsive to stimuli associated with danger such as invasive pathogens. The inflammasomes of FBGCs also sense phagocytized large particulate matter as a danger signal and respond to that danger by triggering the activation of caspase 1, a proteolytic enzyme. This catalyzes the maturation of the highly inflammatory cytokine IL-1β, which is then released and recruits large numbers of macrophages that infiltrate and further inflame the tissue. The result is continuing chronic inflammation, tissue damage and necrosis. Inflammasomes and their role in particle-induced inflammation are discussed further in Chapter 21.

The fusion of macrophages to form FBGCs is triggered by IL-4 and requires the participation of the mannose receptor plus several additional receptors. A particle that is ~10–100 μm in diameter will be internalized via an actin-based mechanism into a phagosome, which fuses with lysosomes to form a phagolysosome (Chapter 15). Then lysosomal enzymes within the phagolysosome attempt to degrade the material hydrolytically. Particles or implants greater than ~100 μm in diameter, a size above the threshold for internalization, attract macrophages that fuse to form FBGCs at the particle or implant surface. In addition, the implant often becomes enclosed within a fibrous avascular capsule, resulting in the formation of a special microenvironment at the FBGC-implant interface. Into this environment, the FBGCs release degradative lysosomal enzymes, reactive oxygen and nitric oxide radicals, and acids that reduce the pH to a value as low as 4. Due to the release of these agents, the biomaterial within this microenvironment is subjected to high concentrations of degradative agents, which engage in bulk extracellular surface degradation as represented in Figure 18.13 (Sheikh et al., 2015). Surface erosion due to the oxidative action of these radicals released by FBGCs has been observed with degradable polymers implanted *in vivo*.

It has been suggested that there might be an underlying biological purpose for the formation of FBGCs as part of the inflammatory response to large particulate foreign material, including material that is nondegradable. For example, a recent suggestion is that the formation of FBGCs might be a possible mechanism for complement-mediated phagocytosis and destruction of large targets (Milde et al., 2015). This suggestion is based on a study involving an *in vitro* model in which sheep red blood cells (RBCs) were used as large targets for phagocytosis. Bone

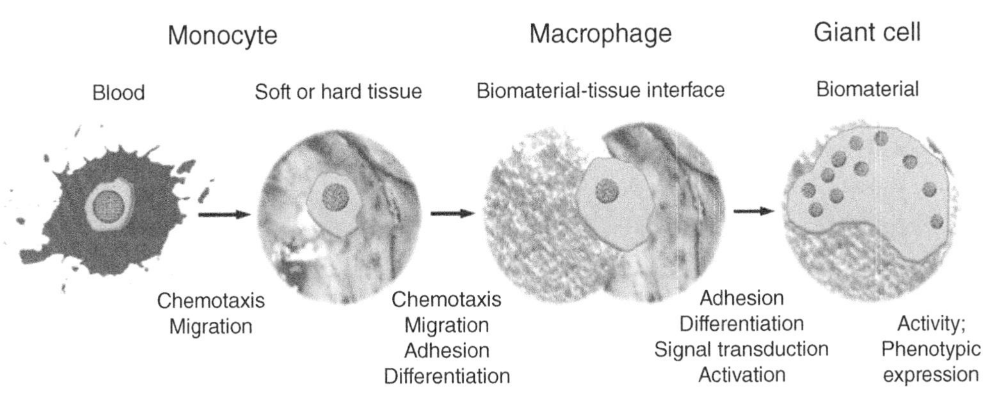

Figure 18.11 Depiction of transition from circulating monocyte to tissue macrophage to foreign body giant cell. *Source:* Sheikh et al. (2015).

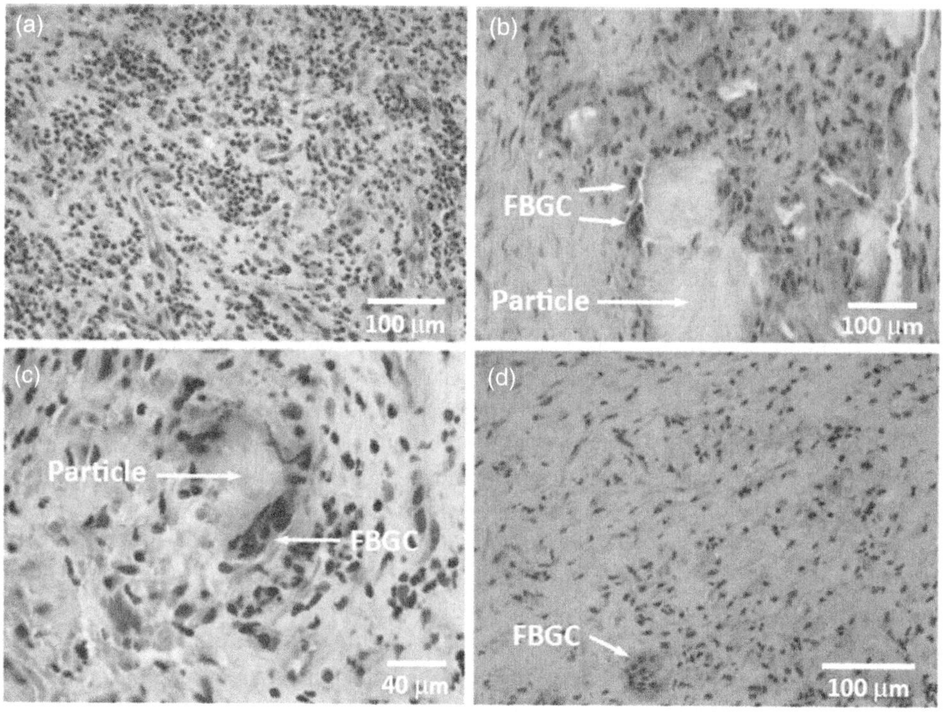

Figure 18.12 (a) Accumulation of acute and chronic inflammatory cells in tissue containing non-degradable particles. (b, c) Macrophages subsequently fuse to form foreign body giant cells (FBGCs) around nondegradable particles (foreign body) in tissue. (d) Late inflammatory response stage showing extensive fibrosis and FBGCs in tissue containing nondegradable particles. *Source:* Sheikh et al. (2015).

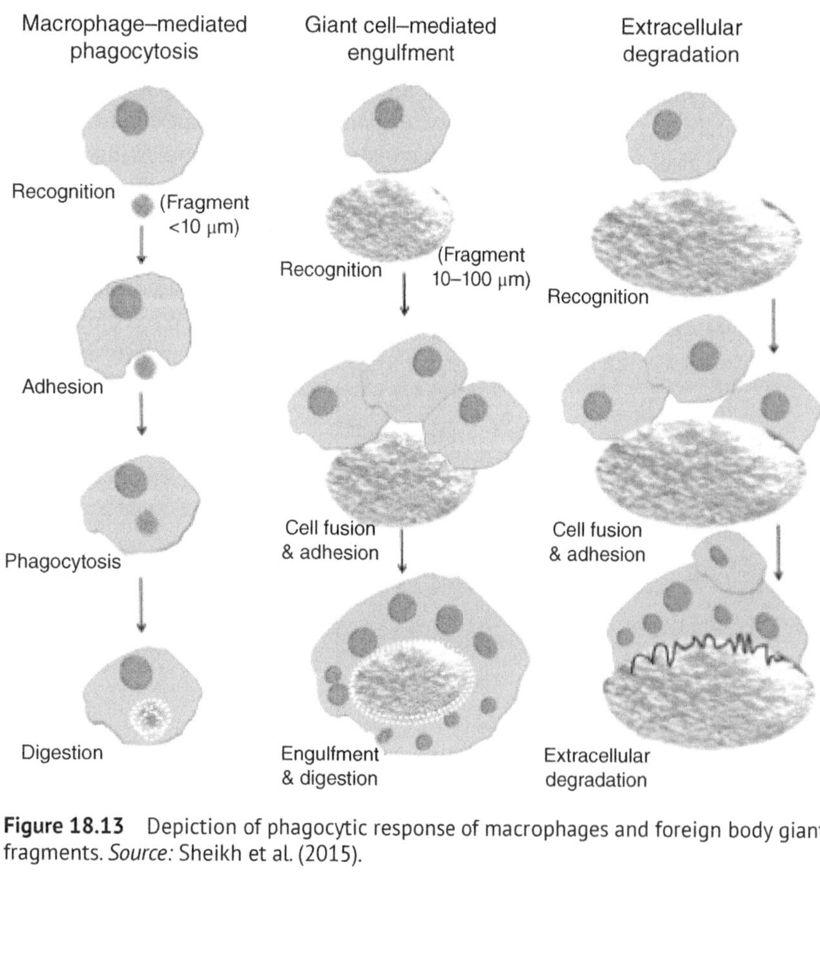

Figure 18.13 Depiction of phagocytic response of macrophages and foreign body giant cells to different size particles and fragments. *Source:* Sheikh et al. (2015).

marrow tissue collected from mice was used as a source of macrophages (Mϕ) and multinucleated giant cells (MGCs) which were induced by treatment with IL-4. The RBCs were opsonized with mouse complement in preparation for targeting, and then added to Mϕ and MGCs on plastic slides. Following incubation for 1 hour, the number of RBCs internalized by Mϕ and MGCs was quantitated. It was found that the number of RBCs internalized by the MGCs was significantly higher than that internalized by Mϕ (Figure 18.14a). This was also found when the results were normalized on the basis of RBCs per nucleus (Figure 18.14b). It was also found that the MGCs developed extensive membrane ruffles, with high amounts of complement receptor 3 being present. Collectively, these results suggest that the formation of multinucleated FBGCs coupled with target opsonization with complement C3 could be mechanisms for the phagocytosis of large particles that enter tissue spaces.

In addition to their degradative effects on implanted biomaterials, FBGCs and their products released into the extracellular space can cause extensive tissue damage. An important example is a foreign body response associated with prosthetic hip implants. Overall, the average lifetime of prosthetic hip implants with a bearing couple consisting of Co–Cr and ultra-high molecular weight polyethylene (UHMWPE) is ~10–15 years, often necessitating revision surgery to replace the failed implant. The primary cause of this failure is osteolysis accompanied by aseptic loosening, caused by wear debris from the articulating surfaces of the implant (Chapters 19 and 21). Over time, wear debris from the UHMWPE acetabular liner collects in the tissue surrounding the implant and is phagocytized by macrophages. Because the UHMWPE particles cannot be degraded, the macrophage lysosomes rupture and release lysosomal enzymes into the extracellular space. These degradative enzymes begin digesting tissue surrounding the implant, resulting in necrosis, loosening and, eventually, failure of the implant. Revision surgery is often necessary, but this is associated with significant trauma and generally leads to poorer outcomes when compared to the original implantation because of the poor condition of the osteolytic host bone.

The foreign body reaction and FBGCs that form at the surface of an implanted biomaterial can persist for months or even years, particularly with long-term implantable devices such as fracture fixation plates and intravascular catheters. During the long periods of implantation, FBGCs remain on the implant surface where they continue to generate and release lysosomal enzymes, ROS and nitric oxide radicals, as well as acids that can have significant degradative effects on both the implant and the surrounding tissues. The foreign body reaction and FBGCs are quickly resolved when the implant is either removed or fully degraded, and the degradation products removed from the implant site.

18.8 Fibrosis and Fibrous Encapsulation

The final stage of the inflammatory response to implanted biomaterials often involves the formation of a sheath of avascular fibrous connective tissue that encapsulates the implant, such as the example shown in Figure 18.15. The fibrous sheath confines the implanted biomaterial, separating it from the surrounding normal tissue. While many implanted biomaterials are susceptible to fibrous

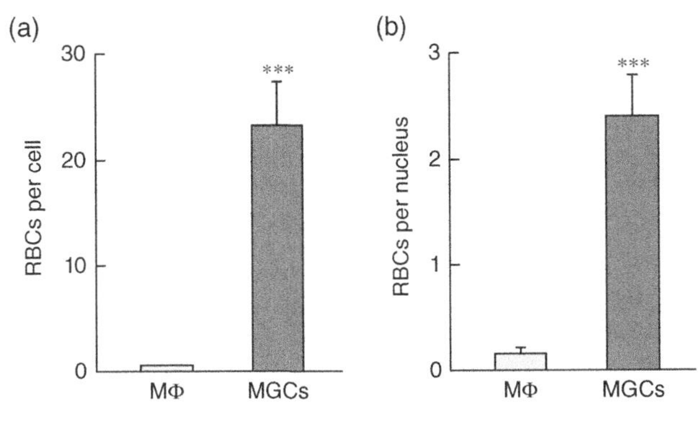

Figure 18.14 (a) Comparison of internalization of C3 opsonized red blood cells (RBCs) by bone marrow derived macrophages (Mϕ) and multinucleated giant cells (MGCs). (b) Comparison of internalization of C3 opsonized RBCs by Mϕ and MGCs with results expressed on per nucleus basis. ***$p < 0.001$. *Source:* Milde et al. (2015).

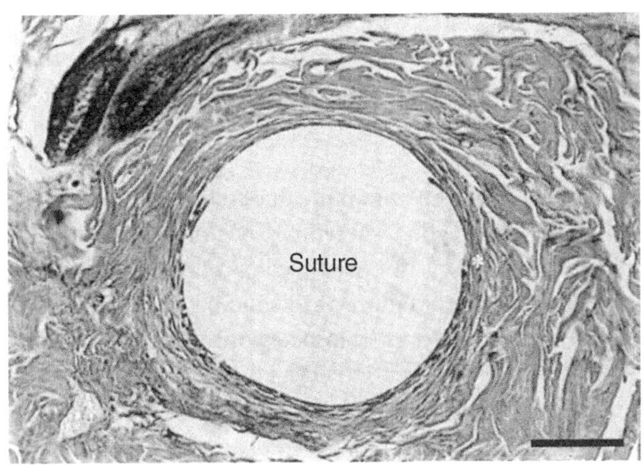

Figure 18.15 Photomicrograph of fibrous encapsulation around polydioxanone suture at 14 days of subcutaneous implantation in the dorsum of a rat. (Bar = 200 μm) *Source:* Andrade et al. (2006).

encapsulation, some biomaterials, such as porous materials and bioactive materials designed for implantation into bone, are normally not encapsulated with fibrous tissue. In general, however, there is greater risk of implant failure when an avascular fibrotic tissue encapsulates the biomaterial.

The biological basis for fibrous encapsulation becomes more understandable in the context of comparing the events of normal and abnormal wound healing. Neutrophils and macrophages that migrate to a wound or an implant site release several different cytokines, including TGFβ1, IL-13, and PDGF. These three cytokines have multiple effects, which include the activation of fibroblasts that migrate into the wound site where they undergo further differentiation to myofibroblasts, a specialized cell type that combines the traits of a fibroblast and a smooth muscle cell. Myofibroblasts appear in the early stages of wound healing and perform multiple tasks that continue until the completion of the final stage (fibrosis and fibrous encapsulation). A major task of myofibroblasts is the synthesis of collagen fibers needed to rebuild the supportive extracellular matrix required for tissue repair and regeneration. An important part of this process is the degradative remodeling of collagen, catalyzed by matrix metalloproteinases (MMPs) released from tissue macrophages. Healing proceeds normally when there is an appropriate balance between the rate of collagen synthesis and the rate of collagen degradation. When collagen synthesis exceeds collagen degradation, the result is fibrosis and the formation of scar tissue that is characteristic of abnormal wound healing.

Multiple factors appear to be linked to the accumulation of collagen that leads to the formation of fibrotic scar tissue. Deceased levels of MMPs result in collagen excess, leading to fibrosis and scarring, while elevated levels of TGFβ1 are associated with the overproduction of collagen in a number of fibrotic diseases. Another factor that appears to promote fibrosis and scarring is the elevated level of NADPH oxidase-derived H_2O_2 within the wound microenvironment.

An example of implant failure due to fibrous encapsulation is the failure that sometimes occurs with silicone-based implants used in breast reconstruction procedures. The primary cause of failure of silicone-based implants is a condition referred to as "capsular contracture," the contraction of a fibrous sheath that can develop as a result of a foreign body reaction to the silicone implant. Another example is the fibrous encapsulation that is occasionally observed with titanium dental implants. When fibrous encapsulation occurs, it normally hinders osseointegration of the titanium implant with the surrounding bone, often resulting in implant failure. Overall, several types of implanted devices can become encapsulated with a fibrous sheath, often leading to implant failure.

The realization that fibrosis and fibrous encapsulation are risk factors associated with many types of implanted materials has focused attention on new materials and strategies designed to inhibit or control fibrosis. A technique that is now widely used to reduce fibrosis and improve osseointegration of titanium implants involves the addition of surface roughness features at the microscale (Chapter 21). Several studies have shown that the addition of nanoscale surface features to smooth (machined) surfaces can also enhance osseointegration. Although the mechanisms of surface roughness alone are not clear, the microscale surface roughness features are thought to provide microenvironmental cues required by osteoblasts for initiating bone formation. A recently developed approach for limiting fibrosis involves coating the surface of titanium implants with the anti-fibrotic protein decorin (He et al., 2017). The effectiveness of the coating was evaluated using *in vitro* tests with fibroblasts seeded on decorin-coated titanium substrates and on uncoated substrates (control). Fibroblasts seeded onto the decorin-coated titanium showed dramatically reduced proliferation and collagen synthesis, which indicates that the coating procedure could potentially be effective in reducing failure caused by fibrous encapsulation of some titanium implants.

Although fibrosis and fibrous encapsulation are often correlated with very serious impairments to the health and longevity of patients, a recent study has shown that fibrous encapsulation could potentially be utilized for *in situ* engineering of autologous vascular grafts (Geelhoed et al., 2017). The basic approach used in the study starts with the subcutaneous implantation of a sterile, rod-shaped, nondegradable material such as a stainless steel, which would normally be performed several weeks prior to the scheduled surgical replacement of the diseased or damaged vascular graft. The expectation is that the stainless steel rod would become heavily encapsulated within a sheath of avascular fibrous tissue, as an outcome of predictable fibrous response to the nondegradable material. Then the stainless steel rod with its surrounding fibrous capsule would be excised, the rod removed, and the fibrous capsule used as an autologous vascular graft. This approach has shown success in animal models, but it has not yet been evaluated in human clinical trials.

18.9 Resolution of Inflammation

Resolution is the cessation of inflammation and the return to a state of normal tissue homeostasis. When a biomaterial is the cause of the inflammation, the complete cessation of inflammation is dependent on either accommodating or eliminating the agent provoking irritation and injury. Accommodation can occur by means of the integration of the biomaterial with host tissue. A well-known example of this is the osseointegration of bioactive glass (45S5) particles and implants by direct bonding between the surface of the bioactive glass and host bone. The bioactive glass can remain in place without the formation of inflammatory tissue between the implant and bone. Elimination of irritation and injury can occur by complete degradation of the biomaterial. An example is the complete degradation of sutures composed of degradable polymers such as PGA and PLGA (Chapter 24). In the absence of complete degradation, resolution may require surgical removal of the implant. Devices such as stainless steel fracture fixation plates used to stabilize bone fractures are normally left in place only temporarily. After satisfactory healing of the fracture, followed by surgical removal of the device, inflammation gradually subsides and a return to normal tissue homeostasis can begin.

Following complete degradation or removal of an implanted biomaterial, the biological process of resolution and a return to normal homeostasis depends on the completion of the following steps (Ortega-Gomez et al., 2013):

- The recruitment of neutrophils to the implantation site should halt. Continued arrival of neutrophils will result in continued tissue injury due to their release of oxidative radicals and proteolytic enzymes.
- Neutrophils at the implant site should undergo apoptosis for the purpose of eliminating injurious neutrophil products and for apoptotic neutrophil-induced transition of macrophages to the anti-inflammatory M2 phenotype (Figure 18.16).
- Macrophages should then also leave the implantation site. The route of macrophage exit from the tissue is via the lymphatic vessel network.
- Finally, there should be satisfactory restoration of the supportive extracellular framework. This requires remodeling and reorganization of the collagen fibers at the implantation site for the tissue to regain its uninjured strength. As part of this process, the myofibroblasts should undergo apoptosis to avoid formation of extensive scar tissue.

Upon completion of these steps, the tissue can satisfactorily return to a state of normal homeostasis.

The inflammatory response to an implanted biomaterial involves the participation and interaction of several different types of cells, biomolecules, and other entities. Table 18.1 provides a brief description of the majority of these cells and biomolecules, along with their major role in the inflammatory response.

18.10 Inflammation and Biocompatibility

Up to this point, the inflammatory response to an implanted biomaterial has been described in terms of a sequence of seven reactions (Section 20.1), beginning with implant interaction with the plasma proteins and progressing through

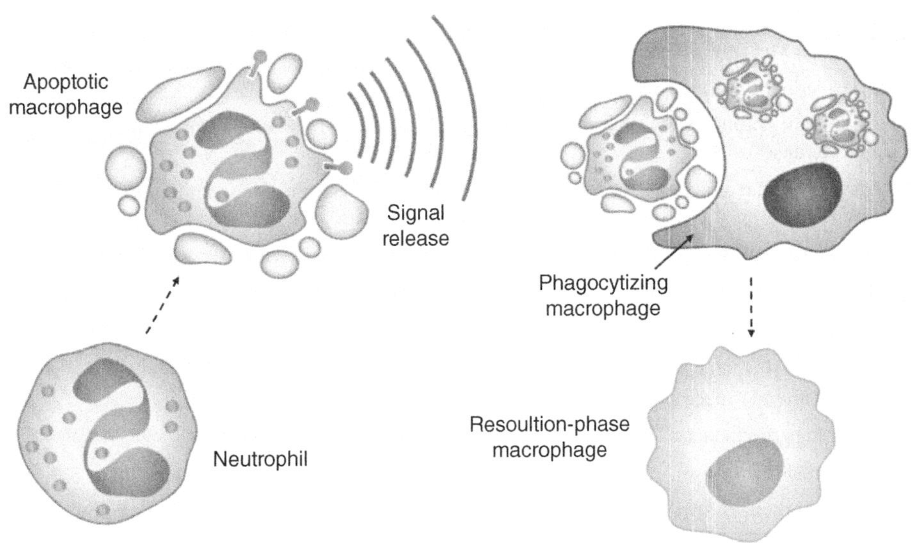

Figure 18.16 Depiction of removal of apoptotic neutrophils during the resolution phase of inflammation. Apoptotic neutrophils release signals that attract M2 macrophages for the purpose of phagocytic removal of neutrophilic debris that would otherwise cause tissue injury. *Source:* Ortega-Gomez et al. (2013).

Table 18.1 Cells, molecules, and other entities involved in the inflammatory response to biomaterials.

Entity	Description	Role in inflammatory response
Collagen type I	Mature collagen type; most abundant protein in the body	Source of fibrosis; main component of scar tissue
Collagen type III	Collagen prominent in fetal tissue and developing bone	Component of provisional matrix
Complement C3	Plasma protein with importance in innate defense	Opsonin labeling of microbes for phagocytosis
Eosinophil	Leukocyte with acidophilic granules	Source of various inflammatory cytokines
Fibrinogen	A plasma protein; inactive precursor of fibrin mesh of blood clot	Adsorbed fibrinogen layer attracts inflammatory cells
Fibroblast	Most abundant cell type in multiple connective tissues	Forms collagen fibers and other components of the ECM
Frustrated macrophage	Macrophage in pathologic state due to inability to ingest large particles	Source of foreign body giant cells (FBGCs)
Foreign body giant cell (FBGC)	Multinucleated cell formed by fusion of frustrated macrophages	Agent of foreign body response; harmful to tissue
Inflammatory cytokines[a]		
IFN-γ	Interferon-γ	Promotes M1 macrophage differentiation
IL-1	Interleukin-1	Potent pro-inflammatory cytokine
IL-4	Interleukin-4	Induces M2 macrophage phenotype
IL-8	Interleukin-8	Monocyte chemoattractant
IL-10	Interleukin-10	Anti-inflammatory effects
IL-13	Interleiukin-13	Promotes formation of myofibroblasts
Lysosomal proteases	Released from neutrophil granules	Degrade tissue ECM; cause tissue damage
M1 macrophage	Pro-inflammatory macrophage	Prominent in chronic phase; high phagocytic activity
M2 macrophage	Anti-inflammatory macrophage	Tissue repair and resolution; removes apoptotic cells
Mast cell	Basophil-like cell present in tissues; derives from basophil	Source of vasodilator histamine
Monocyte	Largest leukocyte; agranulocytic	Origin of macrophages
Myofibroblast	Fibroblast-smooth muscle cell hybrid	Forms fibrotic scar tissue
Neutrophil	Most abundant granulocyte with multiple roles in innate defense	Defense cell in acute phase; recruits leukocytes
Opsonin	Protein marker that attaches to invasive microbe	Labels invasive microbe for phagocytic destruction
P-selectin	Endothelial cell adhesion molecule	Binds neutrophils for extravasation
PDGF	Platelet-derived growth factor	Formation of provisional matrix; resolution
Platelet	Smallest circulating blood cell	Promotes formation of provisional matrix
Reactive oxygen species (ROS)	$O_2^{-}\bullet$, $H_2O_2\bullet$, $HO\bullet$, HOCl, generated by neutrophils and M1 macrophages	Oxidants; important bactericidal effects
TGFβ1	Transforming growth factor-β1	Induces reepithelialization
TNF-α	Tumor necrosis factor-α	Induces P-selection expression
VEGF	Vascular endothelial growth factor	Promotes vasculogenesis in granulation tissue

[a] See also Table 16.2.

additional reactions until the completion of the final stage (fibrosis and fibrous encapsulation). This sequence of seven reactions is the classic response described by several investigators in the literature. In general, many implanted devices trigger all seven of these inflammatory reactions. For example, stainless steel fracture fixation plates and cardiac pacemakers trigger all seven reactions, culminating in the formation of an encapsulating fibrous sheath that persists for as long as the implanted device remains in place.

On the other hand, implanted devices composed of highly biocompatible materials elicit little more than an acute inflammatory response. The inflammatory response of the tissue surrounding these highly biocompatible implants could perhaps be resolved within as little as a week, with the tissue starting to return to a state of normal homeostasis. An example of a highly biocompatible material is chitosan. Investigations involving the subcutaneous implantation of chitosan hydrogels into rats have shown little or no inflammatory response of surrounding tissue. This shows that some highly biocompatible materials elicit only an acute inflammatory response rather than the aforementioned sequence of seven reactions.

Stated somewhat differently, the extent of implant-induced progression through the sequence of inflammatory reactions is directly related to the level of biocompatibility of the implanted material. An illustration of this relationship is presented in Figure 18.17. Materials with a high level of biocompatibility would normally elicit little more than an acute inflammatory response, followed by relatively rapid resolution and a return to normal homeostasis. On the other hand, implanted biomaterials with a moderate level of biocompatibility would normally elicit reactions that also include some extent of chronic inflammation and granulation. With these materials, a longer duration (on the order of weeks) would likely be required for resolution and return to normal homeostasis. In comparison, implanted biomaterials with a low level of biocompatibility would normally trigger the complete sequence of seven inflammatory reactions, thus requiring a much longer time for resolution and a return to normal homeostasis.

18.11 Concluding Remarks

In this chapter, we discussed biocompatibility phenomena related to inflammatory responses of tissues to an implanted biomaterial. Any implanted biomaterial will elicit an inflammatory response but the extent of the response is directly related to the biomaterial's level of biocompatibility. In general, the biocompatibility of an implanted material is described in terms of acute and chronic inflammatory responses. Biomaterials with a high level of biocompatibility may elicit only an acute inflammatory response whereas biomaterials with a lower level of biocompatibility will elicit more severe chronic inflammatory response.

- The acute phase of inflammation is associated with neutrophils that exit the blood vessel by extravasation and migrate to the wound site, where they form bactericidal oxygen radicals and actively phagocytize invading microbes and biomaterial particles. Following acute inflammation, neutrophils undergo apoptosis and clearance from the tissue. The continued presence of neutrophils results in tissue damage that can lead to multiple pathological conditions.
- Chronic inflammation is a longer and more serious phase of inflammation, which particularly involves the activity of two types of macrophages derived from circulating monocytes. Pro-inflammatory M1 macrophages aggressively phagocytize microbes and particulate biomaterials, and generate oxidative radicals such as nitric oxide, whereas anti-inflammatory M2 macrophages remove old collagen and perform other tasks of tissue repair.
- The granulation phase of inflammation involves formation of granulation tissue, a highly vascularized temporary tissue with an extracellular matrix of type III collagen. The presence of granulation tissue usually indicates that the wound tissue is undergoing rebuilding and beginning return to normal.
- The foreign body response occurs when macrophages with phagocytized particles resistant to digestion become

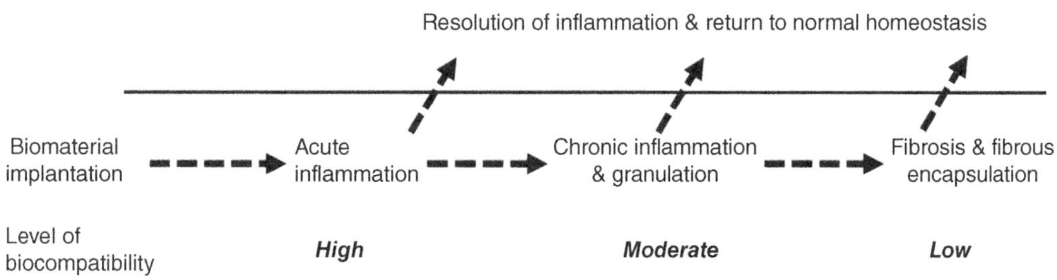

Figure 18.17 Extent of progression through inflammatory reactions to biomaterials as a function of relative biocompatibility of material with resolution of inflammation and return to normal homeostasis.

"frustrated" and begin fusing to form much larger multinucleated FBGCs. Accumulation of FBGCs normally occurs, for example, at tissue sites containing the accumulating wear debris of prosthetic hip implants, resulting in the release of lysosomal enzymes that cause osteolytic bone loss, the primary cause of hip implant failure.

- The last stage of the inflammatory response to many implanted biomaterials, particularly those with a low level of biocompatibility, involves the formation of a sheath of fibrous connective tissue that encapsulates the implant.

Problems

18.1 An experiment is performed with rodents to evaluate *in vivo* responses to subcutaneous implantation of newly developed biodegradable microspheres designed for potential use as a drug delivery vehicle. The experiment includes 21 animals randomly assigned to one of three test groups with 7 animals each. Animals in Group A receive the microspheres only. Group B receives microspheres loaded with the cytokine TGFβ. Group C animals receive the same surgical incision but no implantation. What is the purpose of Group C and why is inclusion of this test group necessary?

18.2 What long-term difference might be expected in tissue responses to a degradable implant material compared to an inert, nondegradable material?

18.3 Investigations of implanted biomaterials focus mainly on assessing possible damage to host tissues inflicted by the implant. Describe one example of damage to the implant inflicted by the host tissue, and include the mechanism(s) by which the damage is inflicted.

18.4 Describe two ways in which biomaterials can be modified to reduce the extent of inflammation when implanted in the body.

18.5 A very small segment of the population has a rare inherited disorder known as chronic granulomatous disease characterized by phagocytes that are unable to produce superoxide or other ROS. What would be the outcome of this ROS deficiency?

18.6 M1 macrophages produce nitric oxide radicals in a metabolic process catalyzed by inducible nitric oxide synthase. Why is it beneficial to have this reaction catalyzed by the inducible isoform rather than the constitutive isoform of nitric oxide synthase? How could it be harmful if M1 macrophages contained the constitutive isoform instead?

18.7 Fibrosis and fibrous encapsulation are two related tissue responses to some types of materials. The fibrotic response can be particularly limiting or harmful to some tissues or organs. Describe one example of a fibrotic response to a specific biomaterial in a specific tissue or organ. Explain how the fibrotic response occurs and why it is harmful or limiting.

18.8 Explain how it might be beneficial to perform a skin sensitivity test before placing a cobalt-chromium alloy implant in a patient. What could possibly happen if a skin sensitivity test is not performed before implantation?

18.9 Cell death by the process of apoptosis causes no damage to surrounding tissues. In comparison, cells that die without going through the process of apoptosis do cause destruction and damage to surrounding tissues. Why does the damage occur in the absence of apoptosis?

18.10 Explain the cellular mechanism by which wear debris from hip implants cause osteolytic bone erosion and implant failure. Indicate how macrophages are involved.

References

Alasmari, S.Z. (2020). in vivo imaging of neutrophil extracellular traps (NETs): visualization methods and outcomes. *BioMed Research International* 2020: 4192745.

Andrade, M.G.S., Weissman, R., and Reis, S.R.A. (2006). Tissue reaction and surface morphology of absorbable sutures after in vivo exposure. *Journal of Materials Science. Materials in Medicine* 17: 949–961.

Geelhoed, W.J., Moroni, L., and Rotmans, J.I. (2017). Utilizing the foreign body response to grow tissue engineered blood vessels in vivo. *Journal of Cardiovascular Translational Research* 10: 167–179.

He, R., Lu, Y., Ren, J. et al. (2017). Decreased fibrous encapsulation and enhanced osseointegration *in vitro* by decorin-modified titanium surface. *Colloids and Surfaces, B: Biointerfaces* 155: 17–24.

Jay, S.M., Skokos, E.A., Zeng, J. et al. (2010). Macrophage fusion leading to foreign body giant cell formation persists under phagocytic stimulation by microspheres in *in vitro* and *in vivo* in mouse models. *Journal of Biomedical Materials Research. Part A* 93: 189–199.

Jhunjhunwala, S., Aresta-DaSilva, S., Tang, K. et al. (2015). Neutrophil responses to sterile implant materials. *PLoS One* 10: e0137550.

Milde, R., Ritter, J., Tennent, G.A. et al. (2015). Multinucleated giant cells are specialized for complement-mediated phagocytosis and large target destruction. *Cell Reports* 13: 1937–1948.

Ortega-Gomez, A., Perretti, M., and Soehnlein, O. (2013). Resolution of inflammation: an integrated view. *EMBO Molecular Medicine* 5: 661–674.

Sheikh, Z., Brooks, P.J., Barzilay, O. et al. (2015). Macrophages, foreign body giant cells and their response to implantable biomaterials. *Materials (Basel)* 8: 5671–5701.

Velnar, T., Bunc, G., Klobucar, R. et al. (2016). Biomaterials and host versus graft response: a short review. *Bosnian Journal of Basic Medical Sciences* 16: 82–90.

Vitkov, L., Krautgartner, W.D., Obermayer, A. et al. (2015). The initial inflammatory response to bioactive implants is characterized by NETosis. *PLoS One* 10: e0121359.

Wood, M.H., Payagalage, C.G., and Geue, T. (2018). Bovine serum albumin and fibrinogen adsorption at the 316L stainless steel/aqueous interface. *The Journal of Physical Chemistry B* 122: 5057–5065.

Zdolsek, J., Eaton, J.W., and Tang, L. (2007). Histamine release and fibrinogen adsorption mediates acute inflammatory responses to biomaterial implants in humans. *Journal of Translational Medicine* 5: 31.

Zhou, J., Tsai, Y.T., Weng, H. et al. (2012). Real-time detection of implant-associated neutrophil responses using a formyl peptide receptor-targeting NIR nanoprobe. *International Journal of Nanomedicine* 7: 2057–2068.

Further Reading

Anderson, J.M. (2001). Biological responses to materials. *Annual Review of Materials Research* 31: 81–110.

Klopfleisch, R. and Jung, F. (2017). The pathology of the foreign body reaction against biomaterials. *Journal of Biomedical Materials Research. Part A* 105: 927–940.

19

Immune Responses to Biomaterials

19.1 Introduction

The preceding chapter focused on implant-induced inflammatory reactions and how such reactions are manifestations of innate, nonspecific immune defense mechanisms, the second of the body's three lines of defense. This chapter will address possible effects of biomaterials on the adaptive, specific immune system, the body's third and final line of defense. Before addressing this topic, it is necessary to describe the nature of the adaptive immune system (also referred to as the acquired immune system) and how it differs from the innate immune system.

As discussed in Chapter 18, the components of the innate immune system may be grouped into three categories: the sequence of inflammatory reactions, the various phagocytic cells such as neutrophils, monocytes, and macrophages, and the family of plasma proteins known as the complement system. The first two categories of innate immune defense components were discussed in Chapter 18. The functions of the complement proteins, which are involved in both the innate immune system and the adaptive immune system, will be addressed in this chapter.

The name of the adaptive (acquired) system provides insight into the underlying major differences between the immune systems. Unlike the innate defenses that rapidly respond to anything identified as not belonging to the body, the acquired immune defenses are adaptable defenses that can be tailored to confront a specific threat among millions of different pathogens and foreign cells. In addition, the adaptive immune defenses are initiated more slowly than the innate defenses, which are inherently present and ready to be activated or mobilized immediately on exposure to an invading agent (typically a microorganism). The more slowly initiated adaptive defense system is antigen-specific and has immunological memory that allows a greatly enhanced response to a second exposure to the same antigen. It is noted that the adaptive immune defense system consists of two major divisions: a division responsible for "humoral" (antibody-based) immunity and a separate division responsible for cell-mediated immunity.

As discussed in Chapter 18, all biomaterials initiate inflammatory reactions upon implantation in the body, and the extent of inflammation that occurs is directly related to the degree of biocompatibility of the biomaterial. In comparison, an adaptive immune response will occur when the biomaterial contains antigens that are recognized by immune cells or releases ions that trigger hypersensitivity. As noted in Chapter 1, one way of classifying biomaterials is in terms of their source, whether they are natural or synthetic in origin. Typically, synthetic biomaterials and most polysaccharides do not elicit an adaptive immune response, whereas biomaterials containing allogeneic or xenogeneic proteins, or one of the few antigenic polysaccharides (such as *N*-glycolylneuraminic acid) can do so.

Overall, then, this chapter will discuss the following topics:

- The components of antibody-based and cell-based adaptive immune defences
- Role of the complement system in adaptive immune defense
- Types of immune responses triggered by some implanted biomaterials.

19.2 Adaptive Immune System

The two divisions of the adaptive immune system, noted in the preceding section, are functional outcomes of two separate groups of lymphocytes. Humoral immunity is a function of B lymphocytes that produce antigen-specific antibodies circulated via blood and lymph. In comparison, cell-mediated immunity is provided by millions of subpopulations of T lymphocytes.

Materials for Biomedical Engineering: Fundamentals and Applications, First Edition. Mohamed N. Rahaman and Roger F. Brown.

Companion website: www.wiley.com/go/rahaman/biomedicalengineering

19.2.1 Lymphocyte Origins of Two Types of Adaptive Immune Defense

The two functional divisions of the adaptive immune system are maintained by the continual, lifelong development of two classes of lymphocytes: B lymphocytes and T lymphocytes. Stem cells in the bone marrow give rise to immature lymphoid cells that go through multiple cell divisions and then either migrate to the thymus gland where differentiation to T cells occurs or remain in the bone marrow where differentiation to B cells takes place. As development in the thymus and bone marrow continues, millions of populations of T lymphocytes and B lymphocyte are formed, with each population exhibiting specificity for a single antigen. This predetermined specificity occurs before any antigen encounter, resulting in naive B cells and T cells during a phase of development referred to as the antigen-independent phase of differentiation. The primary mechanism by which humans are capable of producing many millions of different populations of B and T lymphocytes is by extensive rearrangement of immunoglobulin genes during this phase of differentiation.

The millions of naive B cells and T cells produced are continuously released into the blood for circulation to tissues, with return flow through lymphatic vessels. The majority of naive lymphocytes released are retained by lymphoid organs, especially the lymph nodes, as lymph flows through those organs. The retention of lymphocytes in these organs serves as a strategic pre-positioning of these cells at sites where an encounter with an antigen is most likely. Upon arrival in a peripheral lymphoid organ, an antigen present in the lymph binds to specific receptors on the lymphocyte surface, an event that triggers the antigen-dependent phase of differentiation.

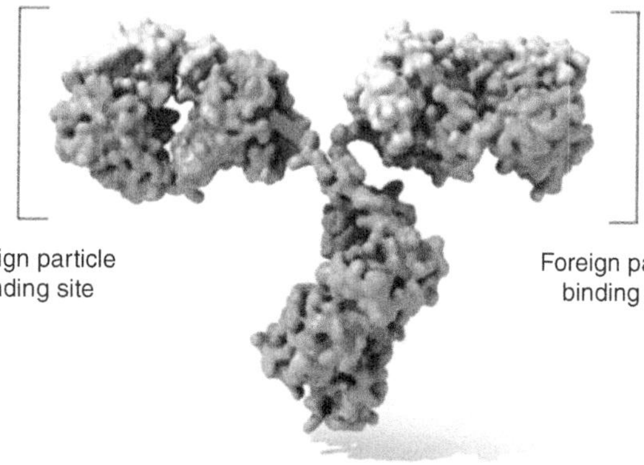

Figure 19.1 Artist model of the three-dimensional structure of an immunoglobulin. The specific example depicted is an IgG class immunoglobin. *Source:* From U.S. National Library of Medicine.

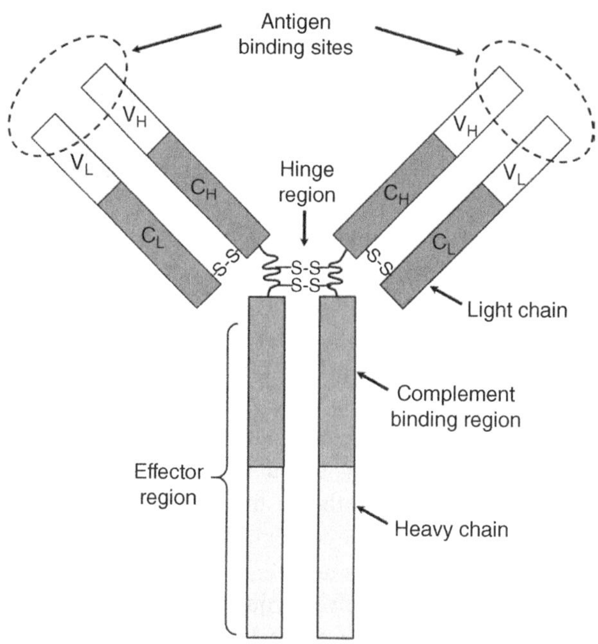

Figure 19.2 Depiction of Y-shaped structure of immunoglobulin IgG composed of two light and two heavy polypeptide chains with antigen binding regions at tips of splayed arms.

19.2.2 Antibody Characteristics and Classes

Circulating antibodies are globular-shaped proteins (Figure 19.1) referred to as immunoglobulins, a group comprising ~20% of the plasma proteins. All immunoglobulins are composed of four polypeptide chains, two identical light chains (molecular weight ~23 kDa) and two identical heavy chains (molecular weight greater than ~53 kDa). Disulfide bonds link the four polypeptide chains in a Y-shaped molecular configuration (Figure 19.2), which is common to all immunoglobulins. Most of the amino acid sequence of each antibody molecule is identical for immunoglobulins of the same class. The regions of identical sequence are referred to as constant regions, a portion of which is the Fc region that binds to Fc receptors on various leukocytes. When compared to the constant regions, the variable regions at the ends of the two splayed arms of each Y-shaped immunoglobulin have amino acid sequences that differ greatly from one antibody to another and function as binding sites for specific antigens. Most antigenic entities are either proteins or large polysaccharide molecules, many of which are found on the surfaces of bacterial invaders or errant, precancerous cells of the body. The antigen binding site of the typical antibody recognizes only a small portion of the antigenic molecule, a localized region referred to as the epitope or antigenic determinant.

Structure	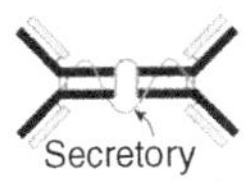				
Class and form	IgA dimer	IgD monomer	IgE monomer	IgG monomer	IgM pentamer
Molecular weight (kDa)	385	180	200	150	900
Percent of total Ig in serum	15	0.2	0.002	75	10
Approximate half-life (day)	6	3	2.5	23	5
Binds complement	No	No	No	Yes	Yes
Physiological function	Protects mucous membranes; in saliva and tears	B cell receptor	Allergy response; hypersensitivity; combats parasites	Main serum antibody; secondary response; can cross placenta	Antibody of initial response to infection; B cell receptor

Figure 19.3 Properties of the five major classes of immunoglobulins.

There are five different classes of immunoglobulins: IgG, IgM, IgA, IgD, and IgE (Figure 19.3). Differences in the heavy chain account for most of the differences between these five classes. IgG antibodies comprise ~75% of circulating antibodies and, among the five immunoglobulin classes, have the longest retention in the blood. Another important effect of IgG antibodies is the activation of the complement system. IgM immunoglobulins, the largest antibodies, are each composed of five Y-shaped monomers linked by disulfide bonds in a pentamer configuration that provides 10 antigen-binding sites. IgM antibodies are the first immunoglobulin type to appear at an infection site, a trait that accounts for their use as a diagnostic indicator of recent infection. IgA antibodies are found mainly at mucosal surfaces and are particularly important in protecting mucosal epithelium from bacterial toxins and binding. IgA is generally a monomer in circulating blood while, at the mucosal surfaces, it is present as a dimer. IgD is found on the exterior membrane of B cells, leading to speculation that IgD may have a role in B cell differentiation. IgE is a highly potent immunoglobulin associated with hypersensitivity, allergic reactions, and response to parasitic worm infections.

19.2.3 Major Histocompatibility Complex and Self-Tolerance

Self-tolerance is largely a manifestation of the presence of highly variable self-marker proteins on all nucleated cells of the body. These self-marker proteins are a family of glycoproteins referred to as the major histocompatibility complex (MHC). There are two classes of MHC that label cells as "self." Class I MHC proteins are present on almost all nucleated cells of the body except those that function as "antigen presentation" cells. In comparison, Class II MHC are present on the cells involved in antigen presentation including macrophages, B lymphocytes and specialized cells known as dendritic cells.

Another aspect of self-tolerance, the lack of immune reaction against one's own cells, is the absence of lymphocyte populations that have receptors for self-antigens. These lymphocyte populations are not present because they are detected and eliminated before birth. The lymphocyte populations that remain and continue to develop are those with receptors for antigens of foreign cells and molecules.

MHC proteins play a major role in antigen presentation. The process of antigen presentation will be described here as it occurs with macrophages (Figure 19.4). As shown, the antigen is initially phagocytized by the antigen-presenting cell (APC), internalized into a phagosome and partially degraded into fragments by lysosomal enzymes. As processing continues, an antigen fragment containing the epitope region (the bonding site recognized by the corresponding antibody) is complexed with a class II MHC protein. The antigen fragment-class II MHC complex is transported to the periphery of the APC where it is displayed (presented) to a type of T cell known as a helper T (T_H) cell. Then the antigen fragment-class II MHC complex interacts with the T_H cell by binding to the T-cell receptor and an accessory protein referred to as CD4. The coupling of this complex with the receptor triggers T-cell activation,

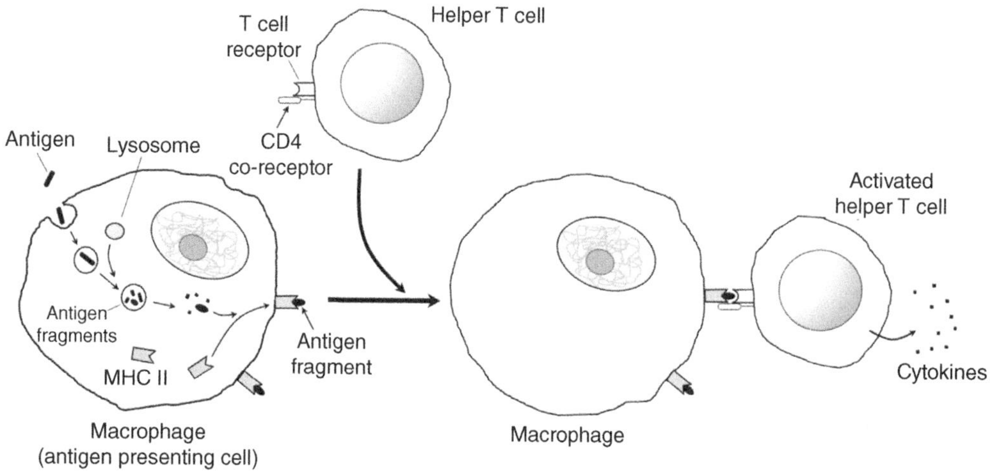

Figure 19.4 Depiction of steps by which macrophages phagocytize and process antigen for presentation to helper T cells.

followed by the release of cytokines which attract additional lymphocytes for subsequent events of the immune response.

19.2.4 B Cell Activation and Release of Antibodies

As noted in Section 19.2.1, immature B lymphocytes progress through the antigen-independent phase of differentiation during which they undergo development into millions of different populations of naive B cells with predetermined antigen specificity (Figure 19.5a). A subsequent initial encounter of naive B cells with free antigens present in lymph initiates the antigen-dependent phase of differentiation (Figure 19.5b). Binding of an antigen to a specific receptor on the surface of the B cell is the first of two steps in the activation process. The second step required for B cell activation involves the participation of a T_H cell sensitized to the same antigen by prior interaction with an APC carrying the antigen fragment-MHC class II complex. The interaction with the T_H cell triggers the B cell to go through multiple cell divisions to form a clone, a population of cells identical to the original cell. The activated cell clone subsequently differentiates into two functional groups: a group of effector cells that launch immune defense actions immediately and a long-lived group of memory cells.

The effector cells of the B cell clone are referred to as plasma cells, short-lived cells that become greatly enlarged and have an extensive rough endoplasmic reticulum (Figure 19.6). This structural specialization allows plasma cells to quickly develop and secrete large quantities of antigen-specific antibodies destined for circulation via the blood and lymph. The cells referred to as memory cells are those that can quickly differentiate into additional effector cells in response to a subsequent encounter with the same antigen. Memory cells can be retained in the circulation for the full lifetime of an individual without additional antigenic stimulation.

The initial antigen encounter that triggers a specific clone of B cells to differentiate into plasma cells is soon followed by an increasing concentration of the antigen specific antibody in the circulating blood. Over a period of ~1–2 weeks, the concentration increases to a peak referred to as the primary immune response (Figure 19.7). The immunoglobulin class secreted into the blood during the primary response is mainly IgM. After reaching the peak, the concentration of this antigen-specific antibody quickly decreases to its normally low level. A second exposure to the same antigen causes the memory B cells of the clone to rapidly give rise to an expanded population of plasma cells resulting in a secondary immune response. The secondary response increases to a peak more rapidly, and it is of greater magnitude and of greater duration than the primary response. It is also important to note that the immunoglobulin class secreted into the circulating blood during the secondary response is mainly IgG. The immunoglobulin (Ig) class switching of the secondary response is an outcome of a process referred to as somatic hypermutation, which affects genes encoding the immunoglobulin heavy chain and light chain.

19.2.5 T Cell Development and Cell-Mediated Immunity

The initial phase of T-cell development, antigen-independent differentiation, occurs in the thymus as noted in Section 19.2.1. This initial phase results in the formation of many different populations of committed naive T cells that express co-receptors for immune cell

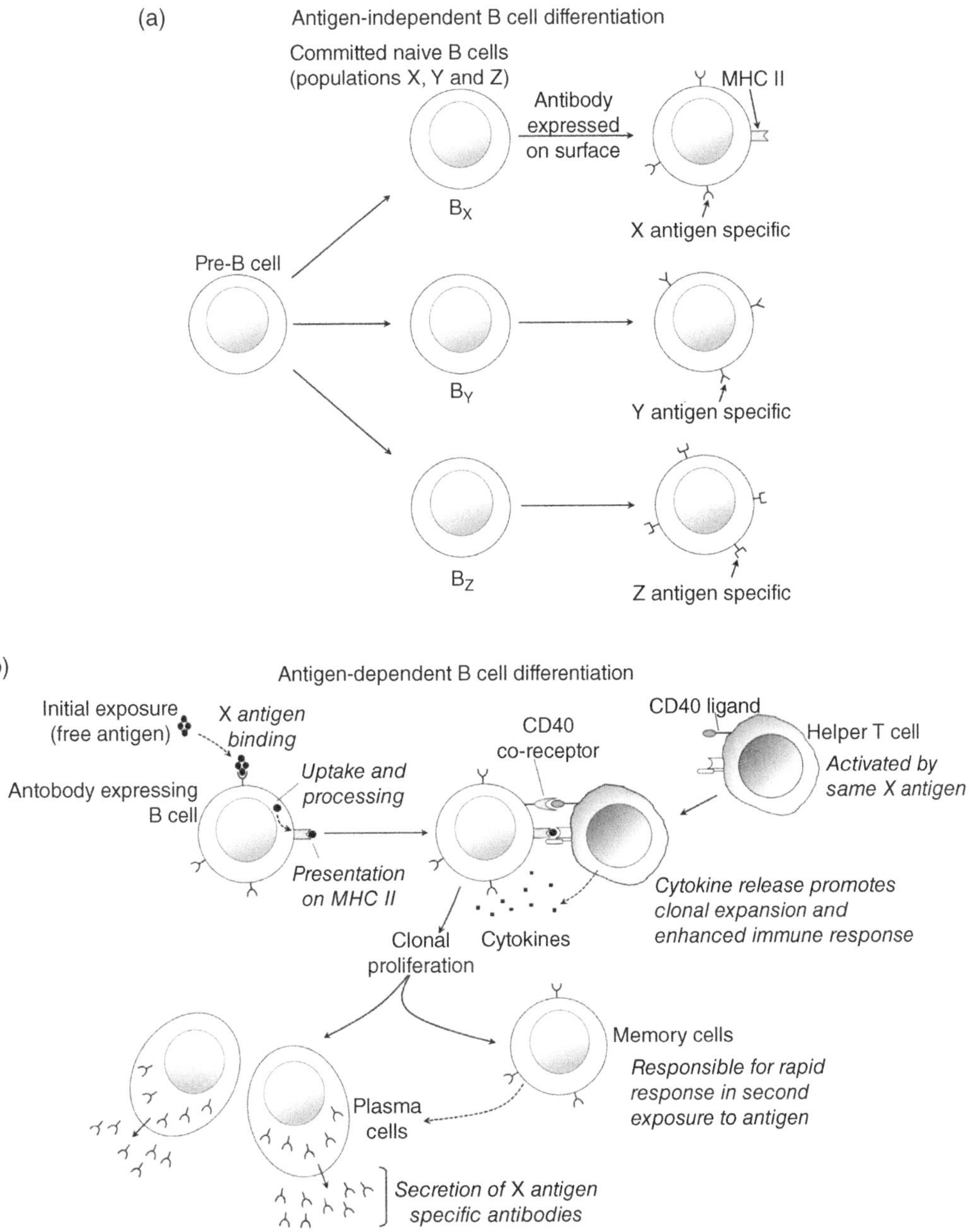

Figure 19.5 Two-step process of differentiation and activation of B cells. (a) The first step, antigen-independent differentiation, occurs in the bone marrow and results in formation of multiple populations of committed B cells. (b) The second step, antigen-dependent differentiation, occurs in peripheral lymphoid organs and is initiated by exposure to free antigen. The activated B cell interacts with a helper T cell which triggers differentiation to antibody-secreting plasma cells and memory B cells.

glycoproteins referred to as CD4 or CD8 (Figure 19.8a). The second phase of T cell development, antigen-dependent differentiation, proceeds in a manner distinctly different from antigen-dependent B cell activation. Unlike B cell activation, which is triggered by free antigens circulated in the blood or lymph, the activation of T cells occurs only in response to the display of a fragment of a specific antigen by an antigen-presenting dendritic cell (Figure 19.8b). Class I MHCs are located at the tip of some of the dendritic arms of the dendritic cell and Class II MHCs at the tip of other dendritic arms. The antigen fragment-Class I MHC complexes interact with $CD8^+$ T cells, while antigen fragment-Class II MHC complexes interact with $CD4^+$ T cells. The display of the antigen fragment-MHC class I complex induces antigen-dependent differentiation of $CD8^+$ T cells into cytotoxic T (T_C) cells, whose mode of action is described in the next paragraph. The display of the antigen fragment-MHC

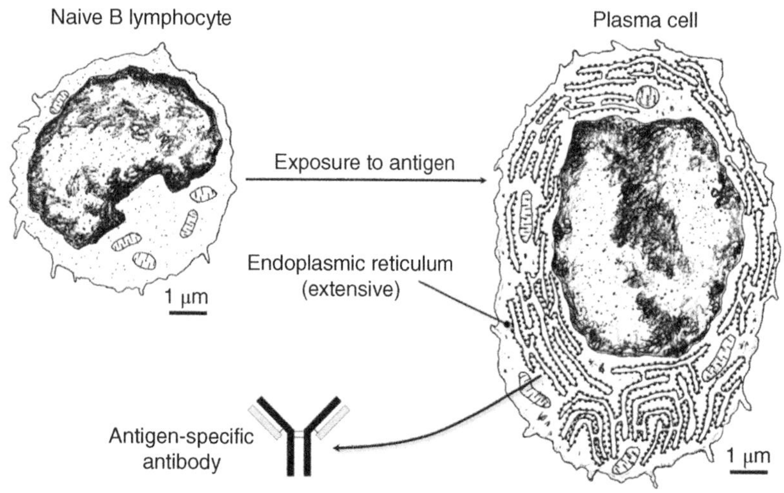

Figure 19.6 Depiction of the differentiation of a B lymphocyte to a plasma cell. The extensive rough endoplasmic reticulum of the plasma cell enables copious production and secretion of antigen-specific antibody.

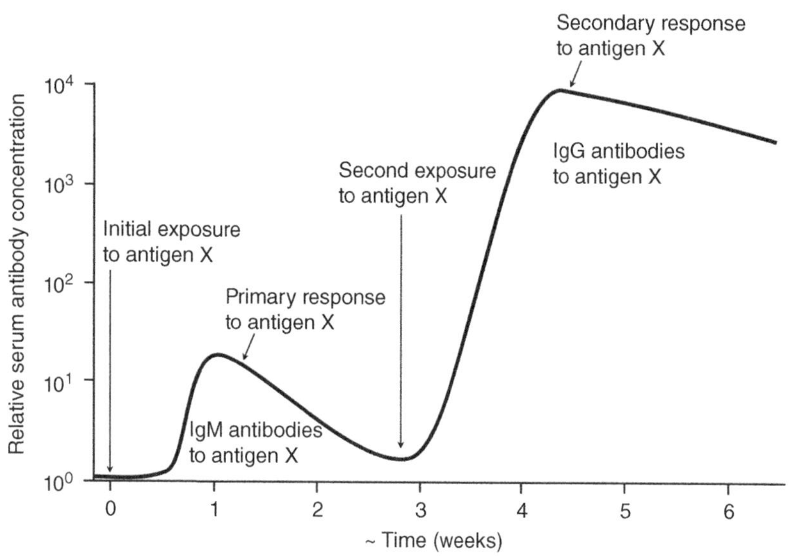

Figure 19.7 Relative levels of specific antibody during primary and secondary immune responses of B cell clone to initial exposure and repeat exposure to antigen. IgM class immunoglobulin is released during the primary response and IgG class immunoglobulin during the secondary response.

class II complex induces antigen-dependent differentiation of $CD4^+$ T cells into T_H cells, regulatory T cells and memory cells. Regulatory T cells modulate T_C cell proliferation and, thereby, prevent autoimmune diseases. Memory T cells, the least abundant T-cell type, remain in circulation for a long duration and can quickly differentiate into a large number of effector T cells in response to subsequent encounter with the same antigen.

A major function of T_C cells is the eradication of cells infected by viruses or bacterial pathogens. Soon after infection occurs, virus-infected cells transport viral antigen to the cell surface for display and recognition by specific T_C cells. The activation of T_C cells that endows them with the ability to recognize and destroy virus-infected cells is a direct result of the process of presentation of a fragment of the viral antigen to T_C cells by antigen-presenting dendritic cells (Figure 19.8b). After attaching to the infected cell, the T_C cell undergoes degranulation to release perforin, a 70 kDa protein with a hydrophobic domain compatible with the interior (middle layer) of cell membranes. Approximately 20 perforin molecules spontaneously assemble as a cylindrically shaped channel in the plasma membrane of target cells, similar to the mechanism of assembly of ion channels described in Section 19.3. The pore that forms results in the nonselective flow of ions and water followed by osmotic lysis of the target cell.

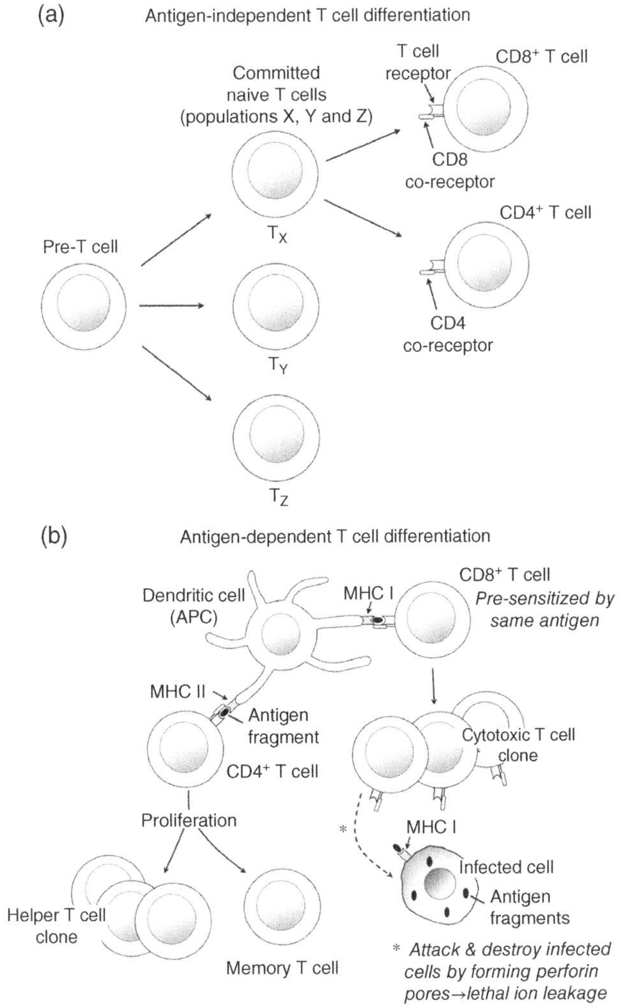

Figure 19.8 Two-step process of differentiation and activation of T cells. (a) The first step, antigen-independent differentiation, which occurs in the thymus, results in formation of multiple populations of committed T cells which express either CD4 or CD8 co-receptor. (b) The second step, antigen-dependent differentiation, which also occurs in the thymus, is initiated by interaction with an antigen presenting dendritic cell. This interaction triggers differentiation $CD4^+$ T cells to T helper cells and memory cells and $CD8^+$ T cells to cytotoxic T cells.

T_C cells also perform a process referred to as immune surveillance, which involves identifying and destroying precancerous cells that appear periodically. The precancerous cells express surface markers that differ from those of normal cells, which form the basis by which the precancerous cells are detected and eradicated by T_C cells.

19.3 The Complement System

The complement system consists of a group of more than 20 plasma proteins that assist or "complement" the functions of both the innate immune defense system and the adaptive immune defense system. These complement proteins are constantly circulated in the blood in an inactive state and are only activated in a complex reaction cascade when needed to assist the innate or adaptive immune defense systems. Many of the complement protein components C1–C9 are proteases which become activated by proteolytic cleavage. Component C3 is the most abundant of the various complement system proteins and arguably the most important. Proteolytic activation of C3 by the enzyme C3 convertase triggers activation of other complement proteins that function in three areas of innate immune defense: chemotaxis and inflammation, opsonization of pathogens for destruction by phagocytic leukocytes (Chapter 18), and formation of a membrane attack complex (MAC) in pathogenic microbes. Some of the complement proteins also contribute to adaptive immune defense. Complement fragments C3a and C5a promote T cell activation and proliferation for immune defense. Attachment of complement fragment C3d to complement receptor on B lymphocytes promotes increased synthesis of antigen-specific antibody.

Formation of the MAC, which occurs in a manner similar to the formation of the perforin pore, involves insertion of complement proteins with a lipophilic core into the plasma membrane of target cells (Figure 19.9). The binding of antibodies to the surface of the pathogen or foreign cell activates complement proteins which bind to the antibody-antigen complex at the cell surface. This triggers the complement cascade, leading to formation of C5b which complexes with C6, C7, and C8. This C5b,6,7,8 complex is very hydrophobic, allowing it to become inserted into the lipid bilayer, after which it is joined by 12–15 C9 molecules that spontaneously polymerize to create a MAC pore in the plasma membrane. This results in leakage of ions through the pore followed by osmotic lysis of the target cell.

Although the complement proteins play important roles in both innate and adaptive immune defense, they also produce detrimental effects with implanted biomaterials. As described in Chapter 18, implanted biomaterials quickly acquire a coating of adsorbed proteins, including complement C3 proteins, which are a potent chemoattractant for neutrophils. The resulting accumulation of neutrophils then promotes inflammation that can cause localized tissue damage.

19.4 Adaptive Immune Responses to Biomaterials

In general, adaptive immune responses to the implantation of biomaterials occur much less frequently when compared to inflammatory responses. The adaptive immune

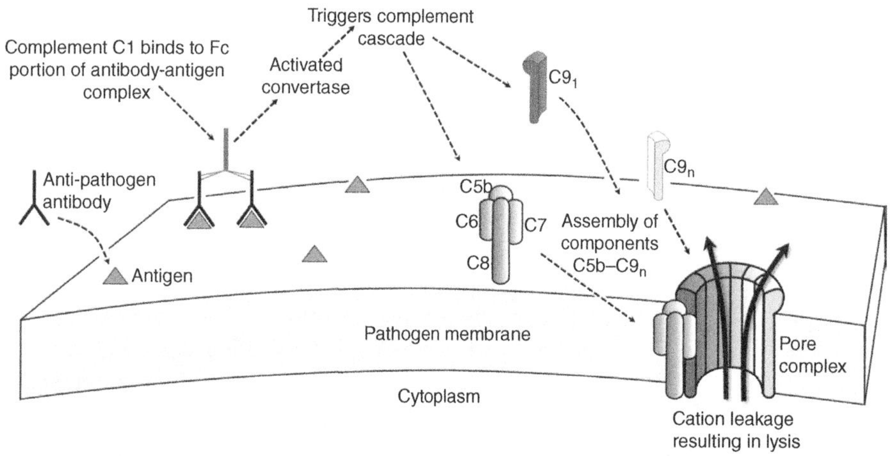

Figure 19.9 Depiction of process by which complement proteins interact with an invasive pathogen and then spontaneously assemble to form a membrane attack complex that causes osmotic lysis of target pathogen.

responses to biomaterials that do occur tend to be of two main types: Type IV hypersensitivity reactions and cell-mediated immune responses to biomaterials containing allogeneic or xenogeneic proteins, or one of the few antigenic polysaccharides.

19.4.1 Hypersensitivity Responses

Hypersensitivity reactions are adverse allergic reactions that occur because of an excessive immune response following exposure to some type of antigen. Hypersensitivity reactions are generally considered to be of four types, referred to as Type I, Type II, Type III, and Type IV, based on the underlying mechanisms by which they trigger the reactions. Types I–III are induced by antibodies or antibody-antigen complexes, whereas Type IV is a cell-mediated response. Overall, Type I and Type IV reactions can be seen with biomaterials, but there is no definitive evidence for the involvement of biomaterials with Type II and Type III reactions.

Type I hypersensitivity reactions, also known as immediate hypersensitivity reactions, occur when plasma cells release allergen-specific IgE antibodies. The IgE antibodies bind to Fc receptors on the surfaces of basophils and mast cells, inducing sensitization of those cells. A repeat exposure to antigens causes the sensitized basophils and mast cells to undergo degranulation, triggering rapid release of vasoactive mediators (histamine, heparin, and cytokines) that stimulate vasodilation and smooth muscle contraction within minutes. These immediate hypersensitivity reactions are usually triggered by skin contact (e.g., contact with latex products such as gloves), inhalation (e.g., pollen), and ingested food (e.g., peanuts). Type I hypersensitivity can vary from mild symptoms such as hay fever and hives to serious, life-threatening emergencies such as systemic anaphylaxis.

Type II hypersensitivity reactions, also described as cytotoxic hypersensitivity reactions, are mediated by IgG or IgM antibodies binding to cell surface antigens, leading to T_C cell destruction of target cells displaying those antigens. The most common example is host response to transfusion of an incorrect blood type. Type III hypersensitivity reactions are triggered by accumulation of small immune complexes at sites such as synovial joints, renal glomeruli, and small blood vessels, which tend to attract inflammatory cells that provoke prolonged inflammation, resulting in localized tissue damage.

Type IV hypersensitivity reactions, also referred to as delayed hypersensitivity reactions because the response may not be seen for several hours or days, are mediated by the interaction of T lymphocytes with antigens. The cells that mediate Type IV reactions are memory T cells, derived from T lymphocytes sensitized by prior exposure to the same antigen. In response to a repeat exposure to the antigen, a population of sensitized T cells exit the circulation and follow the chemotactic trail to the site of antigen location. Upon arrival, the T cells secrete cytokines that attract and activate other effector cells including macrophages. These macrophages can persist for many months at the antigen site and often fuse with each other to form multinucleated giant cells that can cause extensive tissue necrosis. The most common example of type IV hypersensitivity is the response that occurs a few days after contact with poison ivy antigen.

With reference to implanted biomaterials, Type IV hypersensitivity reactions have most commonly been associated with orthopedic devices, particularly devices used in total joint replacement, which normally contain one or more metallic components (Chapter 23). These metallic components can undergo electrochemical corrosion in the physiological fluid (Chapter 14), leading to the release of several types of metal ions, particularly nickel, cobalt, and chromium, which are known to induce hypersensitivity reactions in other situations. Dissolution of fine metallic wear particles can further generate ions in the physiological fluid. Overall, these metal ions can complex with endogenous proteins, which can activate T lymphocytes and lead to cell-mediated immune responses (Figure 19.10).

T cell-mediated metal ion hypersensitivity is particularly problematic as a contributing factor to the failure of prosthetic implants used in total joint replacement. It is estimated that ~15% of total hip joint replacement failures are associated with some degree of metal ion hypersensitivity. Additionally, the incidence of metal ion sensitivity in patients who suffered total hip implant failures due primarily to aseptic loosening has been reported to be as high as 60% (Hallab et al. 2001). An increased incidence of metal ion sensitization has also been observed in some patients who received nickel-containing dental alloy prostheses. Although metal ion-induced hypersensitivity is the most common immune response to implanted biomaterials, rare occurrences of Type IV immune response to nonmetallic biomaterials have been reported, particularly to acrylate components of bone cement.

19.4.2 Immune Responses to Protein-Based Biomaterials and Complexes

As noted in Chapter 9, collagen (type I), prepared mainly from animal tissues such as bovine, porcine, and equine skin and tendons, is the most widely used natural polymer in biomedical applications. While biomaterials containing allogeneic or xenogeneic proteins can elicit an immune response, collagen derived from human tissues have an inherently low antigenicity because these macromolecules are highly conserved proteins, that is, they have nearly identical amino acid sequence across multiple species. However, despite widespread acceptance of the safety of these collagen-based biomaterials, there is always a potential risk of an adverse immune response in a small fraction of the population. Clinical data have shown that ~2–4% of the general population developed an immune response to bovine type I collagen, which appears to be T-cell dependent.

Bioprosthetic heart valves are widely used to replace defective heart valves in humans (Chapter 24). These heart valves are xenografts, commonly made from tissues derived from porcine aortic valves or bovine pericardium

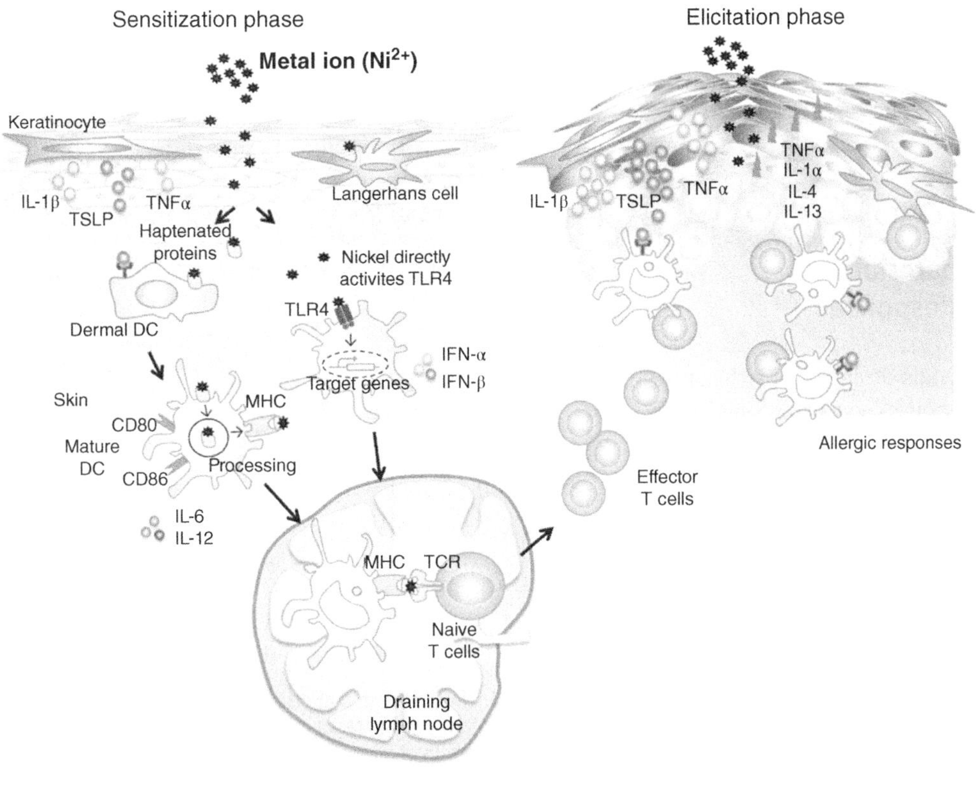

Figure 19.10 Representation of type IV hypersensitivity response to local release of nickel ions. The sensitization phase begins as nickel ions penetrate the skin and promote production of pro-inflammatory cytokines which trigger activation of dermal dendritic cells (DCs). The DCs migrate to draining lymph nodes where they activate effector T cells which in turn migrate to the skin. In response to repeat exposures to nickel ions the pre-sensitized effector T cells release inflammatory cytokines that cause development of skin lesions. *Source:* From Saito et al. (2016).

by a process involving decellularization and glutaraldehyde treatment, procedures intended to remove all porcine or bovine antigens from the tissues. Nevertheless, some antigenicity remains, as evidenced by the presence of antibodies to xenogeneic antigens in patients following implantation of bioprosthetic heart valves derived from porcine aortic valves. This immune response is mainly induced by the epitope galactose-alpha-1-3 galactose (α-gal), a cell surface antigen on porcine valves. However, it has been found that treatment of freshly harvested porcine valves with α-galactosidase removes the α-gal antigen without adversely affecting the *in vivo* performance of the bioprosthetic heart valves.

Haptens are small, non-immunogenic molecules that can indirectly promote an immune response by attaching to a larger carrier molecule, which, by itself, is also non-immunogenic. As an example, human serum albumin is a non-immunogenic protein but when combined with the hapten dinitrophenol, it forms a hapten–protein complex that is highly immunogenic. A similar phenomenon can occur with metal ions released from metallic implants. As noted earlier, metallic implants can undergo electrolytic reaction in the physiological fluid, resulting in the release of metal ions. Many of the released ions, which can be considered small haptens, immediately react with proteins of the interstitial fluid and plasma to form metal ion–protein complexes, which provoke the synthesis of specific antibodies against these complexes. These specific antibodies can lead to the attack of host tissues, resulting in autoimmune disease.

19.5 Designing Biomaterials to Modulate Immune Responses

The development of biomaterials designed to either promote or suppress immune responses at specific sites in the body has attracted interest as a potential therapeutic approach in some applications (Gardner et al. 2013). One area has been the development of suitable biomaterials to deliver protein, peptide, and antibody-based drugs, as well as other therapeutic biomolecules, such as short interfering ribonucleic acid (siRNA) to their site of action (Chapter 26). For example, after administration to the body, siRNA must overcome various physiological and intracellular barriers before it can enter the cytoplasm or nucleus of target cells for it to exert the desired effect. Consequently, suitable biomaterials are required to not only protect siRNA from degradation by serum proteases but also provide it with the ability to evade the immune system. As discussed in Chapter 26, a variety of nanoparticle-based delivery systems, including liposomes, lipid nanoparticles, polymer conjugates, and polymer nanoparticles have been used to deliver siRNA and other therapeutic biomolecules.

An example of a biomaterial designed for a specific type of immunomodulation involves the use of a hydrogel microparticle-based system to promote host acceptance of grafts of allogeneic pancreatic islets (Headen et al. 2018). Immunosuppressive drugs commonly used to control the rejection of allogeneic transplants often have adverse effects on islet function. To alleviate this problem, a system was created in which Fas-ligand, a transmembrane tumor necrosis protein that induces apoptosis, was tethered to hydrogel microparticles composed of maleimide-terminated polyethylene glycol (PEG). A key objective of the approach was to deliver Fas-ligand to T effector cells, the agents responsible for islet allograft rejection. Injection of microparticles carrying the Fas ligand into diabetic mice was found to be effective in promoting apoptosis of T effector cells and prolong the survival of allogeneic islet grafts (Figure 19.11). This approach could potentially be applied to the development of new delivery systems to modulate immune responses in human patients.

A gene targeting procedure (Chapter 25) combined with a technique known as somatic cell nuclear transfer was recently used to prepare genetically altered swine as a source for bioprosthetic heart valves with reduced immunogenicity (Zhang et al. 2018). The objective of the gene targeting was to inactivate three genes encoding enzymes for synthesis of three major xenoantigens: α-gal, *N*-glycolylneuraminic acid, and a blood group antigen known as Sda. Piglet offspring were successfully produced which

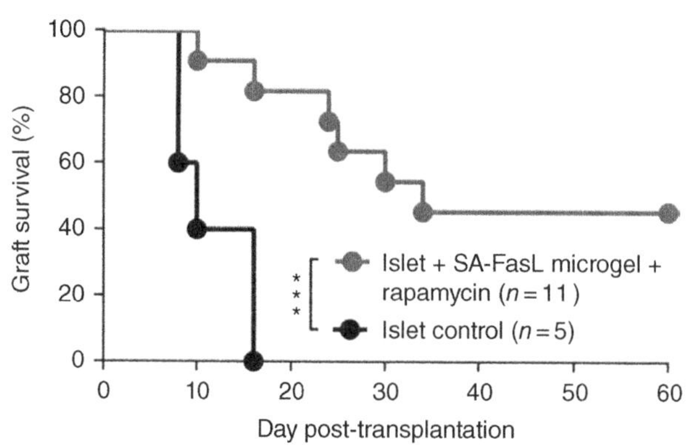

Figure 19.11 Response in diabetic mice to an immunomodulatory biomaterial designed to promote host acceptance of grafts of allogeneic pancreatic islets. The test involved Fas-ligand delivery via hydrogel microparticles synthesized from a polyethylene glycol material with maleimide moieties. The immunomodulatory biomaterial prolonged survival of the allogeneic islet grafts compared to injection of control islets without the immunomodulatory material. ***$p < 0.001$. *Source:* From Headen et al. (2018) / with permission of Springer Nature.

tested negative for the target xenoantigens. In addition, there was minimal binding of human IgG and IgM antibodies to pericardium removed from the test animals. These results suggest that the genetically altered animals could be an ideal source for bioprosthetic heart valves with minimal immunogenicity.

19.6 Concluding Remarks

In this chapter, we discussed the adaptive immune defense system and the immune responses that occur with some types of biomaterials. The immune response defenses are adaptable defenses tailored to millions of different pathogens and foreign cells. They recognize "self" versus "non-self" and possess immunological memory.

- The adaptive immune defense system consists of two major divisions: "humoral" (antibody-based) immunity (function of B lymphocytes) and cell-mediated immunity (function of T lymphocytes). We discussed the components of antibody-based and cell-based immune defenses, and the role of the complement system in immune defense.
- Immune responses to implanted biomaterials can generally be of two types: Type IV hypersensitivity reactions due to metal ions released from orthopedic implants, particularly implants used in total joint replacement which normally contain metal components, and cell-mediated immune responses due to biomaterials containing allogeneic or xenogeneic proteins
- We discussed various approaches that have been investigated for designing biomaterials to modulate immune responses.

Problems

19.1 Explain why the innate defenses can be deployed instantly while the adaptive immune defenses are initiated much more slowly.

19.2 Explain what is meant by the statement that an immune system has adaptability. What is the underlying basis of adaptability and how is adaptability important?

19.3 Explain the benefit of having "memory" cells as major components of the adaptive immune system.

19.4 Explain any advantage that might result from IgM as antibody type of the primary immune response and IgG as antibody type of the secondary immune response.

19.5 Discuss whether or not the testing for metal ion sensitivity should be required before implantation of hip implants containing titanium alloy components.

19.6 Inflammatory reactions occur in response to the implantation of any and all biomaterials. In comparison, relatively few biomaterials provoke immune responses. Explain the physical, biochemical, and/or cellular entities that account for why most biomaterials are not inherently immunogenic.

19.7 Describe any expected immune response to each of the following devices:
A a hip implant with a nickel-containing metal component;
B a heart valve created from porcine aortic valve;
C a degradable suture composed of a poly(α-hydroxy ester) such as polylactic-*co*-glycolic acid;
D a porous bioactive glass scaffold. In each case, explain the basis for the device eliciting or not eliciting an immune response.

References

Gardner, A.B., Lee, S.K.C., Woods, E.C., and Acharya, A.P. (2013). Biomaterials-based modulation of the immune system. *BioMed Research International* 2013 (5): 732182.

Hallab, N.J., Merritt, K., and Jacobs, J.J. (2001). Metal sensitivity in patients with orthopedic implants. *The Journal of Bone and Joint Surgery. American Volume* 83: 428–436.

Headen, D.M., Woodward, K.B., Coronel, M.M. et al. (2018). Local immunomodulation with Fas ligand-engineered biomaterials achieves allogeneic islet graft acceptance. *Nature Materials* 17: 732–739.

Saito, M., Arakaki, R., Yamada, A. et al. (2016). Molecular mechanisms of nickel allergy. *International Journal of Molecular Sciences* 17: 202.

Zhang, R., Wang, Y., Chen, L. et al. (2018). Reducing immunoreactivity of porcine bioprosthetic heart valves by genetically-deleting three major glycan antigens, GGTA1/β4GalNT2/CMAH. *Acta Biomaterialia* 72: 196–205.

Glossary

Antigen A molecular entity that can provoke an immune response; most antigens are proteins.

Antigen presentation A process of display of a specific antigen (or antigen fragment) to a B lymphocyte triggering production of antigen-specific antibody or to a T lymphocyte triggering a cell-mediated immune response.

Complement proteins Proteins of the plasma that function to assist or "complement" processes of both the innate defense system and the immune defense system.

Cytotoxic T cell A type of effector T-cell capable of killing invasive pathogens and other cells of the body that have been invaded by viruses or have undergone transformation to a precancerous state.

Epitope A small portion of an antigen that is the binding site recognized by an antibody against that antigen; a localized region also referred to as antigenic determinant.

Fc receptor The stem portion of Y-shaped immunoglobulins; the portion of an antibody that interacts with surface receptors on macrophages; antibody region recognized by complement proteins.

Hypersensitivity reaction An excessive or uncontrolled immune response following exposure to some types of antigen; there are four types of hypersensitivity reactions, three of which are induced by antibody or antibody-antigen complex, and fourth type is a cell-mediated response.

Immunoglobulins Globular proteins present in the circulating blood that function as antibodies; immunoglobulins constitute about 20% of the plasma proteins.

Major histocompatibility complex A very specific glycoprotein that is part of a family of self-marker proteins present on the surface of all nucleated cells of the body; a marker of "self" unique to each individual.

Membrane attack complex A tubular channel formed in the plasma membrane of a target cell by spontaneous assembly of about 12–15 complement C9 molecules; channel allows ion leakage resulting in osmotic lysis.

Opsonization A process by which a pathogen or other foreign entity is tagged by marker proteins to attract macrophages that will then phagocytize and destroy the entity.

Plasma cell A short-lived cell derived from a unique B lymphocyte clone that is specialized for synthesis of an antigen specific antibody.

Primary immune response A rise in the level of an antigen-specific antibody that is the result of an initial exposure to the antigen.

Secondary response An accentuated rise in the level of an antigen-specific antibody that is the result of a second exposure to the antigen; the secondary response rises more rapidly with greater magnitude and duration.

Self-tolerance Ability of the immune system to recognize and avoid the body's own cells as a result of recognition of "self" MHC and other antigens produced within the body.

Thymus gland A lymphoid organ located within the upper portion of the thorax; the site of maturation of lymphocytes to become T cells.

20

Implant-Associated Infections

20.1 Introduction

The processes and mechanisms of biocompatibility of an implanted biomaterial, as determined by interactions between the biomaterial and the host, have been a major theme so far in this part of the book. However, other factors can have a profound effect on these interactions, one of the main ones being microbial infection, which forms the subject of this chapter.

Bacterial infections are a frequent and especially serious problem associated with the implantation of biomaterials and medical devices. Bacterial contamination of medical devices occurring before or during surgical implantation accounts for approximately two-thirds of hospital-acquired infections. These contaminations are the result of encounters with bacteria in one of two phenotypic states: their free-living (planktonic) state or their sessile adherent form referred to as a biofilm.

Although planktonic bacteria initiate most contaminations of wound sites and implant surfaces, they soon transform into a biofilm, the phenotype that inflicts most of the harm that results from infection. A major outcome of the difference in the harmfulness of these two phenotypes is evidenced by the observation that biofilms are associated with less than 6% of acute wounds but greater than 90% of nonhealing chronic wounds. The nonhealing nature of chronic wounds is due to the inhibitory effects of microbial biofilms on the wound healing process.

Approximately 99% of the bacterial biomass found in nature is surface-attached biofilm. In addition to their prevalence, biofilm contaminations are often of greater concern than contamination by planktonic bacteria. This is because bacteria within a biofilm are protected from the action of antibiotics and host immune defenses by the cocoon-like shielding afforded by the extracellular polymeric substance ("slime") that makes up most of the biofilm mass. Because of the important effects it can have, biofilm contamination of implanted biomaterials will be the main type of bacterial contamination addressed in this chapter.

The most commonly identified microbial agents of implant infections are bacterial species present on the skin of most humans as part of the "commensal" microbiota plus the mildly pathogenic bacterial species. Commensal relationships occur when microbes derive benefit from their human host without inflicting harm. Infections by commensal human microbiota are usually not pathogenic except in those individuals with a compromised immune system. The common microbial inhabitants of the skin and mucous membranes, however, can become opportunistic pathogens due to their ability to form biofilms. This is a particular concern for patients with indwelling catheters or other implanted medical devices due to the ease with which biofilms can form on these devices and thereby cause a high fraction of hospital-acquired infections.

Because the surface characteristics of an implant have a major impact on the extent to which bacteria can adhere to an implant and form a biofilm, modification of these characteristics forms one of the strategies often employed to reduce the incidence of implant infections. Alternative strategies for the prevention or treatment of infections involve the development of new approaches that are not dependent on the use of antibiotics. These new approaches are dictated by the emergence of many types of antibiotic-resistant strains of bacteria and other pathogens.

Overall, then, this chapter will focus on the following topics related to various aspects of bacterial infection of implanted biomaterials and devices:

- The steps by which bacterial biofilms develop on the surface of biomaterials
- Biomaterial properties that favor bacterial attachment and biofilm formation
- How biofilms shield bacteria from antibiotics and impair host defenses against infection
- Strategies to prevent biofilm formation on biomaterial surfaces.

Materials for Biomedical Engineering: Fundamentals and Applications, First Edition. Mohamed N. Rahaman and Roger F. Brown.

Companion website: www.wiley.com/go/rahaman/biomedicalengineering

20.2 Bacteria Associated with Implant Infections

When bacterial contaminations leading to implant failure occur, the majority of these infections are mainly attributable to *Staphylococcus aureus* (*S. aureus*), *Staphylococcus epidermidis* (*S. epidermidis*), and *Pseudomonas aeruginosa* (*P. aeruginosa*), three bacterial species that are common inhabitants of the skin and mucous membranes. *S. aureus* and *S. epidermidis* are both round-shaped bacteria that stain positive with Gram stain, a stain commonly used to identify a bacterial species whose cell wall contains a high amount of the polysaccharide peptidoglycan. *S. aureus* is a mild pathogen of low virulence associated with many common skin infections that can become a more serious opportunistic pathogen. Methicillin-resistant *S. aureus* (MRSA) is a public health risk and the predominant cause of infections that develop within a relatively short duration (<3 months) after surgical implantation. *S. epidermidis* is a nonvirulent saprophyte (that is, it lives on dead organic matter) that inhabits the skin. It usually only becomes a problem within tissues of immune-compromised patients and tissues harboring a foreign material, including an implanted biomaterial. *P. aeruginosa* is a rod-shaped, gram-negative bacterium that is an opportunistic pathogen having multidrug resistance. It is often identified as a contaminant of polydimethylsiloxane (silicone) hydrogel soft contact lenses, and it frequently colonizes moist surfaces such as urinary catheters and pulmonary ventilation machines.

Infections caused by the aforementioned three bacterial species and by other microbial agents can develop in several ways. In addition, there are differences between the times when infections become symptomatic. For example, infections that develop after orthopedic implant surgery are classified as "early" if symptoms appear within 3 months post-surgery, whereas they are said to be "delayed" when symptoms develop between 3 and 12 months post-surgery, and "late" when symptoms develop later than 12 months post-surgery. The main sources of early and delayed infections associated with orthopedic implant surgery are direct contact with the wound and airborne contamination. Although late infections following orthopedic implant surgery rarely occur, these incidents appear to be blood-borne contaminations from a distant site of infection.

In addition to contamination during surgery, contamination can be introduced via indwelling catheters and related devices. Indwelling intravenous cannulae, urinary catheters and endotracheal tubes provide microbial contaminants with readily accessible routes of entry. It is especially relevant to note that infections introduced by means of these types of indwelling devices account for ~45% of hospital-acquired infections, also referred to as nosocomial infections.

20.3 Biofilms and their Characteristics

Although biofilms are formed by many different types of microorganisms, such as bacteria, algae, fungi, and slime molds, most biofilms found in nature are produced by bacteria. Consequently, an understanding of the composition and characteristics of biofilms is essential for understanding the manner in which bacteria behave in their biofilm phenotype. A familiar example of a bacterial biofilm is the plaque that continually forms on teeth. The slime commonly found on the surface of smooth rocks in shallow riverbeds and ponds is also a bacterial biofilm.

The mature biofilm has a dynamic structure that changes in size and shape as environmental conditions and availability of nutrients change. It has a typical architecture consisting of a mushroom-like tower structure with several three-dimensional streamers extending from the cap and numerous voids near the base of the tower (Figure 20.1). The voids function as channels through which liquids flow

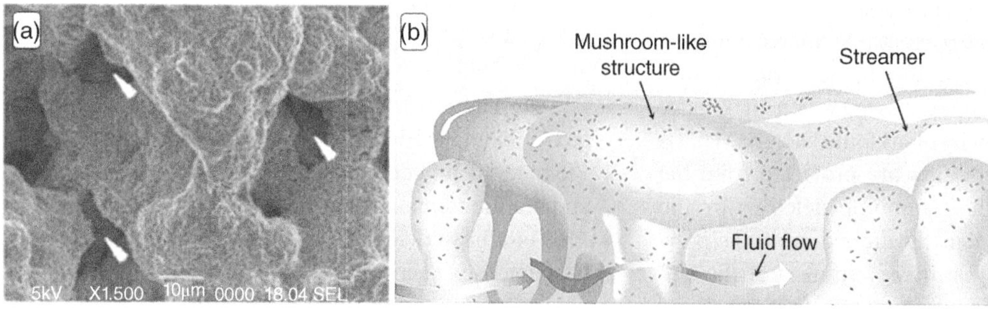

Figure 20.1 Tower-like architecture of a mature biofilm as seen in (a) scanning electron microscope image and (b) illustrative drawing. The drawing shows a mushroom-like structure with streamers plus fluid flow through voids or channels of the biofilm. *Source:* (a) From Asahi et al. (2015); (b) From Montana State University, Center for Biofilm Engineering, with permission.

for transport of oxygen and nutrients, and the removal of metabolic waste products.

Approximately 85–90% of the mass of a mature biofilm consists of a slime-like matrix of extracellular polymeric substances (EPSs) that are synthesized and secreted by the bacteria residing within the biofilm matrix. The EPS components include polysaccharides, proteins, extracellular DNA (eDNA) and, in the case of gram-positive bacteria, teichoic, or lipoteichoic acid. The most abundant components of the EPS produced by most bacterial biofilms are polysaccharides, the components that are the main determinant of the primary conformation of most bacterial biofilms. The biofilm polysaccharides include alginate, the most abundant polysaccharide, plus lesser amounts of polysaccharide intercellular adhesion (PIA), Psl, and Pel. The EPS of some bacterial biofilms contain 1,3-β-linked or 1,4-β-linked hexose residues that make the biofilm more rigid. The eDNA has a structural role in which it contributes to mechanical strength and a genetic role as an agent of gene transfer between cells. Teichoic acid, a copolymer of glycerol phosphate or ribitol phosphate and carbohydrates linked by phosphodiester bonds, strengthens attachment of the gram-positive bacterium *S. epidermidis* to medical devices through stronger attachment to fibronectin.

An important characteristic of biofilms is a multicellular-type behavior of bacterial microcolonies enclosed within the EPS matrix. This cooperative multicellular behavior is the result of a communal activity referred to as quorum sensing, an activity that allows the microbial residents of the biofilm to adapt quickly to environmental changes such as changes in concentrations of oxygen and nutrients. The quorum sensing process involves the exchange of cell-to-cell signals, which causes the biofilm community to behave in a coordinated manner that mimics the behavior of some multicellular organisms. This coordination enables a massive collection of bacteria to conduct a task that would be too large for individual bacteria. Some examples of tasks resulting from the quorum sensing behavior processes include bioluminescence, motility and coordinated dispersal activity. The cell-to-cell signaling mechanism of quorum sensing particularly involves the cyclic dimeric guanosine monophosphate (di-GMP) nucleotide.

20.4 Sequence of Biofilm Formation on Implant Surfaces

The biofilm phenotype allows a bacterium to exist temporarily as a multicellular entity in which "group behavior" is utilized to enhance survival in an adverse environment. The sequence of biofilm formation on an implant surface preconditioned by protein adsorption includes the following steps (Figure 20.2):

- Reversible adsorption of bacterial cells to the implant surface
- Irreversible attachment of bacterial cells to the implant surface

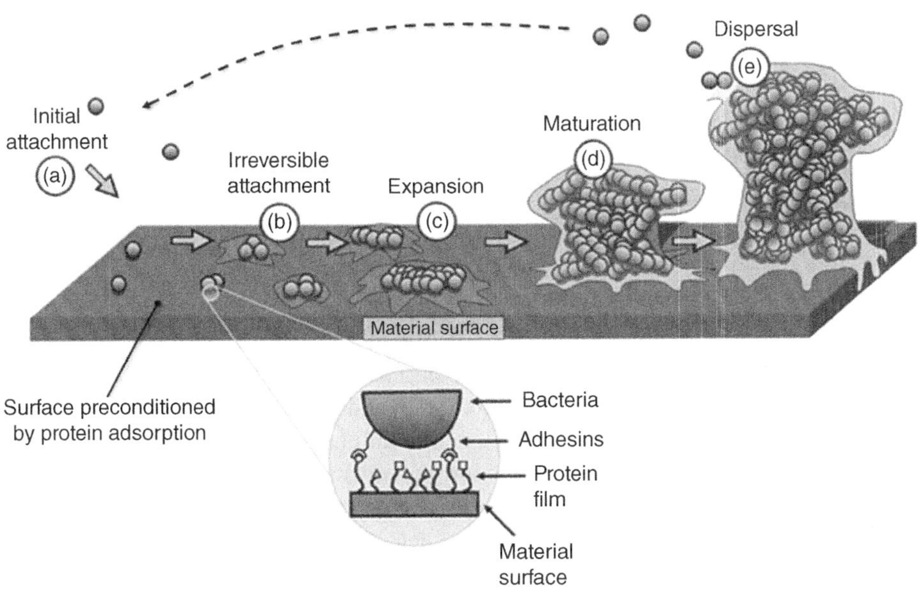

Figure 20.2 Sequence of biofilm formation on an implant surface preconditioned by protein adsorption. Steps include: (a) initial reversible attachment of bacteria; (b) irreversible attachment of bacteria; (c) microcolony expansion; (d) colony maturation with extensive synthesis of biofilm matrix exopolysaccharides and tower formation; and (e) dispersal with return to planktonic state. *Source:* From Bresco et al. (2017).

- Microcolony expansion and synthesis of biofilm EPS matrix
- Biofilm maturation and tower formation
- Dispersal and return to the planktonic state

It is noted that prior to initiation of this sequence of events of biofilm formation, the surface of an implanted biomaterial will be modified by adsorption of biomolecules, particularly proteins present in the plasma, interstitial fluids, and extracellur matrix (Section 16.3). This coating of biomolecules will, in turn, influence the interaction between bacterial cells and the implant surface (Section 20.5).

20.4.1 Passive Reversible Adhesion of Bacteria to Implant Surface

The initial step in the reversible nonspecific attachment of bacteria to an implanted biomaterial mainly involves long-range interactive forces such as van der Waals and electrostatic interactions. Hydrophobic interactions provide the dominant attractive force while bacteria are still tens of nanometers from the surface. The elements that contribute to these hydrophobic interactions include the numerous hydrophobic amino acid residues within the membrane proteins of fimbriae, the microscopic appendages on the surface of many bacteria (Figure 20.3). Other factors that affect passive adhesion of bacteria to the implant include hydrodynamic boundary layer conditions, temperature, and ionic strength. Additionally, the physiochemical characteristics of the implant surface have a strong effect on bacterial adhesion (Section 20.5).

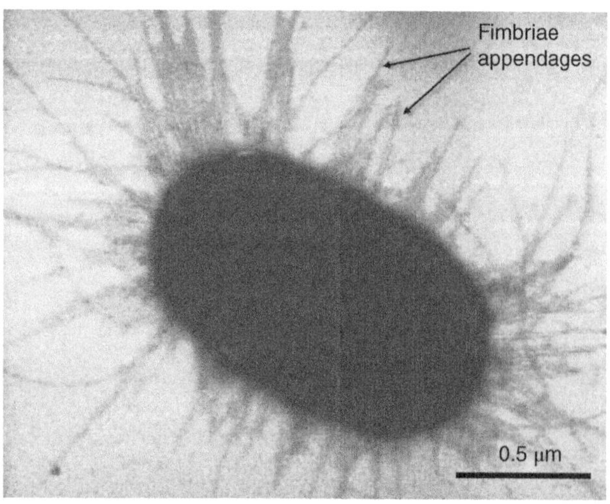

Figure 20.3 Scanning electron microscope image of fimbriae appendages of the bacterium *Escherichia coli* which mediate adhesion to cells and surfaces. *Source:* From Gross (2006).

20.4.2 Specific Irreversible Attachment of Bacteria to Implant Surface

As bacteria are drawn closer to an implant surface (<5 nm), they will be held sufficiently tightly to allow more specific irreversible attachments, the second stage of bacterial adhesion. The irreversible attachments that occur are mediated mainly by adhesins, a term that refers to various types of fine fibril appendages that facilitate adhesion of bacteria to surfaces and to other cells. Most adhesins are lectin-like proteins specific for various biomolecular ligands, including collagens, fibronectins, and other extracellular matrix components. The many different adhesin molecules provide the primary mechanism for specific attachment of bacterial contaminants to implanted biomaterials. Other bacterial cell surface structures and components that participate in the attachment process include flagella, lipopolysaccharides, EPS components and additional proteins. Cell surface structures and components that have a high content of nonpolar amino acid residues are particularly involved in attachment to hydrophobic substrata. In comparison hydrophilic substances such as lipopolysaccharides and many EPS are more important for attachment to hydrophilic materials. The irreversible specific attachment to an implant surface is accompanied by the transition of bacteria from the planktonic phenotype to the biofilm phenotype.

20.4.3 Microcolony Expansion and Formation of Biofilm Matrix

Very quickly after they become firmly attached to the surface of an implanted biomaterial or medical device, bacteria undergo numerous division cycles to form a microcolony. In concert with microcolony formation, the bacteria become more metabolically active and commence synthesis of exopolysaccharides and other components of the EPS matrix or slime. The synthesis of these components continues until the EPS matrix ultimately comprises ~85–90% of the biofilm mass. The EPS becomes highly hydrated, which determines the immediate microenvironment of cells within the biofilm. In addition to water content, the many microenvironment conditions include mechanical stability, porosity, sorption characteristics, hydrophobicity, and biofilm density. As described in Section 20.6, the biofilm very effectively shields bacterial microcolonies from host immune defenses and the action of antibiotics. As an example, Figure 20.4 shows a fluorescence microscope image of polymicrobic biofilm with the bacterial inhabitants enveloped within the EPS matrix for protection from host immune defenses.

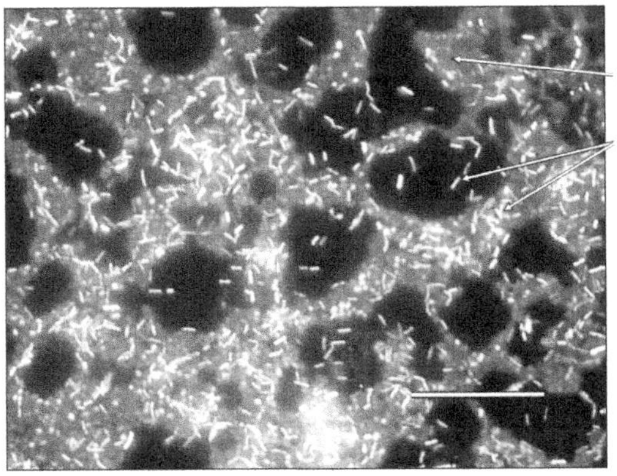

Figure 20.4 Fluorescence microscope image of polymicrobic biofilm grown on a stainless steel surface. Most of the rod-shaped bacteria (visible by staining with 4,6-diamidino-2-phenylindole fluorochrome) are encased in the protective matrix of extracellular polymeric substances (EPS) that is more than half of biofilm mass. Bar = 20 μm. *Source:* From Donlan (2002).

20.4.4 Biofilm Maturation and Tower Formation

As the components of the matrix are synthesized, the EPS expands in height and develops the characteristic tower shape with streamers and interstitial voids (Figure 20.1). Another important feature of biofilm maturation is that upon reaching a critical cell density, the cells of the mature biofilm engage in quorum sensing, the process by which the cells of the biofilm exchange cell-to-cell signals that allow the biofilm community to behave in a coordinated manner mimicking the behavior of some multicellular organisms.

20.4.5 Dispersal and Return to Planktonic State

Dispersal typically involves detachment from the implant surface by a small portion of the biofilm community and movement via active or passive means to a new environment for attachment and initiation of a new biofilm community. Passive dispersal can occur by a physical disruptive force such as hydrostatic shear or physical displacement. Dispersal can also occur via active bacterial responses to environmental changes, such as changes in the concentrations of oxygen and nutrients, temperature, pH, and other factors. The dispersal that occurs via active bacterial responses typically involves bacteria reverting from a biofilm phenotype back to the planktonic phenotype. Many bacteria in the planktonic form possess flagella and other appendages that provide active movement. The cyclic nucleotide cyclic-di-GMP is an especially important cell signal involved in altered gene expression to effect dispersal. Increased cyclic-di-GMP leads to down-regulation of EPS and the sessile state, whereas reduced cyclic-di-GMP leads to up-regulation of swimming and swarming motility. To effect dispersal, the biofilm bacteria often release an enzyme such as alginate lyase, one of several enzymes that catalyze degradation of the EPS matrix. Upon release from the substrate, the dispersing bacteria move until they encounter a suitable environment where they attach to a surface and reinitiate biofilm formation.

20.5 Effect of Biomaterial Characteristics on Bacterial Adhesion

The adhesion of bacteria to an implant surface and subsequent formation of a biofilm are both strongly influenced by the physicochemical characteristics of the implant surface, such as its topography (roughness or smoothness), surface energy, surface charge and, particularly, its surface chemistry (Chapter 5). As noted, upon implantation into the physiological environment, the surface of a biomaterial acquires a surface charge due to adsorption of ions from the aqueous physiological medium. Subsequently, the charged surface is coated with a layer of adsorbed proteins, which will modify the surface charge of the biomaterial. In general, cell attachment is promoted when the energy barrier of interaction is low enough to allow cells to approach the protein-coated surface close enough for adhesion to take place (Section 16.3).

Several studies have reported that bacteria adhere to a lower extent to surfaces with a negative charge, often measured *in vitro* in an aqueous medium (without an adsorbed layer of proteins). Because many bacterial cells have a negative surface charge, this has been attributed to a greater degree of repulsion between the cells and the negatively charged surface. However, the effect of the adsorbed protein layer in modifying the surface charge of the implanted biomaterial is often ignored.

The effects of surface topographical features such as roughness, porosity, grooves and texture on cell and bacterial response have been widely investigated. For example, as discussed in Chapter 21, the introduction of certain microscale surface roughness features on the titanium (Ti) and Ti6Al4V implants is a widely used approach for enhancing the osseointegration of dental and orthopedic devices. The addition of certain nanoscale surface features can further improve the positive outcomes seen with microscale topography. A problem is that the topographical features that enhance osseointegration can often enhance bacterial adhesion and biofilm formation as well. These conflicting outcomes illustrate the dilemma of balancing the benefit of enhanced osseointegration with the risk of enhanced biofilm formation. The conflicting outcomes also provide an example of what is known as the "race for the surface," a term often used in the biomaterials literature to describe the competition between the host tissue and microbial contaminants for colonization of the surface of an implanted biomaterial.

20.6 Biofilm Shielding of Infection from Host Defenses and Antibiotics

The ability to successfully combat the formation of bacterial biofilms is a long-standing and challenging problem, which continues to confront most living organisms, including humans. Both of the two phenotypic forms of bacteria, the planktonic phenotype and the biofilm phenotype, have been present on our planet and within its many ecosystems since bacteria first appeared approximately three billion years ago. Bacteria evolved to use the biofilm lifestyle as an adaptive strategy to enhance their survival in adverse environments, including survival within the human body. The EPS matrix within which bacterial microcolonies of mature biofilms reside has evolved to be a very effective shield against the actions of host defense systems and the actions of most modern antibiotics.

The EPS matrix formed by bacteria after attaching to an implant surface makes the microbes less accessible to the host defense system due in part to the difficulty neutrophils and macrophages encounter in penetrating the thick polysaccharide barrier. Additionally, the dense polymeric matrix of a mature biofilm can be very difficult for macrophages to engulf, resulting in "frustrated" phagocytosis and foreign body giant cell formation (Chapter 18). Biofilms also release various toxins and other factors that provide additional bacterial shielding against neutrophils and macrophages. As an example, *S. aureus* biofilms release a chemotaxis inhibitory protein, as well as a separate factor that blocks complement activation. *S. aureus* is also known to release a toxin that causes neutrophil lysis and another toxin that impedes phagocytosis by macrophages.

Staphylococcal biofilms are also capable of modifying the path of monocyte differentiation toward one of two main macrophage phenotypes (Chapter 18). Staphylococcal biofilms can shift monocyte differentiation away from the M1 pro-inflammatory bactericidal phenotype toward the M2 anti-inflammatory pro-fibrotic phenotype. This shift to the M2 macrophage phenotype reduces the portion of the tissue macrophage population with bactericidal activity.

In addition to their shielding against host defender cells, biofilms also provide very effective shielding against antibiotics. Biofilm bacteria are capable of withstanding antibiotic concentrations that are ~10^3 times greater than the lethal concentration for planktonic bacteria. A combination of factors account for this difference in antibiotic susceptibility, such as limited diffusion through the EPS, antibiotic inactivation by the low pH of the biofilm and removal by EPS matrix components that act as adsorbents. A particularly important factor that accounts for much of this difference in antibiotic susceptibility is the altered metabolic characteristics of bacteria deep within the biofilm matrix. The bacterial cells deep within the biofilm have reduced metabolic and growth rates that make them much less susceptible to antibiotics. This is the basis of their recognition as "persister" cells that cause reinitiation of infection when antibiotic treatment is discontinued.

The problem of low susceptibility of biofilm bacteria to antibiotics is compounded by the widespread use of modern antibiotics. This widespread use and misuse of antibiotics serves as a selection pressure that enhances the appearance of antibiotic resistant strains of bacteria. MRSA, which has become rampant and a widespread public health risk, is particularly challenging. A very high percentage of nosocomial pathogens cultured from patients in intensive care units have been found to be methicillin resistant.

20.7 Effects of Biofilm on Host Tissues and Biomaterial Interactions

Biofilm infections, as we have mentioned (Section 20.1), account for a high fraction of all human infections. Estimates from the US National Institutes of Health and Centers for Disease Control and Prevention indicate that biofilm infections account for ~90% of all infections in humans and approximately two-thirds of all hospital-acquired infections. Biofilms are implicated in many common infectious situations such as dental plaque, contact lens infections, middle-ear infections, and infections of indwelling catheters including intravenous catheters and urethral catheters.

Catheters are frequent sites of biofilm infection, especially urethral catheters. Almost 100% of patients who undergo urethral catheterization for more than four weeks experience biofilm infection of the catheter, which, in some cases results in complete blockage. An example of complete blockage of a urethral catheter resulting from crystalline biofilm formation by the bacterium *Proteus mirabilis* is shown in Figure 20.5. It has also been reported that ~95% of patients that require endotracheal ventilation exhibit biofilm formation on endotracheal tube surfaces (Figure 20.6). The biofilm pathogens are the principal cause of ventilator-associated pneumonia, which occurs as a result of impairment of cough reflex by the endotracheal tube.

Biofilms have also been implicated in much more serious and potentially deadly infections of permanent indwelling

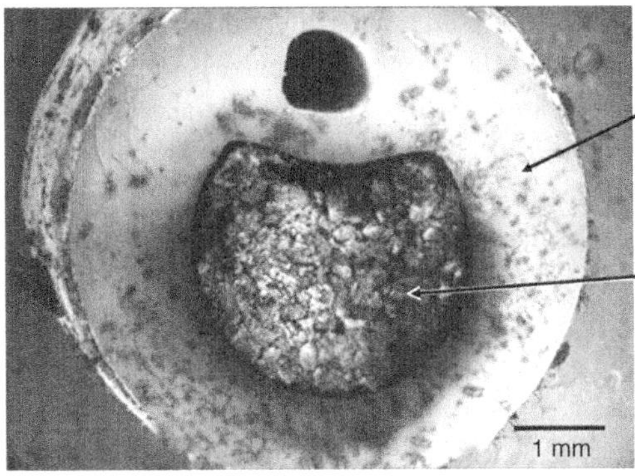

Figure 20.5 Scanning electron micrograph of complete blockage of urethral catheter by crystalline biofilm formed by the bacterium *Proteus mirabilis*. *Source:* From Stickler (2014).

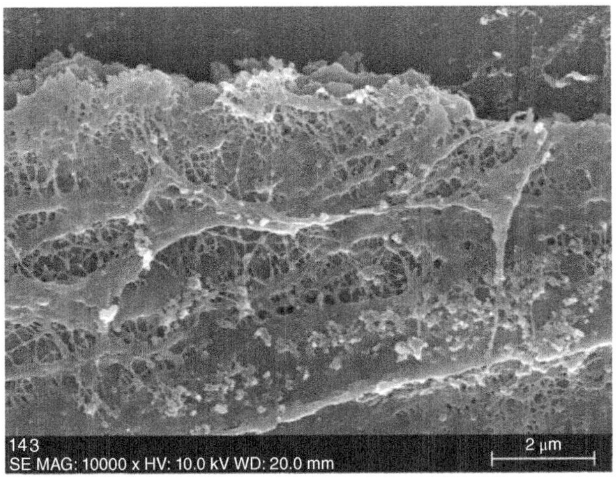

Figure 20.6 Scanning electron microscope image of polymicrobic biofilm on an endotracheal tube from a mechanically ventilated patient. Biofilm formation is the main source of ventilator-associated pneumonia common to these patients. *Source:* From Gil-Perotin et al. (2012).

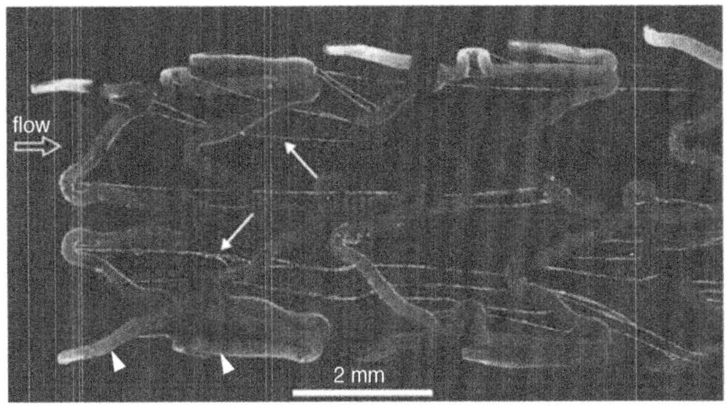

Figure 20.7 Laser scanning confocal microscope image of *P. aeruginosa* biofilm streamers growing from bare-metal stents in a microfluidic flow channel in vitro. Arrows point to streamers of *P. aeruginosa* cells expressing green fluorescent protein; arrowheads point to the bare metal stent. *Source:* From Drescher et al. (2013).

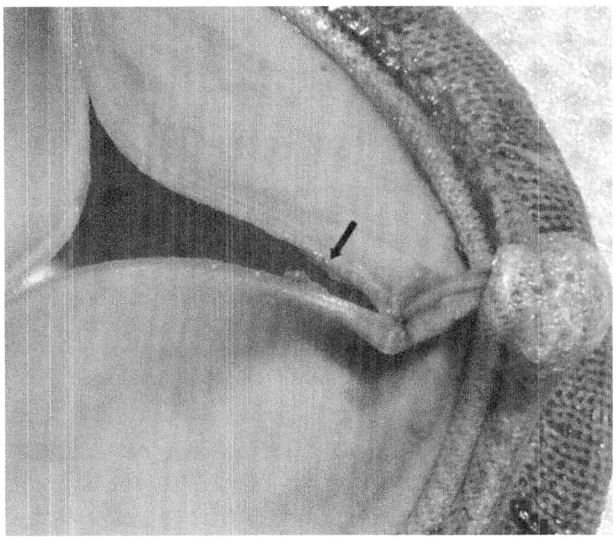

Figure 20.8 Macroscopic photo image of a *Candida* fungal biofilm on a prosthetic aortic valve removed at autopsy from a human patient. Black arrow at edge of prosthetic valve leaflet indicates biofilm that led to multi-organ failure. *Source:* From Michel et al. (2009).

devices such as arterial stents, prosthetic heart valves and total hip implants. This is sometimes due to the attachment and relatively rapid growth of biofilm "streamers" (Figure 20.1). Streamers that detach from a distant infection site and disperse in the circulating blood can sometimes reattach to a bare metal arterial stent (Figure 20.7), which could lead to occlusion of the stent. Infective endocarditis, a frequently fatal disease, is an outcome of biofilm formation on structural components of the circulatory system, which can include implanted devices such as coronary stents and prosthetic heart valves. It is noteworthy that biofilms are the principal cause of most fatal infections within the circulatory system. As an example, Figure 20.8 shows the biofilm-induced failure of a prosthetic heart valve.

Biofilms have several detrimental effects that harm tissues, impair host defenses and contribute to infectious disease processes. One example of direct harm to tissues is the degenerative erosion of bone caused by the biofilm from hematogenous osteomyelitis, a blood borne staphylococcal infection that can develop in one of the long bones of the leg or upper arm. The bacterial microcolonies in biofilms can exchange resistance plasmids, one of the events that lead to reduced susceptibility to antibiotics and the emergence of serious public health risks such as MRSA. An example of severe implant-associated osteomyelitis resulting from MRSA-infected total joint replacement (TJR) is shown in Figure 20.9. Staphylococcal pathogens are responsible for ~75% of osteomyelitis cases, more than half of which are due to MRSA strains.

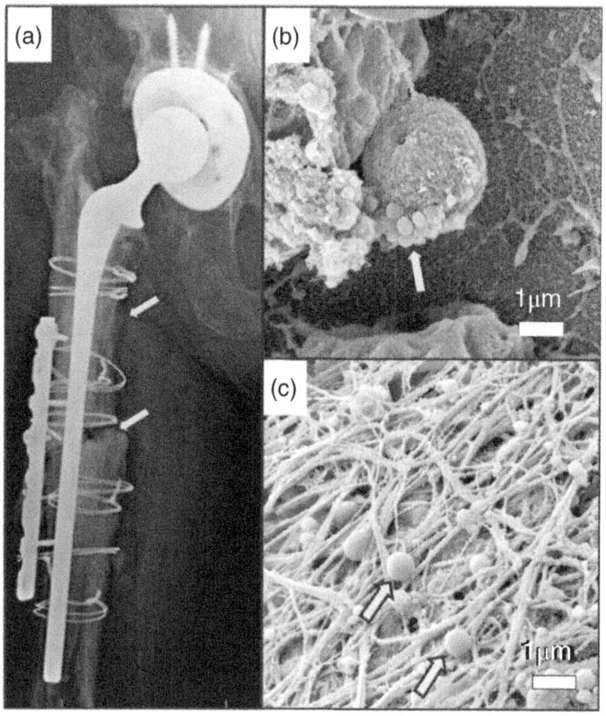

Figure 20.9 (a) Radiographic image prior to revision surgery for MRSA-infected total joint replacement (TJR). White arrows indicate an area of periosteal reaction (upper arrow) and a non-united femoral fracture (lower arrow). Scanning electron micrograph images indicate: (b) staphylococcal bacteria (white arrow) as part of biofilm on surface of explanted TJR; and (c) staphylococcal bacteria (white arrow) attached to fibrin on surface of explanted device. *Source:* From Masters et al. (2019).

It is known that the tissue harm caused by bacterial biofilm is exacerbated when the tissue also contains an implanted biomaterial or medical device. This effect is the result of implant interference with neutrophil and macrophage phagocytosis and other host defense functions. Decreased bactericidal activity of neutrophils after the exposure to biomaterial surfaces has been documented with *in vitro* cell cultures and in rodent models. The host defenses of rodents can normally prevent formation of soft tissue abscesses by *S. aureus* at inoculum densities as high as 10^8 colony forming units (CFU) per milliliter (ml). When injections of *S. aureus* are delivered at tissue sites containing an implanted biomaterial or medical device, however, soft tissue abscesses are formed with inoculum densities as low as 10^2 CFU/ml. In a similar manner, the presence of silk sutures was found to reduce the number of microbes needed for infection by a factor of 10^3 in guinea pigs. The interference that occurs in these cases is an outcome of tissue recognition of an implanted biomaterial as a foreign body and the resultant triggering of a foreign body response, followed by release of lysosomal enzymes that cause serious tissue damage (Chapter 18).

The basis of the increased risk of infection in tissues with an implanted biomaterial or medical device is the chemical inertness of many implant devices. The surface of chemically inert biomaterials is a favorable site for the attachment and colonization of bacteria but it is much less favorable for the attachment of mammalian tissues. The bacterial microcolony and biofilm that develop soon after attachment can soon commence production of exotoxins, which, as noted earlier, will have deleterious effects on host tissue. *S. aureus* and *S. epidermidis* adhere particularly strongly to orthopedic implants with a rough or porous surface, thereby fostering vigorous biofilm infection. In addition, these infections and the resulting tissue damage enhance the likelihood of yet more infection, thus escalating the problem. This escalation frequently includes secondary infections that greatly complicate diagnosis and identification of infectious agents.

20.8 Strategies for Controlling Implant Infections

There is a great need for innovative approaches to control all types of microbial infections, particularly the infections resulting from biofilms that frequently form on the surfaces of implanted biomaterials and medical devices. As noted earlier, upon implantation of a biomaterial, local tissues, and any microbes that may have invaded the implant site compete for colonization of the implant surface. This competition, often referred to as "race for the surface," is usually initiated by a planktonic bacterial contamination. The growing realization that society might soon enter a "post-antibiotic" era when antibiotics will likely no longer be effective coupled with the emergence of antibiotic resistant microbes, intensifies the need for innovative approaches to control microbial infections, particularly biofilm infections. Some examples of strategies for controlling infection without the use of antibiotics are described below.

20.8.1 Orthopedic Implants Designed for Rapid Tissue Integration

One approach for developing orthopedic materials that are less susceptible to bacterial infection is to modify the implant to give the host bone tissue a competitive advantage in the race for the surface. The objective in this case is to modify the implant surface to inhibit bacterial attachment to prevent establishment of a microbial biofilm that overwhelms host tissue while simultaneously encouraging integration of osteogenic cells. One strategy that has been followed in the pursuit of this objective involves application of passive coatings that interfere with microbial attachment as opposed to the use of coatings that releases bactericidal agents. The rationale behind this strategy is that nonbacterial coatings that passively impede bacterial attachment are less likely to promote bacterial resistance to the mode of action of the coating.

An investigation based on this passive coating strategy was undertaken recently to assess the anti-biofilm effectiveness of a coating composed of 5-aryl-2-aminoimidazole (designated LC0024) attached to the surface of Ti discs (Figure 20.10). Molecules of the imidazole compound were covalently bonded to the Ti discs (5 mm in diameter) using carbodiimide chemistry (Chapter 12). *In vitro*, the use of a live/dead staining technique (Chapter 22) confirmed that the LC0024 coating did not have a bactericidal effect *in vitro*. Then the anti-biofilm effectiveness of the coating was assessed in an *in vivo* test, which involved subcutaneous implantation of untreated discs (control-Ti) and LC0024-coated discs (LC0024-Ti) in the dorsum of mice. One day later, actively growing *S. aureus* was subcutaneously injected into the area around each implant and, after an additional four days, the control and coated Ti discs were removed for evaluation of biofilm formation. The measured average *S. aureus* cell content, which was used as an indicator of biofilm size, was ~24 times lower for the coated discs (LC0024-Ti) when compared to the control-Ti discs (Figure 20.10a). However, approximately the same number of planktonic *S. aureus* were recovered from tissues surrounding the control and coated Ti discs (Figure 20.10b), indicating that the LC0024 coating was devoid of bactericidal activity. Additionally, a separate test was conducted to assess the possible effects of the LC0024 coating on osteogenesis, in which uncoated and LC0024-coated porous Ti6Al4V rods (1.5 mm diameter) were implanted in the distal femora of rats. Eight weeks postimplantation, evaluation of the retrieved implants showed that the extent of mineralization was essentially the same on both the uncoated and LC0024-coated discs, indicating the coating did not impair osseointegration. Overall, these results suggest that treatment of Ti implants with this type of anti-biofilm coating could potentially be an effective approach for tipping the balance in favor of the host in the race for the surface.

20.8.2 Surface Nanotopography

Another approach that has been investigated for preventing biofilm formation has been the introduction of nanotopographical features into the implant surface to reduce bacterial adhesion (Chapter 21). As an example, in one study (Feng et al., 2015), aluminum oxide (Al_2O_3) specimens were subjected to a two-step anodization process to produce surface pores of size approximately 15, 25, 50, and 100 nm. Specimens having a smooth surface (at a nanoscale level) were used as controls. Then planktonic cultures of *Listeria monocytogenes*, *S. aureus*, and *S. epidermidis* were inoculated onto the smooth and nanoporous Al_2O_3 specimens. After incubation for 48 hours, confocal laser scanning microscopy showed significantly lower biofilm formation on Al_2O_3 specimens with 15 and 25 nm pores when compared to smooth specimens and specimens with 50 and 100 nm pores (Figure 20.11). The lower biofilm formation on specimens with 15 and 25 nm pores was rationalized in

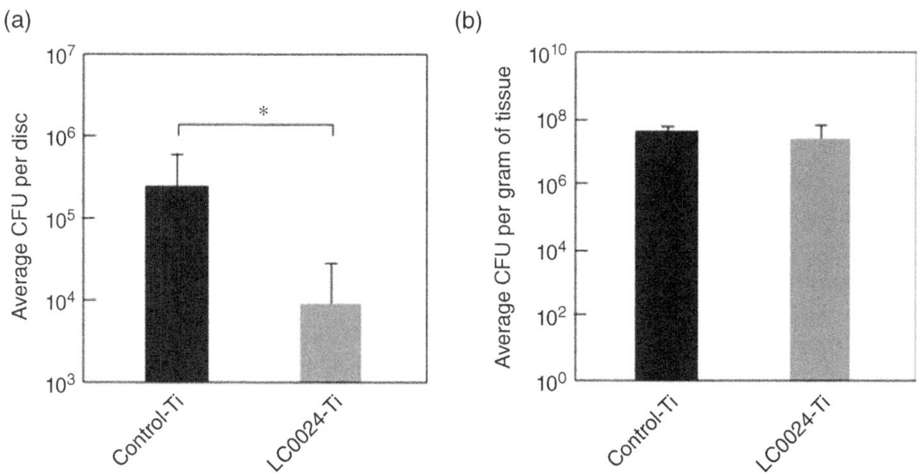

Figure 20.10 (a) Comparison of *in vivo* formation of *Staphylococcus aureus* biofilm on 5 mm diameter untreated titanium (Ti) discs (designated control-Ti) and Ti discs coated with 5-aryl-2-aminoimidazole (LC0024-Ti). Discs were implanted subcutaneously in the dorsum of mice and after 1 day, an inoculum of 10^7 *S. aureus* was injected into the area around the discs. Animals were euthanized after 4 days, and the number of bacteria in the biofilm on each explanted disc was determined by measuring the colony forming units (CFU) (* = $p < 0.05$). (b) Comparison of *S. aureus* colonization of tissue surrounding untreated control-Ti and LC0024-Ti discs after euthanasia. *Source:* From Peeters et al. (2019) / with permission of John Wiley & Sons.

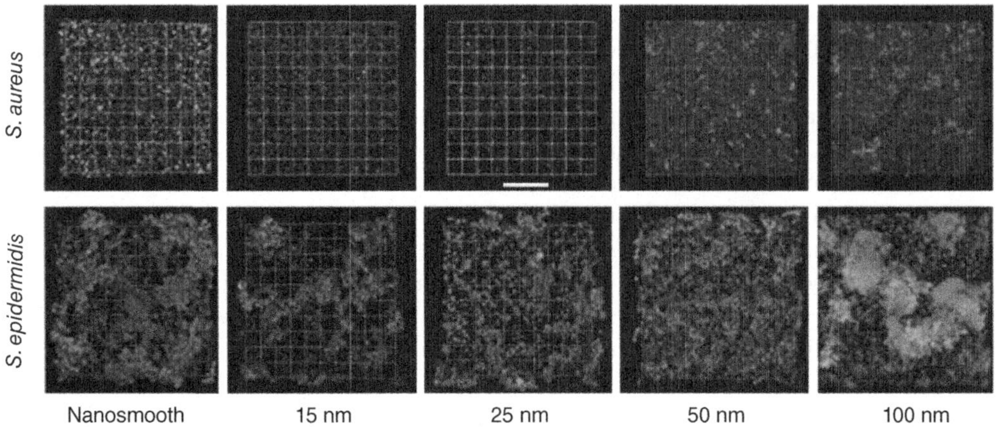

Figure 20.11 Representative confocal microscopy images of biofilms of *S. aureus* and *S. epidermidis* formed during 48-hour incubation on smooth aluminum oxide (control) and anodized surfaces with electrochemically etched 15, 25, 50, and 100 nm diameter pores. Scale bar = 100 μm. *Source:* From Feng et al. (2015).

terms of calculations of interactive forces between the bacterial cells and the nanoporous surface using an extended DLVO theory (Chapter 16). These calculations showed enhanced electrostatic and acid–base repulsive forces associated with the 15 and 25 nm pores.

These results for nanoporous Al_2O_3 are consistent with the results of other studies that show the ability of certain microscale and nanoscale surface topographical features to exert antimicrobial effects (Chapter 21). Overall, the ability of nanoscale surface topographical features to produce conditions that impede bacterial attachment and subsequent biofilm formation represents a potentially attractive method for reducing bacterial contamination of implanted biomaterials.

20.8.3 Silver Nanoparticles

Silver (Ag) ions and nanoparticles have well-known bactericidal properties. The bactericidal effects of Ag nanoparticles appear to involve multiple modes of action including inhibition of DNA replication, unrepairable changes in bacterial cell membranes, blockage of electron transport, and other metabolic pathways. The bactericidal effect of Ag nanoparticles has been observed to be size dependent, with most of the bactericidal effect restricted to Ag particles (or clusters) within a size range ~1–10 nm. These properties provide Ag nanoparticles with the ability to function as a broad-spectrum agent that is not prone to inducing antibiotic resistance. This would potentially permit usage of Ag nanoparticles against a wide variety of bacterial contaminants. For example, the addition of Ag nanoparticles to the surface of titanium-based biomaterials used in orthopedic surgery has been shown to be very effective in enhancing their capacity to resist the growth of *S. aureus*.

20.8.4 Anti-biofilm Polysaccharides

Various polysaccharides in biofilms produced by some bacterial species are known to inhibit the formation of biofilm by other bacterial species. A particularly notable example of this effect occurs with polysaccharides extracted from the biofilm of the uropathogenic *Escherichia coli* strain CFT073, which prevents biofilm formation by a broad variety of bacterial species, including both gram-negative and gram-positive bacteria. The effectiveness of a polysaccharide preparation extracted from biofilm formed by the *E. coli* strain CFT073 for preventing biofilm formation on abiotic surfaces has been evaluated in a recent study. Glass coverslips were inoculated with a nonpathogenic strain of *E. coli* (strain K12) transfected with a vector for green fluorescent protein. After incubation for two hours, a saline suspension of the CFT073 polysaccharide extract or saline alone (control) was added to the coverslip cultures and the incubation continued for another 18 hours. After the incubation, confocal microscopy showed extensive biofilm formation on the saline control surfaces, whereas the surfaces coated with the CFT073 polysaccharide extract were almost devoid of biofilm formation (Figure 20.12). These results suggest that coatings composed of a suitable anti-biofilm polysaccharide extract could potentially provide an effective method for passively preventing biofilm formation on implant surfaces.

20.8.5 Bacteriophage Therapy

Another approach that has been investigated to destroy biofilms involves the use of bacteriophages, that is, viruses that specifically target and kill bacteria. Bacteriophages are harmless to humans because they are each specific for only one type of bacteria, which they destroy by causing the bacterium to lyse. The infection and destruction of the target bacterium occurs by a lytic cycle in which the bacteriophage multiplies within the bacterium until they finally cause lysis of the bacterial cell and release of the bacteriophage progeny (Figure 20.13). Bacteriophage therapy would be much more specific than antibiotic therapy and would not be impeded by the polysaccharide barrier of biofilms as occurs with antibiotic therapy. A disadvantage of bacteriophage therapy would be the high specificity of the phage but this could be overcome by using a bacteriophage mixture to permit targeting of a broader spectrum of bacterial infectious agents. One study has shown that bacteriophage treatment can successfully remove *P. aeruginosa* biofilm contamination from an aortic vascular graft.

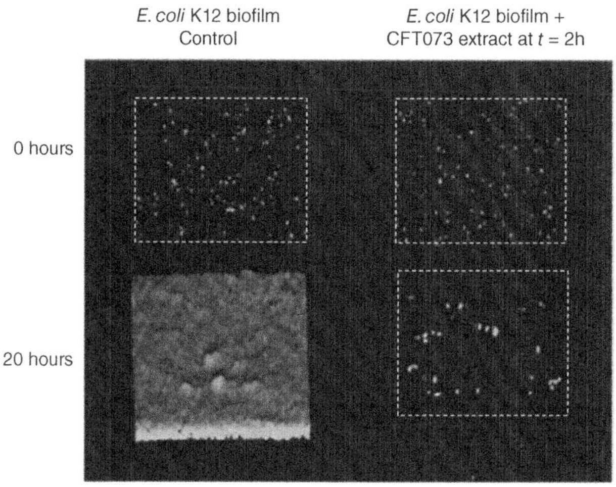

Figure 20.12 Antibiofilm effect of cell wall polysaccharide (serotype K2) of uropathogenic *Escherichia coli* strain CFT073. Glass coverslips were inoculated with a nonpathogenic strain of *Escherichia coli* (strain K12) transfected with a vector for green fluorescent protein. After incubation for 2 h, the coverslips received addition of the K2 capsular polysaccharide extract or saline vehicle only (control) followed by continuation of incubation to 20 h total when relative biofilm size was assessed by confocal microscopy. Dashed white line represents boundary of glass coverslips. *Source:* From Rendueles et al. (2013).

20.8.6 Mechanical Disruption

Biofilm infections are an especially problematic challenge associated with the widespread clinical usage of urinary catheters. Indeed, nearly a third of all hospital-acquired infections are due to biofilm infections of urinary catheters. A novel approach to prevent catheter blockage by biofilm formation has been tested using a polydimethylsiloxane (silicone) elastomer prototype catheter formed by three-dimensional (3D) printing and replica molding. The prototype includes an inflatable inner surface designed to mechanically disrupt and dislodge biofilms after they have formed. Experiments have indicated that this type of inflatable prototype is effective in dislodging crystalline biofilms formed by *Candida albicans*, an opportunistic pathogenic yeast that is a leading cause of urinary catheter blockage (Figure 20.14). The promising outcome of this design suggests that inflatable elastomeric catheters could potentially be effective in significantly reducing the incidence of hospital acquired urinary tract infections.

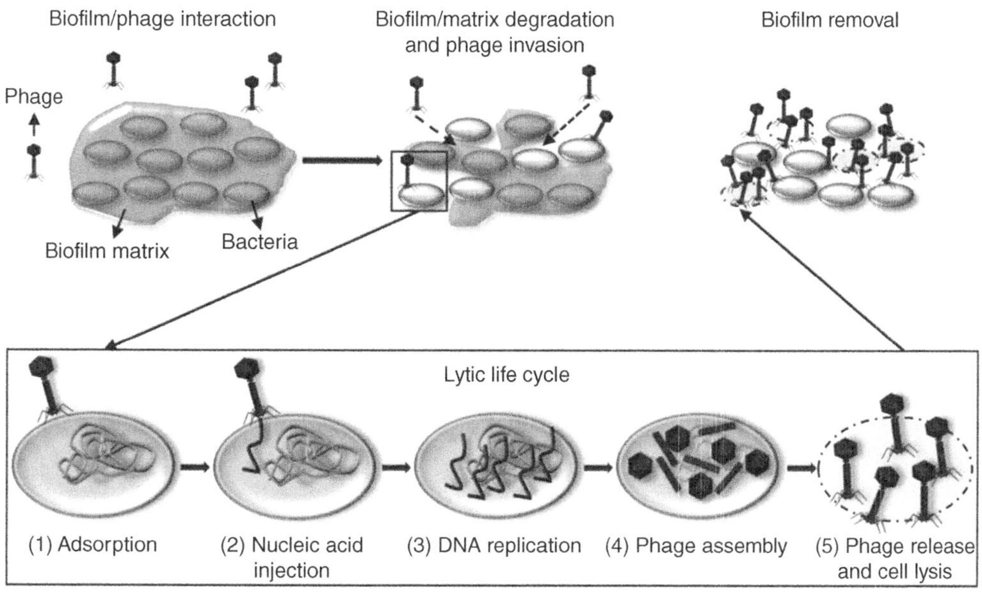

Figure 20.13 Depiction of bacteriophage therapy for degradation and removal of biofilm contamination. Inset enlargement shows steps of the bacteriophage lytic cycle. *Source:* From Gutierrez et al. (2016).

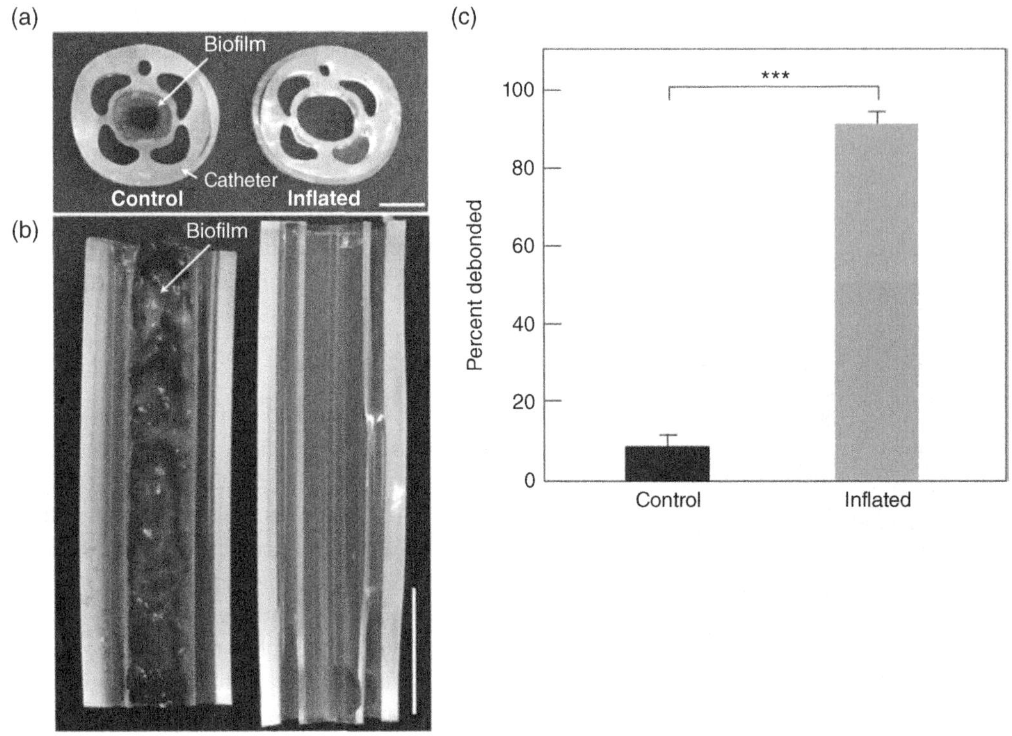

Figure 20.14 Analysis of deformable prototype urinary catheter for controlled removal of *Candida albicans* biofilm. A simulated urine inoculated with *C. albicans* was circulated through the lumen of the prototype catheters for 48 hours prior to test of pressure-driven inflation for removal of adherent biofilm. Macroscopic images show (a) cross sectional and (b) longitudinal sections of prototype catheter before (control) and after inflation. Biofilms were stained with crystal violet for visualization. Scale bars = 2 mm (a) and 5 mm (b). (c) Comparison of percentages of *C. albicans* biofilm mass removed from control and inflated catheters. *** = $p < 0.001$. *Source:* From Maskarinec et al. (2018).

20.9 Concluding Remarks

This chapter provided a discussion of bacterial infection of implanted biomaterials and medical devices, a serious complication that often requires prolonged antibiotic treatment or major surgery to remove the infected implant, particularly for permanently implantable devices such as hip and knee implants. Bacterial infection of a biomaterial surface is caused by contact with bacterial species, particularly *S. aureus*, *S epidermis*, and *P aeruginosa*, in their free-living (planktonic) state or as a biofilm. Biofilms are particularly problematic because they shield bacteria from host immune defenses and from antibiotics.

- We discussed the basic characteristics of biofilms, the steps in their formation on implanted biomaterials, and the mechanisms by which they shield bacteria against defender cells and even against high doses of antibiotics.
- The basic physicochemical properties of biomaterials that influence bacterial adhesion to their surfaces are their surface chemistry, surface charge, surface energy, and surface topography. The effects of topography are discussed further in the next chapter devoted to a discussion of biocompatibility phenomena related to surface topography and particulate biomaterials.
- We discussed various strategies that have been studied to impede or reduce bacterial infection of implanted biomaterials without the use of antibiotics, such as passive coatings, introduction of nanosize porosity, addition of silver nanoparticles, bacteriophage therapy and the use of certain polysaccharides.

Problems

20.1 Briefly describe two ways in which nonantibiotic strategies might be superior to the use of antibiotics for controlling infections of implanted biomaterials.

20.2 Briefly explain why the presence of an implanted biomaterial exacerbates infection. Include the mechanism(s) by which the implanted biomaterial impairs one or more host defenses.

20.3 Describe what is meant by the phrase "race for the surface" that occurs between host tissues and microbial contaminants. Include in your answer the "race for the surface" that often occurs with one specific type of biomaterial implant. In addition, briefly describe one approach that might tip the race in favor of the host tissue.

20.4 Briefly explain why "persister" cells often account for delayed infection of an implanted biomaterial.

20.5 Briefly explain how the improvement in osseointegration by the addition of porosity to the surface of titanium femoral stems of hip implants might be undone by the increased risk of biofilm formation resulting from greater adhesion of bacteria to the porous surface.

20.6 Suggest one approach that might make urinary catheters less susceptible to biofilm formation. Briefly explain the mechanism of action of the catheter modification.

References

Asahi, Y., Miura, J., Tsuda, T. et al. (2015). Simple observation of *Streptococcus mutans* biofilm by scanning electron microscopy using ionic liquids. *AMB Express* 5: 6.

Bresco, M.S., Harris, L.G., Thompson, K. et al. (2017). Pathogenic mechanisms and host interactions in *Staphylococcus epidermidis* device-related infection. *Frontiers in Microbiology* 8 (1401).

Donlan, R.M. (2002). Biofilms and device-associated infections. *Emerging Infectious Diseases* 7: 277–281.

Drescher, K., Shen, Y., Bassler, B.L. et al. (2013). Biofilm streamers cause catastrophic disruption of flow with consequences for environmental and medical systems. *Proceedings of the National Academy of Sciences of the United States of America* 10: 4345–4350.

Feng, G., Cheng, Y., Wang, S.Y. et al. (2015). Bacterial attachment and biofilm formation on surfaces are reduced by small-diameter nanoscale pores: how small is small enough? *NPJ Biofilms and Microbiomes* 1: 15022.

Gil-Perotin, S., Ramirez, P., Marti, V. et al. (2012). Implications of endotracheal tube biofilm in ventilator-associated pneumonia response: a state of concept. *Critical Care* 16: R93.

Gross, L. (2006). Bacterial fimbriae designed to stay with the flow. *PLoS Biology* 4: e314.

Gutierrez, D., Rodriguez-Rubio, L., Martinez, B. et al. (2016). Bacteriophages as weapons against bacterial biofilms in the food industry. *Frontiers in Microbiology* 7: 825.

Maskarinec, S., Parlak, Z., Tu, Q. et al. (2018). On-demand release of *Candida albicans* biofilms from urinary catheters by mechanical surface deformation. *Biofouling* 34: 595–604.

Masters, E.A., Trombetta, R.P., de Mesy Bentley, K.L. et al. (2019). Evolving concepts in bone infection: redefining "biofilm", "acute vs. chronic osteomyelitis", "the immune proteome" and "local antibiotic therapy". *Bone Research* 7: 20.

Michel, R.G., Kinasewitz, G.T., Drevets, D.A. et al. (2009). Prosthetic valve endocarditis caused by *Candida lusitaniae*, an uncommon pathogen: a case report. *Journal of Medical Case Reports* 3: 7611.

Peeters, E., Hooyberghs, G., Robijns, S. et al. (2019). An antibiofilm coating of 5-aryl-2-aminoimidazole covalently attached to a titanium surface. *Journal of Biomedical Materials Research B* 107: 1908–1919.

Rendueles, O., Kaplan, J.B., Ghigo, J.-M. et al. (2013). Antibiofilm polysaccharides. *Environmental Microbiology* 15: 334–346.

Stickler, D.J. (2014). Clinical complications of urinary catheters caused by crystalline biofilms: something needs to be done. *Journal of Internal Medicine* 276: 120–129.

Further Reading

Gristina, A.G., Naylor, P., and Myrvik, Q. (1988-1989). Infections from biomaterials and implants: a race for the surface. *Medical Progress through Technology* 14: 205–224.

Raphel, J., Holodniy, M., Goodman, S.B., and Heilshorn, S.C. (2016). Multifunctional coatings to simultaneously promote osseointegration and prevent infection of orthopaedic implants. *Biomaterials* 84: 301–314.

Schierholz, J.M. and Beuth, J. (2001). Implant infections: a haven for opportunistic bacteria. *The Journal of Hospital Infection* 49: 87–93.

21

Response to Surface Topography and Particulate Materials

21.1 Introduction

The earlier chapters on biocompatibility phenomena discussed several physicochemical properties of biomaterials that influence their biocompatibility, such as chemical composition, hydrophobicity (or hydrophilicity), and stiffness. These chapters were concerned mainly with macroscale biomaterials and did not generally take into account how features at the microscale and the nanoscale influence their biocompatibility. In this chapter, we will consider the effects of these additional features on biocompatibility phenomena, particularly features related to the surface topography of biomaterials and the particulate form of biomaterials. Overall, except for a difference in emphasis, the biocompatibility phenomena associated with surface topography and particulate biomaterials are consistent with those discussed in earlier chapters for macroscale biomaterials.

Surface topographical features relevant to three-dimensional biomaterials, such as roughness, porosity, grooves, and texture, and ways to characterize them were discussed in Chapter 5. In this chapter, we will consider how such topographical features of a biomaterial influence the response of cells and tissues. Overall, certain topographical features at both the microscale and the nanoscale can have a beneficial effect on cell and tissue response, and their addition is now a key design feature to improve outcomes associated with a variety of implanted biomaterials. For example, certain roughness features can enhance the interaction between titanium-based implants and bone cells, leading to better integration of the implant with the surrounding bone. In comparison, a smooth surface normally inhibits the interaction between similar implants and bone cells, and leads to the formation of a fibrous sheath around the implant, which limits integration and proper healing of a bone defect.

Certain surface topographical features at a microscale, such as an array of pillars, grooves, and pores have also been shown to improve the antibacterial activity of some implants, while surfaces with certain nanoscale topographical features can offer additional modalities for combatting bacterial infection, such as bactericidal activity. Bacterial infection of implantable devices, as noted in Chapter 20, is a serious complication, particularly for devices such as in-dwelling catheters and orthopedic prostheses used in long-term implantation. Consequently, the reliance primarily on a biomaterial's surface topographical features could provide an alternative approach to current methods, such as persistent antibiotic therapy or other chemical methods, to treat or prevent bacterial infection. This could be particularly significant in light of the increasing prevalence of antibiotic resistant microbes *in vivo*.

Particulate biomaterials are relevant to biocompatibility phenomena in two main ways. First, particles are accidentally or inadvertently introduced into the physiological environment due to wear, degradation, or corrosion of implanted biomaterials. A well-known example is the release of wear particles from the articulating surfaces of hip and knee implants (Chapter 1). Second, particles are introduced deliberately into the physiological environment as devices to deliver therapeutic and diagnostic agents, or to track or label cells. Overall, particulate biomaterials can have positive or negative effects on cell and tissue response. For example, wear particles released from hip and knee implants can promote local inflammatory responses, which may lead to detrimental effects such as loosening of the implant. In such situations, revision (repeat) surgery is normally required to replace the loosened implant, a procedure that is associated with considerable trauma. In comparison, deliberately introduced particles can provide positive effects, such as controlled local delivery of therapeutic agents or targeted delivery of therapeutic agents to specific cells such as tumor cells.

The normal objective is to optimize the desirable effects of the surface topography of biomaterials and the particulate form of biomaterials, while reducing or avoiding the biocompatibility phenomena that could be harmful. An understanding of the mechanisms by which various topographical and particulate features influence cell response is

Materials for Biomedical Engineering: Fundamentals and Applications, First Edition. Mohamed N. Rahaman and Roger F. Brown.

Companion website: www.wiley.com/go/rahaman/biomedicalengineering

important in achieving this objective. The topographical and particulate features discussed in this chapter are at the microscale, taken as ~0.1 μm to a few tens of micrometers, and at the nanoscale, less than ~50–100 nm. Overall, the topics we shall discuss are in the following broad areas:

- Host responses to microscale and nanoscale surface topographical features of a biomaterial
- Influence of the surface topography of a biomaterial on its antibacterial activity
- Host responses to accidentally or deliberately introduced microparticles and nanoparticles (NPs).

21.2 Effect of Biomaterial Surface Topography on Cell Response

As discussed in Chapter 5, the important surface properties of a three-dimensional biomaterial are its chemical composition, free energy, and topography. Regardless of size, upon introduction of a biomaterial into the aqueous physiological environment, its surface immediately acquires an electrostatic charge by adsorption of ions from the environment. Then a layer of biomolecules, composed mainly of proteins, rapidly coats the surface by preferential adsorption from the physiological fluid (Chapter 16). The surface properties of the original surface control the conformation, concentration, and nature of the adsorbed biomolecules. However, it is the surface covered with a layer of adsorbed biomolecules, and not the bare surface of the biomaterial, which initially interacts with cells in the physiological environment.

Numerous studies have established that cells respond differently to the surface topographical features of a biomaterial depending on the scale and geometry of the surface topography and on the type of cell. Although the surface topographical features that elicit the optimal response of a given cell type are difficult to determine or predict, the features that have the most significant effect on cell response are on the scale of the cell microenvironment, normally smaller than the cell length (see Figure 5.19). It is thought that topographical features, such as roughness, larger than the cell length will generally be perceived as being smooth by the cell. Certain topographical features at the microscale, as noted earlier, are known to improve the outcomes associated with some implanted biomaterials. The addition of certain nanoscale surface features can further improve the positive outcomes seen with microscale topography. Cell response is dependent not only on the scale of the surface topography but also on its geometry. For example, cells show a different response to a biomaterial having a spiky surface when compared to a biomaterial with a smoothly undulating surface, even though the two surfaces may have the same average roughness parameter R_a (Section 5.5).

Different cells respond differently to the same surface topography. For example, osteoblasts (bone cells) proliferate better on rough surfaces when compared to smooth surfaces, whereas fibroblasts, the most common cell type found in connective tissue, proliferate better on smooth surfaces. This has been clearly demonstrated in experiments in which rat calvarial osteoblasts and human gingival fibroblasts were cultured on titanium-coated epoxy resin discs, each having a controlled gradient in roughness over a length of ~10 mm. The measured R_a values of the disc surface varied continuously from ~5.7 μm at one end to ~1.1 μm at the other end. After seven days of incubation, the discs with attached cells were rinsed and stained for fluorescent assessment of cell density (cell number per unit area) over the length of the roughness gradient. The osteoblasts showed a preference for the rougher part of the surface gradient, as indicated by the osteoblast cell density at the rough end being approximately twice that at the smooth end of the gradient (Figure 21.1). In comparison, the fibroblasts showed an opposite trend, in which the cell density at the smooth end was approximately twice that at the rough end.

Cell morphology is seen to change continuously along the roughness gradient (Figure 21.2). Near the smooth end of the gradient, the individual osteoblasts spread well and adopt an approximately round shape with a distinctive actin network. In comparison, near the rough end, there is less spreading, with cells showing a smaller size and an approximately cuboidal shape. The fibroblasts show trends in morphology that are similar in nature to those of the osteoblasts. Near the smooth end of the gradient, the fibroblasts are also well spread, adopting an approximately round shape with a distinctive actin network. In comparison, near the rough end, they spread less, adopting an elongated shape with long filamentous extensions.

Surface topographical features of real biomaterials, such as Ti and Ti6Al4V implants, produced by techniques such as polishing, machining, grit blasting, chemical etching, plasma spraying, or some combination of these methods, normally have a random arrangement. They generally have an R_a value that varies from ~0.2 μm for polished or machined surfaces to ~3.5 μm for grit blasted surfaces, while surfaces produced by a combination of grit blasting and chemical etching, or by plasma spraying may have an R_a value of ~5 μm. On the other hand, the use of more controllable methods, such as electrochemical and lithographical techniques, allows the deliberate creation of surface topographical features such as grooves, pillars, pores, and lines of controllable geometry and

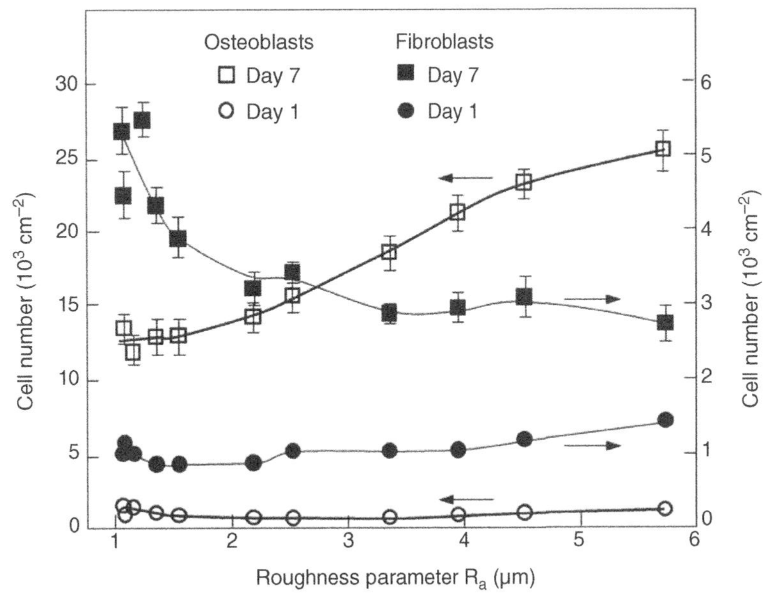

Figure 21.1 Number of rat calvarial osteoblasts and human gingival fibroblasts at different positions on the roughness gradient of titanium-coated epoxy discs at day 1 and day 7 of culture. *Source:* Kunzler et al. (2007) / with permission of Elsevier.

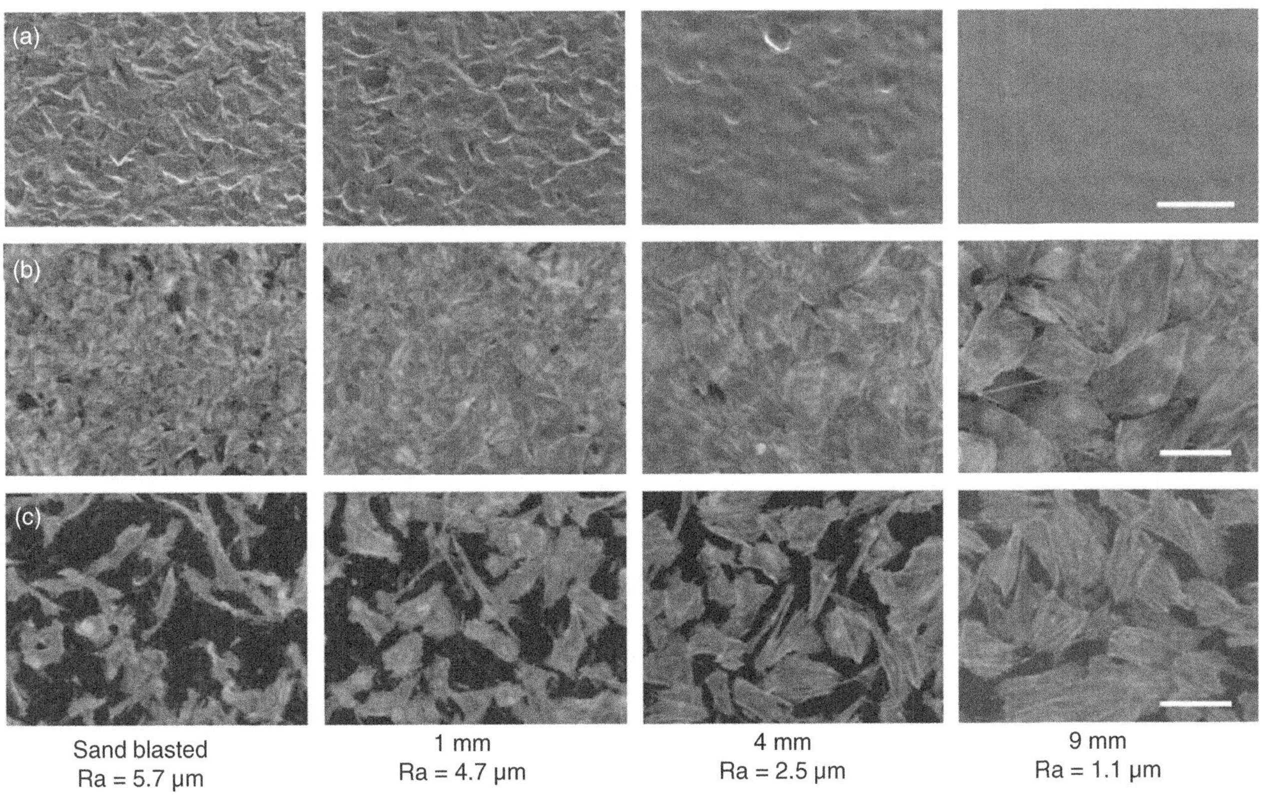

Figure 21.2 (a) SEM images of different positions (sand blasted, 1, 4, and 9 mm) and their corresponding roughness value R_a on the roughness gradient of titanium-coated epoxy discs. Morphology of rat calvarial osteoblasts (b) and human gingival fibroblasts (c) at different positions on the roughness gradient at day 7 of culture. The nucleus was stained with DAPI and the actin network with Alexa Fluor 488 phalloidin. Scale bar = 200 μm. *Source:* From Kunzler et al. (2007).

spacing, and an ordered surface texture (Chapter 5). These methods also allow control of the geometry and spacing of the topographical feature from the microscale (microtopography) down to the nanoscale (nanotopography). Much information for understanding cell response to surface topography has been achieved from studies using these textured surfaces. These studies show that specific topographical features can influence various cellular activities, such as cell adhesion, migration, proliferation, gene expression, differentiation, and apoptosis. For example, a micro-patterned surface topography can guide the alignment and behavior of certain cells such as macrophages, osteoblasts, and neural cells.

Examples of the role that surface topography can play in modifying cell and tissue response in areas such as osteogenesis, neurogenesis, and macrophage differentiation are discussed in the following sections.

21.2.1 Microscale Surface Topography in Osseointegration

Certain surface roughness features at a scale smaller than the dimension of a cell, as noted earlier, can improve the osseointegration of implants, such as Ti and Ti6Al4V implants. A relevant question relates to why such surface roughness features of a biomaterial should influence osteogenesis because this is not self-evident. Although a clear explanation is not available, surface roughness at a microscale is known to be a key requirement for osteogenesis during the natural process of endochondral bone formation (Chambers et al. 1984). The surfaces onto which osteoblasts migrate and deposit a mineralized layer of newly formed bone often contain microscale cavities resulting from the bone degradative activity of osteoclasts, large multinucleated cells responsible for bone remodeling and preparation of the bone bed for mineralization. The acid phosphatase and other degradative enzymes secreted by osteoclasts produce concavities of diameter ~30–100 μm that have a rough surface. These microscale cavities and other surface roughness features are thought to provide microenvironmental cues required for initiation of bone formation by osteoblasts.

Several studies have shown the effectiveness of surface roughening for promoting osseointegration of Ti and Ti6Al4V implants in animals. For example, microcomputed tomography scans of Ti6Al4V pedicle screws implanted for 12 weeks in the lumbar vertebrae of sheep have shown a greater amount of bone-implant contact around Ti6Al4V screws roughened by grit blasting (~74%) when compared smooth (machined) screws (~60%) (Figure 21.3). Additionally, less fibrous tissue formed around the grit-blasted screws, and the torque needed to remove the rough screws from the bone (5.3 ± 0.4 N m) was more than twice that required to remove the smooth screws (2.3 ± 0.3 N m).

The addition of nanoscale surface features to smooth (machined) surfaces using surface modification techniques has been shown to enhance the beneficial osteogenic effects observed for microrough surfaces. In one study, vertical TiO_2 nanotubes (outer and inner diameter ~100 and ~80 nm, respectively, and height ~250 nm) were formed on the surface of commercial purity Ti substrates by an electrochemical anodization process (Figure 21.4a). Then the bone-bonding capacity of these surfaces was compared with that for similar Ti substrates having a microrough surface (~6 μm diameter by ~2 μm deep) produced by grit-blasting with Ti particles. After implantation for four weeks in rabbit tibiae, the Ti specimens with the nanotubular

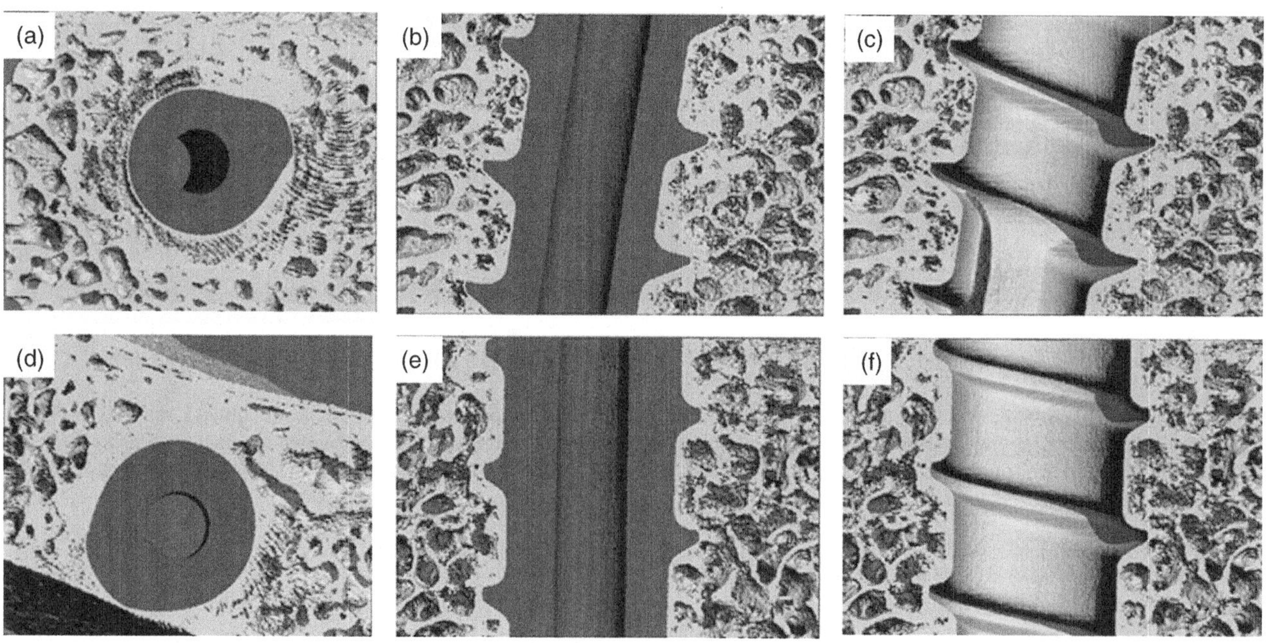

Figure 21.3 Microcomputed tomography scans of machined (a–c) and grit-blasted (d–f) Ti6Al4V pedicle screws in the L4 vertebral bone of sheep. *Source:* From Schwartz et al. (2008).

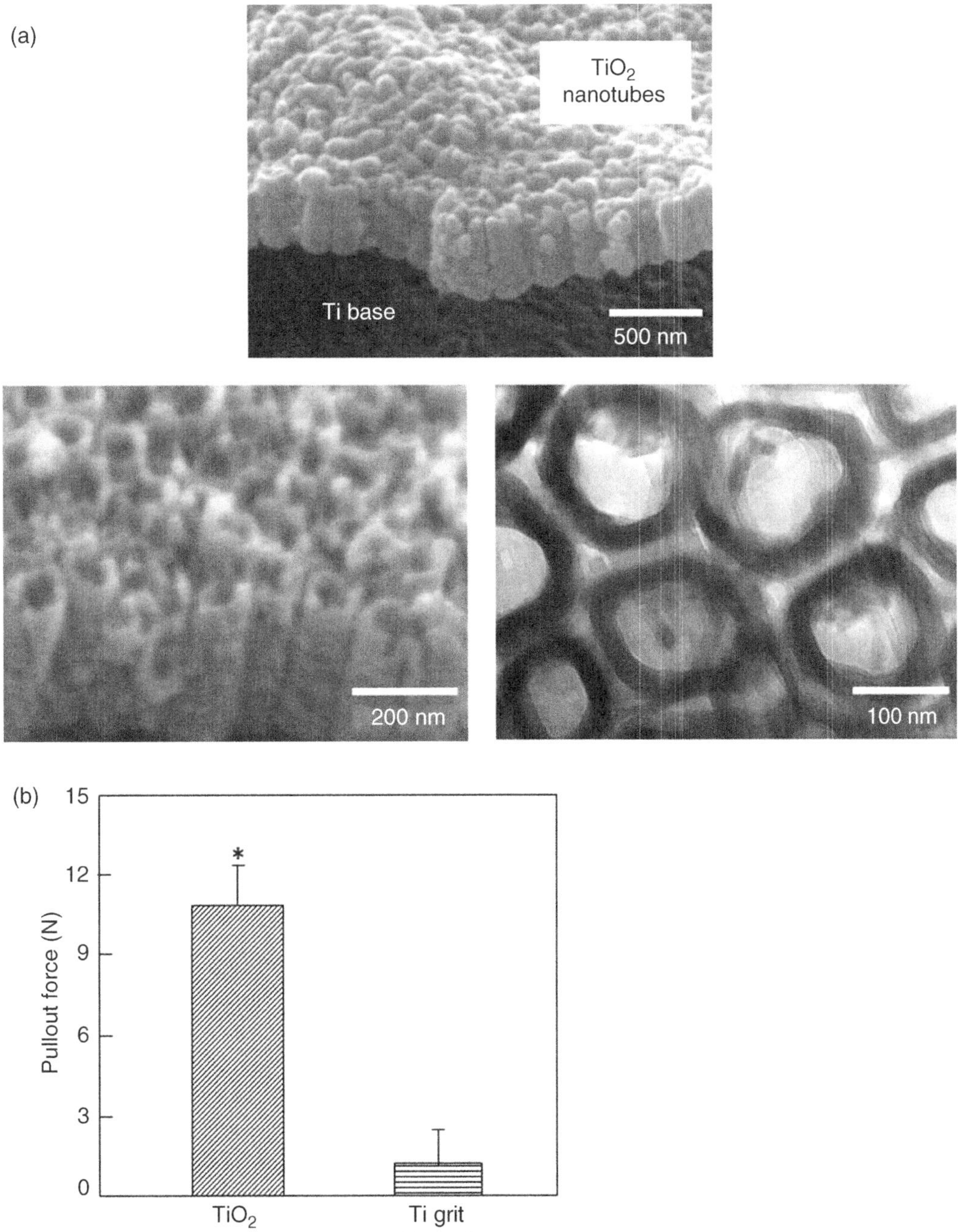

Figure 21.4 (a) Electron microscope images at different magnification showing an array of titanium dioxide (TiO_2) nanotubes deposited on the surface of a titanium (Ti) implant. *Source:* From Oh et al. (2005). (b) Pullout force for Ti specimens with a TiO_2 nanotubular surface and with a grit-blasted surface implanted for four weeks in rabbit tibiae. *Source:* Bjursten et al. (2010) / with permission of John Wiley & Sons.

surface showed a significantly greater amount of bone-implant contact area (~80%) when compared to the micro-rough specimens (~20%). In pullout tests used to measure the strength of the interfacial bonding between the implanted specimens and the surrounding bone, specimens with the nanotubular surface required a significantly higher pullout force (~11 N) when compared to the specimens with the microrough surface (~1 N) (Figure 21.4b).

Overall, there is clear evidence that osteogenesis is improved by the introduction of microscale and nanoscale surface roughness to a smooth surface, particularly for Ti and Ti6Al4V. However, as noted in Chapter 5, for the same material, the introduction of surface roughness can lead to changes in other surface characteristics, such as surface chemistry and surface energy (contact angle and wettability) and not just in the surface topography. This creates an additional level of complexity in understanding the mechanism by which surface topography influences the response of cells and in designing the optimal surface topography for a specific application.

Much of what is known about bone-implant bonding has been obtained from *in vitro* studies with cultured osteoblasts and osteoblastic progenitor cells. These studies have been especially useful in elucidating the initial events that occur when mesenchymal stem cells (MSCs) and progenitor cells colonize an implantation site and respond to microenvironment cues that initiate osteoid deposition and mineralization. In one series of studies, human osteoblast-like (MG63) cells were cultured on Ti substrates having cavities of controlled diameter (10, 30, and 100 μm) created by electrochemical micromachining (EMM), or having a microrough surface composed of pits and peaks (diameter ~1–3 μm) formed by acid etching.

After five days of incubation, examination in the scanning electron microscope (SEM) showed that the cells were anchored within the 30 and 100 μm diameter cavities rather than on the flat surface between the cavities. Enumeration of cells released by trypsinization showed a higher number of cells on the substrates having cavities of these two diameters. These two observations have biological relevance in light of the fact that the average diameter of the human osteoblast is ~20 μm. Assessment of cell response, by quantifying the expression of alkaline phosphatase (ALP), osteocalcin, and other markers of osteoblast differentiation, showed greater expression of these markers on the microrough Ti substrates when compared to the smooth substrates. Overall, these results indicate that the microrough Ti substrates are more favorable for osteoblast differentiation than those with the smooth surface.

Enhancement of osteoblast differentiation of a similar nature has also been observed for osteoprogenitor cells cultured on other biomaterials having a microrough surface. For example, in one study, osteoprogenitor cells derived from fetal rat calvariae were seeded on tissue culture polystyrene (PS) cover strips roughened along a given direction using different grades of grinding paper coated with silicon carbide (SiC) particles to form grooves of average surface roughness R_a of ~0.4, 0.8, 1.7, or 2.9 μm (Figure 21.5a). After incubation for 10 days, cells seeded on the PS strips having grooves with R_a ~ 0.8 μm showed

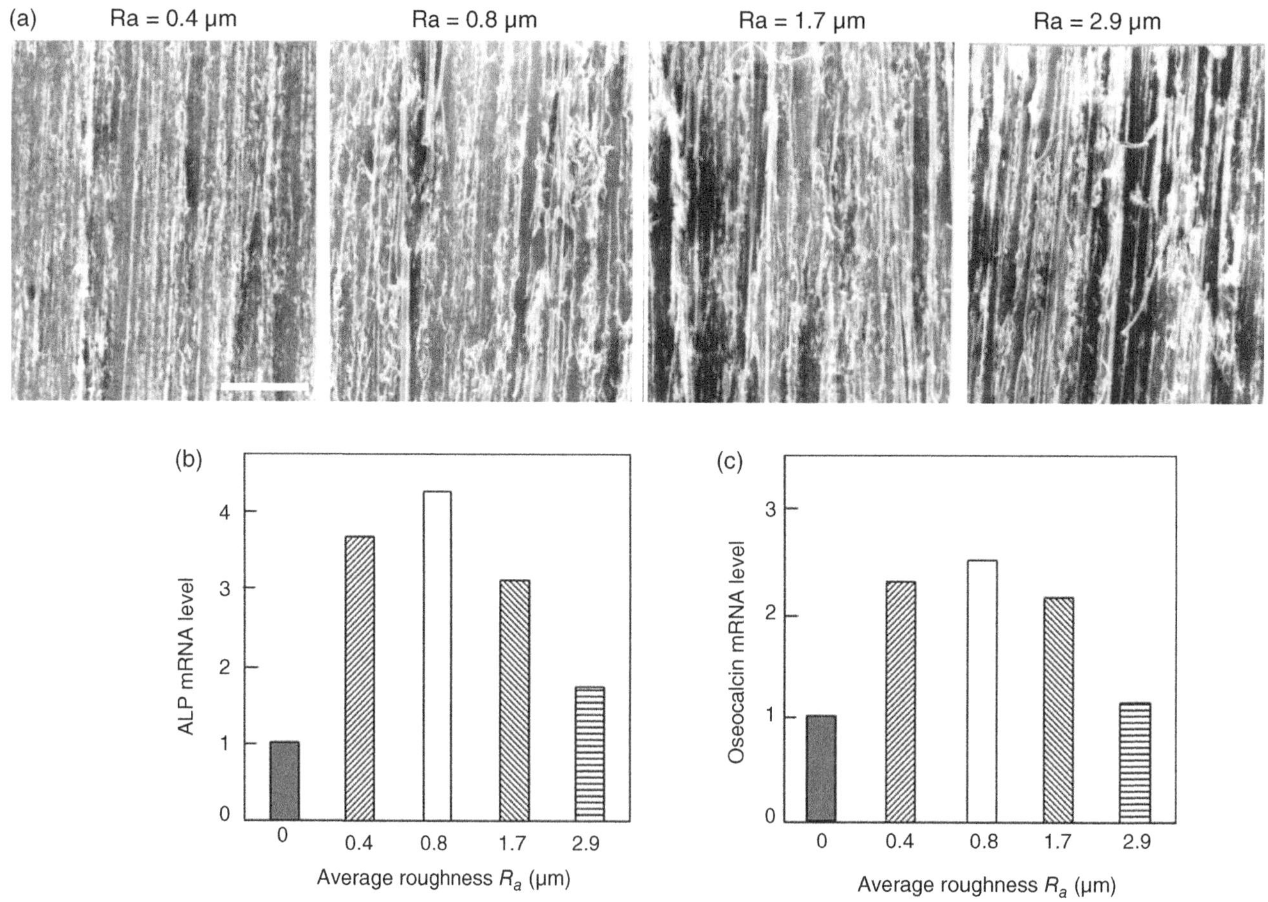

Figure 21.5 (a) SEM images of polystyrene (PS) cover strips of various surface roughness R_a = 0.4, 0.8, 1.7, and 2.9 μm (Bar = 100 μm). (b, c) Effect of surface roughness on alkaline phosphatase (ALP) mRNA and osteocalcin mRNA expression on day 10 of culture with fetal rat calvariae osteoprogenitor cells. *Source:* From Hatano et al. (1999) / with permission of Elsevier.

ALP messenger ribonucleic acid (mRNA) and osteocalcin mRNA expressions that were ~3.5 and ~2.5 times higher, respectively, when compared to cells seeded on smooth PS strips, denoted by $R_a = 0$ (Figure 21.5b).

21.2.2 Microscale and Nanoscale Patterned Surfaces in Macrophage Differentiation

The creation of controlled microscale and nanoscale topographical features on the surface of a biomaterial can be used to guide the differentiation of immune defense cells toward phenotypes that are more favorable for wound healing and less likely to promote inflammatory responses. This is particularly applicable to macrophages (immune defense cells), which respond to microenvironmental cues that influence their differentiation toward the M2 macrophage phenotype known to be favorable for wound healing (Chapter 18).

In vitro studies have shown that the morphology of bone marrow-derived macrophages (BMDMs) can be modulated by culturing them on micropatterned substrates. In one study (McWhorter et al. 2013), the surfaces of polydimethylsiloxane (PDMS) substrates were printed with fibronectin lines, 20 or 50 μm in width with a separation of 20 μm, and coated with Pluronic 127, a nonionic surfactant that blocks cell adhesion (Chapter 11). BMDMs cultured on the micropatterned surfaces with 20 μm fibronectin lines adopted an elongated morphology characteristic of the less inflammatory pro-healing (M2) phenotype, whereas the cells on the surfaces with 50 μm lines were only partially elongated. In comparison, BMDMs cultured on non-patterned surfaces adopted a pancake-like morphology characteristic of the pro-inflammatory (M1) phenotype. Additionally, BMDMs seeded on the micropatterned surfaces with 20 μm lines and 50 μm lines showed an increase in the expression of arginase-1, an established M2 phenotype marker, which was respectively ~3.8 and 1.3 times that of the non-patterned surfaces. In comparison, BMDMs cultured on the micropatterned surfaces and the non-patterned surfaces expressed similar low levels of inducible nitric oxide synthase (iNOS), an M1 phenotype maker.

Similar guidance of macrophage morphology has been achieved by a pattern of fine grooves (150 nm to 50 μm wide) on Ti plates, created using lithographic and ion etching techniques, to obtain cues that influence differentiation via surface topography without relying on chemical signaling. BMDMs cultured on such surfaces were also found to undergo dramatic changes in morphology and gene expression, which were dependent on the width of the grooves rather than the presence of fibronectin. These cells also adopted an elongated morphology characteristic of the anti-inflammatory M2 phenotype macrophages, whereas BMDMs cultured on smooth Ti surfaces showed a pancake-like morphology characteristics of pro-inflammatory M1 phenotype macrophages (Figure 21.6a). The greatest extent of cell elongation occurred with BMDMs cultured on Ti surfaces having grooves of width ~450 nm. Additionally, immunohistochemical techniques showed that BMDMs seeded on grooved Ti surfaces exhibited increased expression of the M2 phenotype marker arginase-1 (Figure 21.6b). In comparison, the cells seeded on smooth Ti plates expressed the M1 phenotype marker iNOS. These findings are significant in the sense that they illustrate the potential utility of intentional patterning of Ti and other biomaterials to promote wound healing and avoid microenvironmental cues that promote inflammation.

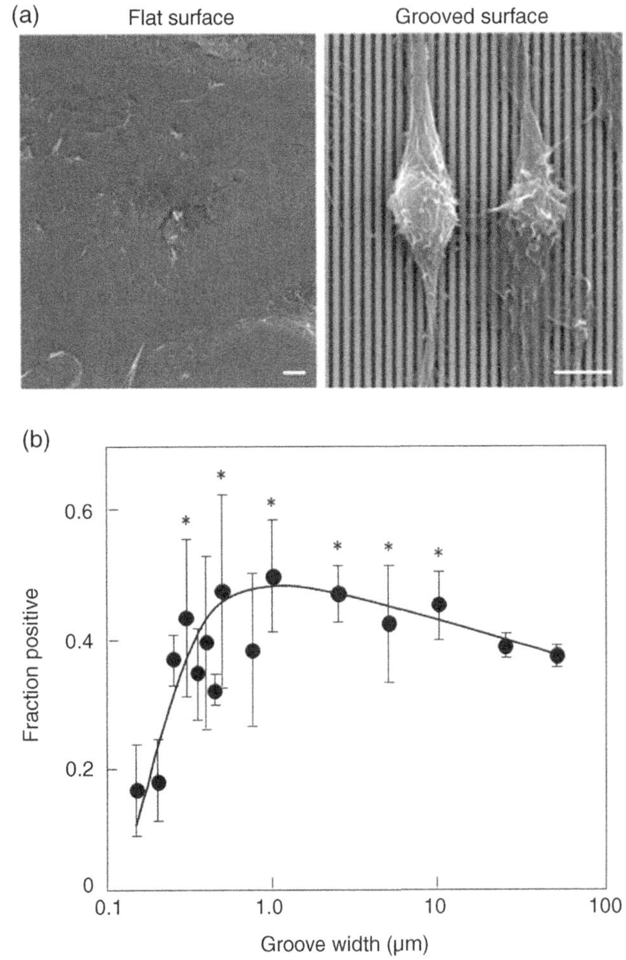

Figure 21.6 Effects of microgroove patterning on phenotype expression of bone marrow-derived macrophages (BMDMs) cultured on titanium (Ti) having a grooved surface of width varying from ~150 nm to ~50 μm. (a) BMDMs on smooth Ti surface exhibit a flat morphology of M1 macrophages, whereas those on grooved substrates show pro-healing M2 phenotype. (b) Expression of arginase-1 (Arg1) characteristic of the M2 phenotype occurs in a manner corresponding to groove width. *Source:* From Luu et al. (2015) / with permission of American Chemical Society.

21.2.3 Microscale Patterned Surfaces in Neural Regeneration

Another area in which the introduction of micropatterning on the surface of a biomaterial may have application is in the construction of nerve guidance conduits (NGCs) for the repair of peripheral nerve injuries. Severe injuries causing segmental defects in peripheral nerves often result in long-term or permanent loss of muscle function, loss of sensation, and debilitation. Currently, the only effective approach for repairing peripheral nerve defects involves the use of nerve autografts, but this approach suffers from limited supply and adverse effects on the body areas innervated by the donor nerve segment.

The development of synthetic NGCs for guided differentiation of neural progenitor cells could potentially provide an effective alternative for repairing peripheral nerve defects in humans. While a variety of designs has been created from several degradable polymers, testing of NGCs for their ability to heal nerve defects has so far been performed only in laboratory studies using animal models and *in vitro* cell culture models. For example, in a recent study, porous tubular NGCs composed of polylactic-*co*-glycolic acid (PLGA), whose inner surface consisted of uniform longitudinal grooves (depth and spacing ~5 μm), were created by a combination of salt leaching and lithographic methods (Figure 21.7a). In order to improve cell adhesion, the inner micropatterned surface of some specimens was coated

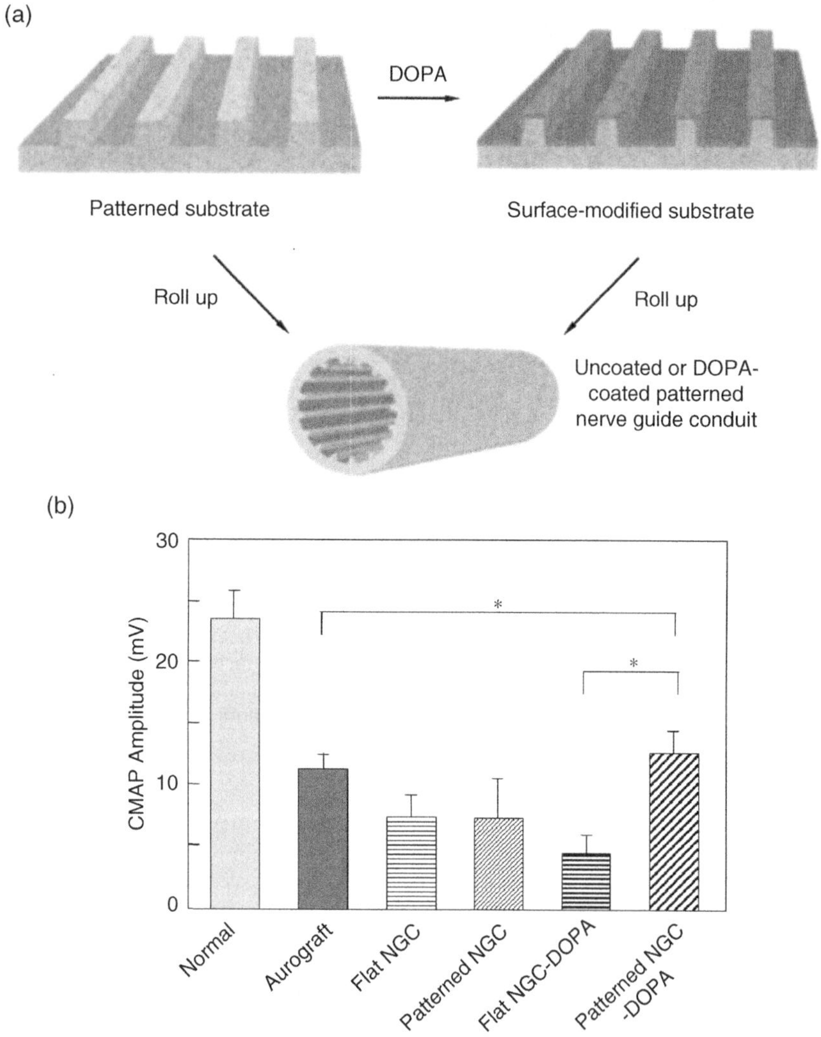

Figure 21.7 (a) Schematic illustration of the creation of patterned PLGA nerve guidance conduit (NGC) with or without DOPA surface modification for sciatic nerve regeneration. (b) Comparison of the electrophysical recordings of the compound muscle action potential (CMAP) obtained from sciatic nerve stimulation used to assess the effectiveness of various NGCs to repair 10 mm long sciatic nerve defects in rats at four weeks postimplantation. CMAP amplitudes for the normal sciatic nerve and defects implanted with autografts are shown for comparison (*significant difference between groups; $p < 0.05$). *Source:* From Kim et al. (2018) / with permission of John Wiley & Sons.

with dihydroxy-L-phenylalanine (DOPA), a substance responsible for the unique adhesion properties of mussels. Then specimens of these DOPA-coated micropatterned NGCs, along with micropatterned NGCs without the DOPA coating, DOPA-coated non-patterned NGCs (flat inner surface), and autografts were implanted in sciatic nerve segmental defects created in adult male rats by surgically excising a 10 mm long segment of the left sciatic nerve near the left hip joint. Electrophysical recordings of the compound muscle action potential (CMAP) obtained from sciatic nerve stimulation were used to assess the effectiveness of the NGCs to repair the nerve defects at four weeks postimplantation. These measurements showed more extensive repair of the defects implanted with micropatterned DOPA-coated NGCs (CMAP ~55% of the value for the normal sciatic nerve used as a control) when compared to the DOPA-coated NGCs with a flat surface (CMAP ~20%) (Figure 21.7b). The micropatterned DOPA-coated NGCs also showed better repair of the defect when compared to the autografts. In comparison, the CMAP for the nerve defects implanted with the micropatterned NGCs without the DOPA coating was ~33% of the value for the normal sciatic nerve, showing the beneficial effect of the micropatterned surface alone.

Immunohistochemical analyses of the nerve defects at four weeks postimplantation provided additional evidence for the effectiveness of the micropatterned NGCs in promoting repair of the nerve defects. High positive signals for the expression of neurofilament markers were detected through the entire length of the nerve defects containing micropatterned DOPA-coated NGCs. In comparison, a much weaker signal for expression of neurofilament markers was detected through much of the length of the nerve defects containing the non-patterned DOPA-coated NGCs.

21.3 Biomaterial Surface Topography for Antimicrobial Activity

Bacterial infection of implanted biomaterials, as noted earlier, is a serious complication, particularly for devices such as intravascular and urinary catheters and for orthopedic devices used in long-term implantation, such as hip, knee, and spinal implants. Bacterial cells can generally adhere and grow on an implant surface over time, producing biofilms that contain tightly attached bacteria in complex communities that resist antimicrobial agents and host defenses (Chapter 20). Persistent antibiotic therapy is generally used to treat infection, but failure of this treatment to cure the infection often necessitates surgical intervention, which can be traumatic.

Since the formation of antibiotic-resistant biofilms can lessen the effectiveness of antibiotic therapy, other approaches based on the creation of antimicrobial implant surfaces have been studied for their effectiveness in treating infection. One approach has been to coat or functionalize the implant surface with a substance known to kill or repel the target bacteria, such as silver or hydrophobic polymers containing cationic functional groups. This approach relies on a chemical mechanism for its bactericidal effect.

Another approach, which has gained significant attention in recent years for inhibiting bacterial adhesion and certain types of biofilm formation, relies on biomaterials having a controlled surface topography. This approach relies on a purely physical mechanism for its bactericidal effect. It often involves the creation of a special surface topography to mimic the topography of various animal species that show bactericidal effects. The topographical features involved range from the microscale down to the nanoscale. A few examples of this approach based on surface topography are discussed in the following sections.

21.3.1 Microscale Topography with Antimicrobial Activity

The surface of shark skin provides an example of a natural textured surface that shows antibacterial activity. It has long been known that shark skin resists the attachment of aquatic organisms such as algae and barnacles, a property that reduces drag and keeps the surface clean. A distinctive feature of shark skin is the presence of microscale dermal denticles arranged as overlapping diamond-shaped plates (Figure 21.8a). The surface of each dermal denticle has a spatial patterning of parallel ridges and microscale grooves spaced ~30–100 μm apart. These microscale grooves greatly reduce bacterial attachment, which is the basis of the antibacterial effect.

The surfaces of synthetic nondegradable polymers such as polymethyl methacrylate (PMMA) and PDMS have been engineered using lithographic techniques to mimic the natural textured surface of shark skin, in order to test their potential to resist bacterial attachment and biofilm formation-based solely on their surface topography. One surface pattern, referred to as Sharklet AF™, consists of pillars of various height and spacing designed to deter the attachment of specific organisms. For example, a pattern designed to resist biofilm formation of *Staphylococcus aureus* (*S. aureus*) (diameter ~1 μm), the most common organism responsible for bacterial infection of implanted orthopedic devices, consists of features ~2 μm in width and spacing, 4–16 μm in length, and ~3 μm in height (Figure 21.8b). After inoculation *in vitro* with broth cultures of *S. aureus*, PDMS sheets having the Sharklet AF

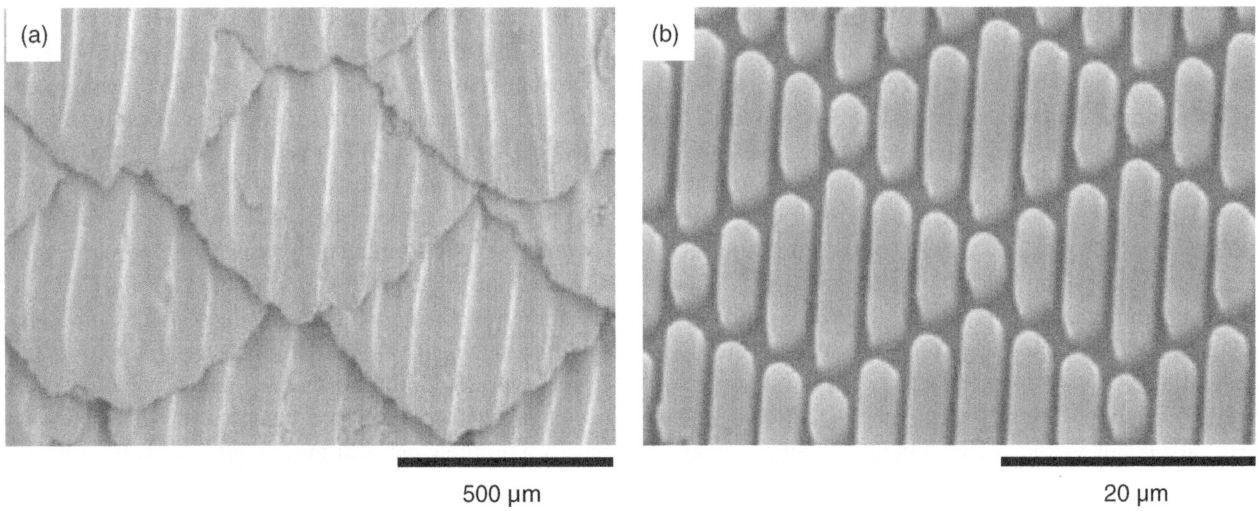

Figure 21.8 Scanning electron microscope images of (a) spinner shark skin and (b) polydimethylsiloxane (PDMS) sheet with the Sharklet AF surface topography. *Source:* From Magin et al. (2010).

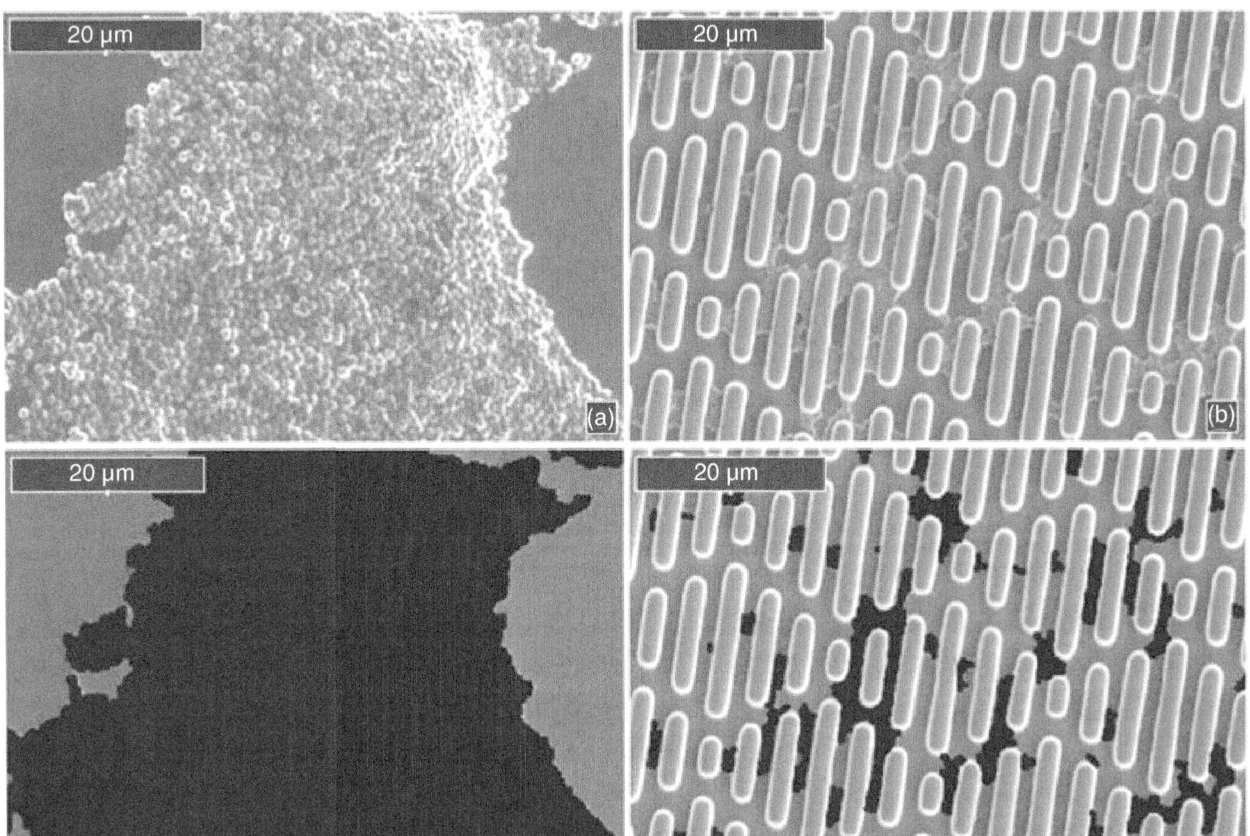

Figure 21.9 SEM images of *Staphylococcus aureus* (*S. aureus*) on polydimethylsiloxane (PDMS) sheets with (a) a smooth surface and (b) with the Sharklet AF surface topography after incubation for seven days. Bacteria covered areas were blackened to compare relative bacterial densities (c, d). *Source:* From Chung et al. (2007).

microtopography showed a significantly better capacity to resist bacterial growth on their surfaces when compared to smooth sheets (Figure 21.9). The average bacterial coverage of the Sharklet AF and smooth surfaces were 35 and 77%, respectively, at 21 days of culture. It is only after 21 days that multilayered bacterial colonies grew above and over the protruding ridges, an indication that the biofilm had expanded. Despite this eventual appearance of the biofilm, the delay in its appearance represents a significant improvement in the control of biofilm formation.

The Sharklet AF surface appears to resist bacterial adhesion, thereby delaying biofilm formation and growth significantly. The mechanism by which it does so is not clear but one suggestion is that it is related, in part, to a change in wettability (contact angle) of the Sharklet AF surface when compared to a smooth surface. It is noted that the topographical features of the Sharklet AF surface are larger than the diameter of *S. aureus* cells. The water contact angle of a smooth PDMS is 115–120°, and the introduction of topographical features can make the surface more hydrophobic (higher contact angle) unless there is heterogeneous wetting (Chapter 5).

Related studies have evaluated the effectiveness of the Sharklet AF surface in limiting bacterial migration through PDMS tubing. This has been done by measuring the incidence of migration of a highly motile uropathogenic strain of *Escherichia coli* (*E. coli*), a strain responsible for urinary tract infection, through PDMS tubing (1 cm long) having the Sharklet AF surface or a smooth surface, positioned as a bridge between two agar blocks. Two variations of the Sharklet AF surface pattern were tested, consisting of pillars oriented parallel or perpendicular to the long axis of the tube. In these experiments, aliquots of log phase *E. coli* (that is, the *E. coli* from a population growing in a logarithmic manner with time) were used to inoculate one agar block, followed by incubation for 24 hours. Then an examination for growth of bacterial colonies was performed on the agar block on the other side of the bridge. The results of these experiments showed that the Sharklet AF pattern significantly inhibited migration of *E. coli*. Tubing segments with the transverse and parallel orientation of the pillars reduced the incidence of bacterial migration by 81% and 50%, respectively, when compared to the smooth tubing. This type of surface topography could potentially be incorporated into the design of urinary catheters to reduce the incidence of catheter-associated urinary tract infections that seriously affect patient recovery and quality of life.

Patterned surfaces with finer topological features, at a scale of a fraction of a micrometer, referred to as a submicron scale to distinguish it from the nanoscale regime, also show antibacterial activity. It is noted that this submicron scale is smaller than the size of bacteria commonly associated with infections of implanted cardiovascular and orthopedic devices. As an example, polyurethane (PU) specimens having a surface pattern consisting of a square array of circular pillars at two different scales were created using a lithographic technique. One pattern, designated 400/400, consisted of pillars of diameter ~400 nm and a spacing of ~400 nm along each row, whereas the other, designated 500/500, consisted of ~500 nm diameter pillars of spacing ~500 nm along each row (Figure 21.10). Thin sheets having these two surface patterns were tested separately under dynamic flow conditions for their effectiveness in preventing bacterial adhesion and biofilm formation. The bacteria used in these tests were *S. aureus* and *Staphylococcus epidermidis* (*S. epidermidis*), two gram-positive bacteria responsible for some of the most commonly encountered infections of implanted cardiovascular and orthopedic devices. PU specimens were mounted on a steel disc, lowered into a suspension of *S. aureus* or *S. epidermidis*, and rotated at a controlled speed to generate shear stresses in the range 0–1.3 Pa across the film surface (Figure 21.11a). This range of shear stresses was intended to simulate the lower part of the physiological range in blood vessels (0 to ~5 Pa).

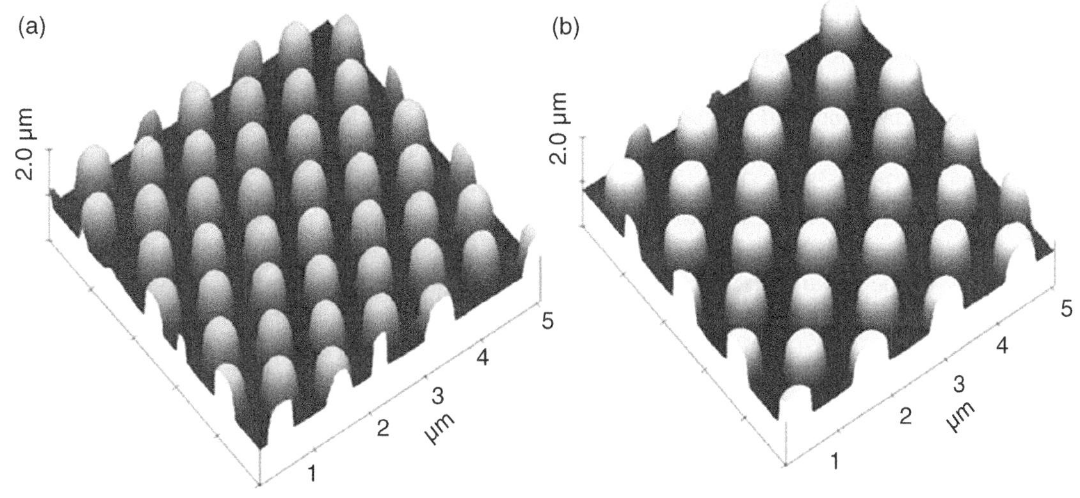

Figure 21.10 Atomic force microscope images of polyurethane specimens with a surface topography consisting of (a) 400 nm and (b) 500 nm pillars. *Source:* From Xu and Siedlecki (2012) / with permission of Elsevier.

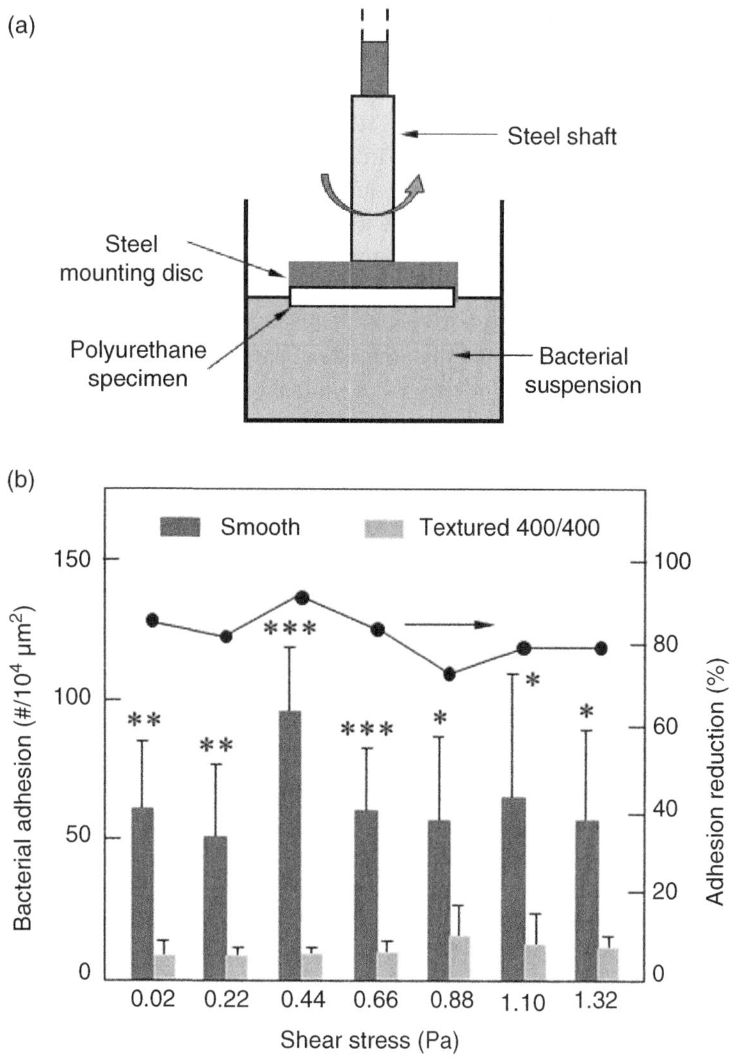

Figure 21.11 (a) Illustration of rotating disc system used for testing of bacterial attachment and biofilm formation. (b) Bacterial adhesion on smooth and textured (400/400) polyurethane specimens after exposure to suspensions of *S. epidermidis* in phosphate-buffered saline for one hour at different shear stresses. *Source:* Xu and Siedlecki (2012) / with permission of Elsevier.

Both PU surface patterns were effective in reducing *S. aureus* or *S. epidermidis* adhesion when compared to smooth PU films. For example, films with the 400/400 patterns reduced *S. epidermidis* adhesion at one hour by an average of ~80% over the entire stress range (Figure 21.1b). In comparison, the reduction in adhesion was lower for *S. aureus*, being ~40% at near static conditions and increasing to ~60% at higher stresses. Additionally, fluorescence imaging showed that the patterned PU surfaces strongly inhibited biofilm formation when compared to the smooth PU surface. High amounts of biofilm were present across the entire surface of the smooth PU surface at two or five days, even at the outer edge where the shear stress was highest. In comparison, only individual bacterial cells were present over most of the 400/400 patterned surface, and little biofilm formation was present even at the center of the disc where shear stress was close to static conditions. As the total surface area of the ends of the pillars is much lower than that of the flat surface, this inhibition of bacterial adhesion and biofilm formation may be related to the lower surface contact area that the patterned surfaces provide for bacterial attachment by the oval shaped staphylococcal microbe.

Silicon nitride (Si_3N_4), a ceramic biomaterial used in spinal repair (Chapter 7), shows not only strong antimicrobial activity but also attractive osseointegrative properties. This combination of properties has been attributed primarily to the unique microrough surface topography of Si_3N_4. This topography is created not by a deliberate process such as roughening or lithography but, instead, by a combination of the inherent crystal structure of Si_3N_4 and the processing conditions used to fabricate the Si_3N_4 implant. Implants for spinal repair typically have a topography consisting of elongated (acicular) grains, 0.2–2 μm in cross section, that protrude out randomly from the surface of the material to various lengths (less than ~5 μm) (Figure 5.25). The surface of these implants has a roughness parameter R_a of ~0.6 μm (as measured using white light interferometry), a composition corresponding to a silicon oxynitride, and a zeta potential of approximately −60 mV.

Several studies have shown the antimicrobial and osseointegrative properties of Si_3N_4 implants. In one study to assess the contribution of surface topography to these properties, *S. aureus* biofilm formation on two forms of Si_3N_4 implants was evaluated *in vitro*. One form had the normal surface microtopography of the as-fabricated material, while the other had a machined (smooth) surface of the same fabricated material. For comparison, machined specimens of Ti and polyether ether ketone (PEEK), two biomaterials also used in spinal repair, and stainless steel were also evaluated. The tests for antimicrobial activity using a methicillin-resistant strain of *S. aureus* (MRSA) involved incubation of the Si_3N_4 specimens for 20 minutes in an overnight broth culture of MRSA, followed by air drying for SEM examination, and vortex mixing in saline, with a 10-fold dilution for colony forming unit (CFU) assays.

Examination of the air-dried specimens in the SEM showed that large numbers of bacteria adhered to the Ti and PEEK (Figure 21.12), with smaller numbers on stainless steel. In comparison, while some bacteria adhered to the machined surface of Si_3N_4, there was none on the microrough as-fabricated surface. The CFU data validated these observations (Figure 21.13), showing a significant reduction in CFU for the as-fabricated Si_3N_4 when compared Ti, PEEK, and stainless steel. Additionally, there was a significant reduction in CFU for the as-fabricated Si_3N_4, by a factor ~3, when compared to the machined Si_3N_4. These results may be seen to show the critical importance of surface microtopography for the inherent antimicrobial properties of Si_3N_4 spinal implants. On the other hand, it is worth noting that the physicochemical properties of the as-fabricated Si_3N_4 surface are not identical to those of the machined surface. Machining the as-fabricated surface can modify its wetting behavior (contact angle) and the chemical composition of its surface oxide layer.

In order to assess the role of surface microtopography on the osseointegrative properties of Si_3N_4, the behavior of *S. aureus*, macrophages, MSCs, and osteoblasts on the surface of as-fabricated and machined forms of Si_3N_4 was studied in an established murine model *in vivo*. Tests involved insertion of sterile or MRSA-contaminated specimens of as-fabricated and machined Si_3N_4 into the tibiae of female Balb/c mice.

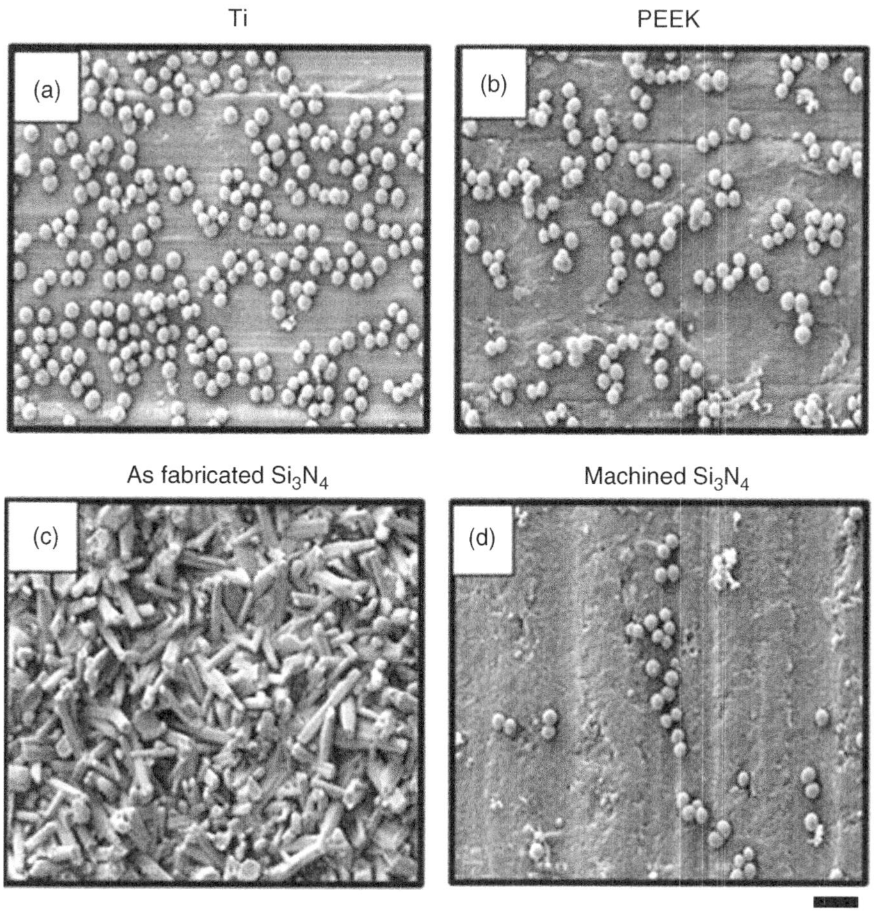

Figure 21.12 SEM images of *S. aureus* adherence to the surface of (a) titanium (Ti), (b) polyether ether ketone (PEEK), (c) as-fabricated silicon nitride (Si_3N_4), and (d) machined Si_3N_4, after overnight culture (Bar = 2 μm). *Source:* From Ishikawa et al. (2017).

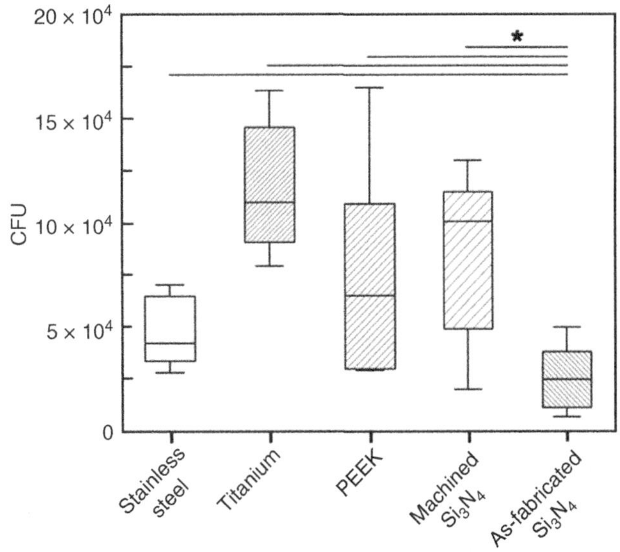

Figure 21.13 Colony forming unit (CFU) of *S. aureus* on the surface of stainless steel, titanium, polyether ether ketone (PEEK), machined silicon nitride (Si_3N_4), and as-fabricated Si_3N_4 after overnight culture. *Source:* Ishikawa et al. (2017) / with permission of John Wiley & Sons.

After euthanizing the animals at specific periods (up to 14 days postimplantation), the tibiae were surgically excised with the implants in place and examined in the SEM.

Images of the sterile as-fabricated Si_3N_4 implants showed evidence of a progressive transformation toward osseointegration as a function of time (Figure 21.14). For example, the image at day 1 showed the presence of leukocytes and red blood cells adhering to a fibrin coating on the surface. At day 3, cells with an oval shape morphologically consistent with migrating MSCs were observed. In comparison, the cells that are prominent on the surface at day 5 had a polygonal appearance consistent with the morphology of differentiating MSCs. The image at day 10 showed a nearly confluent layer of cells with a rounded morphology typical of osteoblasts. These observations suggest that the as-fabricated Si_3N_4 surface supports osteoblastic differentiation. In comparison, the machined surface was almost unreactive with the host cells and showed little change over the same period.

SEM examination of the MRSA-contaminated as-fabricated Si_3N_4 specimens at similar postimplantation times showed a progressive transformation from an MRSA-infected condition to an apparent pre-osseous interface over

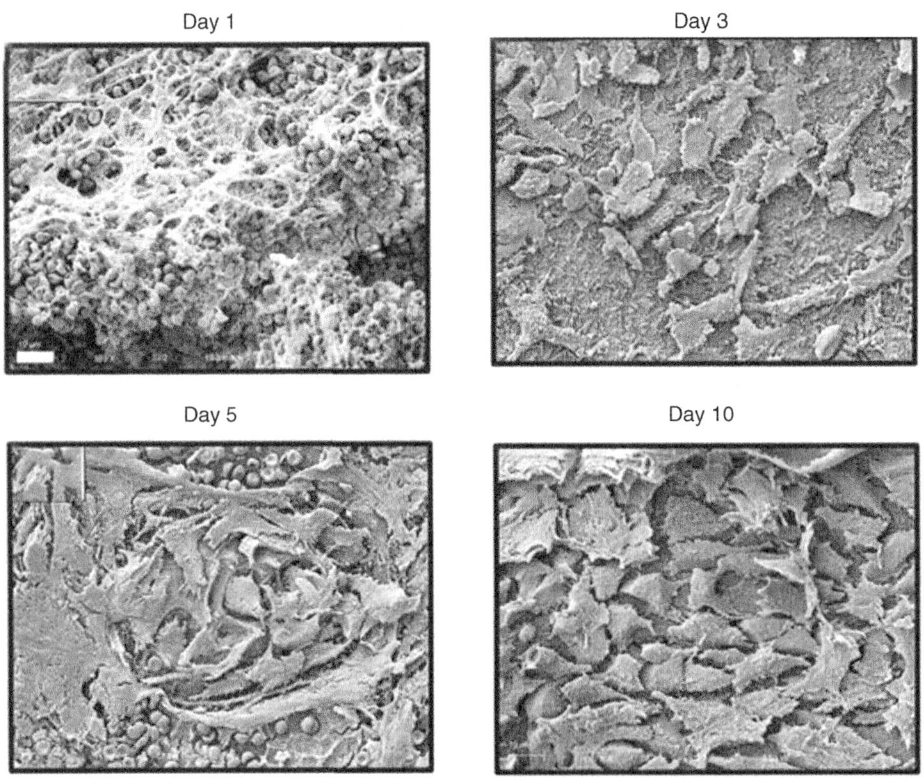

Figure 21.14 SEM images of the surface of as-fabricated Si_3N_4 specimens implanted in the tibiae of mice, showing progressive transformation toward osseous integration over time. Day 1: white and red blood cells on the implant surface; day 3: rounded or oval cells morphologically consistent with mesenchymal stem cells (MSCs); day 5: prominent cells on the implant surface are rounded to polygonal in appearance, phenotypically consistent with differentiating MSCs; day 10: osteoblasts organizing to form a dense coating (Bar = 2 μm). *Source:* From Ishikawa et al. (2017).

time. In comparison, the machined surface showed less activity toward both the bacteria and the host cells. The mechanism by which the machined surface repels the bacteria also appears to limit the adherence of host cells, thereby limiting osteogenesis. Overall, these findings provide clear evidence that as-fabricated Si_3N_4 implants with their unique surface microtopography have a clinically valuable combination of antibacterial and osseointegrative properties.

21.3.2 Nanoscale Topography with Antimicrobial Activity

Several recent studies have shown that surfaces with certain topographical features of dimensions within or near the nanoscale regime (less than ~50–100 nm) can offer new modalities for combatting bacterial infection. The antimicrobial activity of such surfaces depends primarily on their ability to kill bacterial cells that may adhere to them and not on their ability to repel bacterial cells, that is, such surfaces are bactericidal. Overall, studies show that the nanoscale topography of these surfaces, not their surface composition, is directly responsible for their bactericidal activity. Relevant questions relate to the types of nanoscale topographical features that have a significant influence on this bactericidal activity and their mechanisms of action. Overall, an understanding of the major topographical features that influence bactericidal activity and the mechanisms of bactericidal activity is important for designing biomaterials having an optimal nanoscale surface topography for the desired bactericidal activity.

Nanoscale Surface Parameters

Figure 21.15 illustrates the cross section of a simple model of a patterned surface, taken as consisting of conical pillar-like protrusions in a square array, which is used to define various geometrical parameters of a patterned surface. These parameters are the height h of the pillars, the diameter of the pillars at their base D and at their tip d, the aspect ratio of the pillars (h/D), and the distance of separation between adjacent pillars. This separation distance, normally taken along a row, is either the distance l between adjacent pillar surfaces or the center-to-center distance (pitch) P between adjacent pillars. Another useful parameter is the areal density of A_d of the pillars, that is, the number of pillars per unit area of the surface. For this square array, A_d is equal to $1/P^2$ or $1/(l+D)^2$.

In practice, the pillars may have a cylindrical shape rather than a conical shape, and they may have a nonuniform height or shape. For a surface with pillars of nonuniform height, the height of the pillars may be fitted to a distribution function, such as a sigmoidal distribution or a bimodal distribution. Additionally, the tips of the pillars may have various shapes, such as a hemisphere or a pointed shape.

Two natural topographies that show bactericidal activity against certain bacterial strains are the surface of the wings of the cicada and the dragonfly. Figure 21.16 shows the nanoscale topographical characteristics of the wings of one species of cicada, the dog day (DD) annual cicada (*Tibicen* ssp.) and one species of dragonfly, the common sanddragon dragonfly (*Progomphus* ssp.). These and other natural surfaces have been used as models in designing the surface topography of some synthetic biomaterials. The wing surface of the DD cicada consists of a nearly uniform array of hemispherical-capped conical pillars of nearly uniform height $h \sim 180$ nm and base diameter $D \sim 100$ nm, which tapers to a cap diameter $d \sim 60$ nm. The center-to-center distance P of the pillars is ~170 nm. In comparison, the wing surface of the sanddragon dragonfly consists of high aspect ratio hemispherical-capped cylindrical protrusions of varying height and non-uniform shape, which appear to be bundles of three to five smaller

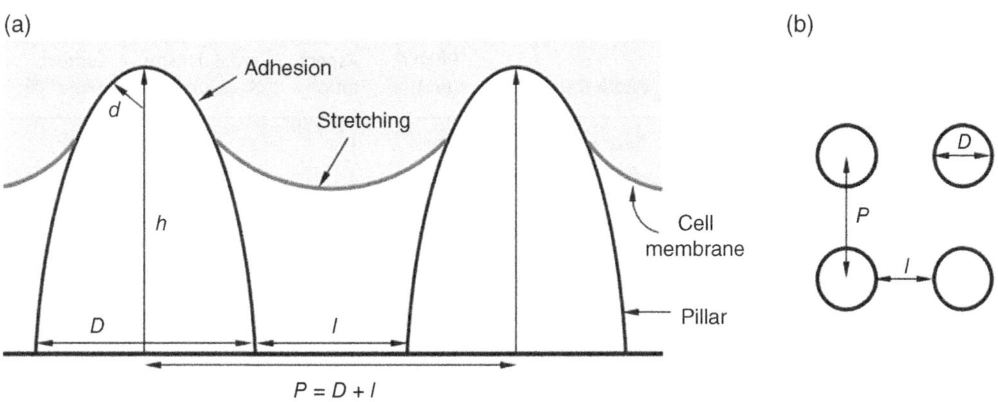

Figure 21.15 (a) Schematic illustration of the cross section of a bacterial cell membrane adsorbed on a patterned surface consisting of pillars, where h is the height, D is the base diameter, and d the tip diameter of the pillars. The pillars have a center-to-center spacing (pitch) P and distance l between adjacent surfaces along a row. (b) Pillars in a square arrangement, showing the parameters D, P, and l.

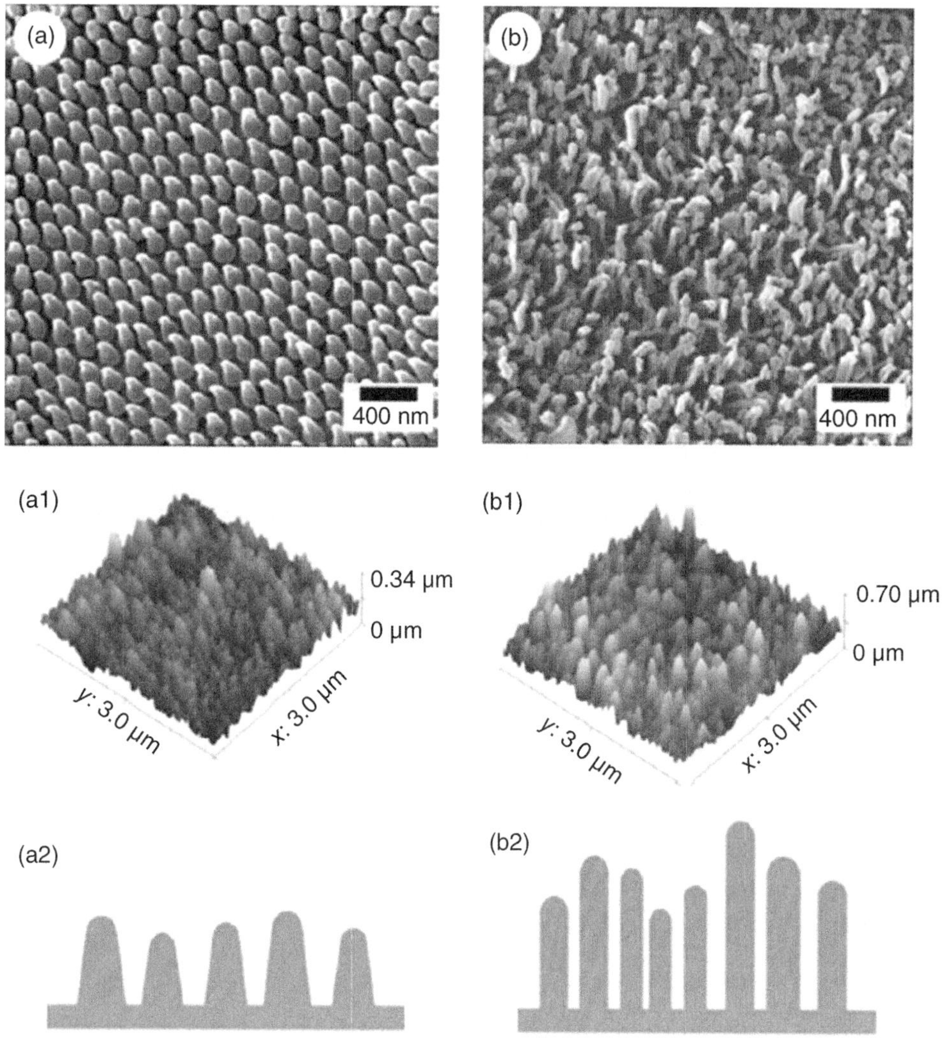

Figure 21.16 Topographical characteristics of the cicada (*Tibicen* ssp.) wing surface (a, a1, a2) and the dragonfly (*Progomphus* ssp.) wing surface (b, b1, b2). (a, b) Helium ion microscope images; (a1, b1) atomic force microscope (AFM) images; (a2, b2) edge-view illustration of topography. *Source:* From Nowlin et al. (2014) / with permission of Royal Society.

Table 21.1 Topographical characteristics of the cicada (*Tibicen* ssp.) and the dragonfly (*Progomphus* ssp.) wing surface.

Species	Pillar geometry	Height h (nm)	Width D (nm)	Pitch P (nm)	Aspect ratio[a]	Areal density A_d (1/μm^2)	Contact angle[b] (°)
Tibicen ssp.	Spherically capped cones	180	Base 100; cap 60	170	1.8	35	130
Progomphus ssp.	Spherically capped cylinders	240	50	120	4.8	70	120

[a] Based on the base diameter.
[b] With water.
Source: Nowlin et al. (2014) / with permission of Royal Society,

protrusions. The protrusions have an average height $h \sim 240$ nm, an average diameter $D \sim 50$ nm at their tip, and a center-to-center distance $P \sim 120$ nm. Table 21.1 summarizes the parameters of these two natural surfaces.

Mechanism of Bactericidal Activity

Because bactericidal action of certain nanoscale topographies is a more recently observed phenomenon, studies are still ongoing to elucidate the mechanisms involved. However, experimental studies have generally shown the following characteristics:

- The nanoscale topography is directly responsible for the bactericidal activity, not the surface chemical composition. For example, coating the natural surface of the cicada wing with a 10 nm film of gold (which produces a negligible topographical change) preserves its bactericidal activity despite the difference in chemical composition and change in measured water contact angle, from ~160° for the cicada wing surface to ~105° for the gold-coated surface.

- The bactericidal activity of a given nanoscale topography is dependent on the type of bacterial cell. For example, the aforementioned cicada wing topology is efficient in killing certain gram-negative bacteria such as *E. coli* and *Pseudomonas aeruginosa* (*P. aeruginosa*) but not gram-positive bacteria such as *S. aureus* and *S. epidermidis*.

Upon encountering the pillars of a nanoscale surface topography, bacterial cells can adhere to them. The cell wall (membrane) may be penetrated, thereby causing cell death, if the tips of the pillars are sharp enough, the cell wall is not robust enough, or a combination of the two. Alternatively, the cell wall can deform over the contours of the surface, causing the cell membrane to stretch. If the stretching is sufficiently large, it can cause the cell membrane to rupture, killing the cell, as exemplified by *P. aeruginosa* cells on the surface of a cicada wing in Figure 21.17. Some experimental studies show that for a given surface topography, bactericidal effectiveness increases with the height h of the pillars but only up to a certain point. Increasing h above a certain value appears to provide little additional benefit because the cell membrane is already stretched to a sufficient degree to cause rupture. On the other hand, upon decreasing h below this threshold value, the pillars eventually become sufficiently short such that they provide little stretching of the cell membrane. Consequently, depending on the biomaterial and the bacterial cell, there is a lower limit on h for effective bactericidal activity.

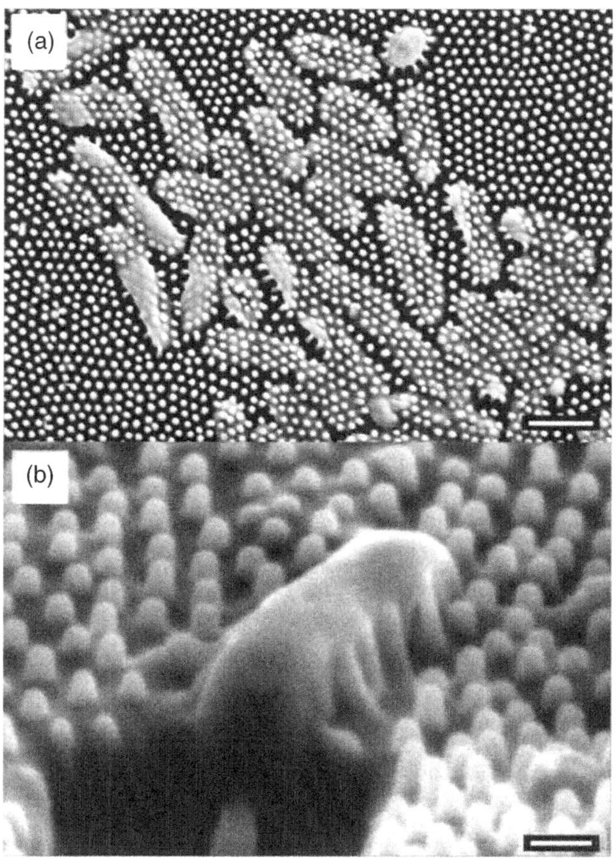

Figure 21.17 Bactericidal effect of cicada wing surface on *Pseudomonas aeruginosa* (*P. aeruginosa*). (a) *P. aeruginosa* cells on the surface of a cicada wing. The pillars on the wing surface penetrate the cells (Bar = 1 μm). (b) SEM image of a *P. aeruginosa* cell sinking between the pillars on the cicada wing surface (Bar = 200 nm). *Source:* From Ivanova et al. (2012).

Additionally, some experiments show an increase in bactericidal activity with increasing areal density A_d of the pillars, indicating that the bactericidal activity increases as the number of pillars with which the bacterial cell comes into contact increases. Often it is difficult to assess the effect of areal density of the pillars alone from experiments that also vary the pillar diameter or aspect ratios, which can also influence bactericidal activity, or from experiments covering only a limited range of areal densities.

Certain nanoscale topographies, as noted earlier, are more efficient in killing gram-negative bacteria than gram-positive bacteria, which has been attributed to a difference in cell wall (membrane) thickness or rigidity. For example, the peptidoglycan layer composed of amino acids and sugars that forms the cell wall of the majority of gram-positive bacteria is ~20–100 nm thick, whereas the peptidoglycan layer of most gram-negative cell walls is much thinner (~5–10 nm). For a given stretching force, the cell wall of gram-negative bacteria may deform more easily and, thus, may be more susceptible to rupturing on the nanoscale surface pillars. It has been suggested that bacterial cell motility may also play a role in the bactericidal activity of nanoscale surface topography. For example, the higher motility of *E. coli* has been suggested as one of the factors responsible for their higher death rate when compared to less motile bacteria.

Another factor that has been suggested to influence bactericidal activity is the adhesion between the cell wall and the pillars of the nanoscale surface topography. This suggestion is based, in part, on experimental observations of the rupturing of two strains of yeast cells (*Saccharomyces cerevisiae*) on natural nanoscale surfaces such as cicada and dragonfly wings. Although the cell wall of these single-celled (eukaryotic) microorganisms are structurally and chemically different from the cell wall of bacteria, rupturing of these yeast cells on the surface of cicada and dragonfly wings shows features that are similar to those observed for bacterial cells. The weaker adhering yeast cells show a loss in viability only on cicada wing surfaces of higher aspect ratio (~2). In comparison, stronger adhering cells show a loss in viability on both lower aspect ratio (~0.5) and higher aspect ratio surfaces.

A variety of biophysical models has been developed to analyze the stretching of an adsorbed bacterial cell on a surface consisting of nanoscale pillars. The cell is normally approximated as a deformable spheroidal mass enclosed within an elastic membrane of a certain biaxial rigidity, referred to as a stretching rigidity (or modulus) but whose structure and composition are neglected. Upon contacting the surface, the cell membrane is assumed to adopt a geometry approximating that shown in Figure 21.15a. A fraction of the membrane area adheres to the pillars whereas the remaining fraction suspended between the pillars deforms due to the force of gravity and other forces that may be present. Some models determine the degree of stretching of the membrane because a larger degree of stretching increases the potential that the cell membrane will rupture and cause cell death.

The predictions of these biophysical models, like models in general, depend on the type of model and on the numerical parameters assumed in the model, such as the biaxial stretching rigidity and the adhesion energy of the membrane on the surface, and the height, radius, and areal density of the pillars. Figure 21.18 illustrates the predictions of one model for the degree of stretching, defined as the change in surface area ΔS normalized to the original surface area S_0, as a function of various parameters. According to these trends, the potential for cell death increases with decreasing biaxial modulus of the cell membrane and increasing adhesive energy of the membrane on the pillars. Bactericidal activity is also predicted to increase with the height h, diameter D, and areal density A_d of the pillars, but only up to a certain point above which bactericidal activity saturates.

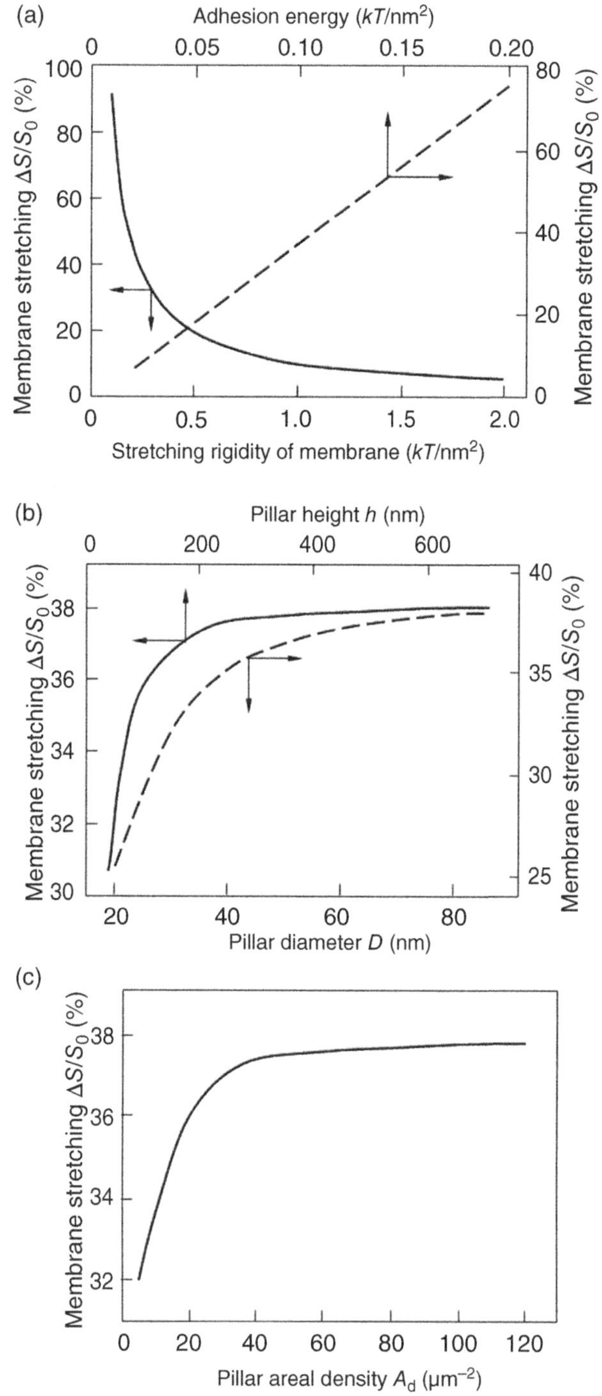

Figure 21.18 Predictions for the degree of membrane stretching as a function of (a) the stretching rigidity of the membrane and the adhesive energy between the cell membrane and the pillars, (b) the pillar diameter D and pillar height h, and (c) the areal density A_d of the pillars. The unchanged parameters in the calculations are cell diameter = 2 μm, stretching rigidity = $0.25kT$ 1/nm^2, bending rigidity = $10kT$ 1/nm^2, adhesive energy = $0.1kT$ 1/nm^2, pillar areal density = 100 μm^2, pillar diameter = 40 nm, pillar height = 300 nm, k is the Boltzmann constant, and T is the absolute temperature. *Source:* Li (2016) / with permission of Royal Society of Chemistry.

Effect of Nanoscale Surface Parameters on Bactericidal Activity

The cicada wing topography, as noted earlier, is efficient in killing gram-negative bacteria such as *E. coli* and *P. aeruginosa* but not gram-positive bacteria such as *S. aureus* and *S. epidermidis*. This natural topography has been mimicked on various synthetic polymers, such as PMMA, PMDS, PS, and PU. In one study, the cicada wing topography (designated P200) and two modified versions of this topography were imprinted on PMMA films spin-coated on glass cover slips using lithographic techniques. One topography consisted of pillars with height $h \sim 300$ nm, diameter $D \sim 190$ nm, and a center-to center distance $P \sim 320$ nm (P300), while the other had pillars with $h \sim 300$ nm, $D \sim 215$ nm, and $P \sim 600$ nm (P600). Then these surface-patterned PMMA films and flat PMMA films (control) were inoculated with log phase suspensions of the gram-negative bacteria *E. coli* and incubated overnight. The antibacterial effectiveness of the patterned films was assessed from the colony forming ability of the bacterial suspensions when compared to the suspensions on the flat film.

Assays showed that exposure of the *E. coli* suspensions to the P200 film decreased their colony forming ability to ~50% when compared to the flat film, whereas the amount of dead bacterial cells on the P200 surface increased by ~100%. Examination in the SEM showed a dramatically altered cell morphology. Bacteria on the P200 film appeared deflated due to stretching over several pillars (Figure 21.19), when compared to the normal, rod-shaped morphology of *E. coli* on the flat surface. These P200 films showed a better ability to inhibit the colony forming ability of *E. coli* and a higher amount of dead cells on their surface when compared to the P300 and P600 films having a coarser topography. Although these patterned films have different height and diameter, their bactericidal activity increased with the areal density of their pillars, in agreement with the trend in Figure 21.18. Overall, these results indicate that the patterned PMMA films are bactericidal, that is, they have the ability to kill *E. coli* cells. On the other hand, additional testing showed little effect of these PMMA films on the viability of cultures composed of gram-positive bacteria, which have a more robust cell wall.

Another natural surface that shows potent bactericidal effects, as noted earlier, is the wing of the dragonfly. The wing surface of the dragonfly *Diplacodes bipunctata* has been shown to be effective against gram-negative bacteria such as *E. coli* and *P. aeruginosa*, and against gram-positive bacteria such as *S. aureus*. The reason for this broader range of bactericidal behavior, when compared to the cicada wing topology, is not clear but one suggestion is the occurrence of a higher degree of stretching due to the random arrangement of pillars of nonuniform height and variable shape. Because of its potent bactericidal activity, the dragonfly wing topography has been used as a model to create a variety of synthetic materials that are potentially bactericidal. One such material is black silicon (Si), named for its black color, which has been developed for photovoltaic applications but has also been used recently for biomedical sensing applications. Black Si, normally created using reactive ion etching in a plasma, has a surface consisting of a thin oxide layer approximating the composition of SiO_2.

In one study, black Si having pillars of height ~0.5 μm and the dragonfly wing were evaluated for their bactericidal activity by measuring the colony forming ability of bacterial cultures exposed to their surfaces for various times. The bacteria used were gram-negative *P. aeruginosa*, gram-positive *S. aureus*, and *Bacillus subtilis* (*B. subtilis*) endospores. Both surfaces were highly effective in promoting the killing of the gram-negative and gram-positive bacteria, as well as the bacterial endospores, which are particularly resistant to most techniques for eradication of bacterial infections.

Another study used black Si with uncoated surfaces consisting of pillars of three different heights (~0.5, 2.5, and 20 μm), tip diameters and areal densities, as well as similar surfaces coated with a thin layer of diamond (referred to as black diamond). The diamond coating served to strengthen the black Si pillars, in addition to making their tips less sharp and increasing their tip diameter. Both types of surfaces were effective in killing gram-negative *E. coli*. In comparison, these surfaces showed little bactericidal activity against gram-positive *Staphylococcus gordonii* (*S. gordonii*), which have a smaller size (~0.7 μm by ~0.5 μm) than *E. coli* (~2 μm by ~0.5 μm), a thicker cell membrane, and a lower motility. Despite

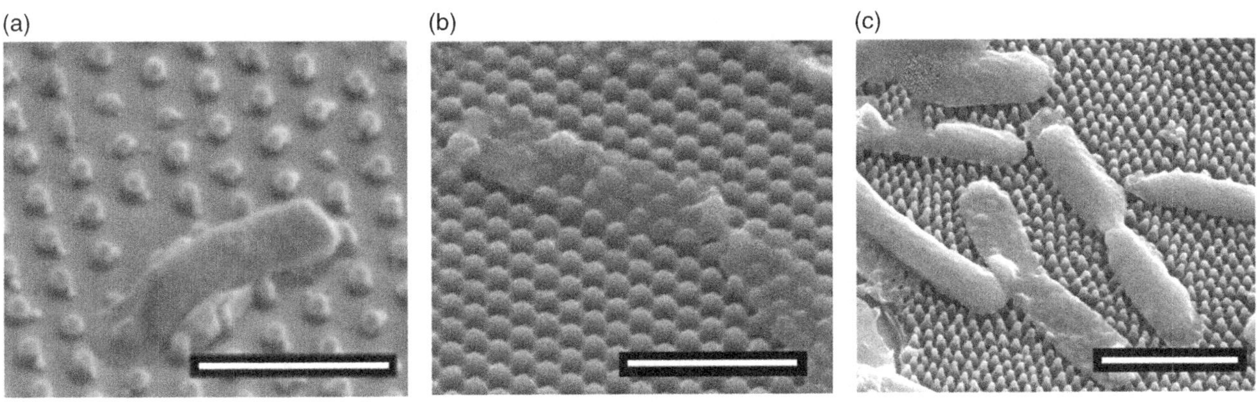

Figure 21.19 Bactericidal effect of polymethyl methacrylate (PMMA) films on glass cover slips, which were imprinted with a cicada-like surface topology using lithographic techniques. SEM images show collapsed *E. coli* cells on PMMA surfaces consisting of pillars with (a) height $h \sim 300$ nm, base diameter $D \sim 215$ nm, and periodicity $P \sim 595$ nm (P600); (b) $h \sim 300$ nm, $D \sim 190$ nm, and $P \sim 320$ nm (P300); (c) $h \sim 210$ nm, cap diameter $d \sim 100$ nm, and $P \sim 170$ nm (P200). (Scale bar = 2 μm) *Source:* From Dickson et al. (2015).

having pillars of larger tip diameter, the bactericidal activity of black diamond was similar to that of black Si, indicating that the sharper tip diameter of black Si was not a significant factor in its bactericidal activity.

There was little correlation between the bactericidal effectiveness of black Si surfaces and an increase in their pillar height. Black Si with the medium pillar height (~2.5 μm) showed a significantly better ability to kill *E. coli* when compared to the surface with the long pillar height (~20 μm), but little difference when compared to the surface of short pillar height (~0.5 μm). A better correlation could be achieved by considering the areal density of the pillars. Short and medium pillars, with a higher areal density (~65 and ~81/μm^2, respectively), showed a significantly better bactericidal activity when compared to long pillars (areal density ~ 1.51/μm^2).

Examination in the SEM showed many of the *E. coli* cells settling between the gaps of the long needles of black Si and appearing to show an undamaged morphology (Figure 21.20). However, where the sharp tips of the long needles appeared to penetrate the cell, there was a loss of turgor pressure, lysis, and flattening of the cell. While cells of *E. coli* showed a damaged morphology on the shorter pillars, *S. gordonii* cells showed their characteristic morphology and signs that they were beginning to form chain-like structures.

21.4 Microparticle-Induced Host Responses

As noted earlier, the surface of a biomaterial that cells and tissues initially encounter consists of a layer of biomolecules composed mainly of serum proteins, which has properties that depend on the particle surface characteristics and the adsorbed biomolecules. The initial interaction of a microparticle with a cell occurs at the cell membrane, and some of the cell and tissue responses that subsequently occur are set in motion by interactions that occur at this membrane. In many cases, however, cell interactions with a microparticle occur because of internalization of the particle into the cell. The mechanisms of these interactions become more clearly discernable with an understanding of the process of endocytosis in which particles are taken into the cell and internalized. Cells also release particulate materials, which occurs by exocytosis, a process that is particularly important for particles used for drug delivery and diagnostic imaging.

21.4.1 Mechanisms of Microparticle Endocytosis

Phagocytosis is generally the main mechanism by which particles of size larger than ~0.5 μm, that is, most microparticles, are

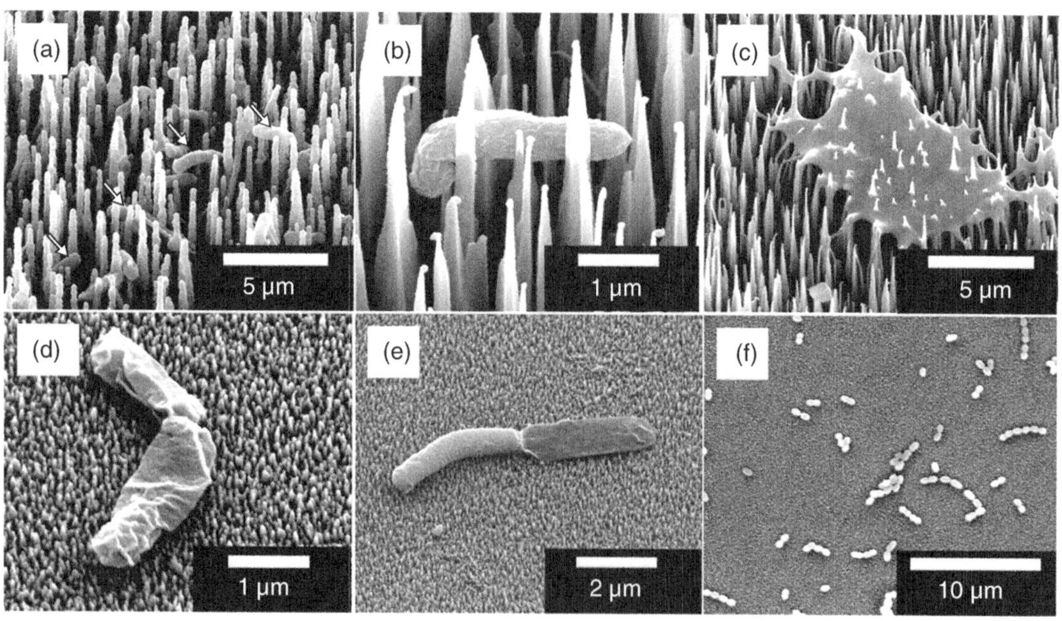

Figure 21.20 SEM images for *E. coli* after incubation for one hour on black silicon (bSi) or black diamond (bD) (bSi coated with diamond). (a) bD long needles with bacteria (arrows) lying in the gaps between the sparsely spaced needles. (b) Close-up image of a single bacterium trapped in the gaps between the bSi long needles. (c) bSi long needles with a lysed bacterium impaled by the needles. (d) Short bD needles with unhealthy bacteria. (e) Short bD needles showing one viable and one lysed bacterium. (f) Seemingly viable *S. gordonii* on short-needle bSi surfaces after incubation for one hour. *Source:* From Hazell et al. (2018).

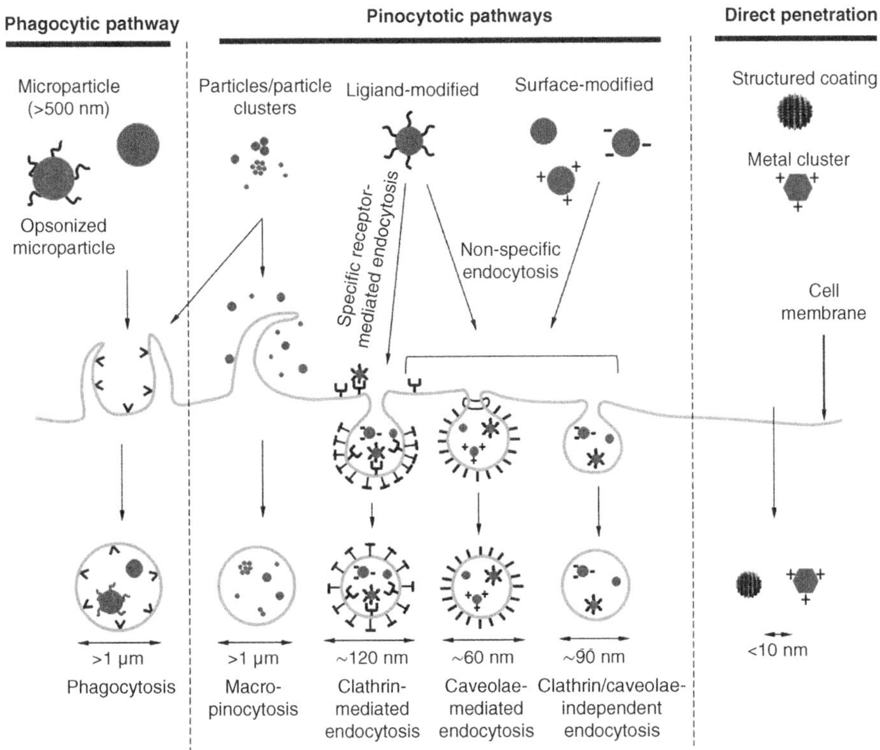

Figure 21.21 Mechanisms of microparticle and nanoparticle endocytosis. *Source:* Adapted from Zhu et al. (2013).

taken into cells (Figure 21.21). Phagocytosis has the following characteristics:

- It is a combination of two separable processes, attachment of a particle to the macrophage cell membrane and internalization of the particle into the cell.
- It is restricted to professional phagocytic cells, of which macrophages and monocytes are principal examples. These phagocytic cells are equipped with receptors that allow recognition of bacteria and other harmful entities (Chapter 18).

The process is normally initiated by opsonization, in which opsonins such as immunoglobulins (for example, immunoglobulin G, abbreviated IgG), various complement proteins and other proteins from the blood are adsorbed on the particle surface. The opsonized particle can now be recognized by and attached to the membrane of a phagocyte via specific ligand–receptor interactions. Binding of the particle to the cell membrane activates intracellular signaling, which triggers actin polymerization, the formation of cell-surface extensions, and subsequent engulfment and internalization of the particle. The particle is now inside the cell, contained within a membrane-bound vesicle called a phagosome.

21.4.2 Response to Microparticles

The ability of a macrophage to take up a particle depends on the availability of membrane receptors for different particle surfaces and on the characteristics of the particle. Much of our understanding of the factors that control phagocytic uptake of microparticles has been obtained from *in vitro* investigations with macrophage cell cultures. These studies show that the microparticle characteristics that have a significant influence on phagocytosis are its size, shape, and chemical composition. Other particle characteristics, such as its Young's modulus, have also been shown to have an effect, but the role of surface topography is difficult to identify.

Effect of Microparticle Size

Several studies to elucidate the effect of microparticle size on phagocytosis have been performed using PS microspheres, generally considered a type of standard nondegradable particle for assessing phagocytosis. Experiments using rat alveolar macrophages and IgG-opsonized and non-opsonized PS microspheres show that phagocytosis apparently consists of two separate and distinct steps, attachment and internalization. While the internalization step is not affected significantly by particle size, the highest amount of attachment

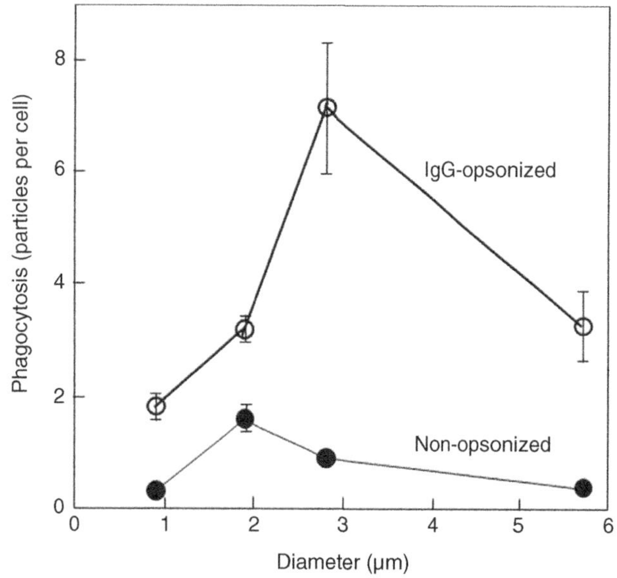

Figure 21.22 Overall phagocytosis uptake (combination of attachment and internalization) for IgG-opsonized and non-opsonized polystyrene microspheres by rat alveolar macrophages as a function of microsphere diameter. *Source:* From Champion et al. (2008) / with permission of Springer Nature.

and phagocytosis occurs at a particle diameter of ~2–3 μm (Figure 21.22). This is noteworthy because this diameter coincides approximately with the size (diameter or width) of most bacteria. Figure 21.22 also shows that pre-adsorption of IgG onto PS microspheres, giving IgG-opsonized particles, enhances attachment of the microspheres to the macrophage surface via binding to its opsonin receptors.

It has been hypothesized that this optimal size for phagocytosis seen for PS microspheres is an outcome of the attachment step, which is influenced by the characteristic features of the membrane ruffles in active macrophages (Champion et al. 2008). Support for this hypothesis comes from the observation that the elimination of these ruffles, via osmotic swelling of the macrophages to give a smooth membrane, eliminates this size dependence of phagocytosis. Figure 21.23 shows a simple model of particle-ruffled membrane interaction to account for the particle size dependence of phagocytosis and for the maximum interaction at a particle diameter matching the dimensions of macrophage ruffle features seen in the SEM.

In general, the inflammatory response of tissues to microscale particles is much more intense when compared to that for nanoscale particles. Much of this difference in tissue response is an outcome of the phagocytosis of particles by macrophages. The level of tissue inflammation triggered by macrophage phagocytosis of particles is size-dependent, with the highest level of inflammation usually occurring with particles ~1 μm in diameter. For example, the release of the pro-inflammatory cytokines tumor necrosis factor-α (TNF-α) and interleukin-1β (IL-1β) by macrophages in response to internalization of PS microspheres of diameter ~1–2 μm is greater than that triggered by smaller or larger diameter PS microspheres. Correlated with this, the highest levels of superoxide radicals released by macrophages in response to internalization of PS microspheres occurs in response to uptake of microspheres of diameter ~1–2 μm. Another example of this size-dependent effect is the macrophage phagocytosis of SiO_2 particles, inorganic particles that show a large negative charge (zeta potential) in aqueous media without serum proteins at the physiological pH (~7.4), which have their strongest cytotoxic effect on macrophages at a size of ~1 μm.

Effect of Microparticle Shape

Phagocytosis studies on microparticles have normally been performed using spherical particles, making the precise role of microparticle shape largely unknown for many years. However, recent studies have shown that microparticle shape can have a strong effect on phagocytosis (Sharma et al. 2010). These studies employed PS microparticles of the same surface composition and volume, but

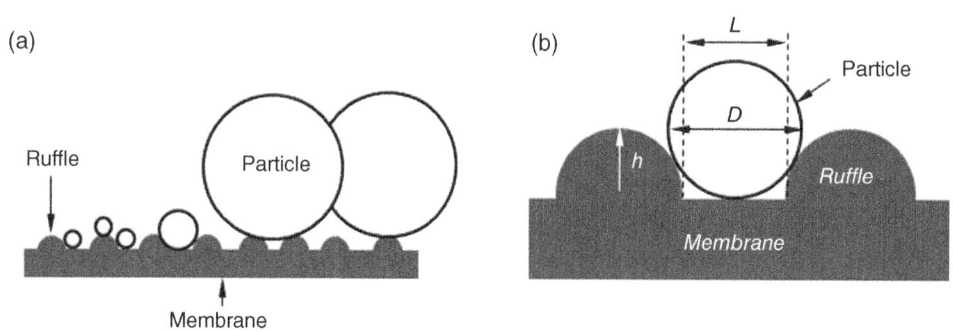

Figure 21.23 Model of particle-ruffled membrane interaction to account for the particle size dependence of phagocytosis. (a) Small particles ($D < L$) and large particles ($D > L$) make fewer contacts with the macrophage membrane when compared to particles of size $D \sim L$. (b) Ruffle and particle length scales used to determine the particle diameter capable of making the most membrane contact. *Source:* Champion et al. (2008) / with permission of Springer Nature.

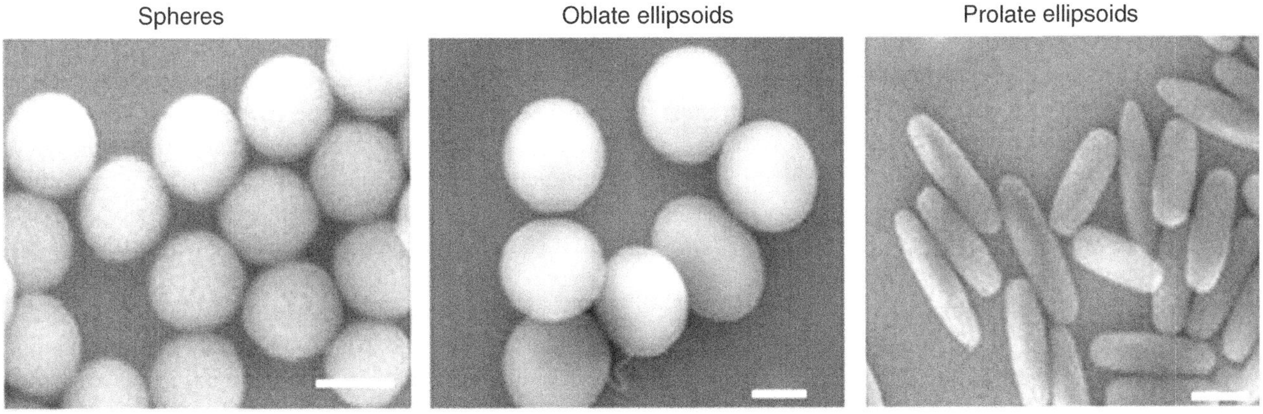

Figure 21.24 SEM images of polystyrene (PS) particles having the shape of spheres, oblate ellipsoids and prolate ellipsoids (Bar = 1 μm). *Source:* From Sharma et al.(2010).

different regular shapes, such as spheres, oblate ellipsoids, prolate ellipsoids, and elliptical discs of aspect ratio (major axis to minor axis) up to ~5. Oblate ellipsoids may be visualized as having the shape of an American football, whereas prolate ellipsoids have a shape corresponding to that of a lentil (Figure 21.24).

Incubation with rat macrophage cells show that, at a single particle level, the particle shape at the point of first contact by macrophages, not the actual particle size, determines whether cells will proceed with phagocytosis or simply spread on the particle surface (Champion and Mitragotri 2006). Internalization is highest when the particle binds vertically to the cell surface (that is, in the direction of its long axis), decreases as the angle between the long axis and the cell surface increases, and becomes zero above a critical angle (~45°). For example, macrophages that attach to elliptical discs along the major axis, internalize them in less than ~6 minutes. In comparison, macrophages that attach to similar elliptical discs along the minor axis or on the flat side did not internalize them even after two hours but, instead, spread on the particle surface, a process that may approximate frustrated phagocytosis (Chapter 18).

For an ensemble of many PS microparticles, consisting of prolate ellipsoids, oblate ellipsoids, and spheres (sphere diameter ~1 μm) of the same volume, attachment of the particles to the macrophages is in the order prolate ellipsoids > oblate ellipsoids > spheres (Figure 21.25). In comparison, internalization of the attached particles is in the order oblate ellipsoids > spheres > prolate ellipsoids. The extent of overall phagocytosis is the product of attachment and internalization. Oblate ellipsoids have a high attachment and the highest internalization, and, consequently, they have the highest phagocytosis. In comparison, the limited phagocytosis of prolate ellipsoids can be explained by the fact that their superior attachment is offset by their poor internalization. It is noted that although the ellipsoids and spheres have the same volume, the ellipsoids have a higher surface area than the spheres because a spherical shape has the lowest surface area for a given volume. Smaller ellipsoids

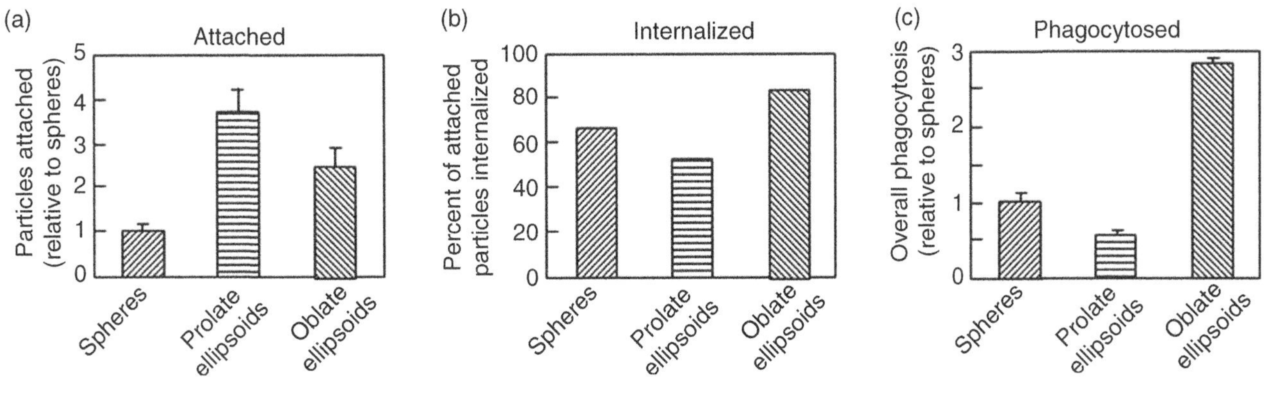

Figure 21.25 Comparison of attachment and internalization of particles of various shapes prepared from spheres of diameter 1 μm to cells of a monocyte/macrophage cell line. (a) Number of particles attached to cells; (b) percent of attached particles that are internalized by cells; (c) overall phagocytosis. *Source:* From Sharma et al. (2010) / with permission of Elsevier.

and spheres (sphere diameter ~0.5 μm) of the same volume show a trend similar to that in Figure 21.25c, but little effect of shape is seen from larger ellipsoids and spheres (sphere diameter ~3.6 μm) of the same volume. Overall, these results for PS microparticles show that, where the particle volume is less than that of a sphere of ~3.6 μm, the shape of the particle independently influences its attachment to macrophages and its internalization.

Effect of Microparticle Composition

As noted for macroscopic biomaterials, the chemical composition of a material influences its physicochemical properties. Two aspects of composition relevant to the biocompatibility phenomena of microparticles and NPs as well, are the chemical composition of the particle itself and its surface chemistry. Depending on their chemical composition, microparticles internalized by both phagocytic cells and non-phagocytic cells may be broadly classified as degradable or nondegradable. Whether a microparticle is degradable or nondegradable directly influences the cellular responses that occur upon its internalization. Some internalized particles degrade partially or fully through the action of one or more of the many different types of lysosomal enzymes or enzymes of other cellular compartments. Internalization of degradable particles, such as phagocytized microbes, apoptotic cell debris, and degradable polymer particles, will normally be adequately metabolized and cleared (by physiological actions of the kidneys and liver), causing no adverse effects to the body.

In comparison, internalization of nondegradable foreign particles often leads to adverse effects, which may include serious pathological manifestations. For example, the internalization of large nondegradable particles (larger than ~3 μm) by macrophages causes these phagocytic cells to go into a state of frustrated phagocytosis, which leads to the formation of foreign body giant cells (FBGCs). Although this represents an attempt to wall off and minimize the effects of the nondegradable foreign body, the formation of FBGCs often causes serious tissue injury (Section 21.4.4). Macrophage internalization of nondegradable particles of size ~0.5–2 μm normally causes lysosomes to swell and then rupture, resulting in the release of lysosomal enzymes that damage neighboring tissues. Figure 21.26 summarizes potential interactions that can occur between internalized microparticles and cells and tissues depending on whether the particle is degradable or nondegradable.

Surface chemistry influences various surface properties of a particle, such as its electrostatic charge (or zeta potential) and hydrophobicity (or hydrophilicity), which influence cell response. In general, *in vitro* studies show that an increase in hydrophobicity leads to an increase in phagocytosis. In one study, the surface of cellulose microspheres (diameter ~1.5 nm) was modified with a variety of functional groups, giving microspheres having a range of hydrophobicity (as

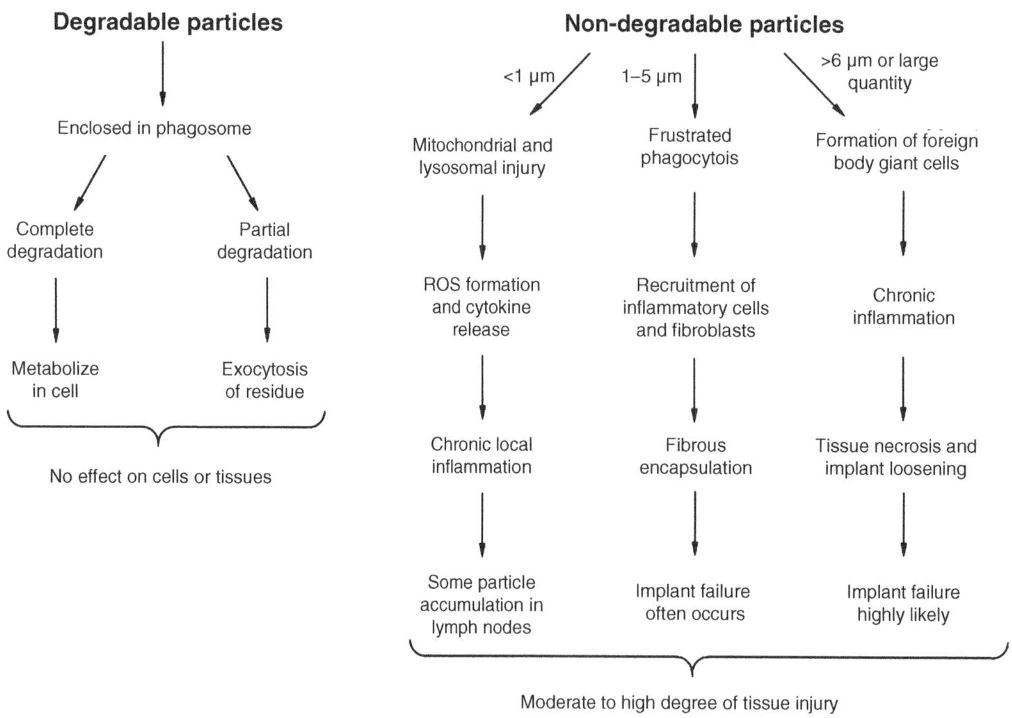

Figure 21.26 Potential interactions that can occur between internalized particles and cells and tissues depending on whether the particles are degradable or nondegradable. (ROS, reactive oxygen species)

determined by the water contact angle) and zeta potential (determined in phosphate-buffered solution at pH ~7.4). Incubation with mouse peritoneal macrophages showed that the microspheres with hydrophobic surfaces were more susceptible to phagocytosis when compared to those with hydrophilic surfaces. Additionally, unmodified cellulose microspheres, characterized by an almost electrostatically neutral hydrophilic surface, were least susceptible to phagocytosis. The greater susceptibility of hydrophobic microparticles to phagocytosis has normally been associated with their better ability to support adsorption of proteins, which could provide a higher concentration of ligands for opsonization.

Some studies have shown that negatively charged microparticles are more easily phagocytosed than microparticles with a positively charged surface but this is not generally so. As most bacteria have a negative charge, this preference of phagocytic cells for negatively charged particles over positively charged particles has been associated with their major functional role of bactericidal activity. However, in the aforementioned study involving surface-modified cellulose microspheres, phagocytosis was found to be lowest for particles with a zeta potential near zero and increased as the absolute value of the zeta potential increased, for both negatively charged and positively charged particles. No significant difference in phagocytosis was found between the positively charged and negatively charged microspheres when considered at a given magnitude of the zeta potential. Overall, a degree of caution should be exercised when assessing the individual effects of hydrophobicity and surface charge because one cannot normally be changed independently of the other (Chapter 5).

Particles having a hydrophilic electrostatically neutral surface, as noted earlier, are less susceptible to internalization by macrophages. This property is taken advantage of practically in one approach to delay phagocytosis of microparticles and NPs used as drug delivery devices (Chapter 26). Recognition and removal of these devices after their introduction into the physiological environment would severely restrict their effectiveness. Modification of the particle surface with physically adsorbed or chemically bonded polymer molecules is used to shield the particle from the hydrophobic and electrostatic interactions that assist opsonins to bond to the surface, thereby delaying phagocytosis. Well-known examples of such shielding molecules are polyethylene glycol (PEG) and PEG-containing copolymers (Chapter 26).

Effect of Microparticle Mechanical Properties

The mechanical properties of a macroscopic biomaterial, such as its Young's modulus, are known to have a significant influence on its biocompatibility (Chapter 17), and they have been shown to influence the phagocytosis of microparticles by macrophages as well. In one study (Beningo and Wang 2002), polyacrylamide (PAAm) microspheres of diameter 1–6 μm were created with the same chemical composition but with different stiffness by varying the extent of their molecular crosslinking. Mechanical testing of thin PAAm sheets prepared by a similar procedure showed that one group of microspheres with a higher degree of crosslinking had a Young's modulus that was more than three times that of the other group. Following covalent bonding of bovine serum albumin (BSA) to their surface, the microspheres were opsonized with the same areal density of rabbit anti-BSA IgG and incubated with BMDMs for 45 minutes at 37 °C. Opsonized microspheres with the higher Young's modulus showed a significantly higher phagocytic index, that is, average number of beads internalized per cell, when compared to opsonized beads with the lower modulus. Since both groups of microspheres had the same areal density of IgG, and, thus, the same level of surface binding to macrophages, this difference in phagocytosis appears to occur during the internalization step. In comparison, non-opsonized beads were poorly phagocytosed. Additionally, the PAAm sheets with the higher Young's modulus showed a greater tendency to cause frustrated phagocytosis, by a factor of ~6 times when compared to the sheets of lower Young's modulus.

21.4.3 Microparticle Distribution in the Organs

An important aspect of the biological effects associated with the endocytosis of microparticles (and NPs as well) is their eventual fate. Phagocytosis of degradable particles, as noted earlier, may lead to rapid degradation and to resolution with no discernable effect on cells or tissues (Figure 21.26). In comparison, nondegradable particles smaller than ~1 μm will be disseminated through the lymphatic system. Phagocytic cells containing these particles can migrate to lymph vessels and accumulate in the lymph nodes. However, as noted in Chapter 14, no material is completely resistant to degradation. Slowly degrading particles may release monomers, oligomers, ions, or other products, which will be cleared through the liver or kidneys. During this process, some of these products, particularly metal ions, may be distributed to distant organs or tissues. Overall, nondegradable particles or remnants of slowly degrading particles will remain in the lymph nodes.

After injection into the peritoneal cavity of mice, PMMA and PS particles (~1 μm or ~5–6 μm) showed an organ distribution that was not dependent on the size of the particles but strongly dependent on the composition of the particles. Peritoneal macrophages with phagocytized PS particles remained near the site of injection, in the fatty-fibrous tissues surrounding the spleen, pancreas, and liver, where they triggered a localized inflammatory response. In comparison, macrophages with internalized PMMA particles tended to travel to distant organs, particularly the spleen, showing that for microparticles of the same size, particle composition has a strong effect on their organ distribution.

21.4.4 The Inflammasome and Microparticle-Induced Inflammation

Introduction of foreign particles into a tissue invariably triggers inflammation and various protective responses that are part of the innate immune defense system. The components of the innate immune defense system responsible for particle-induced inflammation are known as inflammasomes, large multi-protein complexes within the cytoplasm of macrophages and neutrophils (Chapter 18). These inflammasomes detect proteins associated with bacterial pathogens such as the protein flagellin and damaging organic molecules such as gout-causing uric acid crystals. In response to the danger detected, the inflammasome triggers the activation of the proteolytic enzyme caspase 1, which causes extensive damage to the plasma membrane of the macrophage and catalyzes maturation of the highly inflammatory cytokine IL-1β. The dying macrophage releases IL-1β, which attracts additional macrophages and promotes pyroptosis as well, a localized form of programmed cell death that typically occurs in cells harboring intracellular pathogens. The localized pyroptosis, which allows degradation and clearance of an infectious pathogen, may be viewed as a protective response that limits the spread of the pathogen. Pyroptosis becomes harmful, however, when the primary stimulus of inflammation is a nondegradable inorganic substance such as wear particles from implants. The ensuing chronic inflammation will result in fibrosis, tissue damage, and necrosis.

One of the most important inflammasomes is NLRP3 (NOD-like receptor protein 3). There is evidence that the NLRP3 inflammasome signals the adverse tissue responses to particles such as orthopedic wear particles. An important feature that distinguishes NLRP3 is its responsiveness to activation by a broad variety of stimuli such as inorganic particles and bacterial and viral pathogens.

21.4.5 Wear Debris-Induced Osteolysis

Although total hip and knee joint arthroplasty is an effective treatment for most patients suffering from degenerative diseases of the hip and knee joint, ~20% of the hip and knee joint replacement surgeries performed each year are actually revision (repeat) surgeries due to premature failure of the prosthetic joints by osteolysis and aseptic loosening. It is well recognized that microscale particulate debris in the joint space produced by the articulating components of the prosthetic joints lead to periprosthetic inflammation and loosening of the implant (Chapter 1). This particulate debris, having a range of sizes and shapes, is composed mainly of ultrahigh molecular weight polyethylene (UHMWPE) for metal-on-UHMWPE articulating couples or alumina (Al_2O_3) for the less commonly used ceramic-on-ceramic couples (Chapter 23). In addition, incubation of cultured macrophages in the presence of UHMWPE microparticles has been shown to promote the expression of RANKL, IL-1, and other cytokines that stimulate the activity of osteoclasts (bone degrading cells).

Wear debris induced osteolysis and implant loosening occurs by a series of steps, as summarized in Figure 21.27. The wear debris released into the periprosthetic space encounters the synovial fluid lubricating the articular surfaces, which spreads the UHMWPE particles into additional areas of the bone–implant interface where they are quickly recognized and phagocytosed by macrophages. A common feature of the fibrous membrane surrounding failed implants removed during revision surgery is extensive macrophage infiltration, with many of the macrophages containing intracellular UHMWPE microparticles, and the presence of

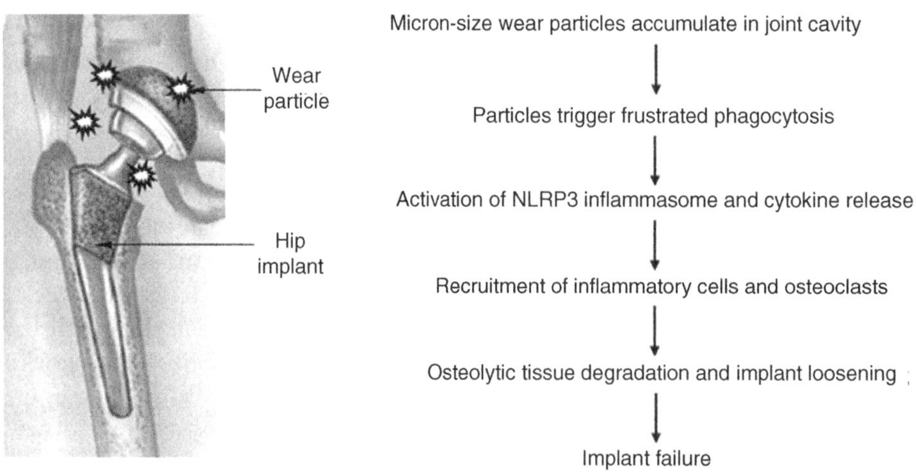

Figure 21.27 Sequence of events following release of wear particles from a hip replacement prosthesis, leading to osteolysis and implant failure. (The wear particles are enlarged for clarity.)

FBGCs, an indication of the frustrated phagocytosis syndrome (Chapter 18). The internalized microparticles then initiate a pro-inflammatory response in the macrophages.

Recent studies show that the underlying mechanism of the macrophage response to internalized wear particles is the activation of the NLRP3 inflammasome, which triggers the release of the highly inflammatory cytokine IL-1β (Manzano et al. 2018). This, in turn, leads to the release of additional cytokines that recruit progenitor osteoclasts and inhibit osteoblasts. The resulting osteoclasts cause osteolysis around the prosthesis. The progressive bone erosion is typically asymptomatic until implant loosening causes catastrophic failure, often accompanied by fracture of the severely weakened bone. Osteolysis and implant loosening caused by wear debris continue to be major challenges in orthopedic surgery.

21.5 Nanoparticle-Induced Host Responses

The foregoing discussion on microparticle biocompatibility phenomena can serve as a basis for understanding the interaction of NPs with cells and tissues but there are significant differences. As with microparticles, the surface of an NP that initially interacts with cells in the physiological environment consists of an adsorbed layer of biomolecules composed mainly of proteins (sometimes called a protein corona). On the other hand, when compared to a microparticle, the much greater (sharper) curvature of an NP surface could lead to different conformations of the adsorbed proteins, which will influence cell response differently. The difference in size could also influence the scale of the interaction between an NP and a cell. The surface of a cell is heterogeneous spatially in both chemical composition and electrostatic charge. Whereas a microparticle tends to experience an average effect of these heterogeneities, an NP of size comparable to the scale of these heterogeneities would interact differently because it experiences a specific localized cell surface environment.

21.5.1 Mechanisms of Nanoparticle Endocytosis

The mechanisms of NP endocytosis are different from those for microparticles and when compared to phagocytosis, these mechanisms are more common and occur in many cell types. Overall, an NP (with its protein corona) interacts with the cell membrane by adhesion to the cell surface lipid bilayer. This interaction can involve both specific and nonspecific interactions. Specific interactions are those between ligands on the particle surface and cell membrane receptors, which can potentially lead to receptor-medicated endocytosis. In comparison, nonspecific interactions involve the particle charge and hydrophobicity, such as electrostatic attraction between a charged particle and the charged phospholipid head groups of the cell membrane, or van der Waals attraction between neutral particles and the cell membrane.

Endocytosis of NPs can occur via different pathways. The primary pathway is clathrin-mediated endocytosis, which can occur via receptor-specific uptake (receptor-mediated endocytosis), or via nonspecific adsorption (receptor-independent endocytosis). Clathrin-mediated endocytosis takes place at small patches of the cell membrane that are rich in the fibrous protein clathrin, covering ~0.5–2% of the total cell surface. Clathrin is responsible for inducing and stabilizing invagination of the cell membrane, resulting in a pit that engulfs the particle (Figure 21.21). Then the pit becomes pinched off by enzymatic activity, forming a vesicle of diameter ~100–150 nm in the cell. A secondary pathway is caveolae-mediated endocytosis. Caveolae lipid raft areas are present on the cell surface. They are characterized by bulb-shaped invaginations of size ~50–80 nm in the cell membrane of epithelial and nonepithelial cells. Caveolin proteins present in the membrane of the caveolae are involved in the formation and stabilization of their characteristic shape. Caveolae enclose the NP and pinch off, forming a vesicle within the cell.

A limited amount of endocytosis may also occur by phagocytosis (Section 21.4.1), particularly if the particles are larger than ~100–150 nm, and by macropinocytosis. Macropinocytosis does not involve pit-forming proteins or lipid rafts. Instead, large membrane extensions or ruffles form as a result of cytoskeleton rearrangement, which then fuse back onto the plasma membrane, creating a vesicle (~0.2–5 μm) that traps a large amount of extracellular fluid. As all particles and molecules in the fluid are taken into the endocytic vesicle, regardless of their specific receptors, macropinocytosis is a form of nonspecific bulk fluid uptake. Clathrin-mediated and caveolae-mediated endocytosis and macropinocytosis are sometimes considered as subsets of pinocytosis. Overall, endocytosed NPs are confined in membrane-bound vesicles derived from the cell membrane and are not capable of reaching the cytosol.

21.5.2 Response to Nanoparticles

As endocytosis initially involves specific and nonspecific interactions between the NP (with its protein corona) and the cell membrane, it is dependent on the characteristics of the particle and on the type of cell. As with microparticles, the particle characteristics that have a significant effect on endocytosis of NPs are their size, shape, and surface chemistry.

Effect of Nanoparticle Size

Several experimental studies show that cellular uptake of NPs is strongly dependent on particle size and that there is an optimal particle size for cellular uptake. This optimal particle size depends on the particle surface composition and the cell type. Gold (Au) NPs have often been used as a type of standard nondegradable particle for assessing endocytosis because of the ease in controlling their size and shape during synthesis, modifying their surface chemistry, and quantifying their concentration using spectroscopy. Experiments with spherical Au NPs of various diameters in the range ~14–100 nm, which were surface modified with citric acid ligands and incubated with HeLa cells (ovarian cancer cells) in the presence of serum proteins, show a maximum cellular uptake for particles of diameter ~50 nm (Figure 21.28). Au NPs coated with transferrin (a blood protein), for which endocytosis occurs primarily by a clathrin-mediated (receptor-mediated) mechanism, also show optimal uptake at a diameter of ~50 nm. In comparison, maximum uptake of spherical PS NPs incubated with Caco-2 cells (human epithelial intestinal cells) has been found to occur at a particle diameter of ~100 nm.

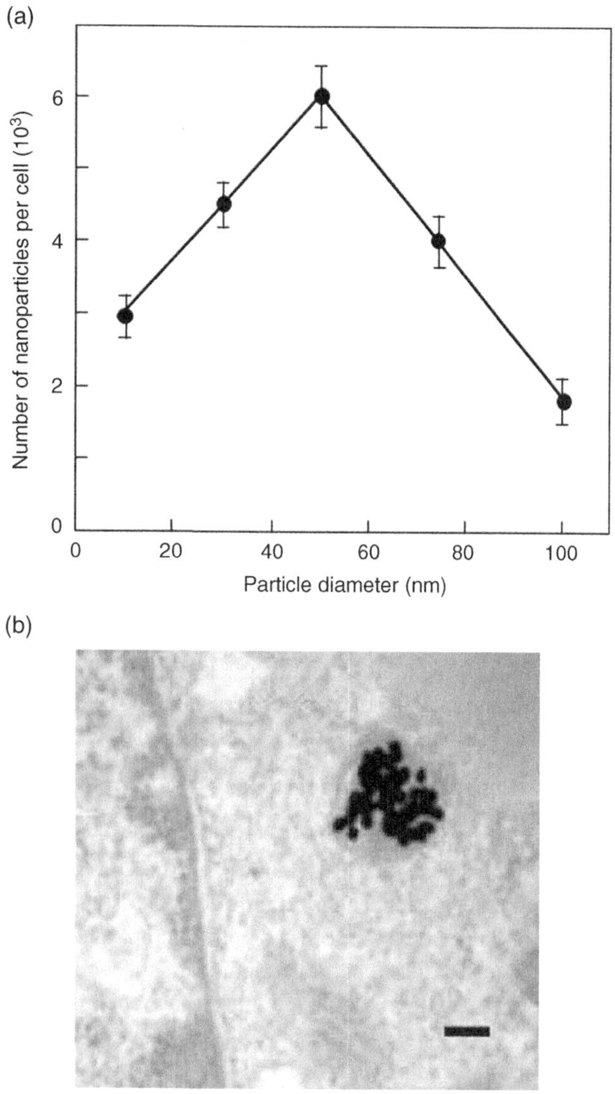

Figure 21.28 (a) Dependence of cellular uptake of spherical gold nanoparticles as a function of particle diameter; (b) TEM image of 50 nm particles within vesicle. (Bar = 100 nm). *Source:* From Chithrani et al. (2006) / with permission of American Chemical Society.

Several theoretical models have been developed to predict the size-dependent endocytosis of NPs. Some models of receptor-mediated endocytosis are based on dynamic and kinetic considerations, such as the rate at which an NP is internalized, whereas other models are based on thermodynamic considerations, such as the total number of NPs internalized at equilibrium. In the thermodynamic models of receptor-mediated endocytosis, whether an NP is internalized or not depends on the adhesion energy between the particle and the cell membrane and the deformation energy due to the bending and stretching of the cell membrane to engulf the particle. The predictions of these models depends on the assumed parameters for the ligand-receptor binding energy and the rigidity of the cell membrane but, overall, using the results of Figure 21.29 as an example, the thermodynamic models predict the following:

- A lower particle diameter of ~40–50 nm below which internalization of individual particles cannot occur
- An optimal diameter for internalization of single particles, equal to ~50 nm
- An upper particle diameter of ~100 nm above which internalization is very low.

The prediction of an optimal diameter for internalization is consistent with the aforementioned experimental results for transferrin-coated Au NPs endocytosed by a receptor-mediated mechanism. These thermodynamic models also predict a lower particle diameter limit (~40–50 nm) below

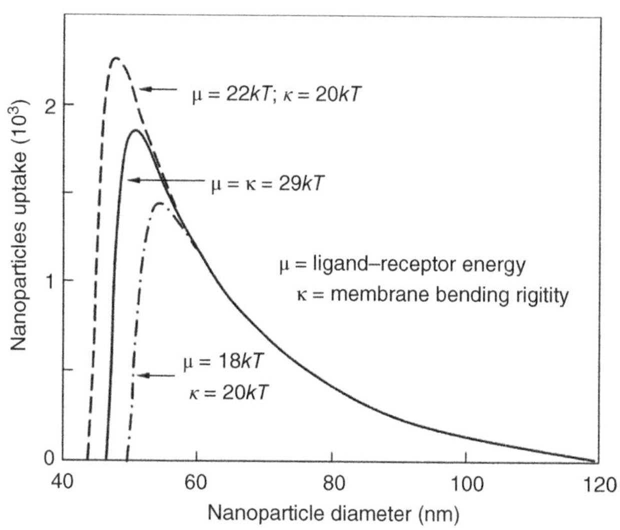

Figure 21.29 Model predictions for the endocytosis of nanoparticles by a ligand-receptor mechanism as a function of particle size, for the assumed values of μ and κ, where k is the Boltzmann constant and T the absolute temperature. *Source:* Zhang et al. (2009) / with permission of John Wiley & Sons.

which internalization of individual particles cannot occur, but as seen from Figure 21.28, a significant amount of endocytosis can occur for a much smaller diameter. This deviation from the model predictions has been explained in terms of clustering (agglomeration) of very fine NPs, whereby several particles adhere together to form clusters (agglomerates) by attractive forces such as van der Walls forces or other forces. Because surface area increases with decreasing particle size, finer particles are more susceptible to clustering when compared to larger particles because this leads to a larger reduction in surface area (surface free energy). In support of this explanation, it was observed that the aforementioned endocytosis of ~14 nm diameter transferrin-coated Au NPs by HeLa cells required the clustering of least six particles before uptake could take place. The diameter of a cluster composed of six of these NPs is ~40 nm, which provides better agreement with the model predictions. Overall, clustering of NPs is a common occurrence, particularly for finer NPs, unless repulsive forces between the particles can overcome the attractive van der Walls forces that are always present.

Effect of Nanoparticle Shape

As found for the phagocytosis of microparticles, particle shape can also have a strong effect on the endocytosis of NPs. When compared to microparticles, the effects of NP shape on endocytosis has received greater attention because the use of nonspherical NPs can provide improved performance in certain therapeutic and diagnostic devices. Elongated shapes relevant to NPs used in such devices are normally solid rods or ellipsoids of aspect ratio (major axis to minor axis) less than ~5, although carbon nanotubes have a considerably higher aspect ratio (Chapter 3). For solid objects with a smooth surface, it is noted that for a given volume, a spherical shape has the lowest surface area.

Overall, there are conflicting results for the effect of particle shape on NP endocytosis. Several reports show that when compared to spherical NPs (nanospheres), elongated NPs such as rod-shaped NPs (nanorods) have a lower uptake in certain cells. For example, in one study, transferrin-coated Au nanospheres and nanorods of varying aspect ratio (~1.5–6) were incubated with HeLa cells, a system for which endocytosis occurs primarily by a clathrin-mediated (receptor-mediated) mechanism. The nanorods showed a significantly lower uptake when compared to the spherical NPs (diameter ~50 nm) (Figure 21.30). On the other hand, one study has shown that for PS particles of a larger size, with the shape of rods or ellipsoids, endocytosis was enhanced when compared to spherical particles. PS spheres of diameter ~200 nm, rod-shaped particles (aspect ratio ~2) and disk-shaped particles (diameter to thickness ~2) of the same volume were surface-modified with carboxyl (C=O)OH groups or with

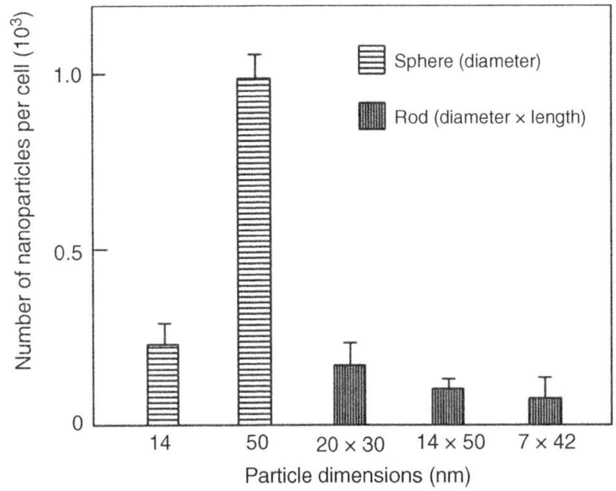

Figure 21.30 Effect of particle shape on the endocytosis of transferrin-coated gold nanoparticles by a clathrin-mediated pathway. *Source:* Chithrani and Chan (2007) / with permission of American Chemical Society.

carboxyl groups conjugated with biotin (a targeting ligand). Then the particles were incubated with Caco-2 cells. The results showed a higher uptake of rod-shaped particles when compared to spherical particles regardless of whether the particles were coated with biotin or not, and that the presence of biotin enhanced endocytosis.

The conflicting results for the effect of particle shape may be due to a variety of factors, such as difference in particle size and surface area and the mechanism of endocytosis. For example, in the aforementioned studies, the PS particles used had a much larger size than the Au particles. Additionally, for the same type of particles, there was a difference in surface area between the elongated particles and the spheres. For example, the surface area of the Au nanorods was considerably lower (by a factor of three times or more) than that of the Au nanospheres. In comparison, for the same volume, a rod-shaped PS particle had ~2 times the surface area of a PS sphere. A higher surface area, for example, presents more sites for ligand–membrane interaction. Another factor, as noted earlier for elongated microparticles, is the angle at which the elongated particle contacts the cell surface.

Effect of Nanoparticle Surface Composition

The composition of most relevance to biocompatibility phenomena of an NP is its surface composition, which controls its protein corona and, thus, its interaction with the cell membrane by specific or nonspecific interactions. It is noted that the phospholipid bilayer making up the membrane of mammalian cells is negatively charged. As the surface composition of a material directly influences its surface charge (or zeta potential), several experimental studies have focused on the relation between the surface charge of NPs and their cellular interaction.

Several studies show that as-synthesized or surface functionalized NPs with a positive surface charge have a higher cellular uptake than similar particles with a negative surface charge. This is normally attributed to the higher electrostatic interaction between the positively charged particles and the negatively charged mammalian cell membrane. For example, in one study, Au nanorods (18 nm in diameter × 40 nm) were synthesized using the cetyltrimethylammonium bromide (CTAB) method and their surface was functionalized with a variety of polyelectrolytes to give zeta potentials in the range +52 to −70 mV in a serum-free aqueous medium. When incubated with HeLa cells in media containing fetal bovine serum, the as-synthesized nanorods with the CTAB surface, which have the highest positive zeta potential, showed the highest uptake. In comparison, the nanorods, surface-modified with the polyelectrolyte poly-4-styrene sulfonic acid (PSS), which have the highest negative zeta potential, showed the lowest uptake.

In another study, nearly spherical Au NPs (~30 nm in diameter) and nanorods (14 nm in diameter × 55 nm in length) were synthesized by a similar CTAB route and surface-modified with PSS or with the polyelectrolyte polydiallyldimethyl ammonium chloride (PDDAC) (Figure 21.31). As synthesized, the nanospheres and nanorods coated with CTAB and PDDAC showed a positive zeta potential, whereas the PSS-coated nanospheres and nanorods showed a negative zeta potential in an aqueous medium. On the other hand, after incubation for 30 minutes in media containing newborn calf serum, the nanospheres and nanorods all showed a negative zeta potential regardless of the composition of the surface coating. An important aspect of these results is that they show how the adsorption of serum proteins can modify the zeta potential normally determined in an aqueous medium without proteins.

When incubated with human breast adenocarcinoma (MCF-7) cells in media containing serum, PDDAC-coated nanospheres and nanorods showed a significantly higher uptake by the cells when compared to CTAB-coated and PSS-coated nanospheres and nanorods. Measurement of the protein adsorption in an aqueous media containing serum showed that the PDDAC-coated nanospheres and nanorods with the highest uptake also had the highest amount of adsorbed proteins. In comparison, the PSS-coated nanospheres and nanorods with the lowest uptake also had the lowest amount of adsorbed proteins. Overall, these results indicate that in addition to the surface charge (zeta potential), which is normally measured in serum-free aqueous media, protein adsorption to the NP surface plays a direct role in cellular uptake.

As noted for the phagocytosis of microparticles, hydrophobic NPs are more susceptible to endocytosis than hydrophilic NPs. This preference of cells for hydrophobic NPs may be attributed to the greater adsorption of blood serum proteins to hydrophobic surfaces and the lipid raft mechanisms of endocytosis. Additionally, as noted earlier, surface modification with hydrophilic polymer molecules such as PEG and PEG-containing copolymers forms an important approach to shield NPs from the interactions that lead to internalization.

Experimental studies show that certain NPs (and macromolecules) penetrate the plasma membrane of cells not by endocytosis but, instead, by generating transient holes in the membrane (Leroueil et al. 2007). The degree of disruption is related to the particle charge and size, and the cell membrane properties. Positively charged NPs of size less than ~10–20 nm have a greater tendency for internalization by this mechanism. The nature of the transient holes generated in the cell membrane may vary from a literal hole to a region where material such as protein, lipid, and cholesterol are reduced. It is thought that strong electrostatic interaction between positively charged particles and negatively charged phospholipid groups of the cell membrane is mainly responsible for the generation of the transient holes.

Internalization of NPs by a mechanism that involves the creation of transient holes in the cell membrane may lead to cytotoxic effects such as cell death or leakage of ions and molecules into, or out of, the cell. On the other hand, some macromolecules (such as cell-penetrating peptides) and synthetic nanomaterials of very small dimensions (such as silver nanoclusters) may be capable of penetrating membranes without disruption of the lipid bilayer. Experiments have shown that Au NPs (diameter 4–5 nm), when coated with a shell composed of an ordered arrangement of alternating anionic and hydrophobic ligands, penetrated the membrane of mouse dendritic cells without evidence of membrane disruption. In comparison, similar Au NPs having an identical shell but one lacking structural order of the same ligands did not penetrate the cell membrane but, instead, were largely trapped in endosomes (Figure 21.32). These results provide additional insight into understanding the molecular interactions that influence the penetration of cell membranes by NPs and into designing efficient low-toxicity NP drug-delivery devices.

21.5.3 Cytotoxicity Effects of Nanoparticles

The expanding range of biomedical applications envisioned for NPs has raised concerns about their potential for causing unintended adverse biocompatibility phenomena. As noted for microparticles, internalization of NPs composed of degradable biomaterials will normally lead to rapid degradation and to resolution with no discernible effect on cells (Figure 21.21). Because of their smaller size when compared to microparticles, for the same composition, degradation of NPs will be faster. Nondegradable or slowly degrading NPs can lead to a variety of cytotoxicity effects, although these effects may be less severe when compared to those caused by microparticles.

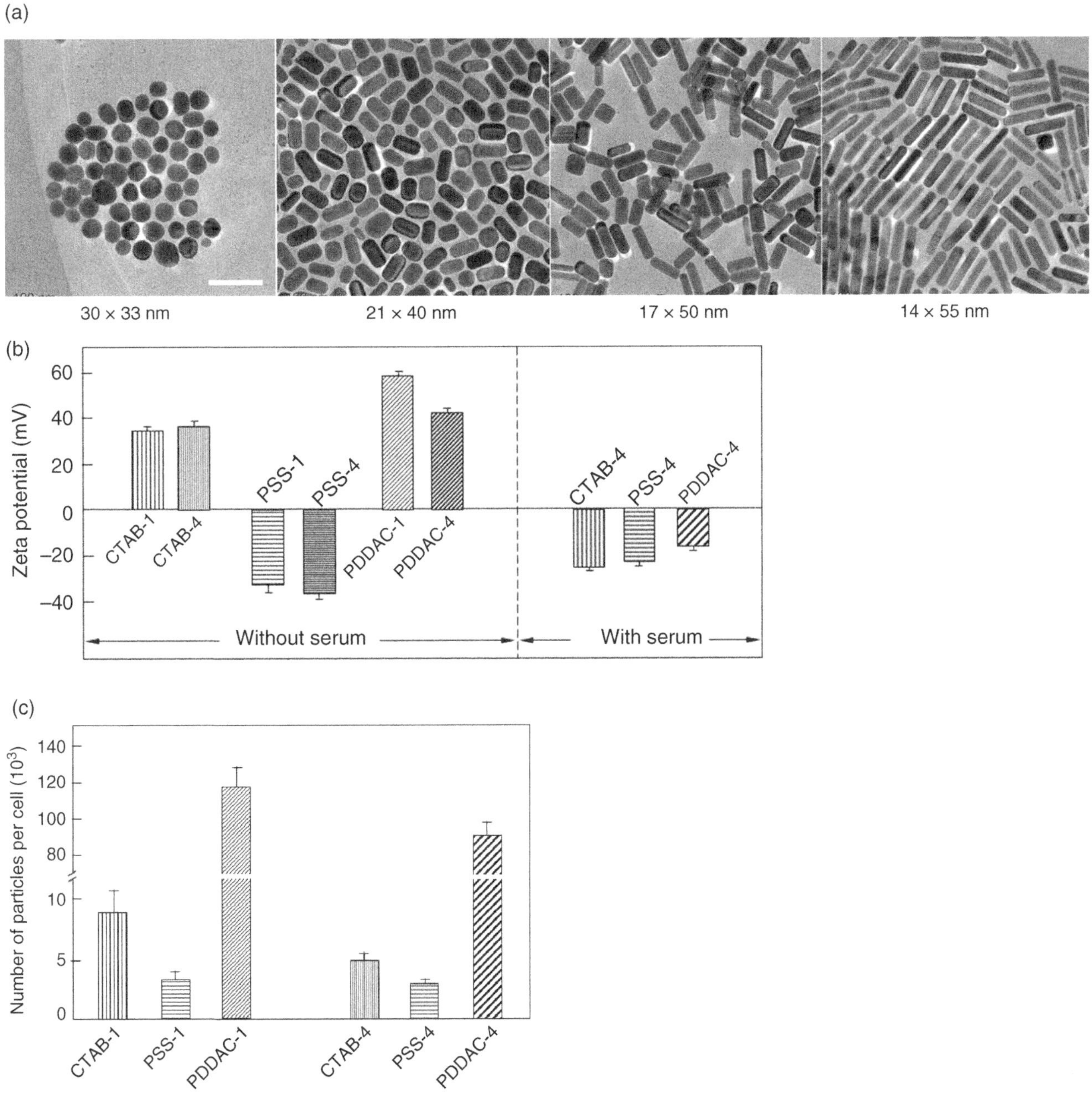

Figure 21.31 (a) TEM images of cetyltrimethyl-ammonium bromide (CTAB)-coated Au nanoparticles. (Bar = 100 nm). (b) Zeta potential of Au nanoparticles (aspect ratio ~1) (labeled 1) and nanorods (aspect ratio ~4) (labeled 4) coated with CTAB, poly-4-styrene sulfonic acid (PSS), or polydiallyldimethyl ammonium chloride (PDDAC). Zeta potentials were measured in aqueous media after the particles were incubated for 30 minutes in media without or with newborn calf serum. (c) Uptake of CTAB, PSS or PDDAC nanoparticles and nanorods upon incubation with human breast adenocarcinoma (MCF-7) cells in media containing serum. *Source:* From Qiu et al. (2010) / with permission of Elsevier.

NPs with a hydrophobic surface or a positively charged surface, as noted earlier, are normally internalized to a larger extent when compared to NPs with a negatively-vcharged surface or a hydrophilic surface. However, this comes at the expense of increased cytotoxicity. The effect of NP surface hydrophobicity on the immune response of cells has been investigated using Au NPs (diameter ~2 nm) whose surfaces were functionalized with head groups of varying hydrophobicity (Figure 21.33a). To bring out the sole effects of surface hydrophobicity, the gold particles with their hydrophobic surface ligands were coated with a passivating layer of spacer molecules composed of tetraethylene glycol (TEG) to remove the background effects of the Au particle core. Then functional groups of varying hydrophobicity (designated NP1 to NP8) were chemically bonded to the spacer molecules. Numerical values for the hydrophobicity of these functional groups were determined from their predicted partition coefficient in the

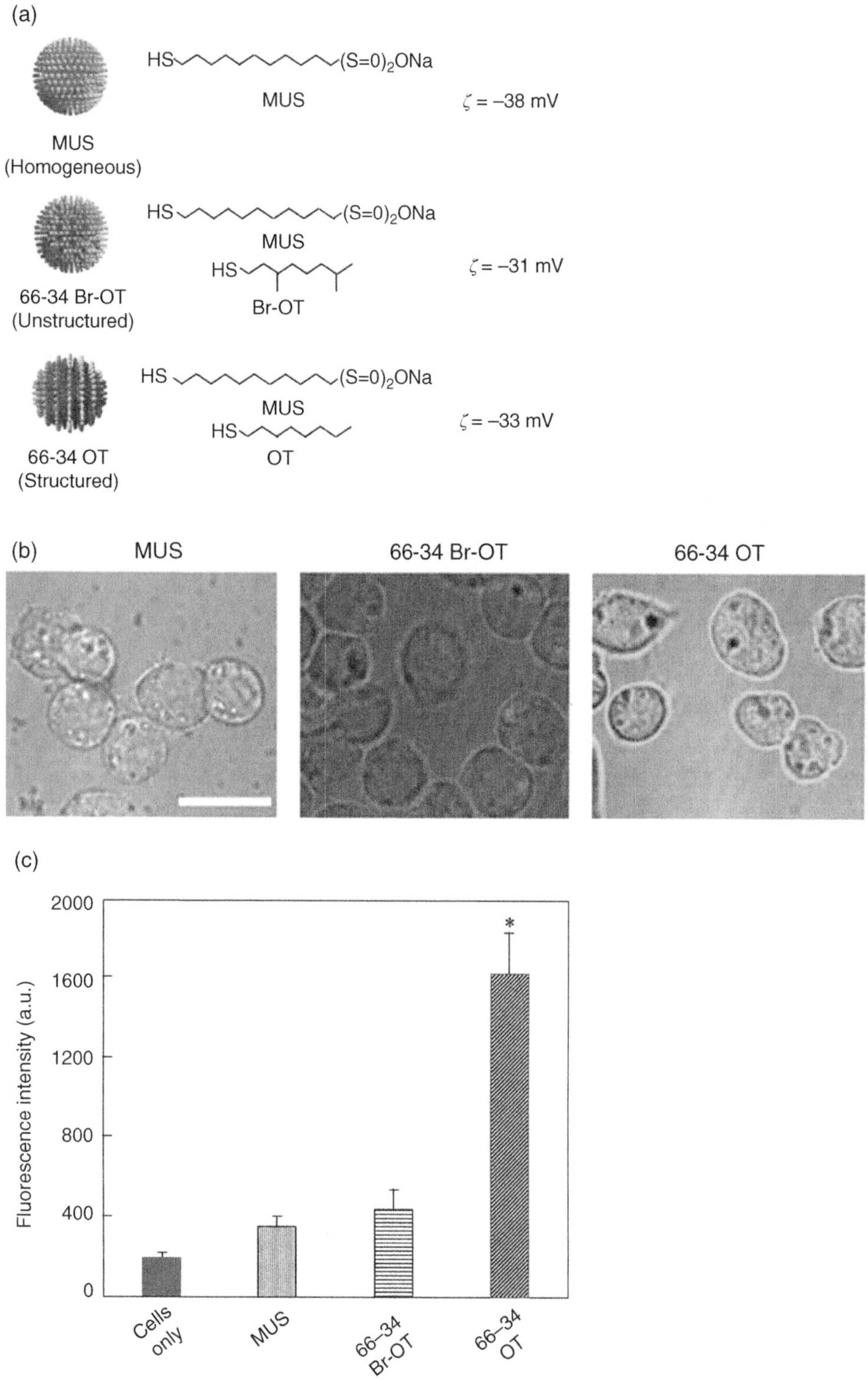

Figure 21.32 Gold nanoparticles with ordered arrangements of hydrophilic and hydrophobic surface functional groups exhibit altered patterns of subcellular localization in cells. (a) Schematic diagrams of nanoparticles with homogeneous ligand shell composed of 11-mercapto-1-undecanesulfonate (MUS), unstructured ligand shell composed of a 2:1 M mixture of MUS and 3,7 dimethyl octane 1-thiol (Br-OT), designated 66-34 Br-OT, and a structured ligand shell composed of 2:1 M mixture of MUS and 1-octanethiol (OT), designated 66-34 OT. The zeta potential (ζ) of the particles with their ligand shell are for measurements in serum-free medium. (b) Bright field fluorescent images of mouse dendritic cells incubated with the nanoparticles at 4 °C. (c) Mean fluorescence intensities (in arbitrary units) from confocal images at 4 °C for the different nanoparticles. *Source:* From Verma et al. (2008) / with permission of Springer Nature.

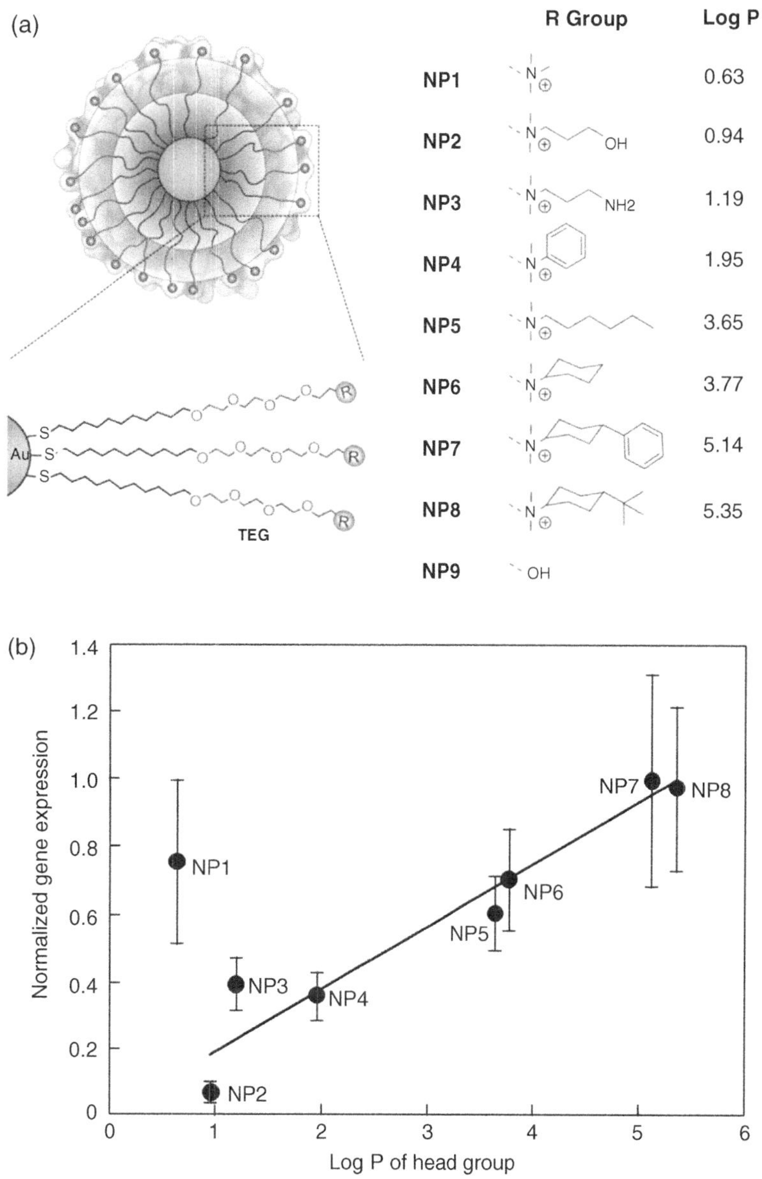

Figure 21.33 Effect of functionalization of gold nanoparticles (diameter ~2 nm) with residues of increasing hydrophobicity. (a) Illustration of nanoparticles functionalized by R groups with increasing hydrophobicity (log *P*) determined from their predicted partition coefficient in the *n*-octanol–water system. (b) Normalized expression of pro-inflammatory TNFα by mouse splenocytes incubated for two hours in the presence of gold nanoparticles functionalized with R groups of increasing hydrophobicity. TEG, tetraethylene glycol. *Source:* Moyano et al. (2012) / with permission of American Chemical Society.

n-octanol–water system. B-lymphocytes harvested from mice were exposed to the surface-modified NPs for two hours and then assayed for the expression of the cytokines IL-2, IL-6, IL-10, TNFα, and interferon-γ (IFNγ) as an indicator of relative immunotoxicity.

As shown in Figure 21.33b, except for the particles terminated with the NP1 group, the expression of TNFα increases with increasing hydrophobicity of the NP surface, which indicates increasing immunotoxicity with hydrophobicity. In comparison, the immune response *in vivo*, determined from injecting the NPs into the tail vein of mice, is more complex. The gene expression of IL-10 at 1.5 hours postinjection increases with hydrophobicity from NP1 to NP4 but decreases slowly thereafter. Additionally, at six hours, correlation between gene expression and hydrophobicity is lost.

NPs can potentially interact with multiple intracellular targets resulting in a variety of adverse effects (Figure 21.34). As noted earlier (Section 21.5.2), positively charged NPs (diameter less than 10–20 nm) can disrupt the plasma membrane by causing transient holes in the membrane. This can permit transient ion flux, which could lead to additional disruption, especially in the case of cellular entry of calcium ions, which trigger unwanted activation of various intracellular signaling pathways. NPs can also interact with membrane proteins resulting in membrane disruption.

Exposures to quantum dots and Ag and TiO_2 NPs are also known to cause mitochondrial swelling and disruption of mitochondrial metabolism. The mitochondrial disruption can lead to generation of reactive oxygen species (ROS) that cause indiscriminate oxidative damage at multiple sites within the cell. Exposures to quantum dots have been shown

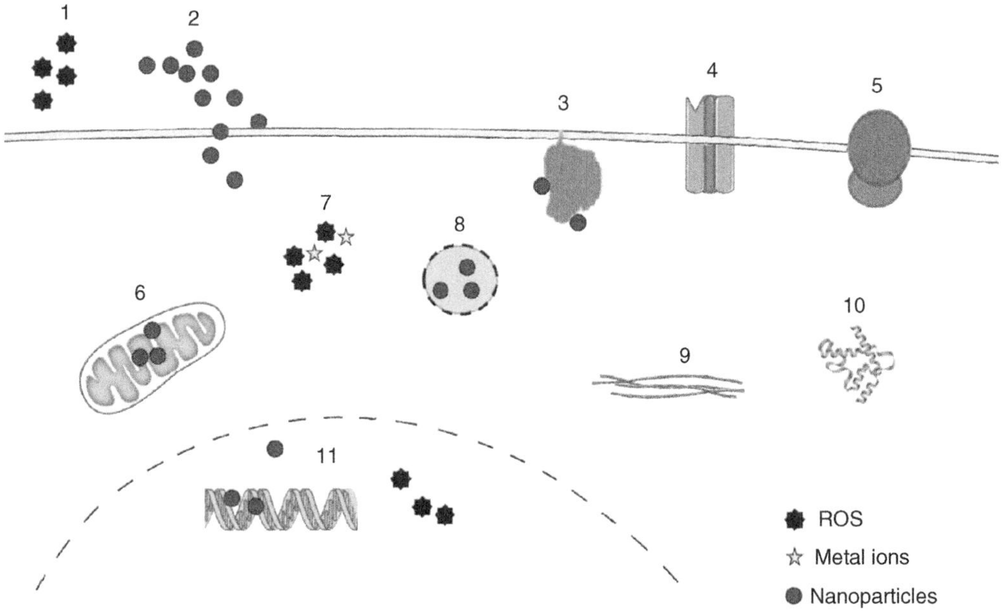

Figure 21.34 Intracellular targets for cytotoxicity of nanoparticles. Nanoparticles may act through extracellular generation of reactive oxygen species (ROS) (1), can damage the plasma membrane by creating transient holes (2), or bind to membrane proteins (3), various membrane ion channels (4), or receptors (5). Internalized nanoparticles can interfere with mitochondrial metabolism (6), causing the generation of radicals and the induction of apoptosis. Intracellular ROS generation by nanoparticles or by metals from lysosomal degradation (7), lysosomal disruption (8), direct binding to components of the cytoskeleton (9), and the induction of structural alterations of cytoplasmic proteins (10) are additional modes of toxic actions. In the nucleus, interference with the transcription machinery and oxidative damage of the DNA (11) may occur. *Source:* Fröhlich (2012) / with permission of Dove Press Ltd.

to cause swelling of lysosomes. Additionally, lysosomes are highly sensitive to oxidative stress, which renders these organelles especially vulnerable to damage by ROS resulting from NP disruption of mitochondrial metabolism.

The nucleus is a target for interaction with intracellular NPs. Electrostatically neutral SiO_2 NPs (less than ~50 nm in diameter) can diffuse into the nucleus and interact with deoxyribose nucleic acid (DNA), resulting in possible genotoxic effect, disruption of transcription, or a combination of both. Charged intracellular NPs can also encounter DNA when the nuclear membrane breaks down during mitosis. Additional DNA damage indirectly attributable to NPs can result from NP-induced formation of ROS. Other intracellular targets for NP-induced cytotoxicity include the cytoskeleton and intracellular proteins. Because of their affinity for proteins, NPs could interfere with protein function. This affinity for proteins could also result in disruption of the cytoskeleton, which could then lead to effects on mobility, cell attachment, morphology, differentiation, and other cytoskeletal-dependent processes.

21.6 Concluding Remarks

In this chapter, we discussed biocompatibility effects related to the surface topography of three-dimensional biomaterials and to particulate biomaterials at the microscale and at the nanoscale. Surface topography and the particulate form can strongly influence the interaction of a biomaterial with cells and tissues, either positively or negatively. Consequently, an understanding of the mechanisms and factors that influence this interaction is useful for optimizing the desirable effects while reducing or avoiding harmful biocompatibility events.

- Surface topography at the microscale or nanoscale is now a major design feature to improve outcomes associated with a variety of implanted biomaterials. We discussed the role that topography can play in modifying cell and tissue response, and the topographical features that influence this response, using examples from osteogenesis, neurogenesis, and macrophage differentiation.
- Certain topographical features at the microscale and nanoscale can improve the antibacterial activity of some implants, while certain features at the nanoscale can provide bactericidal activity. The design and creation of biomaterials having an optimal topography could provide an alternative approach to current methods to treat or prevent bacterial infection.
- We discussed the mechanisms by which microparticles and NPs interact with cells, and the significant particulate characteristics such as chemical composition, size, and shape that influence this interaction. Particulate biomaterials are being used or envisioned for use as devices in a variety of biomedical applications and, consequently, an understanding of these mechanisms is useful for creating devices with optimal functionality.

Problems

21.1 Briefly explain why the type of titanium alloy stems now most commonly used in total hip replacement surgeries are those with a rough surface.

21.2 Briefly explain the difference between antibacterial activity and bactericidal activity. Describe the basis by which the growth of bacterial biofilms on biomaterials with the Sharklet AF surface topology is inhibited, indicating whether the inhibition is an antibacterial effect or a bactericidal effect.

21.3 Consider a surface topography consisting of cylindrical pillars arranged in a square array on a flat surface. Given that the diameter of the pillars is 100 nm and the center-to-center distance between the pillars along a row is 170 nm, determine the areal density of the pillars (number of pillars per unit area).

21.4 A synthetic biomaterial has a surface topology mimicking the pattern of the cicada wing topology. Would the height of the pillars have an effect on the bactericidal activity of the biomaterial? Explain.

21.5 PMMA substrates having a topography mimicking the dragonfly wing topography show a wider range of bactericidal activity when compared to similar substrates mimicking the cicada wing topology. Describe the mechanism and topographical features responsible for this difference in bactericidal activity.

21.6 Particles with a hydrophobic surface are more susceptible to phagocytosis by macrophages when compared to particles with a hydrophilic surface. Explain why.

21.7 Explain why particles with a neutral hydrophilic surface are less susceptible to phagocytosis when compared particles with an electrostatically charged surface.

21.8 For microparticles of diameter larger than ~0.5 μm, the highest amount of phagocytosis by macrophages occurs for particles of diameter ~2-3 μm. Suggest an explanation for this observation.

21.9 Briefly explain why opsonization is not involved in the process of particle uptake by alveolar macrophages.

21.10 Explain the mechanism by which wear debris from hip implants trigger the release of IL-1β and why this poses a serious clinical risk.

21.11 Explain why non-phagocytic cells are attracted to, and interact more intensely with positively charged particles when compared to negatively charged particles.

21.12 Some NPs used as drug delivery devices are coated with a layer of hydrophilic electrostatically neutral polymer molecules. Explain why.

21.13 Explain why NPs with a positive surface charge are considered to have potentially much greater applicability for clinical usage when compared to NPs with a negative charge.

21.14 Gold NPs synthesized by the CTAB method have a positive zeta potential when measured in an aqueous medium without blood serum. Will the zeta potential of the particles change when they are placed in the same aqueous medium containing serum proteins? Explain.

21.15 Compare the surface area of a spherical gold particle of diameter 50 nm with that of a cylindrical gold particle of diameter 14 nm and length 50 nm.

21.16 After a consideration of the various physicochemical properties of NPs, select one property you believe will have a significant effect on the interaction of nanoparticles with host cells and explain briefly what effect this property will have.

References

Beningo, K.A. and Wang, Y. (2002). Fc-receptor-mediated phagocytosis is regulated by mechanical properties of the target. *Journal of Cell Science* 115: 849–856.

Bjursten, L.M., Rasmussen, L., Oh, S. et al. (2010). Titanium dioxide nanotubes enhance bone bonding *in vivo*. *Journal of Biomedical Materials Research. Part A* 92: 1218–1224.

Chambers, T.J., Reveell, P.A., Fuller, K., and Athanasou, N.A. (1984). Resorption of bone by rabbit osteoclasts. *Journal of Cell Science* 66: 383–399.

Champion, J.A. and Mitragotri, S. (2006). Role of target geometry in phagocytosis. *Proceedings of the National Academy of Sciences of the United States of America* 103: 4930–4934.

Champion, J.A., Walker, A., and Mitragotri, S. (2008). Role of particle size in phagocytosis of polymeric microspheres. *Pharmaceutical Research* 25: 1815–1821.

Chithrani, B.D. and Chan, W.C.W. (2007). Elucidating the mechanism of cellular uptake and removal of protein-coated gold nanoparticles of different sizes and shapes. *Nano Letters* 7: 1542–1550.

Chithrani, B.D., Ghazani, A.A., and Chan, W.C.W. (2006). Determining the size and shape dependence of gold nanoparticle uptake into mammalian cells. *Nano Letters* 6: 662–668.

Chung, K.K., Schumacher, J.F., Sampson, E.M. et al. (2007). Impact of engineered surface microtopography on biofilm formation of Staphylococcus aureus. *Biointerphases* 2: 89–94.

Dickson, M.N. Liang, E.I., Rodriguez, L.A. et al. (2015). Nanopatterned polymer surfaces with bactericidal properties.*Biointerphases* 10: 021010-1–02110-8.

Fröhlich, E. (2012). The role of surface charge in cellular uptake and cytotoxicity of medical nanoparticles. *International Journal of Nanomedicine* 7: 5577–5591.

Hatano, K., Inoue, K., Kojo, T. et al. (1999). Effect of surface roughness on proliferation and alkaline phosphatase expression of rat calvarial cells cultured on polystyrene. *Bone* 25: 439–445.

Hazell, G., May, P.W., Taylor, P. et al. (2018). Studies of black silicon and black diamond as materials for antibacterial surfaces. *Biomaterials Science* 6: 1424–1432.

Ishikawa, M., de Mesy Bently, K.L., McEntire, B.J. et al. (2017). Surface topography of silicon nitride affects antimicrobial and osseointegrative properties of tibial implants in a murine model. *Journal of Biomedical Materials Research. Part A* 105: 3413–3421.

Ivanova, E.P., Hasan, J., Webb, H.K. et al. (2012). Natural bactericidal surfaces: mechanical rupture of *Pseudomonas aeruginosa* cells by cicada wings. *Small* 8: 2489–2494.

Kim, S.M., Lee, M.S., Jeon, J. et al. (2018). Biodegradable nerve guidance conduit with microporous and micropatterned poly(lactic-co-glycolic acid)-accelerated sciatic nerve regeneration. *Macromolecular Bioscience* 18: 1800290; 14 pages.

Kunzler, T.P., Drobek, T., Schuler M., and Spencer, N.D. (2007). Systemic study of osteoblast and fibroblast response to roughness by means of surface-morphology gradients. *Biomaterials* 28: 2175–2182.

Leroueil, P.R., Hong, S., Mecke, A. et al. (2007). Nanoparticle interaction with biological membranes: does nanotechnology present a Janus face? *Accounts of Chemical Research* 40: 335–342.

Li, X. (2016). Bactericidal mechanism of nanopatterned surfaces. *Physical Chemistry Chemical Physics* 18: 1311–1316.

Luu, T.U., Gott, S.C., Woo, B.W.K. et al. (2015). Micro- and nanopatterned topographical cues for regulating macrophage cell shape and phenotype. *ACS Applied Materials & Interfaces* 7: 28665–28672.

Magin, C.M., Cooper, S.P., and Brennan, A.B. (2010). Non-toxic antifouling strategies. *Materials Today* 13: 36–44.

Manzano, G.W., Fort, B.P., Dubyak, G.R., and Greenfield, E.M. (2018). Wear particle-induced priming of the NLRP3 inflammasome depends on adherent pathogen associated molecular patterns and their cognate toll-like receptors: an *in vitro* study. *Clinical Orthopaedics and Related Research* 476: 2442–2453.

McWhorter, F.Y., Wang, T., Nguyen, P. et al. (2013). Modulation of macrophage phenotype by cell shape. *Proceedings of the National Academy of Sciences of the United States of America* 110: 17253–17258.

Moyano, D.F., Goldsmith M., Solfiell, D.J. et al. (2012). Nanoparticle hydrophobicity dictates immune response. *Journal of the American Chemical Society* 134: 3965–3967.

Nowlin, K., Boseman, A., Covell, A., and LaJeunesse, D. (2014). Adhesion-dependent rupturing of Saccharomyces cerevisiae on biological antimicrobial nanostructured surfaces. *Journal of the Royal Society Interface* 12: 20140999.

Oh, S.-H., Finones, R.R., Daraio, C. et al. (2005). Growth of nanoscale hydroxyapatite using chemically treated titanium dioxide nanotubes. *Biomaterials* 26: 4938–4943.

Qiu, Y., Liu, Y., Wang, L. et al. et al. (2010). Surface chemistry and aspect ratio mediated cellular uptake of Au nanorods. *Biomaterials* 31: 7606–7619.

Schwartz, Z., Raz, P., Zhao, G. et al. (2008). Effect of micrometer-scale roughness on the surface of Ti6Al4V pedicle screws *in vitro* and *in vivo*. *The Journal of Bone and Joint Surgery. American Volume* 90: 2485–2498.

Sharma, G., Valenta, D.T., Altman, Y. et al. (2010). Polymer particle shape independently influences binding and internalization by macrophages. *Journal of Controlled Release* 147: 408–412.

Verma, A., Uzun, O., Hu, Y. et al. (2008). Surface-structure-regulated cell-membrane penetration by monolayer protected nanoparticles. *Nature Materials* 7: 588–595.

Xu, L.-C. and Siedlecki, C.A. (2012). Submicron-textured biomaterial surface reduces staphylococcal bacterial adhesion and biofilm formation. *Acta Biomaterialia* 8: 72–81.

Zhang, S., Li, J., Lykotrafitis, G. et al. (2009). Size-dependent endocytosis of nanoparticles. *Advanced Materials* 21: 419–424.

Zhu, M., Nie, G., Meng, H. et al. (2013). Physicochemical properties determine nanomaterial cellular uptake, transport and fate. *Accounts of Chemical Research* 46: 622–631.

Further Readings

Behzadi, S., Serpooshan, V., Tao, W. et al. (2017). Cellular uptake of nanoparticles: journey inside the cell. *Chemical Society Reviews* 46: 4218–4244.

Gittens, R.A., Olivares-Navarrete, R., Schwartz, Z., and Boyan, B.D. (2014). Implant osseointegration and the role of microroughness and nanostructures: lessons for spine implants. *Acta Biomaterialia* 10: 3363–3371.

22

Tests of Biocompatibility of Prospective Implant Materials

22.1 Introduction

Biomaterials designed for use in medical devices, as noted in previous chapters, should have the requisite level of biocompatibility over the duration of their application. In this chapter, we will address the topic of biocompatibility phenomena within the context of the testing and regulation required to assure the safety and effectiveness of medical devices that contain biomaterials. After providing an overview of the agencies and organizations that regulate health care products, the process of medical device regulation will be covered. This process will normally begin with an assessment of the potential risks of the proposed device, such as whether it would be used in an invasive or noninvasive manner, after which a plan is developed for evaluating the toxicity of the device to address the various identified risks.

Overall, the level of risk, ranging from low to high, determines the level and complexity of various toxicity tests. The lowest level of complexity normally involves *in vitro* toxicity tests with cultured cells, while tests at the intermediate level involve *in vivo* testing with small animal and large animal models. The highest level of testing, consisting of clinical trials with human subjects, is required when the potential risks of adverse biological responses are assessed as being moderate to high. Various categories of toxicity tests, such as tests focused on hemocompatiblilty, genotoxicity, and other major categories of potentially adverse responses are covered. The principles of these tests, the way in which they work, and the type of results that they provide are discussed. However, the detailed protocols for performing these tests are not covered because they are described more appropriately elsewhere. The final step of the evaluation process involves summarizing the outcomes of all the tests performed into a report to validate the safety and efficacy of the device.

Overall, the major topics that will be covered in this chapter are:

- Standards and agencies for regulating medical devices
- *In vitro* toxicity testing of biomaterials with mammalian cell cultures
- *In vivo* toxicity testing using small and large animal models.

22.2 Biocompatibility Standards and Regulations

Before a new health care product such as a medical device containing biomaterials can be marketed in the United States, the European Union (EU) countries and other major countries, it must be thoroughly evaluated and approved by an appropriate regulatory body for assurance of safety and effectiveness. In the United States, for example, this body is the Food and Drug Administration (FDA), whereas in the EU, biomedical devices are regulated by a network of agencies in the member countries under the coordination of the European Commission. The evaluation processes followed by medical device regulatory bodies, including the FDA and agencies of the EU countries, conform to standards developed by the International Organization for Standardization (ISO), which are contained in a series of documents known collectively as ISO 10993.

22.2.1 ISO 10993

ISO 10993 is a set of standards that address the biological evaluation of medical devices. The evaluation takes into consideration the physical and chemical nature of the component materials making up the device, and the frequency and duration of exposure to the body. The set of standards within the ISO 10993 undergo review every 5 years, with additional revision as needed. Reflecting this revision process, the 2018 version of the ISO 10993 includes a new section specific to guidance standards related to nanomaterials. ISO 10993 is subdivided into twenty parts, with Part 1 defining and describing the applicability of the

Materials for Biomedical Engineering: Fundamentals and Applications, First Edition. Mohamed N. Rahaman and Roger F. Brown.

Companion website: www.wiley.com/go/rahaman/biomedicalengineering

remaining nineteen parts of the document. ISO 10993-1, titled "Biological evaluation of medical devices – Part 1: Evaluation and testing within a risk management process" is, at the time of writing, the most recently updated guidance. Among the various separate ISO standards for biological evaluation of medical devices are tests for in vitro cytotoxicity, genotoxicity, carcinogenicity and reproductive toxicity, interactions with blood, irritation and skin sensitization, and systemic toxicity.

22.2.2 FDA Guidelines and Requirements

The standards recognized by the FDA for evaluating and approving a medical device are based on the standards promulgated in the aforementioned ISO 10993-1. It is noted that the FDA evaluates and approves devices containing biomaterials, and not biomaterials per se. The initial requirement within the FDA evaluation process for a new medical device is the formation of a "Biological Evaluation Plan" consisting of two main parts. The first part is an assessment of potential risks of adverse physiological response to a device based on the extent of direct or indirect contact with the body. The potential risk of adverse response is classified in one of three categories, Class I (low risk), Class II (intermediate risk), or Class III (high risk). The second part of the plan is a detailed list of appropriate biological tests and procedures to address the potential risks of adverse responses identified in the initial assessment. The level of risk, ranging from low to high, determines the stringency of testing required by the FDA to validate that the proposed device is both safe and effective. This validation is required to secure FDA approval for marketing the device.

The process of testing a medical device to assess its potential risk for eliciting adverse biological responses normally consists of a series of increasingly complex steps. As noted earlier, these tests are performed in an order of increasing complexity, from *in vitro* testing with mammalian cell cultures, to *in vivo* testing with small and large animal models, and to clinical trials with human patients.

22.3 *In vitro* Biocompatibility Test Procedures

22.3.1 Cytotoxicity Tests

The procedures utilized in standard *in vitro* cytotoxicity testing of biomaterials involve the use of mammalian cell cultures grown as monolayers in polystyrene (PS) culture vessels, normally after the monolayers reach a confluency of ~80%. The monolayer culture is contacted directly by the biomaterial or by fluid extracts of the biomaterial. An alternative approach involves covering the cell monolayer with an over layer of agarose to cushion the cells from physical injury when they are contacted with the biomaterial. After an appropriate incubation time, the monolayers are examined, which may involve simple qualitative scoring to assess cellular effects beneath and surrounding the biomaterial. Some *in vitro* cytotoxicity assays employ spectrophotometry or other quantitative measurements to evaluate cellular responses to the biomaterial.

Direct Contact Assay

The direct contact assay is a simple and rapid test designed to mimic contact of the biomaterial with cells in the physiological environment. The procedure involves a qualitative assessment of the effect of a biomaterial placed directly on the surface of a sub-confluent monolayer of cultured mammalian cells in a standard 100 mm PS culture plate. After incubation for 24 hours, the culture is examined under an inverted phase contrast microscope for evidence of morphological changes (Figure 22.1). Some of the observable indicators of cytotoxicity include detached dead cells, cell swelling, vacuolization, rounded morphology, and loss of attachment tendrils. The degree of cytotoxic response is assessed qualitatively using a score ranging from 0 (no damage) to 4 (severe). Although the direct contact assay is useful for testing of some biomaterials, it is not appropriate for low-density materials that would float in the culture medium or high-density materials that could crush the cells. A more appropriate qualitative cytotoxicity test for these materials is the agarose overlay (or agarose diffusion) test method.

Agarose Overlay Test

Similar to the direct contact assay, the agarose overlay test is performed with a sub-confluent monolayer of cultured mammalian cells in a standard 100 mm PS culture plate. The liquid growth medium is temporarily removed and replaced by an equal volume of warm medium containing agarose, which is poured over the cells and allowed to solidify. Then the biomaterial is placed on the soft agarose layer, which acts as a cushion to protect the cells from high-density biomaterials or to prevent low-density biomaterials from floating in the liquid growth medium. After incubation for 24 hours, the culture is examined under an inverted phase contrast microscope for evidence of morphological changes in the cells. The indicators of cytotoxicity and the assessment of the degree of cytotoxic response are similar to those for the direct contact assay. The agarose overlay assay is a simple and rapid qualitative test that partially mimics physiological conditions but it also suffers from limitations. For example, the agarose over layer is a simplified simulation of the

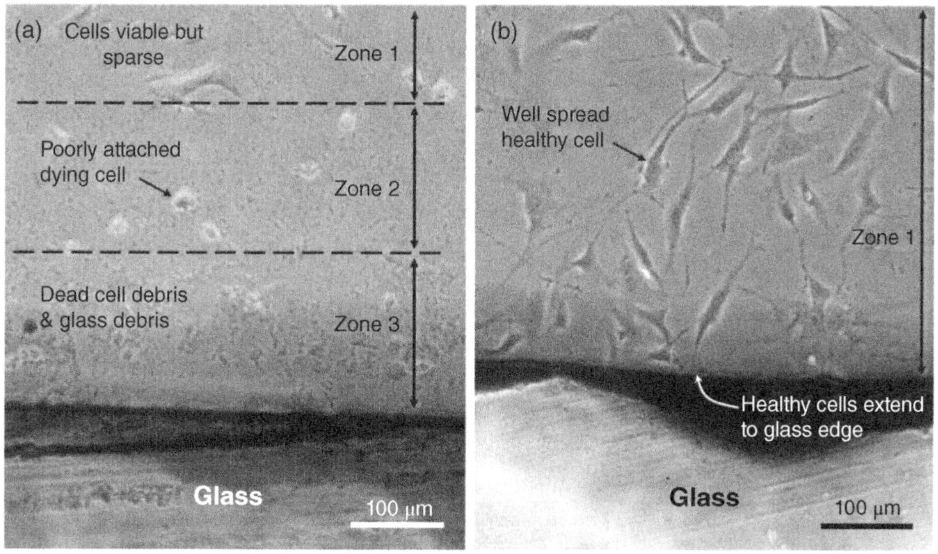

Figure 22.1 Direct contact assay of sub-confluent MC3T3-E1 cells after incubation for 24 hours in contact with two bioactive borate glasses of differing degradation rate: (a) high degradation rate and (b) low degradation rate. Incubation with the high degradation rate glass resulted in a zone of cell death (zone 3) immediately adjacent the glass, while incubation with the low degradation rate glass allowed the zone of well attached, healthy cells to extend to the glass edge.

extracellular matrix, which can allow some water-soluble extracts too easy an access to the cell monolayer.

MTT Assay

The MTT assay is currently the most commonly used *in vitro* method for assessing the potential cytotoxicity of biomaterials. It involves monitoring the reduction of the water-soluble tetrazolium salt 3-(4,5-dimethylthiazol-2-yl)-2,5-diphenyl tetrazolium bromide (MTT) to water-insoluble purple formazan crystals. Because the reduction of MTT to formazan is only catalyzed by succinate dehydrogenase present in functional mitochondria, the amount of MTT activity is directly proportional to the number of live, metabolically active cells. MTT assays are normally performed using an appropriate line of cultured mammalian cells, which are seeded into 96-well tissue culture plates and allowed to attach. Then extracts or leachates of the biomaterial are added to the wells and incubation is carried out for 48–72 hours, with MTT added in the last four hours of the incubation. After completion of the incubation, the plate is washed, blotted dry and ethyl alcohol added for solubilization of the purple formazan crystals, followed by spectrophotometric analysis at a wavelength of 550 nm for quantitation of MTT metabolic activity. The MTT assay is quantitative, rapid, and easy to perform. Additionally, because 96-well plates are used, it allows testing of multiple samples and comparison of their effects with that of known standard cytotoxic substances.

An advantage of the MTT assay is that it is quantitative and does not involve subjective scoring. Another advantage is that the MTT assay can be performed with extracts of the biomaterial being tested or by direct contact with the biomaterial. As an example of the direct contact approach, Figure 22.2a shows relative amounts of MTT reduction by osteoblastic cells seeded on borate glass fiber specimens and incubated for 2, 4, and 6 days. After placing these specimens in ethyl alcohol to solubilize the purple formazan, the amount of formazan, quantitated using spectrophotometry, shows an increase with culture time (Figure 22.2b).

Live/Dead Cell Staining

The live/dead cell staining assay is another widely used *in vitro* cytotoxicity test. The basic principle of the assay is the use of the compounds calcein AM and ethidium homodimer-1 that selectively stain live cells and dead cells, respectively. Calcein AM is a permeable nonfluorescent compound that upon entering live cells, is hydrolyzed by intracellular esterases, enzymes that are present in almost all live cells. The esterase removal of the acetomethoxy (AM) group results in the formation of calcein, a green fluorescent compound that is retained within the cytoplasm of live cells. In comparison, ethidium homodimer-1 is excluded from healthy cells with an intact plasma membrane but diffuses easily into dying cells with damaged membranes, binding to DNA and producing a bright red fluorescence that identifies dead cells.

A typical protocol for the live/dead staining technique involves the use of mammalian cells that are seeded on glass cover discs in 12-well plates. After a duration of ~2 hours to permit cell attachment, culture medium containing an extract of the biomaterial is added to the multi-well plate and incubation is continued for 3–4 days. Upon conclusion of the incubation, the cover discs with attached cells are gently washed with physiological saline and incubated for ~30 minutes in a medium containing a mixture of calcein AM and ethidium homodimer-1. After washing, the cover discs are inverted onto glass slides for examination under an epifluorescence microscope. An example of results obtained by the live/dead staining technique is shown in Figure 22.3. The results of this example compare the response of pig periosteal osteoblasts to culturing on the surface of anodized and non-anodized Ti6Al4V alloy discs. As shown, there is a much higher fraction of live cells on the anodized Ti6Al4V discs and a higher fraction of dead cells on the non-anodized Ti6Al4V discs.

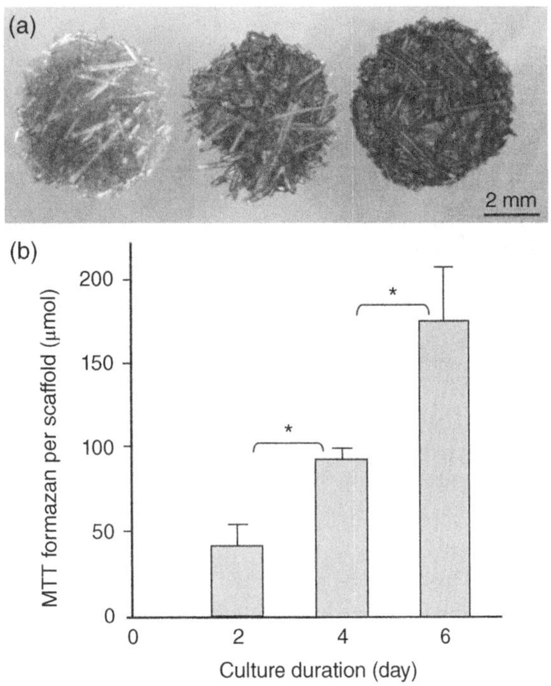

Figure 22.2 (a) MTT assay of proliferation of MLO-A5 osteoblasts seeded on 13-93 glass fiber scaffolds and cultured for 2, 4, and 6 days (left to right, respectively). (b) Spectrophotometric measurement of MTT formazan extracted from 13-93 glass fiber scaffolds. * = $p < 0.05$. *Source:* From Modglin et al. (2012) / with permission of John Wiley & Sons.

22.3.2 Genotoxicity Tests

The Ames Test

The most commonly utilized test for mutagenicity is the Ames test, which is a short-term test for identifying chemical carcinogens based on their ability to trigger reverse mutations in histidine-requiring strains of the bacterium *Salmonella typhimurium*. A chemical suspected of carcinogenic activity is mixed in soft agar with a rat liver supernatant (S9 fraction) plus an inoculum of the *Salmonella* strain and poured onto a Petri dish with solidified minimal agar. The S9 fraction, prepared by homogenization of rat liver in an isotonic KCl solution followed by sedimentation by centrifugation at 9000g, contains cytochrome P450 isozymes that catalyze metabolic activation of chemical pro-mutagens to their mutagenic form. Control plates are prepared in the same manner but without the S9. After incubation for 48 hours, the number of histidine positive revertant colonies is scored (Figure 22.4). Only a limited number of

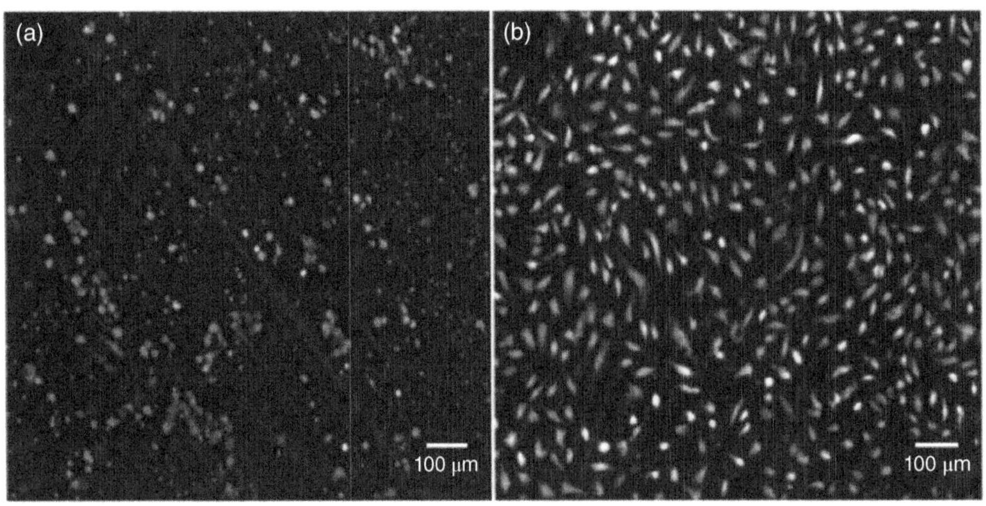

Figure 22.3 Live/dead staining of pig periosteal osteoblasts after three days on (a) non-anodized Ti6Al4V alloy and (b) anodized Ti6Al4V alloy. Staining with calcein AM and ethidium homodimer-1 followed by examination under epifluorescence allows detection of green-fluorescing live cells and red-fluorescing dead cells, respectively. Scale bar = 100 μm. *Source:* From Beltrán-Partida et al. (2015).

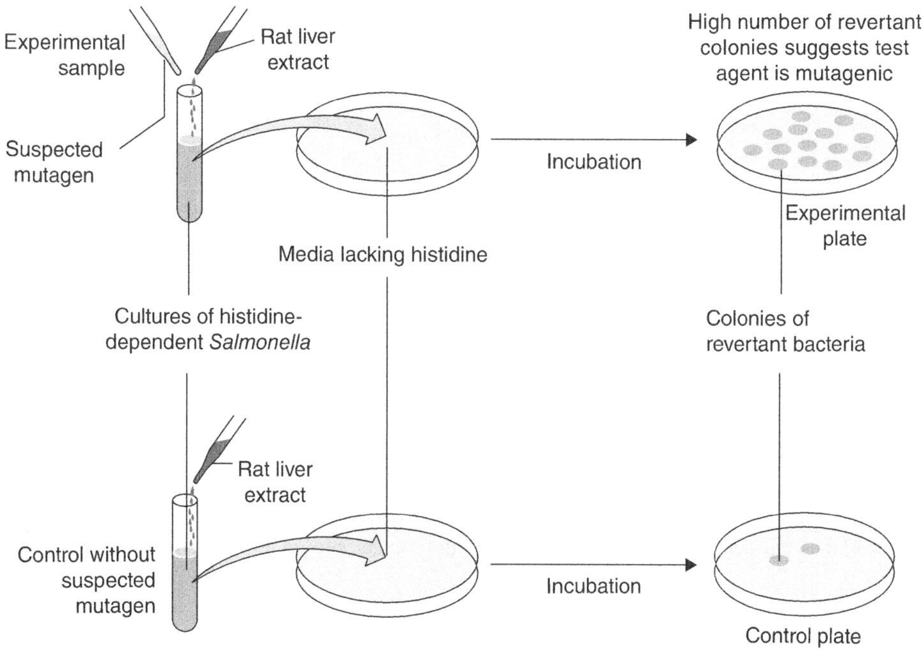

Figure 22.4 Ames test for mutagenicity. Test chemical suspected of carcinogenicity is mixed in soft agar with a rat liver supernatant plus inoculum of a histidine-requiring *Salmonella* strain and poured onto dish with solidified minimal agar. Formation of colonies resulting from reverse mutation to histidine-independent phenotype is positive indication of mutagenic activity in test chemical. *Source:* From Creative Proteomics; with permission.

spontaneous revertant colonies will normally form in the control Petri dishes. In comparison, the presence of a larger number of colonies in dishes with the S9 fraction is an indication that the test chemical causes reverse mutation from the histidine-requiring phenotype to a histidine-independent phenotype.

The formation of histidine-independent colonies is an indication that the test chemical is very likely a carcinogen because ~70% of known carcinogens are detectable as mutagens in the Ames test. However, such results should be interpreted with caution. For example, some chemicals such as nitroglycerin and the amino acid arginine produce false positive results in the Ames test. False negative results can also occur. For example, the potent carcinogen diethylstilbestrol is not detected as a mutagen in the Ames test. Consequently, it is prudent to utilize different types of tests to assess the possible genotoxicity of new biomaterials.

Mammalian Genotoxicity Tests

Several different types of genotoxicity tests using mammalian cell cultures are available for the detection of chemicals that induce gene alterations. Among these, the mouse lymphoma assay (MLA) is currently recognized by the FDA as the assay of choice based on the extensive body of literature indicating that it is effective in detecting many types of genetic alterations. The MLA is performed with the mouse lymphoma cell line L5178Y $TK^{+/-}$, a cell line that is heterozygous for thymidine kinase (TK). Heterozygous refers to an individual having two different alleles for a specific trait, an allele being a version of a gene or specific DNA sequence on a chromosome.

Similar to the Ames assay, the MLA procedure includes culturing the $TK^{+/-}$ cells with the test chemical substance in the presence of liver S9 fraction to achieve metabolic activation of the test agent. However, the MLA is more complex, labor intensive, and time-consuming. The principle of the assay is that cells lacking thymidine kinase due to mutation of the $TK^{+/-}$ phenotype to $TK^{-/-}$ will be unaffected by the cytotoxic effects of trifluorothymidine (TFT), a nucleoside analogue. The mutant $TK^{-/-}$ cells will proliferate and form colonies in the presence of TFT while normal $TK^{+/-}$ cells will not (Figure 22.5).

The standard MLA protocol involves growth and treatment of log phase $TK^{+/-}$ cells in suspension culture. Treatments with the test chemical are done with and without the addition of liver S9 fraction prepared from the livers of rats treated with Aroclor 1254, an agent that induces formation of the cytochrome P450. The typical duration for treatment with the test chemical is 3–6 hours. Then the cells are washed with physiological saline to remove the test agent, suspended in complete growth medium RM10 (regenerative medium 10), and incubated for ~3 days or longer to permit phenotypic expression of the induced mutations.

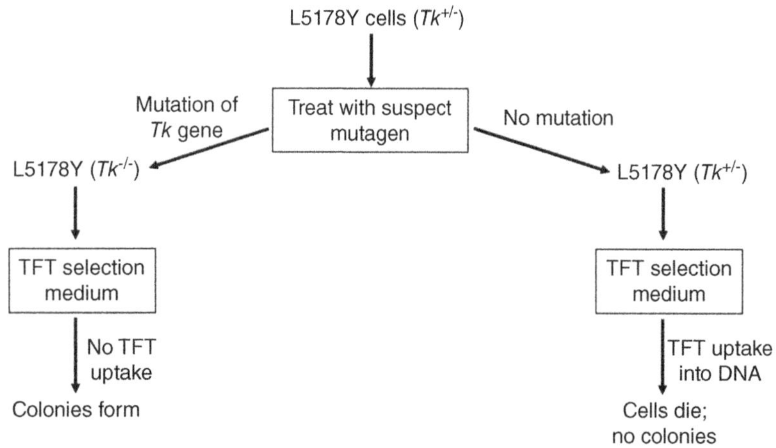

Figure 22.5 Mouse lymphoma assay performed with the mouse lymphoma cell line L5178Y $TK^{+/-}$. The basic principle of the assay is that cells lacking thymidine kinase due to mutation of the $TK^{+/-}$ phenotype to $TK^{-/-}$ will be unaffected by the cytotoxic effects of the nucleoside analogue trifluorothymidine (TFT). The mutant $TK^{-/-}$ cells will proliferate and form colonies in the presence of TFT while normal $TK^{+/-}$ cells will not.

At the conclusion of the three-day expression period, the cells are suspended in selection medium (such as RM10 medium with TFT) and seeded into 96-well plates at a density of ~500–2000 cells/well for detection of mutant cell clones. Companion control 96-well plates are prepared with cells suspended in normal medium and seeded at a density of 1 cell/well for the determination of cloning efficiency. Then the control 96-well plates seeded at 1 cell/well and test plates with ~500–2000 cells/well are incubated for 10–14 days to permit growth of colonies to sizes adequate for visualization. The number of mutant colonies that form in the TFT selection medium is counted and expressed relative to the number of colonies that form in normal medium without TFT. The MLA procedure is an effective and sensitive test for the detection of chemicals that induce gene alterations but, as noted earlier, it is a difficult, labor-intensive, and expensive assay due to the many steps involved.

Several additional types of *in vitro* assays are available for the detection of genotoxicity based on the use of Chinese hamster CHO-K1 and V79 cells, human lymphoblastoid TK6 cells and other cell lines. These assays detect genotoxicity by a variety of mechanisms such as specific gene mutations, damage and subsequent repair of DNA, and detection of chromosomal aberrations. For example, the HPRT test is an *in vitro* assay designed to detect mutagenic materials via induction of mutations in the hypoxanthine-guanine phosphoribosyl transferase (HPRT) gene. The protocol for the HPRT test, which is usually performed with Chinese hamster V79 cells or human lymphoblastoid TK6 cells, is quite similar to the MLA. The unscheduled DNA synthesis (UDS) assay is designed to monitor DNA synthesis due to repair of a damaged segment of DNA by the excision repair pathway. The UDS designation refers to synthesis of DNA at a time during the cell cycle other than the S phase when DNA is replicated in preparation for cell division. The UDS assay typically involves autoradiography to track the incorporation of radioactive thymidine in the nuclei of mammalian cells that are cultured on glass slides, including a period of exposure to a test material.

Another approach to the detection of genotoxic agents involves testing for potential induction of chromosomal aberrations and sister chromatid exchanges (SCEs) in cultured mammalian cells treated with biomaterials. Direct visualization of SCEs in Chinese hamster V79 cells is an effective method of detecting genotoxic agents. The basic SCE procedure involves incubation of cells in the presence of the biomaterial and the thymidine analog 5′-bromodeoxyuridine (BrdUrd) for ~30 hours. This duration is sufficient for two rounds of DNA replication, resulting in unifilarly BrdUrd-substituted DNA in one chromatid of each chromosome and bifilarly BrdUrd-substituted DNA in the sister chromatid. Then mitotic cells are collected, swollen in hypotonic saline and applied to glass slides for spread release of metaphase chromosomes. Staining with the dye Hoechst 33258 is followed by exposure to long-wave UV light, a procedure that produces differential staining of the chromatids, thereby allowing visualization of SCEs (Figure 22.6).

22.3.3 Hemocompatibility Tests

A large number of devices such as catheters, vascular grafts and stents interact directly with whole blood (blood drawn directly from the body) or with components of blood. Consequently, hemocompatibility is one of the most important categories of mandated testing for assurance of device biocompatibility and efficacy. It is particularly important to test new biomaterials to establish with high confidence that they do not damage any of the various blood cells or promote undesirable effects such as hemostasis or activation of the complement system. Consistent with this requirement, it is noteworthy that the guidelines of ISO 10993-4 specify five separate categories of recommended tests for the interaction of devices with blood, including tests that address thrombosis, coagulation, platelets, hematology, and the complement system.

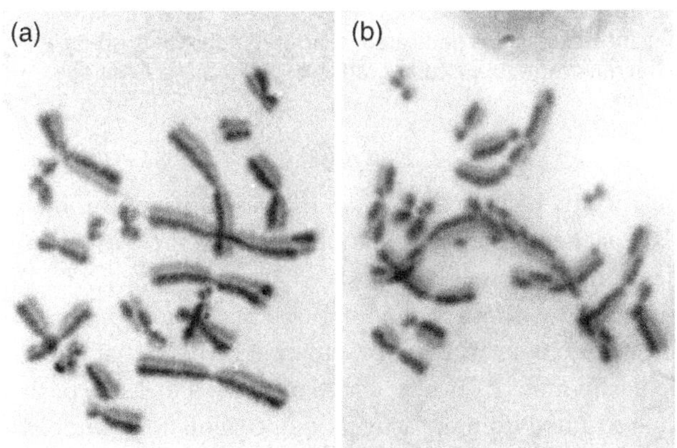

Figure 22.6 Detection of genotoxicity by visualization of sister chromatid exchanges (SCEs). The SCE procedure involves incubation in the presence of test material plus thymidine analog 5′-bromodeoxyuridine (BrdUrd) for a duration sufficient for two rounds of DNA replication resulting in unifilarly BrdUrd-substituted DNA in one chromatid and bifilarly BrdUrd-substituted DNA in the sister chromatid. Staining with Hoechst 33258 followed by UV exposure with the examples shown allow comparison of SCEs in (a) V79 control cells and (b) V79 cells treated with the known mutagen ethyl methanesulfonate, 100 μg/ml.

Complement Activation C3a and SC5b Assay

Biomaterials introduced into the body, as noted in Chapter 16, are rapidly coated with a layer of adsorbed biomolecules composed mainly of proteins. This protein layer promotes the activation of complement system proteins such as C3a and C5a, which recruit inflammatory cells to the site, causing potentially damaging effects. Because complement activation is one of the most prevalent immune defense responses triggered by the introduction of a foreign material into the body, assays for complement activation are particularly important for the reliable assessment of potential immunological effects of devices that contact circulating blood. In this regard, assays specific for the complement proteins C3a and sC5b-9 (the soluble form of the C5b-9 membrane attack complex) are recognized as valid indicators of complement activation. As an example, an early increase of soluble sC5b-9 in circulating blood has been found to be a predictive marker for later development of renal thrombotic microangiopathy following hematopoietic stem cell transplantation. In a similar manner, testing performed with radioimmunoassay techniques to assess the response to titanium alloy dental implants revealed significant activation of C3a and C5a by the implants. However, caution should be exercised in interpreting such results because complement activation is an important part of healing and not always a predictor of a poor outcome.

A widely used procedure for *in vitro* detection of C3a and sC5b-9 in plasma involves the use of an enzyme-linked immunosorbent assay (ELISA) specific for either C3a or sC5b-9. The common form of commercially available test kits consists of a solid-phase sandwich ELISA in polystyrene 96-well plates, illustrated schematically in Figure 22.7. The procedure shown for the detection of C3a involves incubation of the biomaterial in human whole blood followed by withdrawal of plasma. Then, aliquots of the

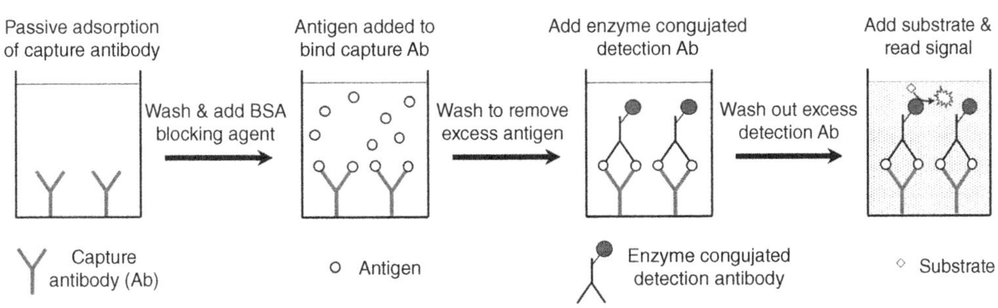

Figure 22.7 Enzyme-linked immunosorbent assay (ELISA) such as for detection of activated complement in human blood exposed to test material. Wells of 96-well plate are coated with capture antibody (e.g., rabbit anti-human C3a antibody). Diluted plasma from the exposed human blood is then added to each well. After excess antigen is removed, detection antibody (e.g., mouse anti-rabbit antibody conjugated with marker enzyme) is added followed by marker enzyme substrate to detect presence of human anti-complement antibody.

diluted plasma are pipetted into wells pre-coated with a primary capture antibody, such as rabbit antihuman C3a antibody. Following incubation for a suitable duration, the wells are washed and albumin is added to block nonspecific binding. Then a secondary antibody, such as mouse anti-rabbit antibody conjugated with the enzyme horseradish peroxidase, is added followed by further incubation. Finally, the plate is washed and a solution containing a chromogenic peroxidase substrate is added. The resultant signal produced by the peroxidase reaction is measured by spectrophotometry. Assays of this type are fast, reliable and yield quantitative results.

Platelets

Another type of undesirable effect that could potentially occur due to the interaction of a biomaterial with blood is the activation of platelets, which could lead to hemostasis or thrombosis. Non-activated circulating platelets have a disc-like shape, a morphology that permits their unimpaired transport within the systemic blood flow. During the process of activation, platelets undergo a change in morphology characterized by the extension of numerous membrane projections that cause these activated cells to adhere to the endothelium and other surfaces (Figure 16.21). The formation of these membrane projections causes platelet clumping, a major factor in the physiological sequence that leads to hemostasis (Chapter 16). The potential for this type of adverse outcome is a major reason for including the analysis of platelet activation as an important test of hemocompatibility.

A variety of methods is available to assess platelet activation. One method involves the use of solid-phase sandwich ELISA procedures to detect a specific biological mediator known as platelet activating factor, a biologically active phospholipid mediator that promotes platelet activation. Another method, which also employs spectrophotometry with 96-well microtiter plates, involves the detection of platelet aggregation by a reduction of turbidity that occurs when a turbid suspension of platelets undergoes clumping. Examination in the scanning electron microscope (SEM) can be used to directly visualize morphological changes of platelets after exposure and adherence to various biomaterials. As an example, the SEM images in Figure 22.8 show the formation of dendritic membrane extensions by platelets adhered to the surface of Dacron synthetic vascular grafts.

Figure 22.8 SEM visualization of activated platelets with dendritic extensions (indicated by arrows) adhered to surface of a Dacron synthetic vascular graft. *Source:* From Weber et al. (2018).

Coagulation

An increase or decrease in blood coagulation time caused by a biomaterial or its degradation product is often viewed as a potentially adverse outcome. This is because an increase in coagulation time would be indicative of an increased risk of excessive bleeding, whereas a decrease in coagulation time would be indicative of an increased risk of thrombosis. For this reason, coagulation is among the five categories of hemocompatibility tests.

One of the most commonly used assays for testing the effects of biomaterials on coagulation is the partial thromboplastin time (PTT) test, which provides a measure of the time required from the activation of coagulation Factor XII to the formation of a stable fibrin clot (Chapter 16). The PTT test is also known as the kaolin-cephalin clotting time (KCCT) test because of the use of kaolin (a clay composed of hydrated aluminosilicate crystals) and cephalin in the standard protocol. In preparation for performing the PTT test, a sample of whole blood is mixed with citrate ions to prevent the blood from clotting, which would invalidate the test. Then the citrate-treated plasma is incubated with the biomaterial for a specific time (approximately minutes). After this incubation, cephalin is added, which reverses the anticoagulant action of the citrate. Formation of a clot, which is the basis for the measurement of PTT, is initiated by the addition of kaolin, which acts as a substitute for platelets that normally serve as the site on which a clot will form (Chapter 16). A deviation from a normal PTT clotting time (~35 seconds) would create the need for additional tests to assess hemocompatibility. The importance of the PTT test is seen from its routine use to detect coagulation abnormalities in patients prior to surgical procedures.

Blood Cell Numbers and Hemolysis

Another important parameter for assessing hemocompatibility is the assurance that contact of blood with the biomaterial does not cause a decrease in the number of circulating leukocytes and erythrocytes, or cause the release of the contents of these blood cells. High levels of free hemoglobin in the plasma due to erythrocyte damage often lead to severe kidney injury, which may lead to kidney failure. It is therefore necessary to ensure that new biomaterials do not have any potentially damaging effects on leukocytes or erythrocytes.

The procedure for assessing the effect of a biomaterial on erythrocyte integrity commonly involves the use of diluted rabbit blood with an added anticoagulant. The biomaterial or an extract of the material is incubated in the diluted blood for a given time, after which chemical tests are performed to assess the presence of free hemoglobin, an indicator of an adverse hemolytic effect. In a similar manner, quantitation of the erythrocyte cell number per microliter and a differential count, that is, the proportion of the different types of white cells in the blood (per microliter) can be performed with blood samples before and after incubation with the biomaterial. Cell counts of this nature provide a method of screening materials for potential adverse effects on specific components of the blood.

Blood Flow Loop

Another procedure for screening biomaterials for hemocompatibility is a procedure referred to as the blood flow loop, which is an *in vitro* shear flow model designed to simulate the dynamic conditions of systemic blood flow in vivo. An example of a blood flow loop apparatus is the Chandler loop, which has seen considerable use in assessing the interaction between various devices and whole blood. In its simplest form, the Chandler loop consists of circular loops of tubing, rotating in a vertical plane in a constant-temperature water bath. The biomaterial or device to be tested forms all or part of the tubing, or is placed tightly within the tubing. Then the tubing is filled nearly completely with fresh blood leaving a small pocket of air, closed with a short piece of tubing that fits tightly around its ends, and is rotated at a constant speed.

As an example, the Chandler loop apparatus has been used to investigate the interaction of human blood with stents coated with various surface-bound bacterial pyrogens such as lipopolysaccharides (LPS) of gram-negative bacteria or lipoteichoic acids (LTA) of gram-positive bacteria. After rotation of the loop system at 30 rpm for 90 minutes at 37 °C, the blood was removed and analyzed for the activation of complement system components and coagulation factors, while the stents were removed and examined in the SEM for adherent cells. The SEM examination showed that a large number of platelets adhered to the pyrogen-coated stents whereas the control stents were free of adherent platelets. Assays of the blood showed high levels of activated complement C5a and C3a, plus neutrophil elastase, a marker of neutrophil activation. The results of this study are valuable in the sense that they focus attention on the importance of not allowing stents to be exposed to pyrogens to assure their continued patency.

22.4 *In vivo* Biocompatibility Test Procedures

In vivo testing in animal models is a critical part of assessing the safety and effectiveness of new materials developed for potential use in the human body for a variety of reasons. Most importantly, *in vitro* tests are, to a large extent, unable to simulate the *in vivo* physiological environment. For example, biomaterials–tissue interactions cannot be simulated to any significant extent by *in vitro* tests. Additionally, most *in vitro* tests with cell cultures are performed under static conditions that do not replicate the dynamic nature of the physiological environment such as the systemic blood flow. Also absent from *in vitro* tests are the effects of hormones, growth factors, and neural control.

22.4.1 Implantation Tests

The first type of *in vivo* testing normally involves surgical implantation of a biomaterial in animals to evaluate its biocompatibility when it comes into direct contact with tissues within the body. All animal experimental procedures should be carried out with prior approval from the relevant institutional animal care and use committee, in compliance with approved guidelines. In the United States, the appropriate guidelines are the National Institutes of Health (NIH) Guide for Care and Use of Laboratory Animals. In most cases, *in vivo* implantation tests are initially performed using small animal models, such as rodents or rabbits. Rodent models provide the advantages of economy in cost and in space for housing the animals, but the implant size is limited. As far as possible, the selected implantation site should be similar to the intended application site for which the biomaterial is designed. It is also important to select an appropriate species for the testing. Assuming favorable biocompatibility, the next type of tests would normally involve large animals, such as dogs, pigs, or sheep. For example, pigs are normally used for evaluating cardiovascular devices such as prosthetic heart valves and vascular grafts because of the close similarity of the porcine cardiovascular system with the human system.

Subcutaneous Implantation Tests

In vivo testing of biomaterials in small animal models often starts with subcutaneous implantation in rodents, for periods ranging from a few days to a few months. At appropriate intervals, the animals are euthanized according to the aforementioned institutional guidelines and examined for local effects on tissues surrounding the implant and, in some cases, for effects at distant sites, as well as macroscopically observable effects. The types of evaluation performed to detect possible adverse effects often include histopathological examination, biochemical assays, and immunological assays for inflammatory effects. Additionally, SEM can be performed to identify cells and tissues attached to the implanted biomaterial.

An example of a subcutaneous implantation test is an investigation designed to evaluate soft tissue responses to ultra-high molecular weight polyethylene (UHMWPE) with or without a covalently bonded low molecular weight macrocyclic ligand that mimics superoxide dismutase (SOD). The objective of this investigation was to determine whether the addition of the SOD mimic would reduce the inflammatory response to UHMWPE, a material known to elicit moderately inflammatory responses *in vivo*. Disc shaped specimens of UHMWPE with or without the SOD mimic were implanted in subcutaneous sites in the dorsum of adult female rats for periods ranging from 3 to 28 days. Histopathological analysis of the soft tissue response at three days postimplantation showed a neutrophil-rich dense layer of tissue surrounding the control UHMWPE discs without the SOD mimic. In comparison, the soft tissue surrounding the discs coated with the SOD mimic was devoid of neutrophils (Figure 22.9). Implantation tests of this type are helpful in identifying procedures for improving the biocompatibility of widely used biomaterials such as UHMWPE.

Intramuscular Implantation Tests

One of the recognized limitations of subcutaneous implantation tests is the low vascular density of the subcutis. Intramuscular implantation is an alternative to subcutaneous implantation in situations where the higher vascular density of muscle tissue would better simulate responses that could occur in well-vascularized tissues. The specific muscle and animal model used for the test should be selected appropriately. For example, the easy accessibility of the thigh muscle of the rat could form the basis of selecting this model for intramuscular implantation of some biomaterials. Alternatively, a muscle such as the erector spinae adjacent to the spine of rabbits could be selected to minimize movement of the implanted biomaterial.

An example of an intramuscular implantation test is an investigation of tissue response to the injection of a hydrogel into the thigh muscle of adult female rats. The biomaterial consisted of a mixture composed of poly(*N*-isopropyl acrylamide) (PNIPAAm), a temperature-responsive hydrogel that gels upon warming to the physiological temperature (Chapter 11), and a decellularized tissue composed of urinary bladder matrix, a substance known to enhance wound healing. The objective of the investigation was to determine whether the injectable mixture could provide a scaffold for cell ingrowth and repair of soft tissue defects in surgical resections and traumatic injuries. Histological examination of the muscle tissue 21 days after the injection showed extensive cell ingrowth into the hydrogel containing the decellularized tissue, along with little fibrous encapsulation (Figure 22.10). In comparison, the hydrogel without the tissue showed little cell infiltration and a thick fibrous encapsulation.

Bone Implantation Tests *In Vivo*

A widely used implantation test to evaluate the potential adverse effects of biomaterials involves implantation into bone, initially using small animal models such as rodents

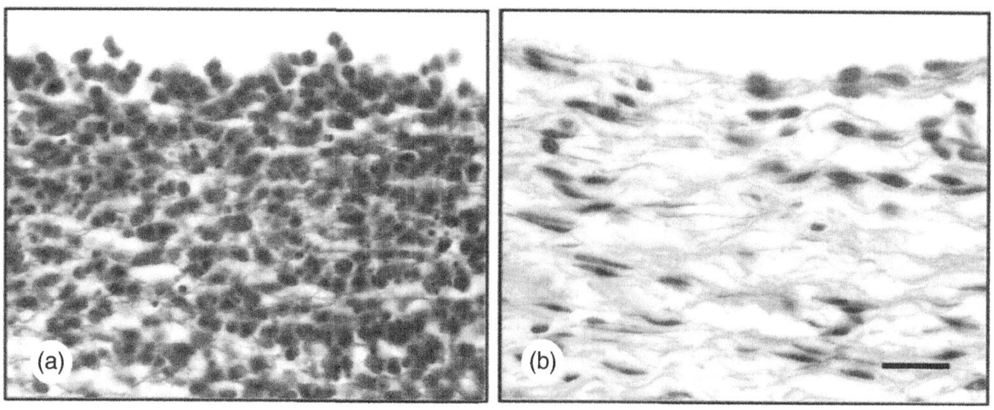

Figure 22.9 Histological analysis of tissue response to subcutaneous implantation of UHMWPE samples with and without covalently bonded low molecular weight ligand mimicking superoxide dismutase (SODm). Analyses at three days reveals neutrophil-rich dense layer formed around (a) control UHMWPE discs without SODm while the soft tissue around samples with SODm (b) was devoid of neutrophils. Bar = 50 μm. *Source:* From Udipi et al. (2000).

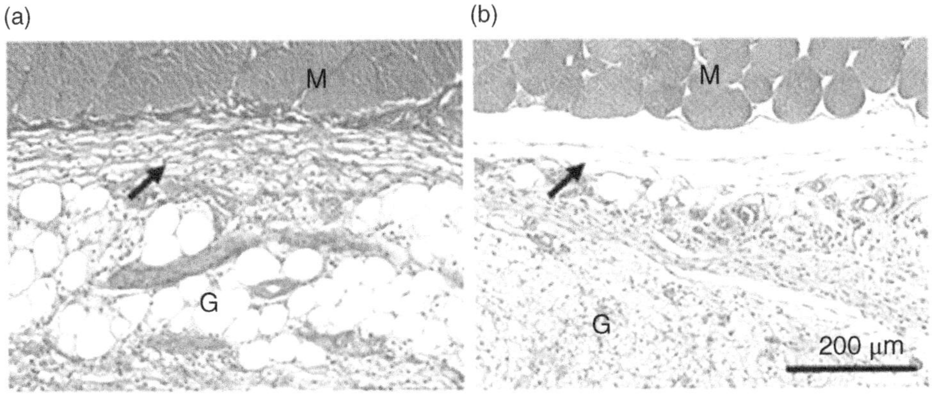

Figure 22.10 Histological analysis of intramuscular response to implantation of a biodegradable porous hydrogel (G) with or without decellularized urinary bladder matrix in the thigh muscle (M) of female rats. Examination at 21 days revealed (a) control hydrogels without the matrix digest developed a thick fibrous encapsulation (arrow) while hydrogels with the matrix digest (b) had minimal fibrous encapsulation (arrow). *Source:* From Zhu et al. (2018).

or rabbits. Depending on the intended application of the biomaterial, the implantation site could be selected based on bone type, such as periosteal or cancellous bone, or based on the load-bearing capacity of the bone, such as structural (load-bearing) bone or nonstructural (non-loaded) bone.

Because bone is a hard, mineralized tissue, evaluating its response to an implanted biomaterial can be more challenging when compared to a soft tissue. For example, conventional microtome techniques cannot be used to prepare standard 5 μm thin sections of bone for histopathological examination. Instead, thin sections are normally prepared by microtome techniques after demineralization by soaking the bone specimens in a calcium-chelating solution such as ethylenediaminetetraacetic acid (EDTA) or in an appropriate acidic solution. Then the specimens are dehydrated, embedded in paraffin and sectioned using a microtome to a thickness of ~5 μm (see Figure 23.9).

Alternatively, the undecalcified bone specimen can be embedded in polymethyl methacrylate (PMMA) to form a resin block that is subsequently sectioned with a diamond-coated blade. Then the sections are ground to a thickness of ~50–70 μm, polished, and stained with an appropriate histological stain to visualize the soft tissue and the mineralized tissue using light microscopy. Polished sections of these PMMA-embedded specimens can also be examined in the SEM using the backscattered electron imaging (BSE) mode. This imaging mode can be used to distinguish between different phases in the specimen cross-section, such as mineralized tissue and vascularized tissue, because it is sensitive to changes in the average atomic number of each phase.

X-ray radiography, computed tomography (CT) and micro-computed tomography (micro-CT/μCT) provide complementary techniques to histological analysis in evaluating biomaterials implanted into bone. CT and micro-CT, like radiography, use X-rays to create cross-sections of the implanted biomaterial and surrounding bone without having to euthanize the animal. Another important feature of CT and micro-CT is that they can be used to recreate a three-dimensional image of the implant within the bone. Consequently, they provide complementary information in evaluating new bone formation and integration of the implant with host bone. Micro-CT is similar to CT but it provides much better resolution, in the micrometer range. It has become very popular in analyzing biomaterials implanted into the bones of small animals, such as calvarial or tibial sites, particularly to evaluate bone ingrowth into the implants and their integration with host bone (see Figure 23.15).

22.4.2 Thrombogenicity Tests

Thrombogenicity tests are particularly important in the evaluation of implantable biomaterials because of concerns that devices in contact with flowing blood could potentially cause adverse biological responses in vivo, such as thrombosis. The types of devices that elicit the highest levels of concern are synthetic and natural vascular grafts because the functionality of these devices depends critically on their ability to maintain their patency and avoid the potential for thrombogenic effects.

In vivo tests for thrombogenicity can be performed with small or large animal models. Testing of vascular grafts in rodents usually involve placement of the graft in the abdominal aorta, while tests in rabbits, dogs, sheep, and pigs can include placement in medium size vessels, the abdominal aorta or the carotid arteries. Potential thrombogenic effects can be evaluated using physiologic monitoring of blood flow by Doppler ultrasound or CT angiography, as well as by gross examination of vessel occlusion or the use of various tests to assess thrombosis and blood cell adhesion to the surface of the graft material.

As an example of *in vivo* testing of vascular grafts for patency and thrombogenicity, grafts created from decellularized human amniotic membrane (3.6 cm in length) were placed in a carotid artery of male rabbits. After four weeks, Doppler ultrasound confirmed the patency of the grafts, while histological analyses showed that the grafts had a very low neutrophil content, which is an indication that they elicited little inflammation. These analyses also showed that the initially acellular grafts became completely filled with cells in organized, circularly aligned layers with an abundance of newly formed collagen (Figure 22.11). The cells of the middle layer contained α-actin, a smooth muscle marker, indicating that this layer is a physiologically functional tunica media (middle layer of an artery or vein).

22.4.3 Irritation and Sensitization Tests

In vivo testing of medical devices for potential adverse responses also includes tests for irritation and sensitization. Irritation is a localized physiological response to various chemical and physical irritants, which damage cells of the epidermis resulting in erythema and edema (redness and swelling) of the skin. These irritants are basically universal irritants, such as capsaicin that provoke essentially the same response in all individuals. The erythema and edema reactions to an irritant occur immediately without involvement of the immune system and quickly subside when the irritant is removed.

Irritation tests are normally performed mainly with albino rabbits using one of three versions of the irritation test known as primary skin irritation, intracutaneous irritation, or ocular irritation. Primary skin irritation and intracutaneous irritation tests require shaving the skin of the animals. The primary skin test involves a patch test procedure in which irritants and suitable controls are applied directly on the skin surface. After exposure for 24 hours, the extent of erythema and edema at the irritant application site is examined and qualitatively scored. The intracutaneous test involves preparation of a fresh saline or vegetable oil extract of the biomaterial and injecting it by means of a small-bore needle into the dermal layer of the skin at three sites, while the carrier liquid (control) is injected at three additional sites. After 24 hours, erythema and edema above the sites of intradermal injection of the material and the control are qualitatively scored and compared to assess possible irritation effects of the biomaterial. The ocular irritation test (also called the Draize test) involves applying the test material or an extract of it directly to the surface of the eye, after which the animals are observed over a period of 14 days to assess the extent of any erythema and edema induced.

The most widely recognized *in vivo* sensitization test involves assessment of dermal sensitization reactions in guinea pigs. The rationale for this animal model is that the dermal sensitization reactions of the guinea pig are similar to those of humans. Although there are several variations of the guinea pig sensitization tests, they all consist of two main parts, induction and challenge. A commonly used variation of the guinea pig sensitization test is the "adjuvant and patch" test (APT). The induction phase of the APT protocol is initiated by administering several intradermal injections of Freund's adjuvant (an agent that enhances the immune response) at multiple sites on the back of the animal. Then the skin above each injection site is abraded, followed by application of a sealed (occluded) patch containing an extract of the biomaterial. This step of skin abrasion

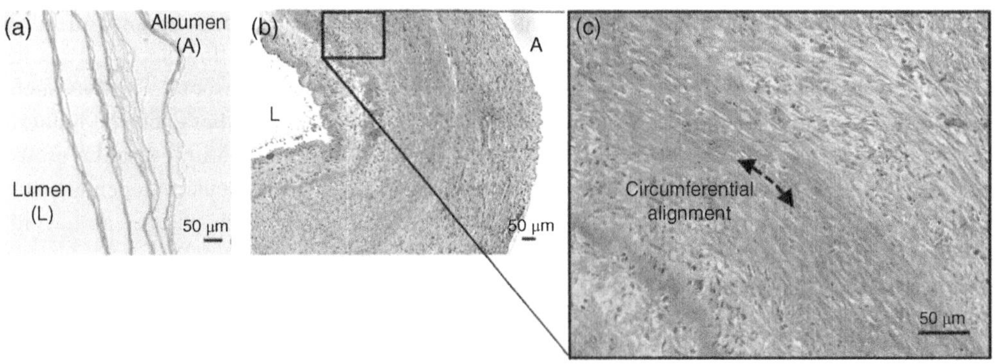

Figure 22.11 Histological analysis of response to graft conduit fabricated from decellularized human amniotic membrane implanted in carotid artery of male rabbits. Histological analyses revealed the initially acellular grafts (a) became completely filled with cells (b) in organized, circularly aligned layers (c) including a middle layer with smooth muscle. *Source:* From Amensag et al. (2017).

and application of an occluded patch with fresh extract is repeated at 24 hours and at 48 hours.

One week later, an occluded patch with fresh extract is applied at each injection site to boost sensitization, followed by a two-week recovery period during which the delayed sensitivity develops. Finally, at 21 days after the initial induction, an extract of the test material is applied to the skin at a distant site for the challenge, which forms the second part of the APT. The challenge site is examined for evidence of erythema and edema, which is evaluated using essentially the same qualitative scoring system used in the in vivo irritation test. It should be noted that any erythema and edema that occurs at the challenge site is an immune response rather than a direct response to the irritant applied to the challenge site. The guinea pig sensitization test performed using a protocol such as the APT is recognized as having a high sensitivity and good reproducibility.

22.4.4 Systemic Toxicity Tests

The *in vivo* toxicity tests discussed so far have focused mainly on assessing the effects of a biomaterial near its implantation site. It is important to note that an implanted biomaterial or substances leached from the biomaterial could potentially elicit adverse effects at remote tissue sites or throughout the body as a whole. These are systemic toxic effects that could manifest within the first 24 hours of exposure (acute systemic toxicity) or later. Toxic systemic effects that manifest themselves between 24 and 28 days, 28 and 90 days, and over 90 days are referred to as sub-acute, sub-chronic, and chronic, respectively.

Systemic toxicity could be an outcome of exposures via various routes, such as intraperitoneal or intravenous injection, oral intake, inhalation, and subcutaneous implantation. in vivo testing of biomaterials or biomaterial extracts for potential adverse systemic effects is often performed with animal models, in most cases, rodents. The tests require careful attention of the mode of exposure. For example, inhalation of gold nanoparticles by rats has been used in a systemic toxicity study designed to simulate occupational exposures to gold nanoparticles in humans.

One investigation of systemic toxicity involved exposure of lactating rats and their suckling offspring to quantum dots administered to the lactating rats via intravenous injection. It consisted of a combination of a sub-acute toxicity test in the lactating rats and a long-term toxicity test in the offspring. The quantum dots consisted of CdSe/ZnS (CdSe with a passivating ZnS shell) having a diameter of ~20 nm, which were coated with amphiphilic polymer molecules. Two groups of lactating rats, a high-dose group and a low-dose group, received tail-vein injections of 5 and 1 nmol quantum dots per rat, respectively, in saline. Control animals received tail-vein injections of saline only. The effects of the quantum dots were monitored over a period of 18 days for the lactating rats and up to 180 days for the surviving offspring, using gross observations (such as behavior, food intake, and body weight), hematology and serum biochemistry measurements, and histopathological examination of various organs.

Measurements of the Cd content in the blood and major organs of the rats showed that the fraction of the injected dose recovered in the liver, kidneys, spleen, and blood was 39%, 17%, 5%, and 0.3%, respectively, for the high-dose group, and 64%, 33%, 10%, and 0%, respectively, for the low-dose group. These measurements indicate that most of the injected quantum dots remained within the rats, with only a small fraction transmitted via breast milk to their offspring.

Multiple indicators of subacute systemic toxicity were present in the rats of the high-dose group, such as a mortality of ~33% and decreased body weight. Serum measurements for rats in the high-dose group showed evidence of severe hepatocellular injury as indicated by high concentrations of alanine transaminase and other markers of liver damage. There was also evidence of kidney damage, as indicated by elevated serum levels of creatinine and blood urea nitrogen. Histopathological examination revealed inflammation and cellular damage in various organs, particularly the liver, kidneys, spleen, and lungs of the rats.

Additionally, the surviving offspring of the high-dose group showed multiple indicators of sub-acute toxicity, such as a survival rate of less than one-third when compared to the control group, decreased body weight, and high concentrations of Cd in maternal milk. The offspring of the high-dose group also showed a few indicators of long-term systemic toxicity, such as a high concentration of Cd in the intestine and histological evidence of edema in intestinal mucosa. In comparison, there were no indications of long-term systemic toxicity in the offspring of the low-dose group. Overall, these observations show several indications of subacute systemic toxicity in the lactating rats caused by the intravenous administration of high doses of the CdSe/ZnS quantum dots, as well as indications of both sub-acute and long-term systemic effects in their surviving offspring.

22.5 Clinical Trials of Biomaterials

Two major factors determine whether a medical device would require testing in human clinical trials prior to application for FDA approval. One factor is whether the new device was substantially similar to previous (predicate) devices, in which case clinical trials would not normally be required. The other factor is whether the device posed a potential risk for unacceptable adverse biologic responses. In this case, the device would be classified as a Class III (high-risk) device by the FDA, requiring clinical trials with human subjects to determine whether it was safe and effective. Essentially all devices exposed to blood or implanted *in vivo* are classified as Class III devices.

22.6 FDA Review and Approval

The FDA requires submission of a Biological Evaluation Report at the conclusion of the testing for toxicological risk, which includes summaries of the results of all *in vivo* and *in vitro* nonclinical tests and, if required, the results of clinical trials. This report is reviewed by the FDA to determine whether the toxicological risks associated with the device have been thoroughly and appropriately investigated. Review of a "non-predicate" device that is, a device substantially different from previously approved (predicate) devices, is particularly rigorous. Assuming a positive outcome of the review process, the FDA gives approval for marketing and clinical use of the device.

22.7 Case Study: The Proplast Temporomandibular Joint

Regulatory approval of a medical device can be very profitable for a company but the consequences of failure of an implantable medical device due to inadequate testing or poor regulation can be disastrous for the patient and the company, and costly for the company as well. There are several examples of such types of failure, particularly dating back to a few decades ago, which provide the need for caution. One example is the Proplast temporomandibular joint (TMJ) disc. The TMJ connects the mandible to the rest of the facial skeleton and is essential for eating and talking. It contains a fibrocartilage disc, called the TMJ disc, between the mandible and temporal bone, which distributes the stresses that develop within the joint as the TMJ articulates. However, trauma and age-related wear can lead to thinning and perforation of the TMJ disc, leading to temporomandibular disorders characterized by pain and limitation of jaw movement. Consequently, attempts were made in the 1970s to develop prosthetic devices to replace the TMJ disc.

One prosthetic TMJ disc device, marketed by a small company (Vitek) under the name Proplast, consisted of polytetrafluoroethylene (PTFE) filled with alumina or vitreous carbon particles. In 1983, Vitek petitioned the FDA for approval to market the Proplast device as a replacement for dysfunctional TMJ discs. In the petition, Vitek asserted equivalence of the Proplast device to an existing TMJ disc replacement device, except that polydimethoxysilane (PDMS) was replaced by PTFE. The petition argued that instead of classification as a Class III device, which has far more stringent requirement for approval, the Proplast should qualify for approval as a Class II device. This argument was accepted by the FDA, which approved the Proplast device for marketing and clinical use based on "substantial equivalence."

As the physiological forces exerted on a joint are complex, inadequate attention to the design of a prosthetic joint device to withstand these forces due to repetitive motion of the joint can lead to disastrous consequences. Within a few years of Proplast being marketed and implanted, many patients suffered significant tissue destruction due to uneven mechanical wear of the Proplast device, resulting in fragmentation of the device and the release of particulate debris. Considerable litigation ensued and, because of limited resources, the small company Vitek filed for bankruptcy protection in 1990. Three years later, the FDA rescinded approval of the Proplast device and reclassified TMJ devices as Class III devices, which require the submission of several years of data on patient outcomes before a new TMJ device can be marketed. It is worth noting that during the ten years that the Proplast devices were marketed, approximately 26 000 patients were implanted with these devices. Many of these patients suffered considerable tissue destruction.

A retrospective study of the failure of the Proplast device identified several questionable aspects in the testing of the device and in the FDA approval process. For example, it was found that the company (Vitek) had used mainly *in vitro* test procedures to predict the potential mechanical changes in the Proplast disc when used in soft-tissue implantation *in vivo*. Additionally, the supplier of the PTFE, DuPont de Nemours, Inc., claimed that they did not knowingly supply the PTFE to Vitek and that the PTFE was marketed as an industrial grade material not intended for use in humans. Related to this, the FDA did not adequately investigate the disclaimers issued by the DuPont company, and apparently did not fully consider the complex biomechanical forces that the Proplast device is subjected to within the joint.

It is worth noting that approximately 20 years before the marketing and clinical use of the Proplast device, Sir John Charnley, the inventor of the original prosthetic hip joint design, had reported on the unsuitability of PTFE as the acetabular liner in hip implants (Chapter 1). He warned that adverse tissue reactions could follow the implantation of PTFE due to abrasion of the liner. Overall, the events associated with the Proplast device provide an example of the disastrous consequences that can arise due to inadequate testing and poor regulatory procedures.

22.8 Concluding Remarks

In this chapter, the topic of biocompatibility phenomena was discussed from the point of view of testing and regulation to assure the safety and effectiveness of medical devices that contain biomaterials.

- We provided an overview of the agencies and organizations that regulate health care products and the standards and procedures that govern the regulation of medical devices containing biomaterials.
- A variety of toxicity tests, ranging from the lowest level involving *in vitro* tests with mammalian cell cultures to more complex *in vivo* tests involving small and large animal models, were discussed.
- *In vitro* toxicity tests are a useful starting point in evaluating the potential risks of a biomaterial to elicit adverse biological responses but *in vivo* testing with animal models is necessary to obtain a more accurate assessment.
- We discussed the events associated with the Proplast temporomandibular device to provide an example of the disastrous consequences that can arise due to inadequate testing and poor regulation of medical devices.

Problems

22.1 Describe one example of a situation in which testing of potential toxicological effect of a biomaterial should be performed with an extract of the biomaterial and describe an appropriate type of toxicological test in this situation.

22.2 Describe a situation in which testing of a biomaterial for potential toxicological effects should be performed with the biomaterial itself rather than an extract of the material, and describe an appropriate type of test in this situation.

22.3 Explain why it would be important to perform a skin sensitivity test before implanting a cobalt–chromium implant in a patient, and describe what could potentially happen in some patients if a skin sensitivity test were not performed.

22.4 Explain how skin irritation and skin sensitization are two distinctly different biological reactions despite the fact that erythema and edema are symptoms of both conditions.

22.5 Describe the most likely risk that would occur in response to implantation of a Dacron synthetic vascular graft in a patient.

A Indicate the tests (*in vitro*, *in vivo*, or both) that would be most appropriate for assessing this risk and give a description of cellular and tissue mechanisms.

22.6 Briefly explain how thrombosis differs from coagulation, and describe an appropriate test that could be performed to assess the risk of thrombosis.

22.7 Describe why assessment of the potential toxicological risk of a biomaterial designed for orthopedic applications should be tested by *in vivo* implantation in an appropriate bone site rather than in a subcutaneous site.

22.8 The use of nanoparticles as biomaterials poses potential toxicological risks that are different from those of macroscale biomaterials.

A Explain one potential toxicological risk of using nanoparticles in the body.

B Describe an appropriate type of test to assess this toxicological risk.

References

Amensag, S., Goldberg, L.A., O'Malley, K.A. et al. (2017). Pilot assessment of a human extracellular matrix-based vascular graft in a rabbit model. *Journal of Vascular Surgery* 65: 839–848.

Beltrán-Partida, E., Moreno-Ulloa, A., Valdez-Salas, B. et al. (2015). Improved osteoblast and chondrocyte adhesion and viability by surface-modified Ti_6Al_4V alloy with anodized TiO_2 nanotubes using a super-oxidative solution. *Materials* 8: 867–883.

Modglin, V.C., Brown, R., Fu, Q. et al. (2012). *In vitro* performance of 13-93 bioactive glass fiber and trabecular scaffolds with MLO-A5 osteogenic cells. *Journal of Biomedical Materials Research. Part A* 100: 2593–2601.

Udipi, K., Ornberg, R.L., Bruce Thurmond, K. II et al. (2000). Modification of inflammatory response to implanted biomedical materials *in vivo* by surface bound superoxide dismutase mimics. *Journal of Biomedical Materials Research* 51: 549–560.

Weber, M., Steinle, H., Golombek, S. et al. (2018). Blood-contacting biomaterials: *In vitro* evaluation of the hemocompatibility. *Frontiers in Bioengineering and Biotechnology* 6: 99.

Zhu, Y., Hideyoshi, S., Jiang, H. et al. (2018). Injectable, porous, biohybrid hydrogels incorporating decellularized tissue components for soft tissue applications. *Acta Biomaterialia* 73: 112–126.

Further Reading

Anderson, J.A. (2016). Future challenges in the *in vitro* and *in vivo* evaluation of biomaterial biocompatibility. *Regenerative Biomaterials* 3: 73–77.

Black, J. (1984). Systemic effects of biomaterials. *Biomaterials* 5: 11–18.

DeJong, W.H., Carraway, J.W., and Geertsma, R.E. (2020). *In vivo* and *in vitro* testing for the biological safety evaluation of biomaterials and medical devices. In: *Biocompatibility and Performance of Medical Devices*, Woodhead Publishing Series in Biomaterials, 2e (ed. J.P. Boutrand), 123–166. Woodhead Publishing.

Duncan, E., Lemons, J.E., Mayesh, J.P. et al. (2004). New products and standards. In: *Biomaterials Science: An Introduction to Materials in Medicine*, 2e (eds. B.D. Ratner, A.S. Hoffman, F.J. Schoen and J.E. Lemons), 784–804. San Diego, CA: Elsevier Academic Press.

Lee, K.C., Eisiq, S.Q., and Perrino, M.A. (2018). Foreign body giant cell reaction to a Proplast/Teflon interpositional implant: a case report and literature review. *Journal of Oral and Maxillofacial Surgery* 76: 1719–1724.

Van Norman, G.A. (2016). Drugs and devices: comparison of European and U.S. approval processes. *JACC: Basic to Translational Science* 1: 399–412.

Part VI

Applications of Biomaterials

23

Biomaterials for Hard Tissue Repair

23.1 Introduction

This chapter and the remaining chapters, making up the final part of the book, are devoted to discussing the applications of biomaterials. We will see how the principles of materials science, engineering, and biology covered in the earlier chapters are used practically to modify and adapt the various classes of materials for use as biomaterials in a variety of biomedical applications. These applications are broadly grouped into four areas representing a significant cross section of the biomedical field and the biomedical industry: repair of hard tissues, repair of soft tissues, tissue engineering and regenerative medicine, and drug delivery. The effectiveness of the biomaterials in these applications will also be discussed using examples taken from a variety of studies.

In this chapter, we will discuss the application of biomaterials in healing defective regions of various hard tissues such as bone, articular joints, the spine, and teeth in order to restore the structure and function of these tissues. Healing of bone fractures and bone loss is a common occurrence in orthopedic surgery. While small fractures can normally heal on their own, often with the aid of external or internal fracture fixation devices, there are many cases where bone damage or bone loss is too severe for self-healing. Clinically, these bone defects are reconstructed using various types of implants called bone grafts or bone graft substitutes. Second to blood, bone is the most widely transplanted tissue. Autologous bone grafts (bone taken from the patient) are the ideal bone grafts, while allografts (bone from a donor or a cadaver) are the most widely used bone graft, but they also have limitations. Consequently, there have been significant efforts to develop alternative implants, referred to as bone graft substitutes, to avoid the limitations associated with bone autografts or allografts. Bone graft substitutes cover a wide variety but a characteristic feature is that they are normally composed of biomaterials, synthetic or natural. While these biomaterials offer significant advantages, they are often not very effective when used alone. On the other hand, when combined with the appropriate cells or osteoinductive growth factors, they can show a significantly better ability to heal bone defects.

Articular joints are composite tissues in which a thin layer of cartilage (a soft avascular tissue) covers the articulating surfaces of the bones where they meet to form a joint. Deterioration of these joints can occur, due to arthritis or trauma, but it is often difficult to detect this deterioration until it has reached an advanced stage. At this point, total joint replacement may be the only viable solution. Total hip joint and total knee joint replacement are common procedures in orthopedic surgery, which have served to improve the quality of life for millions of patients. However, they also suffer from significant limitations, one of which is their limited lifetime *in vivo* due to loosening of the implant. A major cause of this loosening, as noted in Chapter 21, is the adverse response of tissues surrounding the implant to particulate debris produced by wear of the implant's articulating surfaces. Implant loosening often necessitates revision surgery to replace the failed implant, a procedure that is very traumatic. One measure of the significance of implant loosening is that ~20% of the total hip and knee joint replacement surgeries performed annually are actually revision surgeries.

Deterioration of one or more of the intervertebral discs in the lower part of the spine is a major cause of low back pain, a condition that affects millions of people. A common treatment involves the placement of an implant (graft) in the intervertebral space to achieve fusion of adjacent vertebrae. While there is some loss in mobility of the spine in the region of the affected vertebrae, this is more than compensated for by the reduction in low back pain. As spinal fusion implants are generally composed of bone autografts, allografts, or biomaterials, some aspects of bone defect healing are also relevant to spinal fusion implants. Spinal fusion is effective but it is not an ideal long-term solution for many patients in that it provides relatively pain-free back movements over a period of only ~10 years.

Materials for Biomedical Engineering: Fundamentals and Applications, First Edition. Mohamed N. Rahaman and Roger F. Brown.

Companion website: www.wiley.com/go/rahaman/biomedicalengineering

Dental implants and dental restorations are devices composed of biomaterials used to restore the structure, function, and appearance of damaged or lost teeth. Dental implants are implanted into bone to provide structural support for restorative devices such as crowns, bridges, and dentures, while composite resins are used to fill cavities in teeth. An area of significant advances over the last few decades has been the development of metal-free restorations composed of glass-ceramics and polycrystalline ceramics. These biomaterials provide the ability to restore not only the existing tooth structure but also its optical properties and aesthetic appearance.

Overall, the topics discussed in this chapter are in the following broad areas:

- Mechanisms by which bone fractures heal
- Bone grafts and bone graft substitutes for healing bone fractures and bone loss
- Biomaterials for total joint replacement
- Spinal fusion implants
- Dental implants and restorations.

23.2 Healing of Bone Fracture

Treatment of bone fracture, as noted earlier, is a common procedure in orthopedic surgery, requiring different procedures depending on the severity of the fracture. Inherently stable small fractures, in which the bones undergo micromotion only, often heal with nonsurgical treatments such as the use of a plaster or resin cast, or a plastic brace. In comparison, large fractures that are inherently unstable normally require surgical intervention, such as the use of external or internal fixation devices to stabilize the fracture. External fixation does not require opening the fracture site but, instead, bone fragments are held in position by pins placed through the skin onto the skeleton, which are supported structurally by external bars. Internal fixation requires opening the fracture site, reassembling the broken bones into position and stabilizing them in place until the facture heals. Depending on the type of fracture and on the fracture site, a large variety of plates, screws, wires, and intramedullary devices are available to stabilize the fractured bone (Figure 23.1).

23.2.1 Mechanism of Fracture Healing

A relevant question in bone fracture healing is the mechanism by which the fracture heals. Overall, the healing of bone fractures differs from the healing of wounds of the skin and most other tissues in one major way in that healing occurs without the formation of fibrous scar tissue.

Figure 23.1 (a) Examples of internal fracture fixation plates for stabilizing fractures in the long bones of humans. *Source:* Courtesy of G. Richards, AO Foundation; (b) illustration of broken long bone stabilized with a plate and screws.

A distinctive aspect of fracture healing is that it occurs in essentially the same manner as bone development and, in that regard, it is actually a process of tissue regeneration. The facture healing process is generally considered to occur in four overlapping stages, namely hematoma formation, soft callus formation, bony callus formation, and remodeling (Figure 23.2).

Hematoma Formation

Within minutes after a bone fracture, blood released from broken peripheral and intramedullary blood vessels begins to coagulate, forming a relatively large hematoma that fills the fracture site and bulges into the subperiosteal space, forming a raised area encircling the bone (Figure 23.2a). The hematoma serves as an initial scaffold within which subsequent callus formation can begin. Cells trapped within the hematoma also trigger the release of several pro-inflammatory cytokines, such as TNFα, IL-1, and IL-6 (Table 16.2). These cytokines initiate angiogenesis and attract neutrophils and monocytes to the fracture site, which quickly differentiate into macrophages. The TNFα output, which reaches a peak within the first day post-injury, results in macrophage accumulation and promotes the differentiation of mesenchymal stem cells (MSCs) to osteoblasts and osteoclasts. In comparison, the cytokines IL-1 and IL-6 appear to be important in promoting the differentiation of osteoblasts and osteoclasts, angiogenesis, and the initiation of soft callus formation.

The macrophages recruited to the fracture site have multiple important roles in this early stage of fracture healing. Macrophages debride the injury site by phagocytizing necrotic tissue debris to create space needed for ingrowth of progenitor cells for osteogenesis and angiogenesis. Macrophages also release the cytokines TNFα and IL-1 that

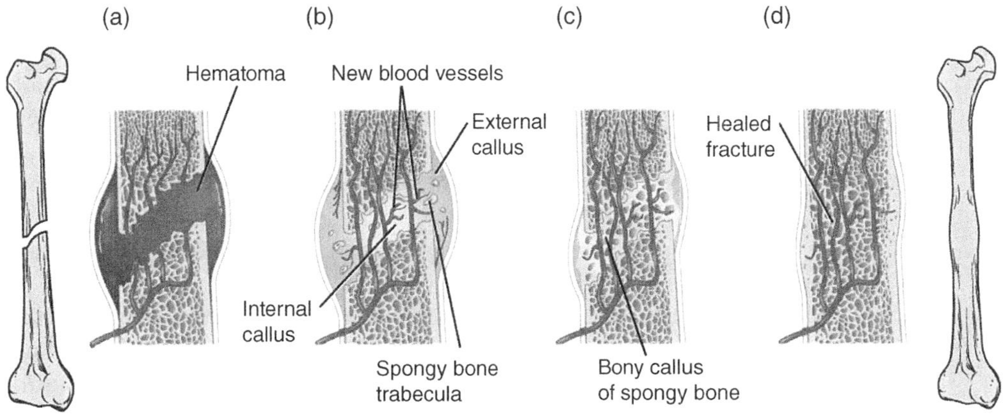

Figure 23.2 Four stages of bone fracture healing: (a) hematoma formation, (b) soft callus formation, (c) bony callus formation, and (d) bone remodeling. The stages overlap to varying extent. *Source:* From: Access for free at https://openstax.org/details/books/anatomy-and-physiology.

amplify the inflammatory reactions and attract more immune cells. Additionally, macrophages appear to have a particularly important role in recruiting MSCs from the circulating blood, marrow tissue and surrounding mesodermal tissue. Although the exact mechanism of macrophage recruitment of MSCs is not clear, activated macrophages are known to secrete specific homing chemokines identified by the acronyms SDF-1, CCL2, and CXCL-8. Pro-inflammatory M1 phenotype macrophages (Chapter 18) are the main type of macrophages present during the early stages of fracture healing. These macrophages promote osteogenic differentiation in MSCs by activation of the COX-2-PGE2 pathway, a metabolic pathway that produces prostaglandin E2. Overall, the events of the hematoma formation stage are mostly completed in ~7 days.

Soft Callus Formation

The transition to the second stage of fracture healing, the formation of a soft callus, is marked by the conversion of the hematoma to granulation tissue (Figure 23.2b). The soft callus, a structure consisting largely of fibrocartilage, forms within the vascular-rich granulation tissue, and completely fills the internal region between the ends of the fractured bone and the external callus outside the periosteum. Because the soft callus serves as a framework that will later undergo mineralization to become hard bony tissue, its formation is arguably the most important stage of fracture healing. In this sense, formation of the soft callus replicates the same basic steps that occur in intramembranous and endochondral bone formation in the developing fetal skeleton.

Chondroblasts and fibroblasts, the cells responsible for forming the fibrocartilage of the soft callus, are derived from MSCs recruited to the fracture site by M1 macrophages in the preceding hematoma stage. They produce an extracellular matrix composed mainly of proteoglycans and collagen type I and type II. The signaling factors that promote chondrogenesis and endochondral ossification (which begin in the later part of this stage) particularly involve several cytokines in the transforming growth factor-beta (TGFβ) superfamily, such as TGFβ2, TGFβ3, and growth/differentiation factor-5 (GDF5).

This soft callus stage of fracture healing also includes angiogenesis and neovascularization, processes essential for bone repair. The main sources of the mesodermal tissue for tissue ingrowth that forms the new vasculature are the ingrowth of vascular buds from the outer periosteum and mesenchymal stem tissue of the bone marrow. Degradation of the fibrocartilage matrix to form cavities needed for the ingrowth of endothelium and vascular buds is a function of the chondroclasts. The process of vascularization is controlled mainly by two signaling cytokines, angiopoietin and vascular endothelial growth factor (VEGF). Angiopoietin stimulates ingrowth of vascular tissue from existing vessels of the periosteum. In comparison, VEGF promotes differentiation of MSCs to microvessel buds that form the rich vascular network of the interior. The soft callus is pliable and can deform easily if the fracture lacks adequate support. The duration of the soft callus formation stage is ~2 weeks.

Bony Callus Formation

Formation of a bony callus, which occurs during the third stage of fracture healing, involves a progressive degradation of the cartilaginous matrix of the soft callus and its replacement by hard, mineralized tissue (Figure 23.2c). This mineralized tissue, referred to as woven bone, is a type of bone that has randomly oriented trabeculae, and it is weaker and less rigid than lamellar bone. The steps that occur in bony callus formation, as noted earlier, are identical to the steps

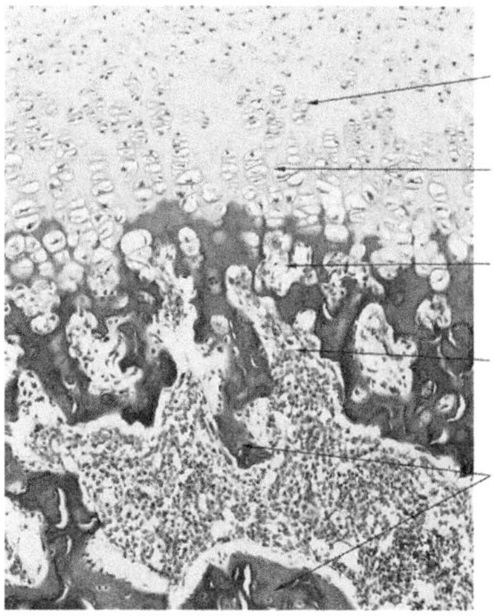

Figure 23.3 Process of endochondral bone formation that occurs in fetal development and normal repair of adult bone. The process particularly includes chondrocyte hypertrophy and apoptosis, cavity formation, and ingrowth of osteogenic progenitor cells. *Source:* M. A. Hill, University of New South Wales, Australia. Downloaded from https://embryology.med.unsw.edu.au/embryology/index.php?title=File:Bone_histology_016.jpg.

of endochondral ossification. Toward the end of the soft callus stage, hypertrophied chondrocytes secrete alkaline phosphatase, which causes the surrounding matrix to mineralize, creating a barrier for nutrient diffusion (Figure 23.3). This reduction in nutrient causes the hypertrophied chondrocytes to begin dying, leaving cavities in the cartilage matrix. Then progenitor cells derived from MSCs begin to invade the cavities and differentiate into osteoblasts. The cytokine TNFα appears to promote the recruitment of MSCs and their differentiation to bone-forming osteoblasts. Other progenitor cells thought to be of hemopoietic origin give rise to osteoclasts. The formation of osteoclasts, which degrade the mineralized matrix to create cavities for migrating ingrowth of osteoprogenitor cells, is stimulated by a factor known by the acronym RANKL.

As the calcified cartilage undergoes gradual degradation and replacement with woven bone, the bony callus becomes progressively more rigid and, thus, more capable of withstanding physiological forces. Conversion of the soft callus to a bony callus usually takes 6–12 weeks.

Bone Remodeling

Although the woven bone of the bony callus provides some mechanical stability to the healing fracture site, it lacks the mechanical strength and, thus, the load-bearing capacity of normal bone. The process of bone remodeling involves the gradual replacement of woven bone of the bony callus with lamellar bone to allow a return to near-normal mechanical strength. This remodeling involves degradation of the woven bone matrix by osteoclasts, which release the enzymes acid phosphatase and matrix metalloproteinase that degrade the calcified matrix and collagen, respectively. Degradation is tightly controlled to occur at a rate that is approximately equivalent to the formation of lamellar bone by osteoblasts. The cytokines TNFα and IL-1 appear to have roles in controlling the biochemical events of bone remodeling, while some bone morphogenetic proteins, such as bone morphogenetic protein-2 (BMP2), also appear to be involved.

During the remodeling stage, the external callus that had provided temporary support is removed, and the fracture site becomes smoother and returns to a normal or near-normal appearance (Figure 23.2d). Bone remodeling is the longest stage of fracture healing and may last for ~2 years or more.

23.2.2 Internal Fracture Fixation Devices

Bone fractures, as noted earlier, often require stabilization for proper healing using various external or internal fixation devices. Because they are inserted *in vivo*, biomaterials used for internal fracture fixation devices should have the following combination of desirable properties and characteristics:

- Biocompatibility, particularly over a long-term duration for permanently implantable devices
- Mechanical properties to support the requisite physiological stresses over the duration of the fracture healing process
- Corrosion resistance when used for permanently implantable devices, or a controlled degradation rate when used for short-term fixation
- Ability to be formed economically into a variety of geometries to provide stable fracture fixation.

Overall, the corrosion-resistant metals such as Ti and its alloy Ti6Al4V, stainless steel (316L), and Co-Cr alloys provide a reasonable approximation to most of these properties and, thus, they find considerable use as internal fracture fixation devices. However, a major drawback of these metallic devices is the adverse mechanical biocompatibility arising from the large difference between their mechanical properties and those of bone (Chapter 17). Because their Young's modulus (or stiffness) is much higher than that of cortical bone (Table 4.1), these plates shield the adjacent bone from the physiological stresses, which can lead to bone resorption, and eventually, to bone fracture. Nylon is used to a limited extent, often as a cerclage band because it has sufficiently high tensile strength and flexibility, while polylactic acid (PLA) or PLA reinforced with hydroxyapatite particles is used as degradable pins and screws to stabilize fractures that are not subjected to significant physiological stresses.

Table 23.1 Applications of biomaterials in fracture fixation.

Biomaterial	Properties and characteristics	Application
Stainless steel (316L)	High yield strength; high Young's modulus; low cost	Surgical wire; pin; plate; screw; intramedullary nail
Ti6Al4V	Lower yield strength and Young's modulus than stainless steel; lower density; good osseointegration	Surgical wire; pin; plate; screw; intramedullary nail
Co-Cr	High yield strength and Young's modulus; more difficult to fabricate than stainless steel	Surgical wire; intramedullary nail
Polylactic acid (PLA)	Biodegradable; low strength and Young's modulus; often reinforced with hydroxyapatite to improve modulus and impart bioactivity	Pin; screw
Nylon	Resistant to degradation; moderate strength and Young's modulus; ease of fabrication	Cerclage band

Table 23.1 summarizes the general properties of these biomaterials and their main fracture fixation applications.

23.3 Healing of Bone Defects

Bone defects are a common occurrence in craniofacial and orthopedic surgery, resulting from bone loss due to trauma, malignancy, infection, and congenital disease. Although bone has a self-repairing capacity, there are many cases where bone damage or bone loss is too severe for self-healing. Clinically, these bone defects are commonly reconstructed using various types of implants consisting of bone grafts or bone graft substitutes. As noted earlier, second to blood, bone grafting is the most common transplantation procedure used to replace tissues in the human body. Worldwide, a few million bone graft procedures are performed each year and this number is projected to grow in the future.

A bone graft (or bone graft substitute) provides an implant on which or into which new bone can grow from the surrounding host bone to heal the defect and to achieve integration of the graft with the host bone (osseointegration). Bone grafts and bone graft substitutes promote bone defect healing by a variety of osteoinductive, osteoconductive, and osteogenic mechanisms.

- Osteoinductivity is the process by which MSCs from the surrounding (host) tissue are recruited to differentiate into chondroblasts and osteoblasts, which, in turn, form new bone through the process of endosteal ossification. Growth factors, the most important of which belong to the TGFβ superfamily, are the primary mediators of this process. When used in relation to biomaterials, osteoinductivity normally refers to the ability of a biomaterial to form new bone within its pores when implanted in a non-bony site, such as a subcutaneous or muscle site.
- Osteoconductivity is the process in which a porous material, implanted in or adjacent to bone, supports the ingrowth of MSCs, perivascular tissues, and capillaries, which form new bone within the porous structure.
- Osteogenicity refers to the ability to synthesize new bone by cells such as MSCs, osteoblasts, and osteocytes present within the implant.

23.3.1 Bone Defects

Bone is a tissue of the skeletal system in which various parts of the skeleton are subjected to different levels of physiological stress. On this basis, we may broadly divide bone defects into two types, depending on the level of physiological stress that the bone is subjected to:

- Nonstructural (nonloaded) defects are defects in bones that are subjected to little or limited physiological stresses, such as defects in the craniofacial bones and small (contained) defects in various anatomic sites
- Structural (or loaded) defects are defects in structural bones that are subjected to significant physiological stresses, such large defects in the long bones of the limbs. Bone grafts (or bone graft substitutes) implanted in such defects normally lack stability, necessitating the use of appropriate fracture fixation devices to stabilize them.

It is noted that the type of bone defect (structural or nonstructural) determines the biomechanical properties of the bone graft (or bone graft substitute) to be used in its healing.

23.3.2 Bone Grafts

Bone grafts may be divided into three major types, referred to as autografts, allografts, and bone graft substitutes.

Bone Autografts

An autograft is bone taken from another part of the patient and transplanted into the defect site. Autografts are the ideal bone graft (often referred to as the gold standard) because they have the potential for healing the defect by a combination of all three mechanisms, that is, osteoinductive, osteoconductive, and osteogenic mechanisms. Cancellous bone, such as bone from the rim of the pelvis (iliac crest) is the most commonly used autologous bone graft. It contains a high concentration of osteoblasts and osteocytes, which provides it with superior osteogenic potential. Cortical bone provides significant structural support but its dense, highly organized structure lowers the amount of cells present within it, thus lowering its osteogenic potential when compared to cancellous bone autografts. Overall, the biological events that occur at the defect site leading to integration (union) between an autograft and host bone generally follow those described earlier for fracture healing.

The use of bone autografts suffers from a few limitations, mainly associated with their harvesting. For example, the surgical procedure takes longer because it requires two operations, the harvesting operation at the donor site and the implantation operation at the defect site. Additionally, the harvesting operation leads to complications at the donor site, such as pain, morbidity, blood loss, and the potential for infection. Autografts also suffer from the problem of limited supply, that is, only a limited amount of bone can be harvested from the donor site.

Bone Allografts

An allograft is a type of bone graft substitute prepared from bone taken from a donor or a cadaver. Allografts are the most widely used bone substitute in orthopedic surgery. The allograft preparation process involves freezing or freeze-drying the donor or cadaver bone, treatment to remove cells, and sterilization by gamma radiation, electron beam radiation or ethylene oxide treatment to eliminate disease transmission. However, it cannot be said that the risk of disease transmission is eliminated completely. There is also the potential problem of an adverse immune response to an allograft. Like bone autografts, allografts can be prepared from cortical bone or trabecular bone, providing implants having a range of mechanical properties and a variety of shapes and sizes.

Because they are prepared from bone taken from a donor or cadaver, allografts avoid the drawbacks of bone autografts due to harvesting, such as the length of surgical procedure, limited supply, and donor site morbidity. On the other hand, because the cells and most of the growth factors are removed during their preparation, allografts are osteoconductive but they lack osteogenicity and they may possess only limited osteoinductive potential (Table 23.2). Thus, allografts are an imperfect substitute for autografts. Allografts are also expensive due to the stringent conditions that are used in their preparation.

Another type of bone substitute is demineralized bone matrix (DBM). DBM is prepared by treating bone from a donor or cadaver with a demineralizing agent such as hydrochloric acid, which removes the mineral phase of the bone but leaves many of the osteoinductive agents such as growth factors and other proteins in the demineralized matrix. When compared to bone allografts, DBM has much better osteoinductivity, much lower osteoconductivity, and much weaker mechanical properties due to the removal of the bone mineral phase, and it suffers from a similar lack of osteogenicity. DBM is normally used alone, or in combination with biomaterials as a bone graft extender to provide osteoinductive factors.

Table 23.2 Types of bone grafts and their characteristics.

Bone graft	Strength	Osteoinductivity	Osteoconductivity	Osteogenicity
Autograft				
Cortical	Strong	Osteoinductive	Osteoconductive	Osteogenic
Trabecular	Weak	Osteoinductive	Osteoconductive	Osteogenic
Allograft				
Cortical	Strong	Limited	Osteoconductive	–
Trabecular	Weak	Limited	Osteoconductive	–
Demineralized bone matrix (DBM)	–	Osteoinductive	–	–
Bone graft substitutes				
Biomaterials	Variable	–	Osteoconductive	–
+ Cells		Osteoinductive	Osteoconductive	Osteogenic
+ Growth factors		Osteoinductive	Osteoconductive	–

23.3.3 Bone Graft Substitutes

While bone autografts and allografts have shown clear benefits in healing bone defects, their aforementioned drawbacks have heightened the search for alternative implants. Using the mechanisms of bone healing (osteoinductivity, osteoconductivity, and osteogenicity) as a basis, a vast number of bone graft substitutes have been created and evaluated, some of which are already being used clinically. Most of these bone graft substitutes are composed of biomaterials, mainly polymers (natural or synthetic), ceramics, or composites composed of a polymer matrix and dispersed ceramic particles (Table 23.3). They generally take the form of a loose assemblage of coarse particles, granules or beads, porous three-dimensional solids, or cements that set (harden) *in situ*.

Overall, an ideal bone graft substitute should approximate a bone autograft. This means that minimally, the bone graft substitute should be

- Biocompatible
- Osteoinductive, osteoconductive, and osteogenic

Table 23.3 Examples of bone graft substitutes composed of biomaterials.

Class of material	Characteristics
Ceramics (and glasses)	
Calcium phosphates (Hydroxyapatite; β-TCP; BCP)	Osteoconductive; bioactive; degradable
Bioactive glass (45S5; 13-93)	Osteoconductive; bioactive
Calcium sulfate	Osteoconductive; degradable
Polymers	
Poly(α-hydroxy esters) (PLA; PGA; PLGA)	Osteoconductive; degradable
Collagen	Osteoconductive; degradable
Metals	
Titanium; tantalum	Osteoconductive; resistant to degradation
Composites	
Polymers (PLA; PLGA; collagen) filled with bioactive ceramic particles	Osteoconductive; degradable; bioactive
Cements	
Calcium phosphate	Osteoconductive; injectable; degradable
Calcium sulfate	Osteoconductive; injectable; degradable

- Osseointegrative, that is, the graft should have the ability to integrate with host bone.

Additionally, the bone graft substitute should have certain beneficial properties and characteristics that enhance its potential to heal the bone defect:

- *Microstructure*: Because bone is a vascularized tissue, a bone graft substitute should have a microstructure that is conducive to cell and tissue ingrowth, the formation of capillaries (angiogenesis) in the infiltrated tissue, and the supply of nutrients to the cells within the bone graft substitute to maintain their viability. In general, the particles in a loose assemblage should have a size (diameter) of few hundred micrometers to provide sufficiently large pore spaces for infiltration and migration of osteoblasts (diameter ~20–50 μm), the cells responsible for the synthesis and mineralization of bone. On the other hand, the pores within a three-dimensional implant should be interconnected and have a minimum size between adjacent pores of ~100 μm, but studies have shown that a pore size of ~300–500 μm is optimal for new bone formation and angiogenesis within the implant.
- *Mechanical properties*: Because bone is also a skeletal tissue, a bone graft substitute should have appropriate mechanical properties to provide the requisite structural support, which will vary depending on the size and anatomic site of the bone defect. For example, bone graft substitutes for healing defects in the cranium are generally required to provide little structural support. In comparison, those for healing large defects in bones subjected to considerable physiological stresses, such as the long bones of the limbs, are required to provide significant structural support.

Bone graft substitutes should also be capable of being produced economically into a variety shapes and sizes encountered with bone defects, and they should have suitable handling properties for facile manipulation and shaping in a surgical procedure.

Bone graft substitutes composed of biomaterials provide some advantages over bone autografts and allografts but they also suffer from disadvantages. Problems related to limited supply and donor site morbidity are avoided, while bone graft substitutes can normally be well sterilized to avoid disease transmission. Additionally, biomaterials can often be formed economically into a variety of shapes, sizes, and internal architectures, which have a range of mechanical properties. Some bone graft substitutes such as calcium phosphate cements can also be injected into the bone defect as a paste that sets *in situ* (Chapter 7), providing the benefit of implantation by minimally invasive surgery.

On the other hand, a major drawback of biomaterials used as bone graft substitutes is that while they are osteoconductive, they normally lack osteoinductivity and osteogenicity. Although some encouraging results have been obtained, the use of osteoconductive biomaterials by themselves often fails to provide sufficient healing of bone defects within a clinically relevant time (approximately 3–6 months). Consequently, approaches based on adding cells, osteoinductive growth factors, or a combination of cells and growth factors to biomaterials have been developed to alleviate their lack of osteoinductivity and osteogenicity and, therefore, to improve their ability to heal bone defects. Generally, these approaches also have a few drawbacks, a common one being an increase in cost. The addition of cells to bone graft substitutes also makes the surgical procedure more challenging, while it is often difficult to control the release of growth factors within a safe and effective therapeutic concentration over a desirable time *in vivo*.

It is noted that a few bioactive ceramics such as hydroxyapatite and bioactive glasses are osseointegrative, in that they form a direct bond at their surface with host bone (Chapter 7). These biomaterials stimulate the osteogenic differentiation of MSCs and osteoprogenitor cells attracted to the defect site, which leads to the formation of new bone on their surface and to direct bonding to host bone. Osseointegration of titanium-based implants, as noted in Chapter 21, can be enhanced by the presence of certain surface topographical features.

In the following sections, we will discuss in a qualitative manner the main types of biomaterials used as bone graft substitutes. The materials science of these biomaterials have been discussed in earlier chapters and while some of this information will be mentioned briefly when appropriate, our main focus is on how the properties and characteristics of these biomaterials are modified and adapted to make them suitable for applications as bone graft substitutes. We classify bone graft substitutes arbitrarily into various types, depending on the class of material used to create them (Table 23.3) and on whether they are incorporated with cells or osteoinductive growth factors.

Ceramic-Based Bone Graft Substitutes

The calcium phosphates hydroxyapatite, β-tricalcium phosphate (β-TCP) and biphasic calcium phosphate (BCP), have been used in medicine and dentistry to heal bone defects since the 1980s. These biomaterials have a compositional similarity to bone (that is, they are composed of the same ions as the mineral constituent of bone) and share some of the properties of bone. They are osteoconductive, bioactive, and do not elicit a strong inflammatory response at the time of implantation. In the form of granules, porous solids, putties or pastes, hydroxyapatite and β-TCP, make up the majority of the synthetic bone graft substitutes available commercially.

Hydroxyapatite, β-TCP, and BCP provide bone graft substitutes with a range of mechanical properties and degradation rates for healing nonstructural bone defects. Porous hydroxyapatite implants can be created with sufficiently high compressive strength and Young's modulus for use in healing some structural bone defects but there are concerns about the potential for brittle failure due to their low flexural strength and fracture toughness.

Apart from their lack of osteoinductivity and osteogenicity, hydroxyapatite, β-TCP, and BCP suffer from additional limitations. Although particles, granules and beads can be implanted in a facile manner to fill a defect, they have some mobility in the defect site due to the presence of physiological fluids and, consequently, they are susceptible to being transported out of the defect site unless properly contained. Porous three-dimensional solids of hydroxyapatite, β-TCP, and BCP typically have a specific shape, often regular in geometry, and a moderate hardness as well, which make their manipulation and modification in a surgical procedure to fit an irregularly shaped defect challenging.

Calcium phosphate and calcium sulfate cements in the form of putties or pastes provide a better fit to the defect shape because they are amenable to being packed into the defect or implanted by minimally invasive surgery. However, cements that set *in situ* are susceptible to being washed away prior to their hardening due to intraoperative bleeding in the defect site. Additionally, upon implantation, they contain only a small fraction of pores of the requisite size to support bone ingrowth and vascularization, which can initially hinder infiltration of new bone.

Bioactive glasses in the form of coarse particles, granules or porous solids (Chapter 7) are an alternative to hydroxyapatite, β-TCP, and BCP. Upon implantation, bioactive glasses convert *in situ* to hydroxyapatite (or a hydroxyapatite-like material) at varying rates, depending on their chemical composition and microstructure. Certain ions released in this conversion process can stimulate the osteogenic and angiogenic differentiation of MSCs and progenitor cells, and can inhibit bacterial activity as well. In the form of particles, granules or porous solids, bioactive glasses such as the silicate glasses designated 45S5 and 13-93 (Table 7.3) have been used to a limited extent to heal nonstructural bone defects. Overall, however, bioactive glasses have not found wide application in healing bone defects.

Polymer-Based Bone Graft Substitutes

Polymers used as bone graft substitutes are composed of natural or synthetic polymers, degradable or nondegradable.

The most commonly used natural polymer is collagen (type I). Because it is a component of bone, collagen provides a favorable environment for bone formation but because it is a protein-based natural polymer, it can elicit an immune response. Collagen has weak mechanical properties and, thus, when used as a bone graft substitute, it serves mainly as a carrier for cells or growth factors. Incorporation of hydroxyapatite particles improves the ability of collagen to provide some degree of structural support, while also enhancing its bioactivity.

In the form of microspheres or porous three-dimensional solids, degradable polymers such as the poly(α-hydroxy esters) PLA, and polylactic-*co*-glycolic acid (PLGA) have been developed and commercialized as bone graft substitutes but they do not see considerable use in these applications. Like many polymers, PLA and PLGA have the inherent advantages of ease of forming and handling to fit the normally irregular shapes encountered with bone defects. On the other hand, they lack osteoinductivity and osteogenicity, and they may elicit an inflammatory tissue response *in vivo* due to the release of acidic degradation products. PLA and PLGA grafts provide better structural support than collagen but they are still not suitable for healing structural defects in bones that are subjected to significant physiological loads.

Metal-Based Bone Graft Substitutes

Metals used as bone graft substitutes are few in number, the most common being titanium (Ti) and its alloy Ti6Al4V and, to a lesser extent, tantalum (Ta). These metals are corrosion resistant, biocompatible, and osteoconductive (Chapter 6). Dense Ti is widely used as dental implants and, as noted in Chapter 21, certain surface roughness features enhance its osseointegration. In a porous form, Ti can have attractive mechanical properties (Figure 4.13) which makes it a potential bone graft substitute in healing large structural bone defects, such as segmental defects in the long bones of the limbs.

Commercial Ta for use as a bone graft substitute is prepared in a highly porous form with a microstructure mimicking that of trabecular bone (Chapter 6). This porous Ta shows good osteoconductivity attributed to its unique micro-rough surface topography and, although its mechanical properties are inferior to those of cortical bone, it can provide significant structural support. A major application is in revision hip and knee joint surgery, to fill large bone voids caused by the removal of the failed implant and to stabilize the femoral stem or tibial insert of the new implant.

In general, bone graft substitutes composed of porous metals such as Ti and Ta have significant drawbacks as alternatives to cortical bone autografts and allografts in healing structural bone defects. They cannot remodel (convert) into bone, remaining essentially unchanged over their implantation lifetime. Of much greater importance is that the mismatch in mechanical properties between these implants and the surrounding host bone can lead to adverse mechanical biocompatibility phenomena (Chapter 17). For example, stress shielding of the surrounding bone leads to bone resorption over time and, eventually, to bone fracture.

Composite Bone Graft Substitutes

Composites used as bone graft substitutes typically consist of a matrix composed of a natural or synthetic polymer and a dispersed phase of bioactive particles composed of hydroxyapatite, β-TCP, or bioactive glass. The particulate phase generally improves the Young's modulus of the polymer, thus improving its ability to provide structural support (Chapter 12). However, a more important role of the dispersed phase is often to impart some degree of bioactivity to the polymer to improve its osseointegrative potential. As in the case of unreinforced polymers, these composites are normally formed into coarse particles, beads or granules (of diameter of a few hundred micrometers), or into three-dimensional solids (porous or dense).

Injectable cements that set *in situ*, composed not of a calcium phosphate but of a hydrogel containing a dispersed phase of bioactive particles such as β-TCP or bioactive glass, have also been developed for use as bone graft substitutes. An example is a composite composed of a chitosan matrix and bioactive glass particles. As noted in Chapter 11, chitosan shows pH-dependent and temperature-dependent gelling, and it can set at the normal physiological temperature and pH. On the other hand, the bioactive glass particles convert to hydroxyapatite and release inorganic ions that can stimulate osteogenesis and angiogenesis. This type of composite bone graft substitute could potentially provide the benefits of injectability, osteoconductivity, and osteogenicity, based on the constituent biomaterials alone.

Another type of injectable bone graft substitute is a composite consisting of an alginate hydrogel and dispersed β-TCP beads. Because alginate hydrogels are attractive biomaterials for encapsulating and delivering cells and osteoinductive growth factors (Chapter 11), these composites could provide not only the benefit of injectability but also the ability to heal bone defects by a combination of osteoconductive, osteoinductive, and osteogenic mechanisms. For example, when seeded with MSCs and implanted for eight weeks in subcutaneous sites in rodents, an injectable composite composed of an alginate hydrogel and β-TCP beads showed the ability to induce new bone formation on the surface of the β-TCP beads.

Combination of Biomaterials with Cells and Growth Factors

As noted earlier, approaches based on the addition of cells, growth factors or a combination of cells and growth factors to osteoconductive biomaterials have been used to address their lack of osteoinductivity and osteogenicity. Bone graft substitutes created by these approaches could potentially provide an optimal combination of all three mechanisms without the drawbacks associated with bone autografts and allografts.

MSCs have been the most common type of cells incorporated into biomaterials to enhance their ability to heal bone defects. Osteoprogenitor cells and osteoblasts have also been used but to a more limited extent. The source of these MSCs has been the bone marrow, but MSCs derived from other tissues, such as adipose tissue, are also effective. Because the number of MSCs in bone marrow is small, *in vitro* cell culture techniques are normally used to mitotically expand them without the loss of their developmental potency. Incorporation into a porous biomaterial can be achieved by adding a cell suspension onto the top of the biomaterial (Figure 23.4a). Uniform seeding of the cells within the pores of the biomaterial depends on the ability of the suspension to wet the surface of the biomaterial and on the biomaterial having the aforementioned pore characteristics for infiltration of the suspension throughout its interior. These cell-seeded biomaterials may be cultivated *in vitro* to allow differentiation into the osteoblastic lineage prior to implantation or implanted directly into the bone defect. Several studies have shown the ability to obtain MSCs from human or animal bone marrow. As the MSCs can be obtained percutaneously from a patient's own bone marrow, this approach would avoid the donor site morbidity associated with bone autografts, and the potential for immune rejection and disease transfer associated with allografts. However, the successful application of this approach depends critically on maintaining the survival of the cells within the implant for a sufficiently long period, which can often be difficult.

Healing of a bone defect by an osteoinductive mechanism depends on the presence of growth factors locally at the site of the defect. The osteoconductive biomaterial of a bone graft substitute serves as a delivery device for the growth factor and may serve as a surface on which cells can attach. The growth factor is often incorporated by adsorption to the pore surface of the biomaterial (such as porous hydroxyapatite) or by absorption into the biomaterial (such as a freeze-dried crosslinked collagen sponge) (Figure 23.4b). The most commonly used growth factors are from the TGFβ superfamily, such as BMP2, BMP7, and TGFβ1 (Chapter 16), of which BMP2 has been the most widely investigated. BMP is now produced commercially in a pure state using recombinant expression and other techniques.

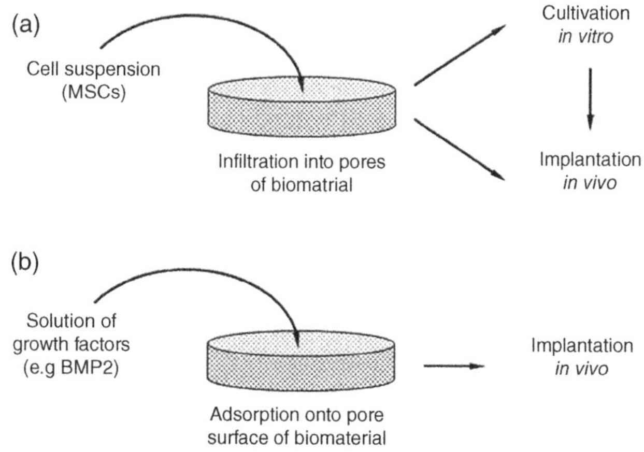

Figure 23.4 Illustration of (a) cell-based and (b) growth factor-based approaches to bone defect healing by bone graft substitutes composed of biomaterials.

Several studies have shown that the addition of BMP2 to osteoconductive biomaterials can significantly enhance their ability to heal bone defects in animal models. The effectiveness of BMP2 depends on a variety of factors such as its dose and duration of release. BMP2, like other growth factors, is very susceptible to losing its biological activity under nonoptimal conditions, and its controlled release from biomaterials at a therapeutically safe and effective concentration over a sustained period is often difficult. Consequently, to ensure efficient healing of bone defects, one approach has been to add BMP2 to bone graft substitutes in doses that are much higher than that normally present in the body (sometimes referred to as supra-physiological doses). However, the use of such high doses has heightened concerns about potential adverse biological effects, such as a higher risk of malignancy, radiculitis, ectopic bone formation, and osteolysis.

Overall, when compared to the use of cells, BMP2 is easier to incorporate into osteoconductive biomaterials and to use in a clinical procedure. However, the aforementioned difficulties related to loss of biological activity, inadequate control over its release over a desirable duration, and the potential for causing adverse tissue responses when released in supraphysiological doses are drawbacks.

Examples of the use of biomaterials as bone graft substitutes to heal various bone defects, either alone or in conjunction with cells or growth factors, are discussed in the following sections. Overall, particulate biomaterials are suitable for healing nonstructural bone defects, whereas porous three-dimensional solids and cements can be used in healing nonstructural and some structural defects. The ability of porous implants to heal structural defects depends on having the requisite combination of pore characteristics to support bone ingrowth and angiogenesis, and the mechanical properties to support the physiological stresses.

Evaluation of a bone graft substitute for its effectiveness in healing bone defects will normally involve *in vitro* biocompatibility tests with cultured cells, followed by *in vivo* testing with small animal models (normally rodents or rabbits) and then with large animal models (such as dogs, pigs, or sheep), assuming that the results from the preceding stage of testing have been satisfactory (Chapter 22). Bone defects created *in vivo* for testing bone graft substitutes should normally have a critical size, that is, a size above which the defect will not heal if left on its own for an appropriate time (longer than ~6–12 months). This critical size depends on the bone site and on the animal model. For example, the critical size defect in rodent or rabbit calvariae is a full-thickness defect of diameter ~5 mm, while critical size defects in the midsection (diaphysis) of rodent femora and rabbit femora are ~6 and ~10 mm long, respectively (Figure 23.5).

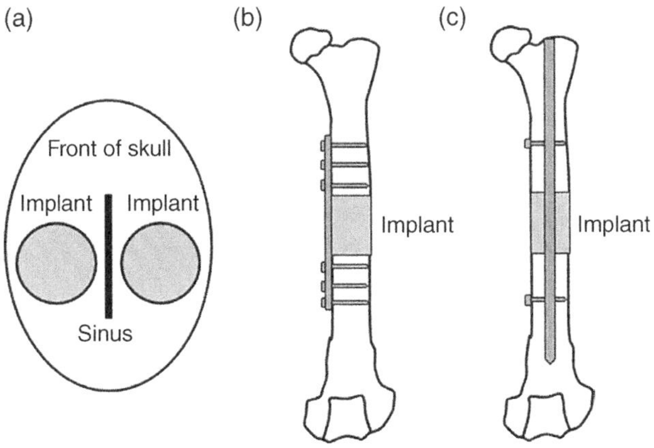

Figure 23.5 Illustration of common bone defect sites in small animals. (a) Bilateral defects in rodent calvarium; (b) segmental defect in rodent (or rabbit) femora and implant stabilization using a plate and screws; (c) segmental defect in rodent (or rabbit) femora and implant stabilization using a locked intramedullary nail.

23.3.4 Healing of Nonstructural Bone Defects

Coarse particles and beads of hydroxyapatite, β-TCP and BCP, as noted earlier, have been widely investigated and applied in healing nonstructural defects. In one study (Alam et al. 2001), microporous granules (diameter = 200–500 μm; pore size = 0.5–2 μm) of hydroxyapatite, β-TCP or BCP were impregnated with 1, 5, and 10 μg recombinant human BMP2, and bonded together using collagen to form disc-shaped specimens (8 mm in diameter × 2 mm). Then these BMP2-containing specimens and specimens without BMP2 (control group) were implanted for eight weeks in rodent calvariae. Histomorphometric analysis of stained sections showed little new bone growth into the implants without BMP2 (Figure 23.6). In comparison, the

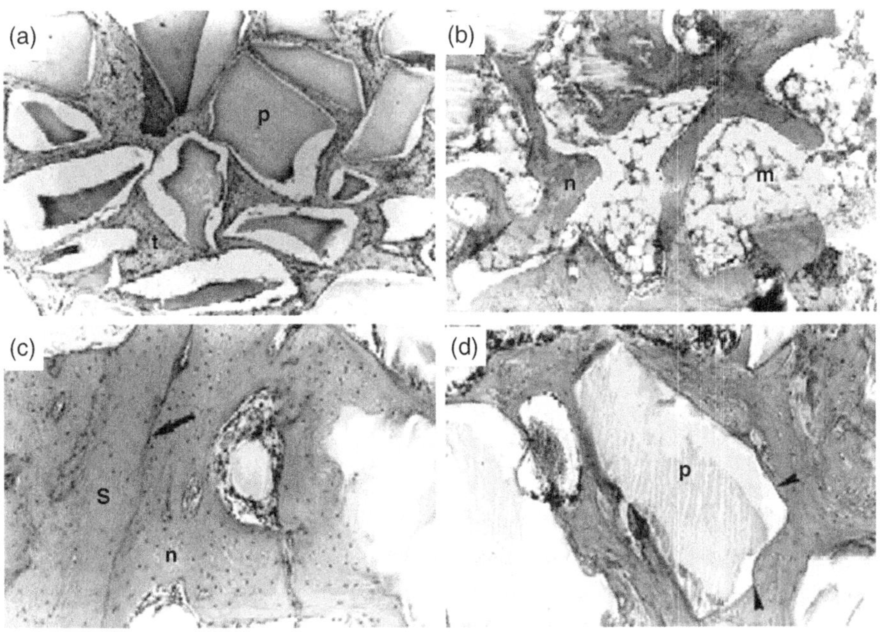

Figure 23.6 Optical images of hematoxylin and eosin (H&E) stained images of rat calvarial defects implanted for 8 weeks with hydroxyapatite particles (200–500 μm) impregnated with BMP2. (a) Implant without BMP2 (control) showing only fibrous connective tissue between the particles; (b) implant with 5 μg of BMP2 showing infiltration of new bone and bone marrow; (c) osseointegration (arrow) between the new bone and host bone of the calvarium; (d) deposition of new bone (arrowhead) occurring directly on the hydroxyapatite particle surface. (Original magnification ×50; p: particle; t, connective tissue; n, new bone; m, bone marrow; s, calvarial bone). *Source:* From Alam et al. (2001).

BMP2-containing implants showed considerable new bone and marrow infiltration into the spaces between the granules, formation of new bone on the granules and integration of the granules with host bone. Figure 23.7 shows that the amount of new bone and bone marrow that infiltrated the implants and the amount of residual calcium phosphate biomaterial in the defects at eight weeks depend on the composition of the implants and on the BMP2 dose. Overall, a BCP composition consisting of 25 wt% hydroxyapatite and 75 wt% beta-TCP, and a BMP2 dose of 5 μg (per defect) shows the best combination of new bone infiltration and resorption.

Collagen sponge is another biomaterial that has been widely used to deliver BMP2 in healing nonstructural bone defects, particularly defects in the cranium. Although collagen is mechanically weaker than synthetic biomaterials such as hydroxyapatite or degradable synthetic polymers such as the poly(α-hydroxy esters), collagen sponge impregnated with BMP2 has shown its effectiveness in healing nonstructural bone defects.

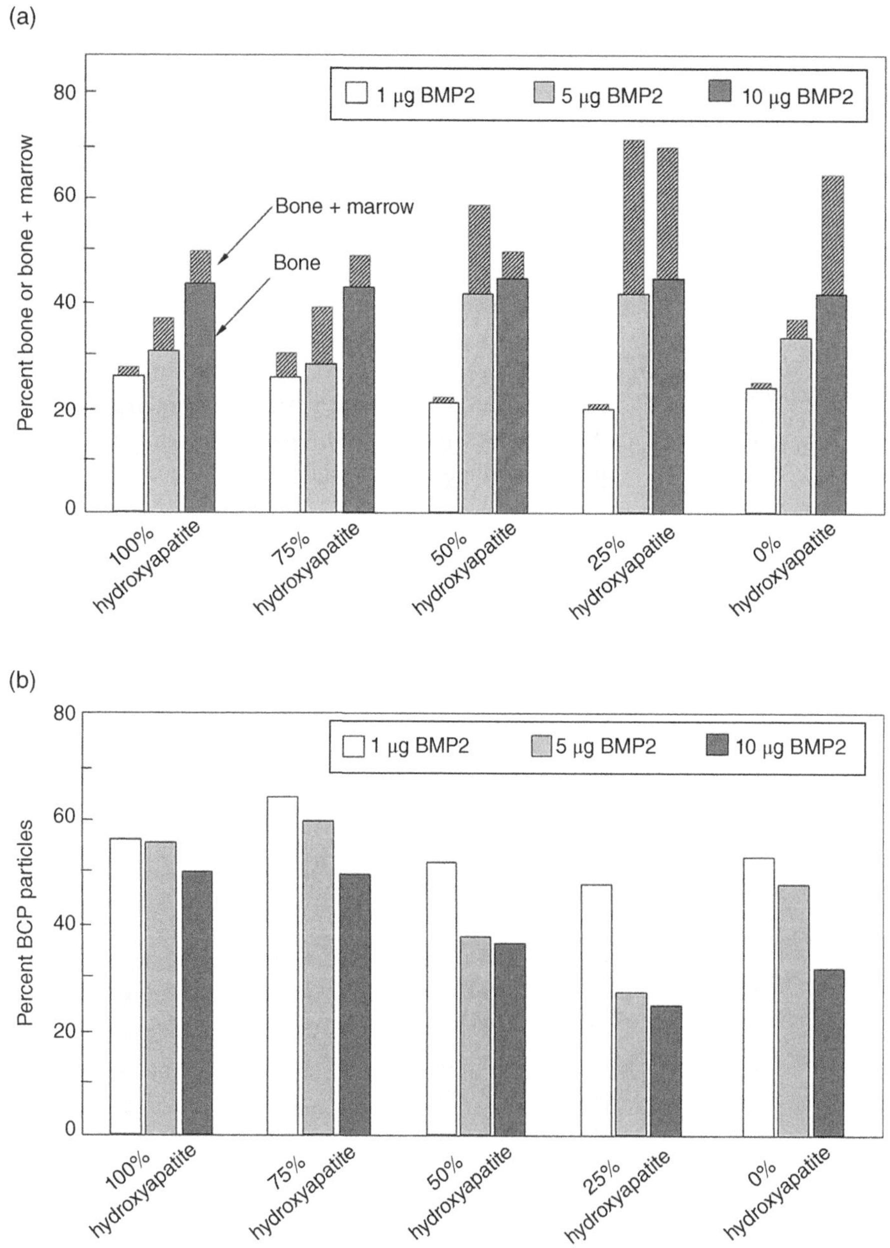

Figure 23.7 (a) Average percentage of bone and bone marrow in stained sections of defects implanted for eight weeks with biphasic calcium phosphate (BCP) particles (200–500 μm) incorporated with 1, 5, or 10 μg BMP2; (b) Average percentage of BCP particles remaining in the stained cross sections. *Source:* From Alam et al. (2001) / with permission of John Wiley & Sons.

Degradable synthetic polymers such as PLA, PLGA and polycaprolactone (PCL) are another type of biomaterial that has been used for healing bone defects. In one study (Cowan et al. 2007), porous implants of PLGA (porosity ~92%; pore size 200–300 μm) were prepared by a solvent casting and particulate leaching method (Chapter 25), and coated with a layer of collagen type I containing varying concentrations of BMP2 (0–5 μg), which released the BMP2 upon degradation. When seeded with osteoblastic MC3T3-E1 cells and incubated in osteogenic differentiation media *in vitro*, these BMP2-containing PLGA implants showed increased cellular density, extracellular matrix production, and mineralization on the PLGA, which increased with the BMP2 dose (30–240 ng/mm^3).

Upon implantation of these BMP2-containing specimens into critical size defects (5 mm diameter) in rat calvariae, microcomputed tomography (micro-CT) imaging showed increased bone formation in the defects in a dose-dependent manner, with higher doses of BMP2 inducing greater bone area, volume, and density (Figure 23.8). Histological sections confirmed the presence of new bone in the defects implanted with the BMP2-containing specimens (Figure 23.9). While the specimens without BMP2 were infiltrated mainly with fibrous tissue, those specimens containing 30–120 ng/mm^3 BMP2 were infiltrated with (immature) woven bone, whereas (mature) lamellar bone infiltrated the specimen with 240 ng/mm^3. An important feature of this study is the combination of micro-CT with more conventional techniques such as histomorphometric analysis in evaluating the healing of bone defects using bone graft substitutes.

Cell-based approaches used in healing bone defects have generally employed biomaterials similar to those used in growth factor-based approaches. For example, BCP particles (100–250 μm) in combination with human bone marrow-derived MSCs have shown the ability to heal critical size defects in rodent calvariae and to provide good integration between the implant and host bone (Mankani et al. 2006). Stained sections of the calvarial defects filled with the MSC-containing implant showed significant bone formation within the implant and union between the implant and the calvarium after only six weeks (Figure 23.10). In comparison, the implants without MSCs

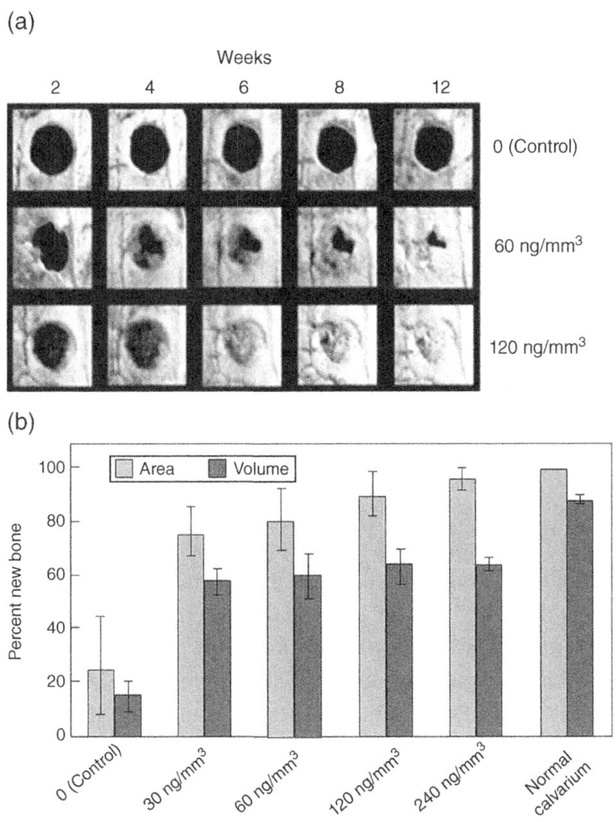

Figure 23.8 (a) Micro-CT images of rat calvarial defects implanted with porous PLGA specimens containing different doses of BMP2 at different times postimplantation. (b) Percentage of bone area and bone volume formed in rat calvarial defects implanted with PLGA specimens containing different doses on BMP2 at 12 weeks postimplantation. *Source:* Courtesy of Benjamin Wu, University of California-Los Angeles.

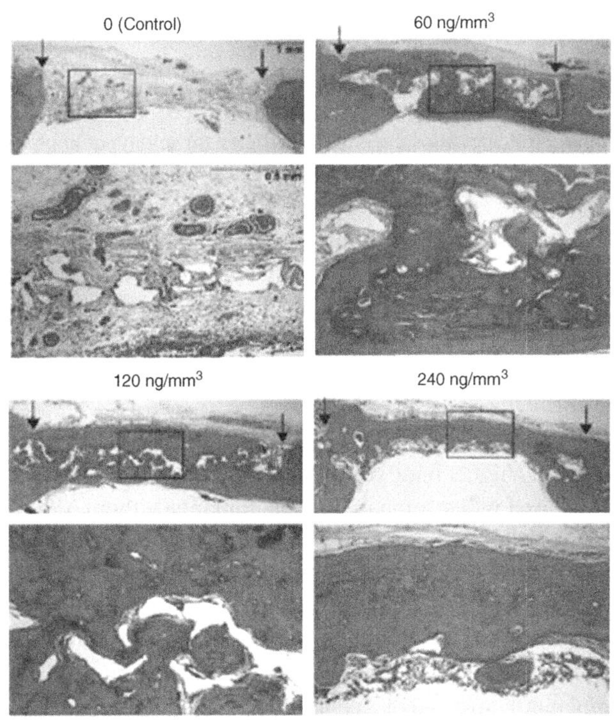

Figure 23.9 Transmitted light images of H&E stained sections of rat calvarial defects filled with porous PLGA implants containing different doses of BMP2 at 12 weeks post-implantation. Arrows indicate the defect margins and the bottom images correspond to the area delineated by the boxes. (Original magnification: top images ×24; bottom images ×400). *Source:* Courtesy of Benjamin Wu, University of California-Los Angeles.

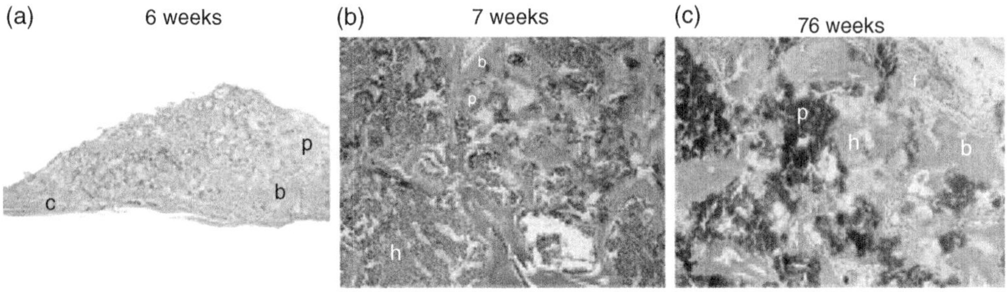

Figure 23.10 Transmitted light images of H&E stained sections of mouse calvarial defects filled with MSC-containing biphasic calcium phosphate (BCP) particles at (a) 6 weeks, (b) 7 weeks, and (c) 76 weeks postimplantation. (Original magnification: (a) ×4; (b), (c) ×10; b: bone within implant; c, calvarium; f, fibrous tissue; h, hematopoietic tissue; p, BCP particle). *Source:* From Mankani et al. (2006).

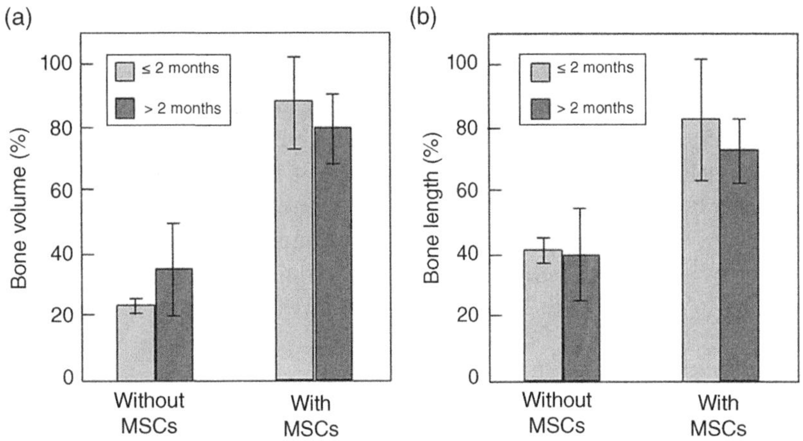

Figure 23.11 Quantitation of bone formation from H&E stained sections of mouse calvarial defects filled with MSC-containing biphasic calcium phosphate (BCP) particles. (a) Volume of implant containing bone; (b) length of the defect closed with bone-containing implant. *Source:* From Mankani et al. (2006) / with permission of John Wiley & Sons.

showed little bone infiltration even after 86 weeks, and were separated from the calvarium by fibrovascular tissue.

Bone continued to form with time in the defects filled with the BMP2-containing implants, which was accompanied by an increase in hematopoietic tissue and a reduction in the volume of the BCP particles. At 96 weeks postimplantation, the MSC-containing implants showed a significantly higher bone volume and length of the defect that was closed when compared to the implants without MSCs (Figure 23.11). Additionally, the hardness and Young's modulus of the bone within the implants at 96 weeks, measured using atomic force microscopy (AFM), were similar to those of the native calvarium. Based on their effectiveness in healing defects in rodent calvariae, these implants, composed of a combination of BCP particles and human bone marrow-derived MSCs, could potentially provide suitable bone graft substitutes for healing nonstructural defects in humans.

As noted earlier, biomaterials used in bone graft substitutes should have an appropriate combination of microstructure and mechanical properties. While a wide variety of techniques can be used to form these porous solids, additive manufacturing (3D printing) techniques have become very popular because of their ability to provide precise control of the microstructure and shape of the biomaterial. In one study, BCP implants (~65 wt% hydroxyapatite) were created with a grid-like microstructure (similar to that shown in Figure 12.9), composed of filaments of diameter ~200 μm and pores of width varying from 100 to 1000 μm. Following the addition of human bone marrow-derived MSCs, they were implanted in subcutaneous sites of immunocompromised female mice for up to 50 weeks to test their ability to support bone formation. These MSC-containing implants showed osteogenic activity in the subcutaneous sites, with pores of width 500 μm supporting the highest new bone formation within the implants (Figure 23.12a). Extensive lamellar bone was present in close proximity to the implant, indicating good bonding between the implant surface and new bone, along with abundant hematopoietic tissue, particularly at longer implantation times (Figure 23.13).

The mechanical properties of the implants showed a complex relationship as a function of the pore size and

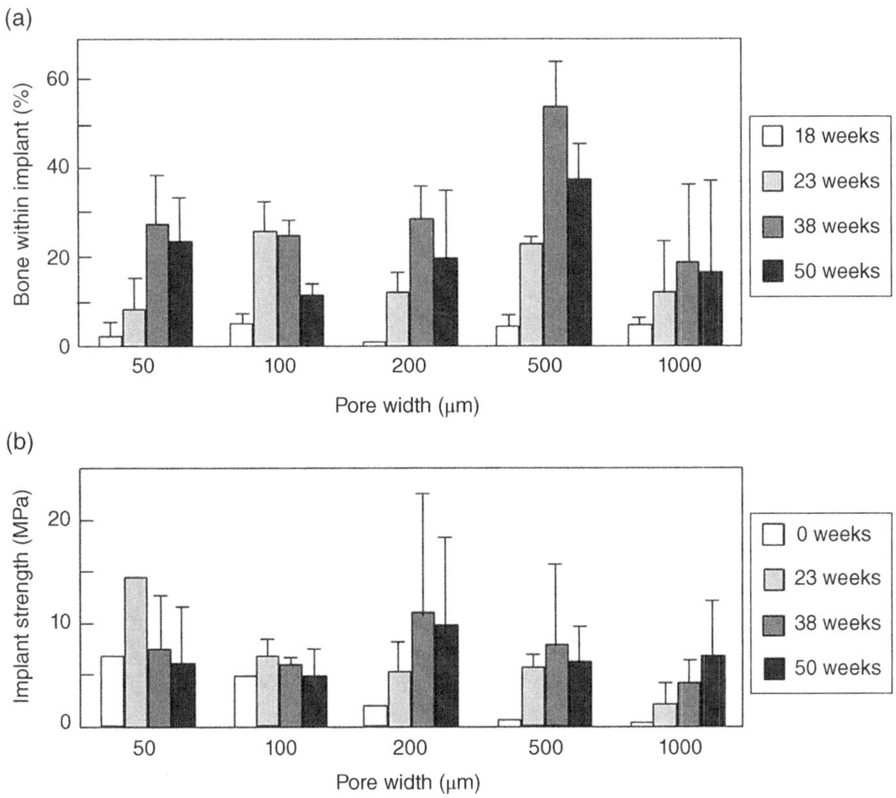

Figure 23.12 (a) Percentage of bone, as a fraction of the total tissue within the pores of the implant, as a function of the pore width of biphasic calcium phosphate (BCP) implants with a grid-like microstructure at various implantation times shown. (b) Compressive strength of the BCP implants, as-fabricated (0 weeks) and at various times postimplantation, as a function of pore width. *Source:* Courtesy of A. P. Tomsia, Lawrence Berkeley National Laboratory, Berkeley, California.

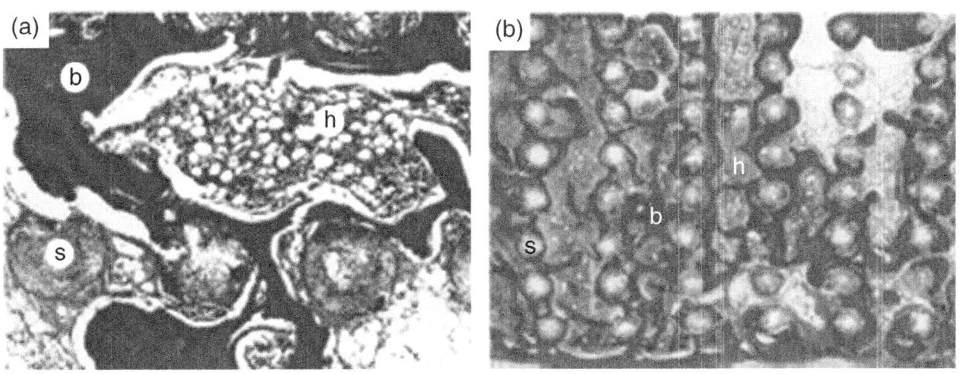

Figure 23.13 Optical images of H&E stained sections of MSC-containing BCP implants with a grid-like microstructure (pore width = 500 μm) after implantation for (a) 23 weeks (magnification ×20) and (b) 38 weeks (magnification ×5) in subcutaneous sites of immunocompromised female mice. (b, bone; h, hematopoietic tissue; s, filament cross section of BCP implant). *Source:* Courtesy of A. P. Tomsia, Lawrence Berkeley National Laboratory, Berkeley, California.

implantation time (Figure 23.12b). As fabricated, implants without cells showed a strong increase in compressive strength as the pore width decreased from 1000 to 50 μm. Because the solid filaments of the grid-like microstructure had the same diameter, the porosity of the as-fabricated implants decreased with a decrease in pore width. Consequently, the observed increase in compressive strength with decrease in pore width is consistent with a decrease in porosity. Upon implantation, the MSC-containing implants all showed an increase in compressive strength up to 23 weeks, matching the increase in new bone formation. Thereafter, the compressive strength showed a complex relationship, which may be due to the competing effects of bone formation and implant degradation. Additionally, while the as-fabricated implants showed a brittle mechanical response typical of most ceramics, those implanted for a long duration (50 weeks), containing a significant amount of bone, marrow, and fibrovascular tissue, showed a more ductile response. Overall, in addition to showing the effectiveness of MSC-containing BCP implants to support bone formation and cell survival within its pores, these results illustrate the complexity of designing porous implants with optimal properties and characteristics for use in bone healing.

While bone marrow-derived MSCs are the most commonly used cells, MSCs derived from other tissues, such as adipose tissue, have also shown their effectiveness in enhancing the ability of bone graft substitutes to heal bone defects. For example, in one study, porous PLGA implants (porosity ~90%; pore size 200–300 μm) were created using a solvent casting and particulate leaching method, coated with a layer of hydroxyapatite, and infiltrated with rodent adipose-derived MSCs. When implanted in critical size defects in rodents, these implants showed an ability to heal the defects in a manner similar to that of implants infiltrated with MSCs derived from rodent bone marrow.

23.3.5 Healing of Structural Bone Defects

The repair of large defects in structural bone, such as segmental bone loss in the midsection (diaphysis) of the long bones of the limbs is a common and challenging clinical problem. Although stabilized by fracture fixation devices, implants are normally required to provide the ability to support physiological stresses starting from the time of their initial implantation and to heal the defect in a clinically relevant time. Bone allograft and porous metals have been used with variable success to heal structural bone defects but, as noted earlier, they suffer from limitations. Consequently, there have been various attempts over the last few decades to develop suitable alternatives. Overall, most of these alternative bone graft substitutes have only been tested in animal models and many of them lack sufficient load-bearing capacity.

In one study (Ohgushi et al. 1989), porous implants of a commercial BCP (~60 wt% hydroxyapatite), with dimensions of 3×3 mm in cross section and 5 mm in length, and containing pores of size ~400 μm, were seeded with rat bone marrow-derived MSCs. Then they were implanted into segmental defects created in the femurs of rats and stabilized with polyethylene fracture fixation plates and Kirschner wires (Figure 23.14a). Defects were also implanted with

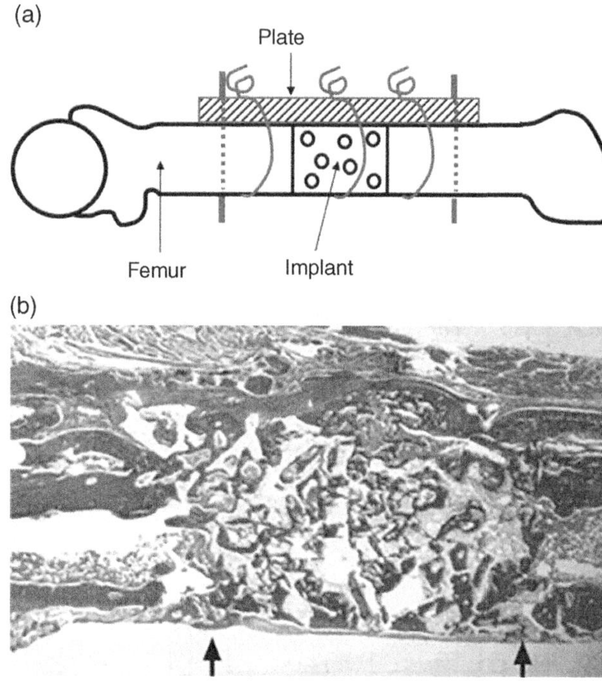

Figure 23.14 Repair of nonunion rat femoral defects with implants composed of biphasic calcium phosphate (BCP) seeded with rat bone marrow-derived mesenchymal stem cells (MSCs). (a) Schematic representation of the BCP implant in the rat femoral defect (5 mm in length). (b) Decalcified histologic section of the femur two months after surgery (arrows indicate the bone-implant junction). The pores of the implant were filled with bone and a bony continuity was observed at both bone-implant junctions. *Source:* From Ohgushi and Caplan (1999) / with permission of John Wiley & Sons.

BCP specimens without MSCs (positive control), or left empty (negative control). Two months postimplantation, the empty defects showed little ability to heal, whereas the defects implanted with the BCP without cells showed little new bone formation, mainly at the ends of the host bone in contact with the implant. In comparison, the MSC-containing implants showed good integration with host bone, brought about by the formation of new bone in and around the implants (Figure 23.14b).

Despite being effective in healing the critical size defects in rodent femurs, these MSC-containing BCP implants required considerable structural support by means of internal fracture fixation plates over their entire period of implantation. Additionally, the weak mechanical properties of BCP are a limitation. Consequently, these MSC-containing BCP implants are not well suited for healing defects in bones where the implant is required to provide significant structural support.

In another study (Chu et al. 2007), tubular implants (outer and inner diameter of 4 and 2 mm, respectively; length = 5 mm) were created from a composite composed of polypropylene fumarate (PPF) and β-TCP particles

(66 wt%) using an additive manufacturing (3D printing) technique. These implants contained four holes (diameter = 0.8 mm) through their sides, into which dicalcium phosphate dihydrate (DCPD) cement, mixed with 10 μg rhBMP2 or containing no BMP2 (control), were inserted. Then they were implanted for 6 or 15 weeks into segmental defects in rat femurs and stabilized with intramedullary wires (see Figure 23.5c).

Whereas the control implants failed to induce union (bridging) between the ends of the implant and host bone and collapsed within the 15-week implantation period, the implants containing BMP2 bridged the defect and maintained the length of the femur throughout the implantation period (Figure 23.15). Additionally, testing of the rat femurs containing their implants at 15 weeks showed that the BMP2-containing implants largely restored the mechanical properties of the host bone. Both the stiffness and ultimate force (to failure) in flexural loading for the femurs implanted with the BMP2-containing implants were approximately equal to those of the intact femurs. Overall, these results indicate that BMP2-containing PPF composites could potentially provide adequate load-bearing capacity to serve as bone graft substitutes in healing some structural bone defects.

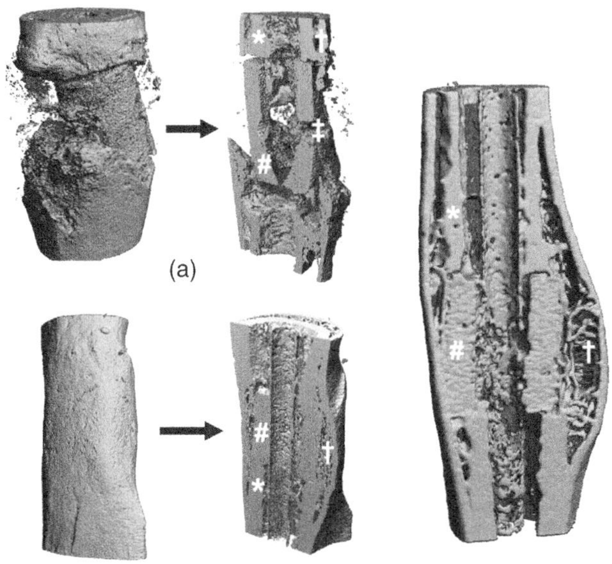

Figure 23.15 Micro-CT images showing external view and cross-section of segmental defects in rat femurs filled with tubular implants composed of polypropylene fumarate (PPF) reinforced with β-TCP particles (a) without BMP2 (control) or (b) with BMP2 at six weeks post-implantation. While the defects filled with the control implants showed minimal bone surrounding the implant and failure of the reparative callus to bridge the defect, the defects filled with the BMP2-containing implants showed a continuous mineralized callus surrounding the implant and integration of the implant with the host bone and mineralized callus. (c) At 16 weeks postimplantation, bone formation occurred on the surface of the BMP2-containing implant leading to enhanced integration. (*, original cortex of the femoral diaphysis; #, implant; †, mineralized callus; ‡, side-hole within the implant). *Source:* From Chu et al. (2007).

Bioactive glasses have also been studied for use as bone graft substitutes to heal structural bone defects. As noted earlier, bioactive glasses form a direct bond to host bone and release certain inorganic ions that have been shown to stimulate osteogenesis and angiogenesis. Consequently, this approach relies on the biomaterial alone, without the addition of cells or growth factors. Strong porous bioactive glass implants have been created and evaluated for their ability to heal structural bone defects in small animal models. For example, bioactive glasses with two different compositions, referred to as 13–93 and 2B6Sr (Table 7.3), were formed into porous implants (porosity ~50%) with a grid-like microstructure (pore width ~200 μm) (similar to that shown in Figure 12.9), and implanted for three months or nine months into segmental defects (10 mm long) created in rabbit femurs (Jia et al. 2015). For comparison, autologous bone grafts, taken from the iliac crest of the rabbits, were also implanted and used as a positive control group. The implants were stabilized in the defects using Ti plates and screws (see Figure 23.5b) and, thus, they were subjected to little physiological load during the implantation period.

Both groups of bioactive glass implants showed the ability to heal the defect, in a manner comparable to that of the autologous bone grafts. X-ray radiographs (Figure 23.16) and micro-CT images at three months showed partial integration between the glass implants and the host bone, and the formation of extensive callus and peri-implant bone. At nine months, there was complete integration with no visible interface between all three groups of implants and host bone, with mature bone bridging the junctions between the implants and host bone. Histomorphometric analysis showed an increasing amount of new bone within the implants and gradual degradation of the bioactive glass with time (Figure 23.16), along with the formation of blood vessels (angiogenesis) within and adjacent to the new bone. At nine months, the amount of new bone and blood vessels within both groups of bioactive glass implants were comparable to those within the autologous bone grafts (Figure 23.17). However, because bioactive glass 13–93 degrades more slowly than 2B6Sr, the amount of residual 13–93 glass implant within the defects was higher.

Overall, these results show the ability of strong porous implants of 13–93 and 2B6Sr glass to heal large defects in structural bone. On the other hand, because the implants were stabilized with Ti plates and screws, they were subjected to minimal physiological stresses.

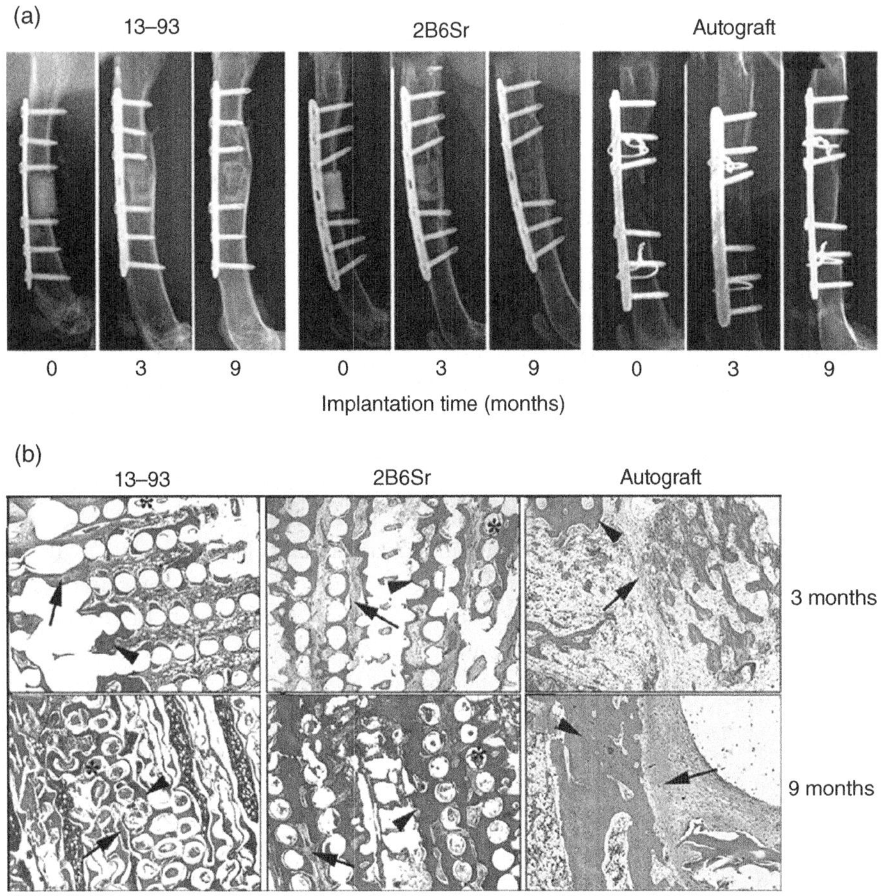

Figure 23.16 (a): X-ray radiographs of rabbit femoral segmental defects filled with porous implants composed of bioactive glass 13–93, bioactive glass 2B6Sr, or autologous bone grafts at 0, 3 months, and 9 months postimplantation. (b): Optical images of H&E stained sections of the defects filled with bioactive glass 13–93 implants, bioactive glass 2B6Sr implants, and autologous bone grafts at 3 months and 9 months postimplantation. (Original magnification × 50) (Arrowheads, mature bone; arrows, collagen fiber; *, cross section of bioactive glass filament.) *Source:* From Jia et al. (2015).

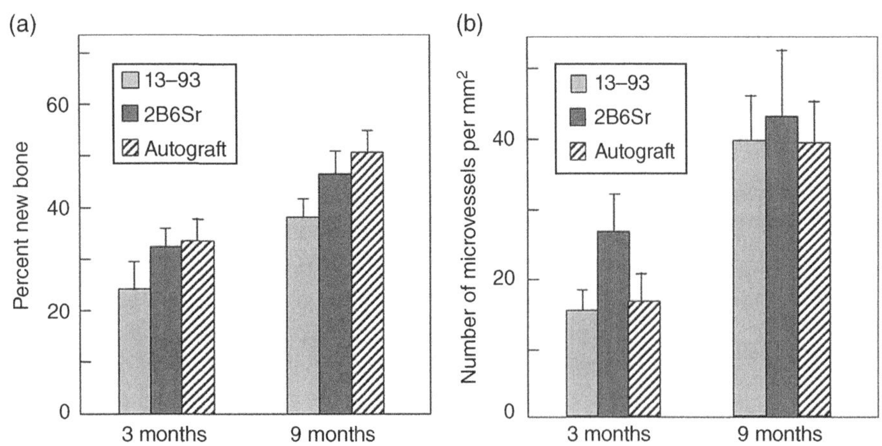

Figure 23.17 (a) Percent new bone (as a fraction of the total defect area) and (b) number of microvessel per mm^2 determined from H&E stained sections, for rabbit femoral segmental defects implanted with bioactive glass 13–93, bioactive glass 2B6Sr, and autologous bone grafts at 3 months and 9 months postimplantation. *Source:* From Jia et al. (2015) / with permission of John Wiley & Sons.

23.4 Total Joint Replacement

Skeletal motion in terrestrial mammals is made possible by synovial joints. While synovial joints in the human body are numerous and vary in structure, they have certain common features, as illustrated by the hip joint shown in Figure 1.6a. For example, a thin layer of cartilage covers the articulating surfaces of the bones where they meet to form a joint, while a very thin layer of a slippery viscous fluid (synovial fluid) separates and lubricates the two cartilage-covered bone surfaces. The joint is an example of a composite tissue, in which two distinct adult tissue phenotypes, bone, and cartilage, with little in common, integrate and function in synchrony to withstand repeated cyclic physiological stresses due to human activity.

Overall, the cartilage layer can deteriorate, due to osteoarthritis, rheumatoid arthritis or trauma, leading to debilitating disorders. Cartilage is an avascular tissue and, thus, it has little regenerative (self-healing) capacity. While small (focal) areas of articular cartilage can be healed by treatments such as chondrocyte transfer or transplantation of cartilage plugs (Chapter 24), cartilage degeneration is often difficult to detect until the underlying (subchondral) bone is reached, resulting in considerable pain. At this late stage, total joint replacement is often the only solution.

Total replacement of hip and knee joints (Figure 23.18) are among the most commonly performed orthopedic procedures in the United States and in some other countries. According to a study published in 2014, over 4.7 million people in the United States have undergone total knee arthroplasty (TKA) and over 2.5 million have undergone total hip arthroplasty (THA), making a total of ~7.2 million people living with the implants. Devices composed of biomaterials are also used to replace other joints, such as the mandibular joint of the jaw and the shoulder joint.

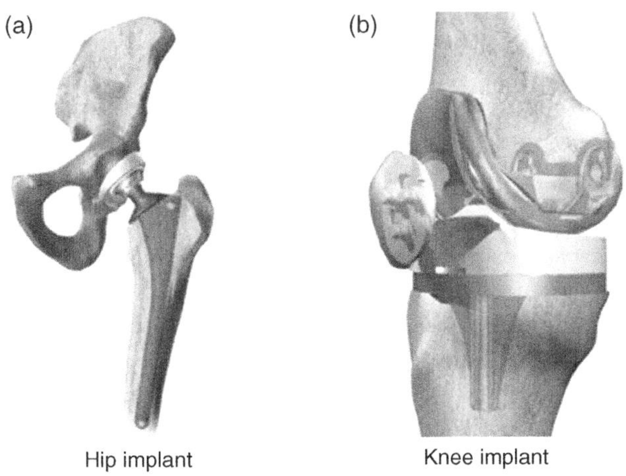

Figure 23.18 Illustration of the placement of prosthetic implants in (a) total hip arthroplasty and (b) total knee arthroplasty.

23.4.1 Total Hip Arthroplasty

In Chapter 1, we discussed the hip implant as a case study in biomaterials design and selection, including the variety of biomaterials and the properties of these biomaterials that determine their effectiveness when used as components in this device. As noted, a critical property is the wear resistance of the biomaterials that make up the articulating components of the device. When used in relation to total joint replacement, the term wear normally refers to the removal of material, typically in the form of particles, by mechanisms such as abrasion, adhesion, and fatigue, due to relative motion between two opposing surfaces under load, such as the articulating bearing surfaces of an implant. Low wear rates are desirable and clinically relevant because excessive generation of wear particles in an otherwise well-functioning THA or TKA normally leads to inflammatory responses in the tissues surrounding the bearings. This, in turn, leads to osteolysis and premature failure of the prosthetic joint by aseptic loosening (Chapter 21). Revision (repeat) surgery to salvage a failed THA or TKA is a major undertaking that is associated with significant morbidity and risks.

Screening tests for determining wear rates are often performed using simplified specimen geometries and methods, but the information obtained from them is often unreliable because of the many variables involved and the limited ability to control these variables under conditions relevant to real implants *in vivo*. In order to address this problem, hip simulators and knee simulators have been developed and used extensively over the last few decades to replicate *in vivo* conditions, such as the type of motion and the forces transmitted to the articulating components of the implant during the repeated walking cycle of a human (Figure 23.19).

Figure 23.20 shows the range of reported wear rates of various articulating couples that have been used or developed for use in prosthetic hip implants. The most widely used bearing couple, as noted in Chapter 1, is a Co–Cr femoral head articulating against an acetabular liner composed of ultra-high molecular weight polyethylene (UHMWPE) or cross-linked UHMWPE. Cross-linked UHMWPE has better wear resistance than UHMWPE but it is more susceptible to brittle fracture because of its lower strain to failure and lower resistance to fatigue crack propagation. When rotating against UHMWPE or cross-linked UHMWPE liners, ceramic femoral heads composed of ZrO_2-toughened Al_2O_3 (ZTA) show much lower wear rates than Co–Cr, while ZTA articulating against ZTA shows still lower wear

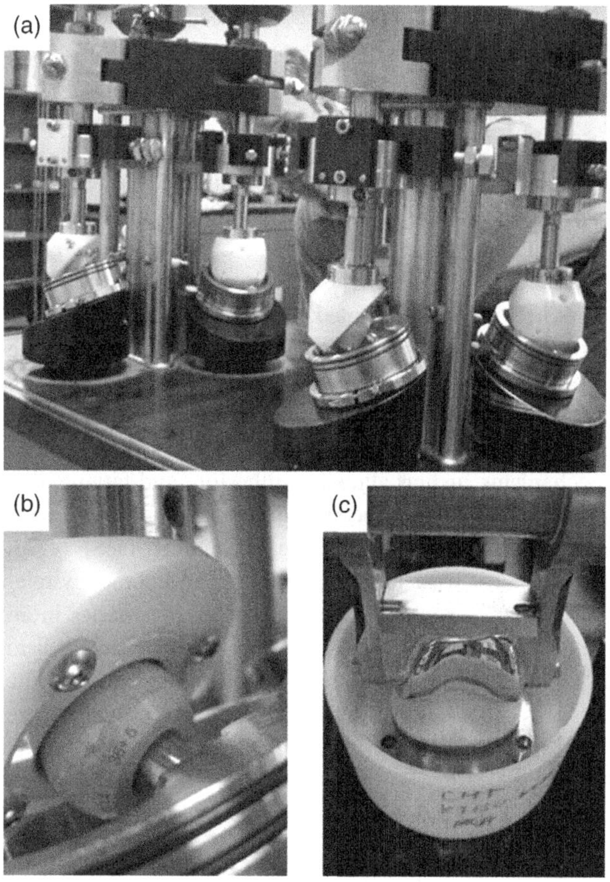

Figure 23.19 (a) View of four stations of an eight-station hip joint simulator with the hip joint bearings in the anatomical (upright) configuration; (b) close-up view of ZrO_2-toughened Al_2O_3 (ZTA) bearing couple in a hip simulator; (c) close-up of Co-Cr–UHMWPE bearing couple in a knee simulator. (A lubricating fluid, normally present in the specimen chambers, was removed to show the bearing couples more clearly.) *Source:* From Rahaman et al. (2007).

rates. Silicon nitride (Si_3N_4) has been developed for use as articulating bearings in hip implants but the US Food and Drug Administration has not yet approved these devices for use in humans. Overall, Si_3N_4 femoral heads show wear rates comparable to ZTA femoral heads when articulating against UHMWPE, cross-linked UHMWPE or Si_3N_4 liners.

The average lifetime of prosthetic hip implants with a bearing couple consisting of Co–Cr and UHMWPE is ~10–15 years. Replacing the UHMWPE liner in this bearing couple with cross-linked UHMWPE leads to an increase in the average lifetime of the implant, but it is still lower than that for bearing couples composed of ceramics. For example, the average lifetime of Al_2O_3–Al_2O_3 bearing couples has been reported to be over 20 years. However, as noted in Chapter 1, this longer lifetime is offset by an increase in cost, the risk of a higher incidence of catastrophic failure due to brittle fracture of Al_2O_3, and the potential for the generation of a squeaking noise if the implant were not properly aligned during implantation.

THA involves the removal of a considerable amount of subchondral bone that may normally be healthy. To avoid this, particularly for younger patients with healthy bones, a hip resurfacing procedure can be used, in which only the surface region of the femoral head bone is trimmed and replaced by a metal cap, often composed of Co–Cr. This cap articulates within an acetabular cup similar to that used in THA, which is inserted into the acetabulum.

23.4.2 Total Knee Arthroplasty

When compared to the hip implant where the femoral head is largely constrained in the acetabular cup, forming a ball and socket arrangement, the physiological stresses on the articulating bearings of a knee implant are more complex due to the greater freedom of movement and the more complex geometry of the articulating bearing couple (Figure 23.18). In general, the biomaterials used for the components of the knee implant are identical to those for the hip implant. The femoral component, commonly composed of Co–Cr, is bonded to the femur using polymethyl methacrylate (PMMA) bone cement and articulates against a tibial plateau composed of UHMWPE (or crosslinked UHMWPE) attached to a tibial support composed typically of Ti6Al4V, which is inserted into the tibia (Figure 23.18). Oxidized zirconium (designated Oxinium™) or Y_2O_3-stabilized ZrO_2 (YSZ) femoral components are used to a limited extent as alternatives to Co-Cr. However, as noted for hip implants, prosthetic knee joints containing YSZ articulating components are not approved by the US Food and Drug Administration (FDA) for use in humans. Silicon nitride has also been developed for use as femoral components and tibial plateaus in TKA but, as noted for THA, these devices have not yet received approval from the US FDA for commercialization.

23.5 Spinal Fusion

Low back pain is a common health problem, affecting millions of people and requiring significant healthcare expenditure. It is caused by the degeneration of one or more of the intervertebral discs in the lower part of the spine. The intervertebral disc is a flat cylindrical piece of cartilage, composed of an outer tough ring, called the annulus fibrosus, which encloses a viscous gel-like mass, the nucleus pulposus. It separates adjacent vertebrae and allows for bending and rotational movement of the spine (Figure 23.21). However, the intervertebral disc can

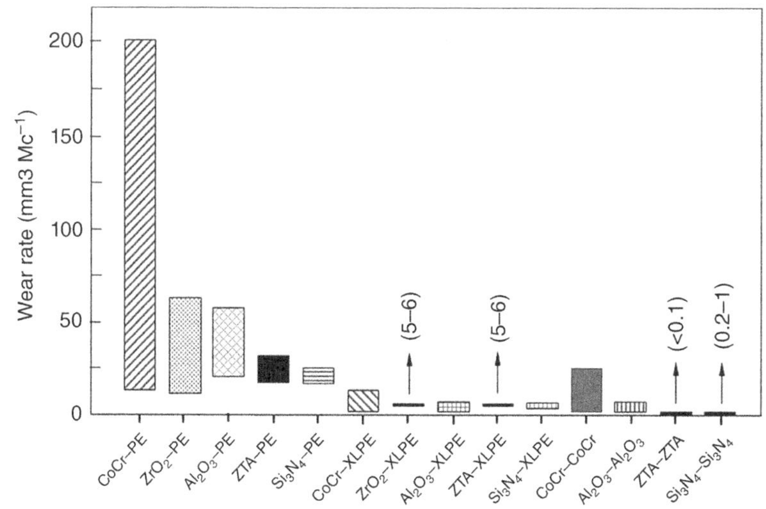

Figure 23.20 Range of wear rates (mm^3 per million cycles) measured in hip simulator tests for various bearing couples. (CoCr, cobalt–chromium alloy; ZrO_2, zirconia stabilized with CeO_2 or Y_2O_3; ZTA, ZrO_2-toughened Al_2O_3; PE, ultrahigh molecular weight polyethylene; XLPE, cross-linked ultrahigh molecular weight polyethylene.) *Source:* From Rahaman and Xiao (2018) / with permission of John Wiley & Sons.

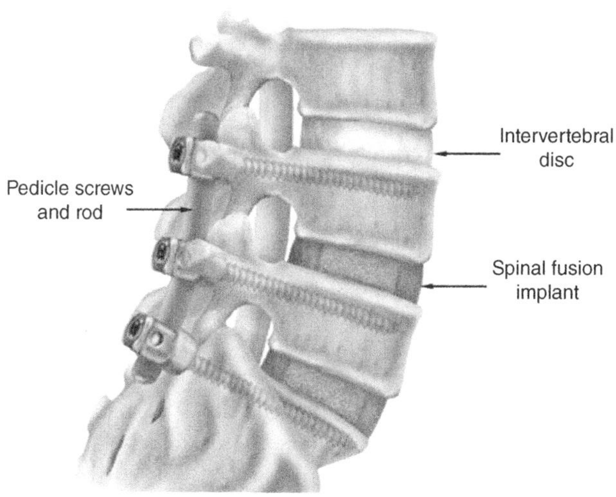

Figure 23.21 Illustration of stabilization of the human spine using implants composed of a bone graft or a biomaterial stabilized by a rod and pedicle screws. *Source:* From Kurtz and Devine (2007) / with permission of Elsevier.

degenerate over time or may be displaced from the intervertebral space, causing intense pain as the vertebrae come into contact and interfere with the nerves.

Cartilage, as noted earlier, has little self-healing capacity. Consequently, one treatment that has been developed to alleviate low back pain involves removal of the degenerated disc and the placement of a graft in the intervertebral space to achieve fusion of adjacent vertebrae. This treatment produces a small loss in mobility of the spine in the region of the affected vertebrae but this is more than offset by a significant reduction in low back pain. Spinal fusion is now a widely used procedure in orthopedic surgery, with an estimated half a million procedures performed annually in the United States, and providing relatively pain-free back movements over ~10 years.

All spinal fusion procedures use some type of bone graft or biomaterial. Because osseointegration between the graft and the adjacent vertebrae is required for fusion, much of our earlier discussion on bone grafts and bone graft substitutes are relevant to the selection of biomaterials for use as spinal fusion implants. Initially, autologous bone grafts harvested from the iliac crest and bone allografts were used to replace the degenerated intervertebral disc, but these grafts suffer from the drawbacks discussed earlier for bone grafts. Subsequently, devices called spinal fusion cages composed of biomaterials were developed for implantation into the intervertebral space. These cages have a variety of designs, a few of which are shown in Figure 23.22. They provide the requisite mechanical properties to support physiological stresses and restore the separation between the adjacent vertebrae. These cages are often attached to the vertebral bone by screws called pedicle screws that are connected by rods to provide good stability during the spinal fusion period. Some cages are designed to contain an autologous bone graft or a bone graft substitute, to enhance fusion or to allow fusion by new bone growth through the cage between the adjacent vertebrae. BMP2 has been introduced into some cages, usually in the form of a collagen sponge impregnated with BMP2, but this design is controversial. As noted earlier for bone graft substitutes, controlling the release of the BMP2 within a desired range of therapeutic doses over an appropriate period is difficult, while the uncontrolled release of the BMP2 in high doses can lead to adverse biological effects.

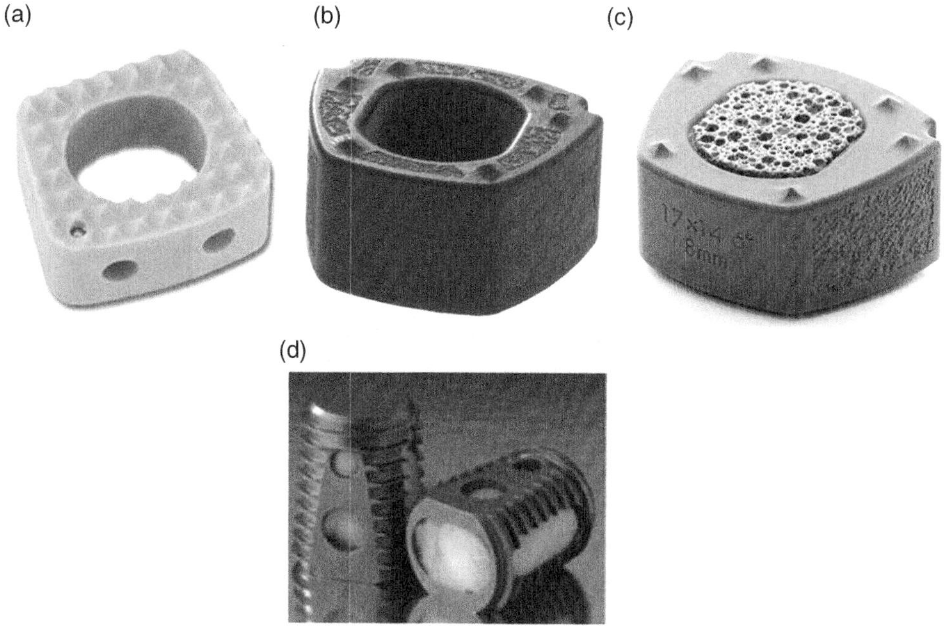

Figure 23.22 Examples of spinal fusion cages composed of biomaterials: (a) polyether ether ketone (PEEK); (b) silicon nitride (Si_3N_4). *Source:* From Ball et al. (2017). (c) Si_3N_4 cage with a porous interior region composed of the same material. *Source:* From Arts et al. (2017). (d) A titanium cage containing a bovine collagen type I sponge from which bone morphogenetic protein-2 (BMP2) is eluted. *Source:* From Huebsch and Mooney (2009).

23.5.1 Biomaterials for Spinal Fusion

Several biomaterials have been evaluated for use as spinal fusion cages, but cages made from Ti or its alloy Ti6Al4V, polyether ether ketone (PEEK), and silicon nitride (Si_3N_4) have been most widely used clinically, with PEEK cages now the most popular. Based on the long history of Ti (and Ti6Al4V) as implants in orthopedic and dental surgery, Ti cages were commercialized and became available for clinical applications around the early 1990s. Titanium has well-known advantages, such as biocompatibility, high corrosion resistance, low density, and the ability to support cell adhesion and osseointegration with host bone, particularly when its surface has certain roughness features. While Ti cages show favorable fusion rates, a major shortcoming is their tendency to subside into the adjacent vertebrae with time, thereby not maintaining the required intervertebral separation. This has been attributed to the high Young's modulus of Ti when compared to bone, which leads to adverse mechanical biocompatibility phenomena such as stress shielding and resorption of the surrounding bone (Chapter 17). Ti is also radiopaque (i.e., opaque to X-rays), making it difficult to visualize new bone formation on or within the cages from radiographic images.

PEEK spinal fusion cages became available for clinical use in the late 1990s and, as noted earlier, are now the most popular. Because PEEK has a Young's modulus comparable to that of human cortical bone (Chapter 8), the use of PEEK cages leads to a reduction in the aforementioned adverse mechanical biocompatibility phenomena associated with Ti cages. Additionally, PEEK has a tensile yield strength comparable to the strength of cortical bone and good ductility. PEEK is radiolucent (transparent to X-rays), which makes it difficult to visualize the cage position and fusion using X-ray imaging techniques, but this drawback is normally overcome by the incorporation of radiopaque markers such as tantalum (Ta) particles. The major drawbacks of PEEK are related to its hydrophobicity and lack of bioactivity. Normally, a thin fibrous tissue surrounds a PEEK cage *in vivo*, limiting osseointegration between the cage and the adjacent vertebrae. This often leads to micromotion between the cage and the vertebrae, which may lead eventually to failure of the implant. Various approaches have been used to reduce this problem, such as coating the cage with hydroxyapatite or incorporating hydroxyapatite particles into the PEEK to improve its bioactivity.

More recently, spinal fusion cages composed of Si_3N_4 have been commercialized and applied clinically. While Si_3N_4 has excellent mechanical properties, of significant importance for long-term implantation applications such as spinal repair are its osseointegrative and antimicrobial properties due largely to its unique surface topography (Chapter 21). Si_3N_4 is also partially radiopaque, allowing visualization of spinal fusion cages *in vivo* by X-ray imaging techniques. Cages of various designs have been produced, such as cages consisting of a dense outer shell and cages consisting of a dense outer shell and a center region consisting of a porous form of the same material (Figure 23.22).

Overall, clinical studies have shown that cages created from Ti, PEEK, or Si_3N_4 are effective in spinal fusion despite their shortcomings. Although PEEK cages are currently the most popular, there is a paucity of data to support the selection of cages created from one biomaterial over the others.

23.6 Dental Implants and Restorations

Biomaterials used in dental applications may be grouped into three major types:

- Dental implants are biomaterials that are implanted into bone, such as the mandible and maxilla of the mouth, to provide support for prosthetic dental devices such as crowns, bridges, and dentures.
- Dental restorations are devices composed of biomaterials, which are used to restore the structure, function and appearance of damaged or lost teeth. They are often divided into direct and indirect restorations. Direct restorations are inserted directly into the prepared site of the tooth, such as gold or composite resins used to fill cavities. In comparison, indirect restorations, such as crowns, bridges, and dentures, are created outside the mouth using a model of the prepared tooth and later fixed into the existing tooth structure.
- Dental cements are specially formulated materials, such as adhesives, used to bond dental restorations to the existing tooth structure.

23.6.1 Dental Implants

Titanium and its alloy Ti6Al4V are the most common biomaterials used in creating dental implants, based on their attractive combination of properties, such as mechanical properties to support indirect restorations and to withstand the stresses arising from mastication, biocompatibility, and corrosion resistance. Dental implants are available in a wide variety of designs (Figure 23.23). The surfaces of these implants are normally modified to improve their potential for bonding and osseointegration with host bone. Common surface modifications involve sandblasting (or grit blasting) to create a desirable surface roughness (Chapter 21), or sandblasting followed by coating with a bioactive material such as hydroxyapatite.

Hydroxyapatite coatings on Ti (or Ti6Al4V) implants are commonly deposited using plasma spraying, in which molten particles are sprayed onto the surface of the

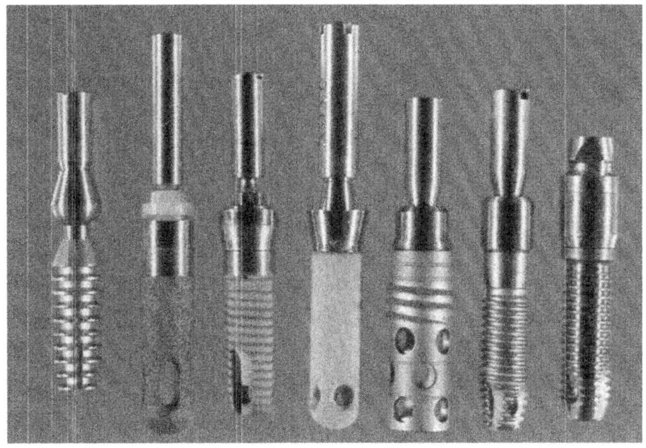

Figure 23.23 Dental implants of various designs and with different surface modification. *Source:* From Ratner (2013).

implant. Although these plasma-sprayed coatings have shown the ability to enhance osseointegration and, thus, to improve the interfacial strength between the implant and host bone in the short term, the long-term performance of these coatings is questionable. Common problems are the weak bonding between the coating and the implant itself, which can lead to delamination of the coating (separation from the surface of the implant), and the susceptibility of the coating to cracking, which can lead to the release of particulate debris, resulting in inflammatory tissue responses (Chapter 21). These problems have their origins in the plasma spraying process itself. As the molten particles impact onto the implant surface and condense to form a layer, they cool rapidly, which leads to interfacial stresses due to a difference in thermal expansion coefficient between hydroxyapatite and Ti. Delamination and cracking can occur when the interfacial region becomes too weak to support these stresses.

23.6.2 Direct Dental Restorations

Dental restorations, direct or indirect, should be biocompatible in the oral environment, but unlike dental implants, they are not required to be osseointegrative or bioactive. Gold is very effective in filling cavities because it bonds strongly to the prepared tooth surface. However, it finds limited use in filling cavities, mainly in the posterior teeth where esthetics are not a major requirement. In comparison, resin composites find widespread use in filling cavities in the anterior or the posterior teeth because they have an adequate combination of physicochemical properties and esthetic characteristics. The technology associated with resin composites is well established. Commonly used resins are based on the monomer Bis-GMA (bisphenol A-glycidyl methacrylate), a diester derived from methacrylic acid and

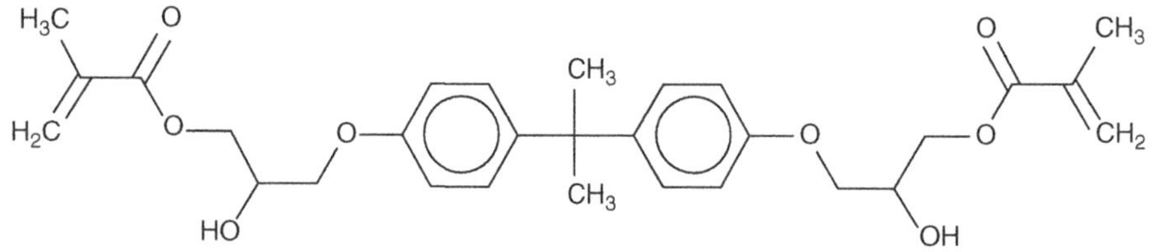

Figure 23.24 Chemical structure of Bis-GMA (bisphenol A-glycidyl methacrylate).

the bisphenol A-diglycidyl ether (Figure 23.24). The presence of two ($C{=}CH_2$) groups in the monomer allows for easy crosslinking (curing) by visible light to form a polymer (cured resin). Bis-GMA has a high viscosity on its own, and it has to be incorporated with a particulate phase to achieve a cured resin with suitable physical properties. Consequently, it is normally diluted with a low molecular weight, low viscosity methacrylate monomer, such as triethylene glycol dimethacrylate (TEGDMA), to achieve a suitable viscosity for filling a cavity. Particles incorporated into Bis-GMA normally consist of chemically inert ceramics of low thermal expansion coefficient, such as coarse quartz or fine (colloidal) silica. Their volume fraction is controlled to achieve a composite having an optimal combination of properties, such as a low shrinkage during curing, a thermal expansion coefficient and a thermal conductivity not vastly different from those of the tooth, adequate mechanical properties to withstand the complex stresses associated with mastication, and suitable abrasive wear resistance to match that of the teeth.

23.6.3 Indirect Dental Restorations

Indirect restorations take a variety of forms, such as crowns, bridges, dentures, inlays, onlays, and veneers. In general, the biomaterials used in creating these dental restorations should ideally have the following combination of properties and characteristics:

- Adequate corrosion resistance to withstand the highly corrosive oral environment
- Mechanical properties and reliability to withstand the complex stresses encountered during mastication without deformation, bulk fracture, or breakage at the margins over a considerable time (preferably the lifetime of the patient but at least ~10 years), and abrasive wear resistance to match that of the natural teeth
- Thermal conductivity comparable to that of the tooth to reduce the potential for large temperature gradients between the restoration and the existing tooth structure
- Ability to be formed economically into the shape of a restoration and once formed, to be capable of being manipulated for placement, shaping, and polishing to a smooth finish in the mouth
- Optical properties to match the color of the existing teeth for aesthetic purposes
- Ability to be visualized in the mouth by standard techniques used in a dental examination, such as X-ray radiography
- Low susceptibility to the accumulation of plaque.

While no material ideally satisfies all of these properties and characteristics, advances in materials science over the last several decades have steadily led to the development of dental restorative materials that provide a reasonable approximation to them. Overall, a wide variety of materials such as porcelain, porcelain–veneered metals, glass-ceramics, and polycrystalline ceramics have been used to create indirect dental restorations. Porcelain–veneered metal frameworks have been used in dentistry since the 1960s, and they still make up a significant fraction of the market for dental restorative materials. However, in the last few decades, the use of metal-free restorative materials composed of glass-ceramics and polycrystalline ceramics has been growing considerably.

Metal-Ceramic Dental Restorations

Metal frameworks composed of corrosion resistant and biocompatible metals such as Au, Pd–Ag alloys and Co–Cr alloys have excellent mechanical properties but they are generally opaque, lacking the optical properties and aesthetic characteristics of teeth. Consequently, a veneer is applied to them to mask their opacity and improve their appearance. Most veneers have a composition in the K_2O–Al_2O_3–SiO_2 phase diagram, a system that also forms the basis in creating various porcelains that have long been used in household and industrial appliances. Because of this general compositional relationship, porcelains developed for use in dental restorations are sometimes referred to as dental porcelains. Dental porcelains are particularly effective for veneering the metal frameworks of crowns and long-span posterior bridges.

Dental porcelains normally have a composition in the range 9–11% K_2O, 11–16% Al_2O_3, 52–62% SiO_2 (wt%), plus 5–7 wt% Na_2O, and smaller amounts of Li_2O and B_2O_3. A layer composed of a particulate mixture with the desired composition is formed on the metal framework. Then the system is sintered (heated) at ~800–900 °C to form a dense porcelain layer that is well bonded to the framework. Reaction between the different particulate materials in the layer during the sintering step leads to the development of a microstructure consisting predominantly of a glass matrix and a small amount of the (crystalline) minerals potassium feldspar ($KAlSi_3O_8$) (also written $K_2O{\bullet}Al_2O_3{\bullet}6SiO_2$) and leucite ($KAlSi_2O_6$) ($K_2O{\bullet}Al_2O_3{\bullet}4SiO_2$). These porcelains are sometimes referred to as feldspatic porcelains because of the presence of potassium feldspar in them.

When compared to glass-ceramics, the crystalline phase in feldspatic porcelains is not formed in a separate step by controlled nucleation and crystal growth of a glass (Chapter 7). Instead, rapid and uncontrolled crystallization occurs during the sintering step, which leads to a disordered arrangement of a low volume fraction of crystals within the glass phase. These feldspatic porcelains also have a coefficient of thermal expansion (~$8–9\times10^{-6}$ °C^{-1}) that is very different from those of the aforementioned metals (alloys) making up the framework (~$14–16\times10^{-6}$ °C^{-1}). This large difference in thermal expansion coefficient leads to mechanical stresses (referred to as thermal stresses) within the porcelain layer upon cooling from the sintering temperature. Because of their uncontrolled microstructure and the presence of thermal stresses within them, feldspatic porcelains have weak mechanical properties, which make them unsuitable for use as veneers for metal frameworks.

Subsequently, feldspatic porcelains with improved properties were develop by combining appropriate areas of feldspatic porcelain technology and glass-ceramic technology. During the sintering process, feldspathic porcelain compositions are subjected to a controlled heat treatment, which leads to the formation of a higher concentration (15–25% by volume) and a more controlled distribution of leucite crystals by a surface crystallization process. Below ~600 °C, leucite exists in a tetragonal crystal structure, which has a high thermal expansion coefficient. The higher concentration of leucite crystals, coupled with their high thermal expansion coefficient, makes the overall thermal expansion coefficient of the porcelain (~$12–13\times10^{-6}$ °C^{-1}) closer to that of the metal framework and, thus, reduces the magnitude of the thermal stresses. Together, the improvement in microstructure and the reduction in thermal stresses make leucite-based porcelains sufficiently strong (flexural strength of ~80 MPa) for successful veneering of metal frameworks such as crowns and bridges.

The production of metal-ceramic dental restorations often involves a few process steps. After casting of the metal framework, a first layer is applied to mask the color of the metal. This layer also serves as a suitable bonding surface for the subsequent layers, which normally consist of two layers to form the dentin and incisal parts of the tooth. Thus, three layers of leucite-based porcelains are bonded together to form a veneer on the metal framework.

As noted below for glass-ceramic restorations, recent advances in dental technology have provided an alternative method for forming the porcelain layer on the metal framework. Instead of sintering, the porcelain layers can be pressed on to the metal framework at a suitably high temperature comparable to that in sintering (a process described as heat pressing). This process makes use of the viscous flow characteristics of the glass phase of the porcelain but it requires a special furnace for heating the material and applying a stress to it as well.

Glass-Ceramic Dental Restorations

Metal-ceramic restorations have high strength and are very effective, but the opacity of the metal framework is a drawback. Consequently, there have been heightened efforts to research and develop metal-free restorations. One type of material that has received considerable attention in such efforts has been glass-ceramics. Advances in glass-ceramic science and technology over the last few decades have led to the development of lithium disilicate ($Li_2Si_2O_5$) glass-ceramics, which are now one of the most widely used metal-free dental restorative materials (Figure 23.25). A range of lithium disilicate glass-ceramics have been developed but the more recently developed materials show a combination of properties and characteristics that are a good approximation to those of an ideal dental restoration noted earlier. Consequently, these recently developed lithium disilicate glass-ceramics are very effective as dental restorative materials. However, when used as bridges, they should be restricted to the fabrication of three-unit bridges for the anterior region of the mouth up to the second

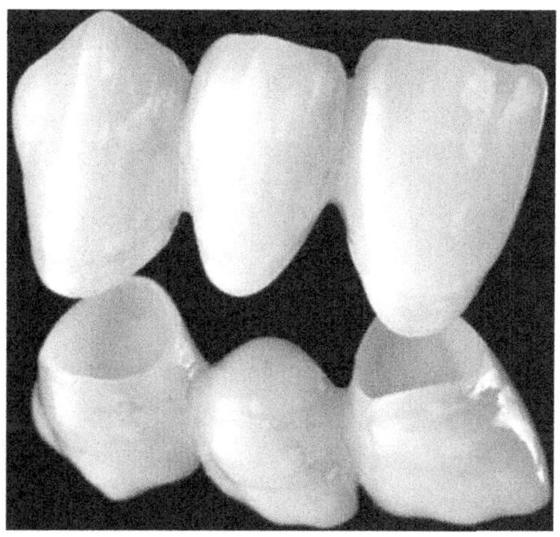

Figure 23.25 Dental bridge reflected on a mirror to show the interior. The bridge consists of a framework of lithium disilicate glass-ceramic veneered with fluorapatite glass-ceramic. *Source:* From Hench et al. (2010).

premolars, where the stresses due to mastication are lower than those in the posterior region.

The principles of glass-ceramic science and technology and their application to the creation of lithium disilicate glass-ceramics were discussed in Chapter 7. In addition to their chemical durability, the most significant features of the recently developed lithium disilicate glass-ceramics that are largely responsible for their effectiveness as dental restorative materials include:

- A microstructure consisting of a high volume fraction of interlocking fine acicular $Li_2Si_2O_5$ crystals homogeneously dispersed with the glass matrix (Figure 7.27b), which is responsible for their attractive mechanical properties (Table 7.4)
- The ability to convert the parent glass to a glass-ceramic in two separate steps, in which a softer lithium metasilicate (Li_2SiO_3) glass-ceramic is formed in the first step, while the required $Li_2Si_2O_5$ glass-ceramic with much better mechanical properties is formed in the second step (Figure 7.24)
- The ability to take advantage of the viscous flow properties of the glass phase of the glass-ceramic to shape the dental restoration
- The ability to take advantage of the optical properties of a glass and the ability to modify these properties through small additions of certain metal oxides to the parent glass composition to achieve a dental restoration having the optical properties and aesthetic characteristics of the existing tooth structure.

Additionally, these properties and characteristics make it possible to produce lithium disilicate dental restorations in a facile manner using two different methods:

- *Heat pressing*: This method makes use of what is described as molding technology, whereby the lithium disilicate glass-ceramic, already with the desired microstructure, is heated in a special furnace to a sufficiently high temperature (~900 °C) and shaped in a mold by applying a pressure to it. This method takes advantage of the viscous flow characteristics of the glass phase to shape the glass-ceramic. Generally, the pressed product maintains the microstructure of the glass-ceramic, but the $Li_2Si_2O_5$ crystals may undergo a limited amount of orientation caused by the viscous flow deformation.
- Computer-aided design (CAD) technology, whereby a block (or ingot) of the parent glass is heated for ~15–30 minutes at ~650 °C to form a lithium metasilicate glass-ceramic, which is softer than the final lithium disilicate glass-ceramic and, thus, easier to machine. Then this glass-ceramic is machined to the shape of the dental restoration from a model of the tooth structure using CAD methods. Thereafter, the shaped lithium metasilicate glass-ceramic is heated for ~30–60 minutes at ~800–850 °C to convert it, with little or no change in volume, to a lithium disilicate glass-ceramic having the requisite microstructure and properties (Figure 23.26).

Overall, the parent lithium disilicate glass is available commercially in a wide variety of shades (colors), such that upon fabrication, the restoration can closely match the color and optical properties of the existing teeth. Because of the simplicity of these two fabrication methods, large variations in the quality of the dental restoration are avoided, even without significant technical expertise. Additionally, the method employing CAD technology allows for repairing the tooth structure in a single visit to the dentist.

Polycrystalline Ceramic Dental Restorations

Although alumina (Al_2O_3), either dense or in a porous form infiltrated with a glass, has been used in creating metal-free dental restorations, one of the most effective polycrystalline ceramics for use in these applications is ZrO_2 stabilized with ~3 mol % Y_2O_3. As noted in Chapter 7,

(a) (b)

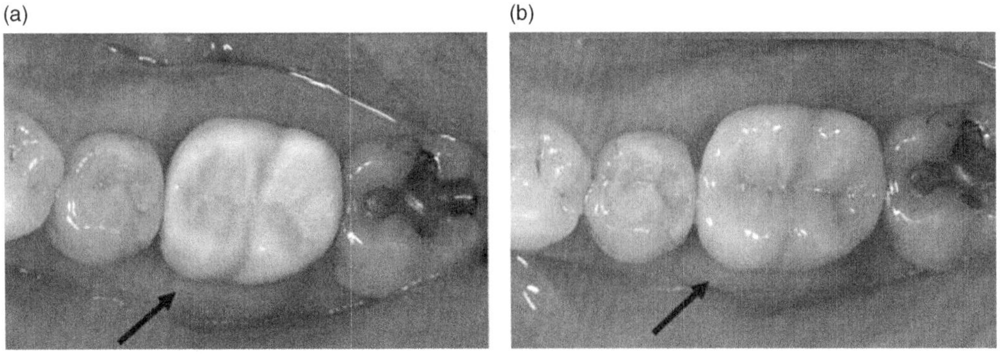

Figure 23.26 Lithium disilicate dental restoration fabricated using computer-aided design (CAD) technology: (a) try-in of a full crown in the state of a lithium metasilicate glass-ceramic; (b) the same crown after heating to form a lithium disilicate glass-ceramic (shaded and glazed) and cemented to the existing tooth structure. *Source:* Courtesy of Wolfram Höland and Arnd Peschke, Ivoclar Vivadent Inc.

this material is referred to as yttria-stabilized tetragonal zirconia polycrystals (Y-TZP) or simply yttria-stabilized zirconia (YSZ). When fabricated under optimal conditions, Y-TZP has a combination of flexural strength (~1 GPa) and a fracture toughness (6–10 MPa $m^{1/2}$) that is among the highest for ceramics. Additionally, Y-TZP has a color that is not very different from that of teeth and it can be veneered with glass-ceramics to provide aesthetic characteristics. Consequently, Y-TZP is well suited for use as restorations such as crowns and bridges, particularly for the posterior teeth where, unlike lithium disilicate glass-ceramics, it is capable of withstanding the higher forces of mastication. However, fabrication of Y-TZP into shapes relevant to dental restorations is more difficult when compared to lithium disilicate glass-ceramics. Additionally, as noted in Chapter 7, there are concerns related to the tendency of Y-TZP to transform to the undesirable monoclinic crystal structure when subjected to a mechanical stress in an aqueous environment, which can lead to a reduction in its mechanical properties.

There are two methods for fabricating Y-TZP dental restorations, which do not differ markedly in difficulty or cost of fabrication. One method employs a single-step sintering process, commonly used for most polycrystalline ceramics. Fine particles of YSZ are formed into a porous specimen of the desired dimensions and shape, which is sintered at ~1400–1500 °C. It is noted that this temperature is considerably higher than that used for fabricating lithium disilicate glass-ceramics. Additionally, the porous specimen shrinks by a large amount (~30 vol%) during the sintering step, and the dense fabricated Y-TZP is difficult to machine because of its high hardness. Overall, proper control of the shape and dimensions of the porous specimen and of the sintering conditions is necessary not only to reduce the amount of costly machining necessary to produce a dental restoration with the right shape and dimensions but also to retain the requisite tetragonal crystal structure of ZrO_2 upon cooling from the sintering temperature.

The other fabrication method involves two separate sintering steps, which are separated by a machining step. First, the as-formed porous specimen (which is very weak) is sintered partially to improve its strength to a sufficient level to withstand the forces encountered in machining. Then this partially sintered specimen is machined to the desired dimensions and shape using CAD technology, after which it is sintered under the aforementioned conditions (~1400–1500 °C) to produce a dense Y-TZP having the right microstructure. When compared to the aforementioned one-step sintering method, the amount of costly machining required to convert the (hard) fabricated specimen to a dental restoration of the right shape and dimensions is reduced. However, the conditions used in the second sintering step still need to be well controlled to achieve the right microstructure and the tetragonal crystal structure.

23.7 Concluding Remarks

In this chapter, we discussed the practical use of biomaterials in healing defective regions in certain hard tissues such as bone, articular joints, the spine, and teeth with the goal of restoring their structure and function. Emphasis was placed on how the principles of materials science, engineering, and biology discussed in earlier chapters are used to modify and adapt biomaterials for use in these applications. We also considered the effectiveness and drawbacks of biomaterials in healing defects in these hard tissues using examples taken from these four application areas.

- Overall, biomaterials have shown a significant level of clinical success in repairing defective regions in the hard tissues discussed in this chapter, which has served to improve the quality of life for millions of people. On the other hand, biomaterials also have their drawbacks. Many of the commonly used biomaterials that we discussed are highly durable, both chemically and mechanically, or suffer from intrinsic limitations. They cannot remodel and convert into the native tissue and, thus, they are limited in their ability to restore the structure and function of the native tissue.
- As noted in Chapter 1, in order to avoid or limit the drawbacks of durable biomaterials, a rapidly growing trend in the last few decades has been the use of tissue engineering (or regenerative medicine) approaches to regenerate native tissues and organs, rather than just repairing them with durable biomaterials. These approaches will be discussed in Chapter 25.

Problems

23.1 List the major stages of bone fracture healing and, for each stage, briefly explain the major physiological processes, the cells involved, and the main function of the cells.

23.2 Select one metal and one synthetic degradable polymer, and compare their advantages and disadvantages for use as internal bone fracture fixation devices.

23.3 Explain the meaning of the terms osteoinductivity, osteoconductivity, and osteogenicity as applied to biomaterials used in healing bone defects.

23.4 Bone autografts, bone allografts, and synthetic bone graft substitutes are the three major types of implants used in healing bone defects. Briefly describe the advantages and disadvantages of each type of implant.

23.5 List and briefly explain what may be the ideal properties and characteristics of a synthetic bone graft substitute for healing bone defects.

23.6 Discuss the major limitations of biomaterials used as synthetic bone graft substitutes to heal bone defects, and describe what approaches might be available for overcoming these limitations.

23.7 Select one calcium phosphate bioceramic and one synthetic degradable polymer and compare their advantages and disadvantages for use as implants to heal bone defects.

23.8 Estimate the compressive stress on a bone graft substitute that is implanted into a segmental defect in the mid-section of the tibia (diameter ~2.2 cm) of a human of body mass 90 kg, during normal walking when the compressive load on the femur is ~three times the body weight.

23.9 Select a synthetic biomaterial that could be used as implants to heal large (critical size) defects in the long bones of the human limbs, and briefly describe its advantages and disadvantages.

23.10 Compare the advantages and disadvantages of porous titanium and porous bioactive glass for use as implants to heal large (critical size) segmental defects in the long bones of human limbs.

23.11 Two types of articulating components used in total hip replacement are metal (Co–Cr)-on-polymer (UHMWPE) and ceramic (ZTA)-on-ceramic (ZTA). Compare the advantages and disadvantages of each combination.

23.12 Briefly describe the desirable properties and characteristics of an implant for use in spinal fusion.

23.13 Titanium, polyether ether ketone (PEEK), and silicon nitride (Si_3N_4) are three biomaterials commonly used as implants in spinal fusion. Compare the advantages and disadvantages of each type of implant in this application.

23.14 Two approaches used for enhancing the osseointegration of titanium dental implants with host bone are surface roughening and coating with a layer of hydroxyapatite. Briefly discuss the advantages and limitations of each approach.

23.15 Compare the advantages and disadvantages of lithium disilicate glass-ceramics and yttria-stabilized tetragonal zirconia polycrystals (Y-TZP) for use as indirect dental restorations such as crowns and bridges.

References

Alam, M.I., Asahina, I., Ohmamiuda, K., and Enomoto, S. (2001). Comparative study of biphasic calcium phosphate ceramics impregnated with rhBMP-2 as bone substitutes. *Journal of Biomedical Materials Research* 54: 129–138.

Arts, M.P., JFC, W., and Corbin, T.P. (2017). Porous silicon nitride spacers versus PEEK cages for anterior cervical discectomy and fusion: clinical and radiological results of a single-blinded randomized controlled trial. *European Spine Journal* 26: 2372–2379.

Ball, H.T., BJ, M.E., and Bal, B.S. (2017). Accelerated cervical fusion of silicon nitride versus PEEK spacers: a comparative clinical study. *Journal of Spine* 6: 1000396.

Chu, T.-M.G., Warden, S.J., Turner, C.H., and Stewart, R.L. (2007). Segmental bone regeneration using a load-bearing biodegradable carrier of bone morphogenetic protein-2. *Biomaterials* 28: 459–467.

Cowan, C.M., Aghaloo, T., Chou, Y.-F. et al. (2007). MicroCT evaluation of three-dimensional mineralization in response to BMP-2 doses *in vitro* and in critical sized rat calvarial defects. *Tissue Engineering* 13: 501–512.

Hench, L.L., Day, D.E., Höland, W. et al. (2010). Glass and medicine. *International Journal of Applied Glass Science* 1: 104–117.

Huebsch, N. and Mooney, D.J. (2009). Inspiration and application in the evolution of biomaterials. *Nature* 462: 426–432.

Jia, W., Lau, G.Y., Huang, W. et al. (2015). Bioactive glass for large bone repair. *Advanced Healthcare Materials* 4: 2842–2848.

Kurtz, S.M. and Devine, J.N. (2007). PEEK biomaterials in trauma, orthopedic, and spinal implants. *Biomaterials* 28: 4845–4869.

Mankani, M.H., Kuznetsov, S.A., Wolfe, R.M. et al. (2006). *in vivo* bone formation by human bone marrow stromal cells: reconstruction of the mouse calvarium and mandible. *Stem Cells* 24: 2140–2149.

Ohgushi, H. and Caplan, A.I. (1999). Stem cell technology and bioceramics: from cell to gene engineering. *Journal of Biomedical Materials Research* 48: 913–927.

Ohgushi, H., Goldberg, V.M., and Caplan, A.I. (1989). Repair of bone defects with marrow cells and porous ceramics: experiments in rats. *Acta Orthopaedica Scandinavia* 60: 334–339.

Rahaman, M.N. and Xiao, W. (2018). Silicon nitride bioceramics in healthcare. *International Journal of Applied Ceramic Technology* 15: 861–872.

Rahaman, M.N., Yao, A., Bal, B.S. et al. (2007). Ceramics for prosthetic hip and knee joint replacement. *Journal of the American Ceramic Society* 90: 1965–1988.

Ratner, B.D. (2013). A history of biomaterials. In: *Biomaterials Science: An Introduction to Materials in Medicine*, 3e (eds. B.D. Ratner, A.S. Hofmann, F.J. Schoen and J.E. Lemons), xli–liii. New York: Elsevier.

Further Reading

Höland, W., Schweiger, M., Peschke, A., and Kappert, H. (2008). Ceramics as materials for dental restoration. *Expert Review of Medical Devices* 5: 729–745.

Karageorgiou, V. and Kaplan, D. (2005). Potential of 3D biomaterial scaffolds and osteogenesis. *Biomaterials* 26: 5474–5491.

Laurencin, C., Khan, Y., and El-Amin, S.F. (2006). Bone graft substitutes. *Expert Review of Medical Devices* 3: 49–57.

Mankani, M.H., Afghani, S., Franco, J. et al. (2011). Lamellar spacing of cuboid hydroxyapatite scaffolds regulates bone formation by human bone marrow stromal cells. *Tissue Engineering Part A* 17: 1615–1623.

24

Biomaterials for Soft Tissue Repair

24.1 Introduction

This chapter discusses the applications of biomaterials in repairing soft tissues. The number of medical devices containing biomaterials that are used in these applications annually is considerable (Table 1.1). The major areas that will be considered are biomaterials used as sutures and adhesives in a variety of surgical procedures, devices to heal diseased or damaged arteries and heart valves in the cardiovascular system, contact lenses and intraocular lenses (IOLs) to improve vision, and treatments to heal skin wounds. Similar to the previous chapter devoted to hard tissue repair, a major focus is on the principles underlying how various biomaterials are modified and adapted for use in these applications. The effectiveness of biomaterials in these applications will also be addressed using examples taken from a variety of studies.

Sutures are devices used in many surgical procedures, ranging from the closure of skin lacerations to more delicate procedures in cardiovascular surgeries, such as the replacement of diseased or damaged arteries and heart valves. They are used more widely and in greater quantity than any other kind of surgical device. Biomaterials used as sutures should be capable of consistently meeting not only a range of physicochemical properties but also a variety of subjective qualities allowing for facile handling and use by clinicians in surgical procedures.

Diseases of the cardiovascular system can lead to a variety of problems, which can be grouped into three major areas. First, arteries can become narrowed or occluded due to atherosclerosis, the formation of plaque or deposits on the luminal (inner) surface of the arteries, thereby decreasing the normal blood flow. Second, the artery walls may become thinner and unable to sustain the pressure of the flowing blood, in which case the artery wall may bulge out and eventually rupture. Third, the functions of the valves responsible for controlling the blood flow into and out of the chambers of the heart are susceptible to impairment by disease, resulting in reduced blood flow, which can lead to congestive heart failure. Devices composed of biomaterials, such as vascular grafts, stents, and prosthetic heart valves find considerable use in repairing or replacing degenerated tissues of the cardiovascular system. These devices deteriorate over time and, consequently, an understanding of the mechanisms and processes by which they deteriorate is useful in developing biomaterials with improved performance in these cardiovascular applications.

Over 100 million people worldwide regularly wear contact lenses to improve their vision, covering daily use to extended use, while over 7 million IOLs are implanted annually to replace the lens of the eye and to restore vision in patients suffering from age-related lens deterioration such as the formation of cataracts. Although these devices are very successful, there is a risk that they could introduce adverse biocompatibility phenomena. For example, insufficient oxygen transport through contact lenses can lead to swelling of the cornea, while protein adsorption and cell adhesion onto the surface of IOLs can lead to opacification and a reduction in visual acuity. An understanding of the interaction of biomaterials with the ocular environment is useful in modifying and adapting biomaterials to make them more effective in these ophthalmologic applications.

Cutaneous wounds are the type of injuries most frequently encountered in life. Although effective treatments are available for healing acute and small-area skin wounds, significant problems can arise in the treatment of chronic wounds, such as large-area burns, infected wounds, and wounds in diabetic patients. Approximately 100 million patients are treated worldwide for traumatic skin wounds, with approximately another 300 million treated for chronic skin wounds. The treatment of skin wounds places a considerable financial burden on healthcare systems worldwide. In addition, the incidence of chronic skin wounds has been increasing due to the increasing prevalence of diseases such as diabetes and peripheral vascular disease. Although they have played an important role in skin wound healing, biomaterials-based strategies vary in their effectiveness, particularly in treating chronic wounds. It is

Materials for Biomedical Engineering: Fundamentals and Applications, First Edition. Mohamed N. Rahaman and Roger F. Brown.

Companion website: www.wiley.com/go/rahaman/biomedicalengineering

therefore vital to understand the wound healing process in order to develop better biomaterials-based strategies to address the challenges in healing chronic wounds.

Cartilage has a limited capacity for self-regeneration and, to make matters worse, cartilage defects are not amenable to repair using biomaterials. As noted in Chapter 23, cartilage degeneration in joints, such as the hip and knee joints often reach an advanced stage before it is detected, which often necessitates total joint replacement using prosthetic devices. The healing of cartilage defects along with the repair of various tissues and organs will be discussed in the next chapter within the context of tissue engineering and regenerative medicine.

Overall, the topics discussed in this chapter are related to the application of biomaterials in the following areas of soft tissue repair:

- Sutures and adhesives for use in surgical procedures.
- Vascular grafts, stents, and heart valves for repairing diseased or damages tissues in the cardiovascular system.
- Contact lenses and IOLs to improve vision.
- Biomaterials-based therapies for healing skin wounds.

24.2 Surgical Sutures and Adhesives

Sutures, as noted earlier, are used more widely and in greater quantity than any other surgical device. The production of sutures is a well-established area of biomedical technology, resulting in a wide range of commercial products that have shown their effectiveness in a variety of surgical procedures. In recent years, there has been a growing use of metal staples, particularly in minimally invasive surgery and in the closure of large incisions such as those used in Caesarian sections and intestinal surgery. Metal staples are commonly made from biocompatible metals such as tantalum, stainless steel (316L), and nickel–titanium alloys. Degradable staples, such as staples composed of polylactic-*co*-glycolic acid (PLGA), find use in skin and internal wound closures. Adhesives are substances used to bond tissues together, particularly very soft tissues where the use of sutures might cause substantial mechanical damage.

24.2.1 Sutures

In addition to being biocompatible and sterile, sutures used in surgical procedures should have the following properties and characteristics:

- Adequate tensile strength and the ability to maintain sufficient strength to remain functional over the required duration of the clinical application.
- Ability to be knotted securely and to provide a strong enough knot.
- A range of diameters for use in a variety of clinical circumstances.
- Adequate compliance (flexibility).
- A variety of subjective qualities allowing for facile use by clinicians in a surgical procedure. Such qualities include suppleness, ease of knotting, low drag or friction between the suture and the tissue to reduce abrasion particularly in delicate cardiovascular, ophthalmic, and neurological procedures, and the overall feel in the hands of the clinician.

Types of Sutures

Sutures may be classified (Table 24.1) in terms of:

- *The type of biomaterial*: This is based on whether the biomaterial is of natural or synthetic origin. Except for fine metal sutures, synthetic sutures are composed of polymeric biomaterials. Natural sutures are composed mainly of silk, cotton, or a collagenous material produced from animal intestine such as bovine, sheep or goat intestine.
- *The permanence of the biomaterial*: This is based on whether the suture is degradable or resistant to degradation (nondegradable within the period of application). Most degradable sutures are produced from synthetic polymers such as polyglycolic acid (PGA), copolymers of glycolic acid with lactic acid or with caprolactone, and polydioxanone (Chapter 9). These synthetic sutures degrade predominantly by hydrolysis. In comparison, degradable sutures produced from animal intestine degrade predominantly by enzyme-mediated hydrolysis. Nondegradable sutures are produced from synthetic polymers such as polypropylene (PP), nylon 6.6, polyethylene terephthalate (PET), and polyethylene (PE) (Chapter 8), or from natural polymers such as silk and cotton (Chapter 10). Metal sutures, normally composed of stainless steel (316L), are used in surgical procedures such as tendon repair requiring strong mechanical properties.
- *The construction of the suture*: This is normally based on whether the suture is in the form of a monofilament or in a braided form.

Overall, sutures composed of synthetic polymers are used more commonly when compared to natural polymers, of which ~50% are degradable and ~50% nondegradable.

Degradable Sutures

An important property of a suture, as noted earlier, is the ability to retain a sufficient fraction of its initial strength in order to remain functional over its intended period of

Table 24.1 Types of materials used for common sutures.

Type of material	Composition[a] or origin	Form[b]	Designation
Synthetic			
Degradable	Polyglycolic acid (PGA)	B	Dexon™
	Polylactic-*co*-glycolic acid (PLGA): 90% glycolic acid + 10% lactic acid	B	Vicryl™
	Polylactic-co-glycolic acid (PLGA)[c]: 90% glycolic acid + 10% lactic acid	B	Vicryl™ Rapide
	Copolymer of glycolic acid (75%) and ε-caprolactone (25%)	M	Monocryl™
	Polydioxanone	M	PDS II™
	Copolymer of glycolic acid (67%) and methylene carbonate (33%)	M	Maxon™
Nondegradable	Nylon 6	M	Ethilon™
	Nylon 6.6	M	Monosof™
	Polypropylene (PP)	M	Propene™, Surgipro™
	Polyethylene terephthalate (PET)	B	Ethibond™
	Ultrahigh molecular weight polyethylene (UHMWPE)	B	Force Fiber™ FiberWire™
	Stainless steel (316L)	M	Surgical steel suture
Natural			
Degradable	Bovine, sheep or goat intestine (predominantly collagen)	M	Plain gut
	Bovine, sheep or goat intestine treated in chromium salt solution	M	Chromic gut
Nondegradable	Silk	B	Perma-hand™
	Cotton		Cotton

[a] Composition of copolymers in wt%.
[b] Braided (B) or monofilament (M).
[c] Lower molecular weight than Vicryl™

application. An approximate measure of the functional lifetime of a degradable suture is the time taken for its tensile strength to decrease by ~50%. Another relevant property of a degradable suture is its degradation time, that is, the time taken for it to degrade completely postoperatively. Overall, these two properties depend on the composition and the structure of the polymer (Chapter 9).

Based on information available in published data sheets, Figure 24.1 shows the approximate average functional lifetime (time for 50% reduction in tensile strength) and the average degradation time for several common commercial monofilament sutures listed in Table 24.1. Monofilament sutures produced from animal intestine degrade rapidly, remaining functional for 1–2 weeks, and degrading completely in ~10 weeks. However, treatment in a chromium salt solution can increase their functional lifetime to 2–3 weeks and their degradation time to 12–13 weeks.

Monofilament sutures produced from degradable polymers provide a wider range of functional lifetimes. Dexon™, the first synthetic degradable suture produced commercially, composed of PGA (a semicrystalline polymer), has a functional lifetime of ~2–3 weeks, and a degradation time of ~13 weeks. Vicyl™ sutures, composed of a copolymer of 90 wt% glycolic acid (glycolide) and 10 wt% lactic acid (lactide), have a similar functional lifetime but a shorter time for complete degradation (8–10 weeks). A more rapidly degrading form of Vicryl™ sutures (designated Vicryl™ Rapide), consisting of a braided suture of lower molecular weight, has a functional lifetime of approximately one week and a degradation time of approximately six weeks. These Vicryl™ Rapide sutures could provide an alternative to natural sutures produced from animal intestine. Monocryl™ sutures, composed of a copolymer of 75 wt% glycolic acid and 25 wt% ε-caprolactone, have a functional lifetime of ~1–2 weeks and a degradation time of ~13–17 weeks. Polydioxanone monofilament sutures (designated PDS II™) have a functional lifetime of ~4 weeks and degrade completely in 25–30 weeks, making it one of the longest lasting degradable sutures.

24.2.2 Soft Tissue Adhesives

Substances used as adhesives to bond soft tissues together should be biocompatible, have the ability to wet tissues and bond to them, be capable of being formed *in situ* by rapid polymerization without producing excessive heat or toxic

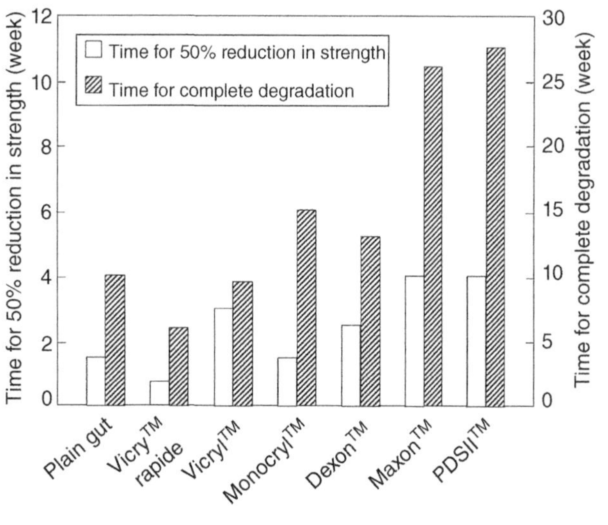

Figure 24.1 Time for 50% reduction in tensile strength and for complete degradation for common sutures listed in Table 24.1. *Source:* Data from Ethicon Wound Closure Suture Products, 2004; Taylor and Shalaby (2013).

byproducts, and be degradable. Additionally, the adhesive should be capable of being easily applied during a surgical procedure and it should not interfere with the normal healing process.

Substances composed of alkyl α-cyanoacrylates such as methyl, isobutyl, or *n*-butyl cyanoacrylate have been used as soft tissue adhesives but based on observations from laboratory studies involving animals, concerns have been expressed about the potential for adverse tissue responses and the formation of tumors. These cyanoacrylates have a chemical structure similar to the corresponding methacrylate (Figure 8.14a), except that a cyano group (C≡N) replaces the CH_3 side group attached to the carbon atom in the methacrylate.

Fibrin is a natural adhesive and sealant, made by mixing separate preparations containing fibrinogen and thrombin, just prior to application at the surgical site (Chapter 10). Being a natural material, fibrin is biocompatible, and it has several additional advantages, such as being hemostatic (able to stop bleeding), able to degrade into nontoxic products, able to adhere to connective tissue and able to promote wound healing. Although the strength of fibrin is lower than that of the cyanoacrylates, it is generally adequate for several clinical applications, particularly when delicate tissues are involved.

24.3 The Cardiovascular System

Diseases of the cardiovascular system, as noted earlier, can lead to a variety of problems. The repair or replacement of diseased tissues in the cardiovascular system requires the consideration of a range of critical factors, which depend not only on the selection of biomaterials having appropriate physicochemical properties but also on converting the biomaterial into a desirable architecture. In order to understand how biomaterials can be modified and adapted to function as devices in the cardiovascular system, it is useful to consider first the major anatomical features of the system itself.

Overall, the cardiovascular system of the human body consists of the heart and the circulatory system. The heart pumps blood through a network of blood vessels in order to provide oxygen, nutrients, and critical biomolecules to all parts of the body and remove waste products and exhaled CO_2. Arteries carry blood away from the heart while veins carry blood back to the heart. Diseases of the cardiovascular system are related to the pumping mechanism of the heart itself, the systems that control blood flow, and the network of blood vessels.

24.3.1 The Heart

The heart can be viewed as a four-chambered pump (Figure 24.2) without which a human could not survive. It is estimated that over an average lifetime, the heart will beat ~3 billion (~3×10^9) times. The upper two chambers, the right and left atria receive returning blood from the systemic circulation and from the pulmonary circulation, respectively, while the two lower chambers, the right and left

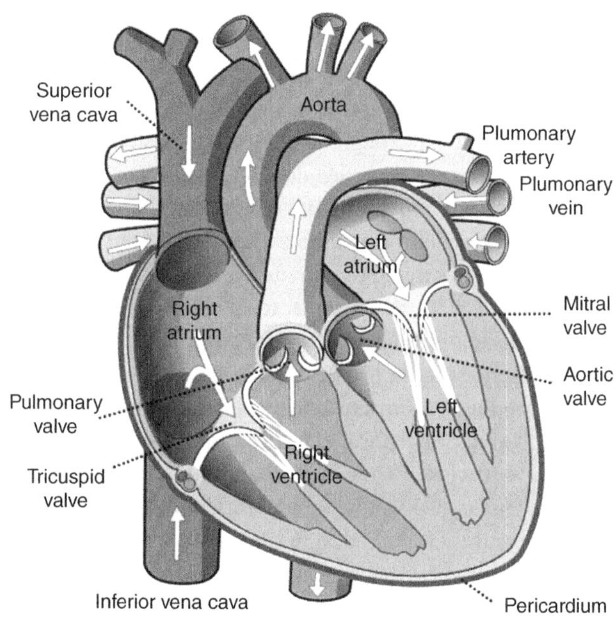

Figure 24.2 Image showing the major anatomical components of the human heart. *Source:* Wapcaplet, Diagram of the human heart (2006), Wikimedia Commons. Licensed Under CC BY Share Alike 3.0 Unported.

ventricles, pump blood into these two circulatory systems. Oxygen-depleted blood enters on the right side of the heart, is then reoxygenated via the lungs and fed back into the left side of the heart, where it is pumped back out into the body by the systemic circulation. The heart contains two inlet and two outlet valves, each essential for maintaining the direction of flow. Blood flow between the left atrium and the left ventricle is controlled by the mitral valve, while the tricuspid valve controls flow between the right atrium and the right ventricle. The pulmonary valve controls flow to the pulmonary artery, while the aortic valve controls flow from the left ventricle into the aorta and then to the systemic circulation. Any of these valves can become diseased, but the mitral valve and particularly the aortic valve are more susceptible because they are subjected to higher pressures.

The wall of the heart consists of three layers, the endocardium (inner layer), which contacts the blood, the myocardium (middle layer of cardiac muscle), and the epicardium (outer layer), and it is protected by the pericardium, a membrane that surrounds it. The pumping action of the heart is provided by the myocardium, a muscle tissue that forms the thickest part of the heart wall. The heart wall receives its own supply of blood from the coronary arteries, consisting of two major arteries that branch off from the aorta near the point where the aorta and the left ventricle meet. Coronary veins take oxygen-depleted blood that has been used by the heart muscles and return it to the right atrium. The coronary arteries are susceptible to disease. If they become narrow or blocked, the myocardium does not receive sufficient oxygenated blood, which can lead to damage to the myocardium and, ultimately, to a heart attack, referred to as myocardial infarction.

The pumping action of the heart is directly initiated by the sinoatrial node, a small body of specialized muscle tissue in the right atrium wall at the junction with the superior vena cava, which acts as a natural pacemaker by producing electrical impulses to stimulate contractile motion of the heart muscle. Irregularities in this contractile motion lead to arrhythmia, a condition in which the heart beats with an irregular or abnormal rhythm. This condition is normally treated by the use of devices such as pacemakers.

24.3.2 The Circulatory System

Oxygenated blood is pumped from the left ventricle through the aortic valve into the aorta, the largest artery in the body (~20 mm in diameter). As they go further away from the heart, the arteries branch and rebranch, becoming smaller and smaller, eventually becoming arterioles (small diameter arteries) that feed into capillaries. The walls of all arteries have a structure consisting of three layers, an inner lining (called the tunica intima), a middle layer (the tunica media), and an outer layer (the tunica adventitia). The tunica intima of small arteries consists of just the endothelium, a one-cell-thick layer of endothelial cells. In comparison, the tunica media, the thickest part of the artery wall, is composed of circumferentially arranged collagen and elastin fibers, and smooth muscle cells, while the tunica adventitia is composed of collagen fibers arranged longitudinally.

The relative amounts of collagen, elastin and smooth muscle present in an artery wall varies with the size and location of the artery. Since the walls of arteries are always under high pressure as blood is forced through them by the heart, they require the ability to expand and contract, thereby changing the diameter of the artery in order to maintain a relatively constant blood pressure. This ability is provided by the elastin fibers present in the tunica media. As would be expected, then, the largest arteries nearest to the heart (the aorta and the pulmonary arteries) have much more elastin and much less smooth muscle in their walls than other arteries. In comparison, even the major arteries branching off from the aorta have walls with much more smooth muscle and much less elastin than the aorta does.

A large number of arteries are present in the circulatory system, examples of which are the femoral and popliteal arteries of the legs, the iliac artery of the pelvis, the radial and ulnar arteries of the forearm, and arteries to the organs, such as the renal artery leading to the kidneys. As noted earlier, arteries can suffer from diseases, such as atherosclerosis, the formation of plaque or deposits on the luminal (inner) surface of the arteries, mainly in the legs, or they may become thinner and unable to sustain the pressure of the flowing blood, leading to aneurisms, mainly in the aorta.

Arterioles feed oxygenated blood to the capillaries, which have very thin walls, allowing them to distribute oxygen and nutrients to tissues and remove oxygen-depleted blood and waste. Capillaries also serve to connect the arterioles with the venules (very small veins), which merge into veins that carry blood back to the heart. The walls of veins are similar to those of arteries, except that they are thinner and less elastic. Veins carry oxygen-depleted blood toward the lungs, where oxygen is received via the pulmonary capillaries, and the pulmonary veins then carry oxygenated blood back to the heart. In general, veins are far less susceptible to disease when compared to arteries.

24.4 Vascular Grafts

Arteries, as noted earlier, can become narrowed or occluded due to atherosclerosis, the formation of plaque or deposits on the luminal (inner) surface of the arteries, thereby decreasing or blocking the normal oxygenated blood flow. A common solution is the replacement or circumvention of the narrowed or blocked part of the artery using a device called a vascular graft (Figure 1.1f), composed of a biological or synthetic material. Synthetic grafts are widely and successfully used for repairing large diameter arteries (greater than ~6–7 mm in diameter) such as the aortic, iliac, and femoral arteries characterized by a high blood flow, but they have not shown similar success with small diameter arteries (diameter less than ~6–7 mm). In comparison, biological grafts are more effective for small diameter arteries with low blood flow, such as the peripheral arteries below the knee and the coronary arteries. The most common treatment for these small diameter arteries involves using an autologous saphenous vein, a subcutaneous vein of the leg. An allograft composed of glutaraldehyde-treated umbilical cord vein can also be used but it is much less effective than the saphenous vein.

24.4.1 Desirable Properties and Characteristics of Synthetic Vascular Grafts

Ideally, synthetic vascular grafts should replicate the structure and function of the natural artery as far as possible. Minimally, they should have the following properties and characteristics:

- Biocompatibility and resistance to infection.
- Mechanical properties to match those of the natural artery. Studies have shown that the mechanical properties of the graft wall have a strong influence on the graft's patency (ability to remain open). The graft should be mechanically durable but should also be compliant to undergo radial expansion and contraction following the systolic and diastolic cycles of the heart. Additionally, synthetic grafts should be mechanically flexible to accommodate repeated bending without kinking, because some arteries, such as the femoropopliteal artery at the back of the knee, undergo repeated bending.
- Resistance to degradation in order to maintain properties, such as mechanical properties, over a long-term duration (normally years).
- Ability to be sutured securely to the healthy artery using fine strong sutures: The anastomosis (join) between the graft and the healthy artery is crucial to the performance of the graft because differences in mechanical properties are normally enhanced at the anastomoses.
- Non-thrombogenic luminal (interior) surface to prevent clotting of the blood.

24.4.2 Synthetic Vascular Graft Materials

It might be apparent that a synthetic graft consisting of a solid impervious tube would not be suitable for use as a vascular graft because it would lack the requisite mechanical compliance and flexibility. Two synthetic materials, developed previously for use as textiles, have been most widely used in creating large diameter vascular grafts. One is a knitted or woven polyester fabric composed of PET, commonly known as Dacron®. The other is a fabric composed of expanded polytetrafluoroethylene (ePTFE), which is an expanded version of a material used for waterproof clothing. Dacron® and ePTFE are biocompatible and resistant to degradation, and the walls of these grafts have a porous architecture designed to provide mechanical compliance and flexibility, as well as ease of suturing to the healthy artery. The porous architecture of an ePTFE graft is shown in Figure 8.18. As discussed in the next section, the patency of synthetic grafts shows a strong correlation with their mechanical compliance. Polyurethane (PU) grafts with a polycarbonate urethane composition can provide a better match to the mechanical compliance of natural arteries when compared to Dacron® and ePTFE grafts (Figure 24.3a), but they have been plagued by two limitations: inadequate chemical stability in the physiological environment and a loss in mechanical compliance following implantation. Consequently, PU vascular grafts have seen little use to date.

While a porous architecture allows for better matching of the mechanical properties of the graft with those of the natural artery, the use of porous synthetic grafts may appear counter-intuitive because the graft should prevent the flowing blood from escaping into the surrounding tissue. However, the porous architecture is actually advantageous in improving the performance of the graft. It leads to the formation of a tissue seal at the luminal surface of the graft, which is more biocompatible than the surface of the synthetic graft itself and which prevents the flowing blood from escaping. As blood enters the porous wall of the graft after implantation, it clots rapidly. Proteins adsorb to the luminal surface of the graft, followed by platelets, macrophages, endothelial cells, and smooth muscle cells. The clot subsequently organizes itself during the healing process to form an inner fibrin layer, a middle layer of foreign body giant cells and an outer capsule of connective tissue. The formation of a new endothelial layer at the luminal surface of the synthetic graft to match that of the healthy artery would be ideal but this does not occur. A layer of compacted fibrin covers the luminal surface of the porous synthetic graft.

24.4.3 Patency of Vascular Grafts

The ability to remain patent (unoccluded) over a long-term duration (normally years) is a critical criterion that governs the performance of vascular grafts, biological, or synthetic. Although certain grafts perform better than others in repairing arteries in different positions of the body, all

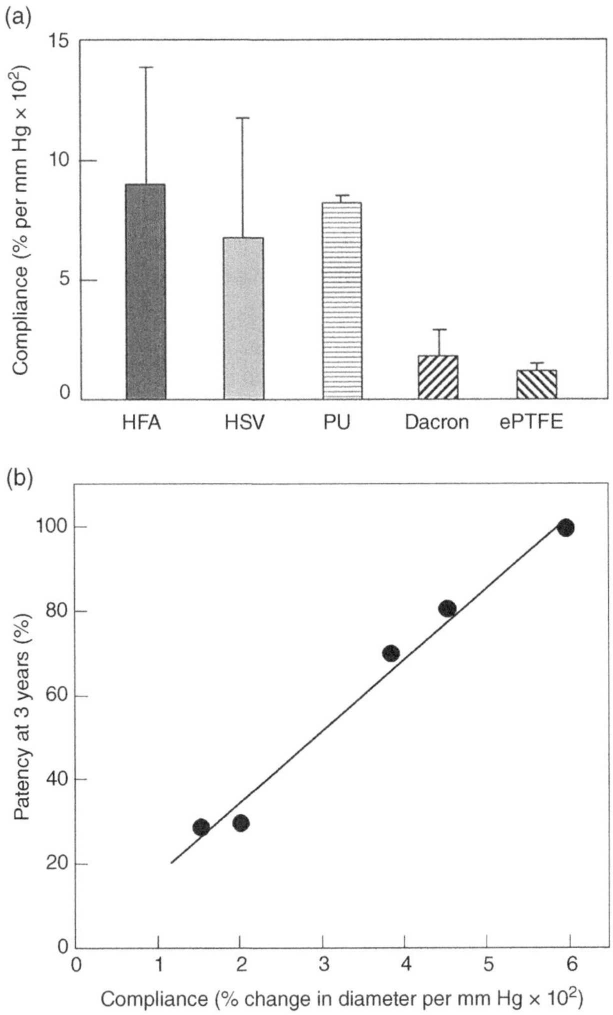

Figure 24.3 (a) Compliance (given as percentage change in diameter per mm Hg × 10^2) in the pressure range 30–100 mm Hg, for various grafts: human femoral artery (HFA), human saphenous vein (HSV), polycarbonate urethane (PU) (Myolink™), Dacron® and expanded polytetrafluoroethylene (ePTFE). *Source:* Data from Walden et al. (1980); Tiwari et al. (2002). (b) Patency at three years as a function of compliance for various grafts in the femoropopliteal position. *Source:* Data from Walden et al. (1980).

grafts deteriorate at varying rates with time. A common measure of the effectiveness of a graft is its patency rate, which is not really a rate but, instead, gives the fractional number of grafts that remain open or unobstructed at some specific time post-surgery. Autologous saphenous vein grafts are much more effective than synthetic grafts and are considered the gold standard for repairing small diameter arteries, but they have a patency rate of ~50% after 10 years. Dacron® and ePTFE grafts show a patency rate of ~90% at five years when used to repair larger diameter aortoilliac arteries (arteries between the aortal and iliac positions), but only ~40–50% for femoropopliteal arteries (arteries between the femoral and tibial positions).

The use of saphenous vein grafts suffers from the limitations of donor site morbidity and the need for an additional surgery (at the donor site). Despite their limited success, synthetic grafts may therefore be required for repairing small diameter arteries when the saphenous vein cannot be used. Consequently, an understanding of the processes and mechanisms by which grafts become occluded and the factors that control their long-term patency is useful in developing improved synthetic grafts that are more successful in a wider range of applications.

In general, occlusion of vascular grafts can occur by the following mechanisms:

- Thrombosis, the formation of a blood clot within the graft. This is a particularly important mechanism in synthetic grafts because their luminal surface lacks the non-thrombogenic endothelial layer of a healthy artery.
- Intimal hyperplasia, the thickening or overgrowth of the luminal surface of the graft, which is relevant to both biological grafts and synthetic grafts. Intimal hyperplasia is a complex process, which begins with injury to the endothelial lining of a biological graft or by platelet aggregation on the (non-thrombogenic) luminal surface of a synthetic graft. Proliferation and migration of smooth muscle cells from the media into the intima is then set in motion (Chapter 17), with subsequent synthesis of new extracellular matrix (ECM) proteins and other ECM material. Together, these two processes lead to intimal hyperplasia.
- Atherosclerosis, the deposition of plaque and fatty substances on the luminal surface, particularly in saphenous vein grafts.

A relevant question relates to the factors that control the long-term patency of vascular grafts. In general, two major factors have received most attention:

- The mechanical properties of the grafts, particularly the mismatch in mechanical properties between the graft and the healthy artery.
- The composition and nature of the graft, particularly in the luminal surface region.

Influence of the Mechanical Properties of the Graft

Natural arteries are viscoelastic but for deformations not much larger than those occurring *in vivo*, they may be considered perfectly elastic with a nonlinear stress versus strain response. Because the graft wall is subjected to both longitudinal stresses (along the length of the graft) and radial stresses (in the radial direction), the mechanical response of the graft in both directions is important. However, far more attention has been paid to the mechanical compliance of the graft, defined by its ability to dilate when subjected to a pressure. The compliance C is defined as the circumferential strain divided by the increase in pressure:

$$C = \left(\frac{\Delta D}{D}\right)\frac{1}{\Delta P} \tag{24.1}$$

where ΔD is the change in diameter D, and ΔP is the change in pressure (often measured in mm Hg).

Despite the interaction of the stresses in the longitudinal and radial directions, studies have shown that the compliance on its own (that is, as a single variable) has a strong effect on the long-term patency of vascular grafts. As shown in Figure 24.3b for the repair of the femoropopliteal arteries in humans, there is a strong correlation between the compliance of the vascular grafts and their long-term patency.

Differences in compliance are enhanced at the anastomoses between the graft and the healthy artery, particularly for small diameter arteries. In general, stiff grafts lead to a higher degree of compliance mismatch when compared to grafts that are more compliant, and the mismatch is greater at the distal anastomosis than at the proximal anastomosis. These differences in compliance lead to hemodynamic changes such as turbulent flow and shear stresses in the direction of the blood flow, resulting in intimal hyperplasia, particularly at the distal anastomosis.

Influence of the Composition and Nature of the Graft

Considerable effort has been made to improve the long-term patency of synthetic vascular grafts through modification of the graft itself using a variety of approaches. One of the most common approaches involves modifying the luminal surface of the graft by coating it with proteins or by chemically immobilizing biomolecules to the surface by covalent bonds. Another approach consists of seeding a layer of autologous endothelial cells onto the luminal surface. These two approaches are designed primarily to reduce thrombosis, which starts to occur following implantation due to the lack of a healthy endothelial lining at the luminal surface of synthetic grafts.

Heparin modification of biomaterials has been used for decades to reduce their thrombogenecity. The end-point immobilization technique described in Chapter 13 has been used to create stable heparin coatings on ePTFE and Dacron® grafts. In a large randomized trial involving over 500 patients, heparin-bonded ePTFE grafts showed a significantly better patency rate than the uncoated grafts when used to repair femoropopliteal arteries. Heparin-bonded ePTFE was found to reduce the risk of graft failure by ~50% when compared to uncoated ePTFE grafts at one year post-surgery. Another method that has been shown to improve the patency rate of synthetic grafts has been the covalent bonding of peptides containing the RGD sequence of fibronectin using, for example, the technique discussed in Chapter 11. These immobilized peptides improve the interaction of the biomaterial surface with a variety of cells, including endothelial cells.

Endothelial cells secrete substances that inhibit both thrombosis and intimal hyperplasia. Consequently, seeding of endothelial cells onto the luminal surface of synthetic grafts is another approach that has received attention for improving the long-term patency of small diameter synthetic grafts. Cell seeding has been performed using two techniques, known as single-stage seeding and two-stage seeding. Two-stage seeding typically involves harvesting autologous endothelial cells from a vein, expanding them in culture, and seeding them onto the luminal surface of the graft. This technique has shown success in clinical trials. For example, in one trial, cell-seeded ePTFE grafts showed a nine-year patency rate of ~65%, which was significantly better than the patency rate (~16%) for unseeded ePTFE grafts. However, the two-stage technique suffers from the disadvantage that the patient has to undergo two surgical procedures.

In comparison, the single-stage technique involves harvesting autologous endothelial cells and seeding them onto the luminal surface of the graft within the timeframe of a typical surgical procedure, which avoids the need for cell culture and reduces the time of the overall surgical procedure. However, the single-stage technique has not shown a similar degree of success when compared to the two-stage technique, which has been attributed mainly to low endothelial cell seeding density and limited cell attachment to the graft surface. A method that has been shown to improve the retention of endothelial cells at the graft surface involves covalent bonding of heparin, RGD peptides, or a combination of heparin and RGD peptides to the surface.

Another approach used to improve the long-term patency of synthetic grafts involves introducing nitric oxide (NO) by attaching NO-releasing substances to the luminal surface of the graft to reduce intimal hyperplasia. Uninjured endothelial cells produce NO by a process in which NO synthase produced by the cells converts L-arginine to L-citrulline and NO. NO has been shown to inhibit platelet aggregation and smooth muscle cell proliferation and migration, and to stimulate endothelial cell proliferation. Consequently, the attachment of NO-releasing substances to the luminal surface of a synthetic graft could reduce intimal hyperplasia following implantation and, thus, reduce the risk of restenosis.

One type of NO-generating substances that has received significant attention consists of *S*-nitrosothiols with the chemical structure shown in Figure 24.4, which serve to store and transport NO within biological systems. Several NO-releasing polymers have been synthesized by covalently bonding S-nitrosothiol molecules to the polymer. For example, copolymers with different NO-releasing kinetics were synthesized by reacting polyethylene glycol (PEG) *N*-hydroxysuccinimide monoacrylated molecules with poly-L-lysine (degree of polymerization = 5), diethylenetriamine (DETA), or L-cysteine (see Figure 11.20) (Bohl and West 2000). Then solutions of these copolymers were reacted with NO gas, forming NO–nucleophile complexes with the amine side-groups of the copolymers, and incorporated into photopolymerizable PEG hydrogels. Depending on the composition of the copolymer incorporated, these PEG hydrogels released NO over periods ranging from hours to months (Figure 24.5), and inhibited smooth muscle cell growth *in vitro* (Figure 24.6) and platelet adhesion on collagen-coated surfaces. These NO-releasing hydrogels could be used to coat synthetic vascular grafts to deliver NO locally. However, a problem with these hydrogels is that they contain a finite reservoir of NO.

24.5 Balloon Angioplasty

Another approach used to address the problem of narrowed or occluded arteries is balloon angioplasty. In this procedure, a catheter is used to physically displace or compress plaque or fatty substances that have accumulated on the luminal surface of an artery. Often the catheter is used in conjunction with an inflatable balloon, which is threaded along the lumen and expanded at the site of the plaque (Figure 24.7). This process is minimally invasive and can provide immediate relief, but it suffers from various disadvantages. For example, inflation of the balloon often damages the artery wall, removing the non-thrombogenic lining of endothelial cells and injuring the underlying smooth muscle cells. This can lead to rapid restenosis of the artery due to thrombosis and to intimal hyperplasia caused by rapid proliferation and migration of smooth muscle cells.

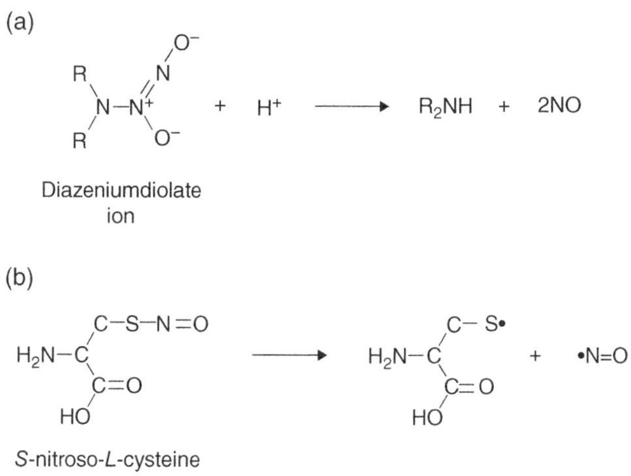

Figure 24.4 Examples of compounds that generate nitric oxide (NO). (a) Generic structure of a diazeniumdiolate; (b) *S*-nitroso-L-cysteine, $C_3H_6N_2O_3S$. (H atoms at two C atoms have been omitted for clarity.).

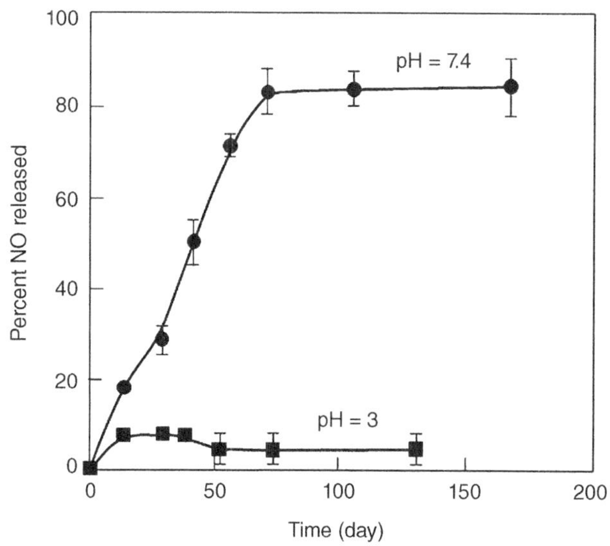

Figure 24.5 Release of NO from polyethylene glycol (PEG) hydrogels containing lysine–nitric oxide (NO) complexes at pH = 7.4 and pH = 3. *Source:* Bohl and West (2000) / with permission of Elsevier.

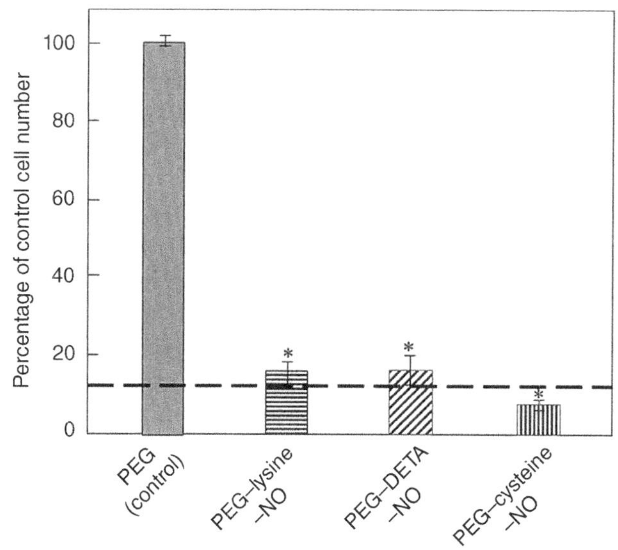

Figure 24.6 Inhibition of smooth muscle cell growth *in vitro* by polyethylene glycol (PEG) hydrogels containing nitric oxide (NO) complexes composed of lysine–NO, diethylenetriamine (DETA)–NO, and cysteine–NO. The control consisted of PEG hydrogel alone. The dashed line represents the initial cell seeding density. (*significant difference when compared to the control). *Source:* Bohl and West (2000) / with permission of Elsevier.

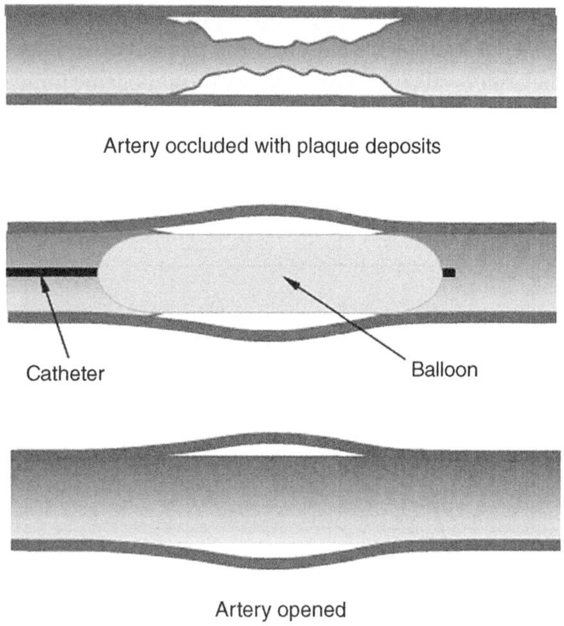

Figure 24.7 Illustration of balloon angioplasty.

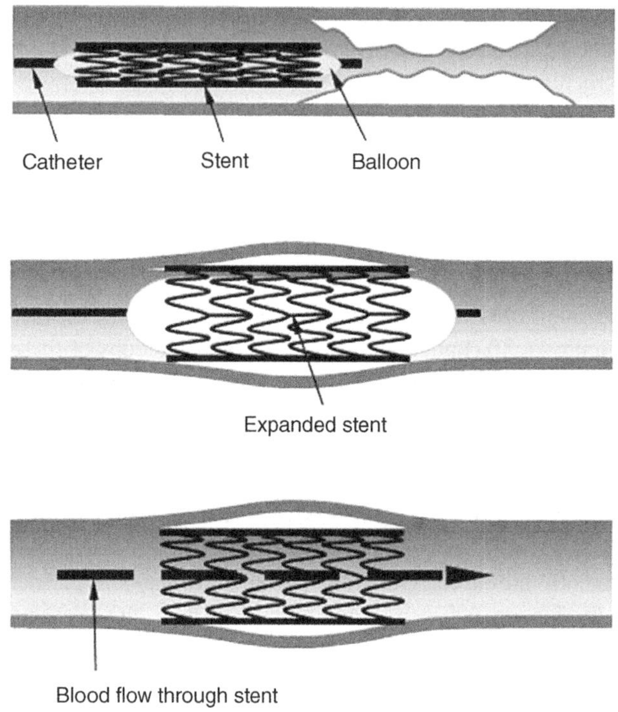

Figure 24.8 Illustration of insertion of a stent using an inflatable balloon.

24.6 Intravascular Stents

A third approach used to address the problem of arteries that have become narrowed or occluded is the use of an intravascular stent to keep the artery open. Intravascular stents are among the most widely used devices of all major medical implantable devices. They are designed to serve as a temporary or permanent implant to maintain or increase the lumen of an artery that has become narrowed due to atherosclerosis. In a surgical procedure, the stent is passed down a catheter and expanded at the site of the narrowed or occluded artery (Figure 24.8).

In Chapter 1, the intravascular stent was discussed in the context of illustrating the shift in biomaterials design over the last few decades, in which advances in biological sciences are playing a larger role in the design of improved biomaterials, often comparable to that of materials science and engineering. This chapter now looks at how biomaterials are actually modified and adapted to improve their performance when used as intravascular stents. In general, a stent should ideally have the following properties and characteristics:

- Biocompatibility and resistance to infection.
- Acceptable mechanical properties, that is, the wall of the stent should have adequate strength in the radial and longitudinal directions (called the hoop strength), radial and torsional elasticity, and fatigue resistance to withstand the pulsating stresses due to the flowing blood.
- Ability to be expanded from a small initial diameter (~50% of the diameter of the artery) prior to implantation to the actual diameter of the artery when placed at the desired arterial site.
- Minimal or no change in diameter following implantation and expansion at the arterial site.
- The stent should cause no damage to the endothelium of the artery.
- The stent should have sufficient longitudinal flexibility for threading through a tortuous artery.
- Thromboresistance in order to remain patent over the desired time.
- Ease of visualization using imaging techniques such as X-rays, computed tomography (CT) or magnetic resonance imaging (MRI), to ensure deployment at the right arterial site.
- Ability to be retrieved from the artery, if needed.

Stent design, as noted in Chapter 1, has evolved over the last few decades. Early stents were designed to consist of durable metals (referred to as bare-metal stents) but, subsequently, a drug-eluting function was introduced (drug-eluting stents). In the last two decades, considerable attention has been paid to the design of degradable stents.

24.6.1 Bare-Metal Stents

It may be apparent that certain corrosion resistant metals with the attractive mechanical properties characteristic of most metals could provide the acceptable biocompatibility, requisite mechanical properties, and ease of visualization for use as stents. Additionally, the metal should be in the form of an open (mesh-like) structure rather than a solid tube to provide sufficient flexibility for deployment. Generally, the thickness of a strut in the mesh-like structure is ~100 μm. Early stent designs employed durable metals that were available off the shelf, such as a stainless steel (316L) or a cobalt–chromium (Co–Cr) alloy (Chapter 6). Stents made of these metals (referred to as bare-metal stents) are deployed by means of a strong balloon inside the stent, which is pressurized once the stent is in the right place and then removed along with the catheter (Figure 24.8). Expansion of the stent to its actual diameter at the desired arterial site relies on plastic deformation of the metal. Some subsequent designs employed shape-memory alloys, such as Nitinol (nickel–titanium alloy), which can expand without the use of a balloon upon warming to the body temperature (Chapter 6).

Although these bare-metal stents work well initially, they themselves can suffer from becoming blocked (restenosis) after a few years due to certain inherent risks associated with the procedure. One risk is that the stent can become occluded by thrombosis because its surface is nonthrombogenic. Another risk is that the stent can irritate the endothelial lining of the artery. This triggers the release of inflammatory cytokines that stimulate proliferation and migration of smooth muscle cells from the media of the artery, which leads to ECM accumulation within the lumen of the stent, often resulting in intimal hyperplasia and restenosis. Consequently, the procedure may have to be repeated at regular intervals.

24.6.2 Drug-Eluting Stents

The problem of restenosis of bare-metal stents has been addressed through the development of drug-eluting stents in which a thin polymer coating incorporated with a drug is applied to the stent surface (Figure 24.9). The design consists of a metal stent composed typically of stainless steel (316L) or a Co–Cr alloy, a nondegradable or a degradable polymer coating incorporated with the drug, and an outer layer of the polymer alone to control the release rate of the drug. Commonly used drugs include paclitaxel and sirolimus, both of which are powerful antiproliferative drugs. These drugs reduce intimal hyperplasia by inhibiting the proliferation of vascular smooth muscle cells and ECM in the lumen of the stent.

Although drug-eluting stents have reduced the incidence of restenosis when compared to bare-metal stents, they have not been entirely successful. The antiproliferative drugs delay the growth of a healthy endothelium over the stent struts, which can lead to the presence of sites for the aggregation of platelets. Consequently, there is a risk of thrombosis, particularly at longer implantation times, necessitating prolonged antiplatelet therapy (longer than ~12 months). Additionally, the stent remains after stenosis has been resolved, which can lead to chronic inflammatory tissue responses (Chapter 18).

24.6.3 Degradable Stents

The need for the stent and the drug delivery function is temporary, that is, their continued presence is unnecessary after stenosis has been resolved (normally a few months after implantation). Degradable stents provide the benefits of bare-metal and drug-eluting stents but because they degrade away, no foreign biomaterial exposed to the blood is left after the resolution of stenosis. Consequently, they provide the potential for a shorter duration of antiplatelet therapy and the avoidance of late-stage thrombosis and chronic inflammatory responses.

The design of degradable stents is similar to that shown in Figure 24.8, except that the stent material is now composed of a degradable polymer or a degradable metal. The polymer stent material is often composed of a high molecular weight poly(L-lactic acid) (l-PLA) or a tyrosine-derived polycarbonate, both of which have reasonably good mechanical properties (Chapter 9). For example, one degradable polymer stent design

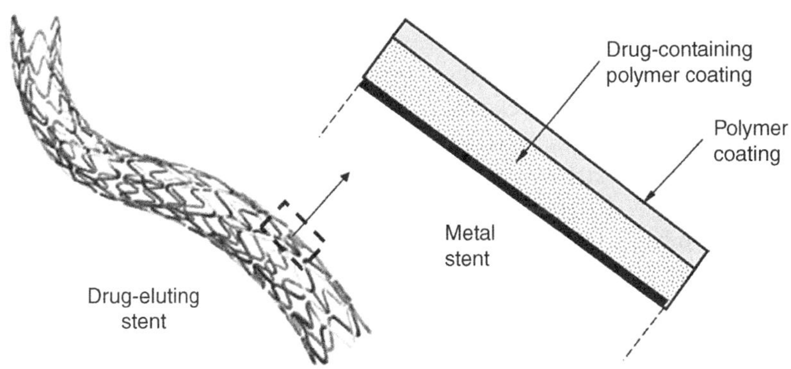

Figure 24.9 Illustration of the structure of a drug-eluting stent.

consists of an L-PLA stent (mesh thickness ~150 μm) and a more rapidly degrading poly(D,L-lactic acid) (d,l-PLA) coating (~1–2 μm thick). Drug elution is completed within ~1 month, while complete degradation of the stent occurs in ~18 months. One type of degradable metal stent consists of a magnesium alloy (Chapter 6) with a mesh thickness of ~100 μm, and a polymer coating composed of a PLGA coating (~1–2 μm thick). Complete drug elution occurs over 2–3 months, while complete degradation of the metal stent occurs within 9–12 months.

Degradable metal stents are an alternative to degradable polymer stents because their mechanical properties and deployment are comparable to those of bare-metal and drug-eluting stents. Overall, the results of clinical trials have shown the feasibility of both degradable polymer stents and degradable metal stents, with promising results.

24.7 Prosthetic Heart Valves

The heart has four valves that are responsible for controlling blood flow into and out of the chambers of the heart (Figure 24.2). All the valves have three cusps except for the mitral valve, which has two. These cusps open to allow the blood to flow forward through the valve and, when the flow slows down and finally stops, a pressure difference across the valve causes the cusps to close, preventing any back flow. Heart valves consist of thin flaps of flexible but tough endothelium-covered fibrous tissue firmly attached at the base to fibrous valve rings. These valves are susceptible to disease, which can be congenital in nature, such as insufficient or malformed leaflets, or the wrong size or shape of the leaflets that open and close, or the disease may be acquired, such as the formation of calcified deposits that cause the leaflets to thicken or become more rigid.

Any of the four valves can become diseased, but the aortic and mitral valves are more susceptible because they are subjected to much higher pressures than the valves on the right side of the heart (~5 times greater with a peak normotensive value of 100–120 mm Hg). A significant fraction of the population with chronic hypertension often have peak systole values of ~160 mm Hg or higher. The consequences of valve disease are reduced blood flow through them or reduced forward blood flow if the valve fails to close properly. Valve disease puts severe stress on the heart and can lead to congestive heart failure. Because the treatment of valve disease by drug therapy is normally ineffective, the most common solution is surgical intervention to repair or replace the defective valve with a prosthetic valve. Worldwide, over 400 000 heart valves are replaced each year.

It is worth examining the requirements of a prosthetic heart valve in order to understand how biomaterials can be modified and adapted to make them effective as devices for repairing heart valves. In general, a prosthetic heart valve should replicate the structure and function of the natural valve. More specifically, this means that ideally, a prosthetic heart valve should have the following properties and characteristics:

- The materials used in creating the valve should be biocompatible and resistant to infection.
- The valve should have a proper opening and closing mechanism, to control the blood flow without too much variation from normal physiological parameters such as flow rate and the energetics of blood flow.
- The valve should not cause damage to blood cells to any significant extent.
- In order to avoid thrombosis, the valve should not interfere with the blood clotting mechanisms.
- The valve should be mechanically reliable for an acceptable length of time. For example, it should have the requisite fatigue resistance to withstand repeated opening and closing and should be resistant to degradation to maintain its mechanical reliability over the desired period.
- The valve should be capable of being attached (sutured) to the heart tissue.

It is unrealistic to expect that any prosthetic heart valve would satisfy this wide combination of ideal properties and characteristics and, thus, the normal objective is to design a device having some optimal combination of properties and characteristics in order to achieve an acceptable performance.

Two types of prosthetic heart valves have been developed and used: mechanical valves created from synthetic biomaterials, and bioprosthetic valves (often referred to as tissue valves) taken from animals or humans.

24.7.1 Mechanical Valves

Mechanical valves have a basic design consisting of an annular component and a circular orifice that contains a mechanism for controlling the blood flow by opening and closing in response to changes in the pressure and blood flow. The annular component is covered with a sewing cuff, composed of Dacron® or Teflon®, to facilitate suturing of the device to the heart muscle. While there are a variety of designs, mechanical valves can be divided into two types depending on the pattern of the blood flow through them (Figure 24.10a). In the earliest designs, introduced in the 1960s, the control mechanism consisted of an occluding ball that moved within a cage structure attached to the annular housing of the valve. In this ball-in-cage design, blood enters the valve in a more peripheral fashion because it has to flow around the outside of the ball, rather than flowing centrally through the valve as in a natural valve. Although mechanically durable, this design suffers

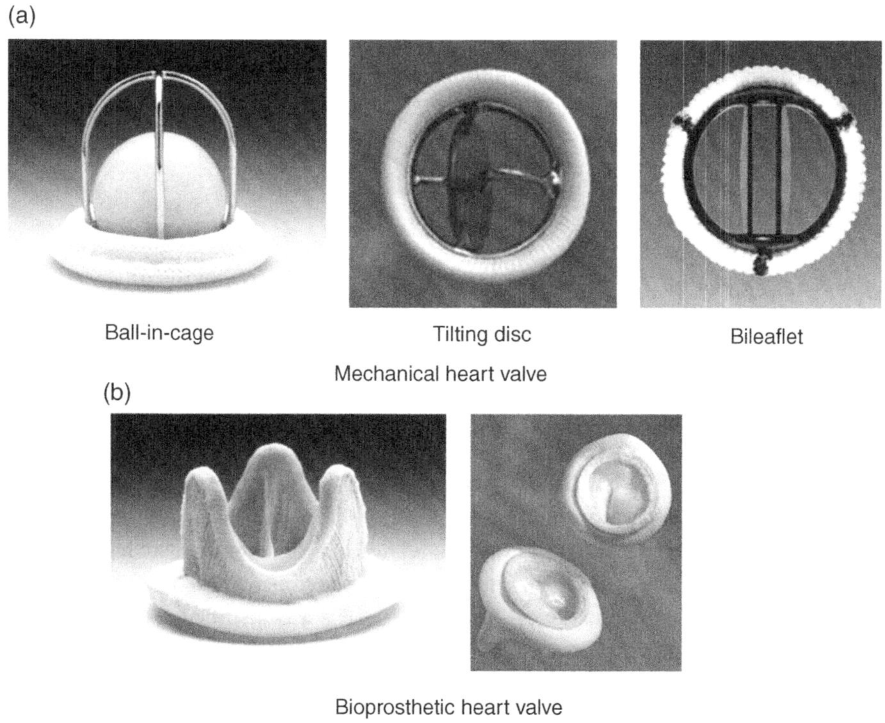

Figure 24.10 (a) Main designs of mechanical heart valves; (b) bioprosthetic heart valve.

from a few drawbacks. The change in direction of the blood flow leads to turbulence. In addition, in order for the valve to close properly, the ball has to be larger than the circular orifice of the valve. Together, these factors place an additional burden on the heart to maintain the blood flow. Also, the ball itself causes damage to the blood cells due to the impact of the cells on its surface, and these damaged cells then release clotting agents, leading to a high risk of thrombosis.

The other type of valve design consists of one or two tilting discs attached to the annular housing, providing a more central flow through the valve. The initial tilting disc design provided an improvement in hemodynamics over the ball-in-cage design by providing a more central flow without allowing back flow. In this way, damage to the blood cells and, thus, the risk of thrombosis was reduced. However, attempts to improve the hemodynamics of the tilting disc design was unsuccessful because of a high incidence of mechanical failure due to fracture of the struts attaching the disc to the annular component. Consequently, the use of this improved design was discontinued.

Introduced in ~1977, the bileaflet valve is now the most commonly used design. It consists of two semi-circular discs composed of pyrolytic carbon-coated graphite attached to a pyrolytic graphite housing. The leaflets swing open into a position nearly parallel to the blood flow, providing an essentially central flow. In the closed position, the leaflets do not meet exactly, which causes some back flow to occur. Pyrolytic carbon had good biocompatibility, showing good resistance to platelet adhesion and, thus, to thrombosis, and it causes little damage to blood cells. Additionally, it has adequate mechanical properties, such as good fatigue resistance to withstand the repeated opening and closing of the valve. Consequently, pyrolytic carbon provides a useful combination of the aforementioned properties required of a prosthetic heart valve material.

24.7.2 Bioprosthetic Valves

Bioprosthetic valves (Figure 24.10b) provide a better resemblance to natural heart valves when compared to mechanical valves. They are made entirely or partly from materials of biological origin. The most widely used bioprosthetic valves are xenografts, consisting of three cusps of tissue derived from porcine aortic valves or bovine pericardium, which have been treated with glutaraldehyde. The glutaraldehyde treatment improves the mechanical durability and biological stability of the tissue through crosslinking of the collagen, and it reduces the already limited antigenicity of the tissue. However, the treatment leads eventually to stiffening of the cusps due to calcification, which is a major cause of valve failure due to cuspal tears, and it kills the cells within the tissue, which prevents remodeling of the tissue cusps *in vivo*. The cusps of xenograft tissue are mounted on a metal or plastic ring, which, as in the case of mechanical valves, is covered with a sewing cuff, composed of Dacron® or Teflon®, to facilitate suturing of the device to the heart tissue.

Another type of bioprosthetic valve consists of allografts (also referred to as homografts), obtained from recipients of heart transplants, organ donors whose hearts are not accepted for heart transplants, or from patients who have died from noncardiac complications. They are normally cryopreserved in the vapor of liquid nitrogen or stored at 4 °C after antibiotic sterilization. Allografts have certain advantages, such as better durability than xenogeneic grafts, good hemodynamic properties, low incidence of thromboembolic events, and the avoidance of lifetime use of anticoagulants. While they suffer from problems such as difficulties in obtaining the right size and the requirement of complex implantation procedures, a major problem is their limited availability due to a shortage of donors. Consequently, allografts see very limited use.

Performance of Heart Valves

Mechanical heart valves, such as the bileaflet valve, are very durable, with an average lifetime of over 20 years. However, one of the major drawbacks of mechanical valves is that they require patients to follow a rigorous anticoagulant treatment for the remainder of their lives because of the risk of thrombosis and thromboembolism (blockage of the valve by a blood clot).

Bioprosthetic heart valves (xenografts) have better hemodynamics than mechanical valves but they are less durable, with a failure rate at 15 years of ~30% in aortic valve replacement and ~50% in mitral valve replacement. Major causes of failure are limited mechanical durability due to calcification and mechanical damage independent of calcification, such as tissue tearing due to proteolytic damage of the collagenous tissue with time. Studies have shown that tissue valve calcification is initiated primarily within residual cells that have been devitalized, usually by the glutaraldehyde pretreatment. The mechanism involves the reaction of calcium-containing extracellular fluid with membrane-associated phosphorus to yield calcium phosphate deposits. Calcification is accelerated in young patients, by glutaraldehyde pretreatment and by increased mechanical stress on the valve. The conditions used in the glutaraldehyde pretreatment have been shown to produce structural changes in the collagenous tissue, which cause a reduction in its compliance. This leads to the development of significant and repetitive localized stresses in the collagenous tissue of the valve upon implantation, which reduces its fatigue resistance, eventually causing mechanical damage.

Incremental improvements in bioprosthetic values over the last few decades, such as pretreatments to reduce the rate of calcification, have produced an increase in their long-term durability. This has resulted in a shift toward increasing use of bioprosthetic valves at the expense of mechanical valves. For example, over a 10-year period from 1997 to 2006, the percentage of implanted bioprosthetic valves in North America increased from ~40 to ~80% (Schoen and Padera 2013).

24.8 Ophthalmologic Applications

Contact lenses and intraoptical lenses (IOLs) represent the main use of biomaterials in ophthalmologic applications (Table 1.1). Worldwide, over 100 million people wear contact lenses as an alternative to spectacles to correct their vision, while over 7 million intraoptical lenses are used each year to replace the eye lens and to restore visual acuity in patients suffering from age-related lens deterioration such as the formation of cataracts. These devices provide useful examples illustrating how biomaterials are modified and adapted to make them effective in the ocular environment.

A schematic illustration of the cross section of the eye is shown in Figure 24.11. The eye is approximately spherical in shape (~25 mm in diameter). It functions by means of a layer of photoreceptors on the retina and two lenses, the cornea and the lens, which focus light onto the photoreceptors. Impulses from these receptors are transmitted to the brain by the optic nerve. Light enters the eye through the transparent cornea and, following passage through the cornea, it is focused onto the retinal receptors by means of the lens changing its shape. Refractive vision problems arise when the cornea does not have the appropriate refractive power or when the axial length of the eye is incorrect. In these situations, devices such as spectacles or

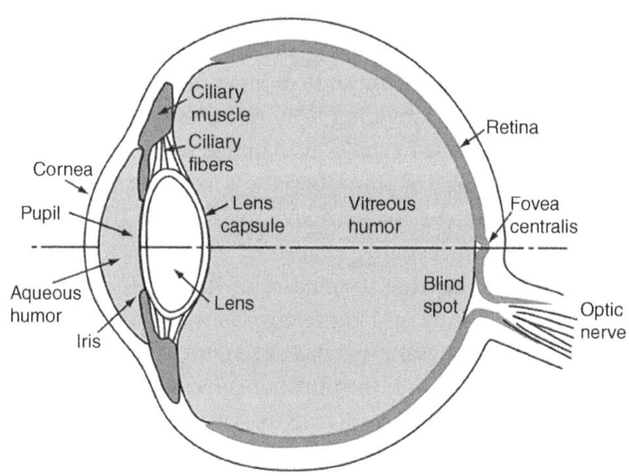

Figure 24.11 Schematic representation of the cross section of the human eye. *Source:* Paul Sherman, WPClipart. Public Domain. Available at https://wpclipart.com/medical/anatomy/eye/eye_cross_section.png.html

contact lenses are required to assist in focusing the light onto the receptors.

Vision problems also arise when the eye lens suffers age-related degeneration, such as the formation of cataracts, reducing the optical transparency of the lens and making it unable to properly transmit and focus the light entering through the cornea. Currently, there is no viable medical treatment for cataracts other than replacing the lens with an implant called an IOL.

24.8.1 Contact Lenses

Because the contact lens rests on the cornea, the contact lens material should minimally have the following properties and characteristics:

- Acceptable biocompatibility in the ocular environment.
- Optical transparency for light transmission.
- Suitable mechanical properties for durability and handling.
- Chemical stability for disinfection and storage.
- Permeability to oxygen transport: Because the cornea receives its oxygen to maintain its clarity, structure and function from the atmosphere and not from the blood supply, corneal swelling can occur if sufficient oxygen does not reach the cornea. Additionally, extended periods of oxygen deprivation can cause blood vessels to grow in the cornea.
- Ease of removal: The contact lens should not adhere strongly to the cornea, and the wearer should be able to remove the lens easily for cleaning, disinfecting, or disposal without discomfort. In addition, movement of the lens on the cornea is required for comfort and to allow the flow of tears between the lens and the cornea.
- Suitable surface characteristics to allow wetting by tears to maintain a continuous tear film between the contact lens and the cornea, to resist protein and lipid adsorption from the ocular environment in order to maintain the optical quality of the lens, and to reduce the risk of allergic and inflammatory responses in the eye.

Contact lenses have been categorized as hard, rigid gas-permeable (RGP), or soft. Hard contact lenses are produced from polymethyl methacrylate (PMMA), a material used to pioneer the development of contact lenses starting in the late 1930s. PMMA has excellent optical transmission, is mechanically rigid and durable, and has good chemical stability in the optical environment. On the other hand, it is rigid and has low permeability for oxygen transport. In general, hard contact lenses are now rarely used because soft contact lenses are more comfortable and better tolerated by wearers.

Polysiloxanes (also referred to as silicones), such as polydimethylsiloxane (PDMS), are rubbery solids (elastomers) (Chapter 8). In addition to having good optical transmission, polysiloxanes have high oxygen permeability attributed to the free volume between the chains in their open chemical structure. On the other hand, polysiloxanes contain no water and, thus, they tend to adhere to the cornea. Polysiloxanes are hydrophobic due to the organic side-groups in their molecular chains and they are lipophilic, giving them low wettability. In addition, they are difficult to machine and polish to the precise shape and surface finish required for contact lenses because of their rubbery nature. Copolymerization of certain siloxane monomers with less hydrophobic monomers such as methyl methacrylate have produced polysiloxanes having better wetting behavior and higher rigidity for ease of machining. These so-called RGP contact lenses, introduced in the early 1980s, currently account for less than ~5% of newly fitted contact lenses. In view of the limited use of rigid PMMA contact lenses and RGP hydrogels, we shall focus mainly on soft contact lens materials.

Soft Contact Lens Materials

In general, soft contact lens materials are composed of hydrogels based on polyhydroxyethyl methacrylate (PHEMA) or polysiloxanes (silicones). These hydrogel lenses are the most commonly worn type of contact lenses because they are comfortable and well tolerated by wearers. In the contact lens literature, hydrogels based on PHEMA are often referred to as conventional hydrogels (or simply hydrogels) to differentiate them from polysiloxane-based hydrogels, referred to variously as silicone hydrogels or silicone-based hydrogels.

PHEMA, credited with being the first synthetic hydrogel created specifically for use in biomedical applications (Chapter 11), has had a significant impact on the evolution of soft contact lens materials since its creation in ~1960. PHEMA has proved to be a very successful contact lens material because it is easy to fabricate economically, is very flexible, and is dimensionally stable to changes in pH and temperature. However, a major disadvantage of PHEMA hydrogels is their low permeability to oxygen transport. This is because they rely on the absorbed water within them for transport of oxygen but water has only a limited capacity to dissolve and transport oxygen.

One approach to improving the oxygen permeability of PHEMA has been to enhance its ability to absorb water by improving its hydrophilicity. This has been achieved by copolymerizing hydroxyethyl methacrylate with an ionic monomer such as methacrylic acid or a more hydrophilic monomer such as *N*-vinylpyrrolidone. For example, at equilibrium, PHEMA hydrogels contain ~40 wt% of absorbed water, whereas a copolymer hydrogel containing ~6 mol% methacrylic acid has a water content of ~60 wt%. However, this approach can increase the oxygen permeability of the hydrogel only up to a certain limit because of the aforementioned low oxygen permeability of water. In general, these PHEMA-based hydrogels do not have a high enough oxygen permeability for them to be worn overnight and, consequently, they are approved for use mainly for daily wear. The higher water content of the hydrogels also has a deleterious effect on certain properties of the contact lens, such as lowering its mechanical durability and increasing its susceptibility to dehydration.

Polysiloxane-based hydrogels were introduced in the late 1990s to address the limitations of PHEMA-based hydrogels, particularly for extended wear, by combining the high oxygen permeability and optical transmission of polysiloxanes with the hydrophilicity, softness, and comfort of hydrogels. One approach used for creating polysiloxane-based hydrogels involves copolymerizing polysiloxane-containing macromolecules with neutral hydrophilic monomers such as hydroxyethyl methacrylate, *N*-vinylpyrrolidone or dimethyl acrylamide. The bulk structure of these polysiloxane-based hydrogels provides high oxygen permeability, high enough water content (greater than ~25 wt%), and adequate mechanical properties. However, because of the hydrophobicity of polysiloxanes, these hydrogels still require further modification in order to improve their wetting behavior and to reduce protein and lipid adsorption.

One type of modification involves surface modification, achieved by treating the polysiloxane-based hydrogel in a gas plasma (Chapter 13). One plasma modification method involves depositing a discrete hydrophilic coating (~25 nm thick) on the hydrogel surface. Another plasma modification method involves oxidizing the hydrophobic siloxane regions of the surface, forming discrete regions (islands) having a silicate composition. Another type of surface modification involves incorporating a wetting agent, such as the electrostatically neutral hydrophilic polymer polyvinylpyrrolidone (PVP) into the hydrogel. When compared to an unmodified siloxane-based hydrogel, these modifications not only improve the wetting behavior of the hydrogel but also lower its susceptibility to protein adsorption (Chapter 16).

Performance of Contact Lenses

The design and development of contact lenses, with improved performance and comfort for a range of lens wear schedules and preferences, have been greatly facilitated through an understanding of the interaction between the contact lens material and the ocular environment. Daily-wear contact lenses require removal each night, cleaning and disinfecting by soaking overnight, and, normally, replacement of the lenses monthly (Table 24.2). Disposable daily-wear lenses do not require cleaning and disinfecting, but instead, they are removed each night and replaced by a new lens the next day. These types of daily-wear contact lenses are generally composed of PHEMA-based hydrogels and do not have the requisite oxygen permeability for overnight use. In the U.S., extended-wear lenses are approved for continuous wear over 7 days and 6 nights, while in some countries they are approved for continuous wear over 15 days or 1 month. These extended-wear lenses are generally composed of polysiloxane-based hydrogels, which, as noted earlier, have a sufficiently high oxygen permeability, while certain types of PHEMA-based hydrogels with adequate oxygen permeability, composed typically of a copolymer of hydroxyethyl methacrylate with methacrylic acid or *N*-vinylpyrrolidone, are also used for continuous wear over seven days and six nights.

The performance of contact lenses in the ocular environment is dependent on a variety of biological issues. For example, oxygen permeability is an important performance issue, which is dependent on the bulk composition of the hydrogel. The lens should also have an appropriate bulk and surface composition to maintain suitable hydration,

Table 24.2 Types of contact lens.

Lens type	Frequency of lens change (day)	Cleaning requirement	Material	Overnight wear
Daily wear	1 (lens changed monthly)	Daily	PHEMA hydrogel	No
Daily disposable	1	None	PHEMA hydrogel	No
Extended wear	7	None	Polysiloxane-based hydrogel[a] or PHEMA-based hydrogel[b]	Yes
Continuous wear	15 or 31	None	Polysiloxane-based hydrogel[a]	Yes

[a] Hydrogel consisting of (i) a copolymer composed of polysiloxane-containing macromolecules with hydroxyethyl methacrylate, *N*-vinylpyrrolidone, or dimethyl acrylamide, or (ii) a polysiloxane-based polymer containing polyvinylpyrrolidone (PVP).
[b] PHEMA-based hydrogel consisting of a copolymer of hydroxyethyl methacrylate with methacrylic acid, *N*-vinylpyrrolidone or dimethyl acrylamide. PHEMA, polyhydroxyethyl methacrylate.
Source: Adapted from McGlinchey et al. (2008).

permit transport of ions, and be wetted by a continuous tear film.

Another important issue relates to protein adsorption from the tear film to the hydrogel surface. Proteins are a major component of the tear film. While they serve a number of important biological functions, protein adsorption to the contact lens surface can also have some negative effects. Adsorbed proteins can not only reduce the optical quality of the contact lens but also lead to negative biocompatibility phenomena, such as a higher risk of allergic and inflammatory responses in the eye, bacterial adhesion and subsequent infection of the eye.

Protein adsorption is determined largely by the contact lens surface composition, which determines its surface physicochemical properties such as wetting behavior and electrostatic charge, and by the chemical composition and concentration of proteins in the tear film. The tear film contains a large number of proteins having a wide range of molecular weights and isoelectric points (IEPs). Approximately 80% of these proteins have a molecular weight less ~10 kDa and their IEPs are in the range ~1–11. The proteins that have been most commonly identified on contacts lenses are lysozyme (molecular weight ~14 kDA; IEP ~11.1), which is the most abundant protein in human tears, constituting one-third of the total protein content, followed by lactoferrin (molecular weight ~80 kDa; IEP ~8.7) and albumin (molecular weight ~66 kDa; IEP ~5.2). As the pH of the tear film is ~7.4, both lysozyme and lactoferrin are positively charged in the ocular environment, whereas albumin is negatively charged.

Similar to biomaterials implanted into the physiological environment, upon placing a contact lens on the cornea, its surface will be rapidly coated with a layer of proteins but the proteins that are subsequently retained on the surface are the ones with higher affinity for the surface (Chapter 16). Commercial contact lenses are large in number, and they have different compositions that are generally not available accurately. However, in general, they will have varying degrees of hydrophobicity (or hydrophilicity) and either an electrostatically neutral or a charged surface. For example, PHEMA-based hydrogels containing methacrylic acid will have a negative surface charge, which will provide a strong tendency for adsorption of lysozyme and lactoferrin by electrostatic attraction. Polysiloxane-based hydrogels have a more complex composition than PHEMA-based hydrogels, due to the various siloxane-based components and other components used in their production. However, a polysiloxane-based hydrogel with a plasma-treated surface composed of silicate islands should have a tendency to support lysozyme and lactoferrin adsorption by electrostatic attraction because silicates normally have an IEP below 7.4 and, thus, will be negatively charged at this pH. On the other hand, a polysiloxane-based hydrogel with an electrostatically neutral surface having some degree of hydrophilicity, such as a polysiloxane containing a wetting agent such as PVP, should provide a better resistance to protein adsorption (Chapter 16).

A risk associated with contact lenses is infection of the cornea, referred to as microbial keratitis, caused by bacterial colonization of the lens surface. Although the risk is low, microbial keratitis is generally the most serious sight-threatening event associated with the wearing of contact lenses. The incidence of microbial keratitis is dependent on a number of factors, such as poor hygiene, inadequate cleaning and disinfection of the lens, overnight wear, and the type of contact lens.

As noted in Chapter 5, the physicochemical properties of the original surface control the conformation, concentration, and nature of the adsorbed biomolecules, but it is the surface covered with a layer of adsorbed biomolecules, and not the bare surface of the biomaterial, which initially interacts with cells. Different microorganisms have been isolated from microbial keratitis associated with contact lenses, consisting mainly of gram-negative bacterial strains such as *Pseudomonas aeruginosa* but also a significant fraction of gram-positive bacteria from the staphylococcal family, such as *Staphylococcus epidermis*. Because bacterial cells are generally negatively charged, contact lenses with a high concentration of adsorbed lysozyme and lactoferrin, such as PHEMA-based hydrogels containing methacrylic acid and polysiloxane-based hydrogels with a plasma-treated surface composed of silicate islands, may be more susceptible to higher amounts of bacterial adhesion. Overall, however, a wide variety of proteins is present in the layer coating the contact lens, differing in composition and concentration, which will influence the amount and type of adherent bacterial cells.

24.8.2 Intraocular Lenses

An IOL is a lens implanted into the eye to replace the natural lens that has suffered age-related deterioration, particularly through the formation of cataracts. The proteins that make up the eye lens may distort, making the lens cloudy, thereby reducing its optical transparency. Currently, there is no effective medical treatment for cataracts except to replace the cloudy eye lens with an IOL. Ideally, an IOL should have the following properties and characteristics:

- Acceptable biocompatibility over the long-term duration of its implantation, normally up to a few decades.
- Retention of optical properties such as transmittance over the duration of its implantation.

- Adequate chemical and mechanical durability to maintain its physicochemical properties.
- Ability to provide certain advantages for the surgeon, such as ease of implantation, and for the patient as well, such as long-term refractive stability.

IOLs have a variety of designs, such as a rigid one-piece design or a foldable one-piece or three-piece design (Figure 24.12). Overall, however, they generally consist of two parts, a central optical part (~5–7 mm in diameter) and side structures, called haptics, for stable fixation of the IOL within the lens capsule. In recent years, there has been a growing tendency toward the use of foldable IOLs because they can be implanted via smaller incisions (~3 mm or smaller) requiring no stitches. In comparison, rigid one-piece IOLs require larger incisions (~5–7 mm or larger) and significant stitching that can lead to problems such as astigmatism.

Intraocular Lens Materials

Biomaterials used for producing the optical part of IOLs can be categorized into two types: acrylic-based polymers and polysiloxane-based polymers (silicones). In general, these polymers have a range of physicochemical properties, which are dependent on their chemical structure (Chapter 8). PMMA, an acrylic polymer with a glass transition temperature T_g ~105 °C, is in the glassy state at the physiological temperature. It is normally used in rigid one-piece designs, because of its stiffness, excellent optical transmission, chemical durability, and high enough refractive index (1.49), and it is well tolerated within the ocular environment. It contains little water (<1 wt%).

Acrylic polymers used in foldable designs are classified in the IOL literature as hydrophobic or hydrophilic. The hydrophobic polymers have a variety of compositions, but generally consist of a copolymer composed of a hydrophobic methacrylate monomer and a hydrophobic acrylate monomer. Foldable hydrophobic IOLs have been very successful since their introduction in the early 1990s. They are softer than PMMA, and have a refractive index in the range 1.45–1.55 and a water content of less than ~1 wt%. For example, one composition, consisting of a copolymer of phenyl ethyl methacrylate and phenyl ethyl acrylate, has a T_g (glass transition temperature) of ~18 °C and, thus, is in a rubbery state at room temperature or above. When compared to PMMA, the bulky nonpolar aromatic side groups lead to lower intermolecular bonding and higher free volume, making it easier for the polymer chains to rotate, while the electron-rich aromatic side groups lead to a higher refractive index (~1.55).

In comparison, hydrophilic acrylic polymers are essentially hydrogels composed of a copolymer of hydrophilic monomers. They have a range of compositions, a water content in the range ~20–40 wt% at the physiological temperature, and a T_g below room temperature. For example, a copolymer of hydroxyethyl methacrylate and hydroxyhexyl methacrylate has a T_g of ~2 °C, a refractive index of 1.47 and a water content of ~20 wt%. Another material that may be grouped with these hydrophilic acrylic polymers is a copolymer composed of a hydrophilic acrylic polymer and porcine collagen. This material is fairly soft and has a water content of ~35 wt%.

The first polysiloxane-based material used for IOLs was PDMS. As noted earlier, PDMS is a rubbery hydrophobic polymer having good optical transmission, and a low water content (less than ~1 wt%). Its contact angle with water is ~100–110°, which is higher than that of the hydrophobic acrylic polymers. The refractive index of PDMS (1.41) is lower than that of the hydrophobic acrylic polymers, but more recently developed polysiloxane-based polymers have a higher refractive index (1.41–1.47). Because of their lower refractive index, polysiloxane-based IOLs tend to have a higher thickness than hydrophobic acrylic IOLs.

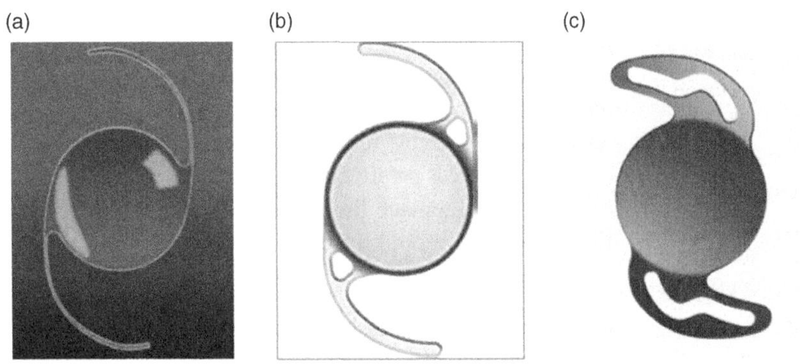

Figure 24.12 Examples of intraocular lens (IOL) designs: (a) rigid one-piece polymethyl methacrylate (PMMA) IOL; (b) one-piece foldable hydrophobic acrylic IOL; (c) foldable hydrophilic acrylic IOL. *Source:* Bellucci (2013) / with permission of Karger Publishers.

Surface modification has been investigated as a means to improve the biocompatibility of IOLs in the ocular environment. As noted in Chapter 13, a variety of techniques is available to modify the surface of biomaterials, such as modifying the outermost molecules of the surface to create new surface functional groups, depositing a discrete coating, or covalent bonding of the desired molecules to the surface. One approach that has been investigated to reduce the susceptibility of hydrophobic acrylic and polysiloxane-based IOLs to protein adsorption and, subsequently, to cell adhesion has been to make their surfaces more hydrophilic. For example, heparin surface modification has been used to make the surface of PMMA IOLs more hydrophilic. These heparin-modified IOLs have shown the ability to reduce inflammatory responses when compared to unmodified IOLs but it is not clear whether they provide any advantage in reducing longer-term adverse biocompatibility phenomena such as IOL opacification.

Other surface modification methods that have been investigated to improve the biocompatibility of IOLs are the covalent bonding of PEG molecules to the surface of hydrogel IOLs and the covalent bonding of 2-methacryloyloxyethyl phosphorylcholine to the surface of hydrophobic acrylic IOLs or polysiloxane-based IOLs. These surface modifications produce no measurable change in the optical transmission of the IOLs but significantly reduce protein adsorption and cell adhesion, which may indicate that they have the potential for reducing the risk of lens opacification. Figure 24.13 shows the reaction scheme for covalently bonding PEG molecules (molecular weight 1100–5000 Da) to the surface of a hydrophobic IOL composed of a copolymer of hydroxyethyl methacrylate and methyl methacrylate.

Additional modifications are commonly made to some IOLs to improve their performance. For example, the incorporation of ultraviolet (UV) absorbing compounds (chromophores) such as benzotriazole into the IOL material is used to protect the retina from UV radiation in the 300–400 nm wavelength range, a protection normally provided by the lens of the eye. Typically, UV-absorbers possess a polymerizable functional group such as a vinyl, acrylate, or methacrylate group in their chemical structure for covalent bonding into the IOL material during polymerization. Blue light chromophores, such as polymerizable yellow dyes, are also added to some IOL materials to absorb harmful blue light radiation. Together these chromophores provide a transmittance curve in the UV and visible range to mimic the protection provided by the eye lens.

Biocompatibility of Intraocular Lenses

The biocompatibility of IOLs is critical to the successful restoration of vision following cataract surgery. Biocompatibility phenomena related to IOLs can be divided into two types (Werner 2008):

- Inflammatory responses of the eye following implantation of the IOL.
- Opacification of the optical component of the IOL, which is often the main and most serious complication.

Cataract surgery with the implantation of an IOL causes a breakdown of the blood-aqueous barrier (the anterior barrier of the eye composed of endothelial cells of blood vessels in the iris and cell layer of the ciliary epithelium), resulting in the release of proteins and inflammatory cells into the anterior chamber (aqueous humor). Protein adsorption to the surface of the IOL is the first event to occur, which influences subsequent interaction of cells with the protein-coated surface of the IOL. Monocytes, lymphocytes, and macrophages migrate through blood vessel walls to the aqueous

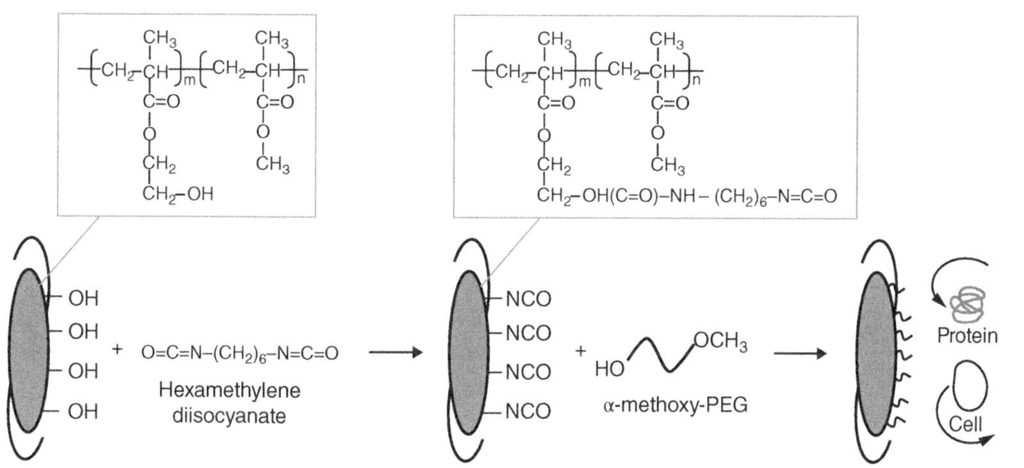

Figure 24.13 Reaction scheme for the covalent bonding of polyethylene glycol (PEG) molecules to the surface of a hydrophobic IOL composed of a copolymer of hydroxyethyl methacrylate (HEMA) and methyl methacrylate (MMA). *Source:* Bozukova et al. (2007) / with permission of American Chemical Society,

humor and eventually to the IOL surface, constituting a foreign-body reaction directed against the IOL. The monocytes transform into small round fibroblast-like cells and, later, the macrophages transform into foreign-body giant cells (Chapter 18). This inflammatory response normally lasts over a period of up to a few months. Subsequently, the giant cells degenerate and detach from the lens surface, usually leaving only an acellular proteinaceous membrane that surrounds the IOL, isolating it from the surrounding ocular tissues.

Because hydrophilic surfaces generally are less susceptible to protein adsorption and, thus, to cell adhesion, hydrophilic acrylic IOLs tend to show lower inflammatory responses. As noted earlier, coating the surface of hydrophobic IOLs with heparin or with certain hydrophilic molecules, such as PEG or 2-methacryloyloxyethyl phosphorylcholine, has the potential to reduce protein adsorption and, thus, cell adhesion. However, these inflammatory reactions are commonly not significant, and IOL implantations generally lead to clinically acceptable outcomes.

Biocompatibility phenomena related to post-operative IOL opacification are due to interactions of the IOL with the eye lens epithelial cells and with the eye lens capsule. The epithelium of the eye lens consists of a sheet of epithelial cells and, upon removal of the eye lens and implantation of an IOL, outgrowth of residual lens epithelial cells from the capsule onto the IOL surface, where they proliferate and differentiate into fibroblast-like cells, leads to opacification of the IOL. Opacification can occur in the anterior capsule of the eye, referred to as anterior capsule opacification (ACO), which generally occurs soon after surgery (often within the first month), but occurs more commonly in the posterior capsule, referred to as posterior capsule opacification (PCO) or secondary cataract formation. PCO in the central region of the IOL is the most common reason for longer-term loss of visual acuity, occurring in ~50% of cases at approximately two to five years after cataract surgery. Neodymium-doped yttrium aluminum garnet (Nd:YAG) laser capsulotomy is the only way to treat PCO but it suffers from significant risks, such as the potential for damage to the retina and to the IOL. Consequently, the prevention of PCO is preferable.

Opacification is dependent on the IOL material but the IOL design can also have a significant effect. The results of some clinical trials appear to suggest that there is a lower incidence of ACO in hydrophilic IOLs than in hydrophobic acrylic IOLs. In addition, it appears that the haptics of hydrophilic IOLs can exert a significant additional effect, depending on whether or not they produce a persistent gap between the IOL optical component and the eye lens capsule. For example, the use of a hydrophilic IOL that does not make persistent contact with the inner surface of the anterior capsule could significantly reduce or prevent ACO. Alternatively, as noted earlier, surface modification of hydrophobic acrylic IOLs or polysiloxane-based IOLs to make their surfaces less hydrophobic can potentially improve their biocompatibility. In general, however, ACO is commonly a short-term effect that interferes with peripheral vision and does not normally have a severe effect on the central vision.

The development of PCO also appears to be dependent on both the IOL material and the IOL design. Results from clinical trials at 1 year and 3 years post-surgery appear to suggest that for the same polysiloxane-based material, the incidence of PCO is lower for IOLs with a sharp-edged optical component when compared to IOLs with a round-edged optical component. Additionally, for IOLs with sharp-edged optical components, there appears to be no significant difference in PCO between hydrophobic acrylic IOLs and polysiloxane-based IOLs. On the other hand, at six years post-surgery, the acrylic IOLs appear to show a higher incidence of PCO when compared to the polysiloxane-based IOLs. As noted earlier, one approach that has been investigated to reduce opacification is covalent bonding of electrically neutral hydrophilic molecules such as PEG to the surface of the IOL.

24.9 Skin Wound Healing

The treatment of cutaneous injuries is a common feature of medical care, with millions of patients treated worldwide each year. While effective treatments are available for healing acute and small-area traumatic wounds, the treatment of patients with large area burns, infected wounds, or chronic skin wounds is challenging. Chronic wounds are generally wounds that fail to proceed through an orderly and timely reparative process to reproduce the anatomic and functional integrity of the injured site. Chronic wounds often occur as pressure ulcers, particularly for patients bedridden for long periods, venous leg and foot ulcers due to poor circulation, and diabetic ulcers in patients suffering from diabetes. Over the last several decades, the incidence of chronic wounds has been increasing due to the prevalence of diseases such as diabetes and peripheral vascular disease. While the treatment of chronic skin wounds places a significant financial burden on healthcare systems, the failure of these wounds to heal can have devastating consequences for the patient, such as lifelong disability or amputation. For example, failure of diabetic foot ulcers to heal is the leading cause of limb amputations in the U.S.

The healing of cutaneous wounds, particularly large area burns and chronic wounds is a complex and intricate process, involving the coordinated interaction of a variety of cell types, such as immune cells, hematopoietic cells and resident cells of the skin, and an appropriate extracellular environment as well. Biomaterials-based strategies have played an important role in healing chronic wounds, but they vary widely in their effectiveness. In the next section, fundamentals of the wound healing process are discussed in order to provide a background for understanding these therapies and for the development of more effective biomaterials-based wound healing strategies.

24.9.1 Skin Wound Healing Fundamentals

The cutaneous wound healing process normally occurs in four overlapping but distinct stages, referred to as hemostasis, inflammation, proliferation, and remodeling (Figure 24.14). In the literature, the process is sometimes also described as occurring in three stages in which the first stage consists of a combination of hemostasis and inflammation.

Hemostasis

Immediately after an injury, the first step is to stop the bleeding and the loss of blood (hemostasis), which is achieved through clotting of the blood at the wound site by platelets. Because blood loss or hemorrhage is a leading cause of death for people with extensive injuries, methods to facilitate hemostasis, such as direct pressure on the wound and the application of hemostatic materials (such as collagen sponge) are very important. Immediately after an injury, the response of tissues at the wound site involves the reflex contraction of smooth muscle cells in the tunica media of severed or damaged blood vessels. This can cause arterial vessels up to a few millimeters in diameter to remain completely closed for a few minutes to prevent excessive blood loss. During this time, a complex sequence of reactions that occur on the surface of platelets is activated (Chapter 16), which leads to the formation of a gelatinous clot composed of fibrin that seals the wound and prevents further blood loss. The coagulation process is normally

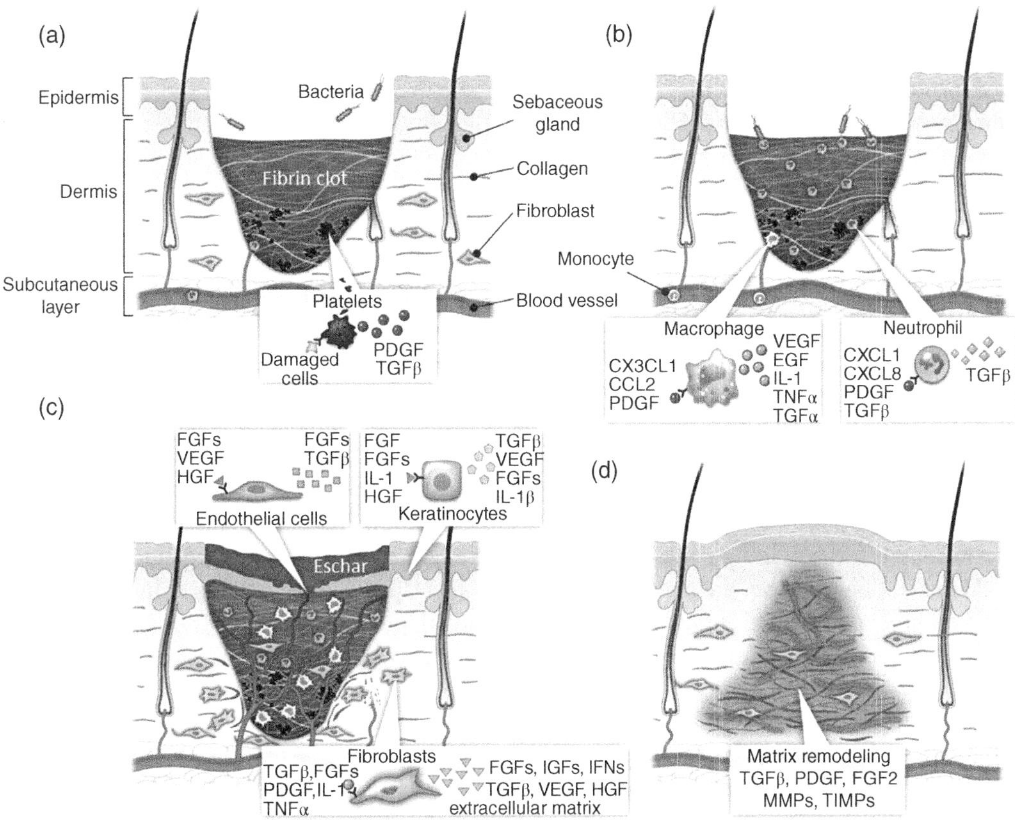

Figure 24.14 Stages of wound healing: (a) hemostasis, (b) inflammation, (c) proliferation, and (d) remodeling. Each stage is characterized by key molecular and cellular events and is coordinated by a host of secreted factors that are recognized and released by the cells of the wounding response. A representative subset of major factors are depicted. PDGF, platelet-derived growth factor; TGF, transforming growth factor; FGFs, fibroblast growth factors; IL-1, interleukin-1; TNF, tumor necrosis factor; KGF, keratinocyte growth factor; IGF, insulin-like growth factor; IFN, interferon; VEGF, vascular endothelial growth factor; HGF, hepatocyte growth factor; MMP, matrix metalloproteinase; TIMP, tissue inhibitor of metalloproteinase. *Source:* Courtesy of Paul A. Khavari, Stanford University and the Veterans Affairs Palo Alto Healthcare System, California.

completed within a few minutes, making hemostasis the shortest stage of the wound healing process. In addition to hemostasis, the fibrin clot serves other important functions. Particularly important in this regard are the platelets entrapped within the gelatinous fibrin clot, which release several growth factors and cytokines that are critically important for the subsequent stages of wound healing.

Inflammation

As discussed in Chapter 18, inflammation is a nonspecific, natural response to injury and an important part of the body's innate immune defense system. Overall, an inflammatory reaction is mounted to attack foreign particles such as bacteria that may be introduced into the wound. This stage begins immediately after the injury and, thus, it is coincident with hemostasis. The inflammatory response is mediated by neutrophils and monocyte-derived macrophages attracted to the wound site by growth factors and cytokines, such as platelet-derived growth factor (PDGF), vascular endothelial growth factor (VEGF), TGFα, TGFβ, TNFα, endothelial growth factor (EGF), and interleukin-1 (IL-1), released by the platelets entrapped within the gelatinous fibrin clot.

Neutrophils are involved mainly in infection control. Upon arrival at the wound site within minutes of the injury, they commence phagocytic ingestion of bacteria and necrotic tissue, an essential activity for cleansing the wound. The arsenal of neutrophil defenses also includes myeloperoxidase and other proteases plus reactive oxygen species (ROS) that have a potent bactericidal effect. Neutrophil infiltration of the wound site normally lasts for ~2 days.

Pro-inflammatory M1macrophages become abundant at the wound site approximately two to three days after the injury. Their main functions are to remove cellular debris and, later, to provide soluble signals to attract cells into the wound site for the third stage of wound healing (proliferation) by releasing various cytokines, proteases, and growth factors into the wound. M1macrophages have greater phagocytic capacity than neutrophils, are better able to survive the increasing acidity of the wound tissue, and they generate ROS and highly reactive NO. The activities of macrophages allow continued destruction of bacteria and removal of necrotic host tissue. Anti-inflammatory M2 macrophages become abundant at the wound site during the latter part of the inflammation stage. The M2 macrophages secrete VEGF, TGFβ, EGF and other growth hormones that promote angiogenesis and regeneration of epithelium, processes that mark the initial formation of highly vascularized granulation tissue and the proliferation stage of wound healing. The release of TGFβ and IL-1 by M2 macrophages dampens the level of tissue inflammation, allowing transition to the proliferation stage. Normally, the inflammation stage is completed within ~7 days.

Proliferation

Proliferation is the rebuilding stage of wound healing, which occurs after satisfactory removal of bacterial and necrotic tissue debris from the wound site and inflammation has subsided. It is characterized by extensive cell proliferation and new tissue formation. As noted earlier, in the later part of the inflammation stage, macrophages release various cytokines, proteases and growth factors to attract cells, predominantly fibroblasts, endothelial cells and keratinocytes, into the wound site.

In response to growth factors such as fibroblast growth factor (FGF), TGFβ, and PDGF, fibroblasts proliferate and secrete various ECM molecules, such as collagen, elastin, proteoglycans, and hyaluronic acid, which provide a provisional matrix, starting at the bottom of the wound bed, for tissue remodeling and angiogenesis. This provisional matrix forms the granulation tissue of the wound, which is essential for proper wound healing. Simultaneously with the provisional matrix deposition, endothelial progenitor cells migrate into the provisional matrix and, in response to VEGF and other growth factors, differentiate to endothelial cells. These endothelial cells initiate angiogenesis and neovascularization, forming a network of capillaries and microvessels in the wound bed to provide nutrients and oxygen to the healing tissue. Angiogenesis and neovascularization are critical components of the wound healing process, particularly for full-thickness skin wounds. The network of capillaries and microvessels formed is responsible for the granular pinkish-red appearance of the provisional matrix, which is the basis for calling it granulation tissue.

As the proliferation of fibroblasts and endothelial cells continues, the granulation tissue expands, filling the wound bed from the bottom upward, giving rise to the reforming dermis. Upon complete filling of the wound site with granulation tissue, the fibroblasts transform to myofibroblasts that begin contracting the margins of the wound, causing the wound to decrease in size. At the wound margin, basal keratinocytes of the epidermis begin migrating inward, commencing the formation of a stratified epithelium. The process continues until the wound is completely covered with a stratified epithelium, at which point the wound has a fully reformed new skin. The proliferation stage can last from ~4 days to ~4 weeks.

Remodeling

The fourth stage of wound healing, which is also the longest, is the remodeling of the wound and surrounding tissue.

It starts at ~3 weeks and may last up to two years, depending on the location and severity of the wound. Fibroblasts and myofibroblasts put down a network of collagen fibers and other ECM molecules in an orderly manner while using proteases to degrade existing disordered tissue. The granulation tissue formed during the proliferation stage consists of immature type III collagen, which is mechanically weak. During the remodeling stage, fibroblasts gradually replace type III collagen with type I collagen oriented parallel to the primary direction of mechanical stress. The enzymes responsible for the degradation of collagen type III are matrix metalloproteinases (MMPs). As the new collagen type I fibers are laid down, they are cross-linked, which improves the strength of the healed tissue. The tissue continues to remodel and mature, gradually gaining strength, but it usually does not return to the full strength of the pre-injured normal tissue. Another feature of the remodeling process is that the wound becomes less vascularized, which causes the coloration of the wound site to change from pinkish-red to pinkish-grey. The wound site may also acquire a shiny white-grey appearance due to the formation of scar tissue, or become darker brown due to hyperpigmentation.

24.9.2 Complicating Factors in Skin Wound Healing

The healing of chronic skin wounds, as noted earlier, is challenging but it can be further complicated by a variety of factors. One factor is the accompanying disease or injury state of the patient. For example, the healing of chronic wounds in diabetic patients is complex and remains a major clinical challenge. An important factor in the failure of a chronic wound to heal is the presence of various species of bacteria, living cooperatively in highly organized biofilms within the wound site (Chapter 20). The biofilm protects the bacteria from antibiotic therapy and the patient's immune response. Cutaneous wounds lend themselves to infection by a variety of bacterial strains but the most common is methicillin-resistant *Staphylococcus aureus* (*S. aureus*), which is also the most common hospital-bone infection affecting millions of patients. Because of the rising incidence of drug-resistant bacterial strains, infection control of cutaneous wounds is very important.

Another complicating problem in cutaneous wound healing, particularly for large area burns and chronic wounds, is scarring, which can have profound functional and aesthetic consequences. A scar represents the sum of the injury, the healing process and subsequent interventions to the reparative process, and subsequent interventions to reduce the extent of scarring as well. Scars can have long-lasting functional, cosmetic, and psychological consequences for the patient. Scarring remains very difficult to treat and difficult to prevent. In humans, problems with scarring in cutaneous skin wounds can arise because of delayed healing or with excessive healing. Delayed healing often occurs in diabetic patients, whereas excessive healing is characterized by the deposition of large amounts of ECM due to alterations in local cell proliferation and vascularization. Excessive healing can produce large disfiguring masses that can physically distort the surface structure of the healed wound site, particularly after major injuries.

24.9.3 Biomaterials-Based Therapies

A wound healing dressing should ideally have a combination of desirable properties and characteristics. In general, the dressing should maintain the wound environment and address the limitations of wound healing specific to the type of wound and the patient's disease or injury state as well. It should be permeable to allow the transport of oxygen, carbon dioxide, and water vapor, and it should be capable of maintaining a balanced moist environment by preventing dehydration of the wound and by absorbing wound exudate. The dressing should also serve as a barrier against infections and provide thermal insulation of the wound. Additionally, it is desirable for the dressing to enhance the aforementioned aspects of the natural wound healing process, such as the recruitment and migration of cells into the healing wound, promotion of angiogenesis and neovascularization, modifying or controlling the influence of immune cells within the wound, and the prevention or treatment of infections.

A variety of biomaterials-based therapies, which vary in their effectiveness, have been developed to treat cutaneous wounds, which vary in their effectiveness (Das and Baker 2016). These therapies may be classified into various types, depending on whether they involve the use of biomaterials alone or biomaterials combined with cells or biomolecules.

Therapies Based on the Use of Biomaterials Alone

Biomaterials for use in the treatment of skin wounds are generally composed of biocompatible polymers, which may be synthetic or natural. More recently, however, certain bioactive glasses in a microfibrous form have shown their effectiveness in healing skin wounds in humans and animals.

One type of dressing consists of a liquid adhesive bandage composed of an amphiphilic polysiloxane-based polymer, with or without an antimicrobial agent, dissolved in a volatile non-stinging solvent such as hexamethyldisiloxane or in a stinging solvent such as ethanol or isopropanol. Upon application to the skin wound by spraying or other methods, the volatile solvent evaporates rapidly, leaving an adherent polymer bandage (coating) on the wound surface. These polysiloxane-based coatings have a composition and, thus, physicochemical properties similar to those of the polysiloxane-based soft contact lenses discussed earlier. They are conformable, transparent, insoluble in aqueous liquids, and permeable to oxygen and water vapor. When combined with an antimicrobial agent, they can also provide a barrier against infection. These liquid adhesive bandages are particularly useful for protecting the surface of skin wounds from damage arising from pressure, friction or shear forces, which may arise, for example, in patients of low mobility or patients confined to bed.

Dextran-based hydrogels have shown promising potential in healing full thickness skin wounds caused by burns in small (murine) and large (porcine) animal models. These hydrogels can be prepared by chemical modification of dextran, as illustrated in Figure 24.15 for an ethylamine-modified dextran. After crosslinking with polyethylene glycol diacrylate (PEGDA), typically in the ratio of ~80 mol% dextran to ~20 mol% PEGDA, ethylamine-modified dextran hydrogels were applied on full thickness burn wounds in a murine burn wound model (Figure 24.16). A thin dressing held the hydrogels in place for the entire healing period. Histological analysis showed that, when compared to wounds treated with the dressing only, the dextran-based hydrogels accelerated healing, resulting in complete skin regeneration after three weeks of treatment. During the treatment, the hydrogels attracted neutrophils to the wound site, which enhanced the hydrogel degradation, and accelerated the recruitment of endothelial progenitors and cells to the wound site, enabling rapid neovascularization and angiogenesis after ~1 week of treatment. Analysis of the regenerated skin structure at three weeks showed that the dextran-based hydrogels promoted skin maturation, resulting in a mature epithelial structure with hair follicles and sebaceous glands.

A recently developed treatment for skin wounds consists of using certain bioactive glasses in a microfibrous form. Although materials such as metals and glasses in a three-dimensional bulk form are very rigid, they generally become soft and pliable when converted into a fibrous form with a small enough fiber diameter. The bioactive glasses used in wound healing have certain special characteristics, which are different from those of the common silicate window and structural glasses. First, they normally have a composition based on the borate glass structure (rather than a silicate glass structure), such as the composition designated 13-93B3 in Table 7.3, which makes them highly degradable in the aqueous physiological environment. Second, they are normally doped with a small amount of metal ions, such as copper ions that are known to stimulate angiogenic gene expression. Thus, upon placement in a wound site, a microfibrous bioactive borate glass dressing starts to degrade, releasing ions such as Ca, Cu, and B (in the form of borate BO_3^{3-} ions), which stimulate cell proliferation and angiogenesis, and provide antimicrobial activity.

An example of a bioactive borate glass dressing used in wound healing is the aforementioned glass designated 13-93B3 doped with less than ~1–3 wt% CuO, composed of fibers with a diameter less than ~1–2 μm, and having a felt-like consistency (Figure 24.17). The microfibers in this dressing degrade completely within ~7–14 days, releasing their constituent ions into the wound site. Bioactive glass dressings have the advantages of low cost and simplicity of application. They are normally applied to the wound and simply held in place by a suitable bandage. Consequently, in the treatment of hard-to-heal wounds, the patient can replace the dressing itself periodically without the need for frequent clinical visits. Bioactive borate glasses have been found to be very effective in healing skin wounds in humans and animals, particularly chronic wounds, with little or no scarring. Figure 24.18 shows an example of the

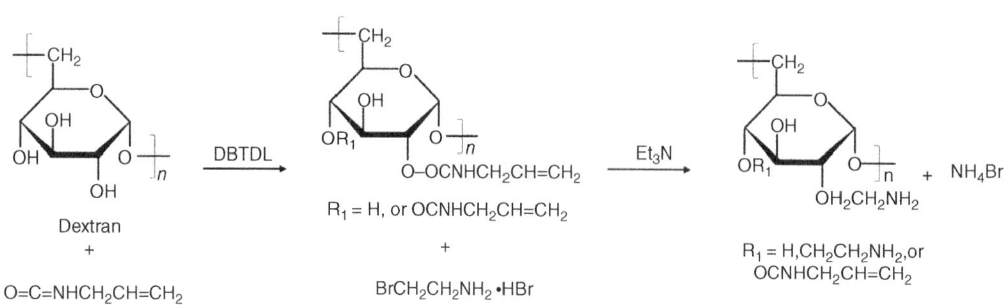

Figure 24.15 Reaction scheme for incorporating ethylamine into dextran. DBTDL, dibutyltin dilaurate; Et_3N, triethylamine. *Source:* Sun et al. (2010) / with permission of John Wiley & Sons.

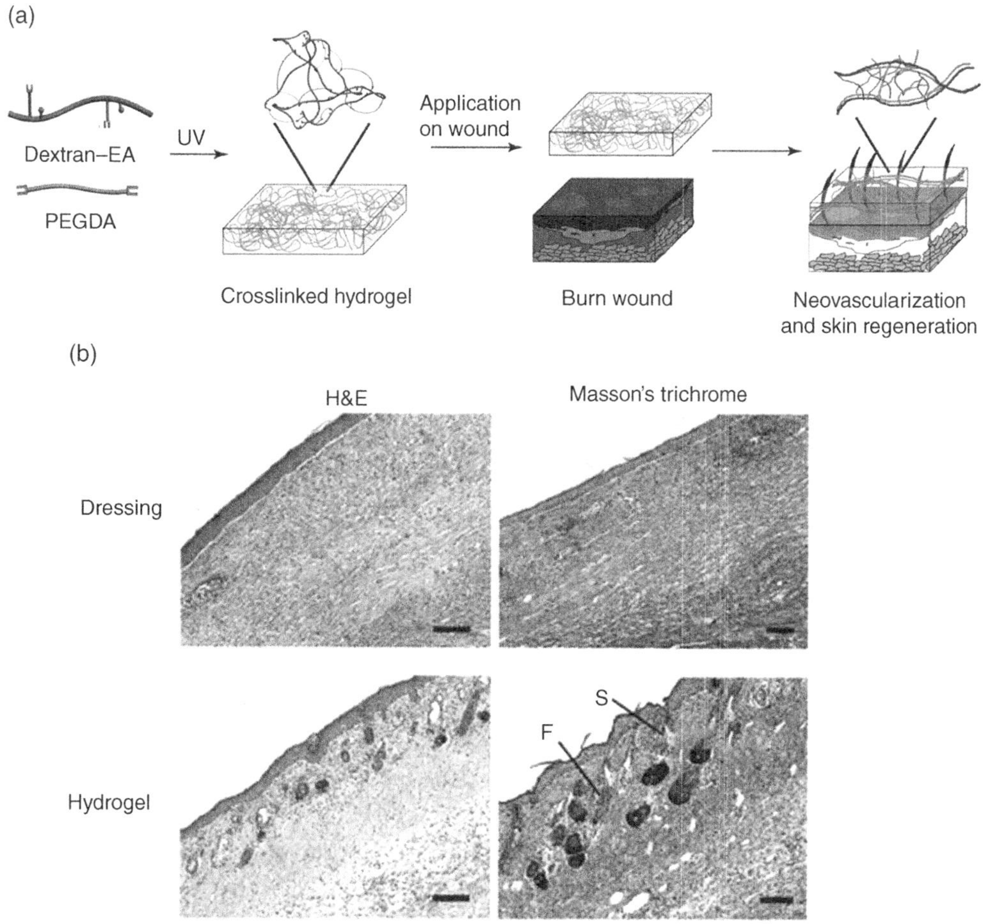

Figure 24.16 Modified dextran hydrogel as a treatment for burn wounds. (a) Illustration of neovascularization and skin regeneration by dextran-based hydrogel; (b) Images of stained histological sections of burn wounds in a murine model treated for 21 days with a dressing alone and a dextran-based hydrogel held in place with the dressing. Hematoxylin and eosin (H&E) stained images show that the dextran hydrogel promoted wound healing with complete skin appendage regeneration, while Masson's trichrome stained images show distinct collagen formed in the dermal layer. F, follicle; S, sebaceous gland. Scale bar = 100 μm. *Source:* Sun et al. (2011) / with permission of National Academy of Sciences.

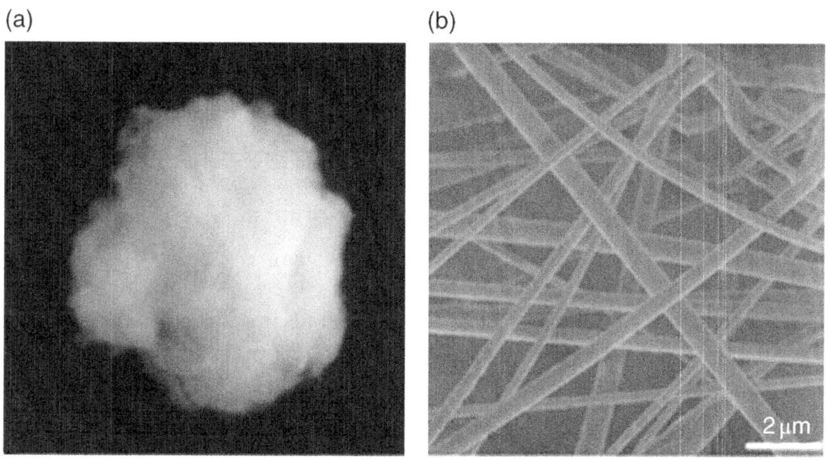

Figure 24.17 (a) Optical image of fibrous bioactive glass dressing for healing skin wounds; (b) higher magnification SEM image showing the fibers in the dressing.

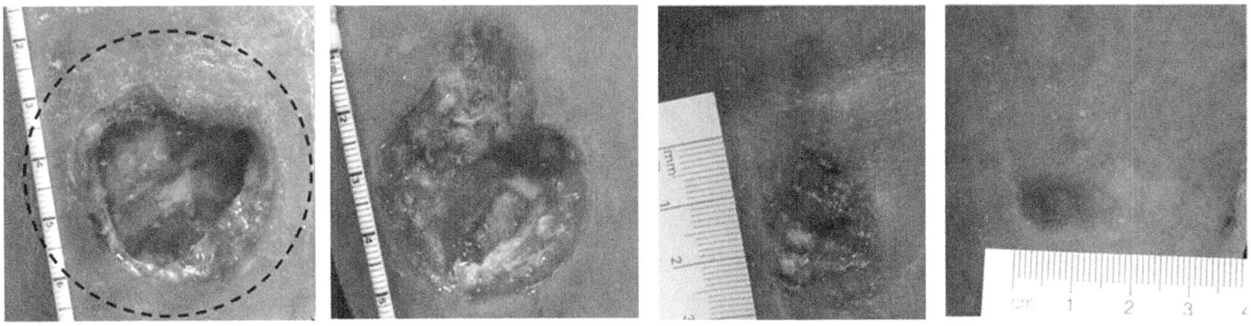

Figure 24.18 Healing of two-month old nonhealing wound (~5 cm in diameter and ~3 cm deep) in a diabetic patient using fibrous bioactive glass dressing. The bioactive glass dressing was applied to the wound, held in place using a bandage and replaced periodically (~7–14 days). Complete healing was eventually achieved without scarring. *Source:* Courtesy of Delbert E. Day, Missouri University of Science and Technology.

healing of a chronic wound in a diabetic patient using a bioactive glass dressing.

Studies designed to understand the role of copper-doped bioactive borate glass dressings in wound healing have shown that *in vitro*, the ionic dissolution product of the microfibrous dressing was not toxic to cells such as human umbilical vein endothelial cells (HUVECs) or fibroblasts (Zhao et al. 2015). Additionally, the ionic dissolution product promoted HUVEC migration, tubule formation and secretion of VEGF, and stimulated the expression of angiogenic-related genes of the fibroblasts. As shown in (Figure 24.19), when used to treat full-thickness skin wounds in rodents, the copper-doped bioactive borate glass dressing accelerated healing when compared to the untreated wounds, and showed a significantly better capacity to stimulate angiogenesis at 7 and 14 days of healing. The defects treated with the copper-doped bioactive borate glass dressing also showed improved collagen deposition, maturity and orientation when compared to the untreated defects.

Biomaterials in Combination with Biomolecules

Approaches based on the delivery of biomolecules or cells by biomaterials have been widely investigated to enhance wound healing. Although various types of biomaterials can be used, hydrogels or sponges are particularly effective in these approaches because of the ease with which biomolecules or cells can be incorporated into them or biomolecules can be covalently bonded to them (Chapter 11).

Because growth factors are known to be important signaling factors in tissue repair and regeneration (Chapter 16) and are well tolerated *in vivo* in physiological doses, their delivery to injured sites has been investigated as a means to enhance wound healing. While the delivery of factors such as platelet-derived growth factor-BB (PDGF-BB), EGF, and fibroblast growth factor-2 (FGF2) has shown promise in animal models of wound healing, the clinical effectiveness of single-factor therapy in humans has been more limited. Recombinant PDGF-BB is the only growth factor currently approved by the U.S. Food and Drug Administration for the treatment of chronic diabetic foot ulcers. However, PDGF-BB therapy is not effective in many patients, and it often requires the use of high PDGF-BB doses, thereby increasing the risk for adverse side effects.

A problem with growth factor therapy is that the factors are administered into a protease-rich, inflamed wound environment, making them susceptible to rapid breakdown and inactivation. This problem can be partly addressed by incorporating the growth factors within an appropriate biomaterial such as a hydrogel, where they can reside and have a better chance of retaining their bioactivity. However, the interactions of growth factors with ECM proteins are particularly important in order to facilitate localized and spatially regulated signaling. Consequently, it was suggested that a major reason underlying the limited success in translating growth factor therapy to clinical applications might be a lack of ECM binding when the growth factors are applied in their clinically used forms (Martino et al. 2014).

Based on the observation that a sequence of twenty-two amino acids in placenta growth factor-2 (PlGF2), designated $PlGF2_{123-144}$, binds strongly to several ECM proteins, this domain was incorporated into the structure of PDGF-BB, VEGF-A, and BMP2, to create engineered growth factor variants having a high affinity for the ECM. In one study, fibrin hydrogels incorporated with low doses of the engineered growth factors PDGF-BB/$PlGF\text{-}2_{123-144}$ or VEGF-A/$PlGF\text{-}2_{123-144}$ were found to significantly enhance the healing of small (6 mm diameter) chronic wounds in a diabetic rodent model, when compared to similar hydrogels incorporated with PDGF-BB or VEGF-A. These results indicate that this approach might have potential for use in diabetic wound healing and in other regenerative medicine applications as well.

Biomaterials in Combination with Cells

Mesenchymal stem cells (MSCs) and progenitor cells have considerable regenerative potential, providing a repair

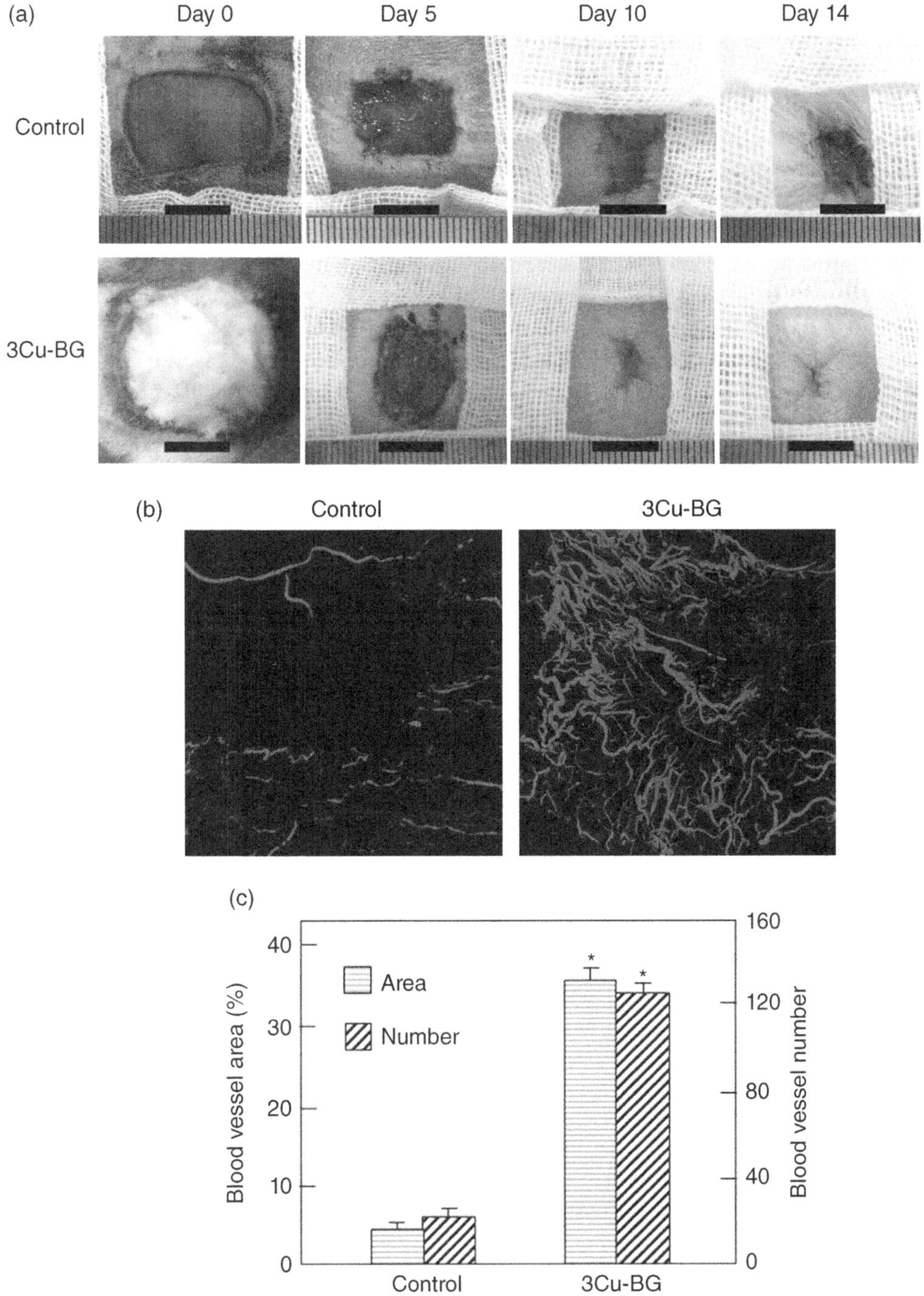

Figure 24.19 Healing of full-thickness skin wounds (2 cm in diameter) in a rodent model using a dressing consisting of fibrous bioactive borate glass doped with 3 wt% Cu (3Cu-BG). (a) Untreated wounds (control) and wounds treated with bioactive glass dressing at various times post-surgery (scale bar = 1 cm). (b) Microcomputed tomography (micro-CT) evaluation of blood vessel formation in skin wounds left untreated (control) or treated with bioactive glass dressing at 14 days post-surgery. (c) Morphometric analysis of new blood vessel area and number of blood vessels at 14 days post-surgery. (*Significant difference when compared to untreated wounds). *Source:* Based on Zhao et al. (2015).

system for healing injuries to tissues and organs. In addition to epidermal progenitor cells, stem cells have shown promising potential in stimulating skin wound healing. A variety of approaches has been developed to incorporate stem cell-based therapies into wound healing. Stems cells can be delivered to the wound site by various approaches, such as direct application or in combination with biomaterials by encapsulation within hydrogels or sponges. For skin wounds, efforts have focused on the use of epidermal progenitor cells, MSCs, adipose tissue-derived stem cells (ASCs) and induced pluripotent stem cells (iPSCs) (Chapter 25).

Although clinical translation is in its early stages, stem cell-based therapies have shown promising potential in healing cutaneous wounds, burns and diabetic ulcers. Studies have shown that MSCs and ASCs secrete factors that attenuate inflammation, stimulate angiogenesis and accelerate wound closure. An advantage of stem cell therapy is that while MSCs and ASCs can both be potentially harvested from autologous sites, ASCs are easy to obtain in a plentiful supply. Autologous ASCs incorporated within a collagen sponge have shown the ability to significantly accelerate wound healing in diabetic healing-impaired mice (Nambu et al. 2009). *In vitro*, cultured ASCs secreted a variety of growth factors and cytokines, such as interleukin 6 (IL-6), granulocyte colony-stimulating factor, granulocyte-macrophage colony-stimulating factor, basic fibroblast growth factor (b-FGF), PDGF-BB, and VEGF. When applied to 1.5 cm diameter full-thickness skin wounds in diabetic mice, ASC-containing collagen sponge significantly enhanced epithelialization of the wound ($58 \pm 13\%$ at one week and $87 \pm 5\%$ at two weeks) when compared to the collagen sponge without ASCs ($13 \pm 9\%$ at one week and $20 \pm 10\%$ at two weeks). In addition, the ASC-containing collagen sponge induced significant formation of granulation tissue and capillaries (Figure 24.20), and significantly accelerated wound healing (Figure 24.21) when compared to the collagen sponge without ASCs.

24.9.4 Nanoparticle-Based Therapies

Several nanoparticle-based therapies are under development for treating skin wounds, particularly chronic wounds. These therapies employ various types of nanoparticles, such as nanoparticles composed of metals, metal oxides, or polymers, antibiotic-loaded nanoparticles, NO-releasing nanoparticles and lipid nanoparticles. Some of these therapies have already reached the stage of human clinical trials and, in general, they have shown promising potential in enhancing the healing of skin wounds. A few examples of these nanoparticle-based therapies are discussed in the remainder of this section.

Metal nanoparticles, particularly silver (Ag) and gold (Au), have shown promising results in healing skin wounds in animal models. Ag and Au nanoparticles can be easily prepared to high purity and with well-controlled sizes and shapes using well-established methods, and they have been among the most studied nanoparticles for use in biomedical and technological applications. Ag is well known for its antimicrobial properties and, in the form of Ag compounds, it has been used clinically as an antimicrobial agent for over a century. For example, a Ag sulfadiazine-containing cream, introduced in the 1960s, has become a widely used topical agent for treating skin burns. However, it is only more recently that Ag nanoparticles have received attention for healing skin wounds.

Ag nanoparticle wound dressings commonly consist of the nanoparticles grafted onto or contained within a dressing. Upon application of the dressing to the skin wound, the Ag nanoparticles dissolve slowly and in a more controlled manner when compared to the aforementioned Ag compounds, forming Ag^+ ions and Ag atomic clusters. The dressings are normally changed regularly to maintain an appropriate Ag concentration at the wound site. Studies involving animal models have shown that Ag nanoparticles promoted skin wound healing through their antimicrobial properties and their ability to reduce inflammation. Additional studies involving full-thickness skin defects in a rodent model showed that Ag nanoparticles promoted wound healing more effectively than Ag compounds in the absence of infectious or inflammatory factors (Liu et al. 2010). Accelerated

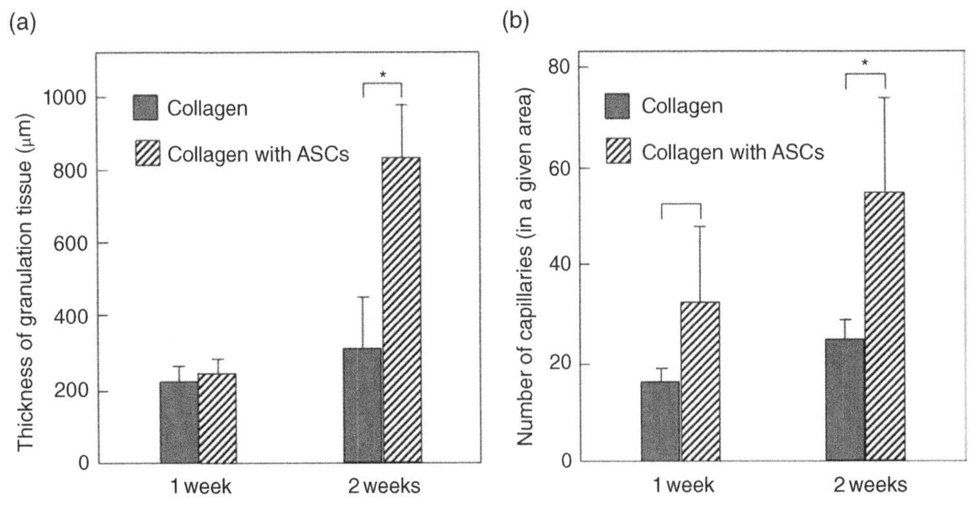

Figure 24.20 Improvement of wound healing in diabetic healing-impaired mice by autologous adipose stem cells (ASCs) incorporated within a collagen sponge. (a) Thickness of granulation tissue and (b) number of capillaries in a micrograph at 1 week and 2 weeks for wounds treated with ASC-containing collagen sponge and collagen sponge without ASCs (control). *Source:* Nambu et al. (2009) / with permission of Wolters Kluwer Health, Inc.

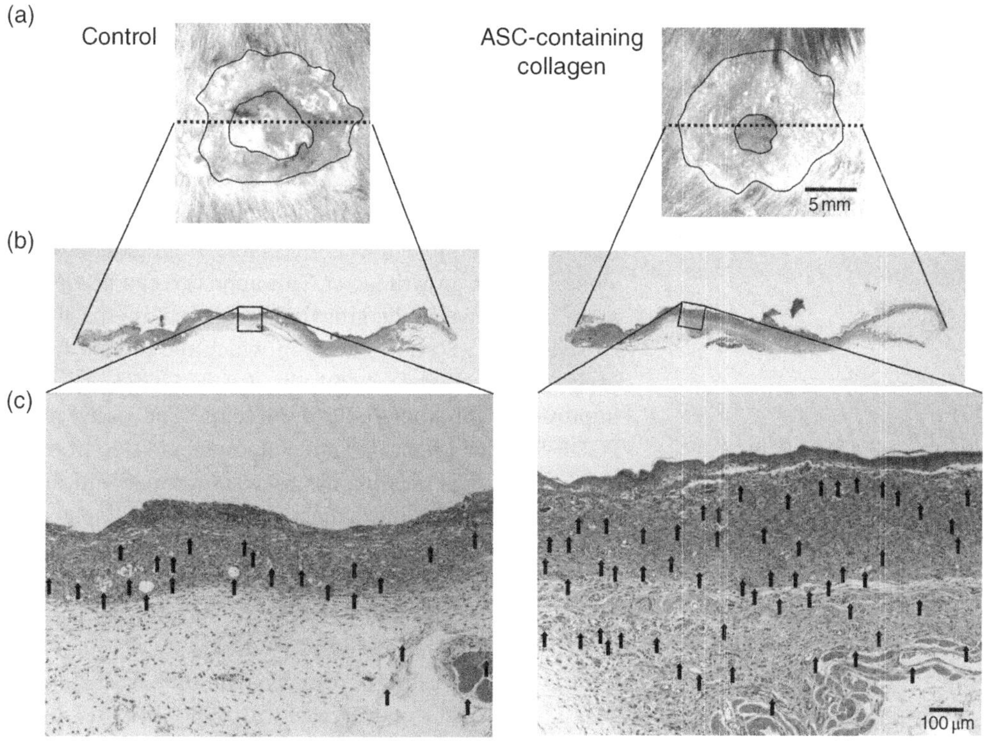

Figure 24.21 (a) Optical micrographs, and (b), (c) hematoxylin and eosin (H&E) stained sections of wounds in diabetic healing-impaired mice after treatment for two weeks with ASC-containing collagen sponge and collagen sponge without ASCs (control). Arrows indicate blood vessels containing erythrocytes. *Source:* From Nambu et al. (2009).

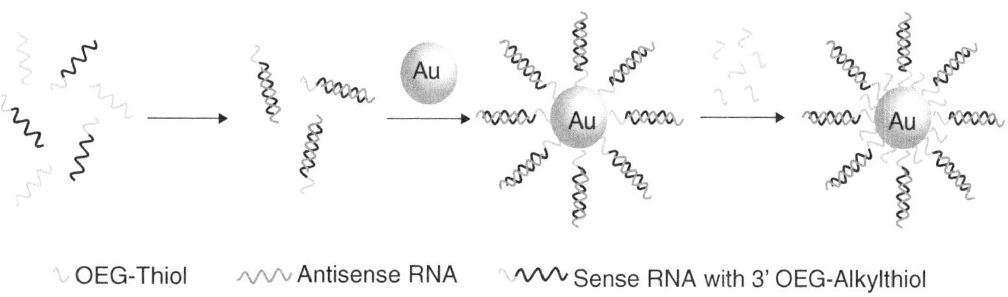

Figure 24.22 Reaction scheme for the preparation of gold nanoparticles functionalized with nucleic acids consisting of short interfering RNA (siRNA) duplexes (OEG-Thiol, oligoethylene glycol-thiol). *Source:* Giljohann et al. (2009) / with permission of American Chemical Society.

wound healing was attributed to the ability of Ag nanoparticles to enhance reepithelialization through their ability to increase proliferation and migration of keratinocytes, and to mediate the differentiation of fibroblast into myofibroblasts, which promoted wound contraction.

In addition to their ease of preparation, Au nanoparticles have another advantage in that their surface can be easily functionalized for delivery of a variety of biomolecules. Nanoparticle conjugates composed of Au nanoparticles (~13 nm in diameter) covalently functionalized with a layer of densely packed and highly oriented nucleic acids (Figure 24.22) were recently prepared and evaluated for their effectiveness in healing skin wounds in diabetic mice. The nucleic acids consisted of short interfering RNA (siRNA) duplexes with the ability to down-regulate ganglioside-monosialic acid 3 synthase (GM3S), an enzyme that is overexpressed in diabetic mice and responsible for causing insulin resistance and impeding wound healing. Topical application of these siRNA-based nanoparticle conjugates, referred to as GM3S spherical nucleic acids (SNAs), to full-thickness skin wounds (6 mm in diameter) in diabetic mice decreased local GM3S expression by over 80% and healed the wounds within 12 days (Figure 24.23). In comparison, control wounds treated with nonsense (NS) SNAs were only ~50% closed. These results show promising potential for the development of oligonucleotide therapies for treating chronic skin wounds.

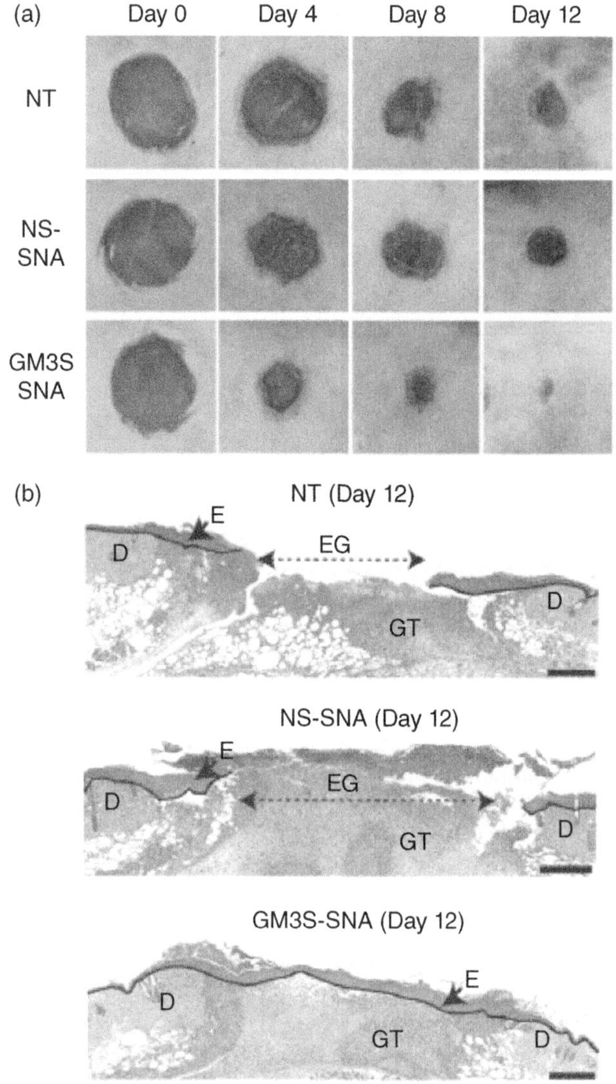

Figure 24.23 Topical application of GM3S (ganglioside-monosialic acid 3 synthase) spherical nucleic acid (SNA) prevents the delayed wound healing in a diabetic diet-induced obese mouse model. (a) Representative clinical images of nontreated (NT) wounds and wounds treated with nonsense SNA (NS-SNA) and with GM3S-SNA. (b) Representative histologic images of NT and NS-SNA- and GM3S-SNA-treated wounds at day 12. D, dermis; E, epidermis; EG, epidermal gap; GT, granulation tissue. Scale bar = 500 μm. *Source:* Courtesy of Amy S. Paller and Chad A Mirkin, Northwestern University.

Lipid-based therapies have shown promising potential in healing skin wounds in small animal models. As noted earlier, growth factors are important signaling factors in tissue repair and regeneration, but in clinical trials the delivery of PDGF-BB has not been particularly successful in healing chronic skin wounds such as diabetic ulcers. Additionally, the delivery of EGF or FGF2 has shown moderate or little benefit. It has been shown that the degradation or loss of cell surface receptors and co-receptors in the diseased diabetic state leads to growth factor resistance, which may reduce or negate the effectiveness of growth factor treatments in healing chronic skin wounds in diabetic patients. For example, a reduction in syndecan-4, a transmembrane heparan sulfate proteoglycan that is a co-receptor for several growth factors, has been observed in the skin of human patients with type 2 diabetes.

One approach to overcoming this resistance to growth factors has been to deliver co-receptors in proteoliposomes along with the growth factor. This approach can be viewed as an alternative to the growth factor-based approach discussed earlier (Section 24.9.3) in which the growth factor itself was modified chemically (with $PlGF2_{123-144}$) to improve its binding to ECM proteins. The delivery of PGDF-BB with syndecan-4 proteoliposomes (S4PL) from an alginate hydrogel dressing has been shown to significantly improve wound healing in diabetic mice when compared to wounds treated with PGDF-BB alone (Figure 24.24). These proteoliposomes were prepared by isolating recombinant syndecan-4 proteins and fusing them into the membranes of liposomes. As illustrated in Figure 24.25, the improvement in wound healing occurred via multiple mechanisms. Wounds in diabetic mice treated with FGF2 and S4PL in an alginate hydrogel have also shown significantly enhanced healing when compared to wounds treated with the hydrogel containing FGF2 alone. Because growth factors at physiologically relevant doses are well tolerated *in vivo*, this lipid-based therapy involving the co-delivery of S4PL and growth factors could potentially provide a useful approach in treating chronic skin wounds.

24.10 Concluding Remarks

In this chapter, we discussed the applications of biomaterials to the healing of a variety of soft tissues of the human body. We focused on the use of sutures used in various surgical procedures, vascular grafts and prosthetic heart valves to repair or replace degenerated tissues of the cardiovascular system, ophthalmological devices consisting of contact lenses and intraocular lenses to improve or restore vision, and biomaterials-based therapies to enhance the healing of skin wounds. Similar to the previous chapter on hard tissue repair, emphasis was placed on using the principles of materials science, engineering, and biology to modify and adapt biomaterials to make them more effective in these applications.

- Biomaterials play a considerable role in improving the quality of life for hundreds of millions of people who suffer from diseased or damaged soft tissues. While they have shown a significant level of clinical success, biomaterials-based therapies also suffer from limitations. Consequently,

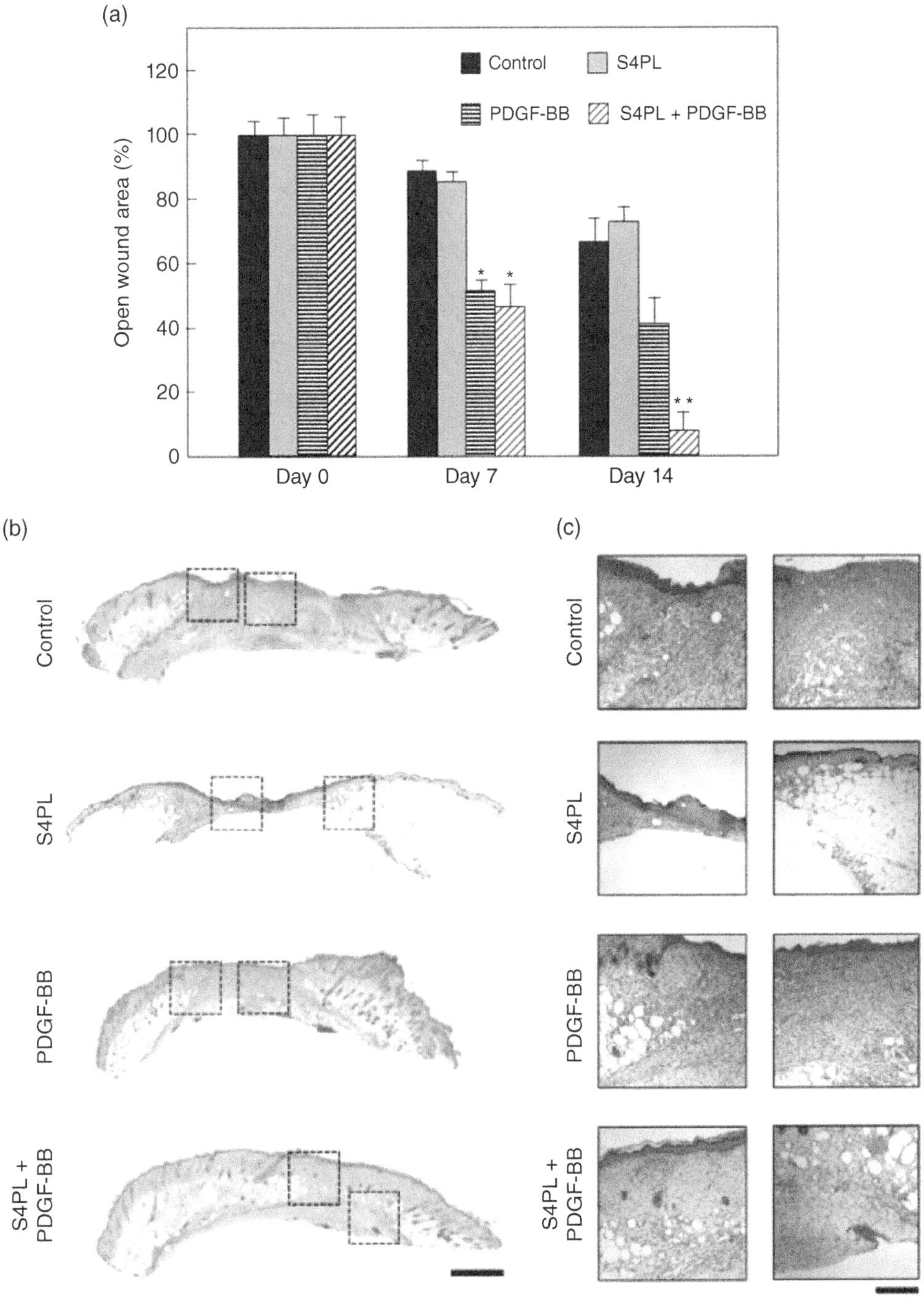

Figure 24.24 Syndecan-4 proteoliposomes enhance wound healing in diabetic mice. (a) Open wound areas calculated from macroscopic wound images. (**Statistically significant difference versus all other groups; *statistically different from the control and S4PL groups). (b) Histological sections through the wound bed at day 14 for different treatments (bar = 1 mm); (c) higher magnification images from the healing wound bed and peripheral regions of the tissue stained with hematoxylin and eosin (bar = 500 μm). Control, untreated wound; S4PL, sandecan-4 proteoliposomes. *Source:* From Das et al. (2016a) / with permission of Elsevier.

an important aspect of biomaterials research and development involves the modification of existing biomaterials or the creation of new biomaterials in order achieve an improved level of effectiveness. This aspect was discussed using a variety of examples taken from the literature.

- In the next chapter, we shall discuss the rapidly growing areas of tissue engineering and regenerative medicine in which biomaterials are used in an alternative approach involving regeneration rather than repair of native tissues and organs.

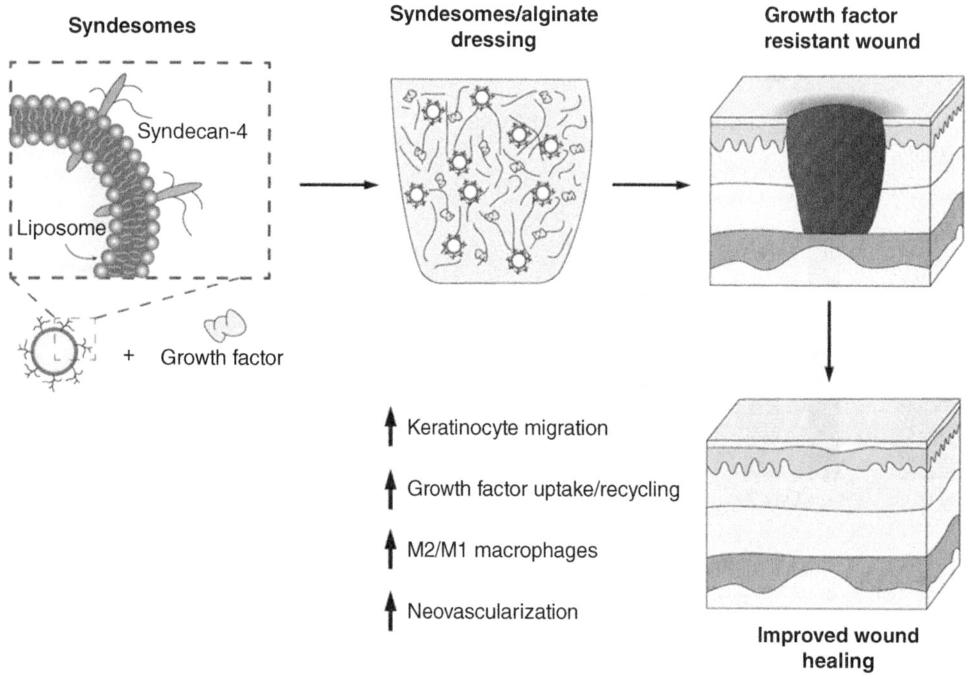

Figure 24.25 Schematic representation of the delivery of sandecan-4 proteoliposomes with a growth factor encapsulated in an alginate hydrogel and the enhancing activity of the proteoliposomes in wound healing. *Source:* Das et al. (2016b) / with permission of John Wiley & Sons.

Problems

24.1 List the desirable properties and characteristics of a suture material. Then select a suture material and briefly discuss how it meets these properties and characteristics.

24.2 Discuss the approaches used clinically to address the problems associated with a blocked or occluded artery.

24.3 Briefly describe the desirable properties and characteristics of a biomaterial for use as a synthetic vascular graft.

24.4 Compare the advantages and disadvantages of synthetic grafts and biological grafts for the repair of blocked or occluded arteries.

24.5 Describe the mechanisms by which an implanted vascular graft can become blocked or occluded and, for each mechanism, briefly explain how it can be reduced or prevented.

24.6 Make a list of issues that should be considered in the design of intravascular stents. Then select a stent composed of a specific material and briefly describe how it compares with your list of design issues.

24.7 Compare the advantages and disadvantages of drug eluting intravascular stents and degradable intravascular stents.

24.8 Discuss the main types of materials that may be used to produce contact lenses, giving their advantages and disadvantages.

24.9 Discuss the main types of materials used to produce IOLs, and briefly compare their advantages and disadvantages.

24.10 Discuss the main problems encountered with implanted IOLs and ways in which these problems could be resolved.

24..11 List the stages of cutaneous wound healing and briefly describe the main processes that occur in each stage.

24.12 Briefly describe why scarring is an important issue in skin wound healing and how it can be reduced or eliminated.

References

Bellucci, R. (2013). An introduction to intraocular lenses: material, optics, haptics, design and aberration. In: *Cataract. ESASO Course Series*, vol. 3 (ed. J.L. Güell), 38–55. Basel: Karger.

Bohl, K.S. and West, J.L. (2000). Nitric oxide-generating polymers reduce platelet adhesion and smooth muscle cells proliferation. *Biomaterials* 21: 2273–2278.

Bozukova, D., Pagnoulle, C., De Pauw-Gillet, M.-C. et al. (2007). Improved performances of intraocular lenses by poly(ethylene glycol) chemical coatings. *Biomacromolecules* 8: 2379–2387.

Das, S. and Baker, A.B. (2016). Biomaterials and nanotherapeutics for enhancing skin wound healing. *Frontiers in Bioengineering and Biotechnology* 4: 82.

Das, S., Majid, M., and Baker, A.B. (2016a). Syndecan-4 enhances PDGF-BB activity in diabetic wound healing. *Acta Biomaterialia* 42: 56–65.

Das, S., Singh, G., Majid, M. et al. (2016b). Syndesome therapeutics for enhancing diabetic wound healing. *Advanced Healthcare Materials* 5 (17): 2248–2260.

Giljohann, D.A., Seferos, D.S., Prigodich, A.E. et al. (2009). Gene regulation with polyvalent siRNA–nanoparticle conjugates. *Journal of the American Chemical Society* 131: 2072–2073.

Liu, X., Lee, P., Ho, C. et al. (2010). Silver nanoparticles mediate differential responses in keratinocytes and fibroblasts during skin wound healing. *ChemMedChem* 5: 468–475.

Martino, M.M., Briquez, P.S., Güç, E. et al. (2014). Growth factors engineered for super-affinity to the extracellular matrix enhance tissue healing. *Science* 343: 885–888.

McGlinchey, S.M., McCoy, C.P., Gorman, S.P., and Jones, D.S. (2008). Key biological issues in contact lens development. *Expert Review of Medical Devices* 5: 581–590.

Nambu, M., Kishimoto, S., Nakamura, S. et al. (2009). Accelerated wound healing in healing-impaired db/db mice by autologous adipose tissue-derived stromal cells combined with atelocollagen matrix. *Annals of Plastic Surgery* 62: 317–321.

Schoen, F.J. and Padera, R.F. Jr. (2013). Substitute heart valves. In: *Biomaterials Science: An Introduction to Materials in Medicine*, 3rde (eds. B.D. Ratner, A.S. Hofmann, F.J. Schoen and J.E. Lemons), 761–771. New York: Elsevier.

Sun, G., Shen, Y.-I., Ho, C.C. et al. (2010). Functional groups affect physical and biological properties of dextran-based hydrogels. *Journal of Biomedical Materials Research. Part A* 93 (3): 1080–1090.

Sun, G., Zhang, X., Shen, Y. et al. (2011). Dextran hydrogel scaffolds enhance angiogenic responses and promote complete skin regeneration during burn wound healing. *Proceedings of the National Academy of Sciences of the United States of America* 108: 20976–20981.

Taylor, M.S. and Shalaby, S.W. (2013). Sutures. In: *Biomaterials Science: An Introduction to Materials in Medicine*, 3rde (eds. B.D. Ratner, A.S. Hofmann, F.J. Schoen and J.E. Lemons), 1010–1024. New York: Elsevier.

Tiwari, A., Salacinski, H., Seifalian, A.M., and Hamilton, G. (2002). New prostheses for use in bypass grafts with special emphasis on polyurethanes. *Cardiovascular Surgery* 10: 191–197.

Walden, R., L'Italien, G.J., Megerman, J. et al. (1980, 1980). Matched elastic properties and successful arterial grafting. *Archives of Surgery* 115 (10): 1166–1169.

Werner, L. (2008). Biocompatibility of intraocular lens materials. *Current Opinion in Ophthalmology* 19: 41–49.

Zhao, S., Li, L., Wang, H. et al. (2015). Wound dressings composed of copper-doped borate bioactive glass microfibers stimulate angiogenesis and heal full-thickness skin defects in a rodent model. *Biomaterials* 53: 379–391.

Further Reading

Gurtner, G.C., Werner, S., Barrandon, Y., and Longaker, M.T. (2008). Wound repair and regeneration. *Nature* 453: 314–321.

Kannan, R.Y., Salacinski, H.J., Butler, P.E. et al. (2005). Current status of prosthetic bypass grafts: a review. *Journal of Biomedical Materials Research* 74: 570–581.

Kapadia, M.R., Popowich, D.A., and Kibbe, M.R. (2008). Modified prosthetic conduits. *Circulation* 117: 1873–1882.

Nicholson, P.C. and Vogt, J. (2001). Soft contact lens polymers: an evolution. *Biomaterials* 22: 3273–3283.

Randeria, P.S., Seeger, M.A., Wang, X.-Q. et al. (2015). siRNA-based spherical nucleic acids reverse impaired wound healing in diabetic mice by ganglioside GM3 synthase knockdown. *Proceedings of the National Academy of Sciences of the United States of America* 112: 5573–5578.

Schoen, F.J. and Levy, R.J. (1999). Tissue heart valves: current challenges and future research perspectives. *Journal of Biomedical Materials Research* 47: 439–465.

25

Biomaterials for Tissue Engineering and Regenerative Medicine

25.1 Introduction

The conventional way to restore diseased or damaged tissues and organs in the human body is to replace them with transplanted tissues and organs, or with prosthetic devices, many of which are composed of durable biomaterials. This approach has been discussed in the previous two chapters for the repair of hard and soft tissues, with the aid of several examples. Transplanted tissues, as we have seen, consist mainly of autografts and allografts, as, for example, in the healing of bone defects using bone autografts or allografts. Commonly used prosthetic devices include hip and knee joints, heart valves, and vascular grafts. Although this approach has worked well and has served to improve the quality of life for millions of patients, it also suffers from limitations. Transplanted tissues and organs are living structures, similar to the tissues and organs to be replaced, but they are generally limited by factors such as availability, donor site morbidity and immune rejection. In comparison, while synthetic biomaterials generally have the advantages of plentiful supply and mechanical durability, they are very different from host tissues and organs. Devices composed of synthetic biomaterials are particularly useful for providing the requisite mechanical and physical functions, but they are normally inadequate for providing the biological functions.

Over the last three decades, there has been a radical shift from the conventional approach of repairing or replacing diseased or damaged human tissues and organs to a new approach involving the regeneration of the patient's own functional tissues and organs. This new approach, now generally known as regenerative medicine, consists of three separate approaches, referred to as tissue engineering, cell therapy, and gene therapy (Figure 25.1). The tissue engineering approach involves harvesting cells, incorporating them into a suitable biomaterial (called a scaffold), stimulating them *ex vivo* for an appropriate time, and implanting the resulting tissue construct into the patient. The principle of this approach is to create living tissues and organs that could potentially replace diseased or damaged tissues and organs in the human body. An alternative tissue engineering approach is to incorporate cells or other biological entities such as growth factors into a suitable biomaterial and implant the construct directly into the body to regenerate the tissue or organ *in situ*.

Cell therapy involves the use of cells, typically stem cells, which are injected or placed at a diseased or damaged tissue site, with the expectation that they would lead to the regeneration of the required tissue. Cells can be delivered without the aid of a biomaterial or by means of a biomaterial. Gene therapy is an experimental technique that uses genes to treat or prevent disease instead of using drugs or surgery. It involves the insertion of specific genes into living cells of a patient to replace defective genes that caused the disease.

Overall, biomaterials form an important component of the tissue engineering approach, whereas they may be involved only minimally in cell therapy and gene therapy. Consequently, the main emphasis of this chapter is on tissue engineering, while cell therapy and gene therapy are discussed only briefly.

Tissue engineering emerged in the early 1990s as a concept for regenerating biological tissues through seeding a patient's own tissue-forming cells into biocompatible materials (Langer and Vacanti 1993). Early efforts provided the proof of principle for the regeneration of single tissues such as skin, bone, and cartilage. Subsequently, the approach has also been applied to the regeneration of more complex structures, including composite tissues composed of two different tissues such as the articular joint and complex organs such as the kidney and liver. These advances would not have been possible without the innovative design and production of biomaterials and the application of principles derived from an improved understanding of biological processes, in areas such as stem cell biology, extracellular matrix (ECM) biology, and the interaction of biomaterials with cells and tissues.

Engineered regeneration of human tissues and organs, other than skin, is not yet used routinely in clinical practice.

Materials for Biomedical Engineering: Fundamentals and Applications, First Edition. Mohamed N. Rahaman and Roger F. Brown.

Companion website: www.wiley.com/go/rahaman/biomedicalengineering

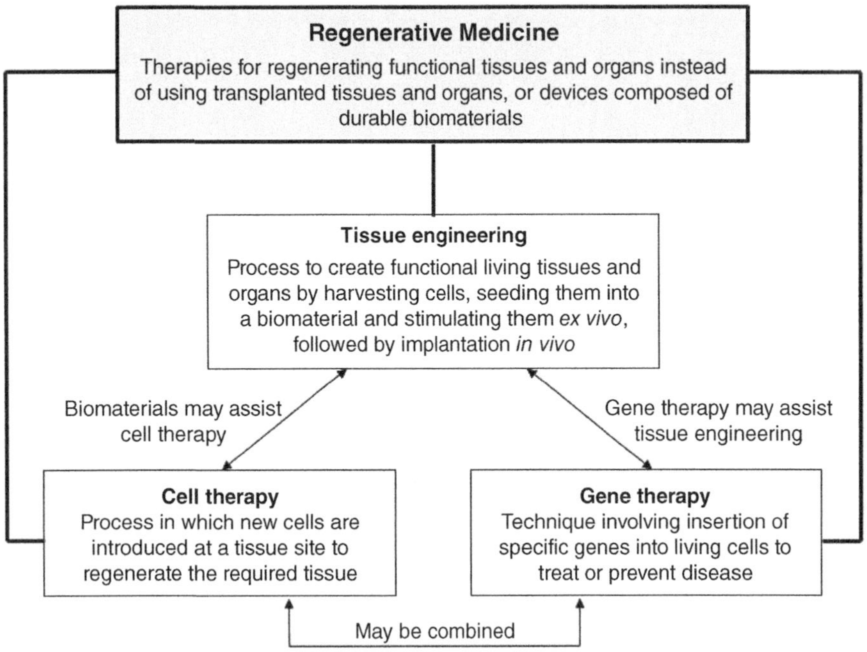

Figure 25.1 Three main approaches to regenerative medicine: tissue engineering, cell therapy and gene therapy. *Source:* Adapted from Williams (2014).

This is related, for the most part, to the complexities involved in regenerating functional tissues and organs and to the relatively short time since the introduction of the tissue engineering concept when compared to the decades-long development of implantable prosthetic devices. Nevertheless, considerable advances have been made in engineering tissues and organs and the approach continues to be a subject of significant research and development.

Overall, the following topics related to engineering functional tissues and organs will be discussed in this chapter:

- Principles of tissue engineering and regenerative medicine.
- Design and production of biomaterials for use in tissue engineering.
- Tissue engineering techniques for the creation of functional tissues and organs.

25.2 Principles of Tissue Engineering and Regenerative Medicine

The tissue engineering approach relies on three core components (Figure 25.2):

- *Biomaterials:* Often having a designed shape and internal architecture (or microstructure) and referred to as scaffolds, biomaterials used in tissue engineering serve as a supporting structure to deliver cells and signaling factors (such as growth factors), and as a template to guide the form and structure of the new tissue.
- *Cells:* Cells are normally required to express or form new tissue.
- *Signaling factors:* Commonly composed of biomolecules such as growth factors, signaling factors provide appropriate chemical signals to stimulate cell response and formation of new tissue.

Various approaches can be used in tissue engineering, depending on whether the regenerated tissue is formed *ex vivo* or *in vivo*. The *ex vivo* approach involves seeding cells, with or without growth factors, into a scaffold, incubating the system in a bioreactor for a sufficiently long time to allow the ECM to build up, after which the tissue construct is implanted into the desired site. In general, this approach consists of the following basic steps:

- *Step 1:* Suitable cells, such as stem cells or differentiated (mature) cells are obtained and manipulated under appropriate culture conditions *in vitro* to optimize their number or phenotype.
- *Step 2:* A biomaterial with a suitable composition is formed into a scaffold having the desired shape, internal architecture, and surface characteristics. Then the requisite number of cultured cells from Step 1 is seeded into the scaffold, with or without the addition of growth factors.
- *Step 3:* The cell-seeded scaffold is placed in a suitable culture medium, into which appropriate growth factors and

other signaling molecules have been added, and the system is incubated in a bioreactor for an appropriate time.

- *Step 4:* The resulting tissue-engineered construct is implanted into the diseased or damaged tissue site.

The *in vivo* (or *in situ*) approach relies on the body itself serving as a bioreactor to regenerate the desired tissue at the implantation site. One *in situ* approach involves seeding cells into a scaffold and, following a brief period of incubation *in vitro*, implanting the cell-seeded scaffold directly into the site of the tissue to be regenerated (Figure 25.3a). The expectation is that the cells would start production of the desired ECM, while the scaffold degrades away and is replaced by new tissue (Figure 25.3b). Another

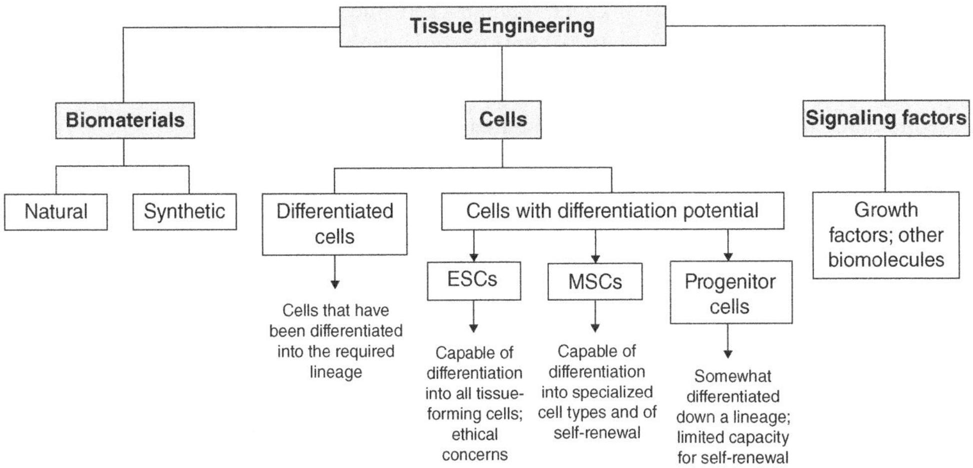

Figure 25.2 Main components of the tissue engineering approach: biomaterials, cells, and signaling factors (ESCs, embryonic stem cells; MSCs, mesenchymal stem cells).

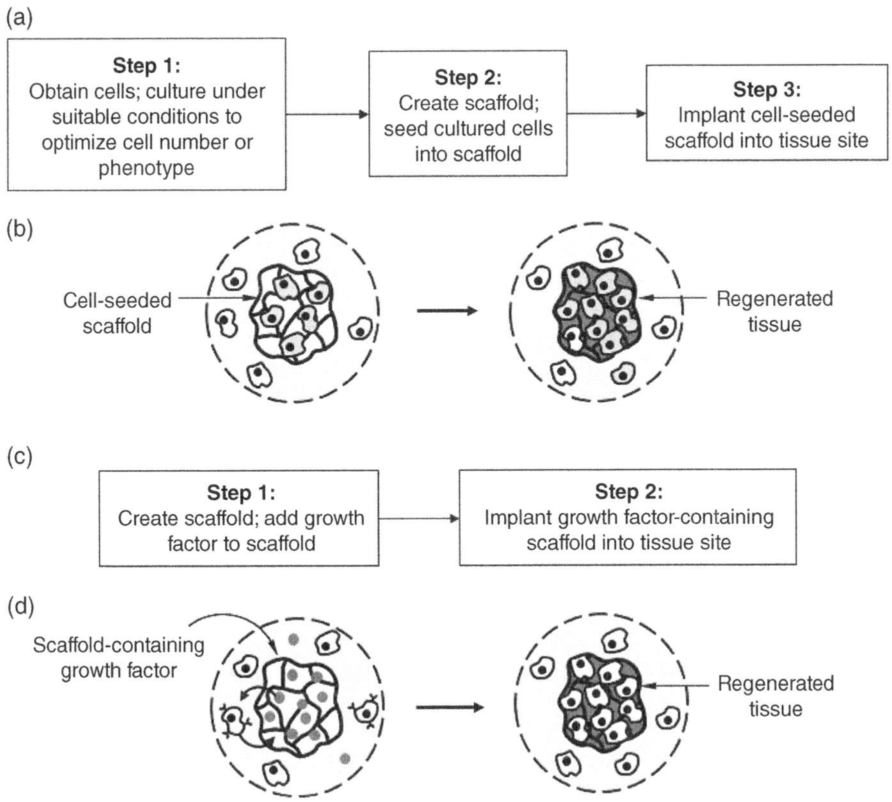

Figure 25.3 Illustration of *in vivo* (or *in situ*) tissue engineering approaches. (a) Main steps in cell-based approach; (b) illustration of tissue regeneration within a defect site by cell-based approach; (c) main steps in growth factor-based approach; (d) illustration of tissue regeneration within a defect site by growth factor-based approach.

in situ approach involves the use of growth factors instead of cells. In this approach, signaling factors (normally growth factors) are incorporated into a cell-free scaffold, either by adsorption, infusion, or covalent bonding to the biomaterial surface, and the construct is implanted into the desired tissue site (Figure 25.3c). The expectation is that the signaling factors would attract endogenous cells from the surrounding tissue to the scaffold to regenerate the desired tissue while the scaffold degrades away and is replaced by new tissue (Figure 25.3d).

Both the *ex vivo* approach and the *in situ* approach includes an expectation that the biomaterial (scaffold) will degrade at an appropriate rate as the ECM is being formed, and that the new tissue will integrate with the surrounding native tissue. Because the *ex vivo* approach allows the ECM to build up prior to the implantation step, the tissue construct could provide mechanical support immediately upon implantation. Additionally, if the tissue construct is mechanically and biologically functional prior to implantation, it has a better chance of remaining so, particularly in tissue sites subjected to physiological loads. In comparison, the *in situ* approach takes advantage of the body's natural regenerative potential, which cannot be adequately reproduced in a bioreactor *in vitro*.

25.2.1 Cells for Tissue Engineering

Cells used in tissue engineering can be divided into two types: differentiated cells such as osteoblasts and chondrocytes, and undifferentiated cells such as stem cells (natural or induced stem cells) and progenitor cells. A general discussion of basic cellular and tissue characteristics was presented in Chapter 16. Here, that discussion is expanded to cover cells and cell lineages that have been particularly useful in tissue engineering.

Differentiated Cells

Table 25.1 provides a list of the main types of differentiated (mature) cells used in tissue engineering of various tissues and organs. Among the cells listed are those within cell lineages that originate from the ectoderm, mesoderm, and endoderm, the three embryonic layers. Ectodermal cells include corneal epithelial cells, mesodermal cells include chondrocytes, fibroblasts, osteoblasts, skeletal myoblasts, tracheal chondrocytes and valvular interstitial cells, and endodermal cells include tracheal epithelial cells and urothelial cells.

The corneal epithelium is composed of four to six layers of non-keratinized squamous epithelial cells and the absence

Table 25.1 Main types of cells used in tissue engineering to regenerate various tissues and organs.

	Cells	
Tissue or organ	**Differentiated cells**	**Stem cells (SCs)/progenitor cells (PCs)**
Bone	Osteoblasts	Bone marrow SCs, peripheral blood SCs
Cartilage	Chondrocytes	Bone marrow SCs
Skeletal muscle	Skeletal myoblasts	—
Skin		
Epidermis	Keratinocytes	Epidermal SCs
Dermis	Fibroblasts	Adipose SCs
Tendon	Tenocytes, dermal fibroblasts	Bone marrow SCs, tendon SCs
Blood vessel endothelium	Vascular endothelial cells	Endothelial PCs
Myocardium	Skeletal myoblasts	Cardiac PCs, endothelial PCs
Heart valve	Valvular interstitial cells	Valvular interstitials PCs
Peripheral nerve	—	Peripheral blood SCs, neural SCs/PCs
Brain	—	Neural SCs/PCs
Liver	Hepatocytes	Amniotic fluid SCs, hepatocyte-like induced pluripotent SCs
Urinary bladder		
Urothelium	Urothelial cells	Urothelial PCs
Muscularis propria	Smooth muscle cells	Bone marrow SCs
Trachea	Chondrocytes, epithelial cells	Adipose SCs
Cornea	Corneal epithelial cells	Corneal epithelial PCs

of keratin renders these cells transparent, allowing unimpeded transmission of light through the cornea. In addition, the corneal epithelium is highly regenerative, a feature of importance due to the exposed nature of the cornea. This high regenerative capacity also enhances the applicability of the corneal epithelium for tissue engineering.

Chondrocytes, osteoblasts, and fibroblasts are three connective tissue cell types that originate from mesenchyme, a primitive mesodermal tissue that has a loose, fluid-like nature. Osteoblasts and fibroblasts are both amenable to *in vitro* cultivation, a feature that makes them attractive for tissue engineering applications. Fibroblasts and osteoblasts have moderate and rapid *in vitro* proliferation rates, respectively, which reflect the moderate rate of fibroblast proliferation in moderately vascularized subdermis and the higher rate of osteoblast proliferation in well-vascularized bone. Chondrocytes are used in tissue engineering applications but less commonly than other connective tissue cells. The slow rate of growth that is characteristic of cells derived from cartilage, a nonvascularized tissue, presents some challenges in tissue engineering. Another group of mesoderm-derived connective cells are valvular interstitial cells (VICs), which comprise most of the mass of aortic heart valves. Present in a quiescent state in the healthy aortic valve, VICs are capable of re-entry into a state of active proliferation by *in vitro* cultivation. VICs have been used for tissue engineering of aortic heart valves, by seeding them on nanofibrous scaffolds mimicking the anatomical geometry of a native heart valve, or by encapsulating them within hydrogels and forming the cell-seeded hydrogel into the anatomical geometry of a native valve by three-dimensional bioprinting.

Skeletal myoblasts are used for tissue engineering of two different muscle types: cardiac muscle and skeletal muscle. Cardiac myoblasts are not suitable for cardiac tissue engineering due to their inherent low capacity for reentry into the cell division cycle. As an alternative approach, skeletal muscle-derived myoblasts, which have the capacity for cell division, have been successfully used for tissue engineering of ventricular myocardium for the repair or replacement of infarcted ventricular myocardium. Skeletal muscle-derived myoblasts have also been used for tissue engineering of skeletal myotubes or myofibers, the multinucleated cells characteristic of mature skeletal muscle. The tissue engineered skeletal myotubes might potentially be utilized in repairing skeletal muscle damaged by muscular dystrophy or related syndromes.

Tracheal epithelial cells and urothelial cells, both of which are derived from the endoderm, are additional examples of cells with high regenerative capacity. The regenerative ability of these cells is a manifestation of their location in organs frequently exposed to hazardous chemicals and particulate materials. The epithelial lining of the trachea often interacts with hazardous materials inhaled into the airways. Urothelial cells, previously known as transitional epithelial cells, form the lining of the urinary bladder, the ureters, and the urethra. The epithelial lining of the bladder is commonly exposed for extended periods to urine and toxic substances frequently present in urine. Because cellular damage caused by these toxic substances often leads to the loss of some epithelial cells, there is a subsequent need for cell replication to replace lost cells. The high regenerative capacity of tracheal epithelial cells and urothelial cells makes these cells attractive candidates for tissue engineering applications.

Cells with Differentiation Potential

The two principal categories of cells with differentiation potential used in tissue engineering include stem cells and progenitor cells. A stem cell is, by definition, an undifferentiated cell capable of unlimited proliferation in which at least one of the two daughter cells of each cell division will be another stem cell of the same type to assure self-renewal. The other daughter cell of each cell division cycle can undergo differentiation to a specific cell type with specialized structure and function. A progenitor cell is also an undifferentiated cell capable of differentiation to a specific cell type but, unlike a stem cell, it has a limited potential for cell division and is slightly differentiated as a progenitor committed to one or more specific cell lineages. The common lymphoid progenitor cell, for example, is committed to the lineages of lymphocytic cells. Some progenitor cells are committed to a single cell lineage, an example of which is the osteoprogenitor cell, which is committed to the bone lineage.

There are two types of natural stem cells: embryonic stem cells and adult stem cells. In addition to these natural stem cells, there are artificially manipulated cells with embryonic stem cell-like traits, examples of which are induced pluripotent stem cells (iPSCs). To understand the two main types of natural stem cells, a basic understanding of various stages of the human embryo, the characteristics of embryonic cell layers, and how these characteristics determine differentiation potential, is useful.

The first stage of the embryo is the zygote, the single cell product of the fusion of a sperm cell with an ovum, which progresses through a series of rapid cell divisions to form a solid 16-cell cluster known as the morula (Figure 25.4), the second stage of the embryo. Because they possess the ability to form all tissues of the body plus the placental membranes that surround, protect, and nourish the fetus, the zygote and the morula are described as totipotent. The next stage of the embryo is marked by the blastocyst, a structure characterized by a fluid-filled cavity and a cluster of cells

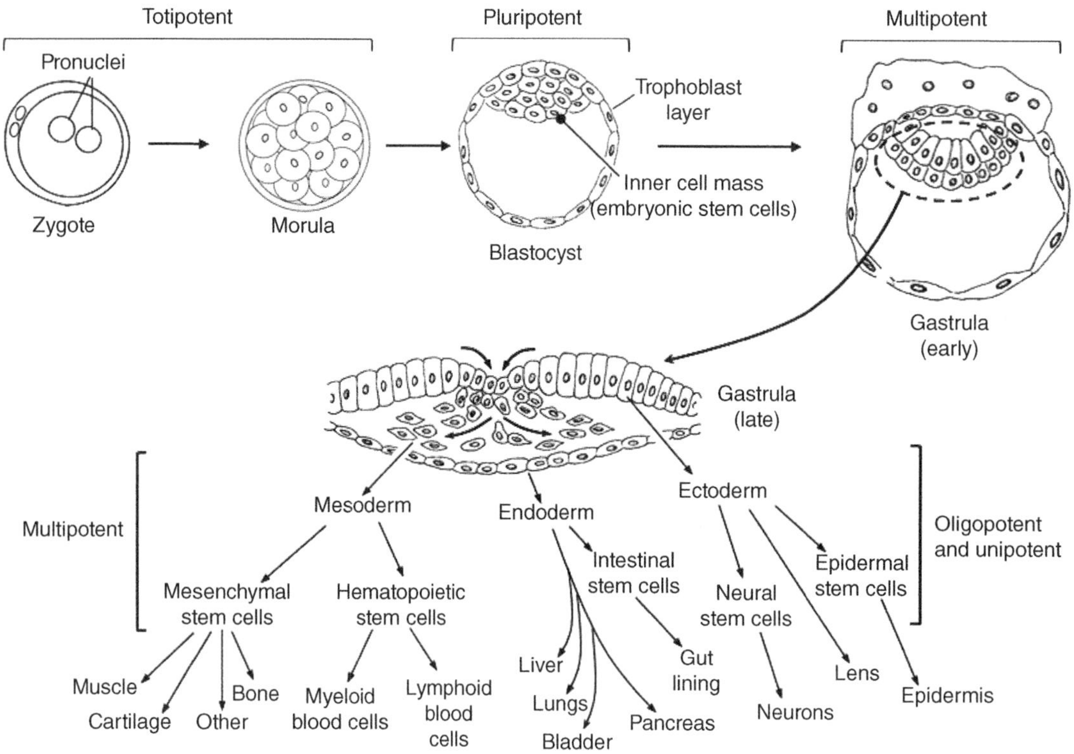

Figure 25.4 Illustration of the stages of human embryo development up to the gastrula stage with formation of three body layers: ectoderm, mesoderm, and endoderm. Included is an indication of differentiation potential ranging from totipotent to pluripotent to multipotent. Also indicated are the various types of adult stem cells and their cell lineages.

known as the inner cell mass enveloped by an outer layer, the trophoblast, from which the placenta develops. The cells of the inner cell mass give rise to the embryo proper and are described as pluripotent because they can form all tissues of the body.

The most significant event of the next stage, the gastrula stage, is the initiation of tissue differentiation, which begins as two separate cell layers form. Because the cells of each of the two layers are capable of forming many but not all tissues of the body, they are referred to as multipotent. With the continuation of development, the two layers of the gastrula partially separate and a new layer of cells forms between the two. There are now three separate layers, which include an upper ectoderm layer, a middle mesoderm layer, and an innermost layer, the endoderm. From these three layers come all of the various tissues and organs of the body (Figure 16.19).

As development continues, the ectoderm, mesoderm, and endoderm give rise to the various types of adult stem cells such as hematopoietic stem cells and mesenchymal stem cells (MSCs). Some adult stem cell types, such as MSCs, are multipotent whereas others, such as hematopoietic stem cells, are referred to as oligopotent because they are limited to forming only a few specific cell types (Figure 25.4). Unipotent stem cells, such as epidermal stem cells, form only a single specific cell type. In addition, there are the progenitor cells committed to specific cell lineages. The combined tightly controlled actions of stem cells and progenitor cells ultimately result in the development of the various tissues and organs of the body (Figure 16.19).

Induced Pluripotent Cells

Ethical concerns related to the use of embryonic stem cells have motivated a search for alternative sources of human cells that show embryonic stem cell-like pluripotent behavior. Successful manipulation of mature mammalian cells to regain the ability to differentiate any cell type of the body was first reported using mouse cells, which formed the initial development of induced pluripotent stem cells (iPSCs) (Takahashi and Yamanaka 2006). One year later, mature human cells were successfully reprogrammed to iPSCs.

The induction of mouse and human pluripotent stem cells from fully differentiated cells normally takes ~3–4 weeks, and the process typically has a low efficiency (less than ~0.1%). IPSCs have been used to produce fully differentiated tissue cells in animals. Particularly notable findings have included iPSC differentiation into functional neurons and into cardiomyocytes. However, despite these promising results, concerns have been expressed about the safety of iPSCs, particularly their propensity to cause the formation of teratomas (non-cancerous tumors). Although non-cancerous, teratomas are of concern because they can impair normal organ function and can potentially become malignant.

Overall, there is considerable interest in finding a safe and workable strategy for reprogramming mature human cells to an embryonic stem cell-like state of pluripotency to permit effective regeneration of damaged or defective tissues in human patients. One approach, involving a procedure referred to as *in situ* cell lineage reprogramming, has been tested in mice as a potential strategy for cardiac regeneration following experimentally induced myocardial infarction (Mohamed et al. 2017). The procedure, which involves the introduction of a cocktail of eight small molecules, has been shown to be effective for *in situ* reprogramming of cardiac fibroblasts to functional cardiomyocyte-like cells.

25.2.2 Biomolecules and Nutrients for *in vitro* Cell Culture

Animal cell culture originated as an established laboratory procedure around 1950, when Harry Eagle described a culture medium of known composition for *in vitro* cultivation and growth of mammalian cells. Eagle studied the role of individual amino acids and vitamins to support cell growth, which led to development of Eagle's minimal essential medium (MEM), a medium of defined composition that is still being used widely. Since that time, a large number of different defined media has been developed for cell culture and tissue culture. In some cases, specific types of differentiation medium have been developed to promote differentiation of stem cells or progenitor cells to specific cell types. Two examples of this are defined culture medium for osteogenic differentiation and defined culture medium for chondrogenic differentiation.

Examples of culture media for osteogenic and chondrogenic differentiation of human bone marrow-derived MSCs are summarized in Table 25.2. The base medium for both of these differentiation media is Dulbecco's modified Eagle's minimal essential medium DMEM. Compared to MEM, DMEM is a more complex medium with increased amino acid and vitamin concentrations. The supplementation with fetal calf serum (FCS) at a concentration of 10% is a standard practice found to be essential for successful cultivation of most mammalian cell lines, including MSCs. Three additional supplements, dexamethasone (Dex), L-ascorbic acid-2-phosphate (AsAP) and β-glycerophosphate (βGP) have been commonly used in many experimental protocols for induction of osteogenic differentiation. Dex is a synthetic glucocorticoid (a type of adrenal steroid hormone) that enhances activity of the transcription factor Runx2, which is essential for osteoblast differentiation. AsAP enhances osteogenic differentiation by increasing the synthesis of collagen type I which is essential for formation of osteoid, the organic portion of bone ECM. βGP helps to promote osteogenic differentiation by functioning as a phosphate source for mineralization. Some osteogenic differentiation media, such as the medium listed in Table 25.2, include bone morphogenetic protein-2 (BMP2), which has been found to further enhance osteogenic differentiation of MSCs, thereby enhancing bone tissue engineering.

Because the developmental process of chondrogenesis is closely related to osteogenesis, it is perhaps not surprising to see that two of the supplements in osteogenic differentiation medium, Dex and AsAP, are also supplements in chondrogenic differentiation medium (Table 25.2). Transforming growth factor-β1 (TGFβ1), the supplement unique to chondrogenic differentiation medium, is a key factor in the promotion of chondrogenic differentiation. TGFβ1 promotes expression of genes for collagen type II and cartilage oligomeric matrix protein, two proteins essential for the cartilage matrix. TGFβ1and Dex synergistically promote expression of genes coding for three additional cartilage matrix proteins: aggrecan, dermatopontin and collagen type XI. AsAP also enhances chondrogenesis, although it does so at a concentration that is ~50 times smaller than that in osteogenic differentiation medium.

25.2.3 Growth Factors for Tissue Engineering

Basic features of growth factors and cytokines, the tissue sources from which they originate, the tissues they affect, and their principal effects have been discussed in Chapter 16.

Table 25.2 Constituents of culture media used for *ex vivo* tissue engineering of human mesenchymal stems cells.

Osteogenic differentiation[a]	Chondrogenic differentiation[b]
Base medium: Dulbecco's modified Eagle's minimal essential medium (DMEM)	Base medium: Dulbecco's modified Eagle's minimal essential medium (DMEM)
Supplements: Fetal calf serum (FCS) (10%)	Supplements: Fetal calf serum (FCS) (10%)
Dexamethasone (Dex) (~100 nM)	Dexamethasone (Dex) (~100 nM)
L-Ascorbic acid-2-phosphate (AsAP) (~50 μM)	L-Ascorbic acid-2-phosphate (AsAP) (~1 μM)
β-Glycerophosphate (βGP) (~10 mM)	Transforming growth factor-β1 (TGFβ1) (~1 μg/ml)
Bone morphogenetic protein-2 (BMP2) (~20 nM)	

[a] Based on Jaiswal et al. (1997).
[b] Based on Solchaga et al. (2011).

Table 25.3 Prominent growth factors used in tissue engineering for the regeneration of various tissues and organs.

Growth factor	Tissues and organs
Angiopoietin 1 and 2 (Ang1; Ang2)	Blood vessel, heart
Bone morphogenetic protein-2 (BMP2)	Bone
Bone morphogenetic protein-7 (BMP7)	Cartilage, kidney
Epidermal growth factor (EGF)	Skin, nerve
Fibroblast growth factor-2 (FGF2)	Blood vessel, bone, skin, nerve, muscle
Hepatocyte growth factor (HGF)	Liver, muscle
Insulin-like growth factor-1 (IGF1)	Lung, kidney, nerve, skin
Nerve growth factor (NGF)	Nerve, brain
Platelet-derived growth factor (PDGF)	Blood vessel, muscle, skin
Transforming growth factor-α (TGFα)	Skin, brain
Transforming growth factor-β (TGFβ)	Cartilage, bone
Vascular endothelial growth factor (VEGF)	Blood vessel

In this chapter, growth factors are further discussed in relation to their importance in tissue engineering. Table 25.3 summarizes the various growth factors now most commonly used in tissue engineering for the regeneration of various tissues and organs. All of these growth factors are peptides, as noted in Chapter 16. Like hormones, growth factors are chemical effectors that interact with specific target tissues possessing surface receptors for the effector chemical. However, unlike hormones, growth factors are produced by many different tissues rather than by just glandular tissues. In addition, the released growth factors typically diffuse only a short distance to affect nearby target tissue and tend to have a short biological half-life. This short biological half-life of growth factors has important implications for the manner in which they are used in tissue engineering. Another item of importance is that a combination of two or more growth factors may be required, sometimes with sequential delivery, to promote specific tissue formation (Lee et al. 2011).

An example of growth factor combinations in tissue engineering is the use of both vascular endothelial growth factor (VEGF) and basic fibroblast growth factor (bFGF) for tissue engineering of blood vessels. VEGF promotes proliferation of endothelial cells (ECs) that form the inner endothelial lining of blood vessels, while bFGF has an angiogenic effect as well as a positive effect on the formation of vascular smooth muscle. The sequential delivery of VEGF and platelet-derived growth factor (PDGF) has also been used for tissue engineering of blood vessels. The delivery of PDGF subsequent to that of VEGF has been reported to promote recruitment of smooth muscle cells (SMCs), thereby enhancing maturation and functionality of the nascent blood vessels. It has been observed that hepatocyte growth factor (HGF) also contributes to neovascularization, a feature that makes the use of HGF potentially attractive for blood vessel tissue engineering.

BMP2 and BMP7 have been utilized for tissue engineering of bone-like tissue by adipose-derived stem cells (ASCs) seeded on β-tricalcium phosphate (β-TCP) scaffolds. A mixture of BMP2 and BMP7 has been shown to have greater osteogenic effectiveness than each of the two growth factors alone. In a similar manner, a mixture of BMP2 and TGFβ3 has been reported to significantly accelerate bone tissue formation by rat MSCs seeded on alginate hydrogels when compared to each of the two growth factors alone. BMPs, which are members of the TGFβ superfamily, also influence the differentiation of tissues other than bone. As an example, BMP7 is used for tissue engineering of nucleus pulposus for intervertebral disc regeneration.

TGFα has a potent stimulatory effect on epidermal tissue, a property that makes it a useful supplement for tissue engineering of cornea-like tissue. In comparison, TGFβ is a potent stimulator of chondroblast proliferation and cartilage matrix formation and, thus, it is commonly used in cartilage tissue engineering. Nerve growth factor (NGF), the first growth factor discovered, has been added to various types of degradable scaffolds for *in situ* regeneration of damaged peripheral nerve tissue. Insulin-like growth factor-1 (IGF1) also has a neurogenic effect, a property that makes it attractive for nerve tissue engineering. In addition, IGF1 stimulates the development of alveolar epithelial cells and, therefore, it is potentially useful for alveolar tissue engineering. The EGF family is the most highly characterized family of growth factors, and EGF finds considerable use in tissue engineering of skin.

25.2.4 Cell Therapy

Cell therapy is a curative procedure in which viable cells of a specific type are injected or implanted in a patient for the purpose of repairing or regenerating diseased or damaged tissue. An area of significant interest is the potential development of a successful cell therapy procedure for regenerating insulin-secreting pancreatic beta cells to treat patients with type I diabetes. Cell therapy is also being investigated as an approach for restoring damaged portions of the left ventricle wall in myocardial infarction patients. An example of this approach is a study involving the use of cardiomyocytes derived from human embryonic stem cells for the repair of infarct-damaged myocardium in athymic rats (Laflamme et al. 2007). An objective of this study was to find an approach for improving the survivability of cardiomyocytes after placement in ischemic infarct-damaged myocardium. To enhance cardiomyocyte survival, a cocktail consisting of the following six components was used: Matrigel, a cell-permeant peptide, cyclosporine A, the drug pinacidil, IGF1 and a caspase inhibitor. The main effect of Matrigel is to prevent anoikis-type apoptosis, a type of programmed cell death that occurs in anchorage-dependent cells when they detach from the surrounding ECM. Cyclosporine A and the cell-permeant peptide both act to attenuate mitochondrial death pathways. Pinacidil is a drug that opens K^+ channels, resulting in peripheral vasodilatation. IGF1 activates a signal transduction pathway that protects cardiomyocyte from apoptosis, while the caspase inhibitor blocks caspase-induced apoptosis.

Myocardial infarction in the rodents was induced by temporary ligation of the left coronary artery to produce a one-hour duration of ischemia followed by reperfusion. Suspensions of cardiomyocytes with or without the cocktail were injected into the damaged left ventricular wall at four days postinfarction, a period known to allow some angiogenesis and initial microvascular growth in the ischemic tissue. The ability of the injected cardiomyocytes to survive and effect repair of the damaged left ventricular wall was evaluated at four weeks postinfarction. Cardiac sites that received the injection of the cocktail-treated cardiomyocytes showed significantly enhanced revascularization and a threefold increase in ventricular wall thickness when compared to sites injected with cardiomyocytes that were not treated with the cocktail. Overall, these results suggest that the use of human embryonic stem cell-derived cardiomyocytes treated with a cocktail of cell-protective agents could potentially be effective as cell therapy for the repair of infarcted left ventricular myocardium in human patients.

25.2.5 Gene Therapy

Gene therapy is an experimental technique for introducing genetic material into cells to fight a disease or to correct a pathological condition. The procedure typically involves replacing a defective gene with a normal gene or inactivating an incorrectly functioning gene. The most commonly used approach for introducing genetic material into cells involves the use of an adenovirus vector. Adenoviruses, double-stranded DNA viruses commonly associated with mild respiratory illnesses, are often used as vehicles for vaccine delivery.

There is also interest in using an adenovirus vector for the delivery of genes for *in situ* reprogramming of pancreatic exocrine cells to insulin-producing beta cells. For example, one study involved the preparation of adenovirus constructs (designated pAD-M3) containing mouse genes coding for three separate transcription factors known to be required for pancreatic beta cell development (Zhou et al. 2008). The mice were treated with the drug streptozotocin to induce a diabetic state, followed one week later by injection of pAD-M3, and their response to the treatment was monitored over an eight-week period. Partial restoration of insulin production and glucose homeostasis was found in animals that received pAD-M3 administration. When compared to the control (normal nondiabetic animals), the serum glucose level in the animals treated with pAD-M3 was ~3 times higher whereas it was ~7 times higher for the animals without pAD-M3 treatment. Examination of recovered tissue showed induced beta cells that were morphologically indistinguishable from native beta cells. On the other hand, the total number of induced beta cells was small when compared to the number in normal nondiabetic animals. Despite the lack of complete reversal of the diabetic state, the results show a potentially promising gene therapy procedure for treating type 1 diabetes.

25.3 Biomaterials and Scaffolds for Tissue Engineering

Biomaterials, as noted earlier, are an essential component of the tissue engineering approach. In order to understand why certain types of biomaterials are preferable for use in specific tissue engineering applications and how these biomaterials are converted into scaffolds with the desired shape and internal architecture, it is useful to first consider the properties and characteristics that a scaffold should have in order to be effective in tissue regeneration. Overall, the scaffold should have certain essential (or required) properties and characteristics but, in addition, certain desirable properties and characteristics which could improve its effectiveness in regenerating the desired tissue or organ.

25.3.1 Properties of Scaffolds for Tissue Engineering

In general, a scaffold should have the following essential properties and characteristics:

- *Biocompatibility:* The biomaterial used in creating a scaffold should be non-cytotoxic and non-immunogenic, and should elicit only a limited inflammatory tissue response.
- *Ability to provide a suitable microenvironment:* The scaffold should provide a suitable microenvironment (or niche) for cells to reside, proliferate, and differentiate

into the required phenotype, which is necessary for the formation of new functional tissue.

- *Surface composition and properties:* The surface of the scaffold should have the right composition and physicochemical properties to facilitate cell adhesion and cell function.
- *Ability to provide molecular signals to target cells:* The scaffold should have the ability to deliver exogenous biomolecules (such as growth factors) or to recruit endogenous biomolecules to provide molecular signals to target cells for optimal cell proliferation, differentiation, and gene expression,
- *Ability to provide mechanical signals to target cells:* As discussed in Chapter 17, the scaffold should have the requisite stiffness to provide mechanical signaling to target cells for optimal cell proliferation, differentiation, and gene expression.
- *Size, shape, and internal architecture:* The biomaterial used in creating the scaffold should be capable of being formed into the size and external shape of the tissue to be regenerated. In addition, the biomaterial should also have the ability to be formed into an optimal internal architecture to facilitate cell infiltration, transport of essential biomolecules and nutrients, and blood vessel formation to support and maintain newly formed tissue within the scaffold.
- *Sterilizable:* The scaffold should be capable of being sterilized by common methods without any reduction in its physicochemical properties.

The desirable properties and characteristics of a scaffold include the following:

- *Degradability and bioactivity:* The ability of a scaffold to degrade is desirable because it provides the potential for the scaffold to be replaced completely by new tissue having the properties and function of the native tissue. A nondegradable scaffold could potentially lead to adverse effects when implanted *in vivo*, such as long-term mechanical stress-induced biocompatibility phenomena (Chapter 17), and a greater risk of infection by providing a surface on which bacteria could adhere. Bioactivity is desirable for stimulating favorable cell and tissue responses to the scaffold. A well-known example is the cell response that leads to the direct bonding of hydroxyapatite or bioactive glass implants to bone (Chapter 23). In comparison, a chemically inert implant is susceptible to encapsulation by fibrous tissue, which will normally limit integration of the implant with surrounding tissue.
- *Degradation rate:* The scaffold should degrade at a rate that is comparable to the rate of tissue regeneration. A rapidly degrading scaffold is undesirable because its ability to guide or stimulate tissue regeneration would be quickly lost. On the other hand, a slowly degrading scaffold could prolong the time for complete tissue regeneration beyond a clinically relevant time.
- *Mechanical properties to support physiological loads:* Scaffolds used for regenerating load-bearing tissues should have adequate mechanical properties, such as strength, elastic modulus, and fatigue resistance, to support repeated physiological loads due to normal activity. In general, the scaffold, at least initially, should have mechanical properties comparable to those of the native tissue to be regenerated. Additionally, in the case of degradable scaffolds, there should be an appropriate balance between the loss in mechanical strength of the scaffold with time and the rate of tissue formation within the scaffold. As illustrated in Figure 25.5, the loss in strength of the scaffold should ideally be compensated largely by an increase in strength, due to the new tissue formed within the scaffold, in order to maintain adequate strength to support physiological loads.
- *Injectability:* An injectable scaffold provides the benefits of ease of implantation and the ability to provide a better fit to the tissue defect to be regenerated, particularly for a complex-shaped defect. Consequently, if necessary for the intended application, the biomaterial used for creating the scaffold should have the ability to be processed into a form with optimal rheological properties and setting time for injection into the tissue site.
- *Processability:* As noted earlier, tissue engineering approaches that rely on the use of cell-seeded scaffolds

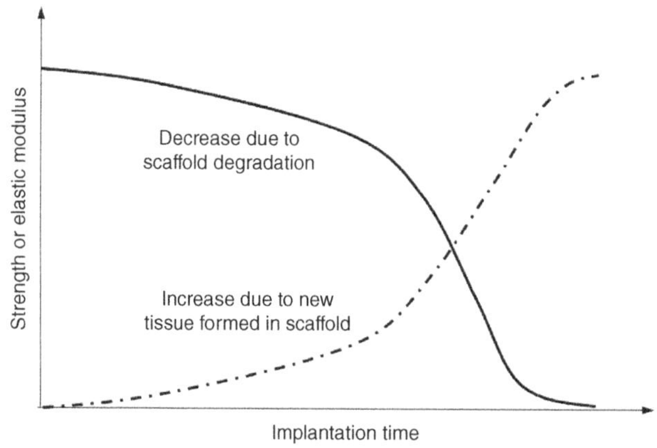

Figure 25.5 Schematic diagram showing trends in the strength of a tissue construct as a function of time *in vivo* due to the separate processes of scaffold degradation and tissue formation within the scaffold. The resultant strength at a given time is the sum of the two strengths. The scaffold provides the main load-bearing function at shorter times, whereas the new tissue formed within the scaffold provides the load-bearing function at longer times. The intersection of the two curves is arbitrary but, in general, a clinically relevant time is no longer than ~3–6 months.

involve two separate steps, the creation of the scaffold followed by cell seeding. An alternative method is to produce the cell-seeded scaffold in a single fabrication process. In this case, the biomaterial used for creating the scaffold should be capable of being processed using methods that simultaneously create the scaffold and incorporate the cells.

25.3.2 Biomaterials for Tissue Engineering Scaffolds

A vast number of biomaterials has been used in the creation of scaffolds for tissue engineering. These biomaterials consist predominantly of polymers and hydrogels, either synthetic or natural in origin, but ceramics such as hydroxyapatite, β-TCP, biphasic calcium phosphate (BCP) and, to a more limited extent, bioactive glasses are also used, mainly as scaffolds for bone regeneration. Degradable metals is an area of considerable recent research and development (Chapter 6) but, currently, most useful biocompatible metals suffer from the disadvantages of being resistant to degradation and of having mechanical properties (such as Young's modulus and tensile strength) that are far superior to those of natural tissues. This difference in mechanical properties normally leads to adverse mechanical stress-induced biocompatibility phenomena (Chapter 17).

The basis for the widespread use of polymers and hydrogels as biomaterials for tissue engineering scaffolds, when compared to ceramics and metals, can be understood in terms of a combination of factors, such as the properties of the tissues to be regenerated, the aforementioned required and desirable properties of tissue engineering scaffolds, and the intrinsic properties of the biomaterials themselves. Except for bone and teeth, the human body is composed of soft tissues and, therefore, based on their intrinsic properties, polymers and hydrogels can be more easily designed to approximate the mechanical properties of soft tissues. Polymers and hydrogels can also be easily synthesized in a wide range of compositions and their surface physicochemical properties can be easily modified using a variety of available techniques. In addition, polymers and hydrogels can also be easily formed into a wide range of shapes and internal architectures using a variety of techniques.

Over the last few decades, scaffolds derived from the ECM have attracted significant attention, and they are already being used in preclinical studies and clinical applications for the regeneration of tissues and organs. These scaffolds normally consist of decellularized allogeneic or xenogeneic tissues and, therefore, they retain to a large extent the structural and biochemical properties of the native ECM from which they are derived. Consequently, ECM-derived scaffolds have the potential to recapitulate the native microenvironment and to provide optimal mechanical and molecular signals for regenerating the desired tissue.

Although a vast number of biomaterials have been used as scaffolds for tissue engineering, the scaffolds themselves generally have three distinct forms: porous three-dimensional solids, hydrogels, and ECM-derived materials. Consequently, one way of classifying this vast number of biomaterials is to group them according to these three distinct forms. In general, significant overlap can exist between these groups. For example, polymers, particularly those of natural origin, can be used in the form of both porous solids and hydrogels. There can also be subdivisions within each group. For example, biomaterials used for creating porous solids or hydrogels can be either synthetic or natural in origin, while some hydrogels can be in a preformed state (having the desired shape and size) or they can be injectable. Additionally, porous solids can be subdivided into hard or soft scaffolds.

Many of the biomaterials used for creating tissue engineering scaffolds in the form of porous solids and hydrogels have been discussed in earlier chapters of this book, in terms of their composition, structure, and properties. Those chapters provide a useful background for the following section in which prominent types of biomaterials are discussed within the context of their ability to approximate the required and desirable properties of tissue engineering scaffolds (Section 25.3.1). Methods used for creating scaffolds with the desired size, shape, and internal architecture are discussed in Section 25.4.

25.3.3 Porous Solids

Table 25.4 shows some examples of polymers used for creating tissue engineering scaffolds in the form of porous solids, subdivided according to their origin (natural or synthetic), along with examples of their applications. Pores provide the means for incorporating cells and biomolecules into the scaffold, for modifying certain physicochemical properties of the scaffold, such as its mechanical properties and degradation rate, and for providing a pathway for cells, biomolecules and nutrients to migrate into or out of the scaffold. Various advantages and disadvantages of synthetic polymers and natural polymers for use as biomaterials were discussed in Section 10.2, in relation to their biocompatibility, mechanical properties, chemical stability in the physiological environment, and processability.

Porous Solids Composed of Synthetic Polymers

Synthetic polymers for tissue engineering scaffolds were initially selected from the degradable polymers or copolymers already approved by the US Food and Drug

Table 25.4 Examples of polymers used as scaffolds in the form of porous solids, and various tissue engineering applications for which they have been investigated.

Polymer	Tissue engineering applications
A. Synthetic polymers	
Polyglycolic acid (PGA)	Bone, cartilage, tendon, nerve, spine
Poly(L-lactic acid) (L-PLA)	Bone, cartilage, tendon, blood vessels, nerve
Polylactic-*co*-glycolic acid (PLGA)	Bone, cartilage, tendon, skin, liver, nerve
Polycaprolactone (PCL)	Bone, cartilage, ligament, skin, nerve, blood vessel
Poly(3-hydroxybutyrate) (P3HB)[a]	Bone
Poly(4-hydroxybutyrate) (P4HB)[a]	Myocardium
Polyglycerol sebacate (PGS)	Myocardium, blood vessel, cartilage, retina
B. Natural polymers: protein-based	
Collagen	Bone, cartilage, skin, tendon, ligament
Silk	Ligaments, blood vessel, nerve, skin, bone, and cartilage
Elastin	Skin, blood vessel
C. Natural polymers: polysaccharide-based	
Hyaluronic acid	Bone, cartilage
Chitosan	Cartilage, bone, skin, nerve
Alginate	Bone, cartilage
Cellulose	Bone, cartilage

[a] P3HB and P4HB can be synthetic or natural in origin.

Administration for use as implantable medical devices. The more commonly used polymers include the poly(α-hydroxy esters) polyglycolic acid (PGA), polylactic acid (PLA) and their copolymer polylactic-co-glycolic acid (PLGA), and polycaprolactone (PCL) (Chapter 9). These polymers generally have adequate biocompatibility but their acidic degradation products can sometimes lead to adverse biocompatibility phenomena such as an inflammatory tissue response, particularly for scaffolds of large size implanted into bone. Subsequently, porous solids of other degradable polymers, such as poly (3-hydroxybutyrate) (P3HB), and polyglycerol sebacate (PGS), have also been studied for use as tissue engineering scaffolds.

In relation to the required properties and characteristics of a tissue engineering scaffold, synthetic polymers generally have adequate biocompatibility, can be easily formed into the size and shape of the tissue to be regenerated, and can be sterilized using common techniques. However, a major limitation is that they are generally unable to provide an optimal microenvironment (or niche) in which cells can reside, proliferate, and differentiate into the required phenotype. Scaffolds composed of synthetic polymers have a composition, structure, bulk physicochemical properties and surface properties that are generally very different from those of the native tissue microenvironment. Consequently, they have a limited ability to deliver optimal mechanical and molecular signals for tissue regeneration.

Various approaches, based on mimicking certain attributes of the ECM, have been used to address this limitation of synthetic polymeric scaffolds. Overall, as discussed in Chapter 16, the ECM consists of fibrous structural proteins, specialized proteins, and glycosaminoglycans (GAGs) composed of polysaccharides. One approach involves modifying the structural features of the scaffold to mimic those of the ECM. An example is the use of the electrospinning technique to create fibrous scaffolds with fiber diameters on the approximate scale of the fibrillar components of the ECM (Section 25.4.2). Another approach involves the use of scaffolds in the form of hydrogels in order to mimic the aqueous gel-like nature of the ECM. A third approach involves the attachment of biomolecules such as growth factors or peptides containing the RGD sequence of fibronectin (Chapter 13) to mimic the molecular signaling characteristics of the ECM. While these approaches have shown varying degrees of success, in general, it is difficult to replicate the native microenvironment of the ECM with its mechanical and molecular signaling characteristics using scaffolds composed of synthetic polymers.

Synthetic polymers used for tissue engineering scaffolds generally lack bioactivity, but they provide a wide range of degradation rates, which can be further modified by copolymerization. Porous solids created from synthetic polymers are generally used in applications where they are subjected to low mechanical loads but, with the right composition and microstructure, they show mechanical properties that are adequate for some load-bearing applications. As discussed in Chapter 13, both the bioactivity and mechanical properties of porous synthetic polymers can be improved through the formation of composites in which the polymer matrix is reinforced with bioactive particles such as hydroxyapatite.

Porous Solids Composed of Natural Polymers

Natural polymers used for creating tissue engineering scaffolds are generally composed of proteins, polysaccharides, or a blend composed of two polymers, such as a protein and a polysaccharide (Table 25.4). Because they are obtained from living species, natural polymers used as biomaterials are normally biocompatible, unless they are processed inadequately or toxic substances are introduced during their processing. Additionally, protein-based polymers should be restricted to those proteins that are highly conserved across species because of the inherent antigenicity of most allogeneic and xenogeneic proteins.

When compared to synthetic polymers, natural polymers have the potential to provide a more attractive environment for tissue regeneration because they have a structural and compositional similarity to individual components of the ECM. Consequently, natural polymers have the potential to provide improved mechanical and molecular signals to target cells *in vivo*. As with synthetic polymers, natural polymers can be easily formed into scaffolds with the requisite size, shape, and internal architecture, and they can be easily sterilized using common methods. However, natural polymers need to be processed under mild conditions, which can reduce the number of available techniques used to form them into scaffolds.

In relation to the desirable properties of scaffolds, porous solids composed of natural polymers, even when crosslinked, show mechanical properties (such as strength and elastic modulus) which are normally weaker than those of the aforementioned synthetic polymers for a similar microstructure. Such scaffolds therefore have a limited ability to support physiological loads and, thus, they are generally more suitable for regenerating nonstructural tissues. On the other hand, in the form of bundles of aligned filaments, collagen has a sufficiently high tensile strength and Young's modulus for use as scaffolds in the regeneration of tendons and ligaments. Because natural polymers used as biomaterials normally degrade in the physiological environment by enzyme-mediated hydrolysis, their degradation *in vivo* is often less controllable when compared to synthetic polymers (which degrade predominantly by hydrolysis). However, better control of their degradation rate can be achieved by the extent to which they are chemically crosslinked (Chapter 10).

Collagen (type I), the most abundant protein in the ECM, is the most commonly used protein-based polymer for tissue engineering applications, as well as for other biomedical applications. Other protein-based polymers used in the form of porous solids include elastin, a component of the ECM, and silk, but they find limited use when compared to collagen. Because of their hydrophilicity, polysaccharides are commonly used in the form of hydrogels, but some polysaccharides such as hyaluronic acid (HA), a component of the ECM, alginate and chitosan have also been used in the form of porous solids.

Copolymers, Blends, and Composites

Copolymerization is a widely used method for modifying the physicochemical properties of solid polymers and hydrogels. In fact, many solid polymers and hydrogels used in tissue engineering are copolymers composed of two or more types of monomers. Blending of two or more polymers provides another method for modifying the physicochemical properties of solid polymers and hydrogels but it is less widely used than copolymerization. Overall, copolymerization and blending are important methods for creating scaffolds with improved mechanical and molecular signaling to target cells *in vivo*. For example, as noted earlier, the ECM can be considered to consist of structural proteins such as collagen and elastin, which are augmented in their biomechanical and biochemical functions by GAGs composed of polysaccharides such as HA and chondroitin sulfate. Consequently, when compared to the individual polymers, copolymers or blends of two natural polymers, for example, a protein-based polymer and a polysaccharide-based polymer, could provide a better approximation to the components of the ECM and, thus, potentially provide a more optimal microenvironment for tissue regeneration. As discussed in Chapter 12, reinforcing polymers with a strong bioactive phase, such as hydroxyapatite or bioactive glass particles provides an approach for creating polymeric scaffolds with improved mechanical properties and bioactivity. Prominent examples are PGA, PLA, and PCL reinforced with hydroxyapatite or bioactive glass, and collagen reinforced with hydroxyapatite, a composition containing the organic and inorganic constituents of bone.

Porous Solids Composed of Ceramics

Porous ceramic solids used as tissue engineering scaffolds are generally composed of the calcium phosphate materials hydroxyapatite, β-TCP, and BCP. These porous solids have been used for decades to heal nonstructural and structural bone defects. As discussed in Chapter 23, when incorporated with cells or growth factors, they are effective in healing bone defects. Hydroxyapatite has high compressive strength, much higher than the compressive strength of synthetic polymers of a similar microstructure, but it degrades very slowly. For example, porous hydroxyapatite implants used for healing large bone defects can take up to a few years to degrade completely *in vivo*. In comparison, porous β-TCP implants show much weaker mechanical properties but degrade considerably faster (a few months for complete degradation). In general, porous solids of BCP can provide some optimal combination of mechanical properties and degradation rate. Porous solids of bioactive glasses have also been used as scaffolds for regenerating bone (Chapter 23), but to a more limited extent when compared to the calcium phosphate bioceramics.

25.3.4 Hydrogels

Hydrogels are water-filled polymers, in which the solid phase consists of a crosslinked three-dimensional network of hydrophilic polymer chains (Chapter 11). An attractive feature of hydrogel scaffolds is that they approximate the aqueous gel-like nature of the ECM. In relation to the required properties and characteristics of a tissue engineering scaffold, hydrogels provide an attractive microenvironment for tissue regeneration because the regions between the molecular chains of the hydrogel provide spaces for cells to reside, proliferate and differentiate, and for the migration of oxygen, nutrients, and essential biomolecules. Hydrogels can vary from being soft and pliable to extremely rigid, providing a range of stiffness and viscoelastic properties for optimal mechanical signaling. In addition, hydrogels can be created from natural polymers that have a compositional similarity to ECM components, or they can be created from synthetic polymers modified with biomolecules, such as growth factors or peptides containing the RGD sequence of fibronectin. Consequently, hydrogels also have the potential to deliver molecular signals to target cells.

The polymers used for creating natural and synthetic hydrogels are normally nontoxic. However, there is a potential risk of toxic effects due to chemical agents used for crosslinking the polymer chains to form the hydrogel. For example, glutaraldehyde is an effective and widely used agent for crosslinking natural polymers such as collagen and HA, but there are concerns about the potential for undesirable side effects due to elution of glutaraldehyde molecules from the implant during its degradation. Additionally, cells are incorporated into the precursor of the hydrogel (that is, prior to the gelation step), which means that the crosslinking technique should not be toxic or detrimental to the cells. As noted in Chapter 11, because of the presence of an aqueous liquid within their pores, hydrogels are generally more difficult to sterilize when compared to solids. However, most hydrogels can be adequately sterilized by aseptic processing of presterilized components, or by terminal sterilization using techniques such as ethylene oxide treatment.

In relation to the desirable properties and characteristics, hydrogels used as tissue engineering scaffolds are normally degradable in the physiological environment, and their degradation rate can be controlled by copolymerization or degree of crosslinking as, for example, hydrogels composed of polyethylene glycol (PEG), collagen or HA (Chapter 11). Overall, hydrogels have weak mechanical properties and, therefore, they are not suitable for regenerating structural tissues. Some hydrogels are injectable as, for example, gelatin hydrogels which show temperature dependent gelling, and chitosan which shows temperature-dependent and pH-dependent gelling.

Table 25.5 Examples of polymers used as scaffolds in the form of hydrogels, and various tissue engineering applications for which they have investigated.

Hydrogel	Tissue engineering applications
A. Synthetic polymers	
Polyethylene glycol (PEG)	Bone, cartilage, skin, myocardium, bladder, blood vessel
Polylactic acid (PLA)	Cartilage
Polyvinyl alcohol (PVA)	Cartilage, blood vessel, intervertebral disc
Polyhydroxyethyl methacrylate (PHEMA)	Skeletal muscle, spinal cord
PEG-PLA	Bone
PEG-PLA-PVA	Cartilage
B. Natural polymers: protein-based	
Collagen	Cartilage, skin, blood vessel, nerve, spinal cord, vocal cord
Fibrin	Cartilage, skin, myocardium
C. Natural polymers: polysaccharide-based	
Hyaluronic acid	Cartilage, skin
Chitosan	Cartilage, bone, skin, nerve, liver, blood vessel
Alginate	Cartilage, muscle, nerve, pancreas, liver
Agarose	Cartilage

Examples of synthetic and natural hydrogels and their applications in tissue engineering are given in Table 25.5. While a wide variety of hydrogels have been used in tissue engineering, the most commonly used compositions are based on those of natural polymers such as collagen, HA, alginate and chitosan, and on the synthetic polymer PEG.

25.3.5 Extracellular Matrix (ECM) Scaffolds

As noted earlier, ECM scaffolds, sometimes referred to as biological scaffolds, are normally prepared by decellularization of intact mammalian tissues followed by sterilization. Unlike scaffolds of some natural polymers, such as collagen or HA, which generally mimic the structure and composition of individual ECM components, ECM scaffolds, when properly processed, generally retain the three-dimensional organization of the structural and functional biomolecules of the tissue from which the ECM was derived. ECM scaffolds have been produced from a variety of tissues, such as dermis, small intestine, urinary bladder, heart, trachea and lung, and from a variety of species, including human, porcine, bovine, and equine species. Decellularization and sterilization methods used in the production of ECM scaffolds are designed to retain maximum functionality of the biologically active molecules, while removing cellular antigens that may cause an adverse host immune response. ECM scaffolds have been used successfully in a variety of tissue engineering and regenerative medicine applications, in preclinical studies and in clinical applications.

Compositionally, because ECM scaffolds consist of the structural and functional molecules of the tissue from which they were derived, the specific composition and distribution of the ECM components will vary, depending on the tissue source. However, a limiting factor is that except for scaffolds derived from porcine small intestine submucosa (SIS) and from urinary bladder, two of the most widely studied and applied ECM scaffolds, the compositions of ECM scaffolds are not known in sufficient detail. SIS scaffolds are composed of >90% collagen, the large majority of which consists of collagen type I, with small amounts of collagen types III, IV, V, and VI (Badylak et al. 2009). Figure 25.6 shows the arrangement of the collagen fibers in a layer of porcine SIS. In comparison, scaffolds derived from urinary bladder contain the same collagen types as SIS, but greater amounts of collagen type III and collagen type VII, are present. The amount of GAGs remaining in ECM scaffolds depends on the decellularization method but, overall, SIS scaffolds contain a variety of GAGs, including heparin, heparan sulfate, chondroitin sulfate, and HA. SIS scaffolds also contain cell adhesion molecules such as fibronectin and laminin, the proteoglycan decorin, the glycoproteins biglycan, and entactin, as well as various growth factors, including TGFβ, and VEGF.

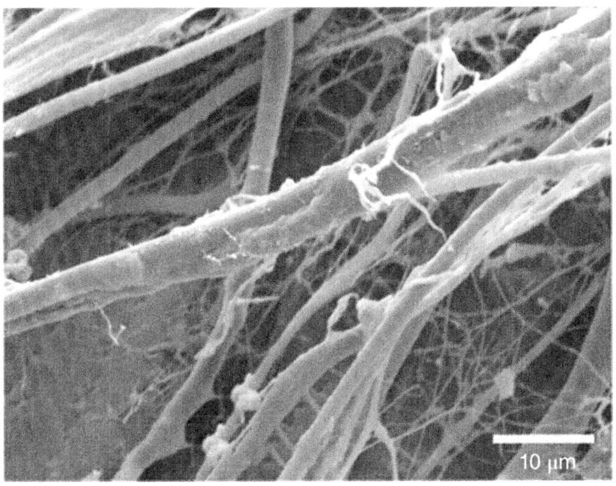

Figure 25.6 SEM image of fibrous collagen network of ECM scaffold derived from porcine small intestine submucosa (SIS). *Source:* From Chang et al. (2016).

As noted earlier, with proper decellularization and sterilization techniques, ECM scaffolds retain the structure and three-dimensional architecture of the corresponding tissue source. It is known that the collagen fiber architecture of the scaffold plays an important role in determining its mechanical behavior. The alignment and organization of collagen fibers are dependent on the function of the tissue from which the scaffold is derived (Chapter 10). For example, the collagen fibers within a tendon or ligament are aligned along the long axis of the tissue to provide the greatest resistance to an applied mechanical strain. Consequently, the use of ECM scaffolds derived from tendon and ligaments is a logical choice for repairing tissues such as the anterior cruciate ligament (ACL).

Based on the aforementioned compositional and structural characteristics, ECM scaffolds have the potential to provide an optimal microenvironment and to deliver optimal mechanical and molecular signals to regenerate the tissue from which the scaffold was derived. On the other hand, decellularization methods generally employ harsh chemical solutions, which can have a deleterious effect on the growth factors within the ECM, the structure of the ECM components, and the surface topology of the ECM scaffold. Effective removal of antigens associated with cell membranes and intracellular components of tissues is necessary to minimize or avoid adverse immune responses by allogeneic and xenogeneic recipients of the ECM scaffold. ECM scaffolds can be easily and effectively sterilized by methods such as ethylene oxide, gamma radiation, and electron beam radiation. Chemical crosslinking can increase the strength of the ECM scaffold but often at the expense of less desirable tissue reconstruction when compared to non-crosslinked forms of the scaffold.

ECM scaffolds can be processed into several forms, such as a hydrated sheet, several of which can be vacuum-pressed to form a (thicker) laminated sheet, and used directly as a scaffold. In addition, hydrated sheets can be freeze-dried, ground into a powder, enzymatically digested and converted to a gel or mixed with a polymer to form a hybrid scaffold. Figure 25.7 gives an overview of the tissue regeneration (or remodeling) process associated with ECM scaffolds.

25.4 Creation of Scaffolds for Tissue Engineering

Scaffolds for use in tissue engineering, as noted earlier, should consist of biocompatible materials with the requisite shape, size, and internal architecture for the desired application. In the formation of hydrogel scaffolds, techniques for crosslinking natural polymer molecules were

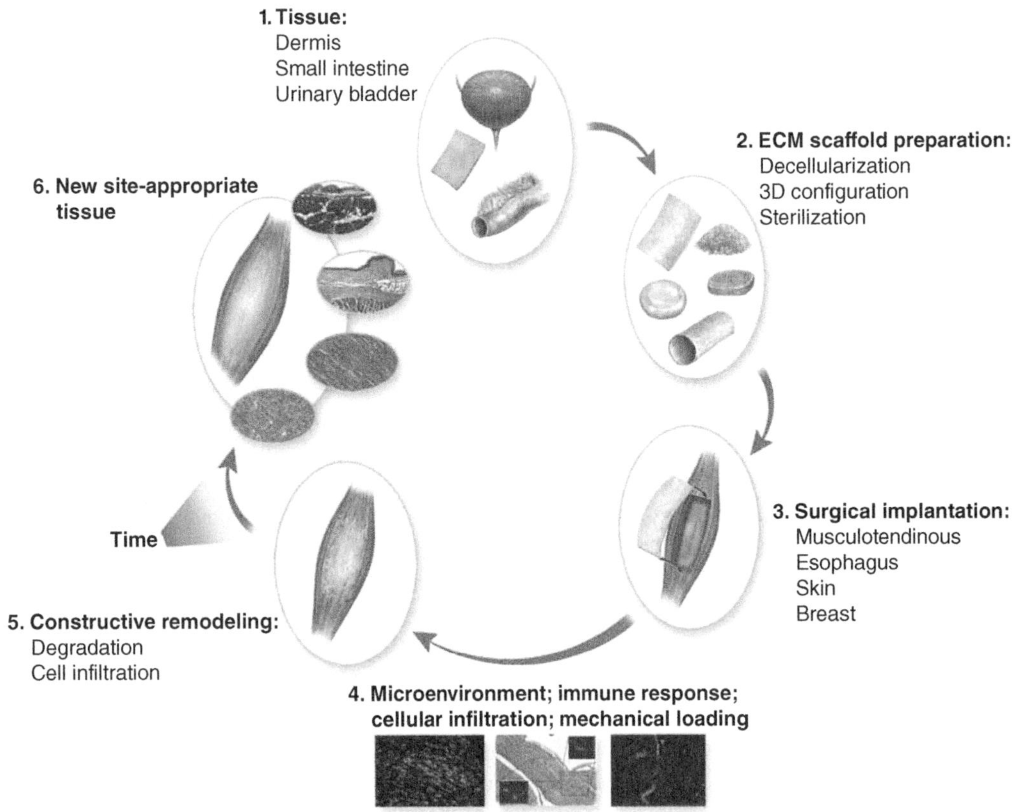

Figure 25.7 Schematic showing an overview of the tissue regeneration process associated with ECM scaffolds. *Source:* From Badylak et al. (2011).

discussed in Chapter 10, while techniques for crosslinking synthetic polymers were discussed in Chapter 11. Consequently, this section will focus mainly on techniques for creating scaffolds in the form of porous solids. A brief discussion of techniques for preparing ECM scaffolds by decellularization of intact mammalian tissues will also be presented.

25.4.1 Creation of Scaffolds in the Form of Porous Solids

A wide variety of techniques is available for creating scaffolds in the form of porous solids, differing in terms of their simplicity, ability to control the scaffold internal architecture, and production cost. A major trend in recent years has been toward a greater use of additive manufacturing techniques, also referred to as three-dimensional (3D) printing, because they provide the ability to precisely control the external shape and internal architecture of the scaffold, often at both the microscale and the nanoscale. Another trend has been the growing use of the electrospinning technique, based largely on its ability to create fibrous structures that mimic the fibrillar components of ECM components. A more recent trend has been the growing use of a technique referred to as 3D bioprinting to create cell-seeded scaffolds in a single fabrication process by incorporating cells into the scaffold during the 3D printing process. Salient features of a variety of techniques are discussed in the following sections, along with their advantages and drawbacks.

Thermal Bonding of Particles

One of the simplest techniques is thermal bonding of solid particles to form a porous three-dimensional solid. It involves placing particles such as microspheres, irregular-shaped particles, or short fibers of the right size into a shaped mold and heating the system to a suitable temperature to bond the particles at their contact points. This technique has been used to create scaffolds composed of degradable polymers such as PGA, PLA, and PLGA, as well as bioactive glass scaffolds (Figure 25.8a) but it now finds little use. Although simple, it is limited in its ability to provide adequate control of the external shape of the scaffold and its internal architecture. The shape of the scaffold is limited to simple shapes (often a cylinder) determined by the shape of the mold, while the size, shape, and packing of the particles determine the architecture of the scaffold. In general, the porosity of the scaffold is limited to a narrow range (~40–60%), while the pore sizes cover a wide range. For example, for a scaffold composed of spherical particles of the same diameter, the distribution of pores sizes in the scaffold covers a range of ~0.15 to ~0.4 of the sphere diameter, and it can become even wider for a scaffold formed from irregular-shaped particles of different sizes. Of greater importance is the constriction of the necks between adjacent pores, resulting in a neck size that is normally much smaller than the actual pore size. This can adversely affect the interconnectivity of the pores, making cell seeding and cell migration through the scaffold difficult.

Use of a Pore-forming Agent

Another simple technique involves the use of a dissolvable substance such as salt (NaCl) particles to aid the formation of pores and to control the porosity and pore size of the scaffold. The originally developed process, often referred to as solvent casting and particulate leaching, involves dissolving a degradable polymer such as PLA or PLGA in an organic solvent such as chloroform or methylene chloride and dispersing the right amount of salt particles of the desired size in the organic solution (in which they are insoluble). Then the suspension is cast onto a surface and dried give a solid polymer–salt construct, which is immersed in water to dissolve out the salt particles, leaving a porous solid polymer, typically in the form of a sheet (Figure 25.8b). When compared to the thermal bonding technique, this technique provides the ability to create scaffolds with a wider range of porosity (~30–90%) and to better control the porosity and pore size of the scaffold. Overall, the scaffold porosity is determined by the volume of the salt particles relative to the volume of the polymer,

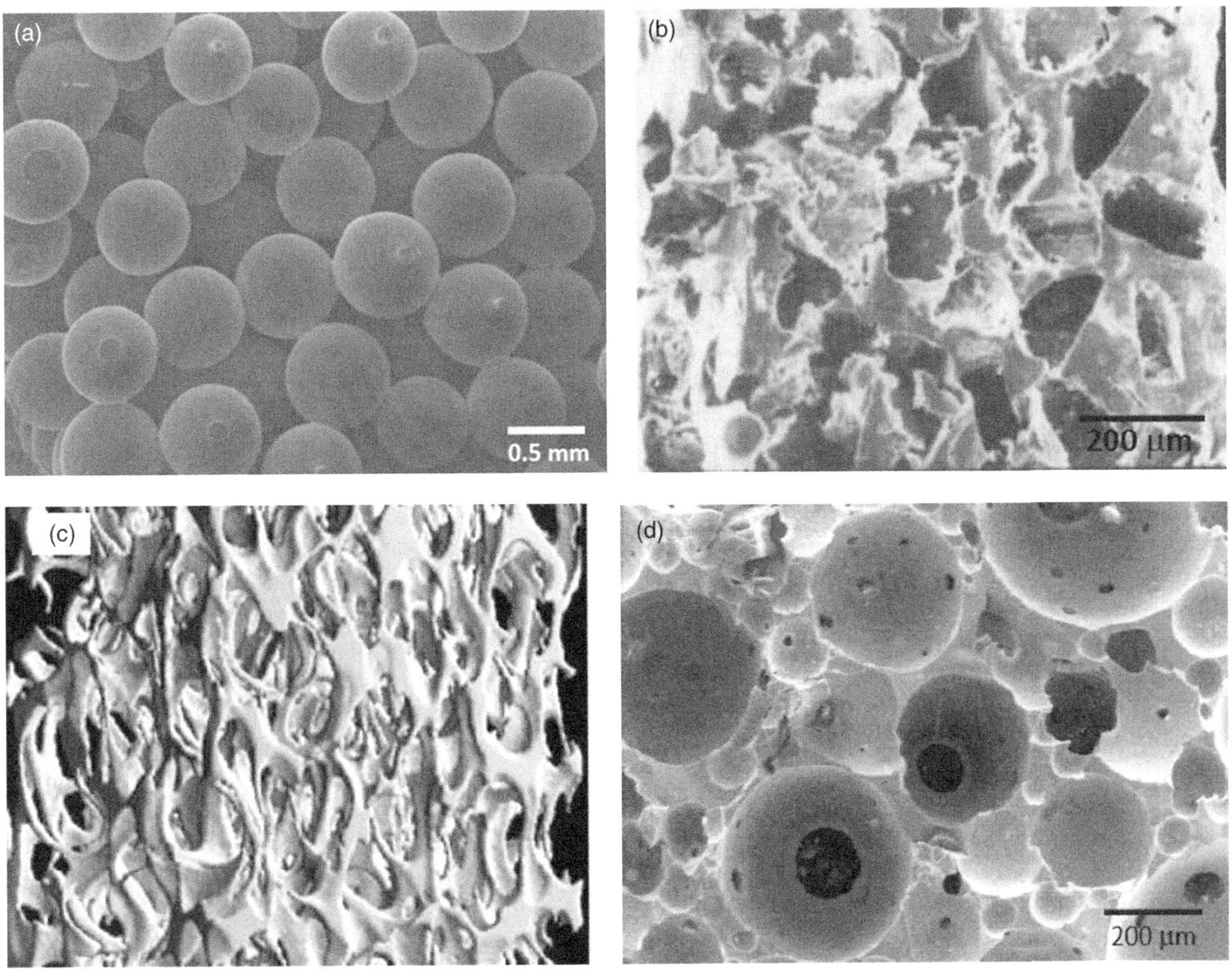

Figure 25.8 SEM images of scaffolds in the form of porous solids prepared by various methods. (a) Scaffold consisting of thermally bonded polylactic-*co*-glycolic acid (PLGA) microspheres. *Source:* From Borden et al. (2003); (b) polylactic acid (PLA) scaffold prepared by solvent casting and particulate leaching. *Source:* From Mikos et al. (1994); (c) PLA scaffold prepared by high pressure CO_2 gas foaming. *Source:* From Mathieu et al. (2006); (d) bioactive glass scaffold prepared by foaming a sol. *Source:* From Sepulveda et al. (2002).

while the pore size is determined by the size of the salt particles, which are commonly sieved to give a narrow size range. On the other hand, as with the thermal bonding technique, achieving good interconnectivity of the pores can be a problem, particularly for scaffolds having a porosity that is not very high. The necks between adjacent pores are constricted, which adversely affects the ability to seed cells into the scaffold and the migration of cells through the scaffold.

The as-formed porous scaffold (after dissolution of the salt particles) is typically in the form of a sheet but thicker sections can be formed by laminating several sheets together. In one variation of the technique, the as-formed polymer–salt construct can be fractured into small pieces, which are compression molded or extruded at a high enough temperature to give a thicker construct. However, complete dissolution of the salt particles can be problematic due to the higher thickness of the constructs. Overall, this solvent casting and particulate leaching was popular upon its introduction in the 1990s but, like the thermal bonding technique and a few other techniques used in forming porous solids, it has been superseded largely by additive manufacturing (3D printing) techniques.

Another technique, used for creating scaffolds of inorganic biomaterials such as hydroxyapatite and bioactive glass involves the use of a pore-forming material composed of organic polymer particles (such as starch particles). Fine hydroxyapatite (or bioactive glass) particles, of size smaller than a few microns, are mixed homogeneously with starch particles that have been sieved to give a desired size range (such as 150–250 μm) and consolidated in a mold to form the desired external shape. Then the as-formed material is heated to ~300–500 °C to decompose the starch particles and sintered to a high enough temperature (~1200 °C for hydroxyapatite) to bond the particles together into a strong network containing pores of approximately the same size as the starch particles. Scaffolds formed by this technique have pore characteristics that are similar to those formed by the aforementioned salt leaching technique and suffer from similar limitations.

Foaming Techniques

Two types of foaming techniques have been used to create scaffolds, one using high-pressure CO_2 gas to create pores in a solid polymer, and the other using a foaming agent to create pores in a suspension, which is used for creating scaffolds of inorganic biomaterials such as hydroxyapatite and bioactive glass. The gas foaming process starts with the formation of solid discs of a polymer, such as PGA, PLA, or PLGA, which are placed in a chamber, warmed, and exposed to CO_2 gas at a pressure of ~5 MPa to saturate the polymer with a significant quantity of absorbed CO_2 molecules. Then the temperature and gas pressure are lowered to the ambient values, when the gas molecules cluster together to reduce their free energy, resulting in the nucleation of fine pores, which grow as more CO_2 molecules diffuse into them. This causes a significant expansion of the polymer, giving a highly porous solid (Figure 25.8c). Porosities of up to ~90% and pore sizes up to ~100 μm can be achieved using this technique, but a drawback is that a significant fraction of the pores, particularly near the surface, can be unconnected.

Another type of foaming technique, which has been used to create scaffolds of ceramics such as hydroxyapatite and of bioactive glasses, starts with the preparation of a suspension of particles, aided by the use of a crosslinking agent such as polyethylenimine. Then the suspension is mixed with a foaming agent (such as polyoxyethylene lauryl ether) and stirred vigorously to give a foamy appearance, after which the system is gelatinized by adding a crosslinking agent to stabilize the foam structure. After casting in a mold and drying to remove the solvent, the resulting material is heated at ~300–500 °C to decompose the organic additives used in the forming process, and sintered at a higher temperature (~1200 °C for hydroxyapatite) to bond the solid phase into a network of adequate strength. Depending on the degree of foaming, scaffolds having porosities of up to ~90% and pore diameters in the range ~10–500 μm are obtained using this technique (Figure 25.8d). A common drawback is that the necks between adjacent pores are normally much smaller than the pore diameter, which can limit cell seeding and cell migration through the scaffold.

Polymer Foam Replication

Polymer foam replication is a popular technique for creating scaffolds of inorganic biomaterials such as hydroxyapatite and bioactive glass. With proper selection of the foam, it can be used to create a highly porous scaffold that mimics the internal architecture of human trabecular bone (Figure 25.9). The process starts with the selection of a commercial polymer foam having the desired pore characteristics, which is infiltrated with a colloidally stable slurry (concentrated suspension) of fine particles (of size less than a few microns) composed of hydroxyapatite (or bioactive glass). This infiltration process involves a few repeated steps of dipping the foam into the slurry, followed by removal and drying, to coat the internal surface of the polymer foam with the desired thickness of particles. Then the particle-coated foam is heated at ~300–500 °C to decompose the foam and sintered at a higher temperature (~1200 °C for hydroxyapatite) to form a network of dense struts, which retains the large pores of the polymer foam.

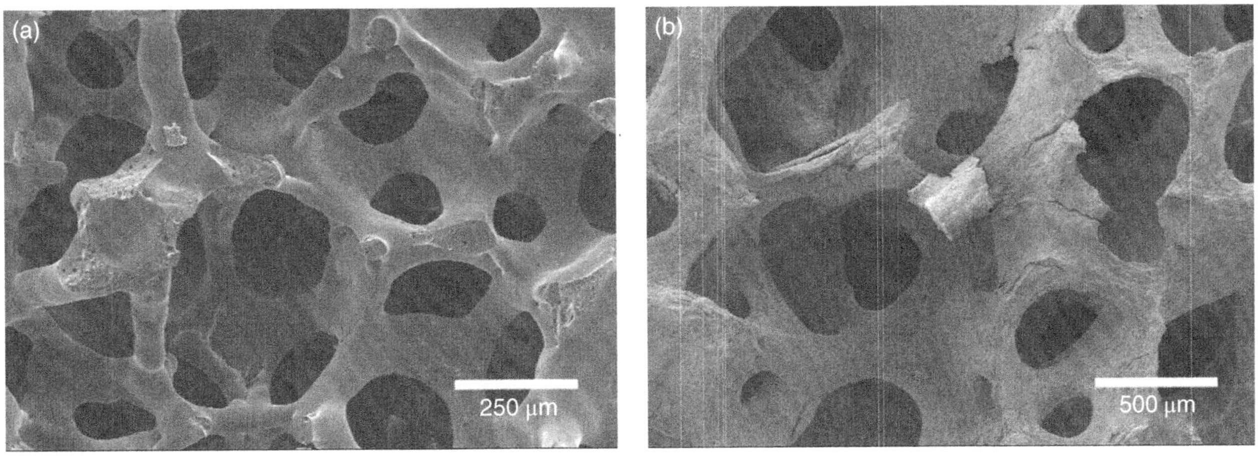

Figure 25.9 SEM images of (a) bioactive glass (13–93) scaffold prepared by a polymer foam replication method, and (b) dry human trabecular bone.

This technique essentially replicates the architecture of the polymer foam, but with the polymer phase replaced by a dense solid. It can be used to created scaffolds having a porosity in the range ~50–80% and good pore interconnectivity. For a given microstructure, the compressive strength and Young's modulus of the scaffold decrease with increasing porosity, but they are normally in the range of values for human trabecular bone (Table 4.1).

Freezing Techniques

Freezing techniques are widely used for creating scaffolds composed of polymers, particularly natural polymers, which normally require processing temperatures that are not far above the physiological temperature (~37 °C). A simple and commonly used technique is freeze-drying, which is an example of a more general phenomenon called thermally induced phase separation (TIPS). A change in temperature (due to freezing, for example) leads to demixing (or phase separation) of a homogeneous solution (or suspension), resulting in the creation a multiphase system (such as a polymer phase and ice). The freeze-drying process starts with the preparation of a homogeneous solution of the polymer molecules. For the creation of collagen scaffolds, this step might involve the preparation of a slightly acidic suspension by adding dilute acetic acid to an aqueous suspension of bovine collagen type I. Then the suspension, placed in a suitable container, is frozen, causing the liquid to freeze (or crystallize), and to separate out from polymer phase. The resulting solid is freeze-dried to remove the frozen liquid by sublimation. For collagen, the freeze-dried material could then be crosslinked, if necessary, and sterilized.

As noted in Chapter 10 for collagen, the starting solution (or suspension) can be frozen isotropically, giving a foam-like structure with a random pore arrangement, or it can be frozen undirectionally, giving a structure in which the pores are oriented in the direction of freezing (Figure 10.2). Overall, the pore size (or diameter) increases with a decrease in the concentration of the suspension and with a decrease in the rate of freezing, while the porosity increases with a decrease in the suspension concentration. For collagen, pores of size (diameter) in the range ~20–200 μm can be obtained using this technique.

The freeze-drying technique has also been applied to the creation of scaffolds composed of inorganic materials such as hydroxyapatite and bioactive glass. The steps leading to the preparation of the freeze-dried material are essentially similar to those for a polymer except that the polymer molecules dispersed in a liquid are replaced by fine particles of the inorganic material (less than a few micrometers in size). While the isotropic freezing route has been used for decades to prepare porous ceramics, it is only more recently that the unidirectional freezing route has attracted significant interest for the creation of bioceramic and bioactive glass scaffolds. Unidirectional freezing, as noted earlier, leads to an anisotropic microstructure of oriented pores and, thus to anisotropic mechanical properties, that is, the mechanical properties in the pore orientation direction are superior to those in the perpendicular direction. This approach could be used to design bioceramic scaffolds with mechanical properties that better approximate those of human cortical bone (Table 4.1), which could potentially be used in healing structural bone defects, such as defects in the long bones of the limbs.

Figure 25.10 shows a schematic of the unidirectional freezing technique for an aqueous slurry of fine particles. Nucleation of ice at the cold surface and crystal growth down the temperature gradient lead to a frozen construct

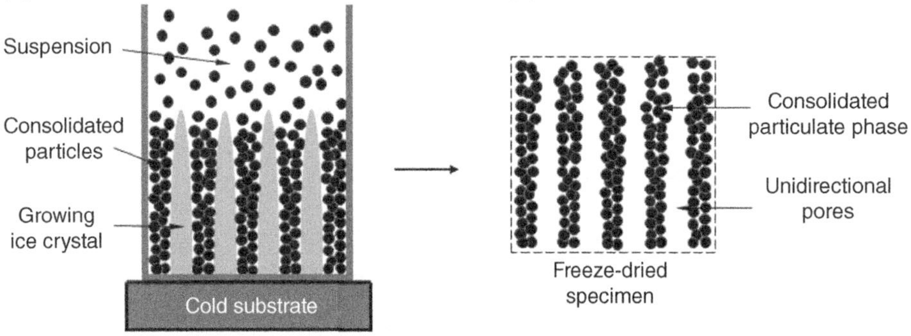

Figure 25.10 Schematic diagram illustrating the unidirectional freezing of suspensions. (a) Illustration of growth of ice crystals and phase separation and (b) illustration of the unidirectional pores in a solid particulate phase after freeze drying.

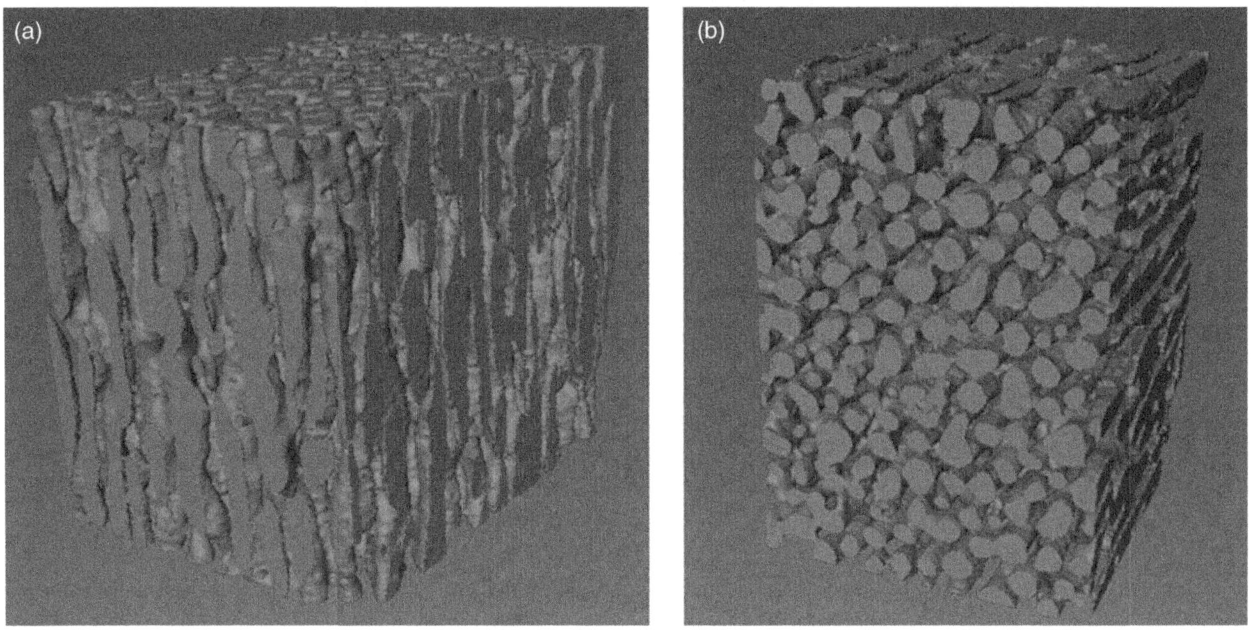

Figure 25.11 Three-dimensional micro-computed tomography (micro-CT) images of bioactive glass (13–93) scaffold (porosity ~50%; pore diameter ~150 μm; compressive strength ~50 MPa) created by unidirectional freezing of suspensions, showing columnar pores with a nearly circular cross section: (a) pores oriented in the direction of freezing; (b) cross section of pores in a plane perpendicular to the freezing direction. (The solid color represents the actual pore phase.)

consisting of ice and a particulate phase. As the pores formed in the construct after freeze drying are a replica of the ice, modifying the concentration of particles in the slurry and the shape of the ice crystals will influence the amount, shape, and size of the pores in the scaffold. For example, an increase in the cooling rate of the slurry will lead to pores of smaller cross section, due to an increase in the nucleation rate of the ice. Additionally, while the use of water as the dispersion medium normally leads to lamellar-shaped pores, the use of other liquids, such as a mixture of water and dioxane or camphene, can give columnar-shaped pores with a nearly circular cross section (Figure 25.11).

Hydroxyapatite and bioactive glass scaffolds created using the unidirectional freezing technique generally have a porosity in the range ~45–65%. Depending on the porosity and freezing rate, lamellar-shapes pore can have dimensions of up to ~500 μm × ~50 μm in cross-section, whereas columnar shapes can have a diameter up to ~150 μm in diameter. Bioactive glass scaffolds of porosity ~50–60% have shown compressive strengths up to ~50–100 MPa in the pore orientation direction, which are ~40–70% of the compressive strength of human cortical bone. These scaffolds have shown the ability to support bone infiltration when implanted in rodent calvariae but, overall, their evaluation in animal models has been limited.

25.4.2 Electrospinning

Electrospinning has been used for decades in a variety of technological areas, but it has attracted significant interest as a technique for creating tissue engineering scaffolds only over the last two decades. This heightened interest in electrospinning stems largely from its ability to create highly porous fibrous scaffolds with structural features, such as fiber diameter, which mimic those of the fibrillar components of the ECM. These structural features provide the potential to deliver optimal mechanical signals for tissue regeneration and, for some natural polymers that have a compositional similarity to ECM components, molecular signals as well.

Electrospinning is used mainly to create polymer scaffolds, but it has also been used to create scaffolds of polymer matrix composites, typically composites containing a dispersed phase of nanoparticles. In addition, the technique has been used to a limited extent to create fibrous scaffolds of some inorganic materials, such as titanium dioxide (TiO_2) and a few bioactive glasses. These inorganic materials have a processing flexibility in that they can be prepared from solutions of starting materials, a route referred to as sol–gel processing, and not only from the more conventional routes employing the sintering of particles or casting a molten material.

Principles of the Electrospinning Process

Electrospinning is a simple and inexpensive technique. Except for the way in which the fibers are created (by means of electrostatic forces), the technique is similar to conventional methods that employ mechanical forces for drawing long continuous fibers, such as glass fibers or polymer fibers, from a molten material. A typical set-up consists of a metallic needle (called a spinneret) connected to a syringe, a high voltage power supply (normally a direct current power supply of range ~1–30 kV), and a collector (a grounded conductor such as a metal) (Figure 25.12a). The syringe contains the polymer solution (or molten polymer), which is fed through the spinneret at a constant and controllable rate using a syringe pump. When the voltage source is turned on, the pendant drop of polymer solution at the tip of the spinneret nozzle becomes electrified, with the induced charges evenly distributed over its surface. As a result, the drop experiences two major types of electrostatic forces: a repulsion between the surface charges of the liquid, and an attraction between the oppositely charged drop and the collector, which causes the drop to elongate.

As the electric field is increased, a point is reached where the electrostatic forces balance the surface tension of the drop, which leads to the development of a shape called the Taylor cone. Upon increasing the electric field further, the electrostatic forces can overcome the surface tension of

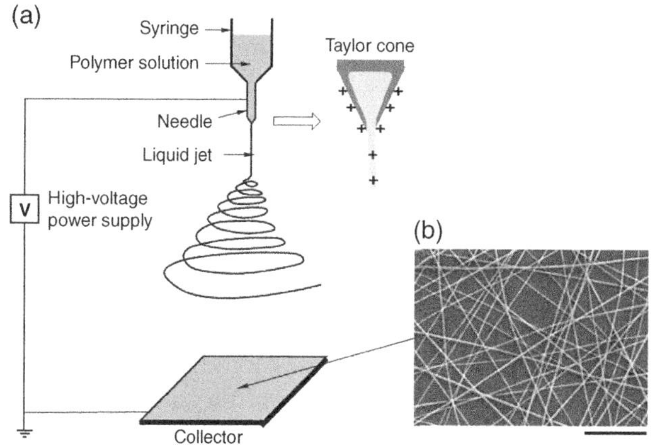

Figure 25.12 (a) Schematic illustration of the basic setup for electrospinning. The inset shows a drawing of the electrified Taylor cone. (b) SEM image of a nonwoven mat of polyvinylpyrrolidone (PVP) deposited on the collector (Bar = 10 μm). *Source:* From Li and Xia (2004) / with permission of John Wiley & Sons.

the liquid, causing a liquid jet in the form of a fiber to be ejected from the apex of the cone and accelerated toward the collector. The electrified liquid jet undergoes stretching, and it experiences a whipping instability that causes it to bend and turn rapidly. Continuous elongation along with evaporation of the solvent leads to considerable shrinkage of the liquid jet and the formation of solid fibers, usually having a smooth surface and a unimodal distribution of fine diameters. Attracted by the electrically grounded collector, often in the shape of a flat plate, the fibers are often deposited in a random arrangement to form a highly porous nonwoven mat (Figure 25.12b).

Parameters and Variations of the Electrospinning Process

The diameter of the electrospun fibers and their arrangement in the deposited material are dependent on a variety of parameters, including:

- The properties of the solution such as its viscosity, which can be controlled by the polymer concentration and its electrical conductivity, the composition and molecular weight of the dissolved polymer, and the polarity and surface tension of the solvent.
- The operational conditions, such as the applied voltage (or electric field), the distance between the spinneret and the collector, the flow rate of the solution through the spinneret, the spinneret tip design, and the collector geometry and composition.
- The ambient conditions, such as the temperature and humidity of the electrospinning environment.

Theoretical models have been developed to analyze the mechanisms underlying the electrospinning process, which could provide an improved understanding in designing and modifying set-ups to better control the fiber diameter and structure of the deposited mats. Additionally, several studies have examined how each of the aforementioned parameters influences the fiber morphology. However, it is difficult to isolate the effect of individual parameters because several of them are interrelated. For example, changing the solution viscosity (or its concentration) affects other properties, such as the surface tension and conductivity of the solution. In practice, a trial-and-error approach is often employed in which the solution properties, the operational conditions, and the ambient conditions are varied until a suitable structure of uniform defect-free fibers is obtained.

In general, scaffolds consisting of a highly porous mat of randomly oriented solid fibers may not be optimal for certain tissue engineering applications, such as the regeneration of blood vessels, tendons, or ligaments. Consequently, several variations have been introduced into the aforementioned technique to modify the arrangement of the fibers and their geometry. For example the use of a collector consisting of a cylinder rotating at high speed has been used to produce aligned fibers (Figure 25.13a), while the use of a spinneret consisting of two coaxial capillaries has been used to create hollow fibers (Figure 25.13b).

Electrospun Scaffolds for Tissue Engineering

Electrospinning has been used to create a large number of scaffolds, composed of natural or synthetic polymers, degradable or nondegradable, which can be homopolymers, copolymers, blends of two or more polymers, or matrices for composites. These scaffolds have been studied and evaluated for engineering a variety of tissues, including tendon, ligament, bone, cartilage, smooth and skeletal muscle, vascular graft, and neural tissue. Electrospinning has also been studied and evaluated for creating drug delivery devices.

As noted earlier, electrospun scaffolds formed by deposition on a flat plate consist of a porous mat of randomly oriented fibers. The characteristic structural features of these mats are high porosity, good pore interconnectivity, fine fibers and high surface area to volume ratio. Several studies have shown porosities in the range ~0.7–0.9 and an average fiber diameter in the range ~10 nm to 10 μm, but some studies have reported values outside these ranges.

As with other scaffolds for tissue engineering, the pore size and pore interconnectivity of electrospun scaffolds are also important because they control the ability of cells to migrate through the scaffold. Normally, the fiber axes in an electrospun mat are inclined only a few degrees to the areal plane of the mat. Consequently, these mats are not truly three-dimensional networks in which the fibers are randomly oriented (in three dimensions) but, instead, they can be thought of as layers of two-dimensional networks stacked upon one another, with their thickness determined by the number of layers. Such networks have been analyzed theoretically to determine the average pore size of and its dependence on various characteristics of the network, such as porosity (in the range 70–90%), fiber radius (1–100 nm), areal density (density per unit area) (1–20 g/cm^2), and fiber density (0.9–1.7 g/cm^3) (Eichhorn and Sampson 2005). Overall, the analysis predicts that the average pore size increases with increasing porosity and decreases with increasing areal density and decreasing fiber density, predictions that are expected intuitively.

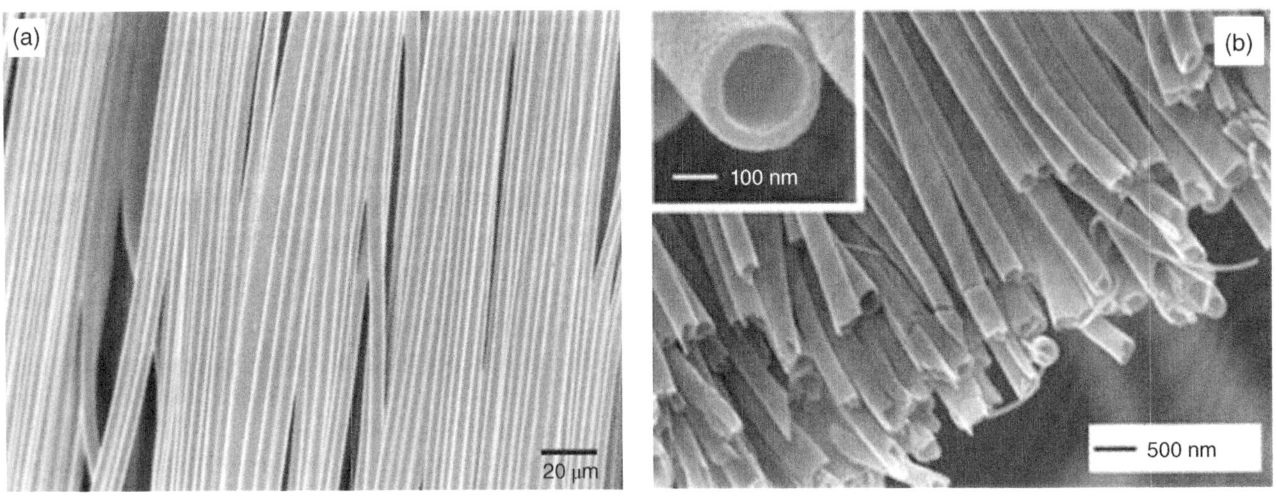

Figure 25.13 Modifications of the basic electrospinning technique to produce mats with various morphologies. (a) SEM image of aligned fibers deposited on a rotating drum collector. *Source:* From Chew et al. (2005). (b) SEM image of hollow fibers produced using a coaxial two-capillary spinneret. *Source:* From Li and Xia (2004).

For a network of porosity = 80% and areal density = 2 g/cm^2, consisting of fibers of diameter = 100 nm and density = 1 g/cm^3, approximating the parameters for an electrospun mat composed of PCL, the average pore size is predicted to be only ~10 nm. Additionally, for a given fiber diameter and fiber density, the mean pore radius is predicted to decrease with increasing areal density (up to ~10 g/cm^2). As the areal density of the network increases with an increasing number of layers, the average pore size will decrease further with an increase in thickness of the network. An interesting prediction of the analysis is that for a given areal density and porosity, the average pore size of the network decreases with decreasing fiber diameter.

Cells relevant to tissue engineering have an average diameter greater than ~10 μm and, consequently, simply making the diameter of the fibers smaller, with the aim of increasing the surface area of the scaffold to improve cell adhesion and proliferation, will severely limit the ability of cells to infiltrate the scaffold. Because of this limitation, electrospun scaffolds with pores in the nanoscale range were initially thought to present a hindrance to certain tissue engineering applications. Overall, however, such scaffolds can provide an advantage in some applications where cell infiltration deep into the scaffold is not a requirement, such as in the tissue-engineering of vascular grafts. Although the majority of the ECM is three-dimensional in nature, the endothelium of blood vessels can be considered as approximating a two-dimensional structure (Chapter 24). An electrospun scaffold with nanoscale pores can provide superior EC attachment because of its high surface area, but its fine pores could potentially prevent SMC migration into the lumen of the vessel, while still allowing transport of biomolecules and nutrients, and removal of waste products.

On the other hand, electrospun scaffolds with nanoscale pores are not advantageous for tissue engineering applications, such as the regeneration of bone or cartilage, which require cells to infiltrate the entire scaffold and create new tissue. Several approaches have been used to alleviate this problem, based mainly on modifying the electrospinning parameters or modifying the deposited structure after the electrospinning process. Increasing the fiber diameter of the electrospun network, as noted earlier, is predicted to result in an increase in average pore size. Thicker fibers and, thus, larger average pore sizes can be obtained by modifying the solution and operational parameters of the electrospinning process, such as polymer concentration, flow rate, applied voltage, and distance between the spinneret and the collector. For example, using this approach, PCL scaffolds of fiber diameter ~2–10 μm and average pore size of ~20–45 μm have been prepared. One approach based on modifying the scaffold structure after electrospinning involves incorporating a sacrificial material into the polymer of interest, which is removed after the electrospinning process. For example, scaffolds composed of a blend of PCL, a polymer that dissolves slowly in water, and PEG, a more soluble polymer, can be formed by electrospinning and subsequently immersed in water to dissolve out the PEG phase, giving PCL scaffolds with larger pores.

25.4.3 Additive Manufacturing (3D Printing) Techniques

Additive manufacturing, commonly referred to in the non-engineering literature as three-dimensional (3D) printing, has become a widely used technique for creating scaffolds in the form of porous solids. As noted earlier, it allows scaffolds to be formed robotically with precise control of their external shape, customized to the shape of the tissue or organ to be regenerated, and their internal architecture, often both at the microscale and the nanoscale, which cannot be easily achieved by traditional methods. Cells or biomolecules are incorporated into the scaffold after its production. Additive manufacturing techniques cover a wide range, depending on the method used in forming the shaped material, and they can be used to create shaped forms of polymers, ceramics and metals, as well as composites. However, some techniques that rely on the use of harsh conditions, such as temperatures well above the physiological temperature (~37 °C), are not appropriate for natural polymers. For ceramics (and glasses), the techniques cannot normally be used to directly produce the final scaffold. Instead, they are used to convert ceramic particles into the desired shaped form, which is subsequently converted into the final scaffold using conventional methods such as sintering to bond the particles into a network of adequate strength (Chapter 7). Composite scaffolds composed of a ceramic particle-filed polymer matrix can often be formed using methods similar to those for the polymer alone but with a starting material composed of a homogeneous mixture of the two materials.

Figure 25.14 illustrates the basic approach of additive manufacturing. The physical object to be built is modeled on a computer-aided design file or an image of the object is obtained using a suitable method, such as a computed tomography (CT) or magnetic resonance imaging (MRI). The model (or image) is converted into a format that is analyzed by a computer, which slices the model into cross sections of the appropriate thickness. Then the cross sections are systematically recreated layer-by-layer in a computer-aided manufacturing step that employs suitable equipment and materials to produce an actual shaped object.

Salient features of some common additive manufacturing techniques for creating polymeric and ceramic scaffolds are discussed in the following sections. However,

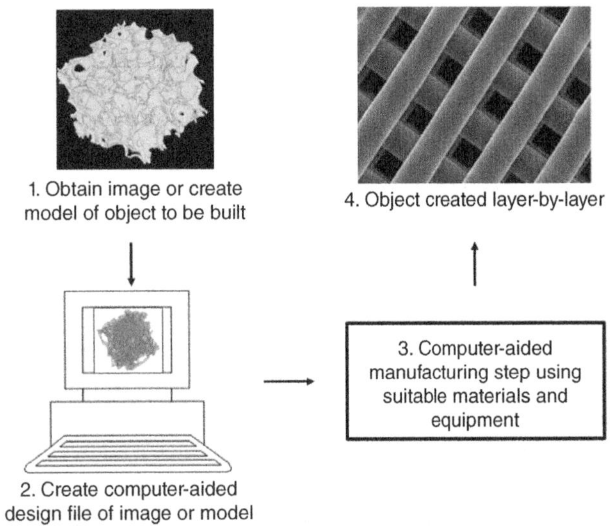

Figure 25.14 Illustration of basic approach to additive manufacturing (3D printing).

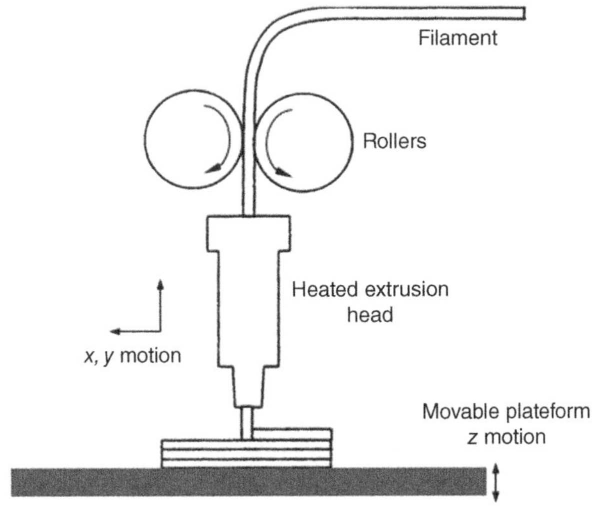

Figure 25.15 Schematic illustration of the fused deposition modeling (FDM) technique.

equipment and methods used in these techniques are not described in detail because they are well covered by the equipment manufacturers or in laboratory manuals. In general, selection of the appropriate technique to be used for a given tissue engineering application will depend on various factors, particularly the ease with which scaffolds having the required shape and internal architecture can be produced and the cost of the equipment.

Fused Deposition Modeling

Fused deposition modeling (FDM) (Figure 25.15) is essentially a computer-controlled hot extrusion process. A coarse polymer filament (~2 mm in diameter) is continuously fed into a computer-controlled hot extrusion head (~100–150 °C), where it is softened and extruded through a nozzle as a continuous finer filament (diameter in the range ~0.25–1.0 mm) to form a three-dimensional structure layer-by-layer according to a computer-controlled pattern. In FDM, extrusion is achieved by means of the force provided by the coarse solid filament fed into the hot extrusion head. As in most extrusion-based forming techniques, the rheology (deformation and flow behavior under an applied stress) of the material being extruded is an important property. The softened polymer should have a low enough viscosity for it to flow and be extruded as a continuous filament and, upon deposition, the printed filament should cool at an appropriate rate to develop sufficient rigidity to form a mechanically stable structure. The diameter of the extruded filament determines the scale of the solid features of the scaffold, while the porosity and pore size of the scaffold are determined by the diameter and spacing of the extruded filament and the directions of the filaments in successive layers.

Selective Laser Sintering

Selective laser sintering (SLS) of polymeric scaffolds uses a starting material in the form of particles (Figure 25.16). A thin layer of particles is rolled onto a platform from a reservoir and a computer directs a laser to raster onto the powder bed, fusing particles in certain areas of the bed into the required pattern. After one layer is formed, the platform is lowered by the appropriate distance, another layer of particles is rolled onto the platform, and the computer directs the laser to form a second layer that bonds to the previous layer. This process is repeated until the entire object is formed. An additional step, such as mechanical vibration of the as-formed object, is required to remove loose particles within the interior of the object in order to obtain the actual porous scaffold. Overall, SLS is not a very useful technique for producing tissue engineering scaffolds because it is limited in its ability to produce scaffolds with fine-scale structural features, plus the equipment is normally expensive, often consisting of industrial scale equipment.

Stereolithography

A schematic of the stereolithography technique is shown in Figure 25.17a. An ultraviolet (UV) laser beam is scanned on the surface of a photopolymerizable monomer solution according to a computer-controlled pattern, curing a thin layer (~100–200 μm) of the monomer solution to form a polymer layer on a platform. Upon forming the first layer, the platform is lowered by a depth equal to the formed layer thickness, and the monomer solution flows over the polymerized layer. The laser is scanned over the new surface to form a second layer and the process is repeated until the scaffold is formed. The photopolymerizable solution

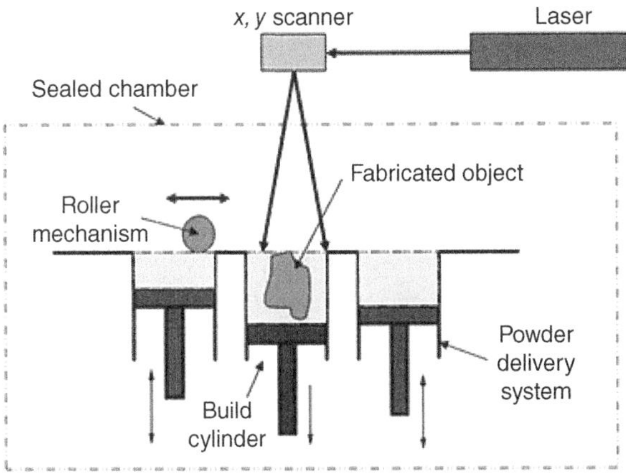

Figure 25.16 Schematic illustration of the selective laser sintering (SLS) technique.

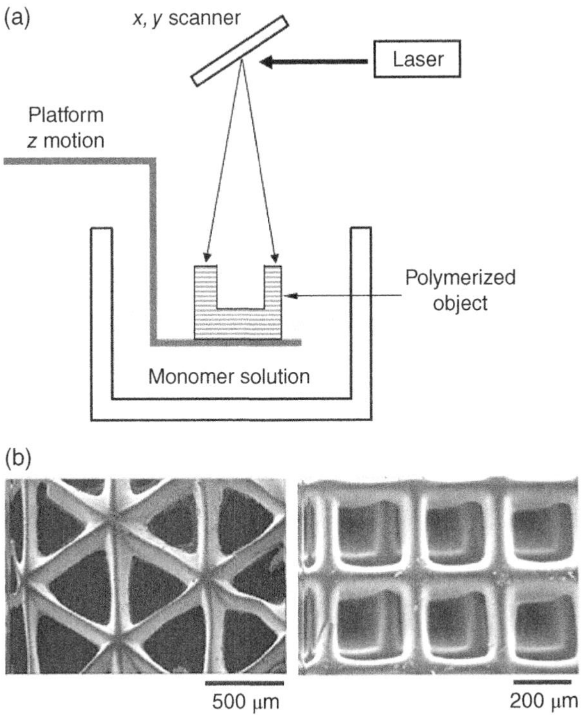

Figure 25.17 (a) Schematic illustration of the stereolithography technique; (b) examples of structures with fine-scale features formed by micro-stereolithography.

typically consists of the monomer and a nontoxic photoinitiator, plus a chemically unreactive solvent to control the viscosity of the solution. A related technique is microstereolithography, which is similar to stereolithography, except that the UV laser is focused to a narrower width (~1–2 μm) to cure a layer of thickness ~1–10 μm. Resolution in the *x*, *y*, and *z* directions at the submicron scale, coupled with the fine UV laser beam, enables precise microfabrication of scaffolds with fine-scale features (Figure 25.17b).

Robocasting

Robocasting can be used to create polymer scaffolds but it has been more widely used to create scaffolds composed of bioceramics such as such as hydroxyapatite, β-TCP and BCP, and bioactive glasses. It has also been used to create scaffolds of composites composed of degradable synthetic polymers such as PLA and PCL reinforced with hydroxyapatite particles. The technique involves computer-controlled, layer-by-layer extrusion of a concentrated polymer solution (for polymer scaffolds) or a concentrated slurry of ceramic particles (for ceramic scaffolds) through a nozzle of diameter in the range ~50–200 μm (Figure 25.18a).

As with fused deposition modeling, robocasting is essentially an extrusion-based process and, thus, the rheological behavior of the paste is important. The paste should have a low enough viscosity to allow flow as a continuous filament of uniform cross section but, upon deposition, it should attain sufficient rigidity to form a mechanically stable structure. This means that the viscosity of the paste should be low at high shear stresses (or stain rates) to allow flow but should increase rapidly as the shear stress gets close to zero. In rheological studies, this type of behavior is referred to as shear thinning, that is, the paste thins out at high shear stresses. The use of a viscometer to characterize the development of a suitable paste having the right rheological properties is often useful. Scaffolds composed of polymers or ceramic particle-reinforced polymers are produced directly by this technique but, as noted earlier, the production of ceramic scaffolds requires an additional sintering step following robocasting. Robocasting is one of the more useful techniques for creating scaffolds of bioceramics and bioactive glasses, providing the flexibility for forming scaffolds with a range of controlled architectures, two examples of which are shown in Figure 25.18b.

Inkjet Printing

In inkjet printing of polymer scaffolds, the polymer is dissolved in a suitable solvent to form a solution (called an ink), which is contained in a chamber and subsequently deposited on a surface as droplets (~10–100 μm in diameter) through a printer nozzle (Figure 25.19). The inkjet printing technique for creating scaffolds and, in general, materials with well-defined structural features, has its roots in commercial inkjet printing technology developed several decades ago for printing an ink on paper. Inkjet printing consists of two main types, referred to as continuous inkjet printing and drop-on-demand inkjet printing. Continuous inkjet printers generate a continuous stream of droplets, even when no printing is required, and the unwanted droplets are often deflected and recycled. Because the ink is forced through a nozzle under pressure, it breaks up into droplets due to a phenomenon known as

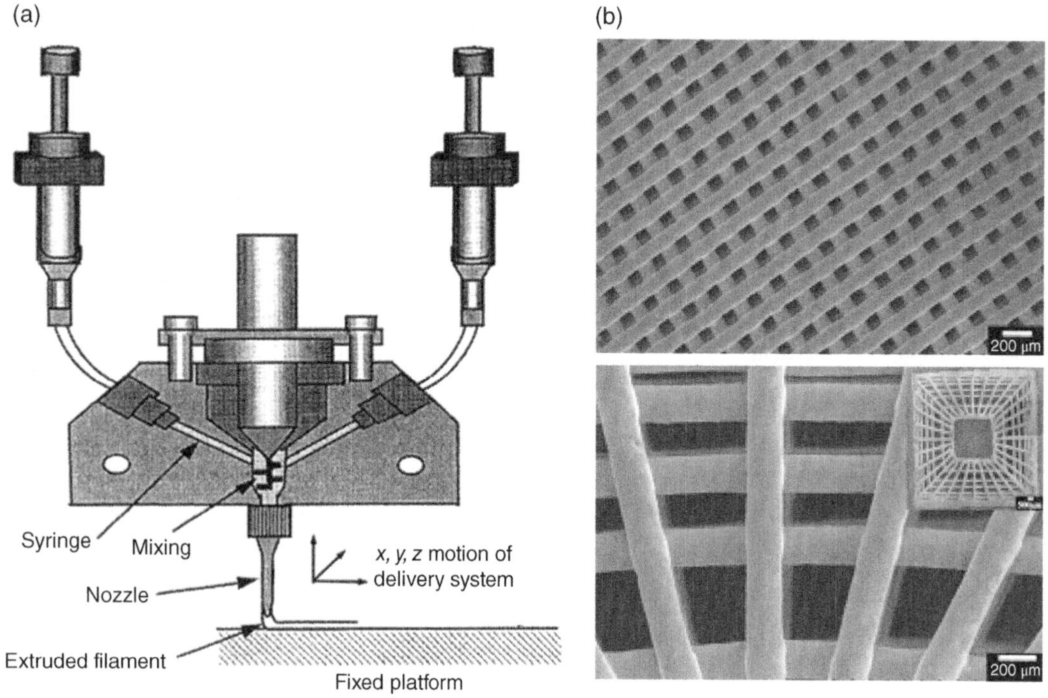

Figure 25.18 (a) Schematic of a two-nozzle delivery system in a robocasting apparatus. *Source:* From Mosissette et al. (2000) / with permission of John Wiley & Sons. (b) SEM images of bioactive glass (13–93) scaffolds with different architectures created by robocasting.

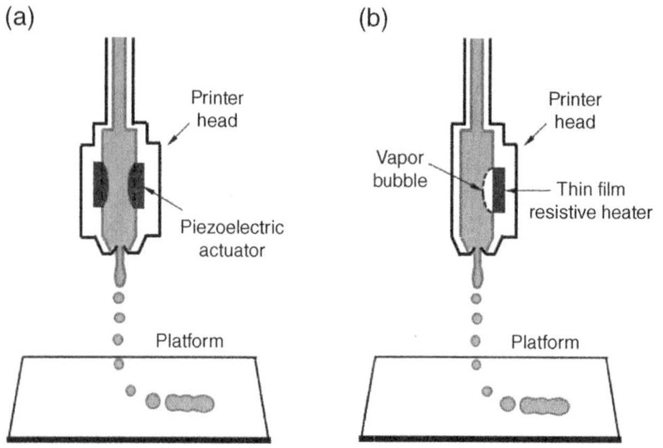

Figure 25.19 Schematic illustration of droplet formation in drop-on-demand inkjet printing by means of (a) a piezoelectric actuator and (b) a thermal actuator.

the Rayleigh instability, which governs the break-up of a cylindrical fluid into a line of spheres.

Drop-on-demand inkjet printers generate droplets only when required, making them more economical and easier to control. Because they are more widely used when compared to continuous inkjet printers, drop-on-demand printers will form the focus of our discussion in the remainder of this section. Drop-on-demand printers consist of single or multiple printheads, with each printhead consisting of a chamber in which a fluid (an ink) is placed and a single nozzle or multiple nozzles. The ink, contained in the chamber and held in place by the surface tension of its meniscus at the nozzle, is subjected to pressure pulses often generated by means of a thermal or piezoelectric actuator, which cause individual droplets to be ejected when the surface tension of the ink is overcome. Upon deposition on a solid surface, contact and adhesion between neighboring droplets, followed by a liquid to solid phase change, lead to the formation of a solid continuous layer according to a computer-controlled pattern. The liquid–solid phase change can occur through solvent evaporation, a change in temperature or gelling. In the creation of scaffolds with three-dimensional architectures, droplets for the second and subsequent layers are deposited on top of the preceding solidified layer. Inkjet printing can also be used to form shaped objects of ceramics and glasses, in which case the ink consists of a colloidally stable suspension consisting of fine ceramic or glass particles.

Inkjet printing can be considered to consist of two distinct processes: the formation of individual droplets that are directed at a specific location on a target surface, and the interaction of the drop with the surface during and following impact (Saunders and Derby 2014). In order to form individual drops, an ink should have physical properties within a certain range. Numerical analysis has shown that droplet formation from a fluid can be defined in terms of a dimensionless parameter Z, which is related to group of physical parameters of the fluid according to the equation

$$Z = \frac{\eta}{(\gamma d a)^{1/2}} = \frac{(\mathrm{We})^{1/2}}{\mathrm{Re}} = \frac{1}{\mathrm{Oh}} \tag{25.1}$$

where η is the dynamic viscosity, γ the surface tension, and d the density of the fluid, and a is a characteristic length, taken as the diameter of the printer nozzle. As noted in Eq. (25.1), Z can also be expressed in terms of the well-known dimensionless parameters called the Weber number We, the Reynolds number Re, and the Ohnesorge number Oh. Individual droplets of uniform size are formed when $1 < Z < 14$. In comparison, for $Z < 1$, viscous dissipation prevents droplet ejection whereas for $Z > 14$, the primary droplet is accompanied by a large number of smaller (satellite) droplets. In addition to the physical properties that govern the formation of uniform individual droplets, it was proposed that a minimum drop velocity is required to overcome the surface tension of the liquid meniscus at the base of the printer nozzle.

The impact of a droplet on a solid surface is also important in inkjet printing because it determines the form that the drop will assume following impact. Analysis has shown that lower droplet velocities lead to spreading on the surface, whereas higher velocities lead to splashing. One characterization of a droplet splashing upon impacting a solid surface is in terms of its Weber number, We, given by

$$\mathrm{We} = \frac{d a U^2}{\gamma} \tag{25.2}$$

where U is the velocity of the droplet, and the other terms are defined from Eq. (25.1). In general, lower We leads to spreading while higher We leads to splashing.

Based on the aforementioned properties and factors, a stability map can be constructed to show the fluid properties that are necessary for drop-on-demand inkjet printing (Figure 25.20). Although the numerical analysis used in constructing this stability map is dependent on the dynamic viscosity, surface tension, and density of the ink, in practice the most critical parameter is often the dynamic viscosity. Many polymers used as scaffold materials have a density comparable to that of the aqueous liquid used as the medium for the ink and, consequently, their dissolution in the liquid causes little change in the density of the ink. The surface tension of the ink changes over a limited range with an upper limit equal to that of water.

The transformation of a droplet to a solid deposit is also important because it has an effect on the overall uniformity

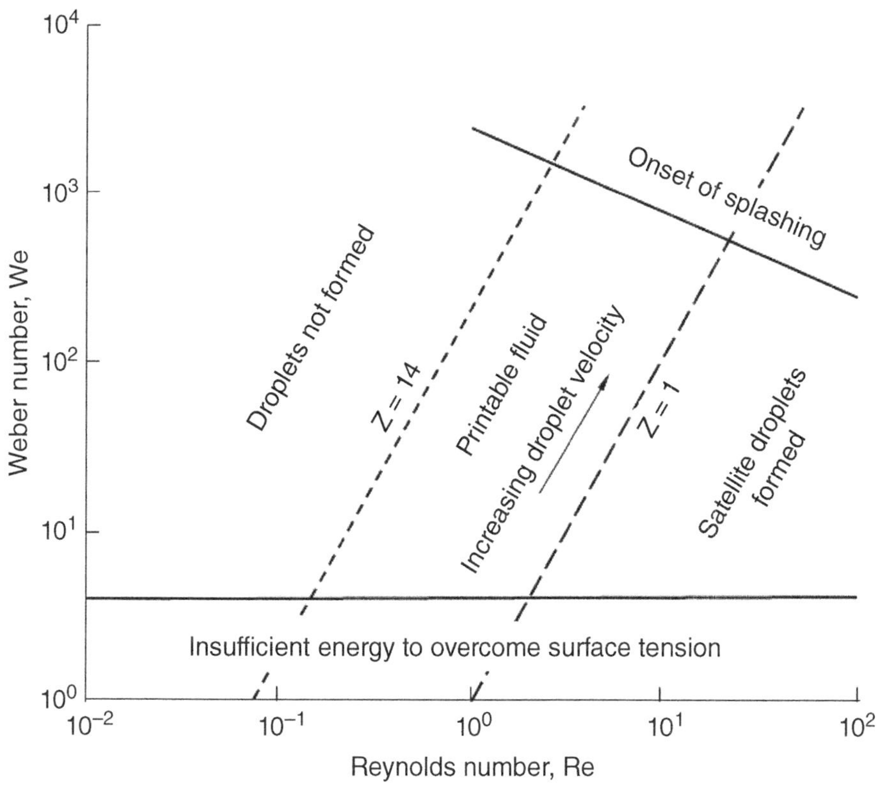

Figure 25.20 Chart showing the region of fluid properties where inkjet droplet formation is optimized. *Source:* Based on Derby (2011).

of the printed layer. This can be seen by considering first an isolated droplet on a solid surface. Upon impact on a surface, the droplet spreads and assumes a shape dictated by the contact angle of the droplet with the surface. When transformation of a droplet to a solid deposit occurs by drying, evaporation of the solvent is fastest at the contact line between the circumference of the droplet and the solid surface. Because of this fast evaporation, deposition of solid occurs first at the contact line, forming a ring that pins the outer edge of the droplet and defines the circumference of the deposit. Thereafter, solvent flows from the interior of the droplet outward toward the edge where it evaporates. This outward flow can also carry solute from the interior of the droplet to the edge, resulting in accumulation of a solid deposit preferentially at the edge of the droplet, not unlike the pattern of a coffee stain on a surface. Significant redistribution of solute can therefore occur during drying, which can lead to a nonuniform distribution of the solid phase. Tissue engineering scaffolds are formed not from isolated droplets but by contact and adhesion between neighboring droplets, followed by their solidification, leading to the formation of a solid continuous layer. However, similar effects apply, which can give a nonuniform thickness over the width of the layer. As the second and subsequent layers are deposited on the preceding solid layer, uniform deposition of the solute over the entire contact area of the droplet with the solid surface or with the preceding solid layer is desirable, rather than the coffee-stain pattern. This requires control of the properties of the ink and the operating conditions such as the drying rate.

An advantage of drop-on-demand inkjet printers is their low cost. In addition, the size of the droplets and their deposition rates can be controlled with high precision. Because the droplets normally have a diameter in the range ~10–100 μm, which defines the scale of resolution of the structural features, another advantage of inkjet printing is the ability to create scaffolds with fine-scale structural features. On the other hand, the biomaterial to be printed should be capable of being formed into a low-viscosity solution to enable the formation of droplets. Additionally, it is not easy to create scaffolds having truly three-dimensional architectures by inkjet printing because a droplet has to be deposited on an already existing (solid) surface.

25.4.4 Formation of Hydrogel Scaffolds

Various physical and chemical methods for creating hydrogels were discussed in Chapter 11, along with the methods used for specific hydrogels composed of synthetic polymer molecules. Methods for specific hydrogels composed of natural polymer molecules were discussed in Chapter 10. When compared to scaffolds composed of porous solids, a key difference is the way in which cells are incorporated into hydrogel scaffolds. Because the mesh size of hydrogels is normally much smaller than a cell diameter, cells have to be incorporated within a hydrogel scaffold prior to its gelation rather than into the as-formed hydrogel. Consequently, methods used for creating hydrogel scaffolds should be consistent with the toxicological limitations of cells.

25.4.5 Preparation of Extracellular Matrix (ECM) Scaffolds

ECM scaffolds, as noted earlier, are normally prepared by decellularization and sterilization of intact mammalian tissues or organs, normally xenogeneic or allogeneic in origin. Because cells and cellular debris remaining in the scaffold can induce an inflammatory response or an immune-mediated rejection of the scaffold, decellularization is a critical step in preparing an ECM scaffold. In general, the goal of the decellularization protocol is to remove all cellular material while minimizing any deleterious effect on the composition, biological activity and mechanical integrity of the remaining ECM scaffold. However, all decellularization protocols have, to some degree, a deleterious effect on the growth factors present in the ECM, the structure of the ECM components, and the topology of the ECM scaffold.

Decellularization protocols vary widely, depending on the tissue from which the ECM is derived and the species of origin, but the most effective decellularization methods often involve a combination of physical, chemical, and enzymatic treatments (Gilbert et al. 2006). Physical treatments, normally the first stage of the decellularization process, include mechanical agitation or sonication, application of mechanical pressure, or freezing and thawing. These treatments generally disrupt the cell membrane, causing the release of the cell contents, and they facilitate subsequent rinsing and removal of cell contents from the ECM. However, the use of physical treatments alone is not effective because some cells and cell debris normally remain in the ECM.

Chemical treatments involve washing the ECM in acidic or alkaline solutions, such as acetic acid, peracetic acid, hydrochloric acid, or ammonium hydroxide. These solutions can effectively disrupt cell membranes and intracellular organelles, but they can also dissociate important biomolecules such as GAGs from collagenous tissues. However, for thin ECM structures, such as those derived from porcine SIS, treatment in dilute peracetic acid has been shown to be effective in preserving the structure and function of various growth factors present in the ECM, such as TGFβ, bFGF, and VEGF, without having any significant deleterious effect on the mechanical behavior of the ECM scaffold. A variety of milder solutions, composed of detergents such as Triton X-100, Triton X-200, or sodium dodecyl sulfate (SDS), have also been shown to be effective as chemical treatments.

Enzymatic treatments often involve protease digestion using trypsin, a highly specific enzyme that cleaves the peptide bond on the carbon side of arginine and lysine if the next amino acid residue is not proline. However, some studies show efficient removal of cellular material from some tissues (such as porcine pulmonary valves) following treatment with a solution composed of trypsin and ethylenediamine tetraaceticacid, whereas other studies show less efficient decellularization.

Overall, the effectiveness of decellularization and alterations to the ECM vary, depending on the source of the tissue and the type of tissue, which control factors such as cell density, matrix density, and thickness and morphology of the tissue. Consequently, depending on the tissue being studied, it will be necessary to optimize the decellularization protocol to achieve acceptable cell removal.

25.5 Three-dimensional Bioprinting

When compared to additive manufacturing (3D printing), three-dimensional bioprinting (3D bioprinting) represents an advance in the tissue engineering approach. Instead of creating scaffolds and subsequently seeding them with cells in a separate step, 3D bioprinting involves the creation of cell-seeded constructs through a single fabrication process involving the printing of biomaterials, seeding them with living cells, and placing the cells at specific locations to mimic their distribution in native tissues and organs.

In general, human tissues contain multiple cell types and extracellular matrices arranged in specific patterns but it is difficult to approximate this complexity of human tissues through the use of additive manufacturing alone or the more conventional fabrication techniques described earlier in this chapter. Those techniques suffer from drawbacks, which include low density and nonuniform cell seeding of the scaffold, limited ability to place cells at specific locations, and limited ability to allow seeding of multiple cell types within the scaffold. 3D bioprinting provides the potential to overcome these drawbacks.

On the other hand, 3D bioprinting involves additional complexities related to selection of biomaterials, cell types, and biomolecules, and additional technical challenges such as using aseptic conditions and maintaining the viability of living cells and the activity of biomolecules present during the fabrication process. For example, cultured mammalian cells should be kept in a 5% CO_2 atmosphere for them to remain viable. In addition, they can be removed from the 5% CO_2 atmosphere for only a few minutes because the medium quickly becomes alkaline, resulting in cell death. Because the duration of the bioprinting process would normally be considerably longer than a few minutes, it is necessary that the cell-containing mixture used for bioprinting and other components of the process be held within a narrow environmental range required to maintain cell viability.

3D bioprinting has been an area of considerable research and development in tissue engineering and regenerative medicine over the last decade or so, and it has already been used for the creation and transplantation of various tissue constructs, including those of skin, bone, cartilage and vascular grafts, and of liver, lung, nerve, heart, and pancreas tissues. A variety of 3D bioprinting techniques are available for creating functional cell-incorporated tissue constructs with the desired shape and architecture, but the most widely used techniques are based on inkjet printing and microextrusion. Microstereolithography has also been used for forming three-dimensional structures with intricate shapes but, except for the ability to form fine-scale features, it is similar to conventional stereolithography. In addition, there are concerns about the toxicity of the chemicals used in the technique, such as the crosslinking agents. The following sections provide a discussion of the principles underlying inkjet and microextrusion bioprinting. As hydrogels are widely used in these two techniques, a review of Chapter 11 is recommended as a background for this discussion.

25.5.1 Inkjet-Based Bioprinting

Except for a difference in the constitution of the ink, the inkjet bioprinting technique is similar to that of inkjet printing described earlier (Section 25.4.3). As noted, droplets are generated, commonly by means of a thermal or piezoelectric actuator, and deposited on a surface to form a three-dimensional structure (Figure 25.19). In thermal inkjet printers, a thin-film resistor heats the ink locally when a voltage pulse is applied to it. This causes a bubble to form, expand and collapse rapidly, generating a pressure pulse within the ink to cause ejection of droplets. In piezoelectric inkjet printers, a piezoelectric actuator rapidly changes its shape when a voltage pulse is applied to it, causing the ink chamber to deform rapidly, which generates a pressure wave that leads to the ejection of droplets.

Inkjet bioprinting, in common with inkjet printing, has the advantages of economical equipment cost and the ability to create constructs having fine-scale structural features due to the small droplet sizes (~10–100 μm). A concern in thermal inkjet bioprinting has been the elevated temperature of the ink close to the thin-film heating element, which may reach ~200–300 °C. However, *in vitro* studies have shown that this elevated temperature does not have a significant impact on the viability of cells or the activity of growth factors encapsulated within the as-formed construct. In general, the duration of each heating cycle is very short (a few microseconds), which is thought to cause little heating of the ink in regions remote from the heater. Concerns have been expressed about the frequencies (~10–25 kHz) used in piezoelectric inkjet printers because they can potentially disrupt cell membranes and the structure of biomolecules, by, for example, causing proteins to unfold. However, *in vitro* studies have shown high average cell viability (greater than ~85–90%) for a variety of cells encapsulated within scaffolds formed by piezoelectric inkjet printers.

A commonly encountered limitation of inkjet bioprinting is the difficulty in achieving a sufficiently high cell density within the scaffold. A low cell concentration in the ink is necessary to maintain the ink viscosity low enough to enable droplet formation and to avoid clogging of the printer nozzle. Consequently, the average number of cells within a droplet of diameter ~100 μm is low (approximately 1–2), which gives a cell density within the scaffold of less than ~10^6 cells/cm^3. This cell density is sometimes too low for efficient tissue formation.

As with inkjet printing, the biomaterial to be printed to form the scaffold should be capable of being converted into a low-viscosity solution to enable formation of droplets and, following deposition, the droplets should be capable of being transformed to a solid phase. However, the presence of cells and other biological entities in inkjet bioprinting introduces additional complexities related to cytotoxicity and biocompatibility. For example, the biomaterial should have a composition such that it has good solubility in a nontoxic aqueous solvent. Additionally, transformation of the droplets to form a solid phase should be achieved under mild conditions to maintain cell viability and avoid deactivation of biomolecules.

A common method to convert droplets to a solid phase is to employ a sol–gel transition. As discussed in Chapter 11, this involves converting a low-viscosity solution of polymer molecules to a high-viscosity gel composed of crosslinked polymer molecules, which can have a consistency ranging from a soft pliable solid to a rigid solid. There are two main approaches to forming a three-dimensional scaffold by means of the sol–gel transition using inkjet bioprinting. In one approach, cell-containing droplets of a polymer solution are deposited from an inkjet printer on a surface to form a layer (Figure 25.21a). Then the polymer molecules in the layer are crosslinked to form a solid network that encapsulates the cells by depositing droplets of a chemical crosslinking agent from another inkjet printer. This process is repeated layer-by-layer until the entire cell-containing scaffold is formed. A variation of this approach is to deposit droplets of the crosslinking agent from an inkjet printer at the desired location on a platform at the surface of a cell-containing polymer solution located in a chamber below the printer cartridge. In this way, a layer of the polymer molecules at the surface of the solution is crosslinked to form a gel. Then the platform is lowered by the appropriate distance (equal to the thickness of the gelled layer) into the cell-containing polymer solution and the crosslinking process is repeated layer-by-layer. Because of the ease of crosslinking sodium alginate solutions under mild conditions using a calcium chloride ($CaCl_2$) solution, this approach has been widely used to create cell-containing alginate scaffolds. Droplets of $CaCl_2$ solution can be deposited on inkjet-printed layers of a cell-containing sodium alginate solution (Figure 25.21a) or on the surface of a cell-containing sodium alginate solution contained in a chamber.

In the other approach (Figure 25.21b), cell-containing droplets are deposited by inkjet printing on a surface, after which the polymer molecules in each layer are crosslinked to form a solid network that encapsulates the cells and other biological entities. Crosslinking can be achieved through the formation of covalent or ionic bonds by a variety of mechanisms, including the use of a chemical

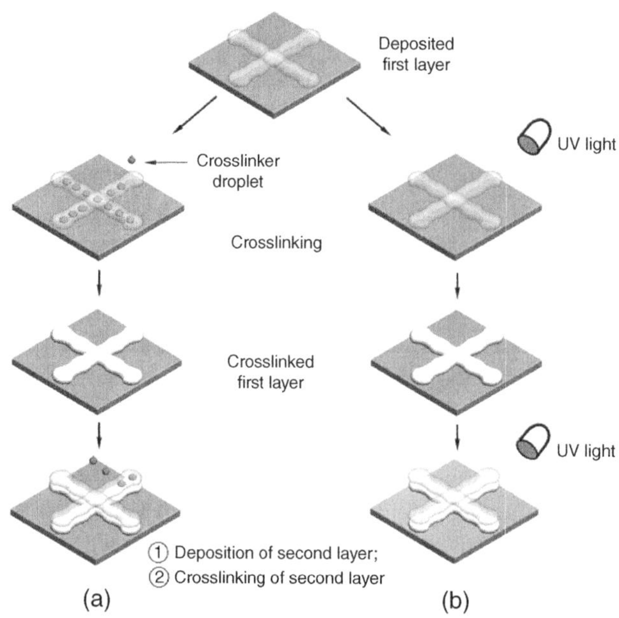

Figure 25.21 Illustration of two approaches for the creation of cell-containing three-dimensional hydrogel constructs using inkjet-based bioprinting by sequential layer-by-layer deposition and crosslinking. (a) Gelation of a solution (sol) by means of droplets of a crosslinking solution; (b) gelation of a solution (sol) by means of photopolymerization. *Source:* Adapted from Gundapati et al. (2016).

crosslinking agent, a change in pH, or electromagnetic radiation such as UV light (Chapter 11). This approach has been widely used to create cell-containing scaffolds of PEG from solutions composed of photopolymerizable molecules such as polyethylene glycol diacrylate (PEGDA).

As noted earlier, one of the advantages of 3D bioprinting is the ability to incorporate multiple cell types within a scaffold and to place them at precise locations to mimic the complexity of native tissue. An example is the creation of tissue constructs composed of an alginate matrix containing three different cell types, human amniotic fluid stem cells (AFSCs), canine SMCs and bovine aortic ECs (Xu et al. 2013). These cells were separately mixed with a $CaCl_2$ solution and loaded into separate printer cartridges. Then the cell suspension was deposited in the form of droplets at predetermined locations onto a solution composed of a mixture of sodium alginate and collagen located in a chamber below the cartridge. Reaction between $CaCl_2$ and sodium alginate resulted in the formation of a composite (alginate-collagen) gel scaffold in which the cells were anchored at specific locations within the gel. Repetition of this process resulted in layer-by-layer formation of a three-dimensional multicell construct (Figure 25.22a).

Fluorescent microscopy showed that the cells were located at their respective predetermined areas within the scaffold, while scanning electron microscopy showed that the construct had a structure resembling numerous randomly dispersed micro-shells, with cells located inside or around these micro-shells (Figure 25.22b). The overall tissue structure of the cell-seeded scaffold was maintained after incubation for three days *in vitro* followed by implantation for two weeks in rodent subcutaneous sites *in vivo*. In addition, staining protocols for SMCs and ECs showed that these cells remained viable in their predetermined locations *in vivo*.

For printed scaffolds containing individual cell types, assays showed that the cells maintained their viability, normal proliferation rates, phenotype expression, and function during *in vitro* culture over a seven-day period. Gross examination of the superficial area of EC-containing scaffolds after implantation in rodent subcutaneous sites for 8 weeks showed more abundant blood vessels when compared to controls (printed scaffolds without ECs) (Figure 25.23). In addition, immunohistochemical analysis showed that the interior of the EC-containing scaffolds contained a number of blood vessels whereas there were no observable blood vessels in the controls. Bone formation was observed within AFSC-containing scaffolds following implantation in rodent subcutaneous sites for 18 weeks (Figure 25.24), whereas no bone formation was observed in the controls (scaffolds without AFSCs). Overall, these results show the feasibility of creating complex tissue constructs containing multiple cell types using inkjet bioprinting and the ability of these constructs to survive and mature into vascularized tissues when implanted *in vivo*.

25.5.2 Microextrusion-Based Bioprinting

Microextrusion-based bioprinting is similar in principle to conventional extrusion techniques used for the production of solid or hollow tubes for industrial applications, except that the extruded filaments are fine (~50–200 μm in diameter) and are deposited robotically to form a three-dimensional structure. Figure 25.25 illustrates the basic

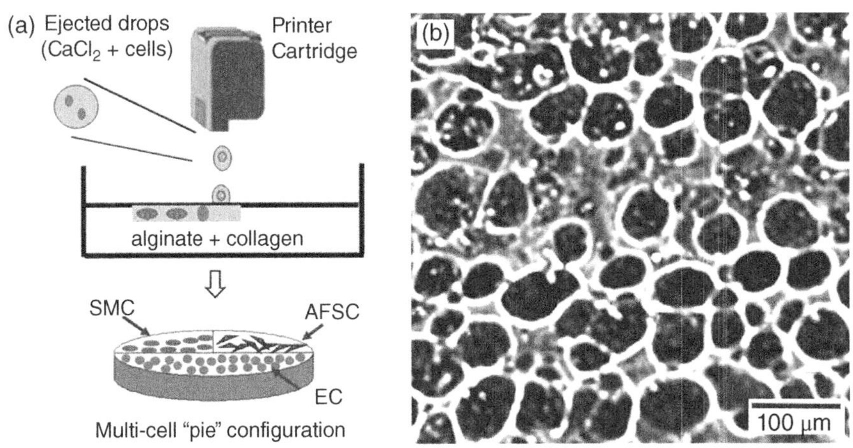

Figure 25.22 (a) Schematic illustration of a 3D bioprinting method to create multicell heterogeneous tissue constructs; (b) optical microscope image of the surface of the as-fabricated construct showing a structure resembling numerous micro-shells dispersed randomly within the construct, with cells (small bright dots) localized inside or around the micro-shells. *Source:* From Xu et al. (2013) / with permission of Elsevier. SMC, canine smooth muscle cells; AFSC, human amniotic fluid-derived stem cells; EC, bovine aortic endothelial cells.

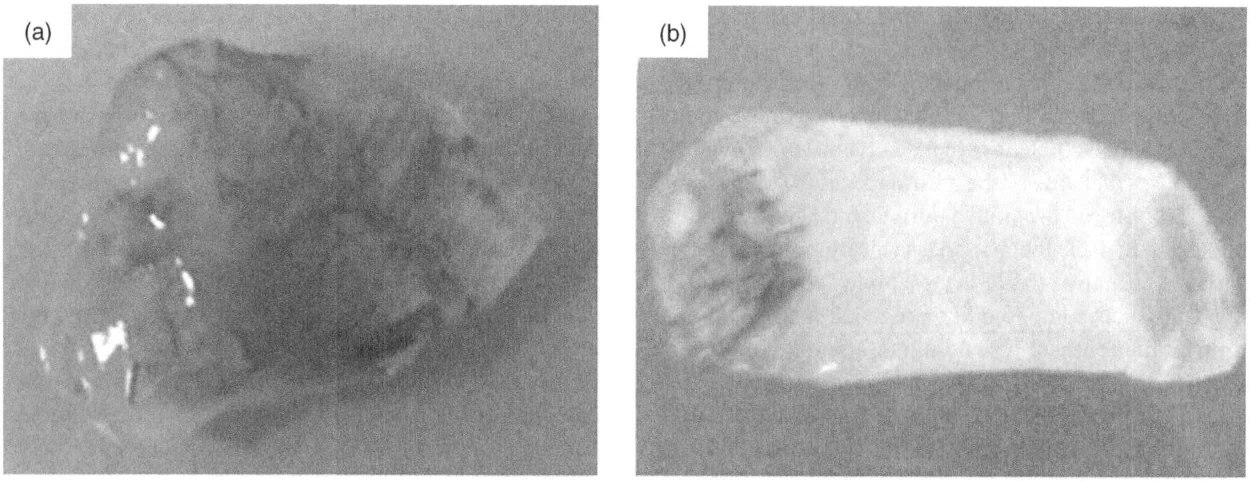

Figure 25.23 Vascularization of endothelial cell (EC)-containing printed tissue constructs after subcutaneous implantation for 8 weeks in athymic mice. Gross examination showed that the superficial area of the EC-containing implants (a) contained more abundant blood vessels than the controls (printed constructs without ECs) (b). *Source:* From Xu et al. (2013).

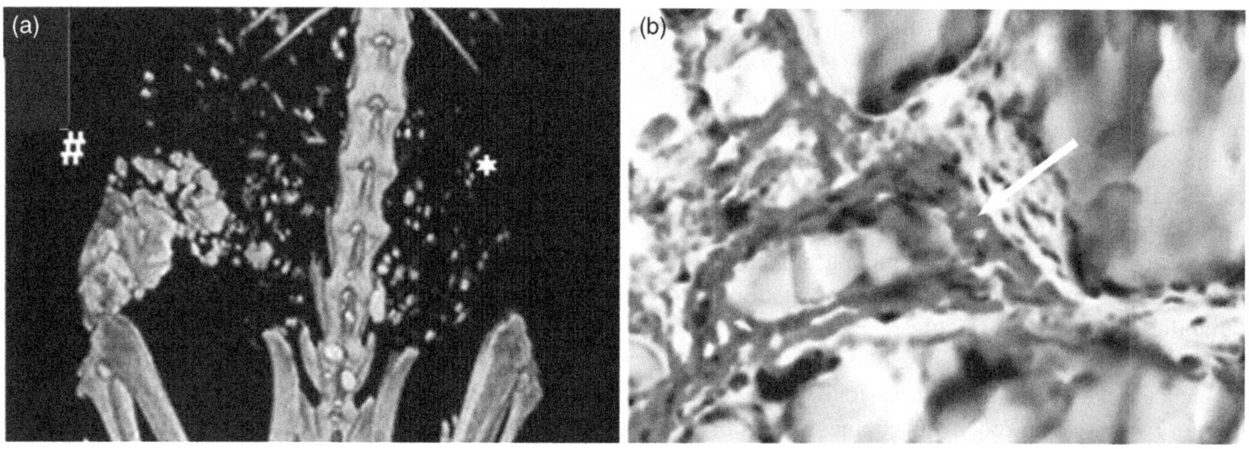

Figure 25.24 Bone formation in amniotic fluid stem cell (AFSC)-containing printed tissue constructs after subcutaneous implantation in athymic mice. (a) Micro-CT images of the printed tissue constructs after subcutaneous implantation for 16 weeks in athymic mice showed bone formation in the AFSC-containing implants (#) but not in the implants without AFSCs (*). (b) Immunohistochemical analysis of the AFSC-containing implants after subcutaneous implantation for 18 weeks showed that the differentiated AFSCs expressed osteocalcin, a typical bone cell marker (arrow). *Source:* From Xu et al. (2013).

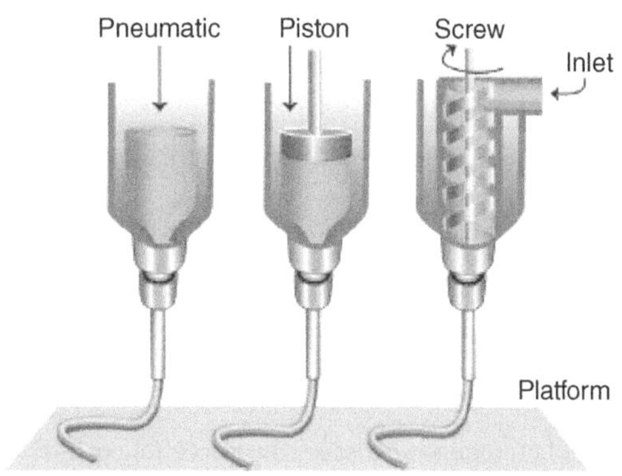

Figure 25.25 Schematic illustration of microextrusion-based three-dimensional bioprinting, showing different methods of extrusion. *Source:* From Malda et al. (2013) / with permission of John Wiley & Sons.

elements of the microextrusion technique for bioprinting cell-containing scaffolds. Typically, an aqueous suspension with the appropriate viscosity, composed of polymer molecules, cells and other biological entities, is forced out through the nozzle of an extruder head and deposited as filaments on a platform to form a three-dimensional structure according to a computer-controlled program. Commonly, the suspension is forced out by means of a pressure, applied either pneumatically (by means of a pressurized gas) or mechanically (using a piston or a screw), while the extrusion head or the platform is capable of movement in the x, y, and z directions. Some equipment may also have more than one extrusion head, each capable of containing different mixtures and of being moved independently.

Microextrusion is a popular bioprinting technique, which has the benefits of economical equipment cost and simplicity of the deposition process. Whereas inkjet bioprinting requires that the biomaterial be converted into a low-viscosity solution suitable for drop formation (viscosity up to ~4–30 mPa s), suspensions with a much wider range of viscosity (~30 to >10^7 mPa s) can be deposited by microextrusion because they are forced out by a controllable applied pressure. Consequently, microextrusion provides the ability to process a wider range of biomaterials and to form scaffolds containing a higher density of cells. As the achievement of a sufficiently high cell density is important, particularly in the tissue engineering of organs, this ability to form constructs with high cell density is a major advantage of microextrusion-based bioprinting when compared to inkjet bioprinting.

On the other hand, microextrusion-based bioprinting also has its drawbacks. High pressures used in extruding the cell-containing suspension can damage the membrane of cells, resulting in a lower viability of the cells encapsulated within the as-formed scaffold. For example, in one study, the viability of HepG2 liver cells encapsulated within alginate hydrogels was observed to decrease with increasing applied pressure (in the range ~35–275 kPa) for nozzle diameters of 100, 250, and 400 μm. Cell viabilities in the range ~50–85% have been reported for constructs formed by microextrusion, which are lower than the viabilities for constructs deposited by inkjet bioprinting (greater than ~85–90%). The scale of the structural features formed by microextrusion (~20–200 μm), corresponding approximately to the width of the extruded filament, is usually coarser than that for inkjet bioprinting (~10–100 μm).

While a variety of biomaterials can be used as scaffolds to deliver cells in microextrusion-based bioprinting, the most commonly used are hydrogels, composed of homopolymer or copolymer molecules, natural or synthetic in origin. In common with the material requirements in other extrusion-based forming techniques, the hydrogel should have a low enough viscosity to allow flow during extrusion but, upon deposition, the printed filaments should develop sufficient rigidity to form a mechanically stable three-dimensional structure. One way of achieving this behavior involves the use of thermoresponsive hydrogels. As discussed in Chapter 11, thermoresponsive hydrogels are normally physical hydrogels in which the mechanism of gelling involves cooperative bonding interactions involving secondary (weak) bonds, such as hydrogen bonds and van der Waals bonds, to form a three-dimensional network. For example, some thermoresponsive hydrogels, such as gelatin, are composed of polymers that have an upper critical solution temperature (UCST). The polymer molecules are in solution above the UCST but, on cooling below the UCST, they phase-separate, forming a hydrogel composed of a three-dimensional solid network. In comparison, thermoresponsive hydrogels composed of polymers that have a lower critical solution temperature (LCST) show the reverse behavior with temperature.

A problem with thermoresponsive hydrogels is that their mechanical properties are often too weak for some tissue engineering applications due to the weak secondary bonds that crosslink their molecular chains. A way to create thermoresponsive hydrogels with better mechanical properties is to start with chemically modified polymer molecules that also allow covalent crosslinking after deposition of the cell-containing hydrogel. As an example, aqueous solutions of gelatin form a thermoreversible hydrogel on cooling below the UCST of gelatin (~30 °C), which can be used to stabilize the structure of the printed filaments. However, the use of a chemically modified gelatin, formed by reacting methacrylic anhydride with gelatin solution, allows formation of a stable structure in a manner similar to that for the unmodified gelatin as well as chemical crosslinking of the molecular chains using UV radiation after deposition of the cell-containing scaffold (Schuurman et al. 2013). The mechanical properties such as Young's modulus of the scaffold can be varied over a wide range by controlling not only the concentration of the chemically modified gelatin in the starting mixture but also the degree of crosslinking after formation of the scaffold (Figure 25.26).

As discussed in Chapter 11, certain triblock copolymers containing PEG or polyethylene oxide (PEO), show thermoresponsive behavior with an LCST, and this type of behavior can also be used to retain the shape of cell-containing hydrogels formed by microextrusion. For example, triblock copolymers of the form A-*b*-PEG-*b*-A, where A is poly(hydroxypropyl methacrylamide lactate), synthesized by a free radical reaction, have an LCST of ~15–20 °C, are degradable by hydrolysis, and are cytocompatible with MSCs (Censi et al. 2011). Hydrogels composed of such triblock copolymers can potentially be used as biomaterials in microextrusion and following microextrusion, they can be chemically crosslinked by UV radiation to control their mechanical properties.

Alginate hydrogels, as noted earlier, have been widely used in tissue engineering because of the ease of encapsulating cells within them under mild crosslinking conditions, their ability to maintain the encapsulated cell phenotype, their low toxicity and their good biocompatibility. On the other hand, alginate solutions generally do not have the requisite rheological behavior for microextrusion into a stable three-dimensional structure. One way to alleviate this problem is to use a mixture composed of alginate and a thermoresponsive polymer such as gelatin. For example, a solution of sodium alginate with an optimal amount of gelatin can be formed by microextrusion into a stable three-dimensional hydrogel with intricate features

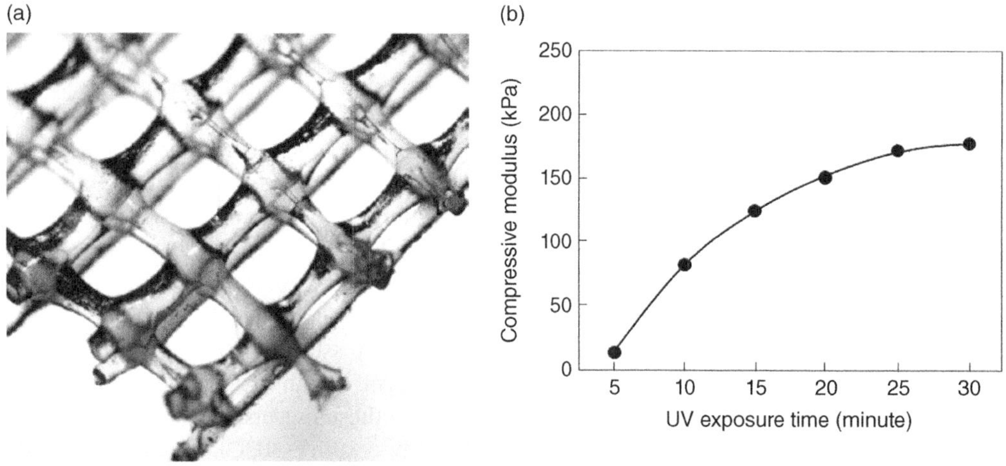

Figure 25.26 (a) Example of the structure of a gelatin-methacrylic anhydride construct formed by thermoreversible gelling followed by chemical crosslinking through exposure to UV light; (b) compressive modulus of the gelatin-methacrylic anhydride construct as a function of UV exposure time, showing that the modulus can be controlled by the UV exposure time. *Source:* From Schuurman et al. (2013).

by taking advantage of the thermoresponsive gelatin phase. Then the as-formed hydrogels can be strengthened by treating them in a $CaCl_2$ solution to chemically crosslink the alginate chains.

Using this approach, alginate–gelatin solutions were separately mixed with two different cell types and extruded to form a three-dimensional cell-containing scaffold with a complex geometry closely matching that of a porcine aortic heart valve (Duan et al. 2013). Heart valves, as noted in Chapter 24, consist of thin flaps of flexible fibrous tissue (the leaflets) firmly attached at their base to a more rigid fibrous valve ring. Consequently, the creation of a tissue-engineered heart valve depends on the ability to mimic not only the complex anatomical geometry of a natural heart valve but also the heterogeneous tissue biomechanics. Experimentally, alginate–gelatin solutions were separately mixed with two different cell types, porcine aortic valve interstitial cells (VICs) and human aortic root SMCs (SMCs) and formed into a tissue construct with the requisite anatomical shape and with the different cell types confined to their respective regions (Figure 25.27). The encapsulated cells both showed good viability (>80%) and phenotype retention over a seven-day culture period *in vitro*. Overall, these results show promising potential for the use of this 3D bioprinting approach for creating tissue-engineered heart valves.

Another approach to forming a mechanically stable three-dimensional structure by microextrusion is to control the rheology of the material being extruded. This is a common approach in conventional extrusion techniques and extrusion-based additive manufacturing techniques such as fused deposition modeling and robocasting. For such techniques, a common requirement is a material that shows shear-thinning rheological behavior (Section 25.4.3). Overall, this approach has received little attention in the development of hydrogels for microextrusion-based 3D bioprinting.

25.6 Tissue Engineering Techniques for the Regeneration of Functional Tissues and Organs

In this section, we will discuss various tissue engineering techniques that have been used in regenerating various tissues and organs. Emphasis is placed on advances made in the areas discussed through a combination of innovative design and production of biomaterials and an improved understanding of biological processes. As noted earlier, while the number of studies has been considerable, except for skin, engineered regeneration of human tissues and organs is not yet used routinely in clinical practice. Consequently, a comparison of tissue engineering techniques with currently used clinical treatments also forms part of the discussion.

25.6.1 Bone Tissue Engineering

Bone tissue engineering techniques were discussed in detail in Chapter 23 within the context of healing bone defects. As noted, autografts are the ideal bone graft because they can promote bone healing by osteoinductive, osteoconductive, and osteogenic mechanisms, while bone allografts, bone from a donor or a cadaver, are the most widely used bone graft, despite their lack of osteogenicity due to the removal of the cells. A variety of biomaterials has been developed

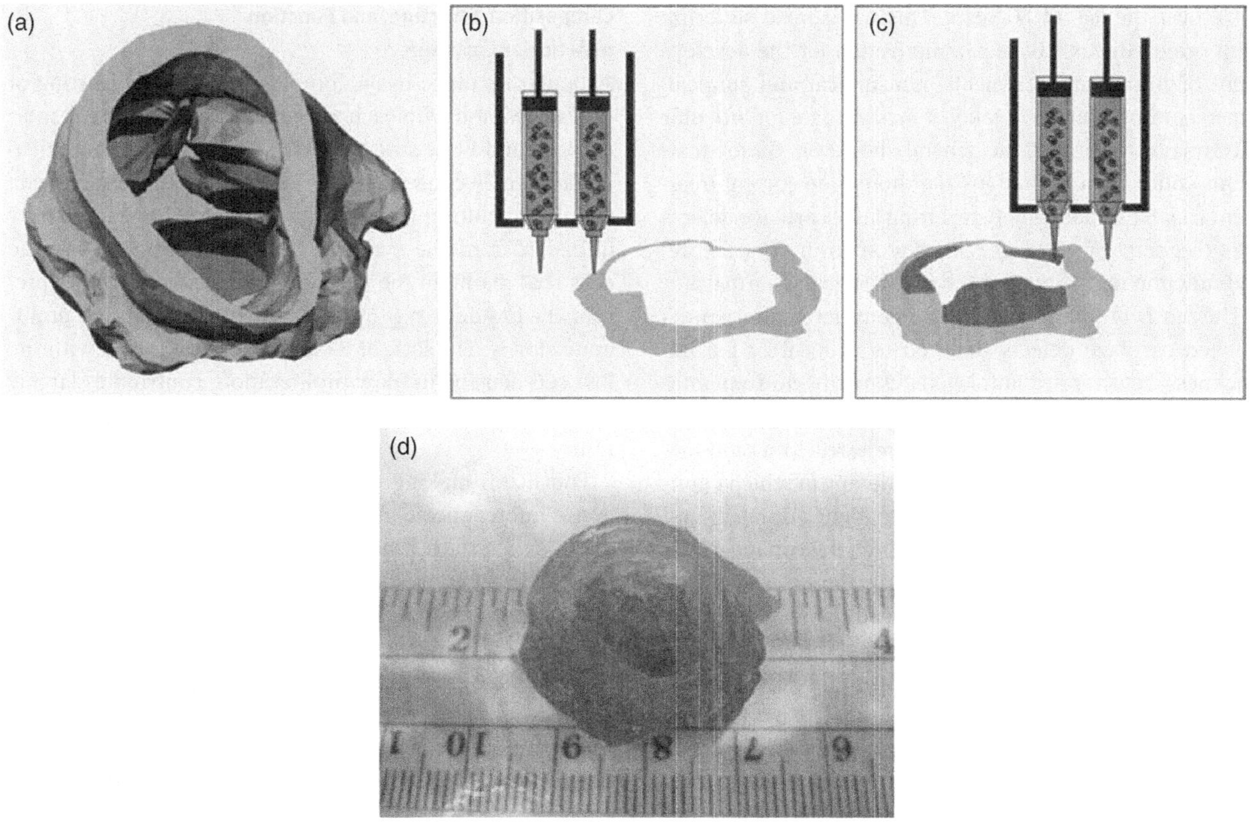

Figure 25.27 Bioprinting of aortic valve conduit using microextrusion technique. (a) Aortic valve model reconstructed from micro-CT images. (b, c) Schematic illustration of the bioprinting process with dual cell types and dual syringes; root region of first layer generated by hydrogel containing smooth muscle cells (SMCs) (b) and leaflet region of first layer generated by hydrogel containing valve leaflet interstitial cells (VICs) (c). (d) As-printed aortic valve conduit. *Source:* From Duan et al. (2013) / with permission of John Wiley & Sons.

and commercialized as bone graft substitutes for healing bone defects. These biomaterials include the calcium phosphate ceramics such as hydroxyapatite, β-TCP, and BCP, calcium phosphate cements, bioactive glasses, and degradable polymers such as PLA and PLGA. Because these biomaterials are osteoconductive and generally lack osteoinductivity and osteogenicity, they often fail to provide sufficient healing of bone defects within a clinically relevant time.

Tissue engineering approaches to healing bone defects have relied mainly on adding cells (osteoblasts or MSCs) or growth factors (commonly BMP2), to scaffolds composed normally of the aforementioned calcium phosphate ceramics or degradable polymers. Constructs created by these approaches could potentially provide an optimal combination of all three mechanisms (osteoconductivity, osteoinductivity, and osteogenicity) without the drawbacks associated with bone autografts and allografts. As discussed in Chapter 23, several studies involving animals have shown the ability of tissue engineering strategies to promote faster healing of bone defects when compared to the use of scaffolds consisting of a biomaterial alone. However, these approaches have been slow to enter clinical practice.

25.6.2 Articular Cartilage Tissue Engineering

Cartilage in the human body is normally divided into three types, hyaline cartilage, fibrous cartilage (fibrocartilage), and elastic cartilage. Here, our focus is on the hyaline cartilage, which makes up the cartilage at the surface of articulating (or synovial) joints such as joints of the limbs, of which the hip joint and the knee joint are examples. Cartilage that joins native tendon (or ligament) to bone is an example of fibrocartilage. The surface of articular joints consists of a layer of cartilage (~2–5 mm thick) that provides frictionless movement while often being subjected to physiological loads as high as several times the body weight. However, articular cartilage can deteriorate from diseases such as rheumatoid arthritis and osteoarthritis, and from trauma due to single or repetitive overloading of the joint, which are a leading cause of physical disability in humans.

Cartilage deterioration due to osteoarthritis is the most common cause of physical disability, estimated to affect ~10% of the US population over the age of 30 years and over

~60% over the age of 60 years. This widespread suffering from osteoarthritis has been a motivation for the development of numerous treatments, nonsurgical and surgical, aimed at improving the quality of medical care for articular joint-related disorders. In general, however, these treatments suffer from limitations and there is no current treatment capable of adequately repairing large cartilage defects or of regenerating cartilage with the composition, structure, and function matching those of native articular cartilage.

The most common surgical treatments for healing small, isolated cartilage defects (referred to as focal or partial-thickness defects) are autologous chondrocyte implantation (ACI), transplantation of osteochondral grafts, and transplantation of cartilage plugs, referred to as mosaicplasty. ACI is a cell transplantation therapy in which chondrocytes are harvested from the patient, isolated and expanded in *in vitro* culture, after which the expanded cell population is implanted into the defect. Limitations of ACI include donor site morbidity, the need for two surgeries, loss of chondrocyte phenotype during culture, leakage of the cells from the implantation site, and the formation of fibrous cartilage tissue. An advance in ACI has been to deliver the cells by means of a hydrogel, such as an agarose-alginate hydrogel capable of maintaining the viability and phenotype of chondrocytes.

Transplantation of osteochondral grafts (grafts consisting of cartilage plus a layer of underlying bone) can involve autografts or allografts. Autografting involves harvesting an osteochondral graft from an area subjected to low load and transplanting it into a load-bearing defect site. This procedure suffers from several drawbacks, such as donor site morbidity, limited supply, incongruity of restored surfaces which can lead to pressure build-up, lack of integration of the transplanted graft, and altered joint biomechanics. Except for donor site morbidity, allografts suffer from the same drawbacks as autografts, plus the potential of an immune response and the risk of disease transmission. Mosaicplasty involves harvesting small cartilage plugs from areas subjected to low loads and transplanting them into a single defect in a load-bearing defect site. In general, mosaicplasty suffers from drawbacks similar to those of osteochondral autograft transplantation. It has been most successful for patients under 50 years of age who do not suffer from joint misalignment.

Because of the limitations of current treatments, considerable efforts have been devoted to the regeneration of articular cartilage using tissue engineering. While significant advances have been made, engineering cartilage with the composition, structure, and function comparable to those of native cartilage has been challenging. In order to understand the challenges involved, a basic understanding of cartilage composition, structure, and function is useful.

Composition, Structure, and Function of Articular Cartilage

Articular cartilage has a complex structure, consisting of four zones, and with each zone having a different cell morphology and ECM structure with distance from the articular surface. Because cartilage is avascular (lacking a blood supply), all nutrient, oxygen, and waste exchange occurs by diffusion from the synovial fluid. Chondrocytes are the cells that maintain the ECM of cartilage, but they are present at a low density (~5% of the tissue volume), and proliferate slowly. The lack of a blood supply, coupled with the low cell density and low proliferation, contributes largely to the low ability of cartilage to repair itself following an injury.

The ECM, making up ~95% of cartilage, consists of a dense interconnected structure of collagen and proteoglycans. Cell-surface binding proteins link chondrocytes to the ECM and are responsible for transmitting mechanical and molecular signals to the chondrocytes. Collagen makes up ~10–20% of the wet weight of cartilage, consisting of a network composed mainly of collagen type II fibers, which account for up to ~90% of the total collagen content, plus small amounts of collagen type IX, X, and XI. The collagen type II fibers are composed of identical polypeptide triple-helical molecules (Chapter 2), which are crosslinked by covalent bonds throughout the tissue.

In addition to collagen, cartilage ECM consists of major and minor proteoglycans, which make up ~10% of the wet weight of cartilage. Proteoglycans consist of a central core protein to which one or more GAGs are covalently bonded. Aggrecan is the major proteoglycan in cartilage, consisting of many branched GAGs, primarily keratan sulfate and chondroitin sulfate, attached to a core protein. The repeating sulfate groups in these GAGs gives each aggrecan molecule a large negative charge at the physiological pH, enabling cartilage to absorb a large amount of water and to support physiological stresses. Aggrecan molecules are bound to long unbranched HA chains by link proteins. This network of aggrecan molecules linked to HA chains is entrapped within the collagen network to give an intricately organized ECM structure (Figure 25.28).

At a larger structural scale, as noted earlier, articular cartilage consists of four zones. The superficial zone contains the articulating surface and extends to ~10% of the cartilage thickness. When compared to the next two zones, it is marked by chondrocytes and collagen fibers that are oriented parallel to the articulating surface, smaller cells with a flattened or ellipsoidal shape, and the highest cell density (number of cells per unit mass of tissue). The superficial zone also has the lowest proteoglycan content (as a percentage of dry weight), and the highest water content (as a percentage of wet weight). The middle zone contains larger

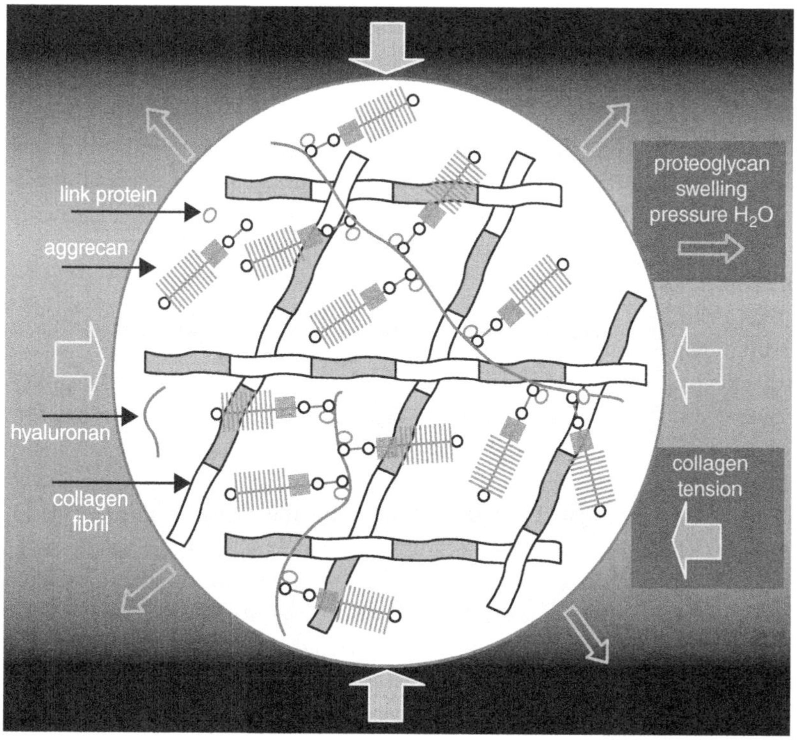

Figure 25.28 Illustration showing the combined properties of collagens and aggrecan in articular cartilage. *Source:* From Kiani et al. (2002) / with permission of Springer Nature.

more rounded cells which, along with the collagen fibers, are randomly oriented, and can often be found in clusters. This zone has the highest collagen content (as a percentage of dry weight), followed by the upper and lower zones. Chondrocytes in the deep zone have an oval shape and, along with the collagen fibers, are oriented in approximately vertical columns perpendicular to the articulating surface. This zone has the highest proteoglycan content, followed by the middle and upper zones. Below the three outer zones is a zone of calcified cartilage connecting the deep zone to the subchondral bone.

Biomaterials and Strategies in Cartilage Tissue Engineering

Both cell-based and growth factor-based tissue engineering strategies have been used for regenerating cartilage in focal defects, but cell-based strategies have been more commonly used. Cell-based strategies, which essentially represent an advance in ACI treatment, involve encapsulating cells (chondrocytes or MSCs) within a scaffold (normally a hydrogel) and after incubation *in vitro* for the appropriate time, implanting the tissue construct into the cartilage defect. Growth factor-based strategies rely on incorporating a growth factor (normally TGFβ1 or TGFβ3) within a scaffold (normally a hydrogel) and incorporating the construct into the cartilage defect.

A difficulty in cartilage tissue engineering is obtaining a sufficiently large and phenotypically stable autologous chondrocyte population. Autologous chondrocytes have been commonly used in studies involving animals, but donor site morbidity makes harvesting a large amount of cartilage impractical in humans. Because of the low amount of harvested chondrocytes, *in vitro* expansion is necessary but retaining the native cell phenotype can be difficult. Chondrocyte proliferation, as noted earlier, is also slow. The use of MSCs can reduce the need for an invasive harvesting procedure. While donor site morbidity is still present for autologous MSCs derived from bone marrow, the use of adipose-derived MSCs is associated with a lower level of donor site morbidity. However, some studies have shown that differentiation of adipose-derived MSCs into chondrocytes can be more difficult.

The scaffold in cartilage tissue engineering should ensure retention of the chondrocyte phenotype, promote chondrocyte proliferation, and provide mechanical stability, in addition to being biocompatible and degradable. Natural polymers are more commonly used when compared to synthetic polymers because they provide a compositional similarity to the ECM components. Additionally, hydrogels have been used to a much greater extent when compared to porous three-dimensional solids because they can provide several advantages (Section 25.3.4). Some natural hydrogels used in cartilage tissue engineering include collagen, alginate, agarose, chitosan, HA and fibrin. Alginate and agarose have shown the ability to maintain the chondrocyte phenotype for long periods (up to several months) but they suffer from limited degradability. PEG-based hydrogels, such as copolymers of PEG with PLA, can be formed by photopolymerization, are degradable and can maintain the chondrocyte phenotype.

The zonal organization of cartilage is important for its physiological function but it appears unlikely that it could be recreated in tissue engineering strategies employing cells from a single source that are homogenously encapsulated within a single homogeneous scaffold. Several studies have attempted to engineer cartilage with the zonal organization of native cartilage using approaches based on the origin of the cells or on depth-dependent design of the scaffold. For example, in a model experimental study (Kim et al. 2003), chondrocytes were harvested from the superficial, middle, and deep zones of bovine articular cartilage and encapsulated separately in a trilayered scaffold formed by sequential photopolymerization of PEG hydrogel layers. Each layer of the scaffold was photopolymerized for the same time and, consequently, had approximately the same crosslink density. After three weeks of *in vitro* culture, each zone of the trilayered tissue construct showed histological and biochemical characteristics that were consistent with the zones of native bovine articular cartilage (Figure 25.29).

Another model study used a bilayered scaffold technique, composed of hydrogel layers with two different agarose concentrations and, thus, two different Young's modulus (or stiffness) (Ng et al. 2005). After encapsulation of chondrocytes isolated from the same source (bovine carpometacarpal joint) these bilayered tissue constructs showed depth-dependent mechanical and biochemical properties during a four-week *in vitro* culture period. Overall, these model studies are instructive for engineering cartilage with a zonal organization resembling that of native cartilage.

25.6.3 Tissue Engineering of Articular Joints

As noted earlier, current surgical treatments for healing articular cartilage defects (such as ACI, mosaicplasty, transplantation of osteochondral grafts) are based on healing focal defects. A problem with tissue engineering approaches for regenerating the cartilage layer in focal defects is the lack of significant bonding between the engineered cartilage and the host tissue surrounding the defect. This leads to limited cartilage-to-cartilage integration and to instability of the engineered cartilage in the defect. One approach to addressing this problem has been the use of tissue-engineered osteochondral (or composite tissue) constructs, composed of a cartilage-like layer firmly attached to an osseous-like layer that can be press-fitted and secured within the subchondral bone of the defect. This approach is based on the aforementioned surgical treatment involving transplantation of osteochondral grafts but without its disadvantages.

Overall, a challenging problem in tissue engineering of osteochondral tissues is due to the contrasting needs

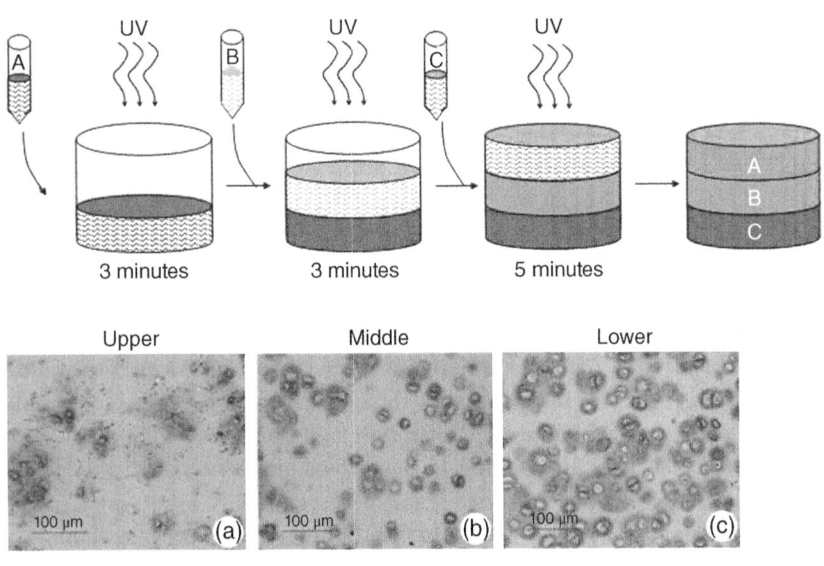

Figure 25.29 (Top): Schematic illustration of technique used to create multilayered constructs to regenerate the zonal organization of articular cartilage. (Bottom): Immunohistochemical staining for collagen type II in top (a), middle (b), and lower (c) layers of the constructs after 3 weeks of *in vitro* culture. Each layer showed histological and biochemical characteristics that were consistent with the zones of native bovine articular cartilage. *Source:* From Kim et al. (2003) / with permission of Elsevier.

for the growth and function of two different adult tissue phenotypes (articular cartilage and bone). These two tissues, with little in common, must integrate structurally and function in synchrony to withstand physiological loads of up to several times the body weight. Despite the fact that bone and cartilage derive from the same progenitor of MSCs (Figure 25.4), bone is a drastically different tissue phenotype when compared to cartilage. For example, articular cartilage is avascular and consists of chondrocytes in a hydrated ECM composed essentially of collagen type II and proteoglycans (Section 25.6.2). In comparison, bone has a rich blood supply and well-organized islands of mineralized collagen type I matrix with osteoblasts residing on the mineralized matrix surface and osteocytes within the matrix.

Osteochondral constructs, typically in the shape of a cylinder (Figure 25.30a), have been created from a variety of biomaterials to address the need for regenerating focal cartilage defects. For example, in one study, a chondral layer consisting of autologous bone marrow-derived MSCs encapsulated within a photopolymerizable PEG hydrogel was attached to an osseous construct composed of porous tantalum (Trabecular Metal™). Following encapsulation of MSCs (~10^7 cells/cm^3), the hydrogel was attached to the osseous layer, and the composite constructs were implanted directly (without *in vitro* culture) into defects created in rabbit knees. At 12 weeks post-implantation, histological and immunohistochemical analyses showed that the regenerated cartilage had integrated with the surrounding cartilage (Figure 25.30b) and closely resembled the native cartilage.

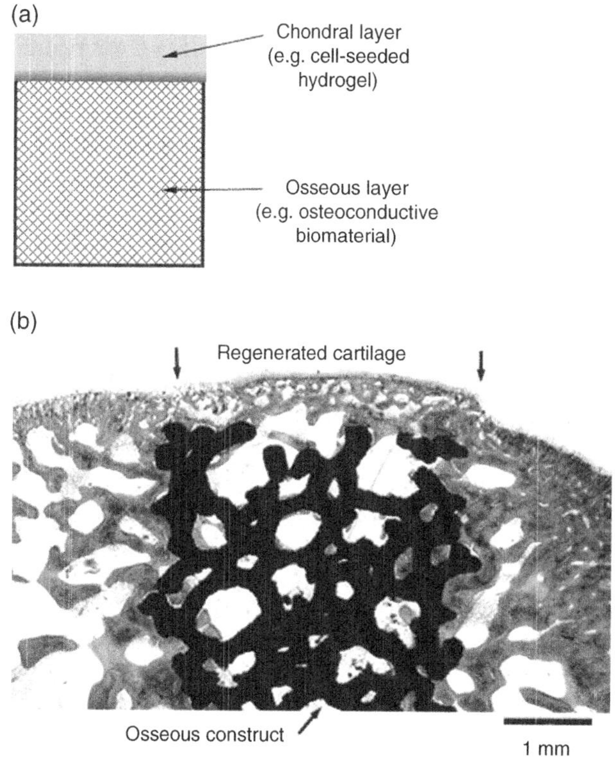

Figure 25.30 (a) Schematic illustration of the cross section of an osteochondral construct in the shape of a cylindrical plug consisting of a cell-seeded chondral layer attached to an osseous substrate. (b) Stained section of an osteochondral construct after implantation for 12 weeks in defects created in rabbit knees. The construct was composed of autologous bone marrow-derived MSCs encapsulated within a photopolymerizable PEG hydrogel (chondral layer) attached to an osseous construct composed of porous tantalum.

Focal defects can deteriorate, with or without surgical treatment, into severe osteoarthritis, eventually necessitating total joint arthroplasty using durable biomaterials. As discussed in Chapter 23, total joint arthroplasty has served to improve the quality of life for millions of patients but the procedure also suffers from limitations, a major one being the limited lifetime of the implant *in vivo* due to aseptic loosening. The limitations of total joint arthroplasty have motivated several efforts aimed at regenerating entire articular joints through tissue engineering of osteochondral constructs with the anatomical shape and surface topology of natural joints. These efforts have so far shown the ability to engineer osteochondral constructs applicable to small joints but not for large joints such as the human knee joint and hip joint.

Biomaterials, cells, and tissue engineering approaches for regenerating articular joints generally follow those described earlier for cylindrical osteochondral constructs, except for a difference in size, shape, and surface topology of the construct. As an example, in one approach (Lee et al. 2009), which is a cell-based approach (Figure 25.3a), hydroxyapatite particle-reinforced PCL was formed into a scaffold with a grid-like architecture and with the shape and surface topology of a human proximal tibial joint using an additive manufacturing (3D printing) technique. The scaffold, of size ~25% that of an actual human tibial joint, was subsequently infiltrated with human MSCs or with human MSC-derived osteoblasts, constituting the osseous portion of the osteochondral construct. Then a PEG layer (~1 mm thick) encapsulating human MSCs or human MSC-derived chondrocytes, was molded onto the scaffold to form an articulating layer (the chondral layer) having the right surface topology.

After subcutaneous implantation for six weeks in rodents, the MSCs in the osseous portion of the construct generated a significantly higher number of blood vessels and larger blood vessels when compared to the MSC-derived osteoblasts. On the other hand, the MSC-derived osteoblasts promoted mineralization within the scaffold. A significantly higher number of cells was present in the chondral layer seeded with MSCs when compared to the chondral layer seeded with MSC-derived chondrocytes. On the other hand, chondrocytes within a GAG matrix was present within the chondral layer seeded with MSC-derived chondrocytes. Overall, MSCs promoted more angiogenesis in the osseous layer, whereas MSC-derived cells produced the corresponding tissue phenotypes. These results are instructive for regenerating human-size articular joints through scale-up of the tissue engineering approach.

In another approach, which can be considered a growth factor-based approach, a scaffold mimicking the shape of a rabbit proximal humeral joint was created from hydroxyapatite-reinforced PCL using an additive manufacturing technique similar to that in the aforementioned cell-based study. However, instead of molding a cell-containing PEG hydrogel layer to the surface of the scaffold to form a chondral layer, the surface of the scaffold was infused with a collagen hydrogel containing TGFβ3 (Figure 25.31). Then the entire articulating surface of proximal humeral joints in skeletally mature rabbits was excised surgically and replaced with these scaffolds infused with TGFβ3-containing hydrogel. Four months postimplantation, TGFβ3-infused scaffolds were fully covered with hyaline cartilage in the articulating surface. In comparison, TGFβ3-free scaffolds had only isolated cartilage formation, while unfilled defects showed no cartilage formation. The TGFβ3-infused scaffolds also recruited a significantly higher number of chondrocytes in the regenerated cartilage, produced uniformly distributed chondrocytes in a matrix containing collagen type II and aggrecan, and a thicker cartilage layer when compared to TGFβ3-free scaffolds (Figure 25.32). In addition, the elastic modulus in compression and in shear for the cartilage produced by the TGFβ3-infused scaffolds was not significantly different from that for the native cartilage. The results of this study provide a proof of principle for the regeneration of the entire articulating surface of a synovial joint without cell transplantation.

Figure 25.31 India ink staining of (a) as-fabricated scaffold, scaffold without TGFβ3 (b) and scaffold infused with TGFβ3 (c) after implantation for 4 months in rabbit proximal humeral joints, and native cartilage (d). *Source:* From Lee et al. (2010).

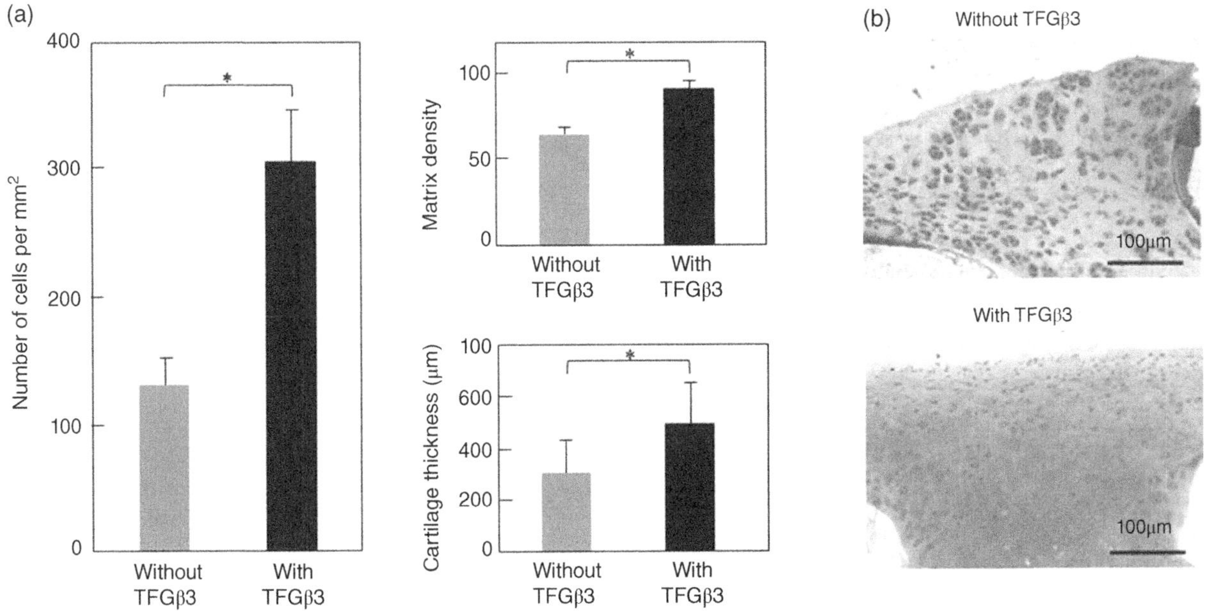

Figure 25.32 (a) Number of chondrocytes in cartilage layer, matrix density, and thickness of cartilage layer for hydroxyapatite-reinforced polycaprolactone scaffolds without and with TGFβ3 after implantation for 4 months in rabbit proximal humeral joints. (b) Safranin O staining of articular cartilage in scaffolds without and with TGFβ3. *Source:* From Lee et al. (2010) / with permission of Elsevier.

25.6.4 Tissue Engineering of Tendons and Ligaments

Tendon and ligament tissues are dense connective tissues composed of approximately parallel collagen fiber bundles connecting bone to muscle and bone to bone, respectively. The primary function of tendons and ligaments is to transmit mechanical forces (primarily tensile forces) generated by skeletal muscles to enable the movement of bones at joints. When these forces exceed a critical threshold, tissue damage can occur, resulting in impaired function and mobility. Tendon and ligament injuries, ranging from chronic degeneration to acute ruptures, are a common clinical problem, representing ~50% of the approximately 30 million musculoskeletal injuries reported in the United States annually. The most commonly injured ligaments are the ACL and the medial collateral ligament (MCL), while the most common tendon injuries involve the tendons of the four muscles of the rotator cuff group, the Achilles tendon, and the flexor tendons of the hand. Chronic degeneration and partial tears of tendons and ligaments are often treated conservatively with physical therapy and corticosteroid injections. On the other hand, because adult tendon and ligament tissues have low regenerative capacity, due to their low cell density and low vascularity, surgical intervention is often the only recourse to healing acute ruptures, representing over ~0.3 million procedures performed in the United States annually.

Current surgical procedures for treating acute injuries rely on replacing the damaged tendons and ligaments with autografts, allografts, and synthetic prostheses, but such techniques suffer from drawbacks and clinical outcomes are generally unsatisfactory. The use of transplanted grafts, as noted earlier, is limited by factors such as donor site morbidity, limited supply, and the risk of disease transmission, while the use of synthetic prostheses generally leads to inadequate long-term functional recovery. Overall, these procedures often fail to restore the structural, functional, and biochemical properties of the repaired tendon or ligament to those of the native (pre-injured) tissue. Consequently, there has been considerable interest in tissue engineering approaches in an effort to improve tendon and ligament tissue healing.

Structure and Function of Tendons and Ligaments

While tendons and ligaments have similar physiology and nanostructural features, they differ in morphology, biochemical composition, and mechanical properties. In tendon development, collagen molecules self-assemble to form collagen fibrils that, in turn, aggregate to form collagen fibers (Figure 10.1). Tendon fibroblasts (tenocytes) reside between the fibers and maintain the ECM. The number and diameter of collagen fibers are variable in the same species and they vary among species. For example, collagen fibers in rat tail tendons have a diameter less than ~30 μm, while collagen fibers in human tendons have a diameter larger than ~300 μm. The collagen fibers in tendons are wrapped with a layer of connective tissue, known as endotenon, which contains blood vessels, lymphatics, and nerves, forming larger-scale structures called fascicles, which are surrounded by another connective tissue layer (the epitenon), to form the tendon tissue (Figure 25.33). Tendons are composed of water (~55% of the wet weight), proteoglycans (less than ~1% of the wet weight), and collagens, with collagen type I being the dominant type (~85% of the dry weight), and smaller amounts of collagen types III, V, XII, and XIV. The dominant proteoglycan is decorin. Adult human tendons are composed of a bimodal distribution of collagen fibrils with average diameters of ~60 and ~175 nm. When compared to ligaments, tendons are less active biologically, and have fewer cells and a smaller number of reducible crosslinks between collagen fibers.

As noted earlier, ligaments differ substantially from tendons in their morphology, biochemical composition and mechanical properties. In addition, ligaments from the same joint, such as the MCL and the ACL have different properties and intrinsic healing responses. Ligaments are composed of water (~60% or more of the wet weight), and four different types of ECM macromolecules, proteins, proteoglycans, elastin and glycoproteins. Approximately 75% of the dry weight of ligaments consists of collagens, predominantly type I (~90%) and type III (~10%). Fibroblasts make up ~20% of the tissue volume, with the remainder being the ECM. As with tendons, collagen molecules are assembled into fibrils, which aggregate to form fibers. Human ligaments contain a bimodal distribution of collagen fibrils, with average diameters of ~40–75 nm, and ~100–150 nm.

Tendons and ligaments can be either fibrous tissue or, near their attachment to bone, fibrocartilage. Fibrocartilage serves to distribute longitudinal and shear forces from soft tissue to subchondral bone. It consists of four tissues at the enthesis (where a tendon or ligament inserts into bone): regular fibrous connective tissue, uncalcified fibrocartilage, calcified fibrocartilage, and bone.

Biomaterials and Strategies for Tendon and Ligament Tissue Engineering

Our focus will be on tissue engineering of tendons but, overall, the principles should be applicable to ligament tissue engineering. Additionally, our focus will be on strategies for regenerating fibrous tendon tissue.

Based on the architecture of tendons (Figure 25.33) and their function (transmission of tensile forces), it may be apparent that porous three-dimensional solids or hydrogels are not well suited as scaffolds for tendon tissue engineering. Commonly used scaffolds consist of an oriented arrangement of fibers composed of synthetic or natural polymers, or decellularized tendon matrices. A number of degradable synthetic polymers, particularly the poly(α-hydroxy esters)

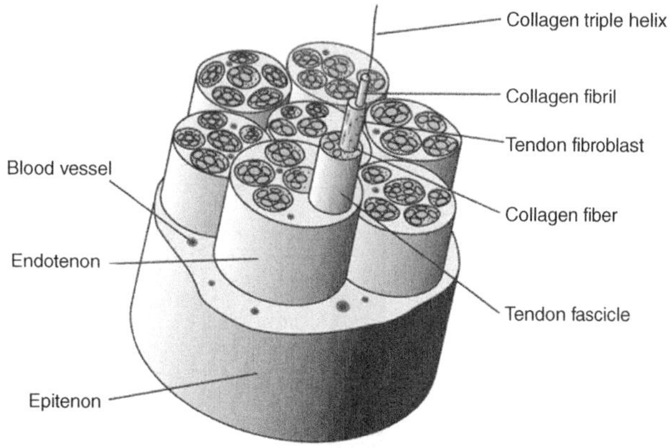

Figure 25.33 The hierarchical architecture of tendon. Collagen triple-helices self-assemble into fibrils. Bundles of fibrils form fibers, which constitute tendon fascicles. Tendon fibroblasts (tenocytes) reside between collagen fibers. Fascicles are wrapped by endotenon, a layer of connective tissue containing blood vessels, nerves, and lymphatics. Multiple fascicles are further surrounded by another connective tissue layer, epitenon, to form the tendon tissue. *Source:* From Yang et al. (2013) / with permission of John Wiley & Sons.

PGA, PLA, and PLGA, in the form of woven or braided fibers, have been used in tendon tissue engineering. However, scaffolds composed of these synthetic biomaterials tend to degrade too rapidly for tendon and ligament tissue engineering and they are generally unable to provide an optimal microenvironment in which cells can reside, proliferate, and differentiate into the required phenotype.

Scaffolds composed of natural polymers and their derivatives may alleviate the limitations of synthetic biomaterials. Collagen-based scaffolds provide a compositional similarity to tendon ECM (composed mainly of collagen type I) and, thus, can provide a more favorable microenvironment for cell adhesion, proliferation, and differentiation. Collagen scaffolds can be crosslinked to reduce their degradation rate and to enhance their mechanical properties, but these properties are generally inadequate for tendon and ligament tissue engineering. When compared to collagen, scaffolds composed of another protein-based natural polymer, silk, have shown better mechanical properties and improved stability against degradation, as well as the ability to support the attachment, proliferation, and differentiation of MSCs.

Topographical cues provided by scaffolds play an important role in tendon tissue engineering. Because tendon tissue is composed primarily of nearly parallel collagen fibers, the ability to mimic this aligned fibrous arrangement is important. For example, tendon-derived stem cells (TDSCs) seeded onto electrospun poly(L-lactic acid) (L-PLA) scaffolds composed of aligned fibers (diameter 400-500 nm) showed significantly higher expression of tendon-specific genes *in vitro* and suppressed osteogenesis when compared to scaffolds with a random fiber arrangement (Yin et al. 2010). Upon implantation for 1 week in rodent subcutaneous sites, the TDSC-seeded scaffolds with aligned fibers induced the formation of spindle-shaped cells, and showed a higher collagen content and a denser matrix when compared to scaffolds with a random arrangement of fibers.

Several studies have shown the ability of ECM scaffolds derived from porcine SIS to enhance the healing of tendons and ligaments in animal models. The positive effects of SIS scaffolds on healing are attributed to their natural structural and biological characteristics, among which is a structure consisting primarily of aligned collagen type I fibers and containing biomolecules such as growth factors. ECM scaffolds prepared from human flexor tendons using an optimized decellularization protocol showed no significant decrease in tensile strength, collagen content, or GAG content when compared to native flexor tendons. Upon seeding with human dermal fibroblasts, these ECM scaffolds showed the ability to support the attachment of viable cells to their surface. Overall, decellularized ECM have attractive properties and characteristics for use as scaffolds in tendon and ligament tissue engineering.

Cell-based strategies have been widely employed for regenerating tendons and ligaments. Cells used in tendon tissue engineering include tendon fibroblasts (tenocytes), dermal fibroblasts, and MSCs. Tenocytes synthesize the constituents of tendon ECM, such as collagen, proteoglycans, and glycoproteins and influence collagen fiber formation. Although tenocyte-based strategies are effective in tendon tissue engineering, harvesting autologous tenocytes can cause donor site morbidity. Dermal fibroblasts have been considered as an alternative cell source to autologous tenocytes because they are more accessible and do not cause major donor site morbidity. While dermal fibroblasts have shown promising potential in some tendon tissue engineering studies, they have been reported to possess limited differentiation potential.

MSCs have been widely studied as a cell source for tendon tissue engineering, and they have shown the ability to enhance tendon and ligament healing. For example, in one study, bone-marrow-derived MSCs were seeded on knitted PLGA scaffolds and used to repair Achilles tendon gaps in rabbits. At 2 and 4 weeks postimplantation, scaffolds seeded with MSCs showed a higher rate of tissue formation and remodeling when compared to unseeded scaffolds. Immunohistochemical analysis showed that the regenerated tendons were composed of collagen type I and type III. At 12 weeks, the regenerated tendon integrated with native tendon and showed microscopic features that were similar to native tendon (Figure 25.34). On the other hand, the elastic modulus of the regenerated tendon was only ~60% of the value for the native tendon but this lower modulus may be due primarily to the physicochemical properties of the scaffold material used.

Bone marrow-derived MSCs have been widely studied for tendon tissue engineering while less attention has been paid to adipose-derived MSCs despite their advantages such as greater availability and ability to be harvested by less invasive procedures. Induced pluripotent stem cells (iPSCs) are another cell source and, in some studies, they have been reported to have greater proliferation and differentiation potential when compared to MSCs. TDSCs, a cell population identified within tendon tissues of several species (Bi et al. 2007), have been shown to have stem cell characteristics such as multipotency and regenerative potential comparable to those of MSCs. As tissue-specific stem cells often show a higher regenerative potential for the tissue in which

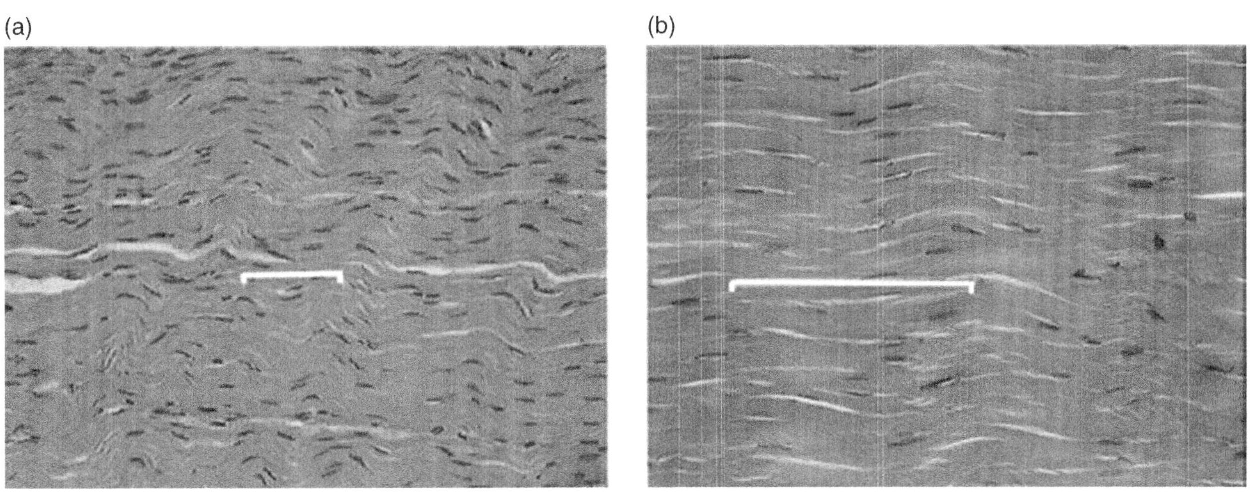

Figure 25.34 Histology (hematoxylin and eosin stained sections) of (a) neotendon formed after implantation of MSC-seeded knitted PLGA scaffolds for 12 weeks in rabbit Achilles tendon gaps, and (b) native tendon. The brackets indicate the length of the crimp periods. (Original magnification 200×). *Source:* From Ouyang et al. (2002).

they reside, TDSCs may provide an attractive cell source for tendon tissue engineering. In one study, allogeneic TDSCs seeded in sponge-like scaffolds composed of knitted silk and collagen showed the ability to enhance the healing of rotator cuffs in rabbits. However, despite the advantages of TDSCs, harvesting of autologous TDSCs would present the same donor site morbidity as tenocytes.

Although growth factor-based strategies have showed success in tendon tissue engineering, they appear to be more challenging when compared to cell-based strategies. Growth factors commonly used, such as IGF1, TGFβ, VEGF, bFGF, and PDGF-BB, are among those identified in the developmental biology of tendons. For example, one study showed that when delivered by means of fibrin glue, PDGF-BB improved the healing of rabbit MCL. In another study, bFGF-releasing scaffolds were created by coating knitted silk microfibers with electrospun PLGA nanofibers incorporated with bFGF. When seeded with rabbit bone marrow-derived MSCs and incubated *in vitro*, these scaffolds showed the ability to enhance cell proliferation when compared to scaffolds without bFGF. The bFGF scaffolds also enhanced production and deposition of collagen type I and type III, and gene expression levels of collagen type I and type III, which are indicators of tenogenic differentiation.

As discussed in Chapter 17, cells and tissues of the body are continuously being subjected to a variety of mechanical forces, and these forces alter cellular responses through mechanotransduction mechanisms. As the primary function of tendons and ligaments is the transmission of tensile mechanical forces, mechanical stimulation would be expected to play an important role in promoting tendon healing *in vivo*. Tissue engineering strategies have relied on mechanical stimulation to pre-condition cell-seeded constructs *in vitro* prior to implantation. This pre-conditioning, in turn, can lead to improved *in vivo* outcomes. For example, studies have shown that cell-seeded ECM scaffolds have higher tensile strength and elastic modulus when subjected to cyclic tensile stretching in a bioreactor. The improved mechanical properties have been explained in part by cell-mediated remodeling of the ECM, with cyclic stretching producing greater deposition of collagen fibrils in the direction of applied strain when compared to a uniform strain. Another study has shown that matching the mechanical properties of MSC-seeded collagen scaffolds with those of native tendon, through mechanical stimulation for example, could improve the mechanical and histological characteristics of healed tendons *in vivo*.

25.6.5 Skin Tissue Engineering

The skin is the largest organ in humans, serving as a protective barrier between the human body and the surrounding environment. Skin wounds due to burns, trauma, and disease are a common occurrence and often require significant medical attention. In Chapter 24, we discussed the fundamentals of skin wound healing and various biomaterials-based wound healing strategies. However, the reparative sequence of events and treatment strategies discussed in that chapter were based on healing small-area wounds. A common clinical problem is the treatment of large-area skin defects such as burn wounds. For such wounds, particularly wounds that involve destruction of the dermis, the healing process generally does not take place efficiently, often resulting in functionally and aesthetically compromised fibrous scar tissue for the remainder of the patient's life. Over 2 million burn wounds require medical attention in the United States alone each year, of which nearly 100,000 require hospitalization.

Autologous skin grafts are the gold standard for treating large-area burn wounds but it suffers from limited availability, particularly for patients with extensive burns because there is little undamaged skin from which autologous grafts can be harvested. Skin allografts are a possibility but they provide only a temporary solution because they undergo immunogenic rejection within a few weeks and must be replaced by autografts.

Because of a variety of factors, including a significant need for skin substitutes, limitations associated with skin autografts and allografts, and the fact that skin therapies are noninvasive, the development of functional skin substitutes received more attention than other areas in the early years of tissue engineering. As a result, a wide range of commercial skin substitute products is now available for clinical treatment of burn wounds. However, skin tissue engineering has yet to provide a clinically and economically effective solution in critical areas such as large-area burn wounds that involve destruction of the dermis.

Burn wounds are often classified into three types, based on the depth of the wound:

- *First degree burn:* A first degree burn, also referred to as an epidermal burn, involves loss of the top layer of the epidermis (see Figure 24.14).
- *Second degree burn:* A second degree burn involves loss of the epidermis and part of the dermis, but it may also involve complete loss of the dermis. Second degree burns are sometimes divided into superficial partial-thickness burns when they involve partial loss of the dermis and deep partial-thickness burns when there is complete loss of the dermis, extending into the subcutaneous layer (called the subdermis or hypodermis).
- *Third degree burn:* A third degree burn extends through the subcutaneous tissue to involve tendon, muscle, bone and other deep structures. It is sometimes called a deep full-thickness burn.

Transplantation of skin autografts is the gold standard for healing large-area first and second degree burns. Grafts are harvested using a dermatome, which detaches the epidermis and a superficial part of the dermis. Such autografts are also called split-thickness skin grafts or split skin grafts. Remaining epidermal cells in the residual dermis of the donor site enables regrowth of an epidermis, and the donor site normally heals within ~1 week. After conversion into a mesh-like form to cover a wider area, the skin graft is applied to the burn where its capillaries merge with the capillary network of the wound site. This so-called "graft take" is essential for proper supply of nutrients and for survival of the graft. As noted earlier, for patients with extensive burns, donor sites are limited as the patient will have little undamaged skin to allow harvesting.

Allografts (cadaver skin) provide a temporary solution, preventing fluid loss and contamination of the wound site. However, allografts elicit an immune response, leading to rejection within ~7–21 days and necessitating replacement with autografts. Allografts also suffer from limited supply and the risk of disease transmission. Porcine xenografts can be used if necessary, but they suffer from disadvantages that are similar to those of allografts.

Tissue-engineered Skin Substitutes

A wide range of tissue-engineered skin products is available commercially or has been used in clinical trials for treating first or second degree burns. Many of these products have been referred to by various names, such as skin substitutes or skin equivalents, sometimes qualified (or preceded) by the term engineered, bioengineered, or biologic, while some have been referred to as living skin equivalents. Skin substitutes available commercially or used in clinical trials can be either cell-free (acellular) or seeded with cells. Cell-free skin substitutes have commonly been derived from allogenic or xenogeneic sources such as decellularized human dermis or porcine SIS, or they may be composed of natural polymers such as bovine collagen and elastin, bovine collagen and chondroitin sulfate, or HA. In addition to protecting the wound, these cell-free skin substitutes stimulate ingrowth of fibrovascular tissue. While autologous cells have been used in a few skin substitutes, most cell-seeded skin substitutes contain allogeneic cells (fibroblasts and keratinocytes) because the use of autologous cells is impractical due to the time taken from harvesting the cells to producing the actual skin substitute. Allogeneic cell-seeded skin substitutes are considered temporary skin substitutes because they suffer from immune rejection within a period of days to weeks due to the presence of the allogeneic cells. They deliver growth factors and ECM to promote healing before rejection.

Another way of classifying the wide range of commercial skin substitutes is in terms of whether they are applicable to the treatment of the epidermal layer (in which case they are referred to as epidermal substitutes), the dermal layer (dermal substitutes), or both the epidermal and dermal layers (dermo-epidermal substitutes). As illustrated in Figure 25.35, allogeneic cell-seeded skin substitutes involve the combination of human fibroblasts, keratinocytes, or a combination of fibroblasts and keratinocytes with a degradable scaffold. Primary keratinocytes and fibroblasts are isolated from human donor sites and expanded *in vitro* prior to seeding onto degradable scaffolds. For example, for a dermo-epidermal skin substitute (DESS), fibroblasts and the scaffold (composed of bovine collagen, for example) are used first to create the dermal part. Then keratinocytes are seeded on top of the dermis to form the epidermal part.

Dermo-epidermal substitutes are the most advanced and sophisticated skin substitutes that are available for clinical use, but an off-the-shelf skin substitute capable of healing deep partial-thickness burns is not yet available. A major problem is the absence of initial vascularization of such skin substitutes, which can impair healing of the wound due to a lack of oxygen and nutrients. One approach to addressing this limitation has been to stimulate the formation of a vascular network in the skin substitute through the addition of growth factors such as VEGF. However, this approach has not been successful because of the short life span of VEGF in the tissue.

Another approach, involving the use of autologous adipose tissue, has been investigated for tissue engineering of pre-vascularized autologous DESSs (Klar et al. 2014). Autologous adipose tissue has two properties that make it a particularly attractive source for tissue engineering applications: it is easily obtainable from most patients by surgical excision or liposuction and it contains a heterogeneous population of cells referred to as the stromal vascular fraction (SVF). Although the SVF consists mostly of mature adipocytes, it also contains ECs, fibroblasts, progenitor cells and stem cells commonly referred to as ASCs. It is noted that progenitor cells and ASCs constitute ~3% of the cell number of the SVF, a relative abundance which is approximately three orders of magnitude higher than that of MSCs in fresh bone marrow. In addition to the ease with which it can be obtained, the SVF has two additional advantages. First, the SVF appears to contain vasculogenic cells and pre-fibroblasts in an optimal ratio and, second, it has a high vasculogenic potential in all age groups including elderly patients and patients with burn injuries.

Experimentally, three types of dermo-epidermal constructs were created for transplantation onto deep partial-thickness skin defects on the dorsum of immune-deficient rats. The three types of constructs, referred to as vascular DESS, nonvascular DESS and skin-derived fibroblast DESS, were prepared by seeding fibrin hydrogels with SVF

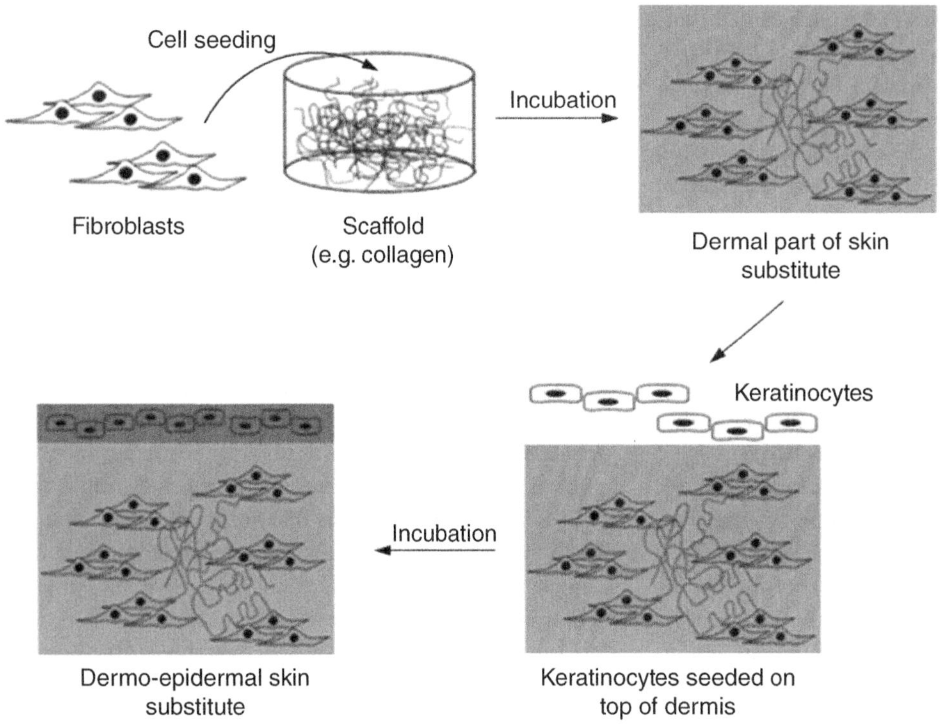

Figure 25.35 Schematic illustration of steps in tissue engineering for the creation of dermo-epidermal skin substitutes. *Source:* Adapted from Groeber et al. (2011).

cells, adipose stromal cells and skin-derived fibroblasts, respectively. Then the hydrogels were incubated for 1 week or 3 weeks in α-modified Eagle's medium (α-MEM) supplemented with 10% fetal bovine serum to permit vascular network formation, and transplanted onto the skin defects. At 4, 7, and 14 days, the transplant sites were excised for histological analysis, immunohistochemical staining and electron microscope examination.

With their pre-formed vascular networks, the vascular DESS showed immune staining for capillary-specific antigenic markers, anastomosed with the host vasculature within four days, and achieved sustained coverage of the wound site. In comparison, the nonvascular DESS and skin-derived fibroblast DESS showed a lower degree of vasculogenesis and lower sustained coverage at the wound sites. A key advantage of the use of SVF in the preparation of vascular DESS is that it is less labor-intensive than DESSs prepared by co-culturing of endothelial progenitor cells and dermal progenitor cells. Overall, the results of this investigation show that SVF-based tissue engineering of prevascularized DESSs is a potentially promising approach for the treatment of large-area skin wounds in human patients.

25.6.6 Bladder Tissue Engineering

The urinary bladder (or simply bladder) in humans is a hollow muscular and distensible organ that stores urine from the kidneys before disposal by urination. Congenital disorders, cancer, trauma, infection, or other conditions can lead to bladder damage, which often requires eventual reconstructive procedures, called bladder augmentation or augmentation cystoplasty, to repair or replace the bladder. Segments of autologous gastrointestinal tissues (such as bowel tissues) are commonly used for augmentation cystoplasty but they often lead to complications such as infection, increased mucous production, and malignant disease. Materials such as skin, decellularized bladder or SIS, natural polymers such as bovine collagen, or synthetic polymers such as polydimethylsiloxane (PDMS), polyurethane or polytetrafluoroethylene (PTFE) have also been used. However, they have not been successful due to structural, mechanical, functional, or biocompatibility problems. Consequently, despite the complications, gastrointestinal tissues remain the gold standard for reconstructing bladder damage.

Over the last few decades, there has been a growing interest in the use of tissue engineering approaches to create functional bladder tissues in an attempt to improve bladder reconstruction. These approaches have led to the creation of functional bladder tissues that have shown promising results in human patients requiring augmentation cystoplasty. Overall, however, the creation of engineered bladder tissues is challenging. In order to understand the challenges involved, it is useful to start with a brief discussion of the composition, structure, and function of the bladder.

The bladder is a hollow organ in the lower abdomen, which provides a storage vessel for urine, the liquid waste produced by the kidneys. The urine stays in the bladder until urination. An outer layer of muscle surrounds the inner lining of the bladder and when the bladder is full, the muscles in the bladder wall tighten to allow urination. After urination, the bladder shrinks in size. The bladder wall is usually considered to consist of three layers, plus an outer layer of fatty connective tissue that separates the bladder from other organs (Figure 25.36). The innermost layer, called the urothelium, consists of a layer of cells, called urothelial cells, which line the inside of the bladder. Under the urothelium is a layer of collagen-rich connective tissue called the lamina propria. The outermost layer, the muscularis propria, consists of muscle tissue. The urothelium serves as a barrier that prevents the contents of the bladder (urine) from permeating into the body cavity, while the collagen-rich layer and muscle tissue maintain the integrity of the bladder. The cells present in these layers interact with each other and other proteins to regulate cellular differentiation and function.

As noted earlier, an important requirement for biomaterials used in tissue engineering is that the biomaterial should provide an optimal environment to stimulate adhesion, proliferation, and differentiation of the relevant cell types. It is apparent that multiple cell types are required to create bladder tissue with a layered structure to mimic that of the native tissue. Each cell type requires different conditions for optimal proliferation and differentiation, which should be taken into account in designing an appropriate tissue engineering strategy. Biomaterials for engineering bladder tissues should also provide structural support for the different cell layers, including a suitable surface for stable cell attachment. In addition, biomaterials used for the inner layer should support the formation of a suitable barrier to retain the bladder contents and, for the outer layer, they should provide adequate biomechanical support to accommodate a high density of SMCs to prevent collapse of the bladder tissue.

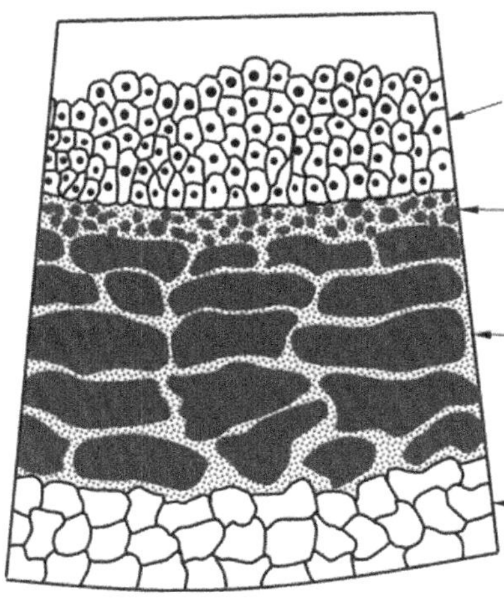

Figure 25.36 Illustration of layers in human bladder tissue.

Engineered bladder tissues created in recent tissue engineering approaches have shown promising results for reconstructing the bladder in human patients needing augmentation cytoplasty. In one approach (Atala et al. 2006), engineered bladder tissues were created by harvesting autologous urothelial cells and muscle cells using bladder biopsy, culturing the cells *in vitro* and expanding them to large quantities, and seeding the cells on bladder-shaped degradable composite scaffolds. Then the cell-seeded constructs were implanted and used for bladder reconstruction in humans. In the design of the engineered bladder tissues, autologous cells were used to avoid rejection of the cell-seeded construct upon implantation *in vivo*. In addition, the scaffold consisted of a bilayer composite created from two different biomaterials that were stitched together. Collagen derived from decellularized bladder submucosa formed the inner layer of the scaffold while the outer layer was composed of a porous synthetic polymer (PGA). The collagen inner layer served as a barrier to prevent loss of the bladder contents while also providing an optimal matrix for urothelial cell proliferation and differentiation. On the other hand, the porous PGA layer provided a matrix to accommodate large numbers of SMCs and to maintain structural integrity of the scaffold.

Experimentally, the porous outer layer of the composite scaffold was formed by first configuring PGA sutures (~15 μm in diameter) into a bladder-shaped mold created specifically for each patient and stabilizing the structure using PLGA sutures. The resulting flexible structure was coated with a solution of PLGA and dried to achieve adequate mechanical strength. Then the collagen inner layer was stitched to the PGA outer layer. Overall, the as-fabricated size of the composite scaffolds ranged from 70 to 150 cm^2, with a total layer thickness of ~2 mm. Cell seeding of the composite scaffold was performed in two steps. The exterior surface of the scaffold was first seeded with SMCs (~50×10^6 cells/cm^3) and following incubation for 48 hours, the inner surface of the scaffold was seeded with urothelial cells (~50×10^6 cells/cm^3). After incubation for 3–4 days, the cell-seeded scaffold was coated with fibrin glue, wrapped in omentum to improve vascularization *in vivo*, and anastamosed to the native bladder of human patients (aged 4–19 years) using PGA sutures (Figure 25.37). Overall, the time between bladder biopsy to harvest the cells and implantation of the cell-seeded scaffolds was ~7–8 weeks.

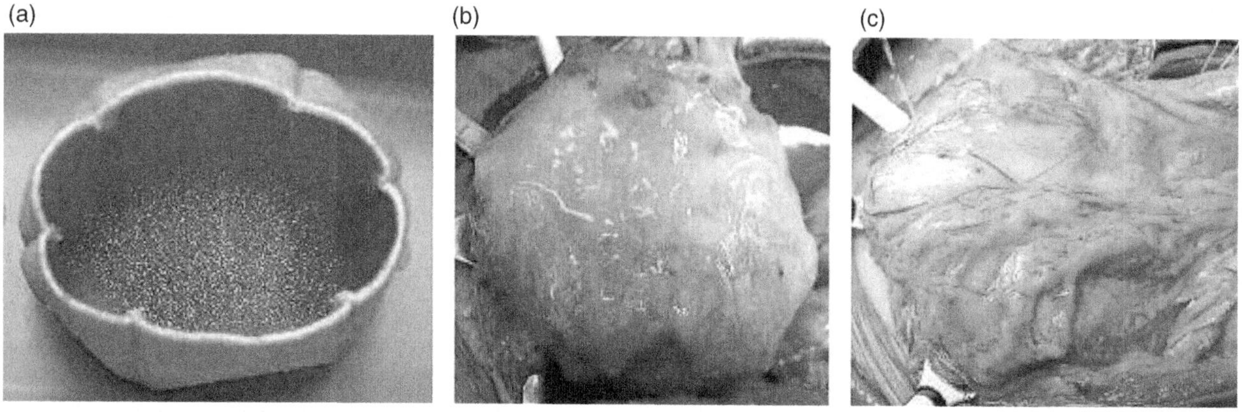

Figure 25.37 Construction of engineered bladder tissue. (a) Scaffold seeded with cells; (b) engineered bladder tissue anastamosed to native bladder with polyglycolic acid sutures; (c) implant covered with fibrin glue and omentum. *Source:* From Atala et al. (2006).

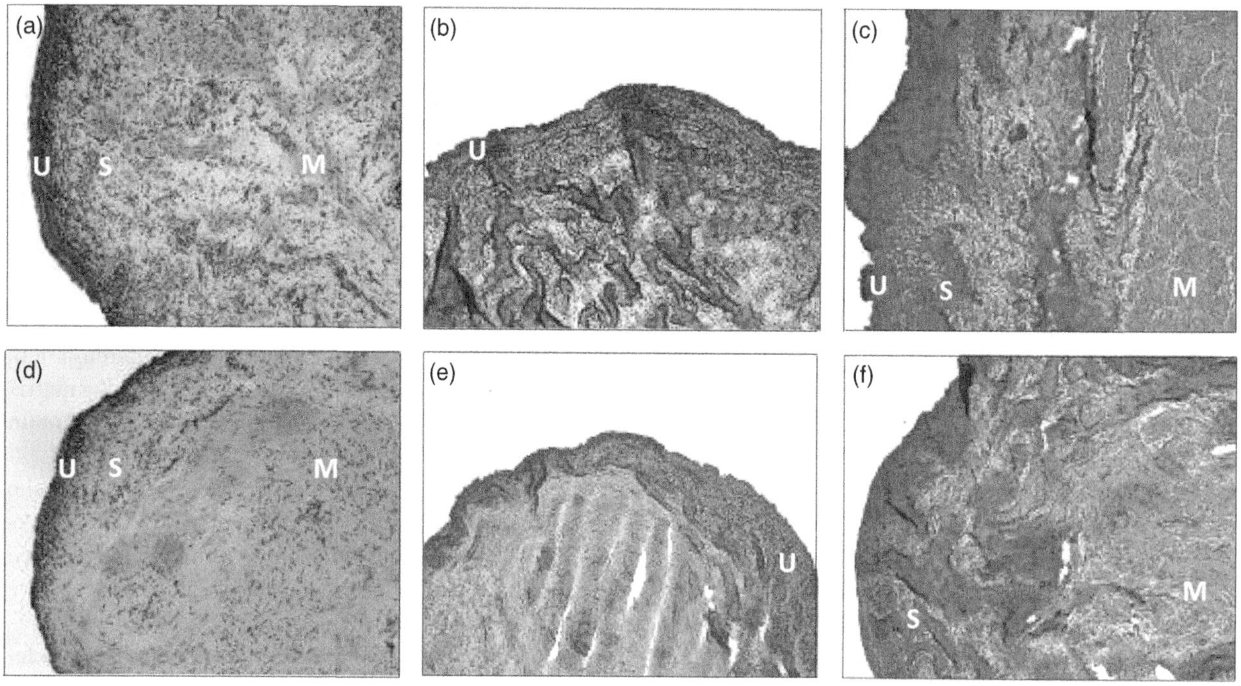

Figure 25.38 Morphological analysis of implanted engineered bladder. (a–c) Cystoscopic biopsies of implanted engineered bladders 31 months after augmentation shows extent of regeneration. Engineered bladder tissue showed trilayered structure, consisting of lumen lined with urothelial cells (U) surrounded by submucosa (S) and muscle (M). Hematoxylin and eosin (a), immunocytochemical analysis with anti-pancytokeratin AE1/AE3 antibodies (b), and anti-α smooth muscle actin antibodies (c) showed presence of phenotypically normal urothelium and smooth muscle. (d–f) Native bladder tissue. (Original magnification: 100×). *Source:* From Atala et al. (2006).

Postoperatively, the bladders reconstructed with the engineered tissues showed improved functional parameters over a follow-up period of 22–61 months. Urodynamic studies showed that bladder compliance was improved and bladder capacity was increased, while urinary continence improved markedly. The urodynamic results were similar to those that would be expected with the use of gastrointestinal tissue but without the aforementioned complications. Histologically and histochemically, full thickness tissue biopsies of the engineered bladder tissue showed an adequate structural architecture and phenotype (Figure 25.38).

Overall, the results of this study show that tissue engineering techniques have promising potential for generating bladder tissues that can be used in patients requiring augmentation cystoplasty. The principles involved in this study are also applicable to the engineering of other hollow organs such as blood vessels and the esophagus.

25.7 Concluding Remarks

In this chapter, we discussed the application of biomaterials in tissue engineering and regenerative medicine approaches to regenerate functional tissues and organs. The overall principle of tissue engineering and regenerative medicine is to regenerate a patient's own tissues and organs instead of the more conventional approaches involving repair or replacement of the patient's tissues and organs using durable biomaterials or transplanted tissues.

- Regenerative medicine encompasses three separate approaches, tissue engineering, cell therapy, and gene therapy. The focus of this chapter was on tissue engineering because of its significant reliance on the use of biomaterials.
- Building on detailed information presented in earlier chapters, the three main components of the tissue engineering approach, biomaterials, cells and signaling factors, were discussed at a basic level.
- Biomaterials with an appropriate external shape and internal architecture (or microstructure), called scaffolds, are an important aspect of tissue engineering. We discussed the required and desirable properties and characteristics of tissue engineering scaffolds.
- Techniques for forming biomaterials into scaffolds were discussed, with emphasis placed on additive manufacturing (3D printing) techniques because of their ability to provide precise control of the scaffold shape and architecture, and on electrospinning based on its ability to create fibrous scaffolds that can mimic structural features of the ECM.
- We discussed a recent advance in the additive manufacturing (3D printing) techniques, called three-dimensional bioprinting, which provides the ability to create scaffolds and seed them with living cells as part of a single fabrication process.
- Applications of the tissue engineering approach to the creation of tissues and organs, including bone, cartilage, articular joints, tendons and ligaments, skin and the urinary bladder were discussed. Emphasis was placed on advances made in these areas through innovative biomaterial design and production, and incorporation of principles derived from improved understanding of biological processes.

Problems

25.1 Discuss the distinction between tissue engineering and regenerative medicine.

25.2 Describe the basic tissue engineering approach to regenerating tissues and organs and variations in this basic approach that have been introduced over time.

25.3 Discuss and compare the advantages and disadvantages of using fully differentiated cells and adult stem cells in tissue engineering.

25.4 Discuss the distinction between stem cells and progenitor cells.

25.5 Compare the advantages and disadvantages of using MSCs derived from bone marrow and from adipose tissue in tissue engineering.

25.6 Briefly describe the advantages and disadvantages associated with induced pluripotent stem cells in tissue engineering applications.

25.7 Cellular and gene manipulations involved in some tissue engineering applications present an element of risk for the development of tumors known as teratomas. Briefly describe one type of cellular or gene manipulation that presents no risk of tumor development and briefly explain the basis for the absence of risk.

25.8 Discuss the essential (required) properties and characteristics of scaffolds for tissue engineering.

25.9 One of the requirements of scaffolds in tissue engineering is the ability to provide a suitable microenvironment (or niche) for cells to reside, proliferate, and differentiate into the required phenotype. Discuss limitations of porous solids composed of synthetic polymers in satisfying this requirement, and what approaches, if any, can be used to address these limitations.

25.10 Discuss and compare the advantages of hydrogel scaffolds and decellularized ECM scaffolds in tissue engineering.

25.11 Discuss the advantages and limitations of electrospinning as a technique for creating scaffolds for tissue engineering.

25.12 Discuss and compare the advantages and disadvantages of 3D printing and 3D bioprinting in tissue engineering.

25.13 Discuss and compare the advantages and disadvantages of inkjet-based 3D bioprinting and microextrusion-based 3D bioprinting in tissue engineering.

25.14 Explain why it is difficult to regenerate articular cartilage. Briefly describe current surgical treatments for healing focal articular cartilage defects in humans.

25.15 You are responsible for a project to engineer a tissue for regenerating full-thickness articular cartilage defects in humans. (a) What criteria should the scaffold material(s) satisfy? (b) What synthetic scaffold material(s) would you choose for this application? Explain why. (c) Briefly describe the procedure you would use to engineer the tissue construct.

25.16 You are responsible for a project to engineer a skin substitute to treat large-area second-degree burns in humans. (a) What criteria should the scaffold material(s) satisfy? (b) What scaffold material(s) would you choose for this application? Explain why. (c) Briefly describe the procedure you would use to engineer the tissue construct.

25.17 You are responsible for a project to engineer a tissue (vascular graft) for reconstructing diseased blood-vessels. (a) What criteria should the engineered tissue satisfy? (b) What scaffold materials would you choose for this application? Explain why. (c) What are the anticipated advantages and disadvantages of the engineered tissue? (d) Briefly describe the procedure you would use to create the engineered tissue.

References

Atala, A., Bauer, S.B., Soker, S. et al. (2006). Tissue-engineered autologous bladders for patients needing cytoplasty. *Lancet* 367: 1241–1246.

Badylak, S.F., Brown, B.N., Gilbert, T.W. et al. (2011). Biologic scaffolds for constructive tissue remodeling. *Biomaterials* 32: 316–319.

Badylak, S.F., Freytes, D.O., and Gilbert, T.W. (2009). Extracellular matrix as a biological scaffold material: Structure and function. *Acta Biomaterialia* 5: 1–13.

Bi, Y., Ehirchiou, D., Kilts, T.M. et al. (2007). Identification of tendon stem/progenitor cells and the role of the extracellular matrix in their niche. *Nature Medicine* 13: 1219–1227.

Borden, M., El-Amin, S.F., Attawia, M. et al. (2003). Structural and human cellular assessment of a novel microsphere-based tissue engineered scaffold for bone repair. *Biomaterials* 24: 597–609.

Censi, R., van Putten, S., Vermondèn, T. et al. (2011). The tissue response to photopolymerized PEG-p(HPMAm-lactate)-based hydrogels. *Journal of Biomedical Materials Research. Part A* 97: 219–229.

Chang, C.W., Petrie, T., Clark, A. et al. (2016). Mesenchymal stem cell seeding of porcine small intestinal submucosal extracellular matrix for cardiovascular applications. *PLoS ONE* 11: e0153412.

Chew, S.Y., Wen, J., EKF, Y. et al. (2005). Sustained release of proteins from electrospun biodegradable fibers. *Biomacromolecules* 6: 2017–2024.

Derby, B. (2011). Inkjet printing ceramics: From drops to solid. *Journal of the European Ceramic Society* 31: 2543–2550.

Duan, B., Hockaday, L.A., Kang, K.H., and Butcher, J.T. (2013). 3D bioprinting of heterogeneous aortic valve conduits with alginate/gelatin hydrogels. *Journal of Biomedical Materials Research. Part A* 101: 1255–1264.

Eichhorn, S.J. and Sampson, W.W. (2005). Statistical geometry of pores and statistics of porous nanofibrous assemblies. *Journal of the Royal Society Interface* 2: 309–318.

Gilbert, T.W., Sellaro, T.L., and Badylak, S.F. (2006). Decellularization of tissues and organs. *Biomaterials* 27: 3675–3683.

Groeber, F., Holeiter, M., Hampel, M. et al. (2011). Skin tissue engineering - *in vivo* and *in vitro* applications. *Advanced Drug Delivery Reviews* 128: 352–366.

Gundapati, H., Dey, M., Ozbolat, I. et al. (2016). A comprehensive review on droplet-based bioprinting: past, present and future. *Biomaterials* 102: 20–42.

Jaiswal, N., Haynesworth, S.E., Caplan, A.I. et al. (1997). Osteogenic differentiation of purified, culture-expanded human mesenchymal stem cells in vitro. *Journal of Cellular Biochemistry* 64: 295–312.

Kiani, C., Chen, L., Wu, Y.J. et al. (2002). Structure and function of aggrecan. *Cell Research* 12: 19–32.

Kim, T.-K., Sharma, B., Williams, C.G. et al. (2003). Experimental model for cartilage tissue engineering to regenerate the zonal organization of articular cartilage. *Osteoarthritis and Cartilage* 11: 653–664.

Klar, A.S., Güven, S., Biedermann, T. et al. (2014). Tissue-engineered dermo-epidermal skin grafts prevascularized with adipose-derived cells. *Biomaterials* 35: 5065–5078.

Laflamme, M.A., Chen, K.Y., Naumova, A.V. et al. (2007). Cardiomyocytes derived from human embryonic stem cells in pro-survival factors enhance function of infarcted rat hearts. *Nature Biotechnology* 25: 1015–1024.

Langer, R. and Vacanti, J.P. (1993). Tissue engineering. *Science* 260: 920–926.

Lee, C.H., Marion, N.W., Hollister, S. et al. (2009). Tissue Formation and vascularization in anatomically shaped human joint condyle ectopically *in vivo*. *Tissue Engineering Part A* 15: 3923–3930.

Lee, C.H., Cook, J.L., Mendelson, A. et al. (2010). Regeneration of the articular surface of the rabbit synovial joint by cell homing: a proof of concept study. *Lancet* 376: 440–448.

Lee, K., Silva, E.A., and Mooney, D.J. (2011). Growth factor delivery-based tissue engineering: general approaches and a review of recent developments. *Journal of the Royal Society Interface* 8: 153–170.

Li, D. and Xia, Y. (2004). Direct fabrication of composite and ceramic hollow nanofibers by electrospinning. *Nano Letters* 4: 933–938.

Malda, J., Visser, J., Melchels, F.P. et al. (2013). 25th anniversary article: engineering hydrogels for biofabrication. *Advanced Materials* 25: 5011–5028.

Mathieu, L.M., Mueller, T.L., Bourban, P.-E. et al. (2006). Architecture and properties of anisotropic polymer composite scaffolds for bone tissue engineering. *Biomaterials* 27: 905–916.

Mikos, A.G., Thorsen, A.J., Czerwonka, L.A. et al. (1994). Preparation and characterization of poly(l-lactic acid) foams. *Polymer* 35: 1068–1077.

Mohamed, T.M.A., Stone, N.R., Berry, E.C. et al. (2017). Chemical enhancement of *in vitro* and *in vivo* direct cardiac reprogramming. *Circulation* 135: 978–995.

Mosissette, S.L., Lewis, J.A., Cesarano, J. et al. (2000). Solid freeform fabrication of aqueous alumina–poly(vinyl alcohol) gelcasting suspensions. *Journal of the American Ceramic Society* 83: 2409–2416.

Ng, K.W., Wang, C.C.-B., Mauck, R.L. et al. (2005). A layered agarose approach to fabricate depth-dependent inhomogeneity in chondrocyte-seeded constructs. *Journal of Orthopaedic Research* 23: 134–141.

Ouyang, H.W., Goh, J.C.H., Mo, X.M. et al. (2002). The efficacy of bone marrow stromal cell-seeded knitted PLGA fiber scaffold for Achilles tendon repair. *Annals of the New York Academy of Sciences* 961: 126–129.

Saunders, R.E. and Derby, B. (2014). Inkjet printing of biomaterials for tissue engineering: bioprinting. *International Materials Review* 59: 430–448.

Schuurman, W., Levett, P.A., Pot, M.W. et al. (2013). Gelatin-methacrylamide hydrogels as potential biomaterials for fabrication of tissue-engineered cartilage constructs. *Macromolecular Bioscience* 13: 551–561.

Sepulveda, P., Jones, J.R., and Hench, L.L. (2002). Bioactive sol-gel foams for tissue repair. *Journal of Biomedical Materials Research* 59: 340–348.

Solchaga LA, Penick KJ, Welter JF Chondrogenic differentiation of bone marrow-derived mesenchymal stem cells: tips and tricks *Methods in Molecular Biology* 698, 253–278, 2011

Takahashi, K. and Yamanaka, S. (2006). Induction of pluripotent stem cells from mouse embryonic and adult fibroblast cultures by defined factors. *Cell* 126: 663–676.

Williams, D. (2014). *Essential Biomaterials Science*, 314. Cambridge University Press.

Xu, T., Zhao, W., Zhu, J.-M. et al. (2013). Complex heterogeneous tissue constructs containing multiple cell types prepared by inkjet printing technology. *Biomaterials* 34: 130–139.

Yang, G., Rothrauff, B.B., and Tuan, R.S. (2013). Tendon and ligament regeneration and repair: Clinical relevance and developmental paradigm. *Birth Defects Research Part C* 99: 203–222.

Yin, Z., Chen, X., Chen, J.L. et al. (2010). The regulation of tendon stem cell differentiation by the alignment of nanofibers. *Biomaterials* 31: 2163–2175.

Zhou, Q., Brown, J., Kanarek, A. et al. (2008). *In vivo* reprogramming of adult pancreatic exocrine cells to beta-cells. *Nature* 455: 627–632.

Further Reading

Brown, B.N. and Badylak, S.F. (2014). Extracellular matrix as an inductive scaffold for functional tissue reconstruction. *Translational Research* 163: 268–285.

Lee, C.H., Marion, N.W., Hollister, S., and Mao, J.J. (2009). Tissue formation and vascularization in anatomically shaped human joint condyle ectopically *in vivo*. *Tissue Engineering Part A* 15: 3923–3930.

Li, D. and Xia, Y. (2004). Electrospinning of nanofibers: reinventing the wheel? *Advanced Materials* 16: 1151–1170.

Sill, T.J. and von Recum, H.A. (2008). Electrospinning: Applications in drug delivery and tissue engineering. *Biomaterials* 29: 1989–2006.

Slaughter, B.V., Khurshid, S.S., Fisher, O.Z. et al. (2009). Hydrogels in regenerative medicine. *Advanced Materials* 21: 3307–3329.

26

Biomaterials for Drug Delivery

26.1 Introduction

Drugs are pharmacologically active substances, which, when delivered (administered) to the body, may have a beneficial therapeutic effect. Drugs can be used, for example, to manage, treat or cure disease or discomfort, facilitate healing of injuries or improve physiological performance. However, the ability of a drug to have its desired therapeutic effect depends not only on the drug itself but also on other factors, including the method by which the drug is delivered to the body, the mechanism by which the drug acts in the body and the ease with which the body responds to the presence of the drug.

The delivery method affects several factors that contribute to the effectiveness of a drug, including its pharmacokinetics and toxicity. Pharmacokinetics is a general term that refers to the activity of a drug in the body over an appropriate period, including the processes by which it is absorbed and distributed in the body, localized in tissues, and excreted from the body. Often, the best mode of delivery is achieved by combining a suitable biomaterial with the drug to form a drug delivery system. For example, this could involve physically dispersing the drug within a degradable polymer, which is implanted or injected into the desired tissue site, where it is released over the desired treatment period by diffusion from the polymer or by erosion of the polymer. Alternatively, the drug molecules could be covalently bonded by means of a linker molecule to the macromolecular chains of a degradable polymer, forming a polymer–drug conjugate. Upon implantation, drug release occurs by cleavage of the linker molecule under appropriate conditions.

As drugs have become more complex over the last few decades to treat serious conditions, the potential to enhance not only their therapeutic benefits but also their harmful side effects has increased. An example is cancer therapy, where many chemotherapy drugs can kill cancerous cells and healthy cells as well. As a result, patients often suffer from adverse side effects, such as nausea, hair loss, and fatigue, resulting, in most cases, in a temporary reduction in quality of life. Advances in genetic engineering and biotechnology have led to the development of new classes of drugs, including drugs composed of nucleic acids, antibodies, proteins and complex macromolecules, which have presented new challenges in designing and creating new biomaterials for use in drug delivery systems. Often, the biomaterial may be required to provide enhanced therapeutic protection for the drug to maintain its pharmacological activity, enhanced specificity of the drug delivery system to target specific cells, and efficient delivery of the drug to its site of action within the interior of targeted cells. Once the drug has reached its site of action, the biomaterial may be required, in addition, to control its release. Delivery systems used in such advanced therapeutic applications are often referred to as controlled drug delivery systems or targeted drug delivery systems. Overall, the clinical use of drug delivery systems is considerable, with a global market estimated at over 100 billion dollars (US) in 2014.

Biomaterials have been designed and used in drug delivery systems since about 1960, but the field of drug delivery and controlled drug release has evolved considerably since that time. Building on early developments, including observations that biomaterials could be used to release small-molecule drugs (typically of molecular weight less than ~1 kDa) over a prolonged period, and an understanding of drug release mechanisms, biomaterials have been designed to release such drugs at a controlled rate. With the subsequent creation of protein and other large-molecule drugs, several strategies have been developed to provide controlled release of large-molecule drugs. These strategies have resulted in the creation of microparticle-based drug delivery systems from which large-molecule drugs could be released at a controlled rate over a prolonged period at specific sites within the body. Additional advances included the development of systems composed of degradable polymers and hydrogels, which could also release their drugs in response to environmental stimuli, such as changes in temperature or pH, or by remote physical stimulation using light, ultrasound waves, or electric or magnetic fields.

Materials for Biomedical Engineering: Fundamentals and Applications, First Edition. Mohamed N. Rahaman and Roger F. Brown.

Companion website: www.wiley.com/go/rahaman/biomedicalengineering

In the last 20 years or so, nanotechnology has had a dramatic effect on drug delivery and drug release, resulting in the design and development of a variety of nanoparticle-based delivery systems, including systems based on degradable polymers, lipids, dendrimers, and inorganic biomaterials. These nanoparticle-based systems are capable of not only encapsulating the drug and maintaining its pharmacological activity, but also targeting its delivery to specific sites, such as tumors, and controlling its release upon reaching its site of action. Nanoparticle-based technologies have already made significant inroads into the creation of drug delivery systems for clinical applications.

Since the identification nearly two decades ago of ribonucleic acid interference (RNAi) as a fundamental pathway to target and induce gene silencing, there has been significant interest in the development of biomaterials to deliver RNA as an alternative class of biomolecules to treat a variety of diseases. A considerable number of RNAi-based therapeutics have already reached clinical studies and many more have been under development. Another area of significant recent interest has been the development of biological delivery systems composed of living systems, such as vesicles, non-pathogenic viruses and cells, to deliver drugs and treat disease because they naturally possess many desirable features for the delivery of biomolecules.

Overall, this chapter is devoted to a discussion of biomaterials for use in drug delivery and controlled release of drugs. The topics to be discussed are in the following broad areas:

- Principles and mechanisms of controlled drug release.
- Designing biomaterials for use in drug delivery systems, particularly systems based on microparticles, hydrogels and nanoparticles.
- Examples of the applications of such delivery systems in the treatment of diseases, particularly serious diseases such as cancer.

26.2 Controlled Drug Release

Drugs can be administered to the body by means of a variety of routes, including the oral route, injections, inhalation, transdermal delivery, and surgical implantation, each of which has its advantages and disadvantages (Table 26.1). The most common routes today are the oral route, by swallowing a pill, and injections. Biomaterials have played a role in improving oral and injectable drug delivery, as well as the other means of drug delivery. Oral pills are used to deliver a majority of small-molecule drugs as evidenced, for example, by tens of billions of aspirin tablets being consumed worldwide each year. Injections administered intravenously, intramuscularly, or subcutaneously, are the primary mode of delivering free (unencapsulated) drugs composed of peptides and proteins, with over 10 billion injections administered worldwide each year. Oral pills offer the convenience of pre-determined and measured doses, portability, defined dosing times and an overall non-invasive nature of administration. However, they are unable to deliver large-molecule drugs composed of proteins and other macromolecules. While injections are able to deliver large-molecule drugs, they are invasive in nature.

In general, oral pills and injections suffer from drawbacks that make them unsuitable for meeting the needs of advanced therapeutic applications. These delivery routes have a limited ability to control the release of the drug at a desirable rate over a prolonged period, to allow direct administration within the desired tissue to reduce systemic side effects, or to deliver high molecular weight biomolecules. Drug administration by oral pills and injections normally requires frequent doses, resulting in high variability in the drug concentration over the treatment period (Figure 26.1). Upon administration of pills and injections, the drug concentration increases to a desirable therapeutic level but, in some cases, toxic side effects can arise when the concentration increases above a maximum safe level. Drug concentrations can also decrease rapidly to levels that are below the effective therapeutic level due to factors resulting from metabolism, degradation of the drug, or transport of the drug away from the target site. This type of release profile, whereby the drug concentration increases above the toxic level and decreases below the effective level, results in increased risk to the patient. Drug concentrations above the toxic level and below the effective level also lead to waste of the drug. Additionally, the effect of the drug is systemic and not localized at a desired tissue site. These limitations can have serious consequences, particularly in the treatment of serious conditions such as cancer.

By combining the drug with a suitable biomaterial, drug delivery systems have been designed to control the release of the drug over a prolonged period, localize its effect and, for some applications, such as cancer treatment, improve its specificity to target certain cells, such as tumor cells. These drug delivery systems, when compared to oral pills, injections of free drugs, and other earlier approaches, provide several advantages, including the following:

- The drug can be continuously released at a concentration within a desirable therapeutic range for a prolonged period (in the range of days to years), thereby reducing the need for frequent administration of the drug (Figure 26.1), improving patient comfort, and improving adherence of the patient to the recommended drug treatment regimen.

Table 26.1 Different routes of drug delivery, their main advantages and disadvantages, tissue targets and examples of delivery systems.

Drug delivery route	Advantages	Disadvantages	Targets	Examples of delivery systems
Injections: intravenous (IV); intramuscular (IM); subcutaneous (SQ); depot	Applicable to a large number of drugs; rapid onset (IV); controlled release (IM, SQ); up to 100% bioavailability	Rapidly cleared from the body (IV); difficult for patient to administer (IV); frequent injections required (IV); greater noncompliance by patients (IV); greater risk of infection	Tissues with blood access (IV); systemic; muscle (IM)	Chemotherapy (IV); vaccines (IM); hormones (depot); hydrogels; nanoparticles
Oral	Convenient for patients; high patient compliance	Low bioavailability; variable adsorption in the presence of food; generally non-targeted; not viable for large-molecule drugs; degradation of drug in stomach and liver	Systemic	Liquid medications; capsules; pills; microparticles; hydrogels; nanoparticles
Transdermal	Painless administration; sustained and controlled release; reduced side effects; high patient compliance	Low bioavailability; variable absorption depending on skin condition and location; risk of incorrect dosage (creams)	Systemic; skin	Patches; microneedles; creams; hydrogels; nanoparticles
Pulmonary (e.g., inhalation)	Ease of use; high bioavailability; rapid absorption and systemic take up; direct access to lungs	Inconsistent delivery due to variation in patient technique; local lung toxicity and immunogenicity; large/bulky administration devices	Lungs; systemic; brain	Aerosols; dry powder; nanoparticles
Surgical implantation	Direct access to a range of diseased tissues; lower off-target toxicity	Requires surgical intervention; time and burden of procedure; risk of infection due to surgery; foreign body response and rejection; may require immunosuppressing drugs	Local delivery to a range of diseased tissues	Polymer implants; microparticles; hydrogels

Source: Based on Fenton et al. (2018) and Anselmo and Mitragotri (2014).

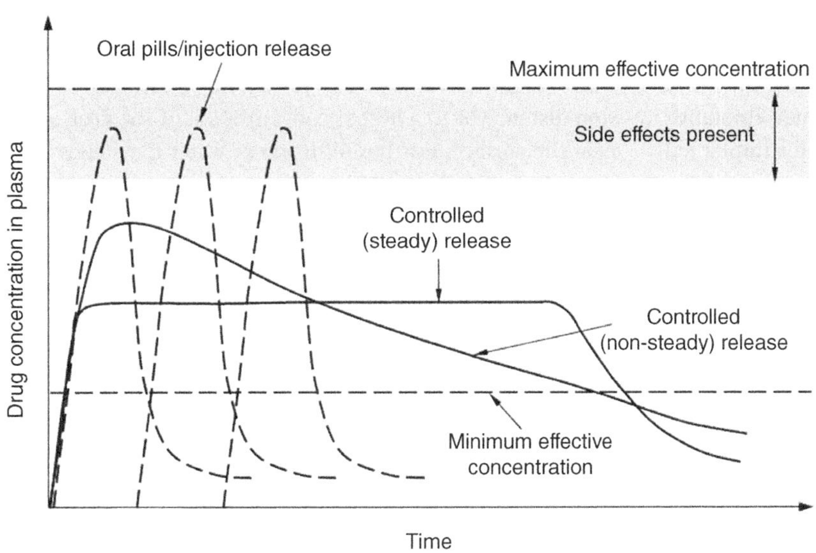

Figure 26.1 Schematic illustration of drug concentration in plasma as a function of time after drug administration by oral tablets and injections, and by controlled drug delivery systems.

- By avoiding peaks and valleys in the released drug concentration outside the desirable therapeutic range, risks to the patient related to reduced effectiveness of the drug or to potential toxic side effects are reduced while, at the same time, waste of the drug is avoided.
- Patient-to-patient variations in drug release can be reduced because the rate-limiting step in release is often the removal of the drug from the biomaterial itself.
- By varying the composition of the biomaterial or altering the procedure by which the drug is combined with the biomaterial, controlled drug delivery systems can provide a wide range of drug release rates for different applications.
- Because drug release can be controlled within a desirable therapeutic range for a prolonged period, controlled drug delivery systems can be implanted or injected within a desired tissue site, providing a drug release depot, which can reduce or eliminate harmful side effects caused by systemic drug administration.

26.2.1 Drug Delivery Systems

As noted earlier, drugs can be delivered by a variety of routes, with each route having its own advantages and limitations. Consequently, each route has its own requirements for designing and creating suitable biomaterials that can be used in drug delivery systems to improve the administration and efficacy of the drugs. Several categories of drug delivery systems have been studied or developed for clinical applications, including systems based on oral pills, liquid medications, transdermal patches, aerosols, microparticles, hydrogels, and nanoparticles. For example, microparticle-based delivery systems often consist of drugs encapsulated within solid degradable polymer microspheres, which can be injected at the desired tissue site to serve as a depot for prolonged drug delivery. On the other hand, delivery systems composed of polymer implants, microparticles, or hydrogels can be surgically implanted into a range of tissue sites, allowing prolonged drug delivery and often reducing the number of invasive procedures required to maintain a similar therapeutic effect.

We shall focus mainly on three types of drug delivery systems chosen because of their large reliance on the use of biomaterials, the considerable research and development interest which they have attracted over the last few decades, and their clinical and commercial potential or impact. These systems consist of polymer-based microparticle delivery systems, hydrogel-based delivery systems, particularly systems that can release their incorporated drugs in response to environmental or external stimuli, and nanoparticle-based delivery systems. Our emphasis will be on the principles of designing biomaterials for use in these delivery systems, mechanisms of controlled drug release, and applications of these systems to the treatment of diseases, particularly cancer.

26.2.2 Mechanisms of Drug Release

Polymers find broad applicability in drug delivery systems and, therefore, it is useful to consider the possible mechanisms by which a drug can be released from them. In general, the basic mechanisms of release are diffusion of the drug molecules from the polymer, erosion of the polymer, cleavage of certain chemical bonds, and osmosis (Figure 26.2).

Release by Diffusion

Diffusion-controlled drug release systems normally rely on diffusion of the drug molecules through water-filled pores of the polymer, which is the common pathway for hydrophilic drugs. Diffusion of small molecule hydrophobic drugs through the polymer phase itself may be possible, but this is most effective when the polymer has a lower molecular weight or is in a rubbery state, such that the free volume between the polymer chains is sufficiently high. Here we shall be concerned mainly with diffusion of hydrophilic drugs through water-filled pores.

In general, the drug is physically dispersed within a degradable polymer matrix. Upon delivery of the drug-containing polymer to the desired tissue site, water molecules will normally start to diffuse into the polymer, forming a network of water-filled pores, starting from the surface of the polymer. These water-filled pores provide a favorable pathway for drug molecules to diffuse outward, driven by a concentration gradient between the polymer and the surrounding tissue. Diffusion of drug molecules starts to occur from a surface layer of the polymer, which may sometimes lead to an initial rapid (burst) release due to the short diffusion distance or to a higher concentration of the drug at or near the surface. Further diffusion of water molecules into the polymer leads to an interconnected network of water-filled pores throughout the polymer, enabling diffusion of the drug from the interior of the polymer. With pores of sufficiently large size and interconnectivity, most of the drug could be released prior to any significant erosion of the polymer. On the other hand, for smaller size pores of lower interconnectivity, release of the drug may slow down after an initial burst release. Subsequently, degradation and bulk erosion of the polymer can lead to larger pores of better interconnectivity. This can lead to an acceleration of drug release until almost all of the drug has been released, often resulting in an overall release profile that approximates a sigmoidal shape. Following release of the drug, the polymer matrix eventually erodes away completely.

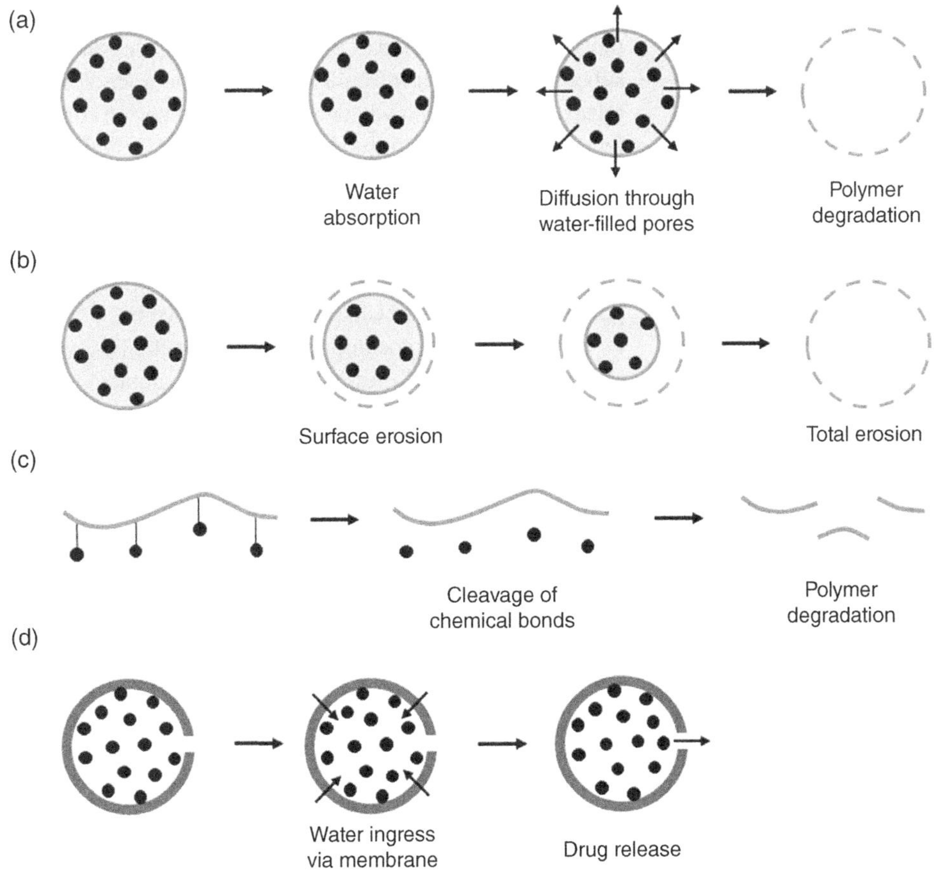

Figure 26.2 Mechanisms of drug release from solid polymeric systems: (a) diffusion, (b) erosion, (c) cleavage of covalent bonds, and (d) osmosis.

Overall, the general trend in the drug release profile as a function of time can vary markedly, depending on a variety of factors, particularly the composition and molecular weight of the polymer, the geometry of the polymer matrix, and the molecular weight (size) of the drug molecules. For example, the composition of the polymer will determine the rate at which water molecules diffuse into the polymer to form an interconnected network of pores, as well as the rate of bulk erosion (Chapter 9). Based on their size alone, smaller drug molecules generally diffuse faster than larger molecules.

Release by Erosion

Generally, in release by erosion, the drug is physically dispersed within a degradable polymer matrix and is released as the polymer erodes in the aqueous physiological environment. As noted in Chapter 9, erosion is a physical process that accompanies degradation of the polymer, whereby oligomers or monomers produced by degradation of the polymer chains dissolve into the aqueous physiological environment. This leads to a loss of mass, and the drug within this mass is released into the physiological environment. Eventually, all of the drug is released upon complete erosion of the polymer matrix. There are two types of erosion, bulk erosion and surface erosion. Mass loss as a function of time by bulk erosion often shows a complicated behavior, making it difficult to control drug release (Figure 9.5). In comparison, the mass loss per unit area in surface erosion ideally increases uniformly with time, giving a constant drug release rate (called zero-order release).

Release by Cleavage of Chemical Bonds

Instead of physically dispersing a drug within a polymer matrix, drug molecules may be covalently bonded by means of linker molecules to the macromolecular chains of a degradable polymer, forming a polymer–drug conjugate. Upon delivery of the conjugate to the desired tissue site, release of the drug occurs by degradation of the linker molecules under appropriate conditions. This release mechanism may be particularly useful for drugs that do not have adequate solubility in the physiological environment. Following release of the drug, the polymer chains normally degrade by hydrolysis.

Release by Osmosis

In osmotic release, the drug is encapsulated within a semi-permeable membrane through which a small hole has been drilled. Some devices may contain more than one hole. The membrane itself is permeable to water but impermeable to the drug. Upon implantation of the delivery system, water enters the drug reservoir by osmosis due to the presence of the drug or to a co-encapsulated salt, and the drug is forced out through the hole by osmotic pressure.

26.3 Designing Biomaterials for Drug Delivery Systems

The design of biomaterials for use in drug delivery systems requires proper consideration of a combination of properties and characteristics of the biomaterial. Minimally, a drug delivery system should normally provide controlled drug release at a therapeutically effective concentration over the desired treatment period at the intended site of action. Consequently, biomaterials used in drug delivery systems should:

- Have adequate biocompatibility and, if degradable, should degrade into nontoxic products.
- Be capable of being incorporated with an adequate amount of a drug in order to provide prolonged release and to achieve the desired therapeutic effectiveness.
- Protect the drug from degradation *in vivo* and provide predictable release of the drug over the entire duration of the treatment, while avoiding peaks and valleys in the drug concentration outside the desired therapeutic range.
- Have the ability to be injected intravenously or to be implanted directly within the desired tissue site to reduce harmful side effects.
- Be capable of delivering the drug efficiently to its intended site of action, such as targeted cells or within the interior of the targeted cells, to enhance the specificity of the drug.

26.4 Microparticle-based Delivery Systems

Microparticle-based drug delivery systems normally consist of degradable polymer microspheres in which the drug is physically dispersed or encapsulated. The polymer protects the drug and controls its release, while the size of the particles, in the range ~1–100 μm, allows for a collection of particles to be easily administered by injection through a syringe needle. Because of their size, microparticles tend to remain at the tissue site where they are injected and, thus, they are often delivered at the desired tissue site, such as a subcutaneous or an intramuscular site. The microspheres release their drugs slowly over a prolonged period (up to a few months), providing an effect locally or systemically. Such delivery systems are often referred to as depot systems or matrix systems.

Polymers used in microparticle-based drug delivery systems are often selected from certain degradable synthetic polymers discussed in Chapter 9 which have been shown to have adequate biocompatibility. Polylactic-*co*-glycolic acid (PLGA) and, to a lesser extent, polylactic acid (PLA) and polyglycolic acid (PGA), have been the most commonly used polymers, based largely on their history of medical use, generally non-toxic degradation products, and ease of formation into microspheres at the laboratory and industrial scales. Other biocompatible polymers, including polyanhydrides and polyorthoesters, have also been used because they can provide better control of the drug release profile.

Overall, polymer-based microsphere delivery systems are versatile depot systems. They allow control of drug delivery rates over a wide range through proper selection of the polymer composition, such as monomer composition, copolymerization, and polymer molecular weight, and the physical characteristics of the particles, such as the microsphere diameter or distribution of diameters. On the other hand, encapsulation of peptide and protein drugs in PLGA microspheres and in other polymer microsphere-based systems, as well as maintaining the stability of the drugs over a prolonged period of delivery, can often be challenging.

Drug release from PLGA microspheres occurs predominantly by diffusion of the drug molecules through water-filled pores (Figure 26.2a), which, in the later stages of release, can be accelerated by bulk erosion of the polymer. As noted earlier, drug release by diffusion can lead to a range of release profiles. In practice, there is also some contribution from polymer erosion, which will further complicate the release profile. In comparison, release from polyanhydrides occur predominantly by surface erosion, which can lead to more predictable drug release rates. For example, with proper selection of the polymer composition, polyanhydrides can be synthesized to degrade at a nearly constant rate over periods ranging from a few days to several months (Figure 9.16).

For diffusion-controlled release and for a given composition, the size of the polymer microspheres has a strong effect because it determines the diffusion distance of water molecules into the microsphere and of drug molecules outward into the physiological environment. For the same mass of microspheres, larger microspheres release their drug more slowly and over a longer period because, on average, the water molecules and drug molecules have to diffuse over a larger distance. Figure 26.3 shows the release profile of piroxicam from PLGA microspheres of three nearly monodisperse diameters, 10, 50, and 100 μm, in phosphate-buffered saline (PBS) (pH = 7.4) at 37 °C. The microspheres, loaded with 10 wt% piroxicam, were prepared from a copolymer composed of ~50 wt% lactic acid and ~50 wt% glycolic acid. The smallest microspheres showed a rapid initial release rate, with ~40–50% of the drug released within 1 day. In comparison, the initial release rate decreased for the larger diameter microspheres, and the release profiles showed an approximately sigmoidal shape. Additionally, by combining mixtures of drug-loaded microspheres of different monodisperse diameters, nearly constant (zero order) release was obtained over 14 days.

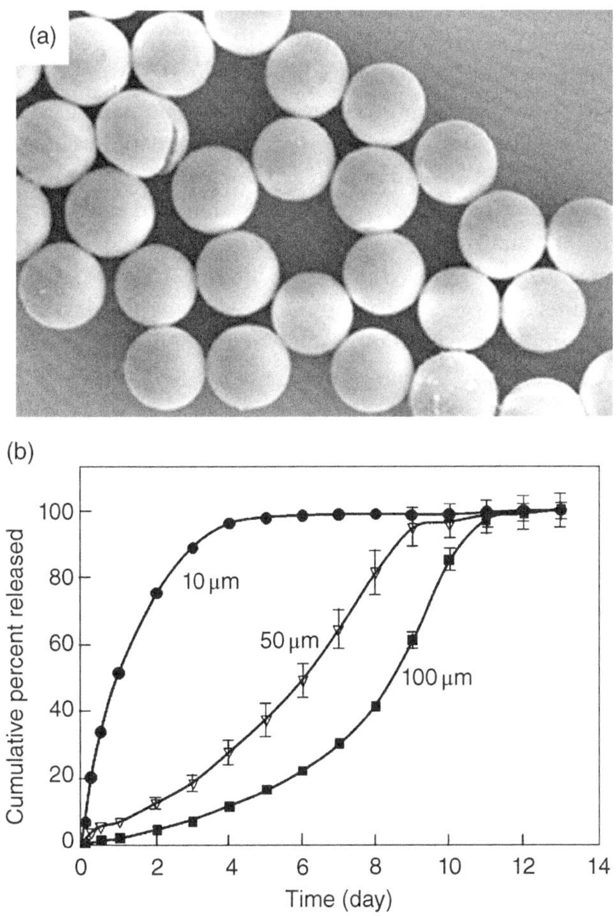

Figure 26.3 (a) Scanning electron microscope (SEM) image of nearly monodisperse polylactic-*co*-glycolic acid (PLGA) microspheres (diameter ~50 μm) incorporated with the drug piroxicam; (b) effect of microsphere size on the release of piroxicam from microspheres incorporated with 10 wt% piroxicam. *Source:* From Berkland et al. (2002) / with permission of Elsevier.

26.4.1 Preparation of Polymer Microsphere Delivery Systems

Microspheres of degradable polymers containing physically dispersed drug molecules can be prepared using a variety of techniques including spray drying of solution droplets, phase separation and emulsion techniques, at both the laboratory and industrial scales. Emulsion techniques have been most commonly used at the laboratory scale because they involve simple procedures and do not require the use of sophisticated equipment. Two common emulsion techniques are the oil-in-water (O/W) technique and the double emulsion (W/O/W) technique, which are widely used to form drug-containing PLGA microspheres. In the O/W technique (Figure 26.4), which is applicable to drugs that have a poor solubility in water, the polymer (PLGA) is dissolved in a volatile nontoxic solvent (such as methylene chloride, also known as dichloromethane or methylene dichloride), after which the drug is dissolved in the polymer solution by stirring. This organic solution, which can be termed an oil (O) phase in relation to common oil–water emulsions, is added to an excess solution of water (W) containing a low concentration of a stabilizer, such as polyvinyl alcohol (PVA). The PVA molecules stabilize the drug-containing microspheres against agglomeration (the tendency to stick together). This mixture is emulsified using a high-speed stirrer (or a sonicator), when the O phase partitions into fine droplets dispersed in the W phase. Thereafter, stirring the mixture gently to keep the droplets dispersed and, if necessary, warming it slightly to accelerate evaporation of methylene chloride, leads to the formation of solid microspheres containing the drug.

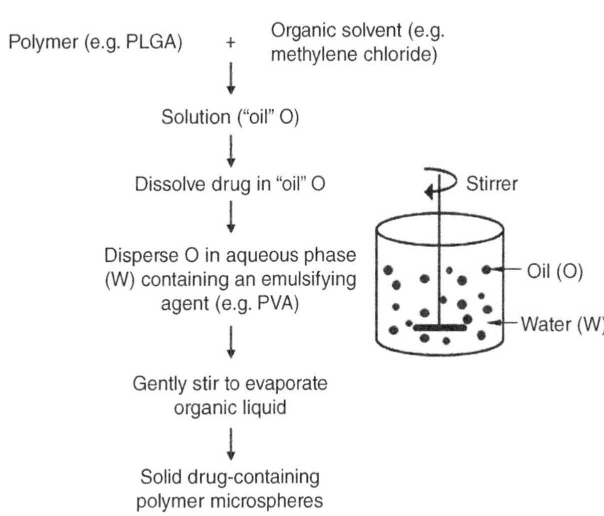

Figure 26.4 Chart showing steps in the oil-in-water (O/W) technique for preparing solid polymer microspheres incorporated with a drug.

In the double emulsion (W/O/W) technique, which is applicable to water-soluble drugs, the drug is first dispersed in an aqueous solvent (forming the first W phase). This solution is added to a solution of the polymer in an organic solvent, such as PLGA dissolved in methylene chloride (the O phase), and the system is emulsified by vigorous stirring using a high-speed stirrer (or a sonicator). Then the resulting W/O emulsion, composed of fine droplets of the W phase in the O phase, is transferred to an excess solution of water (the second W phase) containing a stabilizer, such as PVA. Vigorous stirring of this mixture leads to the formation of a double (W/O/W) emulsion. Thereafter, gently stirring and, if necessary, warming the mixture, as noted for the O/W technique, leads to the formation of solid PLGA microspheres in which the drug is dispersed.

The O/W and W/O/W emulsion techniques, as noted earlier, are easy to perform, and they are capable of producing a high yield of microspheres with efficient encapsulation of the drug. While the microsphere size and distribution of sizes are reproducible, they are often not highly controllable, such that standard deviations of up to 25–50% in the average diameter are common. Modifications to the techniques, such as breaking the O phase in the O/W technique into droplets of well-controlled size prior to addition to the W phase, can lead to the formation of drug-containing microspheres with a narrow distribution of sizes as, for example, the microspheres shown in Figure 26.3.

26.4.2 Applications of Microparticle Delivery Systems

One of the first microparticle-based drug delivery systems approved by the US Food and Drug Administration (FDA) was a product designated Lupron Depot®, composed of PLGA microspheres encapsulating the synthetic hormone leuprolide, which was approved in 1989 as an injectable product for the treatment of prostate cancer. Lupron Depot® provided several advantages of a desirable drug delivery system discussed earlier (Section 26.2). A main advantage was that it drastically reduced the number of injections required for the treatment, thereby improving patient comfort and compliance with the recommended treatment regimen. Whereas treatment with leuprolide alone required daily injections, treatment with Lupron Depot® required injections every 1–6 months depending on the dose. Additionally, the biomaterial used in the delivery system, PLGA, has well-established biocompatibility, can provide a range of release rates by varying its copolymer composition (Chapter 9) and its microsphere diameter (Figure 26.3), and can be formed into microspheres with reproducible characteristics. On the other hand, in common with other microparticle-based delivery systems, maintaining the stability of the encapsulated drug for the duration of the treatment can often be challenging.

Since its approval by the FDA in 1989 for the treatment of prostate cancer, Lupron Depot® has been approved for other treatments, such as management of endometriosis and treatment of central precocious puberty. Many other microparticle-based depot delivery systems have now received FDA approval and are in clinical use for various treatments including schizophrenia, bipolar I disorder, alcohol dependence, opioid dependence and type II diabetes.

Although not microparticle-based, a polyanhydride-based delivery system in the form of dime-sized wafers (designated the Gliadel® wafer) has been used to deliver the chemotherapeutic drug nitrosourea carmustine (BNCU) to treat glioblastoma multiforme (an aggressive type of brain cancer). The matrix of the wafers consists of a copolymer, abbreviated P(CPP-SA), synthesized from 1,3-bis(4-carboxyphenoxy) propane (CPP) and sebacic acid (SA) monomers (Figure 9.15). In the treatment, the tumor is removed as much as possible and up to eight BNCU-containing wafers are surgically implanted into the post-tumor resection, where they release the drug slowly for up to 1 month to kill remaining tumor cells. Because the drug is delivered locally, harmful side effects that normally occur from systemic chemotherapy are minimized. Approved in 1996 by the FDA for use as an adjunct to surgery in patients with recurrent glioblastoma multiforme, the Gliadel® wafer was also approved in 2002 as a first-time treatment.

26.5 Hydrogel-based Delivery Systems

Hydrogels, as we have seen from Chapter 11, have several properties and characteristics that make them attractive biomaterials for use in a variety of applications, including drug delivery. As a prelude to the discussion in this section, it is useful to summarize the salient features of hydrogels relevant to their application in drug delivery systems. In general, hydrogels can be synthesized with a wide range of chemical compositions and macromolecular network structures under mild conditions, making it easy to control their physico-chemical properties. Hydrogels are also generally biocompatible, can be degradable or nondegradable, and can be produced in a variety of forms, such as three-dimensional monoliths, microparticles, nanoparticles, and coatings. The porous three-dimensional network structure of hydrogels, consisting of crosslinked macromolecules, can be easily modified by controlling the density of the crosslinks (or mesh size) and the affinity of the macromolecules for the aqueous environment in which the hydrogel is placed. The network structure permits the incorporation of drugs into the hydrogel matrix under mild conditions and subsequent release of the drugs by diffusion through the network at a rate that is dependent on the size of the drug molecules. Hydrogels can be used as depot delivery systems, maintaining a sufficiently high local concentration of the drug over an extended period, or they can be used for systemic delivery.

On the other hand, hydrogels also have some limitations for use as biomaterials in drug delivery systems. For example, the quantity and homogeneity of drugs incorporated into hydrogels can often be limited. This is particularly the case for hydrophobic drugs because of the hydrophilic nature of hydrogels. Hydrogels are typically applied to the delivery of hydrophilic drugs but the high water content of most hydrogels often leads to rapid release of such drugs, over a period of approximately hours to days. This release is generally much faster when compared to release from systems based on solid polymer microspheres (Section 26.4). The way in which hydrogel-based systems are delivered to the body can also be problematic in clinical applications. Drug-containing hydrogels are usually formed *ex vivo*, before delivery to the body. While some hydrogels are sufficiently deformable to be injected into the desired tissue site, many are not. Consequently, many hydrogels are delivered by surgical implantation rather than by injection, which generally limits their clinical application due, in part, to patient discomfort. However, this problem can sometimes be circumvented by converting the preformed gel into microparticles or into nanoparticles (Section 26.6).

Over the last few decades, significant efforts have been made to improve the way in which drug-containing hydrogels are delivered to the body and to extend the duration of their drug release. For example, hydrogel compositions have been developed that allow injection of drug-containing solutions of polymer molecules, followed by crosslinking and gelling *in situ* through the formation of physical or chemical bonds (Section 11.4). Both physical and chemical strategies have been used to improve the strength of the interaction between the drug molecules and the polymer molecules to prolong drug release. Strategies include electrostatic attraction between oppositely charged groups on the drug and the polymer molecule, and covalent bonding of the drug to the polymer molecule (Figure 26.2c), rather than relying on weak van der Waals bonding or hydrogen bonding.

An important development, particularly for controlled or targeted delivery of drugs, has been the creation of environmentally responsive and stimuli-responsive hydrogels. Environmentally responsive hydrogels respond to specific changes in the local environment, such as temperature, pH, or chemical composition, whereas stimuli responsive hydrogels respond to external physical stimuli such as magnetic or electrical fields, sounds waves, or light. Local environmental changes or the application of external stimuli can be used to trigger drug release from such hydrogels to improve the precision, effectiveness and targeting of the delivered drug.

26.5.1 Environmentally Responsive Drug Delivery Systems

Overall, a wide variety of environmentally responsive hydrogels have been created, some of which were discussed in Chapter 11, particularly hydrogels that respond to local environmental changes in temperature, pH and ionic concentration. In this section, common types of hydrogel drug delivery systems that respond to local changes in pH and temperature are discussed, along with examples of their application in controlled drug delivery.

pH Responsive Hydrogels

pH responsive hydrogels can provide attractive advantages in improving the precision and effectiveness of administered drugs because several tissues, fluids, cells and organelles within the body have different pH values. For example, the gastric environment (pH ~1–3), the interior of lysosomes and endosomes (~4.5–5.5), and inflammatory and tumor tissues (~6.0–7.0), exist at acidic or slightly acidic pH values (<7.0), while others, including the blood (~7.4), the ocular surface (~7.1) and intestines (~6.8) exist at near neutral pH values. Additionally, pH gradients exist across several organ barriers and microenvironments. For example, the tumor microenvironment often exists at a lower pH (~5.7) when compared to surrounding healthy tissues (pH ~7).

The principles of designing pH responsive hydrogels were discussed in Chapter 11. In general, these gels are normally polyelectrolyte hydrogels, composed of chains in which the repeating monomer unit contains one or more ionizable functional groups that can become protonated or deprotonated above or below a certain pH. Consequently, the hydrogel can swell or de-swell (Figure 11.27), resulting in the release of the drug in response to pH changes within targeted tissues or organs. However, a limitation with many swelling-controlled hydrogels is that the response can often be too slow, particularly for hydrogels with a thickness greater than ~1 mm.

As noted earlier, because of the difference in pH between the gastric and enteric (intestinal) environments, pH responsive hydrogels are attractive biomaterials in oral drug delivery. An example is the development of hydrogels for use in delivery systems that remain stable and elastic in the acidic environment of the stomach but can degrade in the pH-neutral environment of the small or large intestines (Zhang et al. 2015). Stability in the stomach provides for prolonged drug delivery whereas dissolution in the intestines prevents intestinal obstruction upon exiting the stomach. To achieve prolonged retention in the gastric cavity without exiting through the pylorus (diameter ~13 mm), the part of the stomach that connects to the duodenum (first part of the small intestine), such devices should ideally have an optimal combination of properties and characteristics. For example, the device should consist, at least in part, of an elastic polymer to allow folding or compacting the whole device into a smaller configuration for delivery through the oesophagus (diameter 15–20 mm). After reaching the stomach, the device should expand to a diameter larger than 20 mm for retention and prolonged drug delivery. However, such large devices, when composed of a nondegradable polymer, can cause serious complications, such as intestinal obstruction upon exiting the stomach, which often necessitates surgical intervention for removal. One solution has been the use of a hydrogel whose macromolecular chains are physically crosslinked by hydrogen bonds, such as hydrogen bonds between carboxyl groups on the chains. As noted in Section 11.10.1, carboxyl groups are electrostatically neutral at lower pH but become deprotonated with increasing pH. Such hydrogels have good elasticity (due to their weak intermolecular bonding), and they also enteric, enabling degradation of the three-dimensional hydrogel network into smaller components that can pass through the intestinal tract.

A schematic representation of the structure and composition of a hydrogel composed of hydrogen-bonded chains is shown in Figure 26.5. This hydrogel is composed of two synthetic macromolecules, one consisting of poly(acryloyl 6-aminocaproic acid), abbreviated PA6ACA (weight-average molecular weight M_w = 347–466 kDa) and the other poly(methacrylic acid-*co*-ethyl acetate) with the commercial designation L 100-55 (M_w = 241 kDa). PA6ACA has side chains of sufficient length for the terminal carboxyl groups to be flexible and accessible for the formation of intermolecular hydrogen bonds. In the acidic gastric environment, where the carboxyl groups are not deprotonated, inter-chain hydrogen bonds between carboxyl groups and amide groups provide a loosely crosslinked supramolecular network with water trapped inside it, which contributes to the elastic property of the hydrogel. On the other hand, in the almost neutral aqueous environment of the small intestine, the carboxyl groups are deprotonated, eliminating the hydrogen bonding and resulting in rapid degradation of the three-dimensional network into soluble chains.

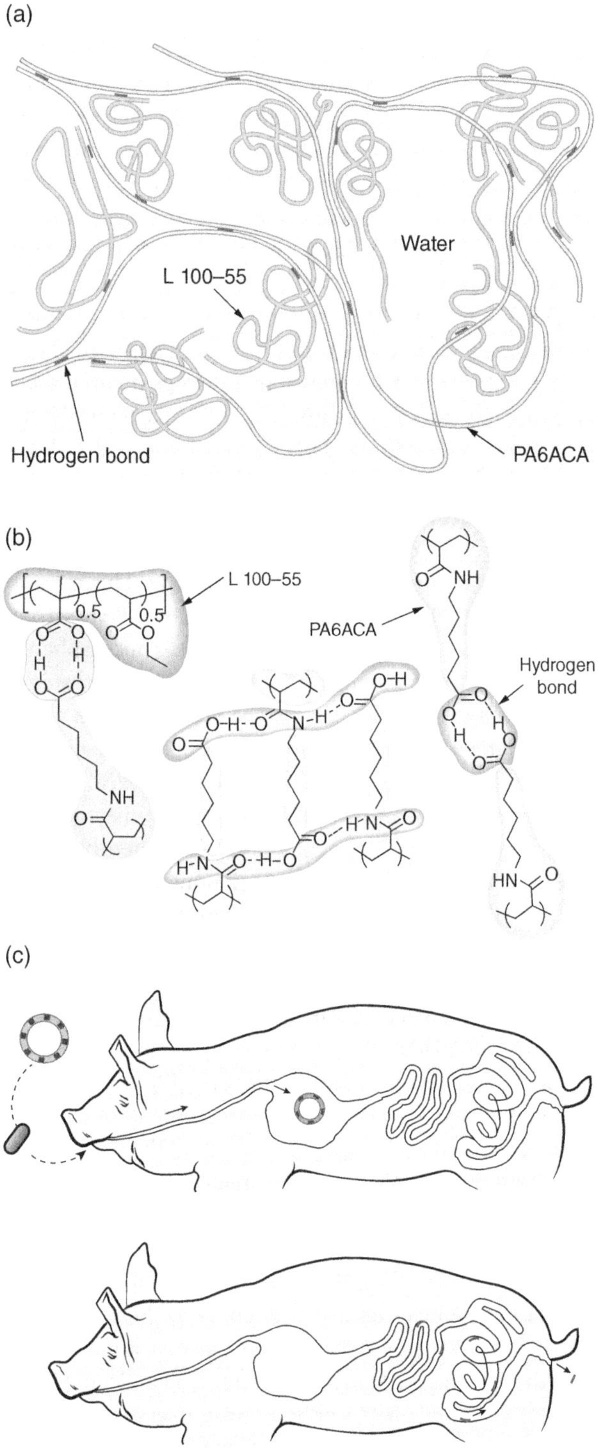

Figure 26.5 (a, b) Proposed structure of supramolecular hydrogel network composed of hydrogen-bonded poly(acryloyl 6-aminocaproic acid), PA6ACA, and poly(methacrylic acid-*co*-ethyl acetate), designated L 100-55. (c) (Top): Schematic illustration of the delivery and gastric retention of a ring-shaped device created by joining polycaprolactone (PCL) arcs to intervening hydrogel linkers composed of the supramolecular hydrogel. (Bottom): Schematic illustration of safe passage of PCL arcs through the small and large intestine on dissociation of the ring-shaped device as a result of the total or partial dissolution of the hydrogel linkers. *Source:* From Zhang et al. (2015) / with permission of Springer Nature.

Prototype gastric devices in the shape of a ring (outer diameter = 32 mm; thickness = 2 mm) were created by joining polycaprolactone (PCL) arcs to intervening hydrogel linkers composed of PA6ACA and L 100-55 in the ratio 1 : 2 by weight. Then the ring-shaped devices were bent and encapsulated separately within standard 000 gelatin capsules (closed length ~26 mm; diameter ~10 mm), and evaluated in a large animal model (Yorkshire pigs) by introduction through the oesophagus (Figure 26.5c). This animal model has a gastric and intestinal anatomy and dimensions similar to those of humans. The devices showed prolonged gastric retention (2–5 days), while loss of the intact device occurred by degradation or rupture of the hydrogel linkers, resulting in disruption of the ring structure into small linear components, which enabled easy and safe passage out of the stomach. These results show that such supramolecular hydrogels with elastic and enteric properties could be used in improving the design of gastric devices for prolonged drug delivery and other applications.

Temperature Responsive Hydrogels

Several temperature responsive hydrogels have been developed for use as biomaterials in drug delivery systems, particularly as depot systems for local delivery. Hydrogels composed of polymers that show a lower critical solution temperature (LCST) near room temperature are particularly useful because they exist as high-viscosity solutions below their LCST but form a gel at the physiological temperature (~37.4 °C). Consequently, drugs can be easily incorporated into such polymer solutions near room temperature by mixing, for example, and be injected at the desired site, where the solution gels and forms a depot for controlled drug release. Common hydrogels formed from such LCST polymers were discussed in Chapter 11, including those formed from polymers based on poly(*N*-isopropyl acrylamide) (PNIPAAm), triblock copolymers composed of polyethylene oxide and polypropylene oxide (PEO-*b*-PPO-*b*-PEO), known as Pluronics, and chitosan.

Another temperature responsive hydrogel is a diblock copolymer composed of polyethylene glycol-*b*-polycaprolactone-*co*-lactic acid (PEG-b-PCLLA), which has an LCST near room temperature and degrades within a few weeks to a few months, depending on its composition. When loaded with a combination of two chemotherapy drugs, doxorubicin and 5-fluorouracil, hydrogels composed of PEG-*b*-PCLLA were found to be effective as injectable depot systems for inhibiting the growth of tumors in a rodent model (Kim et al. 2016). As illustrated in Figure 26.6, the delivery system was prepared by encapsulating doxorubicin within PLGA microshells (diameter ~75 μm), which were dispersed in a viscous solution of PEG-*b*-PCLLA containing 5-fluorouracil. *In vitro* experiments at 37 °C in PBS (pH = 7.4) showed that ~70% of the fluorouracil was released within ~5 days, after which its release slowed considerably. In comparison, doxorubicin was released more slowly, with ~50% released over a period of 35 days.

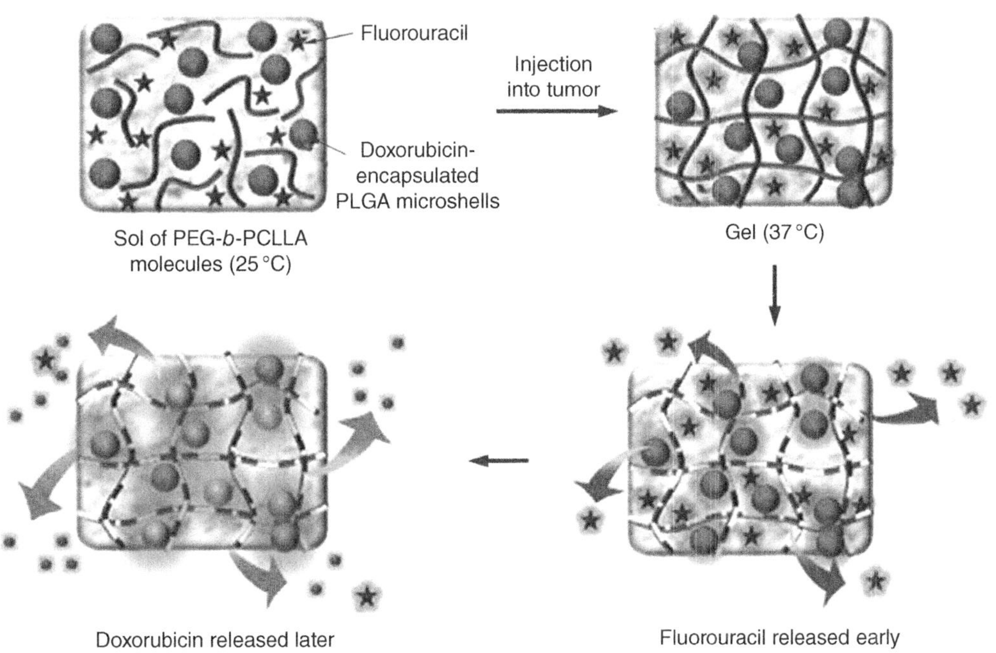

Figure 26.6 Schematic representation of depot drug delivery system composed of a temperature responsive hydrogel, polyethylene glycol-*b*-polycaprolactone-*co*-lactic acid (PEG-*b*-PCLLA), incorporated with a combination of doxorubicin-encapsulated polylactic-*co*-glycolic acid (PLGA) microspheres and 5-fluorouracil. *Source:* Adapted from Kim et al. (2016).

The same delivery system was then injected directly into tumors created by subcutaneous inoculation of B16F10 melanoma cells into the abdomen of six-week-old mice, where it gelled and served as a depot to release the drugs. As shown in Figure 26.7, the PEG-*b*-PCLLA hydrogel delivery system containing the two drugs significantly inhibited the growth of the tumor when compared to the control group (saline injection), microcapsules containing doxorubicin alone, and PEG-*b*-PCLLA containing fluorouracil alone. Additionally, analysis over an eighteen-day period showed that the PEG-*b*-PCLLA hydrogel released its doxorubicin and fluorouracil locally within the tumor, without distribution of the drugs into surrounding normal tissues. Generally, the treatment of tumors by intratumoral injection can only be done when the location of the tumor is known and the tumor can be imaged, which would mean that this treatment may be applicable to relatively late-stage tumors. Nevertheless, these results provide a rationale for potential treatment of such tumors using a temperature responsive hydrogel depot system containing a combination of chemotherapy drugs.

26.5.2 Drug Delivery Systems Responsive to External Physical Stimuli

The majority of depot drug delivery systems, including those based on polymer microparticles and hydrogels discussed earlier, rely on diffusion of the drug out of the depot or on local environmental conditions causing erosion or degradation to control drug release. However, this may not be always appropriate because the local environment of the disease can significantly influence these release processes and can also vary from patient to patient. Additionally, the required release kinetics and dose of the drug may be difficult to predict prior to introduction of the drug. Overall, a problem with depot delivery systems is that the drug release rate cannot usually be modulated or changed once the delivery system has been introduced into the body and drug release has started.

In comparison, drug delivery systems consisting of certain hydrogels and polymers, which are activated by external physical stimuli, such as magnetic or electric fields, sound waves, or light, have the potential to allow drug release at the appropriate time and at the appropriate dose for a specific disease and for a specific patient. Some applications, such as the treatment of diabetes by insulin delivery, for example, may not require continuous (sustained) drug release over a prolonged period but, instead, may require intermittent (or pulsatile) delivery over a prolonged period (Figure 26.8). Some recent studies have shown that pulsatile delivery or the combination of pulsatile delivery and continuous delivery could provide a more effective treatment for some diseases. Overall, delivery systems that are activated by external physical stimuli to provide both spatial and temporal control of drug release are attracting considerable attention for use in clinical applications. In this section, we shall discuss two of the more common types of hydrogel delivery systems that respond to high frequency sound waves (ultrasound) and to magnetic fields, along with examples of their applications.

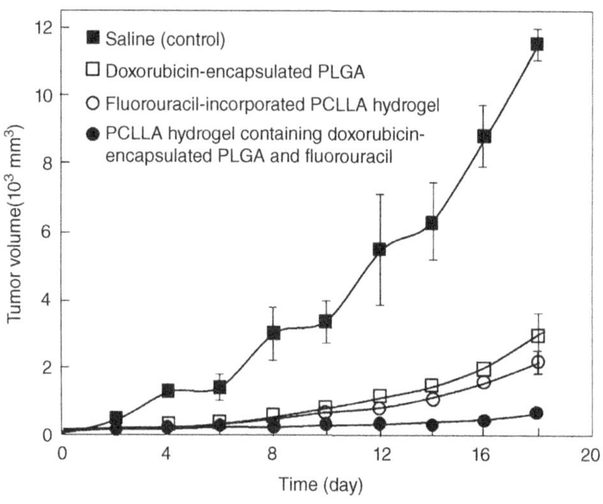

Figure 26.7 Tumor volume versus treatment time for tumors created by subcutaneous inoculation of B16F10 melanoma cells into the abdomen of six-week-old mice and treated with temperature responsive PEG-*b*-PCLLA hydrogel incorporated with a combination of doxorubicin-encapsulated PLGA microspheres and 5-fluorouracil. The tumor volume for tumors treated with saline (control), doxorubicin-encapsulated PLGA microspheres alone and PEG-*b*-PCLLA hydrogel incorporated with fluorouracil alone are shown for comparison. *Source:* From Kim et al. (2016) / with permission of Elsevier.

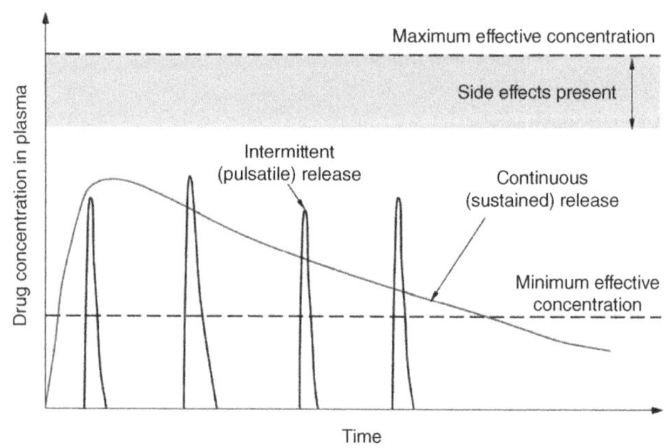

Figure 26.8 Schematic illustration comparing drug concentration in plasma as a function of time due to intermittent (pulsatile) drug delivery systems and continuous (sustained) delivery systems.

Acoustic Stimulated Drug Delivery

Ultrasound (~2–20 MHz) is widely used clinically as an imaging technique to observe tissues and organs within the body. Recently, ultrasound has been studied as a stimulus to provide spatial and temporal control of drug delivery from hydrogels (and polymers) at diseased sites in the body. Ultrasound has been shown to accelerate drug release from hydrogels largely through its ability to degrade (rupture) the crosslinked polymer network. Two mechanisms have been suggested as being responsible for the network degradation. The first mechanism involves strong shear forces arising from bubble formation and collapse within the aqueous phase of the hydrogel, due to high local temperature and pressure produced by the high frequency sound waves. The second mechanism involves the generation of free radicals (H• and HO•) during bubble formation, which persist after bubble collapse and leads to free radical-mediated degradation of the polymer chains (Chapter 15).

In many cases, ultrasound-mediated degradation can lead to permanent damage to the polymer network, altering drug release rate and potentially accelerating release with subsequent application of the ultrasound stimulation (Figure 26.9a). In order to prevent this, efforts have been made to synthesize self-healing polymer networks that are capable of re-forming after removal of the ultrasound stimulation (Figure 26.9b). A variety of hydrogels and polymers composed of such networks has been developed for use in ultrasound-mediated drug delivery, including alginates, hyaluronic acid, polyacrylates, and PLGA. For example, alginates form a hydrogel through ionic crosslinking between divalent ions (typically Ca^{2+}) in aqueous solution and negatively charged (C=O)O$^-$ groups on the alginate chains (Figure 10.20). It has been shown that the aforementioned shear stress resulting from application of ultrasound disrupts the ionic bonds and, after cessation of the ultrasound stimulus, persistence of the Ca^{2+} ions within the hydrogel and the surrounding physiological fluid allows the crosslinks to re-form.

Ultrasound-stimulated release of the chemotherapy drug mitoxantrone from injectable alginate hydrogels has been shown to significantly reduce tumor growth in mice when compared to sustained release of the drug from the hydrogel at a lower dose over a longer duration (without ultrasound stimulation) (Huebsch et al. 2014). Mitoxantrone is used to treat breast cancer but its toxic side effects when administered systemically by injection is a drawback. A possible way to alleviate this problem is local delivery, by means of a depot drug delivery system, for example. Mitoxantrone is composed of ionic molecules that form a complex with alginate chains and, thus, shows a low baseline level of release from alginate hydrogels in the absence of ultrasound stimulation. Treatment of human breast tumors (MDA-MB-231) created in the hind limbs of mice by daily ultrasound-stimulated drug release from alginate hydrogels substantially reduced tumor growth when compared to sustained release alone from the hydrogels (Figure 26.10a). In comparison, tumors in mice receiving no treatment grew rapidly.

Figure 26.10b shows survival curves for animals left untreated between 70 and 180 days. Animals that previously

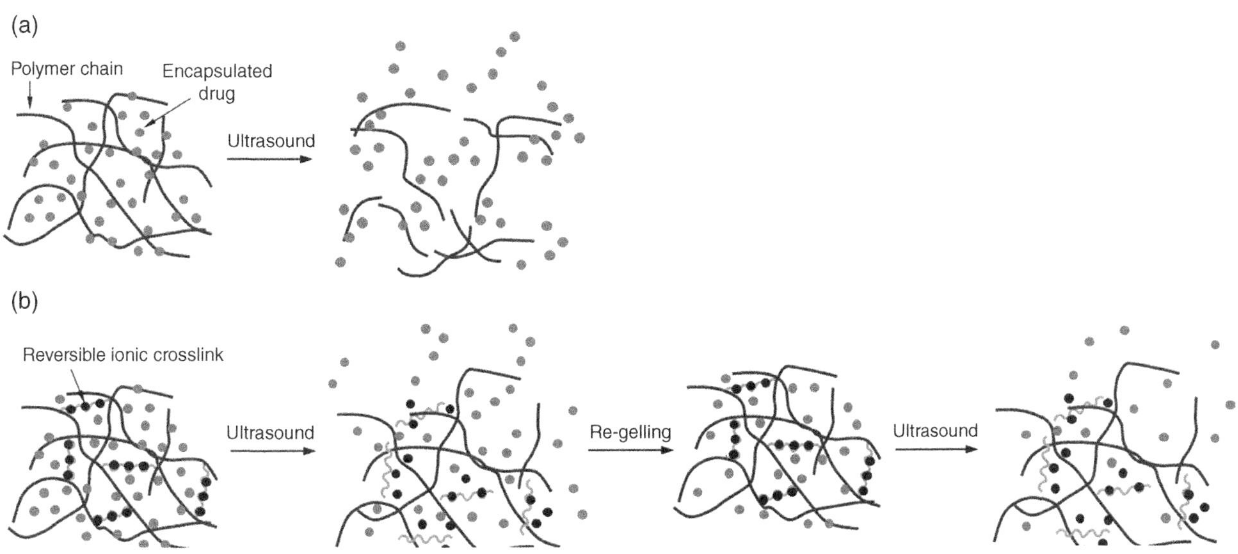

Figure 26.9 Illustration of ultrasound-mediated drug delivery. (a) Ultrasound can mediate the fracture of polymeric depots, releasing entrapped drug. (b) Self-healing gels can release drugs in response to ultrasound and, after removal of the ultrasound, will re-form a gel, trapping unreleased drugs. This allows for repeated stimulation and release without permanent breakdown of the polymeric depot. *Source:* From Brudno and Mooney (2015) / with permission of Elsevier.

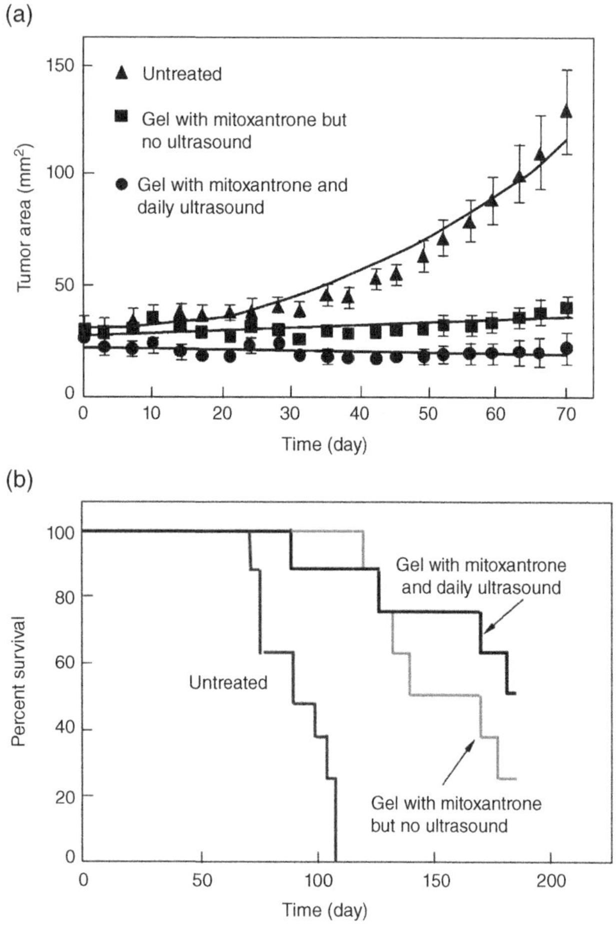

Figure 26.10 (a) Growth curves for human MDA-MB-231 breast tumors in mice. Tumors were left untreated, exposed to a standard alginate hydrogel containing mitoxantrone but without ultrasound, or treated with a combination of mitoxantrone-containing hydrogel and daily ultrasound. (b) Survival curves for mice left untreated between 70 and 180 days following tumor implantation, after receiving no treatment, treatment with a mitoxantrone-containing hydrogel but without ultrasound, or treatment with a combination of mitoxantrone-containing hydrogel and daily ultrasound treatment. *Source:* From Huebsch et al. (2014) / with permission of National Academy of Sciences.

received no treatment succumbed to the effects of the tumors within ~100 days. As would be expected, previous treatment with mitoxantrone-containing hydrogels alone improved survival, with the time to 50% animal mortality increasing from ~100 to ~120 days. In comparison, animals previously treated with daily ultrasound-stimulated release from the mitoxantrone-containing hydrogels showed a time for 50% mortality of ~180 days, which was nearly double that for animals that previously received no treatment. These results suggest that although local mitoxantrone delivery is effective, its effectiveness and, likely, the effectiveness of other drugs, could be improved by local spatial delivery as well as temporal delivery.

Magnetic Stimulated Drug Delivery

The use of a magnetic field to stimulate (or trigger) local drug delivery is another area that is receiving significant attention. Here, we will focus on the application of magnetic fields to stimulate drug release from hydrogel depot systems that contain magnetic responsive constituents. One such system, called ferrogels, consist of magnetic nanoparticles entrapped within a hydrogel. Ferrogels have attracted significant attention for use in drug delivery systems because of the ability to combine a reasonably safe physical stimulus with magnetic nanoparticles that have adequate biocompatibility.

Magnetic nanoparticles used in ferrogel-based drug delivery systems are commonly composed of the ferrimagnetic iron oxide magnetite (Fe_3O_4) or maghemite (γ-Fe_2O_3) (Chapter 4). These iron oxides have been widely studied and applied in other biomedical applications, including magnetic resonance imaging (MRI), cell tracking and hyperthermia treatment of tumors. An advantage of Fe_3O_4 and γ-Fe_2O_3 nanoparticles is that in addition to their adequate biocompatibility, they can be easily synthesized with a wide range of controlled sizes using established chemical techniques. When used in ferrogels, these nanoparticles normally have a size in the range ~5–20 nm, such that each particle consists of a single domain of magnetic dipoles rather than multiple domains. Particles of size in this range show superparamagnetism, a form of magnetism shown by ferromagnetic or ferrimagnetic nanoparticles. Upon application of a magnetic field, the magnetic dipoles within the nanoparticles tend to align with the field and the particles become highly magnetized. However, upon removal of the magnetic field, the magnetic dipoles have sufficient thermal energy to return to a random arrangement, giving the nanoparticles a net zero remanent magnetization. Essentially, the magnetization curve has the sigmoidal shape of multi-domain ferromagnetic or ferrimagnetic materials (Figure 4.21), but there is no hysteresis.

One mechanism for magnetic field stimulated drug release from ferrogels involves local heating of the encapsulated nanoparticles using a high frequency alternating current (AC) magnetic field, which leads to disruption of the polymer network of the hydrogel to enhance diffusion and release of the drug molecules (Figure 26.11a). Hydrogels composed of polymers that show temperature responsive gelling behavior, particularly polymers that undergo a phase or volume transition at a critical temperature near the physiological temperature (~37 °C), have been widely investigated for use in ferrogels. Ideally, the hydrogel should show minimal drug release at the physiological temperature, while the polymer phase should have a lower LCST slightly above the physiological temperature (~40–45 °C). Upon local heating to temperatures above the LCST, the polymer network collapses, forcing out water along with its dispersed drug (Figure 26.11b). An advantage of this type of delivery system it that it offers the potential to combine magnetic stimulated drug delivery with hyperthermia treatment of tumors. However, local heating can also increase the degradation rate of the hydrogel network, thus accelerating rather than actuating drug release.

Polymers based on PNIPAAm have been widely investigated for use in ferrogels because their composition can be easily altered to give an LCST in the range ~35–45 °C. Triblock copolymers composed of PEO-*b*-PPO-*b*-PEO, known as Pluronics, have also received attention, particularly for delivering hydrophobic drugs. As illustrated in Figure 26.11b, local heating of the magnetic nanoparticles due to the applied magnetic field leads to collapse of the polymer network, forcing water along with its dispersed drug out of the system. However, the collapse of the hydrogel network does not always lead to release of the incorporated drug. In some cases, collapse of the network has been shown to suppress drug release due to reduced diffusion of the drug molecules through the hydrogel.

Another mechanism by which a drug can be released from ferrogels depends not on local heating and temperature changes within the system but, instead, on mechanical deformation of the ferrogel itself due to a constant applied magnetic field gradient. When placed in a magnetic field gradient, a ferrogel can expand or contract rapidly along the direction of the magnetic field, brought about by a change in the disposition of the superparamagnetic nanoparticles due to their interaction with the field. Whether the ferrogel expands or contacts depends on a variety of factors, including the initial shape of the ferrogel, the strength of the applied magnetic field, as well as the concentration and the composition of the magnetic particles dispersed within the hydrogel. The deformation of the ferrogel causes water to flow through the interconnected spaces between the polymer network, which can trigger and enhance release of a drug incorporated within the hydrogel (Figure 26.11c).

One problem with most ferrogels is that the elastic modulus of the hydrogel is sufficiently high to resist a large

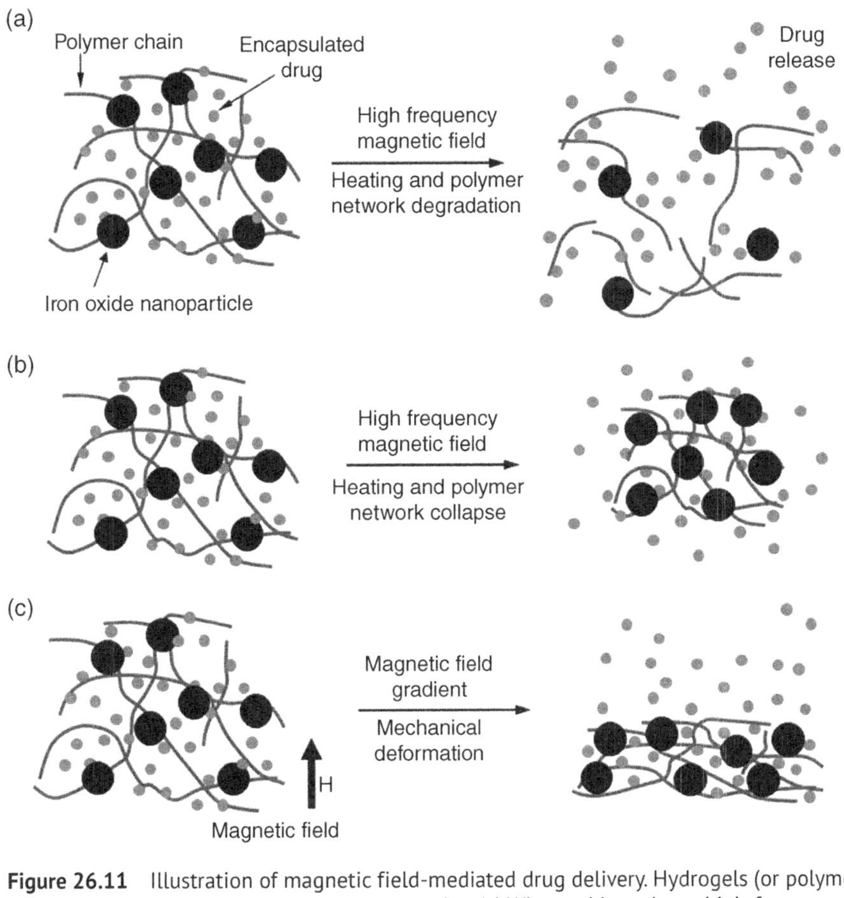

Figure 26.11 Illustration of magnetic field-mediated drug delivery. Hydrogels (or polymers) can be embedded with iron oxide nanoparticles to make them magneto-responsive. (a) When subjected to a high-frequency magnetic field, the system heats up, leading to disruption of the polymer network and drug release. (b) For hydrogels with a lower critical solution temperature (LCST) slightly above the physiological temperature, heating above the LCST leads to collapse of polymer network, forcing out water along with its dispersed drug. (c) A constant magnetic field gradient can induce polymer deformation with expulsion of water and drug molecules with each application of the field. *Source:* Adapted from Brudno and Mooney (2015).

enough deformation under moderate magnetic fields. Another problem is that because the mesh size (average size of the spaces within the crosslinked network) of most hydrogels is small (less than ~50–100 nm), diffusion of large molecules through the hydrogel is slow. One approach used to alleviate these problems involves the creation of macroporous ferrogels in which interconnected large pores (of size varying from a few tens of micrometers to a few hundred micrometers) are incorporated within the hydrogel (Zhao et al. 2011). For example, macroporous ferrogels composed of crosslinked alginate chains (1 wt% alginate) and 13 wt% Fe_3O_4 nanoparticles (~10 nm), prepared by a freeze-drying process, showed interconnected pores of size ~700 μm and an initial elastic modulus (initial slope of compressive stress versus strain curve) of ~2.5 kPa. These hydrogels could be deformed reversibly to a strain of ~80% before fracture (Figure 26.12). In comparison, nanoporous ferrogels of the same composition, prepared without the macropores, showed an initial modulus of ~30 kPa and fractured at a strain of ~20%. When subjected to a moderate magnetic field gradient (~38 A/m^2), cylindrical samples of the nanoporous hydrogel showed a contraction of ~5% in the direction of the applied field, whereas the macroporous hydrogel contacted by ~70%.

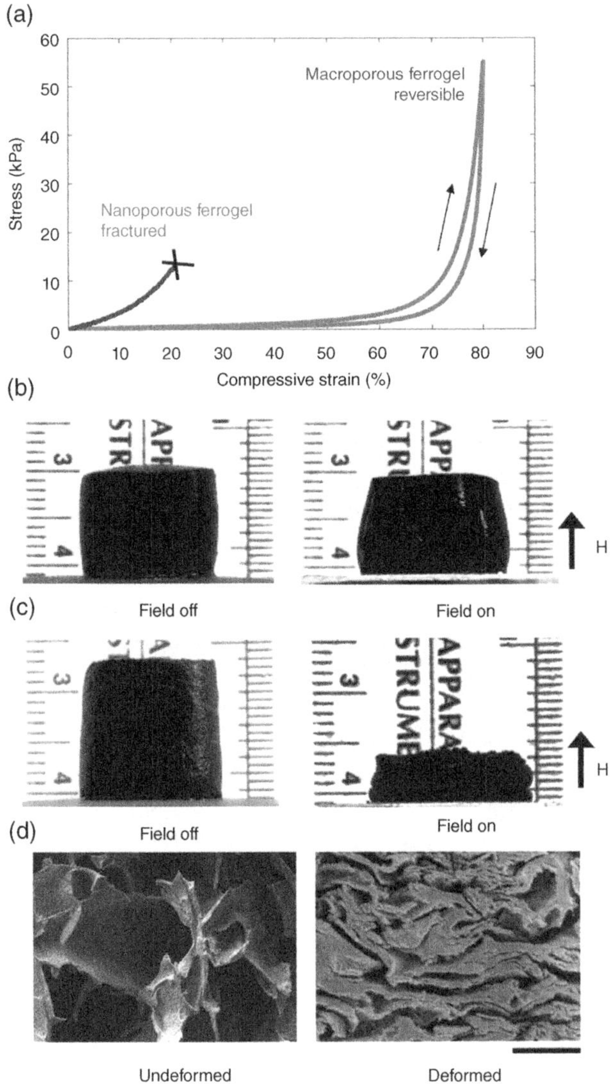

Figure 26.12 (a) Stress versus strain curves for nanoporous ferrogel and macroporous ferrogel tested in compression. (b) The height of a cylindrical nanoporous ferrogel decreased by ~5% when subjected to a magnetic field gradient of ~38 A/m^2. (c) In comparison, the corresponding macroporous ferrogel showed a reduction in height of ~70% when subjected to the same magnetic field. (d) SEM images of a freeze-dried macroporous ferrogel in the undeformed and deformed states (scale bar = 500 μm). *Source:* From Zhao et al. (2011) / with permission of National Academy of Sciences.

Upon stimulation by a moderate magnetic field, these macroporous ferrogels showed the ability to accelerate release of the drug mitoxantrone. As noted earlier, mitoxantrone molecules form an ionic complex with the negatively charged (C=O)O$^-$ groups on alginate chains, which retards their release from alginate hydrogels. Experimentally, mitoxantrone-containing macroporous ferrogels were placed in PBS and, every 30 minutes, they were subjected to 120 on-off cycles of a magnetic field gradient (~38 A/m^2) lasting a total of 2 minutes. Similar macroporous ferrogels not subjected to a magnetic field gradient served as the control group. Cumulative release profiles (Figure 26.13) showed a step-wise increase with magnetic stimulation, and the total amount of mitoxantrone released from the gel was ~7 times more than that released from the control group after three hours. Experiments with higher molecular weight biological agents, plasmid deoxyribose nucleic acid (DNA) and the chemokine SDF-1α, showed release profiles matching the trend for mitoxantrone, indicating the potential utility of this approach with a wide range of drugs.

26.6 Nanoparticle-based Delivery Systems

Nanoparticle-based delivery systems represent the most widely studied drug delivery systems over the last few decades, particularly for the delivery of chemotherapy drugs for cancer treatment. These delivery systems generally consist of three main components (Figure 26.14):

- A core material of nanoscale dimensions (less than ~50–100 nm) capable of carrying drugs, referred to as a nanocarrier, but the core material in some systems may be larger, up to ~200 nm.
- Drugs, often encapsulated within the nanocarrier, but sometimes covalently bonded to the surface of the nanocarrier.

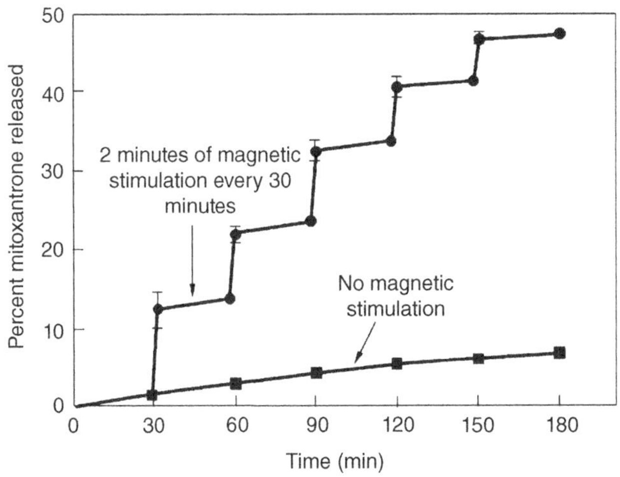

Figure 26.13 Cumulative release profiles of mitoxantrone from macroporous ferrogels subjected to 2 minutes of magnetic stimulation every 30 minutes or no magnetic stimulation. Values represent mean and standard deviation of each increment (n = 3–5). *Source:* From Zhao et al. (2011) / with permission of National Academy of Sciences.

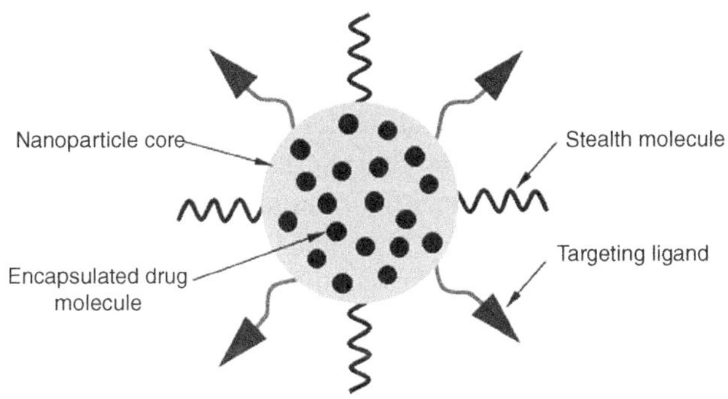

Figure 26.14 Illustration of the basic constituents of nanoparticle-based drug delivery system.

- Surface functional molecules attached to the nanocarrier, normally by covalent bonds, including molecules to endow the delivery system with "stealth" (the ability to avoid rapid clearance from circulation after systemic injection), and targeting ligands to direct the nanoparticles to specific cells or tissues.

Over the last few decades, a large number of nanoparticle-based drug delivery systems have been developed or have been approved for clinical applications, indicating the enormous potential of these systems for improving the treatment of diseases in the future, particularly cancer treatment. As discussed later, these delivery systems include polymer nanoparticles, lipid-based carriers such as liposomes, lipid-polymer hybrid nanoparticles and micelles, polymer-drug conjugates, dendrimers, and inorganic nanoparticles. Among these systems, lipid-based carriers and polymer–drug conjugates have been two dominant types of carriers, accounting for a majority of clinically approved nanoparticle-based drug delivery systems.

Based on their small size, nanoparticle-based drug delivery systems have certain advantages over depot delivery systems such as polymer microparticles and three-dimensional hydrogels discussed earlier. Unlike microparticles, which tend to remain at the tissue site where they are delivered, nanoparticles, when injected intravenously, can extravasate (escape) through the endothelium and epithelium, enabling them to penetrate capillaries, inflammatory tissue sites and tumors. Through passive or active targeting, this provides the potential to more precisely deliver drugs to target cells. Additionally, as the surface area to volume ratio of a sphere varies inversely with its radius, for the same mass, nanoparticles present a higher surface area for functionalization with targeting ligands when compared to microparticles. Consequently, they can provide better binding to specific cell surface receptors. Another advantage of nanoparticles is that unlike microparticles, they should be capable of being delivered by intravenous injection without the risk of embolization in the capillaries.

26.6.1 Distribution and Fate of Nanoparticle-based Drug Delivery Systems

When compared to microparticles, nanoparticles are more easily taken up by cells. Particles can enter cells by a variety of pathways, referred to collectively as endocytosis, which includes phagocytosis and pinocytosis (Figure 21.21). Phagocytosis is a means of taking up microparticles of size less than ~5–10 μm, and can only be done by a few cell types, such as professional phagocytes, including macrophages and monocytes. In comparison, all cell types can take up nanoparticles by pinocytic pathways. This means that cell targeting can be more efficient and more widely applicable for nanoparticles. On the other hand, this also means that nanoparticles can be rapidly taken up by cells and cleared from circulation in blood upon intravenous injection, which will prevent them from reaching their intended site of action. In general, cells of the reticuloendothelial system (RES), particularly cells of the liver, spleen and kidneys, are very effective in rapidly removing nanoparticles. Consequently, an important issue in designing nanoparticle-based drug delivery systems is to avoid rapid take up and clearance by such cells but, at the same time, facilitate entry into target cells.

As discussed in Chapter 21, take up by cells is strongly dependent on the physico-chemical properties of the particle surface. Nanoparticles with a hydrophobic surface will be rapidly cleared from circulation but, as noted, circulation time can be greatly increased by surface modification, such as making the surface hydrophilic and electrostatically neutral. The most commonly used approach is to covalently attach a layer of PEG molecules to the particle surface, a process often referred to as PEGylation. The PEG molecules are said to endow the nanoparticles with "stealth" to avoid take up by cells and, thus, greatly increase their circulation time. It is noted that PEG is a nondegradable polymer but individual PEG chains of molecular weight less than ~10–15 kDa dissolve in the physiological fluid and can be eliminated from the body by excretion through the kidneys (Chapter 11).

26.6.2 Targeting of Nanoparticles to Cells

Two basic strategies are commonly employed for targeted delivery of nanoparticle-based drug delivery systems to treat and eradicate tumors: passive targeting and active targeting. Both strategies exploit general features of tumor tissues, which allow the drug delivery system to accumulate at tumor sites rather than in normal (non-cancerous) tissues, including anatomical features of the tumor vasculature and lymphatic vessel drainage, as well as general metabolic characteristics of tumor tissues.

Tumor Vasculature and Microenvironment

Solid tumors of size less than ~2 mm^3 are capable of receiving an adequate supply of nutrients and oxygen by diffusion. However, as the tumor grows beyond this size, the tumor tissue becomes increasingly hypoxic, resulting in a microenvironment that promotes upregulation of vascular endothelial growth factor (VEGF). The localized expression of VEGF triggers angiogenic budding, expansion of endothelial cells and rapid remodeling of the vasculature to an abnormal architecture. Two of the vascular changes, poorly organized basement membrane (basal lamina) and formation of fenestrations, result in a leaky tumor vasculature, which, in turn, causes high capillary filtration to the tumor interstitial space. In addition to changes in tumor blood vessels, there are also substantial changes in tumor lymphatic vessels, which cause a severe reduction in lymph drainage from tumors.

Because of the low oxygen tension of tumor tissues, anaerobic glycolysis is the dominant metabolic pathway of tumor tissues, resulting in high lactic acid production followed by increased pumping of hydrogen ions to the interstitial space. The principal outcome of this high level of anaerobic metabolism is a tumor microenvironment of pH ~6.0–7.0, making it more acidic than normal tissue in which the interstitial fluid has a pH of ~7.2–7.5. In general, the rapidly growing tumor tissues tend to overexpress receptors for various growth factors and vitamins needed to maintain their growth and accelerated metabolism.

Passive Targeting

The general features of tumors that are particularly exploited as the basis of passive targeting are leaky blood vessels and poor lymphatic drainage. Together, these two features account for what is referred to as the enhanced permeability and retention (EPR) effect. The leaky vasculature allows a nanocarrier to more easily extravasate (escape) into the interstitial space of tumor tissues than into the interstitial space of normal, non-cancerous tissues (Figure 26.15a). In addition, the greatly reduced lymphatic drainage of tumors causes accumulated nanocarriers to be retained for a prolonged period, which allows for greater release of chemotherapy drugs into the interstitial fluid surrounding tumor cells.

A greater therapeutic benefit of the EPR effect is achieved when the nanoparticles can evade immune detection and remain in the systemic circulation for a prolonged period. This is largely dependent on the size and surface physico-chemical characteristics of the nanoparticles. As noted earlier, circulation time can be greatly increased by surface modification, such as making the surface hydrophilic and electrostatically neutral, particularly by PEGylation. An optimal size for prolonging circulation time is in the range ~20–100 nm. Particles smaller than ~10 nm can pass through the filtration pores of the kidneys and be eliminated in the urine. In general, the kidney glomerulus serves as a physical filter, allowing water and small molecules to pass into nascent urine, while larger molecules are retained in circulation. As the pore size of the glomerulus filtration barrier is ~8 nm, delivery systems of size larger than ~20 nm are normally used to limit renal clearance. At the higher end of the optimal size range, studies have shown that particles of size larger than ~200 nm are cleared from circulation more rapidly than particles of smaller size, regardless of whether they are PEGylated or not.

Active Targeting

When compared to passive targeting, active targeting permits much greater specificity to tumors with less likelihood of damage to normal tissues. Active targeting is commonly achieved by conjugating (covalently bonding) a target-specific ligand to the surface of the nanocarrier. Ligands selected for this purpose typically possess specific affinity for a receptor overexpressed by tumor cells or affinity for a receptor overexpressed by endothelial cells of the tumor vasculature (Figure 26.15b). Active targeting ligands can include antibody ligands (the most common type of targeting ligand), aptamers, and non-antibody small molecules.

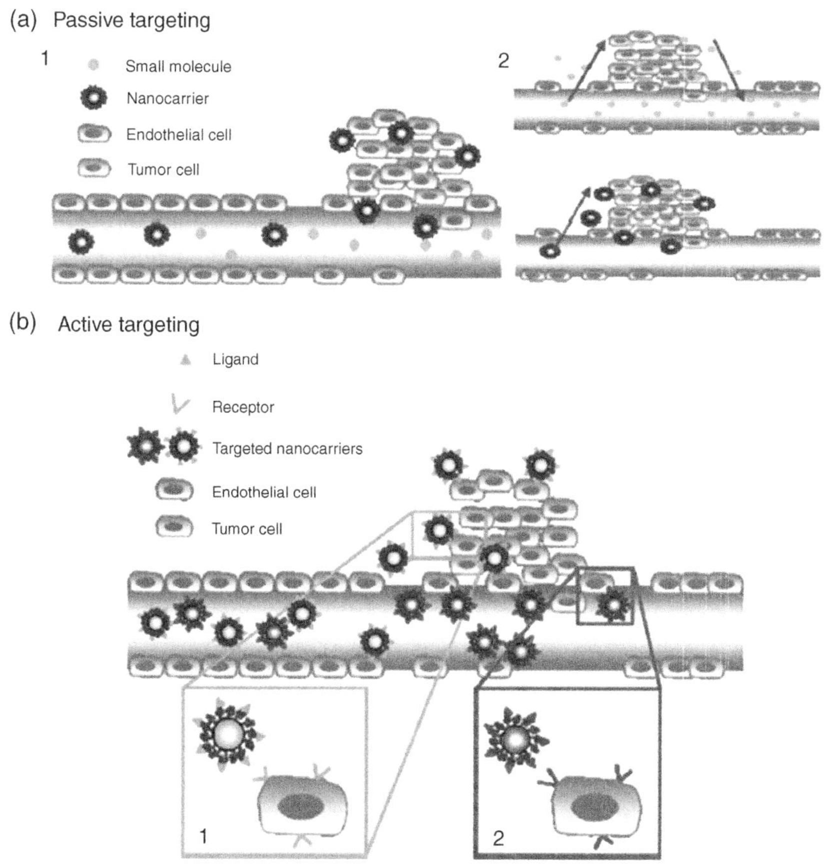

Figure 26.15 (a) Passive targeting of nanocarriers: (1) Nanocarriers reach tumors selectively through the leaky vasculature surrounding tumors. (2) Schematic representation of the influence of the size for retention in the tumor tissue. Drugs alone diffuse freely in and out of the tumor blood vessels because of their small size and, thus, their effective concentration in the tumor tissue decreases rapidly. In comparison, drug-containing nanocarriers cannot diffuse back into the blood stream because of their large size, resulting in progressive accumulation (called the EPR effect). (b) Active targeting strategies. Ligands grafted at the surface of nanocarriers bind to receptors overexpressed by (1) cancer cells or (2) angiogenic endothelial cells. *Source:* From Danhier et al. (2010) / with permission of Elsevier.

Antibody Ligands Antibody ligands used for active targeting include whole immunoglobulin G (IgG) monoclonal antibodies and IgG antibody fragments. Two widely used methods for conjugating whole antibody or antibody fragments to nanocarriers involve thioester linkage via a heterobifunctional crosslinking agent and reaction between a carboxylic acid group and a primary amine group. The conjugated ligand can then direct the nanocarrier to one of several different receptors that are overexpressed by tumor cells but are either not expressed or only minimally expressed by normal cells (Table 26.2).

An example of an overexpressed receptor is epidermal growth factor receptor (EGFR), a member of the ErbB family of tyrosine kinase receptors, which is overexpressed in solid tumors such as breast cancer, colorectal cancer and non-small-cell lung cancer. Humanized anti-ErbB2 monoclonal antibody targeting epidermal growth factor receptor 2 (ErbB2) (also referred to as HER2) was approved by the US FDA in 1998 for chemotherapy in human breast cancer patients.

Another receptor overexpressed by tumor cells and targeted by antibody ligands is the folate receptor, which binds folic acid (folate), a vitamin essential for the biosynthesis of purine and pyrimidine bases. Folate receptor-α (FRα) is overexpressed on ~40% of human cancers. One approach to active targeting of human cancers with overexpressed folate receptors involves the use of folate-conjugated nanocarriers (Figure 26.16). The mode of action with this approach involves target-specific binding of the folate-conjugated nanocarrier to the folate receptor, followed by membrane involution and internalization within an endosome. Metabolic processing of the conjugate within the endosome results in release of the drug, which subsequently exerts a cytotoxic intracellular effect.

The transferrin receptor (TfR), which binds the iron-transporting plasma protein transferrin, is another

Table 26.2 Examples of overexpressed tumor receptors, locations, and targeting antibodies.

Receptor	Tumor or tissue location	Specific antibody
Epidermal growth factor receptor 2 (ErbB2)	Breast, colorectal, pancreatic and non-small-cell lung cancer	anti-ErbB2
Transferrin receptor (TfR)	Breast, ovarian and brain cancers (glioma and glioblastoma)	anti-TfR
Folate receptor-α (FRα)	Breast, lung, kidney, endometrial, colorectal, and pancreatic cancers	anti-FRα
Fibroblast growth factor receptor (FGFR)	Breast, prostate, bladder, and gastric cancers	anti-FGFR
Prostate-specific membrane antigen (PSMA)	Prostate gland	anti-PSMA
Endothelin receptor B (EDNRB)	Metastatic melanomas	anti-EDNRB
Vascular endothelial growth factor receptor-1 (VEGFR1)	Tumor blood vessel endothelial cells	anti-VEGFR1
Vascular endothelial growth factor receptor-2 (VEGFR2)	Tumor blood vessel endothelial cells	anti-VEGFR2
Vascular cell adhesion molecule-1 (VCAM1)	Tumor blood vessel endothelial cells	anti-VCAM1
Integrin $\alpha_V\beta_3$	Tumor blood vessel endothelial cells	anti-$\alpha_V\beta_3$

overexpressed receptor of some human tumors, including pancreatic, colon and lung cancer, as well as glioblastoma, the most aggressive and deadly type of brain cancer. A difficult problem in active targeting of glioblastoma is the identification of strategies that will allow TfR-targeted nanocarriers to successfully cross the blood–brain barrier (BBB). One strategy that appears promising for alleviating this problem involves coating the nanocarrier with the non-ionic, water-soluble surfactant polysorbate 80 (Jain et al. 2015).

Several strategies are utilized for active targeting of overexpressed receptors on the aberrant blood vessels of tumors. As an example, one strategy involves the use of a monoclonal antibody against either vascular endothelial growth factor receptor-1 (VEGFR1) or vascular endothelial growth factor receptor-2 (VEGFR2). Because both of these receptors are overexpressed in aberrant blood vessels of tumors, either can be effectively used in active targeting strategies. Another target of the tumor vasculature is vascular cell adhesion molecule-1 (VCAM1), a transmembrane glycoprotein that is more extensively expressed on the surface of endothelial tumor cells than on the surface of normal endothelial cells. The $\alpha_V\beta_3$ integrin, a cell receptor for extracellular matrix proteins, is another target on endothelial cells of tumor blood vessels. The $\alpha_V\beta_3$ integrin isoform, one of 24 isoforms of integrin transmembrane receptors, is highly expressed on the vascular endothelium of growing tumors such as glioblastoma but poorly expressed on vascular endothelial cells in most normal organs.

Aptamers Aptamers are folded, short oligonucleotide strands or peptides that bind specific biomolecule targets with high affinity. Although oligonucleotide aptamers

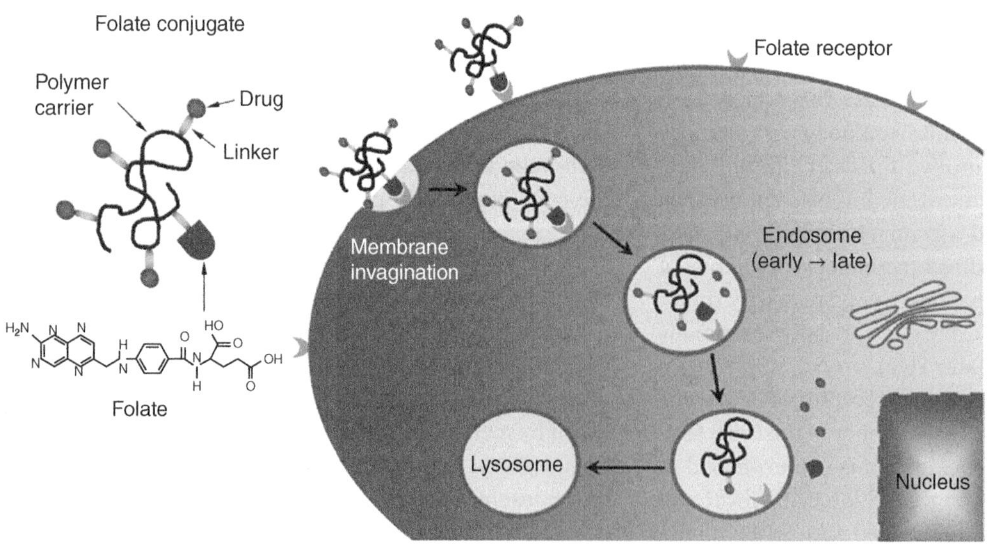

Figure 26.16 Receptor-mediated endocytosis of folate-conjugated drugs. The folate receptors recognize the conjugates, which are subsequently subjected to membrane invagination. As the endosomal compartment acidifies, the conjugate and the drugs are released from the receptor into the cytosol. *Source:* From Park et al. (2008) / with permission of Elsevier.

include both RNA and single-stranded DNA, most oligonucleotide aptamers are RNA sequences. Aptamers have several features, such as the ability to be synthesized in large quantities, the ability to be easily modified, and a lower susceptibility to thermal degradation, which make them potentially more attractive than antibodies as targeting ligands. Aptamers have been prepared which specifically bind a variety of overexpressed receptors of tumors, including prostate-specific membrane antigen (PSMA), VEGFR, and nucleolin, an RNA-binding protein overexpressed on the surface of breast cancer cells and other cancer types. These and other aptamers have been conjugated to various nanocarriers and studied for targeting of tumor cells *in vitro* and *in vivo*.

Small Molecule Targeting Unlike protein antibodies, which have little tendency to permeate cells, certain small molecules are effective as ligands for active targeting because of their tendency to permeate cells. The vitamin folic acid has been shown to be an effective targeting ligand for guiding nanocarriers to the FRα that is overexpressed on many human cancers. An example is the conjugation of folate to cyclodextrin nanoparticles loaded with the anticancer drug paclitaxel. This delivery system has been found to effectively target and destroy human breast cancer cells inoculated in animal models. The mode of action involved high affinity binding to FRα, triggering clathrin-mediated endocytosis, followed by paclitaxel-induced cell death. Two additional examples of small molecules that have attracted interest as targeting ligands are the water-soluble vitamin riboflavin and the compound γ-amino-*n*-butyric acid (GABA). Several studies have shown successful targeting of some tumor types that overexpressed the riboflavin receptor by riboflavin-conjugated nanocarriers. A similar type of targeting has been reported for GABA-conjugated drugs targeted to overexpressed GABA-receptor on pancreatic cancer cells *in vitro*.

In the following sections, we will go on to discuss the design of biomaterials for use as nanocarriers in a variety of drug delivery systems, along with examples of their applications, particularly cancer treatment. Delivery systems for cancer treatment are of particular interest because these systems have been, by far, the most dominant area of research and clinical applications.

26.6.3 Polymer-based Nanoparticle Systems

Polymers have been widely used for the creation of nanoparticle-based drug carriers. Nanoparticles have been created from both synthetic and natural polymers, including PLGA, PLA, and chitosan. An advantage of polymer nanocarriers is that drugs can be incorporated into them without the need for chemical modification of the polymer. However, solid polymer nanoparticles without surface functionalization also have limitations as nanocarriers. In general, drug release, which occurs by diffusion or erosion mechanisms, or a combination of these two mechanisms (Section 26.2.1), is often difficult to control. Because of the shorter diffusion distance when compared to microparticles, a greater proportion of the drug can leave the surface by diffusion, while water can penetrate the nanoparticles more rapidly. This can lead to a greater tendency for a rapid initial (burst) release and an overall faster release of the drug. Nanoparticles composed of polymers that degrade by hydrolysis, such as PLGA, can also deteriorate due to rapid water penetration, which can further accelerate drug release. For nanoparticles formed by emulsion techniques, the amount of a drug that can be incorporated into individual particles generally decreases with decreasing size. This is because during particle formation, drugs dispersed or dissolved in the solvent droplets can partition into the surrounding aqueous phase. The smaller the particle size, the greater the amount of drug that will have access to the surrounding aqueous phase. Consequently, the amount of drug incorporated into individual solid particles can decrease with decreasing particle size.

Several strategies for creating polymer nanoparticle carriers involve preparation of drug-containing nanoparticles followed by their surface functionalization in separate steps. For example, in one strategy, nanocarriers composed of PLGA with a coating of polyethylene glycol (PEG) molecules were prepared from block copolymers of PLGA and PEG (PLGA-*b*-PEG) by an (O/W) emulsion technique (Cheng et al. 2007). The PLGA formed the drug-containing core while the hydrophilic coating of PEG molecules served to endow the nanoparticles with stealth to enhance blood circulation time. Then these PLGA-PEG nanoparticles were functionalized with a targeting ligand (A10 RNA aptamer) in a second step. However, regardless of the preparation strategies used, surface functionalization of the nanoparticles after their preparation often requires the use of an excess amount of reactants, making it difficult to adjust the surface properties of the nanoparticles in a controlled and reproducible manner. Overall, surface modification of the nanoparticles after their preparation offers only a limited ability to control the particle surface properties in a precise manner. Consequently, the nanocarriers produced by such methods often have significant variations in their surface properties, which are not optimal for many drug delivery applications.

One approach that has been used to achieve more precise control of the surface properties of polymer nanoparticles involves the use of pre-functionalized block copolymers possessing all the desired constituents shown in Figure 26.14, after which these copolymers are self-assembled into nanoparticles with the desired physicochemical and biological properties. For example, a model study employed PLGA as the controlled-release polymer system, PEG to endow the PLGA particles with stealth, and the A10 RNA aptamer (Apt) as a hydrophilic targeting molecule (Gu et al. 2008). The Apt molecule binds to the PSMA overexpressed on the surface of some prostate cancer cells. As illustrated in Figure 26.17a, amphiphilic triblock copolymers of PLGA-*b*-PEG-*b*-Apt were synthesized for self-assembly of targeted nanoparticles in a single step.

A key feature of this approach was that by mixing known quantities of PLGA-*b*-PEG-b-Apt and the diblock copolymer PLGA-*b*-PEG prior to self-assembly, nanoparticles with varying Apt surface density can be formed by a one-step self-assembly process (Figure 26.17b). This allowed the optimal Apt surface density for maximum uptake of the nanoparticles by human prostate cancer cells to be determined *in vitro* and *in vivo*. In general, an excess amount of targeting ligands on the surface of the nanoparticles, which do not confer additional targeting benefits, could have an adverse effect in limiting the exposure of the PEG molecules, thereby limiting the ability of the PEG molecules to endow the nanoparticles with stealth.

In vitro experiments showed that nanoparticles with well-defined PLGA-*b*-PEG copolymer composition provided the ability to control the release of the drug docetaxel, which is used clinically in a number of cancer treatments. Additionally, the amount of nanoparticles endocytosed by human prostate cancer cells increased significantly as the fraction of PLGA-*b*-PEG-*b*-Apt in the nanoparticle formulation increased from ~0.5% to 5%. However, a further increase to 10% resulted in only a modest increase in nanoparticle uptake, indicating that the optimal Apt ligand density for endocytosis by the cancer cells was for PLGA-*b*-PEG-b-Apt fractions in the range ~1–10%.

Subsequent *in vivo* experiments, using a model in which human xenograft prostate cancer tumors were induced in

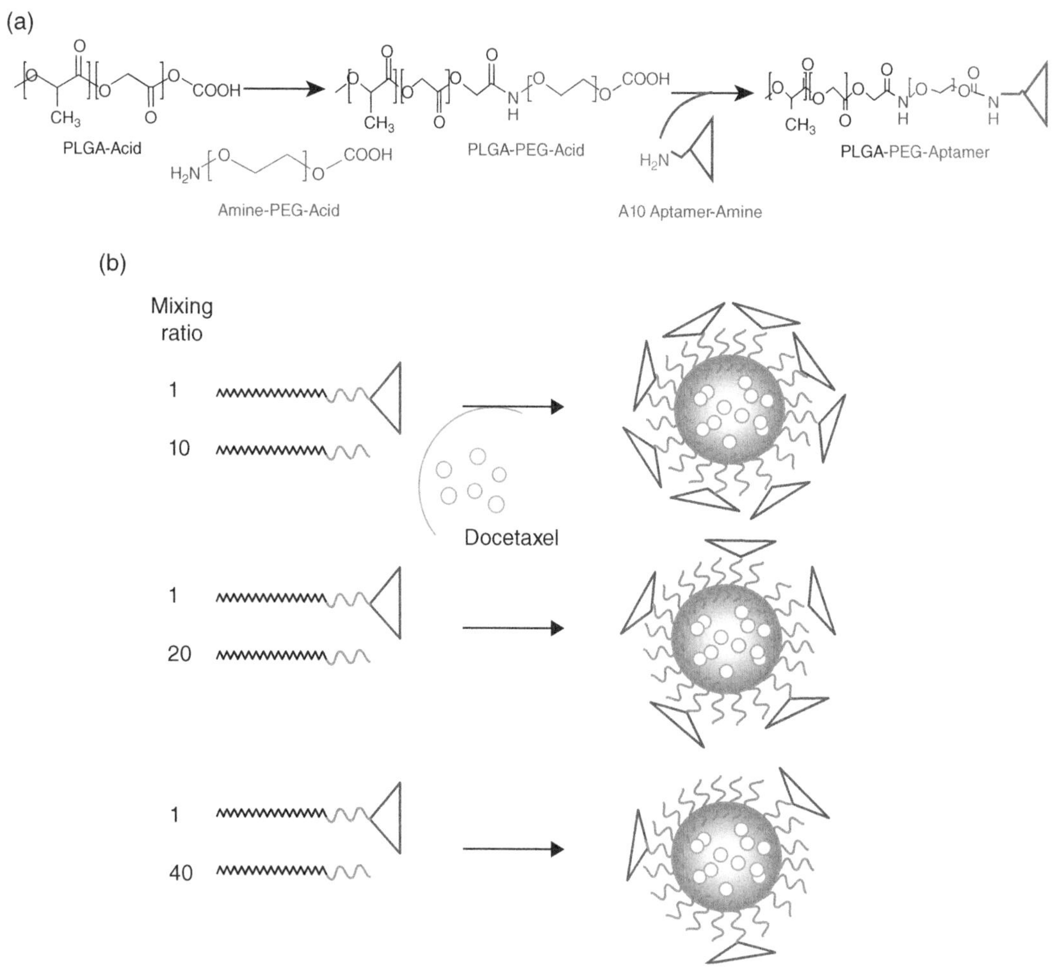

Figure 26.17 Development of prostate-specific membrane antigen (PSMA)-targeted nanoparticles by self-assembly of PLGA-*b*-PEG-*b*-Apt triblock copolymer. (a) Schematic showing synthesis of the PLGA-*b*-PEG-*b*-Apt triblock copolymer. (b) By using distinct ratios of PLGA-*b*-PEG-*b*-Apt triblock copolymer with PLGA-*b*-PEG diblock copolymer lacking the A10 Apt during nanoparticle formulation, the Apt surface density can be precisely and reproducibly changed. *Source:* From Gu et al. (2008) / with permission of National Academy of Sciences.

nude mice, showed a narrow window existed within this range for when the nanoparticles are both maximally targeted and maximally stealth. Nanoparticle formulations containing 1%, 5%, and 10% of PLGA-*b*-PEG-*b*-Apt were administered by systemic injection into the mice. Control groups consisted of nanoparticle formulations containing 10% of the triblock copolymer with mutant Apt (PLGA-*b*-PEG-b-MutApt) and nanoparticle formulations with 0% Apt. Nanoparticle retention in the tumors, as measured by the percent of the injected dose per gram (IDPG) of tissue, increased as the PLGA-*b*-PEG-*b*-Apt in the nanoparticle formulation increased from 1 to 5% (Figure 26.18a). However, with further increase of the PLGA-*b*-PEG-*b*-Apt in the nanoparticles to 10%, nanoparticle accumulation in the tumors decreased. A possible explanation for this decrease is a reduction in the stealth properties of the nanoparticles caused by an increase in Apt surface density, which masks the effect of the PEG molecules, thereby resulting in rapid clearance by the liver. This is supported

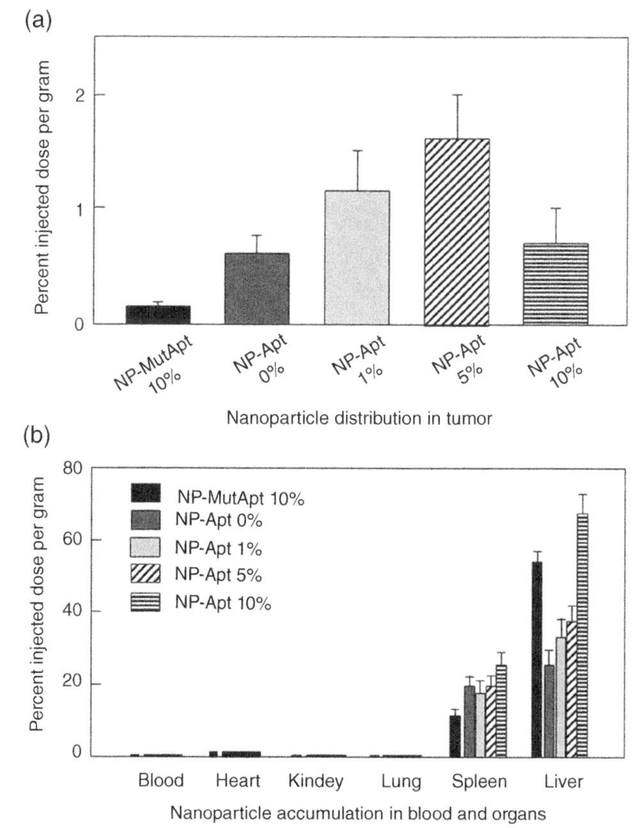

Figure 26.18 Effect of aptamer (Apt) surface density on nanoparticle (NP) distribution *in vivo* for a single systemic administration of NPs containing 1%, 5% and 10% of NP-Apt by retro-orbital injection into tumor-bearing mice. Control groups consisted of NPs without aptamers (NP-APT 0%) and NPs with scrambled nonfunctional aptamer (NP-MutApt 10%). The data for NP-Apt 1% and NP-Apt 5% were statistically significant when compared to NP-Apt 0% and NP-MutApt 10% (mean ± standard deviation; $n = 5$). *Source:* From Gu et al. (2008) / with permission of National Academy of Sciences.

by the observation that the control group of nanoparticles containing 10% of the triblock copolymer with the mutant Apt (PLGA-*b*-PEG-*b*-MutApt) showed less accumulation in the tumors when compared to the other control group of nanoparticles composed of PLGA-*b*-PEG (that is, nanoparticles containing 0% Apt). Nanoparticle formulations containing 10% PLGA-*b*-PEG-*b*-MutApt apparently provide less stealth and are non-targeted, whereas nanoparticle formulations containing 0% Apt provide maximal stealth and are non-targeted, allowing them to reach the tumor by passive targeting alone.

The ligand surface density was also found to affect the accumulation of the nanoparticles in the liver. Nanoparticles containing 10% PLGA-*b*-PEG-*b*-Apt and 10% PLGA-*b*-PEG-*b*-MutApt both showed significantly higher accumulation in the liver when compared to nanoparticles containing a lower Apt surface density (Figure 26.18b). Together, these data suggest that a narrow window exists where the ligand density of targeted nanoparticles can provide an optimal combination of targeting and stealth properties. Consequently, in developing targeted nanoparticles, there should be a balance between the tumor-targeting ligand surface density and the surface properties required for providing stealth. Additionally, the triblock copolymer approach used in this study is instructive for designing polymer nanoparticle delivery systems to target cancer cells.

26.6.4 Lipid-based Nanoparticle Systems

Lipids are composed of short-chain molecules with a hydrophilic (polar or electrostatically charged) head group, which is covalently bonded by means of a linker group to a hydrophobic tail composed of 1 or 2 aliphatic chains. An example is the phospholipid structure shown in Figure 26.19, in which the head group consists of a phosphate moiety. In general, a variety of lipids is available to form structures with controlled physico-chemical and biological properties, which make them attractive nanocarriers for drug delivery applications. Lipids can self-assemble in an aqueous environment to form closed structures, including unilamellar vesicles capable of entrapping and isolating hydrophilic and hydrophobic drugs within their aqueous interior and membrane, respectively, and protecting them from the surrounding environment. The similarity of certain lipid-based structures to cell membranes, particularly structures formed from phospholipids (Figure 16.6), provides the opportunity to deliver drug molecules into cells or their intracellular compartments. In addition, lipid-based structures are generally biocompatible and degradable, and their physico-chemical properties, such as size, charge and surface functionality can be easily modified. These advantages make lipid-based structures a leading delivery system for a wide

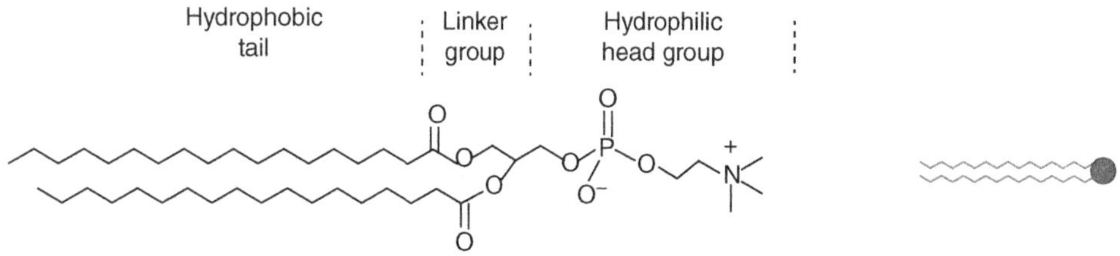

Figure 26.19 (a) Chemical structure showing basic parts of a lipid, shown for the phospholipid 1,2-distearoyl-*sn*-glycero-3-phosphorylcholine (DSPC). (b) Common depiction of a phospholipid.

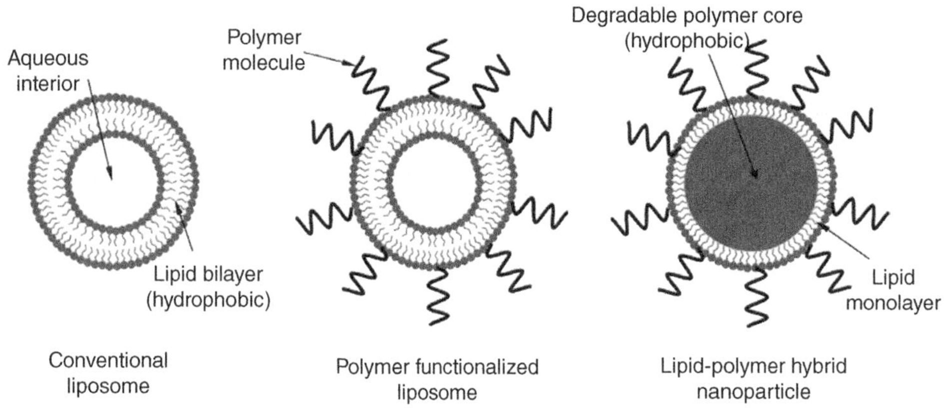

Figure 26.20 Illustration of liposome-like structures: conventional (classical) liposome; polymer-functionalized liposome (commonly with polyethylene glycol molecules), and lipid-polymer hybrid nanoparticle.

range of drugs in clinical applications. Liposomes, lipid–polymer hybrid nanoparticles, polymersomes and micelles are the main types of lipid-based nanocarriers, of which liposomes have become a well-established delivery system.

Liposomes

Liposomes are closed bilayer vesicles, with the shape of a hollow sphere enclosing an aqueous interior (Figure 26.20a), typically formed by self-assembly of natural or synthetic phospholipids in an aqueous environment. These types of liposomes have a similarity to the phospholipid bilayer of eukaryotic cell membranes but without the membrane biomolecules (such as carbohydrates and cholesterol). Drugs can be encapsulated within liposomes, particularly in the aqueous interior (water-soluble, hydrophilic drugs) or within the bilayer (hydrophobic drugs). Typically, liposome delivery systems are designed to be injected intravenously, and release of the drug takes place when the lipid envelope breaks down, which can occur extracellularly or intracellularly. The bilayer thickness of liposomes composed of phospholipids is normally ~3–5 nm. In comparison, the cell membrane is ~8–10 nm thick. While liposomes can be synthesized with a range of diameters, from the nanoscale to the microscale, our emphasis is on liposomes of size less than ~100 nm, sometimes referred to as small unilamellar vesicles, which are relevant to nanocarriers for drug delivery.

Since it was shown nearly 50 years ago that liposomes could entrap drugs and be used as drug delivery systems, there have been a number of important advances in the development of liposomes (or lipidic nanoparticles) to deliver a variety of drugs. These advances have led to numerous clinical trials, which have resulted in liposomes being the first nanoparticle-based drug delivery system to make the transition from concept to clinical application. Liposomes have now been approved by regulatory agencies to deliver a variety of drugs, including small-molecule and large-molecule drugs.

Liposomes initially developed for drug delivery, sometimes referred to as conventional (or classical) liposomes, are formulations typically composed of various phospholipids and cholesterol, which have unaltered surfaces (Figure 26.20a). When compared to the administration of free drugs, conventional liposomal drug delivery systems show improved *in vivo* activity and survival of their entrapped drugs, improved therapeutic efficacy and reduced toxic effects. Several liposomal drugs based on conventional liposomes have been approved for clinical applications, including liposomal daunorubicin for the treatment of Kaposi sarcoma and liposomal doxorubicin for combinatorial therapy of recurrent breast cancer, ovarian cancer and Kaposi sarcoma. Conventional liposomes have been refined over time to improve their effectiveness as drug carriers using approaches designed to make them responsive to various stimuli, such as pH gradients and temperature, for example, and they are being applied to an increasing number of cancer treatments.

Despite these refinements, conventional liposomes still suffer from a few limitations, due in part to their inherent instability. For example, conventional liposomes, particularly those of diameter less than ~100 nm, are susceptible to agglomeration when they collide with each other in order to reduce their high surface area. This typically leads to fusion of their bilayer membranes, often resulting in loss or mixing of their encapsulated drugs. Another limitation of conventional liposomes is their rapid clearance from circulation after systemic injection due to their strong interaction with serum proteins, which leads to opsonization and internalization within cells. This limits their ability to reach their intended site of action. One approach to shielding nanoparticles (and microparticles) from opsonins is to coat their surface with polymer molecules that are hydrophilic and electrostatically neutral (Chapter 21). As noted earlier, PEG is widely used for this purpose and it is also the polymer of choice for liposomes. Typically, the PEG molecules are covalently bonded to the head groups of the phospholipids (Figure 26.20b). In addition to shielding liposomes against opsonization, the PEG molecules provide steric repulsion against liposome fusion and enhances the structural integrity of the phospholipid bilayer. The end groups of the PEG molecules can also be functionalized with specific ligands for targeting to specific sites of cells, tissues, and organs.

When compared to conventional liposomes, PEG–liposome formulations, sometimes referred to as stealth liposomes, have shown prolonged circulation in blood, while encapsulated drugs, such as doxorubicin, for example, have shown increased effectiveness against certain tumors as well as less toxic side effects. A commercial product, designated Doxil®, became the first liposomal nanoparticle formulation approved by the US FDA for the treatment of Kaposi sarcoma, a cancer that causes lesions to grow in the skin, lymph nodes, internal organs, and mucous membranes lining the mouth, nose, and throat. It has since been approved for the treatment of other cancers, including recurrent breast cancer and ovarian cancer.

Lipid–Polymer Hybrid Nanoparticles

Lipid–polymer hybrid nanoparticles, often referred to simply as lipid nanoparticles, generally consist of three functional components (Figure 26.20c):

- A degradable hydrophobic polymer core that can encapsulate poorly water-soluble drugs and release them at a controlled rate.
- A hydrophilic polymer shell that can improve the stability of the nanoparticles and shield them from opsonization, thereby increasing their blood circulation time.
- A lipid monolayer at the interface between the hydrophobic core and the hydrophilic shell, which serves as a molecular barrier to prevent drugs from diffusing rapidly out of the nanoparticles, thereby enhancing drug loading efficiency and yield, and slowing drug release from the polymer nanoparticle core.

Additionally, for targeted delivery of the encapsulated drugs, the surface of the lipid nanoparticles can be functionalized with targeting molecules to direct the nanoparticles to specific cells or tissues.

Various lipid nanoparticle formulations have been developed but nanoparticles composed of a PLGA or a PLA core have received significant attention. This lipid nanoparticle system has the potential to take advantage of the beneficial properties and characteristics of both polymer nanoparticles and liposomes. Another advantage of this system is that the lipid nanoparticles can be prepared by self-assembly using a single-step process that is reproducible and predictable. For example, in one study (Chan et al. 2009), lipid nanoparticles composed of a PLGA core, a soybean phosphatidylcholine (lecithin) lipid monolayer around the core, and a PEG shell composed of 1,2-distearoyl-*sn*-glycero-3-phosphoethanolamine-*N*-carboxy(polyethylene glycol)2000 (DSPE-PEG-COOH), were prepared in a one-step process. The hydrophobic PLGA core is biocompatible and has the ability to encapsulate a sufficiently high amount of hydrophobic drugs, while the PEG shell provides electrostatic and steric stabilization of the lipid nanoparticle, a longer circulation time *in vivo* and functional end groups for targeting ligands.

Figure 26.21 shows a schematic illustration of the one-step process for preparing such lipid nanoparticles encapsulating the chemotherapy drug docetaxel. Overall, the characteristics of the lipid nanoparticles can be controlled by various formulation parameters, such as the lipid to PLGA mass ratio and the lipid to PEG molar ratio. Measured drug release profiles in water showed that drug release from the nanoparticles was controlled mainly by the lipid monolayer, with the PLGA molecular weight and surface density of the PEG layer having only a minor effect. The lipid monolayer appeared to act as a molecular fence, serving to control the drug molecules within the PLGA core, and to limit water penetration into the core, which would increase hydrolysis of the PLGA and enhance drug release, typically by diffusion and bulk erosion.

Additionally, the lipid nanoparticles remained stable in PBS *in vitro* over the five-day measurement period, with no significant change in their size or polydispersity, indicating

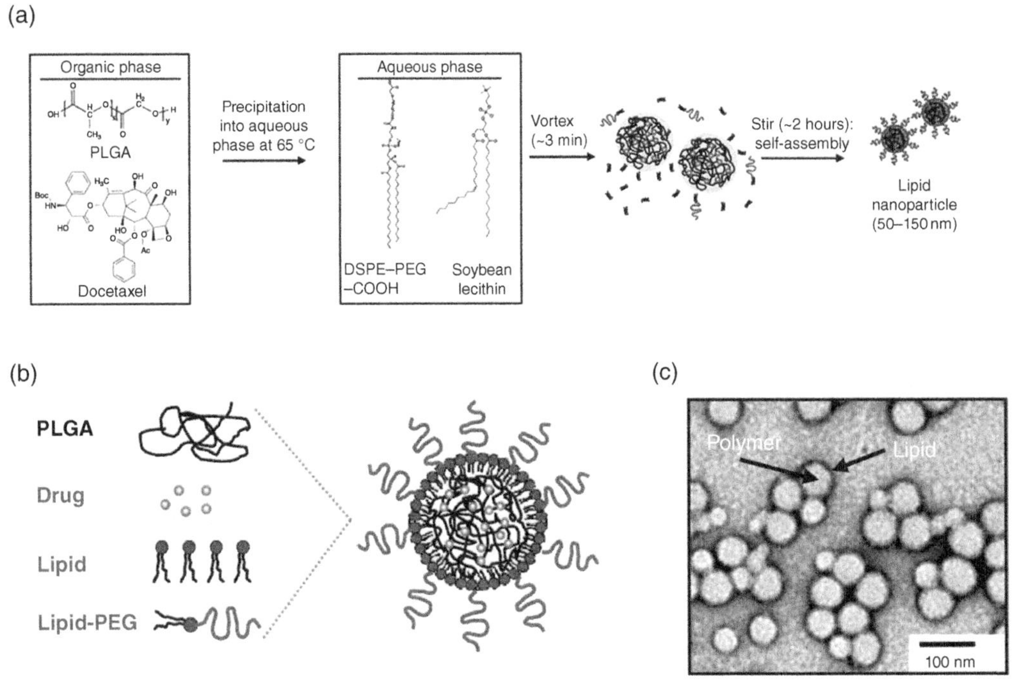

Figure 26.21 (a) Schematic illustration of the preparation of lipid-polymer nanoparticles (NPs) encapsulating a drug (docetaxel) using a self-assembly process. The NPs are composed of a hydrophobic polylactic-*co*-glycolic acid (PLGA) core, a surface layer of hydrophilic polyethylene glycol (PEG) molecules, and a lecithin lipid monolayer at the interface between the hydrophobic core and the hydrophilic surface layer. (b) Illustration of the constituents of the drug-containing lipid-polymer nanoparticles. (c) Transmission electron microscope (TEM) image of the lipid-polymer nanoparticles. *Source:* From Chan et al. (2009) / with permission of Elsevier.

stable bonding between the lipid layer and the PEG molecules. The lipid nanoparticles also remained stable when incubated in human plasma solution or in fetal bovine serum, indicating that plasma protein binding to the nanoparticles was not a significant factor. *In vitro* cytotoxicity experiments showed that that the lipid nanoparticles were well tolerated by the human cell lines HeLa and HepG2. When conjugated to A10 RNA aptamers, by covalent bonding of the aptamers to the carboxyl (C=O)OH group of the PEG surface molecules, the lipid nanoparticles showed their suitability for targeted drug delivery. As noted earlier (Section 26.6.2), A10 aptamer molecules bind to the PSMA overexpressed on the surface of some prostate cancer cells. *In vitro* experiments showed that the aptamer-targeted lipid nanoparticles were selectively delivered to prostate cancer cells that expressed the PSMA antigen but not to prostate cancer cells that did not express the PSMA antigen.

Polymersomes

Polymersomes have been widely investigated over the last few decades for use as drug delivery systems, particularly for the delivery of chemotherapy drugs in cancer treatment. Polymersomes can self-assemble in aqueous media into spherical bilayer vesicles similar to those of liposomes, but a major difference is that they are composed of synthetic amphiphilic block copolymer molecules rather than lipids, which are normally composed of natural molecules, typically phospholipids (Figure 26.22). Overall, lipids and amphiphilic polymer molecules have physical properties that are generally similar, consisting of a polar part covalently bonded to a hydrophobic part. In lipids, the polar part consists of a hydrophilic or charged head group (typically a phosphate moiety) bonded to 1 or 2 aliphatic chains (hydrophobic part), while amphiphilic copolymer molecules consist of 1 or 2 hydrophilic blocks bonded to a hydrophobic block. Consequently, it may not be surprising that such amphiphilic copolymer molecules can also self-assemble into closed spherical bilayer vesicles, called polymersomes, which have a similarity to liposomes.

Self-assembly of amphiphilic block copolymer molecules is driven by van der Waals attraction of the hydrophobic blocks due to the migration of water molecules from between these hydrophobic blocks into the surrounding medium, as discussed in Chapter 11 for PEO-*b*-PPO-*b*-PEO block copolymers (Figure 11.6). In general, amphiphilic block copolymer molecules in aqueous media can self-assemble into a variety of structures, including spheres (micelles), cylinders and closed spherical bilayer vesicles (polymersomes). The main factor that determines the

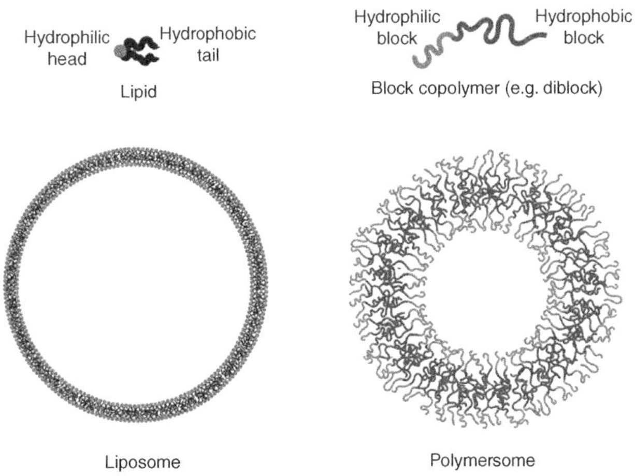

Figure 26.22 Illustration of (*top*) building blocks and (*bottom*) cross sections showing the structures of liposomes and polymersomes. *Source:* From Rideau et al. (2018).

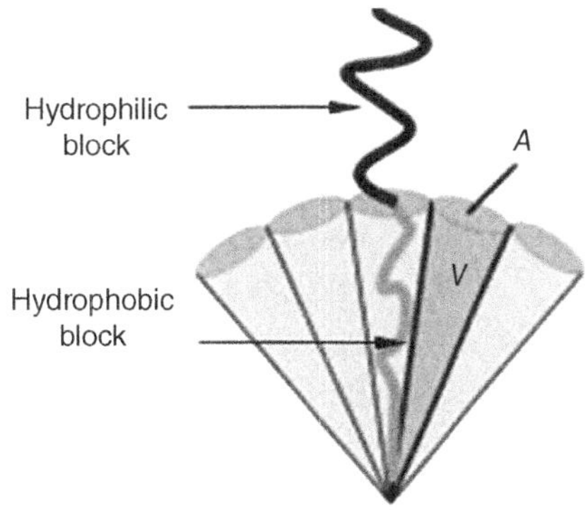

Figure 26.23 Description of the shape formed by amphiphilic block copolymer molecules in terms of a packing parameter equal to the volume V of the hydrophobic block relative to the volume of the hydrophilic block. A is the interfacial area between the hydrophobic block and the hydrophilic block, and l is the hydrophilic chain length normal to the interface.

shape of the structure is the size of the hydrophobic block relative to the size of the hydrophilic block, related through a parameter called the packing parameter f, defined by the equation

$$f = \frac{V}{Al} \tag{26.1}$$

where V is the volume of the hydrophobic block, A is the interfacial area between the hydrophobic block and the hydrophilic block, and l is the hydrophilic chain length normal to the interface (Figure 26.23). Spherical structures (micelles) will form for $f < 1/3$, cylinders will form for $1/3 < f < 1/2$, and polymersomes will form for $1/2 < f < 1$.

Because the molecular weight of block copolymers (>1 kDa) is normally greater than that of phospholipids (~0.1–1 kDa), the bilayer thickness of self-assembled polymersomes (~5–50 nm) is often greater than that in liposomes (~3–5 nm). As in the case of liposomes, polymersomes can be created with a wide range of sizes, but our emphasis will be on small unilamellar vesicles of size in the nanoscale range (diameter less than ~50–100 nm). While both liposomes and polymersomes can encapsulate hydrophilic and hydrophobic drugs within their aqueous interior and membrane, respectively, polymersomes tend to be less permeable. Consequently, polymersomes are generally less susceptible to leakage of their encapsulated drugs and show a greater ability to retain their drugs. Polymersomes also generally have better long-term stability in the aqueous physiological environment when compared to liposomes, which can be further enhanced by crosslinking. Overall, block copolymers have an advantage over phospholipids in that they can be more easily manipulated to achieve the desirable characteristics of the vesicles. Additionally, a wide range of copolymer compositions can be synthesized and their molecular weight and ratio of hydrophilic block to hydrophobic block can be controlled using a variety of polymer chemistry techniques. These properties allow the creation of polymersomes with desirable physico-chemical properties, biocompatibility and responsiveness to environmental stimuli.

Polymersomes can be created by self-assembly of amphiphilic block copolymer molecules using a variety of techniques, the most important of which are referred to as phase inversion and polymer rehydration. In the phase-inversion technique, the copolymer is dissolved in an organic solvent. Then water (or an aqueous solution) is added slowly to the organic solution or the organic solution is injected into water. As the hydrophobic blocks are insoluble in water, the molecules self-assemble in such a way that the hydrophobic blocks minimize their direct exposure to water while the hydrophilic blocks interface with water. Polymer rehydration techniques are based on rehydration of amphiphilic block copolymer films to induce self-assembly. The copolymer is dissolved in an organic solvent, and the solution is cast and dried by evaporation of the organic solvent to form a film. Then the film is hydrated by the addition of water, which permeates the film, leading eventually to the formation of vesicles.

A variety of degradable or non-degradable polymers, composed of diblock or triblock copolymers, have been investigated for the creation of polymersomes (Table 26.3). While other polymers such as polyacrylic acid (PAA) and poly(L-glutamic acid) have been used as the hydrophilic block, PEG has been most often used. This is largely based on the desirable characteristics of PEG, including its adequate biocompatibility, ease of synthesis into macromolecules with controlled molecular weight, ability to reduce opsonization by serum proteins, and ease of functionalization

Table 26.3 Examples of copolymers used to form polymersomes.

Nondegradable polymersomes	Degradable polymersomes
PEG-*b*-PD	PEG-*b*-PLA
PEG-*b*-PEE	PEG-*b*-PCL
PEG-*b*-PS	PEG-*b*-PTMC
PEG-*b*-PDMS	
PAA-*b*-PS	

PEG, polyethylene glycol; PD, polybutadiene; PEE, polyethylethylene; PS, polystyrene; PDMS, polydimethylsiloxane; PAA, polyacrylic acid; PLA, polylactic acid; PCL, polycaprolactone; PTMC, polytrimethylene carbonate.

with biomolecules. For many drug delivery applications, polymersomes are often required to be degradable. For such applications, degradable polymers such as poly(D,L-lactic acid) (D,L-PLA), PCL, and poly(trimethylene carbonate) (PTMC) have often been used as the hydrophobic block, based, in part, on their adequate biocompatibility and ability to degrade at varying rates by hydrolysis, while D,L-PLA and PCL also show pH-responsive degradation behavior.

Similar to cell membranes that incorporate cholesterol and membrane proteins, the membrane of polymersomes can serve as a reservoir for both hydrophobic and amphiphilic drug molecules, while hydrophilic drug molecules can be incorporated into the aqueous interior. For example, hydrophobic chemotherapy drugs can be incorporated into the membrane by dissolving or dispersing them, together with the block copolymer, in an organic solvent, after which the solution (or dispersion) is added to water (or an aqueous solution), as outlined earlier for the phase inversion technique. Several methods have been used to incorporate hydrophilic drug molecules into the aqueous interior of liposomes, commonly by direct encapsulation during formation of the polymersomes.

Drug release from polymersomes occurs mainly by diffusion through the membrane, driven by the drug concentration gradient between the polymersome and the surrounding medium. Consequently, the release kinetics will depend largely on the size of the polymersome and the structure of the membrane because they determine the diffusion distance and the ease of diffusion, respectively. Often, the kinetics, spatial control and preciseness of the drug release cannot be adjusted to the desired level simply by depending on concentration gradient-driven diffusion. As discussed earlier, particularly for hydrogel-based delivery systems, an important approach to alleviating this problem is through the use of delivery systems that respond to environmental or external physical stimuli. While a variety of such systems has been developed, pH-responsive polymersome systems have received significant attention because, as noted earlier, several tissues, fluids, cells and organelles within the body have different pH values

Block copolymers made up of a hydrophilic block composed of PEG and a hydrophobic block composed of D,L-PLA, PCL, or PTMC, which degrade by hydrolysis, have been among the earliest copolymers used for creating pH-responsive polymersomes. Drugs remain encapsulated within the polymersomes during circulation in blood at the physiological pH (~7.4) but are released upon arrival at a site of lower pH, such as a tumors, endosomes or lysosomes (Figure 26.24a). For example, polymersomes composed of PEG-*b*-PLA or PEG-*b*-PCL, incorporated with a combination of paclitaxel and doxorubicin, have been shown to localize in tumor tissue created in mice, release their encapsulated drugs upon entering the acidic tumor environment, and rapidly cause tumor shrinkage (Ahmed et al. 2006). These polymersomes exploit the ability of the bilayer membrane to incorporate a hydrophobic drug (paclitaxel) and the aqueous interior for incorporating a less hydrophobic drug (doxorubicin) (Figure 26.25a). They also exploit the ability to catalyze (accelerate) the hydrolytic degradation of the ester bond in these copolymers by acidic pH environments (Section 9.5.2). When dispersed in HEPES buffer (pH = 5.5) at 37 °C *in vitro*, polymersomes of large size were observed to degrade and develop pores faster than polymersomes dispersed in PBS (pH = 7.4), resulting in faster release of their encapsulated molecules (Figure 26.25b). On the other hand, leakage of

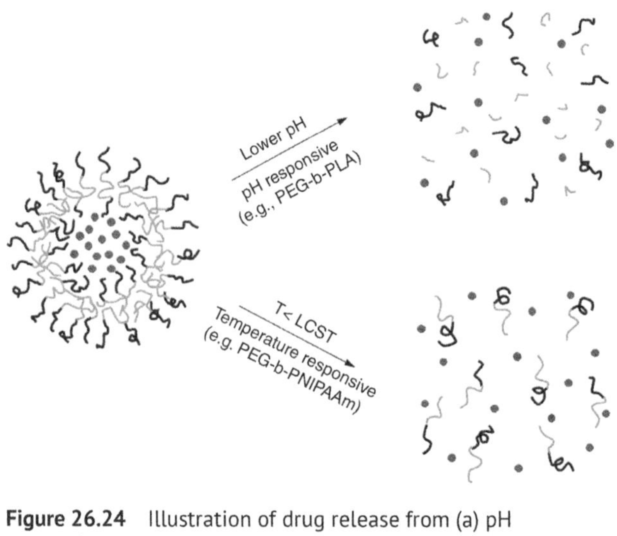

Figure 26.24 Illustration of drug release from (a) pH responsive polymersomes, such as a polymersome composed of the diblock copolymer PEG-*b*-PLA, and (b) temperature responsive polymersomes, such as PEG-*b*-PNIPAAm. PEG = polyethylene glycol; PLA = polylactic acid; PNIPAAm = poly(*N*-isopropyl acrylamide).

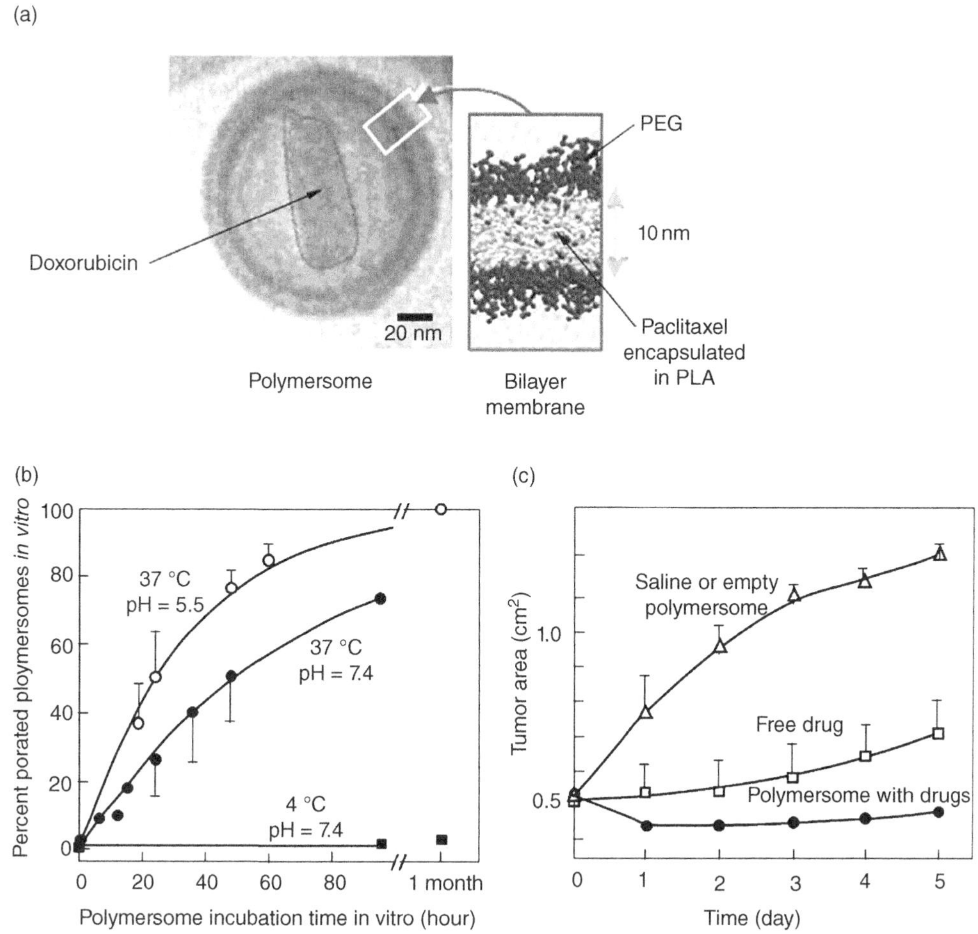

Figure 26.25 Drug incorporation, release and antitumor activity of polymersomes composed of PEG-*b*-PLA block copolymers and incorporated with doxorubicin and paclitaxel. (a) TEM image of polymersome containing doxorubicin and paclitaxel. (b) Polymersomes degraded and developed pores faster in HEPES buffer (pH = 5.5) at 37 °C *in vitro* than in phosphate-buffered saline (PBS) (pH = 7.4), resulting in faster release of their encapsulated drugs. (c) *In vivo*, solid tumors created in mice by injection of MDA-MB-231 human breast cancer cells showed no significant increase in tumor area for up to 5 days following injection of polymersomes containing doxorubicin and paclitaxel. *Source:* From Ahmed et al. (2006) / with permission of American Chemical Society.

encapsulated molecules was limited at 4 °C even after 1 month, which is convenient for storage of drug-containing polymersomes.

In vivo, solid tumors were created in mice by a single subcutaneous injection of MDA-MB-231 human breast cancer cells. Such tumors are known to have a leaky vasculature that allows permeation of nanoparticle-based delivery systems devoid of targeting ligands by the EPR effect (Section 26.6.2). Then single injections of polymersomes incorporating a combination of paclitaxel and doxorubicin or free drugs at their maximum tolerated dose were administered to tumors of approximately the same size by means of tail-vein injections. Controls consisted of systemic injection of buffer or polymersomes without drugs. As shown in Figure 26.25c, whereas the control tumors continued to grow and double in size within ~ 2 days, the tumors treated with the drug-containing polymersomes showed no significant increase in tumor area up to the measurement period of 5 days. In comparison, the area of tumors treated with the free drugs increased by ~50% at 5 days.

Micelles

Micelles are spherical structures, normally of size in the range ~5–50 nm, formed in an aqueous medium by self-assembly of dispersed amphiphilic molecules, composed of a polar (hydrophilic) part covalently bonded to a hydrophobic part (Figure 26.26). At low concentrations, such amphiphilic molecules remain dispersed in the aqueous medium, but as the concentration is increased above a certain value, called the critical micelle concentration, they self-assemble to form micelles. The temperature must also be above a certain value (the critical micelle temperature) for self-assembly to take place. The hydrophobic parts of the molecules form the core of the micelle, which may be

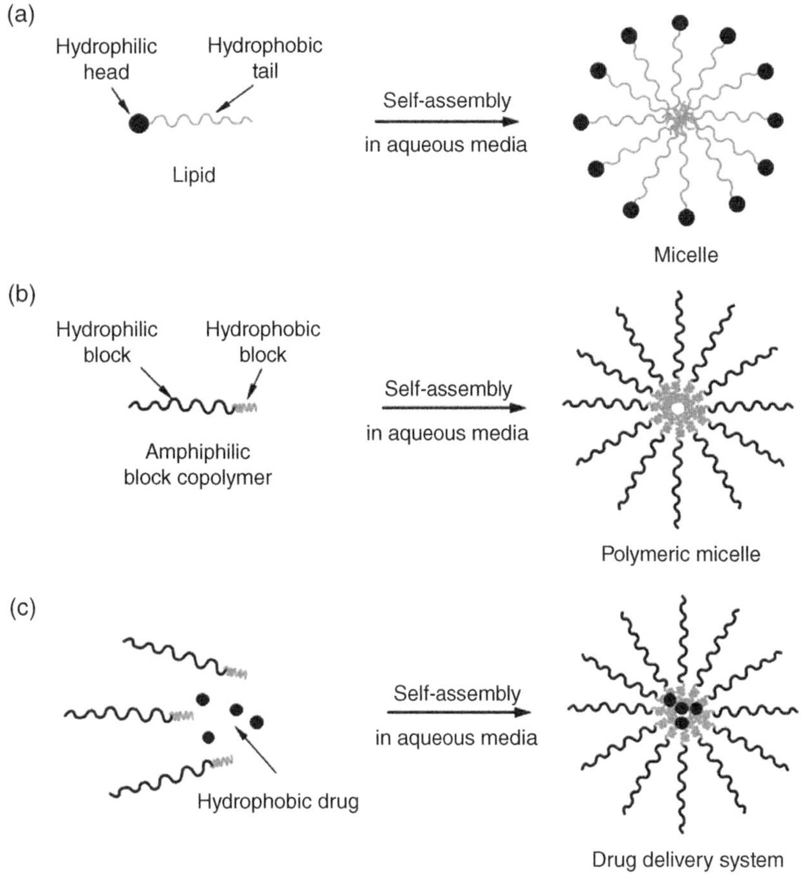

Figure 26.26 Schematic illustration comparing the structure of (a) micelle formed by self-assembly of lipids and (b) polymeric micelle formed by self-assembly of amphiphilic block copolymers. (c) Schematic illustration of drug incorporation into polymeric micelle to form a drug delivery system.

used for incorporating a drug, typically one that is poorly soluble in an aqueous medium. The hydrophilic parts of the molecules form a shell (or corona) surrounding the hydrophobic core, which interfaces with the aqueous medium. Lipids such as fatty acids, for example, composed of a polar (C=O)OH head group and a single short aliphatic chain, are common examples of amphiphilic molecules that self-assemble to form micelles (Figure 26.26a). In general, phospholipids by themselves do not form micelles because their two aliphatic chains attached to the polar head group are too bulky to allow self-assembly into a spherical micellar structure.

Micelles used for drug delivery are commonly composed not of lipids but, instead, of amphiphilic block copolymer molecules (Figure 26.26b). Consequently, our discussion for the remainder of this section will focus on such micelles, often referred to as polymeric micelles. As discussed earlier, amphiphilic block copolymer molecules in aqueous media can self-assemble into structures of different shapes, determined by the parameter f (Equation 26.1), which is related to the size of the hydrophobic block relative to the size of the hydrophilic block. In general, micelles will form when f is less than 1/3. This means that the hydrophobic block will typically be much smaller than the hydrophilic block. Similar to the formation of polymersomes, the driving force for the formation of polymeric micelles is again the van der Waals attraction of the hydrophobic blocks due to migration of water molecules from between the hydrophobic chains into the surrounding medium.

Polymeric micelles provide certain advantages for use as potential drug delivery systems. A major advantage is their ability to deliver hydrophobic drugs (or drugs that are poorly soluble in aqueous media) because the drug is encapsulated within the hydrophobic core of the micelle, as illustrated in Figure 26.26c. In comparison, while liposomes can incorporate poorly water-soluble drugs within their hydrophobic bilayer, as noted earlier, the amount of drugs that can be incorporated is often limited to avoid the risk of disrupting the bilayer. Micelles also protect drugs from degradation in the physiological environment because the drugs are incorporated within the hydrophobic core. In addition, the small size of micelles permits their extravasation and accumulation in physiological sites, such as tumors, where the permeability of the vascular endothelium is increased. This provides the ability of drug-incorporated micelles to target such sites selectively through passive targeting by the EPR effect (Section 26.6.2).

Polymeric micelles investigated for use in drug delivery systems are commonly composed of diblock copolymer molecules, but some micelles composed of triblock copolymer molecules have also been investigated. PEG has been most commonly used as the hydrophilic block in these copolymers, based on its aforementioned properties, including adequate biocompatibility and ability to endow the delivery system with stealth to avoid rapid clearance from circulation in the vasculature. However, other polymers, including polyvinylpyrrolidone (PVP), PVA, and PNIPAAm have been investigated as alternatives in a few studies. In comparison, a variety of polymers have been used as the hydrophobic block, including polypropylene oxide (PPO), D,L-PLA, PCL, poly(L-lysine), and polyaspartic acid.

Another type of polymeric micelle involves the use of lipids as the hydrophobic blocks of the copolymer molecule. Micelles prepared from molecules composed PEG and diacyllipids, particularly phosphatidylethanolamine (abbreviated here as PE, but not to be confused with the abbreviation PE commonly used for polyethylene) have received considerable interest. The presence of two aliphatic chains in PE contributes significantly to an increase in van der Waals attraction within the core of the micelle, resulting in enhanced stability. Polymeric micelles composed of PEG–PE can be incorporated with a variety of chemotherapy drugs, including paclitaxel and tamoxifen, and they have shown long circulating times in blood upon injection into mice (Torchilin et al. 2003). When modified by attaching amino group-containing ligands composed of the antibody mAb 2C5 (Figure 26.27), paclitaxel-containing PEG–PE micelles showed increased accumulation in experimental tumors created in mice and the ability to significantly inhibit tumor growth following administration by tail–vein injection.

26.6.5 Polymer Conjugates

In lipid-based drug delivery systems, as we have seen in the previous section, the nanocarrier and the drug can be considered as separate entities. The drug is encapsulated within the nanocarrier typically by non-covalent bonds and is released mainly by physical mechanisms, particularly diffusion, driven by a concentration gradient between the nanocarrier and the surrounding environment. In comparison, polymer conjugates consist of delivery systems in which the drug is covalently bonded (conjugated) to water-soluble polymer molecules, typically by means of a linker molecule. Upon delivery of the polymer conjugate to the site of action, such as the intracellular compartment, degradation of the linker molecule leads to release of the drug (Figure 26.2c). Polymer conjugates are often divided into two types, polymer-protein conjugates, in which macromolecules such as proteins are

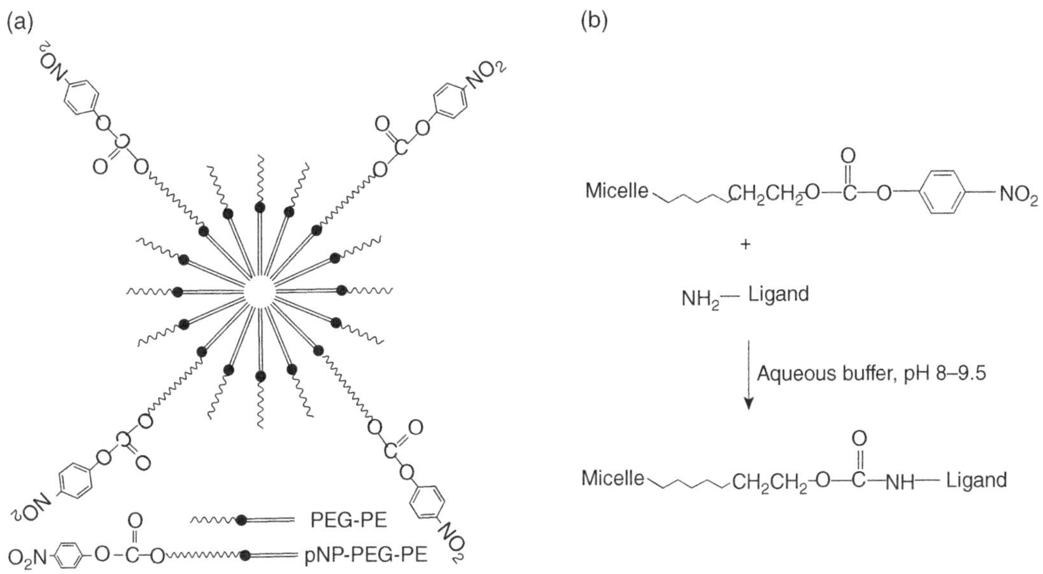

Figure 26.27 (a) Schematic illustration of the structure of micelle formed from molecules composed of polyethylene glycol (PEG) and a diacyllipid, phosphatidylethanolamine (PE), with the free end of PEG terminated with *p*-nitrophenylcarbonyl (pNP) group. (b) Schematic illustrating the coupling of amino group-containing ligands (such as antibodies) with the pNP groups. *Source:* From Torchilin et al. (2003) / with permission of National Academy of Sciences.

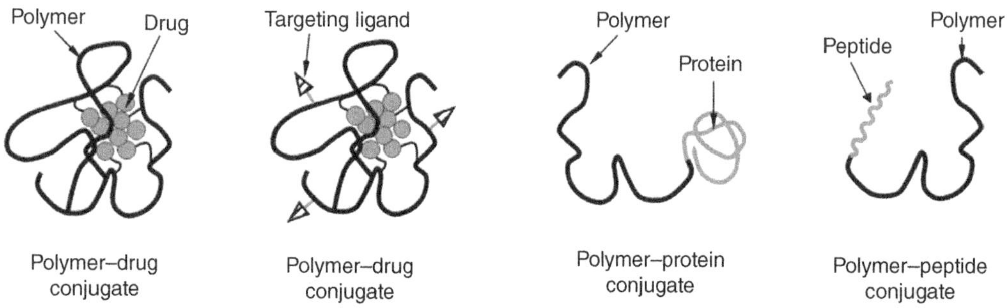

Figure 26.28 Schematic illustrations of some types of polymer conjugates.

bonded to polymer molecules, and polymer–drug conjugates, in which small-molecule drugs, typically chemotherapy drugs, are bonded to polymer molecules (Figure 26.28). These two types of polymer conjugates, along with polymer drugs (drugs composed of polymers alone) and polymer micelles, are sometimes described as polymer therapeutics. Polymer conjugates are normally administered parenterally, that is, directly into the body, typically by subcutaneous or intramuscular injection for polymer–protein conjugates and intravenously for polymer–drug conjugates.

Polymer–Protein Conjugates

A growing number of protein, peptide, and antibody-based drugs have been created by recombinant DNA and other technologies for use as therapeutic agents, but many of them suffer from drawbacks, such as poor solubility, degradation by proteolytic enzymes, rapid clearance from circulation in blood by the kidneys and, in some cases, immunogenicity. An effective approach to alleviating such problems has been to covalently conjugate certain water-soluble polymers, particularly PEG, to them to form polymer-protein conjugates. The first successful application of this type of conjugation technology was the covalent bonding of PEG molecules to proteins several decades ago, a process now often referred to as PEGylation. This approach has since been applied to a variety of proteins, including enzymes, cytokines, and antibodies. Although several water-soluble polymers have been used successfully for protein conjugation, PEG has been used most widely and very successfully used. The first polymer–protein conjugate to receive approval from the US FDA for clinical use was PEG-L-asparaginase in 1994, to treat acute lymphoblastic leukemia. Since that time, a large number of anticancer PEG-protein conjugates have progressed to clinical applications or clinical trials, a few examples of which are listed in Table 26.4.

It is worth recalling the characteristics of PEG that contribute to its success in protein conjugation. Because PEG molecules are highly water-soluble, nontoxic, non-immunogenic, and electrostatically neutral, PEGylation can enhance the solubility and stability of the protein, reduce its immunogenicity, and prolong its circulation time in blood. Although PEG is not degradable in aqueous media, individual molecules of molecular weight less than ~10–15 kDa dissolve in the physiological fluid and can be eliminated from the body by excretion through the kidneys, while PEG of larger molecular weight can be excreted in the feces. Additionally, PEG molecules consist of linear flexible hydrophilic chains, which can extend out from the conjugated protein into the aqueous environment and form a shell to protect the protein. PEG can also be prepared with a single reactive group at one end of the polymer chain, which aids site-specific conjugation to a protein and avoids protein crosslinking during conjugation. The early PEG-protein conjugates were prepared using linear PEG molecules, such as monomethoxy-PEG whose chain ends are terminated by a methyl (CH_3) group and by a hydroxyl (OH) group, and with several chains attached randomly to each protein molecule via the OH group. Subsequently, advances in chemical techniques have resulted in the synthesis of branched PEG molecules that provide better stability to attached proteins when compared to linear PEG molecules of the same molecular weight. Additionally, PEG with different functional groups has been synthesized that improve the site-specific attachment to proteins.

Polymer–Drug Conjugates

Intravenously injected low-molecular-weight drugs, such as chemotherapy drugs, as we have seen, can be cleared rapidly from circulation in blood. However, their attachment to certain polymers, to form polymer–drug conjugates, can enhance their circulation times considerably, facilitating passive tumor targeting by the EPR effect. Additionally, polymer–drug conjugates can be synthesized to contain targeting ligands such as peptides and antibodies to enhance tumor targeting by receptor-mediated endocytosis, thereby limiting cellular uptake to the endocytic route.

Table 26.4 Examples of polymer–protein and polymer–drug conjugates developed or used clinically for cancer therapy.

Polymer conjugate	Cancer therapy
Polymer–protein conjugate	
SMANCS	Hepatocellular carcinoma
PEG-L-asparaginase	Acute lymphoblastic leukemia
PEG-GCSF	Prevention of neutropenia associated with cancer chemotherapy
PEG-IFNα2a	Hepatitis B and C, melanoma, chronic myeloid leukemia and renal-cell carcinoma
PEG-IFNα2b	Hepatitis C, melanoma, multiple myeloma and renal-cell carcinoma
PEG–arginine deiminase	Hepatocellular carcinoma
Polymer–drug conjugate	
Polyglutamate–paclitaxel	Various cancers, particularly non-small-cell lung cancer; ovarian cancer as a single agent or in combination therapy
Polyglutamate–camptothecin	Various cancers
HPMA copolymer–doxorubicin	Various cancers, particularly lung and breast cancer
HPMA copolymer–doxorubicin-galactosamine	Particularly hepatocellular carcinoma
HPMA copolymer–paclitaxel	Various cancers
HPMA copolymer–camptothecin	Various cancers
HPMA copolymer–carboplatin platinate	Various cancers
HPMA copolymer–DACH-platinate	Various cancers
Dextran–doxorubicin	Various cancers
PEG–camptothecin	Various cancers

Source: From Duncan (2006) / with permission of Springer Nature. SMANCS, styrene maleic anhydride-neocarzinostatin; PEG, polyethylene glycol; GCSF, granulocyte colony-stimulating factor; IFNα, interferon-α; HPMA, *N*-(2-hydroxypropyl) methacrylamide; DACH, diaminocyclohexane.

An important issue relates to the features required for designing effective polymer–drug conjugates, which include the following:

- The polymer should have adequate biocompatibility, including nontoxicity and non-immunogenicity.
- The polymer should also have a sufficiently high molecular weight to provide resistance against rapid clearance from circulation but also to ensure renal clearance of the polymer following delivery of the drug. For nondegradable polymers, this means that the molecular weight should be less than ~20–30 kDa, whereas the molecular weight of degradable polymers can be higher, up to ~100 kDa.
- The polymer should be capable of carrying an adequate amount of the drug, which is dependent on the potency of the drug.
- The linker molecule used to attach the drug to the polymer chain should have adequate chemical stability to ensure transportation of the drug to its intended site (such as the intracellular compartment) but, thereafter, it should degrade at an optimal rate to ensure drug release at a therapeutically desired level.
- The size of the polymer–drug conjugate, as measured by its hydrodynamic diameter, for example, should be small enough (less than ~50–100 nm) to ensure extravasation into tumors and endocytic internalization by tumor cells. For small-molecule drugs, the size of the conjugate will depend mainly on the molecular weight of the polymer and its conformation. Approximating the polymer molecule as a freely jointed chain, this means that the molecular weight of the polymer should be less than ~50–100 kDa (see Eq. (8.7)).

While many polymers have been investigated for use in polymer–drug conjugates, only a few of them have progressed to the stage of clinical studies. PEG, as noted in the previous section, has been commonly used in polymer–protein conjugates, and it has been studied for use in polymer–drug conjugates as well. However, its drug-carrying capacity is limited because the PEG chain contains only two terminal end groups which can take part in conjugation. Instead, several clinical studies of polymer–drug conjugates have involved the use of *N*-(2-hydroxypropyl) methacrylamide (HPMA)-based copolymers (Table 26.4). Conjugates consisting of HPMA-based copolymers and chemotherapy drugs involve several of the aforementioned design principles. For example, as the chain backbone of HPMA is nondegradable, polymers of molecular weight ~30 kDa have commonly been used. A peptide composed of Gly-Phe-Leu-Gly has been used as the linker molecule for drug conjugation because it is stable in blood but is degraded by lysosomal proteases once the conjugate has been internalized by endocytosis. While HPMA copolymers conjugated without a targeting ligand have shown passive tumor targeting by the EPR effect and improved anti-tumor activity *in vivo*, the feasibility of receptor-mediated targeting has been shown with a galactose-containing residue.

Figure 26.29 shows an example of an HPMA copolymer-based drug conjugate containing the chemotherapy drug doxorubicin and galactosamine as a targeting ligand. This delivery system progressed to the stage of clinical trials. Recent developments have included strategies to incorporate degradability into the HPMA chain backbone, which allow the use of polymers with higher molecular weight coupled with adequate biocompatibility. Such conjugates containing doxorubicin have shown long circulation times and enhanced antitumor activity when injected intravenously in mice.

26.6.6 Dendrimers

Dendrimers are composed of highly branched macromolecules radiating from a central core, giving a nearly spherical shape (Figure 26.30). They are characterized by a diameter in the range ~1–10 nm, with an almost monodisperse diameter and monodisperse molecular weight (weight-average to number average molecular weight M_w/M_n ~1.0). The intramolecular core of dendrimers has a low structural density, providing a substantial amount of (solvent-filled) voids for entrapping drugs and other biomolecules as well as the potential for their subsequent controlled release. Dendrimers can also be synthesized with a large number of terminal functional groups at their periphery, such as amine NH_2, hydroxyl OH, or carboxyl (C=O)OH groups, which provide the means to attach PEG molecules, drugs, and other biomolecules, such as targeting ligands. Overall, these characteristics distinguish dendrimers from other nanocarriers such as polymer nanoparticles, polymer conjugates, and lipid-based nanoparticles discussed earlier.

Dendrimers have a high solubility in aqueous media. They allow the delivery of drugs by two methods. Drugs can be encapsulated within the aqueous intramolecular spaces of PEGylated dendrimers, in which case targeting of tumors can occur through passive targeting by the EPR effect. Alternatively, drugs and targeting ligands can be chemically attached (conjugated) by means of linker molecules to the large number of functional groups at the periphery of dendrimers. In this case, the drug is released under appropriate conditions by chemical degradation of the linker molecules upon the dendrimers reaching their intended site, while the ligands provide the capacity for active targeting of specific cells. The small size of dendrimers make rapid take up into the intracellular compartment very favorable. Additionally, the small size allows dendrimers to be cleared through the kidneys, which means that while degradability of dendrimers is preferred, it is not a requirement.

Dendrimers can be synthesized in a wide range of compositions using a variety of chemical techniques, although

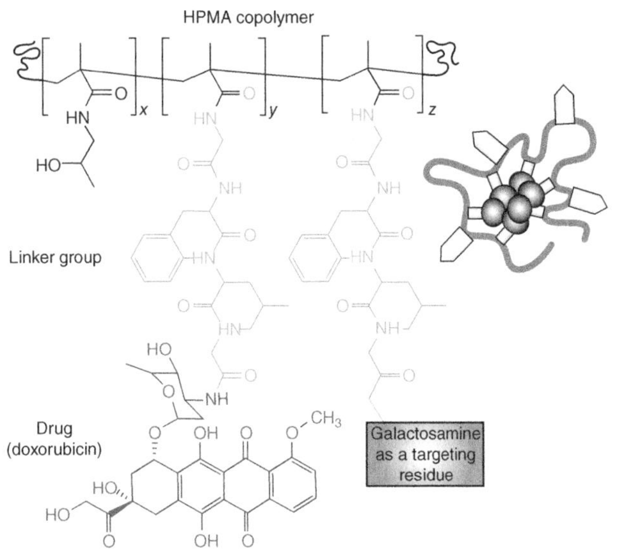

Figure 26.29 Structure of an *N*-(2-hydroxypropyl) methacrylamide (HPMA) copolymer-based drug conjugate containing the chemotherapy drug doxorubicin and galactosamine as a targeting ligand. *Source:* From Duncan (2006) / with permission of Springer Nature.

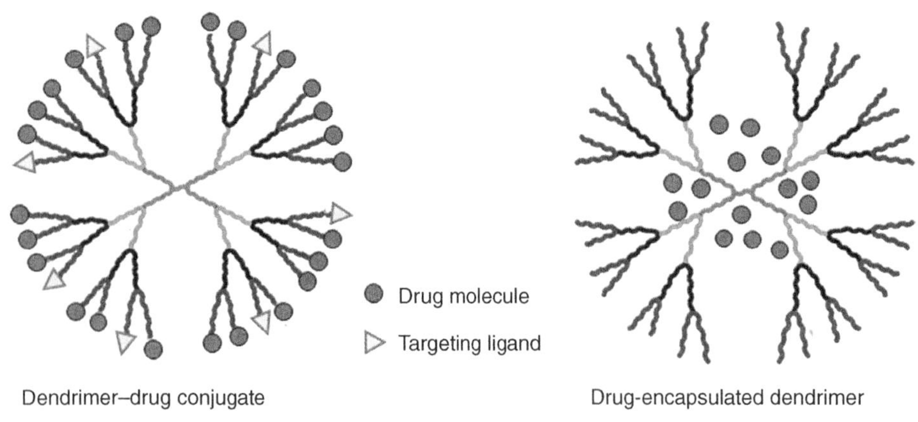

Figure 26.30 Schematic illustration of two types of dendrimer-based drug delivery systems: dendrimer–drug conjugate and drug-encapsulated dendrimer.

many of them are not intended for use in drug delivery systems. One of the first family of dendrimers for drug delivery applications and one of the most widely studied is based on polyamidoamine (PAMAM; Figure 26.31). These dendrimers are synthesized by assembling monomer units in a branch-upon-branch motif according to certain dendritic principles. An iterative reaction sequence leads to concentric shells of branches (called generations) around a central initiator core, whereby the structure grows linearly in diameter with each generation, while the surface groups increase exponentially.

Over the last few decades, a large number of dendrimer compositions and architectures have been synthesized for potential use in drug delivery, diagnostic imaging, and other biomedical applications. However, the transfer of dendrimers to clinical trials has been slow due, in part, to concerns about possible toxicity of some compositions. While PAMAM dendrimers continue to be widely studied, alternative compositions have been synthesized which are non-toxic and, in some cases, degradable. In addition, advances in dendrimer synthesis techniques, coupled with systematic pharmacokinetic studies, are enabling the optimization of dendrimers with improved potential for parenteral drug delivery.

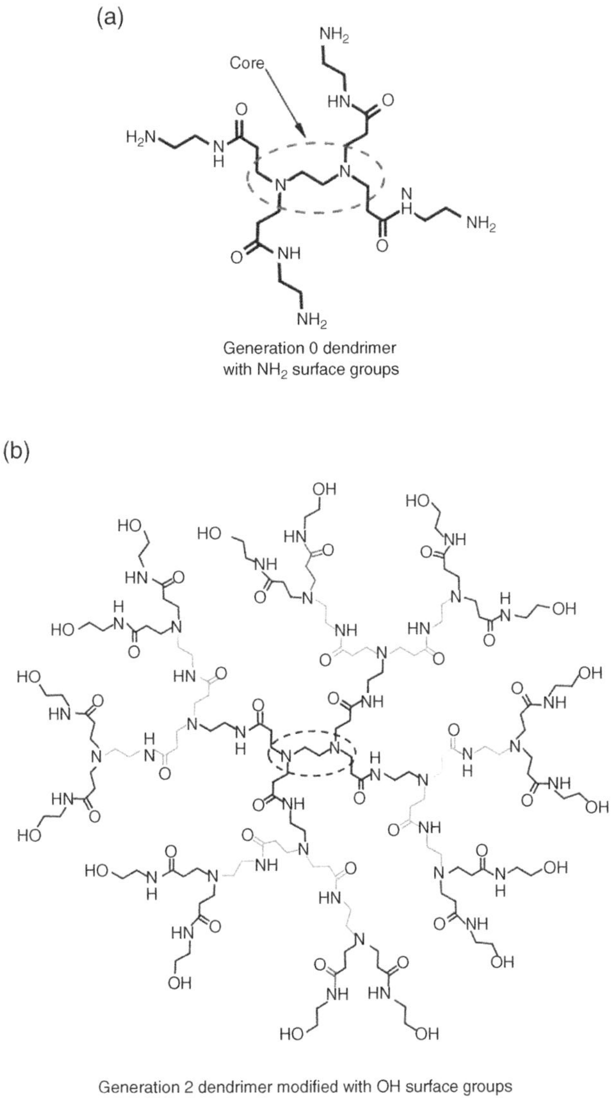

Figure 26.31 Structure of polyamidoamine (PAMAM) dendrimers synthesized from amidoamine, showing (a) a generation 0 dendrimer with NH_2 surface groups and (b) a generation 2 dendrimer with hydroxyl OH surface groups.

26.6.7 Inorganic Nanoparticles

A variety of inorganic nanoparticles has been studied for use in biomedical applications, including gold (Au), quantum dots (for example, cadmium selenide and cadmium sulfide), carbon nanotubes and fullerenes (Chapter 3), iron oxide (Fe_3O_4 and γ-Fe_2O_3) (Chapter 4), and porous silica (SiO_2). These nanoparticles, except for Au, have been studied mainly for diagnostic imaging and other applications, rather than primarily as nanocarriers for drug delivery systems. Overall, these inorganic nanoparticles are nondegradable, while few of them, particularly quantum dots and carbon nanotubes, suffer from concerns related to possible toxicity *in vivo*.

Although non-degradable, Au nanoparticles provide certain advantages for use in drug delivery systems, including ease of preparation into well-controlled sizes and ease of surface functionalization, often via the thiol (–SH) functional group, for the delivery of a variety of biomolecules. Human cells have been shown to take up Au nanoparticles (~20 nm in diameter) without toxic effects following continuous exposure for three days *in vitro*. Upon intravenous injection in mice, spherical gold nanoparticles (10, 50, 100, and 250 nm in diameter) without surface modification were found to be mainly distributed in the liver and spleen, but they also showed a size-dependent distribution in tissues and organs. Whereas the larger particles were found only in the blood, liver and spleen, the 10 nm particles showed a more widespread distribution, in the blood, liver, spleen, kidney, testis, thymus, heart, lung, and brain. As noted earlier, particles of size less than ~10 nm can pass through the filtration pores of the kidneys and be eliminated in the urine. However, unstabilized particles of size in the colloidal range show a greater tendency to agglomerate with decreasing size, forming clusters of varying size, which can account for the more widespread distribution of the 10 nm particles.

Unlike drug delivery systems such as polymeric nanoparticles and lipid-based nanoparticles in which the drug is encapsulated within the nanocarrier, in Au nanoparticle-based delivery systems, the drug is covalently bonded (conjugated) to the surface of the nanoparticles. While Au nanoparticles have been widely investigated for the delivery of chemotherapy drugs in cancer treatment, including small-molecule drugs such as paclitaxel and doxorubicin, and protein-based drugs, very few have progressed to clinical studies. As we have seen, some drug delivery systems consisting of lipid-based nanoparticles or polymer nanoparticles can be self-assembled using pre-functionalized block copolymers that have all the desired components. In comparison, in Au nanoparticle delivery systems, the Au nanoparticles are prepared and functionalized with the desired components in separate steps. Competition among the drug molecules, hydrophilic molecules such as PEG, and targeting ligands for the bonding sites on the Au nanoparticle surface can make it difficult to control the amount of the drug attached to the surface.

One delivery system that has shown promising results in early clinical studies consists of recombinant human tumor necrosis factor-alpha (TNFα) and thiol-terminated PEG molecules (~20 kDa) covalently bonded as separate entities to the surface of Au nanoparticles (~30 nm) (Paciotti et al. 2004). This delivery system was developed through a series of pre-clinical studies, in which the optimal conditions for covalent bonding of controllable amounts of TNFα and thiol-terminated PEG to Au sols were determined. These conditions included the pH (8–10) of the Au sol, and the concentrations of TNFα and thiol-terminated PEG added to the sol to give the desired amount of these two entities bonded to the Au particle surface. Covalent bonding of TNFα apparently occurred via the –SH groups present in its cysteine residues, while PEG was bonded via its terminal –SH group. Intravenous injection in human patients showed that TNFα could be administered at doses higher than those previously shown to be toxic by administration of free TNFα. Additionally, tumor biopsies at 24 hours post-injection showed the presence of Au nanoparticles, indicating the ability of the drug delivery system to reach the tumors by passive targeting.

26.7 Delivery of Ribonucleic Acid (RNA)

In general, aberrant protein production within cells, caused by overexpression of one gene or multiple genes, is a feature of many diseases. For example, several types of cancer are caused by the overexpression of specific genes, which results in rapid and uncontrolled cell proliferation. A common way to treat diseases is to regulate gene expression through the introduction of small-molecule and protein-based drugs into the body, using drug delivery systems, for example, as discussed in the preceding sections of this chapter. However, it has been estimated that only a small fraction of proteins in the body can be inhibited with small-molecule drugs due to the mechanism by which they act.

Since about 1998, when RNAi was identified as a fundamental pathway by which sequence-specific RNA strands are able to target and induce gene silencing (Fire et al. 1998), there has been significant interest in the use of RNAs as an alternative class of therapeutic molecules to treat disease. RNAi is an endogenous pathway for post-transcriptional silencing of gene expression, which is triggered by double-stranded ribonucleic acid (dsRNA), including endogenous microribonucleic acid (miRNA) and short interfering ribonucleic acid (siRNA). In principle, by activating this pathway, siRNA has the potential to silence the expression of almost any gene with high efficiency and specificity, including targets that are untreatable with small-molecule drugs or antibodies. Because of this far-reaching potential, considerable effort has been devoted to the development of siRNA therapeutics for the treatment of a variety of diseases, including viral infections, hereditary disorders, and cancer. A considerable number of RNAi-based therapeutics have already reached clinical studies and many more are under development.

As siRNA is the most common structure used in RNAi-based therapeutic formulations, it will form the focus of our discussion in this section. siRNAs are duplexes of 21–23 nucleotides, with a molecular weight ~13 kDa, a high negative charge (due to the negatively-charged phosphate groups in the chain backbone), and dimensions of ~7.5 nm in length and ~2 nm in diameter. siRNAs can be created intracellularly through the cleavage of long dsRNA by an enzyme called Dicer. However, in practice, siRNAs are commonly produced synthetically and then introduced directly into the cell, thus circumventing the Dicer process. Once within the cytoplasm, siRNA is incorporated into a protein complex called the RNA-induced silencing complex (RISC) (Figure 26.32). A multifunctional protein (Argonaute-2) present in RISC unwinds the siRNA, after which the sense strand (passenger strand) of the siRNA is cleaved. The activated RISC, which contains the antisense strand (guide strand), seeks out and degrades mRNA that is complementary to the antisense strand. Thereafter, it can move on and degrade additional mRNA targets, further propagating gene silencing. This leads to a therapeutic effect lasting ~3–7 days in rapidly dividing cells and several weeks in non-dividing cells. Eventually, siRNAs are depleted below a certain therapeutic level or degraded within the cell, necessitating repeated administration to achieve the desired therapeutic effect.

To activate the RNAi pathway, it is necessary for siRNA to be delivered to the interior of target cells and to be incorporated into the RNAi mechanism. However, various

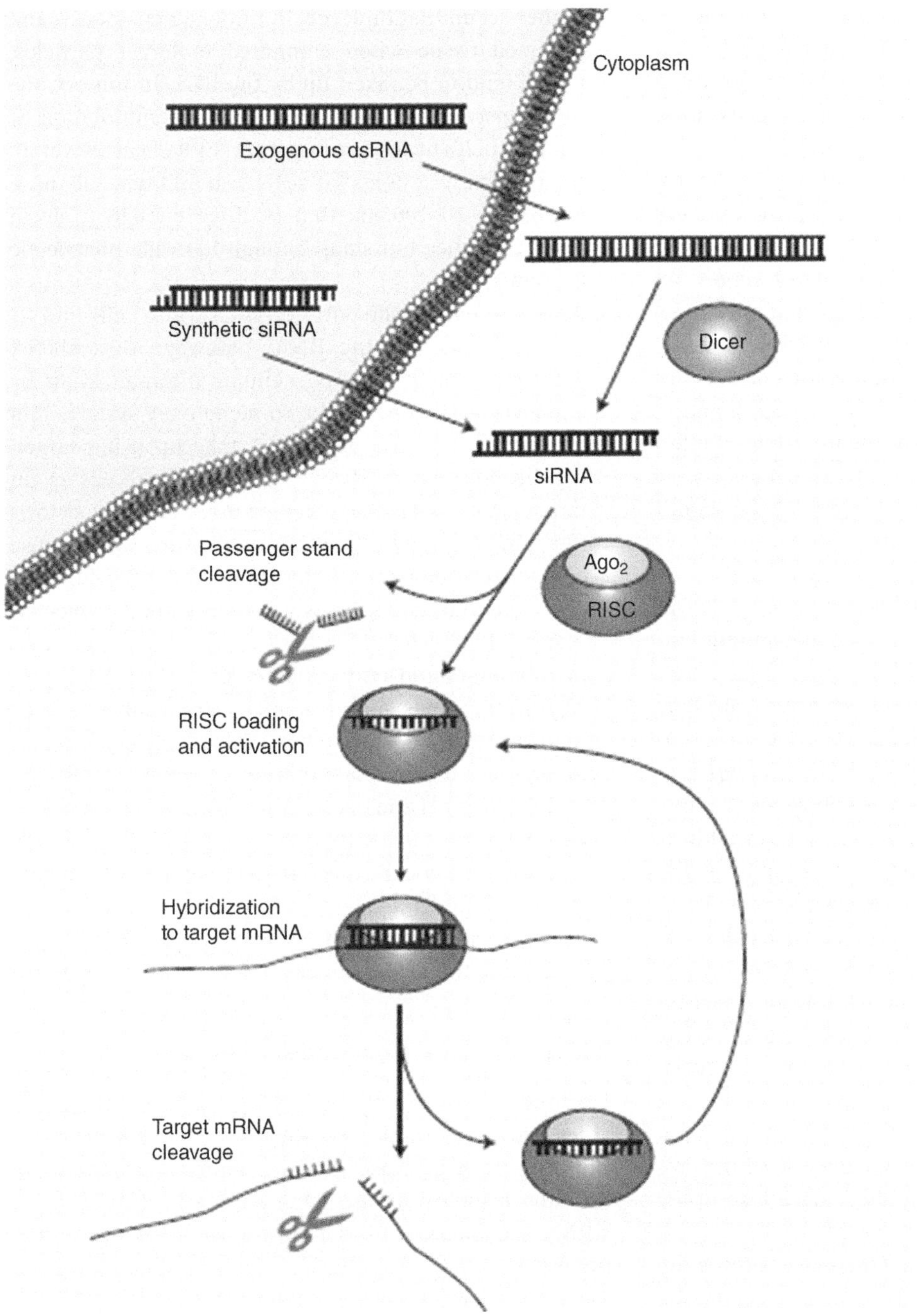

Figure 26.32 RNA interference. Long dsRNA introduced into the cytoplasm is processed by the enzyme Dicer into 22-nt pieces with 2-nt single-stranded overhangs on the 3′ ends. The structure of synthetic siRNA mimics that of Dicer products. The siRNA guide strand is loaded into the RNA-induced silencing complex (RISC), and the passenger strand is cleaved by Argonaute-2 (Ago2). The activated RISC–guide-strand complex identifies and cleaves mRNA that is complementary to the guide strand, preventing translation and thereby silencing gene expression. *Source:* From Kanasty et al. (2013) / with permission of Springer Nature.

physiological barriers must be overcome for safe and effective delivery of siRNA. In general, the ease of siRNA delivery is dependent, in part, on the accessibility of the tissue or organ within the body. Local administration of siRNA, by direct administration to the target tissue, can offer several benefits, including a higher availability of siRNA due to the proximity of the tissue and a reduction of adverse effects typically associated with systemic administration, such as rapid clearance from circulation. However, siRNA must still overcome intracellular barriers before it can enter the cytoplasm or nucleus of target cells for it to exert the desired effect.

Although several tissues and organs are amenable to direct (local) administration of siRNA, including the eye, skin, lung, brain, and isolated tumors, many tissues and organs, including the liver, heart, kidney, and metastasized tumors, can only be reached by systemic administration by means of intravenous injection of delivery systems. However, when administered systemically, siRNA faces additional physiological barriers, when compared to direct administration, before it can reach its site of action. As unmodified siRNA molecules are too large and too hydrophilic (with a high negative charge) to diffuse across (negatively charged) cell membranes alone, the use of a delivery system, chemical modification of siRNA, or a combination of a delivery system and chemical modification, is often required to assist take up of siRNA by target cells.

Overall, systems for intravenous delivery siRNA should be designed to provide the following:

- Ability to protect siRNA from degradation by serum RNAses and provide the ability to evade the immune system: Upon systemic administration, unmodified siRNA can be degraded by serum RNAses (a group of hydrolytic enzymes that degrade RNA molecules), and it can stimulate an innate immune system due to a similarity in molecular structure with invading pathogens. As discussed in the following sections, one way to address these problems involves chemical modification of the siRNA chain, whereas some delivery systems aim to protect siRNA by encapsulating it within nanoparticles.
- Ability to avoid non-specific interactions with serum proteins and avoid delivery to non-target cells: As noted for nanoparticle-based drug delivery systems (Section 26.6.2), the surface of the nanoparticle can become opsonized with serum proteins, resulting in rapid take up by the RES and clearance from circulation. As noted, PEGylation is a common strategy for reducing interaction with serum proteins. However, PEGylation can also reduce take up by cells and silencing efficiency.
- Ability to avoid renal clearance: One main pathway by which siRNA leaves the bloodstream is through the kidneys, where it risks being eliminated in urine. Consequently, the delivery system should also prevent renal clearance of siRNA. As noted for nanoparticle delivery systems, delivery systems of size larger than ~20 nm are normally used to limit renal clearance of siRNA.
- Ability to enable siRNA to leave blood vessels in order to reach target tissues: This occurs in tissues with a leaky vasculature, such as the liver and several tumors (Figure 26.17). The size of the delivery system is important because, as noted for nanoparticle-based delivery systems, particles of size less than ~100 nm tend to have higher accumulation levels in tumors, hepatocytes and inflamed tissue when compared to larger particles, which tend to be taken up by the RES. In tumors, the highly permeable endothelia lead to accumulation of nanoparticles by passive targeting (EPR effect). Overall, then, delivery systems for siRNA should typically have a size of ~20–100 nm, that is, large enough to avoid renal clearance but small enough to evade phagocytic clearance.
- Ability to enable the entry of siRNA into cells and its incorporation into the RNAi pathway: Most siRNA delivery systems undergo cellular internalization by endocytosis (Figure 21.21). Some delivery systems aim to improve the rate of cellular take up using targeting ligands, which bind specifically to receptors on target cells to enable receptor-mediated endocytosis (Section 26.6.2). On the other hand, other systems use cell-penetrating peptides that enable internalization by cells through endocytosis or non-endocytic mechanisms. Endocytosed systems are taken up into membrane-bound endocytic vessels, which fuse with early endosomes and become more acidic as they mature into late endosomes. Some delivery systems incorporate biomaterials that are designed to respond to this lower pH and disrupt the endosomal membrane, thereby releasing siRNA into the cytoplasm. However, in many systems, the mechanism of endosomal release is not well understood. On reaching the cytosol, siRNA must be released from the delivery system for incorporation into the RISC pathway.

26.7.1 Chemical Modification of siRNA

Chemical modification of siRNA, as noted earlier, provides a strategy to address problems related to its degradation by serum proteases and its ability to stimulate an immune response *in vivo*. One approach is to alter the siRNA sequences with chemically modified sugars and linkers. Common modifications include substituting the 2′-hydroxy group of the ribose ring with 2′-fluoro, 2′-*O*-methyl or 2′-deoxy substituent, locked nucleic acids and unlocked nucleic acids, and substitution of the phosphodiester linkages between nucleotides with phosphorothioate linkages (Figure 26.33a). Another approach is the conjugation of siRNA with molecular targeting ligands, including GalNac (a complex galactose derivative), cholesterol, cell-penetrating peptides and antibodies. GalNac (Figure 26.33b) and cholesterol have been shown to be particularly effective. For example, following subcutaneous implantation, siRNA–GalNac conjugates have been found to induce gene silencing in the liver without the need for a delivery system.

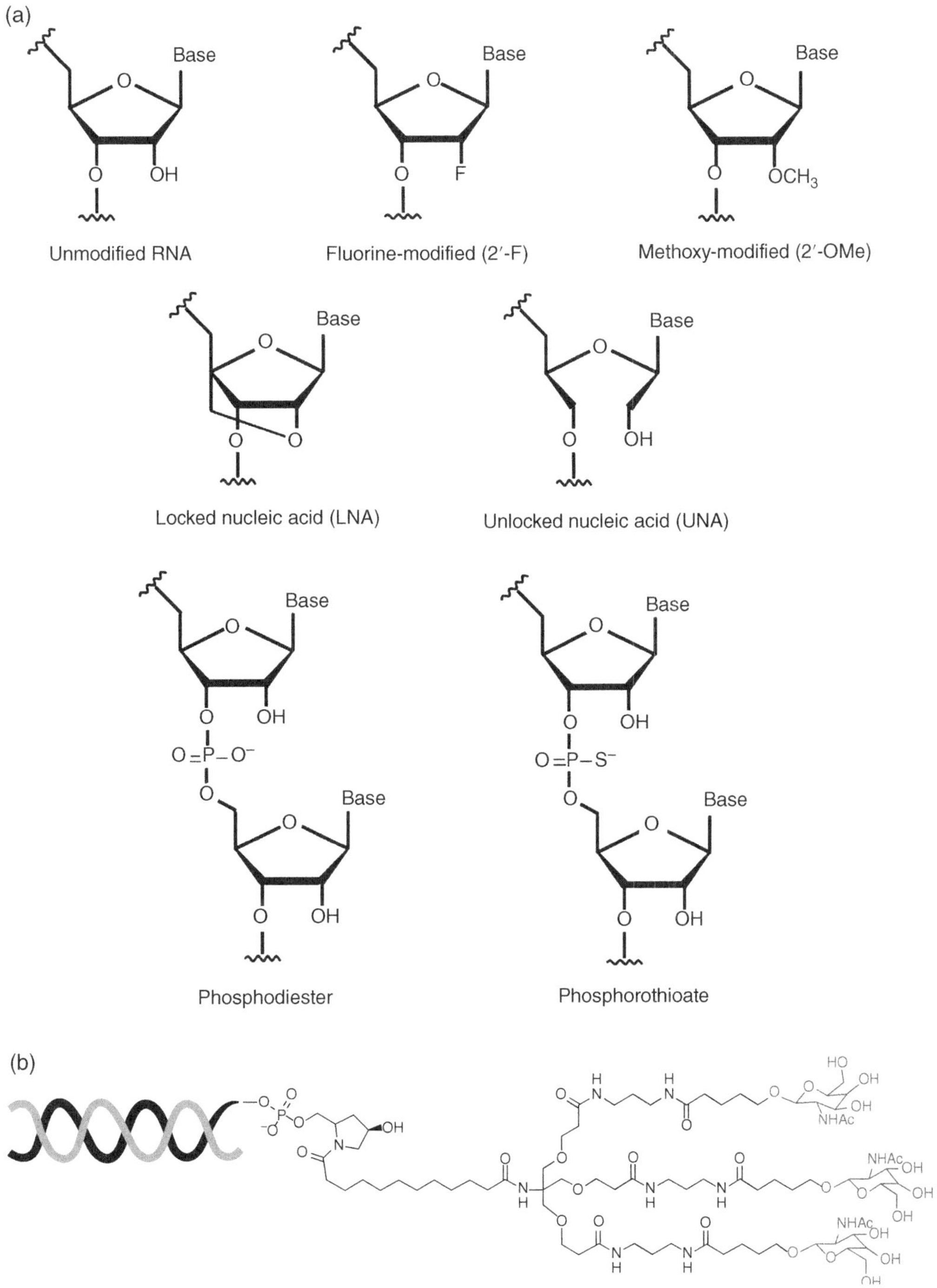

Figure 26.33 Common modifications of the siRNA structure. (a) Modifications to the siRNA chain backbone. (b) Conjugation of siRNA with the molecular targeting ligand referred to as GalNac.

26.7.2 Biomaterials for siRNA Delivery

A variety of biomaterials has been studied and developed to deliver siRNA *in vivo*, including several systems already discussed for use in nanoparticle-based drug delivery systems, such as systems based on lipids, polymer nanoparticles, dendrimers, and inorganic nanoparticles. As noted earlier, siRNA, has a high negative charge due to the phosphate groups in its chain backbone. Consequently, the design of many siRNA delivery systems is centered on its negative charge and the requirement to develop cationic compounds that can form a complex with it, leading to charge neutralization and compaction of the siRNA.

Lipid-based Delivery Systems

One of the most widely studied systems for siRNA delivery is based on lipid building blocks. Liposomes were discussed earlier in this chapter in relation to drug delivery (Section 26.6.4), and they are used for siRNA delivery as well. Prior to being studied for siRNA delivery, liposomes were widely studied and applied for the delivery of DNA, because of their ability to protect oligonucleotides from nuclease degradation and renal clearance, to promote take-up by cells, and to promote endosomal escape.

Many lipid-based systems used for siRNA delivery are based on cationic lipids, whose head groups have a net positive charge in an aqueous environment, or on ionizable lipids, whose head groups become positively charged depending on the pH of the aqueous environment. These lipids interact with negatively charged siRNA through electrostatic interactions, forming liposomal complexes, called lipoplexes. Studies have shown that the structure of the cationic lipid building blocks strongly influences various functional properties of the lipoplexes formed from them. As noted earlier, the three essential structural components of lipids are the hydrophilic head group, linker group and hydrophobic tail (Figure 26.34a).

Several studies have shown that cationic lipids are less efficacious and more toxic when compared to ionizable lipids. Consequently, recent studies have focused on the development of ionizable lipids for siRNA delivery. In general, in order to minimize toxicity without sacrificing effectiveness, the pK_a of an ionizable lipid should be low enough for it to remain unprotonated during circulation in the blood but high enough for it to become protonated in the early or late endosomes. This is because protonation is necessary to promote membrane fusion and lipid mixing with anion lipids in the endosomal membrane. For example, a study involving *in vivo* gene silencing by 53 ionizable lipids showed that a pK_a in the range 5.4–7.6 was efficacious but a pK_a of ~6.4 was optimal. Overall, the pK_a of the liposome appears to be more important for efficacy rather than the pK_a of the individual lipid building blocks. However, despite this strong correlation between pK_a and efficacy,

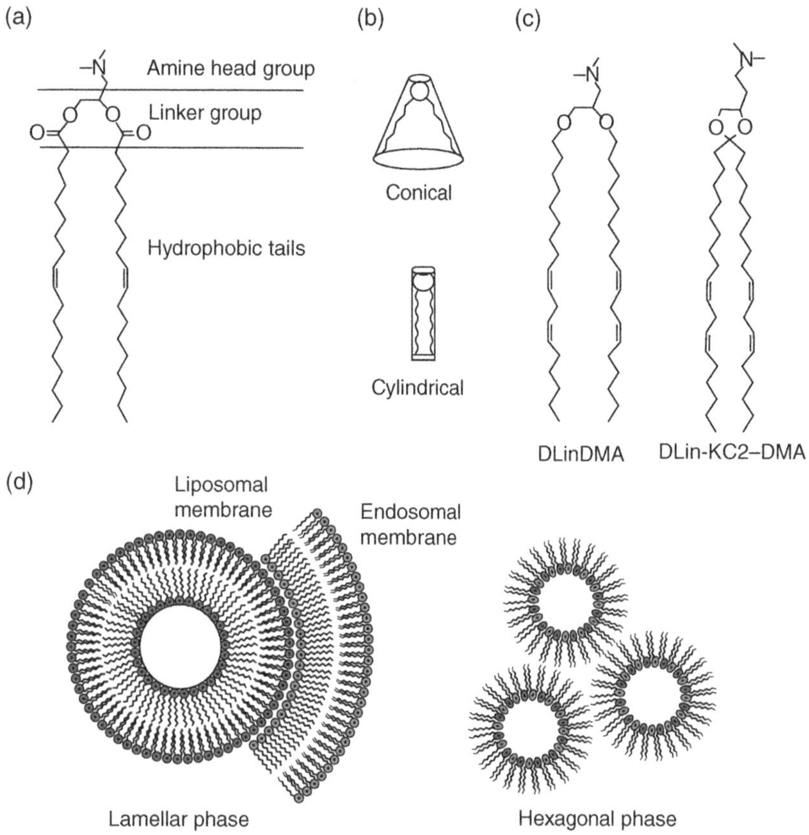

Figure 26.34 Lipid structures and shapes. (a) Ionizable lipids are composed of three sections: the amine head group, the linker group and the hydrophobic tails. (b) Lipids with a small head group and tails composed of unsaturated hydrocarbons tend to adopt a conical structure, whereas lipids with a large head group and saturated tails tend to adopt a cylindrical structure. (c) The structures of siRNA delivery lipids DLinDMA and DLin-KC2-DMA. (d) The mixing of cationic and anionic lipids promotes the transition from the more stable lamellar phase to the less stable hexagonal phase, thus aiding fusion of the liposomal and endosomal membranes. *Source:* From Kanasty et al. (2013) / with permission of Springer Nature.

liposomes with comparable pK_a can have different levels of efficacy *in vivo*.

The structure of the lipid's hydrophobic tail affects not only its pK_a but also its transition temperature and potency. The lipid transition temperature refers to the temperature at which the lipid membranes shift from a more stable lamellar structure to a less stable hexagonal structure (Figure 26.34d). This transition is important because it promotes the destabilization of the endosomal membrane and the release of siRNA from both the lipoplex and the endosome. Lipids with small polar head groups and large unsaturated hydrophobic tails have lower transition temperatures and shift more readily from the lamellar to the hexagonal structure to promote endosomal release (Figure 26.34b). In comparison, lipids with large polar head groups and fully saturated hydrophobic tails are more likely to adopt the stable lamellar structure. In general, successful lipid formulations are designed to remain in the stable lamellar structure during circulation and shift to the hexagonal structure within endosomal compartments.

For lipids with identical head groups, linkers and tail lengths, increased unsaturation of the tails has been found to lead to a decrease in pK_a, a decrease in transition temperature and an enhancement of *in vivo* efficacy. Two of the most efficacious ionizable lipids for *in vivo* delivery of siRNA and other nucleic acids, denoted DLinDMA and DLin-KC2-DMA, have the structures shown in Figure 26.34c.

Altering the structure of the amine head group and the linker component of the lipid building blocks also affects in vivo efficacy. For example, for DLinDMA, replacing the ether linker with an ester group was found to reduce efficacy whereas replacing it with a ketal linker enhanced efficacy. Additionally, altering the length of the carbon chain connecting the amine group to the ketal group resulted in an increase in the pK_a of the lipoplex but the transition temperature remained constant. Optimal efficacy was obtained with DLin-KC2-DMA, which has a pK_a of 6.7.

In order to improve the potency of lipoplexes, many liposomal formulations for siRNA or mRNA include additional constituents, such as cholesterol, lipid-anchored PEG molecules, a phospholipid and targeting ligands, along with the ionizable or cationic lipid to form lipid nanoparticles (Figure 26.35). As noted earlier, PEG molecules serve a variety of functions, including prevention of aggregation and shielding the liposomes from take up by macrophages, thereby increasing their circulation time. Addition of cholesterol serves to lower the transition temperature of liposomal membranes, thereby improving their ease of conversion from the lamellar to the hexagonal structure. Targeting ligands, as we have seen, bind specifically to receptors on target cells, thus improving take up by target cells. Optimal efficacy and distribution of the lipid nanoparticles *in vivo* are achieved through modulating the ratio of these additional constituents and compositional design of the lipid.

Polymer-based Nanoparticle Delivery Systems

A large number of polymeric biomaterials, in the form of nanoparticles, have been investigated for use as delivery systems for siRNA. Most of these polymers contain monomer units that can be protonated under acidic conditions, which, as noted earlier, serves as the driving force for electrostatic complexation with siRNA and may facilitate endosomic escape as well. Examples of such polymers include synthetic polymers such as cyclodextrin, polyethylenimine (PEI), and PAMAM dendrimers, as well as natural polymers such as poly(L-lysine) and chitosan.

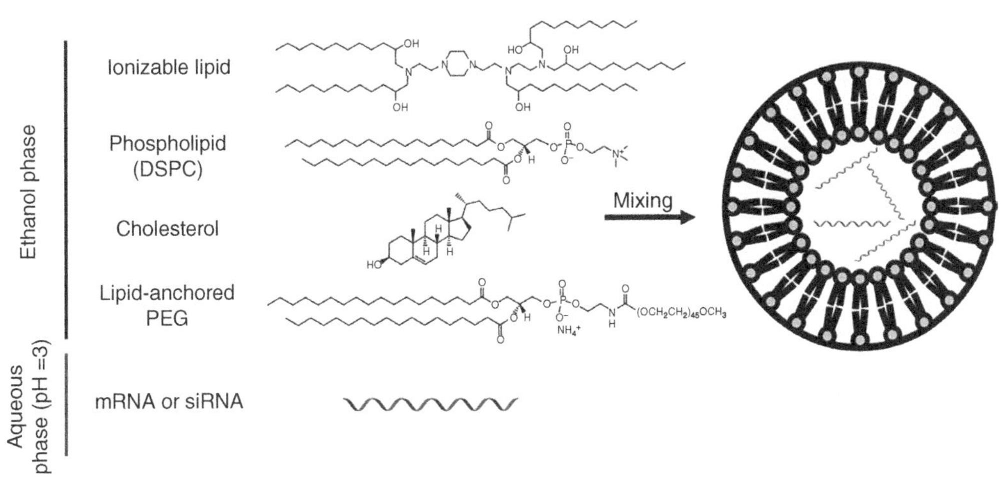

Figure 26.35 Formulation of lipid nanoparticles for the delivery of mRNA or siRNA. Lipid nanoparticles are formed by the mixing of two phases: (1) a four-component ethanol phase containing an ionizable lipid, a phospholipid, cholesterol, and lipid-anchored PEG, and (2) an acidic aqueous phase containing mRNA or siRNA. *Source:* From Kauffman et al. (2015) / with permission of American Chemical Society.

The cyclodextrin-based delivery system was the first targeted nanoparticle delivery system to enter clinical studies for the delivery of siRNA to treat cancer. Cyclodextrin-containing polymers are composed of short polycationic oligomers ($n \approx 5$) containing amidine –C(=NH)NH– functional groups, which can be considered nitrogen analogs of the carboxyl group (Figure 26.36a). In this structure, cyclodextrin forms part of the linear chain backbone of a water-soluble polymer. Protonation at the imino nitrogen (C=N) under acidic conditions leads to positively charged groups that facilitate complexation with siRNA. Additionally, end-capping of the oligomer termini with imidazole functional groups can facilitate endosomal escape, resulting in improved efficacy. Cyclodextrin-based siRNA delivery systems also require additional components in their formulation for stabilization and efficacy *in vivo*. For example, PEG molecules have been attached to the surface of the delivery system to improve stability and circulation times, while a targeting agent, such as transferrin, can be attached to the nanoparticle surface to bind to transferrin receptors that are typically upregulated on cancer cells.

PEI was one of the first cationic polymers investigated as a delivery system for DNA and, having been shown to be efficacious, it was later studied for the delivery of siRNA. PEI is a water-soluble but nondegradable polymer, which can be synthesized in linear or branched chains (Figure 26.36b). When used as a delivery system for siRNA, PEI often has a molecular weight of ~25 kDa. Because the amino nitrogen atom in each monomer unit of its chain backbone can become protonated, PEI molecules can potentially acquire a high positive charge in aqueous media, which provides a large driving force for complexation with siRNA (and other nucleotides). However, data on the protonation characteristics of PEI have varied widely. For example, the pK_a of PEI has been reported over a wide range (~5–9). A certain fraction of the monomer units in PEI are protonated at the physiological pH (~7.4) but this fraction has been reported from as low as 10–20% to as high as 90%. Overall, however, it appears that a significant fraction of the monomer units remain unprotonated at the physiological pH.

PEI has been found to be highly efficacious as a delivery system for siRNA, which been attributed largely to the ability of PEI to protect siRNA from nuclease degradation by a phenomenon called the proton sponge effect. As a significant fraction of its monomer units appears to be unprotonated at the physiological pH, PEI can serve as an effective "sponge" to take up H^+ ions (protons) by adsorption in an

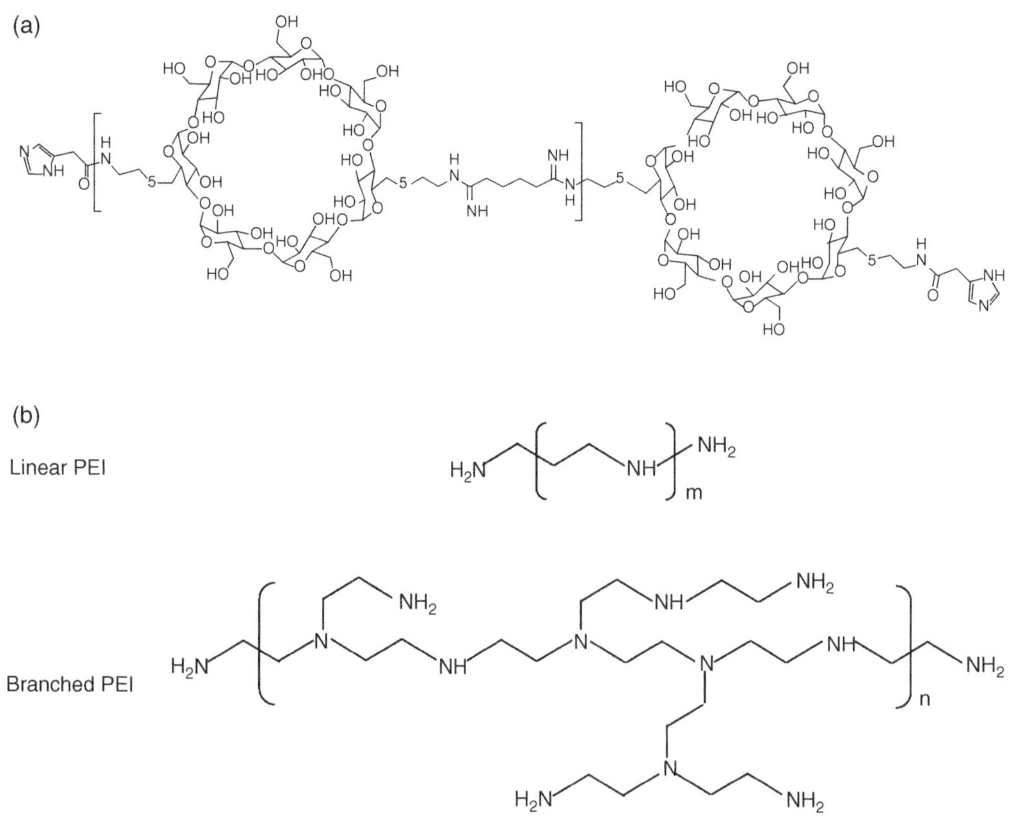

Figure 26.36 (a) Composition of a water-soluble cyclodextrin-based polymer synthesized for use as a nanocarrier to deliver siRNA. *Source:* From Davis (2009) / with permission of American Chemical Society. (b) Structure of linear polyethylenimine (PEI) and a branched PEI.

environment of lower pH, thus serving as a buffer to reduce the change in pH. Upon endocytosis, lysosomal buffering by PEI is thought to protect siRNA from nuclease degradation. Subsequently, a combination of osmotic and PEI swelling due to repulsion between the positively charged PEI chains leads to lysosomal swelling and membrane rupture, and to release of the PEI–siRNA complex into the cytoplasm.

Covalent bonding of PEG to PEI can increase the effectiveness PEI as a delivery system by reducing its interaction with plasma proteins. However, PEI delivery systems have been associated with cytotoxicity issues, which include membrane damage due to free PEI and to apoptosis through mitochondrial-induced pathways.

Inorganic Nanoparticles for siRNA Delivery

As noted earlier, while a variety of inorganic nanoparticles has received attention for use in biomedical applications, including Au, quantum dots, carbon nanotubes, fullerenes, and iron oxide, Au nanoparticles have been most widely studied and applied for drug delivery systems. In recent years, Au nanoparticles have also received significant attention as a delivery system for siRNA and other nucleic acids. One such system, consisting of spherical nucleic acids (SNAs) composed of Au nanoparticles (~13 nm in diameter) covalently functionalized with a layer of densely packed and highly oriented siRNA, was discussed in Chapter 24 for the healing of soft tissue wounds (Figure 24.22). Systemic application of SNAs has also shown potency as an RNAi-based therapy for glioblastoma in mice (Jensen et al. 2013). As glioblastoma multiforme is the most prevalent and lethal form of malignant brain tumors, as well as one of the deadliest human cancers, SNAs could potentially provide an important approach for systemic RNAi therapy for glioblastoma multiforme and other lethal malignancies though systemic injection.

26.8 Biological Drug Delivery Systems

As we have seen from previous sections of this chapter, drug delivery systems composed of biomaterials, including systems based on polymeric microparticles, hydrogels, and nanoparticles, have been widely studied and successfully developed to deliver a variety of therapeutic biomolecules. As noted, the success of these systems relies largely on designing delivery systems to address the physicochemical limitations of free drugs (such as solubility and stability) and to overcome various physiological barriers (such as immune clearance, entry into cells, the BBB and off-target deposition) in order to reach the site of action. More recently, biological particles, including vesicles, pathogens and mammalian cells, have been receiving significant interest for delivering therapeutic molecules. A major reason for this interest is that these biological particles are highly optimized for their specific functions *in vivo* and they possess features that are normally desirable for carriers in the delivery of therapeutic molecules. For example, cell-derived membrane vesicles, which are secreted by most cells in the body and are present in most body fluids, facilitate cell-to-cell communication and naturally transport RNA and proteins. Additionally, they are stable in blood, can confer immune tolerance, and can possess native targeting ligands, which are desirable features for use in the delivery of therapeutic biomolecules. Exosomes are the most widely studied cell-derived membrane vesicles for delivering biomolecules.

26.8.1 Exosomes for Therapeutic Biomolecule Delivery

Exosomes are phospholipid bilayer vesicles of size ~40–100 nm, which are derived from multivesicular bodies, as illustrated schematically in Figure 26.37. They have a bilayer enriched in cholesterol, sphingomyelin, glycosphingolipids, and phosphatidylserine, and their principal content is some type of RNA, including miRNA, mRNA and various noncoding RNAs. However, they can also contain proteins, lipids, DNA and other biomolecules. Exosomes are generated and secreted in a three-step process. The first step involves endocytic invagination of the plasma membrane to form an early endosome, which matures to a late endosome. This maturation to a late endosome includes molecules sorting into multiple smaller intraluminal vesicles, resulting in the structure referred to as a multivesicular body (MVB). The second step involves movement of the MVB to the plasma membrane, while in the third step, fusion of the MVB with the plasma membrane leads to secretion of exosomes. As noted earlier for vesicles, exosomes facilitate intercellular communication, which appears to be achieved by means of their encapsulated miRNA and mRNA. The structure and contents of the exosome varies with the tissue of origin, as well as with the physiological and pathological state of the tissue.

In order for exosomes to be used in the delivery of biomolecules, they must first be isolated efficiently and characterized by valid methods, and then incorporated with the relevant biomolecules. A variety of methods has been used to isolate exosomes from body fluids such as plasma, but the most common method is a differential centrifugation process, involving three or more steps of low-speed centrifugation to remove intact cells and large debris, followed by a

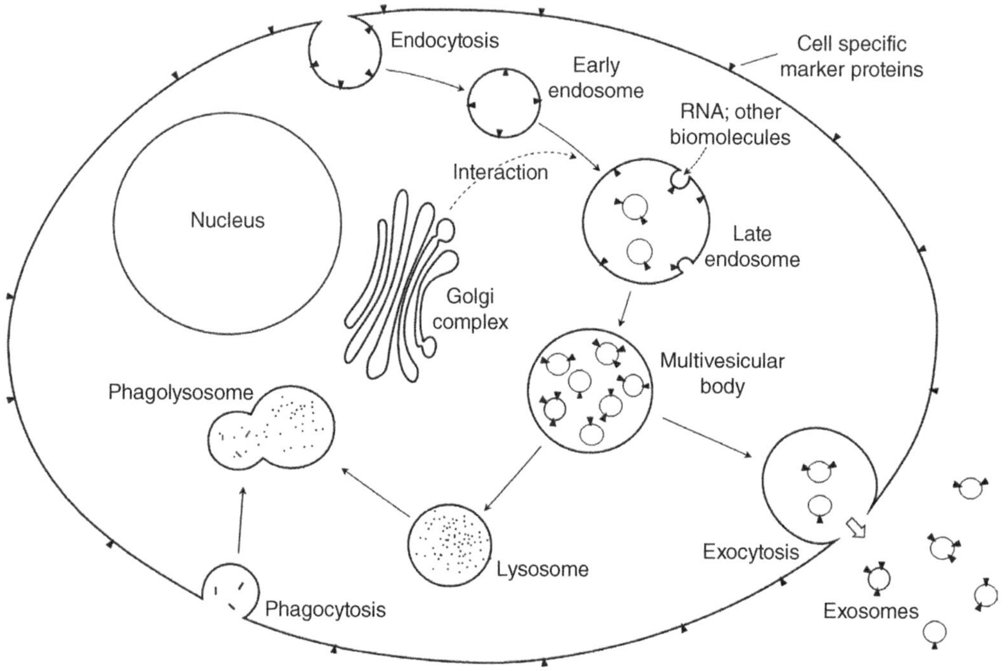

Figure 26.37 Illustration of endosomal pathway for the generation of exosomes.

final high-speed centrifugation. Various methods have been developed for incorporating biomolecules into exosomes, which involve strategies to incorporate the biomolecules either during endosome isolation or after the isolation process. It is also possible to attach targeting ligands to exosomes for targeted delivery to tumors.

Exosomes have been studied as a delivery system for a variety of biomolecules, including nucleic acids and proteins. One of the first proof-of-principle studies involved the use of exosomes to deliver siRNA to the brain in mice to silence expression of BACE1, a protease linked to the development of amyloid plaques in Alzheimer's disease (Alvarez-Erviti et al. 2011). Immature dendritic cells were derived from mouse bone marrow and used as a source of exosomes because they are devoid of lymphocyte stimulatory molecules. Then the exosomes (~80 nm in diameter) were incorporated with exogenous siRNA. The brain was selected as the target tissue because the BBB has been a major obstacle in the delivery of macromolecular drugs to the central nervous system. To ensure that systemically injected exosomes targeted the brain *in vivo*, the dendritic cells were genetically manipulated to express Lamp2b, an exosomal membrane protein, fused with the neuron-specific rabies viral glycoprotein (RVG) peptide. This peptide is known to bind nicotinic acetylcholine receptors present on neurons and the vascular endothelium of the BBB. An electroporation procedure was used to load BACE1 siRNA into the targeted exosomes.

Then BACE1 siRNA encapsulated in RVG exosomes were administered intravenously to a group of normal C57BL/6 mice and compared with four other treatment groups: untreated mice (control), mice given RVG exosomes without siRNA (RVG exosomes); mice given BACE1 siRNA in non-targeted exosomes (siRNA-TR); and mice given BACE1 siRNA coupled to RVG peptide (siRNA-RVG peptide). Brain tissue samples analyzed three days later showed significant reductions in BACE1 protein expression and mRNA expression in animals that were injected with BACE1 siRNA encapsulated in RVG exosomes and BACE1 siRNA coupled to RVG peptide, respectively (Figure 26.38). Overall, these results show that RVG exosomes are capable of delivering siRNA specifically and safely after systemic administration and, therefore, may represent a promising system for gene therapies targeting chronic neurodegenerative disorders.

Despite their attractive features, exosomes suffer from a number of limitations, which have to be overcome prior to possible translation to clinical studies. Overall, there is a need for a better understanding of exosome biology and function, and for economical techniques to isolate and purify exosomes, and to incorporate a variety of biomolecules into them as well.

26.9 Concluding Remarks

In this chapter, we discussed various aspects related to the design and adaptation of biomaterials for use in the delivery of drugs and other therapeutic biomolecules. The ability of

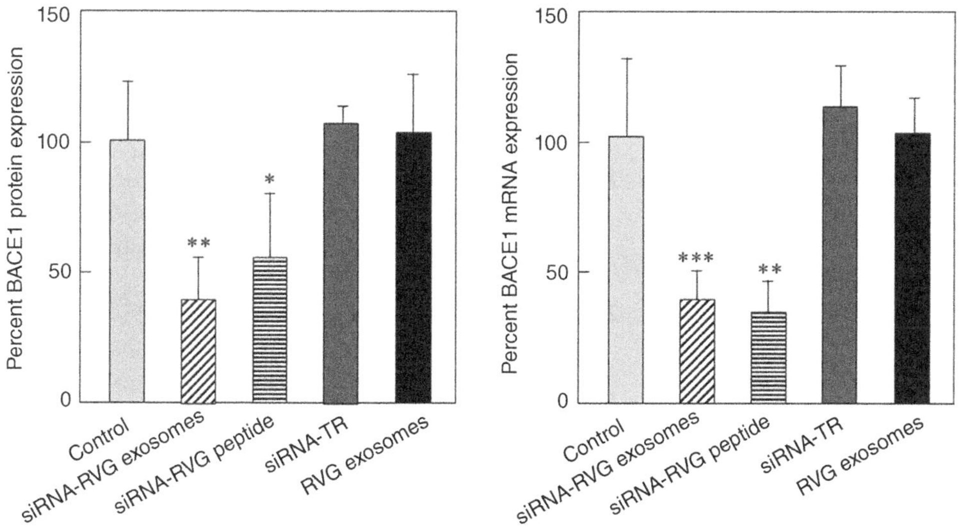

Figure 26.38 *In vivo* delivery of siRNA by exosomes targeted to the brain of mice silences expression of BACE1, a protease linked to the development of amyloid plaques in Alzheimer's disease. BACE1 siRNA encapsulated in exosomes targeted with rabies viral glycoprotein (RVG) peptide (siRNA-RVG exosomes) were administered intravenously to mice and compared with four other treatment groups: untreated mice (control), mice given RVG exosomes without siRNA (RVG exosomes); mice given BACE1 siRNA in non-targeted exosomes (siRNA–TR); and mice given BACE1 siRNA coupled to RVG peptide (siRNA–RVG peptide). Analysis of brain tissue samples three days later showed significant reductions in BACE1 protein expression and mRNA expression in animals treated with siRNA-RVG exosomes and siRNA–RVG peptide. (*, ** and *** indicate $P < 0.05$, $P < 0.01$, and $P < 0.001$, respectively, versus the control group; mean ± standard deviation; $n = 5$). *Source:* From Alvarez-Erviti et al. (2011) / with permission of Springer Nature.

a drug to have its desired therapeutic effect depends not only on the drug itself but also on other factors, including the method by which the drug is delivered to the body. Combining a suitable biomaterial with the drug to form a drug delivery system is often the best mode of delivery. Drug delivery systems used in the treatment of serious conditions are often controlled or targeted systems, in which the biomaterial is required to protect the drug, deliver it to a specific site of action and, once there, control its release at a desirable rate over a prolonged period.

- While drugs are administered using a variety of routes, we focused on drug delivery systems based on polymer microparticles, hydrogels and particularly nanoparticles, because these systems have been the subject of considerable interest and development over the last few decades. These systems are normally administered by intravenous, intramuscular or subcutaneous injections, or by surgical implantation.
- We discussed biomaterials, particularly in the form of nanoparticles, for the delivery of ribonucleic acids, and biological systems, particularly exosomal vesicles, as alternatives to synthetic biomaterials for the delivery of therapeutic biomolecules. These systems have been the subject of numerous recent studies and are likely to have a significant impact on the treatment of serious conditions, such as cancer, in the future.

Problems

26.1 Explain what controlled drug delivery systems are and why they are important.

26.2 List and briefly discuss the main requirements and characteristics of biomaterials for use in controlled drug delivery systems.

26.3 A commonly used microparticle-based drug delivery system consists of PLGA microspheres containing dispersed drug molecules. Discuss the factors that control drug release from this system and the mechanism of drug release.

26.4 Compare the advantages and disadvantages of drug delivery systems based on polymer microspheres and on hydrogels.

26.5 Discuss the role of pH responsive polymers in controlled drug delivery systems.

26.6 Discuss the role of temperature responsive polymers in controlled drug delivery systems.

26.7 Discuss the role of PNIPAAm in controlled drug delivery systems.

26.8 Describe an example where intermittent (pulsatile) release of drugs is more desirable than continuous (sustained) release, and describe a drug delivery system that could be used for intermittent drug release.

26.9 Explain what ferrogels are and discuss the principles underlying their role in drug delivery systems.

26.10 Discuss the factors that control the release of drugs in acoustic (ultrasonically) controlled drug delivery systems and the mechanism of drug release.

26.11 Describe a delivery system from which drug release can be controlled by a magnetic field and discuss the mechanism of drug release.

26.12 Distinguish between passive targeting and active targeting for targeting nanoparticle-based drug delivery systems to cells.

26.13 Is there an optimal size range for nanoparticles for use in drug delivery systems? If so, give the optimal range and explain the reasons for the optimal size range.

26.14 Discuss the use of liposomes in nanoparticle-based drug delivery systems, including the principles underlying their advantages and limitations.

26.15 Distinguish between liposomes and lipid–polymer hybrid nanoparticles, and compare their advantages and disadvantages for use as nanocarriers in drug delivery systems.

26.16 Distinguish between liposomes and polymersomes, and discuss the principles underlying their advantages and limitations for use in drug delivery systems.

26.17 Discuss the requirements of a system that could be used to deliver short interfering ribonucleic acid (siRNA) by intravenous injection. Then select a specific system and explain how it approximates these requirements.

26.18 Explain what exosomes are and discuss the principles underlying their advantages and limitations for use as systems to deliver biomolecules *in vivo*.

References

Ahmed, F., Pakunlu, R.I., Srinivas, G. et al. (2006). Shrinkage of a rapidly growing tumor by drug-loaded polymersomes: pH-triggered release through copolymer degradation. *Molecular Pharmaceutics* 3: 340–350.

Alvarez-Erviti, L., Seow, Y., Yin, H.F. et al. (2011). Delivery of siRNA to the mouse brain by systemic injection of targeted exosomes. *Nature Biotechnology* 29: 341–345.

Anselmo, A.C. and Mitragotri, S. (2014). An overview of clinical and commercial impact of drug delivery systems. *Journal of Controlled Release* 190: 15–28.

Berkland, C., King, M., Cox, A. et al. (2002). Precise control of PLG microsphere size provides enhanced control of drug release rate. *Journal of Controlled Release* 82: 137–147.

Brudno, Y. and Mooney, D.J. (2015). On-demand drug delivery from local depots. *Journal of Controlled Release* 219: 8–17.

Chan, J.M., Zhang, L., Yuet, K.P. et al. (2009). PLGA-lecithin-PEG core-shell nanoparticles for controlled drug delivery. *Biomaterials* 30: 1627–1634.

Cheng, J., Teply, B.A., Sherific, I. et al. (2007). Formulation of functionalized PLGA–PEG nanoparticles for *in vivo* targeted drug delivery. *Biomaterials* 28: 869–876.

Danhier, F., Feron, O., and Préat, V. (2010). To exploit the tumor microenvironment: passive and active tumor targeting of nanocarriers for anti-cancer drug delivery. *Journal of Controlled Release* 148: 135–146.

Davis, M.E. (2009). The first targeted delivery of siRNA in humans via a self-assembling, cyclodextrin polymer-based nanoparticle: from concept to clinic. *Molecular Pharmaceutics* 6: 659–668.

Duncan, R. (2006). Polymer conjugates as anticancer nanomedicines. *Nature Reviews. Cancer* 6: 688–701.

Fenton, O.S., Olafson, K.N., Pillai, P.S. et al. (2018). Advances in biomaterials for drug delivery. *Advanced Materials* 30: 1705328.

Fire, A., Xu, S., Montgomery, M.K. et al. (1998). Potent and specific genetic interference by double-stranded RNA in *Caenorhabditis elegans*. *Nature* 391: 806–811.

Gu, F., Zhang, L., Teply, B.A. et al. (2008). Precise engineering of targeted nanoparticles by using self-assembled biointegrated block copolymers. *Proceedings of the National Academy of Sciences of the United States of America* 105: 2586–2591.

Huebsch, N., Kearney, C.J., Zhao, X. et al. (2014). Ultrasound-triggered disruption and self-healing of reversibly crosslinked hydrogels for drug delivery and enhanced chemotherapy. *Proceedings of the National Academy of Sciences of the United States of America* 111: 9762–9767.

Jain, A., Jain, A., Garg, N.K. et al. (2015). Surface engineered polymeric nanocarriers mediate the delivery of transferrin-methotrexate conjugates for an improved understanding of brain cancer. *Acta Biomaterialia* 24: 140–151.

Jensen, S.A., Day, E.S., Ko, C.A. et al. (2013). Spherical nucleic acid nanoparticle conjugates as an RNAi therapy for glioblastoma. *Science Translational Medicine* 5: 209ra152.

Kanasty, R., Dorkin, J.R., Vegas, A., and Anderson, D. (2013). Delivery materials for siRNA therapeutics. *Nature Materials* 12: 967–977.

Kauffman, K.J., Dorkin, J.R., Yang, J.H. et al. (2015). Optimization of lipid nanoparticle formulations for mRNA delivery in vivo with fractional factorial and definitive screening designs. *Nano Letters* 15: 7300–7306.

Kim, D.Y., Kwon, D.Y., Kwon, J.S. et al. (2016). Synergistic anti-tumor activity through combinational intratumoral injection of an in-situ injectable drug depot. *Biomaterials* 85: 232–245.

Paciotti, G.F., Myer, L., Weinreich, D. et al. (2004). Colloidal gold: a novel nanoparticle vector for tumor directed drug delivery. *Drug Delivery* 11: 169–183.

Park, J.H., Lee, S., Kim, J.H. et al. (2008). Polymeric nanomedicine for cancer therapy. *Progress in Polymer Science* 33: 113–137.

Rideau, E., Dimova, R., Schwille, P. et al. (2018). Liposomes and polymersomes: a comparative review towards cell mimicking. *Chemical Society Reviews* 47: 8572–8610.

Torchilin, V.P., Lukyanov, A.N., Gao, Z., and Papahadjopoulos-Sternberg, B. (2003). Immunomicelles: targeted pharmaceutical carriers for poorly soluble drugs. *Proceedings of the National Academy of Sciences of the United States of America* 100: 6039–6044.

Zhang, S., Bellinger, A.M., Glettig, D.L. et al. (2015). A pH-responsive supramolecular polymer gel as an enteric elastomer for use in gastric devices. *Nature Materials* 14: 1065–1071.

Zhao, X., Kim, J., Cezar, C.A. et al. (2011). Active scaffolds for on-demand drug and cell delivery. *Proceedings of the National Academy of Sciences of the United States of America* 108: 67–72.

Further Reading

Gao, W., CMJ, H., Fang, R.H., and Zhang, L. (2013). Liposome-like nanostructures for drug delivery. *Journal of Materials Chemistry B* 1: 6569–6585.

Lee, J.S. and Feijen, J. (2012). Polymersomes for drug delivery: design, formation and characterization. *Journal of Controlled Release* 161: 473–483.

Li, J. and Mooney, D.J. (2016). Designing hydrogels for controlled drug delivery. *Nature Reviews Materials* 1: 1–17.

Menjoge, A.R., Kannan, R.M., and Tomalia, D.A. (2010). Dendrimer-based drug and imaging conjugates: design considerations for nanomedical applications. *Drug Discovery Today* 15: 171–185.

Index

Materials for Biomedical Engineering: Fundamentals and Applications, First Edition. Mohamed N. Rahaman and Roger F. Brown.

Companion website: www.wiley.com/go/rahaman/biomedicalengineering

d

n

o

q

r

s

t

u

v

/691

Printed and bound by CPI Group (UK) Ltd, Croydon, CR0 4YY